Michael Erdmann
July 1996

Robotics Research
The Seventh International Symposium

Springer
London
Berlin
Heidelberg
New York
Barcelona
Budapest
Hong Kong
Milan
Paris
Santa Clara
Singapore
Tokyo

Georges Giralt and Gerhard Hirzinger (Eds)

Robotics Research

The Seventh International Symposium

With 580 Figures

Springer

Georges Giralt, Docteur es Sciences
Centre National de la Recherche Scientifique
Laboratoire d'Analyse et d'Architecture des Systemes
7 Avenue du Coronel Roche, 31077 Toulouse Cedex, France

Gerhard Hirzinger, Prof. Dr.-Ing.
DLR Oberpfaffenhofen
Institute for Robotics and System Dynamics, 82234 Wessling, Germany

ISBN 3-540-76043-1 Springer-Verlag Berlin Heidelberg New York

British Library Cataloguing in Publication Data
Robotics research : the seventh international symposium
1.Robotics - Research
I.Giralt, Georges II.Hirzinger, Gerd
629.8'92'072
ISBN 3540760431

Library of Congress Cataloging-in-Publication Data
Robotics research : the seventh international symposium / Georges Giralt and Gerhard Hirzinger (eds.).
p. cm.
Includes bibliographic references.
ISBN 3-540-76043-1 (hardcover : alk. paper)
1. Robotics - - Research - - Congresses. I. Giralt, Georges. II. Hirzinger, Gerd. III. International Symposium on Robotics Research (7th : 1995 : Herrsching am Ammersee, Germany)
TJ211.R56817 1996 96-12837
629.8'92 - - dc20 CIP

Printed in Great Britain

Typesetting: Camera ready by editors
Printed and bound by Professional Book Supplies Ltd., Steventon, Oxfordshire
69/3830-543210 Printed on acid-free paper

Contents

Preface

The Seventh International Symposium of Robotics Research was held in Herrsching near Munich, Germany, from October 21 to 24, 1995, the first to be organized following the renewal undergone by the International Foundation of Robotics Research (IFRR) during the preceding Symposium (Hidden Valley, October 1993).

A board of eighteen officers was appointed:

Suguru Arimoto	(Tokyo University)
Ruzena Bajcsy	(University of Pennsylvania)
Robert Bolles	(SRI International)
Mike Brady	(Oxford University)
Paolo Dario	(Pisa University, Italy)
Joris De Schutter	(Katholieke Universiteit Leuven, Belgium)
Olivier Faugeras	(INRIA, France)
Georges Giralt	(LAAS-CNRS, France), Secretary
Gerd Hirzinger	(DFVLR, Germany)
Hirochika Inoue	(Tokyo University), President-elect
Ray Jarvis	(Monash University, Australia)
Takeo Kanade	(Carnegie Mellon University)
Hirofumi Miura	(Tokyo University)
Richard Paul	(University of Pennsylvania), President
Marc Raibert	(Massachusetts Institute of Technology)
Bernie Roth	(Stanford University)
Yoshiaki Shirai	(Osaka University)
Tsuneo Yoshikawa	(Kyoto University)

It was our commitment to bring together active, leading robotics researchers from academia, government, and industry, with the ambitious objective to assess the state of Advanced Robotics and to discuss future research directions.

Papers representing authoritative reviews of established research areas as well as papers reporting on new areas were sought for presentation. A number of leading researchers were asked to submit extended abstracts outlining papers representing their areas of research.

In addition to inviting participants, a call for papers was issued in order to include researchers who had made significant new contributions to robotics. All abstracts and submitted papers were reviewed by the eighteen IFRR officers acting as the Symposium Program Committee before final paper selection was made. Attendance at the Symposium, 63, was limited to presenters and invited participants.

It is worth noting that of the sixty three participants, only thirty were present at Hidden Valley in 1993.

The organization of the Proceedings follows that of the Symposium, with a chapter devoted to each session. Senior members of the robotics research community were asked to chair each session and to write a session summary which introduces the chapter.

During the four-days of sessions, fifty five papers were presented in a single track with emphasis being put in open discussion with a pro-active role of the chair.

We believe to express the general opinion of the attendees in writing that we had a most interesting four-days work with an outstanding set of original results as well as clean, enlightening new formulations of standing problems. Hence, the Symposium endeavoured to contribute with meaningful steps forward to build the emerging body of concepts, methods, scientific and technical knowledge and tools that shape Robotics.

Indeed, the papers cover not only a broad spectrum of subjects, but, most important, they range from pure exploratory research to application-oriented research work, sometimes even directly related to a real-world generic case.

Thus whichever viewpoint we stand for, Robotics revival, birth or just normal evolution and progress we hope to share with the readers what we experienced together : an already quite mature body of knowledge is fostered by novel cutting edge applications and shall respond to the two-folded challenge of scientific and technical soundness as well as economical and technical impact.

We would like to express, on behalf of IFRR our thanks to the following institutions that financially sponsored the 7th ISRR : the Bavarian Ministry of Economics, Traffic and Technology, DLR (German Aerospace Research Establishment), KUKA (German robot manufacturer), LAAS-CNRS (French Research Laboratory, and Secretariat of the International Foundation for Robotics Research), SIEMENS.

Our special thanks also go to the staff of DLR for a very efficient and most pleasant venue organisation, and to the IFRR Secretariat for the continuing logistic support.

Gerhard Hirzinger and Georges Giralt
Co-Chairs
The Seventh International Symposium of Robotics Research

List of Authors

PART 1
MANIPULATION
SESSION SUMMARY

J. De Schutter
Katholieke Universiteit Leuven – Division PMA
Celestijnenlaan 300 B, 3001 Heverlee, Belgium
e-mail: Joris.DeSchutter@mech.kuleuven.ac.be

The session on Manipulation contains four papers on dynamic interaction, design of dextrous hands, object reorientation using grasp gaits and palmar manipulation. The first two papers are review papers in their respective fields; the last two papers present pioneering work in the planning of manipulations. These last papers focus on different types of manipulative actions: the third paper deals with the planning of *prehensile* manipulation combined with a sequence of regrasps as in the case of a multifingered hand; the fourth paper deals with the planning of *nonprehensile* manipulation, such as sliding and dropping, as in the case of manipulation using multiple palms.

The first paper by E. Fasse and N. Hogan, *Control of physical contact and dynamic interaction*, reviews several different approaches to controlling mechanical interaction, including impedance and admittance control and compares them with hybrid force/position control. The problems of physical contact and the strengths and weaknesses of interaction control are considered in view of specification of a desired behavior and implementation.

The second paper by A. Bicchi, *Hands for dextrous manipulation and powerful grasping: a difficult road towards simplicity*, discusses the state of the art in the design of artificial hands, and comments about expected future developments. Bicchi favors a minimalistic attitude in the design of hands for practical applications; designing simple and effective devices for executing non-trivial devices is in fact much more difficult than contriving very complex systems for the same task. He distinguishes between dextrous hands and hands for grasping. Both have opposite requirements in some sense as dexterity means the ability to change the position and orientation of the manipulated object with respect to the hand, while grasping is intended to prevent motion relative to the hand.

The third paper by S. Leveroni and K. Salisbury, *Reorienting objects with a robot hand using grasp gaits*, addresses the planning of robot finger motions to enable continuous and stable reorientation of grasped objects. With a given grasp on an object, the range of reorientation through which the fingers may move the object is limited by the workspace of the fingers and grasp stability. If however the object can be properly regrasped without dropping it, further motion of the object in the desired direction may be possible. Such a sequence of finger/objects motions and regrasps is called a grasp gait. The paper develops tools and methods to find suitable grasp gaits that enable a robot hand to reorient convex objects in two dimensions.

The final paper by M. Erdmann, *An exploration of nonprehensile two-palm manipulation: planning and execution*, presents a system for orienting parts using two palms which consists of a planner and an executive. Based on a geometric description of a part, the coefficients of friction between the part and each of the palms, and a start and goal configuration of the part, the system plans and executes a sequence of nonprehensile palm motions (e.g. purposeful sliding and constrained dropping) which orient the part from the specified start to the specified goal configuration.

Hands for Dexterous Manipulation and Powerful Grasping: A Difficult Road Towards Simplicity

Antonio Bicchi

Centro "E.Piaggio", Università di Pisa, via Diotisalvi, 2. 56126 Pisa (Italia).
Email: bicchi@anaxagoras.piaggio.ccii.unipi.it

Abstract

In this paper, an attempt at summarizing the evolution and the state-of-the-art in the field of robot hands is made. In such exposition, a critical evaluation of what in the author's view are the leading ideas and emerging trends, is privileged with respect to exhaustiveness of citations. A basic distinction is made between hands designed for mimicking the human anatomy and physiology, and hands designed to meet practical, if restricted, task requirements.

The fact that most of the sophisticated robot hands produced so far never found their way into real-world applications is a strong stimulus at re-thinking some of our approaches. In particular, arguments are presented in favour of a "minimalistic" attitude in the design of hands for practical applications, i.e., use the least number of actuators, the simplest set of sensors, etc., for a given task. To achieve this rather obvious engineering goal is a challenge to our community. The paper illustrates some of the new, sometimes difficult theoretical problems that are brought about by building and controlling simpler, practical devices.

1 Introduction

"*'Αναξαγόρας μὲν οὖν φησι διὰ τὸ χεῖρας ἔχειν φρονιμώτατον εἶναι τῶν ζώων ἄνθροπον : εὔλογον δὲ διὰ τὸ φρονιμώτατον εἶναι χεῖρας λαμβάνειν*[1]". In one of his books on nature sciences [2], the greek philosopher Aristotle (384–322 BC) thus argued against the conceptions of his late colleague Anaxagoras (500?–428 BC) regarding the relationship between human hands and mind. As they appear to be the two most distinguished features of humans among animals, the two philosophers debated whether it was because humans had dextrous hands that they grew intelligent, or the other way around. It has been later on confirmed by several findings of paleoanthropologists that the mechanical dexterity of the human hand has been a major factor in allowing homo sapiens to develop a superior brain (a similar role played by the anatomical structure of the human larynx in relation with speech capabilities has been also recognized).

While the dexterity of the human hand has been admired since the oldest times, it still represents an unmatched standard for artificialists, and probably will for good. Nonetheless, there are theoretical and practical intermediate goals that are worth our best efforts to work towards dexterity. This paper discusses in what state the art of building artificial hands is at present times, and argue about possible directions it may take in the future. Excellent surveys are available on robot hand systems and components (see e.g. [44], [49], [111], [42], [100], [122], [53], and [95]), and the reader is referred there for other views on the state of the art.

In Section 2 the dualism between anthropomorphic and non–anthropomorphic design is considered, and issues raised by real–world applications of artificial hands are introduced. The strong need for simplicity in the design of practical robot hands is underscored. In Section 3 these themes are substantiated with examples of manipulation–oriented hands, while in Section 4 robustness of grasping is addressed. Throughout the paper, open problems of both theoretical and technological nature are presented.

[1] "Anaxagoras says that because of having hands, man grew the most intelligent among animals. [I think] it is correct to say that because of his intelligence he has hands"

2 Anthropomorphism in Design

Although stronger or faster artificial hands may be built, performance of human hands are unequalled if a sufficiently broad scope of manipulation tasks is considered. It is therefore natural for a roboticist to take inspiration from such a design success, and set forth for himself the goal of building hands that achieve, albeit partially, such capabilities. However, the toolbox nature can use is extremely different from what we have, in terms of actuators, sensors, and control means. Hence, the question whether artificial hands should ***look like*** those of humans, is not settled.

Good reasons for building anthropomorphic hands are found in applications where a ***replacement*** is needed. In other words, if the system is to use the same interface with the environment that was designed for the human hand (e.g. handles, consoles, tools etc.), then an anthropomorphic hand can save much effort in redesigning the interface. Such is typically the case with prosthetic devices (see e.g. [71], [139], [30], [58]). The same criteria apply in a closer field to robotics, i.e., telemanipulation (see e.g. [35], [8], [131], [37], [46], [130], [79], [62]), although there exist examples of non strictly anthropomorphic hands intended also for remote operation (see e.g. [21]). It should be noted that anthropomorphic design also makes it easier for the human operator to map his natural manipulation behaviours into commands for the device. Future robots for the domestic environment will probably have human–like hands, although at present it seems hard to reasonably forecast how close such future is.

On the other hand, if the control of the robot hand is realized by computer programs, and the environment is at least partially available for design decisions (as it happens in industrial parthandling, for instance), then several reasons may suggest that an anthropomorphic hand is not the best solution. Among the drawbacks of human–like hands, are the complex kinematic structure, the high number of actuators, and the sophistication of sensing systems. Cost–effectiveness and reliability are at a premium in factory applications of robot hands, and make the simplest grippers an optimal solution for most trivial grasping tasks. Manufacture of large enough batches of products justifies the development of specialized grippers for the task (Kato [70] reviewed a very large number of such devices). However, as the life cycle of products decreases in the technological competition, the need for flexibility in parthandling devices becomes more and more important.

In between the completely unstructured world and the perfectly defined environments, there is a whole gray scale of applications where the familiar flexibility/efficiency tradeoffs have to be seeked for actively. This concept is well radicated in the robotics community, see for instance [151] and [25]. Design of devices for this class of problems usually obey the good old engineering principle of minimalism: choose the simplest mechanical structure, the minimum number of actuators, the simplest set of sensors, etc., that will do the job (or class of jobs).

Complexity reduction is especially important in terms of hardware components of the system, as they often make for most of the cost, weight, and failures of robots. On the other hand, it often turns out that sophisticated design, analysis, programming, and control are required to perform difficult tasks by means of simple devices. Designing simple and effective devices for executing non–trivial tasks is actually much more difficult than contriving very complex systems for the same job. This is true both in a technological and theoretical sense, as this paper attempts at illustrating.

3 Design of dextrous hands

"Dexterity" is a rather broad concept in common language, that involves aspects of ability and stability in peforming motions of the manipulated object by means of the hand. We will restrict here to the notion, widely accepted in the robotics manipulation literature, that dexterity means the capability of changing the position and orientation of the manipulated object from a given reference configuration to a different one, arbitrarily chosen within the hand workspace. In this section, we examine several attempts at achieving dexterity by robot hands.

Robot hands are systems comprised of two or more fingers that act on a manipulated object via contacts. The presence of contact phenomena in manipulation makes for the most of the peculiarity of manipulation among other robotic systems, and clearly contact models deeply affect the analysis of manipulation systems. A standard classification of contact models introduced in robotics ([126] [31]) distinguishes point-contact-with-friction (or "hard-finger"), "soft-finger", and complete-constraint contacts (or "very-soft-finger"). Other important aspects of contact modeling regard the visco-elastic behavior (rigid, isotropically elastic, etc.) and the behavior in sliding and rolling conditions, namely, the static and kinetic coefficients of friction, and whether the contact point moves on the contacting surfaces as they rotate with respect to each other ("rolling contact"), or not. The latter case corresponds to an idealized situation of contact between surfaces with infinite relative curvature.

Salisbury ([83]) showed first that the minimum theoretical number of d.o.f.'s to achieve dexterity in a hand with rigid, hard-finger, non-rolling and non-sliding contacts, is 9. As a simple explanation of this fact, consider that at least three hard-fingers are necessary to completely restrain an object. On the other hand, as no rolling nor sliding is allowed, fingers must move so as to track with the contact point on their fingertip the trajectory generated by the corresponding contact point on the object, while this moves in 3D space. Hence, 3 d.o.f. per finger are strictly necessary. The Salisbury Hand ([124]) was accordingly designed to have nine joints, distributed in each finger so as to optimize a measure of individual "manipulability" of the finger ([125]).

Several other hands developed in University or Government research centers have adopted design schemes similar to Salisbury's under this regard, as e.g. the University of Karlsruhe Hand ([Wöhlke, 1990]), the hand developed at the Technical University of Darmstadt ([105], [150]), and that of Delft University ([64]). Hands of this type are not usually strictly anthropomorphic.

Some researchers preferred to introduce redundant degrees of freedom in their hands to achieve more flexibility of use. In one of the earliest successful hand designs, Okada employed two four-joint fingers and one three-joint thumb (see [101]). In the design of the hand of the Technical University of München ([87]), the three-joint three-finger design has been modified by introducing one more joint per finger, the motions of which are however mechanically coupled so that a total of nine d.o.f.'s is maintained. Coupling some of the joints in the hand allows to mechanically introduce grasping behaviors by curling the fingers to wrap the object, which is a feature of the human hand and of some prosthesis-oriented hands ([139]). Other authors introduced more than three fingers in their robot hands, with a basically twofold motivation: four- and five-fingered hands are closer to the anthropomorphic model, and allow to alternate the fingers in contact with the object so as to achieve richer manipulation patterns. Typical examples of anthropomorphic hands are the Utah/MIT Hand ([59]) and the Anthrobot Hand ([1]).

Actuation means are another active area of research. While shape memory alloys once used in the Hitachi hand do not seem to have found their way into practical applications, the hand developed at the New York University by Ebner and Wallace [34] has 15 joints actuated by direct-drive magnetic actuators. A two fingered hand with 12 d.o.f. for miniaturized object handling has been presented by Tanikawa ***et al.*** [137] that is actuated by piezoelectric motors. Hayakawa and Kawamura [47] presented a pneumatic bellows system realizing a soft gripper.

Notwithstanding the great effort spent, and the impressing technological and theoretical results achieved by the robotics community in building and controlling dextrous robot hands, the number of applications in the real-world and the performance of such devices in operative conditions should be frankly acknowledged as not yet satisfactory. In particular, the high degree of sophistication in the mechanical design prevented dextrous robotics hand to succeed in applications where factors such as reliability, weight, small size, or cost, were at a premium. One figure partially representing such complicacy is the number of actuators, that ranges between 9 and 32 for hands considered above. Further reduction of hardware complexity, even below the theoretically minimum number of 9, is certainly one of the avenues for overcoming this impasse.

It should be recalled at this point that Salisbury's analysis of minimal design requirements for dexterity was based on a particular definition of dexterity and a set of assumptions on the contact model. Thus, for instance, it can be easily shown that if soft-finger contacts are considered, fingers with at least four d.o.f.'s are needed to achieve dexterity in the sense above defined. Even the human hand can not be considerd dextrous if soft-finger contacts are enforced at the fingertips.

Other means of achieving dexterity can be devised if we allow some modifications of the concept of dexterity, and of the assumptions on contact models. In most applications, for instance, it is not necessary that the manipulated object can track a given trajectory in position and orientation at every instant during manipulation. Rather, it is sufficient that the object can be brought from the initial to the desired configuration, irrespective of what path it follows in the process (in control theoretic terminology, this is a distinction between ***functional*** and ***asymptotic*** reproducibility).

New possibilities of achieving dexterity in this latter sense are disclosed by considering different contact models. Thus, if one allows contacts between fingers and the object to be detached at some point during manipulation, and a new contact to be established in a different position, manipulation by "regrasping" or by "finger gaiting" can be accomplished.

Manipulation by regrasping involves a sequence of grasps on the object, alternated with phases in which the object is left alone on a support surface (e.g., a work table). Tournassoud ***et al.*** [140] proposed a planner for manipulation by regrasping. It should be noted that such manipulation method may rely on hands as simple as standard parallel–jaw grippers. A simple modification of the parallel–jaw gripper obtained by introducing a passive sliding bearing on one finger, along with a sophisticated planning strategy, may allow efficient part orientation by regrasping ([40]), or part sorting ([66]). Among the disadvantages of manipulation by regrasping, is certainly the need for grasping and releasing the object several times during manipulation, and the consequent time consumption in the process. Also, in the manipulation of irregular 3D objects, there may be a very limited number of stable configurations of the object on the supporting plane, in which the hand can leave the object safely enough during the release phase. The analysis and minimization of execution times for regrasping plans, and the characterization of robustness of such plans, are among the open problems in this area.

On the other hand, finger gaiting involves the use of four or more fingers, whereof one at a time is repositioned on the surface of the object, while the remaining fingers manipulate the object locally. Finger gaiting has been demonstrated for instance by Okada [101] and Fearing [38] to manipulate a sphere and a stick, respectively. Detailed theoretical analyses of some aspects of finger gaiting are reported in [50], [121], [27], [102]. Complexity and robustness of planning and execution, in particular for complex 3D objects, are two of the major open problems for finger gaiting manipulation as well.

A further degree of flexibility in manipulation is gained if one allows some of the contacts to slide during certain intervals of time. This kind of manipulation is actually most often observed in human hands, where controlled slippage is almost ubiquitous. Among the most important contributions towards exploiting slippage for enhancing dexterity, we cite [38], [19], [17], [29], [68], [27], [69], and [153]. In order to control slippage, beeing able to predict its occurrence is instrumental. This implies the need for an accurate analysis of friction and slippage phenomena. In particular, in the case of combined torsion and shear loading, evaluating a "margin of stability" for the contact before slipping is a very important but rather difficult task, of which only partially satisfying solutions are known so far (see e.g. [83], [61], [14], [54]; [41]). A second open problem in this area is the synthesis of sets of contact locations for selectively preventing and allowing slippage motions of grasped objects. Tools for the solution of problems of this sort can be derived from results in the synthesis of mechanical fixtures (see e.g. [3], [149]) and in the analysis of partial form–closure (see e.g. [74], [141], [10]). Also, the close relationship between research in the area of manipulation "at large" (regrasping, finger gaiting, and controlled slippage) and the field of part feeding and orienting by pushing, tilting, or fencing (see e.g. [83], [112], [36], [136], [78]) is brought to the attention of the reader, although it cannot be discussed here.

In the process of accurately analyzing the setup of the manipulation problem, with the aim of reducing the complexity of the hand hardware, a dramatic improvement is achieved if the assumption that bars rolling contacts is removed. In fact, as it will be discussed shortly, considering the effects of rolling contacts is a very effective way of lifting the difficulties of dextrous manipulation from the hardware level to the software's (i.e., to planning and control algorithms).

In most of the literature on dextrous manipulation, the non–rolling contact assumption is motivated by the hypothesis that fingers have very sharp curvature, so that the contact point between a fingertip and an object does not change much if the two roll on each other. However, the high–curvature hypothesis is hardly verified in most hand models, and changes in the contact point position due to rolling deeply affect grasping and manipulation. Effects on manipulation are evident from observing how the fundamental kinematic relations in manipulation, embodied in the hand jacobian $\mathbf{J}$ and the hand grasp matrix $\mathbf{G}$, are affected. In fact, while they usually only depend on finger configurations $\mathbf{q}$ and object posture $\mathbf{u}$ only, respectively (i.e., $\mathbf{J} = \mathbf{J}(\mathbf{q})$, $\mathbf{G} = \mathbf{G}(\mathbf{u})$), if rolling (or slipping) occurs, in general they become complex functions of both, $\mathbf{J} = \mathbf{J}(\mathbf{q}, \mathbf{u})$ and $\mathbf{G} = \mathbf{G}(\mathbf{q}, \mathbf{u})$. This entails that the kinematics, statics, and dynamics of the system are completely changed, and usually appear substantially more complex. A thorough exposition of the analytical and control aspects of manipulation with rolling is reported in [95].

If regarded as an undesired effect, rolling has to be compensated for in manipulation by using real-time feedback from tactile sensors indicating the actual position of the contact point at each time instant. Work in the direction of compensating effects of rolling has been carried out, among others, by [127], [154], [116], [106], [80], [26], [45].

It has been recently underscored that rolling can actually be turned to play in advantage of the design of simpler dextrous hands. In fact, rolling between rigid bodies in three-dimensional space is a well-known example of nonholonomically constrained motion. A knife-edge cutting a sheet of paper and a cat falling onto its feet are examples of natural nonholonomic systems, while bycicles and cars (possibly with trailers) are familiar examples of artificially designed nonholonomic devices. The most notable characteristic of nonholonomic systems is that they can be driven to a desired configuration in a d-dimensional configuration manifold using less than d inputs. Since "inputs" in engineering terms translates into "actuators", devices designed by intentionally introducing nonholonomic mechanisms can spare hardware costs without sacrifying dexterity.

While nonholonomy in a system is often regarded as an annoying side-effect of other design considerations (this is how most people consider e.g. car maneuvering for parallel-parking), purposeful introduction of nonholonomy in robotic system design has been considered previously (see e.g. [18], [104], [133]).

In the case of surfaces rolling on top of each other, the nonholonomy stems from the fact that the rolling-without-slipping constraints are written in terms of the time derivatives of the Lagrangian coordinates of the system, and that such constraints can not be integrated in a form that depends only on the coordinates themselves. As an effect of this, if a fingertip rolls on the surface of the object, not only the contact point moves on both surfaces, but, if a closed loop path is followed in the input space, the final contact point will not in general come back to the starting position. Roughly speaking, concatenating suitable sequences of input loops, it is possible to gain control over more configurations than available inputs (actuators).

To exploit such possibilities, a detailed analytical model of the kinematic laws of rolling contacts is fundamental. Formulas for predicting how the contact points and the relative orientation of the surfaces evolve with rolling, have been investigated first by Cai and Roth [23] and Montana [92], independently. Li and Canny [75] studied the problem of rolling by putting it in the framework of nonlinear control systems theory, and showed that a ball rolling on a plane can be displaced and reoriented at will within its five-dimensional configuration manifold (i.e., is controllable) by only using two inputs. The same was shown to hold for two balls rolling on each other, provided that their radius is different. A geometric algorithm was proposed by these authors to plan motions of the sphere on the plane, leading the sphere to an arbitrary final configuration on the plane. Murray *et al.* [95] report about using controlled rolling for repositioning the fingertips of a hand on the surface of the grasped object.

With such theoretical work as motivation, [12] investigated the possibility of building dextrous hands with a minimal number of actuators by exploiting rolling. Based on Li and Canny's results, it can be conjectured that *almost any* object can be dextrously manipulated by using hands with only three actuators (the third input simply translates the plane finger along its normal direction). A recent result reported in [16], stating that any convex object of revolution rolling on a plane finger is controllable by rolling, can be regarded as a strong support to such conjecture. A schematic diagram of the first prototype of the dextrous hand with a minimal number of actuators built in our laboratory is reported in fig. 1. The lower finger translates in its plane, thus rolling the object. The vertical actuator on the upper finger is used for compensating the change in height of objects during rolling, while maintaining a suitable grasping force. Such actuator could in principle be replaced by a constant-force spring. To achieve full dexterity, a further actuator (not shown in fig 1) is used to translate the whole hand vertically. An intrinsic tactile sensor is used for detecting contact points, measuring

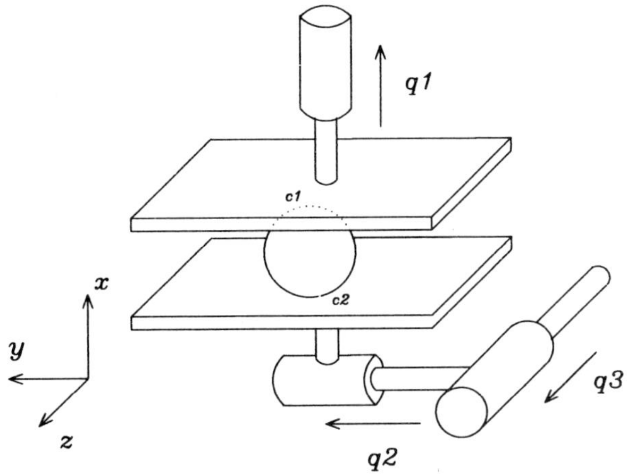

Figure 1: Schematic of the hand developed at Centro "E. Piaggio" for experimenting dextrous manipulation by rolling.

contact forces, and avoid slippage during rolling. The hand is currently capable of dextrously manipulating spheres and objects of revolution, using both Li and Canny's algorithm and an algorithm proposed in [12]. For objects of more general shape, no closed form algorithm for planning is known at the state, and only iterative planners have been realized so far. If the object shape is not known a priori, the hand can manipulate the object by rolling it between its fingers, and data obtained from intrinsic tactile sensors on the fingers ([13]) are used to reconstruct a model of the surface, on which the planner subsequently operates.

Manipulation by rolling is a challenging new area, whose promises in terms of hardware simplification still need much theoretical work to be fully supported. Among the open problems, we mention the further difficulties in planning due to workspace limitations of fingers, and the problem of stabilizing nonholonomic systems in general (due to Brockett's theorem). Also, the generalization to nonholonomic systems of useful notions such as manipulability and dextrous workspace (see e.g. [108] and [28]) seems to be important towards engineering applications of rolling. Finally, extensions to manipulation by rolling of polyhedral or irregular objects is necessary to support applications in most industrial environments. Work in this latter direction is being undertaken with the help of the theoretical framework of [77], and the relationship with work reported in [40] is also to be investigated.

4 Design of hands for grasping

"Grasping" indicates an action of a robot hand on an object consisting in preventing its motions relative to the hand, possibly in the face of disturbance forces acting on the object itself. The task of grasping is therefore, at least in some sense, converse to that of manipulation, and it can be expected that in the design of a hand, tradeoffs between dexterity and grasping robustness have to be seeked.

In order to define what grasping robustness is, the notions of "form–closure" and "force–closure" of grasps are instrumental. These properties concern the capability of the grasp to completely or partially constrain the motions of the manipulated object, and to apply arbitrary contact forces on the object itself, without violating friction constraints at the contacts.

Form–closure is the ability of a hand to prevent motions of the object, relying only on unilateral, frictionless contact constraints. This problem (which also has direct bearing to the design of mechanical fixtures and jigs for manufacturing parts) has been studied since at least 1875. Early results showed that at least four frictionless contacts are necessary for grasping an object in the plane, and seven in the 3D case. An active area of research is the synthesis of force–closure grasps, i.e., given the object geometry, where to place contacts so as to prevent object motions. [4], [91], [128], and [82] successively discussed the possibility of finding form–closure grasps on different surfaces, and showed that there is no such grasp on finite surfaces of revolution. Mishra ***et al.*** [91] and Markenscoff ***et al.*** [82] showed that four and seven contacts are also sufficient for the form–closure grasp of any polyhedron in the 2D and 3D case, respectively. The work of Nguyen [98], and Markenscoff and Papadimitriou [81], generated much interest in constructive procedures for placing contacts on given objects to achieve form–closure.

There is also a form–closure analysis problem, i.e., given an object and a set of contact locations, to decide whether the object has any degree-of-freedom left, and which. Both qualitative (true/false) tests (see e.g. [74], [83], [98], [91], [48]) and quantitative (quality index) tests (see e.g. [73] and [141]) have been proposed for form–closure. As already mentioned, the concept of ***partial*** form–closure proves very useful in an-

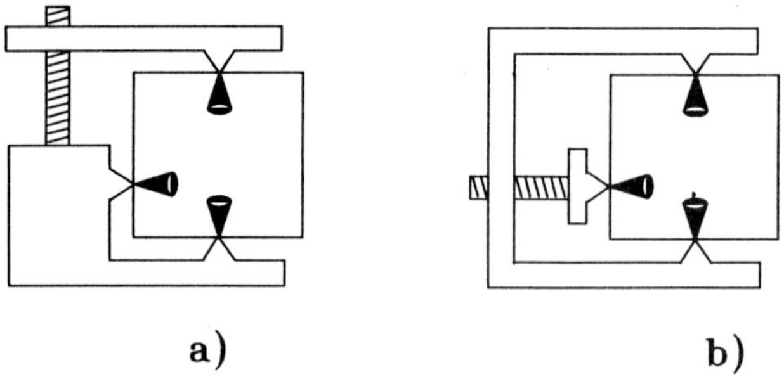

Figure 2: Force–closure depends upon the end–effector.

lyzing and planning manipulation by controlled slippage. A recent extension of the classical notion of form–closure is the so-called immobilization problem, where the secon-order effects of the curvature of the surfaces are taken into account, to provide more detailed results (see e.g. [119], [120], [143]).

The analysis of form–closure is intrinsically geometric, and does not take into account the kinematics and characteristics of the end–effector. On the other hand, the concept of "force–closure" is often used with the intuitive meaning that motions of the grasped object are completely (or partially) restrained despite whatever external disturbance, by virtue of suitably large contact forces that the constraining device (the end-effector) is capable to exert on the object. Frictional contacts are usually considered in force–closure. Literature on the synthesis of force–closure grasps include [98], [99], [109], [114]. The analysis of force–closure has been considered among others by [99], [39], [27], and [97].

Few authors in the literature payed attention to the relations between grasping and the end-effector structure and kinematics. Among those who did, are [142], [147], [113], and [55]. To explain why the end-effector kinematic structure is relevant to force–closure, consider the grasps depicted in fig. 2–a and fig. 2–b, where the same object is held by two different end-effectors through three identical contacts (friction cones are depicted by shaded sectors). It is intuitively clear that, while the grasp in fig. 2–a can resist arbitrary forces externally applied on the object by suitably "squeezing" the object, the grasp in fig. 2–b can not oppose e.g. to forces pulling the object to the right in the horizontal direction, since no "squeezing" is allowed by the end-effector. Similar cases may occur whenever the

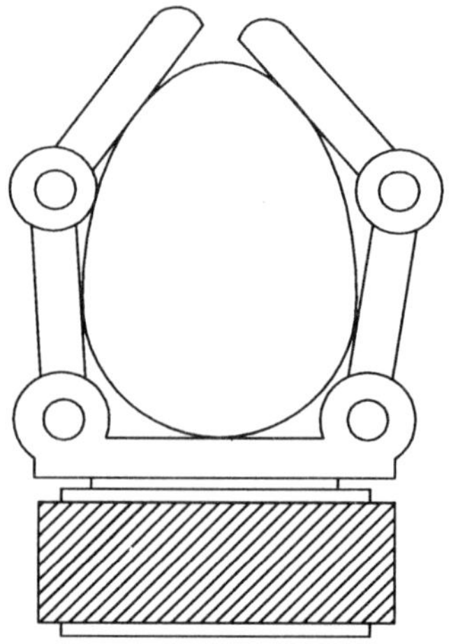

Figure 3: An example of power grasping.

end-effector has fewer degrees–of–freedom than necessary to arbitrarily control contact forces (i.e., it is ***kinematically defective***). Far from being a pathological case, kinematic deficiency is rather a normal condition in simple industry-oriented grippers, as well as in more complex devices such as dextrous robot hands when used in "power grasp" configuration. A definition of force–closure that takes into account the kinematics of the gripping device was proposed in [10], along with an efficient, exact algorithm for testing such property.

By the term "power grasping", or the equivalent expressions "enveloping grasping" and "whole–hand manipulation" ([123]), the action of a hand holding an object by using not only its fingertips, but also the internal phalanges and the palm is denoted. An example of such grasp is depicted in fig. 3. In the observation of human and animal grasping and manipulation, the fundamental role played by the inner parts of the hand (palm and proximal phalanges) to enhance both the stability of the grip and the versatility of operation can be frequently observed (see e.g. [31], [56]). To transfer this enhanced robustness into robotic devices, researchers have conceived hands with the ability of using inner surfaces for contacting the object, and capable of sensing contact interactions. Ulrich ***et al.*** [144] designed a medium–complexity hand capable of several grasp modes, including power grasping. Mirza and Orin [89] showed the largely increased holding capability of a robot hand exploiting its inner links and palm for grasping, given limits on the actuator torques, and built the so-called DIGITS system (for Dexterous Integrated Grasping with Intrinsic Tactile Sensing) to experimentally assess such grasping style. Vassura and Bicchi [145] proposed a hand whose design was integrally thought for whole-hand manipulation. A hand inspired to such philosophy has been realized at the University of Bologna by Bonivento, Vassura, and co-workers (see e.g. [85], [86]). The hand described in [117] was also designed to manipulate objects by using its inner surfaces.

Although both simple industrial grippers and power grasping hands share kinematic defectivity, the attention of the robotics comunity has been drawn only rather recently on the specific problems engendered by this feature. It turns out that most of the analytical studies in dextrous manipulation carried out in the simplifying assumption that objects are manipulated by fingertips only, do not apply directly to whole-hand manipulation and to defective grippers in general. Notice that, based on arguments similar to Salisbury's (see section 3), it can be easily argued that any hand with less than nine actuators is defective. To design and effectively use simpler hands, it is therefore necessary to deeply understand the implications of kinematic defectivity on the kinematics, dynamics, and control of hands.

The use of defective limbs in manipulating an object may imply that the object can not be controlled to move in arbitrary directions, but rather its velocity is constrained within certain subspaces. Tools for the analysis of the kinematics of series-parallel coordinating manipulation systems were provided in [55], [147], [146], and [157], while [60] reported about the analysis of underactuated manipulators. Explicit consideration of the kinematics and manipulability of whole-hand manipulation and of kinematically defective cooperating limbs in general was made in [15]. As a result of such analysis, it can be clearly seen that the more defectivity is introduced to get robust grasping, the less manipulability is left to the object. As already mentioned, one way of avoiding this impasse is to exploit rolling for gaining dexterity without increasing the number of joints.

A crucial problem in robot manipulation is the control of grasping forces to avoid slippage. Grasping, or internal, forces are defined as contact forces lying in the nullspace of the grasp matrix $\mathbf{G}$. Contact forces that are not internal directly affect the equilibrium of the object, and are sometimes referred to as manipulating forces. The problem of choosing joint torques so as to realize manipulating forces required by the task, while imposing grasping forces that guarantee slippage avoidance, is often referred to as the force distribution problem. The force distribution problem is a common problem with other robotic areas, as e.g. legged locomotion, cooperating and/or constrained manipulation, and has attracted a very large attention in the past few years, mostly with regard to non-defective systems (see e.g. [103], [72], [76], [97], [148], [63], [96], [107], [22], [20]).

The author pointed out that in defective systems, i.e., systems whose jacobian matrix is not full row rank, it may not be actually possible to choose grasping forces at will ([9]). Such phenomenon happens everytime the nullspace of the grasp matrix and the nullspace of the transpose of the hand jacobian have nontrivial intersection (i.e., the system is hyperstatic), such as is the case for the simple gripper in fig. 2-b. In the cited paper, the subspace of internal forces that can be actually used for avoiding slippage is evaluated by an algorithm that uses information on the compliance of bodies in contact. Several aspects of the force distribution problem in defective systems have been since studied, as reported e.g. in [6], [155], [84], [57], [90], [156], [65], and [52].

The fact that in the above mentioned case the rigid body model fails, and it is necessary to have an estimate of the bodies' compliance to solve the force distribution problem, shows that in defective manipulation hyperstatic condition may arise. The role of compliance in grasping has been investigated by many authors, beginning with [43]. Cutkosky and Kao [32] discussed how to compute the aggregated compliance matrix of a hand-object system, while [132] reported on contact stress models for manipulation. Relations of compliant (and rolling) contacts with the stability of the grasp were considered in [93], [94], [141], and [51]. Compliant manipulation was also considered in [88]. Bicchi and Prattichizzo [11] considered the dynamic structure of general (including defective) manipulations systems, and presented some standard forms of the linearized dynamics of manipulation systems. Control of power grasping systems has been considered by [129], [118], and [115].

Many open problems remain to be solved in order to be able to design robot hands to effectively exploit defectivity to increase grasp robustness and reduce hardware complexity. Among these, perhaps the most important is the need for a reliable estimate of contact compliance. In fact, it is hardly reasonable in any practical case to assume that such data are known a priori. However, it is conceivable that from the measurement of joint displacements and contact forces, compliance parameters can be identified on-line, in a fashion similar to that used to estimate inertial parameters in adaptive controllers for robots.

5 Conclusions

In this paper, a review of some of the work being done in robotic manipulation has been provided, and trends have been highlighted that, in the author's view, might allow those devices to find larger applications in the real world. Due to space limitations, many other important aspects could not be discussed, such as tactile sensing. It is noted in passing that also in those fields, a trend towards simplification of hardware by application of more sophisticated analysis can be recognized (consider for instance work in "early taction", see e.g. [110], [134], [24]). As a final remark on ways that might be taken in order to simplify manipulation systems, work being done on the synthesis of simple perceptual and grasping strategies that try to capture human skill in synthetic behaviours for the hand should also be considered carefully (see e.g. [33], [135], [5], [67], [7], [138]).

Acknowledgements

The author wishes to thank the organizers and participants to the 7th ISRR for the atmosphere of the meeting, fertile of scientific discussion and constructive comments offered to an earlier version of this paper.

This work was partially supported by EC programs ESPRIT (project W.G. "LEGRO") and HCM (European Robotics Network, ERNET), and by CNR (project "Manipolazione Robotica: Coordinazione e Destrezza").

References

NOTE. The following acronyms are used in references:

- **ACC: Proc. American Control Conf.;**
- **CDC: Proc. IEEE Conf. on Decision and Control;**
- **ICRA: Proc. IEEE Int. Conf. on Robotics and Automation;**
- **JRA: IEEE Jour. on Robotics and Automation;**
- **TRA: IEEE Trans. on Robotics and Automation;**
- **TSMC: IEEE Trans. on Systems, Man, and Cybernetics;**
- **IROS: Proc. IEEE/RSJ Int. Conf. on Int. Robots and Systems;**
- **ISER: Proc. Int. Symp. Experimental Robotics;**
- **ISRR: Proc. Int. Symposium on Robotics Research;**
- **JRAS: Journal of Robotics and Autonomous systems;**
- **JRSAR: Japan Robotics Society – Advanced Robotics;**
- **ROMANSY: CISM-IFToMM Symp. on Theory and Practice of Robots and Manipulators;**
- **ROVISEC: Proc. Int. Conf. on Robot Vision and Sensory Control;**
- **SYROCO: Proc. IFAC Symposium on Robot Control.**

[1] Ali, M.S., Kyriakopoulos, K.J., and Stephanou, H.E.: The kinematics of the Anthrobot-2 dextrous hand. ICRA, pp. 3-705-3-710, 1993.

[2] Aristotle, *De partibus animalium*, 687a 7, ca. 340 BC.

[3] Asada, H., and By, A.B.: Kinematic analysis of workpart fixturing for flexible assembly with automatically reconfigurable fixtures. TRA, vol. 1, no. 2, pp. 86–94, 1985.

[4] Baker, B.S., Fortune, S., and Grosse, E.: Stable prehension with a multifingered hand. ICRA, pp. 570–575, 1985.

[5] Bard, C., and Troccaz, J.: Automatic preshaping for a dextrous hand from a simple description of objects. IROS, 1990.

[6] Barkat, B., Bessonnet, G., and Lallemand, J.P.: Hyperstatic grasping optimization with finger deformability and sliding constraints. ICRA, pp. 1923–1930, 1994.

[7] Bekej, G. A., Liu, H., Tomovic, R., and Karplus, W. J.: Knowledge -based control of grasping in robot hands using heuristics from human motor skills. TRA, vol. 9, no. 6, pp. 709–722, 1993.

[8] Bergamasco, M., Dario, P., Sartini, G.: Design of a Medium Complexity End Effector for Teleoperation Applications in Space," Proc. of the 1st IARP Int. Workshop on Robotics in Space, 1991.

[9] Bicchi, A.: On the problem of decomposing grasp and manipulation forces in multiple whole–limb manipulation. JRAS, vol. 13, pp. 127–147, 1994.

[10] Bicchi, A.: On the closure properties of robotic grasping. IJRR, vol.14, no.4, 1995.

[11] Bicchi, A. and Prattichizzo, D.: A standard form for the dynamics of general manipulation systems. ICRA, pp. 2810–2815, 1995.

[12] Bicchi, A., and Sorrentino, R.: Dexterous manipulation through rolling. ICRA, pp. 452–457, 1995.

[13] Bicchi, A., Salisbury, J.K., and Brock, D.L.: Contact sensing from force and torque measurements. IJRR, Vol.12, no.3, 1993.

[14] Bicchi, A., Salisbury, J.K., and Brock, D.L.: Experimental evaluation of friction data with an articulated hand and intrinsic contact sensors. ISER, 1994b.

[15] Bicchi, A., Melchiorri, C., and Balluchi, D.: On the mobility and manipulability of general multiple limb robotic systems. TRA, vol.11, no.2, pp. 215–228, 1995a.

[16] Bicchi, A., Prattichizzo, D., and Sastry, S.S.: Planning motions of rolling surfaces. CDC, 1995b.

[17] Brock, D.L.: Enhancing the dexterity of a robot hand using controlled slip. ICRA, pp. 249–251, 1988.

[18] Brockett, R.W.: On the rectification of vibratory motion. Sensors and Actuators, vol. 20, pp. 91–96, 1989.

[19] Brost, R.C. 1988. "Automatic grasp planning in the presence of uncertainty. IJRR, vol. 7, no. 1, pp. 3–17, 1988.

[20] Bruyninckx, H., Demey, S., Dutré, S., and DeSchutter, J.: "Kinematic models for model based compliant motion in the presence of uncertainty, IJRR, vol. 14, no.5, pp. 465–482, 1995.

[21] Buss, M., and Hashimoto, H.: Dextrous robot hand experiments. ICRA, pp. 1680–1686, 1995.

[22] Buss, M., Hashimoto, H., and Moore, J.B.: Grasping force optimization for multifingered robot hands. ICRA, pp. 1034–1039, 1995.

[23] Cai, C., and Roth, B.: On the spatial motion of a rigid body with point contact. ICRA, pp. 686–695, 1987.

[24] Caiti, A., Canepa, G., DeRossi, D., Germagnoli, F., Magenes, G., Parisisni, T.: Towards the realization of an artificial tactile system: Fine-form discrimination by a tansorial tactile sensor array and neural inversion algorithms. TSMC, vol. 25, no. 6, pp. 4–15, 1995.

[25] Carlisle, B.: Is U.S. robotics research of any use to U.S. industry? IEEE Robotics and Automation Society Newsletter, 1991.

[26] Chang, D.C., Cutkosky, M. R.: Rolling with deformable fingertips. IROS, pp. 194–199, 1995.

[27] Chen, I-M., and Burdick, J.W.: A qualitative test for n-finger force–closure grasps on planar objects with applications to manipulation and finger gaits. ICRA, pp. 814–820, 1993.

[28] Chen, W., and Kumar, V.: Workspace of planar cooperating robots with rolling contacts. JRSAR, vol.9, no. 4, 1995.

[29] Cole, A. A., Hsu, P., and Sastry, S. S.: Dynamic control of sliding by robot hands for regrasping. TRA, vol. 8, no.1, pp. 42–52, 1992.

[30] Crossley, E.F.R., and Umholtz, F.G.: Design for a three–fingered hand. Mechanism and Machine Theory, vol. 12, pp. 85–93, 1977.

[31] Cutkosky, M.R.: Robotic grasping and fine manipulation. Kluwer Academic, 1985.

[32] Cutkosky, M.R., and Kao, I.: Computing and controlling the compliance of a robotic hand. TRA, vol. 5, no. 2, pp. 151–165, 1989.

[33] Dario, P., and Buttazzo, G.: An anthropomorphic robot finger for investigating artificial tactile perception. IJRR, vol. 6, no. 3, pp. 25–48, 1987.

[34] Ebner, M., and Wallace, R.S.: A direct–drive hand: design, modeling and control. ICRA, pp. 1668–1674, 1995.

[35] Engler, J., and Groover, M.P.: Design of an anthropomorphic electromechanical hand with exoskeletal control to emulate human hand dexterity. Proc. 1989 World Conf. on Robotics Research, pp. 6–41, 1989.

[36] Erdmann, M.A.: On a representation of friction in configuration space. IJRR, vol. 13, no. 3, 1994

[37] Farry, K.A., and Walker, I.D.: Myoelectric teleoperation of a complex robotic hand. ICRA, pp. 502–509, 1993.

[38] Fearing, R.: Simplified Grasping and Manipulation with Dextrous Robot Hands. JRA, vol. 2, no. 4, pp. 188–195, 1986.

[39] Ferrari, C., and Canny, J.: Planning optimal grasps. ICRA, pp. 2290–2295, 1992.

[40] Goldberg, K.: Feeding and sorting algorithms for the parallel -jaw gripper. ISRR, 1993.

[41] Goyal, S.: Planar sliding of a rigid body with dry friction: Limit surfaces and dynamics of motion". Ph.D. Thesis, Cornell University, 1989.

[42] Grupen, R.A., Henderson, T.C., and McCammon, I.D.: A survey of general–purpose manipulation. IJRR, vol. 8, no. 1, pp. 38–62, 1989.

[43] Hanafusa, H., and Asada, H.: Stable prehension by a robot hand with elastic fingers. Robot Motion Planning and Control, M. Brady (ed.), pp. 323–336. MIT Press, 1982.

[44] Harmon, L.D.: Robotic taction for industrial assembly. IJRR, vol. 3, no. 1, pp. 72–76, 1984.

[45] Härtl, H.: Dextrous manipulation with multifingered robot hands including rolling and slipping of the fingertips. JRAS, vol. 14, pp. 29–53, 1995.

[46] Hashimoto, H., Ogawa, H., Umeda, T., Obama, M., Tatsuno, K.: An unilateral master-slave hand system with a force–controlled slave hand. ICRA, pp. 956–961, 1995.

[47] Hayakawa, Y., and Kawamura, S.: Sensing actuators using a pneumatic bellows system and its application to a soft gripper. ICRA, pp. 398–403, 1993.

[48] Hirai, S., and Asada, H.: Kinematics and statics of manipulation using the theory of polyhedral convex cones. IJRR, vol. 12, no. 5, pp. 434–447, 1993.

[49] Hollerbach, J.: Robot hands and tactile sensing. AI in the 1980's and beyond, W.E.L. Grimson and R.S. Patil, eds., MIT Press, pp. 317–343, 1987.

[50] Hong, J., Lafferriere, G., Mishra, B., and Tan, X.: Fine manipulation with multifinger hands. ICRA, pp. 1568–1573, 1990.

[51] Howard, W.S., and Kumar, V.: Stability of planar grasp. ICRA, pp. 2822–2827, 1994.

[52] Howard, W. S., and Kumar, V.: Modeling and analysis of the compliance and stability of enveloping grasps. ICRA, pp. 1367–1372, 1995.

[53] Howe, R. D. and Cutkosky, M. R.: Touch sensing for robotic manipulation and recognition. Robotics Review 2, O. Khatib ed., MIT Press, Cambridge, MA. 1992.

[54] Howe, R.D., Kao, I., and Cutkosky, M.R.: The sliding of robot fingers under combined torsion and shear loading. ICRA, pp. 103–105, 1988.

[55] Hunt, K.H., Samuel, A.E., and McAree, P.R.: Special Configurations of Multi–finger Multi–freedom Grippers — A Kinematic Study. IJRR, vol. 10, no.2, pp.123–134, 1991.

[56] Iberall, T.: The nature of human prehension. ICRA, 1987.

[57] Ismaeil, O.M., and Ellis, R.E.: Grasping using the whole finger. ICRA, pp. 3111–3116, 1994.

[58] Jacobsen, S.C., Meek, S.G., and Fullmer, R.R.: An adaptive myoelectric filter. 6th IEEE Conf. Engineering in Med. & Biol. Soc., 1984a.

[59] Jacobsen, S.C., Wood, J.E., Knutti, D.F., Biggers, K.B.: The Utah-MIT Dextrous Hand: Work in progress. IJRR, vol.3, no.4, pp. 21-50, 1984b.

[60] Jain, A., and Rodriguez, G.: An analysis of kinematics and dynamics of underactuated manipulators. TRA, vol. 9, no. 4, pp. 411–422, 1993.

[61] Jameson, J.W.: Analythic techniques for automated grasp. Ph.D. Thesis, Stanford University, 1985.

[62] Jau, B. M.: Dexterous telemanipulation with four fingered hand system. ICRA, pp. 338–343, 1995.

[63] Joh, J., and Lipkin, H.: Lagrangian wrench distribution for cooperating robotic mechanisms. ICRA, pp.224–229. 1991.

[64] Jongkind, W.: Dextrous gripping in a hazardous environment. Ph.D. Thesis, Delft Univ. of Technology, 1993.

[65] Kaneko, M., Imamura, N., and Honkawa, K.: Contact points detection for inner link based grasps. JRSAR, vol.9, no. 4, 1995.

[66] Kang, D., and Goldberg, K.: Sorting parts by random grasping. TRA, vol. 11, no. 1, pp. 146–152, 1995.

[67] Kang, S.B., and Ikeuchi, K.: Toward automatic robot instruction from perception — Recognizing a grasp from observation. TRA, vol. 9, no. 4, pp. 432–443, 1993.

[68] Kao, I., and Cutkosky, M. R.: Quasistatic manipulation with compliance and sliding. IJRR, vol. 11, no. 1, pp. 20–37, 1992.

[69] Kao, I., and Cutkosky, M.R.: Comparison of theoretical and experimental force/motion trajectories for dextrous manipulation with sliding. IJRR, vol. 12, no. 6, pp. 529–534, 1993.

[70] Kato, I.: Mechanical hands. Survey Japan, 1982.

[71] Kato, I.: Development of WASEDA robot – The study of biomechansms at Kato Laboratory. Third Edition. Dept. of Mech. Eng., Waseda University, Tokyo, 1991.

[72] Kerr, J., and Roth, B.: Analysis of Multifingered Hands. IJRR, vol. 4, no. 4, pp. 3–17, 1986.

[73] Kirkpatrick, D., Kosaraju, S.R., Mishra, B., and Yap, C–K.: "Quantitative Steinitz's Theorem with Applications to Multifingered Grasping. TR460, Dept. of Computer Science, Courant Inst. of Mathematical Sciences, New York University, 1989.

[74] Lakshminarayana, K.: Mechanics of form closure. ASME paper 78-DET-32, 1978

[75] Li, Z., and Canny, J.: Motion of two rigid bodies with rolling constraint. TRA, vol. 6, no.1, pp. 62–72, 1990.

[76] Li, Z., Hsu, P., and Sastry, S.: Grasping and coordinated manipulation by a multifingered robot hand. IJRR, vol.8, no.4, 1989.

[77] Liu, Y., and Popplestone, R.: A group theoretic formalization of surface contact. IJRR, vol. 13, no. 2, pp. 148–161, 1994.

[78] Lynch, K. M., and Mason, M. T.: Pulling by pushing, slip with infinite friction, and perfectly rough surfaces. IJRR, vol. 14, no. 2, pp. 174–183, 1995.

[79] Machida, K., Toda, Y., Murase, Y., Komada, S.: Precise space telerobotic system using 3–finger multisensory hand. ICRA, pp. 32–38, 1995.

[80] Maekawa, H., Tanie, K., and Komoriya, K.: Tactile sensor based manipulation of an unknown object by a multifingered hand with rolling contact. ICRA, pp. 743–750, 1995.

[81] Markenscoff, X., and Papadimitriou, C.H.: Optimum grip of a polygon. IJRR, vol. 8, no. 2, pp. 17–29, 1989.

[82] Markenscoff, X., Ni, L., and Papadimitriou, C.H.: The geometry of grasping. IJRR, vol. 9, no. 1, pp. 61–74, 1990.

[83] Mason, M.T. and Salisbury, J.K.: Robot hands and the mechanics of manipulation. MIT Press, Cambridge, MA, USA. 1985.

[84] Melchiorri, C.: Static force analysis for general cooperating manipulators. ICRA, pp. 888–893, 1994.

[85] Melchiorri, C., and Vassura, G. "Mechanical and control features of the University of Bologna Hand Version 2. IROS, pp. 187-193, 1992.

[86] Melchiorri, C., and Vassura, G.: Implementation of whole–hand manipulation capability in the U.B. Hand system design. JRSAR, vol.9, no. 4, 1995.

[87] Menzel, R., Woelfl, K., and Pfeiffer, F.: Grasping with a dextrous robot hand. SYROCO, pp. 303–308, 1994.

[88] Michelmann, P., and Allen, P.: Compliant manipulation with dextrous robot hand. ICRA, pp. , 1993.

[89] Mirza, K., and Orin, D.E.: Force distribution for power grasp in the Digits system. ROMANSY, 1990.

[90] Mirza, K., and Orin, D.E.: General formulation for force distribution in power grasp. ICRA, pp. 880–887, 1994.

[91] Mishra, B., Schwartz, J.T., and Sharir, M.: On the existence and synthesis of multifinger positive grips. Algorithmica, vol. 2, no. 4, pp. 541–558, 1987.

[92] Montana, D. J.: The kinematics of contact and grasp. IJRR, 7(3), pp. 17–32, 1988.

[93] Montana, D.J.: The condition for contact grasp stability. ICRA, pp. 412–417. 1991.

[94] Montana, D.J.: Contact stability for two–fingered grasps. TRA, vol. 8, no. 4, pp. 421–430, 1992.

[95] Murray, R.M., Li, Z., and Sastry, S.S.: A mathematical introduction to robotic manipulation. Boca Raton: CRC Press. 1994.

[96] Nagai, K., and Yoshikawa, T.: Dynamic manipulation/grasping control of multifingered robot hands. ICRA, pp. 1027–1032, 1993.

[97] Nakamura, Y., Nagai, K., and Yoshikawa, T.: Dynamics and stability in coordination of multiple robotic systems. IJRR, vol. 8, no. 2, pp. 44–61, 1989.

[98] Nguyen, V.D.: The synthesis of stable force-closure grasps". TR 905. Cambridge (MA): MIT A.I. Lab., 1986.

[99] Nguyen, V.D.: Constructing force–closure grasps. IJRR, vol. 7, no. 3, pp. 3–16, 1988.

[100] Nicholls, H.R. and Lee, M.H.: A survey of robot tactile sensing technology. IJRR, Vol 8, No. 3, pp. 3-30, 1989.

[101] Okada, T.: Object handling system for manual industry. TSMC, vol. SMC–9, no. 2, 1979.

[102] Omata, T, and Nagata, K.: Planning reorientation of an object with a multifingered hand. ICRA, pp. 3104–3109, 1994.

[103] Orin, D.E., and Oh, S.Y. : Control of Force Distribution in Robotic Mechanisms Containing Closed Kinematic Chains. J. Dyn. Sist. Meas. Contr., vol. 102, 1981.

[104] Ostrowski, J., Lewis, A., Murray, R., and Burdick, J.: Nonholonomic mechanics and locomotion: the snakeboard example. ICRA, pp. 2391–2396, 1994.

[105] Paetsch, W., and Kaneko, M.: A three-finger multijointed gripper for experimental use. IROS 1990.

[106] Paljug, E., Yun, X., and Kumar, V.: Control of rolling contacts in multi–arm manipulation. TRA, vol. 10, no. 4, pp. 441–452, 1994.

[107] Panagiotopoulos, P.D., and Al–Fahed, A.M.: Robot hand grasping and related problems: optimal control and identification. IJRR, vol. 13, no. 2, pp. 127–136, 1994.

[108] Park, F.C., and Brockett, R.W.: Kinematic dexterity of robotic mechanisms. IJRR, vol. 13, no. 1, pp. 1–15, 1994.

[109] Park, Y.C., and Starr, G.P. : Grasp synthesis of polygonal objects using a three-fingered robot hand. IJRR, vol. 11, no. 3, pp. 163–184, 1992.

[110] Pati, Y.C., Krishnaprasad, P.S., and Peckerar, M.C.: An analog neural network solution to the inverse problem of "early taction" . TRA, vol. 8, no. 2, pp. 196–212, 1992.

[111] Pertin–Troccaz, J.: Grasping: a state of the art. The Robotics Review I, MIT Press, pp. 71–98, 1989.

[112] Peshkin, M.A., and Sanderson, A.C.: Planning robotic manipulation strategies for workpieces that slide. JRA, vol. 4, no. 5, pp. , 1988.

[113] Pollard, N.S., and Lozano–Perez, T.: Grasp stability and feasibility for an arm with an articulated hand. ICRA, pp. 581–585, 1990.

[114] Ponce, J., Stam, D., and Faverjon, B.: On computing two-finger force-closure grasps of curved 2D objects. IJRR, vol. 12, no. 3, pp. 263–273, 1993.

[115] Prattichizzo, D., Salisbury, J.K., and Bicchi, A.: Contact and grasp robustness measures: Analysis and experiments. ISER 1995.

[116] Reynaerts, D., and Van Brussel, H.: Design and transputer based control of a two fingered robot hand. SYROCO, pp. 309–314, 1994.

[117] Reynaerts, D., and Van Brussel, H.: Whole–finger manipulation with a two–fingered robot hand. JRSAR, vol.9, no. 4, 1995.

[118] Reznik, D., and Lumelsky, V.: Multi–finger "hugging": a robust approach to sensor based grasp planning. ICRA, pp. 754–760, 1994.

[119] Rimon, E., and Burdick, J.W.: Towards planning with force constraints: On the mobility of bodies in contact. ICRA, pp. 994–1000, 1993.

[120] Rimon, E., and Burdick, J. W.: New bounds on the number of frictionless fingers required to immobilize 2D objects. ICRA, pp. 751–757, 1995.

[121] Rus, D.: Dextrous rotations of polyhedra. ICRA, pp. 2758–2763, 1992.

[122] Russell, A.: Robot tactile sensing. Prentice Hall, 1990.

[123] Salisbury, J. K.: Whole-arm manipulation. ISRR, 1987.

[124] Salisbury, J.K., Brock, D.L., Chiu, S.L.: Integrated language, sensing and control for a robot hand. ISRR 1985.

[125] Salisbury, J.K., and Craig, J.J.: Articulated hands, force control and kinematic issues. IJRR, Vol. 1, No. 1, 1982.

[126] Salisbury, J. K., and Roth, B.: Kinematic and force analysis of articulated mechanical hands. J. Mech. Transm. Automat. in Des., vol. 105, 1983.

[127] Sarkar, N., Yun, X., and Kumar, V.: Dynamic control of 3–D rolling contacts in two–arm manipulation. ICRA, pp. 978–983, 1993.

[128] Selig, A.J., and Rooney, P.K.: Reuleaux pairs and surfaces that cannot be gripped. IJRR, vol. 8, no. 5, pp. 79–87, 1989.

[129] Sheridan, M.J., Ahalt, S.C., and Orin, D.E.: Fuzzy control fro robotic power grasp. JRSAR, vol.9, no. 4 , 1995.

[130] Shimojo, M., Sato, S., Seki, Y., and Takahashi, A.: A system for simultaneous measuring grasping posture and pressure distribution. ICRA, pp. 831–836, 1995.

[131] Singh, L., Silverthorne, A., Gottschlich, S., and Stephanou, H.: "Variably–autonomous manipulation. ICRA, pp. 3217–3222, 1994.

[132] Sinha, P.R., and Abel, J. M.: A contact stress model for multifingered grasps of rough objects. TRA, vol. 8, no. 1, pp. 7–22. February 1992.

[133] Sordalen, O. J., and Nakamura, Y.: Design of a nonholonomic manipulator. ICRA, pp. 8–13, 1994.

[134] Speeter, T.H.: Three–dimensional finite element analysis of elastic continua for tactile sensing. IJRR, vol. 11, no. 1, pp. 1–19, 1992.

[135] Stansfield, S.A.: Robotic grasping of unknown objects: A knowledge–based approach. IJRR, vol. 10, no. 4, pp. 314–326, 1991.

[136] Swanson, P.J., Burridge, R.R., and Koditscheck, D.E.: Global asymptotic stability of a passive juggler. ICRA, pp. 1983–1988, 1995.

[137] Tanikawa, T., Arai, T., Ojala, P., Saeki, M.: Two–finger micro hand. ICRA, pp. 1674–1679, 1995.

[138] Teichmann, M., and Mishra, B.: Reactive algorithms for grasping using a modified parallel jaw gripper. ICRA, pp. 1931–1936, 1994.

[139] Tomovic, R., and Boni, G.: An adaptive artificial hand. IRE Transactions Automatic Control, vol AC-7, no.3, pp.3–10, 1962.

[140] Tournassoud, P., Lozano–Perez, T., and Mazer, E.: Regrasping. ICRA, pp. 1924–1928, 1987.

[141] Trinkle, J.C.: On the stability and instantaneous velocity of grasped frictionless objects. TRA, vol.8, no.5, 1992.

[142] Trinkle, J.C., Abel, J.M., and Paul, R.P.: An investigation of frictionless enveloping grasping in the plane. IJRR, vol. 7, no. 3, pp. 33-51, 1987.

[143] Trinkle, J.C., Farahat, A.O., and Stiller, P.F.: Second order stability cells of a frictionless rigid body grasped by rigid fingers. ICRA, pp. 2815–2821, 1994.

[144] Ulrich, N., Paul, R., and Bajcsy, R.: A medium–complexity compliant end effector. ICRA, pp. 434–436, 1989.

[145] Vassura, G. and Bicchi, A.: Whole hand manipulation: design of an articulated hand exploiting all its parts to increase dexterity. Robots and Biological Systems, NATO-ASI Series, Springer-Verlag, 1989.

[146] Waldron, K. J., and Hunt, K. H.: Series–parallel dualities in actively coordinated mechanisms. IJRR, vol. 10, no. 5, pp. 473–480. 1991

[147] Waldron, K.J., Raghavan, M., and Roth, B.: Kinematics of a hybrid series–parallel manipulation system. J. Dyn. Sys. Meas. Control, vol. 111, no. 2, pp. 211–221, 1989.

[148] Walker, I.D., Freeman, R.A., and Marcus, S.I.: Analysis of motion and internal loading of objects grasped by multiple cooperating manipulators. IJRR, Vol. 10, No. 4, Aug. 1991.

[149] Wallack, A., and Canny, J.: Planning for modular and hybrid fixtures. ICRA, pp. 520–525, 1994.

[150] Weigl, A., and Seitz, M.: Vision assisted disassembly using a dextrous hand–arm system: An example and experimental results. SYROCO, pp. 314–322, 1994.

[151] Whitney, D.E.: Real robots don't need jigs. ICRA, 1986.

[152] Wöhlke, G.: A programming and simulation environmnet for the Karlsruhe dextrous hand. JRAS, vol. 9, pp. 243–263, 1990.

[153] Xue, Y., and Kao, I.: Dextrous sliding manipulation using soft fingertips. ICRA, pp. 3397–3402, 1994.

[154] Yun, X.: Object handling using two arms without grasping. IJRR, vol. 12, no. 1, pp. 99–106, 1993.

[155] Zhang, X.Y., Nakamura, Y., Goda, K., and Yoshimoto, K.: Robustness of power grasp. ICRA, pp. 2828–2836, 1994.

[156] Zhang, Y., and Gruver, W. A.: Definition and force distribution of power grasp. ICRA, pp. 1373–1378, 1995.

[157] Zheng, Y. F., and Luh, J.Y.S.: On the inertia duality of parallel–series connections of two robots in operational space. TRA, vol. 9, no. 6, pp.846–854. 1993.

An Exploration of Nonprehensile Two-Palm Manipulation: Planning and Execution

Michael Erdmann
School of Computer Science and the Robotics Institute
Carnegie Mellon University
5000 Forbes Avenue
Pittsburgh, PA 15213-3891

1 Introduction

This paper describes our current research into *nonprehensile palm manipulation*. The term "palm" refers to the use of the entire device surface during manipulation, as opposed to use of the fingertips alone. The term "nonprehensile" means that the palms hold the object without wrapping themselves around it, as distinguished from a force/from closure grasp often employed by a fingered hand. Indeed, nonprehensile operations such as purposeful sliding and constrained dropping constitute important palm primitives.

We have implemented a system for orienting parts using two palms. The system consists of a planner and an executive. As input, the system expects a geometric description of a part, its center of mass, the coefficients of friction between the part and each of the palms, and a start and goal configuration of the part in stable contact with one of the palms. As output, the system computes and executes a sequence of palm motions designed to reorient the part from the specified start to the specified goal configuration.

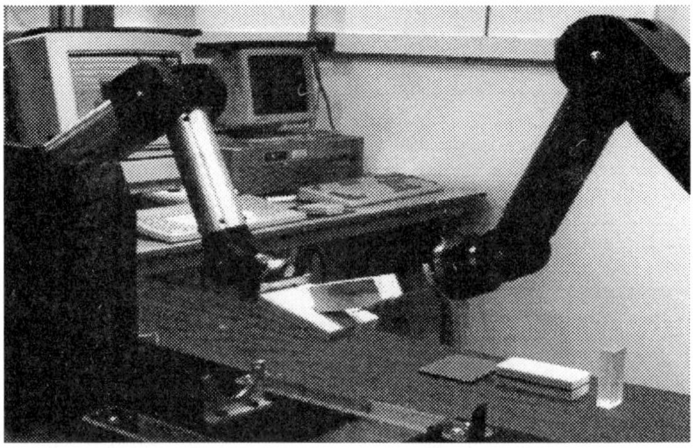

Figure 1: The experimental setup consists of two Zebra robots, with flat palms attached to each wrist. The Zebras are shown manipulating an irregular block.

2 System Specifics

For hardware we are using two Zebra Zero robots. We replaced the standard grippers on these arms with two flat surfaces; these are the "palms". Figure 1 shows the experimental setup. We used three different materials for the surfaces of these palms, namely aluminum, aluminum covered with duct tape, and aluminum covered with a hard impact-absorbing foam. These three surfaces provided us with a range of friction coefficients.

The system expects 2.5-dimensional polyhedral parts, specifically parts that are extruded convex polygonal shapes. The actual parts oriented consisted of several rectangular blocks as well as one irregularly-shaped block. The irregular block's two-dimensional cross-section consisted of a five-sided polygon with three stable edges, one unstable edge, and one marginally stable edge. Again, we used a variety of materials for the parts, namely hard foam, hard plastic, and wood.

The planner is planar, that is, it assumes that the parts have two translational and one rotational degrees of freedom. Hence the 3D parts oriented were extruded 2D shapes. The current version of the planner focuses on Type-B contacts, to use the classification of [Lozano-Pérez 1983, Section V]. In other words, the planner focuses on contacts involving the palm surfaces but not the boundaries of those surfaces. Viewed in 2D, this class of contacts includes both vertex and line contacts of the part with the palm lines. It excludes Type-A contacts, that is, again viewed in 2D, it excludes contacts between a line segment of the part and a vertex of the palms.

Finally, a word about sensing. The plans are sensorless. The planner takes careful account of object geometry, friction, and gravity, including knowledge of the object's center of mass. The resulting plans execute without any sensing of the part by the robots.

3 Examples

Figures 2 and 3 show two plans produced by our planner. We will refer to the left palm as "Palm 1" and the right palm as "Palm 2". The first plan reorients a rectangular block. The block was plastic, Palm 1 was aluminum, and Palm 2 was hard foam. The second plan reorients the irregular block mentioned earlier. The block was wood, Palm 1 was aluminum, and Palm 2 was aluminum covered with duct tape. The exact coefficients of friction may be found in the figure captions. The frames correspond to snapshots before and after actions planned by our planner.

Both plans have a similar overall character, but the precise actions differ. Both plans start with a TILT of Palm 1 to an angle that causes the block to slide to the right. The planner takes the tilt angle in each case to be 3 degrees greater than the angle of friction for that case. As part of the TILT, Palm 2 catches the block. With that, the reorienting process starts. Here are some highlights:

- **Slide.** A SLIDE operation consists of some action by the palms in which one palm slides relative to the part. Each plan happens to include one such SLIDE operation, beyond the initial tilt. In Figure 2, Palm 2 drags the block to the right between Frames 5 and 6, moving tangentially to Palm 1. In Figure 3, Palm 2 drags the block down and to the right between Frames 2 and 3, again moving tangentially to Palm 1. The basic operation is the same in both cases, but the angles of the palms are different, in order to account for the different coefficients of friction.

- **Regrip.** Both plans contain a number of REGRIPS. A REGRIP is an action in which a palm rotates about its contact point with the part. The plans call for such REGRIP operations in order to strategically configure the palms for some subsequent action. For instance, between Frames 4 and 5 of Figure 2, Palm 2 reorients itself to a shallow angle in order to then drag the block to the right. Having dragged the block to the right, Palm 2 then reorients itself to a steeper angle in order to perform the rotation from Frame 7 to Frame 8. Similarly, the actions between Frames 3 and 15 of Figure 3 basically constitute a dance of REGRIP and ROTATE operations required to turn the block upright without permitting it to slip at the contacts or otherwise lose equilibrium.

- **Rotate.** No matter what else the palms do, ultimately they must ROTATE the blocks from their start to their goal configurations. In the plans shown, this is generally accomplished by moving Palm 2 so that its contact point traverses an arc of a circle about Palm 1's contact point. It is up to the planner to choose the amount of rotation so the part does not slip. Both plans contain a number of ROTATE operations. For instance, in Figure 2 there is a small ROTATE operation between Frames 2 and 3, and a large one between Frames 7 and 8. In Figure 3, there is a tiny ROTATE between Frames 4 and 5, just enough to bring the small unstable edge into flush contact with Palm 1. Two other rotations occur between Frames 6 and 7 and between Frames 10 and 11, each of which brings Palm 2 into flush edge contact with the block. (There are several more ROTATE actions subsequently.)

- **Release.** The two plans exhibit two different release modes. In Figure 2, Palm 2 releases the plastic block by doing a very careful dance with Palm 1. Before Palm 2 lets go of the block and moves away, the block and Palm 1 are in their final configurations (see Frame 14). In contrast, in Figure 3, Palm 2 actually pushes the block past its stable point, and allows the block to rotate to a new stable equilibrium. This happens between Frames 16 and 17. The action between these frames is a ROTATE. Palm 2 rotates the block about its contact with Palm 1. As the center of mass passes over the contact point, the block rotates on its own under the influence of gravity; Palm 2 eventually catches up. This operation is very delicate. The planner had to choose the angle of Palm 1 sufficiently steep to prevent the block from rotating too far and tipping over. After all, the goal edge is barely stable. On the other hand, choosing the tilt of Palm 1 too steeply would have caused the block to slide and thus slip between the two palms, again falling over. Finally, once the block rotates to its goal edge, Palm 2 can let go (Frame 18), and Palm 1 can then tilt to horizontal (Frame 19).

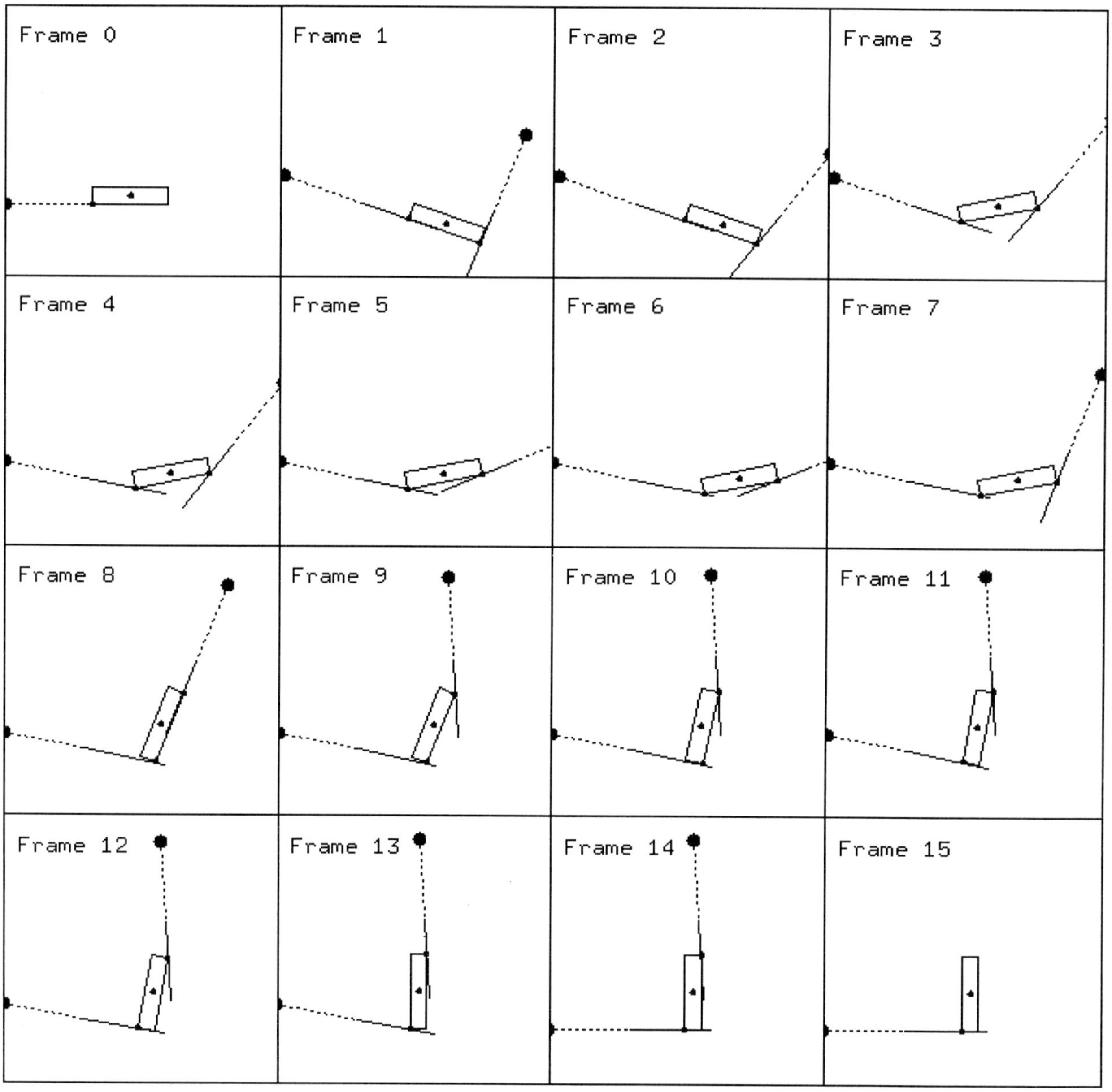

Figure 2: Plan for reorienting a rectangular block using two robot palms. The plan was generated automatically, given a geometric description of the part and the coefficients of contact friction. Two Zebra Zero robots executed this plan, each acting as one palm. — Friction is 0.287 at the left palm and 0.810 at the right palm. The block dimensions are 83mm by 18mm. The palm lengths are drawn to scale. The big black dots are the locations of the robot wrist joints; the dashed lines depict unreachable parts of the robot hands, such as the force sensor housings. The small dots indicate the primary vertex of contact for each palm. — The friction coefficients were computed by measuring the friction angles to about one degree of accuracy, then taking tangents of those angles. The three digits shown do not reflect three digit accuracy.

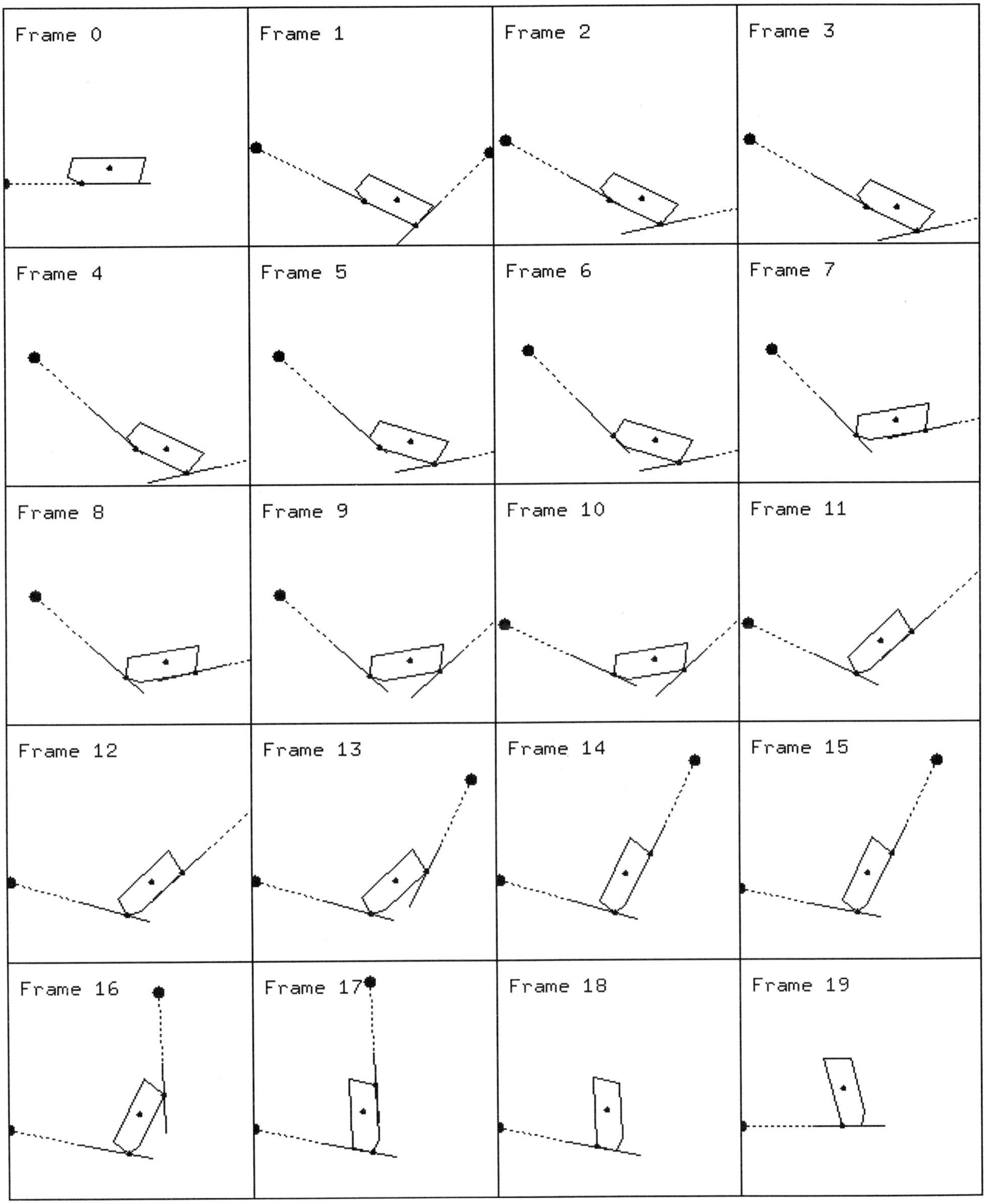

Figure 3: Plan for reorienting a five-sided block using two palms. Friction at the left palm is 0.445 and friction at the right palm is 0.554.

4 Related Work

Nonprehensile Manipulation

The importance of nonprehensile manipulation in robotics first makes its written appearance in Mason's Ph.D. thesis [Mason 1982] and the paper [Mason 1986]. Mason in turn points to work by Pingle, Paul, and Bolles from 1974, a conversation with Inoue, and general parts feeding approaches. The basic idea is for a robot to manipulate an object purposefully even though the robot does not have full control over the object. Mason looked at the problem of pushing objects in the plane. He analyzed the mechanics of pushing, including the classification of sliding and rolling contacts, and showed how pushing could be used to reduce an object's uncertainty. Mason's research spawned over a decade of further work on pushing. Peshkin analyzed the pushing problem [Peshkin 1986] and showed how to design a series of fences along a conveyor belt to orient parts on the conveyor [Peshkin and Sanderson 1988]. Peshkin's work was later extended by [Brokowski, Peshkin, and Goldberg 1993] to design curved fences that smoothly engaged and reoriented the parts on the conveyor. [Brost 1988] implemented a system that could orient parts through a series of pushing and squeezing operations. [Akella and Mason 1992] showed how to pose any polygon using only pushing strategies. [Lynch 1992] showed how to pose objects by pushing without relative slip, and [Lynch and Mason 1995] investigated various non-intuitive properties of pushing strategies. Finally, [Donald, Jennings, and Rus 1994] implemented a variety of pushing and reorienting strategies, using cooperating mobile robots to move large objects such as boxes and couches between rooms.

A number of researchers have investigated and implemented other ingenious forms of nonprehensile manipulation. [Sawasaki, Inaba, and Inoue 1993] considered the task of moving an object over a planar surface by tumbling it about various edges of contact. [Aiyama, Inaba, and Inoue 1993] considered the task of moving an object over a planar surface by pivoting it about various vertices, much as one might move a large refrigerator. These researchers observe that pushing, tumbling, and pivoting form a geometrically complete class of motions. Specifically, these operations preserve, respectively, plane, line, and vertex contact with a support surface. From this perspective, palm manipulation is a simple generalization of these operations, in which the support surface itself becomes moveable.

Palm-Like Manipulation

Probably the first to examine palm-like manipulation explicitly were Salisbury in one setting and Trinkle in another. [Salisbury 1987] proposed the idea of whole arm manipulation, in which an arm manipulates an object by making contact with the object anywhere on the arm's surface, as needed, rather than restricting contact to the fingertips of the hand. Around the same time, in a series of papers Trinkle and colleagues analyzed the mechanics of strategies for scooping up frictionless objects resting on a table, by enveloping them with two fingers [Trinkle 1992], [Trinkle, Abel, and Paul 1988], [Trinkle and Paul 1990].

Yun and colleagues looked at the problem of manipulating large objects by supporting them with two palms [Yun 1993], [Paljug and Yun 1993], [Paljug, Yun, and Kumar 1994]. These researchers developed and implemented control laws for stably moving and reorienting objects with non-sliding rolling contacts. Bicchi and colleagues analyzed the kinematics and manipulability of pairs of cooperating robots that manipulate objects with flat palm-like surfaces [Bicchi, Melchiorri, and Balluchi 1995], [Bicchi and Sorrentino 1995].

Slip

An important aspect of our system is the ability to purposefully slip one palm relative to the object being held, or, equivalently, to use one palm to drag or push the object relative to the other palm. There has been considerable prior work on slip. [Fearing 1984] and [Nguyen 1988] used slip to attain stable grasps on objects of unknown shape or pose. The works by Mason and Trinkle and their colleagues cited earlier made extensive use of slip to orient objects and reduce uncertainty. [Brock 1988] used slip to reorient grasped objects in the presence of gravity. [Carlisle, Goldberg, Rao, and Wiegley 1994] used slip to reorient grasped objects by pushing them against other objects. [Cole, Hsu, and Sastry 1992] developed a dynamic control law for regrasping objects using sliding. Perhaps the most extensive investigation of slip originated with work by Kao and Cutkosky [Kao and Cutkosky 1989, 1992, 1993]. In one example, a robot rotates a card lying on a table, by slipping two fingers over the top surface of the card. Finally, [Yoshikawa, Yokokohji, and Nagayama 1993] analyze three-fingered grasping of three-dimensional objects using slip motions.

Friction

A key component of most grasping strategies, including our palm manipulation system, is the ability to account for friction. The heart of our planner is based on earlier work on friction found in [Erdmann 1994]. That paper also contains an extensive review of prior work on friction. Finally, the work reported in this paper has benefited greatly from the planar friction cone model, in particular the ideas underlying the work in the paper [Brost and Mason 1989].

5 Slide-Release

When two hands manipulate an object, there are three basic operations the hands perform: holding or rotating the object in an equilibrium grasp, allowing the object to fall in disequilibrium, and sliding part of one hand against the object. In this section we focus on the sliding operation.

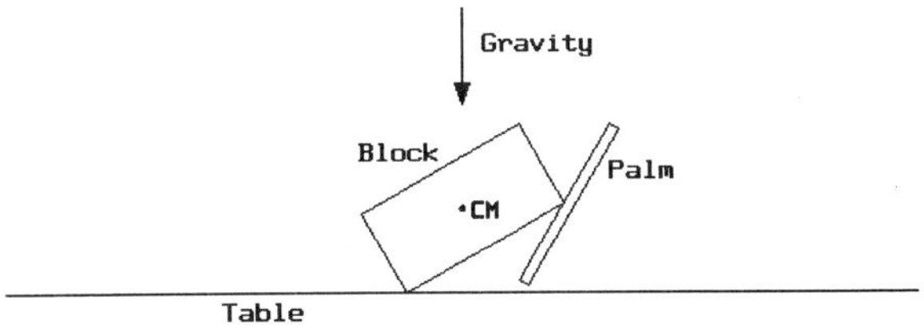

Figure 4: Block in contact with a table and a palm.

5.1 Single-Handed Manipulation

Even single hand manipulation gains scope and competence by judicious use of sliding. Imagine for instance placing a large dictionary on a table, using one hand. Shortly before release, one edge of the book will be in contact with the table, while one's hand is holding an opposite edge. In all likelihood, one's fingers are supporting the bottom face of the book, while one's thumb is wrapped around the top side. In fact, the thumb is superfluous. Sooner or later one will release one's grip on the book, slide the fingers out from underneath, and allow the book to fall into place. This is a classic *slide-release*. It demonstrates one form of relative sliding, in which the hand slides relative to the object, while the object remains stationary in the world frame. In the example, the fingers slide out from under the book, while the book does not slide relative to the table. Of course, once the hand has slid all the way out from under the book, the book drops into place on the table, which is the whole point of a slide-release.

There is another form of sliding, dual to the one described, in which the hand moves and the object remains fixed to the hand, sliding in the world frame while tracking the hand. Such sliding can be used prior to a release to position the object in the world frame. In the example, we might move our hand so as to position the dictionary precisely on the table, and only thereafter release the book. Since we have a thumb, we would probably perform this fine positioning by grasping the raised end of the dictionary. However, again, the thumb is really unnecessary. One can both push and pull the dictionary simply by choosing carefully the angle with which one's palm or fingers support the raised end of the dictionary. Friction can substitute for the force closure provided by a thumb.

5.2 The Book-Placing Task

In the remainder of this section we will examine the book placing example in detail. We will develop a simple test for determining when sliding is possible and what type of sliding is possible. This test constitutes the central tool used by our two-palm planner for carving up its search space.

The basic problem is shown Figure 4. The figure depicts a block in contact with a horizontal table and a tilted palm. Gravity acts straight down. We assume that friction arises from dry Coulomb friction. We assume further that there is no difference between the static and dynamic coefficients of friction.

5.3 Equilibrium Conditions

Let us look at the conditions under which equilibrium is possible. Figure 5 again shows the block in contact with the table and the palm, but now with the contact friction cones depicted as well. We have depicted the cones as two-sided friction cones in order to emphasize the lines of action of the reaction forces. We have also drawn a dashed vertical line through the center of mass of the block. This line depicts the line of action of the force of gravity.

First, we ask whether the block will fall or remain stationary. It turns out the block will fall; Figure 6 shows the result after 8 centiseconds, as determined by our configuration space friction cone simulator.

In contrast, consider Figure 7. In that figure, the coefficient of friction between the block and the table is now $\mu_1 = 0.5$, larger than it was in Figure 5. This time the block remains stationary; it is in equilibrium. See Figure 8.

Mechanically, the difference between the examples of Figure 5 and Figure 7 is that the contact reaction forces can balance gravity in the second case but not the first. Geometrically, two conditions must be satisfied. First, the negative of the gravity vector must lie in the positive-span of the two contact reaction forces. Second, the lines of action of the two contact reaction forces and the gravity vector must all intersect

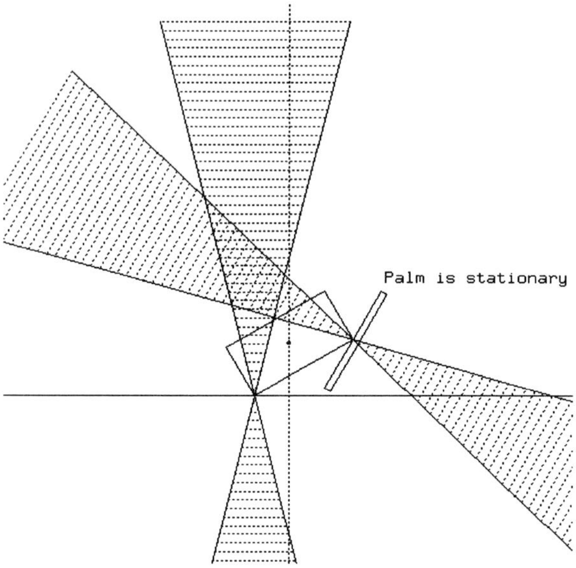

Figure 5: Initial configuration of block, table, and palm, along with the two two-sided contact friction cones and the vertical center-of-mass line. Friction is 0.25 at each contact. Observe that the friction cones *do not* intersect anywhere along the center-of-mass vertical.

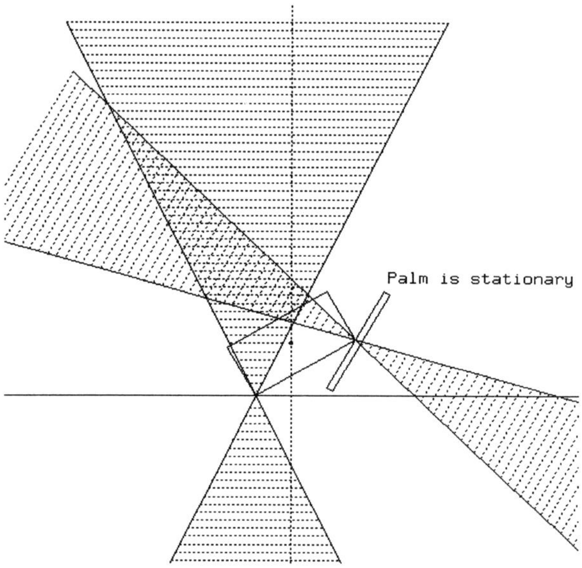

Figure 7: Another initial block-table-palm configuration, along with the two-sided friction cones and the vertical center-of-mass line. In this case friction is still 0.25 at the palm contact, but it is now 0.5 at the table contact. Observe that in this example the friction cones *do* intersect along the center-of-mass vertical.

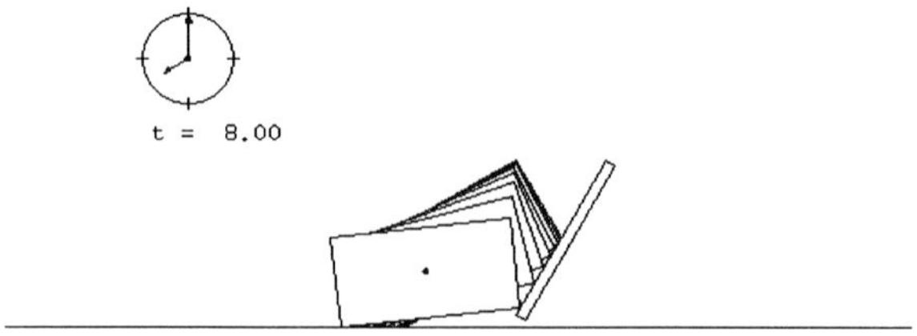

Figure 6: Simulation of motion starting from the configuration in Figure 5. The block falls.

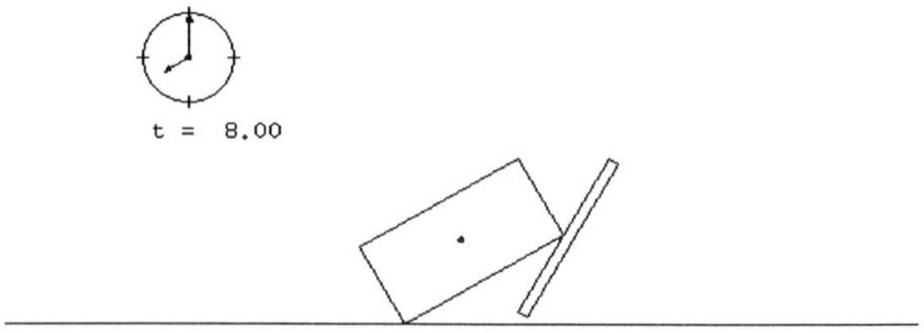

Figure 8: Simulation of motion starting from the configuration in Figure 7. The block remains stationary.

at a common point. In other words, to use Brock's terminology [Brock 1988], the *force focus* of the three forces must exist. Together, the two conditions ensure that force-torque balance is physically possible.

We will focus mainly on the second condition in the remainder of this paper, that is, the *force focus test*. Nonetheless, we caution the reader not to forget the first condition, that is, the *positive-span test*.

Next, we ask ourselves whether we can tell instantly from Figures 5 and 7 if the force focus test is satisfied. The answer is yes. We need merely intersect the two two-sided friction cones and compare the resulting geometric object to the vertical line drawn through the object's center of mass. If the intersection of the two two-sided friction cones overlaps the vertical line, then it is possible to satisfy the force focus test. Conversely, if the intersection of the friction cones does not overlap the vertical line then it is not possible to satisfy the force focus test.

5.4 Sliding Contact Conditions

A robot expands its effective workspace when it can slide its hands relative to the objects it is manipulating. By strategically sliding the hands, the robot can reposition the objects in collision-free and mechanically advantageous locations on the hands.

Consider Figure 9. The figure again shows the vertical line through the object's center of mass and the two two-sided friction cones. Suppose the palm moves to the right, tangential to the table. Will the block track the palm or fall? Figure 10 shows the simulation results after 8 centiseconds. We see that the block falls. In contrast, consider Figures 11 and 12. In that example the block tracks the hand.

The difference between these two examples may be explained and predicted using similar reasoning as we used with the equilibrium examples of Section 5.3, only we must be more careful in describing the source of the reaction forces. The reaction force at the sliding contact cannot arise arbitrarily from within its contact friction cone. Instead, the reaction force must be on an edge of the friction cone, namely the edge that opposes the sliding motion. The reaction force at the other contact, the non-sliding contact, can of course potentially arise from anywhere within its contact friction cone.

For the example of Figures 9 and 11, the reaction force at the block-table contact must lie on the left edge of its friction cone. In deciding whether quasi-static dragging is possible, we must therefore intersect the line through the left edge of the friction cone with the vertical line through the object's center of mass. If this intersection point lies within the two-sided friction

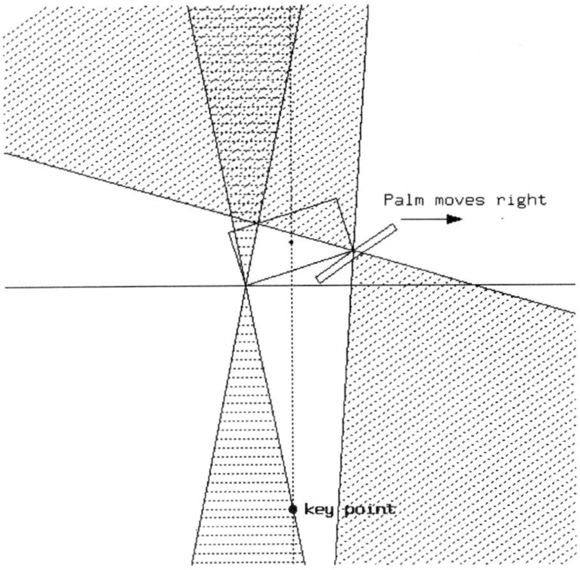

Figure 9: Initial configuration of block, table, and palm. The palm is about to move straight to the right. Friction is 0.2 at the table contact and 0.8 at the palm contact. The "key point" is the intersection of the table's left friction cone edge with the center-of-mass vertical. Observe that the key point lies *outside* the palm's friction cone.

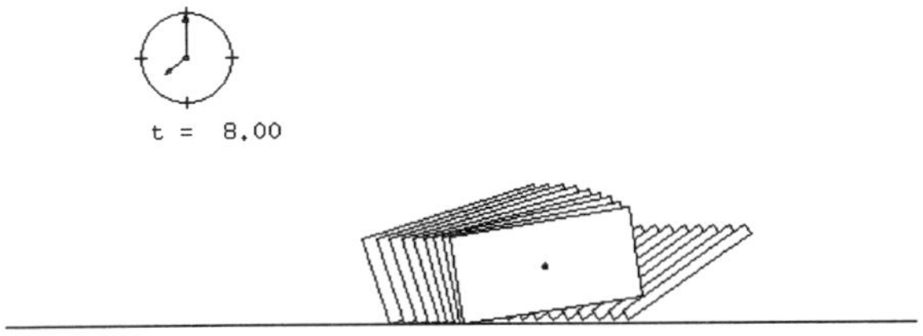

Figure 10: Simulation of motion starting from the configuration in Figure 9. As the palm moves, the block slips and rotates down.

cone anchored at the block-palm contact, then quasi-static sliding is indeed possible. Otherwise it is not.

The conditions for pushing the block to the left, or for sliding the palm relative to the block without disturbing the block, are similar.

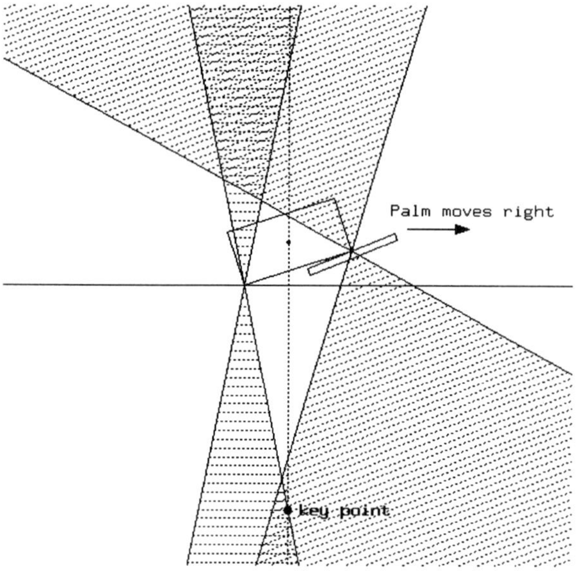

Figure 11: Initial configuration of block, table, and palm. The palm is about to move straight to the right. Friction is 0.2 at the table contact and 0.8 at the palm contact. Observe that the palm is at a shallower angle than it was in Figure 9. Consequently, the "key point" now lies *inside* the palm's friction cone.

5.5 Summary

To decide whether static equilibrium is possible one checks whether the two friction cones have a common overlap with the vertical line drawn through the center of mass. To decide whether quasi-static sliding is possible one checks whether an edge of one friction cone has a common overlap with the other friction cone on the vertical. The specific edge used depends on the particular sliding mode. The two tests are really the same, merely constrained differently to account for the different object motions.

5.6 Planning Tools

It is easy to generalize these equilibrium and sliding tests into planning tools. For a fixed block configuration and a fixed set of contacts, the truth value of an equilibrium or sliding test can only change at certain discrete orientations of the palm. These critical orientations are those angles at which a friction cone edge of one friction cone intersects a friction cone edge of the other friction cone at a point lying on the center-of-mass vertical (or lying at infinity). In short, the mechanics of the task splits the orientation space of the palm into a finite set of sectors yielding qualitatively similar object behaviors.

6 The Planner

6.1 Carving up the Configuration Space

The fundamental planning stage for the planner is a two-dimensional space that encodes the possible orientations ϕ_1 and ϕ_2 of the two palms for a fixed orientation θ of the object. The basic role of the two-palm planner is to carve up (ϕ_1, ϕ_2) space into regions within which the permissible palm motions are qualitatively identical. The planner does this by first computing certain critical curves. The critical curves describe the palm orientations (ϕ_1, ϕ_2) at which two friction cone edges intersect at the same spot on the vertical center-of-mass line, or at which the positive-span test might change truth value. Finally, the planner creates a cylindrical decomposition based on these critical curves. The permissible behavior of the palms and the object are invariant across palm orientations within any region thus constructed.

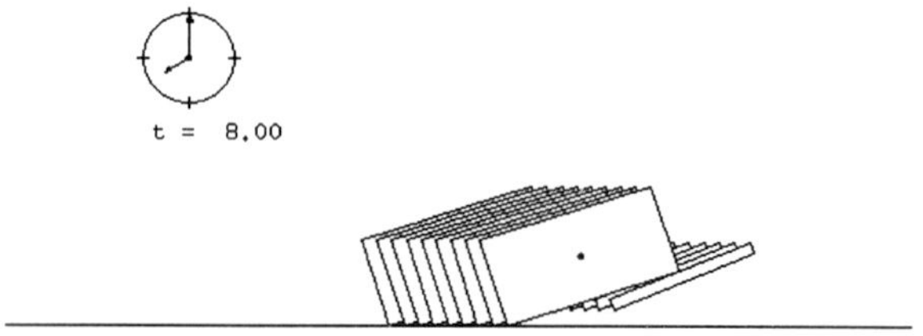

Figure 12: Simulation of motion starting from the configuration in Figure 11. As the palm moves, the block tracks the palm without rotating.

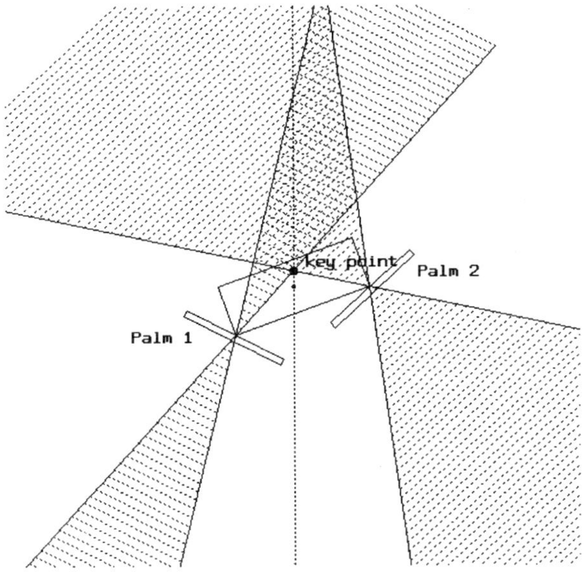

Figure 13: Two palms holding a block. Friction at the left contact (Palm 1) is 0.25; friction at the right contact (Palm 2) is 0.70. The figure shows a singular configuration, in which Palm 1's right friction cone edge intersects Palm 2's left friction cone edge smack dab on the center-of-mass vertical.

Figure 13 shows two palms supporting a block in the presence of gravity and contact friction. The configuration is a critical configuration; the right edge of friction cone 1 intersects the vertical line at the same point as does the left edge of friction cone 2. If we rotate either palm slightly the resulting contact would become non-critical. Separating such generic contact states is a critical curve in (ϕ_1, ϕ_2) space. At each point on this curve, the right edge of Palm 1's friction cone meets the left edge of Palm 2's friction cone smack dab on the vertical through the block's center of mass. The planner constructs all such critical curves, for all possible contacts of the two palms with vertices of the block.

6.2 Planning Motions

The planner uses labelled subdivisions of the form shown in Figure 14 to plan motions of the palms to move the part from the start to the goal. The planner uses yet another critical event analysis, this time in θ, to construct (ϕ_1, ϕ_2) subdivisions at a small number of generic part orientations θ.

The planner searches the (θ, ϕ_1, ϕ_2) configuration space from start to goal by searching its collection of (ϕ_1, ϕ_2) subdivisions. Moving between different subdivisions is equivalent to rotating the part from one θ orientation to another. For a fixed part orientation, the palms can perform two types of motions: (i) they can rotate about their contact points, or (ii) they can slide left or right. The labelled subdivisions tell the planner which of these motions are legal. For instance, the region boundaries in the subdivision tell the planner how far it can reorient a palm before either changing region classification or making palm-edge contact with the part. And the region classification itself tells the planner whether it can slide a given palm left or right.

7 Future Directions

Although reasonably automatic, the system is far from truly autonomous. It is easy to design tasks for which the planner produces unexecutable plans. Low friction tasks are prime candidates. One way to increase the competence of our planner is to expand its mastery of mechanics. There are three directions future work should pursue:

1. Incorporate principled tests for contact stability.
2. Distinguish between static and dynamic coefficients of friction.
3. Build planners that can manipulate parts without requiring continuous stability or equilibrium.

Item (3) is probably the most interesting direction to explore. Our current planner has a modicum of nonequilibrium motion, namely the release operation discussed in Section 3. However, much more is possible. [Zumel 1995] is currently pursuing this direction, also in the domain of palm manipulation, focusing on tasks with very low friction. Additionally, [Mason and Lynch 1993] are pursuing this direction with their work on dynamic manipulation.

8 Acknowledgments

Many thanks to Craig Johnson for his excellent work with the Zebra Robots. This research benefited greatly from discussions with Tamara Abell, Bruce Donald, Kevin Lynch, Matt Mason, and Nina Zumel.

Financial support for this research was provided in part by Carnegie Mellon University through a Faculty Development Award, in part by an equipment grant from AT&T, and in part by the National Science Foundation through the following grants: NSF Presidential Young Investigator award IRI-9157643 and NSF Grant IRI-9213993.

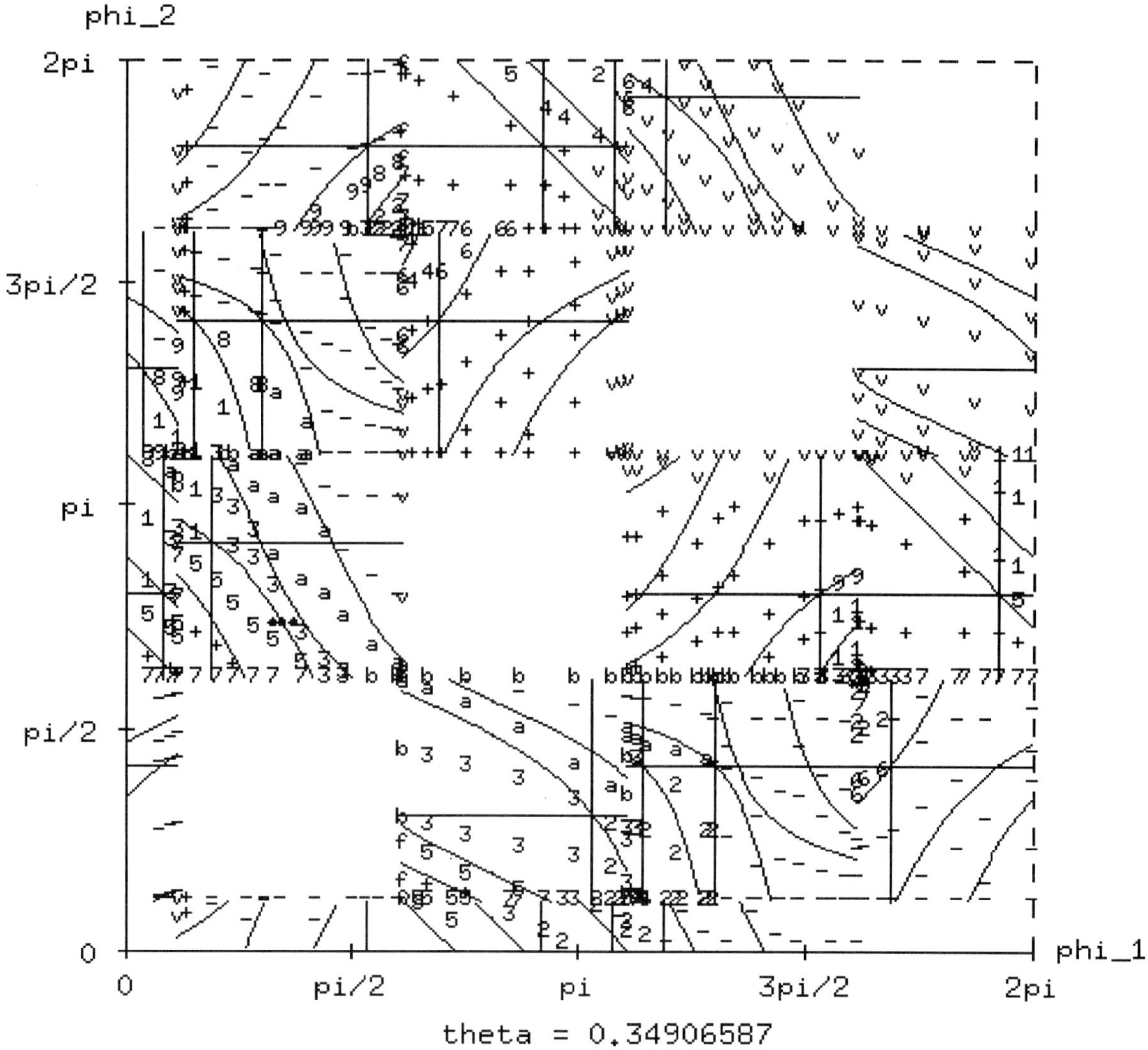

Figure 14: The planner subdivides the space (ϕ_1, ϕ_2) of palm orientations for a fixed object orientation θ into regions of invariant contact mode, using a cylindrical decomposition. This figure shows the location of the representative state of each invariant region. The block orientation is fixed at 20 degrees, that is, $\theta = 0.349$ radians. The palms can make contact with any vertex of the block. Of course, not all contacts yield equilibrium states. — Each state is depicted by a letter. The letters are internal codes used by the planner to describe the types of motions possible at that state. For instance, the planner uses a one letter hexadecimal representation to encode the sliding modes possible when the block is in equilibrium. And a "v" means that the block is falling without rotating. — Note the three dots in the left central area. The middle dot corresponds to the critical configuration depicted in Figure 13. The underlying curve describes the orientations (ϕ_1, ϕ_2) at which the right edge of Palm 1's friction cone meets the left edge of Palm 2's friction cone smack dab on the center-of-mass vertical.

9 Bibliography

Aiyama, Y., Inaba, M., and Inoue, H. 1993. Pivoting: A New Method of Graspless Manipulation of Object by Robot Fingers. *Proceedings of the 1993 IEEE/RSJ International Workshop on Intelligent Robots and Systems*, Yokohama, Japan, pp. 136–143.

Akella, S., and Mason, M. T. 1992. Posing Polygonal Objects in the Plane by Pushing. *Proceedings of the 1992 IEEE International Conference on Robotics and Automation*, Nice, France, pp. 2255–2262.

Bicchi, A., Melchiorri, C., and Balluchi, D. 1995. On the Mobility and Manipulability of General Multiple Limb Robots. *IEEE Transactions on Robotics and Automation.* **11**(2):215–228.

Bicchi, A., and Sorrentino, R. 1995. Dexterous Manipulation Through Rolling. *Proceedings of the 1995 IEEE International Conference on Robotics and Automation*, Nagoya, Japan, Vol 1, pp. 452–457.

Brock, D. L. 1988. Enhancing the Dexterity of a Robot Hand Using Controlled Slip. *Proceedings of the 1988 IEEE International Conference on Robotics and Automation*, Philadelphia, Pennsylvania, pp. 249–251.

Brokowski, M., Peshkin, M., and Goldberg, K. 1993. Curved Fences for Part Alignment. *Proceedings of the 1993 IEEE International Conference on Robotics and Automation*, Atlanta, Georgia, Vol 3, pp. 467–473.

Brost, R. C. 1988. Automatic Grasp Planning in the Presence of Uncertainty. *International Journal of Robotics Research.* **7**(1):3–17.

Brost, R. C., and Mason, M. T. 1989. Graphical Analysis of Planar Rigid-Body Dynamics. *Fifth International Symposium on Robotics Research*, Tokyo.

Carlisle, B., Goldberg, K., Rao, A., and Wiegley, J. 1994. A Pivoting Gripper for Feeding Industrial Parts. *Proc. 1994 IEEE Intl. Conference on Robotics and Automation*, San Diego, California, pp. 1650–1655.

Cole, A. A., Hsu, P., Sastry, S. S. 1992. Dynamic Control of Sliding by Robot Hands for Regrasping. *IEEE Transactions on Robotics and Automation.* **8**(1):42–52.

Donald, B. R., Jennings, J., and Rus, D. 1994. Analyzing Teams of Cooperating Mobile Robots. *Proceedings of the 1994 IEEE International Conference on Robotics and Automation*, San Diego, California, pp. 1896–1903.

Erdmann, M. A. 1994. On a Representation of Friction in Configuration Space. *International Journal of Robotics Research.* **13**(3):240–271.

Fearing, R. S. 1984. Simplified Grasping and Manipulation with Dextrous Robot Hands. *Proceedings of the 1984 American Control Conference*, pp. 32–38.

Kao, I., and Cutkosky, M. R. 1989. Dextrous Manipulation with Compliance and Sliding. *Fifth International Symposium on Robotics Research*, August 28–31, 1989, Tokyo.

Kao, I., and Cutkosky, M. R. 1992. Quasistatic Manipulation with Compliance and Sliding. *International Journal of Robotics Research.* **11**(1):20–40.

Kao, I., and Cutkosky, M. R. 1993. Comparison of Theoretical and Experimental Force/Motion Trajectories for Dextrous Manipulation With Sliding. *International Journal of Robotics Research.* **12**(6):529–534.

Lozano-Pérez, T. 1983. Spatial Planning: A Configuration Space Approach. *IEEE Transactions on Computers.* **C-32**(2):108–120.

Lynch, K. M. 1992. The Mechanics of Fine Manipulation by Pushing. *Proceedings of the 1992 IEEE International Conference on Robotics and Automation*, Nice, France, pp. 2269–2276.

Lynch, K. M., and Mason, M. T. 1995. Pulling by Pushing, Slip with Infinite Friction, and Perfectly Rough Surfaces. *International Journal of Robotics Research.* **14**(2):174–183.

Mason, M. T. 1982. Manipulator Grasping and Pushing Operations. AI-TR-690. Ph.D. thesis. Massachusetts Institute of Technology, Artificial Intelligence Laboratory.

Mason, M. T. 1986. Mechanics and Planning of Manipulator Pushing Operations. *International Journal of Robotics Research.* **5**(3):53–71.

Mason, M. T., and Lynch, K. M. 1993. Dynamic Manipulation. *Proceedings of the 1993 IEEE/RSJ International Workshop on Intelligent Robots and Systems*, Yokohama, Japan, pp. 152–159.

Nguyen, V-D. 1988. Constructing Force-Closure Grasps. *International Journal of Robotics Research.* **7**(3):3–16.

Paljug, E., and Yun, X. 1993. Experimental Results of Two Robot Arms Manipulating Large Objects. *Proceedings of the 1993 IEEE International Conference on Robotics and Automation*, Atlanta, Georgia, Vol 1, pp. 517–522.

Paljug, E., Yun, X., and Kumar, V. 1994. Control of Rolling Contacts in Multi-Arm Manipulation. *IEEE Transactions on Robotics and Automation.* **10**(4):441–452.

Peshkin, M. A. 1986. Planning Robotic Manipulation Strategies for Sliding Objects. Ph.D. thesis. Carnegie Mellon University, Physics Department.

Peshkin, M. A., and Sanderson, A. C. 1988. Planning Robotic Manipulation Strategies for Workpieces that Slide. *IEEE Journal of Robotics and Automation.* **4**(5): 524–531.

Pingle, K., Paul, R., and Bolles, R. 1974. *Programmable Assembly, Three Short Examples*. Film. Stanford Artificial Intelligence Laboratory.

Salisbury, J. K. 1987. Whole Arm Manipulation. *Proceedings 4th International Symposium of Robotics Research*, pp. 183–189.

Sawasaki, N., Inaba, M., and Inoue, H. 1989. Tumbling Objects Using Multi-Fingered Robot. *Proceedings of the 20th ISIR*, pp. 609-616.

Trinkle, J. C. 1992. On the Stability and Instantaneous Velocity of Grasped Frictionless Objects. *IEEE Transactions on Robotics and Automation.* **8**(5):560–572.

Trinkle, J. C., Abel, J. M., and Paul, R. P. 1988. An Investigation of Frictionless Enveloping Grasping in the Plane. *International Journal of Robotics Research.* **7**(3):33–51.

Trinkle, J. C., and Paul, R. P. 1990. Planning for Dexterous Manipulation with Sliding Contacts. *International Journal of Robotics Research.* **9**(3):24–48.

Yun, X. 1993. Object Handling Using Two Arms Without Grasping. *International Journal of Robotics Research.* **12**(1):99–106.

Yoshikawa, T., Yokokohji, Y., and Nagayama, A. 1993. Object Handling by Three-Fingered Hands Using Slip Motion. *Proceedings of the 1993 IEEE/RSJ International Workshop on Intelligent Robots and Systems*, Yokohama, Japan, pp. 99–105.

Zumel, N. B. 1995. Analysis and Planning for Nonprehensile Two Palm Manipulation. Ph.D. thesis proposal. The Robotics Institute, Carnegie Mellon University.

Control of Physical Contact and Dynamic Interaction

Ernest D. Fasse*
University of Twente, Dept. of Electrical Engineering
7500 AE Enschede, the Netherlands
e-mail: fasse@u.arizona.edu

Neville Hogan
M.I.T., Dept.'s of Mechanical Engineering and Brain & Cognitive Sciences
Cambridge, MA 02139, USA
e-mail: neville@mit.edu

1 Motivation

This paper considers the influence of physical contact and mechanical interaction on the dynamics and control of manipulators. Manipulation fundamentally requires contact with the object(s) being manipulated; contact implies mechanical interaction; and mechanical interaction can have a profound influence on manipulator dynamics and control – for example, controller stability is easily jeopardized. In many present robot applications the dynamic effects of physical contact may be neglected either because contact forces are relatively small or because they need not be controlled accurately. However, the dynamics of physical contact are likely to become more prominent as newer robot designs permit relatively larger payloads and robots are applied to tasks requiring more precise control of mechanical interaction, e.g., applications involving intimate physical interaction with humans such as haptic virtual environments or personal-care robots.

This paper will review several different approaches to controlling mechanical interaction, including impedance and admittance control, and compare them to hybrid force-position control. The problems of physical contact and the strengths and weaknesses of interaction control will be considered under two broad headings: (1) implementation and (2) specification of a desired behavior for a given task.

*E.D. Fasse is currently with the University of Arizona, Dept. of Aerospace & Mechanical Engineering, Tucson, AZ 85721, USA.

2 Current Approaches

2.1 Regulator Design

The majority of the vast literature on control system design has been dominated by the problem of regulation or tracking. Control authority is exerted to ensure that some variable of interest is kept close (in some appropriately defined sense) to a desired value despite uncertainties and hardware limitations. From that perspective, contact tasks are often assumed to require an appropriately coordinated hybrid of position control and force control. In many applications regulator design has been remarkably effective and it has been refined to a high degree of sophistication but it is not necessarily the best approach to all problems. Applied to physically interacting systems, it can be misleading.

From a regulator design perspective, interaction between a control system and its environment causes a "disturbance" that should be counteracted. Disturbances are usually assumed to be small in some sense (e.g., have less influence on system behavior than the available control effort); are typically assumed to have a frequency content distinct from that of the variable to be tracked; and are typically assumed to be independent of the state of the system. The consequences of physical contact commonly satisfy none of these conditions. Consider two persons shaking hands: each can exert comparable forces, so the "disturbance" due to contact is unlikely to be small; each has a comparable bandwidth for voluntary movement, so the frequency content of the "disturbance" is likely to be the same as that of the desired behavior, not distinct from it; and far from being independent, the "disturbance" experienced by each hand is largely due to its own actions. It may seem that with a sufficiently detailed model of the environment these problems could be avoided but the behavior of the environment may be arbitrarily complex and very poorly known. For example, in its sensory role, the human hand is frequently used to explore objects and identify their dynamic behavior. By definition these objects have not been modeled a priori, yet the hand remains well-controlled during physical contact and exhibits no pathological behavior. As reviewed below, design constraints have been identified that permit a robot controller to perform satisfactorily during physical contact with objects of arbitrarily complex and almost completely unknown dynamic behavior.

2.2 Interactive Behavior

Perhaps the most important limitation of the regulator/tracker design approach is its focus on the "forward-path" response to a command or reference input. In situations involving physical contact, response to the environment may be at least as important as response to a reference input, perhaps more so. An example may be found in "haptic virtual environments", computer-driven, human-interactive displays to synthesize the illusion of a (software- generated) environment. Visual and acoustic displays dominate current virtual environment technologies, but there is a clear need for *haptic* displays to synthesize the touch and feel of contact with objects. A robot with a suitable means of coupling to a human can serve as a programmable haptic display and some sophisticated designs have recently emerged.

The design of the controller for a haptic display device presents an interesting challenge. If it is approached as a regulator design problem, what should be the reference input? While the hand is not in contact with a virtual object, zero force may be required. A force regulator might therefore seem appropriate, but how can the boundary of a virtual object be described in terms of a reference force? Within the region occupied by a virtual object, hand motion should be opposed or even prevented. This might seem to call for a position regulator, but how can the arbitrary motions permissible when not in contact be described in terms of a reference position? It is simpler to regard the controller as implementing a (specified) relation between force and position (or motion), a mechanical impedance, than as regulating force or position. When not in contact, impedance should be low, probably zero. Within the region occupied by a virtual object, impedance should increase to reflect how the object resists deformation.

2.3 An Interaction-Port Perspective

Given that a manipulator may contact a wide variety of objects, some with largely unknown dynamics, it is useful to direct the controller design toward those properties of a manipulator that do not change when it contacts objects. One such property is how it feels from the "outside"; the behavior of the manipulator exhibited at the interaction port by which it couples to its environment.

We define an *interaction port* by a set of variables that describe the exchange of energy between a system and its environment. Typically these are the forces and positions at points of contact but the idea is readily extended to other variables and to non- contact forms of energetic interaction. The definition of the variables requires care: each force or moment (more generally, *effort*) must be associated with a corresponding position or angle (more generally, *displacement*) and its rate of change (velocity, angular rate or, more generally, *flow*) such that each effort-flow pair (known as *power conjugate variables*) properly defines a power flow by which the system energy may change.[1]

The *interaction port behavior* determines a map relating the port variables. Borrowing terms that originated in electrical network theory, *impedance* relates motion to force and *admittance* relates force to motion. Impedance and admittance are commonly used as frequency domain descriptions of linear systems, but they are readily generalized to nonlinear systems. More precisely, impedance is a causal dynamic operator that maps an input motion time function $\dot{x}(t)$ to an output force time function $F(t)$; admittance is the dual operator relating $F(t)$ to $\dot{x}(t)$. In the linear case one is the inverse of the other; in the nonlinear case the inverse operators may not be definable. For example, assuming a state-determined repre-

[1] The generalized efforts and generalized velocities used in mechanics are an example of an effort-flow pair for which a power flow is defined. Less commonly used wrenches and screws provide another example.

sentation of system dynamics, impedance is described by rate equations $\dot{z} = Z_r(z, \dot{x}, t)$ and output equations $F = Z_o(z, \dot{x}, t)$, where z is a finite-dimensional state vector and $Z_r()$ and $Z_o()$ are algebraic (memoryless) functions. Impedance (or admittance) should not be confused with its more common parameterizations. It is more general than a combination of mass, friction and elasticity; an impedance may be described by arbitrarily complex, nonlinear dynamics and there are many alternative parameterizations of the map from motion to force.

Impedance control is a generic term like motion control or force control. Impedance determines interactive behavior by definition; the impedance control approach is to characterize, modulate and (to the extent possible) determine the impedance at the port(s) of interaction with the environment. Impedance control should not be confused with its particular implementations. It is not confined to position and velocity feedback control of rigid-member inertial mechanisms actuated by current-controlled torque motors; in its most general form it is not simply emulation of mass-spring-damper systems. The main distinction of impedance control is that it attempts to govern the interactive behavior of a manipulator that is unaffected by contact with objects. It is often convenient to think of mechanical impedance as a dynamic generalization of stiffness. Stiffness at a point is a system property independent of any objects it may contact. In contrast, the force at that point is strongly affected by contact with objects. Force control (and likewise motion control) is therefore fundamentally sensitive to object contact; impedance control need not be.

3 Contact Instability

Physical contact and mechanical interaction can have a profound affect on control system stability. Attempts to control the force exerted by a robot have been thwarted by the phenomenon of contact instability. A machine that can stably execute unrestrained motions may become unstable on contact with an object. Whitney [29], [30] first reported the problem with robot force feedback control. He observed that the stiffness of the contacted surface plays a role analogous to a feedback gain, hence stable contact imposed a limit on the stiffness of the surface or the stiffness of the robot at the point of contact. Despite the work of numerous researchers, the problem proved to be remarkably refractory and was regarded as a fundamental challenge of robotics by Paul [22].

Several workable solutions emerged from subsequent work. One important factor is that in a typical configuration, significant dynamics (e.g., due to flexibility of the robot joints or members or to sensor dynamics) are interposed between the force feedback sensors and the robot actuators; the sensors and actuators are *non-collocated.* A force feedback control loop (which is not closed until contact) excites the interposed dynamics and its maximum stable gain is severely limited by their presence. One way to avoid this problem is to minimize robot structural dynamics and significantly improved designs (e.g., using composite materials) have been developed. Another class of solutions is to minimize the non-collocation problem using small, fast end- effectors near the point(s) of contact [27], [24], [25].

However, it is important to recognize that contact instability is not solely due to non-collocation of sensors and actuators. Indeed, contrary to common misperception, contact instability is not confined to force feedback controllers contacting stiff surfaces. It is a more general problem that can re-surface despite improvements in robot design as new and more demanding applications are attempted. For example, most of the devices presently being used as haptic displays are prone to contact instability for certain classes of emulated (virtual) objects or as the human operator changes grip strength.

3.1 Interactive Stability and Passivity

An interaction port perspective facilitates a general approach to the contact instability problem and has been used to identify conditions for preserving stability on contact that are both sufficient and necessary (for an appropriately defined class of objects). A large class of objects a manipulator may encounter are passive. They can store and return energy, but cannot generate energy. This is important because passivity is widely used in control theory for stability analysis. The stability of adaptive controllers can be established using a passivity argument in which an energy storage function is used as a candidate Lyapunov function [19], [21]. Using an input-output analysis, Cho and Narendra [3] showed that a negative feedback interconnection of strictly passive operators is sufficient to guarantee stability.[2]

When two objects are coupled at an interaction port they may exchange energy but the connection itself cannot generate energy. This means that the mathematical operators describing their interaction port behavior are restricted in the ways they may be combined. If the interaction port is described using force and velocity, the interaction port behavior of each object is an impedance or admittance. If two objects are coupled their interaction is equivalent to a unity gain negative feedback connection of their respective impedance and admittance. For example if the velocity at a point of contact is common to both objects, then by Newtons third law, the forces on each object at the contact point are equal and opposite. As a result, a *sufficient* condition to preserve stability on contact with a passive object is that the manipulators impedance (or admittance) should appear to be passive.

Using a linear analysis, the necessity of this condition may be deduced by considering the class of all passive objects. A sketch of the argument is presented here; details are in [8], [9]. Linearizing the dynamic equations about the nominal contact conditions, at steady state in response to sinusoidal excitation both the force and velocity at the contact point will be sinusoids. Considering the entire class of passive objects, the relative magnitudes of the two sinusoids are unrestricted but the relative phase angle between the two may not exceed $\pm 90°$ because passivity requires that the average power transmitted to the object must be non-negative.

Because physical contact of two systems is equivalent to a feedback connection of their impedance and admittance operators, the Nyquist criterion may be used for stability analysis. The necessary and sufficient condition for stable interaction is that the Nyquist plot of the product of the manipulator impedance and object admittance (or vice versa) must not encircle the -1 point on the complex plane. The Nyquist plot of the product may be obtained by rotating the Nyquist plot of the manipulator impedance through the angle determined by the Nyquist plot of the object admittance, which can be no more than $\pm 90°$, and multiplying it by the magnitude of the Nyquist plot of the object admittance. Because the magnitude of the object admittance can take on any value the necessary and sufficient condition for stability of the coupled system is that the Nyquist plot of the manipulator impedance must lie exclusively in the right half of the complex plane; but that means that the manipulator impedance must be a positive-real function, i.e., its interaction port behavior must be indistinguishable from that of a passive object.

[2] A rigorous definition and detailed discussion of the several definitions passivity may be found in [31], [7].

3.2 Conservatism and the Practicality of Passivity

A common objection to passivity-based analysis of control system stability is that it yields sufficient conditions that may be excessively conservative. Therefore the claim that a passive interaction port behavior is necessary as well as sufficient should be examined carefully. First of all, the necessary condition was established assuming a linearized representation of interaction port behavior and consequently it is only as true as the fidelity of the linearized model, i.e., it is a local result valid in the vicinity of the nominal operating conditions. For nonlinear systems, passive interaction port behavior is sufficient to ensure contact stability (e.g., [8], [12], [13], [14]) but a global necessary condition has yet to be established.

Second, the necessity of passive interaction port behavior for linearized systems arises from the assumed class of objects: all possible passive objects. That is an extremely broad class that includes infinitely stiff and infinitely massive objects. In any realistic application the class of objects is likely to be more restricted (e.g., the stiffness or mass of objects contacted may be bounded) and it may be possible to use knowledge about that restricted class of objects to relax the coupled stability condition. Despite these caveats the contact stability condition provides useful insight for practical applications. In one early experiment using a simple two- link robot as a test platform, Colgate [5], [9] designed a control system to make its end-point follow a commanded trajectory in a small region in the workspace center. The controller used only position and velocity feedback and was designed using the LQG/LTR technique, a widely-recognized modern approach to multi-variable controller design. That technique does not constrain the interaction port behavior to be passive and, in fact, both analytical prediction and direct measurement on the hardware showed that the impedance due to the LQG/LTR controller violated the condition for contact stability. The passivity analysis predicted that instability would occur on contact with certain classes of objects including both sufficiently stiff springs and sufficiently large masses. Experimental observation confirmed these predictions; although position was well-controlled when unconstrained, the robot was unstable on contact with a stiffness of 420 ($\pm$20) N/m and mass of 1.4 ($\pm$0.1) kg.

Note that in addition to confirming the value of the passivity analysis, this experiment clearly demonstrated that contact instability is not merely a quirk of force-feedback controlled robots contacting rigid surfaces. It is a fundamental problem of contact; it may occur in the absence of force feedback and while contacting unconstrained objects. Nor is this a peculiarity of the LQG/LTR technique: PID controllers are commonly used to control robot joint position but they violate the passivity condition by producing excessive phase lag between (disturbance) force and motion at low frequencies. The analysis predicts that these controllers will exhibit instability when coupled to sufficiently large unconstrained masses and this was also confirmed experimentally [9]. This kind of instability may become an important consideration if robots are to be used to manipulate large unconstrained masses (e.g., in space or underwater applications).

3.3 Passivity of Discrete-Time Controllers

An important limitation of the passivity-based condition for contact stability is that it constrains the phase of the interaction port behavior at all frequencies. If a discrete-time implementation is used then the delay due to data sampling introduces a phase lag that grows without bound at high frequency and this violates the passivity condition. Given the overwhelming advantages of digital control this would seem to limit the usefulness of a contact stability condition that requires passivity. This motivated Chapel and Su [2] to investigate "nearly passive" robots.

In fact, recent work by Colgate [6] has dramatically reduced the conservativeness of the coupled stability criteria and Colgate and Schenkel [7] have identified conditions that guarantee the passivity of machines under discrete control. An elegant solution is to include a physical energy dissipator (e.g., a linear damper) at the interaction port that serves to remove energy added by the active behavior of the discrete-time controller. Loosely speaking, this physical energy dissipator may be considered analogous to the use of an anti-aliasing filter in an analog-to-digital converter. It ensures that the interaction between the continuous domain of physical variables and the discrete domain of digital computation does not lead to pathological behavior.

3.4 A Small-Gain Approach to Interactive Stability

The problem of contact instability can be approached from a different perspective. Kazerooni [18] (see also [16], [17], [28]) presents an analysis (with corroborating experiments) that uses the small-gain theorem to derive a sufficient condition for stable compliant motion using force feedback and a controller design procedure is developed from that sufficient condition. The permissible force feedback gain is upper-bounded by a quantity that decreases as either the robot or the contact surface become stiffer, a result consistent with widespread experimental observation.

This analysis has certain features that limit its generality. First, it is assumed that a force feedback loop is closed around a position-controlled robot. The analysis does not apply to alternative control architectures, e.g., motion feedback around a high- gain force/torque controller. Second, it is assumed that, in the absence of force feedback, the position- controlled robot is stable when in dynamic contact with its environment. Position control is certainly common in robot applications, but because stability is assumed at the outset, the analysis cannot identify any constraints on the position controller to ensure stable contact.

Third, the approach describes both the displacement response of the robot to contact forces and the environmental forces evoked by robot displacements using operators that have magnitude-bounded gains. This means that certain classes of robot position control cannot be considered, including the commonplace PID position control which does not have a bounded magnitude at DC. Also, infinitely stiff surfaces cannot be considered, even though that may be a practical description of the kinematic constraint due to contact.

In contrast, the passivity-based analysis restricts only the *phase* of the operators used to relate contact forces and motions; their magnitudes may be arbitrary. Given that a manipulator may have to contact completely unfamiliar objects a description that applies to infinitely stiff or infinitely compliant objects seems preferable. Given that some applications may require the application of large forces a stability condition that does not restrict the magnitude of the robot reaction seems preferable.

Finally, note that the result derived from the application of the small-gain theorem to a force controlled robot yields a gain constraint that changes with the stiffness of the contacted object. This is because contact force is not a property of the manipulator nor of the object. On the other hand, the passivity analysis yields a constraint on a manipulators interaction port behavior, and that is a property of the manipulator independent of any objects that are contacted.

4 Computational Complexity

One of the main impediments to using interaction controllers (e.g., impedance, admittance, compliance or stiffness controllers [4], [15], [20], [23]) is that force or motion at the interface are not explicitly controlled but implicitly determined by the interaction port behavior. In practice it has proven difficult to select appropriate impedance parameters to execute useful tasks. One important reason is that the relation between interaction port behavior (e.g. at a robots end-effector) and the corresponding actuator behavior (e.g., in robot joint coordinates) is extremely complex, a highly nonlinear function of the geometric and kinematic properties of the manipulator. This problem is particularly acute for spatial manipulators which involve the complex kinematics of spatial rotations.

The stiffness of a manipulator at its end-effector defines a static relation between end-effector configuration and wrenches (torques and forces). Let ϕ be a chart mapping end-effector configurations to generalized coordinates with $\phi(q) = x_r$. Translation is commonly represented using Cartesian coordinates; orientation is commonly represented using angles such as roll, pitch and yaw angles. Associated with generalized coordinates, x_r, are generalized velocities, $\dot{x}_r$, and generalized forces, F. Stiffness is then a static relation between x_r and F. A commonly used relation is

$$F = K_\phi(x_r - x_v), \tag{1}$$

where K_ϕ is a stiffness matrix, x_r is an array of coordinates of the robot end-effector configuration and x_v is an array of coordinates of the so-called virtual equilibrium configuration. The configuration-wrench behavior of this apparently simple stiffness is remarkably complex. It is highly dependent on the manipulator and equilibrium configurations and as a result it is inordinately difficult to select appropriate stiffness parameters.

This problem is further compounded by the transformation of interaction port behavior into a coordinate frame relevant to a task, which is similarly complex and nonlinear, especially for spatial tasks. That means that the already complex mapping of task-relevant behavior into manipulator behavior may change significantly in different phases of a task and that makes it difficult to break a complex application down into manageable parts.

Recent work [10], [11] has shown how the interaction port behavior may be decomposed into "spatial" and "nonspatial" parameters. "Spatial" parameters, e.g., principal directions of translational stiffness, have well-defined, intuitive, spatial transformation properties and can be selected depending on the configuration of objects with which the robot must interact. "Nonspatial" parameters, e.g., principal translational stiffnesses, can be selected independently of object configurations. This decomposition dramatically simplifies the specification of impedance parameters.

The approach is summarized in the following sections which analyze the spatial transformation of manipulator compliance, part of the impedance selection problem, and describe a *spatial compliance controller* for a serial, rigid manipulator. The results generalize readily to other impedance parameters [11].

5 A Spatial Compliance Controller

The configuration of a rigid body can be represented by a frame attached to the body. A *frame* is a quadruplet (p, e_1, e_2, e_3), where p is a displacement vector representing position and vectors e_1, e_2 and e_3 are an orthonormal triplet representing orientation. The vectors are described with respect to a fixed, orthonormal reference frame. Each frame can be identified with a homogeneous matrix:

$$H = \begin{bmatrix} R & p \\ 0^t & 1 \end{bmatrix}, \tag{2}$$

where $R = [\; e_1 \quad e_2 \quad e_3 \;]$ is an orthonormal matrix.

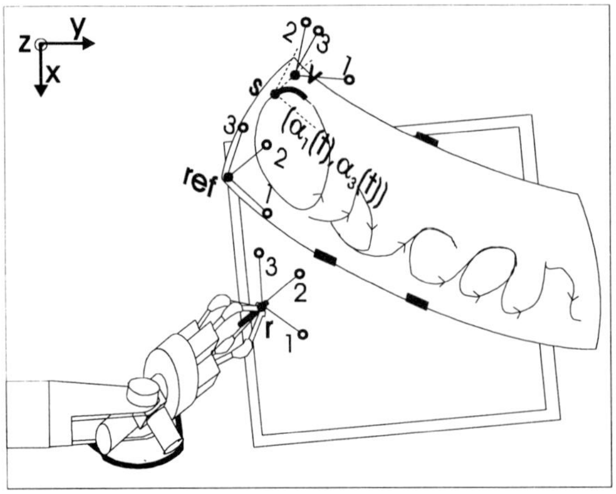

Figure 1: Interaction with a spherical surface. Compliance acts to align the robot and virtual equilibrium frames.

Shown in Fig. 1 are three frames. Frame $(p, e_1, e_2, e_3)_r$ is attached to the end-effector of the robot. Frame $(p, e_1, e_2, e_3)_s$ is attached to a body of interest, in this case a spherical surface. Frame $(p, e_1, e_2, e_3)_v$ represents the virtual equilibrium configuration of compliance.

"Spatial" compliance acts to align the end-effector frame and the virtual equilibrium frame, and is simply described in terms of its potential energy:

$$\begin{aligned} U &= \tfrac{1}{2}\Delta p^t S \Lambda_t S^t \Delta p - \operatorname{tr}(\Lambda_o R_v^t R_r) \\ &= \tfrac{1}{2}\Delta p^t K \delta p - \textstyle\sum_{i=1}^{3} \kappa_i e_{ir}^t e_{iv}; \end{aligned} \tag{3}$$

where $\Delta p = p_r - p_v$ and where $\operatorname{tr}(A)$ is the trace of matrix A. This energy function defines a family of potential energy functions parameterized by S, H_v, Λ_t and Λ_o. Parameter $S = [\begin{array}{ccc} v_1 & v_2 & v_3 \end{array}]$ is an orthonormal matrix, parameter Λ_t is a diagonal matrix with elements k_1, k_2 and k_3, and parameter Λ_o is a diagonal matrix with elements κ_1, κ_2 and κ_3.

The first energy term is that of a linear, translational spring of stiffness $K = S\Lambda_t S^t$, which acts to coincide points p_r and p_v. The principal directions of stiffness are v_1, v_2 and v_3. The corresponding principal stiffnesses are k_1, k_2 and k_3. The second energy term is that of a rotational spring that acts to align R_r and R_v.

The compliance associated with each $-\kappa_i e_{ir}^t e_{iv}$ term acts to align corresponding vectors e_{ir} and e_{iv}. Energy $-\kappa_i e_{ir}^t e_{iv}$ is minimized when vectors e_{ir} and e_{iv} are aligned ($e_{ir}^t e_{iv} = 1$), and maximized when they are anti-aligned ($e_{ir}^t e_{iv} = -1$). Parameters κ_1, κ_2 and κ_3 are orientational compliance parameters referred to here as orientational stiffnesses. In the sequel it is shown that these parameters determine the effective rotational stiffnesses for small displacements.

5.1 Control Law Based on Kinestatic Robot Model

It is straightforward to derive a corresponding control law for serial manipulators. Assume that the generalized coordinates representing robot joint configuration are $\theta = \{\theta_i\}$ and that the corresponding generalized forces are $\tau = \{\tau_i\}$. If static friction, gravity, link compliance, etc., can be neglected in the manipulator kinestatic model, taking the partial derivative of energy (Eqn. 3) with respect to the generalized coordinates yields a suitable control law:

$$\tau = \left(\frac{\partial p_r}{\partial \theta}\right)^t K(p_r - p_v) - \sum_{i=1}^{3} \kappa_i \left(\frac{\partial e_{ir}}{\partial \theta}\right)^t e_{iv}, \tag{4}$$

in which each $\partial a/\partial \theta$ for $a \in \{p_r, e_{1r}, e_{2r}, e_{3r}\}$ is a 3×6 Jacobian-like matrix. This assumes that the generalized efforts are positive when work is done on the robot. In practice it is necessary to add other control terms to compensate friction and gravity, and to dissipate energy. This control law was simulated and implemented by Bonnes and Colard [1] and Tigchelaar [26]. The stability properties of the controlled robot are investigated in [11].

5.2 Dependency of Wrenches on End-Effector Configuration

The end-effector wrenches for a particular end-effector configuration are determined by the differential of the energy function at that configuration. Let δH_r be an arbitrary, infinitesimal displacement from a particular configuration, H_r, so that

$$H_r + \delta H_r = \left[\begin{array}{cc} R_r + R_r\delta\tilde{\theta} & p_r + R_r\delta p \\ 0^t & 1 \end{array}\right]. \tag{5}$$

In this expression, $\delta\theta$ is an infinitesimal rotation expressed in the end-effector frame and δp is an infinitesimal translation expressed in the end-effector frame. Associated with $\delta\theta$ is an antisymmetric matrix:

$$\delta\tilde{\theta} = \left[\begin{array}{ccc} 0 & -\delta\theta_3 & \delta\theta_2 \\ \delta\theta_3 & 0 & -\delta\theta_1 \\ -\delta\theta_2 & \delta\theta_1 & 0 \end{array}\right] \tag{6}$$

given $\delta\theta = [\ \delta\theta_1 \quad \delta\theta_2 \quad \delta\theta_3\ ]^t$. This notation shall be used to associate antisymmetric matrices and vectors. The infinitesimal work given δH_r can be computed from Eqn. 3, yielding

$$\delta W = \Delta p^t S \Lambda_t S^t R_r \delta p + \frac{1}{2}\mathrm{tr}\left(2\mathrm{as}\left(\Lambda_o R_v^t R_r\right)^t \delta\tilde{\theta}\right), \tag{7}$$

where as(A) is the antisymmetric part of matrix A. The resultant wrench expressed in the end-effector frame is

$$\tilde{m} = 2\mathrm{as}\left(\Lambda_o R_v^t R_r\right) \text{ and } f = R_r^t S \Lambda_t S^t (p_r - p_v), \tag{8}$$

because the infinitesimal work given an arbitrary, infinitesimal displacement δH_r is

$$\delta W = f^t \delta p + \frac{1}{2}\mathrm{tr}\left(\tilde{m}^t \delta\tilde{\theta}\right). \tag{9}$$

5.3 Wrenches Resulting from Small Displacements

For small displacements of the end-effector from the virtual equilibrium configuration, the wrench-configuration relation of Eqn. 8 can be approximated by a linear relation

$$\begin{bmatrix} m \\ f \end{bmatrix} \approx K \begin{bmatrix} \delta\theta \\ \delta p \end{bmatrix}, \text{ assuming} \tag{10}$$

$$H_r \approx \begin{bmatrix} R_v + R_v\delta\tilde{\theta} & p_v + R_v\delta p \\ 0^t & 1 \end{bmatrix}. \tag{11}$$

The linear approximation is computed by substituting Eqn. 11 into Eqn. 8 and discarding high-order terms, resulting in

$$f \approx R_v^t S \Lambda_t S^t R_v \delta p, \text{ and } m \approx (\mathrm{tr}(\Lambda_o) I - \Lambda_o)\, \delta\theta. \tag{12}$$

The corresponding stiffness matrix is

$$K = \begin{bmatrix} \mathrm{tr}(\Lambda_o) I - \Lambda_o & 0 \\ 0 & R_v^t S \Lambda_t S^t R_v \end{bmatrix}. \tag{13}$$

Expansion of Eqn. 13 shows that for H_r near H_v, $\kappa_2 + \kappa_3$ is the effective rotational stiffness about axis e_{1v}, $\kappa_3 + \kappa_1$ is the effective rotational stiffness about axis e_{2v}, and $\kappa_1 + \kappa_2$ is the effective rotational stiffness about axis e_{3v}.

5.4 Spatial Transformation of Compliance

Compliance is parameterized by two kinds of parameters. *Spatial parameters* are parameters for which the action of a rigid body transformation is defined. In this case the spatial parameters are the principal directions of translational stiffness, S, and the virtual equilibrium configuration, H_v. *Nonspatial parameters* have no special properties. In this case the nonspatial parameters are the translational and orientational stiffnesses, Λ_t and Λ_o. The compliance family has a useful transformation property. Let

$$H_\sigma = \begin{bmatrix} R_\sigma & t_\sigma \\ 0^t & 1 \end{bmatrix} \tag{14}$$

represent an arbitrary rigid body transformation. Substitution of $H_\sigma H_r$, $H_\sigma H_v$ and $R_\sigma S$ for H_r, H_v, and S, respectively, in Eqn. 8 yields

$$\tilde{m} = 2\mathrm{as}(\Lambda_o R_v^t R_\sigma^t R_\sigma R_r) = 2\mathrm{as}(\Lambda_o R_v^t R_r) \text{ and} \tag{15}$$

and

$$\begin{aligned} f &= R_r^t R_\sigma^t R_\sigma S \Lambda_t S^t R_\sigma^t (R_\sigma p_r + t_\sigma - R_\sigma p_v - t_\sigma) \\ &= R_r^t S \Lambda_t S^t (p_r - p_v). \end{aligned} \tag{16}$$

This shows that if the robot configuration and all spatial parameters are subject to a rigid body transformation, then the compliant wrenches are unchanged. This transformation property greatly simplifies compliance selection. Its value is demonstrated in the following by considering the task of writing on a spherical surface.

6 Example: Writing on a Spherical Surface

Figure 1 shows a robot end-effector and a spherical surface with a known radius of curvature, r. Shown are (1) the robot end-effector frame, $(p, e_1, e_2, e_3)_r$, labelled **r** in the figure, (2) a frame normal to the surface at a point of interest, $(p, e_1, e_2, e_3)_s$, labelled **s** in the figure, (3) a fixed, reference frame on the surface, $(p, e_1, e_2, e_3)_{ref}$, labelled **ref** in the figure, and (4) the virtual equilibrium configuration frame, $(p, e_1, e_2, e_3)_v$, labelled **v** in the figure. The robot is shown holding a writing tool. The tip of the tool is assumed to coincide with p_r; the axis of the tool is assumed to coincide with e_{2r}. The reference frame on the surface is assumed either to be known priori or to be perceptible from sensory information. Point p_{ref} lies in the surface; e_{2ref} is normal to the surface.

The task is to make a stroke on the surface. This task can be decomposed into a number of subtasks: (1) moving to a point above the surface near the start of the stroke, (2) contacting the surface at the start of the stroke, (3) making the actual stroke, and (4) moving away from the surface if necessary. Consider only the subtask of making the actual stroke, assuming contact has already been made at the start of the stroke.

The stroke is described by real-valued displacement functions $\alpha_1(t)$ and $\alpha_3(t)$ having units of distance. The spherical surface is identified with a flat surface by means of a mercatorial projection, the familiar projection used in cartography. Vectors e_{1ref} and e_{3ref} can be used to generate local directions of latitude and longitude. The displacement functions describe the stroke using coordinates of latitude and longitude.

The frame normal to the surface at the desired point of contact, $(p, e_1, e_2, e_3)_s$, can be thought of as the result of first transporting frame $(p, e_1, e_2, e_3)_{ref}$ along a latitudinal segment of arclength $\alpha_1(t)$, and then along a longitudinal segment of arclength $\alpha_3(t)$. Let $c_i(t) = \cos(\alpha_i(t)/r)$ and $s_i(t) = \sin(\alpha_i(t)/r)$. The orientation of frame $(p, e_1, e_2, e_3)_s$ is

$$R_s = R_{ref} \begin{bmatrix} c_1(t) & -s_1(t)c_3(t) & -s_1(t)s_3(t) \\ s_1(t) & c_1(t)c_3(t) & c_1(t)s_3(t) \\ 0 & -s_3(t) & c_3(t) \end{bmatrix}. \tag{17}$$

The position of this frame is

$$p_s = p_{ref} + re_{2ref} - re_{2s}, \tag{18}$$

where e_{s2} is determined by Eqn. 17. Point $p_{ref} + re_{2ref}$ is the center of the sphere.

Intuitively, the principal directions of stiffness can be chosen to be aligned with the frame normal to the surface at the desired point of contact:

$$S = R_s(t). \tag{19}$$

The translational stiffnesses tangential to the surface, k_1 and k_3, are chosen to be equal and high. The translational stiffness normal to the surface, k_2, is chosen to be low. Assume that the writing tool is to be held at a constant angle of tilt with respect to the local surface, which corresponds to a common human writing strategy. The tool is to be tilted at an angle β around the e_{3s} axis. The corresponding virtual equilibrium orientation is

$$R_v = R_s \begin{bmatrix} \cos(\beta) & -\sin(\beta) & 0 \\ \sin(\beta) & \cos(\beta) & 0 \\ 0 & 0 & 1 \end{bmatrix}. \tag{20}$$

Orientational stiffnesses κ_1 and κ_3 are not important in this task as near-point contact is assumed. They are assumed to be equal and low. Orientational stiffness κ_2 is assumed to have a higher value. A simple strategy for writing on a surface is to displace the virtual equilibrium position from the desired position by a distance d normal and into the surface. Using this strategy the virtual equilibrium position is

$$p_v(t) = p_{ref} + re_{2ref} - (r - d)e_{2s}(t), \tag{21}$$

where d is the desired distance of displacement. This is a reasonable strategy if both the friction of the instrument-surface interface and the inertia of the robot can be neglected. The actual equilibrium position is then $p_s(t)$.

Equations 17-21 are sufficient to describe the task independently of the surface configuration, $(p, e_1, e_2, e_3)_{ref}$. It is not claimed that this is the best nor even a good strategy for writing; the point is to show that the spatial compliance family is parameterized in such a way that it can be used to describe complex spatial interaction with bodies in the environment in a simple way. This eliminates one of the computational barriers to applying interaction controllers to interesting tasks. An assembly example is given in [11].

7 Discussion

Consider again "conventional" stiffness or compliance control. One could call the eigenvectors of K_ϕ and the coordinates of the virtual configuration, x_v, "directional parameters". One could call the eigenvalues of K_ϕ "nondirectional parameters", but as will be shown this would not be useful.

Compliance selection is often simplified by introducing "task coordinates". Let ψ be a second chart mapping end-effector configurations to generalized coordinates with $\psi(q) = y_r$. These coordinates might be three Cartesian coordinates and three angles with respect to a frame attached to a body of interest, such as the $(p, e_1, e_2, e_3)_{ref}$ frame in the example. We then have $x_r = \phi \circ \psi^{-1}(y_r)$ and $\delta y_r = J\delta x_r$, where J is a Jacobian matrix. Using chart ψ compliance can be expressed as

$$G = K_\psi(y_r - y_v) \tag{22}$$

where G is the generalized force corresponding to the generalized velocity $\dot{y}_r$. Selection of the virtual equilibrium coordinate and stiffness matrix with respect to chart ψ, i.e., selecting y_v and K_ψ, is assumed to be straightforward. The virtual equilibrium coordinate and stiffness with respect to chart ϕ are then

$$x_v = \phi \circ \psi^{-1}(y_v) \text{ and } K_\phi = J^t K_\psi J. \qquad (23)$$

If Eqn. 23 is satisfied then the compliance relations of Eqn.'s 1 and 22 generate the same configuration-wrench behavior for small displacements of the end-effector from the virtual equilibrium configuration. Matrices K_ψ and K_ϕ do not necessarily have the same eigenvalues; the eigenvalues of K_ϕ will in general depend on the configuration of the objects with which the robot interacts. In the example we were able to choose nonspatial parameters, k_1, k_2, k_3, κ_1, κ_2, κ_3, independently of the configuration of the surface, $(p, e_1, e_2, e_3)_{ref}$. The preceding discussion shows that one could not choose the eigenvalues of K_ϕ independently of the coordinates of the surface, x_s. The spatial parameters, v_1, v_2, v_3 and $(p, e_1, e_2, e_3)_v$, were related geometrically to the configuration of the surface, $(p, e_1, e_2, e_3)_{ref}$. One could not relate x_v or the eigenvectors of K_ϕ so simply to the coordinates of the surface, x_{ref}. The compliance of Eqn. 1 is not described using the relevant geometrical structure of the configuration manifold of a rigid body.

8 Future Directions

Recent developments have shown how to design and implement practical robot controllers that are robust to physical contact and dynamic interaction with a broad class of passive objects with almost arbitrarily complex dynamic behavior. A new parameterization of interaction port behavior that explicitly represents its geometrical structure simplifies the deployment of these controllers in complex, realistic tasks. Of course, numerous unsolved problems remain. One is the case of dynamic interaction with active objects, in which the actions of the manipulator may be equivalent to coupling a power source to the interaction port. An example is the use (or misuse) of certain classes of power tools. How to control physical contact and dynamic interaction in this situation is an open question. Interestingly, humans seem remarkably adept at this class of manipulation.

Acknowledgments. E.D. Fasse was supported by an NSF-NATO postdoctoral fellowship, and by the Mechatronics Research Centre Twente. N. Hogan was supported in part by NIH grant AR40029. The authors would like to thank Justin Won for his invaluable contributions. Portions of the work summarized in this paper have been published elsewhere.

References

[1] Bonnes A., Colard G.J.B., *Peg-in-hole assembly with the OSCAR-6 robot using impedance control*, joint M.Sc. thesis reportnr. 021R94, University of Twente, Enschede, the Netherlands, 1994.

[2] Chapel J.D., Su R., "Coupled Stability Characteristics of Nearly Passive Robots", *Proc. IEEE Int. Conf. on Robotics and Automation*, 1992, pp. 1342-1347.

[3] Cho Y.S., Narendra K.S., "Stability of nonlinear time-varying feedback systems", *Automatica*, vol. 4, 1968, pp. 309-322.

[4] Colbaugh R., Seraji H., Glass K., "Direct Adaptive Impedance Control of Robot Manipulators," *J. of Robotic Systems*, vol. 10, 1992, pp. 217-248.

[5] Colgate, J.E., "Coupled Stability of Multiport Systems-Theory and Experiments", *J. Dynamic Systems Measurement and Control*, vol. 16, 1994, pp. 419-428.

[6] Colgate J.E., "Coordinate Transformations and Logical Operations for Minimizing Conservativeness in Coupled Stability Criteria", *J. Dynamic Systems, Measurement and Control*, vol. 116, 1994, pp. 643-649.

[7] Colgate J.E., Schenkel G., "Passivity of a Class of Sampled-Data Systems: Application to Haptic Interfaces", submitted to *J. Dynamic Systems, Measurement and Control*, 1993.

[8] Colgate J.E., Hogan N., "Robust Control of Dynamically Interacting Systems", *Int. J. of Control*, vol. 48, 1988, pp. 65-88.

[9] Colgate J.E., *The Control of Dynamically Interacting Systems*, Ph.D. Thesis, M.I.T., 1988.

[10] Fasse E.D., "Simplification of Compliance Selection using Spatial Compliance Control", *Proc. of the ASME Dynamic Systems and Control Division*, vol. 57, 1995, pp. 193-198.

[11] Fasse E.D., Broenink J.F., "A Spatial Impedance Controller for Robotic Manipulation," manuscript conditionally accepted for publication by *IEEE Trans. on Robotics and Automation*, 1995.

[12] Fasse E.D., Hogan N., "Coupled Stability of Impedance-Controlled Robots", in Youcef-Toumi K., Kazerooni H. (eds.) *Robotics Research*, ASME, New York, 1989, pp. 5-11.

[13] Fasse E.D., Hogan N., "A Lyapunov-Based Approach to Designing Manipulator Controllers Robust to Interaction with Dynamic Environments", in Youcef-Toumi K., Kazerooni H. (eds.) *Symposium on Robotics*, ASME, New York, 1988, pp. 115-124.

[14] Hogan N., "On the Stability of Manipulators Performing Contact Tasks", *IEEE Trans. on Robotics and Automation*, vol. 4, 1988, pp. 677-686.

[15] Hogan N., "Impedance control: An approach to manipulation", *J. of Dynamic Systems, Measurement and Control*, vol. 107, 1985, pp. 1-24.

[16] Kazerooni H., Waibel B., Kim S., "On the Stability of Compliant Motion Control: Theory and Experiments", *J. of Dynamic Systems, Measurement and Control*, vol. 112, 1990, 417-426.

[17] Kazerooni H., "Contact Stability of the Direct Drive Robot when Constrained by a Rigid Environment", *IEEE Trans. on Automatic Control*, vol. 34, 1990, pp. 710-714.

[18] Kazerooni H., "On the Robot Compliant Motion", J. of Dynamic Systems, Measurement and Control," vol. 111, 1989, pp. 416-425.

[19] Landau Y.D., *Adaptive Control–The Model Reference Approach*, Marcel Dekker, New York, 1979.

[20] Newman W.S., "Stability and Performance Limits of Interaction Controllers", *ASME J. of Dynamic Systems, Measurement and Control*, vol. 114, 1992, pp. 563-570.

[21] Ortega R., Spong M., "Adaptive Motion Control of Rigid Robots: A tutorial", *Proc. Int. Conf. on Decision and Control*, 1988.

[22] Paul R.P., "Problems and Research Issues Associated with the Hybrid Control of Force and Displacement", *Proc. IEEE Conf. on Robotics and Automation*, 1987, pp. 1966-1971.

[23] Salisbury K., "Active Stiffness Control of a Manipulator in Cartesian Coordinates", *Proc. IEEE Conf. on Decision and Control*, 1980.

[24] Sharon A., Hogan N., Hardt D.E., "High Bandwidth Force Regulation and Inertia Reduction Using a Macro/Micro Manipulator System", *Proc. IEEE Conf. on Robotics and Automation*, 1988.

[25] Sharon A., Hogan N., Hardt D.E., "The Macro/Micro Manipulator: An Improved Architecture for Robot Control", *Robotics & Computer-Integrated Manufacturing*, vol. 10, 1993, pp. 209-222.

[26] Tigchelaar T., *Enhancement of the OSCAR-6 robot with vision*, M.Sc. thesis reportnr. 020R94, University of Twente, Enschede, the Netherlands, 1994.

[27] Tilley S. W., Cannon R.H. Jr., "Experiments on Endpoint Position and Force Control of a Flexible Arm With a Fast Wrist", *Proc. AIAA Guidance and Control Conf.*, 1986.

[28] Waibel B.J., Kazerooni H., "Theory and Experiments on the Stability of Robot Compliance Control", *IEEE Trans. on Robotics and Automation*, vol. 7., 1991, pp. 95-104.

[29] Whitney D.E., "Historical Perspectives and State of the Art in Robot Force Control", *IEEE Int. Conf. on Robotics and Automation*, 1985, pp. 262-268.

[30] Whitney, D.E., "Force Feedback Control of Manipulator Fine Motions", *J. of Dynamic Systems, Measurement and Control*, 1977, pp. 91-97.

[31] Wyatt J.L., Chua L.O., Gannett J.W., Göknar I.C., Green D.N., "Energy Concepts in the State-Space Theory of Nonlinear n-Ports: Part I – Passivity", *IEEE Trans. on Circuits and Systems*, vol. 28, 1981, pp. 48-61.

Reorienting Objects with a Robot Hand Using Grasp Gaits

Susanna Leveroni
Kenneth Salisbury
AI Laboratory and Dept. of Mechanical Engineering
MIT, Cambridge 02139, MA, USA
srl@ai.mit.edu, jks@ai.mit.edu

Abstract

This paper addresses the planning of robot finger motions to enable continuous and stable reorientation of grasped objects. With a given grasp on an object, the range of reorientation through which the fingers may move the object is limited by the workspace of the fingers and grasp stability. If however the object can be properly regrasped without dropping it, further motion of the object in the desired direction may be possible. If we use a hand with more than the minimum number of fingers required for a stable grasp, then it is possible to remove one (or more) of the fingers from the grasp without dropping it. This finger (or these fingers) may be then be replaced on the surface to form a new grasp with which a further increment of reorientation may be achieved. We call the resulting sequences of finger/object motions and regrasps *grasp gaits.* To develop an approach to planning such gaits we have addressed the problem of reorienting objects which are constrained to lie in the plane. Since a stable grasp in this environment can be achieved with only two fingers, a hand with three fingers may be able to execute a series of regrasps and object motions such that continuous reorientation is possible. Our results show that for certain types of objects, simple stereotypic gaits may be found which achieve this goal. For more restrictive cases, such as low friction at the contacts or oddly shaped objects, we have developed methods which enable us to find more complex gait patterns if they exist.

1 Why Grasp Gaits?

Robotics researchers have long sought to build and utilize articulated end-effectors, or hands, to extend robotic grasping and manipulative capability. Simple one-degree-of-freedom grippers are limited to grasping a small range of objects and cannot impart any motion to the object. By employing multiple, articulated fingers in the end-effector design it becomes possible to stably grasp a wider range of objects. Furthermore, if sufficient freedom in the hand/object system exists, then the fingers may be used to impart controlled motions to the grasped object. The potential increase in robot capability such designs promise has motivated an enormous amount of research into how to build and use robot hands. Traditionally such research has focused on how to build the hands [1],[2], how to plan grasps [3],[4],[5], how to adjust the internal forces within a grasp [6], how to sense and interpret tactile and proprioceptive information [7],[8], and how to make small moves once an object has been grasped [1],[9]. In addition, a fairly general framework for motion planning with an articulated hand, using coordinated finger motions, as well as rolling and sliding motions has been proposed in [10].

Traditional approaches to moving objects held by multiple fingers have focused on utilizing small coordinated motions of the fingers (in a grasp configuration fixed with respect to the moved object) to impart motion to the grasped object. The extent to which a reorientation can be achieved within a single grasp configuration is limited by the workspace of the fingers and the fact that inner links of the finger will collide with the object if large motions occur. Other approaches have included considering objects slipping between between two or more fingers to a new position [11],[12],[13], organizing the fingers so that a finite reorientation is possible [14], or using one grasp configuration with finger substitutions [15],[16]. However, in general, these methods reach either a workspace or a stable grasp limit, and therefore are not suited to a large or continuous reorientation.

When more than the minimum number of fingers needed for a secure grasp are present then *grasp gaits* that permit continuous reorientation may be possible. When a given grasp reaches its limit for rotating the object, an extra finger may be placed on the object to permit removal of one of the other fingers. If the finger removed was the one at its motion limit for the desired object rotation, and if the finger placed can be located so that it forms a new stable grasp with the remaining fingers, then the object may be further rotated in the desired direction until yet another finger reaches its workspace limit. In other cases, the object may need to be only rotated partially in order to ensure that a new grasp exists that will allow rotation to continue. By repeating the sequence of regrasping and rotating the object, continuous rotation may be achieved. The problem of planning such grasp gaits is essentially a search for a way to walk the fingers around the object without loosing grasp stability, not unlike the problem of finding multi-legged vehicle gaits [18]. In performing this search many questions arise: How far should a given grasp configuration be used to rotate the object before a new grasp is taken? Which of the many new grasp configurations possible should be used? Are you guaranteed to even find a new grasp? Is there something that should have been done earlier to allow that a new grasp could be found? These are questions that we seek to answer to enable a robot to reorient objects through large or continuous motions.

1.1 Problem Scope

We have selected a subset of the general grasp gait planning problem in order to gain insight on the more general problem with this more tractable, but none-the-less instructive case. In particular we look at objects which are allowed to move only in the plane. We assume these object are convex. We consider all finger, body and disturbance forces to act in the plane of the object. We assume a robot hand whose fingers are point contacts with friction. Therefore, the minimum number of fingers required to achieve a force closure grasp in this planar case is two. We assume a three-fingered robot hand, which allows the object to be regrasped from an initial two-fingered grasp by placing a third finger at a new point on the object and then by removing one of the original fingers.

We assume the fingertips to be points, so that when the object is rotated, the complexity of rolling contacts is not present. We assume that the fingers have some known workspace limits and that their workspaces are mutually exclusive. We also assume that the fingers do not slip along the object, so that the only way a finger moves to a new position on the object is by being removed from the object and then replaced at a new location. Our goal is to find a series of two-fingered grasps and intermediate rotations, together called a grasp gait, which allows the object to be continually rotated.

2 Tools For Generating Grasp Gaits

2.1 Grasp Maps

The first tool we use to generate grasp gaits is the grasp map. This is similar to the contact configuration space used by [17]. For a given convex object, any point along its contour can be specified by one parameter, the angle that its position vector from a reference frame f makes with the positive horizontal axis. The reference frame f is constrained to lie within the object's contour and is fixed to the object (See Figure 1). Any two-fingered grasp for that object can then be represented with respect to the object by a pair of angles. All possible two-fingered grasps for that object can be represented in a two-dimensional space, where the horizontal coordinate gives the position on the object of a finger i, and the vertical coordinate gives the position on the object of a finger j. A grasp map can be constructed by numerically solving (point by point) for the regions in the map which represent stable grasps. The regions which correspond to stable grasps are represented as shaded areas, while the unstable regions are white. (See Figure 2.)

A few general qualities of grasp maps are worth noting. First, there is never a shaded region crossing the $y = x$ line, since such a point would correspond to a grasp where both fingers were at the same position on the object. Second, the grasp maps are symmetric about the $y = x$ line, since the interchanging of the coordinates simply corresponds to the interchanging of fingers i and j. Third, these maps repeat every 360°for both coordinates and therefore the space wraps around like a torus.

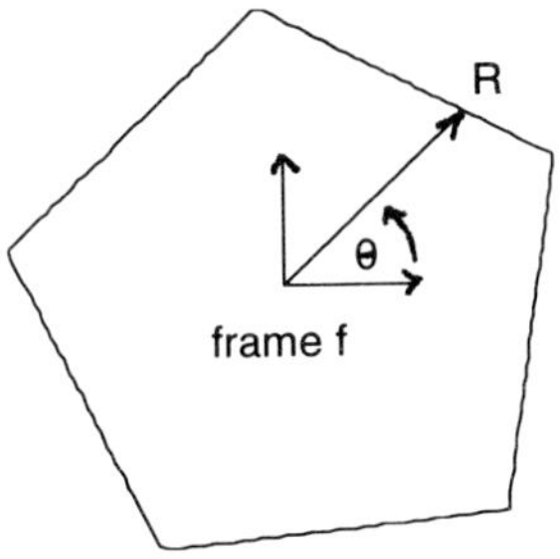

Figure 1: A convex object defined from reference frame f.

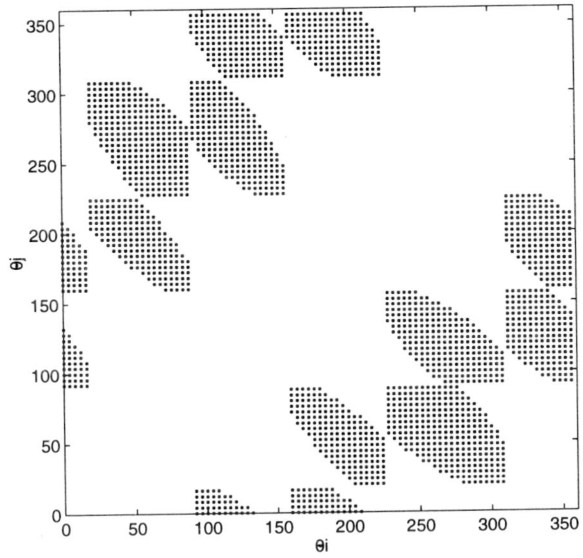

Figure 2: A grasp map for the object in figure 1.

Using the grasp map, we can then look for grasp gaits which consist of a series of regrasps represented by points which always lie within the shaded regions of the map. Since object rotations occur without the fingers moving relative to the object, the grasp map is invariant with respect to object orientation.

2.2 Finger Workspace Maps

The second tool we use to generate grasp gaits is the finger workspace map. For any robot hand, each robot finger has real workspace limits, beyond which the finger cannot move. When manipulating an object, each finger may, in addition, have 'virtual' workspace limits, where finger motion is further limited due to the presence of the object or the other fingers. We define a finger's workspace by the range of angles through which it can move, measured in a reference frame F. Reference frame F is fixed to ground and shares an origin with reference frame f, the object reference frame. We assume the finger workspaces to be equal in size and separation and define them to be non-overlapping to avoid finger crossover problems. Figure 3 shows sample workspaces for fingers that can each move through a 90°range.

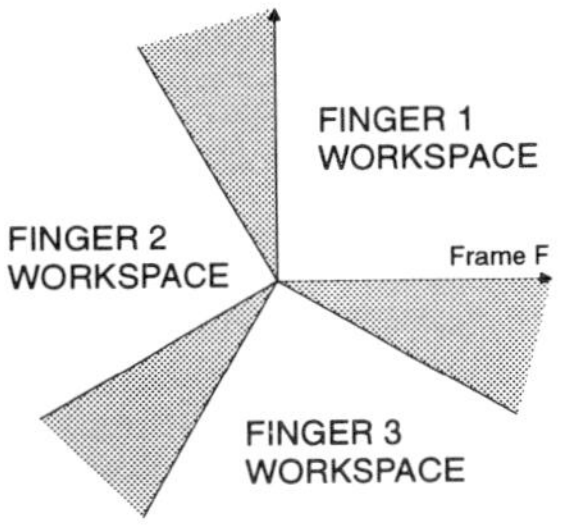

Figure 3: Finger workspaces defined from reference frame F.

We can represent these workspace limits in a two-dimensional space similar to the grasp map representation. The horizontal coordinate represents the angle of finger i with respect to frame F, and the vertical coordinate represents the angle of finger j with respect to the same frame. However, now the value of either coordinate implies a specific finger, since for every angle in reference frame F there is only one finger that can move to that position. In addition, there may be angles to which no finger can move. Therefore any point in this space corresponds to either a grasp of a particular pair of fingers, or a grasp which cannot be achieved due to the specific workspace limits. A workspace map can be constructed by defining the boundaries of regions which contain valid finger pairings. These boundaries are squares for the symmetric and non-overlapping workspace case. We call these "finger pair boxes". For a grasp to satisfy workspace constraints, it must be in within one of the finger pair boxes. For example, using the workspace limits shown in figure 3, where finger 1 can move between 0°and 90°, finger 2 between 120° and 210°and finger 3 between 240°and 330°, the open box in the top left hand corner of figure 4 corresponds to grasps between fingers 1 and 3. Again, the workspace map is symmetric about the $y = x$ line, and it repeats every 360°. We refer to the finger pair boxes that are furthest from the $y = x$ line as "outer" boxes, and those closest to the $y = x$ line as "inner" boxes. If the workspace map is extended beyond 360°, the inner and outer boxes are bound by two lines with a slope of positive 1.

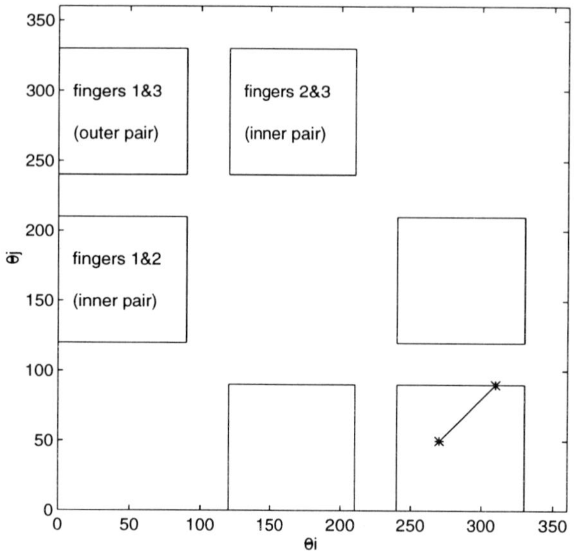

Figure 4: A workspace map for workspaces in figure 3.

Using the workspace map, we then look for grasp gaits whose regrasps occur according to the following rules. First, it is assumed that when switching from one two-fingered grasp to another, there is always one finger and its corresponding position which is common to the two grasps. (It is assumed that the new finger is placed on the object before the old finger is removed, so that there are always at least two fingers on the object.) Therefore, to switch to a new grasp and corresponding point on the map, only vertical or horizontal moves are permitted, since these are the only moves which change only one finger at a time. Further, a new grasp cannot be found within the same finger pair box, since this would require that the same finger be removed from the object and replaced at another position, during which time there would be only one finger remaining on the object.

When the object is rotated about the origin of both reference frames f and F, both of the fingers in the grasp increase their angles by the same amount as seen in reference frame F. Therefore, in the workspace map, rotations are represented by the grasp point moving along a diagonal line with a slope of 1. For counterclockwise rotations, both finger positions increase, and for clockwise rotations, both decrease. If the object continues to rotate, eventually the grasp point will hit one of the edges of the finger pair box, corresponding to one of the fingers hitting an edge of its workspace. In figure 4, a two fingered grasp defined by the finger positions (270, 50) is represented by the lower left hand star in the lower right hand finger pair box. The line traces the change of coordinates with respect to reference frame F as the object is rotated 40°counterclockwise. The upper right hand star represents the final finger positions in workspace frame of (310,90) and shows that the object can no longer be rotated using such a grasp.

2.3 Combining the Two Maps

Since reference frames F and f share a common origin, before the object is rotated the two frames are coincident and thus the two maps can be overlayed. (See Figure 5 (a).) Feasible grasps can now be seen to be those which satisfy both the grasp stability and workspace constraints. When looking for a new grasp, the new grasp point must be in a region shaded in the grasp map, and it must also be either a vertical or horizontal move from the original point, as well as in a new finger pair box. As the object is rotated, not only the point representing the present grasp travels along a diagonal line as seen from the workspace map, but every point in the grasp map travels along a diagonal line, since every point on the object rotates by the same amount relative to the workspace reference frame. Therefore a rotation of the object is represented by a sliding of the grasp map along the diagonal of the workspace map. For a counterclockwise rotation, the grasp map would slide up and to the right relative to the workspace map; for a clockwise rotation, it would slide down and to the left. In figure 5 (b) we show a counterclockwise rotation of 45°as seen from the object frame. The finger pair boxes have slid along the diagonal in the negative direction, from the upper right corner to the lower left. Since the workspace map is repeating, the box edges that have disappeared from the lower left corner have reappeared in the upper right corner.

The problem of finding a gait can now be posed in map space. The goal of rotating an object continually is equivalent to the goal of continually sliding the finger pair boxes of the workspace map along the positive diagonal of the grasp map. The constraint that the object must be held at

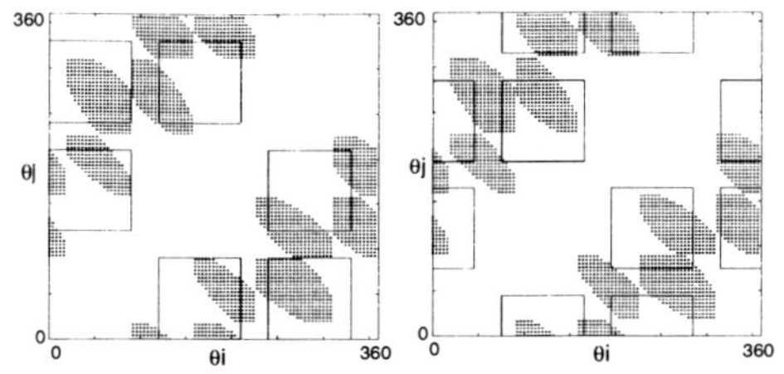

Figure 5: Series showing effect of rotation on grasp and workspace maps.

all times by a stable two fingered grasp is equivalent to the constraint that there must always be a point which is both in the stable region of the grasp map and inside one of the moving finger pair boxes of the workspace map. The problem of finding a gait is equivalent to finding a series of points (1) which satisfy the above constraints, (2) which are connected in grasp map space by only vertical or horizontal moves, and (3) where no two consecutive points are in the same finger pair box in workspace map space.

3 Brute Force Search

One way to look for a gait that could successfully rotate an object is to do a brute force search through all of the possible motions. Any step in a gait, or state of the system, can be completely described by three variables: the angle of the object relative to reference frame F (the workspace reference frame) and the angle of the two fingers grasping the object relative to reference frame f (the object reference frame). Then for any state, all the possible actions which can occur next can be enumerated using the grasp map and workspace map. At a given state two actions are possible, rotating the object or regrasping the object. The object can be rotated by an angle just greater than zero to an upper limit given by the minimum distance from the grasp point to the right or upper edge of the finger pair box in the workspace map. Alternatively, the present grasp can be replaced by another grasp, by moving horizontally or vertically to a new finger pair box in the workspace map to any point in a shaded region in the grasp map. Although both these actions involve a choice among a multiple infinity of choices, we discretize them to permit searching among distinct choices. The number of alternative moves from each state depends on the state, the object shape, the coefficient of friction and the discretization chosen for the object and finger angles. In general, the number of alternative motions, or the branching factor of the search, is large. We have chosen a discretization of 2°for both finger and object positions, yielding a typical branching factor of 35.

3.1 Breadth First Search

With branching factors on the order of 35, a breadth first search would be expected to be infeasible. (A breadth first search explores every possible action from a given grasp as a possible path before exploring the next level of possible motions in the search tree.) We ran a breadth first search for a number of different objects, with a goal of rotating the object 360°. For none of the cases was the search anywhere near a solution after using 500Mb of disk swap space. Thus a breadth first search was considered intractable.

3.2 Depth First Search

Alternatively, a depth first search was tried for a variety of objects. (A depth first search explores only one possible motion from a given grasp until either the goal is met or a dead end is reached. If a dead end is reached, the search continues from a possible motion from the closest previous node.) Since object rotation is the goal of the search, the nodes which involved rotation were put at the head of the queue, in front of the nodes involving regrasps. Therefore, regrasps were only sought if rotation was no longer possible. Using this approach, a solution was found for some objects, but was again far from a solution (after using 500Mb of disk swap space) for others. For example, using a pentagon similar to the one shown in figure 7 and a coefficient of friction of .7, a path was found after creating over 2,000,000 nodes, expanding over 2,000 of them, and resulting in a path length of over 100. (Each node represents a state in the system defined by the particular grasp and object orientation. A path length is the number of nodes in the final path which leads to the goal.) For the ellipse shown in figure 8 with the same coefficient of friction, however, the search was still not near a solution after using the available disk swap space. Therefore, a depth first search was not considered to be sufficient for finding gaits if they existed for general object shapes.

3.3 Best First Search

One of the reasons that a depth first search creates and expands so many nodes while trying to find a gait is that many regrasps are investigated which do not lead directly to continued rotation. A second reason is is that the graph of possible paths is highly interconnected. (Since the number of partial paths leading to the same node can influence the number of times that each path leading away from that node is investigated, a highly interconnected graph may lead to a lot of redundant searching.) This second problem is easily solved by considering only the first path found leading to a given node and terminating all other paths leading to that same node. The first problem can be addressed by using a metric to evaluate each partial path and then choosing the node with the highest value to be expanded next. To avoid unnecessary regrasps, the value of the path should be decreased each time a new grasp is used. To guide the path towards rotation, the value of the path should be increased as the object is rotated. To avoid a rotation occuring as the sum of a number of smaller rotations, the value of the path is should also decreased when the object is rotated. Therefore, we choose a metric $M = \alpha\theta - \beta\eta - \gamma r$, where θ = the object angle in degrees, η = the number of regrasps, and r = the number of rotations. A range of α, β and γ were used and all of them equaling 1 gave satisfactory results. If $\alpha/\beta \gg 1$, the object was often rotated too far and many unnecessary regrasps were tried before backing up to find an alternate path. If $\gamma = 0$, all combinations of rotations adding up to the maximum rotation was tried before backing up if a dead-end had been reached.

This *best first* approach was used to search for a gait to rotate many different object shapes by 360°. In every case the search was able to determine whether or not a path existed. Figure 6 shows 4 objects for which successful gaits existed using a workspace size for each finger of 90 degrees. In each case, 10 different initial grasps were used. Table 1 and 2 in the Comparing Methods section shows the results for the best first search under the column heading *general.* Table 1 shows results for a coefficient of friction of .8, and table 2 shows them for a coefficient of friction of .7. Three different data types are used: average nodes opened, average path length and average number of regrasps required by the final path. (Nodes opened are nodes which have been explored as a possible path. The path length includes both regrasps and rotations, since a state change includes either a regrasp or a rotation.) The difference between the number of nodes expanded and the length of the path can be used as a measure of the efficiency of the search, since the number of nodes expanded that were not in the final path led to dead ends.

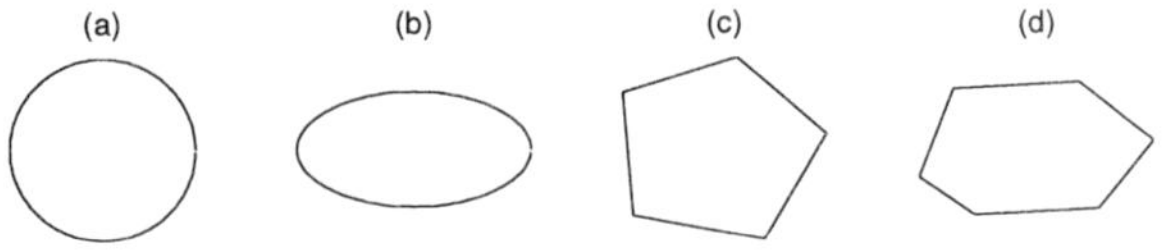

Figure 6: Objects used to test manipulation methods.

Thus, using the tools developed in Section 2 along with a metric to direct the search toward a reasonable path, we were able to construct a search which finds sufficient gaits to rotate objects. However, finding a sufficient solution using a search method does not in itself lead to an understanding of the grasp gait problem. There are still questions we would like to answer. Why does a gait exist in some cases and not in others? Why is a gait harder to find in some cases than in others? Is there an instructive way to classify different types of gaits? Answers to these questions would be interesting in their own right as well as helpful in finding more efficient solutions than the best first search. In addition, they would be helpful in guiding methods to manipulate objects when only general information about the object is known, and a robot hand with sensors is searching for the next stable grasp instead of a computer, in which case maximum efficiency is essential. To answer these questions, we take a closer look at a few of the gaits generated.

4 A Closer Look At Grasp Gaits

In looking at gaits which successfully rotate objects, we note that for certain objects and grasps, the object can be rotated until a finger reaches the end of a workspace, at which point a new grasp can be found which will allow rotation to continue. Figure 7 (1)-(4) shows a sequence of rotations and regrasps for a polygon with its representation in grasp map space to the right. Figure 7 (1) shows an initial grasp of (50,270) in grasp map coordinates, between fingers 1 and 3,

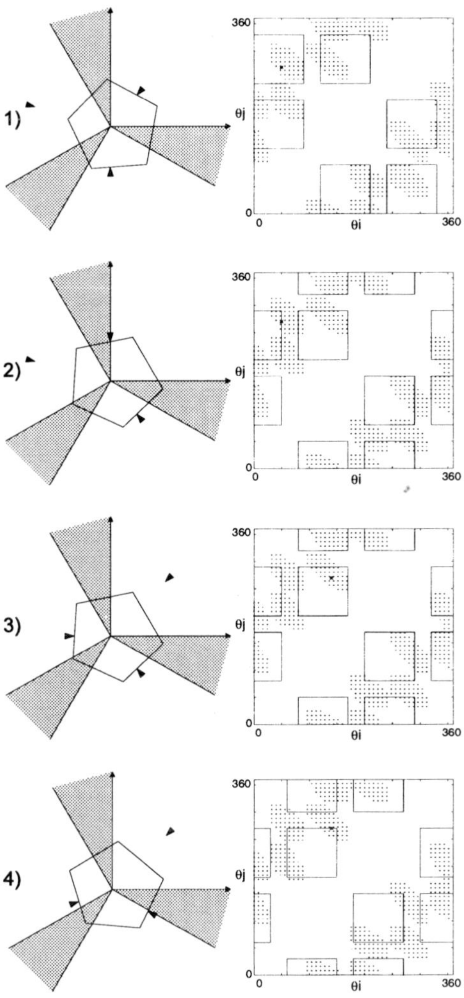

Figure 7: Series showing pentagon rotating successfully.

represented by the x in the outer finger pair box in map space. Figure 7 (2) shows the object after being rotated counterclockwise by 40°. Finger 1 has reached the end of its workspace and thus the object can no longer be rotated. Figure 7 (3) shows that a grasp between finger 1 and 3 has been replaced by a grasp between fingers 2 and 3, which is a positive horizontal move in the grasp map space to the coordinates (140,270). Figure 7 (4) shows the state after the object has been rotated an additional 20°. Looking at the map we see that a positive vertical move in map space will allow the object to be rotated further.

For other objects or other grasps, a new grasp cannot always be found if the object is rotated until a finger reaches a workspace limit. However, in some cases, a new grasp *could* have been found if it was sought at an earlier point in the rotation. Figure 8 and 9 illustrate this point. Figure 8 (1) shows an ellipse with an initial grasp of (135,245) in grasp map coordinates, using fingers 2 and 3. Figure 8 (2) shows the ellipse after it is rotated counterclockwise by 75°. Finger 2 has reached the end of its workspace. The corresponding map shows that although a vertical move in the map is possible, the second finger will still be at the edge of its workspace, and no additional moves to a stable region are possible which at the same time will allow the object to be rotated further. Figure 9 (1)-(4) shows an alternative path that will allow the object to be rotated beyond this point. Figure 9 (1) shows the object after it has been rotated only 45°from its original position. Figure 9 (2) shows a negative horizontal move in the grasp map to a grasp of (45,245) in grasp map coordinates. Figure 9 (3) shows a subsequent positive horizontal move to a grasp of (75,245) which now enables the object to be rotated further. Figure 9 (4) shows the object after being rotated to a position of 85°, at which point finger 3 has reached the end of its workspace. The corresponding map shows that a positive vertical move will allow the object to be rotated further still.

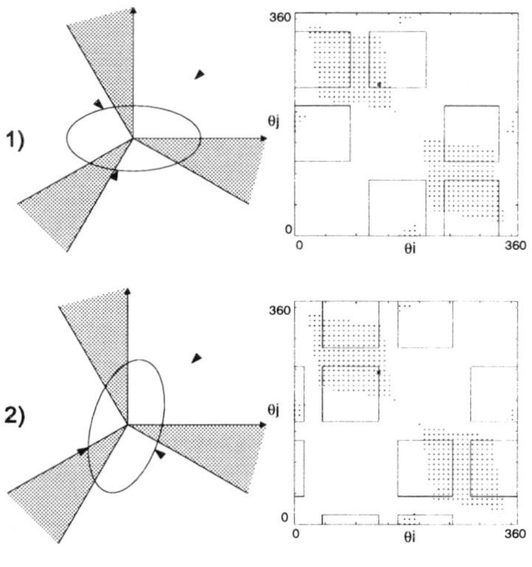

Figure 8: Series showing ellipse unable to rotate further.

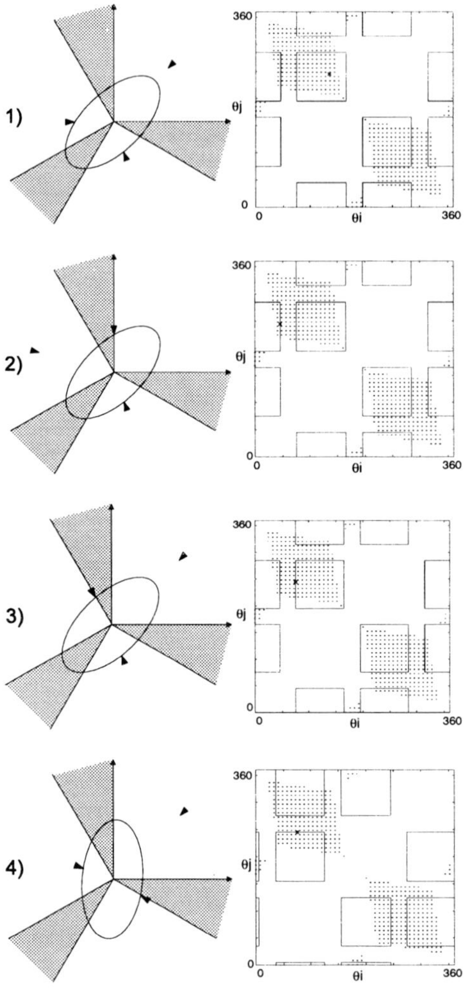

Figure 9: Series showing ellipse rotating successfully.

5 A Closer Look at Grasp Maps

To gain a better understanding of the different types of grasp gaits as those shown above, we need to take a closer look at the different types of grasp maps that can arise for convex objects. In general, grasp maps are a function of the the shape of the object and the coefficient of friction. With regard to shape, there are two parameters that we have used to characterize object shape, where varying these parameters leads to distinctive types of grasp maps. The first parameter we use is the eccentricity of the object. For our purposes, we define eccentricity, E to be equal to the ratio of the largest radius to the smallest radius, where the radius is measured from the surface of the object to the object's centroid. The second parameter we use is the smoothness of the object. We define smoothness, S, to be equal to the number of sides of the object. A smooth object (no discontinuities in slope) is defined to have infinite smoothness.

5.1 Varying Eccentricity and Smoothness

To show the effect of eccentricity on a grasp map, we vary eccentricity while keeping the coefficient of friction and smoothness constant. Figure 10 shows grasp maps for objects with eccentricity of 1, 1.33, 2 and 4. Smoothness is infinite and the coefficient of friction = .7. We see that for an object

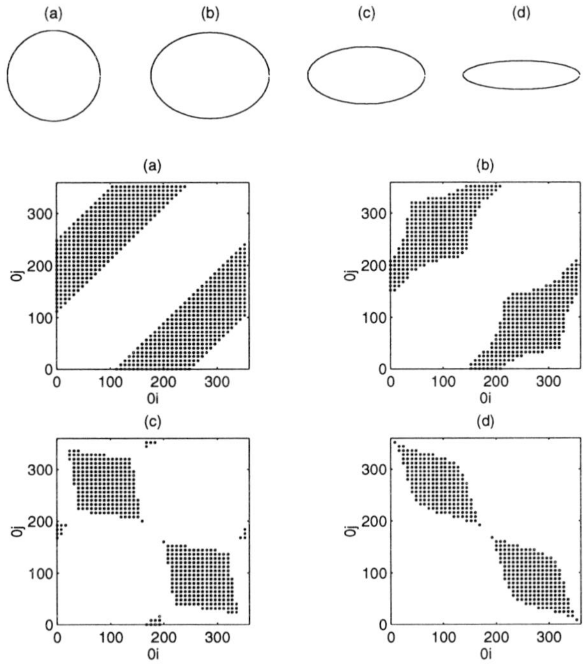

Figure 10: Objects of varying eccentricity and their grasp maps.

with an eccentricity of 1, the stable grasp region in the grasp map is a diagonal band with a slope of positive 1. Because of the repeating nature of grasp maps, as a grasp point moves along the stable region of the map in the direction of increasing angles, the grasp point wraps around to the bottom of the lower stable region. We refer to this type of map as a continuous map. In contrast, an object with an eccentricity of 4 has a stable grasp region that is a band with a slope of negative 1. This band is a discontinuous region, in that if a grasp point moves out in any direction from a point in the stable region, eventually it moves into an unstable region. As eccentricity varies between the two extremes, the stable region varies from the positive diagonal band to the negative diagonal band, passing through a stage where it is a region of points not correlated in either direction. For these intermediate eccentricities, the region is discontinuous as well.

To show the effect of smoothness on a grasp map, we vary smoothness while keeping the coefficient of friction and eccentricity as constant as possible. (If the smoothness is not infinite, the eccentricity must be greater than 1 since all radii cannot remain equal by definition.) Figure 11 shows grasp maps for objects with varying smoothness. We see that as smoothness decreases, the stable region in the grasp map becomes less smooth as well, and finally breaks up into isolated regions. However, these isolated regions still reveal information about the object's eccentricity, as they still fall along the positive diagonal.

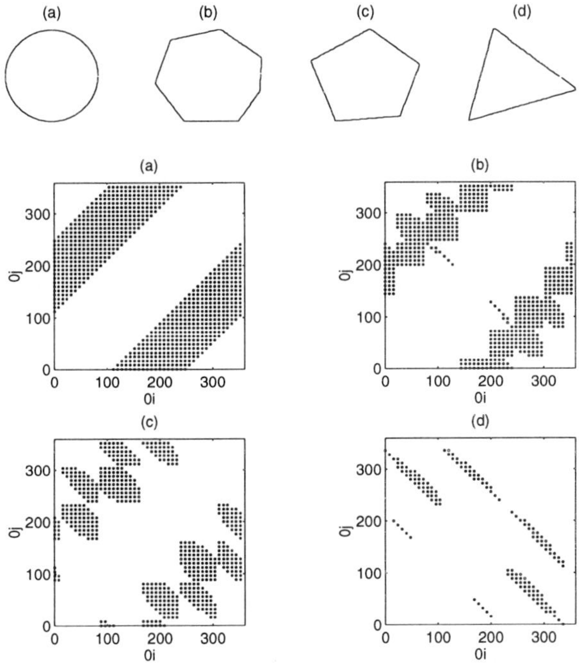

Figure 11: Objects of varying smoothness and their grasp maps.

5.2 Finding Simply Shaped Regions in Complex Maps

One major difference between grasp maps is whether the stable regions are more or less continuous and lying along the positive diagonal or discontinuous and lying along a different orientation. This difference is related to the eccentricity of the object. A useful goal is to seek a simply shaped region for each type of map which tends to fit within the stable region of that map type. From there we can seek a rule-based algorithm that will generate a grasp gait for that simpler prototype region. In addition, we parameterize these regions, so that for a given set of parameters for a specific map, the rule-based algorithm generates a unique gait for that specific object.

For the continuous grasp map, the stable region can be approximated as a solid band of varying width and placement, lying along the positive diagonal (where the inner and outer edge have a slope of 1). This prototype holds for many continuous grasp maps. Figure 12 (a) shows an example grasp map (from Figure 10 (a)) with the band prototype within. For the discontinuous grasp map, the stable region can be approximated as a box of varying height, width and placement. This prototype fits within many discontinuous grasp maps. Figure 12 shows objects an example grasp map (from Figure 10 (c)) with the box prototype within.

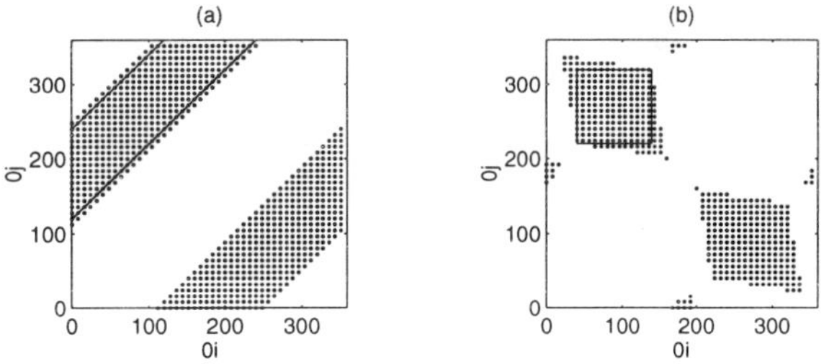

Figure 12: Prototype grasp regions within grasp maps.

6 Grasp Gaits for Prototype Grasp Maps

For a given object, we can now use a grasp map modified to include only the simpler prototype pattern along with a workspace map to seek rule-based algorithms to rotate the object. (Further discussion of these methods are to be presented in [19]).

6.1 Reorienting Low-Eccentricity Objects Using Band Method

For many low eccentricity objects, a band readily fits within its more complex grasp map. Thus far the only constraint on the band is that its inner and outer edge have a slope of 1. Since object rotations are represented in map space by a relative sliding along the 45°line of the grasp and workspace maps and the workspace boxes are aligned along the 45°line, the grasp map with this sort of band is invariant with respect to the workspace map. To ensure that regrasps can be achieved, the first requirement of the band is simply that it be at least as wide as the space between the finger workspaces (which we shall call ω). The second requirement is that the band be located such that it intersects both the inner and the outer finger pair boxes. It can be shown that these are the only requirements of the band to ensure that a gait exists to continually rotate the object. In addition, it can be shown that there is a requirement of the initial grasp to ensure that a gait can be implemented. The requirement is that if the grasp is in an outer box it must be at least ω from the inner edge of the band, and if it is in an inner box it must be at least ω from the outer edge. If these conditions are met, then a successful gait rotating the object counterclockwise (clockwise) can be generated using the following rules in map space:

1. Rotate until the grasp reaches the edge of its finger pair box
2. If the grasp is in an outer box, take a positive horizontal (negative vertical) step to the left (top) edge of the inner box
3. If the grasp is in an inner box, take a positive vertical (negative horizontal) step to the bottom (right) edge of the outer box
4. Return to step 1

Grasp gaits following these rules create a staircase-like pattern in the grasp map, as they are made by always taking alternate positive (negative) horizontal and vertical steps in the grasp map. (See figure 13 (a).)

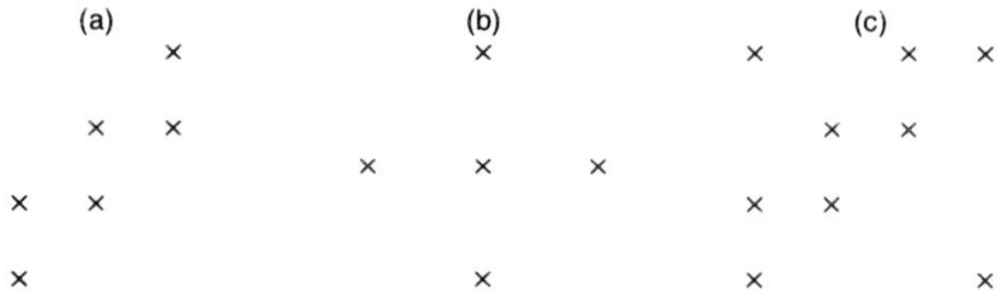

Figure 13: Typical gait patterns as seen from the grasp map: (a) staircase, (b) plus (+), (c) combined.

6.2 Reorienting High-Eccentricity Objects Using Box Method

As the eccentricity of an object increases, the grasp map becomes discontinuous and is no longer invariant as seen from the workspace map during reorientations. Therefore, as the object is rotated and the fingers reach the end of their workspaces, at some object orientations there is not going to be a new grasp moving up along the positive diagonal in the grasp map (which is the direction along which the fingerpair boxes lie). Some regrasps will have to jump backwards in the grasp map. However, since the stable grasp region is discontinuous, at some object orientations there is not going to be a new grasp moving backwards in the grasp map. It can be shown that if there is a box at least 2ω by 2ω within the grasp map, and its center passes through both the inner and outer finger pair boxes in the workspace map during rotations, then a gait exists for the object to be rotated continually. The following rules will generate such a gait for counterclockwise rotation. (Clockwise rotations can be generated by substituting right for left, top for bottom, positive for negative and visa versa.)

1. If the grasp is at least a distance ω from both the top and right edge of the grasp map box, rotate until the grasp reaches the edge of its finger pair box. (We refer to this distance as the rotation limit.)
2. If the grasp is less than a distance ω from the top of the grasp map box, calculate the distance that the object can be rotated before the bottom of the grasp map box will be a distance ω from the bottom of the present finger pair box. (We refer to this distance as the upper limit.) Then rotate either the rotation limit or the upper limit, whichever is less.

3. If the grasp is less than a distance ω from the right of the grasp map box, calculate the distance that the object can be rotated before the left of the grasp map box will be a distance ω from the left of the present finger pair box. (We refer to this distance as the right limit.) Then rotate either the rotation limit or the right limit, whichever is less.

4. If the grasp is less than a distance ω from both the right and top of the grasp map box, rotate either the rotation limit, the upper limit, or the right limit, whichever is least of the three.

5. If the object was rotated by the rotation limit, take either a positive horizontal or vertical step of size ω, whichever is possible. Return to step 1.

6. If the grasp was rotated by the upper limit, take either a negative horizontal or vertical step, whichever is possible. Take the largest step possible.

7. If step 6 was just performed using a horizontal step, take the largest possible negative vertical step.

8. If step 6 or 7 was just performed, take a positive vertical step equal to ω. Return to step 1.

9. If the grasp was rotated by the right limit, take either a negative vertical or horizontal step, whichever is possible. Take the largest step possible.

10. If step 9 was just performed using a vertical step, take the largest possible negative horizontal step.

11. Take a positive horizontal step equal to ω. Return to step 1.

Using this method, if the conditions for the box size and placement are met, there is no initial condition that would preclude a successful gait. However, if, for a counterclockwise (clockwise) rotation, the grasp was less than ω from the top or right (bottom or left) edge, and the opposing edge was in the same finger pair box, the object would first have to rotate backwards to get the grasp in a position that it could rotate continually in the desired direction. For a grasp map in which only a box of the required size fits, the above rules generate a plus (+) pattern in the grasp map, requiring exactly five points in the grasp map to rotate the object continually. For maps which include larger boxes, these rules generate a hybrid of + and staircase patterns. (See figure 13 (b) and (c).)

7 Extending Rule-Based Solutions Using Modified Search

As the smoothness of an object decreases, the grasp map becomes less regular but still follows the general trend for its given eccentricity. Therefore some grasp maps may not contain a band or box of sufficient size, but the missing pieces of the map may not be required for a gait to exist for that object. In that case, the object could be rotated using the same general gait pattern (either staircase or +), but the step sizes would need to be modified to fit the specific grasp map. For these cases a modified search is used which looks only for the general structure in the grasp map (either staircase or +) already known to work for that type of object, as opposed to the general search, which looks at any viable finger motion or rotation. As explored in the next section, this modified search is much more efficient than the general search method at finding a path, and in general it is equally as efficient as the rule-based method.

8 Comparing Methods

We now compare results of the different approaches to generating gaits investigated in this paper. The first approach is the general search, made feasible by using a best first technique and by not considering multiple paths through a node. The second approach is either the rule-based method if the sufficient conditions for such a method are met, or the modified search based on the rule-based method. The objects and initial conditions used for the best first search in section 3.3 were used again to compare methods for achieving a goal of reorienting the object 360 degrees. (The rule-based methods will rotate an object continually, but a finite goal is needed for comparison.) Figure 6 shows the objects used and table 2 and 3 compare the results from the two methods. Table 2 shows results using a coefficient of friction of .8 and table 3 shows results using a coefficient of friction of .7. The three types of data used for comparisons are (1) the number of nodes opened, (2) the path length, and (3) the number of regrasps required. For the modified rule-based method, the path length always equals the number of regrasps, since object rotations are implicit in the structure of the gait in the grasp map. For the rule-based search, all three measures are always equal, since no extra nodes ever need to be opened. Interesting comparisons between the two groups are (1) nodes opened - path length (showing an efficiency of the method) and (2) number of regrasps (showing an efficiency of the path found). Objects (a), (b) and (d) were successfully rotated using the rule-based methods, and object (c) required a modified search. Object (c) is a typical case in that, even though a modified search was used, its efficiency was 100% (the number of nodes equalled the path length). For the many examples we tried which used the modified search, the number of nodes opened tended to be equal or close to equal to the path length. These results show that the rule-based or modified rule-based method is much more efficient at finding a path than the best-first search. In addition, the number of regrasps required tend to be less for the rule-based approach.

DATA TYPE	circle (a)	
	general	rule-based
nodes opened	500.2	9.0
path length	44.8	9.0
regrasps	22.4	9.0
	ellipse (b)	
	general	rule-based
nodes opened	5,746.5	12.7
path length	48.2	12.7
regrasps	31.5	12.7
	pentagon (c)	
	general	rule-based
nodes opened	315.0	14.5
path length	37.4	14.5
regrasps	21.9	14.5
	hexagon (d)	
	general	rule-based
nodes opened	439.5	17.8
path length	42.7	17.8
regrasps	24.4	17.8

Table 1: Results of method comparison using coefficient of friction of .8.

DATA TYPE	circle (a)	
	general	rule-based
nodes opened	957.7	9.0
path length	82.0	9.0
regrasps	41.1	9.0
	ellipse (b)	
	general	rule-based
nodes opened	14,486.3	13.8
path length	30.4	13.8
regrasps	20.3	13.8
	pentagon (c)	
	general	rule-based
nodes opened	208.8	14.5
path length	29.5	14.5
regrasps	18.0	14.5
	hexagon (d)	
	general	rule-based
nodes opened	1,421.2	18.6
path length	83.4	18.6
regrasps	58.0	18.6

Table 2: Results of method comparison using coefficient of friction of .7.

9 Conclusions

In this paper we address the problem of generating grasp gaits that will enable a robot hand to reorient convex objects in two dimensions. We begin with a grasp map and further develop the analytical tools used to pose the problem in a mathematically tractable way. We then use a general search method to find sufficient solutions in map space. As an alternative approach, we find general patterns within grasp maps which are related to general parameters of objects. We then find rule-based solutions which generate gaits deterministically for these patterns. We use these deterministic solutions to modify the general search for cases where the deterministic solutions do not fit. We find that the deterministic solutions and the modified search solutions are significantly more efficient than the general search methods. In addition, we have not yet found a convex object where the rule-based or modified rule-based solutions have failed while the general search has not. We conjecture that the analytical solutions may be complete as well as more efficient, although this remains to be rigorously tested as part of our future work. It is ultimately desirable to plan gaits for non-planar motions, including rotations about an arbitrary axis and translation. This will require at least 4 fingers to permit sequences of 3 finger grasps and an extension of our representation and search techniques to this higher dimensional problem.

10 Acknowledgments

This work was supported by the Office of Naval Research, University Initiative Program Grant N00014-92-J-1814, and the National Science Foundation Graduate Student Fellowship Program. We would also like to gratefully acknowledge the assistance of Prof. Tomás Lozano-Pérez.

References

[1] Mason, MT, Salisbury, JK. Robot Hands and the Mechanics of Manipulation, MIT Press, Cambridge, MA, 1985.

[2] Jacobsen, SC, et al. Design of the Utah/MIT Dextrous Hand, Proc. 1987 IEEE International Conference on Robotics and Automation, San Francisco, Apr. 1987.

[3] Nguyen, V-D. Constructing Force-closure Grasps. IJRR 1988; 7(3):3-16.

[4] Coelho Jr., J. Grupen, R. Optimal Multifingered Grasp Synthesis. In Proc. 1994 IEEE Int Conf. on Robotics and Aumation, San Diego, 1994.

[5] Pollard, NS. Parallel Methods for Synthesizing Whole-Hand Grasps from Generalized Prototypes, MIT AI Lab. TR-1464, January 1994

[6] Bicchi, A. "Optimization of Robotic Grasping Forces," in Proc. of the American Control Conference, 1992.

[7] Howe, RD, Cutkosky, MR. Touch Sensing for Robotic Manipulation and Recognition, in The Robotics Review 2, MIT Press, 1992.

[8] Allen, P. Robotic Object Recognition Using Vision and Touch, Kluwer Academic Publishers, 1987.

[9] Kerr, J. Roth, B. Analysis of Multifingered Hands, The Int. J. of Robotics Research, 4(4):3-17,1986.

[10] Li, Z, Canny, J, Sastry, S. On Motion Planning for Dextrous Manipulation, Part I;: The Problem Formulation. Proceedings of IEEE Conf. on Robotics and Automation, Scottsdale AZ, 1989.

[11] Brock, D. Enhancing the Dexterity of a Robot Hand Using Controlled Slip. TR992, Artificial Intelligence Laboratory, Massachusetts Institute of Technology, May 1987.

[12] Rus, D. Coordinated Manipulation of Polygonal Objects, IEEE/RSJ Int. Conf. on Intelligent Robots and Systems, 1993.

[13] Abell, T, Erdmann, M. Stably Supported Rotations of a Planar Polygon with Two Frictionless Contacts. IEEE/RSJ International Conference on Intelligence Robots and Systems, vol. 3, pp.411-418, Pittsburgh, PA, August 1995.

[14] Omata, T and Nagata, K. Planning Reorientation of an Object with a Multi-Fingered Hand, Proceedings of the 1994 IEEE International Conference on Robotics and Automation, May, 1994, San Diego, CA volume 4, pp3104-3110

[15] Fearing, R. Simplified Grasping and Manipulation with Dexterous Robot Hands. IEEE International Journal of Robotics and Automation 1986; 2(4).

[16] Hong, J, Lafferriere, G, Mishra, B, and Tan, X. Fine Manipulation with Multifinger Hands. IEEE International Conference on Robotics and Automation, pp.1568-1573, Cincinatti, OH, May 1990.

[17] Chen, I-M, and Burdick, JW. A Qualitative Test for N-Finger Force-Closure Grasps on Planar Objects with Applications to Manipulation and Finger Gaits. IEEE International Conference on Robotics and Automation, pp.814-820, Atlanta, GA, 1993.

[18] McGhee, R. and Frank, A. On the Stability Properties of Quadruped Creeping Gaits, In Mathematical Biosciences, vol. 3, pp. 331-351, 1968.

[19] Leveroni, S. PhD Thesis, Dept. of Mechanical Engineering, MIT. In preparation.

PART 2

CONTROL

SESSION SUMMARY

Suguru Arimoto
Department of Mathematical Engineering and Information Physics
University of Tokyo, Bunkyo-ku, Tokyo 113, Japan

Research of robot control can be traced back to the origin in 1980 when two remarkable papers proposing fast algorithms for the computed torque method were published by John M. Hollerbach and independently by J.Y.S. Luh, M.W. Walker, and R.P.C. Paul. Since then, with help of the rapid progress of microelectronics and computer technology and belief in computational power, much work has been done in developing special-purpose VLSIs and designing micro-proccesing parallel architechtures that implement and execute such fast computed torque methods. However, there is little of actual robots in industry in which such a fast computed torque method is implemented for carrying out real and practical tasks. It is possible to raise some reasons but the most important one that can not be overlooked is that belief in power of computations has eluded robotists from looking inside the black box of robot dynamics forced them to be blind to physical meanings of dynamics of a robot and physical interactions between the robot and its task environment. To make robots acquire dexterity for execution of versatile tasks, understanding of physical interactions of robots with objects or their environments is indespensable, which must be granted by exploring inside the black box of robot dynamics.

This session consists of the following five papers:

1) S. Arimoto. Another language for describing robot language for describing robot motions: A nonlinear position-dependent circuit theory.
2) B. Siciliano. Parallel force/position control of robot manipulators.
3) M. Uchiyama and A. Konno. Modeling, controllability and vibration suppression of 3D flexible robots.
4) K. Yoshida, D.N. Nenchev, and M. Uchiyama. Moving base robotics and reaction management control.
5) Y. Nakamura, T. Suzuki, and M. Koinuma. Nonlinear behavior and control of underactuated robot mechanisms.

All these papers aim to open the black box of robot dynamics in their own ways in attacking different problems. In the first talk, Arimoto proposed a framework of "nonlinear position-dependent circuits" that can describe Lagrange's equations of robot motions even when they are interacting with environments. The first question from the floor was as to the extensibility of expressing Lagrange's equations of dynamics of flexible-link robots by such a nonlinear circuit and describing a physical phenomenon of impact between a tool endpoint and the environment via such nonlinear circuits. Another question was posed from the attendees, which pointed out a previous literature on the possibility of expressing Lagrange's equations by means of energy-based "bond graph" approach. The passivity as a generalization of impedance concept to cope with nonlinear dynamics was stressed by the presenter. In the second talk, Dr. Siciliano presented a detaild survey on recent advances of hybrid position/force control and showed his own recent results. Questions were concerned with 1) the practical performance of the proposed methods and 2) limitations on the methods due to the assumptions on the shape of concerned constraint surface. Dr. Uchiyama presented an interesting talk on finding a new singularity called "uncontrollable mode" of the configuration of a flexible arm. He also stressed the effective use of Holtzer's method in modeling dynamics of flexible robots. One question was concerned with the possibility of analyzing the uncontrollable mode in a structured way without computing numerically the controllability matrix. Dr. Yoshida talked about moving base robotics in general at the beginning and pointed out a variety of its applications in practical issues of robotics. Then he focused on his own research subject on dynamic analysis for reaction management control using the reaction null space. Finally, Dr. Nakamura presented his talk on nonlinear behaviors of dynamics of underactuated robotic mechanisms. He showed a variety of interesting phenomena like chaotic motions even for a simple mechanism of two degree of freedom with a single actuator. The most important question was concerned with the necessity of using discontinuous or time-varying control in stabilizing the motion of such a mechanism.

The first paper treated only the case that the degree of freedom is equal to the number of actuators. The second paper was concerned with the case that the degree of freedom degrades caused by a geometric constraint. The remaining three papers were all concerned with cases that the number of actuators is less than the degree of freedom, that is, cases of under-actuated dynamics.

In summary, all these papers attempted to open the sophistcated inside of a robot dynamics black box by aiming at exploring and finding physical principles underlying robot dexterity from the viewpoint of robot dynamics and control. It is expected that the session may give rise to a turning point towards physically understanding human dexterity and realizing intelligent robots in a certain future.

Another Language for Describing Robot Motions: A Nonlinear Position-dependent Circuit Theory

Suguru Arimoto
Faculty of Engineering, The University of Tokyo, Bunkyo-ku, Tokyo, 113 (Japan)
e-mail: arimoto@sr3.t.u-tokyo.ac.jp

Abstruct: Dynamics and control of nonlinear mechanical and advanced mechatronic systems can be investigated more vividly and efficiently by using corresponding nonlinear position-dependent circuits that describe Lagrange's equations of motions and interactions with objects or/and task environments. Such expressions of Lagrange's equations via nonlinear circuits are indebted to lumped-parameter discretization of mechanical systems as a set of rigid bodies through equations of motion due to Newton's second law. This observation is quite analogous to validity of electric circuits that can be derived as lumped-parameter versions of Maxwell's equations of electromagnetic waves. Couplings of nonlinear mechanic circuits with electric circuits through actuator dynamcis are also discussed. In such electromechanical circuits the passivity should be a generalization of impedance concept in order to cope with nonlinearities of circuits and play a crucial role in their related motion control problems. In particular, it is shown that the passivity as an input-output property gives rise to a necessary and sufficient characterization of H_∞-tuning for disturbance attenuation of robotic systems, which can give another system-theoretic interpretation of the energy conservation law.

1 Introduction

It is widely known as explained in many textbooks that there is an analogy between dynamcis of a simple linear mechanical system of mass-damper-spring (MCK) and that of a simple lumped-parameter linear electric circuit of inductor-resistor-capacitor (LRC) (see Fig.1). Nevertheless, dynamics of actual mechanical and mechatronics systems such as robot arms and mechanical hands have never been analyzed intensively based on their corresponding circuit-theoretic expressions though we do not know whether such circuits exist in general or not. Indeed, there is a dearth of papers that have discussed directly any possibility of expressing Lagrange's equation of motion for nonlinear mechanical systems via lumped-param- eter circuits like electric circuits. Conversely, there is a well-known approach called "Euler-Lagrange formalism" for analyzing electric circuits (see Meisel [1]) and "Lagrange-Formalism" for discrete elecktromechanical systems (see Maisser and Steigenberger [2]). More classically, there was a proposal for description of mechanical systems via a network called "bond graph" (see Paynter [3]), Karnopp and Rosenberg [4]). However, all these were proposed before opening a new "robot era" and therefore have not yet been extended to cope with dynamics of nonlinear mechanical and mech- atronics systems. Indeed, dynamics of actual mech- atronics systems such as anthropoid robot arms and hands are nonlinear and have strong couplings between joint variables. As a matter of course in electric circuit theory, generalization of electric circuits towards nonlinear circuits has been attempted in

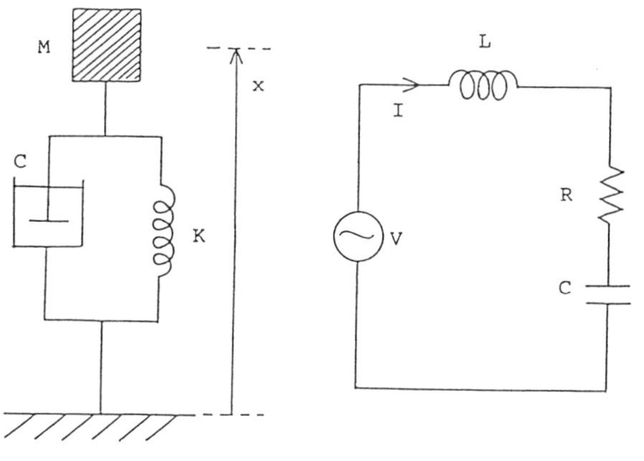

Fig.1 Analogy between a Mass-Damper-Spring mechanical system and a linear Induetor-Resistor-Capacitor electric circuit.

1960's and 1970's (see Brayton and Moser [5]) and Chua and Green [6]), but only a limited class of circuits was investigated by introducing individually a few of single nonlinear elements like a nonlinear resistor, an inductor with nonlinear flux, and a nonlinear capacitor.

This paper shows first that a mechanical system governed by Lagrange's equation of motion implies the exsistence of a nonlinear position-dependent circuit and its dynamics corresponds to that of the circuit. In such a nonlinear circuit the velocity vector can be regarded as the current vector that passes along a bundle of branches and flows into and out of such blocks with n input and output terminals as kinetic inductor, resistor, virtual resistor, and gravity capacitor. These linear or nonlinear blocks are considered to be operators that operate on a domain of velocity vectors and generate voltage drops. A nonlinear kinetic inductor can be regarded as a block with n inputs and n outputs being dependent on the position and retains a kinetic energy. A gravity capacitor is also dependent on the position and retains a potential energy. Both Kirchhoff's current and voltage laws are valid to this class of nonlinear position-dependent circuits, correspondingly to d'Alambert's principle. In addition, a new concept called "displacement capacitor" is introduced in order to cope with set-point control problems of nonlinear mechanical systems. The displacement capacitor can be interpreted as an extension of a mechanical spring. Thus, it is shown that a similar method to the direct-current analysis for electric circuits can be applied for the stability analysis of set-point control without compensation for the gravity force. However, the proposed approach is not a simple extension of this analogy because mechanical circuits as a language for describing dynamcis should have more flexible syntax and general context to cope with various nonlinearities in electromechanical systems and three underlying physical variables "velocity", "torque", and "position" differently from electric circuits that have only two kinds of underlying physical variables "current" (corresponding to "velocity") and "voltage" (corresponding to "torque"). A physical variable "electric charge" appears as an integral of "current" in time in electric circuits like "position" as an integral of "velocity" in time, but a vector of position variables signifies a real configuration of the system in space. Hence, a kinetic inductor is a genuinely new concept of inductor that is dependent on not only "velocity" but also "position", that is, the inductance operator that induces "voltage drop" depends on the present configuration of the system.

In this paper we first introduce definitions of elementary blocks that may constitute a mechanical circuit and show a few examples of nonlinear position-dependent circuits that express Lagrange's equations of motion of mechatronics systems. Secondly we deal with coupling of the motion of a robot arm drived by a set of voltage-controlled DC motors as joint actuators with a set of linear armature electric circuits. Thirdly we treat the most fundamental problems of set-point position control of robot arms with or without compensation of gravity forces from the passivity viewpoint. In particular, we present a necessary and sufficient characterization of H_∞-tuning of a PD control with off-line gravity compensation in the sense of disturbance attenuation. In all cases, the passivity as a system-theoretic input-output property plays a crucial role, which also characterizes the sufficient condition for H_∞-tuning in most cases. This in turn means that the H_∞-tuning for mechatronics systems in the sense of disturbance attenuation can be interpreted as another characterization of "energy preservation law." In the concluding section, we discuss future research directions of the proposed approach based on nonlinear position-dependent circuits.

2 Elementary Blocks of Circuits

Consider a block, into and out of which a bundle of n physical variables $\dot{q} = (\dot{q}_1, \cdots, \dot{q}_n)$ flows like a vector of n electric currents (see Fig.2). For a mechanical system consisting of a series of n rigid links connected through n joints (see Fig.3), $\dot{q}_i$ stands for the angular velocity at joint i if the joint is of revolute-type or does for the velocity if it is prismatic. Accordingly q_i denotes the angle or the linear position and hence $\dot{q}_i = dq_i/dt$. We introduce several kinds of elementary block with corresponding physical properties as follows:

a) Resistance block; this is a block that induces the torque/force vector $R\dot{q}$ like a vector of voltage drops, where R is an $n \times n$ symmetric non-negative definite matrix (see Fig.4(a)).

b) Virtual-Resistance block; this is a block that

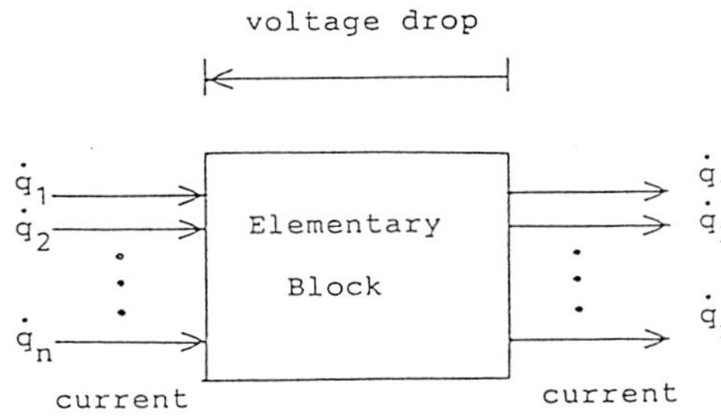

Fig.2 Elementary block of nonlinear position-dependent circuits.

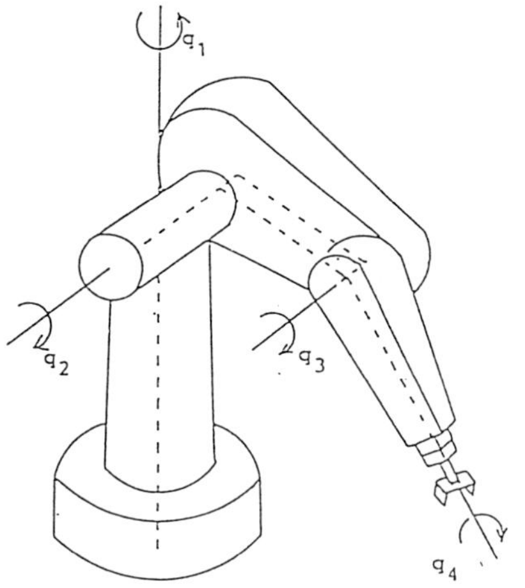

Fig.3 Anthropoid manipulator with all revolute-type joints

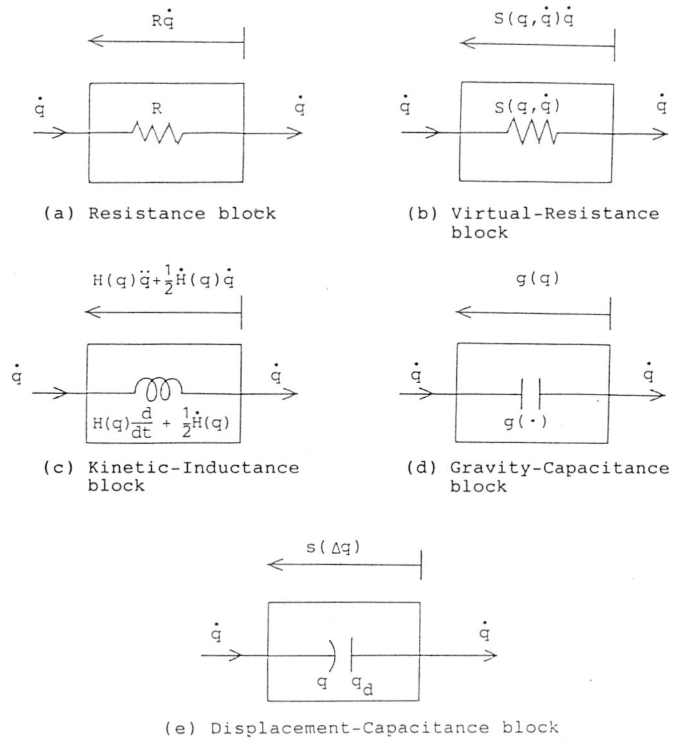

Fig.4 Elementary blocks of nonlinear position-dependent circuits

induces the torque/force vector $S(q,\dot{q})\dot{q}$, where $S(q,\dot{q})$ is an $n \times n$ skew symmetric matrix dependent on q and $\dot{q}$ (see Fig.4 (b)).

c) Kinetic-Inductance block; this induces the torque/force vector

$$\{H(q)\frac{d}{dt} + \frac{1}{2}\dot{H}(q)\}\dot{q} \quad (= H(q)\ddot{q} + \frac{1}{2}\dot{H}(q)\dot{q}) \tag{1}$$

where $H(q)$ is an $n \times n$ symmetric positive definite matrix dependent on position q (see Fig.4(c))

d) Gravity-Capacitance block: This induces the torque/force vector $g(q)$ that is an n-dimensional vector-valuded function of q (Fig.4(d)).

e) Displacement-Capacitance block; this induces the torque/force vector $s(\Delta q) = (s_1(\Delta q_1), \cdots, s_n(\Delta q_n))^T$, where $q_d = (q_{d1}, \cdots, q_{dn})^T$ is a given target position, $\Delta q = q - q_d, \Delta q_i = q_i - q_{di} (i = 1, \cdots, n)$, and each $s_i(\theta)$ is a linear or nonlinear function satisfying

$$s_i(0) = 0 \quad \text{and} \quad \theta s_i(\theta) > 0 \text{ for } \theta \neq 0 \tag{2}$$

(see Fig.4(e) and Fig.5).

f) Current (Angular Velocity)-Source or Voltage (Torque)-Source; this is a source that can generate the specified current (angular velocity) flow (see Fig.6(a)) or maintains the specified voltage (torque/force) between two terminals (see Fig.6(b)).

Furthermore, it should be interpreted that a) a resistance block with resistance R (an $n \times n$ symmetric non-negative definite matrix) induces the energy consumption rate $\dot{q}^T R\dot{q}$, b) a virtual-resistance block induces none of energy consumption because the inner product of $\dot{q}$ with $S(q,\dot{q})\dot{q}$ vanishes due to the skew symmetry of $S(q,\dot{q})$, c) a kinetic-inductance block retains the kinetic energy $K(q,\dot{q}) = \frac{1}{2}\dot{q}^T H(q)\dot{q}$ likely as an inductor retains magnetic energy in an electric circuit, and d) a gravity-capacitance block retains the potential energy $P(q)$ where $g_i(q) = \partial P(q)/\partial q_i, i = 1, \cdots, n$. All nonlinear blocks from b) to d) and a linear resistance block a) can be dependent on the same and common position q if they are connected in series (see Fig.7). To illustrate this, consider again the robotic arm depicted in Fig.3. As was already known, the Lagrange equation of

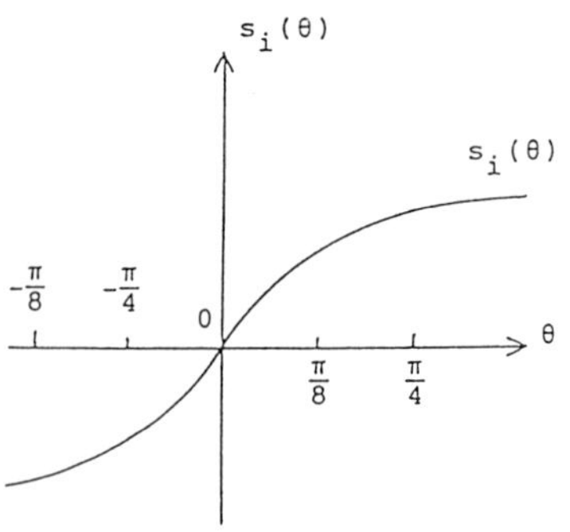

Fig.5 Gradient of a potential.

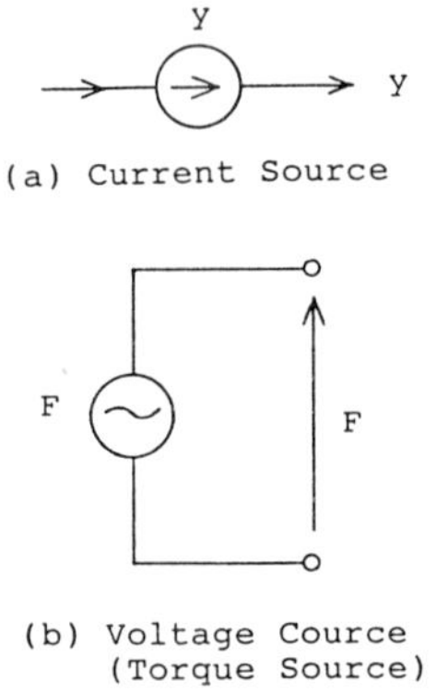

Fig.6 Sources for nonlinear position-dependent circuits

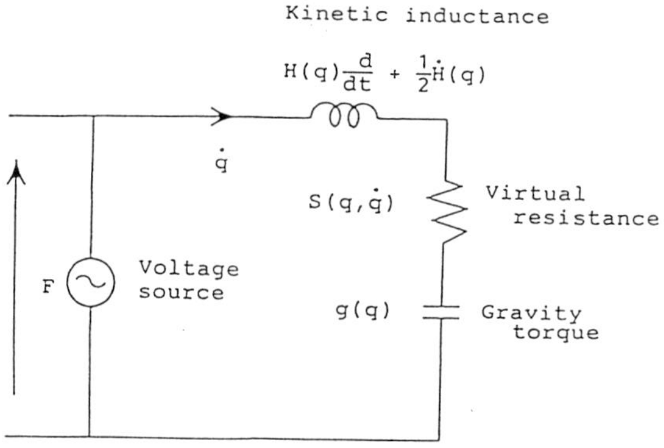

Fig.7 A nonlinear circuit that expresses Lagrange's equation of motion.

motion of the system is described by

$$\frac{d}{dt}\{\frac{\partial}{\partial \dot{q}}L(q,\dot{q})\} - \frac{\partial}{\partial q}L(q,\dot{q}) = F \qquad (3)$$

where $L(q,\dot{q}) = K(q,\dot{q}) - P(q), K(q,\dot{q})$ denotes the kinetic energy, $P(q)$ the potential energy, and F the generalized force vector whose i-th component stands for a torque generated by an actuator at joint i. In detail, (3) can be written as

$$H(q)\ddot{q} + \dot{H}(q)\dot{q} - \frac{\partial}{\partial q}K(q,\dot{q}) + \frac{\partial}{\partial q}P(q) = F. \quad (4)$$

Further, note that (3) can be expressed as

$$\{H(q)\frac{d}{dt} + \frac{1}{2}\dot{H}(q)\}\dot{q} + S(q,\dot{q})\dot{q} + g(q) = F \quad (5)$$

where

$$S(q,\dot{q})\dot{q} = \frac{1}{2}\{\dot{H}(q)\dot{q} - \frac{\partial}{\partial q}\dot{q}^T H(q)\dot{q}\}. \qquad (6)$$

It is well known that $S(q,\dot{q})$ becomes skew-symmetric (see Arimoto and Miyazaki [7]).

Now, applying Kirchhoff's voltage law for the circuit of Fig.7 leads to equation (5). Thus, the Lagrange equation of motion for the system of Fig.3 is now represented by the nonlinear position-dependent circuit depicted in Fig.7 (Corresponding to equation (5)).

Finally it should be remarked that a displacement-capacitor can be interpreted as a generalization of "mechanical spring" since each component $s_i(\Delta q_i)$ of $s(\Delta q)$ plays a role of restoring force, which can be described by a linear or nonlinear function of displacement Δq_i. Further, a vertual resistance block characterized by a skew-symmetric matrix $S(q,\dot{q})$ can be regarded as a generalization of gyrators in electric circuits.

3 Passivity Analysis

Impedance or admittance is an essential concept of linear electric circuits, that is defined as a ratio of Fourier-transformed input and output. As to Fig.1(b), admittance $G(i\omega)$ between input voltage V and output current I is defined as $G(i\omega) = I(i\omega)/V(i\omega) = (Li\omega + R + (Ci\omega)^{-1})^{-1}$, where $I(i\omega)$ and $V(i\omega)$ signify the Fourier transform of $I(t)$ and $V(t)$ respectively, provided that both the inductor L and the capacitor C retain zero energy at $t = 0$. Similarly, mechanical impedance concerning the mass-damper-spring system between input force F and output velocity $\dot{x}(= dx/dt)$ is defined as $G(i\omega) = i\omega(M(i\omega)^2 + C(i\omega) + K)^{-1} = (Mi\omega + C + K(i\omega)^{-1})^{-1}$. However, the concept of impedance or admittance is not effective in general in analyzing nonlinear dynamics like (4), because Fourier and Laplace transforms can not be applied for describing input-output relations of that dynamics. It is however fortunate that, according to Anderson [8], passivity of system dynamics is the alternative expression of impedance or admittance, which is defined as (for example, in the case of Fig.1(b))

$$\begin{aligned}\int_0^t I(\tau)V(\tau)d\tau &= E(t) - E(0) + \int_0^t RI^2(\tau)d\tau \\ &\geq -E(0) = -\gamma_0^2 \qquad (7)\end{aligned}$$

where $E(t) = \{LI^2(t) + C^{-1}q^2(t)\}/2$ and $dq(t)/dt = I(t)$. In the case of nonlinear dynamcis (5), we obtain the following relation by taking an inner product of (5) by $\dot{q}$ and taking an integration of the product over $\tau \in [0,t]$:

$$\int_0^t \dot{q}^T(\tau)F(\tau)d\tau = E(t) - E(0) \geq -E(0) = -\gamma_0^2 \tag{8}$$

where $\dot{q}^T$ stands for the transpose of $\dot{q}$ and

$$E(t) = \frac{1}{2}\dot{q}^T(t)H(q(t))\dot{q}(t) + P(q(t)). \tag{9}$$

Since the constant term of potential $P(q)$ can be taken arbitrarily, we set $inf_q P(q) = 0$ provided that all joints of the system are rotational. Then, $E(t)$ is nonnegative definite and hence $E(0) = \gamma_0^2$, which asserts the second inequality in (8). Suggested by this observation, we generally define the passivity of nonlinear dynamics. Suppose that a nonlinear dynamics with input u is given and output y associated with the dynamcis is specified. If this input-output pair $\{u, y\}$ satisfies

$$\int_0^t y^T(\tau)u(\tau)d\tau \geq -\gamma_0^2 \tag{10}$$

where γ_0^2 depends only on the system state at $t = 0$, then we say that the pair satisfies the passivity.

Now, consider firstly a setpoint position control problem and discuss three types of servo-loop for robotic systems described respectively by

$$F = g(q_d) - A\Delta q - B\dot{q} + u, \tag{11}$$

$$F = -A\Delta q - B\dot{q} - C\int_0^t y(\tau)d\tau, \tag{12}$$

$$F = -A\Delta q - B\dot{q} + Z(q_d)\hat{\Theta} - \beta y + u, \tag{13}$$

where q_d is a given desired target position, $\Delta q = q - q_d$, and y and $\hat{\Theta}$ are defined respectively as follows:

$$y = \dot{q} + \alpha s(\Delta q), \tag{14}$$

$$\hat{\Theta}(t) = \hat{\Theta}(0) - \int_0^t \Gamma^{-1}Z^T(q_d)y(\tau)d\tau, \tag{15}$$

where we intentionally make $s(\Delta q)$ composed of saturated function $s_i(\Delta q_i), i = 1, \cdots, n$, as shown in Fig.5 (see Arimoto [9]). The servo-loop of (11) is called a PD feedback with off-line gravity force compensation, that of (12) an SP-ID feedback (Saturated-Proportional, Integration and Differentiation) since each component of $s(\Delta q)$ is saturated, and that of (13) an adaptive PD control. It is implicitly assumed in all cases that A, B, and C are positive definite diagonal matrices, $\beta \geq 0$ is a scalar constant, and $\hat{\Theta}$ an estimate of Θ, a vector of link masses, i.e., $\Theta = (m_1, \cdots, m_n)^T$. If the gravity term $g(q_d)$ at the desired posture can not be evaluated exactly because of uncertainty of link masses, it is then possible to use the fact that link masses appear linearly in the gravity term $g(q)$ and therefore $g(q)$ can be expressed as

$$g(q) = Z(q)\Theta \tag{16}$$

where $Z(q)$ is an $n \times n$ matrix called the gravity regressor. Note that, instead of real-time computation of $Z(q(t))$ as in (16), the adaptive PD control defined by (13) through (15) uses only a constant regressor $Z(q_d)$ that can be computed off-line. Among these three servo-loops the SP-ID feedback must be the simplest because each servo-loop at ith joint can be independently and distributedly implemented since A, B, and C are diagonal and each component of $s(\Delta q)$ depends only on its corresponding position error, i.e., $s_i(\Delta q) = s_i(\Delta q_i)$ for $i = 1, \cdots, n$.

First consider the PD feedback with off-line gravity compensation. Substituting (11) into (5) yields

$$\{H(q)\frac{d}{dt} + \frac{1}{2}\dot{H}(q)\}\dot{q} + \{S(q,\dot{q}) + B\}\dot{q} + g(q) - g(q_d) + A\Delta q = u. \tag{17}$$

This dynamics is expressed by the circuit depicted in Fig.8. Since $P(q)$ can be expressed as a combination of sinusoidal functions of q, its hessian $G(q) = (\partial^2 P(q)/\partial q_i \partial q_j)$ is also a combination of sinusoidal functions of components of q. Therefore, there exists a constant matrix $A > 0$ that satisfies

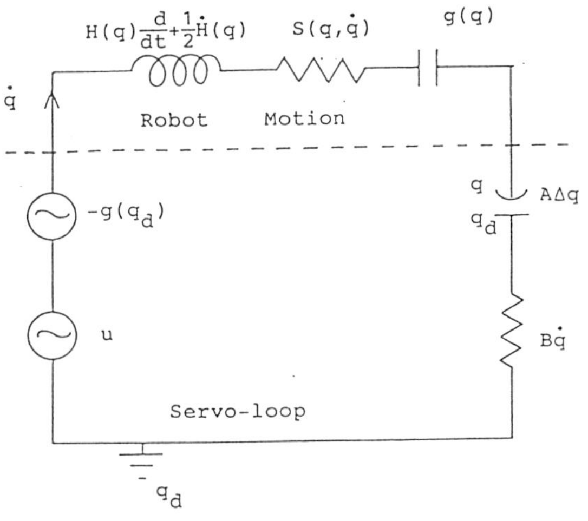

Fig.8 The input-output pair $\{u,\dot{q}\}$ satisfies the passivity.

$$\frac{1}{2}\Delta q^T A \Delta q + P(q) - P(q_d) - \Delta q^T g(q_d) \geq a \parallel \Delta q \parallel^2 \tag{18}$$

for all q and a fixed small constant $a > 0$. Then, as already pointed out by Arimoto [9], it follows from taking an inner product between $\dot{q}$ and (17) and an integration of it over $[0, t]$ that

$$\int_0^t \dot{q}^T(\tau)u(\tau)d\tau = E(t) - E(0) + \int_0^t \dot{q}^T(\tau)B\dot{q}(\tau)d\tau \tag{19}$$

where

$$\begin{aligned} E(t) &= \frac{1}{2}\dot{q}^T(t)H(q(t))\dot{q}(t) + P(q(t)) - P(q_d) \\ &- \Delta\dot{q}^T(t)g(q_d) + \frac{1}{2}\Delta q^T(t)A\Delta q(t). \end{aligned} \tag{20}$$

Since E is positive definite in Δq and $\dot{q}$, (19) means (10) with $y = \dot{q}$. That is,

Property 1 The input-output pair $\{u, \dot{q}\}$ of the circuit depicted in Fig.8 satisfies the passivity.

It is well known (Takegaki and Arimoto [11]) that the target state ($q = q_d$, $\dot{q} = 0$) of the closed-loop system of (17) with $u = 0$ is globally asymptotically stable, which was proved with the aid of LaSalle's invariance theorem. The meaning of this result can be tractable from the direct-current analysis of the circuit depicted in Fig.8. In this circuit, $u = 0$ and the desired target q_d is set at the earth. Since the voltage of the source is constant, the current tends to vanish as $t \to \infty$, which means that the sum of voltage drops through the capacitance blocks and the source, that is, $\{g(q) - g(q_d) + A\Delta q\}$, must tend to vanish. Therefore, If $A > 0$ is large enough satisfying inequality

$$\Delta q^T\{g(q) - g(q_d)\} + \Delta q^T A \Delta q \geq a \parallel \Delta q \parallel^2 \tag{21}$$

with a specified small $a > 0$ as well as inequality (18), then $\Delta q(t)$ must tend to vanish.

Secondly consider the SP-ID feedback defined by (12) that can avoid calculation of the gravity force. In relation to this, we consider the passivity of the pair $\{u, y(= \dot{q} + \alpha s(\Delta q))\}$ concerning the dynamics of (17), which corresponds to the passivity of input torque u (voltage input) and velocity y (current output) of the circuit depicted in Fig.9. According to Arimoto [9][12], it follows that:

Property 2 There is a number $\alpha_0 > 0$ such that the pair $\{u, y(= \dot{q} + \alpha s(\Delta q))\}$ for any fixed α satisfying $\alpha_0 \geq \alpha > 0$ concerning (17) satisfies

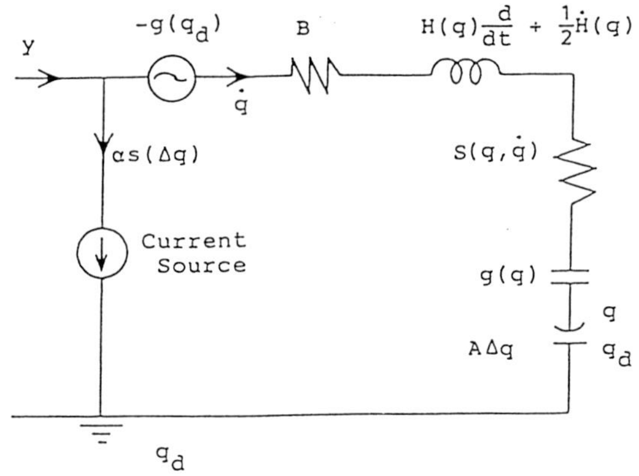

Fig.9 A saturated position feedback is implemented as a current source so that the input-output pair {u,y} satisfies the passivity.

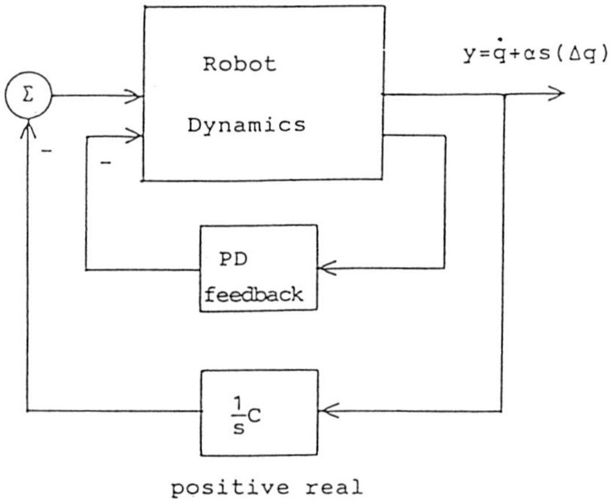

Fig.10(a) A negative feedback connection of a nonlinear hyper-stable system with a linear positive real system.

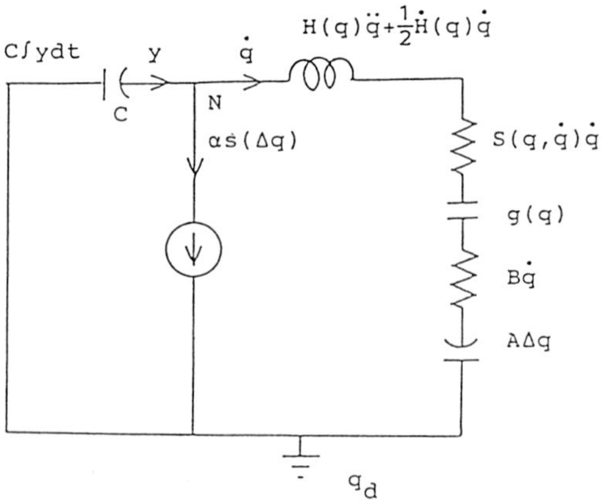

Fig.10(b) A nonlinear position-dependent circuit that exhibits the closed-loop system of equation (22).

the passivity.

Owing to this property, the SP-ID feedback of (12) can be interpreted as a negative hyper-stable feedback connection of the nonlinear passive dynamics with a linear passive system with transfer function $G(s) = (1/s)C$ (see Fig.10(a)). Therefore, Popov's hyper-stability (Popov [13]) suggests the following:

Property 3 There exists a number $\alpha_0 > 0$ such that the SP-ID feedback defined by (12) for any fixed $\alpha(\alpha_0 \geq \alpha > 0)$ leads to the global asymptotic stability of the equilibrium state ($q = q_d$, $\dot{q} = 0$) concerning the closed-loop dynamics:

$$\{H(q)\frac{d}{dt} + \frac{1}{2}\dot{H}(q)\}\dot{q} + \{S(q,\dot{q}) + B\}\dot{q} + g(q) + A\Delta q + C\int_0^t y(\tau)d\tau = 0 \quad (22)$$

where $y = \dot{q} + \alpha s(\Delta q)$.

Note that the above dynamics can be expressed by the circuit depicted in Fig.10(b). A rigorous proof of Property 3 was givin by Arimoto [9]. However, from the circuit-theoretic viewpoint it is easy to see taht as $t \to \infty$ the current through the displacement capacitor accumulates in that block and its integral aggregates the value

$$\lim_{t\to\infty}\int_0^t y(\tau)d\tau = -g(q_d) \quad (23)$$

which just withstands the desired gravity force $g(q_d)$ at the gravity capacitor block.

Thirdly, consider the adaptive PD control defined by (13) through (15). Substituting (13) into (5) yields

$$\{H(q)\frac{d}{dt} + \frac{1}{2}\dot{H}(q)\}\dot{q} + \{S(q,\dot{q}) + B\}\dot{q} + Z(q_d)\Delta\Theta + g(q) - g(q_d) + A\Delta q + \beta y = u \quad (24)$$

where $\Delta\Theta = \Theta - \hat{\Theta}$. Similarly to the argument in Arimoto [14], it is possible to state:

Property 4 There is a number $\alpha_0 > 0$ such that the adaptive PD control defined by (13) for any fixed α satisfying $\alpha_0 \geq \alpha > 0$ gives rise to the passivity of the pair $\{u, y\}$ concerning (24) and to the global asymptotic stability of the equilibruim state ($q = q_d$, $\dot{q} = 0$) concerning (13) with $u = 0$, (14) and (15).

A slightly weaker version of this stability property was firstly presented by Tomei [15] and an-

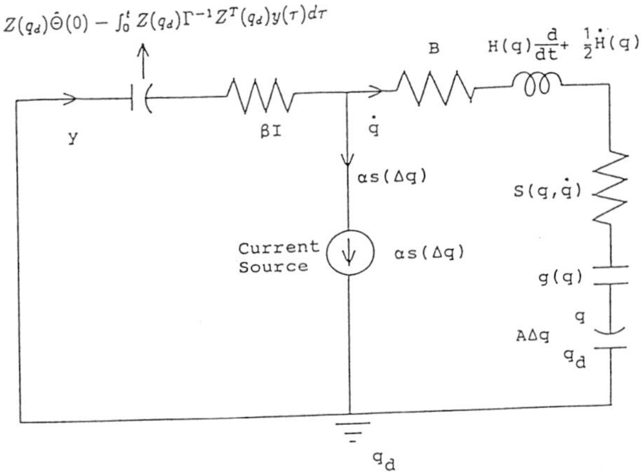

Fig.11 A nonlinear circuit of robot dynamics with a controller satisfying the global asymptotic stability without compensating the gravity term.

alyzed in detail by Kelly [16]. Note that equation (24) can be expressed by a nonlinear circuit depicted in Fig.11. It is interesting to see that the gravity regressor plays a role of displacement capacitor because, according to (15), the term $Z(q_d)\Delta\Theta$ can be expressed as

$$Z(q_d)\hat{\Theta}(t) = Z(q_d)\hat{\Theta}(0) - C\int_0^t y(\tau)d\tau \quad (25)$$

where $C = Z(q_d)\Gamma^{-1}Z^T(q_d)$.

4 Coupling of Mechanical Circuits with Electric Circuits

We consider coupling of a mechanical motion of a robot arm with a set of electric cricuits of joint actuators. Consider a voltage-controlled DC servo-motor depicted in Fig.12(a). The dynamic equation of the armature circuit correponding to the i-th joint actuator can be expressed as

$$L_i\dot{I}_{ai} + R_iI_{ai} + K_{bi}\dot{\theta}_i = v_i, \quad i = 1, \cdots, n \quad (26)$$

where L_i, R_i, I_{ai}, K_{bi}, and v_i denote the armature inductance, armature resistance, armature current, constant of motor back electro-motive force, and armature voltage of the i-th motor armature circuit. We denote the transmission gear ratio $n_i : 1$ from the motor shaft angle to the joint angle by a number $r_i (= 1/n_i)$ and assume that the arm link is sufficiently stiff to be considered as rigid and hence $r_i\dot{\theta}_i = \dot{q}_i$, where q_i denotes the i-th component of position coordinates q of the arm (see Fig.12(b)). On the other hand, the torque

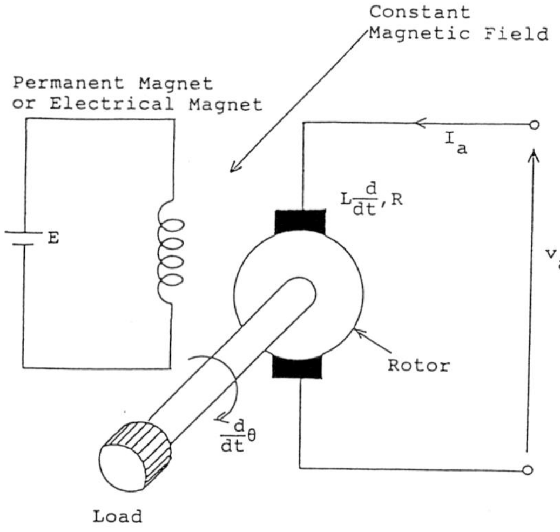

Fig.12(a) A DC servo-motor.

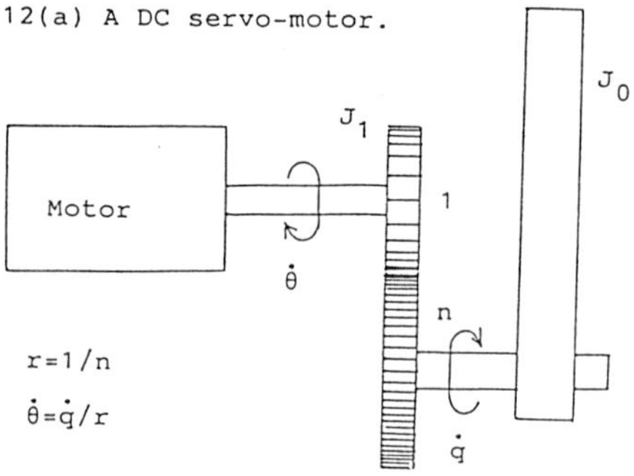

Fig.12(b) A transmission gear with reduction ratio 1:n is attached at the motor shaft.

generated from the i-th joint motor is equivalent to $K_{\tau i}I_{ai}$, where $K_{\tau i}$ denotes the torque constant. According to the d'Alambert principle, the torque $K_{\tau i}I_{ai}$ is equal to the load due to the inertia moment and damping from the motor side itself and the i-th contribution r_iF_i of the motion of the arm, where F_i denotes the i-th component of F in equation (5). This can be expressed as

$$\begin{aligned} NK_\tau I_a &= F + H_0\ddot{q} + B_0\dot{q} \\ &= \{H_0 + H(q)\}\ddot{q} + \{\frac{1}{2}\dot{H}(q) + B_0 \\ &\quad + S(q,\dot{q})\}\dot{q} + g(q) \quad (27) \end{aligned}$$

where $N = diag(r_1^{-1}, \cdots, r_n^{-1})$, $K_\tau = diag(K_{\tau 1}, \cdots, K_{\tau n})$, $H_0 = diag(J_{m1}r_1^{-2}, \cdots, J_{mn}r_n^{-2})$, $B_0 = diag(b_{m1}r_1^{-2}, \cdots, b_{mn}r_n^{-2})$ provided that all joints are driven by such voltage-controlled DC servo-motors.

Now, the nonlinear position-dependent circuit governed by equation (27) can be coupled with the linear electrical circuit governed by equation (26). Since $r_i\dot{\theta}_i = \dot{q}_i$ for $i = 1, \cdots, n$ and in principle the motor torque constant $K_{\tau i}$ is equivalent to the constant K_{bi} of back electro-motive force, the circuit of the total dynamics can be depicted in Fig.13, where coupling between the mecanical nonlinear circuit and electrical armature circuit is realized by a "transformer" at the velocity level. The voltage drop (torque) induced by the transformer at the side of mechanical motion is equivalent to $-NK_\tau I_a$ (proportional to the current vector I_a) and the actual voltage drop induced by the transformer at the armature circuit is considered to be equivalent to the voltage $NK_\tau\dot{q}$ induced by the back electro-motive force. According to equation (26) and $K_{\tau i} = K_{bi}$, the set of armature circuits can be described in the following form:

$$L_a\dot{I}_a + R_aI_a + NK_\tau\dot{q} = v \quad (28)$$

where $I_a = (I_{a1}, \cdots, I_{an})^T$, $v = (v_1, \cdots, v_n)^T$, $L_a = diag(L_1, \cdots, L_n)$, and $R_a = diag(R_1, \cdots, R_n)$. At this stage it is natural to expect that the input and output pair $\{v, I_a\}$ for the total circuit of Fig.13 may satisfy the passivity. In fact, it is possible to show the validity of this property by taking inner products between equation (28) and I_a and between equation (27) and $\dot{q}$ and summing up these two resulting quantities, which results in the form

$$\begin{aligned} I_a^Tv &= \frac{d}{dt}[\frac{1}{2}I_a^TL_aI_a + \frac{1}{2}\dot{q}^T\{H_0 + H(q)\}\dot{q} \\ &\quad + P(q)] + I_a^TR_aI_a + \dot{q}^TB_0\dot{q}. \quad (29) \end{aligned}$$

Since $\min_q P(q) = 0$, it follows from (29) that

$$\int_0^t I_a^T(\tau)v(\tau)d\tau \geq -E(0) = -\gamma_0^2, \quad (30)$$

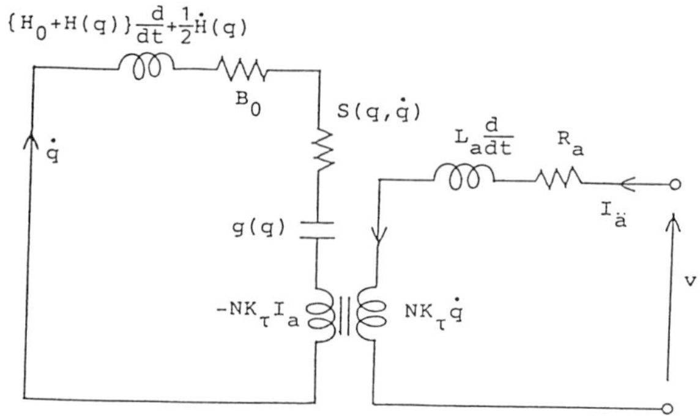

Fig.13 The transformer connects the motion of a nonlinear mechanical system with a set of linear electrical circuits. The input-output pair $\{v, I_a\}$ satisfies the passivity.

where $E(0)$ dentoes a nonnegative quantity at $t = 0$ of the total energy defined by the content inside the square bracket [ ] in (29). A setpoint control problem for the total system can be solved by taking advantage of this passivity problem though the total system seems more complicated with $3n$ state variables $(q, \dot{q}, I_a)$. This has been treated in Arimoto [14].

5 Disturbance Attenuation in a Sense of H_∞ Control

In this section we treat a special case when the inductance matrix L_a is small enough in comparison with R_a so that the first term of (28) can be ignored in the armature circuit dynamics. Then, (28) can be rewritten in the form

$$NK_\tau I_a = F - \bar{B}_0 \dot{q} \tag{31}$$

where $F = NK_\tau R_a^{-1} v$ and $\bar{B}_0 = NK_\tau R_a^{-1} NK_\tau$. Substituting (31) into (27) yields

$$\{H_0 + H(q)\}\ddot{q} + \{\frac{1}{2}\dot{H}(q) + B_1 + S(q,\dot{q})\}\dot{q} + g(q) = F \tag{32}$$

where $B_1 = B_0 + \bar{B}_0$. We implicitly assume the exact knowledge on N, K_τ, and R_a and thereby the input F can be composed of the input voltage v through $F = NK_\tau R_a^{-1} v$. Further, we must bear in mind that there are other nonlinear frictional forces that can not be described and hence can not be included in (32). Therefore, if we consider a PD feedback with gravity compensation defined by (11) and a closed-loop system (obtained by substituting (11) into (32))

$$\{H_0 + H(q)\}\ddot{q} + \{\frac{1}{2}\dot{H}(q) + S(q,\dot{q}) + \bar{B}\}\dot{q}$$
$$+ g(q) - g(q_d) + A\Delta q = u = h(\Delta q, \dot{q}) + \Delta u \tag{33}$$

where $\bar{B} = B_1 + B$, then extra input u must be a sum of unknown but structured frictional forces $h(\Delta q, \dot{q})$ and another genuine disturbance Δu as described in (33). It should be also remarked that a position feedback of time delay $A\Delta q(t - \Delta t)$ induces a positive velocity feedback as shown by

$$-A\Delta q(t - \Delta t) \cong -A\Delta q(t) + \Delta t A \Delta \dot{q}(t). \tag{34}$$

Since such a positive velocity feedback must be taken into account in the structured disturbance $h(\Delta q, \dot{q})$, we suppose a class of structured disturbances satisfying

$$\gamma^2 \int_0^t h(\Delta q(\tau), \dot{q}(\tau))^T \dot{q}(\tau) d\tau \le \int_0^t \| \dot{q}(\tau) \|^2 d\tau \tag{35}$$

for a fixed γ^2 and any $t > 0$.

Definition 1 If the input-output pair $\{\Delta u, \dot{q}\}$ concerning the system of (33) satisfies the passivity for any structured disturbance $h(\Delta q, \dot{q})$ satisfying (35), then it is said that it satisfies the passivity with margin γ^{-2}.

At the same time, we are concerned with the performance of disturbance attenuation in general in the sense of L^2 norm of the output $\dot{q}$ versus that of the total disturbance u in (33) over any time interval $[0, t]$.

Definition 2 Assume that $q(0) = q_d$ and $\dot{q}(0) = 0$. If for a fixed $\gamma^2 > 0$ and any time interval $[0, t]$ it holds that

$$b_0 \int_0^t \| \dot{q}(\tau) \|^2 d\tau \le \gamma^2 \int_0^t \| u(\tau) \|^2 d\tau, \tag{36}$$

then we say that the system described by (33) establishes the H_∞-tuning with level γ^{-2}. Here in (36) b_0 is defined as b_0 =minimum eigenvalue of $\bar{B}$, $(b_0 = \lambda_m(\bar{B}))$.

The most important result concerning these two concepts can be stated in the following:

Theorem 1 A necessary and sufficient condition for the system of (33) to satisfy the passivity with margin γ^{-2} and to establish the H_∞-tuning with level γ^{-2} simultaneously is that

$$\gamma^{-2} \le b_0 \quad (= \lambda_m(\bar{B})), \tag{37}$$

provided that $A > 0$ is chosen to satisfy (18) and (21) with a specified small $a > 0$.

Proof The sufficient part follows from the passivity of pairs $\{u, \dot{q}\}$ and $\{\Delta u, \dot{q}\}$ of (33). In fact, taking an inner product of (33) by $\dot{q}$ and an integral of it over $[0, t]$ yields

$$\int_0^t \dot{q}^T(\tau) u(\tau) d\tau = E(t) - E(0) + \int_0^t \dot{q}^T(\tau) \bar{B} \dot{q}(\tau) d\tau \tag{38}$$

where

$$E(t) = \frac{1}{2}\dot{q}^T(t)\{H_0 + H(q(t))\}\dot{q}(t) + P(q(t))$$
$$- P(q_d) - \Delta q^T(t) g(q_d) + \frac{1}{2}\Delta q^T(t) A \Delta q(t). \tag{39}$$

If $q(0) = q_d$ and $\dot{q}(0) = 0$, then $E(0) = 0$. Hence it follows from (38) that

$$E(t) + \int_0^t \dot{q}^T(\tau)\bar{B}\dot{q}(\tau)d\tau = \int_0^t \dot{q}^T(\tau)u(\tau)d\tau$$
$$\leq \int_0^t \frac{1}{2}\{\gamma^{-2} \parallel \dot{q}(\tau) \parallel^2 + \gamma^2 \parallel u(\tau \parallel^2\}d\tau. \quad (40)$$

Applying (37) for this inequality leads to

$$b_0 \int_0^t \parallel \dot{q}(\tau) \parallel^2 d\tau \leq \gamma^2 \int_0^t \parallel u(\tau) \parallel^2 d\tau - E(t) \quad (41)$$

which shows the establishment of H_∞-tuning with level γ^{-2}. To consider the passivity of the pair $\{\Delta u, \dot{q}\}$ of (33), rewrite (38) in the following way:

$$\int_0^t \dot{q}^T(\tau)\Delta u(\tau)d\tau = \int_0^t \dot{q}^T(\tau)\bar{B}\dot{q}(\tau)d\tau$$
$$+ E(t) - E(0) - \int_0^t h^T(\Delta q(\tau), \dot{q}(\tau))\dot{q}(\tau)d\tau. (42)$$

Substituting inequality (35) into this equation yields

$$\int_0^t \dot{q}^T(\tau)\Delta u(\tau)d\tau \geq E(t) - E(0)$$
$$+ \int_0^t \dot{q}^T(\tau)\{\bar{B} - \gamma^{-2}I\}\dot{q}(\tau)d\tau. \quad (43)$$

Applying (37) to this expression shows the passivity of (33) with margin γ^{-2}. The proof of the necessity part is rather lengthy and therefore omitted in this paper (in detail, see Arimoto [23]).

Theorem 1 presents a physical interpretation that if the energy dissipation rate $\dot{q}^T\bar{B}\dot{q}$ has a margin beyond $\gamma^{-2} \parallel \dot{q} \parallel^2$ for any $\dot{q}$ then the H_∞-tuning is established. This is valid for the nonlinear dynamics of (33) in a global sense. Another noteworthy feature of Theorem 1 is that the sufficient condition for establishment of the H_∞-tuning is expressed in a very simple form, which need neither solve any of Riccati's matrix equations (see Doyle et al. [17]) nor use solutions to a Hamilton-Jacobi-Issacs equation (Van der Shaft [18][19], Isidori and Astolfi [20], and Astolfi and Lanari [29][22]). Readily from the proof of the necessity part of Theorem 1 the following result follows (the proof will be given in Arimoto [23]):

<u>Theorem 2</u> As for the circuit depicted in Fig.14, the input-output pair $\{\Delta u, \dot{q}\}$ satisfies the passivity if and only if $\gamma^{-2} \leq b_0 (= \lambda_m(\bar{B}))$, provided that $A > 0$ is large enough satisfying (18) and (21) simultaneously with a specified $a > 0$.

As for H_∞-tuning problems in the case of set-point control schemes without compensating the gravity forces defined by (12) and (13), it is possible to derive similar results on sufficient conditions. We treat only the adaptive PD control defined by (13). In this case the closed-loop dynamics is described in the form

$$\{H_0 + H(q)\}\ddot{q} + \{\frac{1}{2}\dot{H}(q) + S(q,\dot{q}) + \bar{B}\}\dot{q}$$
$$+ g(q) - Z(q_d)\hat{\Theta} + A\Delta q + \beta y = u \quad (44)$$

which is obtained from substitution of (13) into (32). We assume that each saturated function $s_i(\theta)$ is differentiable with a continuous derivative and has a potential satisfying $S_i(\theta) = \int_0^\theta s_i(\xi)d\xi$ and furthermore there exist constants $c_i > 0$ and $d_i > 0$ such that

$$\theta s_i(\theta) \geq d_i\theta^2 \text{ and } S_i(\theta) \geq c_i s_i^2(\theta), \quad i = 1, \cdots, n. \quad (45)$$

Then, there exists a constant $\bar{c}_0 > 0$ such that

$$s^T(\Delta q)\{-\frac{1}{2}\dot{H}(q) + S(q,\dot{q})\}\dot{q} - \dot{s}^T\{H_0 + H(q)\}\dot{q}$$
$$\geq -\bar{c}_0 \parallel \dot{q} \parallel^2 . \quad (46)$$

At the same time, it is easy to see from (45) that for given $s(\Delta q)$ the gain matrix A can be chosen so as to satisfy (18) and the following instead of (21):

$$s^T(\Delta q)\{g(q) - g(q_d) + A\Delta q\} \geq a \parallel \Delta q \parallel^2 \quad (47)$$

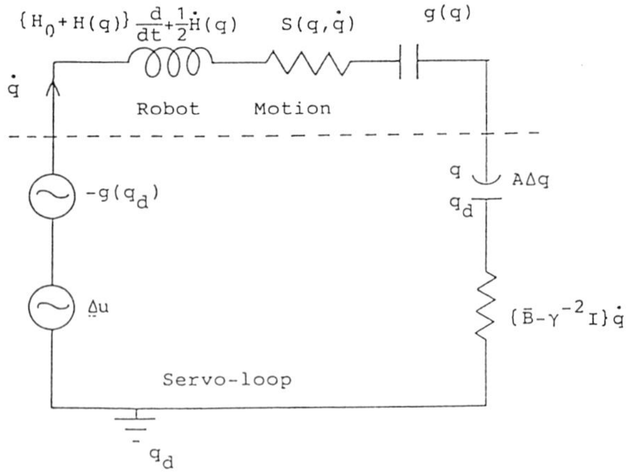

Fig.14 The input-output pair $\{u,\dot{q}\}$ satisfies the passivity if and only if $\gamma^{-2} \leq \lambda m\{\bar{B}\}$, where $\lambda m\{\bar{B}\}$ denotes the minimum eigenvalue of $\bar{B}$.

with a specified small $a > 0$. Then, it has been proved that (see Arimoto [14]):

Theorem 3 If $\alpha > 0$ satisfies

$$\bar{B} > \alpha \bar{c}_0 I \quad \text{and} \quad \bar{B} > \alpha \lambda_M C^{-1} \qquad (48)$$

where λ_M denotes the maximum eigenvalue of $\{H_0 + H(q)\}$ for all q and $C = diag(c_1, \cdots, c_n)$, then the input-output pair $\{u, y\}$ of the closed-loop system (44) satisfies the passivity, that is,

$$\int_0^t y^T(\tau)u(\tau)d\tau = E(t) - E(0)$$
$$+ \int_0^t \{W(\Delta q(\tau), \dot{q}(\tau)) + \beta \parallel y(\tau) \parallel^2\} d\tau \quad (49)$$

where $E(t) \geq 0$ and $W(\Delta q, \dot{q})$ is positive definite in Δq and $\dot{q}$.

The total energy $E(t)$ in the above equation can be described in detail as follows:

$$E(\Delta q, \dot{q}, \Delta\Theta) = \frac{1}{2}\dot{q}^T\{H_0 + H(q)\}\dot{q}$$
$$+ \frac{1}{2}\Delta q^T A \Delta q + P(q) - P(q_d) - \Delta q^T g(q_d)$$
$$+ \alpha \sum_{i=1}^{n} \bar{b}_i S_i(\Delta q_i) + \alpha s^T(\Delta q)\{H_0 + H(q)\}\dot{q}$$
$$+ \frac{1}{2}\Delta\Theta^T \Gamma \Delta\Theta \qquad (50)$$

where $\bar{b}_i$ denotes the ith diagonal component of $\bar{B}$. After several maneuverings of the robot arm the estimate $\hat{\Theta}$ of unknown parameters is expected to converge to the true value Θ. Therefore, if we assume that, at $t = 0$, $q(0) = q_d$, $\dot{q}(0) = 0$, and $\Delta\Theta(0) = 0$, then it is possible to assert the H_∞ tuning in a sense of disturbance attenuation concerning the system described by (44).

Theorem 4 Under the initial conditions mentioned above, a sufficient condition for the closed-loop (44) with output $y(= \dot{q} + \alpha s(\Delta q))$ to establish the H_∞ tuning for any disturbance u, i.e.,

$$\beta \int_0^t \parallel y(\tau) \parallel^2 d\tau \leq \gamma^2 \int_0^t \parallel u(\tau) \parallel^2 d\tau, \quad (51)$$

is that $\gamma^{-2} \leq \beta$.

Since integration of $\parallel y \parallel^2$ over $[0, t]$ can be expressed as

$$\int_0^t \parallel y(\tau) \parallel^2 d\tau = V(t) - V(0)$$
$$+ \int_0^t \{\parallel \dot{q}(\tau) \parallel^2 + \alpha^2 \parallel s(\Delta q(\tau)) \parallel^2\} d\tau \quad (52)$$

where

$$V(t) = 2\alpha \sum_{i=1}^{n} S_i(\Delta q_i(t)), \qquad (53)$$

each component $\dot{q}$ or $\alpha s(\Delta q)$ of y establishes the H_∞-tuning, that is, it holds that

$$\beta \int_0^t \parallel \dot{q}(\tau) \parallel^2 d\tau \leq \gamma^2 \int_0^t \parallel u(\tau) \parallel^2 d\tau, \quad (54)$$

$$\beta \int_0^t \alpha^2 \parallel s(\Delta q(\tau)) \parallel^2 d\tau \leq \gamma^2 \int_0^t \parallel u(\tau) \parallel^2 d\tau \quad (55)$$

6 Conclusions

In this paper we have introduced a framework of nonlinear position dependent circuits as another language for describing nonlinear dynamics of mechanical and advanced mechatronic systems. Instead of Fourier and Laplace transforms that are used effectively in analyzing linear lumped-parameter electric circuits, the concept of passivity becomes fundamental and can be used in effect as a basic tool of characterizing input-output properties of such nonlinear position-dependent circuits. It is shown that the design problems of controllers for nonlinear mechanical systems become tractable on the basis of such visible expressions via physical circuits and their performance analysis becomes also easier by means of examining the input-output passivity. It is also shown that for mechanical systems the H_∞ tuning problem in a sense of disturbance attenuation can be interpreted as another input-output characterization of the energy conservation law. In the case of the simplest PD controller with off-line gravity compensation a necessary and sufficient condition for the H_∞-tuning has been presented, which is expressed in a simple form without using any solutions to Riccati's equations and Hamilton-Jacobi-Issacs equations.

Further developments of nonlinear position-dependent circuits to cope with dynamics under geometric constraints and trajectory tracking dynamics are now investigated intensively (see Arimoto [14][23]).

There are a lot of problems to be attacked on the basis of development of a nonlinear position-dependent circuit theory. 1) Analysis of nonlinear circuits in cases that the system has both rotational and prismatic joints. 2) Analysis of nonlinear circuits when the number of actuators is less than the degree of freedom. 3) Design of controllers via nonlinear circuits for redundant robots or robotic systems under constraints in general. 4) In particular, design of controllers via developments of coupled circuits for coordination of multiple robots. 5) Analysis of couplings between mechanical motions and electric circuits through actuators. 6) How to cope with, by nonlinear circuits, problems of a transition of dynamics that causes a change of the degree of freedom or that of constraints.

References

[1] Meisel J. Principle of Electromechanical Energy Conversion. McGraw-Hill, New York, 1966

[2] Maisser P, Steinberger J. Lagrange-Formalismus fur diskrete electromechanische System, Z Angew Math Mech 1979; 59: 717-730

[3] Paynter HM. Analysis and Design of Engineering Systems. The MIT Press, Cambridge, Massachusetts, 1960.

[4] Karnopp D, Rosenberg RC. Analysis and Simulation of Multiport Systems. The MIT Press, Cambridge, Massachusetts, 1968.

[5] Brayton RK, Moser JK. A theory of nonlinear networks Part I and Part II. Quart Appl Math 1964;22:1-33 and 81-104

[6] Chua LO, Green DN. A qualitative analysis of the behavior of dynamic nonlinear networks: Stability of autonomous networks. IEEE Trans. on Circuits and Systmes 1976; 23:355-279

[7] Arimoto S, Miyazaki F. Stability and robustness of PID feedback control for robot manipulators of sensory capabilities. Robotics Research, First Int Symp (M. Brady and P.P. Paul eds.) MIT Press, Cambridge, MA: 783-799

[8] Anderson BDO. A system theory criterion for positive real matrices, SIAM J Control; 5:171-182

[9] Arimoto S. A class of quasi-natural potentials and hyper-stable PID servo-loops for nonlinear robotic systems. Trans. SICE 1994; 30:1005-1012

[10] Arimoto S. Passivity of robot dynamics implies capability of motor program learning, Proc. of the Workshop on Nonlinear and Adaptive Control Applications to Robotics, Grenoble, No.21-23, 1990, Springer-Verlag, Berlin, pp.49-68, (1991).

[11] Takegaki M, Arimoto S. A new feedback method for dynamic control of manipulators, Trans ASME J of Dynamic Systems, Measurement, and Control 1981;3:119-125

[12] Arimoto S. Fundamental problems of robot control: Part I, Innovation in the realm of robot servo-loops. Robotica 1995; 13:19-27

[13] Popov VM. Hyperstability of Control Systems. Springer-Verlag, New York, 1973

[14] Arimoto S. Nonlinear position-dependent circuits for mechanical motion control. Proc. of IFAC Workshop Motion Control, Munich, Germany Oct. 9-11, 1995

[15] Tomei P. Adaptive PD controller for robots with elastic joints. IEEE Trans. on Robotics and Automation 1991; 7:565-570

[16] Kelly R. Commentson 'Adaptive PD controller for robot manipulators.' IEEE Trans. on Robotics and Automation 1993; 9:117-119

[17] Doyle JC, Glover K, Khargonekar PP et al. State-space solutions to standard H_2 and H_∞ control problems. IEEE Trans.on Automatic Control 1989;34:831-847

[18] Van der Schaft AJ. On a state-space approach to nonlinear H_∞ control,Systems and Control Letters 1991; 16:1-8

[19] Van der Schaft AJ. L_2-gain analysis of nonlinear systems and nonlinear state feedback H_∞ cotrol. IEEE Trans. on Automatic Control 1992; 37:770-784

[20] Isidori A, Astolifi A. Disturbance attenuation and H_∞ control via measurement feedback in nonlinear system. IEEE Trans. on Automatic Control 1992; 37:1283-1293

[21] Astolfi A, Lanari L. Optimal tuning of PD controllers for rigid robots. Proc. of the Asian Control Conf, Tokyo, July 1994, 141-142

[22] Astolfi A, Lanari L. H_∞ control of rigid robot Proc. of IFAC Symp. on Robot Control, Sept. 1994, Capri, Italy, 199-204

[23] Arimoto S. Nonlinear position-dependent circuits: A language for describing motions of nonlinear mechanical systems, to be presented at the IUTAM Symp. on Interaction between Dynamcis and Control in Advanced Mechanical Systems, April 21-26, 1996, Eindhoven, Netherlands.

Nonlinear Behavior and Control of Underactuated Robotic Mechanisms

Y. Nakamura T. Suzuki M. Koinuma
Dept. of Mechano-Informatics, University of Tokyo
7-3-1 Hongo, Bunkyo-ku, Tokyo 113 (Japan)
email: nakamura@mech.t.u-tokyo.ac.jp

Abstract

The nonlinear motion of underactuated mechanisms has drawn recent interests of researchers. The nonlinearity is often referred to as nonholonomy and related to many both theoretical and practical issues. In this paper, we first analyze the nonlinear behavior of a two-joint planer manipulator with the second joint free from nonlinear dynamics point of view. We then discuss the simultaneous positioning of the both joints. We use a time-periodic input and propose an amplitude modulation of the feedback error. The analysis via the Poincaré map shows that the behavior becomes even chaotic with large amplitude. The effectiveness and robustness of the proposed positioning control are verified by experiments.

1 Introduction

Study of the nonlinearity of motion is a major common problem in many subfields of robotics. In particular, the nonlinear motion of underactuated mechanisms has drawn recent interests of researchers. The nonlinearity is often referred to as *nonholonomy* and related to many both theoretical and practical issues of interest, such as planning and control of mobile robots[1][2] [3][4], base-body reorientation of free-flying space robots[5] [6], rolling contacts of multifingered hands[7], controllability of pushing[8], and so on.

The equations of motion of the above-mentioned systems are commonly represented by first-order differential equations with equality constraints on velocities. Many research efforts have been paid on the control aspects such as controllability[9], stabilizability[10], feedback stabilization [11][12], time-periodic control[13], chained form transformation [14][15], and so on. The developed and accumulated knowledge of control even enabled us to design a new nonholonomic mechanism[16][17] that would have been neither controlled nor utilized without it. This clearly illustrates that the study of nonlinear motion of underactuated mechanisms not only allows us to control the already-existing mechanisms with higher speed or better accuracy, but also empowers us to create new robotic mechanisms that have certain advantages, but have never been thought of possible.

The constraints of motion are sometimes due to dynamics of mechanisms. Underactuated mechanisms have the generalized coordinates that are not actuated. This is the origin of dynamic constraints. Nakamura[18] studied the integrability of such constraints. Oriolo and Nakamura[20] clarified that the dynamical constraints of a manipulator with free joints are generally nonintegrable and, therefore, second-order nonholonomic.

On control of a manipulator with free joints, Arai and Tachi[19] studied a path control of a manipulator with electromagnetic brakes at free joints. Nakamura and Iwamoto[21] discussed a space multi-link system with free joints and its shape control. Wichlund et al.[22] discussed the integrability and stabilizability of a class of underactuated dynamical systems. Nakamura et al.[24] proposed a stabilization of a 2DOF free-joint manipulator to its stable node.

In this paper, we investigate the nonlinear behavior of free-joint manipulators. We choose a simple planner two-joint manipulator with the second joint free so that we can get deep mathematical insights of the nonlinearity. The approaches and tools of analysis we develop in this paper will be the basis of further research of free-joint manipulators with higher complexity. The analysis of nonlinear dynamical behavior and the simultaneous positioning of the both joints are the main focus of this paper.

2 A Free-Joint Manipulator

2.1 Dynamics Modeling

The two-joint-planer manipulator in Fig.1 is with the first joint actuated and the second joint free and resides in the horizontal plane. This is the simplest model of nonholonomic free-joint manipulators.

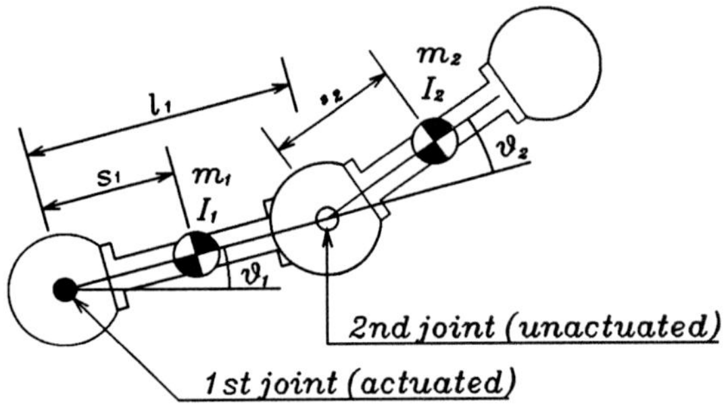

Figure 1: A 2DOF free-joint manipulator

The Hamiltonian is represented by

$$H = \frac{aJ_2{}^2 + 2b\cos\theta_2 \cdot J_2(J_2 - J_1) + c(J_2 - J_1)^2}{2(ac - b^2\cos^2\theta_2)} \tag{1}$$

where $\boldsymbol{\theta} = (\theta_1, \theta_2)^T$ is the generalized coordinate and $\boldsymbol{J} = (J_1, J_2)^T$ is its momentum. $\boldsymbol{J}$ is obtained through the Hamilton's canonical transformation as follows:

$$\boldsymbol{J} = \begin{pmatrix} (a + 2b\cos\theta_2 + c)\dot{\theta}_1 + (b\cos\theta_2 + c)\dot{\theta}_2 \\ (b\cos\theta_2 + c)\dot{\theta}_1 + c\dot{\theta}_2 \end{pmatrix} \tag{2}$$

The a, b, and c are constants defined by

$$\begin{aligned} a &\stackrel{\text{def}}{=} m_1 s_1{}^2 + m_2 l_1{}^2 + I_1 \\ b &\stackrel{\text{def}}{=} m_2 l_2 s_2 \\ c &\stackrel{\text{def}}{=} m_2 s_2{}^2 + I_2 \end{aligned} \tag{3}$$

where m_i and I_i denotes the mass and the inertia of link i. The dynamics of the manipulator is obtained by

$$\begin{aligned} \frac{d\boldsymbol{\theta}}{dt} &= \frac{\partial H}{\partial \boldsymbol{J}} \\ \frac{d\boldsymbol{J}}{dt} &= -\frac{\partial H}{\partial \boldsymbol{\theta}} \end{aligned} \tag{4}$$

Namely,

$$\frac{d}{dt}\begin{pmatrix} J_1 \\ J_2 \\ \theta_1 \\ \theta_2 \end{pmatrix} = \begin{pmatrix} \tau \\ \frac{b\sin\theta_2 \cdot P(\theta_2, J_1, J_2)\cdot Q(\theta_2, J_1, J_2)}{(ac - b^2\cos^2\theta_2)^2} \\ \frac{Q(\theta_2, J_1, J_2)}{ac - b^2\cos^2\theta_2} \\ \frac{-P(\theta_2, J_1, J_2) - Q(\theta_2, J_1, J_2)}{ac - b^2\cos^2\theta_2} \end{pmatrix}$$

$$(J_1, J_2, \theta_1, \theta_2) \in \boldsymbol{R}^2 \times \boldsymbol{T}^2 \tag{5}$$

where

$$\begin{aligned} P(\theta_2, J_1, J_2) &= b\cos\theta_2 \cdot J_1 - (a + b\cos\theta_2)J_2 \\ Q(\theta_2, J_1, J_2) &= cJ_1 - (b\cos\theta_2 + c)J_2 \end{aligned}$$

2.2 The Conserved Motions

When there is no input, namely, $\tau = 0$, we have two obvious conserved quantities. One is the generalized momentum of the 1st joint, and the other is the Hamiltonian that is equivalent to the total energy. With $J_1(0) = J_{10}$ and $H(0) = H_0$ as initial conditions, the conservations are represented by

$$J_1(t) = J_{10} \tag{6}$$

$$\begin{aligned} &(a + 2b\cos\theta_2 + c)J_2{}^2 + 2(b\cos\theta_2 + c)J_{10}J_2 \\ &+ cJ_{10}{}^2 - 2(ac - b^2\cos^2\theta_2)H_0 = 0 \end{aligned}$$

Hence, the possible reachable space in the four-dimensional state space restricts its dimension to two. Since Eq.(1) does not involve θ_1, the possible reachable space is symmetric with respect to θ_1 and has the intersection with a plane perpendicular to θ_1 axis as shown in Fig.2. Figure 2 shows the intersections for $J_{10} = 0.1$[kg·m^2·rad/sec] and various values of H_0. The figure shows that the motions follow ordered closed paths. Note that since the space of θ_2 is not $\boldsymbol{R}^1$ but $\boldsymbol{S}^1$, the upper two and lower two curves in Fig.2 are also connected at $\theta_2 = \pm\pi$ and show closed paths.

The rich behaviors of nonlinear systems are drawing much attention in various fields of research. Chaos is one of such behaviors. It is known[26] that chaos occurs only in 3 or more dimensional space for continuous systems. The two conserved quantities of Eq.(2.2) reduce the dimension of our system and makes the well-ordered behavior as Fig.2.

2.3 Chaos with Periodic Inputs

When the system is subject to time-periodic input of $\tau = \gamma\cos\omega t$, the whole dynamics is represented

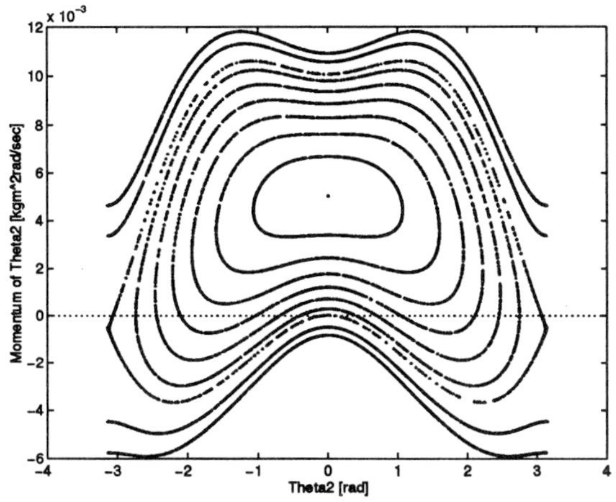

Figure 2: Phase plane at $J_1 = 0.1[\mathrm{kgm}^2\cdot\mathrm{rad/sec}]$

by

$$\frac{d}{dt}\begin{pmatrix} J_1 \\ J_2 \\ \theta_1 \\ \theta_2 \\ \phi \end{pmatrix} = \begin{pmatrix} \gamma\cos\phi \\ \frac{b\sin\theta_2 P(\theta_2,J_1,J_2)\cdot Q(\theta_2,J_1,J_2)}{(ac-b^2\cos^2\theta_2)^2} \\ \frac{Q(\theta_2,J_1,J_2)}{ac-b^2\cos^2\theta_2} \\ \frac{-P(\theta_2,J_1,J_2)-Q(\theta_2,J_1,J_2)}{ac-b^2\cos^2\theta_2} \\ \omega \end{pmatrix}$$

$$(J_1, J_2, \theta_1, \theta_2, \phi) \in \boldsymbol{R}^2 \times \boldsymbol{T}^2 \times \boldsymbol{T}^1 \quad (7)$$

where constants γ and ω denote the amplitude and the angular frequency. Note that a new state ϕ is included in Eq.(7) to represent the dynamics as of an autonomous system. Although J_1 is not conserved any more, it can be easily integrated and solved as

$$J_1 = J_{10} + \frac{\gamma}{\omega}(\sin\phi - \sin\phi_0) \quad (8)$$

which implies a new conserved quantity. Since we neither have the energy conservation, the dimension of the state space is four, which possibly allows chaotic behaviors.

Chaos is characterized by two fundamental natures, namely, the sensitive dependence on initial conditions and the topological transitivity. We briefly summarize them for the readers' convenience in appendix A. Figure 3 shows trajectories starting from four mutually close initial values for the case with $\gamma = 4[\mathrm{rad/s}^2]$ and $\omega = 2\pi[\mathrm{rad/s}]$. The solid line has $(J_1, J_2, \theta_1, \theta_2)|_{t=0} = (0,0,0,0)$ as the initial condition. The other lines have small differences of θ_2 as seen in the figure. The figure illustrates that the difference of trajectories extends largely as time goes on, which implies the sensitive dependence on initial conditions.

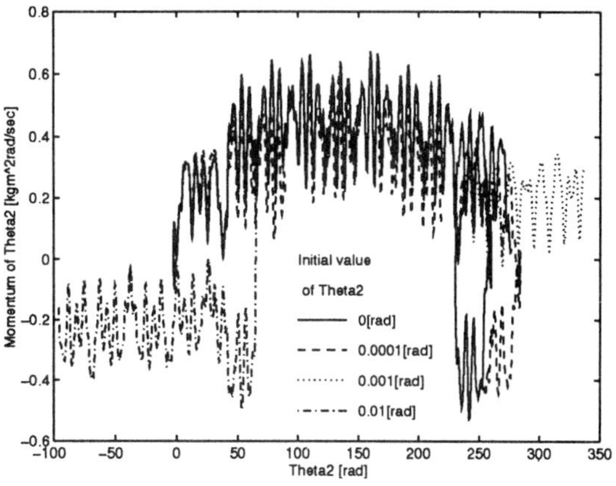

Figure 3: Sensitive dependence on initial conditions

The Poincaré map is a intersection of trajectories in the whole phase space with a certain hyper-plane. It removes at least one variable and makes it easy to understand the system's behavior. Let $\boldsymbol{x}(t) \stackrel{\mathrm{def}}{=} (J_1(t), J_2(t), \theta_1(t), \theta_2(t))$ be the solution of Eq.(7) and the following $\Sigma^{\bar{\phi}_0}$ be the intersection in the phase space.

$$\Sigma^{\bar{\phi}_0} \stackrel{\mathrm{def}}{=} \{(\boldsymbol{x}, \phi) \in \boldsymbol{R}^2 \times \boldsymbol{T}^2 \times \boldsymbol{T}^1 \mid \phi = \bar{\phi}_0\} \quad (9)$$

Then, the Poincaré map is represented as

$$\begin{aligned} P_{\bar{\phi}_0} \quad : \quad & \Sigma^{\bar{\phi}_0} \mapsto \Sigma^{\bar{\phi}_0}, \\ & \left(\boldsymbol{x}\left(\frac{\bar{\phi}_0 - \phi_0}{\omega}\right), \bar{\phi}_0\right) \mapsto \\ & \left(\boldsymbol{x}\left(\frac{\bar{\phi}_0 - \phi_0 + 2\pi}{\omega}\right), \bar{\phi}_0 + 2\pi\right) \quad (10) \end{aligned}$$

or

$$\boldsymbol{x}\left(\frac{\bar{\phi}_0 - \phi_0}{\omega}\right) \mapsto \boldsymbol{x}\left(\frac{\bar{\phi}_0 - \phi_0 + 2\pi}{\omega}\right) \quad (11)$$

The fixed points of $P_{\bar{\phi}_0}$ imply the periodic trajectories of Eq.(7) with the period $2\pi/\omega$, and the k-period points of $P_{\bar{\phi}_0}$ imply the periodic trajectories that pass k times through $\Sigma^{\bar{\phi}_0}$ before returning the initial state.

Figures 4 through 6 show the Poincaré maps being cut at $\phi = 0$ and projected onto the θ_2-H plane. Note that we projected onto the θ_2-H plane rather than the θ_2-J_2 plane, since the former projection provides better physical insights for our current problems. The behaviors in the θ_2-J_2 plane also show the similar characteristics. The initial conditions and the parameters of input were

$$(\theta_1, \theta_2, J_1)|_{t=0} = (0[\mathrm{rad}], 0[\mathrm{rad}], 0[\mathrm{kg}\cdot\mathrm{m}^2\cdot\mathrm{rad/sec}])$$

Fig.4: $\gamma = 0.04$ [rad/sec^2], $\omega = 2\pi$ [rad/sec^2]
Fig.5: $\gamma = 0.4$ [rad/sec^2], $\omega = 2\pi$ [rad/sec^2]
Fig.6: $\gamma = 4$ [rad/sec^2], $\omega = 2\pi$ [rad/sec^2]

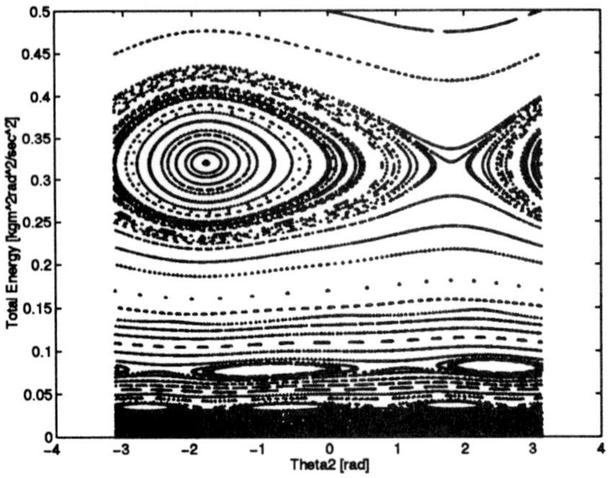

Figure 4: Poincaré map with $\gamma = 0.04, \omega = 2\pi$

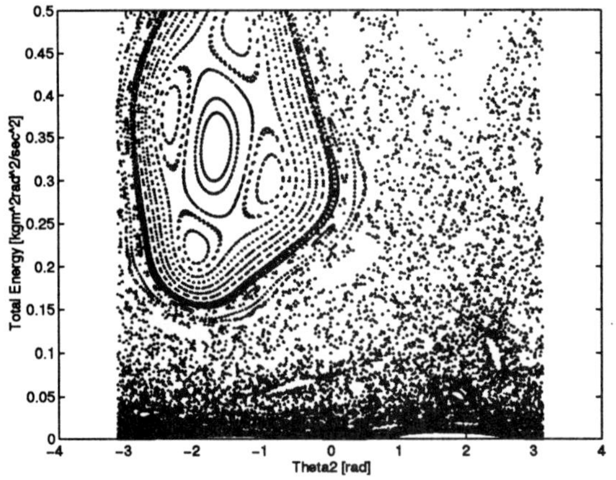

Figure 5: Poincaré map with $\gamma = 0.4, \omega = 2\pi$

Various initial value of J_2 were chosen and their trajectories were plotted in the figures. Only amplitude γ was differently set for each figure, and $\omega = 2\pi$ was used.

The conserved motion of the system without input is represented by a horizontal straight line in the H-θ_2 plane, which means rotation of θ_2. When we have a small time periodic input, the motion is perturbed and shows a slight difference. If we take the Poincaré map, the frequency of the rotation of θ_2 that basically comes from the conserved motion, and the frequency of time-periodic input mutually cause the resonance. The series of cavities or eye balls along lines parallel to the θ_2 axis in Fig.4 show the harmonic resonance of several different orders.

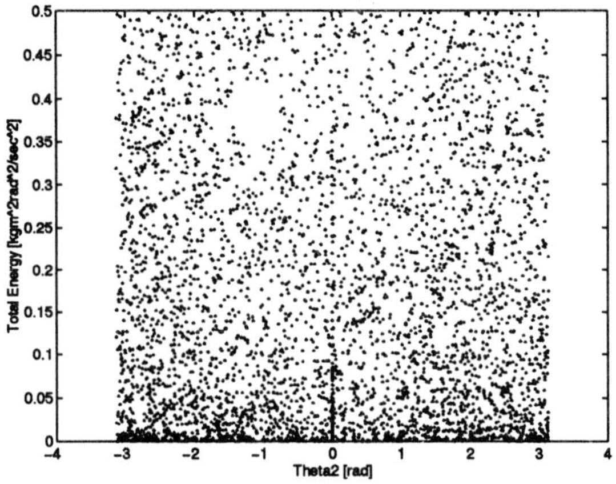

Figure 6: Poincaré map with $\gamma = 4, \omega = 2\pi$

The figures illustrate that as amplitude γ becomes larger, the homoclinic trajectories connecting the saddle points in Fig.4 start collapsed as seen in Fig.5. When the amplitude grows further in Fig.6, the system shows complete topological transitivity. These results conclude that the 2DOF free-joint manipulator driven by a time-periodic input follows a cyclic trajectory in the θ_2-H plane when the amplitude is small, while it behaves chaotic with the large amplitudes.

3 Nonlinear Control

3.1 Control with Periodic Inputs

We discuss positioning of the both joints of the free-joint manipulator. Oriolo and Nakamura[20] concluded that smooth feedback stabilization to a single equilibrium point is not possible, and derived a control law for the asymptotic stabilization to equilibrium manifold as

$$M = \{(\dot{\theta}_1, \dot{\theta}_2, \theta_1, \theta_2) \mid \theta_1 = \theta_{1d}, \dot{\theta}_1 = \dot{\theta}_2 = 0\}$$

In this section, we propose a control law that enables positioning to an arbitrary single equilibrium point from an arbitrary initial point. The control law uses time-periodic inputs and means neither feedback control nor stabilization in the strict sense.

The basic strategy of positioning is (1) $(\dot{\theta}_1, \theta_1) \to (0, \theta_{1d})$ and then (2) $(\dot{\theta}_2, \theta_2) \to (0, \theta_{2d})$ with a time-periodic input of the first joint. The first subgoal can be attained by simple feedback control of the first joint disregarding the motion of θ_2, or by Oriolo and Nakamura's asymptotic stabilization to an equilibrium manifold. In what follows, we mainly focus on control to meet the second subgoal.

The input torque of the first joint is determined such that the acceleration of the joint is time-periodic, namely, $\ddot{\theta}_1 = \gamma \cos \omega t$, which guarantees that $(\dot{\theta}_1, \theta_1)$ always returns to $(0, \theta_{1d})$ after each cycle. We can represent the dynamics of the system with the input as follows:

$$\frac{d}{dt}\begin{pmatrix} \dot{\theta}_1 \\ \dot{\theta}_2 \\ \theta_1 \\ \theta_2 \\ \phi \end{pmatrix} = \begin{pmatrix} \gamma \cos \phi \\ -\gamma \cos \phi (1 + n \cos \theta_2) - n\dot{\theta}_1^2 \sin \theta_2 \\ \dot{\theta}_1 \\ \dot{\theta}_2 \\ \omega \end{pmatrix}$$

$$(\dot{\theta}_1, \dot{\theta}_2, \theta_1, \theta_2, \phi) \in \boldsymbol{R}^2 \times \boldsymbol{T}^2 \times \boldsymbol{T}^1 \qquad (12)$$

where $n = \dfrac{b}{c} = \dfrac{m_2 l_1 s_2}{m_2 {s_2}^2 + I_2}$.

Integrating with initial values of $\dot{\theta}_{10} = 0$ and $\phi_0 = 0$, $\ddot{\theta}_1$ and $\dot{\theta}_1$ can be eliminated from Eq.(12). Nondimensionalizing Eq.(12) offers the following simpler equation:

$$\frac{d}{ds}\begin{pmatrix} \Omega \\ \theta_2 \\ s \end{pmatrix} = \begin{pmatrix} -\varepsilon(1 + n \cos \theta_2) \cos s - n\varepsilon^2 \sin \theta_2 \sin^2 s \\ \Omega \\ 1 \end{pmatrix}$$

$$(\Omega, \theta_2, s) \in \boldsymbol{R}^1 \times \boldsymbol{T}^1 \times \boldsymbol{R}^1 \qquad (13)$$

where

$$s = \omega t,\ \varepsilon = \frac{\gamma}{\omega^2},\ \Omega = \frac{d\theta_2}{ds} = \frac{1}{\omega}\dot{\theta}_2$$

Note that Eq.(13) includes only a constant, ε.

We now investigate the nonlinear behavior of Eq.(13) by using the Poincaré map as in subsection 2.3. With large amplitudes of the input, it is observed that the dynamics shows chaotic behavior similar to what we saw in subsection 2.3 as Fig.7.

On the other hand, with rather small amplitudes of γ, namely, of ε, the system shows well-ordered behavior as seen in the Poincaré map of Fig.8, which illustrates the trajectories with various values of θ_2 and $(\dot{\theta}_2, \gamma, \omega) = (0, 6, 4\pi)$ as the initial condition. From the figure, the system with a small amplitude follows an ellipse-like closed trajectory in the phase plane. It was also observed that increasing the amplitude tends to stretch the ellipses in the direction of $\dot{\theta}_2$ as Fig.9 with $\gamma = 8$.

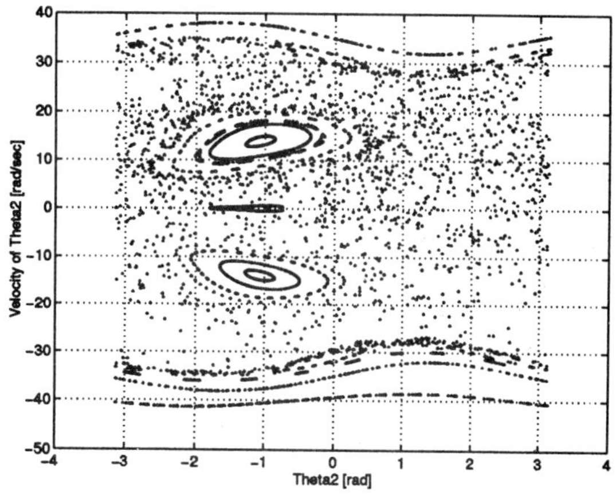

Figure 7: Poincaré map with $\gamma = 60, \omega = 4\pi$

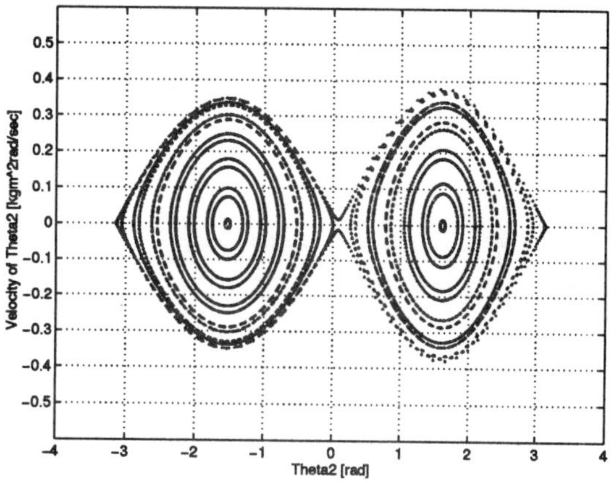

Figure 8: Poincaré map with $\gamma = 6, \omega = 4\pi$

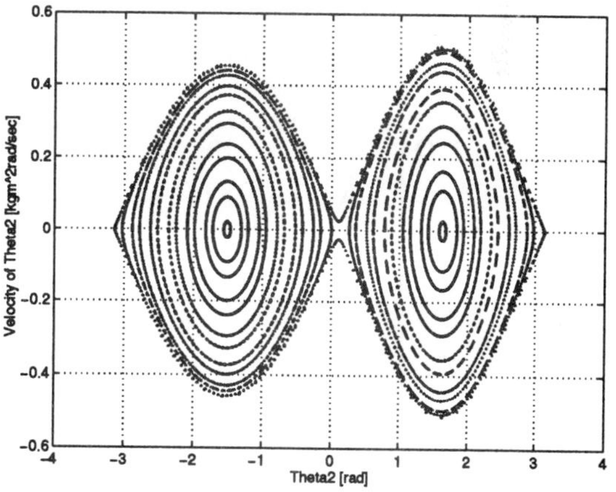

Figure 9: Poincaré map with $\gamma = 8, \omega = 4\pi$

3.2 Feedback Stabilization to an Elliptic Trajectory

In this subsection, we design a feedback stabilization law to an elliptic trajectory that passes through $(\dot{\theta}_2, \theta_2) = (0, \theta_{2d})$. We represent the elliptic approximation of the trajectory by

$$\frac{\left(\frac{\dot{\theta}_2}{\alpha}\right)^2 + (\theta_2 - p)^2}{{r_d}^2} = 1 \qquad (14)$$

where α denotes the eccentricity of ellipse, $r_d = |\theta_{2d} - p|$, and $(0, p)$ implies the center of ellipse.

α and p are determined by the amplitude of time-periodic input γ. Note that in Figs.8 and 9 the two families of ellipses in the left and right half planes are not symmetric in the strict sense, and that p nearly but not precisely equals to $\frac{\pi}{2}$. These are due to the nonlinearity and the finite magnitude of γ.

We now consider γ as an input variable and change it at the beginning of each periodic cycle in response to the error. Namely, we propose the amplitude modulation of feedback error. The modulation is according to the following law:

$$\gamma = \gamma_0 + k_\gamma \delta_r \qquad (15)$$

$$\delta_r = \frac{\left(\frac{\dot{\theta}_2}{\alpha}\right)^2 + (\theta_2 - p)^2}{{r_d}^2} - 1 \qquad (16)$$

where gain k_γ has a constant magnitude and changes its sign depending upon the area of the phase plane as shown in Fig.10. γ_0 denotes The

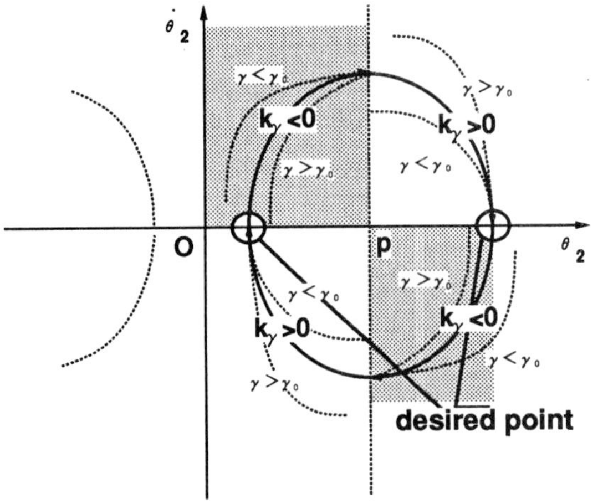

Figure 10: Amplitude modulation of the feedback error

nominal amplitude of the ellipse-like Poincaré map that Eq.(14) approximates. Therefore, if $(\dot{\theta}_2, \theta_2)$ is on the desired ellipse, $(\dot{\theta}_2, \theta_2)$ should almost stay there afterward with this amplitude. If $(\dot{\theta}_2, \theta_2)$ is not on the desired ellipse, with the feedback law of Eqs.(15) and (16) the elliptic trajectory works as an attractor and attracts $(\dot{\theta}_2, \theta_2)$.

Figure 11 shows the result of simulation with the initial condition $(\dot{\theta}_2, \theta_2) = (0, 0)$ and the desired position $\theta_{2d} = 3\pi/4$[rad]. The constants used were $\gamma_0 = 6$[rad/s^2] and $\omega = 4\pi$[rad/s]. The dotted line in Fig.11 indicate the desired ellipse. The solid line implies the trajectory of the system. The small circles in the figure denote the Poincaré map of the trajectory. It is observed from the figure that the desired ellipse attracted $(\dot{\theta}_2, \theta_2)$.

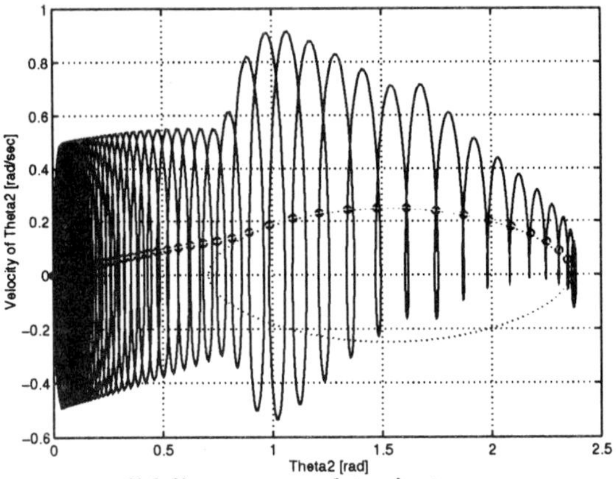

solid line : real trajectory
dotted line : desired trajectory
o : Poincaré map

Figure 11: Simulation : control from 0 to $\frac{3\pi}{4}$[rad]

3.3 Stopping at the Destination

Although the Poincaré map follows the elliptic trajectory by the feedback stabilization discussed in subsection 3.2, it does not imply that we can stop at the destination $(\dot{\theta}_2, \theta_2) = (0, \theta_{2d})$ that the ellipse passes through, as seen in Fig.11. The motion near the destination is enlarged and shown in the left figure of Fig.13. The motion should continue and follow the elliptic path, although in Figs.11 and 13 we terminated the computation when the Poincaré map passed the destination. We can stop at the destination only when the Poincaré map very fortunately makes its footprint (a small circle) exactly there. In this subsection, we propose to determine the amplitude of periodic input in the neighborhood of the destination (namely, the input of the last cycle before the destination) so as to make a footprint on the horizontal axis ($\dot{\theta}_2 = 0$) rather than to use the feedback law of Eqs.(15) and (16). Although this strategy does not guarantee to reach the destination, it does stop the motion in the neighborhood of the destination. The following describes how to compute the amplitude.

Let the state $\boldsymbol{x}(s,\varepsilon)$ denote

$$\boldsymbol{x}(s,\varepsilon)=\begin{pmatrix}\Omega(s,\varepsilon)\\ \theta_2(s,\varepsilon)\end{pmatrix} \tag{17}$$

where s and ε are the normalized time and amplitude as defined below Eq.(13). Also let $\boldsymbol{x}(s)$ imply the s-evolution of the same state when $\varepsilon \equiv 0$, of which initial condition coincides with that of $\boldsymbol{x}(s,\varepsilon)$. Namely, $\boldsymbol{x}(0,\varepsilon)=\boldsymbol{x}(0)$. When ε is small, the s-evolution of $\boldsymbol{x}(s,\varepsilon)$ can be approximated by applying the Taylor series expansion as follows:

$$\begin{aligned}\boldsymbol{x}(s,\varepsilon) &= \boldsymbol{x}(s,0)+\varepsilon\left.\frac{\partial \boldsymbol{x}(s,\varepsilon)}{\partial\varepsilon}\right|_{\varepsilon=0}\\ &\quad+\frac{1}{2}\varepsilon^2\left.\frac{\partial^2 \boldsymbol{x}(s,\varepsilon)}{\partial\varepsilon^2}\right|_{\varepsilon=0}+\boldsymbol{O}(\varepsilon^3)\end{aligned} \tag{18}$$

The Poincaré map of $\boldsymbol{x}(s,\varepsilon)$ can be calculated using Eq.(13) as follows:

$$\begin{pmatrix}\Omega\\ \theta_2\end{pmatrix}\mapsto\begin{pmatrix}\Omega+\frac{1}{2}\varepsilon^2 n^2\pi\sin 2\theta_2+O(\varepsilon^3)\\ \theta_2+2\pi\Omega+\frac{1}{2}\varepsilon^2 n^2\pi^2\sin 2\theta_2+O(\varepsilon^3)\end{pmatrix} \tag{19}$$

In order to make the footprint on the horizontal axis, $\dot{\theta}_2=0$, after the cycle, the upper part of Eq.(19) suggests to determine the amplitude by

$$\varepsilon_d=\sqrt{-\frac{2\Omega}{n^2\pi\sin 2\theta_2}} \tag{20}$$

From Eq.(19), this mapping becomes

$$\begin{pmatrix}\Omega\\ \theta_2\end{pmatrix}\mapsto\begin{pmatrix}O({\varepsilon_d}^3)\\ \theta_2+\pi\Omega+O({\varepsilon_d}^3)\end{pmatrix}=\begin{pmatrix}O({\varepsilon_d}^3)\\ \theta_2+O({\varepsilon_d}^2)\end{pmatrix} \tag{21}$$

which explains that the error of velocity and θ_2 are $O({\varepsilon_d}^3)$ and $O({\varepsilon_d}^2)$ respectively. The corresponding real amplitude is computed by $\gamma_d=\varepsilon_d\omega^2$.

In practice, we set constants $\delta\theta_2$ and ε and apply the proposed termination control when the state enters the following region:

$$\Sigma \stackrel{\text{def}}{=} \left\{(\dot{\theta}_2,\theta_2)\in\boldsymbol{R}^1\times\boldsymbol{T}^1 \,|\, \gamma_d<\gamma_0, |\theta_2-\theta_{2d}|<\delta\theta_2\right\} \tag{22}$$

Note that Eqs.(19) and (20) imply that the change of velocity by ε is very small in the neighborhood of $\theta_2=\pm\frac{N\pi}{2}$ and the strategy in this subsection does not work well. A stabilization control at $\theta_2=\pm\frac{\pi}{2}$ proposed by Nakamura and Iwamoto[24] can be used in this case. $\theta_2=0$ and $\pm\pi$ are saddle points as seen in Fig.8 and can be reached carefully following a stable manifold.

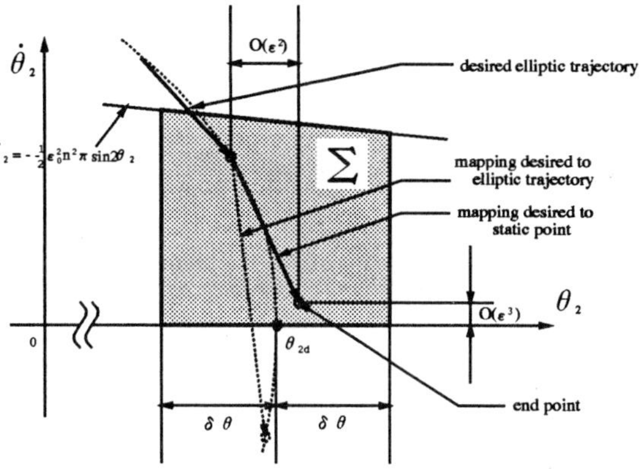

Figure 12: Stopping at the destination

In the control strategy we proposed in subsection 3.2 and this, the state error reflects on the amplitude of time-periodic input and determines it at the beginning of each periodic cycle. Therefore, within a cycle, the control is open-loop, though it is closed-loop in the global view. This problem is the most critical in the very last cycle.

Figure 13 compares the simulation results of the strategies proposed in subsection 3.2 (left) and this subsection (right). They are in the same situation as Fig.11 but that the input amplitude of the last cycle was computed by the different strategies.

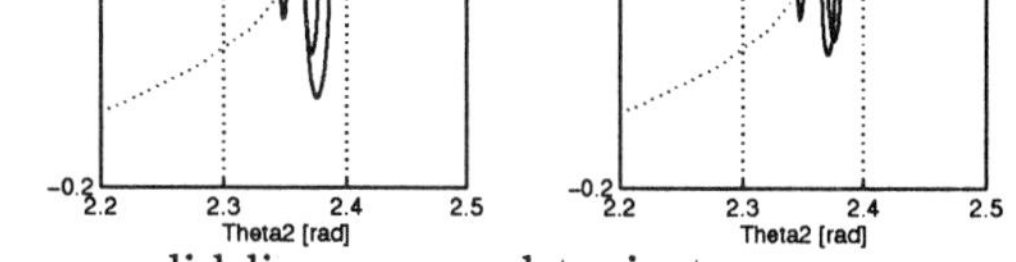

solid line	: real trajectory
dotted line	: desired trajectory
o	: Poincaré map

Figure 13: Simulation : stopping at $\frac{3\pi}{4}$

3.4 Compensating the Modeling Error

In this subsection and the next, we describes some details to improve the accuracy and the global nature of the control strategy developed in subsections 3.2 and 3.3.

We used an ellipse to model a Poincaré map of Fig.8. However, the modeling error becomes large for Poincaré maps passing near $\theta_2 = 0, \pm\pi$. This would result in that the actual Poincaré map does not go through the desired position θ_{2d}, even though the elliptic trajectory is designed to do so and the feedback stabilization of Eqs.(15) and (16) attracts the trajectory. In order to compensate this error, we propose to adaptively modify the desired elliptic trajectory. Namely, whenever the actual Poincaré map passes the θ_2 axis (every half rotation of the Poincaré map), we change constant r_d in Eq.(16) according to the following rule:

$$r_d^{(k+1)} = r_d^{(k)} \left(1 + k_r \cdot \frac{r_d - r_0}{r_0}\right) \tag{23}$$

where $r_d^{(k)}$ denotes the k-th re-evaluation of r_d and r_0 means the intersection of the θ_2 axis and the connected actual Poincaré map. $k_r(> 0)$ is a positive constant. The effect of this compensation was verified by simulation. Figure 14 shows a result of simulation. The initial condition and the destination were $(\dot{\theta}_2, \theta_2) = (0, \pi/3)$ and $(\dot{\theta}_2, \theta_2) = (0, \pi/18)$ respectively. The dotted line in the figure shows the original ellipse that goes through the destination. The connected Poincaré map is represented by the solid line. Although the destination is close to $\theta_2 = 0$ that implies a large modeling error, the actual Poincaré map reached the destination after a few rotations.

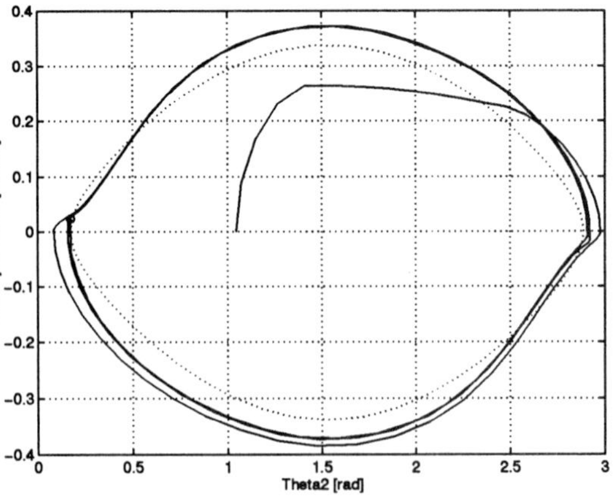

Figure 14: Simimulation with compensating the modeling gap

3.5 Global Attraction

Feedback stabilization makes use of two elliptic families as seen in Fig.8. The family in the right half plane ($0 \leq \theta_2 \leq \pi$) has the half plane as the region of attraction and so does the family in the left. This can be further interpreted as follows. Suppose the current state $(\dot{\theta}_2, \theta_2)$ is above or below the family of ellipse in the right half plane in Fig.8. By increasing the amplitude of periodic input, we can always stretch the family of ellipses, so that one goes through the current state. Thus, changing the amplitude means stretching or shrinking the ellipses in the $\dot{\theta}_2$ direction. The feedback stabilization in subsection 3.2 systematically executes this procedure and stabilizes onto an ellipse passing θ_{2d}.

The above interpretation implies local attraction of the ellipse through θ_{2d}, since nothing is promised for an initial state having its destination in the opposite half plane. In this subsection, we illustrate that the global attraction can be guaranteed by a simple modification.

It is important to note on the Poincaré maps above or below the elliptic families in Fig.8, which are not explicitly shown. The Poincaré maps above the families flow from the left to the right through $\dot{\theta}_2$ axis, while those below the families flow from the right to left. Therefore, if the initial state resides in $\dot{\theta}_2 > 0$ and $\theta_2 < 0$, for example, we can find an appropriate amplitude γ such that the initial state flows from the left plane to the right plane. Accordingly, we can include $\dot{\theta}_2 > 0$ and $\theta_2 < 0$ in the region of attraction.

When $\dot{\theta}_2 \leq 0$ and $\theta_2 < 0$, it is not possible to find a single constant of γ that brings the initial state to the right half plane. However, we can do this by switching γ as follows. First, we apply the feedback stabilization developed in subsection 3.2. The state converges and follows an elliptic trajectory on the left half plane. Then, when the state enters $\dot{\theta}_2 > 0$ and $\theta_2 < 0$, we can apply the strategy in the previous paragraph. With this switching, we now have the whole state space as the region of attraction of an ellipse passing θ_{2d}. Figure 15 shows the result of simulation. The initial and desired positions were $\theta_{20} = -5\pi/12$[rad] and $\theta_{2d} = 3\pi/4$[rad] respectively. The state converged to an ellipse in the left, and then, being switched, followed a flow from the left to the right. Finally, the state was stabilized to an ellipse in the

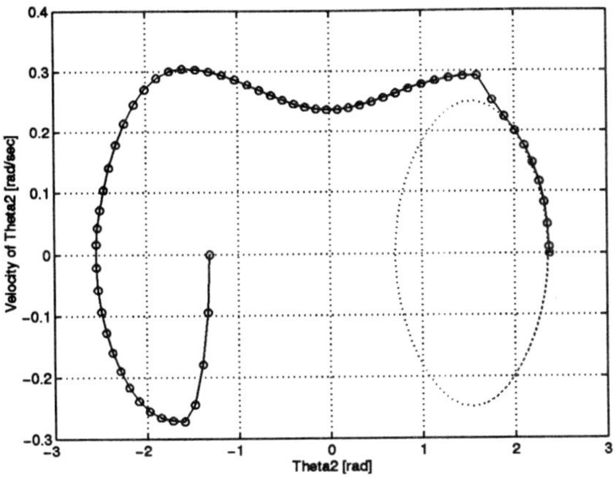

Figure 15: Simulation with global attraction

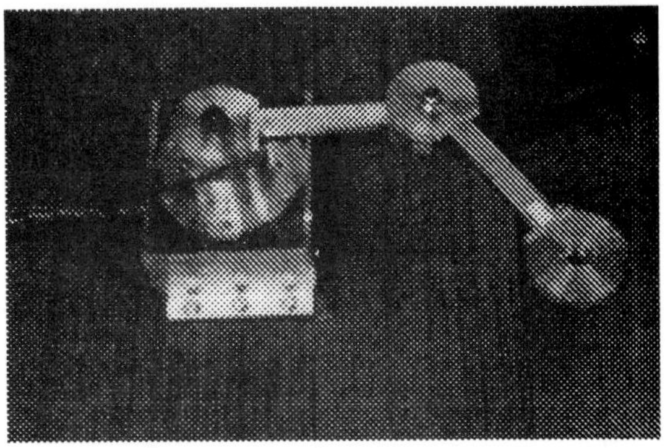

Figure 17: Photo of the free-joint manipulator

right half plane. Note that the switching in the left plane was done on the top of the ellipse, since it is preferable to have a not-small momentum to travel from the left to the right passing through the $\dot{\theta}_2$ axis.

4 Experiments

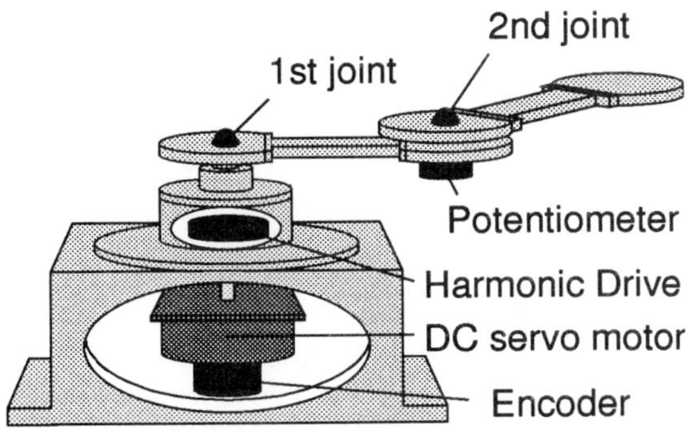

Figure 16: Structure of the free-joint manipulator

Table 1: Dynamic parameters

link	1	2
length [m]	0.240	0.240
gravity center (GC) [m]	--	0.111
mass [kg]	1.00	0.930
inertia about GC [kg·m^2]	–	0.0115

We executed experiments to verify the control strategy proposed in section 3. Figures 16 and 17 show the structure and the photograph of the 2DOF free-joint manipulator developed for the experiments. The both joint axes are vertical. The first joint is actuated by a DC servo motor, while the second joint is free to move. A low-friction type potentiometer is equipped at the second joint. The dynamic parameters necessary for control is summarized in Table 1. For the drive of the first joint, we applied a simple velocity servo loop. We used the following values for the control constants: $\gamma_0 = 6.00[\text{rad/s}^2]$, $\omega = 4\pi[\text{rad/s}]$, $p = \pm 1.53[\text{rad}]$, $\alpha = 0.300$, $k_\gamma = 15.0[\text{rad/s}^2]$, and $k_r = 0.500$. Note that we applied the developed strategy in section 3 as it is, though the real experimental set up involves friction and other uncertainties that we neglected in the modeling.

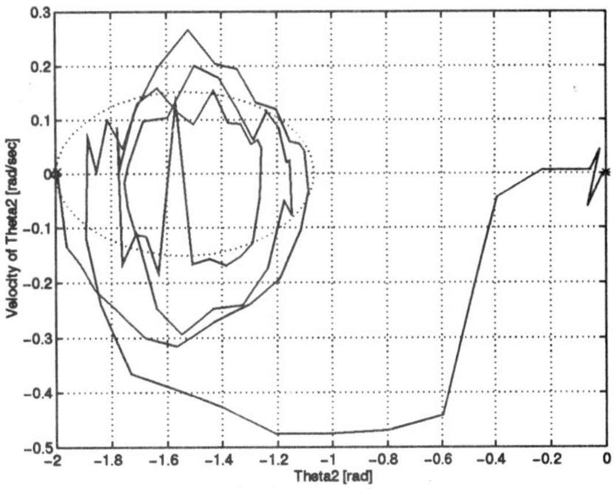

○	: desired point
$*$	: initial point
×	: final point
solid line	: experimental Poincaré map
dotted line	: computed ellipse

Figure 18: Experiment 1: from 0[rad] to −2[rad]

Figures 18 and 19 show the experimental results. Figure 18 is the case with $\theta_{20} = 0[\text{rad}]$ and $\theta_{2d} = -2[\text{rad}]$, and Fig.19 is with $\theta_{20} = -2[\text{rad}]$ and $\theta_{2d} = -0[\text{rad}]$. Due to the friction, $(\dot{\theta}_2, \theta_2) = (0, p)$ becomes a point of attraction. The conver-

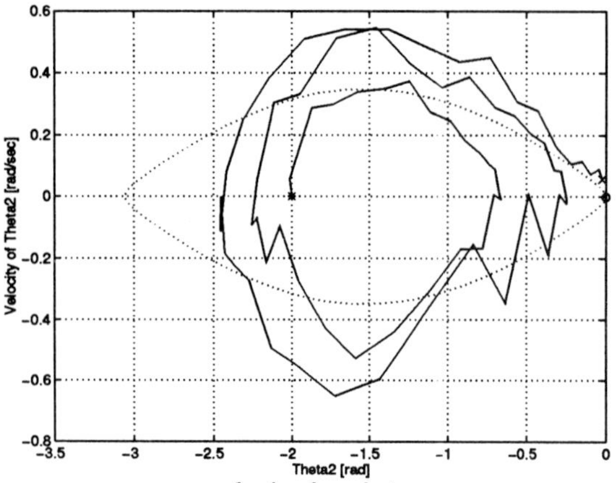

o	: desired point
*	: initial point
×	: final point
solid line	: experimental Poincaré map
dotted line	: computed ellipse

Figure 19: Experiment 2: from −2[rad] to 0[rad]

gence to elliptic trajectory was not as smooth as that of Figs.11 or 14, which is due to the effect of the point attraction. In spite of such difficulty, thanks to the feedback stabilization, the state terminated near the destinations. Note that though the point of termination shows a small residual velocity in Fig.19, it actually stopped in the experiment. It can be explained as (1) the residual velocity was damped out by the friction, or (2) the residual velocity was due to discretization error in computing velocity from the potentiometer signal.

The experimental results clearly show effectiveness and robustness of the developed positioning control law.

5 Conclusion

We investigated the nonlinear behavior of a 2DOF free-joint manipulator and its positioning control.

Choosing a time-periodic input for the first joint was our fundamental strategy. When the amplitude of periodic input remains small, the behavior in the phase space showed an elliptic closed path. As the amplitude grows, the Poincaré map showed chaotic behaviors.

We proposed a series of positioning control strategy where the amplitude modulation of the error signal was used for feedback stabilization to a desired elliptic trajectory.

The experiments were carried out to verify the developed positioning control strategy. The results clearly shows its effectiveness and robustness.

Acknowledgment

Discussions with Prof. Ken'ichi Yoshimoto was helpful for accomplishing this work. The 2DOF free-joint manipulator was developed by Mr. Ryuji Iwamoto. This research was in part supported by the Toyota Physical and Chemical Research Institute.

References

[1] Laumond, J.-P.: Feasible trajectories for mobile robots with kinematic and environment constraints, in *Intelligent Autonomous Systems*, (O.L. Hertzberger, and F.C.A. Groen, eds), pp. 346-354, 1987.

[2] Murray, R.M., and Sastry, S.S.: Steering nonholonomic systems using sinusoids, Proc. Conf. on Decision and Control, pp. 2097-2101, Honolulu, Hawaii, 1990.

[3] Lafferriere, G., and Sussmann, H.J.: A differential geometric approach to motion planning, in *Nonholonomic Motion Planning*, (Z.X. Li and J.F. Canny, eds.), pp. 235-270, 1993.

[4] Bushnell, L., Tilbury, D., and Sastry, S.: Extended Goursat normal form with applications to nonholonomic motion planning, Proc. IEEE Conference on Decision and Control, San Antonio, Texus, 1993.

[5] Nakamura, Y., and Mukherjee, R.: Bi-directional approach for nonholonomic path planning of space robots, in Robotics Research: the Fifth International Symposium (August 1989, Tokyo), MIT press, pp. 405-416, 1990.

[6] Sreenath, N.: Nonlinear control of multibody systems in shape space, Proc. IEEE International Conference on Robotics and Automation, pp. 1776-1781, Cinncinati, OH, 1990.

[7] Li, Z., Canny, J., and Heinzinger, G.: Robot motion planning with nonholonomic constraints, in *Robotics Research: the Fifth International Symposium* (August 1989, Tokyo), MIT press, pp. 309-316, 1990.

[8] Lynch, K.M., and Mason, M.T.: Controllability of pushing, Proc. IEEE International Conference on Robotics and Automation, Nagoya, Japan, pp. 112-119, 1995.

[9] Laumond, J.-P.: Controllability of a multibody mobile robot, Proc. Inter. Conf. on Advanced Robotics, Pisa, Italy, 1991.

[10] Walsh, G.C., and Bushnell, L.G.: Stabilization of multiple input chained form control systems, Proc. IEEE Conference on Decision and Control, San Antonio, Texus, 1993.

[11] Samson, C., and Ait-Abderrahim, K.: Feedback stabilization of a nonholonomic wheeled mobile robot, Proc. IEEE/RSJ Inter. Workshop on Intell. Robots and Systems '91, pp. 1242-1247, Osaka, November, 1991.

[12] Sørdalen, O.J., and Canudas de Wit, C.: Path following and stabilization of a mobile robot, Proc. IFAC Nonlinear Control Systems Design Symp., pp. 513-518, Bordeaux, France, 1992.

[13] Pomet, J.-B.: Explicit design of time-varying stabilizing control laws for a class of controllable systems without drift, System and Control Letters, vol. 18, pp. 139-145, 1992.

[14] Murray, R.M., and Sastry, S.S.: Steering nonholonomic systems in chained form, Proc. Conf. on Decision and Control, pp. 1121-1126, Brighton, England, 1991.

[15] Sørdalen, O.J.: Conversion of the kinematics of a car with n trailers into a chained form, Proc. IEEE Inter. Conf. on Robotics and Automation, pp. 1/382-1/387, Atlanta, Georgia, 1993.

[16] Sørdalen, O.J., Nakamura, Y., and Chung, W.J.: Design of a nonholonomic manipulator, Proc. IEEE International Conference on Robotics and Automation, San Diego, CA, pp. 8-13, 1994.

[17] Chung, W.J., Nakamura, Y., and Sørdalen, O.J.: Prototyping a nonholonomic manipulator, Proc. IEEE International Conference on Robotics and Automation, Nagoya, Japan, pp. 2029-2036, 1995.

[18] Nakamura, Y.: Integrability of dynamic constraints of underactuated mechanisms, Proc. 6th International Symposium of Robotics Research, Pittsburg, PA, pp.99-110, 1993.

[19] Arai, H., and Tachi, S.: Dynamical control of a manipulator with passive joints - psition control experiments by a prototype manipulator, Proc. IEEE/RSJ International Workshop on Intelligent Robots and Systems, 1990.

[20] Oriolo, G., and Nakamura, Y.: Free-joint manipulators: motion control under second-order nonholonomic constraints, Proc. IEEE/RSJ International Workshop on Intelligent Robots and System, Osaka, Japan, pp.1248-1253, 1991.

[21] Nakamura, Y., and Iwamoto, R.: Space multibody structure connected with free joints and its shape control, Proc. IEEE Conference on Decision and Control, San Antonio, Texus, pp.3126-3131, 1993.

[22] Wichlund, K.Y., Sørdalen, O.J., and Egeland, O.: Control properties of underactuated vehicles, Proc. IEEE International Conference on Robotics and Automation, Nagoya, Japan, pp.2009-2014, 1995.

[23] Oriolo, G., and Nakamura, Y.: Control of mechanical systems with se cond-order nonholonomic constraints: underactuated manipulators, Proc. IEEE Conference on Decision and Control, Brighton, Uk, 1991.

[24] Nakamura, Y., Iwamoto, R., and Yoshimoto, K.: Control of nonholonomic mechanisms with drift, The Journal of Robotics Sciety of Japan, Vol.13, No.6, pp.830-837, 1995.

[25] Wiggins, S.: *Introduction to applied nonlinear dynamical systems and chaos*, Splinger-Verlag (1990).

[26] Tabor, M.: *Chaos and integrability in nonlinear dynamics – an introduction*, Wiley Inter-Science (1989).

A Definition of Chaos

The definition of chaos is rigorously presented by Wiggins [25, pp. 608,45] as follows:

Lemma 1 *We consider a* $\boldsymbol{C}^r$ $(r \geq 1)$ *autonomous vector field on* $\boldsymbol{R}^n$ *and represent it by*

$$\dot{\boldsymbol{x}} = \boldsymbol{f}(\boldsymbol{x}) \tag{A.1}$$

We denote the flow generated by Eq.(A.1) by $\phi(t, \boldsymbol{x})$ $(\forall t > 0)$ *and assume that* $\Lambda \subset \boldsymbol{R}^n$ *is a compact set invariant under* $\phi(t, \boldsymbol{x})$, *i.e.,* $\phi(t, \Lambda) \subset \Lambda$ *for all* $t \in \boldsymbol{R}$. *Then* Λ *is said to be* chaotic *if*

(1) *Sensitive dependence on initial conditions (s.d.i.c.) : there exists* $\varepsilon > 0$ *such that, for any* $\boldsymbol{x} \in \Lambda$ *and any neighborhood* U *of* $\boldsymbol{x}$, *there exist* $\boldsymbol{y} \in U$ *and* $t > 0$ *such that*

$$|\phi(t, \boldsymbol{x}) - \phi(t, \boldsymbol{y})| > \varepsilon$$

(2) *Topological transitivity : for any two open sets* $U, V \subset \Lambda$, *there exists* $t \in \boldsymbol{R}$ *such that*

$$\phi(t, U) \cap V \neq \emptyset$$

Parallel Force/Position Control of Robot Manipulators

Bruno Siciliano
Dipartimento di Informatica e Sistemistica
Università degli Studi di Napoli Federico II
Via Claudio 21, 80125 Napoli, Italy
siciliano@na.infn.it

Abstract

This paper is aimed at presenting a survey on a class of parallel force/position control schemes which have been proposed in the latest five years by the author and co-workers. Several control schemes have been devised under this framework for the case of contact with a compliant planar surface. A common feature of such schemes is that, at the equilibrium, the force can be shown to be regulated to a desired constant value at the expense of a position error which depends on environment stiffness. If tracking of end-effector position along the unconstrained task space directions is desired, a passivity-based control scheme can be used which can be naturally made adaptive with respect to manipulator dynamic parameters. On the other hand, if only regulation of position is desired, a simple PID control can be used which can be made adaptive as well in the case of imperfect gravity compensation, and can even avoid velocity measurements.

1 State-of-Art

Control of interaction between a robot manipulator and the environment is crucial for successful execution of a number of practical tasks where the robot's end effector has to manipulate an object or perform some operation on a surface. The specific feature of robot tasks such as polishing, deburring, or assembly, demands control also of the exchanged forces at the contact. The contact force is thus the quantity describing the state of interaction in the most complete fashion, and the interaction control problem [47] has attracted a wide number of researchers in the last decade.

Interaction control strategies can be grouped in two categories; those performing *open-loop force control* and those performing direct *closed-loop force control*. The main difference between the two categories is that the former achieve indirect force control via closed-loop position control, without explicit closure of a force feedback loop; the latter, instead, offer the possibility of controlling the contact force to a desired value, thanks to the closure of a force feedback loop.

To the first category belong *compliance* (or *stiffness*) *control* [31, 33] and *impedance control* schemes [18, 20], where the position error is related to the contact force through a mechanical stiffness or impedance of adjustable parameters. A robot manipulator under impedance control is described by an equivalent mass-spring-damper system with the contact force as input. The resulting impedance may be linear or nonlinear, depending on the fact whether force feedback is used or not.

The most common strategy belonging to the second category is the *hybrid position/force control* which aims at controlling position along the unconstrained task space directions and force along the constrained task space directions. A selection matrix acting on both desired and feedback quantities serves this purpose for typically planar contact surfaces [32], whereas the explicit constraint equations have to be taken into account for general curved contact surfaces [50, 25, 26, 14, 2, 49].

An alternative strategy still in the second category is the *inner/outer position/force control* where, along each constrained task space direction, an outer force control loop is closed around the inner position control loop which is typically available in a robot manipulator [15]. By a suitable design of the force control action (typically an integral term), it is possible to achieve regulation of the contact force to a desired value. The inclusion of an integral action in the force control loop, in fact, guarantees removal of the steady-state force error [46] and may provide robustness with respect to force measurement delays [45, 48].

A new strategy has been recently proposed, that is, the so-called *parallel force/position control* [8]. The key concept of this strategy is to combine the simplicity and robustness of the impedance and inner/outer position/force control schemes with the ability of controlling both position and force typical of the hybrid control schemes. In order to embed the possibility of controlling motion along the unconstrained task space directions, a desired position can be input to the inner loop of an inner/outer position/force control scheme. The result is two control actions, working in parallel; namely, a force control action and a position control action. In order to ensure force control along the constrained task space directions, the force action is designed so as to dominate the position action [6].

2 Modeling

2.1 Manipulator Dynamics

In order to study interaction with the environment, it is worth considering the dynamic model of a robot manipulator in the task (operational) space. This can be written in the form

$$\boldsymbol{B}(\boldsymbol{x})\ddot{\boldsymbol{x}} + \boldsymbol{C}(\boldsymbol{x},\dot{\boldsymbol{x}})\dot{\boldsymbol{x}} + \boldsymbol{g}(\boldsymbol{x}) = \boldsymbol{u} - \boldsymbol{f}, \quad (1)$$

where $\boldsymbol{x}$ is the $(m \times 1)$ vector of end-effector location, $\boldsymbol{B}$ is the $(m \times m)$ symmetric and positive definite inertia matrix, $\boldsymbol{C}\dot{\boldsymbol{x}}$ is the $(m \times 1)$ vector of Coriolis and centrifugal generalized forces, $\boldsymbol{g}$ is the $(m \times 1)$ vector of gravitational generalized forces, $\boldsymbol{u}$ is the $(m \times 1)$ vector of driving generalized forces, and $\boldsymbol{f}$ is the $(m \times 1)$ vector of contact generalized forces exerted by the end effector on the environment; all task space quantities are expressed in a common reference frame.

The end-effector direct and differential kinematics are defined by

$$\boldsymbol{x} = \boldsymbol{k}(\boldsymbol{q}) \quad (2)$$

$$\dot{\boldsymbol{x}} = \frac{\partial \boldsymbol{k}(\boldsymbol{q})}{\partial \boldsymbol{q}}\dot{\boldsymbol{q}} = \boldsymbol{J}_A(\boldsymbol{q})\dot{\boldsymbol{q}} \quad (3)$$

where $\boldsymbol{q}$ is the $(n \times 1)$ vector of joint variables and $\boldsymbol{J}_A$ is the $(m \times n)$ manipulator analytical Jacobian matrix [34]. The $(n \times 1)$ vector $\boldsymbol{\tau}$ of joint actuating generalized forces is computed as

$$\boldsymbol{\tau} = \boldsymbol{J}_A^{\mathrm{T}}(\boldsymbol{q})\boldsymbol{u}. \quad (4)$$

When $m = n$ and the manipulator moves in a singularity-free region of the workspace, the vector of operational variables $\boldsymbol{x}$ constitutes a set of Lagrangian generalized coordinates and $\boldsymbol{B}$ assumes the meaning of a true inertia matrix. Instead, in the case of kinematically redundant manipulators ($m < n$), $\boldsymbol{B}$ is only a pseudo inertia matrix [22].

The following notable properties of the task space dynamic model (1) can be established from similar properties of the joint space dynamic model.

1. The matrix $\boldsymbol{B}$ is symmetric and positive definite. If λ_m (λ_M) denotes the minimum (maximum) eigenvalue of $\boldsymbol{B}$, then

$$\boldsymbol{O} < \lambda_m \boldsymbol{I} \le \boldsymbol{B}(\boldsymbol{x}) \le \lambda_M \boldsymbol{I}, \quad (5)$$

where $\boldsymbol{I}$ is the $(m \times m)$ identity matrix; in the case of all revolute joints, it is $\lambda_M < \infty$.

2. There exists a choice of the matrix $\boldsymbol{C}$ such that the matrix

$$\boldsymbol{N}(\boldsymbol{x},\dot{\boldsymbol{x}}) = \dot{\boldsymbol{B}}(\boldsymbol{x},\dot{\boldsymbol{x}}) - 2\boldsymbol{C}(\boldsymbol{x},\dot{\boldsymbol{x}}) \quad (6)$$

is skew-symmetric [23, 41]. This also implies that

$$\dot{\boldsymbol{B}}(\boldsymbol{x},\dot{\boldsymbol{x}}) = \boldsymbol{C}(\boldsymbol{x},\dot{\boldsymbol{x}}) + \boldsymbol{C}^{\mathrm{T}}(\boldsymbol{x},\dot{\boldsymbol{x}}). \quad (7)$$

Further, the matrix $\boldsymbol{C}$ is upper-bounded in $\boldsymbol{x}$ and linear in $\dot{\boldsymbol{x}}$; hence, a constant $0 < k_C < \infty$ exists such that

$$\|\boldsymbol{C}(\boldsymbol{x},\dot{\boldsymbol{x}})\| \le k_C \|\dot{\boldsymbol{x}}\|. \quad (8)$$

3. The gravity force vector $\boldsymbol{g}$ can be thought of as given by

$$\boldsymbol{g}(\boldsymbol{x}) = \frac{\partial U(\boldsymbol{x})}{\partial \boldsymbol{x}} \quad (9)$$

where $U(\boldsymbol{x})$ is the gravitational energy in the task space that is bounded for any $\boldsymbol{x}$. A helpful property of the gravity term is that $\boldsymbol{g}$ has bounded partial derivatives [44], i.e.,

$$\left\| \frac{\partial \boldsymbol{g}(\boldsymbol{x})}{\partial \boldsymbol{x}} \right\| \le \psi \quad (10)$$

for some $\psi > 0$. This also implies that

$$\|\boldsymbol{g}(\boldsymbol{x}_1) - \boldsymbol{g}(\boldsymbol{x}_2)\| \le \psi \|\boldsymbol{x}_1 - \boldsymbol{x}_2\| \quad (11)$$

for any $\boldsymbol{x}_1$ and $\boldsymbol{x}_2$.

4. The dynamic model (1) is linear in terms of a suitable set of manipulator and load constant parameters [28], i.e.,

$$\boldsymbol{B}(\boldsymbol{x})\ddot{\boldsymbol{x}} + \boldsymbol{C}(\boldsymbol{x}, \dot{\boldsymbol{x}})\dot{\boldsymbol{x}} + \boldsymbol{g}(\boldsymbol{x}) = \boldsymbol{Y}(\boldsymbol{x}, \dot{\boldsymbol{x}}, \ddot{\boldsymbol{x}})\boldsymbol{\pi} \tag{12}$$

where $\boldsymbol{Y}(\boldsymbol{x}, \dot{\boldsymbol{x}}, \ddot{\boldsymbol{x}})$ is an $(m \times p)$ matrix and $\boldsymbol{\pi}$ is a $(p \times 1)$ vector of parameters which depend on link masses, first moments of inertia and inertia tensors. With specific reference to the gravity term, the vector $\boldsymbol{g}$ in (1) can be written in the form

$$\boldsymbol{g}(\boldsymbol{x}) = \boldsymbol{G}(\boldsymbol{x})\boldsymbol{\theta} \tag{13}$$

where $\boldsymbol{G}(\boldsymbol{x})$ is an $(m \times r)$ matrix and $\boldsymbol{\theta}$ is an $(r \times 1)$ vector of manipulator and load parameters; the dimension r is smaller than p, since the gravity term depends only on link masses and first moments of inertia.

The above properties will be used to derive parallel force/position control schemes other than the basic inverse dynamics parallel control.

In the remainder, the attention is restricted to the case of nonredundant nonsingular manipulators with $m = n = 3$, i.e., only translational motion and force components are considered; then, $\boldsymbol{x}$ denotes the end-effector position.

2.2 Environment

Accurate modeling of the contact between the manipulator and the environment is usually difficult to obtain in analytic form, due to complexity of the physical phenomena involved during the interaction. It is then reasonable to resort to a simple but significant model, relying on the robustness of the control system in order to absorb the effects of inaccurate modeling. Following these guidelines, the case of an environment constituted by a frictionless and elastically compliant plane is analyzed. The choice of a planar surface is motivated by noticing that it is locally a good approximation to surfaces of regular curvature [7]. The total elasticity, due to end-effector force sensor and environment, is accounted through the compliance of the plane. Friction effects are neglected within the operational range of interest.

With the above assumptions, the model of the contact force considered takes on the simple form

$$\boldsymbol{f} = \boldsymbol{K}(\boldsymbol{x} - \boldsymbol{x}_0), \tag{14}$$

where $\boldsymbol{x}$ is the position of the contact point, $\boldsymbol{x}_0$ is a point of the plane at rest, and $\boldsymbol{K}$ is the (3×3) constant symmetric stiffness matrix of rank 1; note that Eq. (14) holds only when the end effector is in contact with the environment and all quantities are expressed in the common reference frame. It is worth considering the rotation matrix expressing the orientation of the contact frame with respect to the reference frame

$$\boldsymbol{R} = (\, \boldsymbol{t}_1 \quad \boldsymbol{t}_2 \quad \boldsymbol{n} \,) \tag{15}$$

where $\boldsymbol{n}$ is the unit vector normal to the contact plane, and $\boldsymbol{t}_1, \boldsymbol{t}_2$ are two orthogonal unit vectors lying in the plane. In view of (15), the stiffness matrix can be written as

$$\boldsymbol{K} = \boldsymbol{R}\,\mathrm{diag}\{0, 0, k\}\boldsymbol{R}^{\mathrm{T}} = k\,\boldsymbol{n}\boldsymbol{n}^{\mathrm{T}}, \tag{16}$$

where $k > 0$ is the stiffness coefficient.

The elastic contact model (14) and (16) suggests that a null force error can be obtained only if $\boldsymbol{f}_d = f_d\boldsymbol{n}$. If no information about the geometry of the environment is available, i.e., the direction of $\boldsymbol{n}$ is unknown, the null vector can be assigned to $\boldsymbol{f}_d$ that is anyhow in the range space of any matrix $\boldsymbol{K}$. Analogously, it can be recognized that null position errors can be obtained only on the contact plane, while the component of $\boldsymbol{x}$ along $\boldsymbol{n}$ has to accommodate the force requirement specified by $\boldsymbol{f}_d$; thus, $\boldsymbol{x}_d$ can be freely reached only in the null space of $\boldsymbol{K}$, i.e., along the unconstrained directions of the task space.

3 Parallel Control

The underlying philosophy of the *parallel control* approach is to combine the simplicity and robustness of the impedance control and the inner/outer position/force control with the capability of controlling both force and position of the hybrid control. This is realized by designing two control loops —one in position and one in force— acting in parallel along each task space direction. Conflicts between position and force actions are handled through a rule-based priority strategy.

A physical analysis of the interaction leads to recognizing that dominance of the force control loop over the position control loop should be achieved so as to accommodate unplanned contact forces in any situation. The most natural way to implement the sought dominance is to use a PI force control loop working in parallel to a PD position control loop. In this respect, the scheme can be regarded as an extension of an impedance control scheme (with added direct force control capabilities) and an inner/outer position/force control scheme (with improved position control capabilities). At the same time, force and position controlled directions are not established a priori in the parallel control, as instead in the hybrid control; full sensor measurements can thus be exploited without any task-based filtering action.

The task planning provides force or position references along suitable task space directions, as in the hybrid control case. A perfect planning obviously makes the task successful, but contact is safely handled by the parallel control even in the case of planning errors. Recovery from imperfect planning is made possible thanks to the force dominance rule.

3.1 Inverse Dynamics

According to an *inverse dynamics* concept with contact force sensor measurements, the vector of driving forces in (1) can be synthesized as

$$\boldsymbol{u} = \widehat{\boldsymbol{B}}(\boldsymbol{x})\boldsymbol{a} + \widehat{\boldsymbol{C}}(\boldsymbol{x}, \dot{\boldsymbol{x}})\dot{\boldsymbol{x}} + \widehat{\boldsymbol{g}}(\boldsymbol{x}) + \boldsymbol{f}, \qquad (17)$$

where the hats denote the available estimates of the dynamic terms $\boldsymbol{B}$, $\boldsymbol{C}$, $\boldsymbol{g}$, and $\boldsymbol{f}$ is the measured contact force. Substituting control (17) into model (1), under the assumption of perfect dynamic compensation and exact force cancellation, gives

$$\ddot{\boldsymbol{x}} = \boldsymbol{a} \qquad (18)$$

that is a linear decoupled system expressing a resolved acceleration.

Let $\boldsymbol{x}_d$ and $\boldsymbol{f}_d$ denote the desired values of position and force, respectively. According to the parallel control approach, the new control input is designed as the sum of a position control action and a force control action; namely, as [6]

$$\boldsymbol{a} = \boldsymbol{a}_x + \boldsymbol{a}_f \qquad (19)$$

where

$$\boldsymbol{a}_x = \ddot{\boldsymbol{x}}_d + m_d^{-1} k_D \Delta\dot{\boldsymbol{x}} + m_d^{-1} k_P \Delta\boldsymbol{x} \qquad (20)$$

$$\boldsymbol{a}_f = m_d^{-1} k_F \Delta\boldsymbol{f} + m_d^{-1} k_I \int_0^t \Delta\boldsymbol{f} d\sigma \qquad (21)$$

where $\Delta\boldsymbol{x} = \boldsymbol{x}_d - \boldsymbol{x}$ is the position error, $\Delta\boldsymbol{f} = \boldsymbol{f}_d - \boldsymbol{f}$ is the force error, m_d is a desired mass, and $k_P, k_D, k_F, k_I > 0$ are suitable feedback gains. Substituting (19) with (20) and (21) in (18) yields

$$m_d \Delta\ddot{\boldsymbol{x}} + k_D \Delta\dot{\boldsymbol{x}} + k_P \Delta\boldsymbol{x} + k_F \Delta\boldsymbol{f} + k_I \int_0^t \Delta\boldsymbol{f} d\sigma = \boldsymbol{0} \qquad (22)$$

which reveals that, thanks to the *integral* action, $\Delta\boldsymbol{f}$ is allowed to prevail over $\Delta\boldsymbol{x}$ at steady state.

Assuming that the desired force is aligned with the normal to the contact plane and contact is not lost, the stability analysis of the system (1) under the inverse dynamics parallel control (17), (19), (20) and (21) with the environment (14) and (16) can be developed according to classical linear systems theory. In detail, plugging (14) in (22) gives

$$\begin{aligned} & m_d \ddot{\boldsymbol{x}} + k_D \dot{\boldsymbol{x}} + (k_P \boldsymbol{I} + k_F k \boldsymbol{n}\boldsymbol{n}^{\mathrm{T}})\boldsymbol{x} \\ & \quad + k_I k \boldsymbol{n}\boldsymbol{n}^{\mathrm{T}} \int_0^t \boldsymbol{x} d\sigma \qquad (23) \\ & = m_d \ddot{\boldsymbol{x}}_d + k_D \dot{\boldsymbol{x}}_d + k_P \boldsymbol{x}_d \\ & \quad + k_F(\boldsymbol{f}_d + k \boldsymbol{n}\boldsymbol{n}^{\mathrm{T}} \boldsymbol{x}_0) + k_I \int_0^t (\boldsymbol{f}_d + k \boldsymbol{n}\boldsymbol{n}^{\mathrm{T}} \boldsymbol{x}_0) d\sigma \end{aligned}$$

which represents a third-order linear system, whose stability can be analyzed by referring to the unforced system, i.e., by setting to zero the right-hand side of (23). According to (15), projection of the position vector on the contact frame yields

$$\boldsymbol{R}^{\mathrm{T}} \boldsymbol{x} = \begin{pmatrix} x_1 \\ x_2 \\ x_n \end{pmatrix} \qquad (24)$$

which leads to the system of three scalar decoupled equations

$$m_d \ddot{x}_1 + k_D \dot{x}_1 + k_P x_1 = 0 \qquad (25)$$

$$m_d \ddot{x}_2 + k_D \dot{x}_2 + k_P x_2 = 0 \qquad (26)$$

$$m_d \ddot{x}_n + k_D \dot{x}_n + (k_P + k_F k) x_n + k_I k \int_0^t x_n d\tau = 0 \qquad (27)$$

revealing that a stable behavior is ensured by a proper choice of the feedback gains k_P, k_D, k_F, k_I for the third equation. The following remarks are in order [10]:

- Stability is obtained independently of the actual normal direction to the plane; this essential feature of the parallel approach allows designing the controller based on the contact stiffness coefficient while the actual contact geometry is taken into account only at the planning level.
- The decoupled dynamics of the system (23) derives from structural properties of the parallel control scheme by virtue of the contact force measurement; this is different from the hybrid approach where a decoupled dynamics is imposed by the control law on the basis of the environment model.

4 Force Regulation and Position Tracking

The above parallel force/position control scheme is based on an inverse dynamics concept and, as such, is sensitive to the effects of unmodelled dynamics or disturbances.

4.1 Passivity-based control

For the motion control problem, it has been argued that *passivity-based* controllers are expected to have enhanced robustness with respect to inverse dynamics controllers, since they do not rely on the exact cancellation of nonlinear terms [1].

In the following, a passivity-based parallel control scheme is presented where, like in the case of motion control, the resulting control law is composed of a nonlinear model-based term and a linear compensator action. A functional expression of the reference vector is established to be used in the controller which is related both to the end-effector position error and to the contact force error. This is conceptually different from previous passivity-based hybrid force/position controllers [24] where each component of the reference vector is related either to a position error or to a force error in respect of the task space selection mechanism.

Consider the following control law

$$\boldsymbol{u} = \widehat{\boldsymbol{B}}(\boldsymbol{x})\dot{\boldsymbol{r}} + \widehat{\boldsymbol{C}}(\boldsymbol{x},\dot{\boldsymbol{x}})\boldsymbol{r} + \widehat{\boldsymbol{g}}(\boldsymbol{x}) - k_D(\dot{\boldsymbol{x}} - \boldsymbol{r}) + \boldsymbol{f}, \quad (28)$$

where the hats denote the estimates, $\boldsymbol{r}$ is a (3×1) *reference* vector, and $k_D > 0$.

Assuming that $\widehat{\boldsymbol{B}}$, $\widehat{\boldsymbol{C}}$, $\widehat{\boldsymbol{g}}$ have the same functional form of $\boldsymbol{B}$, $\boldsymbol{C}$, $\boldsymbol{g}$, the control law (28) can be rewritten, in view of (12), as

$$\boldsymbol{u} = \boldsymbol{Y}(\boldsymbol{x},\dot{\boldsymbol{x}},\boldsymbol{r},\dot{\boldsymbol{r}})\widehat{\boldsymbol{\pi}} - k_D(\dot{\boldsymbol{x}} - \boldsymbol{r}) + \boldsymbol{f}. \quad (29)$$

Setting

$$\boldsymbol{r} = \dot{\boldsymbol{x}} - \boldsymbol{e} \quad (30)$$

$$\dot{\boldsymbol{r}} = \ddot{\boldsymbol{x}} - \dot{\boldsymbol{e}} \quad (31)$$

where $\boldsymbol{e}$ is a suitable error vector, and combining (1) with (29) gives

$$\boldsymbol{B}(\boldsymbol{x})\dot{\boldsymbol{e}} + \boldsymbol{C}(\boldsymbol{x},\dot{\boldsymbol{x}})\boldsymbol{e} + k_D\boldsymbol{e} = \boldsymbol{Y}(\cdot)\widetilde{\boldsymbol{\pi}} \quad (32)$$

where $\widetilde{\boldsymbol{\pi}} = \widehat{\boldsymbol{\pi}} - \boldsymbol{\pi}$ is the parameter error vector.

From passivity theory and the skew symmetry of the matrix in (6), it is well known that if the mapping $-\boldsymbol{e} \mapsto \boldsymbol{Y}(\cdot)\widetilde{\boldsymbol{\pi}}$ is passive and $\boldsymbol{Y}(\cdot)\widetilde{\boldsymbol{\pi}}$ is bounded, then $\boldsymbol{e} \to \boldsymbol{0}$ as $t \to \infty$ [29]. Thus, the key point is to find an *error* vector $\boldsymbol{e}$ related to the force and position errors, so that the force and position errors are the outputs of an exponentially stable, strictly proper filter whose input is $\boldsymbol{e}$ [16].

Let then $\boldsymbol{x}_d$ denote the *time-varying* desired end-effector position with bounded derivatives up to the second order, and $\boldsymbol{f}_d$ denote the *constant* desired force. Imposing a constant force is not a limitation from a practical viewpoint, since typical contact tasks require force regulation to a desired amount.

A keen choice for the error vector in (32) is

$$\boldsymbol{e} = \big(s + \boldsymbol{k}_x(s)\big)\Delta\boldsymbol{x} + s^{-1}\boldsymbol{k}_f(s)\Delta\boldsymbol{f} \quad (33)$$

where s is the Laplace variable and $\boldsymbol{k}_x(s)$, $\boldsymbol{k}_f(s)$ denote the transfer functions of the position and force compensators to be designed. It is worth pointing out that, in view of (30) and (31), $\boldsymbol{k}_x(s)$ and $\boldsymbol{k}_f(s)$ in (33) must have relative degree greater than or equal to 0 for the implementation of the control law (28) with only measurements of position, velocity and force.

The first term on the right-hand side of (33) is the same as in typical passivity-based motion control schemes. The presence of an *integral* action in the second term is motivated by the desire to achieve a null steady-state force error at the expense of a finite steady-state position error along the constrained task space direction, on condition that a suitable choice of $\boldsymbol{k}_x(s)$ and $\boldsymbol{k}_f(s)$ is made. It is assumed that $\boldsymbol{k}_x(s)$ does not have poles at the origin.

By virtue of (14) and (16), Eq. (33) can be effectively decomposed into the component along $\boldsymbol{n}$ and the components on the contact plane. Hence, the analysis is reduced to the following three (one + two) equations:

$$e_n = (s + k_x(s))\Delta x_n + s^{-1}k_f(s)\Delta f_n \quad (34)$$

$$\boldsymbol{e}_t = (s + k_x(s))\Delta\boldsymbol{x}_t - s^{-1}k_f(s)\boldsymbol{f}_{dt}, \quad (35)$$

where, with obvious notation, Eq. (34) involves the normal components of the force and position vectors, while Eq. (35) involves the components on the plane. Notice that the contact model (14) and (16) implies $\boldsymbol{f}_t = \boldsymbol{0}$; hence it is reasonable to choose $\boldsymbol{f}_{dt} = \boldsymbol{0}$. As emphasized in the previous section, if the direction of $\boldsymbol{n}$ is unknown, then it is advisable to choose $\boldsymbol{f}_d = \boldsymbol{0}$.

The position along the normal direction is given by

$$x_n = k^{-1}\Delta f_n + k^{-1}f_{dn} + x_{On}. \quad (36)$$

From (34) and (36), it follows that

$$\Delta f_n = k\big(g(s)e_n + h(s)(x_{dn} - k^{-1} f_{dn} - x_{0n})\big) \quad (37)$$

where

$$g(s) = \frac{s}{s^2 + k_x(s)s + kk_f(s)} \quad (38)$$

$$h(s) = \frac{s\big(s + k_x(s)\big)}{s^2 + k_x(s)s + kk_f(s)}. \quad (39)$$

Further, Eq. (35) with $\boldsymbol{f}_{dt} = \mathbf{0}$ gives

$$\Delta \boldsymbol{x}_t = \frac{1}{s + k_x(s)} \boldsymbol{e}_t. \quad (40)$$

Notice that, since $k_x(s)$ and $k_f(s)$ have relative degree greater than or equal to 0, both $g(s)$ and $1/\big(s + k_x(s)\big)$ have relative degree 1, whereas $h(s)$ is proper. Therefore, $k_x(s)$ and $k_f(s)$ must be chosen so that the transfer functions $g(s)$ and $1/(s + k_x(s))$ are both exponentially stable. In that case, from (39), also $h(s)$ is exponentially stable.

On the other hand, the second term on the right-hand side of (37) plays the role of a disturbance on the force error along the normal. Since $h(s)$ possesses at least a zero at the origin, the effect of a constant disturbance is rejected.

It can be demonstrated that, in the *known parameter case* ($\tilde{\boldsymbol{\pi}} = \mathbf{0}$), if $k_x(s)$ and $k_f(s)$ are chosen so that both $g(s)$ and $1/(s + k_x(s))$ are exponentially stable and the poles of $k_x(s)$ and $sk_f(s)$ have all negative real parts, then $\Delta \boldsymbol{x}_t, \Delta \dot{\boldsymbol{x}}_t \to \mathbf{0}$ as $t \to \infty$ and all signals in the system remain bounded. Further, if x_{dn} is a constant, then $\Delta f_n \to 0$ as $t \to \infty$. The proof can be found in [39] and is based on typical passivity arguments.

On the other hand, in the *unknown parameter case* ($\tilde{\boldsymbol{\pi}} \neq \mathbf{0}$), it is sufficient to add to the control law (28), the parameter *adaptive* law

$$\dot{\hat{\boldsymbol{\pi}}} = -\boldsymbol{\Gamma}^{-1} \boldsymbol{Y}^{\mathrm{T}}(\cdot)\boldsymbol{e} \quad (41)$$

where $\boldsymbol{\Gamma}$ is a $(p \times p)$ symmetric, positive definite matrix. It can be shown that the same result is recovered as in the known parameter case.

The above results are based on a suitable choice of $k_x(s)$ and $k_f(s)$. The simplest choice is to take them as constants, i.e., $k_x(s) = \lambda_1$ and $k_f(s) = \lambda_2$ [37] with $\lambda_1, \lambda_2 > 0$. This ensures that the transfer functions (38) and (39) are always exponentially stable independently of k, although the actual value of the stiffness coefficient affects the performance of the system. By virtue of this choice, from (30) and (31) the reference vector $\boldsymbol{r}$ and its derivative become

$$\boldsymbol{r} = \dot{\boldsymbol{x}} - \boldsymbol{e} = \dot{\boldsymbol{x}}_d - \lambda_1 \Delta \boldsymbol{x} - \lambda_2 \int_0^t \Delta \boldsymbol{f} d\sigma \quad (42)$$

$$\dot{\boldsymbol{r}} = \ddot{\boldsymbol{x}} - \dot{\boldsymbol{e}} = \ddot{\boldsymbol{x}}_d - \lambda_1 \Delta \dot{\boldsymbol{x}} - \lambda_2 \Delta \boldsymbol{f}. \quad (43)$$

It can be shown that the control law (28) with (42) and (43) leads to the following equilibrium trajectory

$$\boldsymbol{x}_e = (\boldsymbol{I} - \boldsymbol{n}\boldsymbol{n}^{\mathrm{T}})\boldsymbol{x}_d + \boldsymbol{n}\boldsymbol{n}^{\mathrm{T}}\left(k^{-1}\boldsymbol{f}_d + \boldsymbol{x}_0\right) \quad (44)$$

$$\dot{\boldsymbol{x}}_e = (\boldsymbol{I} - \boldsymbol{n}\boldsymbol{n}^{\mathrm{T}})\dot{\boldsymbol{x}}_d = \dot{\boldsymbol{x}}_d \quad (45)$$

$$\ddot{\boldsymbol{x}}_e = (\boldsymbol{I} - \boldsymbol{n}\boldsymbol{n}^{\mathrm{T}})\ddot{\boldsymbol{x}}_d = \ddot{\boldsymbol{x}}_d \quad (46)$$

$$\boldsymbol{f}_e = k\boldsymbol{n}\boldsymbol{n}^{\mathrm{T}}(\boldsymbol{x}_e - \boldsymbol{x}_0) = \boldsymbol{f}_d. \quad (47)$$

on condition that x_{dn} is a constant. A Lyapunov stability proof of the closed-loop system around the above trajectory can be found in [38]. Notice that the above trajectory is the same that can be obtained with an inverse dynamics control with perfect dynamic compensation.

Notice that the above expressions (42) and (43) are similar in nature to the reference trajectories proposed in [42] and [2] which were derived in the joint space, though. Differently from those works, however, here no selection between force-controlled and position-controlled components of $\boldsymbol{r}$ and $\dot{\boldsymbol{r}}$ is accomplished by the control law (28). Interestingly enough, the control law (28) with (42) and (43) can be regarded as an extension of the impedance controller in [21] in that a desired force $\boldsymbol{f}_d$ different from zero can be specified.

Another simple choice for the force action is $k_f(s) = \lambda_2 + \lambda_3/s$ which determines an additional zero of $h(s)$ at the origin and thus ensures a null steady-state force error along the normal in case of a constant $\dot{x}_{dn}$. Nevertheless, the presence of a double integrator on the force error may lead to an oscillatory behavior in the force response. More complex choices for $k_x(s)$ and $k_f(s)$ are feasible if suitable filtering actions on the position and force errors are sought. In such cases more accurate estimates of the stiffness coefficient are required to tune the coefficients of the compensators.

In summary, both in the known and in the unknown parameter case, it is possible to design a passivity-based control scheme (with adaptive law) which guarantees *tracking* of the end-effector *position* along the unconstrained directions with *regulation* of the contact *force* along the constrained direction.

5 Force and Position Regulation

The parallel control scheme presented in the above section requires complete knowledge of the manipulator dynamic model in order to ensure tracking of the desired end-effector position along the constrained task space directions. Nevertheless, if the actual motion along the contact surface is not relevant but only a desired contact force is to be maintained, a computationally lighter control scheme can be devised which ensures *force regulation* along the constrained task space directions and *position regulation* along the unconstrained task space directions.

5.1 PID control

Consider the *constant* set points $\boldsymbol{x}_d$ and $\boldsymbol{f}_d = f_d\boldsymbol{n}$, and the control law

$$\boldsymbol{u} = k_P\Delta\boldsymbol{x} - k_D\dot{\boldsymbol{x}} + \widehat{\boldsymbol{g}}(\boldsymbol{x}) + \boldsymbol{f}_d + k_F\Delta\boldsymbol{f} + k_I\int_0^t \Delta\boldsymbol{f}d\sigma \tag{48}$$

where $k_P, k_D, k_F, k_I > 0$. This controller corresponds to position PD action + gravity compensation + desired force feedforward + force PI action. Notice that the use of gravity compensation is the only model-based requirement and is inherited from ordinary PD position control to recover steady-state position errors [43].

Remarkably, it can be shown that in the case of perfect gravity compensation ($\widehat{\boldsymbol{g}} = \boldsymbol{g}$) the equilibrium for the system (1) under control (48) is described by the same position in (44) and force in (47) [9].

The study of stability for the force/position regulation case requires methods from nonlinear systems theory. To this purpose, an energy-based argument inspired by the kind of Lyapunov functions used for stability of PID position control [3] can be pursued. It should be emphasized that the Lyapunov method is used only as a means to prove stability of the closed-loop system, and not to derive the control law in a constructive manner; the control law, in fact, has been postulated above on the basis of physical considerations related to the parallel approach in a problem of interaction with an elastically compliant planar surface.

The key point is to find a state description for the system which is suitably augmented to take into account the interaction force in respect of the constraints imposed by the contact. Such a description should lead to a Lyapunov function composed of a potential energy term related to the deviation from the equilibrium contact position, a kinetic energy term related to the system rate of motion, as well as a term related to the energy stored along the normal direction to the plane due to the integral force action. This is accomplished by considering the (7×1) state vector [11].

$$\boldsymbol{z} = (\,\dot{\boldsymbol{x}}^{\mathrm{T}} \quad \boldsymbol{\varepsilon}^{\mathrm{T}} \quad w\,)^{\mathrm{T}}, \tag{49}$$

where

$$\boldsymbol{\varepsilon} = \boldsymbol{x}_e - \boldsymbol{x} = \Delta\boldsymbol{x} + d\boldsymbol{n} \tag{50}$$

$$w = k^{-1}\boldsymbol{n}^{\mathrm{T}}\left(\int_0^t \Delta\boldsymbol{f}d\sigma - k_P k_I^{-1} d\boldsymbol{n}\right), \tag{51}$$

with

$$d = k^{-1}f_d + x_{0n} - x_{dn} \tag{52}$$

being a constant quantity taking into account the effects of the environment contact force and the desired force set point along the constrained task space direction. It is important to remark that $\boldsymbol{z} = \boldsymbol{0}$ corresponds to the equilibrium (44) and (47), as can be easily verified. Also, note that the following relations hold:

$$\boldsymbol{n}^{\mathrm{T}}\boldsymbol{\varepsilon} = k^{-1}\boldsymbol{n}^{\mathrm{T}}\Delta\boldsymbol{f} \tag{53}$$

$$\dot{\boldsymbol{\varepsilon}} = -\dot{\boldsymbol{x}} \tag{54}$$

$$\dot{w} = \boldsymbol{n}^{\mathrm{T}}\boldsymbol{\varepsilon}. \tag{55}$$

The augmented system described by (1), (54), and (55) under the control (48) with $\widehat{\boldsymbol{g}} = \boldsymbol{g}$ can be written in the standard compact homogeneous form

$$\dot{\boldsymbol{z}} = \boldsymbol{F}\boldsymbol{z} \tag{56}$$

where

$$\boldsymbol{F} = \begin{pmatrix} -\boldsymbol{B}^{-1}(\boldsymbol{C} + k_D\boldsymbol{I}) & \boldsymbol{B}^{-1}(k_P\boldsymbol{I} + k_F' k\,\boldsymbol{n}\boldsymbol{n}^{\mathrm{T}}) & k_I k\boldsymbol{B}^{-1}\boldsymbol{n} \\ -\boldsymbol{I} & \boldsymbol{O} & \boldsymbol{0} \\ \boldsymbol{0}^{\mathrm{T}} & \boldsymbol{n}^{\mathrm{T}} & 0 \end{pmatrix} \tag{57}$$

with $k_F' = 1 + k_F$. Notice that some handy reductions —using the structural properties of $\boldsymbol{K}$ in (16) and the definition of w in (51)— have been performed to derive (57).

On the basis of the above augmented state space description, suitable Lyapunov function candidates can be constructed to derive local stability results around the origin of the state space in (49). The key feature of such functions is the introduction of off-diagonal terms and positive constants which are remarkably not used by the control law. These constants serve as additional degrees of freedom to satisfy conditions on the feedback gains guaranteeing stability of the system (56) and (57). Two major results can be established.

- *Local asymptotic stability* can be demonstrated by choosing the following Lyapunov function:
$$V = \frac{1}{2}\boldsymbol{z}^{\mathrm{T}}\boldsymbol{P}\boldsymbol{z}, \tag{58}$$
where
$$\boldsymbol{P} = \begin{pmatrix} \boldsymbol{B} & -\rho\boldsymbol{B} & \boldsymbol{0} \\ -\rho\boldsymbol{B} & (k_P + \rho k_D)\boldsymbol{I} + k'_F k\,\boldsymbol{n}\boldsymbol{n}^{\mathrm{T}} & k_I k\,\boldsymbol{n} \\ \boldsymbol{0}^{\mathrm{T}} & k_I k\,\boldsymbol{n}^{\mathrm{T}} & \rho k_I k \end{pmatrix} \tag{59}$$
with $\rho > 0$.
- *Local exponential stability* can be demonstrated by choosing the following Lyapunov function:
$$W = \frac{1}{2}\boldsymbol{z}^{\mathrm{T}}\boldsymbol{Q}\boldsymbol{z}, \tag{60}$$
where
$$\boldsymbol{Q} = \begin{pmatrix} \boldsymbol{B} & -\delta\boldsymbol{B} & -\gamma\boldsymbol{B}\boldsymbol{n} \\ -\delta\boldsymbol{B} & (k_P + \delta k_D)\boldsymbol{I} + k'_F k\,\boldsymbol{n}\boldsymbol{n}^{\mathrm{T}} & (k_I k + \gamma k_D)\,\boldsymbol{n} \\ -\gamma\boldsymbol{n}^{\mathrm{T}}\boldsymbol{B} & (k_I k + \gamma k_D)\,\boldsymbol{n}^{\mathrm{T}} & \delta k_I k + \gamma(k_P + k'_F k) \end{pmatrix}, \tag{61}$$
with $\delta, \gamma > 0$.

The stability proof exploits the skew symmetry of the matrix in (6) and the properties (5), (7) and (8). Technical details are omitted for brevity and can be found in [12]. It is worth reporting here that k_P is not involved by the conditions on the feedback gains that guarantee local asymptotic stability, and then is available to meet further design requirements during the unconstrained phase of the task. On the other hand, local exponential stability is more demanding and in fact leads to more complex conditions on the feedback gains involving also k_P. Ensuring exponential stability, though, provides a performance measure of the system rate of convergence to the equilibrium.

In the case of imperfect gravity compensation ($\widehat{\boldsymbol{g}} \neq \boldsymbol{g}$), the equations of the system (1) under control (48) become
$$\dot{\boldsymbol{z}} = \boldsymbol{F}\boldsymbol{z} + \boldsymbol{\mu} \tag{62}$$
with $\boldsymbol{z}$ as in (49), $\boldsymbol{F}$ as in (57), and
$$\boldsymbol{\mu} = \begin{pmatrix} -\boldsymbol{B}^{-1}\boldsymbol{G}(\boldsymbol{x})\widetilde{\boldsymbol{\theta}} \\ \boldsymbol{0} \\ 0 \end{pmatrix} \tag{63}$$
where $\widetilde{\boldsymbol{\theta}} = \boldsymbol{\theta} - \widehat{\boldsymbol{\theta}}$ is the parameter error vector in (13).

Remarkably, local asymptotic stability holds also in the case of imperfect gravity compensation [35], and the proof exploits the properties in (9), (10), and (11). It can be shown that a different equilibrium is reached for the system (1) under the control (48) with $\widehat{\boldsymbol{g}} \neq \boldsymbol{g}$, i.e.,
$$\widehat{\boldsymbol{x}}_e = k^{-1}\boldsymbol{n}\boldsymbol{n}^{\mathrm{T}}(\boldsymbol{f}_d + k\boldsymbol{x}_0) \\ +(\boldsymbol{I} - \boldsymbol{n}\boldsymbol{n}^{\mathrm{T}})\big(\boldsymbol{x}_d - k_P^{-1}(\boldsymbol{g}(\widehat{\boldsymbol{x}}_e) - \widehat{\boldsymbol{g}}(\widehat{\boldsymbol{x}}_e)\big) \tag{64}$$
$$\boldsymbol{f}_e = \boldsymbol{f}_d. \tag{65}$$
In this case, the force set point is still attained while a different end-effector equilibrium position is reached compared to the case of perfect gravity compensation. More specifically, a comparison between (44) and (64) reveals that the components of $\boldsymbol{x}$ along the constrained task space direction $\boldsymbol{n}$ coincide, while the imperfect gravity compensation only affects the components of $\boldsymbol{x}$ along the unconstrained task space directions $\boldsymbol{t}_1, \boldsymbol{t}_2$.

In order to counteract imperfect estimation of the gravity term, the control law (48) can be made adaptive with respect to the vector of parameters in (13) by adding the parameter estimate update law
$$\dot{\widehat{\boldsymbol{\theta}}} = -\frac{1}{\nu}\boldsymbol{G}^{\mathrm{T}}(\boldsymbol{x})\left(\dot{\boldsymbol{x}} - \delta\Big(\Delta\boldsymbol{x} + k_I k_P^{-1}\int_0^t \Delta\boldsymbol{f}d\sigma\Big)\right) \tag{66}$$
with $\nu > 0$.

Local asymptotic stability around the original equilibrium (44) and (47) can be proven [36]. It is worth mentioning here that the proof is based on the Lyapunov function candidate
$$W' = \frac{1}{2}\boldsymbol{z}^{\mathrm{T}}\boldsymbol{Q}\boldsymbol{z} + \frac{1}{2}\nu\widetilde{\boldsymbol{\theta}}^{\mathrm{T}}\widetilde{\boldsymbol{\theta}}, \tag{67}$$
in which noticeably $\boldsymbol{Q}$ is the same as in (61). Differently from the local exponential stability

case, γ is not available to satisfy conditions on the feedback gains but it is keenly chosen to render the adaptation law a function of physically measurable quantities already used in the control law (48), i.e., position, velocity and force measurements.

5.2 Output feedback

The implementation of the control law (48) requires joint velocity measurements for the computation of the end-effector velocity vector $\dot{\boldsymbol{x}}$ through the differential kinematics equation (3). A number of robot manipulators are endowed with joint position sensors (encoders or resolvers) only, and joint velocities have to be reconstructed. Hence, it is worth considering the problem of designing a stable force/position regulator *without velocity measurements*.

The regulator presented above can be enhanced in an output feedback setting by relaxing the requirement of joint velocity measurements. Velocity can be reconstructed through a suitable linear filtering action, similarly to the case of motion controllers in [4, 30], on condition that an extra term is introduced in the control law.

It can be shown that the system remains asymptotically stable around the equilibrium (44) and (47) even if $\dot{\boldsymbol{x}}$ in the control law (48) is replaced with a vector $\boldsymbol{v}$ obtained by filtering the position vector $\boldsymbol{x}$, i.e.,

$$\boldsymbol{v} = \mathrm{diag}\left\{\frac{\beta s}{s+\alpha}\right\}\boldsymbol{x}, \tag{68}$$

where $\alpha, \beta > 0$ and s is the Laplace variable.

Consider the control law

$$\begin{aligned}\boldsymbol{u} = {} & k_P \Delta\boldsymbol{x} - k_D \boldsymbol{v} + \boldsymbol{g}(\boldsymbol{x}) \\ & + \boldsymbol{f}_d + k_F \Delta\boldsymbol{f} + k_I \int_0^t \Delta\boldsymbol{f}\, d\sigma - k_\phi \boldsymbol{\phi}\end{aligned} \tag{69}$$

where $k_\phi > 0$ and $\boldsymbol{\phi}$ is obtained as

$$\boldsymbol{\phi} = \mathrm{diag}\left\{\frac{\beta}{s+\alpha}\right\}\Delta\boldsymbol{f}. \tag{70}$$

Assuming $\boldsymbol{v}(0) = \boldsymbol{0}$ in (68) and $\boldsymbol{\phi}(0) = \beta\Delta\boldsymbol{f}(0)$ in (70), by virtue of (14), the following equality holds

$$\dot{\boldsymbol{\phi}} = -\boldsymbol{K}\boldsymbol{v}. \tag{71}$$

In order to study stability of the system (1) with (14) and (68)–(71), the same error vector as in (50) can be considered. Substituting (69) into (1), and accounting for (50) and (54), gives

$$\boldsymbol{B}(\boldsymbol{x})\ddot{\boldsymbol{\varepsilon}} + \boldsymbol{C}(\boldsymbol{x}, \dot{\boldsymbol{x}})\dot{\boldsymbol{\varepsilon}} + k_P\boldsymbol{\varepsilon} - k_D\boldsymbol{v} + y\boldsymbol{n} = \boldsymbol{0} \tag{72}$$

where

$$y = \boldsymbol{n}^{\mathrm{T}}\left(k_F'\Delta\boldsymbol{f} + k_I \int_0^t \Delta\boldsymbol{f}\, d\sigma - k_\phi\boldsymbol{\phi} + k_P d\boldsymbol{n}\right). \tag{73}$$

Differentiating (73) with respect to time, and taking into account (53), (54), (71), and (16), gives

$$\dot{y} = k\,\boldsymbol{n}^{\mathrm{T}}\left(k_F'\dot{\boldsymbol{\varepsilon}} + k_I\boldsymbol{\varepsilon} + k_\phi\boldsymbol{v}\right). \tag{74}$$

In view of (54), the time domain equivalent of (68) is

$$\dot{\boldsymbol{v}} = -\alpha\,\boldsymbol{v} - \beta\,\dot{\boldsymbol{\varepsilon}}. \tag{75}$$

Eqs. (72), (74), and (75) provide a state space representation of the closed-loop system (1), (68), (69), and (71) in terms of the (10×1) state vector

$$\boldsymbol{z}' = \left(\dot{\boldsymbol{\varepsilon}}^{\mathrm{T}} \quad \boldsymbol{\varepsilon}^{\mathrm{T}} \quad \boldsymbol{v}^{\mathrm{T}} \quad y\right)^{\mathrm{T}}. \tag{76}$$

The system equations can be rewritten in the standard compact homogeneous form:

$$\dot{\boldsymbol{z}}' = \boldsymbol{H}\boldsymbol{z}' \tag{77}$$

with

$$\boldsymbol{H} = \begin{pmatrix} -\boldsymbol{B}^{-1}\boldsymbol{C} & -k_P\boldsymbol{B}^{-1} & k_D\boldsymbol{B}^{-1} & -\boldsymbol{B}^{-1}\boldsymbol{n} \\ \boldsymbol{I} & \boldsymbol{O} & \boldsymbol{O} & \boldsymbol{0} \\ -\beta\boldsymbol{I} & \boldsymbol{O} & -\alpha\boldsymbol{I} & \boldsymbol{0} \\ kk_F'\boldsymbol{n}^{\mathrm{T}} & kk_I\boldsymbol{n}^{\mathrm{T}} & kk_\phi\boldsymbol{n}^{\mathrm{T}} & 0 \end{pmatrix}. \tag{78}$$

Local asymptotic stability of the system (77) and (78) around the origin of the state space can be demonstrated by choosing the following Lyapunov function:

$$V' = \frac{1}{2}\boldsymbol{z}'^{\mathrm{T}}\boldsymbol{S}\boldsymbol{z}' \tag{79}$$

where

$$\boldsymbol{S} = \begin{pmatrix} \boldsymbol{B} & \rho_1\boldsymbol{B} & \rho_1\rho_2\boldsymbol{B} & \boldsymbol{0} \\ \rho_1\boldsymbol{B} & k_P\boldsymbol{I} & \boldsymbol{O} & \boldsymbol{0} \\ \rho_1\rho_2\boldsymbol{B} & \boldsymbol{O} & \beta^{-1}k_D\boldsymbol{I} & \boldsymbol{0} \\ \boldsymbol{0}^{\mathrm{T}} & \boldsymbol{0}^{\mathrm{T}} & \boldsymbol{0}^{\mathrm{T}} & (k_F'k)^{-1} \end{pmatrix} \tag{80}$$

with

$$\rho_1 = k_F'^{-1}k_I \tag{81}$$

$$\rho_2 = k_P^{-1}k_D. \tag{82}$$

As in the previous case of full state feedback, the stability proof exploits the skew symmetry of the matrix in (6) and the properties (5), (7) and (8); in particular, the gain of the additional term in (69) is to be chosen as

$$k_\phi = \rho_2 k_I = k_P^{-1}k_D k_I. \tag{83}$$

Technical details are omitted for brevity and can be found in [40]. It is worth reporting here that the the conditions on the feedback gains that guarantee local asymptotic stability can be satisfied independently of the value of the environment stiffness k.

6 Future Work

This paper has surveyed a number of force/position control schemes for a robot manipulator interacting with a compliant planar surface. The schemes have been developed in the so-called parallel framework which exploits full force and position measurements. Contact force regulation along the constrained task space directions and either regulation or tracking of end-effector position along the unconstrained task space directions can be achieved.

It is worth pointing out that the advantageous feature of the presented schemes consists in the simplicity of the control laws which do not contain explicit information on the environment. It should be clear, however, that the geometry and the mechanical characteristics of the environment influence the performance of the manipulator during the contact under such kind of controllers. Extensive simulation results can be found in the papers in the list of References. On the other hand, initial experimental results of parallel control have been performed and will be described in forthcoming papers, e.g., in [13].

Two existing limitations of the approach, which anyhow are common also to the majority of force/position control schemes, are:

- All control schemes have been derived on the assumption that the manipulator's end effector has already come in contact with the environment and, once contact is established, it is not lost. Impact phenomena may occur which deserve careful consideration [19], and there is a need for global analysis of control schemes including the transition from non-contact into contact and vice versa, e.g., [27, 5].
- Only translational motion and linear force have been considered. Hence, there is a need of extending the control schemes to the full-dimensional task space. Preliminary results on a stability analysis of 6-degree-of-freedom parallel force/position control can be found in [17].

Acknowledgments

The contributions of Stefano Chiaverini, Lorenzo Sciavicco and Luigi Villani to the concepts described in this paper are gratefully acknowledged. This work was supported by *Ministero dell'Università e della Ricerca Scientifica e Tecnologica.*

References

[1] R.J. Anderson, "Passive computed torque algorithms for robots," *Proc. 28th IEEE Conf. on Decision and Control*, Tampa, FL, 1989, pp. 1638–1644.

[2] S. Arimoto, Y.H. Liu, and T. Naniwa, "Model-based adaptive hybrid control for geometrically constrained robots," *Proc. 1993 IEEE Int. Conf. on Robotics and Automation*, Atlanta, GA, 1993 pp. 618–623.

[3] S. Arimoto and F. Miyazaki, "Stability and robustness of PID feedback control for robot manipulators of sensory capability," in *Robotics Research: First Int. Symp.*, M. Brady and R.P. Paul (Eds.), MIT Press, Boston, 1984, pp. 783-799.

[4] H. Berghuis and H. Nijmeijer, "Global regulation of robots using only position measurements," *Systems and Control Lett.*, vol. 21, pp. 289–293, 1993.

[5] B. Brogliato and P. Orhant, "On the transition phase in robotics: Impact models, dynamics and control," *Proc. 1994 IEEE Int. Conf. on Robotics and Automation*, San Diego, CA, 1994, pp. 346–351.

[6] S. Chiaverini and L. Sciavicco, "Force/position control of manipulators in task space with dominance in force," *Proc. 2nd IFAC Symp. on Robot Control*, Karlsruhe, D, 1988, pp. 137–143.

[7] S. Chiaverini and L. Sciavicco, "Edge-following strategies using the parallel control formulation," *Proc. 1st IEEE Conf. on Control Applications*, Dayton, OH, 1992, pp. 31–36.

[8] S. Chiaverini and L. Sciavicco, "The parallel approach to force/position control of robotic manipulators," *IEEE Trans. on Robotics and Automation*, vol. 9, pp. 361–373, 1993.

[9] S. Chiaverini and B. Siciliano, "On the stability of a force/position control scheme for robot manipulators," *Proc. 3rd IFAC Symp. on Robot Control*, Vienna, A, 1991, pp. 183–188.

[10] S. Chiaverini and B. Siciliano, "Stability of parallel control," *1993 IEEE International Conference on Robotics and Automation*, Tutorial on 'Force and Contact Control in Robotic Systems: A Historical Perspective and Current Technologies', Atlanta GA, 1993, pp. 127–140.

[11] S. Chiaverini, B. Siciliano, and L. Villani, "A stable force/position controller for robot manipulators," *Proc. 31st IEEE Conf. on Decision and Control*, Tucson, AZ, 1992, pp. 1869–1874.

[12] S. Chiaverini, B. Siciliano, and L. Villani, "Force/position regulation of compliant robot manipulators," *IEEE Trans. on Automatic Control*, vol. 39, pp. 647–652, 1994.

[13] S. Chiaverini, B. Siciliano, and L. Villani, "Parallel force/position control schemes with experiments on an industrial robot manipulator," *13th IFAC World Congress*, San Francisco, CA, 1996, to appear.

[14] A. De Luca and C. Manes, "Hybrid force-position control for robots in contact with dynamic environments," *Proc. 3rd IFAC Symp. on Robot Control*, Vienna, A, 1991, pp. 377–382.

[15] J. De Schutter and H. Van Brussel, "Compliant robot motion II. A control approach based on external control loops," *Int. J. of Robotics Research*, vol. 7, no. 4, pp. 18–33, 1988.

[16] C.A. Desoer and M. Vidyasagar, *Feedback Systems: Input-Output Properties*, Academic Press, New York, NY, 1975.

[17] J.T. Gravdahl, O. Egeland, B. Siciliano, and S. Chiaverini, "Stability analysis of 6-dof force/position control for robot manipulators," *Proc. 33rd IEEE Conf. on Decision and Control*, Lake Buena Vista, FL, 1994, pp. 2408–2409.

[18] N. Hogan, "Impedance control: An approach to manipulation, Parts I–III," *ASME J. of Dynamic Systems, Measurement, and Control*, vol. 107, pp. 1–24, 1985.

[19] N. Hogan, "On the stability of manipulators performing contact tasks," *IEEE J. of Robotics and Automation*, vol. 4, pp. 677–686, 1988.

[20] H. Kazerooni, P.K. Houpt, and T.B. Sheridan, "Robust compliant motion for manipulators," *IEEE J.of Robotics and Automation*, vol. 2, pp. 83–105, 1986.

[21] R. Kelly, R. Carelli, M. Amestegui, and R. Ortega, "Adaptive impedance control of robot manipulators," *IASTED Int. J. of Robotics and Automation*, vol. 4, no. 3, pp. 134–141, 1989.

[22] O. Khatib, "A unified approach for motion and force control of robot manipulators: The operational space formulation," *IEEE J. of Robotics and Automation*, vol. 3, pp. 43–53, 1987.

[23] D. Koditschek, "Natural motion of robot arms," *Proc. 23rd IEEE Conf. on Decision and Control*, Las Vegas, NV, 1984, pp. 733–735.

[24] R. Lozano and B. Brogliato, "Adaptive hybrid force-position control for redundant manipulators," *IEEE Trans. on Automatic Control*, vol. 37, pp. 1501–1505, 1992.

[25] N.H. McClamroch and D. Wang, "Feedback stabilization and tracking of constrained robots," *IEEE Trans. on Automatic Control*, vol. 33, pp. 419–426, 1988.

[26] J.K. Mills and A.A. Goldenberg, "Force and position control of manipulators during constrained motion tasks," *IEEE Trans. on Robotics and Automation*, vol. 5, pp. 30–46, 1989.

[27] J.K. Mills and A.A. Goldenberg, "Control of robotic manipulators during general task execution: A discontinuous control approach," *Int. J. of Robotics Research*, vol. 12, pp. 146–163, 1993.

[28] G. Niemeyer and J.-J.E. Slotine, "Performance in adaptive manipulator control," *Int. J. of Robotics Research*, vol. 10, pp. 149–161, 1991.

[29] R. Ortega and M.W. Spong, "Adaptive motion control of rigid robots: a tutorial," *Automatica*, vol. 25, pp. 877–888, 1989.

[30] R. Ortega, A. Loria, and R. Kelly, "A semiglobal stable output feedback PI^{2}D regulator for robot manipulators," *IEEE Trans. on Automatic Control*, vol. 40, pp. 1432–1436, 1995.

[31] R. Paul and B. Shimano, "Compliance and control," *Proc. 1976 Joint Automatic Control Conf.*, San Francisco, CA, 1976, pp. 694–699.

[32] M.H. Raibert and J.J. Craig, "Hybrid position/force control of manipulators," *ASME J. of Dynamic Systems, Measurement, and Control*, vol. 103, pp. 126–133, 1981.

[33] J.K. Salisbury, "Active stiffness control of a manipulator in Cartesian coordinates," *Proc. 19th IEEE Conf. Decision and Control*, Albuquerque, NM, 1980, pp. 95–100.

[34] L. Sciavicco and B. Siciliano, *Modeling and Control of Robot Manipulators*, McGraw-Hill, New York, NY, 1996.

[35] B. Siciliano and L. Villani, "Force/position regulation of robot manipulators with gravity parameter adaptation," *Prepr. 12th World Congress of IFAC*, Sydney, AUS, 1993, vol. 7, pp. 403–406.

[36] B. Siciliano and L. Villani, "An adaptive force/position regulator for robot manipulators," *Int. J. of Adaptive Control and Signal Processing*, vol. 7, pp. 389–403, 1993.

[37] B. Siciliano and L. Villani, "A passivity-based force/position control scheme for robot manipulators," *Proc. 1994 IEEE Int. Conf. on Robotics and Automation*, San Diego, CA, 1994, pp. 3265–3270.

[38] B. Siciliano and L. Villani, "A Lyapunov-stable adaptive scheme for force regulation and motion control of robot manipulators," *Prepr. 5th IFAC Symp. on Adaptive Systems in Control and Signal Processing*, Budapest, H, 1995, pp. 311–316.

[39] B. Siciliano and L. Villani, "A passivity-based approach to force regulation and motion control of robot manipulators," *Automatica*, vol. 32, no. 3, 1996.

[40] B. Siciliano and L. Villani, "A force/position regulator for robot manipulators without velocity measurements," *1996 IEEE Int. Conf. on Robotics and Automation*, Minneapolis, MN, 1996, submitted.

[41] J.-J.-E. Slotine and W. Li, "Adaptive strategies in constrained manipulation," *Proc. 1987 IEEE Int. Conf. on Robotics and Automation*, Raleigh, NC, 1987, pp. 595–601.

[42] C.-Y. Su, T.-P. Leung, and Q.-J. Zhou, "Force/motion control of constrained robots using sliding mode," *IEEE Trans. on Automatic Control*, vol. 37, pp. 668–672, 1992.

[43] M. Takegaki and S. Arimoto, "A new feedback method for dynamic control of manipulators," *ASME J. of Dynamic Systems, Measurement, and Control*, vol. 102, pp. 119–125, 1981.

[44] P. Tomei, "Adaptive PD controller for robot manipulators," *IEEE Trans. on Robotics and Automation*, vol. 7, pp. 565–570, 1991.

[45] R. Volpe and P. Khosla, "A theoretical and experimental investigation of explicit force control strategies for manipulators," *IEEE Trans. on Automatic Control*, vol. 38, pp. 1634–1650, 1993.

[46] J.T. Wen and S.H. Murphy, "Stability analysis of position and force control for robot arms," *IEEE Trans. on Automatic Control*, vol. 36, pp. 365–371, 1991.

[47] D.E. Whitney, "Historical perspective and state of the art in robot force control," *Int. J. of Robotics Research*, vol. 6, no. 1, pp. 3–14, 1987.

[48] L.S. Wilfinger, J.T. Wen, and S.H. Murphy, "Integral force control with robustness enhancement," *IEEE Control Systems Mag.*, vol. 14, no. 1, pp. 31–40, 1994.

[49] B. Yao and M. Tomizuka, "Adaptive control of robot manipulators in constrained motion," *Proc. 1993 American Control Conf.*, San Francisco, CA, 1993, pp. 1128–1132.

[50] T. Yoshikawa, "Dynamic hybrid position/force control of robot manipulators—Description of hand constraints and calculation of joint driving force," *IEEE J. of Robotics and Automation*, vol. 3, pp. 386–392, 1987.

Modeling, Controllability and Vibration Suppression of 3D Flexible Robots

Masaru Uchiyama* and Atsushi Konno**
*Department of Aeronautics and Space Engineering, Tohoku University
Aramaki-aza-Aoba, Aoba-ku, Sendai 980-77, Japan
e-mail: uchiyama@space.mech.tohoku.ac.jp
**Department of Mechano-Informatics, University of Tokyo
Hongo 7-3-1, Bunkyo-ku, Tokyo 113, Japan
e-mail: konno@jsk.t.u-tokyo.ac.jp

Abstract

A robot with flexible structure called a flexible robot is necessitated. Among various flexible robots 3D (three dimensional) flexible robots are important in practice because most tasks done by robots are in three dimensional space. Modeling, controllability and vibration suppression control of the 3D flexible robot are exclusively discussed in this paper. A lumped-parameter technique using a so-called Holzer's method is presented and how simply the technique can make a practical model for the flexible robot is shown. Using the technique controllability of the flexible robot is analyzed and interesting uncontrollable configurations for an example are found. A model-based, and thus variable-gain, depending on robot configurations, vibration suppression control scheme is derived using the model. In the scheme feedback gains for the control are varied depending on the robot configuration. The results in this paper are all verified experimentally.

1 Introduction

Light-weight robots are attractive because they can be carried easily, move fast with low energy, touch objects softly, and therefore less dangerous. Most of the currently used industrial robots, however, have massive structures, looking very stiff, but are unable to realize quick movements, only requiring superfluous energy just to move themselves. The reason is because the designer wants to avoid the vibration and elastic deformation of the robots. However, this is not an effective solution. The advantages of the light-weight robots could be exploited after controlling the vibration and deformation. The easiness of carrying the robots is very suitable for applications of robots in space or on mobile platforms. The safety of motion is suitable for man-robot coexistence. The robot with light-weight structures is named as a "flexible robot."

In order to solve the vibration and deformation problems, a considerable number of studies have been devoted in the past decade (e.g. [1], [2], [3], [4], [5], [6]), forming a research field of flexible robots. Most papers, however, deal with flexible robots that make only planar motion. We call them a "2D (two dimensional) flexible robot." Only a few papers (e.g. [7], [8], [9], [10]) have dealt with flexible robots to work in three dimensional space, named as a "3D (three dimensional) flexible robot." The 3D flexible robot is practically very important because most tasks to be done by robots are in three dimensional space. The aim of this paper is to give solutions to some of the important problems for the control of 3D flexible robots: modeling, controllability and vibration suppression.

Modeling of 3D flexible robots is highly complicated. Using partial differential equations will be useless in practice. Some of the approximation methods for vibration analysis of flexible systems [11] have been applied to modeling of flexible robots. Sunada and Dubowsky applied finite element techniques to dynamic analysis of industrial robots [12], for example. This kind of modeling, however, being based on a distributed-parameter model, is too complicated to be used in control. Thus, there has been renewal of interest in lumped-parameter models in recent years. Yoshikawa, Murakami and Hosoda developed a very simple lumped-parameter model for a 3-DOF (degrees-of-freedom) 3D flexible robot [9]. Their modeling is simple enough but is specific to their robot. On the other hand, we have presented a more general model in which a so-called Holzer's method is effectively applied ([13] and also in this paper).

An issue which, as long as 2D flexible robots are concerned, has not attracted researchers but is very interesting and also very serious for 3D flexible robots is a controllability issue. The question is if the variables such as end-effector positions can be controlled by the actuators. Uchiyama and Jiang have proposed a concept of compensability which represents if the end-effector position errors due to the elastic deformation can be compensated by joint displacements [14]. They analyzed the compensability of 3D flexible robots and found uncompensable configurations. However, their study is only for static case. Tosunoglu, Lin and Tesar discussed accessibility and tried to relate it to controllability of flexible robots [15]. But their accessibility concept is not successful in doing so. On the other hand, we discuss the controllability using the lumped-parameter model and find interesting uncontrollable configurations in the midst of workspace [16], [17], [18]. Also we propose modal accessibility to explain uncontrollable configurations.

Configuration dependency which is noted in the analysis of controllability also plays an important role in control of structural vibrations of flexible robots. Some researchers have noticed that variable gain control is effective or even mandatory for the control of 3D flexible robots [7], [19]. This is due to the non-collocated actuators to control the vibrations. A vibration control scheme, using linear quadratic regulators, has been proposed as a solution for this problem [9]; however, state feedback gains must be computed for various configurations, so that the gain table must be quite large in the case of robots with many DOF. An alternative solution that we have proposed is a configuration-dependent variable-gain vibration-suppression control scheme using our lumped-parameter model [19], [20], [21], [22]. The scheme computes the feedback gain for vibration suppression depending upon the configuration at each sampling time, therefore, no gain table is needed, and the scheme is very effective for 3D flexible robots. The control scheme is examined in a two-flexible-link seven-joint type robot through experiments.

This paper is organized as follows: First, we present the lumped-parameter modeling technique in Section 2. Exploiting this technique, in Section 3, we present controllability analysis. In Section 4, based on the lumped-parameter model, we present a vibration suppression control scheme with experimental results, before concluding this paper and giving future research issues in Section 5.

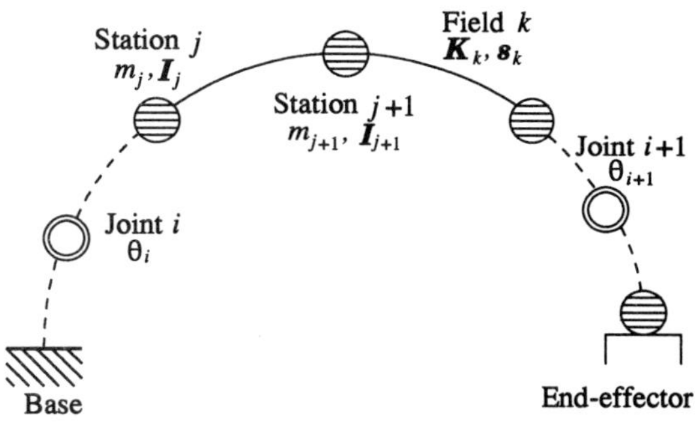

Figure 1: A conceptual sketch of the Holzer's modeling of the flexible robot.

2 Holzer's Modeling

2.1 Modeling

One of the lumped-parameter modeling techniques for the flexible structure is a so-called Holzer's method [11]. The method is simple enough to be applied to modeling of 3D flexible robots. The modeling using other techniques such as an assumed modes method is usually very tough, that is, too complicated to be practical.

In the Holzer's method, a flexible structure is modeled by lumped-masses and massless springs. The lumped-masses are called as "stations," while the massless springs are called as "fields." In applying this method to flexible robots, difference between structures and robots has to be carefully considered. The flexible robot can be regarded, in a sense, as a flexible structure although it is a mechanism like other robots if we disregard its structural flexibility. To deal with this aspect we add "joints" to the conventional Holzer's model. Figure 1 shows a conceptual sketch of the Holzer's modeling of the flexible robot.

Important parameters in this modeling are, as shown in Figure 1, the joint angle θ_i at the i-th joint, the mass m_j and the inertia tensor $\boldsymbol{I}_j$ at the j-th station, and the spring constant $\boldsymbol{K}_k$ and

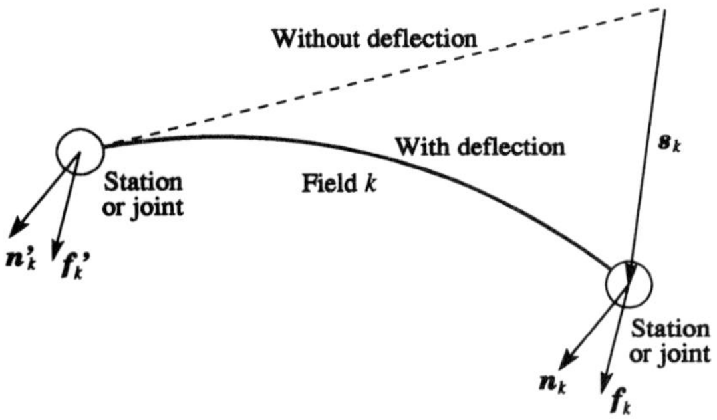

Figure 2: Forces, moments and deflections at the k-th field.

the deflections $\boldsymbol{s}_k$ of the k-th field. We define the joint angle vector $\boldsymbol{\theta}$ as

$$\boldsymbol{\theta} = \left[\begin{array}{cccc} \theta_1 & \theta_2 & \cdots & \theta_n \end{array} \right]^T, \tag{1}$$

and the inertia matrix $\boldsymbol{R}_j$ as

$$\boldsymbol{R}_j = \left[\begin{array}{cc} m_j \boldsymbol{E}_3 & \boldsymbol{0} \\ \boldsymbol{0} & \boldsymbol{I}_j \end{array} \right], \tag{2}$$

where $\boldsymbol{E}_3$ is the 3×3 unit matrix.

Figure 2 shows the forces and moments on the k-th field and its deflection due to them. The field deflection vector $\boldsymbol{s}_k$ generally consists of six components: three translational and three rotational deflections. The force $\boldsymbol{f}_k$ and the moment $\boldsymbol{n}_k$ at one end of the k-th field is given by

$$\left[\begin{array}{c} \boldsymbol{f}_k \\ \boldsymbol{n}_k \end{array} \right] = \boldsymbol{K}_k \boldsymbol{s}_k. \tag{3}$$

The force $\boldsymbol{f}'_k$ and the moment $\boldsymbol{n}'_k$ at the other end can be easily calculated from $\boldsymbol{f}_k$ and $\boldsymbol{n}_k$. The vector $\boldsymbol{e}$ to represent the all components of the field deflections is given by

$$\begin{aligned} \boldsymbol{e} &= \left[\begin{array}{cccc} \boldsymbol{s}_1^T & \boldsymbol{s}_2^T & \cdots & \boldsymbol{s}_q^T \end{array} \right]^T \\ &= \left[\begin{array}{cccc} e_1 & e_2 & \cdots & e_m \end{array} \right]^T. \end{aligned} \tag{4}$$

2.2 Kinematics

In our lumped-parameter model the motion, both flexible and rigid, of the robot is represented by the motion of the all stations. Kinematic equations to describe the motion are, therefore, those to obtain positions, velocities, and accelerations of each station. Let $\boldsymbol{p}_j$ and $\boldsymbol{\nu}_j$ represent position and velocity vectors of the j-th station, respectively. Each vector has three translational and three rotational components and they are represented by

$$\boldsymbol{p}_j = \left[\begin{array}{cccc} \boldsymbol{r}_j^T & \alpha_j & \beta_j & \gamma_j \end{array} \right]^T, \tag{5}$$

$$\boldsymbol{\nu}_j = \left[\begin{array}{c} \boldsymbol{v}_j \\ \boldsymbol{\omega}_j \end{array} \right], \tag{6}$$

respectively, where $\boldsymbol{r}_j$ represents the position of the j-th station, α_j, β_j and γ_j are the orientation angles of the j-th station, and $\boldsymbol{v}_j$, $\boldsymbol{\omega}_j$ are the translational and rotational velocities of the j-th station, respectively. Calculation of $\boldsymbol{p}_j$ from $\boldsymbol{\theta}$ and $\boldsymbol{e}$, and $\boldsymbol{\nu}_j$ from $\boldsymbol{\theta}$, $\boldsymbol{e}$, $\dot{\boldsymbol{\theta}}$ and $\dot{\boldsymbol{e}}$ is not difficult after having common knowledge of kinematic analysis of rigid robots. Particularly, the velocity $\boldsymbol{\nu}_j$ is calculated by the linear equation:

$$\boldsymbol{\nu}_j = \boldsymbol{J}_{\theta j} \dot{\boldsymbol{\theta}} + \boldsymbol{J}_{ej} \dot{\boldsymbol{e}}, \tag{7}$$

using the Jacobian matrices:

$$\begin{aligned} \boldsymbol{J}_{\theta j} &= \left[\begin{array}{c} \boldsymbol{J}_{\theta j}^{v} \\ \boldsymbol{J}_{\theta j}^{\omega} \end{array} \right] \\ &= \left[\begin{array}{cccccc} \boldsymbol{j}_{\theta j1} & \boldsymbol{j}_{\theta j2} & \cdots & \boldsymbol{j}_{\theta jN_j} & \boldsymbol{0} \cdots \boldsymbol{0} \end{array} \right], \end{aligned} \tag{8}$$

$$\begin{aligned} \boldsymbol{J}_{ej} &= \left[\begin{array}{c} \boldsymbol{J}_{ej}^{v} \\ \boldsymbol{J}_{ej}^{\omega} \end{array} \right] \\ &= \left[\begin{array}{cccccc} \boldsymbol{j}_{ej1} & \boldsymbol{j}_{ej2} & \cdots & \boldsymbol{j}_{ejM_j} & \boldsymbol{0} \cdots \boldsymbol{0} \end{array} \right], \end{aligned} \tag{9}$$

where N_j and M_j are the highest joint number and the highest component number of the field deflections before the j-th station. Equation (7) can be more concisely represented as

$$\boldsymbol{\nu}_j = \boldsymbol{J}_j \dot{\boldsymbol{q}}, \tag{10}$$

where

$$\boldsymbol{J}_j = \left[\begin{array}{cc} \boldsymbol{J}_{\theta j} & \boldsymbol{J}_{ej} \end{array} \right], \tag{11}$$

$$\dot{\boldsymbol{q}} = \left[\begin{array}{c} \dot{\boldsymbol{\theta}} \\ \dot{\boldsymbol{e}} \end{array} \right]. \tag{12}$$

2.3 Dynamics

We derive the equations of motion using Lagrange's equations. First, the kinetic energy T and the potential energy U of the robot are given by

$$T = \sum_{i=1}^{n} \frac{1}{2} I_{ai} \dot{\theta}_i^2 + \sum_{j=1}^{p} \frac{1}{2} \boldsymbol{\nu}_j^T \boldsymbol{R}_j \boldsymbol{\nu}_j, \tag{13}$$

$$U = \sum_{k=1}^{q} \frac{1}{2} \boldsymbol{s}_k^T \boldsymbol{K}_k \boldsymbol{s}_k - \sum_{j=1}^{p} \boldsymbol{r}_j^T m_j \boldsymbol{g} = \frac{1}{2} \boldsymbol{e}^T \boldsymbol{K} \boldsymbol{e} - \sum_{j=1}^{p} \boldsymbol{r}_j^T m_j \boldsymbol{g}, \quad (14)$$

respectively, where

$$\boldsymbol{K} = \begin{bmatrix} \boldsymbol{K}_1 & & \boldsymbol{0} \\ & \ddots & \\ \boldsymbol{0} & & \boldsymbol{K}_q \end{bmatrix}, \quad (15)$$

the vector $\boldsymbol{g}$ represents the acceleration of gravity, I_{ai} is the moment of inertia of the i-th motor rotor, and p, q are the numbers of stations and fields, respectively. The kinetic energy of the motor rotor is approximated by the one for the motor which does not change its position and orientation. The Lagrangian is

$$\begin{aligned} L &= T - U \\ &= \sum_{i=1}^{n} \frac{1}{2} I_{ai} \dot{\theta}_i^2 + \sum_{j=1}^{p} \left(\frac{1}{2} \boldsymbol{\nu}_j^T \boldsymbol{R}_j \boldsymbol{\nu}_j + \boldsymbol{r}_j^T m_j \boldsymbol{g} \right) \\ &\quad - \frac{1}{2} \boldsymbol{e}^T \boldsymbol{K} \boldsymbol{e}. \end{aligned} \quad (16)$$

The Lagrange's equations are given by

$$\tau_i = \frac{d}{dt}\left(\frac{\partial L}{\partial \dot{\theta}_i}\right) - \frac{\partial L}{\partial \theta_i} \quad (i = 1,\ 2,\ \cdots,\ n), \quad (17)$$

$$0 = \frac{d}{dt}\left(\frac{\partial L}{\partial \dot{e}_l}\right) - \frac{\partial L}{\partial e_l} \quad (l = 1,\ 2,\ \cdots,\ m). \quad (18)$$

We call the first equation, the equation of link motion and the second, the equation of field motion. The first equation represents the rigid motion of the links and the second the elastic motion of the fields, that is, the link vibration. Substituting Eq. (16) into Eqs. (17) and (18), we obtain

$$\begin{aligned} \boldsymbol{\tau} &= \boldsymbol{R}_a \ddot{\boldsymbol{\theta}} + \left(\sum_{j=1}^{p} \boldsymbol{J}_{\theta j}^T \boldsymbol{R}_j \boldsymbol{J}_{\theta j} \right) \ddot{\boldsymbol{\theta}} \\ &\quad + \left(\sum_{j=1}^{p} \boldsymbol{J}_{\theta j}^T \boldsymbol{R}_j \boldsymbol{J}_{ej} \right) \ddot{\boldsymbol{e}} \\ &\quad + \frac{d}{dt} \left(\sum_{j=1}^{p} \boldsymbol{J}_{\theta j}^T \boldsymbol{R}_j \boldsymbol{J}_j \right) \dot{\boldsymbol{q}} \\ &\quad - \frac{1}{2} \frac{\partial}{\partial \boldsymbol{\theta}} \sum_{j=1}^{p} \left(\dot{\boldsymbol{q}}^T \boldsymbol{J}_j^T \boldsymbol{R}_j \boldsymbol{J}_j \dot{\boldsymbol{q}} \right) \\ &\quad - \sum_{j=1}^{p} \left(\boldsymbol{J}_{\theta j}^T m_j \boldsymbol{g}_6 \right), \end{aligned} \quad (19)$$

$$\begin{aligned} \boldsymbol{0} &= \left(\sum_{j=1}^{p} \boldsymbol{J}_{ej}^T \boldsymbol{R}_j \boldsymbol{J}_{\theta j} \right) \ddot{\boldsymbol{\theta}} + \left(\sum_{j=1}^{p} \boldsymbol{J}_{ej}^T \boldsymbol{R}_j \boldsymbol{J}_{ej} \right) \ddot{\boldsymbol{e}} \\ &\quad + \frac{d}{dt} \left(\sum_{j=1}^{p} \boldsymbol{J}_{ej}^T \boldsymbol{R}_j \boldsymbol{J}_j \right) \dot{\boldsymbol{q}} \\ &\quad - \frac{1}{2} \frac{\partial}{\partial \boldsymbol{e}} \sum_{j=1}^{p} \left(\dot{\boldsymbol{q}}^T \boldsymbol{J}_j^T \boldsymbol{R}_j \boldsymbol{J}_j \dot{\boldsymbol{q}} \right) \\ &\quad - \sum_{j=1}^{p} \left(\boldsymbol{J}_{ej}^T m_j \boldsymbol{g}_6 \right) + \boldsymbol{K} \boldsymbol{e}, \end{aligned} \quad (20)$$

respectively, where

$$\boldsymbol{R}_a = \begin{bmatrix} I_{a1} & & 0 \\ & \ddots & \\ 0 & & I_{an} \end{bmatrix}. \quad (21)$$

Equations (19) and (20) are concisely represented as

$$\begin{aligned} \boldsymbol{\tau} &= \boldsymbol{M}_{11}(\boldsymbol{\theta}, \boldsymbol{e}) \ddot{\boldsymbol{\theta}} + \boldsymbol{M}_{12}(\boldsymbol{\theta}, \boldsymbol{e}) \ddot{\boldsymbol{e}} \\ &\quad + \boldsymbol{h}_1(\boldsymbol{\theta}, \dot{\boldsymbol{\theta}}, \boldsymbol{e}, \dot{\boldsymbol{e}}) + \boldsymbol{g}_1(\boldsymbol{\theta}, \boldsymbol{e}), \end{aligned} \quad (22)$$

$$\begin{aligned} \boldsymbol{0} &= \boldsymbol{M}_{21}(\boldsymbol{\theta}, \boldsymbol{e}) \ddot{\boldsymbol{\theta}} + \boldsymbol{M}_{22}(\boldsymbol{\theta}, \boldsymbol{e}) \ddot{\boldsymbol{e}} \\ &\quad + \boldsymbol{h}_2(\boldsymbol{\theta}, \dot{\boldsymbol{\theta}}, \boldsymbol{e}, \dot{\boldsymbol{e}}) + \boldsymbol{g}_2(\boldsymbol{\theta}, \boldsymbol{e}) + \boldsymbol{K} \boldsymbol{e}, \end{aligned} \quad (23)$$

respectively.

3 Controllability Analysis

3.1 Controllability Matrix

For some 3D flexible robots the structural vibration controllability is configuration-dependent. Such robots, therefore, may have some vibration uncontrollable configurations. To analyze this we derive, first, a controllability matrix using the equations of motion derived in the previous section.

In this analysis we assume that the robot is stationary at a certain point and its structural vibration is excited. Therefore, we have $\boldsymbol{h}_1 = \boldsymbol{0}$ and $\boldsymbol{h}_2 = \boldsymbol{0}$. Also, we assume that $\boldsymbol{e}$ is small enough so that $\boldsymbol{M}_{ij}(\boldsymbol{\theta}, \boldsymbol{e})$ and $\boldsymbol{g}_i(\boldsymbol{\theta}, \boldsymbol{e})$ for $i = 1, 2$ and $j = 1, 2$ can be approximated by $\boldsymbol{M}_{ij}(\boldsymbol{\theta}, \boldsymbol{0}) = \boldsymbol{M}_{ij}(\boldsymbol{\theta})$ and $\boldsymbol{g}_i(\boldsymbol{\theta}, \boldsymbol{0}) = \boldsymbol{g}_i(\boldsymbol{\theta})$, respectively. Hence, from Eqs. (22) and (23)

$$\begin{bmatrix} \boldsymbol{\tau} \\ \boldsymbol{0} \end{bmatrix} = \begin{bmatrix} \boldsymbol{M}_{11}(\boldsymbol{\theta}) & \boldsymbol{M}_{12}(\boldsymbol{\theta}) \\ \boldsymbol{M}_{21}(\boldsymbol{\theta}) & \boldsymbol{M}_{22}(\boldsymbol{\theta}) \end{bmatrix} \begin{bmatrix} \ddot{\boldsymbol{\theta}} \\ \ddot{\boldsymbol{e}} \end{bmatrix}$$

$$+\begin{bmatrix} 0 & 0 \\ 0 & \boldsymbol{K} \end{bmatrix}\begin{bmatrix} \boldsymbol{\theta} \\ \boldsymbol{e} \end{bmatrix}+\begin{bmatrix} \boldsymbol{g}_1(\boldsymbol{\theta}) \\ \boldsymbol{g}_2(\boldsymbol{\theta}) \end{bmatrix} \tag{24}$$

is obtained. Assuming $\ddot{\boldsymbol{\theta}} = \boldsymbol{0}$ and $\ddot{e} = \boldsymbol{0}$ we have an equilibrium point as

$$\begin{bmatrix} \boldsymbol{\tau}_o \\ \boldsymbol{0} \end{bmatrix} = \begin{bmatrix} 0 & 0 \\ 0 & \boldsymbol{K} \end{bmatrix}\begin{bmatrix} \boldsymbol{\theta}_o \\ \boldsymbol{e}_o \end{bmatrix}+\begin{bmatrix} \boldsymbol{g}_1(\boldsymbol{\theta}_o) \\ \boldsymbol{g}_2(\boldsymbol{\theta}_o) \end{bmatrix}. \tag{25}$$

Now defining small deviations from the equilibrium point as

$$\Delta\boldsymbol{\theta} = \boldsymbol{\theta} - \boldsymbol{\theta}_o, \tag{26}$$

$$\Delta\boldsymbol{e} = \boldsymbol{e} - \boldsymbol{e}_o = \boldsymbol{e} + \boldsymbol{K}^{-1}\boldsymbol{g}_2(\boldsymbol{\theta}_o), \tag{27}$$

$$\Delta\boldsymbol{\tau} = \boldsymbol{\tau} - \boldsymbol{\tau}_o = \boldsymbol{\tau} - \boldsymbol{g}_1(\boldsymbol{\theta}_o), \tag{28}$$

and assuming that $\boldsymbol{g}_1(\boldsymbol{\theta}) = \boldsymbol{g}_1(\boldsymbol{\theta}_o)$ and $\boldsymbol{g}_2(\boldsymbol{\theta}) = \boldsymbol{g}_2(\boldsymbol{\theta}_o)$, we have, from Eqs. (24) and (25), configuration-dependent linear equations of motion as

$$\begin{aligned}\begin{bmatrix} \Delta\ddot{\boldsymbol{\theta}} \\ \Delta\ddot{e} \end{bmatrix} &= \begin{bmatrix} \boldsymbol{M}_{11} & \boldsymbol{M}_{12} \\ \boldsymbol{M}_{21} & \boldsymbol{M}_{22} \end{bmatrix}^{-1}\begin{bmatrix} \Delta\boldsymbol{\tau} \\ -\boldsymbol{K}\Delta\boldsymbol{e} \end{bmatrix} \\ &= \begin{bmatrix} \boldsymbol{X}_{11} & \boldsymbol{X}_{12} \\ \boldsymbol{X}_{21} & \boldsymbol{X}_{22} \end{bmatrix}\begin{bmatrix} \Delta\boldsymbol{\tau} \\ -\boldsymbol{K}\Delta\boldsymbol{e} \end{bmatrix},\end{aligned} \tag{29}$$

where $\boldsymbol{X}_{11}, \boldsymbol{X}_{12}, \boldsymbol{X}_{21}, \boldsymbol{X}_{22}$ are $n\times n$, $n\times m$, $m\times n$, $m\times m$ matrices, respectively. From Eq. (29) the equation to represent elastic motion, namely, vibration is separated as

$$\Delta\ddot{e} = -\boldsymbol{X}_{22}\boldsymbol{K}\Delta\boldsymbol{e} + \boldsymbol{X}_{21}\Delta\boldsymbol{\tau}. \tag{30}$$

Rewriting this equation in a state-space form we have

$$\begin{bmatrix} \Delta\dot{e} \\ \Delta\ddot{e} \end{bmatrix} = \begin{bmatrix} \boldsymbol{0} & \boldsymbol{I} \\ -\boldsymbol{X}_{22}\boldsymbol{K} & \boldsymbol{0} \end{bmatrix}\begin{bmatrix} \Delta\boldsymbol{e} \\ \Delta\dot{e} \end{bmatrix}+\begin{bmatrix} \boldsymbol{0} \\ \boldsymbol{X}_{21} \end{bmatrix}\Delta\boldsymbol{\tau}. \tag{31}$$

The controllability matrix for this is

$$\boldsymbol{G}_c = [\ \boldsymbol{B} \quad \boldsymbol{AB} \quad \cdots \quad \boldsymbol{A}^{2m-1}\boldsymbol{B}\ ], \tag{32}$$

where

$$\boldsymbol{A} = \begin{bmatrix} \boldsymbol{0} & \boldsymbol{I} \\ -\boldsymbol{X}_{22}\boldsymbol{K} & \boldsymbol{0} \end{bmatrix}, \quad \boldsymbol{B} = \begin{bmatrix} \boldsymbol{0} \\ \boldsymbol{X}_{21} \end{bmatrix}. \tag{33}$$

If

$$\text{rank } \boldsymbol{G}_c < 2m, \tag{34}$$

the vibration is uncontrollable. It should be noted that the controllability matrix $\boldsymbol{G}_c$ is configuration dependent, that is, a function of $\boldsymbol{\theta}$.

3.2 Modal Accessibility

To see more about the controllability we analyze modal accessibility. To do so we transform Eq. (30) into a set of modal equations. Solving the eigenvalue problem:

$$\boldsymbol{X}_{22}\boldsymbol{K}\boldsymbol{\Phi}_r = \Omega_r\boldsymbol{\Phi}_r \quad (r = 1, \cdots, m), \tag{35}$$

where $\Omega_1 < \Omega_2 < \cdots < \Omega_m$, we have

$$\boldsymbol{\Phi} = [\ \boldsymbol{\Phi}_1 \quad \boldsymbol{\Phi}_2 \quad \cdots \quad \boldsymbol{\Phi}_m\ ]. \tag{36}$$

For simplicity we assume that $\Omega_r \neq \Omega_s$ for $r \neq s$. Using $\boldsymbol{\Phi}$ we consider

$$\Delta\boldsymbol{e}^* = \boldsymbol{\Phi}^{-1}\Delta\boldsymbol{e}. \tag{37}$$

We assume that $\boldsymbol{\Phi}$ is dependent only on $\boldsymbol{\theta}$ and not on time because $\dot{\boldsymbol{\theta}}$ is very small. Then, substituting Eq. (37) into Eq. (30), we have

$$\Delta\ddot{e}^* = \boldsymbol{\Omega}\Delta\boldsymbol{e}^* + \boldsymbol{\Gamma}\Delta\boldsymbol{\tau}, \tag{38}$$

where

$$\begin{aligned}\boldsymbol{\Omega} &= -\boldsymbol{\Phi}^{-1}\boldsymbol{X}_{22}\boldsymbol{K}\boldsymbol{\Phi} \\ &= -\text{diag}[\,\Omega_1, \cdots, \Omega_m], \end{aligned} \tag{39}$$

$$\boldsymbol{\Gamma} = \boldsymbol{\Phi}^{-1}\boldsymbol{X}_{21}. \tag{40}$$

It is obvious from Eq. (38) that if $\boldsymbol{\Gamma}$ has some rows which consist only of 0s, the vibration modes corresponding to these rows are uncontrollable. The modal accessibility is defined using the index

$$a_r(\boldsymbol{\theta}) = \sqrt{\gamma_r\gamma_r^T}, \tag{41}$$

where γ_r is the r-th row vector of $\boldsymbol{\Gamma}$; if $a_r = 0$ the r-th mode is inaccessible.

3.3 Experiments

To clarify our discussion we consider the horizontal motion of our experimental flexible robot called FLEBOT II (FLExible roBOT II) with two elastic links and three actuated joints (Figure 3). Mechanical parameters of the robot are presented in Table 1. The robot has a structure similar to those of many industrial robots (like PUMA) and space robots (like Shuttle Remote Manipulator). Therefore, the results here is of practical value.

FLEBOT II is modeled by lumped-masses and springs as shown in Figure 4. Joint 1, mass 1 and joint 2 are located at the same point while mass 2 and joint 3 are at the same point. Horizontal deflections of link 2 and link 3 are denoted as δ_{z2}

Table 1: Mechanical parameters of FLEBOT Ⅱ.

Parameter	Notation	Value
Length of the links	l_2 [m]	0.50
	l_3 [m]	0.44
Stiffness of the links	E_2I_2 [Nm^2]	41.81
	E_3I_3 [Nm^2]	13.23
Joint masses	m_2 [kg]	1.5
	m_3 [kg]	0.3

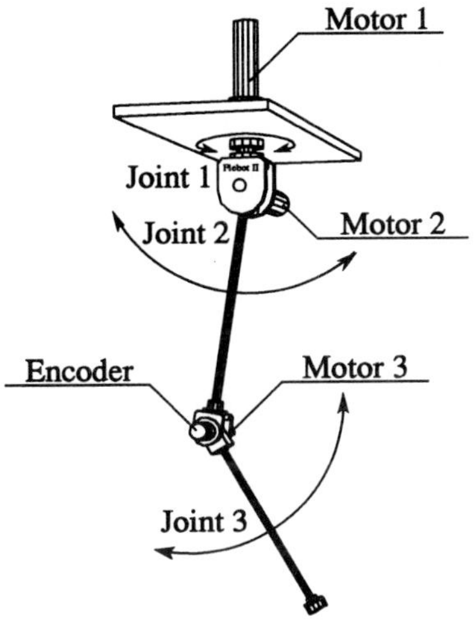

Figure 3: Overview of the experimental flexible robot FLEBOT Ⅱ.

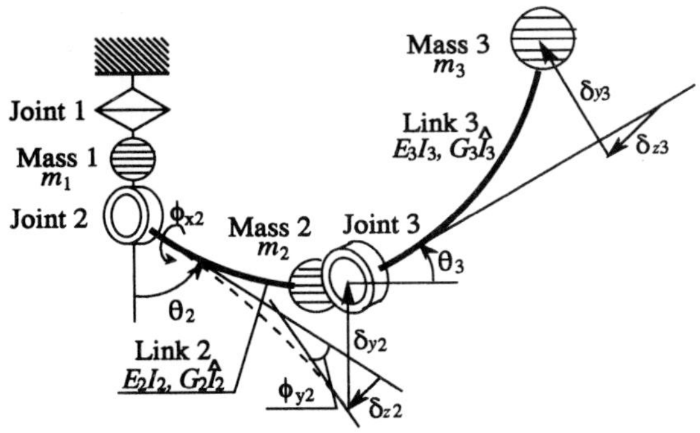

Figure 4: Lumped-mass and spring model of FLEBOT Ⅱ.

and δ_{z3}, respectively. The elastic angular deflections ϕ_{x2} and ϕ_{y2} can also be considered as a part of a deflection vector, but if the moments of inertia of the lumped-masses are small enough to be ignored, ϕ_{x2} and ϕ_{y2} are dependent on δ_{z2} and δ_{z3} [9]. For simplicity, we ignore the moments of inertia of the lumped-masses.

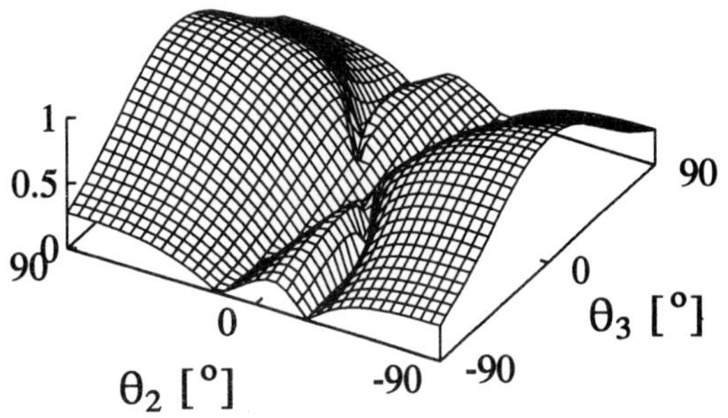

Figure 5: Minimum singular value of the controllability matrix.

Table 2: Modal accessibility indices and natural frequencies.

	Point A	Point B	Point C
a_1 [N^{-1}s^{-2}]	0.314	0.015	0.804
a_2 [N^{-1}s^{-2}]	0.454	0.667	0.003
f_1 [Hz]	3.77	3.60	4.55
f_2 [Hz]	6.44	6.68	6.24

With this model the controllability matrix is a function of the joint angles θ_2 and θ_3. Figure 5 shows the minimum singular value of the controllability matrix. It is seen that there are two continuous lines of uncontrollable configurations in the workspace. One is from the quadrant $\theta_2 > 0$, $\theta_3 < 0$ to the quadrant $\theta_2 < 0$, $\theta_3 > 0$, and the other is from the quadrant $\theta_2 < 0$, $\theta_3 < 0$ to the quadrant $\theta_2 > 0$, $\theta_3 > 0$. The two lines correspond to the first mode inaccessible configuration and the second mode inaccessible configurations, respectively.

Let us consider points A: $(\theta_2, \theta_3) = (0°, 80°)$, B: $(\theta_2, \theta_3) = (-16°, 80°)$, and C: $(\theta_2, \theta_3) = (26°, 80°)$. The point A is picked up from the controllable configurations; points B and C from the configurations which are very close to the uncontrollable configurations. The modal accessibility indices a_r and the natural frequencies f_r $(r = 1, 2)$ for the three points are shown in Table 2. The table shows that the points B and C are in the vicinity of the first and the second mode inaccessible configurations, respectively.

Step response experiments are performed for

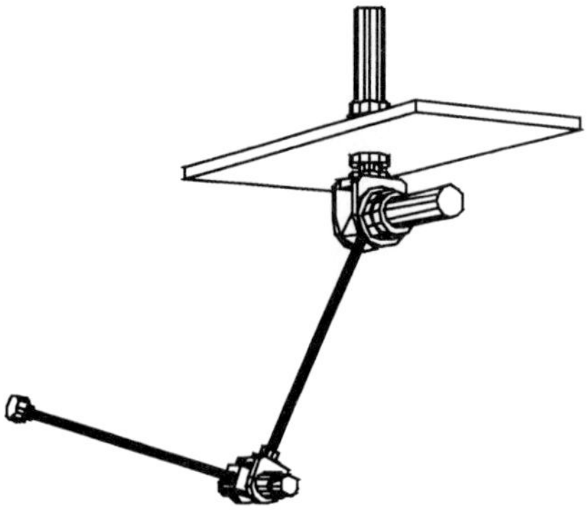

Figure 6: A sketch of point C configuration where $(\theta_2, \theta_3) = (26°, 80°)$.

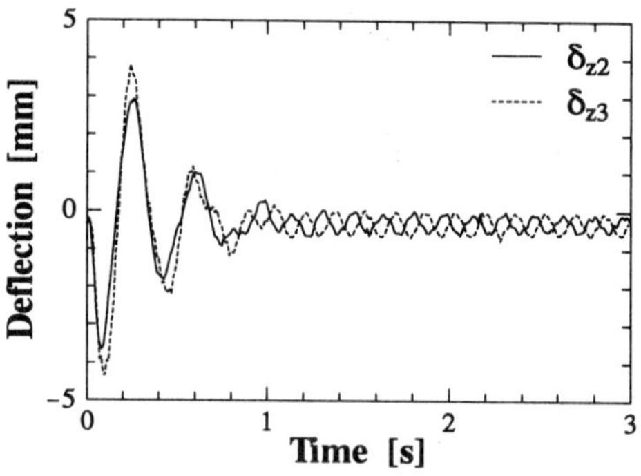

Figure 7: Experimental results at a second mode inaccessible configuration (point C).

the three points of configuration with the joint 1 moving from 0° to 15°. In the experiments the control scheme presented in [17], [22] (also outlined in Section 4) is used to suppress vibration. Good agreement with the theoretical analysis is observed. An example is shown in Figures 6 and 7. Figure 6 is a sketch of the point C configuration. Figure 7 presents experimental results at the point C. It is clearly seen that during the step motion only the first mode is excited while after the motion the second mode can hardly be controlled. This shows that at the point C the second mode is inaccessible and hence the vibration is uncontrollable as expected in the theoretical analysis.

4 Vibration Control

4.1 Velocity Command as Control Input

Looking back to the equations of motion derived in Section 2, we notice that in the field equation of motion (Eq. (23)), the acceleration $\ddot{\boldsymbol{\theta}}$ may be regarded as a control input to the system of the structural vibration. This idea seems attractive because most industrial robots have velocity servo cards in their controllers and the velocity commands to the servo cards are manipulated by a computer as a control input. The motor torques are usually inaccessible by the computer. Assuming that the velocity servo is of perfect performance we ignore the link equation of motion (Eq. (22)) and as equations of motion for controller design we have

$$\ddot{\boldsymbol{\theta}} = \dot{\boldsymbol{v}}_c, \tag{42}$$

$$\begin{aligned} \mathbf{0} = \ & \boldsymbol{M}_{21}(\boldsymbol{\theta}, \boldsymbol{e})\ddot{\boldsymbol{\theta}} + \boldsymbol{M}_{22}(\boldsymbol{\theta}, \boldsymbol{e})\ddot{\boldsymbol{e}} \\ & + \boldsymbol{h}_2(\boldsymbol{\theta}, \dot{\boldsymbol{\theta}}, \boldsymbol{e}, \dot{\boldsymbol{e}}) + \boldsymbol{g}_2(\boldsymbol{\theta}, \boldsymbol{e}) + \boldsymbol{K}\boldsymbol{e}, \end{aligned} \tag{43}$$

where $\boldsymbol{v}_c$ is a velocity command vector. With this framework we have worked on vibration suppression control [19], [20], [21], [22] and have obtained some useful results. It should be noted that Eqs. (42) and (43) describe only an ideal case and analysis and simulation are performed using Eq. (22) too.

4.2 Linearization of Equations of Motion

In designing controllers we make the same assumption as in controllability analysis. The reason is because we are mainly interested in vibration suppression at a certain point after the motion. With the assumption, and using the variables $\Delta\boldsymbol{\theta}$ and $\Delta\boldsymbol{e}$ defined by Eqs. (26) and (27), respectively, we have

$$\boldsymbol{M}_{22}\Delta\ddot{\boldsymbol{e}} + \boldsymbol{K}_{22}\Delta\boldsymbol{e} = -\boldsymbol{M}_{21}\ddot{\boldsymbol{\theta}}. \tag{44}$$

This says that the vibration is undamped and excited by the acceleration of the joint motion.

4.3 Active Damping Method

If we can generate a positive damping term in Eq. (44) by the control input $\ddot{\boldsymbol{\theta}} = \dot{\boldsymbol{v}}_c$ we will have a asymptotically stable system. This can be achieved easily: Letting

$$\dot{\boldsymbol{v}}_c = \boldsymbol{M}_{21}^{+}\boldsymbol{M}_{22}\boldsymbol{K}_{ve}\Delta\dot{\boldsymbol{e}}, \tag{45}$$

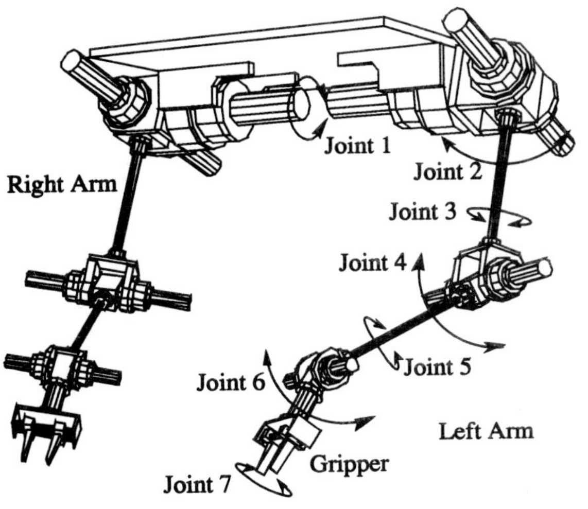

Figure 8: Overview of the experimental flexible robot ADAM.

we have

$$M_{21}^{+}M_{22}\left(\Delta\ddot{e}+K_{ve}\Delta\dot{e}+M_{22}^{-1}K_{22}\Delta e\right)=0. \tag{46}$$

Therefore, if

$$\mathrm{rank}M_{21}=\dim(\Delta e), \tag{47}$$

we have

$$\Delta\ddot{e}+K_{ve}\Delta\dot{e}+M_{22}^{-1}K\Delta e=0. \tag{48}$$

By choosing K_{ve} as positive we have a damped vibration system and hence vibration suppression control is attained. We call this control method an active damping method.

In the above control scheme we should note two points: One is that the feedback control gain in Eq. (45) is configuration dependent, that is a function of θ. This is obvious from the process of deriving M_{21} and M_{22} (see Section 3.1).

The second point is that if Eq. (47) does not hold we cannot have Eq. (48) and hence no guarantee of damping. In this case an alternative solution should be sought. We have proposed to apply an optimal regulator technique to this case. Exploiting the linearized equation (44) we have a state feedback law. Again, the feedback gain is dependent on the configuration. In this case, however, a big gain table is required. Successful experimental results are reported in [22].

4.4 Experiments

To verify the effectiveness of the active damping method we apply the method to a 3D flexible robot shown in Figure 8. The robot has dual arms and is called ADAM (Aerospace Dual-Arm Manipulator). Details of the robot can be found in [8]. The two arms are exactly the same. Each arm has two flexible links (upper and lower arms) and seven active joints.

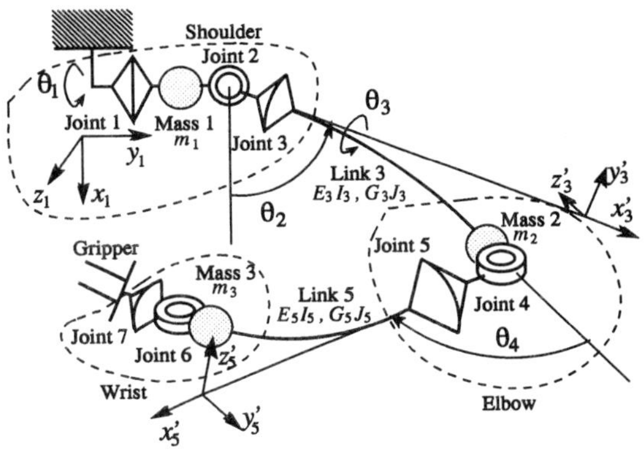

Figure 9: Lumped-mass and spring model of ADAM.

Table 3: Mechanical parameters of ADAM.

	Link 3	Link 5
Length	500 [mm]	500 [mm]
Elastic part	359 [mm]	394 [mm]
Diameter	13.0 [mm]	10.0 [mm]
Material	SUP-6	SUP-6
EI	288.1 [Nm^2]	100.8 [Nm^2]
Mass	700 [g]	500 [g]

Figure 9 shows a lumped-parameter model for the robot. Mechanical parameters of the robot are shown in Table 3. The models for both arms are the same. As is seen in the figure the joints 5, 6 and 7 do not strongly affect the vibrations of the links 3 and 5. Mainly the joints 1 through 4 affect the vibrations. Therefore, we control the vibrations using the joints 1 through 4 and have

$$\theta=\left[\begin{array}{cccc}\theta_1 & \theta_2 & \theta_3 & \theta_4\end{array}\right]^T \tag{49}$$

as the joint space vector in control. The deflections of the links 3 and 5 are δ_{yi}, δ_{zi}, ϕ_{xi}, ϕ_{yi}, and ϕ_{zi} $(i=3,5)$. However, ϕ_{xi}, ϕ_{yi}, and ϕ_{zi} are dependent to δ_{yi} and δ_{zi} by the same reason as in FLEBOT II. Therefore, the link deflection vector e is given by

$$e=\left[\begin{array}{cccc}\delta_{y3} & \delta_{z3} & \delta_{y5} & \delta_{z5}\end{array}\right]^T. \tag{50}$$

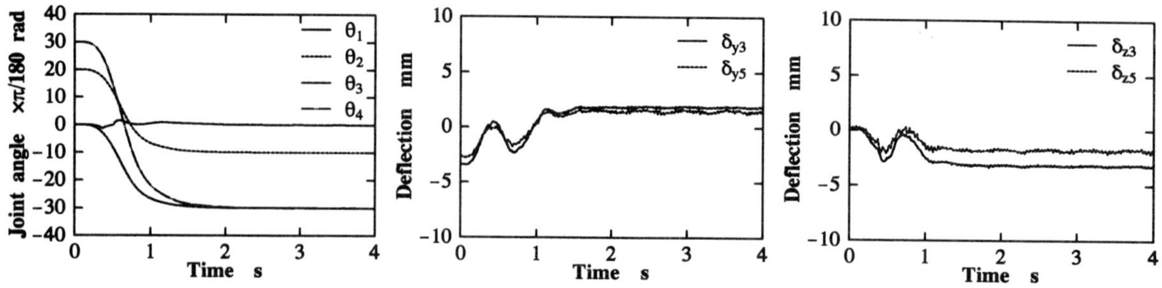

(a) Response of joints. (b) Tip displacement in y direction. (c) Tip displacement in z direction.

Figure 10: Experimental results of vibration suppression control.

In experiment, of course, not only the link vibration but also the joint motion has to be controlled. We control the joint motion by the velocity command $\boldsymbol{v}_c$ calculated by

$$\dot{\boldsymbol{v}}_c = \ddot{\boldsymbol{\theta}}_d + \boldsymbol{K}_{vt}(\dot{\boldsymbol{\theta}}_d - \dot{\boldsymbol{\theta}}) + \boldsymbol{K}_{pt}(\boldsymbol{\theta}_d - \boldsymbol{\theta}), \quad (51)$$

where $\boldsymbol{\theta}_d$ is the desired value of $\boldsymbol{\theta}$, and $\boldsymbol{K}_{vt}$ and $\boldsymbol{K}_{pt}$ are constant feedback gain matrices.

The velocity command to achieve both the vibration suppression and the joint tracking motion is obtained as the sum of the velocity commands calculated by Eqs. (45) and (51).

Experimental results are shown in Figure 10. It is seen that the link deflections are well damped and that at the same time the joint motion is well controlled.

5 Conclusions

In this paper we have discussed modeling, controllability and vibration suppression control of 3D flexible robots. For modeling we have presented a lumped-parameter technique using a so-called Holzer's method and shown how simply the technique can make a practical model for the flexible robot. Using the technique we have found interesting uncontrollable configurations for an elbow-type flexible robot. Finally we have derived a model-based vibration suppression control technique exploiting the lumped-parameter model. In the scheme feedback gains are varied depending on the robot configuration. All the results in this paper have been verified experimentally.

Future research will be directed to controllability analysis with more precise models, stability analysis of the control scheme, and force control of the flexible robot with application to cooperative control of two flexible robots. Also, solving inverse kinematics and dynamics will be a target to challenge.

References

[1] Book WJ, Maizza-Neto O, Whitney DE. Feedback Control of Two Beam, Two Joint System with Distributed Flexibility. *Trans. ASME, J. of Dynamic Systems, Measurement and Control* 1975; 97(4):424–431

[2] Truckenbrodt A. TRUNCATION PROBLEMS IN THE DYNAMICS AND CONTROL OF FLEXIBLE MECHANICAL SYSTEMS. In: *Prepr. of 8th IFAC World Congress*, Kyoto, Japan, July 1981, pp XIV60–XIV65

[3] Cannon, Jr. RH, Schmitz E. Initial Experiments on the End-Point Control of a Flexible One-Link Robot. *Int. J. of Robotics Research* 1984; 3(3):62–75

[4] Pfeiffer F, Gebler B. A Multistage-Approach to the Dynamics and Control of Elastic Robots. In: *Proc. of IEEE Int. Conf. on Robotics and Automation*, Philadelphia, USA, April 1988, pp 2–8

[5] Uchiyama M, Jiang ZH, Hakomori K. Compensating Control of a Flexible Robot Arm. *J. of Robotics & Mechatronics* 1990; 2(2):97–106

[6] Svinin MM, Uchiyama M. Contribution to Inverse Kinematics of Flexible Robot Arms. *JSME Int. J., Series C: Dynamics, Control, Robotics, Design and Manufacturing* 1994; 37(4):755–764

[7] Murayama M, Yoshimoto K. Control of 3 DOF Spatial Manipulator with a Flexible Link. *Trans. of Japan Society of Mechanical Engineers* 1990; 56(522):424–430 (in Japanese)

[8] Uchiyama M, Konno A, Uchiyama T, Kanda S. Development of a Flexible Dual-Arm Manipulator Testbed for Space Robotics. In: *Proc. of IEEE Int. Workshop on Intelligent Robots and Systems '90*, Tsuchiura, Japan, July 1990, pp 375–381

[9] Yoshikawa T, Murakami H, Hosoda K. Modeling and Control of a Three Degree of Freedom Manipulator with Two Flexible Links. In: *Proc. of 29th Conf. on Decision and Control*, Honolulu, USA, December 1990, pp 2532–2537

[10] Jacobus RF, Serna MA. Mechanical Design of the CEIT Flexible Robot. In: *Proc. of 23rd Int. Symp. on Industrial Robots*, Barcelona, Spain, October 1992, pp 329–334

[11] Meirovitch L. *Analytical Methods in Vibrations.* Macmillan Publishing Co., Inc., 1967

[12] Sunada WH, Dubowsky S. On the Dynamic Analysis and Behavior of Industrial Robotic Manipulators With Elastic Members. *Trans. ASME, J. of Mechanisms, Transmissions, and Automation in Design* 1983; 105:42–51

[13] Konno A, Uchiyama M. Modeling of a Flexible Manipulator Dynamics Based on the Holzer's Method. *J. of Robotics Society of Japan* 1994; 12(7):1021–1028 (in Japanese).

[14] Uchiyama M, Jiang ZH. Compensability of End-Effector Position Errors for Flexible Robot Manipulators. In: *Proc. of 1991 American Control Conf.*, Boston, USA, June 1991, pp 1873–1878

[15] Tosunoglu S, Lin SH, Tesar D. Accessibility and Controllability of Flexible Robotic Manipulators. *Trans. ASME, J. of Dynamic Systems, Measurement and Control* 1992; 114:50–58

[16] Konno A, Uchiyama M, Kito Y, Murakami M. Configuration-Dependent Controllability of Flexible Manipulators. In: Yoshikawa T, Miyazaki F (eds) *Experimental Robotics III*. Springer-Verlag London Ltd., 1994, pp 531–544

[17] Konno A, Uchiyama M, Kito Y, Murakami M. Vibration Controllability of Flexible Manipulators. In: *Proc. of IEEE Int. Conf. on Robotics and Automation*, San Diego, USA, 1994, pp 308–314

[18] López-Linares S, Konno A, Uchiyama M. CONTROLLABILITY OF FLEXIBLE MANIPULATORS. In: *Prepr. of Symp. on Robot Control (SY.RO.CO.) '94*, Capri, ITALY, 1994, pp 509–516

[19] Uchiyama M, Konno A. Computed Acceleration Control for Vibration Suppression of Flexible Robot Manipulators. In: *Proc. of 5th Int. Conf. on Advanced Robotics*, Pisa, Italy, 1991, pp 126–131

[20] López-Linares S, Konno A, Uchiyama M. Vibration Suppression Control of Flexible Robots Using Velocity Inputs. In: *Proc. of IEEE/RSJ/GI Int. Conf. on Intelligent Robots and Systems (IROS '94)*, München, Germany, 1994, pp 1429–1437

[21] Konno A, Uchiyama M. VIBRATION SUPPRESSION CONTROL OF SPATIAL FLEXIBLE MANIPULATOR. In: *Prepr. of Symp. on Robot Control (SY.RO.CO.) '94*, Capri, ITALY, 1994, pp 517–523

[22] Konno A, Uchiyama M, Kito Y, Murakami M. Vibration Suppressing Control of 3D Flexible Manipulators Using the Acceleration Command. *J. of Robotics Society of Japan* 1994; 12(8):1166–1174 (in Japanese)

Moving Base Robotics and Reaction Management Control

Kazuya Yoshida, Dragomir N. Nenchev and Masaru Uchiyama
Dept. of Aeronautics and Space Engineering
Tohoku University
Aoba, Sendai 980-77, JAPAN
yoshida@space.mech.tohoku.ac.jp

Abstract

This paper addresses a new paradigm of moving base robotics, for a class of robots which have manipulator systems on a moving base. For such robots, kinematic and dynamic coupling between the manipulator arm and the base degrades positioning accuracy and operational dexterity. Moving base robots are classified into the following four categories: (a) Free Floating Manipulator System, (b) Flexible Structure mounted Manipulator System, (c) Macro-Mini Manipulator System, and (d) Mobile Vehicle mounted Manipulator System. Among them, this paper focuses the Flexible Structure mounted Manipulator System, because this class of system displays inherent dynamic coupling between manipulator operation and base structure flexibility, hence control methods to avoid base vibration are strongly needed. A dynamic analysis for reaction management control using the Reaction Null Space is introduced and applied to yield a reactionless operational path for PTP operation. Force control of this class of system is also addressed.

1 Introduction

Manipulators leap from the ground onto moving bases and expand their application field.

This paper addresses a new paradigm of moving base robotics, for a class of robots which have manipulator systems on a moving base. For such robots, kinematic and dynamic coupling between the manipulators and the base degrades positioning accuracy and operational dexterity. A crude approach to this problem may treat the base motion as unknown disturbance to the manipulator system. However, if we have a proper knowledge about the dynamics of the system including the base motion dynamics, we can develop sophisticated control methods. Our basic motivation regarding moving base robotics is that we should discuss this coupling problem by developing a comprehensive dynamic model, and study manipulator control methods from the point of view of reaction management such as coordinated manipulator control to compensate the base reaction or to yield a minimum reaction to the base.

In this paper, we shall first briefly review the previous work and current application projects of manipulator systems with a moving base, such as free-floating space telerobots, dexterous manipulator systems mounted on a space or terrestrial long-reach manipulator arm, and so on. From them, we abstract a common dynamics problem, such as the "moving base robotics" issue, and categorize the moving base robots into five groups. As an example, we then focus on a class of manipulator systems mounted on a flexible structure, because this class of systems displays inherent dynamic coupling between manipulator operation and base structure flexibility, hence control methods to avoid base vibration are strongly needed. We propose a coordinated manipulator control method paying attention to reaction management based on the "Reaction Null Space" model, and apply it to develop a reactionless operational path for PTP operation. We also discuss the force control issue which is actual when the moving base robot contacts with the environment through its manipulators.

2 Applications of Moving Base Robots and Pioneering Works

Nowadays we can find a number of interesting papers seeing potential applications of moving base robots. Here we shall make a brief overview of the diversity of those applications and pioneering papers.

Space free-flying robots for orbital operation are one typical example of moving base robots. This type of systems comprises one or more manipulator arms mounted on a free-floating or attitude controlled space vehicle. In such systems, dynamic coupling between the manipulator arm and the base satellite is a major control issue. The manipulator reaction disturbs the base satellite attitude, and it is very harmful for the station keeping of the satellite. Therefore, a coordinated control paying attention to the inertial reaction management is necessary. It should be noted that for systems without any external forces (such as from gas-jet thrusters, for example), but with internal forces obtained from reaction wheels, momentum conservation is a major motion constraint.

Earlier important works on the dynamics and control of these space robots are summarized in the book [1]. There are flight projects of free-flying robots currently planned and actively developed both in the US, named Ranger [2], and in Japan, named ETS-VII [3]. Control schemes for motion coordination between the manipulator and the base attitude, paying attention to the reaction management, are discussed in [4], [5].

Another category of space robots are manipulator systems mounted on a long-reach flexible structure. The Special Purpose Dexterous Manipulator (SPDM) mounted at the end tip of the Space Station Remote Manipulator System (SSRMS) [6] and the Japanese Experiment Module Remote Manipulator System (JEMRMS) proposed by Japan's NASDA [7], are examples of this category now being developed. In these systems, the long-reach part of the manipulator (e.g. SSRMS) can work as a passive structure to provide an operational base for the dexterous manipulator (e.g. SPDM) during the dexterous manipulation phase. But due to the inherent characteristics of the long-reach structure, the long-reach base vibrates when excited by the reaction from the manipulator part. Hence vibration dynamics is more important than inertial reaction dynamics or momentum conservation.

This type of systems is used not only in space, but also on earth. Recently, applications in inaccessible environments, such as nuclear power plant sites, building or power line maintenance sites, and so on, attract interests. Some projects have been proposed [8], [9], [10].

A major control issue is vibration suppression of the base to ensure the positioning accuracy and the operational dexterity.

Pioneering work within this category has been done by Dubowsky and his group. The coupling map approach was proposed to find out an operational path that yields smaller base vibration [11]. The inferred endpoint control approach was proposed and successfully demonstrated for end-effector fine positioning despite of base vibration [12]. For quick absorption of the vibrational energy and to increase damping, the pseudo passive energy dissipation control was proposed [13].

The above examples of long-reach systems we consider as a version of macro-mini manipulator systems. The difference is that the long-reach systems assume a passive and flexible structure in the supporting base (macro part). On the other hand, the so-called "macro-mini" systems generally suppose an actively controlled base.

The macro-mini manipulator systems are relatively well studied in the literature. A general frame work for the operational space control of macro-mini manipulators was proposed by Khatib, and applied to various types of macro-mini systems [14]. An approach that uses the mini part to compensate the flexible deformation of the macro part was proposed by Yoshikawa et al [15], [16]. In addition, an impedance control law of the macro-mini manipulator was proposed by Nagai and Yoshikawa [17].

Yet another type of moving base robots are manipulator systems mounted on a terrestrial mobile vehicle, including a wheeled vehicle or a walking machine. There are a number of research projects within this category, which, however, go beyond the scope of this paper. From the view point of manipulator base control, active suspension of the vehicle is of our interest, and considered a version of the general moving base robot framework we develop here. Consideration of static balance and dynamic reaction to control the vehicle's stability and payload's acceleration environment is a major issue [18]-[25].

One interesting example from this category is the KR-II robot proposed by Hirose and Morishima [26]. The robot is designed to mechanically coordinate its locomotion and manipulation segments.

3 Modeling of Moving Base Robots

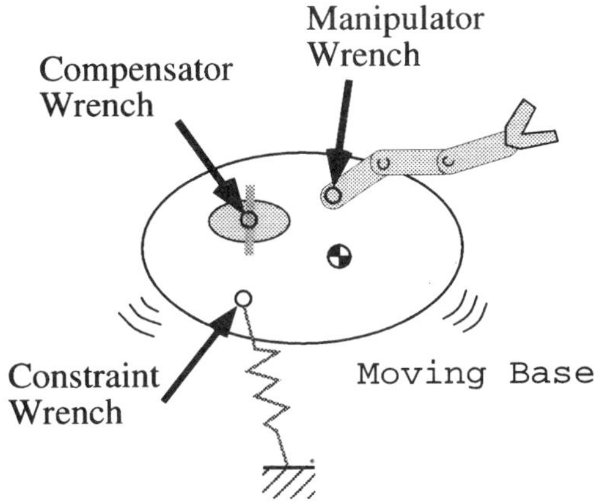

Figure 1 Moving base robot

We propose a general framework to model moving base robots, paying attention to the wrenches applied to the base. Figure 1 shows our model, comprising a free-floating base and a number of mechanisms attached to it. Three types of mechanisms are considered: a) manipulator mechanisms, b) constraint mechanisms, and c) compensator mechanisms. The manipulator mechanisms are usually a target part that we want to operate. The constraint mechanisms are parts which disturb or constrain the base motion. Examples are solar paddles attached on a free-flying space robot, or a flexible structure supporting the base of a long reach manipulator system. Or even for terrestrial mobile vehicle-mounted systems, wheels and suspension mechanisms fall within this category. The compensator mechanisms are such devices as reaction wheels which control and cancel the reaction exerted from the other mechanisms.

The above mechanisms affect the reaction wrenches on the base, and the sum of the wrenches determines its physical motion in inertial space. Obviously, manipulator and compensator mechanisms impose active constraints on the base motion, while the constraint mechanisms impose passive ones. For terrestrial systems, like a vehicle-mounted manipulator, gravity wrenches make a dominant effect and the inertial reaction wrenches may affect very little. In other applications, such as terrestrial long reach manipulator systems or dynamic walking machines or hopping machines, both gravity and inertial reaction wrenches are important. For space borne systems, the inertial reaction becomes critical.

We then classify the moving base robots in terms of the effective wrenches.

- **Free Floating Manipulator System (FFMS)**

An example from this category is a free-flying space robot, where inertial reaction wrenches are dominant. The operation of the manipulator mechanisms produces the reaction wrench on the base. Compensator mechanisms, such as reaction wheels, generate the reaction wrench to control or dynamically counterbalance the manipulator reaction. Solar paddles or flexible appendages mounted on the base work as constraint mechanisms generating a reaction wrench due to the vibrational motion. If gas-jet thrusters are used, they produce an external force wrench directly on the base (Figure 2a).

Underwater robots are also in this category: water current and drag, which are accounted as the external wrenches rather than the inertial reaction wrenches of the manipulator arms, play dominant roll in the system motion dynamics.

- **Flexible Structure mounted Manipulator System (FSMS)**

This can be either a space free-flying system or a terrestrial system. In the latter case, the gravitational wrenches of each mechanism must be accounted additionally. This category is specifically characterized by the elastic wrench resulting from the flexible supporting structure, which we consider to be a constraint mechanism. Vibrations of the flexible structure are excited by reactions from the other mechanisms, then dynamic intercoupling occurs (Figure 2b).

- **Macro-Mini Manipulator System (MMMS)**

This can be a version of some of the other categories, mainly of FSMS. However, if we say Macro-Mini system here, we mean a system that consists of manipulator mechanisms for operation ("Mini" part) and a manipulator-like *active* constraint mechanism ("Macro" part) used to support the base. The supporting constraint mechanism is actively controlled and produces a wrench to the base. It is contrasted with the supporting constraint mechanism of FSMS, because the latter produces a visco-elastic wrench just passively (Figure 2c).

- **Mobile Vehicle mounted Manipulator System (MVMS)**

This category has a special constraint mechanism in the form of a wheel/suspension system or a leg system. The terminal of the constraint mechanism is not fixed to the earth, and we should pay special attention to this terminal condition. If the terminal condition is well defined and the suspension system is controllable, such a system can be considered a version of MMMS (Figure 2d).

- **Ground Based Manipulator System (GBMS)**

This class of robot is not a moving base robot any more. It is considered a special case with a strong constraint mechanism. However, if the base is not fixed to the earth, but just lying on it, then a similar discussion as in the case of MVMS is possible with regard to the terminal condition of the constraint mechanism.

Finally, we would like to note that techniques used typically within one of the above categories, recently are being transferred to solve similar problems within other categories. As an example, we point out the work of Tsujita and Tsuchiya [27], describing the use of reaction wheels for a biped walking gait stabilization.

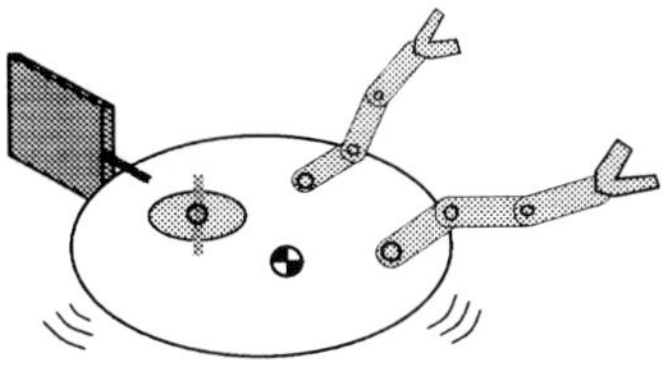

a. Free-floating manipulator system (FFMS)

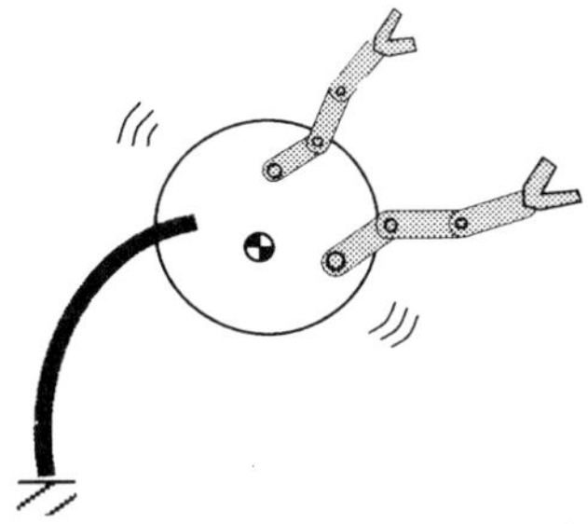

b. Flexible structure mounted manipulator system (FSMS)

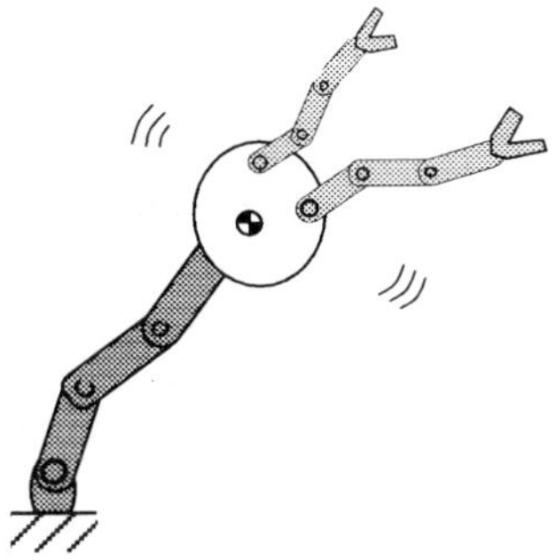

c. Macro-Mini manipulator system (MMMS)

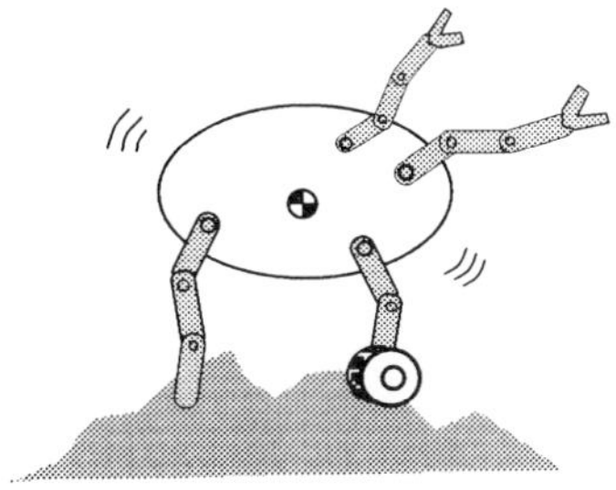

d. Mobile vehicle mounted manipulator system (MVMS)

Figure 2 Possible versions of moving base manipulator systems

4 Equation of Motion

In this section, we derive a general equation of motion of a moving base robot. We assume a moving base robot comprising a single rigid body as a base, and a number of manipulator mechanisms and constraint mechanisms attached to it. Compensator mechanisms will be treated as a version of manipulator mechanisms, and therefore below we shall refer only to the latter ones. Besides these mechanisms, we consider the wrenches of external forces and moments, which can be generated by the contact with the environment.

Figure 3 depicts a model of the system under consideration, where we define the following notation:

l : number of manipulator mechanisms

p : number of constraint mechanisms

q : number of external wrenches resulting from gravity and contact

point O : the origin of the inertial coordinate frame

point B : the mass center of the base

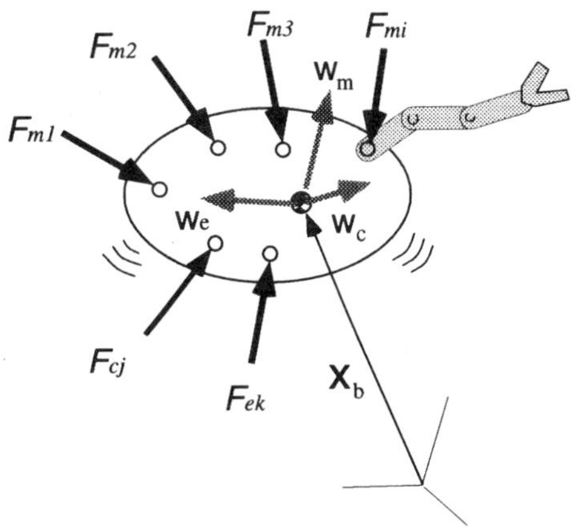

Figure 3 Dynamics model of a moving base robot

point M_i : the attaching point of the i-th manipulator mechanism

point H_i : the end point of the i-th manipulator mechanism

point G_i : center of mass of the i-th manipulator mechanism

point C_j : the attaching point of the j-th constraint mechanism

point E_k : the point where the k-th external wrench is acting

$x_{(*)} \in R^6$: position and orientation variables of a coordinate frame attached at the corresponding point:

$(*) = b, mi, hi, gi, cj, ek$

$\mathcal{F}_{(*)} \in R^6 = \left[f_{(*)}^T, n_{(*)}^T\right]^T$: static/dynamic wrench acting at the corresponding point

$R_{(*)} \in R^{6\times6} = \begin{bmatrix} E & -\tilde{r}_{(*)} \\ O & E \end{bmatrix}$: a transformation matrix where $\tilde{r}_{(*)}$ denotes a 3×3 skew-symmetric matrix

$M_b \in R^{6\times6} = \begin{bmatrix} m_b E & O \\ O & I_b \end{bmatrix}$: inertia matrix of the base in Cartesian space where m_b is mass and I_b is moment of inertia

w_m : manipulator wrench acting at the base centroid

w_c : constraint wrench acting at the base centroid

w_e : external wrench acting at the base centroid

n_i : number of joints in the i-th manipulator mechanism

$n = \sum_{i=1}^{l} n_i$: total DOF in manipulator mechanisms

m_i : mass of the i-th manipulator mechanism

$\phi_i \in R^n$: joint angle vector of the i-th manipulator mechanism

$\tau_i \in R^n$: joint torque vector of the i-th manipulator mechanism

$r_{gi} \in R^3$: a vector pointing from the attachment point to the centroid of the i-th manipulator mechanism

$r_{ij} \in R^3$: a vector pointing from the attachment point to the centroid of the j-th link in the i-th manipulator mechanism

m_{ij} : mass of the j-th link in the i-th manipulator mechanism

I_{ij} : moment of inertia of the j-th link in the i-th manipulator mechanism

ω_{ij} : angular velocity of the j-th link in the i-th manipulator mechanism

$O \in R^{3\times3}$: 3 by 3 null matrix

$E \in R^{3\times3}$: 3 by 3 identity matrix

$I_i \in R^{i\times i}$: i by i identity matrix

4.1 Wrench Dynamics and Dynamic Equilibrium of the Base

We first derive basic dynamics equations in terms of the wrenches exerted on the base. For this purpose, we transform the wrenches acting at different points of the moving base robot $\mathcal{F}_{(*)}$ by means of the transformation matrices $R_{(*)}$ representing a moment arm:

$$w_m = \sum_{i=1}^{l} R_{mi}^T \mathcal{F}_{mi} \tag{1}$$

$$w_c = \sum_{j=1}^{p} R_{cj}^T \mathcal{F}_{cj} \tag{2}$$

$$w_e = \sum_{k=1}^{q} R_{ek}^T \mathcal{F}_{ek} \tag{3}$$

We model each constraint mechanism as a spring-damper system:

$$\mathcal{F}_{cj} = -K_{cj}(x_{cj} - x_{cjref}) - D_{cj}\dot{x}_{cj}, \tag{4}$$

where K_{bj}, D_{bj} are stiffness and damping matrices, respectively.

The equation of motion of the base with respect to point B is:

$$M_b\ddot{x}_b + h_b + g = w_m + w_c + w_e, \tag{5}$$

where h_b represents non-linear dynamics of the base, and g= $[0, 0, g]^T$ is the gravity gradient.

We shall rewrite the last equation in a more compact form. We compose all three wrenches being exerted on the mass centroid of the base (point B):

$$\begin{aligned} w &= R^T\mathcal{F} \\ &= [R_{m1}^T, R_{m2}^T, \ldots, R_{eq}^T] \\ &\quad \bullet[\mathcal{F}_{m1}^T, \mathcal{F}_{m2}^T, \ldots, \mathcal{F}_{eq}^T]^T, \end{aligned} \tag{6}$$

where the dimension of the augmented matrices is $R \in R^{6\times6(l+p+q)}$, $\mathcal{F} \in R^{6(l+p+q)}$. Then we obtain

$$M_b\ddot{x}_b + c_b + g = R^T\mathcal{F}. \quad (7)$$

Note that Equation (7) has the same structure as the equation of an object grasped by multi fingers [28].

Solving Equation (7) for $\mathcal{F}$, using the generalized inverse $[o]^\#$, we obtain:

$$\mathcal{F} = [R^T]^\#(M_b\ddot{x}_b + c_b + g) + (I_6 - [R^T]^\# R^T)\zeta, \quad (8)$$

where $\zeta \in R^6$ is an arbitrary vector.

Equation (8) involves the null-space of the moment-arm matrix R, which is a kinematic expression. The equation thereby indicates the relationship between the base motion and the dynamic balance of acting wrenches. In the special case when the state of the base has not to be changed, the above equation becomes:

$$\mathcal{F} = (I_6 - [R^T]^\# R^T)\zeta. \quad (9)$$

This is the solution for the dynamic equilibrium of wrenches. We here name the above null-space as *"Dynamic Equilibrium Null Space."*

4.2 Static Equilibrium of the Base and the Reaction Null Space

Let us assume that the initial state of the base is static equilibrium, which is assured through the passive constraint wrenches w_c. Initially there are no active and external wrenches exerted on the base ($w_m = w_e = 0$), which means that the whole system is also in static equilibrium.

Next, let us assume that the manipulators have to perform some task, during which also external wrenches might be applied to the system. Our goal is to control the manipulators in such a way that the initial equilibrium state of the base will be not disturbed.

The wrenches acting at each manipulator attachment point are

$$\mathcal{F}_{mi} = \begin{pmatrix} f_{mi} \\ n_{mi} \end{pmatrix} = \begin{pmatrix} m_i(\ddot{r}_{gi} - g) \\ (r_{gi} \times f_{mi}) + \frac{d}{dt}\sum_{j=1}^{n_i}(I_{ij}\omega_{ij} + m_{ij}r_{ij} \times \dot{r}_{ij}) \end{pmatrix}. \quad (10)$$

Rewriting the above equation in terms of joint angles, we obtain

$$H_{bmi}\ddot{\phi}_i + h_{bmi} + g_{bmi} = \mathcal{F}_{mi}, \quad (11)$$

where $H_{bmi} \in R^{6\times n_i}$ denotes an inertia coupling matrix, whereas h_{bmi}, g_{bmi} stand for nonlinear and gravity terms, respectively. This equation describes the contribution of manipulator dynamics to the base disturbance.

For the whole system of manipulators imposing the active constraint on the base, Equation (11) is rewritten as

$$H_{bm}\ddot{\phi} + h_{bm} + g_{bm} = \mathcal{F}_m, \quad (12)$$

where the terms take on the appropriate dimension. The condition for maintaining the static equilibrium of the base is obtained from Equation (5), with $\ddot{x}_b = 0$, $c_b = 0$, $w_c = -g$, and hence,

$$w_m = -w_e,$$

which results in

$$\mathcal{F}_m = -R_m^{-T} w_e. \quad (13)$$

Substituting the last relation into Equation (12) and solving for the joint acceleration, we obtain

$$\begin{aligned} \ddot{\phi} &= H_{bm}^\#(-R_m^{-T} w_e - h_{bm} - g_{bm}) \\ &\quad + (I_n - H_{bm}^\# H_{bm})\zeta. \end{aligned} \quad (14)$$

From the Equation (14) we see that the inertia coupling matrix has a null space. Applying joint accelerations derived from this null space

$$\ddot{\phi} = (I_n - H_{bm}^\# H_{bm})\zeta, \quad (15)$$

the static equilibrium of the base will be maintained. We call this null space *"The Reaction Null Space"*. We must emphasize here that this null space is different from the Dynamic Equilibrium Null Space defined in the previous Subsection. We note also that the concept of Reaction Null Space we introduced earlier with regard to free-floating space robots [29]. In this paper we have generalized the concept to include the whole class of moving base robots. An important issue is the existence of this null space, which must be verified for each particular case. Below, we shall provide an example.

5 Joint Torques

In order to be able to control the manipulator mechanisms, we need to derive the joint torques under the base static equilibrium constraint. This can be done in a straightforward manner, in substituting the joint accelerations from Equation (15) into the following equation describing the dynamics of the composed manipulator mechanism:

$$H_m\ddot{\phi} + h_m + g_m = \tau_m + J_m^T\mathcal{F}_h. \quad (16)$$

where $J_m \in R^{n\times 6q}$ is a manipulator Jacobian and $H_m \in R^{n\times n}$ is a manipulator inertia matrix, which both are in the conventional form but augmented for multiple arms. And $\mathcal{F}_h \in R^{6q}$ is a part of the external wrenches acting at the end-points of the manipulators.

After substitution, we obtain

$$\tau_m = H_m(I_n - H_{bm}^{\#} H_{bm})\zeta + h_m + g_m - J_m^T \mathcal{F}_h \quad (17)$$

which shows that there is a set of joint torques that would guarantee the static equilibrium of the base.

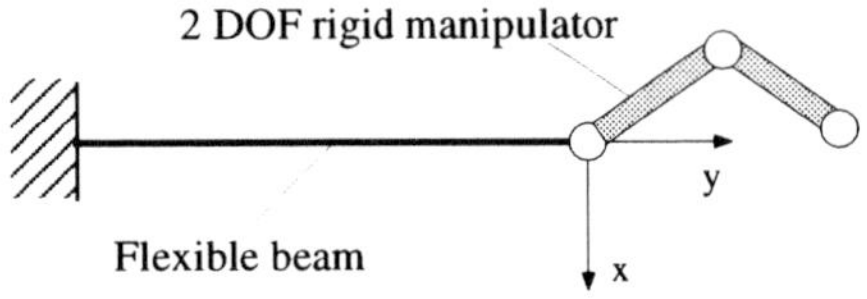

Figure 4 A planar 2 DOF FSMS

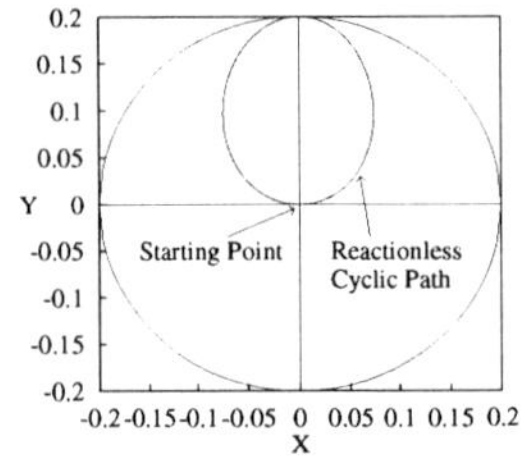

(a)

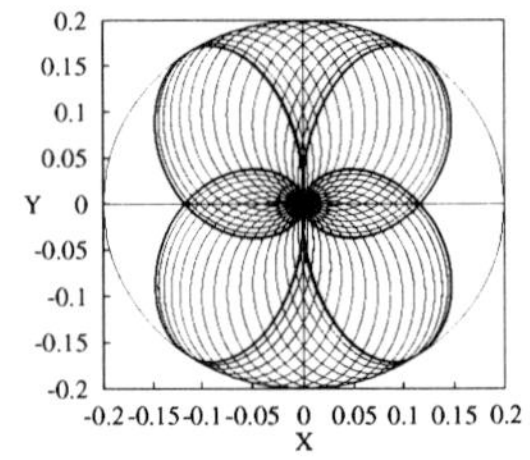

(b)

Figure 5 Reactionless cyclic paths

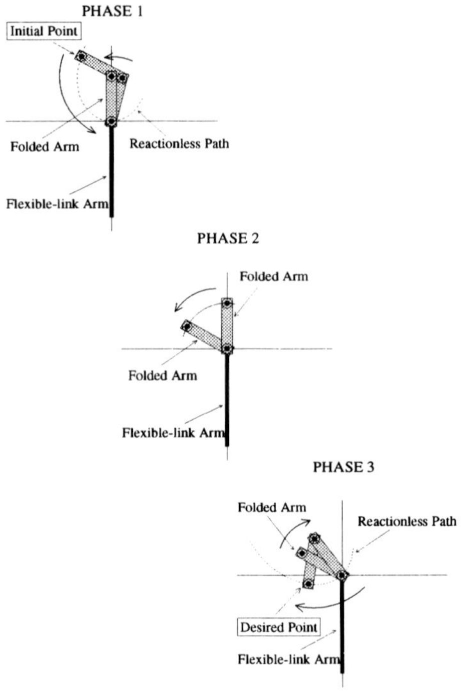

Figure 6 3-phase path planning for FSMS

6 Path Planning Issue with Example

In this section, as a representative moving base robot example, we use a simple model of a Flexible Structure mounted Manipulator System (FSMS) that has a 2 DOF rigid manipulator arm (the upper limb and the lower limb have the same length) mounted on a flexible beam, see Figure 4. We foucus on this FSMS here, because there are a number of potentially useful applications, mainly for space robotics, including SPDM/SSRSM or JEMRMS system, but also for terrestrial applications, and because it imposes interesting dynamic problems to solve.

In this system, the manipulator reaction wrench in y direction does not deform the supporting structure, because the structure has high stiffness in this direction. On the contrary, the wrench in x direction causes deformation and vibration, because of low stiffness. Considering only the x direction of the base motion (1 DOF), we shall obtain the reaction null space in a 2 DOF manipulator system.

6.1 Reactionless paths

Applying solution (15) and integrating it twice from the position $(0, 0)$ (completely folded configuration) and with zero initial velocity, we can obtain an interesting cyclic reactionless path in Cartesian space, as shown in Figure 5a. Varying the initial direction of the folded arm, we obtain multiple paths as shown in Figure 5b. Note that these paths ensure always zero reaction at the base point towards x direction, regardless of the velocity and the acceleration along them. As long as the end-tip travels on any of these paths, the supporting structure does not vibrate. Thus it is possible to perform very fast motion without base disturbance.

It should be noted from Figure 5b that the reactionless paths originating at the folded-arm configuration cover most of the reachable workspace of the rigid arm.

6.2 3-phase path planning

Now we propose a path planning strategy for PTP operation of the FSMS. The above discussion suggests that there is a unique reactionless path between an arbitrary point within the workspace defined from Figure 4b and the base point, where the arm is folded.

Our idea of path planning is based on the following 3-phase motion:

1. From the initial point toward the base point (folded arm), following the corresponding reactionless path (Figure 6, Phase 1).

2. Reconfigure the folded arm in order to switch to another reactionless path (Figure 6, Phase 2).

3. Move to a desired goal point following the new reactionless path (Figure 6, Phase 3).

During phases 1 and 3 we expect zero reaction, whereas during phase 2, base vibration will be excited. However the latter will be relatively small because the center of mass of the folded arm is located close to the base, and the arm does not have to accelerate/decelerate significantly.

The validity of the proposed 3-phase PTP operation is successfully demonstrated using the experimental testbed named "TREP" developed at Tohoku University (Figure 7). As introduced in Section 2, the coupling map approach has been proposed to solve the same task. The coupling map method investigates the dynamic coupling of the manipulator reaction and the base stiffness, then finds a path generating smaller elastic deformation of the base. On the contrary, our method doesn't account for the base stiffness, but finds completely vibration-free paths except its intermediate phase 2.

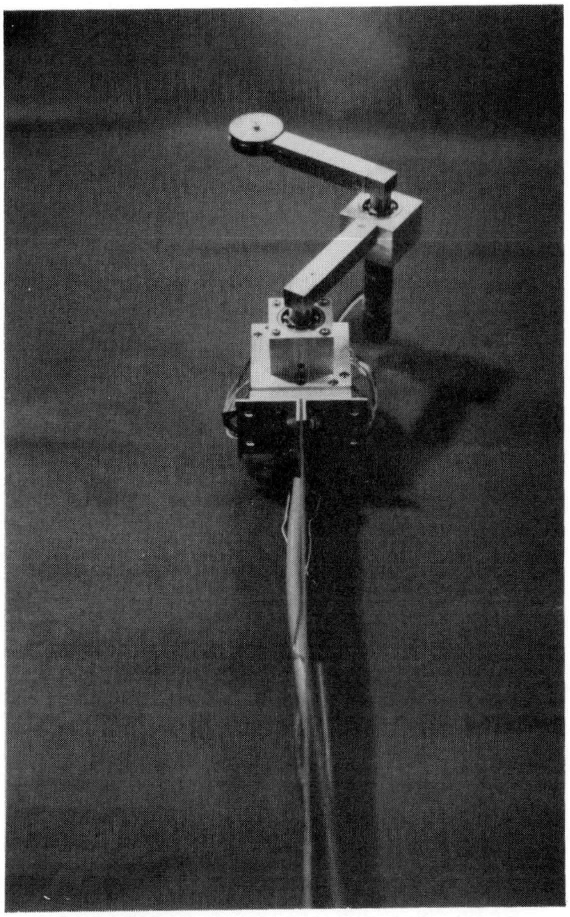

Figure 7 The TREP testbed at Tohoku Univ.

7 Force Control Issue

In this section, we discuss the operation of the FSMS making contact with environment. As a general frame work of the force control issue, we investigate impedance control laws so that the FSMS can perform the dynamic behavior representing a desired impedance despite the given mechanical impedance of the supporting structure. In this section, for simplicity, we assume a FSMS comprising a single elasitc supporting beam and a single n DOF manipulator mechanism on the base. Our motivation is illustrated in Figure 8.

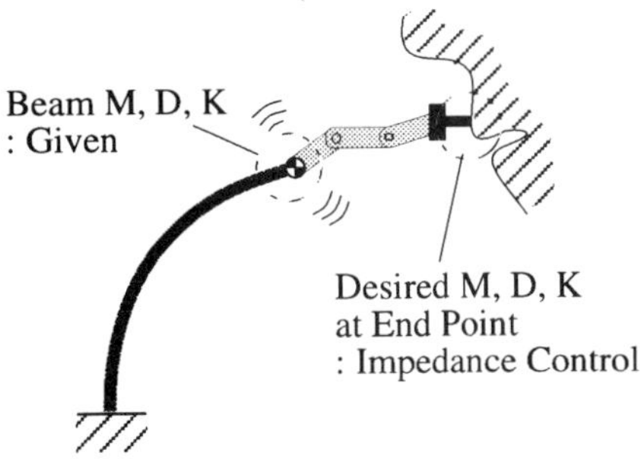

Figure 8 Impedance model of the FSMS

Let us recall the equation of motion of the base (5) and conbine it with that of the constrain flexible beam

(4). We then obtain:

$$M_b\ddot{x}_b + D_b\dot{x}_b + K_b(x_b - x_{bref}) + h_b + g = \mathcal{F}_b, \quad (18)$$

where

$$\mathcal{F}_b = w_m + w_e$$

and

$$D_b = R_c^T D_c$$
$$K_b = R_c^T K_c.$$

The parameters (M_b, D_b, K_b) expresses the inherent impedance of the supporting structure.

On the other hand, the equation of motion (16) is rearranged with the terminal force/torque conditions of the system under consideration as:

$$H_m\ddot{\phi} + h_m + g_m = \tau + J_m^T(\mathcal{F}_h - R_{bh}\mathcal{F}_b), \quad (19)$$

where R_{bh} is a matrix representing a moment arm between points B and H.

Based on the above preparation, we can derive an impedance control law. Define the desired impedance relationship that the manipulator end point should perform as:

$$M\ddot{x}_h + D\dot{x}_h + K(x_h - x_{href}) = \mathcal{F}_h. \quad (20)$$

From the equations (18)(19) (20) and

$$\dot{x}_h = J_m\dot{\phi} + \dot{x}_b$$
$$\ddot{x}_h = J_m\ddot{\phi} + \dot{J}_m\dot{\phi} + \ddot{x}_b,$$

we finally obtain:

$$\begin{aligned}\tau &= -J_m^T(M\ddot{x}_b + D\dot{x}_b + K(x_h - x_{href})) \\ &\quad + J_m^T R_{bh}\mathcal{F}_b + \alpha \quad (21)\end{aligned}$$

where we assume, as is commonly used in impedance control laws, that the desired inertia has the same property of the manipulator arm: i.e. $M = J_m^{-T}H_mJ_m^{-1}$, and α represents non-linear terms.

Using Equation (21), we can control the FSMS so that the system has the desired impedance (M, D, K), knowing the motion of the base $\mathcal{F}_b$, $\ddot{x}_b$, and $\dot{x}_b$ and the position error from the desired end point $x_h - x_{href}$. As an option, substituting (18) into (21) to eliminate $\mathcal{F}_b$, we then obtain:

$$\begin{aligned}\tau &= -J_m^T(M - R_{bh}M_b)\ddot{x}_b \\ &\quad - J_m^T(D - R_{bh}D_b)\dot{x}_b \\ &\quad - J_m^T(K\Delta x_h - R_{bh}K_b\Delta x_b) + \alpha \\ &\quad (22)\end{aligned}$$

In this equation, the impedance transformation from the inherent structure impedence (M_b, D_b, K_b) into the desired one (M, D, K) accommodated by the manipulator operation is clearly seen.

As a control example, highlighting to increase system damping using manipulator operation, we obtain the following control law:

$$\tau = -J_m^T(D - R_{bh}D_b)\dot{x}_b \quad (23)$$

This idea is relevant to the PPED control method proposed in [30].

The discussion on the stability of impedance transformation control will be an interesting future issue.

8 Conclusions

This paper discusses moving base manipulator systems and formulate basic general equations of their dynamics. We define the Dynamic Equilibrium Null Space and the Reaction Null Space, which are useful concepts to discuss the reaction management control of the base of the moving manipulator systems. We propose a path planning method which shall not excite the vibration of the flexible base, and also discuss an impedance control method to realize a desired impedance at the end point in spite of the inherent impedance of the flexible base.

References

[1] *Space Robotics: Dynamics and Cotrol,* ed. by Xu and Kanade, Kluwer Academic Publishers, 1993.

[2] D. Akin et al, "Development of a low-cost free-flying telerobotic space flight vehicle," *43rd Congress of IAF,* Washington, DC, Aug.28–Sept.5, 1992.

[3] M. Oda et al, "The ETS-VII, the world first telerobotic satellite," *in Proc. Artifici al Intelligence, Robotics and Automation in Space'92,* 1992, pp.307–318.

[4] K.Yoshida, "Practical coordination control between satellite attitude and manipulator reac tion dynamics based on computed momentum concept," *in Proc. 1994 IEEE/RSJ Int. Conf. Intelligent Robots and Systems (IROS'94),* Munich, Germany, Sept. 1994, pp.1578–1585.

[5] Y. Ohkami and M. Oda, "The coordinated control of spacecraft attitude and space manipulator," *in Proc. IFAC Int. Conf. on Intelligent Autonomous Control in Aerospace,* Beijing, China, Aug. 1995.

[6] W. Brimley, D. Brown and B. Cox, "Overview of international robot design for space station freedom," *Teleoperation and Robotics in Space, ed. by Skaar and Ruoff,* Progress in Astronautics and Aeronautics Vol.161, AIAA, 1994, pp.411–441.

[7] K. Kuraoka et al, "Design and development status of the JEMRMS," *in Proc. of Int. Symp. on AI, Robotics and Automation in Space (i–SAIRAS'90),* Kobe, Japan, November 1990, pp.23–26.

[8] J.Y. Lew, D.J. Trudnowski, M.S. Evans and D.W. Bennett, "Micro manipulator motion control to suppress macro manipulator structural vibrations," *in Proc. 1995 IEEE Int. Conf. Robotics and Automation*, Nagoya, Japan, May 21–27, 1995, pp.3116–3120.

[9] D-S. Kwon, D-H. Hwang, S.M. Babcock and B.L. Burks, "Input shaping filter methods for the control of structurally flexible, long-reach manipulators," *in Proc. 1994 IEEE Int. Conf. Robotics and Automation*, San Diego, California, May 1994, pp. 3259–3264.

[10] A.A. Goldenberg et al, "A remote manipulator for forestry operation," *IEEE Trans. on Rob. and Automation*, Vol.11, No.2, 1995, pp.185–197.

[11] M.A. Torres and S. Dubowsky, "Path-planning in elastically constrained space manipulator systems," *in Proc. 1993 IEEE Int. Conf. Robotics and Automation*, Atlanta, Georgia, May 2–6 1993, pp.812–817.

[12] C. Mavroidis, P. Rowe and S. Dubowsky, "Inferred endpoint control of long reach manipulator systems," *in Proc. of IEEE/RSJ Int. Symp. on Intelligent Robots and Systems (IROS'95)*, Pittsburg, PA, August 1995.

[13] S. Dubowsky, "Dealing with vibrations in the deployment structures of space robotic systems," *5th International Conference on Adaptive Structures*, Sendai, Japan, December 1994, pp.3–12.

[14] O. Khatib, "Inertial properties in robotic manipulation: an object-level framework," *Int. Journal of Rob. Res.* Vol.14, No.1, 1995, pp.19–36.

[15] T. Yoshikawa, K. Hosoda, T. Doi and H. Murakami, "Quasi-static trajectory tracking control of flexible manipulator by macro-micro manipulator system," *in Proc. 1993 IEEE Int. Conf. Robotics and Automation*, Atlanta, Georgia, May 2–6 1993, pp.210–215.

[16] T. Yoshikawa, K. Hosoda, T. Doi and H. Murakami, "Dynamic trajectory tracking control of flexible manipulator by macro-micro manipulator system," *in Proc. 1994 IEEE Int. Conf. Robotics and Automation*, San Diego, CA, May 1994, pp.1804–1809.

[17] K. Nagai and T. Yoshikawa, "Impedance control of redundant macro-micro manipulators," *in Proc. 1994 IEEE/RSJ Int. Conf. Intelligent Robots and Systems (IROS'94)*, Munich, Germany, Sept. 1994, pp.1438–1445.

[18] S. Dubowsky and A. B. Tanner, "A Study of the Dynamics and Control of Mobile Manipulators Subject to Vehicle Disturbances," *4th Int. Symp. on Robotics Research*, pp.111–117, 1987.

[19] S. Dubowsky and E. E. Evance, "Planning of Mobile Manipulator Motions Considering Vehicle Dynamic Stability Constraints," *Proc. 1989 Int. Conf. on Robotics and Automation*, pp.1271–1276, 1989.

[20] T. Fukuda et al, "Manipulator/Vehicle System for Multi-Robot Cooperation," *Proc. 1992 Int. Conf. on Robotics and Automation*, pp.74–79, 1992.

[21] Y. Yamamoto and X. Yun, "Modeling and Compensation of the Dynamic Interaction of a Mobile Manipulators," *Proc. 1994 Int. Conf. on Robotics and Automation*, pp.2187–2192, 1994.

[22] A. Ghasempoor and N. Sepehri, "A Measure of Machine Stability for Moving Base Manipulators," *Proc. 1994 Int. Conf. on Robotics and Automation*, pp.2249–2254, 1995.

[23] H. Seraji, "Configuration Control of Rover-Mounted Manipulators," *Proc. 1995 Int. Conf. on Robotics and Automation*, pp.2261–2266, 1995.

[24] A. Torige, T. Matsumura, and N. Ishizawa, "Dynamic analysis of gravity center position controlled manipulator," *in Proc. 7th Int. Conf. on Advanced Robotics (ICAR'95)*, Sant Feliu de Guixols, Spain, Sept.20–22, 1995, pp.539–546.

[25] P. Fraisse et al, "Mobile manipulation of a fragile object," *in Preprints of the 4th Int. Symp. on Experimental Robotics*, Stanford, California, June 1995, pp.290–295.

[26] S. Hirose and A. Morishima, "Design and control of a mobile robot with an articulated body," *Int. Journal of Rob. Res.* Vol.9, No.12, 1990.

[27] K. Tsujita and K. Tsuchiya, "Experimental study on motion control of a biped locomotion machine using reaction wheels," JSME, ROBOMEC'95, Kawasaki, Japan, June 1995, pp.202–203.

[28] I.D. Walker, A. Freeman and S.I. Marcus, "Multiple cooperating manipulators: the case of kinematically redundant arms," *in Proc. NASA Conf. on Space Tel erobotics*, JPL, Pasadena, January 1989, Vol.2, pp.15–24.

[29] K. Yoshida and D.N. Nenchev, "Space robot impact analysis and sattelite-base impulse mimization using reaction null-space," *in Proc. IEEE Conf. on Robotics and Automation*, Nagoya, Japan, May 1995, pp.1271–1277.

[30] M. Torres, *Modeling, Path-Planning and Control of Space Manipulators: The Coupling Map Concept*, PhD Thesis, Department of Mechanical Engineering, MIT, February 1993.

PART 3

ROBOT ASSISTANCE FOR SURGERY AND PERSONAL CARE

SESSION SUMMARY

Ruzena Bajcsy
GRASP Laboratory, University of Pennsylvania, Philadelphia, PA (USA)
email: bajcsy@central.cis.upenn.edu

Our session contained three papers, two on applications of robotics in surgery and one pertaining to the design and manufacture one-of-a-kind items for people with disabilities. The motivation for these applications comes from the following observations:

1. Machines may be better than humans for certain tasks. The notion of "better" is defined with respect to accuracy or precision, compensating, for example, for human weakness during eye surgery (i.e., a shaking hand), or otherwise compensating for the fatigue of a human performer.

2. Certain tasks inherently involve remote manipulation, as is the case in laproscopic surgery.

3. In tasks where multiple measurements are available, as they are in different modality imaging systems (CT, MRI, Ultrasound, Positron Emission Tomography, etc.), it is useful to provide the physician with different visualization tools to assist in data analysis.

4. The technologies we have generally come to associate with robotics can be used to automate the design and manufacture of one-of-a-kind devices for patients, such as different prosthetic and orthotic devices, and potentially lower their cost.

The common underlying threads in all these applications are:

1. Machines must cooperate, adapt and adjust to people; hence, there is limited autonomy.

2. Machines must be able to be highly customized and individualized for each user.

3. Machines must be highly reliable, since a performance failure, even if minuscule, can be tragic.

These problems provide a tremendous opportunity, but also a great challenge, for robotic technologies!

During the presentations we saw prototypes and research into these areas. The complexity of the environment in which manipulation takes place is very striking. The usual rigidity assumption does not hold at all since biological muscle and skin are inherently non-rigid objects. The technologist must take into account changes in friction due to different amounts of fluids during manipulation. Sensing is a great challenge as well since, in the laproscopic application for example, one does not have much physical room for sophisticated sensing. The interface between the technology and the user is far more important than in any other application. If the physician does not feel comfortable with the man-machine interface, he/she will either not use it or it will be misused. One almost feels that the man-machine interface must be individualized to the extent the "impedance match", using an analogy, must be achieved between the physician and the machine. This is, of course, the case for robots in surgery and laproscopy, as well as robotic devices for the physically disabled.

We heard over and over again of robot assistants, robotic extendors, and robotic enhancement technologies. Clearly this is the right direction! We believe that one can make very useful devices today, yet open scientific questions remain. Just to mention a few:

1. How does one deliver high quality and real-time sensory observations in a confined space?

2. How does one combine real-time planning and control from high degree-of-freedom manipulators in a non-rigid, highly viscous environment?

3. How does one predict the dynamics of coupled human and complex devices performing tasks such as feeding, grooming, etc.?

4. How does one individualize man-machine interfaces without making each system from scratch?

And many more.

Robotic Aids for Laparoscopic Surgery Problems

Alícia Casals*, Josep Amat* and Enric Laporte**

*Dep. of Automatic Control and Computer Engineering
Universitat Politècnica de Catalunya
Pau Gargallo n° 5, 08028 Barcelona (Spain)
e-mail: casals@esaii.upc.es

** Adrià Surgical Centre -Vall d'Hebron Hospitals
Vallmajó34-36 08021 Barcelona (Spain)

Abstract. Minimally invasive surgery has already become a normal practice. Currently new endoscopical procedures are being investigated in order to increase its benefits, both for the surgeon and for the patient. These benefits are: less damage to surrounding tissues, less post-operative recovery time and no scarring. The extended use of laparoscopic techniques has demonstrated that there are still important limitations in its practical applications. This paper overviews the difficulties appearing in laparoscopic surgery, that comprises the geometrical limitations of the instruments movements and the need to visualize the scene through TV cameras, and analyzes the existing aids as well as the new possible technological improvements.

1. INTRODUCTION

Laparoscopic surgery started in the late eighties and it produced an important revolution in the field of minimally invasive surgery [1]. Laparoscopic surgery requires the introduction in the patient abdomen of some surgical tools, such as: forceps, scissors, graspers, stapling and others, as well as the endoscope, known as the laparoscope to carry on surgical procedures. Only some skin incisions using a trocar are necessary to introduce them in the patient abdomen. Consequently, laparoscopic surgical interventions are in some way a kind of teleoperation, avoiding the need to cut the abdominal wall as it happens in open surgery. The result is that patients can undergo major surgery with short post operative recovery process.

The price of these advantages are the problems associated with teleoperated systems. First, the arms and fingers degrees of freedom that the surgeon use to utilize in the surgical procedure are reduced to only the four DoF of the instruments, when they enter the abdominal wall through the trocar. The second is due to visual perception, there is no direct vision to the working area, only through a TV camera, producing a lack of visual workspace reference frames. And third, the surgeon loos the haptic perception.

In laparoscopy, the main efforts done up to now to aid the surgeon and to decrease intervention risks have been oriented towards both, the increase of the tele-perception capabilities and to the development of robotic systems to hold and move the endoscope. The main efforts in improving the surgeon and assistants perception has been the development of 3D visualization systems [2], to avoid the loss of information and tiredness produced by flat images. To replace haptic perception, some devices have been studied to incorporate force sensors in the instruments [3]. In what refers to robotics aids, a six degrees of freedom manipulator controlled by the surgeon by means of a foot pedal is described in [4]. Other manipulators have already been designed and tested using different interfaces to receive the surgeon orders, either through oral commands or by means of head movements [5] or teleoperated [6]. Additionally, a feasibility study in the use of image processing and other human-machine telerobotic command interfaces have been studied and evaluated [7]. The automation of the camera guidance system by means of computer vision is described in [8].

Based on the RET (Robotic Enhancement Technology) concept, current research is being done in the study and development of robotic manipulators under human control [4] combining the human dexterity and intelligence with the robotics accuracy and control capabilities. Image processing and analysis [9] for diagnosis and surgical planning is also essential for many robot assisted surgical interventions. Virtual reality [10] is also a field to explore to improve visual, touch and even sound perception [11] in minimally invasive surgery and its extension to tele-presence and tele-surgery. These technologies are improving the ergonomics of laparoscopic surgery trying to solve the problems of eye-hand coordination, visualization, sensing feedback and skill acquisition, as described in [12].

2. IMPROVEMENTS ON SCENE VISUALIZATION

The perception of the workspace through a monitor represents to the surgeon an additional handicap to the difficulty of operating inside the body through some holes in the abdomen, that restrict the movements of the surgical tools.

The lack of tridimensional perception using a conventional monitor demands from the surgeon ability and experience to operate efficiently. To compensate this deficiency of the perception system it is necessary to change frequently the TV camera position to look for the optimal points of view. With the aim to improve the scene visualization some tridimensional perception systems are used. Some systems, named pseudo-3D displays, use computer graphics capabilities to generate images, which through shading and changes of the point of view allow to appreciate the scene tridimesionality. Other systems produce true 3D images representation, based on endoscopes provided with two sets. These systems can be classified in two different types, using:

- A monitor with a left-right alternating image and commuted eyeglasses synchronized with the image frame.
- Eyeglasses with micromonitors incorporated on them.

In these systems the image acquisition is based on the use of a laparoscope endowed with two optical sets from which the two images are obtained. The need to minimize the probe size forces to reduce the distance between the two optics to a value of 12 mm, or even less. These short distances are attainable using Hopkins optics.

Even with these short distances between the optics it is possible to produce 3-D perception similar to that obtained through direct vision. This possibility is due to the reduced distances between the optics and the scene, that in these environments are shorter than usual in human vision.

The difference between these two systems is in the way the images are visualized. In the first case, a monitor presents alternatively the right and left image, while the special glasses occlude synchronally the vision to one or the other eye. To assure a visualization without flickering using alternatively one eye after the other, it is not possible to use the standard 50Hz (or 60Hz) TV cameras frequency. This fact forces to double the frequency of the monitor by projecting the two simultaneously obtained images in 20ms (or 16.6ms), alternatively at 10ms per image (or 8.3ms) switching the polarizer. One such system is shown in figure 1.

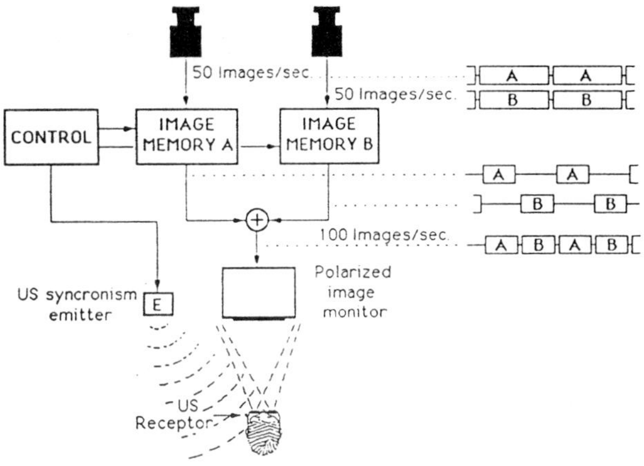

Fig. 1 Acquisition of alternating images at 100Hz from two TV cameras with a 50Hz frame rate

Crystal liquid eyeglasses are used to occlude alternatively the vision of one eye and the other. These glasses allow to change alternatively the polarizing angle at each glass producing the image occlusion if the monitor emits polarized light through a fixed polarizer.

The polarizing commutation is synchronized by an US emitter that transmits the image synchronism signal. This kind of radiation has the advantage over IR beams that there is no need of direct vision between the emitter and the receiver. This situation can frequently appear in a surgical intervention.

The advantage of using a monitor with the alternated images and commuted eyeglasses is that the obtained tridimensional image can be observed by different users, which can at the same time see their environment. This system main drawback is the need to be relatively close to the monitor and that it produces visual tiredness when used for long time periods.

To reduce the visual tiredness produced by this tridimensional visualization the system based on monitored eyeglasses can be used. These eyeglasses incorporate their own mini-monitor in front of each eye. Consequently, each eye sees simultaneously the image obtained from the corresponding camera, at 50Hz (or 60Hz) without the need to commute. The drawback of this system is that the surgeon can only see the scene of the patient interior. He has to take out the eyeglasses to see the environment. For this reason, even its higher visualization quality, this system is rarely used in laparoscopic surgery.

3. THE ROLE OF COMPUTER VISION

The main problems concerning to visual imagery are the reduced field of view, the occlusions produced by the own patient organs due to the reduced operating environment dimensions forced by the minimally invasive techniques and the loss of a visual reference frame. The frequent use of 2D images visualization constitutes, as already mentioned, an additional problem. In order to tackle these problems, research in computer vision techniques is currently leading to the development of some aids for the surgeon to work in much better conditions, and thus reducing the intervention time and risks.

The problem derived from the reduced field of view of the camera requires a great effort from the surgeon. He or she has to mentally locate in position and orientation the patient organs and surgical tools in the workspace and to compensate the perspective distortion of the reduced operating area visualized within the global work environment. This requirement produces a much higher tiredness and tension level than that produced in classical surgery, and interpretation errors can lead to dramatic consequences. Also the camera guidance to track the surgeon zone of interest constitutes a problem to be solved. Robotics can offer great possibilities by means of image processing techniques to control and guide the TV camera movements with 3, 4, 5 or 6 degrees of freedom. That means that the computer vision system must be able to interpret up to some level the image, so as to make possible to automatically stabilize the visualized image.

The use of computer vision can be extended to the control of the electronic zoom. Electronic zoom provides the possibility to approach the image without the need to physically approach too much the endoscope to the operating zone. This control is specially useful in high resolution cameras. With such systems the surgeon won't need to mentally translate or rotate the image with respect to the movements he does and to care much about the selection of the visualization area.

Computer vision is advancing towards the improvement of images visualization, either by choosing the most adequate field of view, by stabilizing the image or by adding to it new information to facilitate the operating space interpretation. In what follows a system developed to automate the camera movements is described and possible future research lines are pointed out.

3. 1 Image enhancement and virtual sensors capabilities

Image processing can be useful in the enhancement of the TV images in laparoscopy and other surgical fields, as it is already used in other application areas. Figure 2 shows (printed in black and white) an example of a color image processed to improve the quality of TV images visualization.

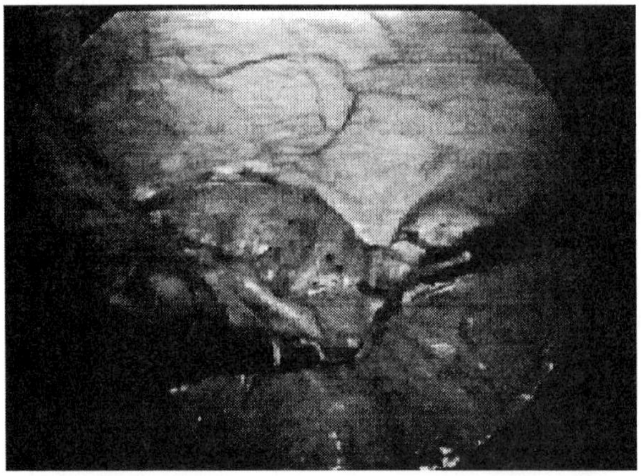

a)

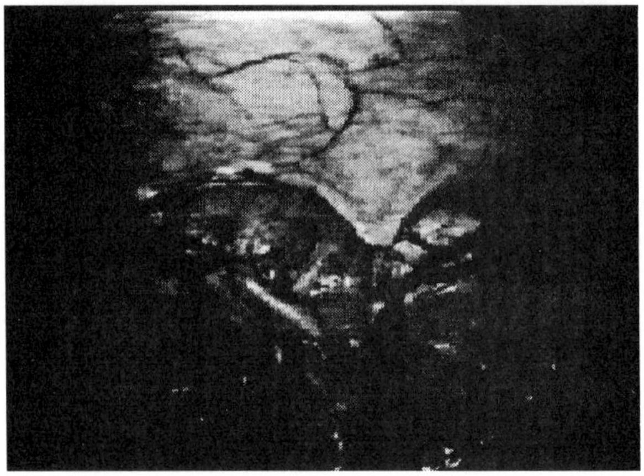

b)

Fig. 2 Image processing to improve scene visualization. a) Original image, b) enhanced image

The extraction of relevant information from the acquired TV image can be used to improve its quality. Thus, image segmentation could be used to define more clearly the boundaries of the different organs and body elements, drawing over the image the boundary lines. Also, the application of pattern recognition techniques could be used to detect an area of interest and to enhance its visualization by means of brightness or superposing a mark.

The use of other imaging devices (USS, RMI...) information could be applied with the same goal.

Computer graphics techniques allow to generate an image from the addition of different views taken with the laparoscope from a sweeping of the scene.

The acquisition of one such panoramic image of the working scene, from the different acquired images requires, on the one hand, to perform the geometric corrections required to minimize the boundary discrepancies appearing between contiguous images. On the other hand, the image shifts which are necessary to obtain the maximum coincidence among them must be found automatically.

This procedure allows to punctually produce panoramic images which could not be obtained by means of the optical sets used in laparoscopy, resulting in a useful aid for the surgeon to locate the laparoscope image within a wider scene environment.

3.2 Visual simulation

With the use of computer graphics it is also possible to generate synthetic images that can be superposed over the real obtained images with the aim to test different strategies and actuation without the need to carry them on physically. An example of its application is shown in figure 3, which visualizes the process of the inginal hernia repair interventions with a polypropylene mesh and the synthetic image generated. In this case, it is possible to simulate the introduction and application of the mesh on the screen, checking the mesh size and its right position and orientation over the desired element to obtain the desired results.

3. 3 A Vision System for Robotic Guidance of the Laparoscope

To provide the surgeon with the visualization of the intra-abdominal working space it is necessary to maintain the laparoscope focusing adequately the

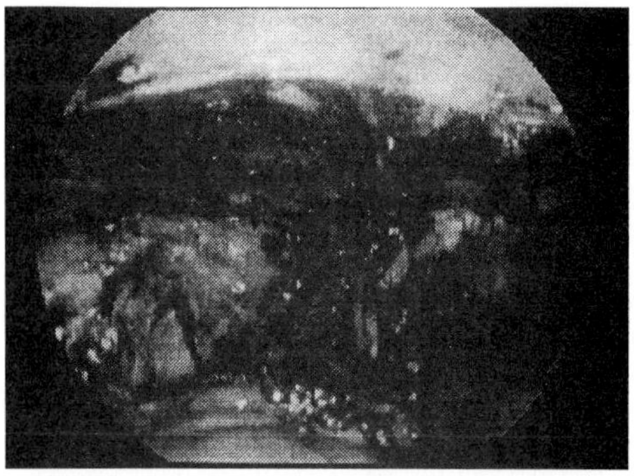

a)

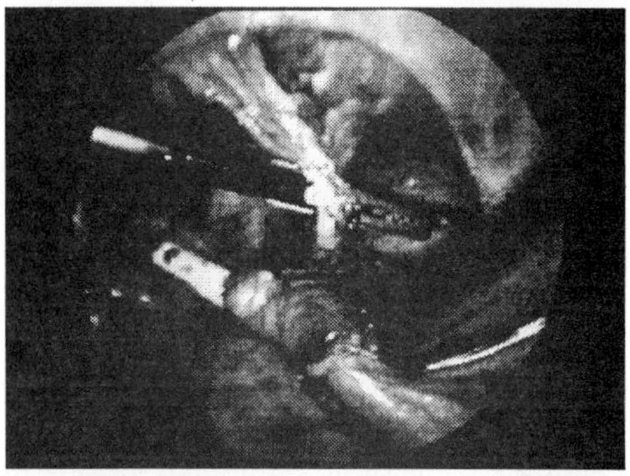

b)

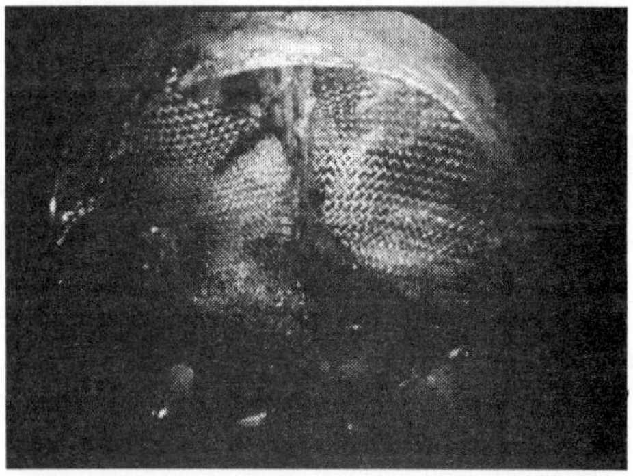

c)

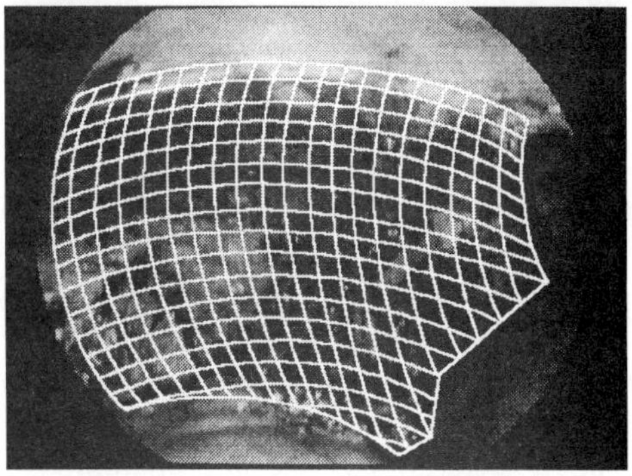

d)

Fig. 3 Inginal hernia repair. a) Image of the inginal zone, b) a step forward in the repairing process, c) after the application of the mesh, d) synthetic image working zone. A good centering of the camera field of view onto the interest zone is essential to operate efficiently. The movement of the camera probe, the endoscope, to adequately track the zone of interest can be done in the following ways:

- Fixed observation
- Manual guidance
- Teleoperated
- Automatic guidance

Fixed observation is based on the use of a mechanical holder that hangs the camera in a predefined position. Punctual manual repositioning by the surgeon or the assistant will be needed. The advantage is the stability of the obtained image and to avoid the assistant tiredness, but on the other hand, it requires to use wider views and consequently providing less scene details. The surgeon himself can change the image point of view when necessary.

The manual guidance done by an assistant, or the surgeon himself, provides a more effective track of the workspace. In this case, a certain assistant experience and a surgeon - assistant mutual understanding is necessary, although usually difficult to obtain. Consequently, some efforts have been done towards the automation of these movements under the surgeon orders. Some teleoperated systems have been built and experimented, as already mentioned, to reduce the problem of the surgeon - assistant mutual understanding, and the adequate interfaces (foot pedal, helmet, voice recognition) are used to teleoperate a robot.

The automatic guidance of the TV camera can be done from the analysis of the visualized image itself. The experimental system developed in our Lab. [8], consists of a robot which holds the camera probe handler and the vision based control system to move the robot. Figure 4 shows the system structure.

The camera movements are controlled from the distance measured by the image processing system between the center of the obtained image and that of the estimated interest center, according to the operation mode indications given by the surgeon.

The guidance system efficiency relies on the kind of scene tracking strategy used to continuously keep the visualized image over the interest zone. Consequently, either the surgeon working tool, an auxiliary tool (as for instance the abdominal wall-lift) or a determined relevant anatomical scene element can be tracked. Additionally to the tracking process, the scene elements tracked have to be adequately filtered to avoid the continuous and too sharp camera movements which would produce tiredness to the surgeon.

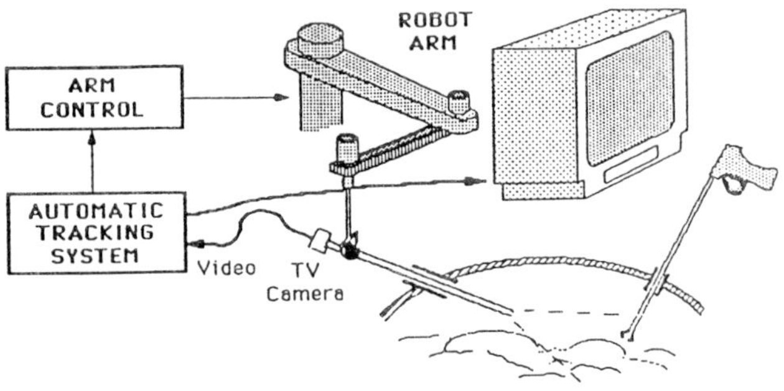

Fig. 4 Structure of the vision based laparoscope system

The system is programmed to track one of the tools according to the surgeon needs or the central point between two of them, as shown in figure 5. It is also possible to select a given relevant scene element.

The computing time required for this identification has to be short enough so as to attain the response speed required to track the scene elements [13]. This vision system is based on the extraction of a very elemental set of features but effective enough to assure the minimum robustness required. Proceeding this way, the obtained operation speed is 12 Hz which allows to perform a very efficient tracking given the dynamics of this type of scenes. A supervisor has been implemented to automatically determine the behavior to be followed by the tracking system and robot guidance. The three different behaviors considered are:

- Fixed observation
- To track the element of interest

The robot can also be used as the support in the fixed observation mode. In this case, the change of the point of view done from time to time by the surgeon, or the approaching and moving away movements, are carried on in a much easier way.

The automatic tracking system is based on the kind of movements observed. In the present state they depend on the stable position of the working tool, on the period and amplitude of the movements centered in a point and on the non repetitive transfer movements.

This behavior is not evaluated as equally satisfactory by all the surgeons, consequently an adaptation is required to be done for each surgeon. In the future, this adaptation will be able to be done not in a totally empirical and manual way, but automatically. That will be possible using a neural controller with learning capability.

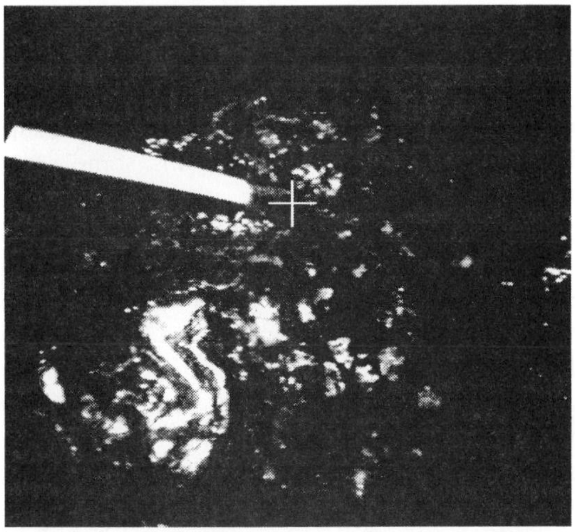

a)

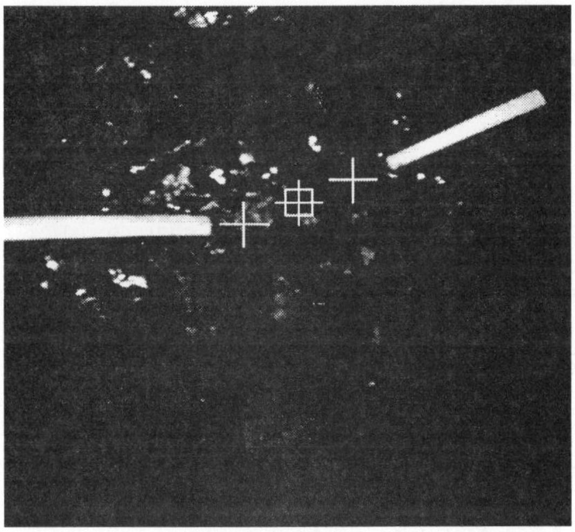

b)

Fig. 5 Tracking of the scene interest points. a)scene with only one tool in operation, b) tracking from two tools

When the operating conditions do not require to track the tools, the surgeon can fix the arm to a point or he can stabilize the image with respect to an element in the scene. In this case, the scene, that usually does not contain elements relevant enough is recognized from the extraction of some given characteristics.

To interact with the vision system some external orders select the tracking and guidance mode. These orders can be given by the surgeon or by an operator. In this phase of the project, these orders are given by the operator.

Thus, the vision system does not completely avoids the requirement of an assistant but the assistant does not need to continuously track the scene and also he does not occupy a work space around the patient. In order to really release more space it will be necessary to design a robot lighter and less invasive than the industrial robot utilized in this experimental phase (Fig. 6).

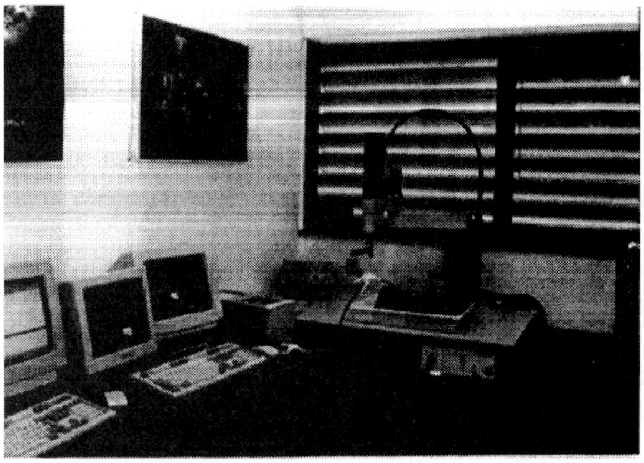

Fig. 6 Visual robot guidance system testing environment

4. ROBOTICS POSSIBILITIES IN LAPAROSCOPIC SURGERY

RET systems, having an increasing use in different application fields today, introduce the combination of sensor based robotic manipulation with human control and supervision. The result is a system that can perform tasks beyond the individual capability of either the human and the robot. The surgeon - robot cooperation can be done in different operating modes.

In a first mode, the robot is an assistant arm that can hold an instrument (fig.7) and can aid in the execution of some actuation such as to make a hole. The advantage of this assistant hand is that the robot can stay fixed, even supporting some efforts. It neither gets tired or vibrates. The robot can also be endowed with force control and then perform some force based control strategies.

In a second mode the surgeon operates through the robot that acts as an improved surgical tool. The operation is done by the surgeon that teleoperates the robot (Fig. 8). This cooperation combines the human intelligence in guiding the task and the robot that can move more precisely, with slighter motions and without vibrations. In this case the robot can be compared as the ruler for the draughtsman, who can not draw straight lines freehand. The design of the adequate interface and the means to improve the surgeon perception are the keys to a good cooperation.

A third means of surgeon - robot cooperation can operate from the robot actions that are preprogrammed and computer controlled. Based on a model of the part of the body to be operated or the corresponding data obtained by different imaging devices such as RMI or tomography a task previously defined by the surgeon can be programmed and executed. With this cooperation the robot appears as an active tool that can decide the movements and even actions based on the model or images and on the programmed task. This kind of tasks can be easily programmable, as to make a hole, can be repetitive, as sewing, or more complex ones in the future. The surgeon in this case supervises the robot operation and corrects the strategy when an eventuality or not foreseen situation appears. The task can be automated just up to the level in which the sequence of actions and sensor based movements can be predicted, and in any case, the surgeon is totally responsible of the previous program definition and the supervision during the programs execution (Fig. 9).

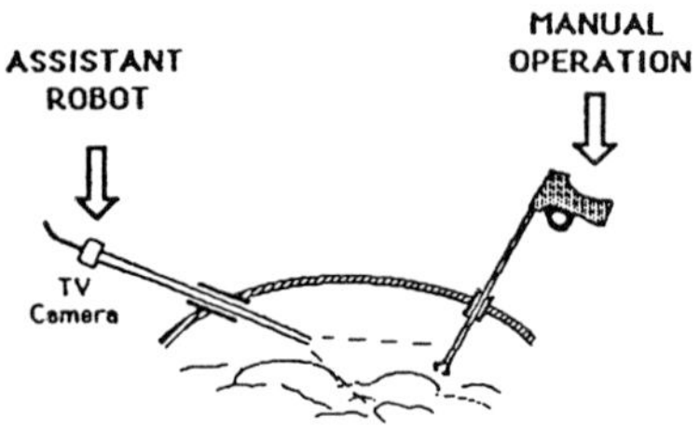

Fig. 7 The robot acting as a surgeon assistant hand.

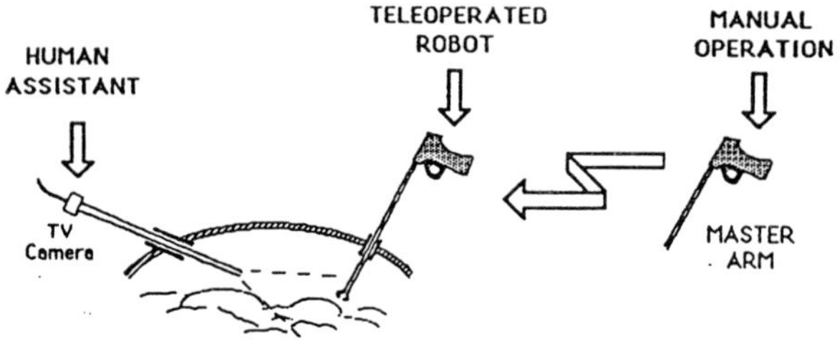

Fig. 8 Structure of a system for teleoperated surgical interventions

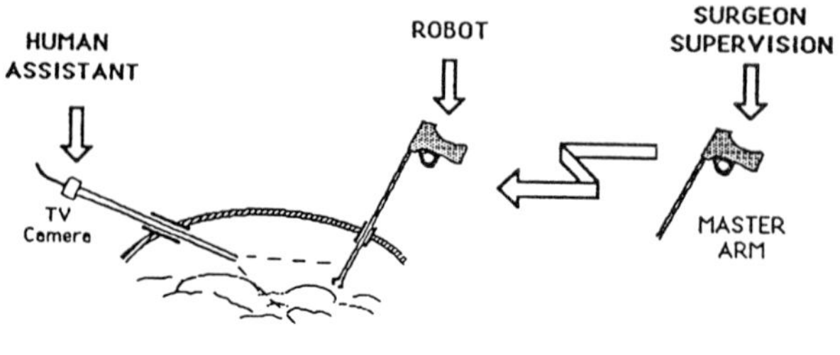

Fig. 9 Computer assisted robot intervention with surgeon supervision

Based on either of these operation modes, robotics aids in laparoscopic surgery can be oriented to: first, the manipulation of the aiding systems such as the holder and mainly the camera holder or laparoscope, and second, the robot could in some cases manipulate the surgical tools.

Current advances in robotics aids for laparoscopic surgery have been done in the first direction, the control and guidance of the TV camera. Different robotic arms have already been built or are under experimentation which are controlled by the surgeon. One such robotic arm moves according to the commands given by the surgeon through a joystick, foot pedal, a strip on the head, a helmet or even a voice recognition system. Among them we can mention the EndoSista robot and Aesop (Automatic Endoscopic System for Optimal Positioning). Their different levels of control have already been discussed including their control by means of vision.

Robotic manipulation of the surgical tools has been investigated and experimented in other fields of minimally invasive surgery, but not yet in laparoscopy. Analyzing the main characteristics of medical robotics it is appreciated that precision is a main feature of human-robot cooperation in this field. The hand tremor problem, appearing in any kind of microsurgery, is much less of a problem in laparoscopy because each instrument is attenuated at the fulcrum when it enters the abdomen. Control of in and out movements is more of a problem and there robotics could help. Also, laparoscopic surgery requires a high level of dexterity in the execution of some operations to be done in very restricted environments and without direct perception of the work in course. Again robotic assistance in such tasks can open new possibilities.

On the other hand, robots programming difficulties due to the amount of uncertainties in surgical environments are decreasing due to the progress going on in medical imagery and sensing. Robot control based on environment sensing can be classified in:

- Geometric control
- Force control
- Imaging model based control

4.1 Geometric control

The eye-hand problem due to the mirror effect in laparoscopic visualization produces different levels of paradoxical movements of the surgeon in his task and can even block his mind specially in very critical moments. Other visual aspects due to the partial vision of the scene can also difficult the right interpretation of the parts relative size, their distance, relative position, etc. The surgeon - robot cooperation could avoid these problems if the robotic system could take absolute reference elements over the patient.

In a robotized or teleoperated task it is possible to define some movement restrictions that in the case of surgery could avoid some damages to the patient. The restriction to move the arm in a predefined direction or to force the robot to move only on a plane, even when the surgeon manual guidance shifts towards a forbidden direction, avoids problems derived from hand tremor or loss of depth information among others. Figure 10 shows an example of a geometrically bounded movement.

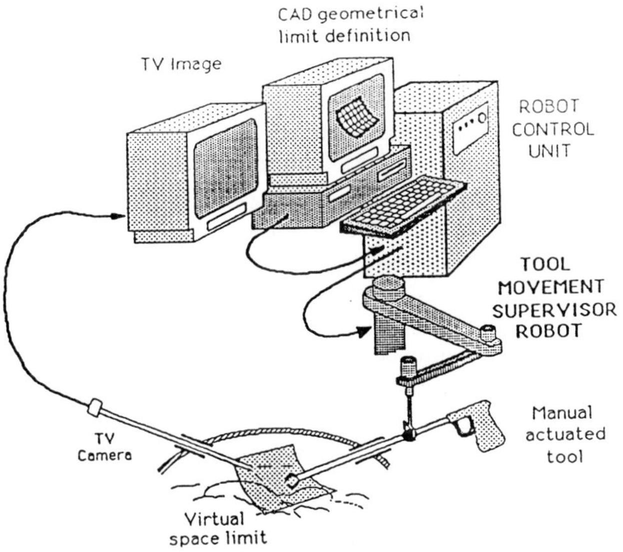

Fig.10 Geometrically bounded movements

When the robot reaches a predefined surface the safety control strategy can force the robot to stop, as if it collides with a virtual obstacle (Fig. 11), or to damp the movement emulating a collision with a virtual pillow.

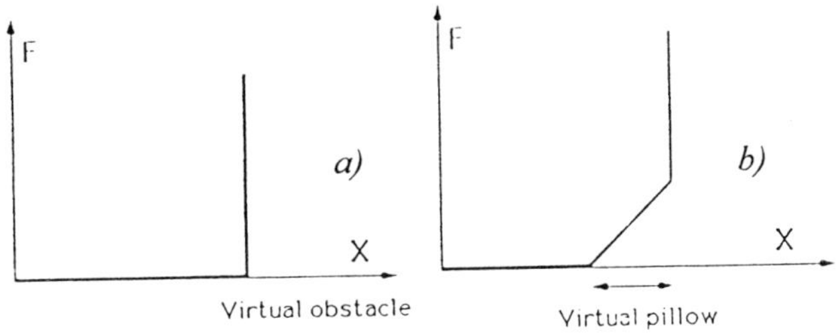

Fig. 11 Geometrical bounded movement strategies. a) Solid virtual obstacle, b) Soft virtual obstacle

4. 2 Force control

The use of force sensors allows to provide the surgeon with the feeling of touch [14] for his better perception of the environment. Force sensing also allows to improve the robot control strategies and to increase the operation safety. A force sensor located on the tool measures the efforts done directly on the operating point, not those done by the surgeon. With this data, the supervision system can limit the tool movements to avoid the application of a force greater than that programmed. A tridimensional measure of the exerted forces and torque's is required.

Figure 12 shows a robotic arm with tool movement supervision based not in geometrical data but on the applied forces. Some experimentation is being done in the introduction of force sensors in the laparoscope. The main problems are the sensor size and the cables to transmit the signals through the laparoscope.

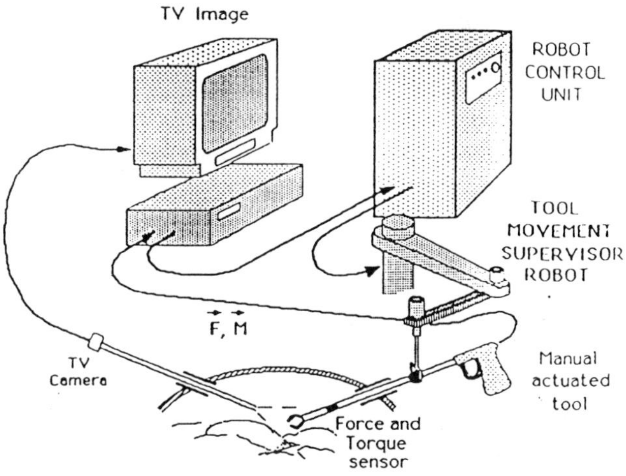

Fig. 12 Supervision based on the applied force

4. 3 Imaging model based control

The use of models of the working area based on the imaging devices usually used in medicine such as RMI, tomography or ecography allows to plan and program safety trajectories. The execution of these trajectories with the aid of robots is essential in surgical interventions where the surgeon precision is not enough, this is the case of neurosurgery. In laparoscopy, where precision is not so critical, the problem to robotize is not the positioning of the robot end effector in a point defined by the imaging model, but the access strategy. The accessibility problem comes from the lack of a sufficient number of degrees of freedom to move inside the patient body. Another difficulty is that the target point coordinates are not known enough and they can change during the operation.

The possible interest of robotics to actuate the surgical tools is in the execution of routine tasks. In this direction and based on the obtained images, the surgeon could, for instance, define the end points and several passing points along a line to be sewed, or define the limits of a surface to be scraped. Based on this image interpreted by the surgeon and the data he generates, the robot could effectively execute a predefined and programmed task.

4. 4 Safety issues

In the field of robotics surgery the patient safety is an essential issue, and different safety aspects have to be considered.

In a first group we can consider all the safety aspects related to the required electrical protections. The lack of an efficient connection of an instrument to ground could produce the Faraday effect with respect to other electrical instruments used. On the other hand it is necessary to verify the efficient protection of the equipment against EMI, specially when using electrical instruments.

A second aspect to consider is that related to the possible physical damages, produced by undesirable movements. In this case, the control based on the use of force sensors protects against improper overefforts. This control is performed either by the fact of providing the surgeon with haptic perception when working manually or through teleoperation, or when the robot operates automatically. An additional safety step is the introduction of mechanical fusses, that is, elastic couplings or self-unplugging devices.

The process supervision constitutes a third safety issue. The introduction of different kind of microsensors to detect presence, flow or level provides the means to minimize the surgeon perception and reaction time, in front to different events.

5. FUTURE TRENDS

The actuation of forceps and other surgical tools through the trocar makes laparoscopic surgery easier to robotize than open surgery. This is due to the fact that these surgical instruments are closer to a robot end effector than the classical tools and the surgeon hands, and their actuation easier to automate. The difficulty, with respect to industrial robot applications, comes from the uncertainties of the working environment that prevents to program a priori a task. Robotics in this case offers the possibility of the control from human gestures, that is, teleoperation.

The requirements in teleoperation tasks do not rely uniquely in providing the desired mobility to the tools needed for a given surgical intervention, which do not implies technological problems, but also in the availability of sensors feedback to the operator, in this case the surgeon. This feedback perception allows to operate safer and with more efficiency.

Therefore, the possibility to introduce telerobotics in laparoscopic surgery is conditioned to the development of the adequate perception systems that allow to manipulate indirectly the tool. Once this problem is solved, that is the tools are mechanically manipulated from the surgeon gestures, he or she can already operate from outside the operating theater. This possibility can mitigate both, accessibility problems that frequently appear, or even to have available in a given situation a specialist that is physically far away [15]. This kind of operation can then be done with the same guarantees than the operations done with the direct presence of the surgeon close to the patient. An on-line interactive ISDN network for telemedicine and telepresence is described in [16].

The requirements in sensorics perception are on the one hand the vision and on the other hand the force. In what refers to vision, the added distance when the surgeon is not present, does not imply any additional obstacle since the sensor is a TV camera that generates a signal that can be transmitted without problems at long distances. The use of the laparoscopic surgical tools through the trocar reduces the force perception in current laparoscopic surgery. Thus, the force feedback required to operate with the same guarantees at some distance must be at least the same level that the surgeon can perceive when operating *in situ*. The introduction of miniaturized force and torque sensors on the surgical tools constitutes the way for the not very far future use of telerobotics in laparoscopic surgery and to extend this technology to the exploration or surgical interventions with the telepresence of an specialist. Both research lines are already alive and some experimentation has been done. The possibility that these specialists operate at any distance and that they are only required in the more critical moments can bring in the future an increase of quality. It will be possible both, to work punctually with an additional support and also to have the availability to remove part of the work equipment outside the operating theater, in larger working areas more comfortable to the surgeon. The additional possibility of introducing force restrictions, of defining boundary surfaces that can not be surpassed by the robot end effector and the definition of magnified geometries of the working environment, would allow the surgeon to operate with much more precision and safety even than that attained in current open surgery. This improvement is due to the actuation and compliance difficulties that are produced many times in minimally invasive surgery.

Another future possibility is the availability to use real time image processing techniques superposing, at the surgeon will, some computer information extracted from other sources over the real images. The sources of additional data could be tomography or RMI from which it is possible to generate synthetic images.

The feedback of vision, force and visually enhanced synthetic images converge towards virtual reality [17], which progress will have also a big influence in tele-surgery. The massive implantation of wide band computer networks, that allow data and images transmission in real time, will facilitate these advances.

Research in microrobotics and micromechanisms has also opened a new expectative in the development of flexible endoscopes, which provide an easier way to move and to operate inside the patient [18]. The development of flexible instruments will provide an easier maneuverability and dexterity in some given operations.

6. REFERENCES

[1] M. H. Ornstein. Robotics in Laparoscopy and Thorascop. In: First European Conference on Medical Robotics, Robomed'94 Barcelona, Spain. June 20-22, 1994, pp. 17-21.

[2] R. M. Satava. 3D Vision Technology Applied to Advanced Minimally Invasive Surgery Systems. In: Surgical Endoscopy, Springer Verlag New York Inc. 1993, pp. 429-431.

[3] R. D. Howe et al. Remote palpation technology for surgical applications. In: IEEE Eng. in Medicine and Biology Magazine, 1995

[4] N. I. Narwell, D. R Weker and Y. Wang. Force Controllable Macro-Micro Manipulator and its Application to medical robotics. In: JPL Computer Motion Inc. report 1994

[5] P. A. Finlay. Small is Beautiful. In: First European Conference on Medical Robotics, Robomed'94, Barcelona, June 20-22, 1994, Spain. pp. 13-15.

[6] R. H. Taylor. An overview of Computer-Assisted surgery at IBM. In: G. Robotics Research: IFRR, 1994, pp. 533-544.

[7] D. Hurteau et al. Laparoscopic Surgery Assisted by a Robotic Cameraman: Concepts and experimental results. In: IEEE International Conference on Robotics and Automation, 1994. pp. 2286-2289.

[8] A. Casals, et al. Vision Guided Robotic System for Laparoscopic Surgery. In: 7th Int. Conference in Advanced Robotics, ICAR'95. Sant Feliu de Guíxols-Spain, September 1995, pp. 33-36

[9] B. Geiger. Three-Dimensional Modelling of Human Organs and its Application to Diagnosis and Surgical Planning. In: Report n° 2105 INRIA, November 1993

[10] R. A. Roob and B. Cameron. VRASP: Virtual Reality Assisted Surgery Program. Journal of Computer Aided Surgery Vol. 1 n. 2, 1995, pp. 33-45.

[11] D. G. Caldwell et al. Tele-Presence: Visual, Audio and Tactile Feedback and Control of a Twin Armed Mobile Robot. In: 1994 Int. Conference on Robotics and Automation, San Diego-USA, May 1994, pp. 244-249.

[12] M. Patkin and L. Isabel. Ergonomics and Laparoscopic General Surgery. In: Annual Scientific meeting, Royal Australasian College of Surgeons. Adelaide-Australia, August 1991, pp. 83-88.

[13] J. Amat et al. Tracking capabilities in Computer Vision. Optical Engineering, vol. 32 n 11, November 1993 pp. 2796-2804.

[14] S.C. Ho et al. Force Control for Robotic Surgery. In: 7th Int. Conference in Advanced Robotics, ICAR'95. Sant Feliu de Guíxols-Spain, September 1995, pp. 21-31

[15] J. W. Hill et al. Telepresence Surgery Demonstration System. In: IEEE International Conference on Robotics and Automation, San Diego-USA, 1994. pp. 2302-2307

[16] A. K. Bejczy and P. Fiorini. An On-Line Interactive ISDN Network Connection for Telemedicine / Telescience. In: Report Jet Propulsion Lab , 1995

[17] A. K. Becjzy. Virtual Reality in Telerobotics. In: 7th Int. Conference in Advanced Robotics, ICAR'95. Sant Feliu de Guíxols-Spain, September 1995, pp. 3-12

[18] A. B. Slatkin and J. B. Burdick. The Development of a Robotic Endoscope. In: International Symposium on Experimental Robotics, ISER'95, Stanford, California, 1995. pp. 100-105

Design of Customized Rehabilitation Aids

Vijay Kumar and Ruzena Bajcsy
GRASP Laboratory, University of Pennsylvania, Philadelphia, PA (USA)
email: kumar/bajcsy@central.cis.upenn.edu

William Harwin
Applied Science and Engineering Laboratories, A. I. duPont Institute
University of Delaware, Wilmington, DE (USA)
email: harwin@asel.udel.edu

Abstract

We present the component technologies that are essential for rapidly designing and producing customized rehabilitation aids for people with motor disabilities. We show that methods traditionally used in robotics and computer vision can be used to formulate and solve many of the problems that are encountered in automating this process.

1 Introduction

There have been many attempts to design robot systems [2, 5, 10, 19, 20] for aiding people with physical disabilities in their activities of daily living. These include robotic aids that operate in such structured environments as offices and act as vocational assistants [20], as well as systems which are designed for operating in unstructured environments [19]. Systems that belong to the first category allow disabled people to pursue a vocation and to attain a certain level of autonomy [32]. The robotic manipulation tasks encountered here are considerably simpler because the environment is structured. In contrast, systems in the second category include such general purpose aids as a six degree-of-freedom robot manipulator mounted on a wheelchair with different end effectors [19] and a wheelchair with two arms that can be used for manipulation as well as for climbing over or stepping across obstacles [34] and are potentially superior for functioning in unstructured environments. However, these versatile aids appear to be be less acceptable to people with disabilities.

There are many reasons for the lack of success of general purpose robotic aids in this community [28]. Such systems tend to be very complex and therefore unreliable. They are also prohibitively expensive. Robot arms move slowly and awkwardly, but at the same time appear to move too quickly in tasks involving contacts with the environment or the user. People with disabilities are often leery of electromechanical and possibly computer controlled aids that rely on sophisticated technology. Finally, one of the main obstacles is the difficulty that the users have in controlling such complex systems [21]. In contrast to a prosthetic limb where the user is in intimate contact with the limb and therefore has proprioceptive feedback (Doubler and Childress [6] have called this extended proprioceptive feedback), the user typically only has visual feedback from a robot system. While haptic interfaces is an active area of research, there appear to be inherent limitations with the technology that preclude simple and cost-effective mechanisms for force and touch sensing [4].

Consider, as an example, the task of feeding for a person who has little or no control over the upper extremities. The person may use intact musculature in the legs or in the head and neck, either directly [10, 27] or by generating myoelectric signals [3], to control a robotic arm to feed with utensils. However, the robot arm is an expensive product and is difficult to use. In contrast, consider the solution shown in Figure 1, called the Magpie [7], in which an articulated passive feeding mechanism allows the user to manipulate a feeding utensil *via* a set of cables. It is reminiscent of the early teleoperators which had no actuators, sensors or computers [33]. The Magpie is an example of a *telethesis*, a device that is closely coupled to the user and acts as an "extension" of the person. This telethesis has four degrees of freedom and is controlled through movements of the leg and foot. Although this device is human

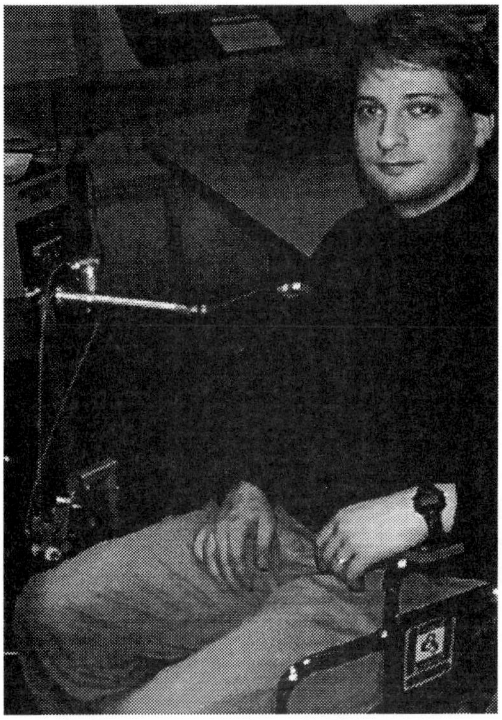

Figure 1: The articulated mechanism for feeding in a foot-controlled feeding device (Magpie) designed at the Nuffield Orthopeadic Center in Oxford [7].

powered, it is possible to use actuators to assist the human so that it remains passive and therefore stable [14]. Finally such devices are simple, reliable and inexpensive. Clinical trials with the Magpie [7] show that there is a high degree of consumer acceptance.

In this paper we consider the customized design and manufacture of a class of rehabilitation aids that is typified by the telethesis shown in Figure 1. The design and manufacture of such products presents a novel problem. Because each person presents a unique neuro-physiological picture, there is considerable variation of performance and function and therefore, it is essential to design tools that are specific to that person. It is necessary to involve the customer in any design process, but this is especially true for rehabilitation aids. Furthermore there are biological changes that occur over time, and it is necessary to allow for adjustments and maintenance or to rapidly redesign and manufacture a new product. Traditional models for product development and manufacturing focus on low-cost, high-volume products. In contrast to this, the manufacture of rehabilitation aids requires the infrastructure and technology to design and produce a wide array of quality products each of which targets specific market needs. While agile manufacturing [29] makes it possible for a designer to move quickly from a preliminary design concept to a prototype, it does not specifically address the need to customize products to individuals.

The primary focus of this paper is on the component technologies that are essential for rapidly designing and prototyping such customized products. Regardless of the specific product class, the first important step in the production of a customized product is the quantitative assessment of the needs of the individual. This involves the acquisition of geometric, kinematic, dynamic and physiological information about the individual which is necessary for developing design specifications and for detailed design. Because the product volume for customized products is likely to be small, the manufacturing cost must be kept low. Thus, there is a need to automate the process of measuring the customer and designing the product from specifications derived from these measurements. In addition, there is always pressure to provide the product quickly and be able to respond to the consumers' needs rapidly.

We consider as a test product a telethesis that could be used by a quadriplegic for feeding, and possibly for typing on a keyboard, for turning pages of a book and for shaving. More than 10,000 spinal cord injuries (SCIs) result in quadriplegia or paraplegia each year in the U.S. alone. Of the population of 300,000 people in the U.S. with SCI, 55% are quadriplegics, and 58% are between 16 and 30 years old [22]. There is often a high level of motivation for disabled people to learn to eat independently [28]. Thus there is a significant population that will benefit from an inexpensive, customized feeding aid.

While most of the discussion in this paper is limited to rehabilitation aids, there are many other examples of customized products. All products that are worn on our bodies, or more generally, products that depend on prolonged contact with human bodies for their functionality, need customization at some level. This class of products includes wrist braces, computer interfaces (keyboards, joysticks), eyeglasses, helmets and sports equipment. The level of customization that is justifiable for each product depends on the cost, time for development and for production, the production volume and ultimately profitability. A discussion of successful product development is beyond the scope of this paper.

Instead, our main objective is to focus on engineering technologies that are required to support the rapid design and production of customized goods and services tailored to the needs of a specific consumer.

2 The Design Process

The design process for rehabilitation products that are customized to a person will involve a number of steps, as shown in Figure 2. Of these, there are three stages that are of central interest here:

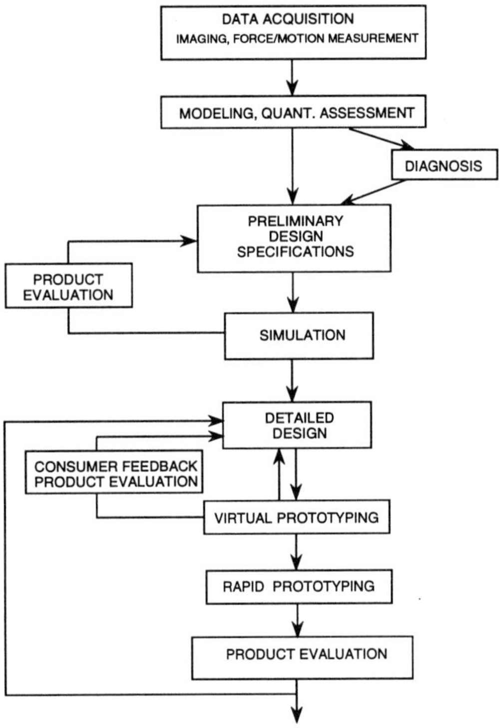

Figure 2: Design and rapid prototyping of one-of-a-kind rehabilitation products.

1. *Data Acquisition*: to measure and observe the user while performing the desired tasks in the appropriate environment and to develop quantitative models for assessment by a therapist or a designer and for providing design specifications;

2. *Virtual Prototyping*: to simulate the user, the product, and their (physical) interaction through powerful workstations with high-resolution graphics throughout the different stages of product design; and

3. *Rapid Design and Prototyping*: to take a preliminary design, convert it into a detailed design and quickly produce prototypes for evaluation and later for production.

In the rest of this paper we describe each of these steps in greater detail, with particular reference to the computer integration. The end target is to design and prototype a successful customized product.

We use the example of a proposed head-controlled feeding device to discuss the basic issues in customized design and prototyping. A head-controlled feeding mechanism is particularly useful for people with quadriplegia. When quadriplegia is the result of a spinal cord injury, the upper extremities and even the torso may be impaired, depending on where the lesion occurs. However, the quadriplegic generally has good control over his/her head and neck movements. In contrast to the foot and ankle control for the telethesis shown in Figure 1, we consider a head-control interface, as shown in the schematic in Figure 3, with which head and neck movements can control the extension and retraction of the cables and therefore the movement of the feeding utensil.

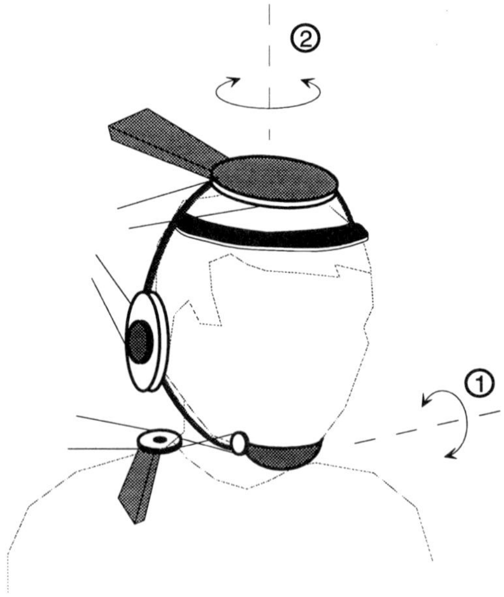

Figure 3: A schematic of a head control interface. The user can move his or her head, neck and chin to control multiple (two are shown in the figure) degrees of freedom.

3 Data Acquisition

It is necessary to measure the capabilities and needs of the individual, his/her environment and to describe the task in quantitative terms in order to generate the specifications for the design problem. For example, the custom design of a head-controlled telethesis for feeding requires the measurement of the geometry of the head, the kinematics of the head and neck, and the forces that the person can apply with his/her head. Similar measurements may also be required for the feeding task (for example, the ranges of motion of the spoon or fork and the forces that are encountered during the task). For customized design we require, in addition to geometric measurements (shape, size), information about the kinematics and dynamics of the individual. Each of these categories is discussed next.

3.1 Geometric measurements

The typical input will be range data from a given human part (shoulder, arm, wrist, neck, head) obtained through the use of two monocular cameras and/or from range finding scanners. Figure 4 illustrates the kind of geometric information that can be obtained from a commercially available optical range scanner. In addition to modeling the human body part, such measurements can also provide information for the design of products that are worn by the user. For example, a helmet or a rigid head band worn on the head, or (as shown in Figure 5) a customized chin strap can be automatically prototyped with such geometric information. Geometric models for such components can be created automatically, and from these models, CAD/CAM models can be easily generated [17]. These models can be used either for prototyping (and manufacturing), as shown in Figure 5, or for inspection and quality control.

The main challenges in obtaining the required measurements turn out to be important research problems in computer vision. First, several range images must be taken of the object from different vantage points and merged into a single surface description. It is necessary to automatically determine where to position the range camera next, making sure all of the range images are perfectly aligned, and merging the range images into a single surface model [25]. Second, robust techniques are needed for concurrently segmenting and fitting geometric models to each segment from data obtained from multiple viewpoints [16]. Finally,

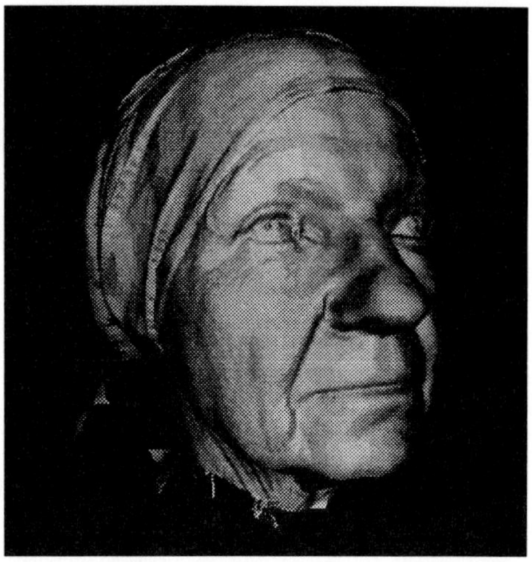

Figure 4: Geometric information of a human head obtained by a Cyberware 3030PS scanner. The information is obtained by scanning the head from different angles and registering the different images. The information of the chin is used to make the customized prototype in Figure 5.

there is a need for methods for representing geometric models at various levels of granularity depending on the desired tolerance and the complexity of the part [16, 17].

3.2 Kinematic measurements

The basic needs are the accurate measurement of the three-dimensional motion of moving body parts (for example, the head and neck) without interfering with the dynamics of the moving bodies and to develop methods for modeling the three dimensional movements. The available measurement systems include optical (accurate to 1-2 mm) and electromagnetic (accurate to 0.3-0.8 mm) sensors [23]. Another possibility is a mechanical manipulandum (an articulated, low-inertia linkage with high-resolution position sensors, one end of which is attached to or moved by the individual) which is accurate to within 0.01 mm [8].

Regardless of the measurement device, it is necessary to automatically determine kinematic models for articulated human limbs with minimal sensing (see, for example, [23]). This problem is similar to the kinematic calibration problem encountered in robotics [12]. While the medical community has developed sophisticated techniques for measuring joint movements, many of these techniques are time consuming and re-

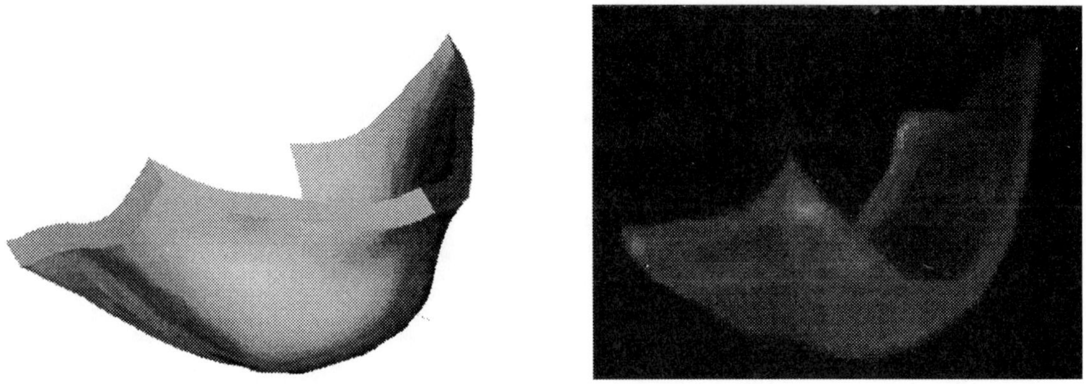

Figure 5: A customized chin strap designed for the subject shown in Figure 4: (a) a ProEngineer solid model; (b) a stereolithography prototype.

quire clinically trained professionals and expensive equipment. Also, in this community, the kinematic models for limb movements are generally limited to ranges of motion at different joints. In contrast, we need fast, non-invasive measurement systems and methods for developing analytical models of limb movements. And frequently, we are not interested in the anatomy of internal features (such as the cervical vertebrae in head/neck movements), but we are only interested in capturing the kinematics of the gross motion.

Our approach to automatically obtaining kinematic models is based on methods developed in a doctoral dissertation [23]. To obtain a kinematic model of, for example, head and neck movements, an electromagnetic sensor is strapped to the user's head. The user is asked to move his/her head while measurements of the head position and orientation relative to a fixed base are recorded. If the kinematic chain connecting the head to the torso has less than six degrees of freedom (less than three degrees of freedom for planar movements), the six (three for the planar case) independent measurements obtained from the sensor must satisfy appropriate closure equations. These consistency equations allow us to infer the structure of the kinematic chain. The cervical spine is incredibly difficult to model because it consists of segments whose relative motions are roll-slide motions (and not pure rotations or translations). However, it turns out that the complex motions of the head and neck are adequately modeled by a kinematic serial chain consisting of only revolute joints.

A model for head and neck movements in the sagittal plane is shown by the planar serial-chain linkage in Figure 6 (a). The movement of the reference frame attached to the sensor (and therefore to the head) with respect to a fixed inertial frame is characterized by a $p \times 1$ vector of constant parameters, x, and a $n \times 1$ vector of joint variables, q. In this case, x consists of the link lengths, a_0, a_1, and a_2, and the angles, θ_0 and θ_3. The variable q consists of the joint variables θ_1 and θ_2 that parameterize the two-parameter motion.

For any position of the head (and the sensor) we can write three (six for spatial models) closure equations of the form:

$$f_i(x, q) = 0, \qquad i \in 1, 2, 3$$

The model is completely described by the vector x and therefore we want to estimate x. Since q varies with the position of the head, each measurement gives us three equations with n additional unknowns. Thus m measurements yield $3m$ nonlinear equations in $p + nm$ unknowns. In this case, $p = 5$ and $n = 2$. If $m > 5$, a nonlinear least squares algorithm can be used to determine the best-fit $2 - R$ linkage. A typical example is shown in Figure 6 (b). While the solution is not unique (the optimization problem is not convex), it is not difficult to obtain very good models. In the example shown in the figure, model matches the experimental data to within 0.01 inches and 0.2 degrees.

Similarly, for general, spatial motions it is possible to estimate best-fit n-parameter models where $n < 6$. For more complicated spatial models, the numerical methods work better if we can eliminate the variables q from the closure equations (which one can always do if $n < 6$). It turns out that a simple variation of the inverse

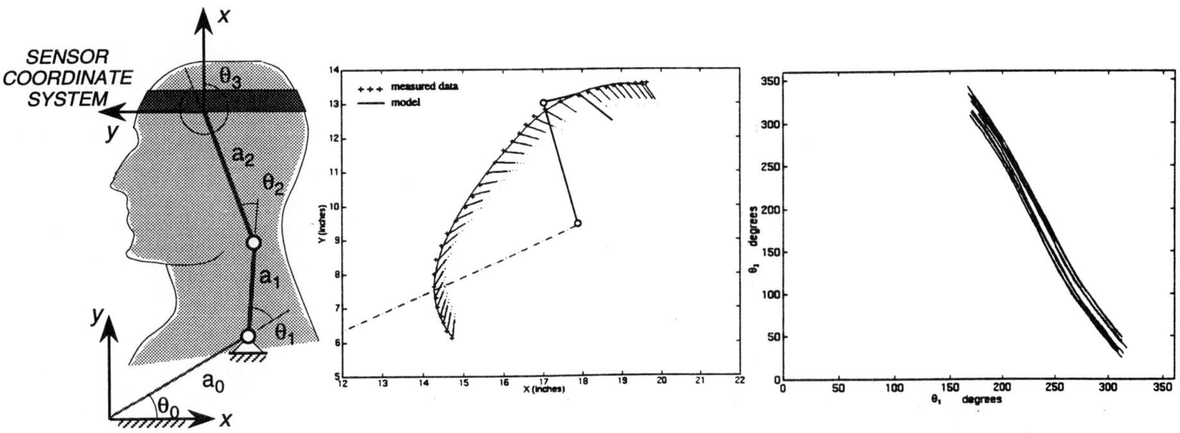

(a) (b) (c)

Figure 6: Head movement in the sagittal plane: (a) a simple 2– model; (b) typical experimental data and the best-fit model; and (c) the movements recorded over multiple trials plotted in joint space. (Figures from [23].)

kinematics procedure for general serial chains [26] allows us to do this. Once the constants x are obtained, the variation of the n joint angles (q) can be obtained by inverse kinematics for each of the measurements.

This procedure is quite general and can be used to construct models for a wide range of n-parameter motions, where n must be less than 6 for three-dimensional movement and less than 3 for planar motions. Frequently, n turns out to be very small for voluntary movements. For example, if we look at the variation of the joint variables in our model with the head movement by plotting the motion in joint space (Figure 6 (c)), we can see that the motions appear to follow the same curve in joint space and further, this curve is closely approximated by a straight line. Our experiments with healthy as well as disabled people (subjects with spinal cord lesions in the cervical spine) indicate that head movements in the sagittal plane are actually 1-parameter motions [23].

Another approach which is not limited by the number of degrees of freedom involves the analysis of the deforming silhouette of moving limbs in a controlled studio-like environment [13]. In this work, no *a priori* model of the human body is assumed. Nor is it necessary to perform the difficult task to segment the image of the human body into limbs. The underlying model is a *composed deformable model* that allows for gross three-dimensional movements as well as local deformations. The basic procedure involves hypothesizing a suitable underlying anatomical structure and then to test this hypothesis against the observed images. The technique applies to humans of any anthropometric dimension. An important direction for future research is the development of algorithms that will perform the data reduction and provide analytical models that can be used for designing the appropriate products.

3.3 Dynamic performance

To measure the performance of an individual, it is often necessary to measure the kinematics of three-dimensional motion while measuring the forces and moments that can be exerted at the end-segment. The motion and force measurements, when appropriately transformed to joint coordinates, reflect joint motion ranges and strength characteristics.

There are very few general-purpose instruments or general techniques to obtain such information. Our approach has been to use a manipulandum [8, 11] or a robot [27] to measure forces and moments that can be exerted by a hand during manipulation. A similar approach has been used to measure forces and moments that can be applied using head and neck movements by quadriplegics and healthy subjects [30]. This measurement combined with a suitable kinematic model for head movement yields a complete model of the static strength characteristics of the individual. Once such a model is available, the design of the feeding linkage can be formulated

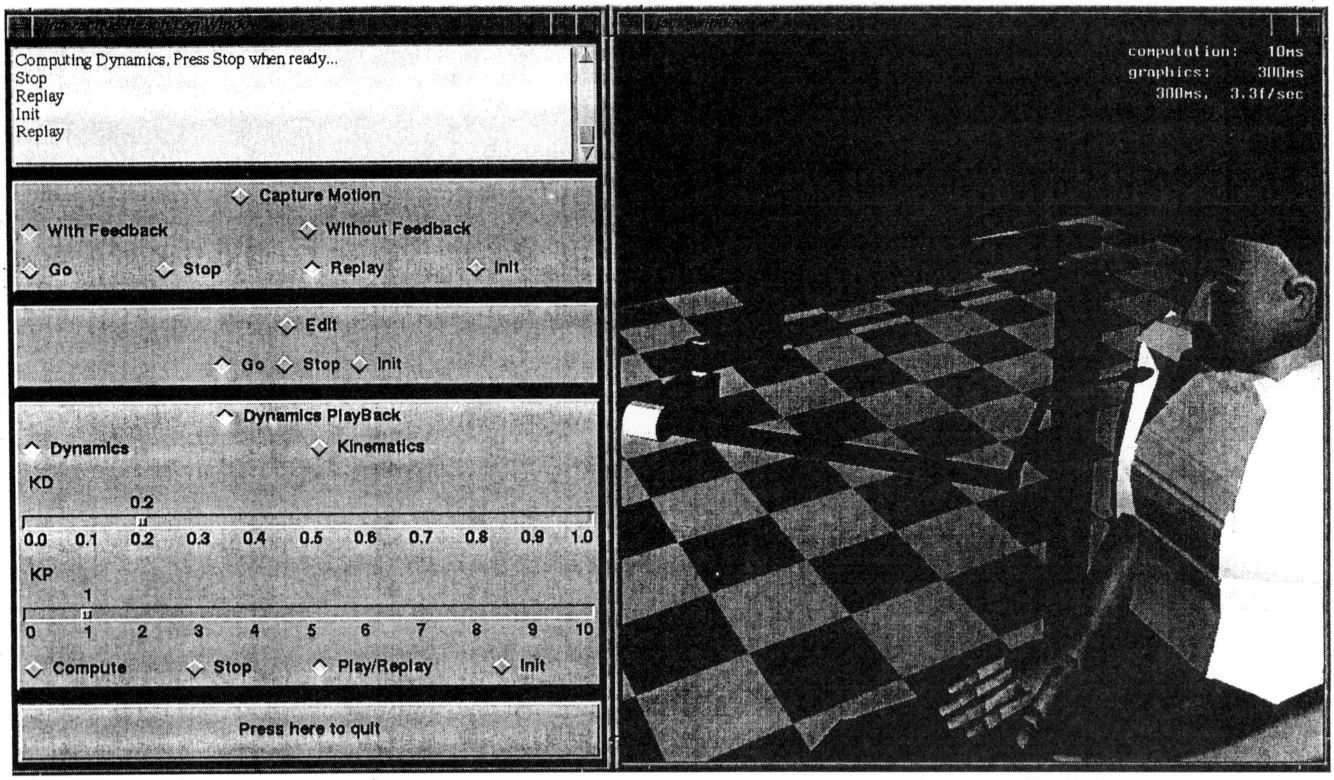

Figure 7: A virtual prototype of a feeding device and the design interface.

fairly precisely.

A natural extension of this work is the dynamic modeling of the head/neck movements where one could obtain impedance models for a range of head positions. However, this is not essential for the application considered here. Other major goals for future research are a general framework for modeling and representing information on human performance, and algorithms for data reduction and representation.

4 Virtual Prototyping

Virtual prototyping is the process of design, analysis, simulation and testing of a product within the computer, and using the results to refine the concept and redesign the product before making a physical prototype. Over the last decade, high speed computer graphics workstations have proven to be very effective in allowing visualization of three-dimensional complex systems [1]. With advances in robotics technology, the potential for developing haptic interfaces that allow the user to feel forces exerted by the virtual environment (in addition to seeing the environment) has been successfully demonstrated [9]. As computers become faster and as more sophisticated actuators and sensors are developed, computer interfaces will enable the user to feel, touch and see the virtual product in a virtual environment.

For customized design and prototyping, it is essential to integrate virtual prototyping with data acquisition. With the measurement of the user, the task and the environment, we can create accurate dynamic models (specific to the user, the task and the environment) and investigate the virtual creation and installation of a customized virtual product on a virtual human user as an integral part of the engineering process. Consider again the example of a feeding device. To evaluate candidate designs, it is useful to create a simulation of the user and the mechanical system as shown in Figure 7.

The mechanism that links the human head to the feeding device is not shown in the figure. The designer can experiment with different kinematic coupling mechanisms and see how the movements of the user are translated into the movement of the end effector or the spoon. Three-dimensional graphics provides visual information about the design, while a real-time dynamics simulation package elicits information about the forces and the velocities that are required of the human head and neck to effectively accomplish feeding. By linking to an appropriate physiological database one can verify the feasibility of the required head and neck motions and also investigate possible sources of discomfort or trauma with the virtual prototype before clinical tests are performed.

Being able to develop a virtual prototype of the product also allows the consumer to use and evaluate the virtual product in an appropriate virtual environment before the designer commits to the expense of creating the physical prototype. In the rehabilitation engineering domain, the designer may miss important constraints due to a lack of affinity with the consumer [28]. As shown in Figure 2, consumer feedback (and evaluation by experts such as therapists) during the virtual prototyping phase and the redesign of the product in response to this feedback at a very early stage can ensure the success of the product and possibly avoid building multiple physical prototypes and incurring the resulting expenses. To this end, in addition to allowing consumers to test the product by testing the virtual prototype, it may be beneficial to provide haptic feedback so that a user can feel the dynamics of the product. Current work is directed to developing a head mounted haptic interface [27] (see Figure 8) in which the user wears a helmet that is attached to a six degree-of-freedom manipulator. For example, the manipulator can then be controlled to simulate the dynamics of the feeding device. But the current state of the art with haptic interfaces [4, 9] shows that much needs to be done before such devices can made safe, reliable and effective.

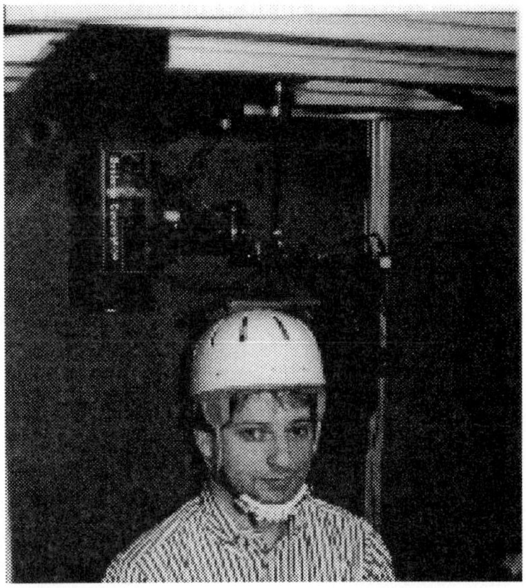

Figure 8: A head-controlled haptic interface for a virtual environment. The user's helmet is attached to the PerForce, a six degree of freedom manipulator (Cybernet Systems, Ann Arbor, MI). The manipulator can be controlled to simulate the dynamics of a virtual prototype.

5 Rapid Design and Prototyping

The design process can be divided into a concept development and system-level design phase and a detail design phase [31]. By rapid design we mainly refer to speeding up of the detail design phase, which includes the specification of the geometry, materials, and the manufacturing process for each component.

A key issue during the detail design phase is to synthesize the coupling mechanism that will take a feasible and comfortable voluntary movement of the user and transform it into the motion required to perform a specified task. We note that selecting the voluntary "input" movement may be a difficult process and involves physiological considerations. The process of "designing" the task is similar to the problem of task planning in robotics and is by no means trivial. However, once the desired input and output motions are identified, the basic problem can be formulated as shown in Figure 9. The telethesis can be decomposed into an input subsystem that is coupled to the user and driven by his/her motions, and an effector subsystem that performs the desired task but is driven by the input subsystem through an appropriate transmission. We outline the general procedure and some important considerations for designing the telethesis when the output motion is a 1-parameter (but possibly complicated and three-dimensional) motion.

The input motion can be mathematically modeled as a multi-parameter motion. In the kinematic model for head/neck movements discussed earlier, the model was a serial chain with axial joints and the parameters corresponded to joint variables. While the input motion is, in general, a complicated three-dimensional motion, in many cases, this motion can be described by a simpler 1-parameter motion. This was seen to be true for head/neck movements (Figure 6 (c)). The input motion is denoted by $\mathbf{u}$ and the parameter by λ in Figure 9. The parameter λ cannot be arbitrarily chosen. It should be possible to design a link-

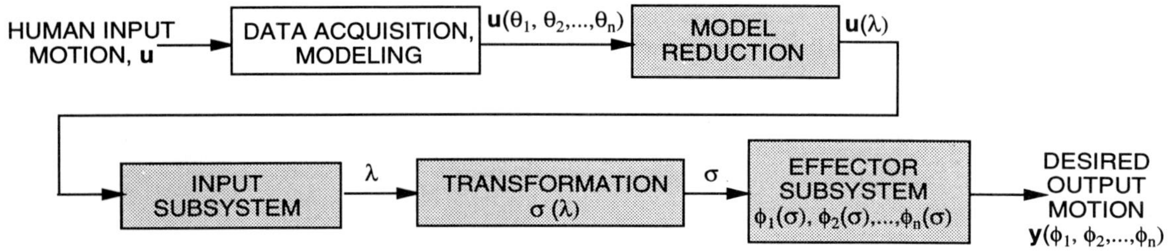

Figure 9: The synthesis of a one degree-of-freedom telethesis for a desired input and output motion.

age for the input subsystem that will "extract" the parameter λ. Physically, this means the input linkage should be such that one of its joint displacements must be proportional to λ so that it is possible to drive the effector subsystem by coupling it to this joint.

The design of the effector linkage is also an interesting problem. While a six-degree of freedom serial chain linkage, such as those seen in robots, will suffice, it is clearly not required. Given the set of tasks that need to be performed, it is necessary to synthesize a linkage that will execute those tasks. By restricting ourselves to serial-chain linkages we can formulate the design problem in the manner of [15]. By allowing for cabled transmissions in serial chain linkages, we obtain a wider class of solutions for the effector subsystem. There are two important design considerations. First, it is beneficial to keep the number of joints in the system to a minimum. Second, it should be possible to drive the effector subsystem with a single joint input, σ. Given the input and output (effector) subsystems, the only remaining task is to design the transformation between λ and σ. This transformation can be designed to provide appropriate mechanical advantage and velocity amplification along the trajectory, $\mathbf{y}(\sigma)$.

A possible effector subsystem for a feeding mechanism is shown in Figure 10 as an example. The linkage can be used to scoop up the food from the plate with any approach angle and bring the food to the user's mouth as he/she pitches his/her head forward. The mechanism has three degrees of freedom driven by cables labeled 1, 2 and 3 in the figure. σ_1, σ_2, and σ_3 are the corresponding cable extensions. The nominal yaw movement of the head, λ_1, causes the linkage to rotate about a vertical axis and translate in a horizontal plane so that the spoon is always in the line of the sight of the user. The nominal pitch movement of the head, λ_2, drives a planar $2-R$ open chain whose joints are coupled linearly so that the spoon performs a planar motion that involves scooping up the food and bringing it to the mouth. The nominal roll movement, λ_3, causes the spoon to pitch about a transverse axis. The three independent parameters $\sigma_1, \sigma_2, \sigma_3$ are coupled to three input parameters $\lambda_1, \lambda_2, \lambda_3$ through a linear transformation (not shown in the figure).

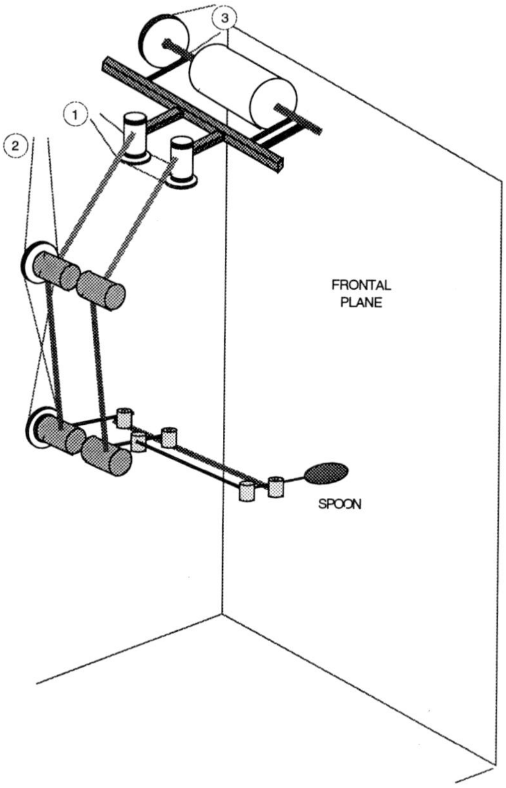

Figure 10: A candidate feeding mechanism that can be mounted to the back of the wheelchair.

While most of the discussion here has been limited to kinematic synthesis, it is clear that there will be many applications in which the task forces will be significant and will impact the design problem. Also, we may not be able to decompose the multi-input, multi-output design problem into several single-input, single-output design subproblems. Even if this decomposition is

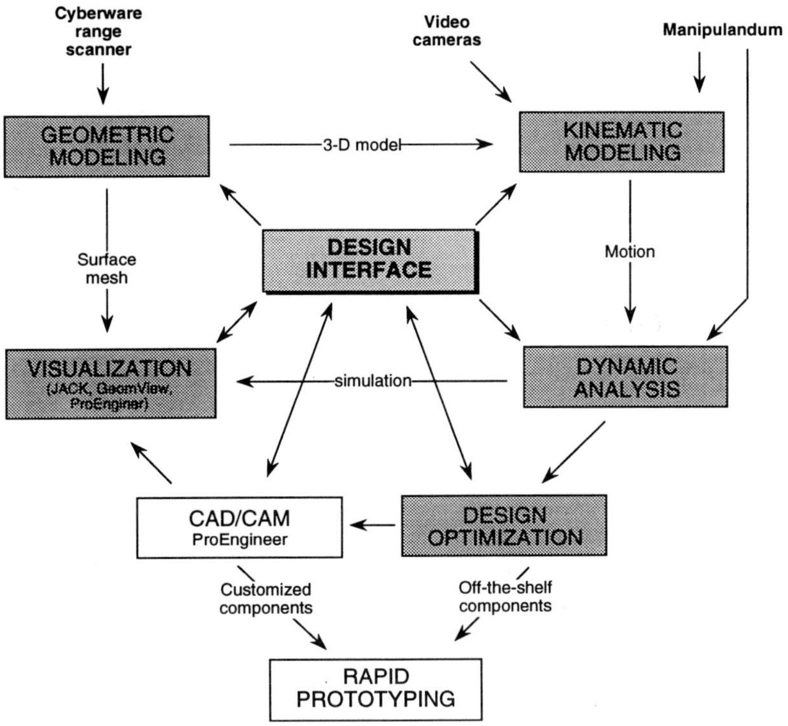

Figure 11: The user interface for the designer.

possible, much of the literature on linkage design is limited to *kinematic* synthesis of *closed chain* linkages and is of very little use here. The synthesis of serial-chain mechanisms with cabled transmissions for specialized tasks, and in particular, the synthesis of multiple degrees of freedom mechanisms, is an important direction for future work.

For rapid design and especially for customized products, it is essential to adopt a computer integrated approach where the designer can access and manipulate various heterogeneous pieces of information pertaining to geometry, kinematics and dynamics. At the heart of our design package is a graphical user interface (see Figure 11) which also acts as a server to support the interactive design and analysis processes. The front end is developed using *Jack*, a package for human body simulation [1]. The key idea is to have a generic request procedure which enables any of the component design/analysis packages or modules to call another package to obtain relevant information. Thus, information from any data acquisition, virtual prototyping or simulation module can be easily displayed on the visualization package. Finally, since the modules operate on different machines/architectures, efficient communication protocols between separate processes (relying on Unix TCP/IP calls) are employed. This graphical server allows a modular approach to software development and enables the designer to interact with each module at different levels. Currently, the modules consist of the two data acquisition modules for geometry [16] and kinematics [13, 23], and the virtual prototyping module described earlier. The user interface allows the designer to create a personalized simulation of a particular individual with the relevant geometric and kinematic characteristics (see Figure 7). The designer can prescribe a desired trajectory (positions and orientations) for the simulated human agent using a mouse or a joystick. It is possible to see the simulated human execute the motions while conforming to the kinematic, dynamic and physiological constraints that are characteristic of the individual and while being subject to the dynamics of the environment (the feeding linkage, in this example).

Our discussion of the design of rehabilitation aids was primarily focussed on the problem of synthesizing customized linkages that take as an input human motion and generate appropriate movement required for specific tasks. Engineering design requires an integrated approach combining such diverse methods as reliability analysis, models for costing [31], and optimization [24]. A detailed discussion of these issues is beyond the scope of this paper. Further, we did not attempt to discuss rapid prototyping, the process of quickly making a physical prototype of a product from a design in order to evaluate or test the product. We refer the reader to [18] for some of the challenges and research problems in this area.

6 Concluding Remarks

We presented the essential components necessary for the computer-integrated manufacture of low-volume customized products such as rehabilitation aids for people with motor disabilities. We also briefly described some of the open research problems in the areas of data acquisition, virtual and physical prototyping, and systems integration. Many of these problems are not very different from problems encountered in robotics, computer vision and databases. The basic concept of customized production and the computer-integrated approach to design and prototyping is applicable to a wide range of products from clothing, sports and recreational equipment to computer workstations and machine interfaces.

Acknowledgments: The support of National Science Foundation, grant number MIP 94-20397, is gratefully acknowledged. Thanks are due to Mohamed Ouerfelli for his work on the modeling of head and neck movements reported in this paper. We also thank Venkat Krovi, Rich Pito, and Ioannis Kakadiaris for their help with the illustrations. We are grateful to Jean-Marc Vezien and Tariq Rahman for discussions on the subject of customized design of rehabilitation aids.

References

[1] N.I. Badler, C.B. Phillips and B.L. Webber (1993), *Simulating Humans: Computer Graphics, Animation, and Control,* Oxford University Press, New York, NY.

[2] M.F. Burrow and T.G. Single (1990), The Georgia Tech Robotic Manipulator: A Six Degree of Freedom Arm for Rehabilitation Applications, *Proc. Int'l. Conf. on Rehabilitation Robotics,* Wilmington, DE, pp. 41-51.

[3] G. Bush, I. Al-Temen, J. Hancock, J. Bishop, E.M. Slack and I. Kurtz (1994), Development of a Myoelectrically Controlled Wheelchair Mounted Object Manipulator for Quadriplegics, *Proc. 4th Int'l. Conf. on Rehabilitation Robotics,* Wilmington, DE, June 14-17, pp. 103-106.

[4] J.E. Colgate and J.M. Brown (1994), Factors affecting the Z-Width of a Haptic Display, *Proc. IEEE Int'l. Conf. on Robotics and Automation,* San Diego, CA, May 8-13, pp. 3205-3210.

[5] J.L. Dallaway and R.D. Jackson (1993), The RAID Workstation for Office Environments, *Proc. RESNA Int'l. '93,* pp. 504-506.

[6] J.A. Doubler and D.S. Childress (1984), An Analysis of Extended Physiological Proprioception as a Prothesis Control Technique, *Jrnl. of Rehabilitation Research and Development,* Vol. 21, No. 1, pp. 5-18.

[7] M. Evans (1991), *Magpie: It's development and evaluation,* Technical report, Nuffield Orthopeadic Center, Headington, Oxford, England OX3 7LD, 1991.

[8] G. Garvin, M. Zefran, E. Henis and V. Kumar (1994), A study on optimality criteria for two hand reaching tasks, *Proc. 13th Southern Biomedical Engineering Conf.,* Washington, DC, April 16-17.

[9] V. Hayward (1995), Performance Measures for Haptic Interfaces, *Proc. 7th Int'l. Symposium on Robotics Research,* Hirshing, Germany, October 21-25.

[10] J.R. Hegarty and M.J. Topping (1991), HANDY 1 — A Low-Cost Robotic Aid to Eating, *Proc. Int'l. Conf. on Rehabilitation Robotics,* pp. 17-25.

[11] N. Hogan (1982), Control and coordination of voluntary arm movements, *Proc. American Control Conf.,* pp. 522-528.

[12] J.M. Hollerbach (1989), A survey of kinematic calibration, *Robotics Review I,* Eds: O. Khatib, J.J. Craig and T. Lozano-Perez, MIT Press, Cambridge, MA, pp. 207-242.

[13] I. Kakadiaris, D. Metaxas and R. Bajcsy (1994), Active part-decomposition, shape and motion estimation of articulated objects: A physics-based approach, *Proc. IEEE Conf. on Computer Vision and Pattern Recognition,* Seattle, WA, June 21-23. pp. 980-984.

[14] H. Kazerooni (1990), Human Machine Interaction via the Transfer of Power and Information Signals, *IEEE Trans. Systems, Man and Cybernetics,* Vol. SMC-20, No. 2, pp. 450-463.

[15] C.J.J. Paredis and P.K. Khosla (1993), Kinematic Design of Serial Link Manipulators from Task Specifications, *Int'l. Jrnl. of Robotics Research,* Vol. 12, No. 3, June, pp. 274-287.

[16] V. Koivunen (1993), *Processing and interpretation of 3-D sensory data with application to geometric modeling,* Ph.D. dissertation, Department of Electrical Engineering, Universitas Ouluensis, Finland.

[17] V. Koivunen and R. Bajcsy (1993), Geometric methods for building CAD models from range data, *Proc. SPIE Geometric Methods in Computer Vision II,* San Diego, CA.

[18] V. Kumar, R. Bajcsy, W. Harwin and P. Harker (1996), Rapid Design and Prototyping of Customized Rehabilitation Aids, *Communications of the ACM,* Special Section on Computers in Manufacturing, (to be published).

[19] H.H. Kwee (1990), Rehabilitation Robotics: Softening the Hardware, *Proc. Int'l. Conf. on Rehabilitation Robotics,* pp. 69-79.

[20] D. Lees, R. Crigler, H.F.M. Van der Loos and L. Leifer (1988), A Third Generation Desktop Robotic Assistant for the Severely Disabled, *Proc. ICAART 88,* Montreal, Canada, June 1988.

[21] L. Leifer (1992), RUI: factoring the robot user interface, *Proc. RESNA Int'l. '92,* RESNA Press.

[22] "National Spinal Cord Injury Statistical Center Annual Report," University of Alabama at Birmingham, 1992.

[23] M. Ouerfelli (1994), *Identification of Open and Closed Kinematic Chains with Applications to Robotics and Biomechanics,* Ph.D. dissertation, Department of Mechanical Engineering, University of Pennsylvania, Philadelphia, PA.

[24] P. Papalambros and D. Wilde (1988), *Principles of Optimal Design,* Cambridge University Press, New York, NY.

[25] R. Pito and R. Bajcsy (1995), A Solution to the Next Best View Problem for Automated CAD Model Acquisition Using Range Cameras, *Modeling, Simulation, and Control Technologies for Manufacturing, SPIE Photonics East*, Philadelphia, PA.

[26] M. Raghavan and B. Roth (1990), Kinematic analysis of the 6-R Manipulator of General Geometry, *Proc. 5th Int'l. Symposium on Robotics Research,* MIT Press, Cambridge, MA, pp. 263-269.

[27] T. Rahman, W. Harwin, S. Chen and R. Mahoney (1994), Rehabilitation robot control with enhanced sensory feedback, *Proc. 4th Int'l. Conf. on Rehabilitation Robotics,* Wilmington, DE, June 14-16, pp. 43-48.

[28] J.B. Reswick (1990), The Moon Over Dubrovnik — A Tale of Worldwide Impact on Persons with Disabilities, *Advances in external control of human extremities,* 4092, 1990.

[29] J.H. Sheridan (1993), Agile manufacturing: stepping beyond lean production, *Industry Week,* Vol. 242, No. 8, pp. 3-46.

[30] C.A. Stanger, A.C. Phalangas and M.F. Cawley (1994), Range of Head Motion and Force of High Cervical Spinal Cord Injured Individuals for the Design of a Test-bed Robotic System, *Proc. 4th Int'l. Conf. on Rehabilitation Robotics,* Wilmington, DE, June 14-17, pp. 37-42.

[31] K.T. Ulrich and S.D. Eppinger (1995), *Product Design and Development,* McGraw-Hill, New York, NY.

[32] G. Verberg, M. Milner, S. Naumann, J. Bishop and O. Sas (1992), An Evaluation of the Manus Wheelchair-mounted Manipulator, *Proc. RESNA Int'l. '92,* RESNA Press.

[33] J. Vertut and P. Coiffet (1986), Teleoperations and Robotics: Evolution and Development, *Robot Technology,* Vol. 3A, pp. 191-194, Prentice Hall.

[34] P. Wellman, V. Krovi, V. Kumar and W. Harwin (1995), A Wheelchair with Legs for People with Motor Disabilities, *IEEE Trans. on Rehabilitation Engineering,* (in press).

PART 4

TIGHT COUPLING OF PERCEPTION WITH ACTION

SESSION SUMMARY

Hirochika Inoue
Department of Mechano-Informatics, University of Tokyo

Recent progress of real time vision technology has freshened and accelerated the study of intelligent behavior for robotic systems. In particular, video-rate processing of visual data enables robots to observe, control, feedback, verify, and estimate not only the dynamic behavior of robots themselves but also dynamic changes in the real world, in real time. This allows experiments on the behavior-related intelligence to be realized more easily, quickly and economically. In such a context, research on perception must be no longer separate from that of action control. Perception can be enhanced by means of supporting actions. Action can be enriched by using adequate perception channels. Challenges for coupling perception with action will open fertile research fields such as behavior based robotic intelligence. This session discusses some problems which appear in such tightly-coupled perception-action behavior approach.

In the first paper, "Toward a System Theory for the Composition of Dynamically Dexterous Robot Behavior", Dan Koditschek reports their efforts to develop robot controller composition techniques for very fast visual guidance in dexterous "batting" maneuvers. In their experiment, a robot with a flat paddle is required to strike repeatedly at a falling ball. The robot's workspace is cluttered with obstacles that disconnect the free space formed when the ball and paddle remain in contact. The controller compositions created will guarantee that a ball introduced in the safe workspace remains there and is ultimately brought to the goal. Their approach to autonomous robot design seeks to achieve user goals that are encoded by state space goal and obstacle sets. Their architecture implements event driven robot policies whose closed loop dynamics drive the coupled robot-environment state toward a goal set and away from obstacles. The theoretical attention on the system as well as the experimental attention on building a distributed computational environment are discussed. In the paper they consider the simplest possible command lexicon – goal cells comprised of singletons and obstacle cells of a particularly simple shape – and hand build an interpretation. In so doing, they become convinced that a wide range of behaviors uncooperative dynamical dexterity may be forthcoming from their present palette of control primitives.

In the second paper, "Perception–Action Network for Sensor Fusion and Planning", Suckhan Lee proposes a perception-action network as a means of efficiently integrating sensing, knowledge, and action for sensor fusion and planning. The network consists of a number of heterogeneous computational units, representing feature transformation and abstraction, data and concept fusion, as well as planning and decision-making for action, which are interconnected as a dynamic system. New input stimuli to the network invoke the evolution of network states to a new equilibrium, through which a real-time integration of sensing, knowledge, and action can be accomplished. Simple simulation experiments are explained. Discussions about generality and effectivity for more complex cases are raised but the issues remain unsolved. Generally, from the viewpoint of implementation for tightly-coupled perception-action behavior study, research of system architecture and software implementation is important, however only a few efforts have been reported so far. Further studies on the structure or principles of such behavior seem necessary to design powerful system implementation tools.

The third paper "Research Program on Mechanisms for Emerging Machine Intelligence" is not a purely technical paper, rather it introduces a research program on intelligent robotics which started in 1995 as a grant from the Ministry of Education of the Japanese Government. The purpose of such "priority research programs" is to promote inter-university research activities in selected areas. "Intelligent robotics" was selected as one of these priority areas and will run for three years. The research program was proposed in order to challenge various issues in the intelligence and autonomy of advanced mechanical systems.

The aim of this project is summed up by the two key phrases: perception-motion coupling, and emergent machine intelligence. The objective of this project is to investigate the mechanism or algorithm that gives rise to machine intelligence. For that purpose, we chose a tightly-coupled perception-motion behavior approach. A machine is a physical entity that behaves in real world. The research interest is focussed on the varieties of intelligence, which are inherently required for robust, efficient, and quick interactions between the machine and real environment. This project aims to find the principles of such mechanical intelligence and demonstrate them by building physical entities. Four major issues were chosen to structure the research. They are: kinematic understanding of manipulation skill, real world understanding through tightly-coupled sensing and action, human-robot cooperative systems, and biologically inspired autonomous systems. All of these issues require a tight coupling of perception with action, and each issue closely relates with the others.

In this program, 19 universities have joined, addressing 43 specific research topics. Abstracts of these research themes are included. The approach and steering principles for this research program are also discussed.

Perception-Action Network for Sensor Fusion and Planning

Sukhan Lee

Jet Propulsion Laboratory
California Institute of Technology
Pasadena, CA 91109

Dept. of EE-Systems and CS
University of Southern California
Los Angeles, CA 90089-0781

Abstract - *Robot intelligence requires a real-time connection between sensing and action. A new computational principle of robotics that efficiently implements such a connection is utmost important for the new generation of robotics. In this paper, a perception-action network is presented as a means of efficiently integrating sensing, knowledge, and action for sensor fusion and planning. The network consists of a number of heterogeneous computational units, representing feature transformation and abstraction, data and concept fusion, constraint satisfaction for consistency, as well as planning and decision-making for action, which are interconnected as a dynamic system. New input stimuli to the network invoke the evolution of network states to a new equilibrium, through which a real-time integration of sensing, knowledge, and action can be accomplished. The network provides a formal, yet general and efficient, method of achieving sensor fusion and planning. This is because the uncertainties of signals, propagated in the network, can be controlled by modifying sensing parameters and robot actions. Algorithms for sensor planning based on the proposed network are established and applied to robot self-localization. Simulation and experimental results are shown.*

1 Introduction

A robotic system is a real-time system which connects sensing and action in such a way as to achieve intelligence in behavior and dexterity in motion. To achieve such intelligence and dexterity often requires integration of distributed sensors and actuators, so as to provide a rich source of sensory and movement patterns that can be clustered into higher levels of concepts and activities. The key to successful integration may be a system architecture in which a certain computational principle of robotics is embedded. Especially, the way how such processes as feature transformation and abstraction, data and concept fusion[1, 2, 3, 4, 5, 6], constraint and knowledge propagation for consistency satisfaction, and cooperative planning and decision-making are interconnected and interacted plays an essential role.

In spite of the fact that a decade of research and development in robotics has produced numerous theoretical and experimental results, robots are yet to acquire the level of intelligence and dexterity required for the autonomous execution of tasks in an unstructured environment. Conventional approaches where a robotic system has been built without an underlying computational principle for integrating sensing, knowledge, and action in real-time seem to suffer from the limitations of task complexity that it can handle. Should the robot intelligence be measured in terms of a power-to-weight ratio, where the power is defined as the product of the complexity and execution speed of a task it can handle and the weight is defined as the product of the volume and cost associated with the required hardware and software, an order of magnitude of improvement in such a power-to-weight ratio may be necessary for the new generation of robotics.

In this paper, we present a perception-action network as a system architecture for integrating sensing, knowledge, and action in real time. The network consists of a number of heterogeneous computational units, representing feature transformation and abstraction, data and concept fusion, constraint satisfaction for consistency, as well as planning and decision-making for action. The computational principle of robotics embedded in the network is a dynamic process by which the state of network is evolved from one equilibrium to another in real time as stimuli are input to the network. Through such a dynamic process, a real-time integration of sensing, knowledge, and action can be accomplished.

It is also shown in this paper that the proposed perception-action network can be incorporated into a formal, yet general and efficient, method for achieving sensor fusion and planning. This can be done by controlling the uncertainties of signals, propagated in the network, based on modifying sensing parameters as well as determining robot actions.

This paper is organized as follows: Sec. 2 describes the proposed perception-action network with the embedded dynamic process for integrating sensing, knowledge, and action. Sec. 3 presents sensor planning with the perception-action network based on both modifying sensing parameters and determining robot actions. Sec. 4 describes conclusions and future directions.

2 Perception-Action Network

The proposed Perception-Action Network (PAN) is formed by the interconnection of four basic computational modules: the data fusion module, the feature transformation module, the constraint satisfaction module, and the action module, as shown in Fig. 1. A data fusion module(DFM) takes one or more data representing an object feature as input and produces the optimal estimate of the feature in conjunction with the current state of the module. A feature transformation module (FTM) extracts a primitive feature from the raw sensory data or transforms a set of primitive features into a more abstract, a higher level of, feature. An action module (AM), as a special case of an FTM, issues an action command based on the predefined laws triggered by a set of feature values associated with. A constraint satisfaction module (CSM) represents system knowledge which imposes a constraint upon a set of feature values associated with, such that feature values should be modified to satisfy the consistency with the corresponding system knowledge[18].

In Fig. 1, the feature values, x_1 and x_2, computed by FTM1 and FTM2 are connected to DFM1 as the input. The output of DFM1 and DFM2, y_1 and y_2, are in turn input to FTM5 to generate the output z_1. The output of FTM5 and FTM6, z_1 and z_2, are then subject to the constraint satisfaction module, CSM1. The dotted line from an external source represents the input from a human engaged in interactive and cooperative sensing and decision-making. The output of each module represents the current estimates of the corresponding feature value or constraint evaluation, and is regarded as the current state of the module. The state of the network is defined as the collection of the states of individual modules.

The network is structured in such a way that a change of the state of a module propagates to adjacent modules and triggers a chain of state changes throughout the network. Note that the propagation of state change is bidirectional, both forward and backward, such that the network achieves the required constraint satisfaction through backward propagation of state changes. This is shown in Fig. 1 as a feedback loop associated with each module. The bidirectional change of network state is governed by the network dynamics described in Sec. 2.1.

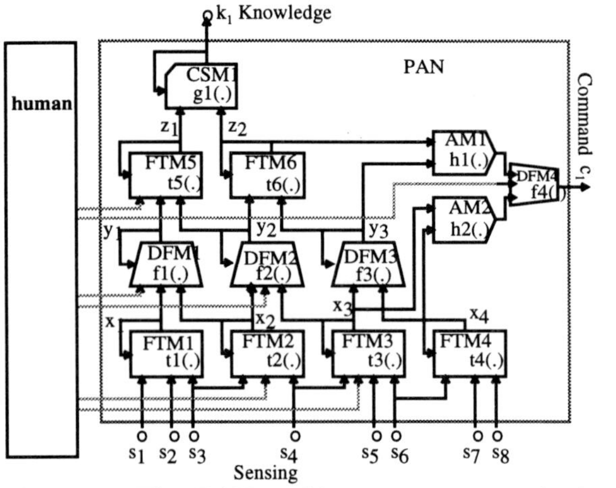

Figure 1: The PAN architecture composed of 4 different modules: FTM, DFM, CSM and AM.

PAN may be considered as a generalized and elaborated form of logical sensor system used for modeling the capabilities of an integrated sensor system. An example of a logical sensor system is shown in Fig. 2, where the capability of extracting features by an integrated sensor system composed of proximity, tactile, and camera sensors are organized in a hierarchy of logical and physical sensors. PAN not only represents a logical sensor system through the organization of FTMs, but also integrates it with the data fusion and constraint satisfaction functions based on network dynamics.

2.1 Bidirectional Signal Propagation

PAN is operated with bidirectional signal propagation: the forward and backward processes. In the forward process, input signals are transformed into output signals based on the transformation functions defined at individual modules. For instance, a lower level of features are transformed into a higher level of features through FTMs, and the fusion of data or concept takes place through DFMs whenever multiple and redundant data are available. In the backward process, CSMs initiate to propagate error signals, if any, backward in such a way as to satisfy network consistency. That is, the feature values connected to CSMs are subject to being updated for consistency satisfaction. The updated feature values of an upper level in turn invoke the update of feature values of a lower level. Through the cycles of forward and backward propagation of signals, the

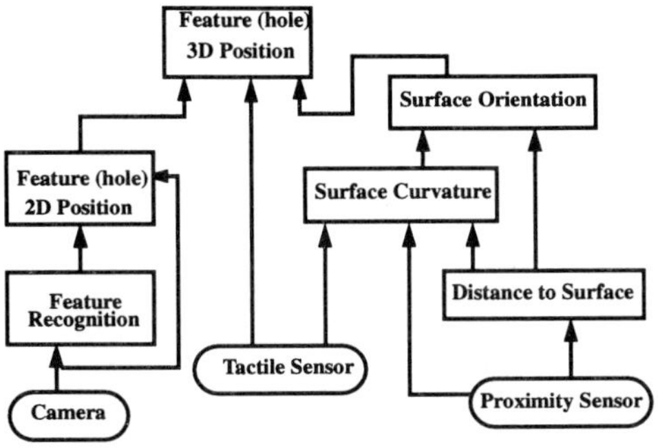

Figure 2: A schematic illustration of a logical sensor system.

network eventually reaches an equilibrium state at which all the features and concepts have optimal and consistent estimates. By optimal, we mean that redundant sources of data are fused under the constraints from system knowledge.

The signals of PAN, represented as sensor data, values associated with features and concepts, and errors from system constraints, are represented by their nominal values and associated uncertainties. Thus, during the cycles of forward and backward signal propagations, not only the nominal values but also their uncertainties need to be updated. Probabilistic modeling and inference can provide a means of achieving such updates.

However the forward and backward signal propagations based on the conventional methods of probabilistic models and inference[7, 8] may not be computationally efficient enough for real-time implementation, besides the difficulty of processing non-Gaussian signals generated by nonlinear transformations. Therefore, in this paper, we present a new approach for accomplishing forward and backward processes by representing PAN as a dynamic system. In the proposed approach, the state of network is evolved from one equilibrium to another in real time as the network is stimulated by input signals, through which the forward and backward processes are accomplished, as shown in Sec. 2.2.

2.2 Network Dynamics

For the clear description of the concept of dynamic evolution of PAN, consider a simple PAN illustrated in Fig. 3. Let us first assume that the network is initially in an equilibrium state, $\{z^0, y_1^0, y_2^0, x_1^0, x_2^0, x_3^0, x_4^0, s_1^0, s_2^0, s_3^0, s_4^0\}$, and satisfies consistency as follows:

$$g(y_1^0, y_2^0) = z^0, f1(x_1^0, x_2^0) = y_1^0, f2(x_3^0, x_4^0) = y_2^0,$$

$$t1(s_1^0, s_2^0) = x_1^0, \quad \text{and} \quad t2(s_3^0, s_4^0) = x_2^0.$$

Now, assume that, upon the stimuli given to the network, FTM1 and FTM2 have new inputs

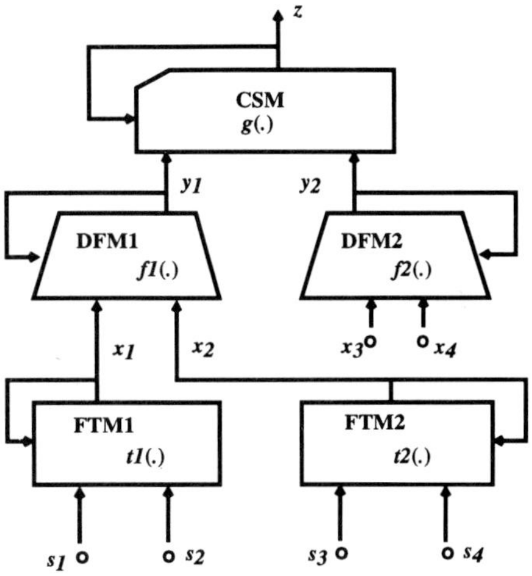

Figure 3: A simple PAN used for the description of network dynamics.

s_1, s_2 and s_3, s_4, respectively. This prompts the states of individual modules simultaneously evolve toward the new equilibrium states that satisfy network consistency. Here, we propose the following as the dynamics governing the evolution of system states:

1. The DFM-CSM dynamics:

$$\begin{aligned} \dot{y_1} = &-\lambda_{y_1}\{y_1 - f1(x_1^t, x_2^t)\} \\ &-\gamma_{y_1}\nabla\phi\{z^0 - g(y_1, y_2)\} \end{aligned} \tag{1}$$

2. the FTM-DFM dynamics:

$$\begin{aligned} \dot{x_1} = &-\lambda_{x_1}\{x_1 - t1(s_1, s_2)\} \\ &-\gamma_{x_1}\nabla\phi\{y_1 - f1(x_1, x_2)\} \\ \dot{x_2} = &-\lambda_{x_2}\{x_2 - t1(s_3, s_4)\} \\ &-\gamma_{x_2}\nabla\phi\{y_1 - f1(x_1, x_2)\} \end{aligned} \tag{2}$$

where the initial states are given at the current network equilibrium.

Eqs. (1) and (2) represent a set of dynamic equations governing the behavior of PAN toward a new network equilibrium state. The first term of a dynamic equation represents the forward process, whereas the second term represents the backward process.

In order to deal with more complex networks, the same form of dynamic equation shown in (1) and (2) needs to be repeated for individual modules with the proper assignment of module functions. In the above example, z_0 of (1) is assumed constant, since it represents the knowledge of absolute certainty. In the case where the knowledge is subject to uncertainty, z_0 should be allowed to be modified in such a way that the uncertainty of z_0 can be reduced by the fusion with the output

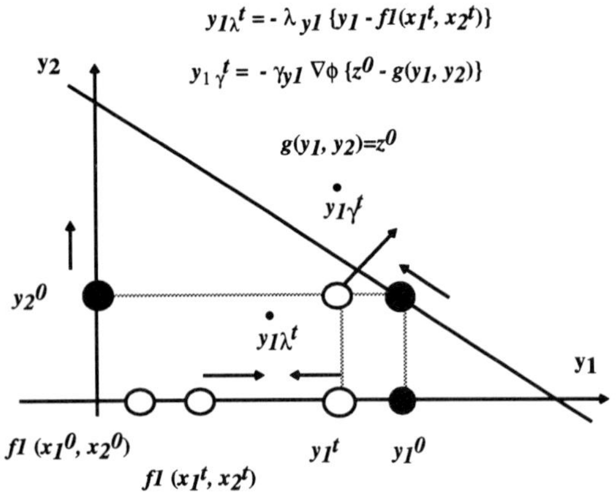

Figure 4: The motions involved in the DFM-CSM dynamics.

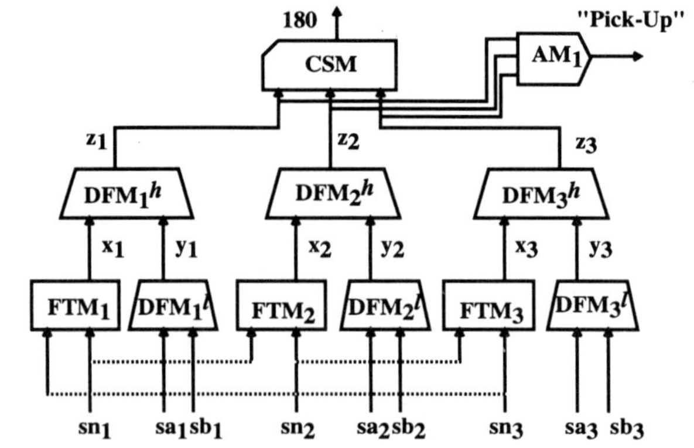

Figure 5: The PAN for the task to pick up the right triangle.

of the corresponding CSM module. This can be done simply by adding

$$\dot{z} = -\lambda_z\{z - g(y_1, y_2)\}$$

to the above dynamic equations.

The above dynamic equations can be interpreted as follows: Assume that the stimuli of s_1, s_2, s_3, and s_4 are occurred at time t_0. The first term of (1) represents the motion of y_1 due to the forward process, starting from y_1^0 toward merging with the output of DFM1, $f1(x_1, x_2)$, as illustrated by $y_1\lambda^t$ in Fig. 4. This is as if the two objects at the locations y_1^t and $f1(x_1, x_2)$ at time t attract each other. The second term of (1) represents the motion of $y_1\gamma^t$ due to the backward process, which is influenced by the attraction force generated by the constraint, as illustrated by $y_1\gamma^t$ in Fig. 4. Note that, when y_1^t moves toward $f1(x_1^t, x_2^t)$, $f1(x_1^t, x_2^t)$ also moves simultaneously toward y_1^t. This is because the backward process of DFM1 enforces to match $f1(x_1^t, x_2^t)$ to be matched with y_1^t.

The amount of variation of y_1 due to the forward and backward process should be determined in proportional to the uncertainties associated with them. This can be done by controlling the ratio between the corresponding coefficients, λ_{y_1} and γ_{y_1}, to be proportional to that of their uncertainties. By the same token, the amounts of variation of y_1 and $f(x_1, x_2)$ in the merging process should be proportional to the uncertainties associated with y_1^0 and $f(x_1, x_2)$, respectively, which can also be accomplished by controlling the ratio between the coefficients, λ_{y_1}, γ_{x_1} and γ_{x_2}. It is possible that the dynamic coefficients be assigned in such a way that the dynamic evolution of states matches the result from a probabilistic model such as the minimum variance estimate described above. If so, the network dynamics defined by the above equations can be considered as a general form of the minimum variance estimate. This allows to establish an optimal setting of parameters of dynamic equations as well as an optimal update of the uncertainties involved in individual states.

2.3 Example

To demonstrate the operation of PAN based on network dynamics given by (1) and (2), we chose the following simple example. A robot is given a task to pick up a right triangle among many different shapes of triangles. The robot is assumed to have two logical sensors [4]: one for the measurement of edge normals(called the "edge-normal sensor") and the other for the measurement of internal angles(called the "angle sensor"). The angle sensor is assumed easier to handle, but has more uncertainty than the edge-normal sensor. For a given triangle, the robot measures each angle twice with the angle sensor and takes the average of two measurements. On the other hand, the robot measures the edge normals with the edge-normal sensor, and computes the internal angles from the measured edge normals. Then, a decision is made on the size of each angle based on data from both sensors. If any of three angles is a right angle, the "pick-up" command is issued. The system constraint is that the sum of three internal angles be 180 degrees. It is assumed that the sensor data have known, independent Gaussian distributions.

For the task described above, we can organize the PAN as shown in Fig. 5. The two sensed data from the angle sensor, sa_1 and sb_1, are fused into y_i, through DFM_i^l. FTM_i takes the edge normals sn_i as its input, and computes the internal angle, x_i, between those two edges. Through the feature transformation equation, these x_i and y_i are in turn fused into the angle estimates, z_i, through the higher level DFM_i^h. CSM checks if z_i satisfy the constraint, i.e., the sum of internal angles is 180 degrees. As mentioned in section

2.1, if an error is detected in CSM, it is used to adjust z_i, and propagated backward to adjust x_i and y_i. The variances of x_i, y_i, and z_i can be computed from those of sensor data, based on the assumption of the independent, Gaussian distribution. The feature transformation function in FTM_i, $t_i(.)$ is defined as:

$$t_i(sn) = (180 - (sn_i - sn_j))\text{MOD}360)$$

where sn_i and sn_j are two edge normals, and we simply use the averaging function as the data fusion function, $f_i(.)$, i.e.,

$$f_i(s) = (\Sigma_{j=1,M} s_j)/M$$

where s is the input vector with dimension M.

Based on (1) and (2), the dynamic equation for z_i is formulated as:

$$\dot{z_i} = -\tau_{z_i}(z_i - (x_i + y_i)/2) + \sigma_{z_i}(180 - \Sigma_{j=1,3} z_j), \quad \text{for i=1,2,3.}$$

The second term of the right hand side comes from the forward process through DFM_i, whereas the third term from the error in CSM. Since the standard deviation sz_i is used for the coefficient of the CSM output error, with a smaller variance of z_i is less affected by the output error of CSM. Similarly, the dynamic equations for x_i and y_i are formulated as:

$$\dot{x_i} = -\tau_{x_i}(x_i - ((180 - (sn_i - sn_j))\text{MOD}360)) + \sigma_{x_i}(z_i - x_i), \quad \text{for i=1,2,3}$$

$$\dot{y_i} = -\tau_{y_i}(y_i - (sa_i + sb_j)/2) + \sigma_{y_i}(z_i - y_i), \quad \text{for i=1,2,3}$$

Note that the error is defined here as the difference of z_i from y_i and x_i, since the input and output of DFM should be equal when an equilibrium state is reached.

Although the PAN shows the different result according to the input, a typical result for a triangle in Fig. 6 is shown in Fig. 7. The initial equilibrium state is chosen as the ideal data for the equilateral triangle in which all the x, y, and z are [60.0 60.0 60.0] without any errors in CSM and DFMs. The sensed edge-normal data vector, sn, is assumed a lot more accurate than the sensed angle data vector.

To explore the effect of the backward process, the backward process is invoked only at time $t = 2$. As shown in Fig. 7, the forward process converges to equilibrium value but only with considerable deviation from accurate value without the backward process. However, with the backward process, it moves to a new equilibrium very close to the accurate value, as shown in a). The error between x and the accurate value is drawn in b) which shows clearly the error correcting effect of the backward process. The graphs in c), d) and e) also show the adjusting effects of the backward process on the errors in CSM and DFMs.

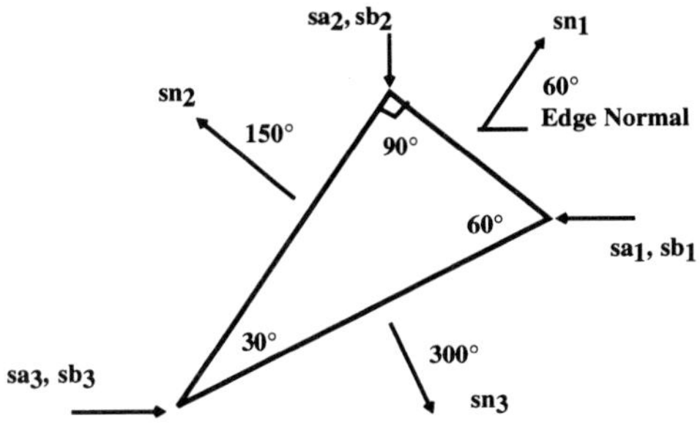

Figure 6: The sample triangle and sensing location.

3 Sensor Planning with PAN

The objective of sensor planning is to dynamically change sensor configurations in such a way as to achieve the given sensing goals. Sensing goals are associated with measuring necessary geometric and physical features of an environment with desired accuracy but at the minimum sensing cost. There has been a recent growth of interest in applying sensor planning (in terms of active or dynamic sensing [11, 14, 15, 16, 17]) to robotics and vision problems[12, 13]. This is because sensor planning can provide the extension of sensing capability as well as the enhancement of accuracy and efficiency in sensing. However, most work on sensor planning currently available are based on ad hoc methods customized to particular problems, lacking formality and generality in problem representation and solution methodology.

In this section, we first represent sensor configuration in terms of PAN and determines sensor poses and parameters under the framework of PAN, while satisfying the system and parameter constraints. By the system constraint, we mean the known relationship among feature values that should be kept true. By the parameter constraint, we mean the ranges of sensor poses and parameters that enable the measurement of targets, and of the inter-sensor geometry that make sensor fusion and integration valid.

3.1 Uncertainty Propagation In PAN

The input and output variables of PAN involve uncertainties. The goal of sensor planning is to minimize the uncertainties involved in the network output variables of interest by modifying the controllable parameters and sensor poses, subject to minimum sensing cost. Let us first describe a method of representing and defining uncertainties in PAN.

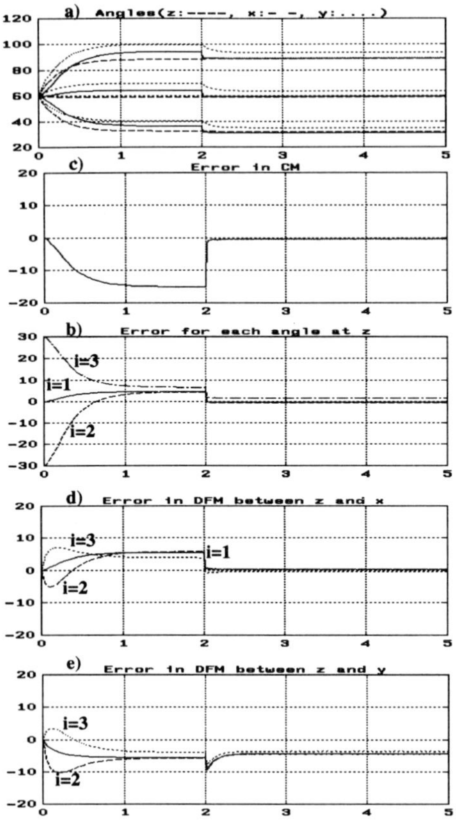

Figure 7: Simulation results for the triangle in Fig. 6 with Backward Process started at t=2.

3.1.1 *Uncertainty Representation*

The uncertainties of PAN are due mainly to the errors involved in the measurement data from physical sensors and to the biases involved in the modeling of feature transformations. Here, we assume that the biases in modeling are not controllable through sensor planning but should be compensated though calibration. The errors in measurement data are due to random noise which can be represented as uncertainty hypervolume or error ellipsoids, assuming that the noise characteristics of sensors are known. For instance, the boundary of an error ellipsoid can be determined either as the maximum error bound, if available. Formally, we represent the uncertainty of a network variable, x, as an ellipsoid of the following form defined in the error space, $\{dx\}$:

$$dx^t W_x dx \leq 1 \qquad (3)$$

where W_x represents the symmetric weight matrix determining the size and shape of the ellipsoid.

3.1.2 *Uncertainty Propagation*

The measurement uncertainties as defined above propagate through the input-output relationships of FTM and DFM modules, under the constraints defined by CSM modules. Let us first represent the relationship between the input vector, x, and the output vector, y, of a FTM or a DFM by

$$y = f(x,p) \qquad (4)$$

where p represents a controllable system parameter vector associated with the corresponding module, if available. Then the error propagation through (4) can be approximated as the first order Jacobian relationship, assuming that f is smooth and the error, dx, of x is small, as follows:

$$y + dy = f(x+dx,p) \approx f(x,p) + \frac{\partial f}{\partial x}\Delta x \qquad (5)$$

Therefore,

$$dy \approx \frac{\partial f}{\partial x}dx = J(x,p)dx \qquad (6)$$

where $J(x,p)$ represents the Jacobian relationship between dy and dx. The uncertainty of x, represented as an error ellipsoid of (3), can now be propagated to the uncertainty of y, represented as an error ellipsoid in terms of dy, through (6). By substituting $dx = J^+(x,p)dy$, obtained from (6), to (3), we have

$$dy^t (J^+)^t W_x J^+ dy \leq 1 \qquad (7)$$

where J^+ represents the pseudo-inverse of J. (7) can be rewritten as

$$dy^t W_y dy \leq 1 \qquad (8)$$

where the symmetric weight matrix, W_y, is defined as $W_y \equiv (J^+)^t W_x J^+$. Eq. (8) is of the same form as (3), and the error propagation can be continued toward the modules of a higher level by properly defining the weight matrices of their input vectors: The weight matrix, W_x, of an input vector composed of x_1 and x_2, can be specified by combining the individual weight matrices W_{x_1} and W_{x_2} of x_1 and x_2, respectively, as follows:

$$W_x = \begin{bmatrix} W_{x_1} & 0 \\ 0 & W_{x_2} \end{bmatrix} \qquad (9)$$

Note that, in (8), W_y is a function of x and p, since J is a function of x and p. For a CSM with its input-output relationship represented as $g(x) - c = 0$, the error propagation implies the constraint on the input error vector, dx:

$$J_g dx = 0 \quad \text{with} \quad J_g = \frac{\partial g}{\partial x} \qquad (10)$$

The propagation of uncertainties in PAN is defined by the error ellipsoids obtained from (8) and (9) with the system constraints given by (10).

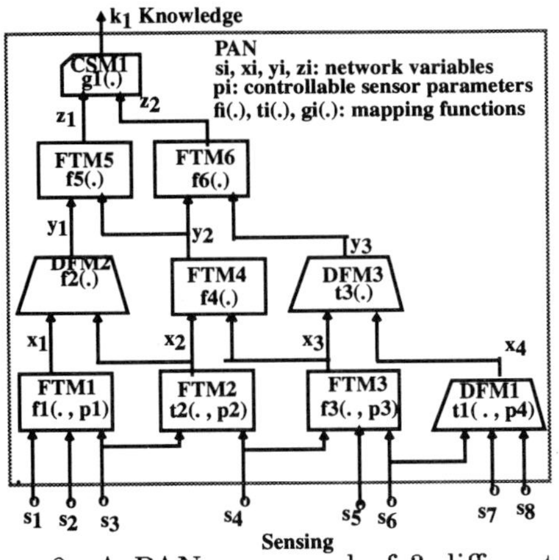

Figure 8: A PAN composed of 3 different modules: FTMs, DFMs, and CSMs.

3.2 Planning Algorithm

Sensor planning here focuses on generating the desired uncertainties for the network output variables of interest by controlling the system parameters subject to minimum sensing cost.

3.2.1 *Representation of Sensing Goals*

The desired accuracy of network output vectors can be represented by the desired error ellipsoids associated with the corresponding output vectors. For example, the desired error ellipsoid, $E_d(y)$, of an output vector, y, can be represented as $dy^t W_{yd} dy \leq 1$, with W_{yd} being the desired weight matrix. Then, the performance of PAN can be defined as a closeness measure between the desired and actual error ellipsoids of the output vector. To be more specific, let us assume that the desired error ellipsoid has principal radius vectors of $(1/\sqrt{\lambda_1})u_1$ and $(1/\sqrt{\lambda_2})u_2$, as shown in Fig. 9.

u_1 and u_2 are the orthonormal singular vectors of the desired weight matrix, W_d, whereas λ_1 and λ_2 are the corresponding eigenvalues. Then, we can obtain the two vectors, $\alpha_1 u_1$ and $\alpha_2 u_2$, from the origin to the boundary of the actual error ellipsoid in the direction defined by each singular vector, u_1 and u_2, of the desired error ellipsoid, as follows:

$$\alpha_i^2 u_i^t W_a u_i = 1, \quad \text{for i = 1,2} \tag{11}$$

where W_a represents the weight matrix of the actual error ellipsoid, which is a function of the controllable system parameter, p. Note that α_i, $i = 1, 2$, can be computed directly from (11). Then, the closeness of two ellipsoids can be defined by comparing between α_i and $(1/\sqrt{\lambda_i})$ for $i = 1, 2$. Finally, comparing the performance of PAN can be defined in terms of given sensing goals by

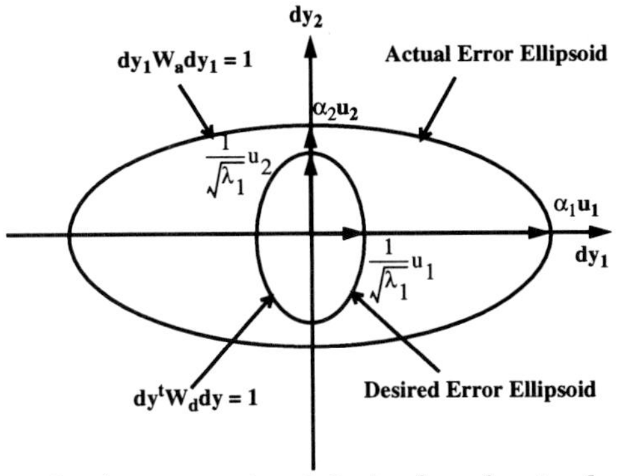

Figure 9: An example of desired and actual error ellipsoids for the representation of network performance.

$(1/\sqrt{\lambda_i})$ and α_i, for individual output vectors, as follows:

$$\frac{1}{2}\sum_k \sum_i (\alpha_{ik} - 1/\sqrt{\lambda_{ik}})^2 \tag{12}$$

where k represents the kth network output vector.

Although the example shown in Fig. 9 is for a 2-D case, the method described here is general and can be applied to any dimensional error space.

3.2.2 *Parametric Sensor Planning*

The problem of parametric sensor planning can be formally stated here as follows: Determine the value of controllable system parameters, p, that minimizes the network performance measure, (12), subject to minimum sensing cost, while satisfying the system and parameter constraints.

We propose to solve the above sensor planning problem based on the iterative parameter update method. In the iterative parameter update method, the parameter vector, p, is updated iteratively by

$$p(t) = p(0) + \int_0^t \dot{p}\, d\tau \tag{13}$$

for a continuous-time system, or

$$p(k+1) = p(k) + dp(k) \tag{14}$$

for a discrete-time system, until an equilibrium at which an optimal solution is found with $\dot{p}$ or dp being 0 is reached.

The parameter update, $\dot{p}$ or dp, can be generated in the parameter space from the direction that best satisfies the following criteria:

1. The satisfaction of system and parameter constraints.

2. The minimization of network performance measure.
3. The minimal sensing cost.
4. The avoidance of local minima.

To determine the parameter update based on the above criteria, let us first derive the three directional vector, $\dot{p_1}$, $\dot{p_2}$ and $\dot{p_3}$ from each of the top three criteria that can be combined into the final parameter update $\dot{p}$. The directional vector, $\dot{p_1}$, changes p in a direction that leads the network to satisfy existing system and parameter constraints. To determine $\dot{p_1}$, we define the following Lyapunov function candidate, V_1:

$$V_1 = \sum_{k=1}^{2} V_{1k} \tag{15}$$

where

$$V_{11} = \frac{1}{2}\sum_{i=1}^{l_1} h_{1i}^2(x,p) \tag{16}$$

$$V_{12} = \frac{1}{2}\sum_{i=1}^{l_2} S^2[h_{2i}(x,p)] \tag{17}$$

V_{11} is defined from the equality constraints of the system and parameter constraints such that $h_{1i}(x,p)$ represents the ith equality constraint, $h_{1i}(x,p) = 0$. On the other hand, V_{12} is defined from the inequality constraints of the system and parameter constraints such that $h_{2i}(x,p)$ represents the ith inequality constraint, $h_{2i}(x,p) > 0$, where

$$S[h_{2i}(x,p)] = \begin{cases} h_{2i}(x,p), & \text{if } h_{2i}(x,p) < 0 \\ 0, & \text{if } h_{2i}(x,p) \geq 0 \end{cases} \tag{18}$$

From (15), (16) and (17), we can see that $V_1 = 0$ only when both the equality and inequality constraints are satisfied, i.e., only when the system and parameter constraints are completely satisfied. Based on (15), (16), and (17), we can derive the parameter update, $\dot{p_1}$, that makes the time derivative of V_1 negative, $\dot{V_1} < 0$, when $V_1 \neq 0$, as follows:

$$\dot{p_1} = -\sum_{k=1}^{2} \frac{dV_{1k}}{dp} / \left\| \frac{dV_{1k}}{dp} \right\| \tag{19}$$

where

$$\frac{dV_{11}}{dp} = \sum_{i=1}^{l_1} h_{1i}(x,p)\left(\frac{\partial}{\partial x}h_{1i}(x,p)\frac{\partial x}{\partial p} + \frac{\partial}{\partial p}h_{1i}(x,p)\right) \tag{20}$$

and

$$\frac{dV_{12}}{dp} = \sum_{i=1}^{l_2} S[h_{2i}(x,p)]\left(\frac{\partial}{\partial x}S[h_{1i}(x,p)]\frac{\partial x}{\partial p} + \frac{\partial}{\partial p}S[h_{1i}(x,p)]\right) \tag{21}$$

Eq. (19) guarantees that , only when $V_1 = 0$, i.e., when both the equality and inequality constraints are satisfied. The term $\left\|\frac{dV_{1k}}{dp}\right\|$ used in the denominator of (19) allows the value of $\dot{p_1}$ to be exploded in the vicinity of local minima, which helps to avoid local minima.

By the same way as we derived $\dot{p_1}$, we can define a directional vector, $\dot{p_2}$, which changes p in the direction to minimize the closeness measure, V_2, defined by

$$V_2 = \frac{1}{2}\sum_k \sum_i (\alpha_{ik} - 1/\sqrt{\lambda_{ik}})^2 \tag{22}$$

where α_{ik} is a function of x and p. Then, from the condition that keeps the time derivative of V_2 negative when $V_2 \neq 0$, we can determine $\dot{p_2}$ as follows:

$$\dot{p_2} = -\sum_k \sum_i (\alpha_{ik} - 1/\sqrt{\lambda_{ik}})\frac{d\alpha_{ik}}{dp} / \left\|\frac{d\alpha_i}{dp}\right\| \tag{23}$$

where

$$\frac{d\alpha_{ik}}{dp} = \frac{\partial \alpha_{ik}}{\partial x} \bullet \frac{dx}{dp} + \frac{\partial \alpha_{ik}}{\partial p} \tag{24}$$

The directional vector, $\dot{p_3}$, that incurs the minimum sensing cost can be searched by

$$\dot{p_3} = \min_{\dot{p}} C(\dot{p}) \tag{25}$$

where $C(\dot{p})$ represents sensing cost in terms of the parameter change, $\dot{p}$, at the current sensor configuration.

With $\dot{p_1}$, $\dot{p_2}$, and $\dot{p_3}$, defined respectively for the constraint satisfaction, the performance optimization, and the minimum sensing cost, the final parameter update, $\dot{p}$, can be determined by combining $\dot{p_1}$, $\dot{p_2}$, and $\dot{p_3}$, as follows:

$$\dot{p} = \dot{p_1} + N(\dot{p_1})\dot{p_2} + N(\dot{p_1} + N(\dot{p_1})\dot{p_2})\dot{p_3} \tag{26}$$

where $N(\dot{p_1})\dot{p_2}$ implies the projection of $\dot{p_2}$ onto the null space of $\dot{p_1}$. Note that, when the system and parameter constraints are satisfied, then $\dot{p_1} = 0$, consequently, $N(\dot{p_1})\dot{p_2} = \dot{p_2}$. Eq. (26) is referred to here as the iterative parametric sensor planning equation. Note that the priority given to $\dot{p_1}$, $\dot{p_2}$, and $\dot{p_3}$, for the computation of $\dot{p}$ can be changed, i.e., a different order of $\dot{p_1}$, $\dot{p_2}$, and $\dot{p_3}$, can be used in (26) for computing $\dot{p}$.

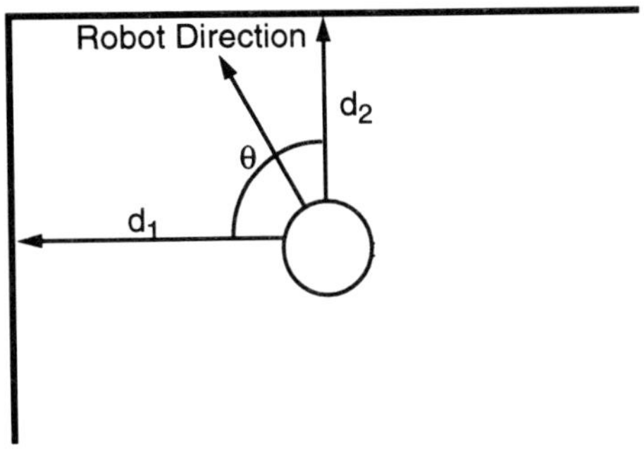

Figure 10: Experimental environment.

3.3 Experimentation

The mobile robot RB5X controlled by a PC notebook and equipped with an ultrasonic sensor is used for the experimentation. The robot is capable of translational and rotational movement. During the experiment, the robot changed its pose by moving to the next optimal location to take another measurement for data fusion. The environment model is assumed available. By considering pose variation as the controllable parameter, the robot repeated the parameter update until it reaches within the desired error bound.

The measurement before and after the movement is fused to help reduce uncertainty. The movement errors are incorporated into the data fusion.

3.3.1 *Environment*

A simple environment, a corner in a rectangular room, is shown in Fig. 10, in which d_1, d_2 are the distances to the walls measured by the ultrasonic sensor, and θ is the angle between two measurement directions of each side of the corner. In this environment, the position of the robot is presented in a 2D Cartesian space based on the (x, y) coordinate and the rotation angle Φ with respect to the world frame.

Because of the ultrasonic nature of the sensor, the distance can only be measured accurately if the measurement direction is within $\pm 10°$ of the normal direction of the wall. When this condition is met, the measurable distance falls between 15 inches and 90 inches with a maximum error 10% of the measured distance. The maximum movement error of the robot is also 10% of the moved distance, but the orientation error is ignorable. As distance measurement takes place for every 10° the robot rotates, for each wall, two or three valid measurements can be obtained. The minimum of them is considered as the perpendicular distance to the wall. In summary, the maximum error bound of the distance measurement is 10% of the measured distance, while the maximum error bound of the wall orientation measurement

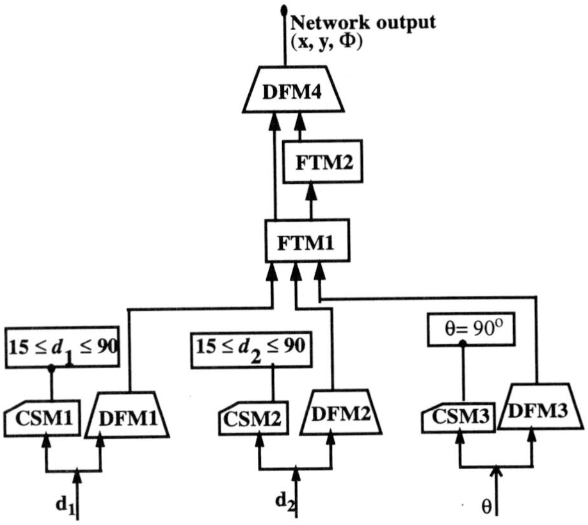

Figure 11: PAN for experiment.

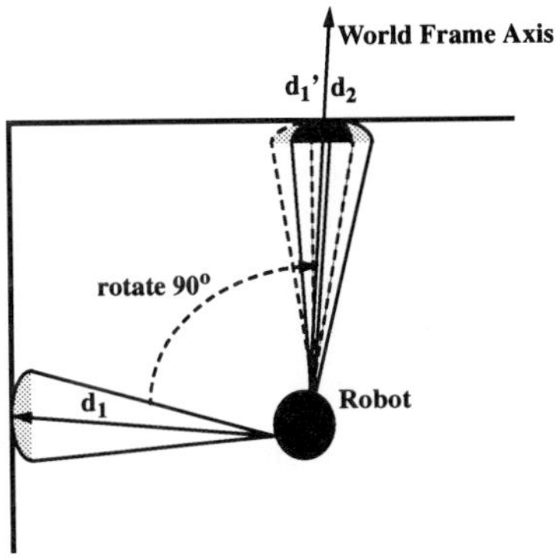

Figure 12: Overlap the orientation errors of the two measurements, the median is taken as one axis of world frame, intersection is the error bound.

is 10°. The controllable parameter is the robot pose variation. The system constraints used in the experiment are the measurement range of the ultrasonic sensor, i.e., $15 \leq d_i \leq 90$ for i = 1, 2 and the consistency of the measurement directions, i.e. $\theta = 90$.

3.3.2 *Procedure*

Fig. 11 illustrates the PAN constructed for the given localization problem. DFM1 and DFM2 modules fuse 10 times of measurement by taking the average. DFM3 performs the intersection of orientation uncertainty of d_1 and d_2. Three CSM modules perform the system constraints given above. FTM1 module transforms the translational and rotational deviation of world frame with respect to the robot frame into that of robot frame with respect to world frame. FTM2 module propagates the error according to the motion vector. DFM4 module fuses the data before and after the movement and gives out the current location and orientation of the robot.

The algorithm performed by the robot during the experiment is as following:

1. Take measurement, S, $S \equiv [d_1, d_2, \theta]^T$, with the uncertainty bound $E_S(\Delta d_1, \Delta d_2)$.

2. From the measurement, the world frame and error range are determined by fusion (Fig. 12):

   - Rotate the orientation error range of the measurement direction of d_1 by 90°,
   - Overlap it with the orientation error range of the measurement direction of d_2 and find the intersection.
   - The median of the intersection is taken as one axis, the range of the intersection is the error bound $\Delta\Phi_w$.
   - From the error bound of measurements E_s, uncertainty bound of world frame $E_w(\Delta x_w, \Delta y_w, \Delta\Phi_w)$ is defined: $\Delta x_w = \Delta d_1, \Delta y_w = \Delta d_2$.
   - The location of the robot with respect to the fixed world frame $L(x, y)$ is $x = d_1, y = d_2$.

3. Fix the world frame as the estimated one, transform the translational $(\Delta x_w, \Delta y_w)$ and rotational deviation $(\Delta\Phi_w)$ of world frame into that of robot frame with respect to the fixed world frame: $E_r^T = JE_w^T$, where $E_r = (\Delta x_r, \Delta y_r, \Delta\Phi_r)$.

   Assume the vector from the origin of the fixed world frame to the robot frame is P, J can be calculated from the following equation:

$$\begin{aligned} \begin{bmatrix} \Delta x_r \\ \Delta y_r \end{bmatrix} &= \Delta\Phi_w \times P + \begin{bmatrix} \Delta x_w \\ \Delta y_w \end{bmatrix} \\ \Delta\Phi_r &= \Delta\Phi_w \end{aligned} \tag{27}$$

4. Calculate the desired relative location ΔL by

$$\Delta L = \left[\frac{\partial E_r}{\partial L}\right]^T (E_d - E_r) \tag{28}$$

$$\frac{\partial E_r}{\partial L} = \frac{\partial E_r}{\partial E_w} \cdot \frac{\partial E_w}{\partial E_S} \cdot \frac{\partial E_S}{\partial S} \cdot \frac{\partial S}{\partial L} \tag{29}$$

   where E_d is desired error bound.

5. Move the robot by $M = \alpha\Delta L$ with uncertainty $E_M(\Delta x_M, \Delta y_M, \Delta\Phi_M)$, where α indicates movement limit.

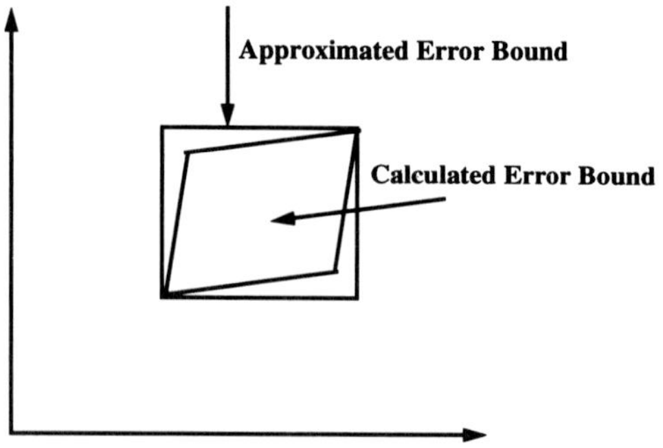

Figure 13: Approximate the error bound by a rectangle.

6. Propagate E_r according to the motion vector M.

$$\begin{bmatrix} \Delta x_m \\ \Delta y_m \end{bmatrix} = \begin{bmatrix} \Delta x_r \\ \Delta y_r \end{bmatrix} + \begin{bmatrix} \Delta x_M \\ \Delta y_M \end{bmatrix} + \Delta\Phi_r \times M$$

$$\Delta\Phi_m = \Delta\Phi_r + \Delta\Phi_M \tag{30}$$

7. Take new measurement S. Calculate $E_r^{'}$ same as calculating E_r in step 3.

8. Intersect the results of step 6 and 7.

$$E_r = E_m \cap E_r^{'} \tag{31}$$

9. If $(E_d < E_r)$ and the robot is not too close to the wall, goto step 4; otherwise stop.

To simplify the intersection process in step 8, we can approximate E_m and $E_r^{'}$ by rectangles. This can be done by finding the leftmost point, the rightmost point, the highest point and the lowest point of the error bound as shown in Fig. 13. We need to note that this method enlarges the error bound, and it finds a conservative solution for the uncertainty of the location. Other methods such as ellipsoid approximation may also be used.

As shown in Fig. 14, for each iteration, the robot tries to move to the desired location defined by (28) with certain movement limit. The error bound reduces by data fusion and/or by higher accuracy measurement.

3.3.3 *Result*

The experiment was performed for several different movement limits of each time of movement. In Fig. 15a, the movement limit for each iteration was set to be 1 inch. The actual location after the last movement was X=16.1 inches, Y=15.8 inches. The location computed by the robot was X=15.9 inches, Y=15.4 inches with the uncertainty EX= 0.67 inch, EY=0.63 inch. We can

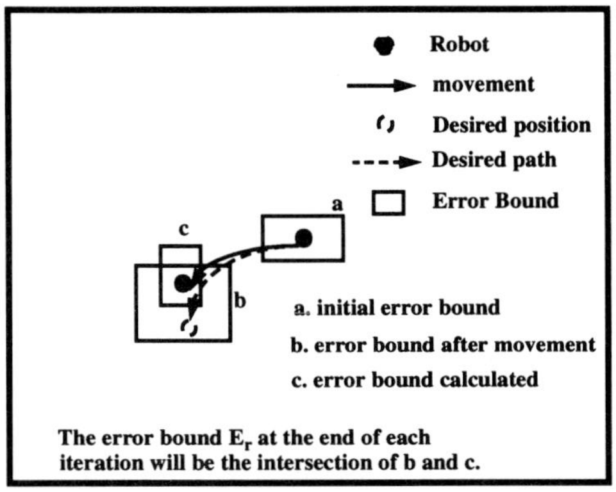

Figure 14: Change of Error bound for each iteration.

see that the maximum error bounds for the position are gradually converged. In Fig. 15b, the movement limit for each iteration was 2 inches. With the same initial position as in Fig. 15a, the convergence is faster than the previous case. However, the worse accuracy was reached for two reasons. First, less iterations were performed before the robot stopped, so that less data fusion were performed. Second, During each movement, it has larger movement error. In Fig. 15c, the movement limit for each iteration was 5 inches. In this case, the movement error was too large compare to the sensor uncertainty. Thus the data fusion procedure (Eq. 13) always had $E_r = E_r^{'}$ after each iteration.

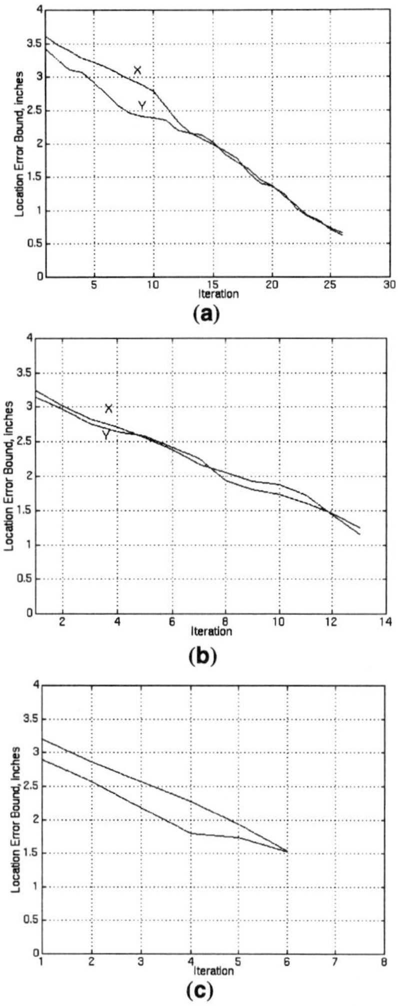

Figure 15: Experiment Results: movement limit for each iteration is (a) 1 inch, (b) 2 inches, (c) 5 inches.

4 Conclusion

The proposed perception-action network provides a mechanism to integrate sensing, knowledge, and action in real time for robotics. The integration is done based on the network dynamics which governs the evolution of the output states of heterogeneous computational units according to input stimuli and system constraints. Also shown is that the network provides a more general but formal way of accomplishing sensor fusion and planning.

However, there still remain many concepts to be extended and many problems to be solved:

1. The proposed network focuses more on perception than on action. It is interesting to see whether we can build a hierarchy of computational units representing action or a logical actuator system and connect it to perception for completion.

2. What would be the best way of implementing the network? For instance, can the network dynamics be implemented as a connection model with analog hardware, while computational units are implemented with digital hardware?

3. What would be the advantage of proposed perception-action network in terms of improving the degree of intelligence defined in the introduction of this paper?

4. The comparison of the proposed network with existing robotic architectures such as subsumption[20], cerebella model articulated computer or NASREM[21], neural network-based architecture, etc.

5. The convergence of network dynamics to correct equilibrium without trapping into local minima.

5 Acknowledgments

The author wishes to express his appreciation to Dr. Jun Park, Mr. Xiaoming Zhao, and Mr. Sookwang Ro of USC for their simulation and experimental works as well as to Prof. Wook Hyun Kwon of Seoul National University for his stimulation and support. This work was Partially supported by the Engineering Research Center for Advanced Control and Instrumentation of Seoul National University.

References

[1] Ren C. Luo, Min-Hsiung Lin, and Ralph S. Scherp, "Dynamic Multi-Sensor Data Fusion System for Intelligent Robots," IEEE Journal of Robotics and Automation, Vol. 4, No. 4, pp. 386-385, August 1988.

[2] Hans P. Moravec, "Sensor Fusion in Certainty Grids for Mobile Robots," AI Magazine, pp. 61-74, Summer 1988.

[3] Scott W. Shaw, Rui J. P. deFigueiredo, and Kumar Krishen, "Fusion of Radar and Optical Sensors For Space Robot Vision," The proceeding of IEEE Conference on Robotics and Automation, pp. 1842-1846, 1988.

[4] Tom Henderson and Esther Shilcrat, "Logical Sensor System," Journal of Robotic System, 1(2), pp. 169-193, 1984.

[5] T. L. Huntsberger and S. N. Jayaramamurthy, "A Frame Work for Multi-Sensor Fusion in the Presence of Uncertainty," Proceedings of Spatial Reasoning and Multi-Sensor Fusion Workshop, pp. 345-350, 1987.

[6] Hugh F. Durrant-Whyte, "Sensor Models and Multi-Sensor Integration," IEEE Journal of Robotics and Automation, 1987.

[7] k. Fukunaga, "Introduction to Statistical Pattern Recognition," Academic Press, 1972.

[8] A. Gelb et al, "Applied Optimal Estimation," The M.I.T. Press, 1974.

[9] S. Lee and R. M. Kil, "Bidirectional continuous associator based on Gaussian potential function network," IEEE International Joint Conference on Neural Networks, 1:45-53, 1989.

[10] S. Lee and R. M. Kil, "Robot Kinematic Control Based on Bidirectional Mapping Neural Network," IEEE International Joint Conference on Neural Networks, 3:327-335, San Diego,1990.

[11] H. F. Duran-White, "A Bayesian approach to optimal sensor placement," International Journal of Robotics Research, 9(5):70-88, 1990.

[12] C. K. Crown and A. Bergman, "Determining the camera and light source location for a visual task," In IEEE International Conference on Robotics and Automation, Scottsdale, (USA), pages 509-514, April 1989.

[13] K.Tarabanis, R. Y. Tsai, and P. k. Allen, "Automated sensor planning for robotic vision tasks," In IEEE International Conference on Robotics and Automation, Sacramento, (USA), pages 76-82, 1991.

[14] S. A. Hutchinson, R. L. Cormwell, and A. C. Kak, "Planning sensing strategies in robotics and automation," Philadelphia (USA), pages 1068-1075, April 1988.

[15] R. Bajcsy, "Active perception vs. passive perception," In IEEE Workshop of Computer Vision: Representation and Control, pages 55-62, Bellaire, MI, 1985.

[16] S. Lacroix, P. Grandjean, and M. Ghllab, "Perception planning for a multi-sensor interpretation machine," In IEEE International Conference on Robotics and Automation, Nice, France, 1992.

[17] G. Hager and M. Mintz, "Task-directed multisensor fusion," In IEEE International Conference on Robotics and Automation, pages 662-667, Scottsdale, AZ, 1989.

[18] S. Lee and X. Zhao, "Sensor Planning with Hierarchically Distributed Perception Net," in IEEE International Conference on Multisensor Fusion and Integration for Intelligent Systems, Oct. 1994, Las Vegas, NV.

[19] A Pugh, Robot sensors, Volume 1, 1986, IFS ltd. UK.

[20] Rodney Brooks, "A Robust layered control system for a mobile robot", IEEE Journal of Robotics and Automation, Vol. RA-2, NO. 1, March 1986

[21] James S. Albus, " Brains, behavior, and robotics", BYTE Books, Peterborough, NH, 1981

Toward A Systems Theory for the Composition of Dynamically Dexterous Robot Behaviors

R. R. Burridge, A. A. Rizzi and D. E. Koditschek
Artificial Intelligence Laboratory and Controls Laboratory
EECS Department, College of Engineering
University of Michigan
1101 Beal Ave, Ann Arbor MI 48109-2110
{burridge,arizzi,kod}@eecs.umich.edu

Abstract

We report on our efforts to develop robot controller composition techniques in the context of dexterous "batting" maneuvers. A robot with a flat paddle is required to strike repeatedly at a falling ball until its height is regulated and it is brought to zero horizontal velocity at a specified position. The robot's workspace is cluttered with obstacles that disconnect the freespace formed when the ball and paddle remain in contact — the machine is forced to "let go" for a time in order to bring the ball to the desired state. The controller compositions that we create will guarantee that a ball introduced in the "safe workspace" remains there and is ultimately brought to the goal. We believe that the developing systems discipline described herein may be extended to build a variety of useful dexterous machines that are similarly single-minded in their pursuit of the user's goal behavior and ability to surmount unanticipated perturbations along the way.

1 Introduction

In this paper we explore empirically a simple approach to the composition of robot behaviors that seems to offer both practical utility as well as analytical tractability. We cast "behaviors" — robotic implementations of user specified tasks — in a form amenable to representation as global regulation via state feedback to a specified set point in the presence of obstacles. Our approach results in a partition of state space induced by a palette of pre-existing feedback controllers. Each cell of this partition is associated with a unique controller. The controllers are ordered in such a fashion that capture in any cell guarantees passage to successively "lower" cells until the "lowest," the goal cell, has been achieved. The volume of the composite domain of attraction around the goal cell is as large as the union of the domains of all constituent controllers.

1.1 Overview of the Approach

In previous work we have paid a great deal of attention to the lower level palette. Our architecture implements event driven robot policies whose closed loop dynamics drive the coupled robot-environment state toward the goal set and away from obstacles. We strive to develop control algorithms that are sufficiently tractable as to allow correctness guarantees with estimates of the domain of attraction as well. Thus, we have focused theoretical attention on "practicable stability mechanisms" — dynamical systems for which effectively computable local tests provide global (or, at least, over a "large" volume) conclusions. At the same time, we have focused experimental attention on building a distributed computational environment that supports flexibly reconfigurable combinations of sensor and actuator hardware controllers, motion estimation and control algorithms, and event driven reference trajectory generators. The result has been a series of laboratory robots that exhibit indefatigable goal seeking behavior, albeit in a very narrow range of tasks. In the present work we explore the "higher level" problem of generalizing the range of tasks by composing the existing controllers.

1.1.1 *Robust Dynamical Dexterity*

In the past work, we have concentrated on tasks requiring dynamical dexterity — the ability to perform work on the environment by effecting changes in its kinetic as well as potential energy. In the present paper, we continue to focus on this same domain of tasks.

Specifically, we want to explore dynamical interactions between robot and environment such as active balancing, hopping, throwing, catching, juggling, and other tasks requiring active regulation of kinetic energy. In doing so, we hope to gain insight into how to design and control machines that exhibit dexterous behavior, robust in the face of disturbances. In particular, we seek a means of exploiting the capabilities of our robot's previously developed dexterous behaviors to achieve new tasks that no single existing controller could have achieved in isolation.

1.1.2 *Feedback Confers Robustness*

Physical disturbances to a system can be classified, broadly speaking, into one of two categories. The first includes the small, persistent disturbances arising from model inaccuracies and sensor and actuator noise. Spin and air drag on a falling ball modelled as a point mass, or spurious changes in camera measurements introduced by lighting variations in an illuminated scene represent typical examples of this first type of disturbance. The second category includes large, significant changes to the world as expected. For example, when a parts tray is severely jostled or arrives unexpectedly empty then an assembly robot's world model is likely to have been grossly violated. Such large potentially catastrophic disturbances typically occur quite rarely (otherwise, they would have been accounted for systematically by the robot's designers!).

A common tradition in robotics and control has been to model these disturbances in the hope of treating them rationally. One typically appeals to stochastic representations of small perturbations and introduces an optimal design (for example a Kalman filter) to handle them. Analogously, recovery from catastrophe is similarly "model" based, typically being addressed by special "exception handling" routines specifically written for each particular failure anticipated. But disturbances are generally left out of nominal models precisely because they are so difficult to model. The systematic origin of the typical small disturbance often defies a stochastic representation and the resulting optimal designs may be overly aggressive. Analogously, if the human system designers haven't imagined a severe failure mode, then the system will be unable to react to it when it occurs, with possibly disastrous results.

In contrast, feedback controlled systems can be designed to confer a level of robustness against both types of disturbance without recourse to explicit disturbance models. On the one hand, asymptotically stable systems are locally generically structurally stable. This means that the first class of disturbances can degrade performance (e.g., biasing the goal set, or coarsening the neighborhood of the goal set to which all initial conditions in its domain of attraction are eventually brought) but, if small enough, cannot destroy stability. On the other hand, if the closed loop system induced by a feedback controller exhibits a global domain of attraction then large disturbances can never take it "by surprise", so long as the original state space and dynamical model used to represent the world (not the disturbance) remain valid following the perturbation. The system will merely continue to pursue the same shepherding function, now applied at the new state, with the same indefatigable progression toward the goal as before.

1.1.3 *Feedback Strategies as Funnels*

Exactly one decade ago, Mason [1] introduced the notion of a funnel as a metaphor for the robust, self-organizing emergent behavior displayed by otherwise unactuated masses in the presence of passive mechanical guideways.[1] He understood that the shape of the guideways constituted an embodied computational mechanism of control that ought to be rendered programmable. In this paper, we embrace Mason's metaphor and seek to extend it to active electromechanical systems (i.e., possessing sensors and actuators) as being more obviously programmable.

Figure 1 depicts the graph of a positive definite scalar valued function centered at a goal point — its only zero and unique minimum. If a feedback control strategy results in a closed loop dynamics on the domain (depicted as the x-y plane in these figures) all of whose motions lead down the fun-

[1] Other researchers in robotics and fine motion planning have introduced comparable metaphors[2, 3].

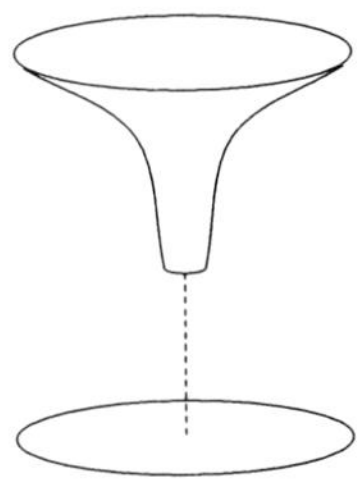

Figure 1: The Lyapunov Function as Funnel: Graph of a positive definite function over its state space.

nel, then this is said to be a Lyapunov function. In such a case, Mason's metaphor of sand falling down a funnel is particularly relevant since all initial conditions on the plane lying beneath the over-spreading shoulders of the funnel (now interpreted as the graph of the Lyapunov function) will be drawn toward its bottom — the goal.

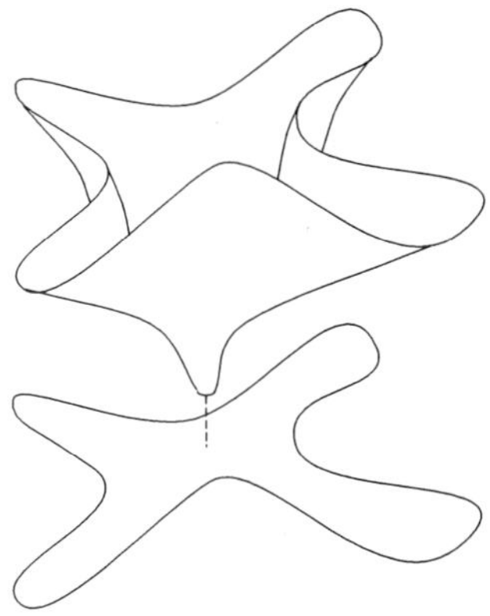

Figure 2: Ideal "funnel" capturing the entire "obstacle free" state space of the system.

In Figure 2 we depict a complicated boundary such as might arise when the user specifies an "obstacle" — here, the set of states outside the bounded area that the user forbids us to enter.[2] Ideally, we would like a control strategy that "funnels" the entire state space of the system to the goal. We suggest this as well in the figure by sketching a complex funnel whose shape is exactly adapted to the domain.[3] Unfortunately, it is usually very difficult or impossible to find globally attractive control laws, so Figure 2 often cannot be realized for practical reasons. Moreover, there is a large and important class of mechanical systems for which no single funnel (continuous stabilizing feedback law) exists [5, 6]. Fortunately, we need not yet abandon the recourse to global attraction.

1.1.4 *Sequential Composition as Preimage Backchaining*

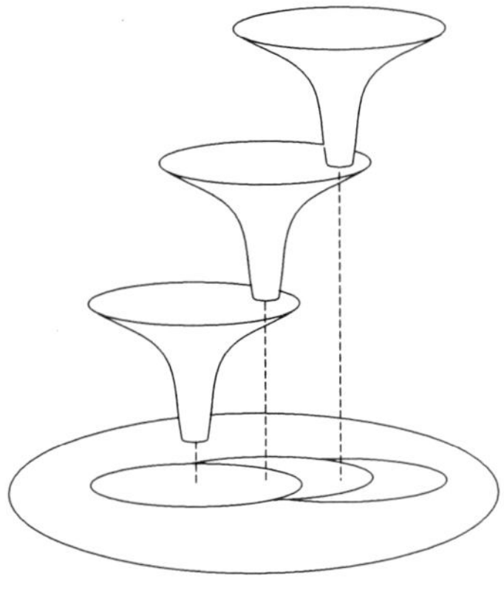

Figure 3: Sequential composition of funnels: the goal point of each previous controller lies within the domain of attraction induced by the next.

Instead, we turn to another tradition introduced within the fine motion planning literature roughly one decade ago by Lozano-Perez, Mason, and Taylor [2] — the notion of preimage backchaining. Assume the availability of a palette of controllers whose induced funnels have targets that can be placed at will. Then we may deploy a set of multiple controllers whose combined (local) domains of attraction cover a large portion of the entire state space of the system. If such a set is rich enough that the domains overlap and goal sets lie within domains of other controllers, then it is possible to "backchain" away from the overall task goal, partitioning the state space into regions where the different controllers will be active, and arrive at a switching scheme between the controllers that will *provably* drive any state in the union of the various domains to the overall goal state.

This method of combining controllers in sequence is depicted in Figure 3. Note that the controller represented by each funnel is only active when the system state is in the appropriate

[2] Of course, this picture dramatically understates the potential complexity of such problems since the free space need not be simply connected as depicted!

[3] In a previous series of papers with Rimon [4], the third author has, addressed the design of such global funnels for purely geometric problems wherein the complicated portion of the domain is limited to the configuration space. In contrast, in this paper, we are concerned with complex boundaries that run through the velocities as well — e.g., see Figure 11.

region of state space, as sketched at the bottom of the figure. As each controller drives the system toward its local goal, the state crosses a boundary into another region of state space where another controller is active. This process is repeated until the state reaches the final partition, which is the only region to contain the goal set of its controller.

We are now in a position to approximate Figure 2 with Figure 4.

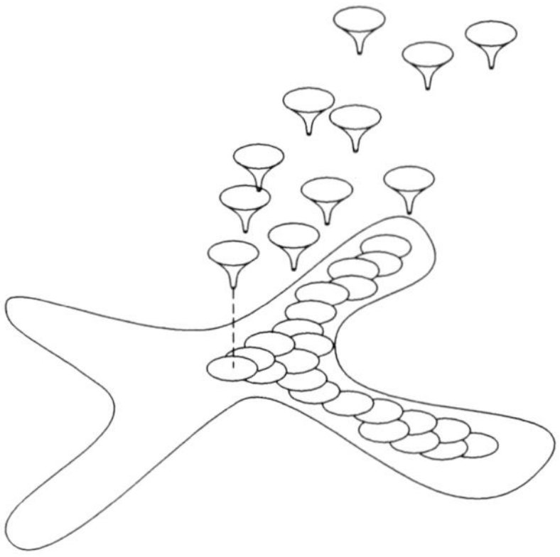

Figure 4: A suitable composition of local funnels partitions (a close approximation to the) free space state space into cells assigned to a unique controller.

This scheme affords the same robustness against disturbances discussed above. Local stability against small disturbances obtains because each funnel represents a Lyapunov function for that particular controller and goal point. It also provides robustness against large, unanticipated disturbances because the control system has no higher-level state, or preconceived plan. If the state of the system shifts to a distant location, then the controller appropriate for *that particular region of state space* automatically activates, and the process begins again. Similarly, if the state takes a favorable leap toward the goal state, or starts in the goal partition, then intermediate controllers will never be activated. Thus as long as the state doesn't jump outside the union of all the domains in the partition, it will be effectively handled.

As we try to suggest in Figure 4 this version of preimage backchaining is particularly convenient when the obstacle free space has a complicated geometry. Because the domains used for the partition of state space must be invariant, there is a natural notion of safety that arises from this approach. If there are obstacles in state space, but they do not intersect any of the constituent domains of the partition of state space, then it follows that the state will never reach an obstacle while under the composite controller.

1.2 Organization and Contributions of Paper

The paper follows a logical development roughly suggested by Figures 1 through 4. In Section 2 we provide a high level overview of the the hardware and software that combine to form the basic entry in the controller pallete suggested by Figure 1. Section 3 we introduce the notion of a safe controller suggested by Figure 2 as well as define in formal terms the sequential composition depicted in Figure 3. In Section 4 we implement the partition depicted in Figure 4 using numerical methods and report on our empirical results to date.

In [7] we introduced the notions of dynamical safety and obstacle avoidance, while in [8] we discussed the controller composition algorithm mentioned in Section 3. In this paper, we further our work by showing how to compose controllers in such a manner as to avoid an obstacle that disconnects the workspace. We make use of a discrete projection of the closed-loop system dynamics to arrive at a complex controller that will provably bring the union of the domains of the constituent controllers on both sides of the obstacle to the overall goal. Empirical results in section 4 demonstrate the efficacy of our approach.

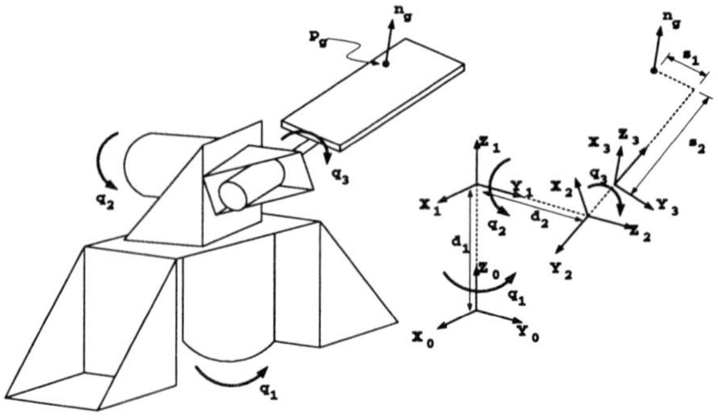

Figure 5: The Bühgler Arm

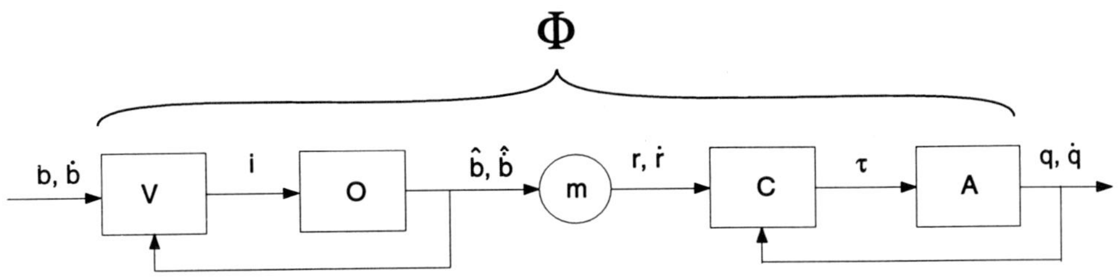

Figure 6: Flow chart showing the various state machines in the system.

2 Experimental Setup and Task

2.1 Hardware

For these experiments we use the "Bühgler," a three degree of freedom direct drive machine pictured in Figure 5, which we have discussed in a variety of other settings [9]. The robot is required to manipulate a ping pong ball from a wide range of starting states to a goal state while avoiding obstacles along the way. Two cameras provide ball images at 60Hz, and all computation is done on a network of approximately twenty transputers, as detailed in [9].

2.2 Software

In Figure 6 we outline the flow of data in the software. Ball states[4], $(b, \dot{b}) \in \mathcal{B}$, are viewed by the vision system, V (cameras, and window management), which produces image plane locations, i. An observer, O, uses i to triangulate ball positions, and produce estimates of the full state of the ball, $(\hat{b}, \hat{\dot{b}})$. These ball state estimates are fed back to V to direct its focus of attention, and also interpolated to produce estimates at 1 kHz for the rest of the system. This V - O sensing loop is discussed in detail in [9].

Estimated ball states, $(\hat{b}, \hat{\dot{b}})$, are passed through a memoryless transformation, m, to produce reference robot states, $(r, \dot{r})$. We refer to m as a *mirror law*. The references, $(r, \dot{r})$, are fed to an inverse dynamics joint space controller [10], C, which produces torques, τ, for the actuator block, A (amplifiers, motors, etc.), which generates true robot joint states, $(q, \dot{q}) \in \mathcal{Q}$. The robot states are fed back to the controller for feedback control.

We refer to the entire filter from ball state to robot state as Φ. Both the sensing loop, V - O, and actuation loop, C - A, are globally asymptotically stable. For the purposes of the work presented in this paper, it will suffice to assume that these loops produce immediately accurate ball states (i.e., $(\hat{b}, \hat{\dot{b}}) \approx (b, \dot{b})$, and $(q, \dot{q}) \approx (r, \dot{r})$). Laboratory experience supports this assumption, which allows us to reduce Φ to m. In the sequel we will be concerned with various choices of m, but will denote the various controllers that arise by Φ.

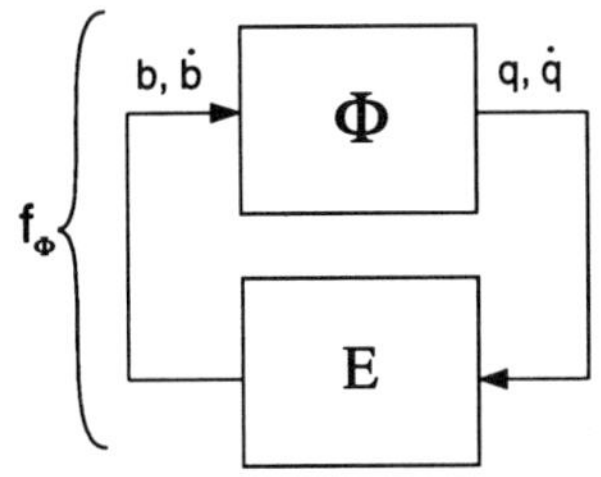

Figure 7: The closed loop dynamics, f, induced by Φ.

In Figure 7 the environment, E, is added to close the loop back from robot to ball. The closed loop dynamics induced by a particular controller, Φ, are denoted f_Φ.

We are interested in encoding and achieving user goals taking the form of a desired set of world states, $\mathcal{G}$. A controller, Φ, achieves those goals if $\mathcal{G}$ is an attracting invariant subset, in which case we are interested in rendering its domain of attraction, $\mathcal{D}(\Phi)$, as large as possible. We have developed over the last five years an approach to robot controller design that accomplishes these goals by essentially reshaping the total energy exchanged between the robot and its environment, as depicted in Figure 1.

[4] At times it will be convenient to further divide the ball state into its vertical and horizontal parts: $b = (h, v)$, $\dot{b} = (\dot{h}, \dot{v})$.

To aid the reader in making sense of the discussion to follow, it seems worth pointing out that there will be two different "time-scales" of interest over which control must be asserted and studied. First, the robot is reacting essentially continuously to the motion of the ball according to the active mirror law. If the ball is resting on the paddle, then the robot has the opportunity for continuous control over the state of the ball. Obviously, however, when the ball is in flight the motion of the robot does not affect it except at impact, at which point it will bounce (we model the bounce using an instantaneous restitution model [11]), and be in flight again. The intermittent nature of the robot's influence on the ball gives rise naturally to a discrete system. The state of the ball at apex is invariant during flight, so we will use apex coordinates when discussing the discrete system.

Thus, by an abuse of notation, the closed loop map, f_Φ, in Figure 7 defines either a continuous flow on $(b, \dot{b})$ or a discrete map on the apex states (i.e., those $(b, \dot{b}) = (h, v, \dot{h}, \dot{v})$ such that $\dot{v} = 0$) (see footnote 4 depending upon the selection of the controller, Φ. Some examples now follow.

2.2.1 *Continuous Closed Loop Dynamics*

A good example of a controller with closed loop dynamics is the Palmer. The palming controller, Φ_P, uses the "waist" of the robot to keep the centerline of the paddle directly under the ball at all times. The other two degrees of freedom use "proportional derivative" style laws to servo about a particular horizontal set point, h^* (see footnote 4). If the ball is displaced to one side of h^*, then the paddle tilts in such a manner as to push it back toward the set point. The set of locations in joint space at which contact occurs is known as the *contact surface*, which is similar to a bowl for this controller.

If the ball is in flight, then it will lose vertical energy every bounce until it comes to rest on the paddle. At the same time, the PD terms cause the ball to bounce toward the set point every impact, although for sufficient vertical energy the initial effect of these terms may be to knock the ball too far, landing further from the goal. Once the ball is on the paddle, the induced closed loop dynamics resemble the classic "ball-on-beam" problem. The robot rolls the ball about on the paddle, bringing it eventually to the horizontal set point, h^*, at zero velocity. Note that h^* is the only attracting goal point for this controller.

2.2.2 *Discrete Closed Loop Dynamics*

As mentioned above, it is possible to create mirror laws that impart vertical energy back to the ball during impact, perhaps more than it loses. In such cases, the ball may never come to rest on the paddle, and a bouncing limit cycle can arise, which we call "batting." When this is true, we consider each flight phase to be a single state in apex coordinates, and work with the (discrete) return map from apex to apex.

In this section we introduce the one degree of freedom vertical mirror law, and then show how we combine it with Φ_P to get juggling behavior.

2.2.2.1 The Vertical Batter: Φ_B

Consider a one degree of freedom "batting" problem: a piston and a bead on a vertical wire, with the task being to provide a control strategy for the piston which brings the bead to a particular limit cycle. Buehler [12] developed a mirror law that globally asymptotically stabilizes the discrete apex dynamics of the ball to a periodic orbit. This mirror law essentially introduces "proportional control" of the total energy of the bead about a desired energy, η^*. Of course, specifying a vertical energy, η^*, is equivalent to specifying an apex height, $(v^*, \dot{v} = 0)$ (see footnote 4).

2.2.2.2 The Juggler : Φ_J

Buehler used Φ_B as the basis for his planar juggler, which employed a rotating bar to bat a puck to a stable limit cycle on a backplane. Rizzi [9] extended Buehler's work to the 3 degree of freedom robot by simply adding Φ_B on the "shoulder" actuator to Φ_P (i.e., $m_J = m_P + m_B$). The contact surface remains the same as for the Palmer, but now the paddle has a vertical velocity at each location on it, varying with the vertical energy of the ball. Instead of settling on the paddle, the ball is now driven toward a set point in apex space positioned horizontally at h^*, with vertical energy η^*.

Note that we can change h^* and η^* independently, and effectively cause the set point to be located anywhere in the cylinder above the workspace (up to the limits of the abilities of the vision system and actuators to accurately perform their functions).

3 Composition and Safety

3.1 Sequential Composition

Say that controller Φ_1 *prepares* controller Φ_2, denoted $\Phi_1 \succeq \Phi_2$, if its goal lies within the domain of the second $\mathcal{G}(\Phi_1) \subset \mathcal{D}(\Phi_2)$. For any set of controllers, $\mathcal{U} = \{\Phi_1, ..., \Phi_N\}$, this relation induces a directed generally cyclic graph.

Assuming that the overall task goal set, $\mathcal{G}$, coincides with the goal set of at least one controller, $\mathcal{G}(\Phi_i) = \mathcal{G}$, then by starting with Φ_i and recursively tracing the relation backwards through the corresponding graph, one arrives at $\mathcal{U}_\mathcal{G} \subset \mathcal{U}$ — the set of all controllers from whose domains the overall goal might eventually be reached by appropriate switching among the controllers. The domain of a properly conceived composite controller, Φ, should then be

$$\mathcal{D}(\Phi) = \bigcup_{\Phi_j \in \mathcal{U}_\mathcal{G}} \mathcal{D}(\Phi_j). \tag{1}$$

Such a sensor based composite results naturally from a partition of the state space of the ball into cells "governed" by the members of $\mathcal{U}_\mathcal{G}$ as follows. The relation, $\succeq$, induces a partial order (possibly more than one) on $\mathcal{U}_\mathcal{G}$ with a minimal element, Φ_i, as may readily be computed by pruning the original graph into an acyclic graph with root at Φ_i (e.g., perform a breadth-first search back from that node) that we call a *preparation graph* for Φ_i and denote by Γ. This partial order now induces a partition on the dynamical workspace, obtained from an arrangement of the domains, $\{\mathcal{D}(\Phi_i)\}_{\Phi_i \in \mathcal{U}_\mathcal{G}}$, of the participating controllers in the obvious manner.[5] The process of constructing a composite controller from a set of "simpler" controllers is denoted by $\Phi = \bigvee \mathcal{U}_\mathcal{G}$.

We have achieved reasonable experimental success with algorithms of this form where the placement and tuning (hereafter referred to as the *deployment*) of the individual controllers is carefully designed by hand [8]. We report here a specific instance of such a "hand-designed" deployment. Ultimately, we expect to build systems where the controller deployment is automatically generated from user-specified task goals.

[5]For example, controller Φ_1 is assigned that portion of its domain, $\mathcal{D}(\Phi_1)$, not contained in that of any "$\succeq$-lesser" controller; controller Φ_{k+1} is assigned the portion of its domain not contained in that of either a $\succeq$-lesser controller one of index $j \leq k$ [8].

3.2 Obstacles and Safety

There are three types of obstacle with which we are concerned. The first, where the ball and robot are in continuous contact, has been studied carefully in the motion planning literature [13]. The second type of obstacle consists of regions of the configuration space "off limits" to the robot, $\mathcal{O}_\mathcal{R}$, whether or not it is in contact with the ball. Robot obstacles include the floor, joint limits, etc., as well as any physical obstacle in the workspace, and have been studied in the kinematics literature. Finally, the third consists of subsets of ball state space which are "off limits" to the ball, even when it is not in contact with the robot. The system is considered to have failed if it allows the ball to enter any of these regions, collectively denoted $\mathcal{O}_\mathcal{B}$. We are most interested in the last type of obstacle in the context of intermittent contact, and thus discrete closed loop dynamics, particularly since it has not been studied methodically to date.

Recall that for a particular mirror law, m, each ball state induces a corresponding robot configuration. We define a workspace obstacle, $\mathcal{O}$, as the union of $\mathcal{O}_\mathcal{B}$ and the set of ball states that induce the robot into an obstacle, $m^{-1}(\mathcal{O}_\mathcal{R})$. The *urgency* of a particular ball state is defined as the time to impact with $\mathcal{O}$ in the absence of the paddle, denoted $\tau_\mathcal{O}$.

Say that a controller Φ is *safe with respect to* $\mathcal{O}$ if there can be found an invariant open subset in $\mathcal{D}_\Phi$ such that the next impact with the robot always occurs before the ball reaches an obstacle, or $\tau_\Phi < \tau_\mathcal{O}$. We denote such a *safe domain* by $\mathcal{D}_\mathcal{O}(\Phi)$.

Notice that for any obstacle and controller, a safe domain is invariant during flight by definition. Thus, once Φ has been selected, only a violation of our models — e.g., a perturbation during flight or impact, an overly large observer error, etc. — can cause a transition out of $\mathcal{D}_\mathcal{O}(\Phi)$.

3.3 Safe Deployments

To achieve the obstacle avoidance outlined in the previous section for Φ_J, we have used numerical methods to compute conservative approximations to $\mathcal{D}_{\mathcal{O}}(\Phi_J)$ for a particular goal point, $\mathcal{G} = (h^*, \eta^*)$. The rotational symmetry of the robot has been exploited to generate a family of safe controllers that differ only in the placement of the goal, $\mathcal{G}$.

If all the constituent controllers, Φ_i, in a deployment are safe with respect to an obstacle, then it follows that the resulting composite controller, $\Phi = \bigvee \Phi_i$, *must* also be safe with respect to the obstacle. This is because any ball state within $\mathcal{D}(\Phi)$ is also within $\mathcal{D}(\Phi_j)$ for some j, and a constituent controller Φ_k will only be active when the ball state lies within $\mathcal{D}(\Phi_k)$.

3.4 Parallel Composition

In as yet unpublished work, we are exploring how to show that $\Phi_J = \Phi_B \times \Phi_P$ implies $f_{\Phi_J} = f_{\Phi_B} \times f_{\Phi_P}$. That is, Φ_J arises from the parallel composition of the batting and palming behaviors as described in Section 2.2.2, and the closed loop behavior seems to decouple as well. In the present paper, we accept this as given by the experimental evidence, and exploit the composition as follows.

The regulation of the vertical subsystem, Φ_B, to the set point, η^*, is very fast relative to that of the horizontal subsystem, Φ_P, to its set point, h^*. Furthermore, the two systems completely decouple at the set point, and are only weakly coupled nearby [14]. Practically, this means that the horizontal cross-sections of domains change only slightly across the useful range of apex heights, so it is effective in practice to "ignore" the vertical subsystem, f_{Φ_B}, when planning or analyzing horizontal motions, f_{Φ_P}. Thus, we study the closed loop four-dimensional horizontal subsystem in isolation, for example as in figures 9,10,12, and 13.

4 Practical Implementation

We introduce into the robot's workspace a flat obstacle at the height of the paddle when held flat. Figure 8 depicts this obstacle's locus within the planar set defined by these horizontal robot con-

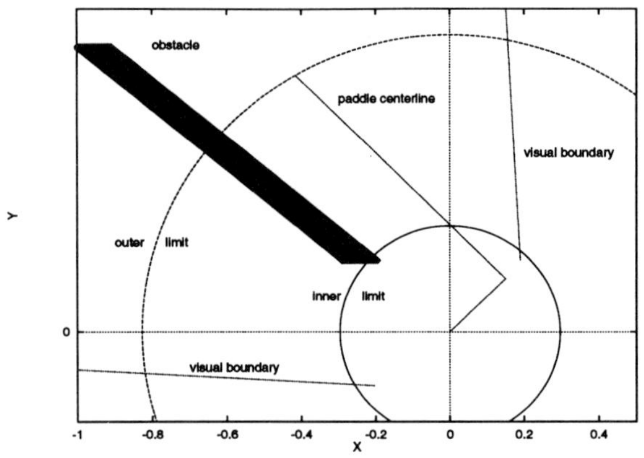

Figure 8: Horizontal configuration space with obstacle and visual boundaries.

figurations: an annular region owing to the paddle's finite length and "shoulder" offset, and the projection of the edges of the visual workspace. The robot may pass under the obstacle, and the ball may pass over it, but it should be clear from the figure that the pure palming controller, Φ_P, sees a disconnected workspace. Thus, in addressing the problem of how to bring a ball from one side of the obstacle to the other, we encounter a simple situation wherein a re-grasp maneuver is necessary to the success of the task.

In section 4.1 we show results of numerical simulation of Φ_J for a generic set point. The resulting domain, $\mathcal{D}(\Phi_J)$, is complex and not easily represented computationally, so in section 4.2 we outline our approach to finding a conservative approximation for a safe domain, as well as our approach for finding an invariant disconnected domain for a controller that crosses the obstacle. Section 4.3 briefly describes the experimental results to date.

4.1 Numerical Approximation of f_{Φ_J}

Due to the complexity of the system, it is difficult to "feel out" the boundaries of the domain of attraction for Φ_J through systematic juggling experiments. Nevertheless, in Figure 9, we display experimental data used to formulate an approximation of the safe domain for the juggling controller operating within the bounded workspace formed from the robot's paddle and visible region — the region of Figure 8 with no flat obstacle inserted [7]. The axes R and Phi refer to the horizontal polar coordinates of the data points in

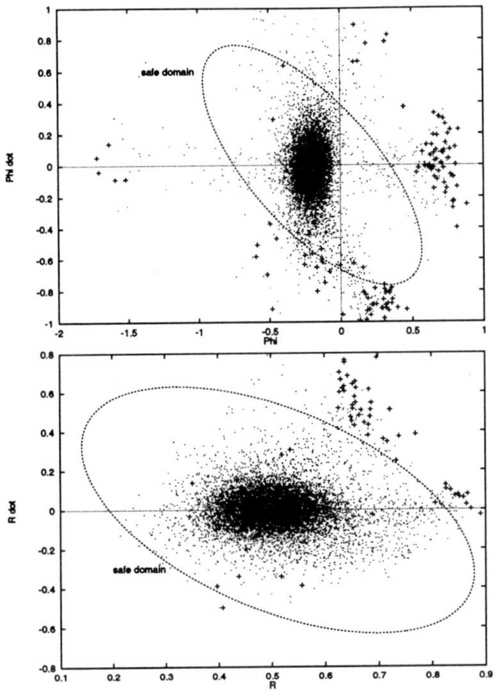

Figure 9: Empirically derived estimate for the horizontally safe juggling domain. The symbol(s) "+" (".") denote the points which failed (succeeded) in remaining within the robot's annular and visual workspace under the action of Φ_J.

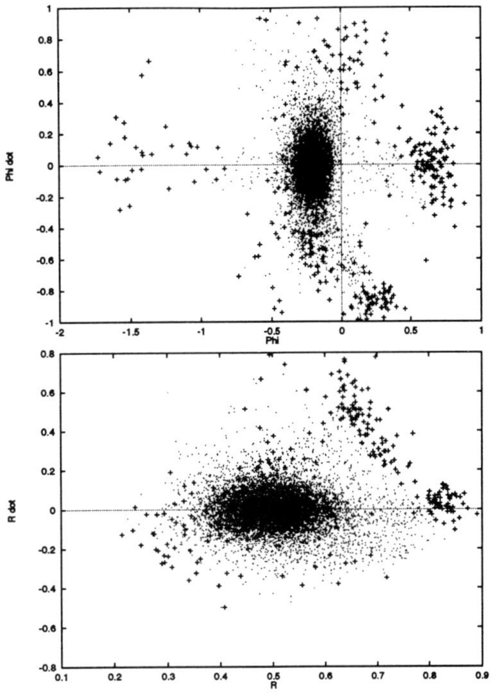

Figure 10: Numerically derived estimate for the horizontally safe juggling domain. The symbol(s) "+" (".") denote(s) the points which failed (succeeded) in remaining within the robot's annular and visual workspace under the action of the numerical simulation of Φ_J.

these plots.

To overcome this difficulty, and to speed up deployment, we created a numerical simulation of the juggler. In Figure 10, we display the same data as Figure 9 run through the numerical simulation of the robot operating under Φ_J. Since all but one of the starting conditions that failed on the real robot failed in simulation, we are persuaded that the simulation provides a conservative approximation to the robot's true actions.

Figure 11 depicts the results of a simulation run to establish the largest safe domain of Φ_J with respect to the disconnected obstacle-cluttered free workspace illustrated in Figure 8. The shaded region consists of free points that reach the goal — the spot labelled "+" in the figure — and remain free (that is, they stay clear of the obstacle, stay within the visible workspace, and remain within reach of the paddle) along the way. This, of course, is merely the zero velocity subset of a four dimensional region that extends with a rather complicated shape in the velocity directions, in a manner we have tried to suggest intuitively in Figure 2.

4.2 A Computationally Derived Safe Deployment

At the present time, for reasons summarized by the discussion surrounding Figure 11, we have no convenient means of parameterizing the largest safe $\mathcal{D}(\Phi_J)$. Yet to build the sequential compositions suggested in this paper we require just such a parameterization. Thus, we are led to find smaller safe domains whose shape is more readily represented, as introduced in Figure 1. Ellipses make an obvious candidate shape and have added appeal from the analytical viewpoint.

4.2.1 *A Conservative Invariant Set*

In Figure 12, we display four slices through the surface of an invariant four-dimensional ellipsoid, along with the same slices of the forward image of that surface computed by the simulation of Φ_J. Although these pictures do not in themselves suffice to guarantee invariance, they provide a useful visualization of what is going on. Indeed, these

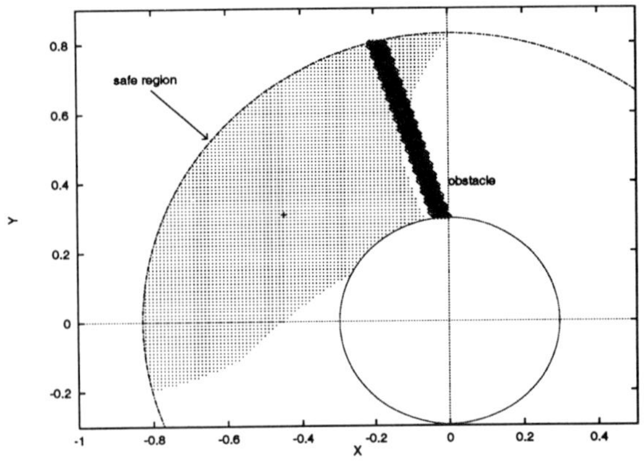

Figure 11: Simulation results: the largest safe subset of the free workspace defined by Φ_J.

four projections were used as the primary tool in a search by hand for an invariant ellipsoid using the simulation. When a promising candidate was created by rotating and elongating the four-dimensional ellipsoid, the forward image was tested for inclusion in the original ellipse. After many trials, the ellipse shown in Figure 12 was found and shown to map entirely within itself.[6]

Once a single invariant ellipsoid is found for a particular set point, several of such ellipsoids may be placed within the workspace by making rotationally symmetric copies (the mirror law, m_J, is rotationally symmetric). This allows us to create a complex controller on each side of the obstacle by straightforward application of the composition algorithm discussed in section 3. The new controller on the goal side draws all states within the copies of the invariant domain toward the task goal point (which is assumed to be the goal of one of the constituent controllers), while the one on the disconnected-from-goal side draws a similar union of domains to a goal point near the obstacle. The next task is to develop a controller that will cross the obstacle.

To get across the obstacle, we start with two controllers and domains of interest – the highest priority pre-obstacle controller, and the lowest priority post-obstacle controller, each with its own invariant domain. The latter will form the basis of the jumping, or trans-obstacle con-

[6]This process will be sped up enormously by access to an analytically tractable return map. Even without analytical results, the search for invariance could be automated by recourse to positive definite programming (a convex nonlinear programming problem) on the parameters (in this case, the entries of a positive definite matrix) defining the shape of the invariant set.

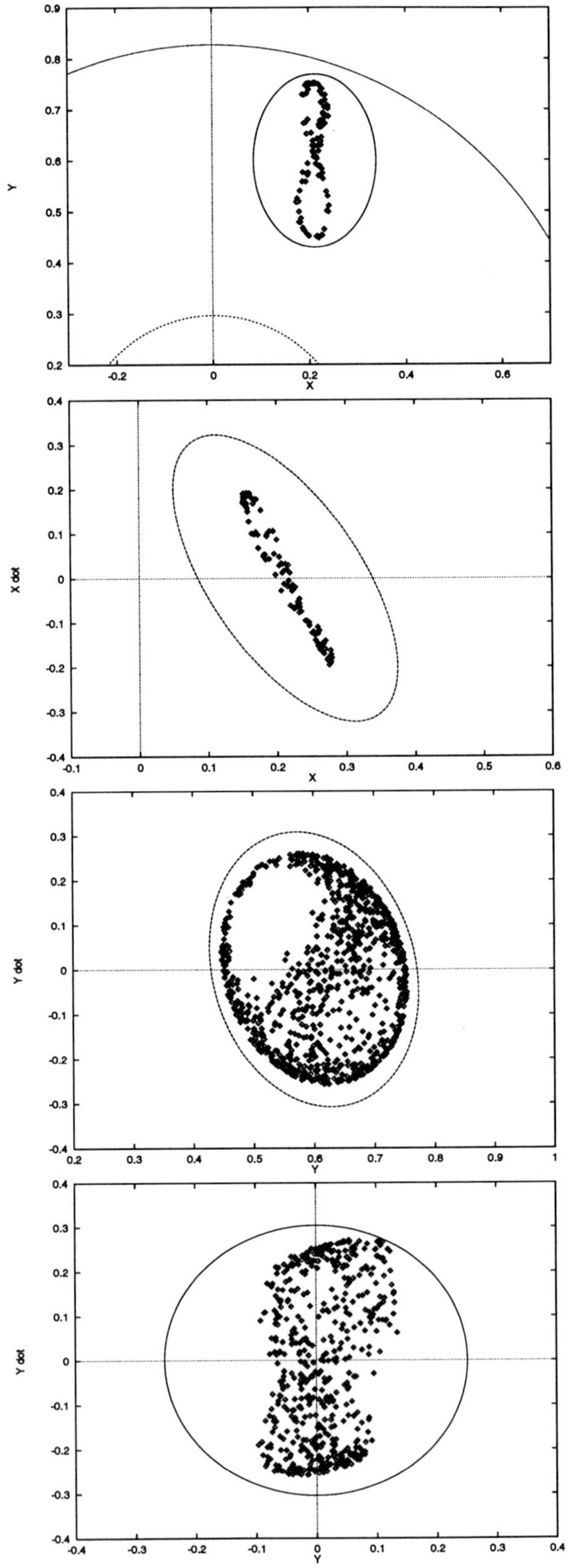

Figure 12: Four 2-D slices of the invariant ellipse.

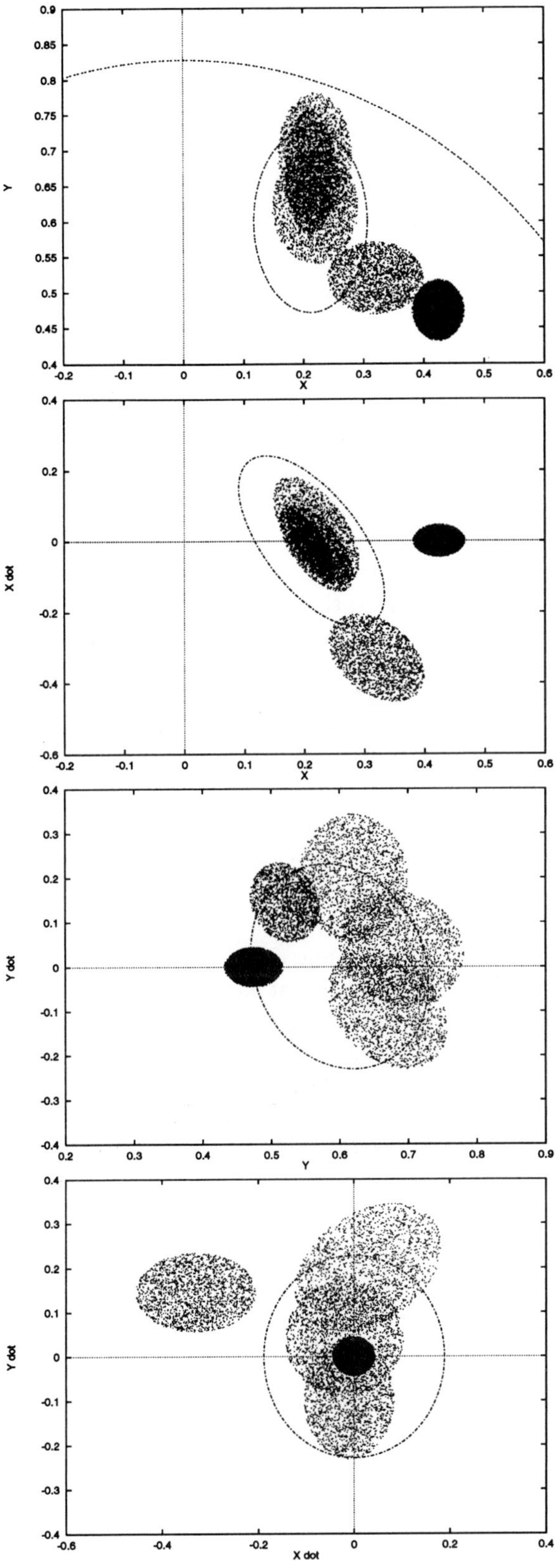

Figure 13: Jump controller invariant union of ellipsoids.

troller. Since the pre-obstacle controller will regulate the ball to arbitrary accuracy (at least in simulation), we can choose a small neighborhood around that set point as a "launching pad" for the trans-obstacle controller. Having chosen such a ball, we take the forward image under the mirror law in simulation. The result may no longer be invariant, but we can readily find an ellipsoid that contains it. The new "containing" ellipsoid is now mapped forward and its image is once again contained in some new ellipsoid. Iterating this process, we eventually find an invariant ellipsoid that contains the forward image of the nth "container" ellipsoid ***and*** lies within the original invariant domain. The union of all these ellipsoids is invariant and safe under the trans-obstacle controller, as the forward image of each member of the union is fully contained within another, and they all lead into the safe invariant domain already found. In Figure 13 we show a union of ellipsoids created by this process projected onto four orthogonal planes.

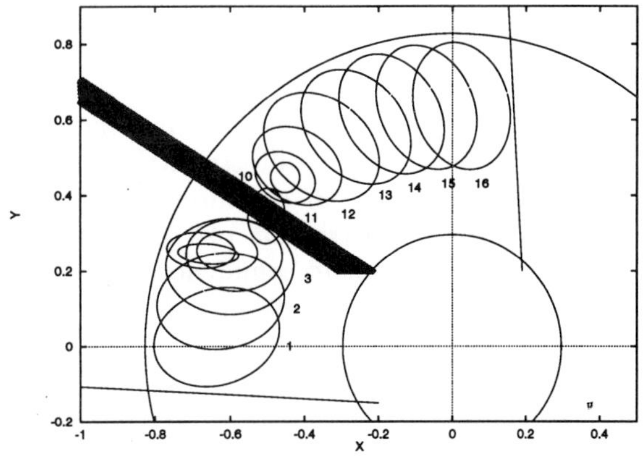

Figure 14: A "theoretical" deployment using the jump controller of Figure 13.

4.2.2 *A "Hand-Crafted" Deployment of Invariant Sets*

In Figure 14, we depict the the entire hand-crafted deployment discussed so far by projecting the domain of each onto the horizontal configuration space introduced in Figure 8. There are a total of fifteen ellipsoids for nine different controllers, the union of six of them associated with the trans-obstacle controller. This is our hand-crafted realization of the concept depicted in Figure 4.

The deployment of Figure 14 gives rise to a

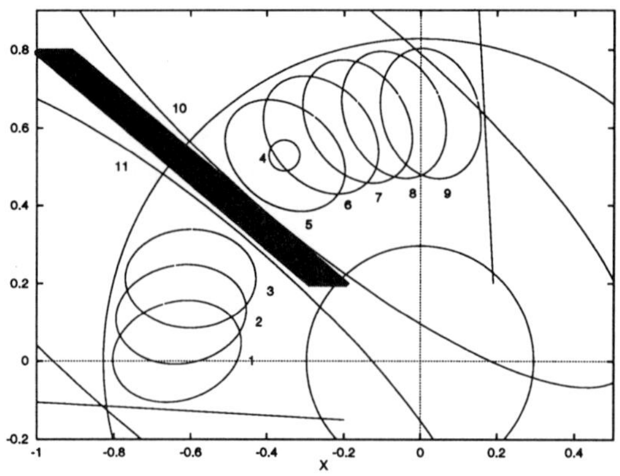

Figure 15: The implemented deployment, adding a "catch-all" (an ellipsoidal approximation to the largest safe domain) to supplement the jump controller.

safe controller under the $\bigvee$ composition described previously. In simulation, all states in the union of all domains (1 - 16 in the figure) converge to the goal associated with domain 1 in Figure 14 – the overall task goal. Unfortunately, due to the conservative nature of our numerical procedure, the domain of attraction of the composite is not large enough to lend confidence that a ball could easily be introduced into it. Worse, given the inevitable infidelities of the robot and ball with respect to the ideal models of ball flight, impacts, and robot kinematics, a physical implementation will remain roughly faithful to the simulation results, but inevitably depart frequently enough and with significant enough magnitude that some further "safety net" is required to bring balls that start (or, "unexpectedly stray") outside the domain of the composite back into it. The "small" disturbances overwhelm the finely tuned trans-obstacle controller.

For purposes of this initial study, we have chosen to use very coarse (and, unfortunately, not necessarily conservative) approximations to the largest safe domain of Φ_J depicted in Figure 11. These take the form of large ellipses that share some of their boundary (at zero velocity) with the disconnecting obstacle as depicted in Figure 15. The robot is assigned the corresponding juggling controller whenever the ball is outside the domain of the sequential composite depicted in Figure 14, yet still within the reachable workspace.

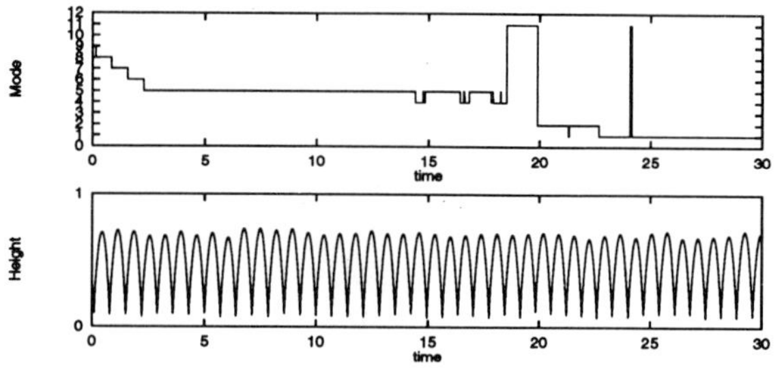

Figure 16: Anecdotal time trace of vertical position and mode number for a typical run.

4.3 Empirical Results to Date

In Figure 16 we show a typical trace of the control mode plotted against time, with the height of the ball also plotted to provide a sense of time scale.[7] In this run, the ball steps smoothly across the domains until it is in mode 5, representing the last one before crossing the obstacle (at approx. $t = 3$). This control mode continues for some time while the system waits for the ball to enter the small "launch window" for mode 4. Finally, this condition holds at an impact and the ball is projected into the domain of mode 11 (one of the safety nets). This controller brings the ball into mode 2, which leads to mode 1, the goal controller.

In Figure 17 we show for this same typical run the horizontal positions at each hit (depicted by a diamond symbol in the plot) to convey a sense of where the domains for these various controllers reside in the workspace. The "launch window" is small relative to the accuracy of the machine, thus it takes several bounces on the far side of the obstacle before finding it, as clearly seen in the figure.

Although the numerical simulation does not perfectly model the real dynamics, it is sufficiently close to allow us to hand-craft a deployment that induces obstacle avoidance while bringing the overall system to a goal state. We are currently focusing attention on methods for automating deployment.

[7] Note that the presently working machine doesn't actually respect the painstakingly hand-crafted trans-obstacle mode because the simulation is not accurate enough in the face of system noise. Instead, the ball is captured after a trans-obstacle controller episode by the large "safety net" controllers depicted in Figure 15.

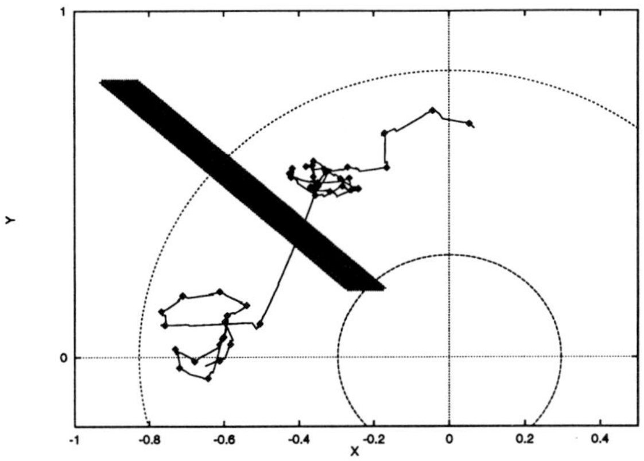

Figure 17: Horizontal projection of the sample trace above.

References

[1] Matthew T. Mason. The mechanics of manipulation. In *Proc. International Conference on Robotics and Automation*, pages 544–548, March 1985.

[2] Tomás Lozano-Perez, Matthew T. Mason, and Russell H. Taylor. Automatic synthesis of fine-motion strategies for robots. *The International Journal of Robotics Research*, 3(1):3–23, 1984.

[3] J. Ish-Shalom. The cs language concept: A new approach to robot motion design. In *Proceedings 23rd IEEE Conference on Decision and Control*, pages 760–767, Las Vegas, Nev., Dec 1984.

[4] Elon Rimon and D. E. Koditschek. Exact robot navigation using artificial potential fields. *IEEE Transactions on Robotics and Automation*, 8(5):501–518, Oct 1992.

[5] Roger W. Brockett. Asymptotic stability and feedback stabilization. In Roger W. Brockett, Richard S. Millman, and Hector J. Sussman, editors, *Differential Geometric Control Theory*, chapter 3, pages 181–191. Birkhäuser, 1983.

[6] D. E. Koditschek. Task encoding: Toward a scientific paradigm for robot planning and control. *Journal of Robotics and Autonomous Systems*, 9:5–39, 1992.

[7] Robert R. Burridge, Alfred A. Rizzi, and Daniel E. Koditschek. Obstacle avoidance in intermittent dynamical environments. In *The Fourth International Symposium on Experimental Robotics*, pages 43–48, 1995.

[8] Robert R. Burridge, Alfred A. Rizzi, and Daniel E. Koditschek. Toward a dynamical pick and place. In *IROS*, pages 2:292–297, August 1995.

[9] Alfred A. Rizzi. *Dexterous Robot Manipulation.* PhD thesis, Yale University, 1994.

[10] Louis L. Whitcomb, Alfred Rizzi, and Daniel E. Koditschek. Comparative experiments with a new adaptive controller for robot arms. *IEEE Transactions on Robotics and Automation*, 9(1):59–70, 1993.

[11] J. L. Synge and B. A. Griffith. *Principles of Mechanics.* McGraw Hill, London, 1959.

[12] M. Bühler, D. E. Koditschek, and P.J. Kindlmann. A family of robot control strategies for intermittent dynamical environments. *IEEE Control Systems Magazine*, 10:16–22, Feb 1990.

[13] Jean-Claude Latombe. *Robot Motion Planning.* Kluwer, Boston, MA, 1991.

[14] A. A. Rizzi and D. E. Koditschek. Further progress in robot juggling: Solvable mirror laws. In *Int. Conf. Rob. and Aut.*, pages 2935–2940, 1994.

Research Program on Mechanisms for Emergent Machine Intelligence

Hirochika Inoue
Department of Mechano-Informatics, The University of Tokyo

Abstract

This paper introduces a research program on intelligent robotics which started in 1995 for three years as a grant from the Japanese Ministry of Education. The title of the program is "Research on Emergent mechanism for Machine Intelligence — A Tightly-Coupled Perception-Motion Behavior Approach". This program aims to study mechanisms for the emergence of robotic intelligence and autonomy. Intelligence which emerges from the physical interaction between a machine and the world is essential for real robots. From this viewpoint, adopting a tightly coupled sensor-action approach, this project will tackle the theoretical foundations of machine intelligence, and experimentally study autonomous systems which recognize and behave in the real world. Research topics include: 1) Kinematic understanding of manipulation skills, 2) Tightly coupled sensor-behavior approaches for real world understanding, 3) Intelligence and autonomy for human-robot cooperation, 4) Biologically inspired autonomous systems. abstract

1 Introduction

Human beings invented various tools and machines for improving our everyday life and these have contributed to the progress of civilization in our society. The robot is one of the most sophisticated machines which is created by building computers into a machine. Since early attempts for computer controlled manipulation and computer vision, research of intelligent machinery has challenges of creating hand-eye systems, industrial robot automations, fundamental developments for robotic component technologies, and theoretical foundations for robotics. Research on intelligence and autonomy becomes important because highly advanced computers will expand their working domain even into the changeable real world environment rather than the closed world of logical symbols. Also, from a social viewpoint, the intelligent autonomous machines are expected to perform various tasks that supports our aging society.

The research program described here is proposed in order to activate the challenges for the intelligence and autonomy of advanced mechanical systems. Intelligence which emerges from the physical interaction between a machine and the world is essential for real robots. From this viewpoint, and adopting a tightly coupled sensor-action approach, this project will tackle the theoretical foundations of machine intelligence, and experimentally build autonomous systems which recognize and act in the real world.

2 Research Issues

Four major issues are chosen to structure the research. They are; kinematic understanding of manipulation skill, real world understanding through tightly-coupled sensing and action, human-robot cooperative systems, and biologically inspired autonomous systems. All of these issues require a tight coupling of perception with action, and each issue closely relates with others. Figure 1 shows the relationship between the four issues. The research program consists framework research topics, which defines the domain and approach, and supplemented research topics which enhance the planned approach.

2.1 Physical Understanding of Dexterity

A hand is sometimes recognized as a physical agency of the brain. Dexterity of the hand is one of the characteristic features that distinguishes humans from apes. Such intelligence that has been acquired through the long history of biological development is considered to be implicit intel-

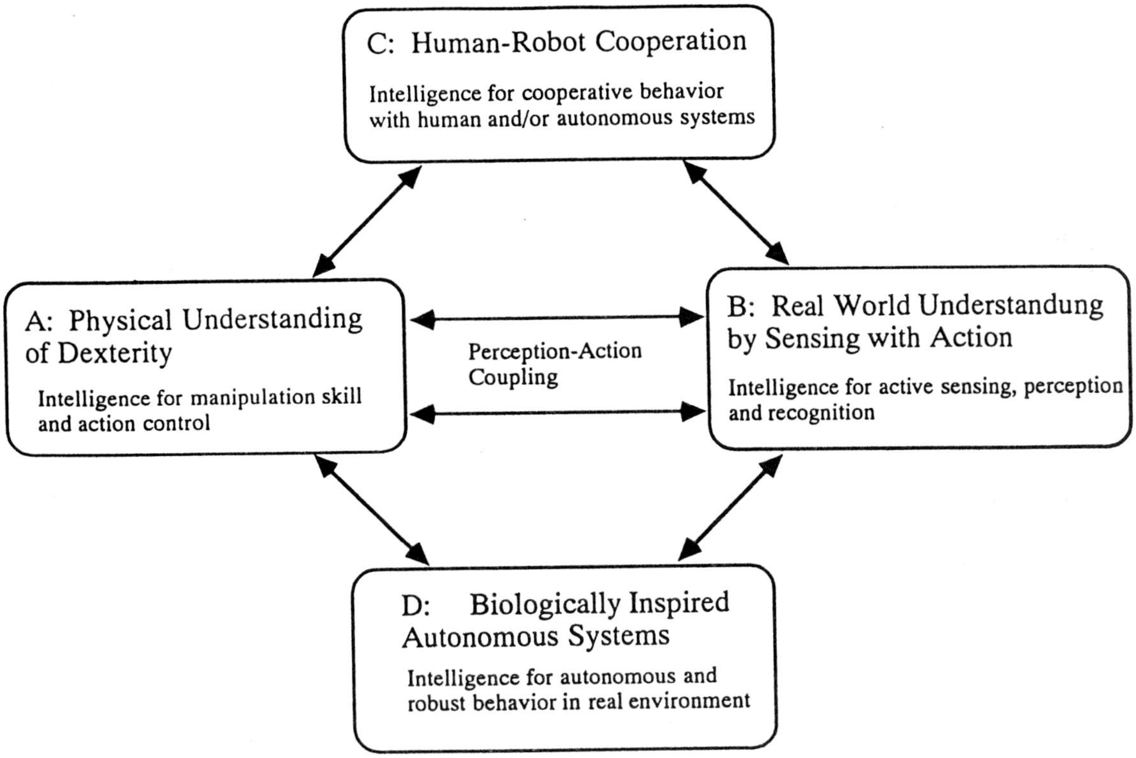

Figure 1: Relationship between the four issues.

ligence that emerges from the physical interaction without consciousness. The manipulation skill emerges from the interactive process between a hand and objects/environment. The principle of the skill seems to be hidden inside the kinematic nature of interactive processes. In this research issue, the kinematic representation of manipulation process, comprehension of human manipulation skill, and implementation of robotic skill, will be studied based upon the kinematic analysis of interactive process for manipulation skills.

2.2 Real world understanding by sensing with action

So far, the phases of recognize-plan-act in robotic behavior are connected in a simple loop if any. However, recent progress of computing performance provides new possibilities for constructing the perception-action coupling. As the speed of visual processing becomes higher, robots can perform very sophisticated perception and perception-action coupling. For instance, according to the partial results of recognition, the other perception agents for the lost or lacking information can be invoked reactively. If those cycles for perception and sensing action become very fast, the way of recognition can configured into more sophisticated constructs for fast, flexible, efficient or robust perception. In this research issue, based upon the paradigm of a tightly coupled perception and action, new framework for optimal understanding of the changing real world, proper action in the real world, and effective memorization of the world will be addressed.

2.3 Intelligence and autonomy for human-robot cooperation

Modern human life is supported by many kinds of machines. However the relationship between those machines and human users is not natural enough. Machines often force humans to adapt to them rather than the machines adapting to humans. Thus, human centered design of human-machine interaction process is strongly required. The purpose of this issue is the study of intelligence and autonomy that is required to provide new cooperative relationships between human and machines. Topics include the machine intelligence to cooperate with humans, machine intelligence to support human activities, and the intelligence to realize cooperative maneuvering.

2.4 Biologically inspired approach to autonomous systems

Biological systems embody a sort of intelligence that enables flexible adaptation to changes in the environment, and robust pursuit of goals, and thus they keep themselves alive. In this issue, through the study of biological intelligence, we will understand the mechanism of intelligence and autonomy of the living things and attempt to build real mechanical systems of such nature. This issue will uncover ways to create reliable, robust and flexible machinery that behave intelligently in the real environment. It includes the topics such as biological understanding of the sensor-motor coupling, artificial insects, group robotics, sensor based behavior of mobile robots, and bio-inspired robots.

3 Approach and Guiding Principles

The aim of this project is summed up by the two keywords: perception-motion coupling, and emergent machine intelligence. The objective of this project is to investigate the mechanism or algorithm that gives rise to machine intelligence. For that purpose, we chose a tightly-coupled perception-motion behavior approach. A machine is a physical entity that behaves in real world. Our research interest is focussed onto some sort of intelligence, which is inherently required for the robust, efficient, quick interactions between the machine and real environment. This project aims to find the principles of such mechanical intelligence and demonstrate them by building physical entities.

3.1 Changing the research framework

So far, usual research of robot intelligence is divided into components such as motion control, environment recognition, task planning, and so on. However in this program we take a perception-motion coupling approach for creating intelligent systems. The approach will be shifted from functional division to multi-function integration. For instance; from the control of hands to the understanding of manipulation skill; from recognition without action to real world understanding by general use of perception and action. Hence this study of system architecture for emerging intelligence in robotic behavior, learning, memorization, and autonomy.

3.2 Interaction with outer world

Robot intelligence must cope with various noise, uncertainty, change, and unsteadiness in the real world. And, intelligence for reaction or symbol grounding is much more important than classical symbol based intelligence. Not only perception but also action can be equally recognized as interaction process with outer world. The essence of real world intelligence can be found in the interaction process between a machine and the environment.

3.3 Tight coupling of perception with action

Recent progress of computer hardware is dramatically changing the prerequisites for robot intelligence research, that is, real time vision, massive memory, enormous computing power by massive parallel computer, and so on. Very fast and very flexible coupling of perception process and action has now become available. Better perception might be supported by proper active sensing, on the other hand, behavior must be supported by high and multimodal sensor integration. Recursive structure of perception and action hints interesting principles for intelligence for mechanical systems.

3.4 Infrastructures for real world robotics

In robotics research, the development of experimental devices or tools are also very important, because process of building robots is also very effective approach for understanding robotic principle. However, as proper platforms for robotic experiment are not available, many researchers build similar but not the same experimental devices. If standard platform like computer hardware is available, software libraries will be accumulated and help to accelerate the speed of research. For such purpose, in our research program, we are planning to develop common research tools for distribution. Our current plan is: real time tracking vision based upon correlation technique, another high performance general color vision system in DSP based multiprocessor architecture, compact low cost legged robot, servo-module and manipulator system design, and so on.

4 Research on Physical Understanding of Dexterity

This project aims to explore physical principles understanding dexterity of human/robot motions in execution of sophisticated tasks from the viewpoint of dynamics and control theory. It also aims to discover implementation algorithms based on the physical principles, which can function in robotic arms and hands and execute tasks with sufficient dexterity even if they are physically interacting with objects in the environment. A novel high-performance robot arm with light weight and a robot finger unit of three degree of freedom with force and touch sensors are designed and will be used commonly as test tools for a variety of experiments. The group leader for this issue is Prof. Suguru Arimoto of University of Tokyo.

(1) Physical Understanding of Adaptation and Learning for Dextrous Robot Motions

This research subject aims to clarify physical conditions of robot dynamics and control that underlie adaptability and learnability of robots in execution of robot tasks with necessary dexterity. In particular, the s ubject attempts to clarify a crucial role of "passivity" in adaptability and learnability of robots that are even subject to external constraints or physical interactions with their environments. Advanced concepts for designing novel controllers realizing dextrous robot motions are introduced and applied by developing and analyzing nonlinear position-dependent circuits that describe Lagrange's equation of robot motions. [S. Arimoto et. al., Univ. of Tokyo]

(2) Mechanics and Control of Skilled Manipulation Process

This study focuses on analysis, modeling and control of such dexterity as exposed in skilled human handwork. First, we make a system which consists of a tracking vision system and force/torque sensors to measure and analyze skilled human handwork. Then, we use the data which is obtained by the system for modeling the mechanics of the process of skilled handwork. Using the model we design a control system for skilled handwork. In the experiment, we adopt a DD parallel manipulator to perform the modeled dexterity for the skilled human handwork. [M. Uchiyama et.al. Tohoku Univ.]

(3) Dynamics and Control in Emerging Mechanism of Skill

This project aims to study intelligent machines through analysis and synthesis of force sensing and force control in skillful manipulation of objects by robotic hands and multiple manipulators. First the role of force sensing and force control is analyzed in grasping and manipulation of objects by robot hands. Then a methodology of constructing cooperative control systems for grasping and manipulation of objects will be established and experimental study of this methodology will be performed. An algorithm for force display system for displaying manipulating feel of dynamic virtual objects with articulation will be developed and experiments on this algorithm will be performed. [T. Yoshikwa et. al. Kyoto Univ.]

(4) Synthesis of Cooperative Behavior through Sensor Integration

The goal of the research is to establish a theory to synthesize a self-organizing mechanism that emerges cooperative behavior of machines through multiple sensor integration. We first build a network of sensors and mechanisms by combining simple and holistic feedback loops. We study to design a self-organizing system that modifies the initial network to fit for a given purpose. The focus of our research will be on how to develop such systems from the viewpoints of nonlinear dynamics and nonlinear control theory. As an example with generality, we pick up a multi-fingered robot hand and study on the synthesis of grasping behaviors. We have already worked on the theory of grasping, and started to build a multi-fingered robot hand system involving vision and force sensors. A parallel computer will be used for real-time control of a manipulator with the multi-fingered hand and for processing of massive sensory information and self-organization. Tracking vision systems are used to provide real-time visual information. [Y. Nakamura, Univ. of Tokyo]

(5) Analysis of Dexterity of Living Things and its Application to Robot

Observing manipulations by human often provides us with a good hint for achieving several dexterous tasks by a robot. The goal of this work is to globally analyze human behaviors, such as active sensing motions, dexterous manipulation, and hand level impedance assignment during tasks, and then to extract essential parts of their characteristics for applying them to a motion planning or an impedance planning for manipulating an object by a robot. We measure the hand level impedance during a simple task by human and also observe the grasp phase transition through the grasp planning by human. Based on these observations, we propose not only an approach on impedance setting depending upon each task required but also a grasp planning learn from human behavior. Also, a multi-fingered hand will be also designed and developed to confirm the grasp planning proposed along with this idea. [M. Kaneko et. al., Hiroshima Univ.]

(6) Study on the Understanding of Manipulation Skills on the basis of Kinematical and Dynamic Feature Extraction of Motion in Contacts

[K. Takase, Univ. of Electro-Communications]

(7) Motion Based Control for Robot Manipulators to Realize Position and Force Patterns with High Accuracy

[S. Kawamura, Ritsumeikan University]

(8) Research on Human and Machine Skills in Manipulation of Deformable Soft Objects

[S. Hirai, Osaka Univ.]

5 Tightly Coupled Sensor and Behavior for Real World Recognition

Recognition of real world is important for intelligent systems. Our approach to the problem is to tightly couple sensing and behavior so that behaviors are improved by sensing and better sensing is performed by proper behaviors. Main research themes are the followings: 1) Interpretation of sensor information by sensing and behavior, 2) Sensing mechanism for immediate adaptive behavior, 3) Action control for a given task based on sensing and a world model, 4) Adaptation mechanism for sensor-behavior systems. This research team includes 5 planned research and 9 proposed research. The group leader for this issue is Prof. Yoshiaki Shirai of Osaka University.

(1) Interpretation of Sensor Information for Real World Understanding

This research aims to study mechanisms of interpretation of sensor information for understanding real world environments. The task of interpretation involves acquisition of various sensor information, interpretation of sensor information for rough understanding of the environment, determination of what information is necessary for compensation of the uncertainty of the interpretation, and determination of the next action for better understanding of the environment. Understanding of real world environments requires a feedback mechanism loop of sensing, interpretation, and action. The features of the research are (a) reliability of interpretation is obtained from that of sensor information and interpretation process, and (b) the optimal plan is made by considering the cost of motions and that of sensor information interpretation. [Y. Shirai et.al., Osaka Univ.]

(2) Tightly Coupled Sensor-Behavior Approach for Real World Recognition

To move in the dynamically changing road environment, the robot should get continuously the sensory information of the environment in very short cycle and perform the sensory behavior action adaptive to the environment change. We study the real time sensory information processing of static objects in the road and study the representation and matching of the sensory information in the digital map. We also study dynamic objects in the road and evaluate their danger estimations by using traffic rules and positions of them. We develop a mobile robot which includes a real time image processing system and a sonar system. We implement the sensory behavior action algorithm on the robot and test it in the sidewalk. [H. Mori et. al., Yamanashi Univ.]

(3) Behavior Control Based on Environment Model and Sensory Information

Recognition of real world is important for intelligent systems. Our approach to the problem is tightly couple sensing and behavior so that behaviors are improved by sensing and better sensing is performed by behaviors. According this conceptual guide-line we develop the strategic algorithms which decide the most suitable observation points by multi-sensor system, and also proper subtask execution series for final task accomplishment. [M. Kakikura, Tokyo Denki Univ.]

(4) Behavior Control Based on Environment Model and Sensory Information

As an actual application, our research focuses on an action control for a given manipulation task based on sensing and a world model. The main thrust of the research consists in developing a method of reasoning: what kind of sensor information is required to complete the task and how to obtain it; how to identify the state of the task using sensor information. The plan of this year is to develop a method of representing a sequence of action and sensing from the description of the environment and that of the task specification for assembly of objects with complex shapes. We also study how to detect the difference between the plan and the actual states. [T. Hasegawa, Kyushu Univ.]

(5) Adaptive Functions in Perception-Action Systems

This research aims to study adaptive functions in perception-action systems that behave in the real environments. In order to perform adaptive behaviors in the real world, the autonomous systems should have control mechanisms to select sensor processing and reactive motions to suit the environments. This research remarks the memory organization to recognize and react and approaches to not only action selection but also automatic acquisition of reactive behaviors. In order to evaluate the architectures for adaptive functions in perception-action systems, the real architecture of the behavior to show adaptive functions will be implemented on real robotics system that recognizes the environment through multi-modal sensors and controls a complex body with multi-degrees of freedom based on massively parallel computers with distributed large memory systems. [H. Inoue et.al., Univ. of Tokyo]

(6) Active Sensory-Motor Integration System Using Vision and Touch
[M. Ishikawa et. al., Univ. of Tokyo]

(7) Active Vision for Robotic Agents in a 3-D Outdoor Environment
[H. Ishiguro et.al., Kyoto Univ.]

(8) Detecting Object Surfaces Using Occlusion Information from Active Binocular Stereo
[F. Miyazaki et. al., Osaka Univ.]

(9) Focus of Attention and Perception of Robot's Environment by Integrated Visual Sensor
[M. Yachida et. al., Osaka Univ.]

(10) Behavior Acquisition and Integration Based on Mapping between Perception and Action
[M. Asada et. al., Osaka Univ.]

(11) Research on Cognition of Changes in Structure of Environment
[S. Tsuji et. al., Wakayama Univ.]

(12) Visual Servoing Combining Environment Recognition and Motion Control
[K. Hashimoto et. al., Okayama Univ.]

(13) Research on Active Recognition System in an Indoor Environment- Determination of Viewpoint Based on Object Recognition
[H. Oki et. al. Oita Univ.]

(14) Shape Recognition by Integrating of Sensing and Manipulation Based on the Aspect Information
[G. Kinoshita et. al., Chuo Univ.]

6 Human-machine Cooperative System

To realize novel human-machine cooperation systems, the following three research topics will be investigated in terms of human-robot cooperation intelligence. 1) Research on machine intelligence to cooperate humans: Multiple robot environment based on agent programming will be realized to supports cooperative works between humans. Also mechanism, sensor, controller and man-machine interface for mobile robot with TV camera will be constructed. 2) Research on machine intelligence to support human activities: A function to grasp most of the object in a room by tracking and updating them as well as a function to express emotional information based on facial expression and gesture of the robot will be realized. 3) Research on machine intelligence to realize cooperative maneuvering: After modeling of human skill, a cooperative maneuvering system will be designed. Abstracts of the 5 planned research and the title of the 4 proposed research are shown below. The group leader for this issue is Prof. Tomomasa Sato of University of Tokyo.

(1) Research on Human Support System

Machine intelligence to support human activities will be investigated. 1) Clarify the necessity of the field on robot system that support human activities. Also clarify the necessary function and research subject. 2) Realize a function to deal room (robotic room) directly as the object of information processing. A function to grasp most of the object in a room by tracking and updating them using real-time vision system will be realized. 3) Understand human condition: Realize a function to understand the condition of human from his/her unconscious reaction. 4) Realize a method for creating Action Score. [T. Sato et. al., Univ. of Tokyo]

(2) Research on Machine Intelligence for Human Cooperative Work

Autonomous robots are essential for supporting human cooperative work in the physical world. To do so, we need new points of view for designing human-robot networks and human-robot interfaces, with which we can use those robots without conscious effort. In exploring this issue, we also expect that many novel research problems would be generated: how communication between human and robots should be, who should be responsible for actions of robots, and so forth. In this research, we investigate cooperation among autonomous agents (human and robots) in human-robot networks, based on a multi-agent computation model. Also we try to explicate the role of robot intelligence for human cooperative work. Autono mous robots for the experiment is to be constructed with CCD cameras, its controllers and a human-robot interface for cooperation among multiple human and robots. [Y. Anzai et.al., Keio Univ.]

(3) Machine Intelligence on Cooperating Motion of Men and Intelligent Mobile Machines

Suppose there exist "office messenger robots" working in an office to deliver some documents to someone. Those robots' destinations sometimes are quite uncertain and the situations are always changing. The main purpose of this study is to clarify how to move under these uncertain circumstances in the mobile action of robots with the aid of the cooperation and dialogue between men and robots. [E. Nakano et. al., Tohoku Univ.]

(4) Machine Intelligence for Human Skill Acquisition

In this project, we try to make system which can realize human complicated performances. The project will lead us to the stage where we can understand the essence of human intelligence and operate the system without feeling incongruous.

The main problem is how to combine human skill and machine high performances. It is a new relation between human and machine. To a human operator the most desirable technical system is intelligent, partly autonomous and easy to operate. Central issues in realization of this future technical system are human-machine interfaces, very human friendly input devices and the problem of realizing intelligence. In this research we are developing virtual reality interfaces to measure human performances to accomplish this aim. We will make skill model based on the measurement and control a dexterous robot hand by using the skill model. And then we will try to make the dexterous robot hand cooperate with people. [H. Hashimoto, Univ. of Tokyo]

(5) Research on Mechanical Intelligence for the Emotional Human-Machine Communication

To open up an entirely new field, emotional-communication-robots, this research aims to realize the human-assistance based on a natural and smooth communication function of a robots which has an emotional emergence model in reference to human, and can express the emotional information. To be more concrete, the hypothesis about the emergence of emotion will be proposed by analyzing the relation between the change of human living information and the emergence of emotion. Based on it, the emotions that the mechanical system should involve will be analyzed. As a result, a robot hardware and the controller which can communicate with human emotionally will be developed and evaluated. Human sensation to the robot will be also evaluated in addition to the emotional function of the robot. [S. Sugano, Waseda Univ.]

(6) Research on Robot Task Teaching and Knowledge Base Constitution Method

[S. Aramaki et. al., Fukuoka Univ.]

(7) Visual Attention Control Utilizing Speech Recognition for the Understanding of Human Behaviors

[Y. Nakamura et. al., Univ. of Tsukuba]

(8) A Study on Using Inference of the Operator's Intention and Information Transformation to Convey Critical Aspects of the Operational Environment in Tele-Micro-Surgery

[M. Mitsuishi et. al., Univ. of Tokyo]

(9) Dynamical Motion Planning for Robots Adaptive to Human Movement

[T. Takamori et. al., Kobe Univ.]

7 Research on Biologically Inspired Autonomous System

The purpose of this research project is to develop methodologies to realize intelligent robots that achieve high autonomy as in a creature, adaptability in the different environments and sociability. Creatures are inherited abilities and intelligence for themselves to be alive autonomously in a real environment. They can adapt themselves to the variations of their environment flexibly and robustly. They can accomplish their wishes to survive. When we consider the design of intelligent machine that works in the real environment, the creatures provide us very noteworthy knowledge. There exit, of course, several studies to develop machines in imitation of human or creatures. However, we can not realize such an intelligent machine that is worth being compared with human or creatures if we only imitate their appearance. This research project stands on the point of view that "Intelligence of machine arises from close connection between behaviors and sensations". We take such insights from intelligence of creatures and make it clear why such intelligence can emerge from them. Finally, we develop methodology to realize intelligence and autonomy of the machine system. The group leader for this issue is Prof. Shin'ichi Yuta of Tsukuba University.

(1) Integration of a Reflexive and Qualitative Method for Autonomous Navigation

The purpose of this study is to improve robustness of navigation and operator interface for map construction, etc. by using a reflexive algorithm. This research theme includes improvement of sensors, development of algorithm and system integration. The research subjects are: a) Long distance navigation in an indoor environment based on the wall-following using ultrasonic sensor. b). Realization of locomotion control of a mobile robot by visual feedback using reference images. c). Integration of a reflexive behavior based on a tactile sensor and an objective behavior in navigation. [S. Yuta et.al., Univ. of Tsukuba]

(2) Research on Insect-Model Based Autonomous System

The insect does not have high level intelligence as the mammal. Their behavior is mainly based on the reflex. But it looks intelligent. They can recognize the environment easily and adapt themselves to it. Often we can see foraging ants behave very intelligently. This may be a good example which shows they have excellent autonomous ability. Studying insects, we are trying to find out the way to design and control the autonomous mechanical system. [H. Miura et. al., Univ. of Tokyo]

(3) Study of Autonomous Robot with Biological Behavior

This research studies the autonomous quadruped walking vehicle in view of leg driving mechanism, sensor system and intelligent gait controllers, and investigates the general concepts of the machine-specific-intelligence of mobile robots. There were several researches on the sensor-based gait control problems of walking vehicle until now in many countries. But it have not been studies at all about the control problems of the walking vehicle when it is walking on extremely rough terrain. This research specially focuses on this walking motion, and clarify the following problems; 1) Specific control problem of the walking vehicle when the vehicle can utilize multiple sensor information and walking on extreme terrain such as on the cliff. 2) Demonstrates the autonomous walking on rough terrain by construct quadruped walking vehicle. [S. Hirose et. al., Tokyo Institute of Technology]

(4) Autonomous System with Learning and Adaptation, Self-evolution and Cooperation Abilities

We aim at application of self-organization, self-evolution, and self-repair of the living things, and study on the robots inspiring the biological structure. To realize learning and adaptation ability, we design the brachiation robot and make its prototype model that has multiple degrees of freedom. The system employs redundant mechanism and its control becomes very difficult. We investigate and study on self-organization, self-evolution, and self-repair of the living things, and consider their application for this control. [T. Fukuda et. al., Nagoya Univ.]

(5) Autonomous and Cooperative Multi-Agents Robot Systems with Some Life-Like Behaviors

The purpose of the research is to establish a method with which multiple robots behave autonomously and cooperatively in order to achieve an overall objective of the system. The research is being made through performing assembly tasks and handling tasks using multiple hands with visual and tactile sensors. How visual and tactile information are integrated in order to make each robot behave autonomously and cooperatively is an important point in the research. We plan to realize a flexible and robust planning and executing system which provides the replannability by monitoring circumstances and detecting accidents through visual agents. Moreover, we are making a study on a cooperative object-handling system using multiple hands with tactile sensors and visual ones. [T. Nagata et. al., Kyushu Univ.]

(6) Development of Autonomous Intelligent Behaviors based on Ecological Computations
[K. Suzuki et. al., Hokkaido Univ.]

(7) Analysis of the Morphology of Snake Movement and Development of the Snake-like Robot
[S. Ma, Ibaraki Univ.]

(8) Group Intelligence Emerging from Individuals Behaving under Majority Rule
[T. Arai et. al., Univ. of Tokyo]

(9) Study on Flight Mechanics of Insects
[K. Kawachi, Univ. of Tokyo]

(10) Generation of Intelligent Behavior by Autonomous Distributed Robots Learned from Grouping Behavior of Creatures
[N. Inou, Tokyo Institute of Technology]

(11) Development of Intelligent Robotic Model from Visually Guided Behavior of the Frog
[N. Matsumoto et. al., Kyushu Instutute of Technology]

8 Concluding remarks

The essential nature of intelligence for real mechanical systems can be found in the process of various interactions with real environment. From this viewpoint, inter-university research program on intelligent robotics sets out to understand the emergent intelligence of mechanical systems, and to create real robots that behave intelligently in the real world based on a tightly-coupled perception-motion approach. The important motivation of this program formation is to restrucure the field. Personal observation on the situation about the robotics field follows. Robotics has become mature. The field is almost established in some sense. The structure of the field is fixing. The number of researchers is increasing. Conferences are held very frequently. And, a large number of research papers are produced every year. But, research funds are steady or declining. The enthusiasm for robotics research in decades ago seems lost. On the other hands, in related areas, several novel approaches for systems are growing very rapidly. They are, for instance, micro machines, virtual reality, A-Life, brain technology, new media technology, and so on. Actually robotics is included in all of those field as examples. On the contrary, can robotics include them together? Competing with such vivid research topics, can robotics service? If so, then how should we envision future robotics? We need a paradigm shift for robotics research. The attempt to sum up the current robotics research and to find new directions, new names, and new paradigms for future robotics is another important points.

PART 5
TELEROBOTICS AND VIRTUAL REALITY
SESSION SUMMARY

Paolo Dario
Scuola Superiore Sant'Anna, Pisa, Italy

Telerobotics and Virtual Reality have evolved somewhat independently for years, but they are now deeply interconnected from the viewpoints of the theoretical problems they raise, the technologies they require and the applications they aim.

The evolution of telemanipulation can be traced back to the mechanical telemanipulator developed in 1948 in the U.S. at the Argonne National Laboratory. The telemanipulator was a structure that "linked" physically the human operator ("master") to the mechanical manipulator ("slave") in a spatial and temporal symbiosis.

The introduction of electric motors and of electronic drive units to actuate and control the manipulator allowed the physical separation between the master and the slave disclosed an entirely new realm of applications to teleoperation (the terms "telepresence" was introduced at this time) and posed to research a number of challenging theoretical and technological problems.

Some degree of autonomous behaviour was sought to solve the problems determined by the temporal delays between the command at the master and the action at the slave. (The term "Telerobotics" was used to identify this approach).

Beginning in years 1985-1990, the field of computer graphics evolved rapidly leading to the impressive growth of the areas of flight simulators and scientific visualization technologies, and ultimately, along with the development of interactive techniques based on progressively more performant user interfaces, to the field of "Virtual Reality" (VR).

The main difference – which may become subtle though – between VR and teleoperation is the control on the environment, that usually lacks in VR. Teleoperation also took advantage from computer graphics by involving the human operator in the control loop again, thus leading to the area known as "Human supervisory control".

Finally, in the 1990's, the two areas became quite deeply interconnected owing to the development of methods and techniques to model and represent the interactions between the operator and the environment. In fact the operating environment can be either remote and "real" (in the field of teleoperation), or immaterial and "virtual" (in the field of VR), and the interactions with such environment(s) are carried on by means of procedures essentially transparent to the operator. (Who may not even know if the environment he/she is manipulating is real or virtual).

This session reflects quite faithfully the state of the art of problems, technologies and intended applications in the areas of teleoperation and VR.

The first paper, entitled "Operator Control of Robotic Systems", by R.P. Paul, C.P. Sayers and J.A. Adams, addresses the problem of how to handle constrained communication channels. After a brief history of the evolution of teleprogramming, the authors propose a method exploiting high bandwidth feedback both at the operator and at the remote sites, while using relatively low bandwidth communication between the two sites. Experimental results are reported which support the efficacy of the proposed approach.

The second paper, entitled "Intelligent Tele-Robotics", by H. Hashimoto, investigates the issue of dynamic force simulation and the use of recording human skills to render a robotic system partly autonomous. An interesting implementation of the concept is described, in which a 1 d.o.f. master-slave system defined "hand-shake device" (HSD) is used, in combination with a relatively high speed communication network (ISDN) to demonstrate the feasibility and investigate the usefulness of transmitting to a remote user not only information and data, but also physical actions.

The third paper, entitled "Haptic Virtual Reality: Display of Operating Feel of Dynamic Virtual Objects", by T. Yoshikawa and H. Ueda, investigates the problem of modelling the dynamic properties of manipulated virtual objects during operation. The paper proposes first a useful analysis of existing force display devices. Then two different approaches are described for impedance display, such as Measuring Force and Display Motion (MF/DM), and Measuring Motion and Display Force (MM/DF). The architecture of an experimental system and results of tests are also presented.

The fourth paper is entitled "Exoskeletons as Haptic Interfaces", by M. Bergamasco. It addresses the basic concepts, the design and the implementation of two different systems for reflecting haptic information detected by a real or virtual slave to the operator. In particular experimental results are presented obtained when using innovative force reflecting haptic interfaces while following the surface of different virtual objects.

The final paper, entitled "Performance Measures for Haptic Interfaces", by V. Hayward and O.R. Astley, presents a useful analysis of haptic channels versus other channels, and proposes criteria for evaluating the performance of different haptic interfaces. As the authors point out, there is at present an equivocal situation about the performance of haptic interfaces given in the literature, which are mostly those suitable for traditional robots not those truly pertinent in telerobotics and VR.

At the end of each paper and of the overall session, many questions were raised addressing technical aspects of each presentation, as well as some fundamental and controversial issues. Among the main research issues that were pointed out as open are the assessment of the advantages and disadvantages of developing very complex haptic interfaces with multiple d.o.f., capable of reproducing faithfully and completely tactile stimuli, and of replicating forces with an accuracy and dynamic response matching human's. This goal still exceeds state of the art knowledge and technology, but some researchers argued that it may be even not desirable to achieve such goal if the resulting interfaces turned out to be too complex, expensive and cumbersome. Other researchers opposed such conclusion, saying that complexity is justified by high performance.

Additional controversial issues related to the lack of generality of current design of haptic interfaces, and to the need – widely recognized – for characterizing different interfaces performance to facilitate comparison and indicate directions for improvements in this area that, although affected sometimes by the fact of being "fashionable", is definitely of major importance for the future of robotics research.

Force Replication to the Human Operator: The Development of Arm and Hand Exoskeletons as Haptic Interfaces

Massimo Bergamasco

PERCRO
Simultaneous Presence, Telepresence and Virtual Presence
Scuola Superiore S.Anna
Via Carducci, 40 Pisa, Italy

Abstract

This paper deals with a particular aspect of the research on teleoperation and on interaction with Virtual Environments (VE), i.e. the design and the development of man-machine interfaces adequate to both record the movements of the human operator's hand and arms and replicate on them forces detected by the end-effector of the remote robot, or modelled in the VE, at the interaction with the external, or virtual environment.

At first the set of requirements needed for the force Display Device (FDD) to allow the human operator to perform a good control manipulative and exploratory tasks are presented. Such issues as maneuverability, fidelity, transparency or passivity of the FDD will be addressed. Two basic topologies of FDDs are analyzed in terms of workspaces and force replication.

The work carried out at Scuola Superiore S.Anna for the development of FDDs is then presented. In particular two prototypes of Arm and Hand Exoskeletons are described in detail together with the experiments carried out up to now by exploiting a virtual environment testing scenario.

The approach to the design of a second prototype of Arm Exoskeleton, devised for application in the rehabilitation field, is presented.

1 Introduction

Tasks related to teleoperation or to the interaction with the Virtual Environments (VE) require the capability for the human operator to acquire knowledge about features of objects belonging to the remote or virtual environment. Common object's features refer to shape, hardness, surface temperature, surface texture, presence of holes on occluded sides, etc. The acquisition of such informations in a real environment is performed by haptic exploration procedures exploiting cutaneous and kinesthetic inputs [1]. The rendering to the human operator of the sensation of a physical interaction with a remote or virtual environment can be achieved by utilizing appropriate interface systems capable of generating adequate sensory *stimuli* to the operator. However, since a strict correlation exists between the typical hand movement pattern related to a specific exploratory procedure and the particular object knowledge associated, the interface system should allow the human operator to execute such a procedure by exploiting correct, natural, movements.

The possibility for the human operator to execute realistic movements and, at the same time, to achieve adequate sensory *stimuli* is then a fundamental requisite for achieving satisfactory control of the operation. *Realism of interaction*, especially for haptic perception tasks, is then strictly related to the capabilities of the interface system to allow natural movements to the human operator's hand as well as to the fidelity in reproducing adequate stimuli on them.

Such a *transparency* of the interface system can be achieved through appropriate design integrating the above functionalities.

The basis for haptic perception is represented by *cutaneous* and *kinesthetic* inputs. During the execution of an exploratory procedure of an object belonging to a virtual environment, the *kinesthetic inputs* that are presented to the operator's perceptual systems are *real* kinesthetic informations he/she achieves from the movements and postures of his/her hands. Adequate kinesthetic inputs should then derive from correct exploratory movements executed by the human operator. This need directly affects the design approach of the interface system.

The other essential issue to be considered for haptic perception is represented by *cutaneous inputs*. Under this type of sensory informations we include *force* and *tactile inputs*. Since now, interface systems devoted to the rendering of force and tactile information during the interaction with the VE, have been separately developed as force and tactile feedback systems. In this paper we discuss the development of force display devices (FDD) capable of replicating forces at the level of the human upper limb and, at the same time, maintaining its correct natural movement pattern. However, we believe that a correct approach should address the design of force and tactile feedback systems as an integrated Haptic Interface system embedding the replication of both functionalities. Recently, an important trend in the design of interface systems has considered the development of "Haptic interfaces" (or manipulandum, or hand controllers) [2][3][4][5]: in these cases the human operator usually interacts with an actuated external device by means of an handle or a stylus. In this way the force stimuli perceived is generated by the device, while the tactile input is completely real (and not realistic) since the user grasp the end-point of the real device. Hand controllers allow only a limited control of manipulative or exploratory procedures that do not require the presence of a tool in the hand; natural movements of the human hand working in free conditions cannot be easily performed.

We are interested in the ultimate concept of haptic interfaces allowing a large spectrum of natural pattern of movements of the human upper limb (Hand included) and capable of replicating adequate cutaneous inputs. The design of the FDD developed at our laboratory has followed this last approach.

2 Force Feedback Systems

The importance of force replication, or force feedback, on the human operator is well recognized since the first telemanipulation experiments carried out in US in late 1950s [6]. From the development of electrical master systems, characterizing the second generation of telemanipulation systems [7], the present level of technology of force feedback interfaces greatly exploits the achievements of the last two decades in the areas of sensor technologies and control architectures. More recently, the possibility of integrating complex end-effectors at the slave level, such as sensorized dexterous robotic hands, brought new emphasis on the design of adequate hand controllers [8][9]. At the present a great boosting action to the design of new man-machine interfaces has been induced by the innovative research in the field of the interaction with the Virtual Environments (VE).

Although conceptually different, the problem of the design of the interface system, especially in terms of afferent sensory information replicated to the human operator's hand, can be considered in VE as very close to the one to be tackled for the control of teleoperators for telemenipulation task [10]. In the course of the paper we assume that the Force Feedback system is composed of:

- one or more Force Display Devices (FDD);
- their control systems;
- a modelling module devoted to generate the force to be replicated.

The set of general features that force feedback systems must possess for allowing a good control of the task by the human operator comprehends issues such as:

- **maneuverability**: the interface system must allow a natural mobility to the human hand and arm. This requirement can be seen in terms of the natural workspace of the human hand-arm complex which must not be restricted from the presence of mechanical structure of the FDD. This condition is difficult to be achieved from the design point of view because the FDD possesses its own workspace which intersection with the human-hand workspace must be optimized;
- **fidelity**: the interface system must be able to reproduce in a faithful way to the human hand-arm the forces modelled in the VE. In previous papers we have analyzed how the forces (wrenches) which can be replicated depend on the kinematics of the FDD [11][12];

- **impedance**: the impedance offered to the human hand-arm complex by the force feedback system must match the same condition of impedance seen by the human hand-arm complex during the performance of a real grasping or manipulative operation. This requirement implies a condition of transparency of the interface system during operation. This condition of passivity of the force feedback system subtends also the need of rendering, for whatever kinematic configuration of the structure, a complete inertial and weight compensation;
- **wearability**: wearability and also portability of the FDD represent two also important features for the synthesis of the final mechanism.

From the topological point of view it is possible to devise two main categories of FDDs [13]:

1. FDD with several DOF which structure is external with respect to the human limb (finger, hand, arm, leg, etc). Only the FDD end-point is connected to the human limb and through this point the forces are transmitted. Since the human limb is constrained to be connected with the FDD end-point (which base frame is fixed to the ground), the resulting maneuverability is limited to the intersection of the FDD's and limb workspaces. In terms of force replication, however, the force exerted by the FDD is perceived by the human operator as an external force, without "side effects"; in the case that multiple contact points should be considered, an equivalent number of FDDs must be utilized;
2. FDD possessing several DOF which structure wraps up the human limb (finger, arm, leg, trunk, etc...). In this case the FDD kinematic reflects the one of the limb since the joint axes are the same (Exoskeletons).

   Exoskeletons can be designed with one or more points of attachment with the human limb as well as, on the other hand, with their base frame fixed to the ground (external to the body) or to the human limb (see Fig.:2). In the last condition side effects phenomena due to the presence of reaction forces at the level of the base frame attachment point can be present.

When the interaction between the human operator and the virtual environment comprehends grasping, manipulative as well as exploratory procedures, the set of virtual forces that can be modelled in the virtual environment can belong to a large spectrum, in terms of magnitude, direction and also point of application. This topic as been analyzed in [14].

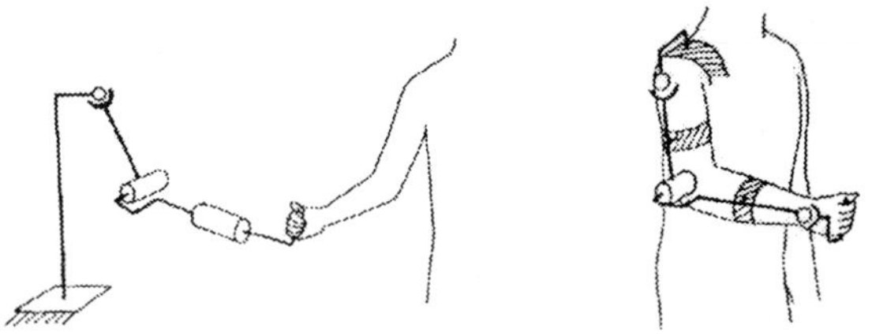

Figure 1: Different Topologies of Force Display Devices

The main assumption we have introduced in our work in order to design a force feedback system capable of replicating the above complete range of forces is the following: "since the proprioceptive information (sensation on force exerted) perceived by humans is based on signals generated by sensors located at the joints, muscles and tendons, we base the concept of force replication by considering forces that can stimulate tracts of the limb comprehensive of the same muscles and joint which will be interested during the performance of the same task in a real condition". These forces will be replicated to the human limbs by different FDDs, possessing different performances in terms of magnitude, orientation and direction of forces to be replicated.

3 Force Display Devices

At PERCRO of the Scuola Superiore S.Anna, a complete Force Feedback System consisting of a) two serially connected FDD components, b) their control systems, c) a modelling module for generating the forces and d) a graphical module for the representation of the VE scenario, has been designed and realized.

The FDD components, represented in Fig. 3, consist of two exoskeleton systems devoted to replicate forces at the level of the arm and of the operator's fingers. In particular the components are:

- a FDD component, called Arm Exoskeleton or External Force Feedback (EFF) system, with 7 DOF which are coincident with the principal joint axes of the human arm.

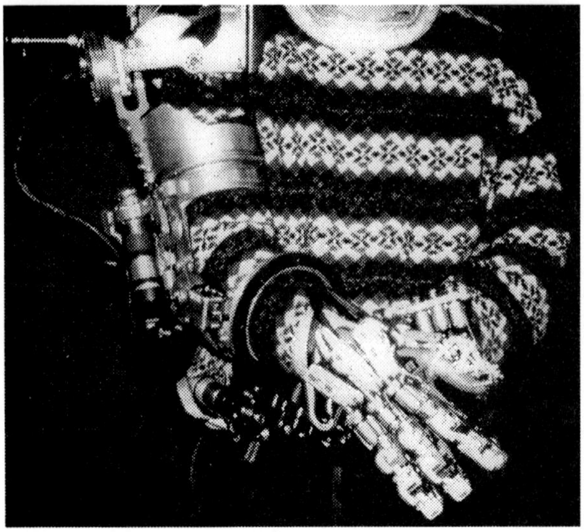

Figure 2: The Arm and the Hand Exoskeletons

The Arm Exoskeleton has been designed to be completely supported by the operator's body by means of a purposely designed trunk structure or by a fixed frame structure which allows easier wearing conditions in case of repeated experimental tests;

- a FDD component, called Hand Exoskeleton or Hand Force Feedback (HFF) system, consisting of 4 parallel FDDs wrapping up four fingers of the human hand (little finger excluded) and each one possessing 3 DOF in correspondence of the finger joint axes (the total number of DOF is then 12). The 4 FDD are connected to a base plate located at the metacarpus and corresponding to the end-point of the Arm Exoskeleton.

Both FDD components have ben designed in order to maximize the resulting workspaces of the human arm and hand during operation. In case the EFF system is supported by the trunk structure, the complete FDD system becomes portable and allows a large physical mobility to the human operator inside the control space. This fact extends its use also for augmented reality applications.

Consequences of the described solution can be summarized as:

- the serial links kinematics of the EFF and HFF exoskeletons is given and cannot be changed to improve its inertial and elastic properties;

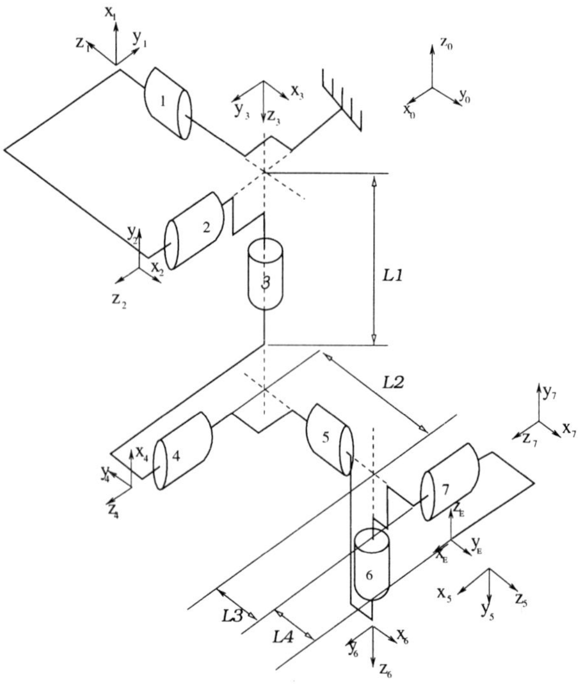

Figure 3: Kinematic Representation of the Arm Exoskeleton

- the mechanical design presents a certain degree of complexity: limitations to the human arm movements must be reduced and for this reason the links should be slender and their surfaces have to be free of protrusions which could hurt the operator limbs;

- the portability of the FDD system on the operator trunk imposes tight constraints in terms of weight of structural parts and actuators and it requires a careful mass distribution to mitigate the operator effort needed to balance the system.

3.1 Arm Exoskeleton

Fig. 3.1 depicts the scheme of the kinematic chain of the complete 7 DOF Arm Exoskeleton [15]. In this configuration (called *Exos7*), the end-point of the mechanical structure consists of a plate where the Hand Exoskeleton can be connected. The Arm Exoskeleton can be also utilized according to a configuration with 5 DOF (called *Exos5*) in which the last 2 DOF, corresponding to the wrist flexion-extension and abduction-adduction movements, have been removed and substituted by a fixed handle.

The Denavit-Hartenberg parameters describing the Arm Exoskeleton kinematics are given in Ta-

Link	a	d	α	θ	*Mass*	
#1	0	0	$-\pi/2$	0	1.5	
#2	0	0	$-\pi/2$	$-\pi/2$	0.53	
#3	0	$-L_{arm}$	$-\pi/2$	0	2.775	
#4	0	0	$-\pi/2$	$-\pi/2$	0.644	
#5	0	L_{farm}	$\pi/2$	$\pi/2$	1.423	0.75
#6	L_w	0	$-\pi/2$	$\pi/2$	0.199	0
#7	H_h	L_h	0	0	0.052	0

Table 1: Denavit-Hartenberg parameters and link masses for the Arm Exoskeleton

ble 1.

The length of the arm L_{arm} is adjustable ($\pm 0.015\ m$) around the nominal value of $0.33\ m$. The values of the other parameters are $L_{farm} = 0.255\ m$, $L_w = 0.01\ m$, $L_h = 0.063\ m$, $H_h = 0.023\ m$ for the *Exos7* configuration and $L_{farm} = 0.33\ m$, $L_w = L_h = H_h = 0.0\ m$ for the *Exos5*.

The actuation and transmission system of each joint of the Arm Exoskeleton is composed of a DC electric motor, a gear box mounted on the motor axis, a cable transmission system transmitting the movement of the gear box output stage

The basic consideration for the choice of the actuation and transmission system has been that of using gear boxes that would increase the maximum output torque of small and light motors while still keeping the joints backdrivable. At the same time, the cable circuits which are routed with idle pulleys allow a further reduction ratio with very high efficiency and, as far as possible, to place the actuators on the more proximal links instead on each joint. The actuators for joints #3, #4, #5 are placed on link #3, while on link #5 the actuators for joint #6 and #7; the great influence of the motor location on the mass distribution of the Arm Exoskeleton can be seen by analyzing Table 1.

Although all the Arm Exoskeleton joints are backdrivable, their actual friction is still far to be negligible. The solution devised to overcome friction problems consists in the integration in the cable transmission system of differential tendon tension sensors. Such devices give an indirect measure of joint torques so that a local joint torque feedback loop can be implemented to reduce the influence of motor , gear box and cable circuit friction on joint behaviour. In the first prototype of Arm Exoskeleton, differential tendon tension sensors are present only on the two wrist joints and on the forearm prono-supination joint; the next version of the Arm Exoskeleton will mount them

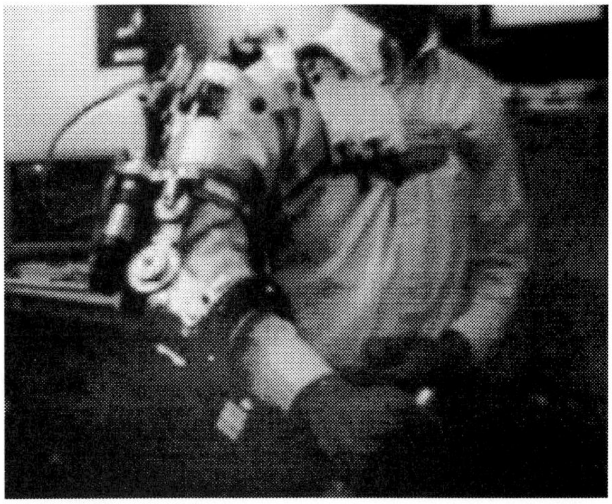

Figure 4: The Arm Exoskeleton in the 5 DOF configuration (*Exos5*)

on all joints. Table 2 gives the joint excursion limits which have been measured with the operator actually wearing the Arm Exoskeleton to obtain values corresponding to the intersection between the human arm workspace and that of the FDD component.

Joint	θ_{min}	θ_{max}
#1	$-25.0°$	$-73.26°$
#2	$-40.38°$	$49.4°$
#3	$-80°$	$72.9°$
#4	$-21.17°$	$74.56°$
#5	$-85.0°$	$90.0°$
#6	$-80.0°$	$80.0°$
#7	$-35.0°$	$25.0°$

Table 2: Joint rotation limits (in degrees) of the operator wearing the EFF

Kinesthetic sensors of the Arm Exoskeleton consists of optical encoders located on the motor shafts; encoders #1 to #5 give out 500 impulses per revolution; encoders #6 and #7 give out 15 impulses per revolution.

3.2 Force Modelling and FDD control strategy

Experiments have been carried out with the aim of studying the behaviour of the human operator during the execution of interaction procedures in the VE.

The implementation of force feedback from the VE is based on evaluating the operator hand position from the Arm Exoskeleton kinesthetic sensors data, performing collision detection procedures, calculating the forces due to collisions with virtual surfaces, and finally controlling the force between the FDD and the operator's hand to the computed value [12][17].

Models of the interaction between the human hand and virtual objects can be very complex and computationally heavy if the detailed geometry of contact is taken into account [16]. The operator hand is modelled with only one control point c so that its inter-penetration into virtual bodies is summarized by a single interpenetration vector d which represents its distance from the surface.

A contact between the operator hand and a virtual body generates a force $\boldsymbol{F}_{react}$ according to a dynamic model of contact which represents the elastic and viscous properties of the virtual body:

$$\boldsymbol{F}_{react} = \begin{cases} -K\,d + B\,v_{op} \cdot \frac{\boldsymbol{d}}{|\boldsymbol{d}|} & \text{at contact} \\ 0 & \text{otherwise} \end{cases}$$

The elastic component of the surface reaction force models a rigid surface of stiffness K. The viscosity coefficient B is introduced to take into account the dissipative phenomena occurring during a real collision in the same way a restitution coefficient of the collision would do [18]. The force-based interaction between the operator and the VE is affected by system sampling time, simulation and communication delays, errors in force replications due to the FDD.

The implementation of the proposed interaction is realized by using as control variables the Arm Exoskeleton join torques [15]. The joint torques to be utilized for the control of the Arm Exoskeleton joints in order to exert on the operator hand a desired wrench $\boldsymbol{F}_{react}$ should compensate the inertia, the Coriolis and centrifugal effects, the friction and the weight. A complete mapping of external wrenches at the hand level into joint torques will thus be dependent on exoskeleton configuration, joint velocities and accelerations:

$$\tau = M(\boldsymbol{q})\ddot{q} + C(\boldsymbol{q}, \dot{\boldsymbol{q}}) + D(\boldsymbol{q}, \dot{\boldsymbol{q}}) + G(\boldsymbol{q}) + J^T(\boldsymbol{q})\boldsymbol{F}_{react}$$

where M is the inertia matrix of the Arm Exoskeleton, C is the vector of Coriolis and centrifugal terms, D is the vector of friction terms, and G is the vector of gravity effects.

In our case we have decided to use only the gravity compensation term; this choice greatly simplifies the control law but it introduces a force error especially when joint velocities and accelerations are high. Since no force/torque sensors are present on joint #1-#4, an open loop control law has been implemented as follows:

$$\tau_{control} = \hat{G} + J^T(\boldsymbol{q})\boldsymbol{F}_{react}$$

where $\hat{G}$ indicates an estimate of G. Such an open loop scheme can be used since all joints are backdrivable. Friction and modelling errors in the estimation of G introduce further force errors. An improvement in the force replication accuracy can be obtained by using a closed loop control law when joint torque sensors are available.

The Arm Exoskeleton controller and the Force Feedback subsystem of the VE have been digitally implemented on a 6 transputer network and run at a sampling frequency $\frac{1}{t_c}$, with t_c the cycle period of sampling, varying in the range $1 - 0.001s$ [12].

Experiments have been carried out by using the Arm Exoskeleton with the aim of reproducing two basic interactions of the human hand with a rigid surface: the collision with a virtual surface and the following of the virtual surface. Experiments have been carried out by varying the parameters of the VE such as the system sampling time t_c, the virtual surface elastic coefficient K and the virtual surface viscosity coefficient B. Results have been extensively reported in [12] and summarized in [17] and [18].

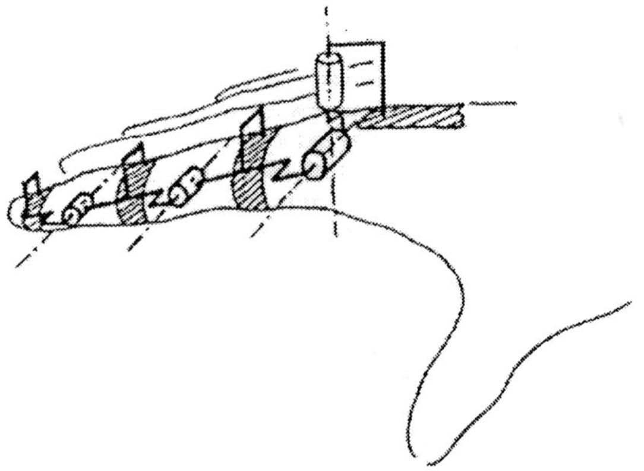

Figure 5: Kinematics scheme of the Hand Exoskeleton

3.3 Hand Exoskeleton

The second component of the complete FDD system is called Hand Exoskeleton and consists of

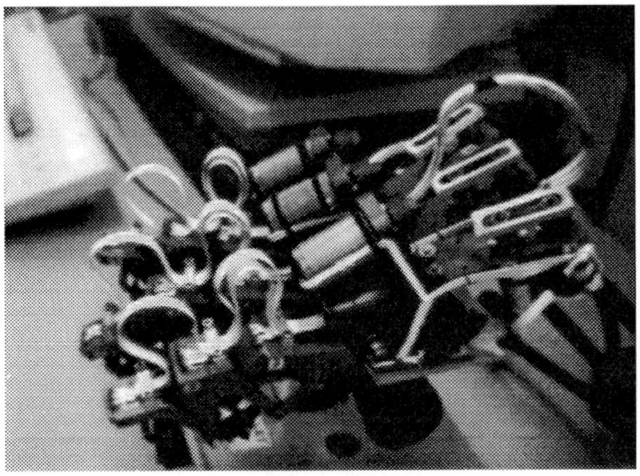

Figure 6: The Hand Exoskeleton

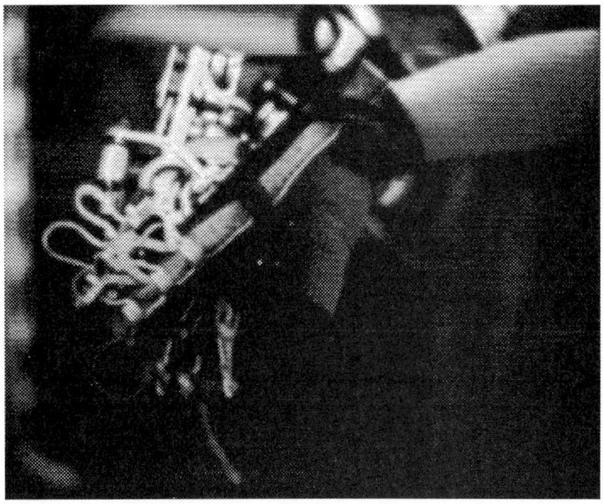

Figure 7: Another view of the Hand Exoskeleton

4 FDD exerting forces to the phalanges of the hand's fingers (little finger excluded). Each finger exoskeleton, which kinematic scheme is represented in Fig.3.2, consists of four links connected by revolute joints disposed as the joints of each finger. For each joint of the finger exoskeleton, the joint axis has been designed in order to approximate the instantaneous position of the flexion-extension axis during operation. At the metacarpo-phalangeal joint a passive abduction-adduction movement has been also integrated.

The actuation system for one finger exoskeleton is based on three DC servomotors and associated tendon tension transmission systems. Each tendon is pulling on the middle point of each phalanx of the finger in order to execute the extension movement; at each joint, the flexion movement is obtained by a passive torsion spring integrated on the joint axis. The three motors are located on a cantilever structure fixed with the base frame of each finger exoskeleton. Rotation sensors, based on conductive plastics technologies, are integrated at each joint while force sensors, capable of recording the interaction force between the exoskeleton structure and each phalanx, are located directly on the dorsal surface of each phalanx link.

A picture of the resulting Hand Exoskeleton is given in Fig.3.3.

The kinematic structure of the thumb exoskeleton is slightly different to the one of the other fingers. In particular, from the construction point of view, the cantilever supporting the three motors of the thumb assumes a completely different aspect with respect to the one of the other fingers.

The Hand Exoskeleton system is controlled by a transputer-based control architecture. One of the critical factors encountered during the design of the system has been that of obtaining a system possessing limited weight and volumes, in such a way to allow good manoeuvrability of the hand. During wearing conditions, the ranges of motion of the fingers can be considered very close to those of the free human hand. In terms of the mechanical performances of the Hand Exoskeleton system, we obtained a maximum extension force of $0.3N$, being the force sensor range of $-0.5N$—$3.0N$. Force resolution is $0.0025N$, while the force feedback bandwidth is $0.5Hz$ with an angular displacement of 90° for all the 3 DOF.

In order to test the performances of each finger exoskeleton, a purposely designed test-bed actuated structure has been realized. Fig.3.3 shows the operating condition of the test-bed system: one finger exoskeleton is fixed to the base plate of the test-bed and connected, by means of appropriate connection particulars, to the links of the test-bed system. The test-bed is controlled in parallel to the finger exoskeleton in order to execute both isometric and isotonic tests. During isotonic tests, the finger exoskeleton is commanded to exert constant forces on its phalanges, while the test-bed is controlled in order to impose appropriate movement patterns to the finger exoskeleton links. The objective of the test is that of monitoring the force error as a function of the frequency of the displacement at each joint. In the framework of isometric tests, sinusoidal force inputs to the Hand Exoskeleton will be provided, while the test-bed will impose on it constant displacements (fixed configurations). In terms of positioning bandwidth, the test-bed

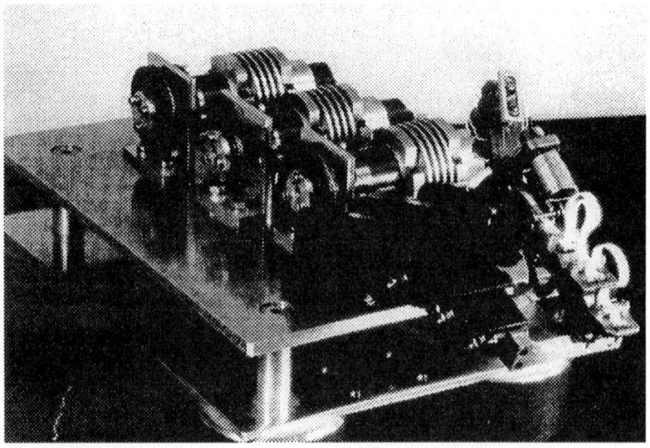

Figure 8: The Hand Exoskeleton Test-bed

system has been designed for $8Hz$ on 90° for all the 3 DOF. The maximum transmittable force to the Hand Exoskeleton is $110N$, and the position resolution has been estimated to $3.5arcmin$.

4 The Design of a New Arm Exoskeleton

In the framework of a project dealing with rehabilitative applications, the design of a new Arm Exoskeleton has been addressed. The objectives of the new FDD system can be summarized as:

1. capability to exert forces not only at the level of the hand but also on the forearm or/and arm (multiple contact point);
2. capability to wear the structure from the lateral side and without inserting the human arm only through the shoulder hollow;
3. general reduction of the masses and inertia by exploiting a remote arrangement of the motors. The objective is the reduction of mobile masses in order to diminish unbalancing moments on the human body;
4. to analyze the influence of a set of design parameters on the accuracy of force replication.

The design of the new Arm Exoskeleton exploits a kinematics similar to the one presented in Section 3 for the prototypical solution. However, several attachment points have been foreseen along the human arm; at present, it has been assumed that the replication of forces at the level of the arm-hand complex will be performed only at one of the possible locations, but, with the purchasing of low-cost force sensors, the control could be contemporarily extended to several points.

In order to allow a lateral wearing of the FDD (for disable people this condition is fundamental) the joints of the arm rotation and forearm prono-supination have been designed by utilizing open circular guides with an angular encumbrance of 210° instead of 360° of common ball bearings. The open circular guides leave the necessary clearance for lateral insertion of the arm.

Structural parts will be analyzed with finite elements methods and built by exploiting composite material technologies: Fig.4 shows the first prototype of the motor case located on the back of the human operator; this prototype has been designed and realized by Ferrari S.p.A., Modena, Italy.

The actuation system has been carefully studied in order to reduce the total weight. Torque motors have been selected to obtain high torque/weight ratios. For the first 4 DOF the actuation system consists of the DC torque motor, the gear box with a transmission ratio of 5 and one resolver as rotation sensor. These 4 motors have been remotely located on the back of the operator, while the 3 motors corresponding to the last 3 distal joints have been integrated on the forearm structure close to the elbow joint.

The yardstick on which verify the performances in terms of force feedback has been devised as the accuracy of force replication. Our objective is to obtain an accuracy of 4% on the whole operative range (force, displacement, temperature). All the design parameters influencing accuracy will be analyzed (stiffness of the arm exoskeleton, backlash, transmission friction, resolution and accuracy of position and torque sensors at the joints, controller design). At present, since the arm exoskeleton structure presents several DOF, the analysis has been approached by assuming simplified hypotheses.

5 Conclusions

The work described in this paper has addressed the description of Force Display Devices developed at PERCRO of Scuola Superiore S.Anna, Pisa, Italy. The design of FDD has been approached by considering the requirements of natural manoeuvrability and fidelity in replicating forces that the interface system should possess in order to satisfy conditions of realism during the

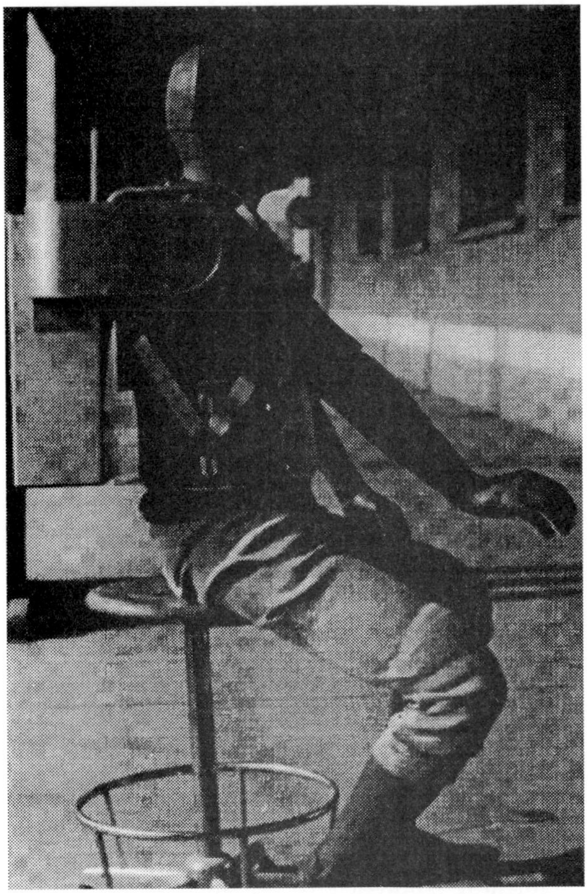

Figure 9: Prototype of structural parts of the new Arm Exoskeleton (By courtesy of Ferrari S.p.A.)

interaction with the virtual environment. A complete FDD system has been presented consisting of two serially connected FDD components: an Arm Exoskeleton system and a Hand Exoskeleton system. Results of preliminary tests related to the control of the interaction with virtual surfaces and to contour following procedures of virtual surfaces have been introduced. The design of a new Arm Exoskeleton has been preliminary presented by focusing on the innovative included features. Until now the approach to the design of FDD systems has been characterized by a great effort in terms of pragmatical work and pure drawings of mechanical solutions. However, on such a basis, PERCRO is now starting a new approach for the design of FDD which exploits analysis and methodological estimations of the foreseen solutions. We believe that a great amount of theoretical work can still be opened inside this research field.

Acknowledgements

The author wishes to thank Carlo Alberto Avizzano and Giuseppe Maria Prisco for their contribution in the preparation of the paper, and Ferrari S.p.A. for the collaboration. The research described in this paper has been supported by the Commission of the European Union in the framework of the ESPRIT n.5363 project GLAD-IN-ART and the TIDE n. 1216 project VETIR.

References

1 Lederman S.J., Kaltzky R. L., "Haptic Exploration and Object Representation" in M.A. Goodale (Ed.), Vision and Action: the control of Grasping, Ablex Publishing Corporation, Norwood, NJ, 1990

2 V. Hayward, "Toward a Seven Axis Haptic Device", Proc. of 1995 IROS, Pittsburgh, US, Aug. 1995.

3 T. H. Massie, J. K. Salisbury, "The PHANToM Haptic Interface: a Device for Probing Virtual Objects", Proceedings of the ASME Winter Annual Meeting, Chicago, November 1994.

4 J. E. Colgate , J. M. Brown, "factors affecting the Z-Width of a Haptic Display", 1994 IEEE International Conference on robotics and Automation, S. Diego, CA, May 1994.

5 D. A. Lawrence, J. D. Chapel, "Performance Trade-offs for a Hand Controller Design", 1994 IEEE International Conference on Robotics and Automation, S. Diego, Ca, May 1994.

6 Goertz R. C., "Manipulator Systems Development at ANL," Proc. of 12th RSTD Conference, 1964.

7 Flatau C. R., "Development of servo Manipulator Systems of High Energy Accelerator Requirements," Proc. of 13th RSTD Conference, 1965.

8 Brooks T. L., Bejczy A.K., "Hand Controllers for Teleoperation," JPL Publication 85-11, NASA, March 1985.

9 Jau B., "Man-Equivalent Telepresence Through Four Fingered Human-like Hand System, Proc. IEEE

Intern. Conf. on Robotics and Automation, Nice, France, 1992.

10 Bergamasco M., "Design of Hand Force Feedback Systems for Glove-like Advanced Interfaces, Proc. of the First IEEE Int. Workshop on Robot and Human Communication ROMAN92, Hosei University, Tokyo, Japan, 1992.

11 B. Allotta, M. Bergamasco, "The Range of Wrenches that can be Statically Exerted by a Force Replication Device", 1994 IEEE ROMAN International Workshop on Robot and Human Communication, Nagoya, Japan, July 1994.

12 G. M. Prisco, "The Control of the Interaction with Virtual Environments: Theoretical and Experimental Analysis of the Replication of Forces on the Human Arm by means of Exoskeleton Systems", Thesis in Electrical Engineering, University of Pisa, July 1994

13 M. Bergamasco, "Theoretical Study and Experiments on Internal and External Force Replication, "Lecture notes workshop on Force Display in Virtual Environments and its Application to Robotic Teleoperation (A. Bejczy, G. Hirzinger, K. Tanie, Eds.) 1993 IEEE Intern . Conf. on Robotics and Automation, May 2, Atlanta, GA, USA

14 M. Bergamasco, "Haptic Interfaces: The Study of Force and Tactile Feedback Systems" Proc. of IEEE Workshop on Robot and Human Communication ROMAN '95, Waseda University, Tokyo, Japan, 1995

15 M. Bergamasco, B. Allotta, L. Bosio, L. Ferretti, G. M. Prisco, G. Parrini, F. Salsedo, G. Sartini, "An Arm Exoskeleton System for Teleoperation and Virtual Environments Applications", 1994 IEEE International Conference on Robotic and Automation, S. Diego, CA, May 1994

16 M. Bergamasco, P. Degl'Innocenti, D. Bucciarelli, "A Realistic Approach for Grasping and Moving Virtual Objects", 1994 IROS International Conference on Intelligent Robot and Systems, Munich, Germany, 1994

17 M. Bergamasco , G. M. Prisco, "Virtual Surfaces Contour Following: An Experimental Approach Exploiting An Arm Exoskeleton As Haptic Interface, "Proc. of 1995 ASME International Mechanical Engineering Congress and Exposition, Session DSC-12A, November 1995, San Francisco.

18 M. Bergamasco, G. M. Prisco , "An Experimental Approach to Virtual Surfaces Exploration exploiting An Arm Exoskeleton as Haptic Interface," Proc. of SPIE 1995, Philadelphia, USA.

Intelligent Tele-Robotics

Hideki HASHIMOTO

Institute of Industrial Science, Univ. of Tokyo
2-22-7 Roppongi,Minato-ku, Tokyo 106, Japan.
Tel:+81-3-3479-2766 Fax:+81-3-3423-1484
E-mail: hashimoto@iis.u-tokyo.ac.jp

1 Introduction

"Tele-Operation" has been already known as a technical word in our daily life. Most people think that "Tele-Operation" is the technology which realizes remote control of robots in space or hazardous environments like a nuclear plant. Of course some others know that it implies more wider technology field. In this case a lot of researchers notice that the "Tele-Operation" is one of the necessary technologies for the information infrastructure because of the most valuable services for people on this infrastructure, require some motions together with motion images and sounds, like in multi-media. Robots, for instance could be good intelligent interfaces between computer networks and people, and on the other hand robots could assist people in such scenes where some motion as moving objects are required. Usually it is thought to be difficult to realize that robots have to possess considerable autonomous intelligence to understand environments and intentions of the users. But now we can use our knowladge to let robots have some similar intelligence as humans have by using tele-operation. Tele-operation with some degree of autonomous intelligence is a key technology to realize truly useful robots in the information infrastructure.

Intelligent Tele-Robotics is a new discipline which main body is tele-operation, intelligent robot and high-speed computer network. The "Intelligent Tele-Robotics" is appearing as an important technology field in the coming new-age society. Key words for research topics could be the Haptic which means how to produce more real sensation, Virtual Reality, Skill, Intelligence and Communication. Each of these topics are so attractive that we can find easily a lot of research papers on them. But quite a few people are trying to use the information infrastructure composed by optical fiber and ATM. We can use very high-speed and big capacity communication channels for our robotic and mechatronic system. This direction might cause a new breakthrough because the communication volume is more than ten times biger compared with the usual ethernet technology. Also this direction is facing to our daily life not only to Industry. Amusement, welfare, medicine and assistance of aged people are the targets of the "Intelligent Tele Robotics".

We focus on two topics in this paper to discuss "Intelligent Tele-Robotics". One is the Haptic Interface itself and the other is the Communication of Haptic Interface. For the Haptic Interface research we present a Dynamic Force Simulator (DFS) for force feedback in human-machine systems. We propose a virtual world model with two force flows: one is the force flow from human to an object, the other is the force flow from an object to human. To use this model, DFS simulates object dynamics, contact model and friction characteristics of the human hand interacting with the object in Virtual Reality. After derivation of kinematic and force relations between hand and object space, we balance the two forces: one from the human and the other from the object in the contact force space in Virtual World and then realize the adequate feedback forces to human operator. Interaction with DFS allows calculation and feedback of appropriate forces to the force controlled actuators of the sensor glove we have developed. In this paper grasping of a cylinder in the virtual world is presented. During object grasping, we measure joint angles and torques using the sensor glove system. In the future we will use this system to analyze human dextrous manipulations called Human Skill. In the other related research we proposed a 1DOF master-slave system called HandShake Device (HSD). HSD is composed of 2 systems: one is a master system, which is grasped by an operator, the other is a slave system, which grasps the operator's palm. Two HSDs are respectively placed at a local site and a remote site so

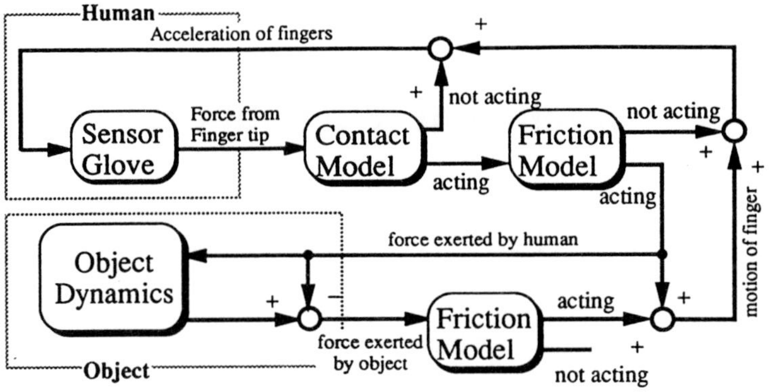

Fig.1 Simplified Force Flow between Human and Object

that people can physically communicate (by shaking hands) each other through HSDs. HSD applies the virtual model method as an algorithm of teleoperation (tele-handshake). HSD was exhibited in SIGGRAPH'95 with Network Neuro Baby (NNB). NNB is a research project in Artificial Life which simulates a living baby acting as a network agent in the computer network. HSD is an interface device which an operator can communicate with NNB physically.

2 Dynamic Force Simulator for Haptic Interface

The extremely high speed development in the field of the computer technology the last decade opened a new gate for a dreamfull and interesting research world called as Virtual Reality or Artificial Reality. Research in Virtual Reality and human-machine interface has so far concentrated on high-quality graphics animation using Data Glove, but the force feedback remained outside of researcher's horizon. The development of effective force feedback devices is just in the beginning phase, where some rather sophisticated devices have emerged[3][9]. We strongly think that in the future an easy-to-use human-machine interface has to include some force exchange as well[1][4]. In this paper we propose a Dynamic Force Simulator(DFS) as an integrated part of a futural human-machine interface which can feedback forces from human operators to the Virtual Reality and vice-versa. The DFS – in the present state of the research – simulates the dynamics of the object, contact model and characteristics of the friction between the human hand and the interacting object in a Virtual Reality. In addition to our system aims at analysis of human motions, for example human skill acquisition as a first step of a more complex Intelligent Assisting System (IAS) proposed previously [2]. To obtain human skill we need object state and parameters, joint torques and angles. A virtual reality system using for skill acquisition offers the following possibilities

1. Iterating operations under identical conditions
2. Creating arbitrary environments
3. Exact knowledge of physical parameters
4. Decreasing the number of Sensors
5. Getting torques of fingers by using force feedback.

The human's motion can be analyzed by the proposed DFS, for example grasping and handling. The knowledge obtained by DFS helps in controlled dextrous hand. The sensor glove device we have developed for force feedback in 11 degrees-of-freedom. We can measure joint angles and torques of human fingers with this device. The operator can interact with the DFS. The DFS calculate the appropriate forces and feed them back by the force controlled actuators, which is a kind of force displays i.e. the sensor glove device.

2.1 Dynamic Force Simulator

2.1.1 Modeling of Virtual World

The proposed DFS has detailed derivation of the force flow during interaction of the human operator and objects in the virtual environment. The DFS environment model in the virtual world is shown in Fig.1. The idea of the DFS considers the force flow between the human operator and the virtual world. Let us imagine the case when there is a force exchange between the human operator and objects in Virtual Reality. In this case the force flow includes two main streams: one from the human operator through the sensor glove and the other is originated from the object in the virtual world. When the force flow from human pass through the contact model and friction model, these submodels divide the force into two flows: one influence object motion, the other doesn't influence. Both force flows are balanced on contact points. We can create the virtual environment with these force flows, some coordinate transformations, contact models and friction models.

The block diagram shown in Fig.1 is implemented on the computing resources as shown in the next section.

2.1.2 Coordinate Transformations

Figure 2 illustrates the transformations between finger-tip, contact and object coordinate frames. Let us define transformation matrices $\boldsymbol{U}_i \in \mathbb{R}^{4\times4}$ and $\boldsymbol{V}_i \in \mathbb{R}^{4\times4}$, where $\boldsymbol{U}_i$ transforms from the finger-tip coordinate frame $\boldsymbol{T}_i \in \mathbb{R}^{4\times4}$ to the contact frame $\boldsymbol{C}_i \in \mathbb{R}^{4\times4}$, and $\boldsymbol{V}_i \in \mathbb{R}^{4\times4}$ from the contact frame to the object frame $\boldsymbol{B} \in \mathbb{R}^{4\times4}$.

The x- and y-axis of the contact coordinate frame $\boldsymbol{C}_i$ are tangents and the z-axis is perpendicular to the object surface curvature(Fig.3).

The transformation matrices $\boldsymbol{\Gamma}_i \in \mathbb{R}^{6\times6}$ and $\boldsymbol{\Lambda}_i \in \mathbb{R}^{6\times6}$ are defined such a way that $\boldsymbol{\Gamma}_i$ transforms finger-tip forces ${}^{T_i}\boldsymbol{f}_i$ into forces ${}^{C_i}\boldsymbol{f}_i$ in the contact coordinate frame and $\boldsymbol{\Lambda}_i$ transforms these to forces ${}^{B}\boldsymbol{f}_i$ in the object reference frame $\boldsymbol{B}$.

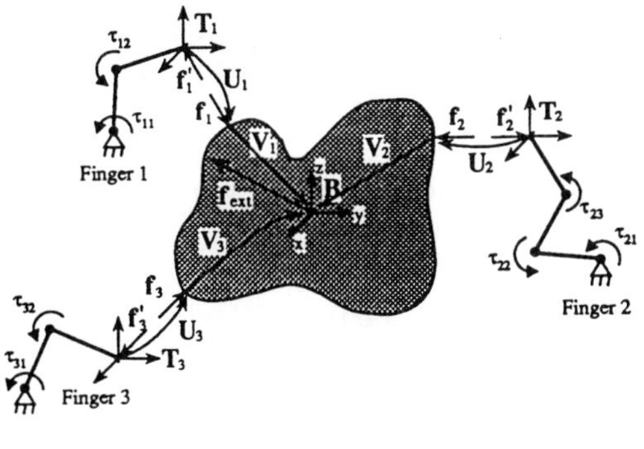

Fig.2 Object Grasped by 3 Fingers

The relationship of coordinate transformations can be written as

$$\boldsymbol{C}_i = \boldsymbol{T}_i\,\boldsymbol{U}_i \tag{1}$$

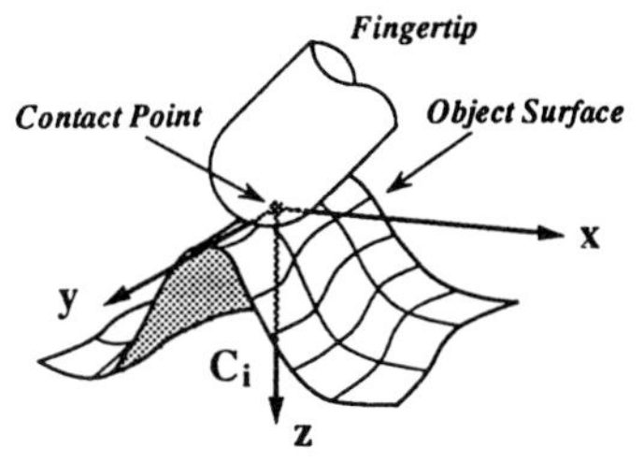

Fig.3 Contact Point and Contact Frame

$$\boldsymbol{B} = \boldsymbol{C}_i\,\boldsymbol{V}_i = \boldsymbol{T}_i\,\boldsymbol{U}_i\,\boldsymbol{V}_i \tag{2}$$

$${}^{C_i}\boldsymbol{f}_i = \boldsymbol{\Gamma}_i\,{}^{T_i}\boldsymbol{f}_i \tag{3}$$

$${}^{B}\boldsymbol{f}_i = \boldsymbol{\Lambda}_i\,{}^{C_i}\boldsymbol{f}_i = \boldsymbol{\Lambda}_i\,\boldsymbol{\Gamma}_i\,{}^{T_i}\boldsymbol{f}_i \tag{4}$$

where $\boldsymbol{U}_i$, $\boldsymbol{V}_i$ are homogeneous transformations and $\boldsymbol{\Gamma}_i$, $\boldsymbol{\Lambda}_i$ are force transformations [7]

The sensor glove has two sensor outputs, the torques actuated by the human operator $\boldsymbol{\tau}_i$ and the finger joint angle values $\boldsymbol{\theta} \in \mathbb{R}^{10}$. $\boldsymbol{T}_i$ are calculated by forward kinematic equations from $\boldsymbol{\theta}$ and distances of the fingers from objects in the virtual world can be calculated. If fingers are in contact with the objects, the transformations $\boldsymbol{U}_i$, $\boldsymbol{V}_i$, $\boldsymbol{\Gamma}_i$ and $\boldsymbol{\Lambda}_i$ are calculated immediately.

2.1.3 Contact Models

A specific contact model has to be chosen (compare [5][6][8] for more details of contact models) to perform the transformation of the contact forces into the object coordinate frame $\boldsymbol{B}$. According to the contact model we have a number of wrenches exertable on the object that we write as unit contact wrenches ${}^{C_i}\boldsymbol{w}_{ij}$ in the contact frame $\boldsymbol{C}_i$.

The contact forces ${}^{C_i}\boldsymbol{f}_i$ can be divided into two parts: one is balanced by other grasping or external forces and it can be written a linear combination of the contact wrenches ${}^{C_i}\boldsymbol{w}_{ij}$. The other part referred as a rest force ${}^{C_i}\hat{\boldsymbol{f}}_i$ is unbalanced and it accelerate the particular finger.

$${}^{C_i}\boldsymbol{f}_i = {}^{C_i}\hat{\boldsymbol{f}}_i + \sum_{j=1}^{k_i} c_{ij}\,{}^{C_i}\boldsymbol{w}_{ij} = {}^{C_i}\hat{\boldsymbol{f}}_i + {}^{C_i}\tilde{\boldsymbol{f}}_i \tag{5}$$

where k_i is the number of contact wrenches ${}^{C_i}\boldsymbol{w}_{ij}$ for finger i that effect the object according to the contact model. The rest force ${}^{C_i}\hat{\boldsymbol{f}}_i$ in (5) yields acceleration of the finger-tip and therewith changes contact configuration because it is not balanced by other grasping forces or the external force.

The force ${}^{C_i}\tilde{\boldsymbol{f}}_i$ is given as [5]

$$\begin{aligned}{}^{C_i}\tilde{\boldsymbol{f}}_i &= [\boldsymbol{w}_{i1}, \boldsymbol{w}_{i2}, \cdots, \boldsymbol{w}_{im}]\,[c_{i1}, c_{i2}, \cdots, c_{im}]^T \\ &= {}^{C_i}\boldsymbol{W}_i \boldsymbol{c}_i\ . \end{aligned} \tag{6}$$

The force acting on the object is given as

$$\begin{aligned}{}^B\tilde{\boldsymbol{f}} &= {}^B\tilde{\boldsymbol{f}}_1 + \cdots + {}^B\tilde{\boldsymbol{f}}_n \\ &= \left[\boldsymbol{\Lambda}_1 {}^{C_1}\boldsymbol{W}_1, \cdots, \boldsymbol{\Lambda}_n {}^{C_n}\boldsymbol{W}_n\right]\left[\boldsymbol{c}_1^T, \cdots, \boldsymbol{c}_n^T\right]^T \\ &= {}^B\boldsymbol{W}\boldsymbol{c}\ . \end{aligned} \tag{7}$$

where the matrix $W \in \mathrm{IR}^{6\times n}$ contains the n contact wrench directions in its columns. In general case, there isn't unique solution of $\boldsymbol{c}$ for a given ${}^B\tilde{\boldsymbol{f}}$. Any solution for $\boldsymbol{c}$ can be split into two vectors, the particular solution $\boldsymbol{c}_p$ and the homogeneous solution $\boldsymbol{c}_h$ such that

$$\boldsymbol{c} = \boldsymbol{c}_h + \boldsymbol{c}_p \tag{8}$$

where $\boldsymbol{c}_p$ is orthogonal to $\boldsymbol{c}_h$. $\boldsymbol{c}_h \in null(\boldsymbol{W})$, $\boldsymbol{c}_h$ lies in the null space of $\boldsymbol{W}$ and corresponds to the internal grasping forces. The internal grasping force vector $\boldsymbol{c}_h$ is lost during transformation of the matrix $\boldsymbol{W}$. So it needs to balance the forces, which come from operator and object, in the contact force space.

Tab. 1 Contact Models

Contact Model With Friction	
Point Contact	Soft Finger
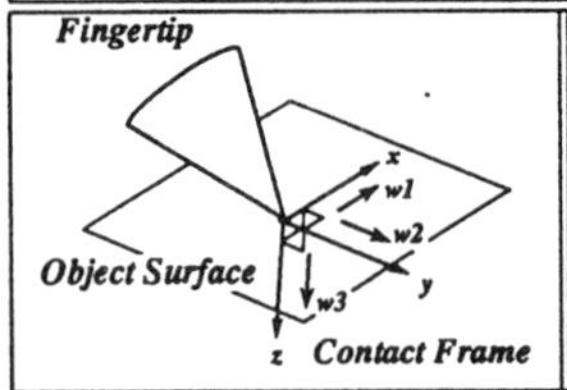	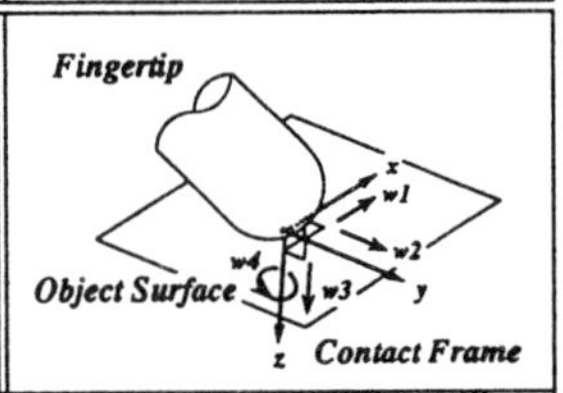
$w_1 = (1\,0\,0\,0\,0\,0)^T$ $w_2 = (0\,1\,0\,0\,0\,0)^T$ $w_3 = (0\,0\,1\,0\,0\,0)^T$	$w_1 = (1\,0\,0\,0\,0\,0)^T$ $w_2 = (0\,1\,0\,0\,0\,0)^T$ $w_3 = (0\,0\,1\,0\,0\,0)^T$ $w_4 = (0\,0\,0\,0\,0\,1)^T$

2.1.4 Friction Model

A specific coulomb friction model is considered which limits the tangential components c_i before transformation in the object frame. According to this model there are two types of friction: one is the statical friction, the other is the dynamic friction. In case of Statical Friction the tangential forces c_{ix} and c_{iy} are limited by the friction cone as

$$\sqrt{c_{ix}^2 + c_{iy}^2} < \mu_s c_{iz}\ , \tag{9}$$

where μ_s is the statical friction coefficient. If eq.(9) is not valid then the particular finger start to move.

On the other hand, if fingertip is moving on the object surface, the dynamic friction force will act on the object. This dynamic friction force $\boldsymbol{f}_{df}$ is given as

$$\boldsymbol{f}_{df} = \mu_d ||{}^C\tilde{\boldsymbol{f}}_{i_n}|| \frac{\boldsymbol{v}_i}{||\boldsymbol{v}_i||}\ , \tag{10}$$

where μ_d is the dynamic friction coefficient and ${}^C\tilde{\boldsymbol{f}}_{i_n}$ is the normal force. This force acts opposide to the direction that of the fingertip velocity vector $\boldsymbol{v}_i$.

Now we need to switch from the dynamic friction to the statical friction models according to the finger-tip velocity vector $\boldsymbol{v}_i$ as follows

$$\begin{cases} ||\boldsymbol{v}_i|| \le \alpha & : \text{Statical Friction} \\ ||\boldsymbol{v}_i|| > \alpha & : \text{Dynamic Friction}\ , \end{cases} \tag{11}$$

where α is a scalar value. Though α is normally zero, it is preferable to employ a small value for α to avoid an unstable switching situation.

2.1.5 Force exerted by Object

In the most case the force exerted by the object is reacting force. Let us assume a force ${}^B\tilde{\boldsymbol{f}}$ acting on the object in the virtual world is exerted by the human hand in the sensor glove. The object is accelerated by ${}^B\tilde{\boldsymbol{f}}$ and gravitation. Describing the object dynamics by the Newton-Euler equation as

$${}^B\boldsymbol{f}_{ext} = \begin{bmatrix} \hat{\boldsymbol{m}} & 0 \\ 0 & \boldsymbol{H} \end{bmatrix} \begin{bmatrix} \dot{\boldsymbol{v}} \\ \dot{\boldsymbol{\omega}} \end{bmatrix} - {}^B\tilde{\boldsymbol{f}} \tag{12}$$

where

$${}^B\boldsymbol{f}_{ext} = \begin{bmatrix} \hat{\boldsymbol{m}} & 0 \\ 0 & \boldsymbol{H} \end{bmatrix} \begin{bmatrix} {}^B\boldsymbol{g} \\ 0 \end{bmatrix} - \begin{bmatrix} \boldsymbol{\omega} \times \hat{\boldsymbol{m}}\boldsymbol{v} \\ \boldsymbol{\omega} \times \boldsymbol{H}\boldsymbol{\omega} \end{bmatrix}\ , \tag{13}$$

where $\dot{\boldsymbol{v}}, \dot{\boldsymbol{\omega}}$ are the body accelerations, $\hat{\boldsymbol{m}} = \mathrm{diag(m, m, m)} \in \mathcal{R}^{3\times 3}$ is a diagonal matrix with the object mass in the diagonal elements, $\boldsymbol{H} \in \mathcal{R}^{3\times 3}$ is the inertia matrix, ${}^B\boldsymbol{g} \in \mathcal{R}^3$ is the vector of gravitation. The force exerted by the object noted as ${}^B\tilde{\boldsymbol{f}}_{ext}$.

2.1.6 Model of Virtual World with Sensor Glove

Now we can redraw Fig.1 as Fig.4 including above mathematical tools. This block diagram consists of 4 main parts: the sensor glove, the object dynamics, the force flow and the calculation of Transfer Matrix. The joint torques $\boldsymbol{\tau}_{ij}$ and angles $\boldsymbol{\theta}_{ij}$ are measured by the Sensor Glove. The force ${}^C\boldsymbol{f}_i$ on finger-tip is

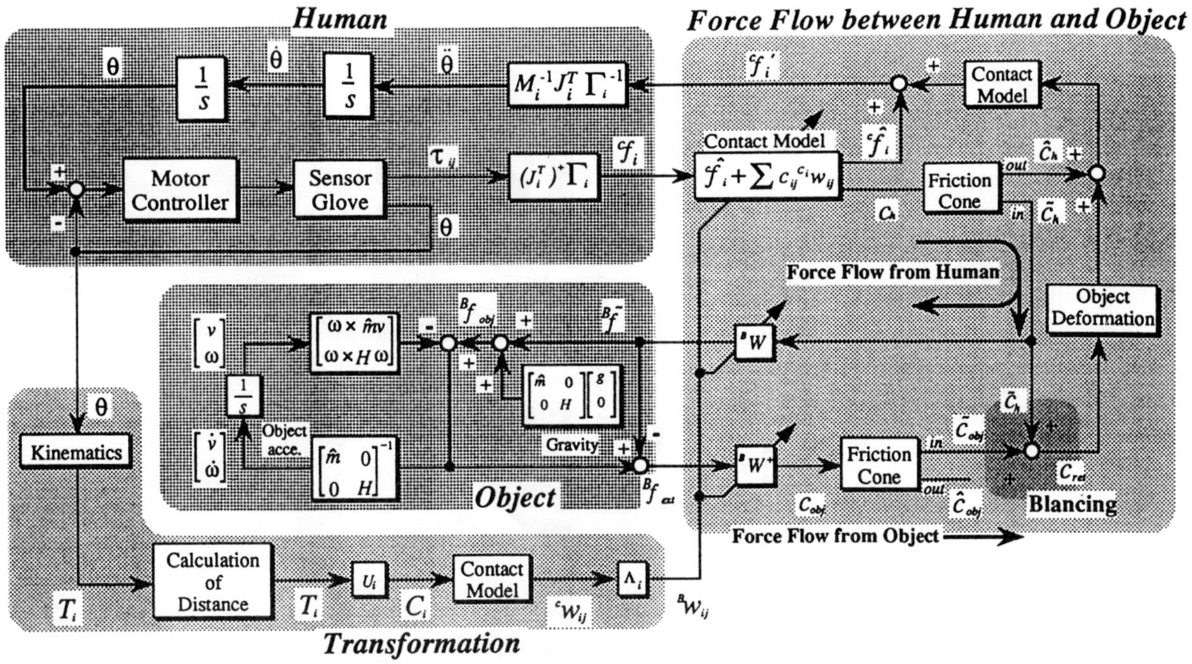

Fig.4 Detailed Force flow of Dynamic Force Simulator

obtained from the manipulator jacobian matrix $\boldsymbol{J}_i$ after the coordinate transformation $\boldsymbol{\Gamma}_i$. The force ${}^C\boldsymbol{f}_i$ will pass through the contact model and friction model, then ${}^C\boldsymbol{f}_i$ is divided into ${}^C\hat{\boldsymbol{f}}_i$, $\hat{\boldsymbol{c}}_h$ and $\tilde{\boldsymbol{c}}_h$. ${}^C\hat{\boldsymbol{f}}_i$ and $\hat{\boldsymbol{c}}_h$ yield acceleration of the finger-tip and $\tilde{\boldsymbol{c}}_h$ acts on the object. The force $\tilde{\boldsymbol{c}}_h$ is transformed into Object coordinate frame by ${}^B\boldsymbol{W}$ matrix and $\tilde{\boldsymbol{c}}_h$ after transformation called as ${}^B\tilde{\boldsymbol{f}}$. The force ${}^B\tilde{\boldsymbol{f}}$ exerted by the human is acting on the object, which is accelerated by sum of this force, gravitation and inertial force. The force ${}^B\boldsymbol{f}_{ext}$ exerted by an object is given by an acceleration of an object subtract by ${}^B\tilde{\boldsymbol{f}}$. This force ${}^B\boldsymbol{f}_{ext}$ is balanced with human's force after consideration the friction model. The balanced force is feedbacked to the operator. Using this diagram we can make a virtual environment considering contact conditions, friction conditions and deformation of an object.

2.2 Experimental System Structure

The experimental system structure of the DFS is shown in Fig. 5 and Fig.6 (a). The maximum frequency with which a typical human hand can transmit motion commands to the hand master is 5-10[Hz] while it demands that the position and force feedback signals be presented to it at a frequency not less than 20-30[Hz] [9][10]. We should construct system as fast as we can.

DFS system is composed of three kinds of computers to get more computational power. The low-level motion control of the sensor glove system is realized by using T-800 transputer system we have developed. This transputer system communicates to both the Intel i860 and the Hewlett Packard Apollo 735 workstation. The i860 receives data of sensors from the sensor glove system (Fig.6 (c)). The sensor glove joint angles are calibrated by T-800 which is the controller of i860. The DFS is realized on the Intel i860 which generates the motor commands for the Sensor Glove systems and these commands are sent to T-800 transputer system. The Hewlett Packard Apollo 735–48Z workstation, which has a graphic accelerator, is used for displaying the state of the hand and an object in the virtual world with real-time graphics animation(Fig.6 (b)). The overall network of these computing resources enables parallel processing of DFS calculations, motion control, and sensor glove calibration in real-time.

2.3 Experimental Results

We show an experimental result. In this example a cylindrical object is grasped by human hand in

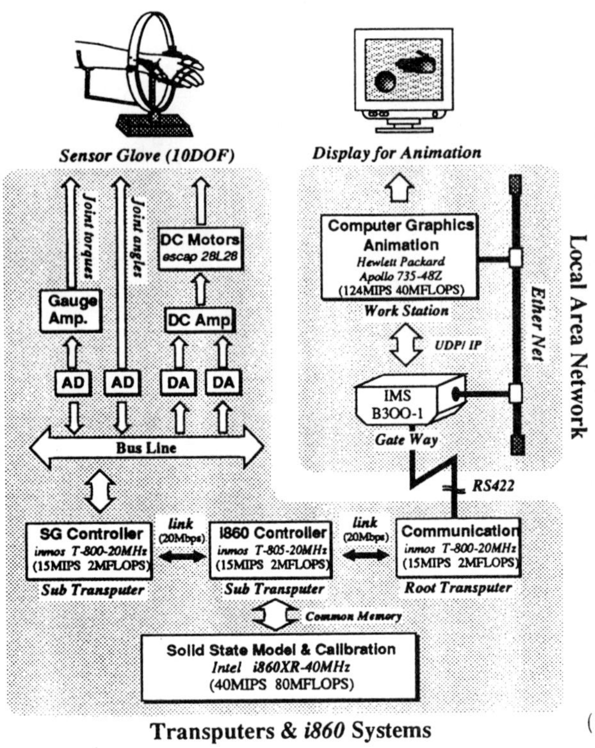

Fig.5 DFS Experimental System Structure

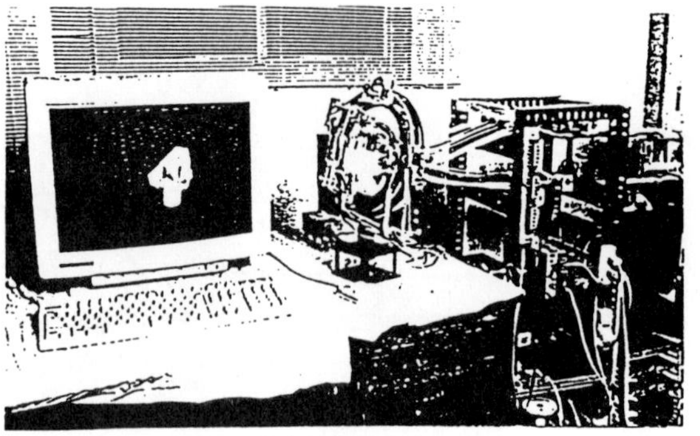

(a) DFS System

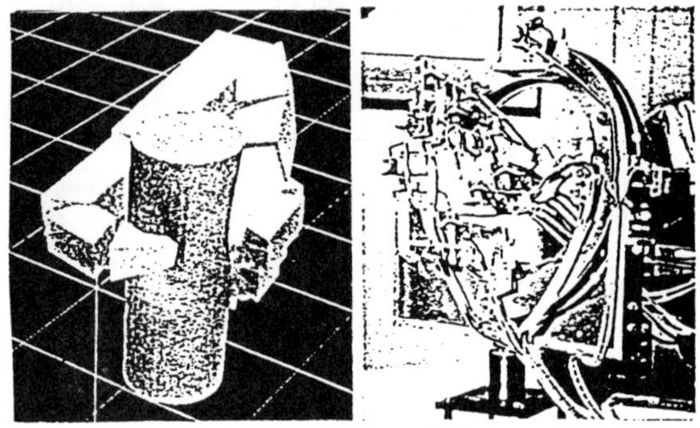

(b) Virtual World in DFS (c) Sensor Glove

Fig.6 View of DFS System

the virtual world using the sensor glove as shown in Fig.6. We assumed a point contact model with friction, where the mass of the object is $1[Kg]$ and the statical friction coefficient μ_s between the object and the finger is ∞. During grasping, the operator sense force feedbacked from the grasped virtual object via the sensor glove system(Fig.6 (b)). Human operator is grasping an object from (a) to (f) in Fig.7. And then he release this object in (g). From (h) to (l), he try to grasp this object again and move it to the right side of his hand in Virtual World. During this operation, we can measure joint torques and angles. Figure 8 shows the measured data of index finger joint angles and torques during object grasping. In the first experience, the finger was moving in the free space to grasp the virtual object. Subsequently, some fingers made contact. For example, the index finger contacted from 1.4[sec] to 4.2[sec] (Fig.8). Finally the fingers released the virtual object and the operator moved in free space.

During grasping, theoretically the finger-tips do not move in the object coordinate frame during above experiment. The forces applied by the operator should be balanced by the torques calculated by inverse kinematics from the reactive forces at the contact points. However, the finger-tip penetrated into the inside of the virtual object due to calculation time delays and effect of wire flexibility.

2.4 Conclusion

In this paper we proposed a Dynamic Force Simulator (DFS) for force feedback to the human operator. Exact simulation of object dynamics, contact model and friction characteristics enables realistic feedback to the human operator. Two influencing flows from the human operator through the sensor glove and from the object to the human were shown in detail. The block diagram of the DFS was realized on the computing resources and an experimental result of a cylindrical object grasping was shown.

In the present stage of the research we use the existing DFS and sensor glove to analyze the human motion. Applying the result of this analysis, we intend to develop a new sensor glove, which is more complex , and to discuss about collision phenomena and its stability further more we will aim at more

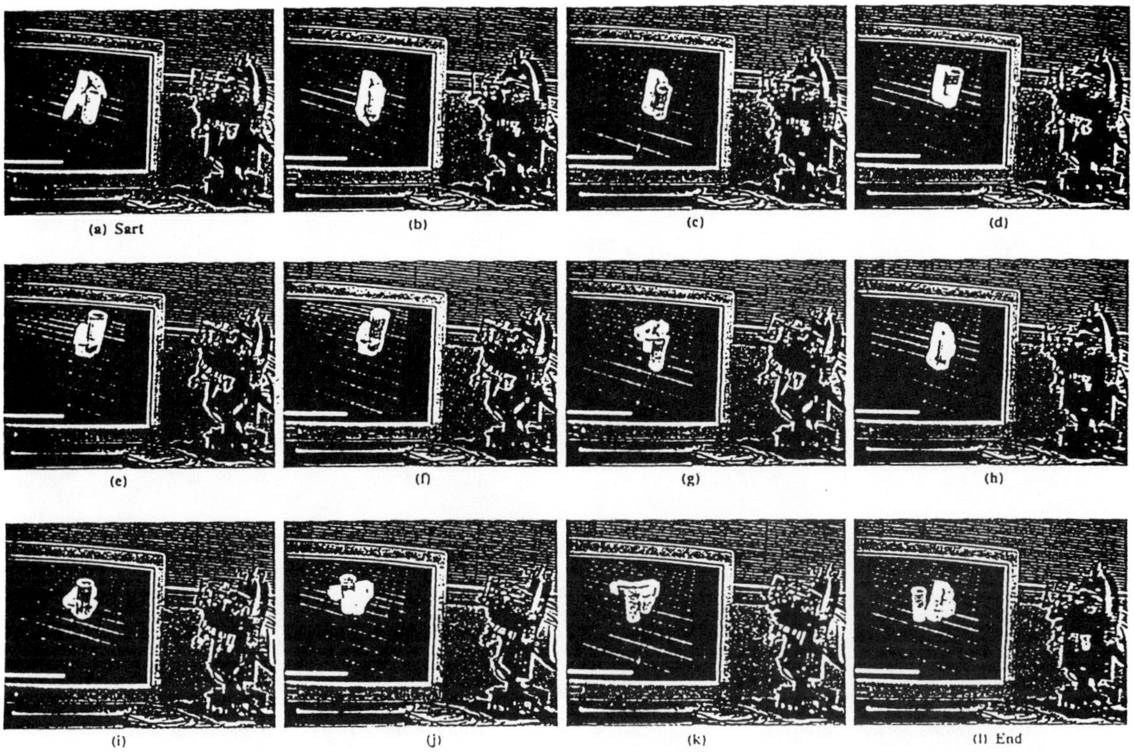

Fig.7 Phases of Object Grasping by Sensor Glove in Virtual World of DFS

dextrous manipulation and getting more detailed data about fingers and object state parameters in the future. In the next step, we will implement this our system to operate a robot manipulator.

3 Development of Haptic Interface: Handshake Device

We have been developing "HandShake Device (HSD)" which is a part of Network Neuro Baby (NNB) project [11][12][13]. NNB is a research project in Artificial Life which simulates a living baby acting as a network agent in the computer network(Fig.9). NNB project was exhibited in SIGGRAPH'95 held in Los Angeles, USA.

HSD is an interface device which an operator can communicate with NNB physically. NNB returns some reactions , for example getting angry. These reactions dipend on a magnitude of a grasping force. The HSDs with NNB are placed in a local site and a remote site so that people can physically communicate each other with NNB through HSDs. The structure of HSD is shown in Fig.10. HSD can make force sensation to an operator and measure forces by force sensors. HSD system has a local mode and a network mode. In the local mode, an operator grasps his or her palm by himself or herself. In the network mode, HSD is assumed as the other's right hand and then the operator can feel existence of human through the force sensation generated by other people. Those HSDs can send and receive force sensation through information network. The virtual dynamics method is applied to realize a teleoperation(tele-handshake)[16].

3.1 HandShake Device (HSD)

Fig.12 shows structure of HSD. The HSD is composed of 2 linear motion motors(AM20: WITTENSTEIN's product), 2 position sensors, and 2 force sensors which are implemented to measure force

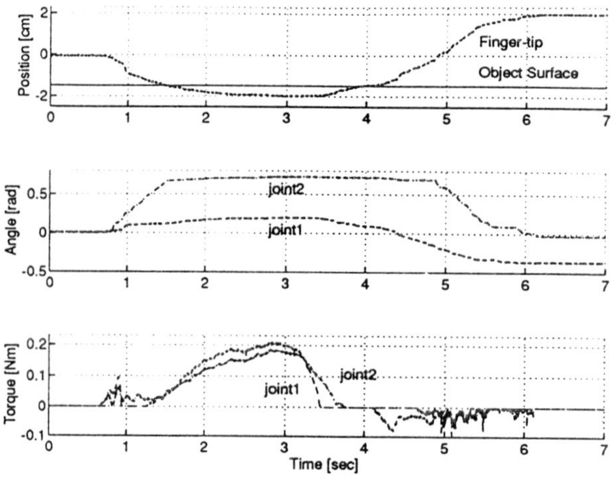

Fig.8: Grasping a cylinder in Virtual World (The angle and torques of Index finger are measured, The position is calculated)

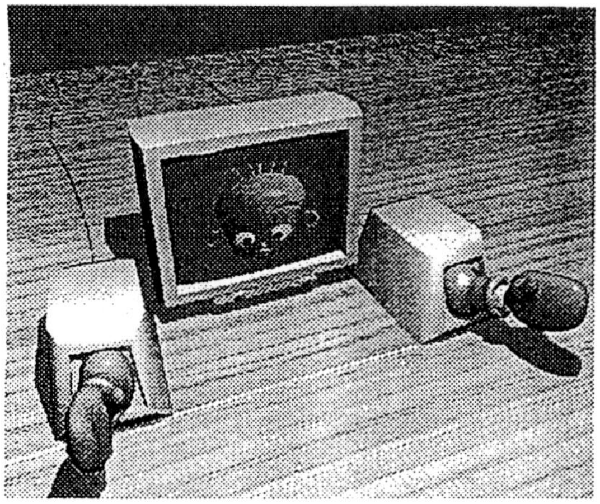

Fig.9 Neuro Baby with Handshake Device

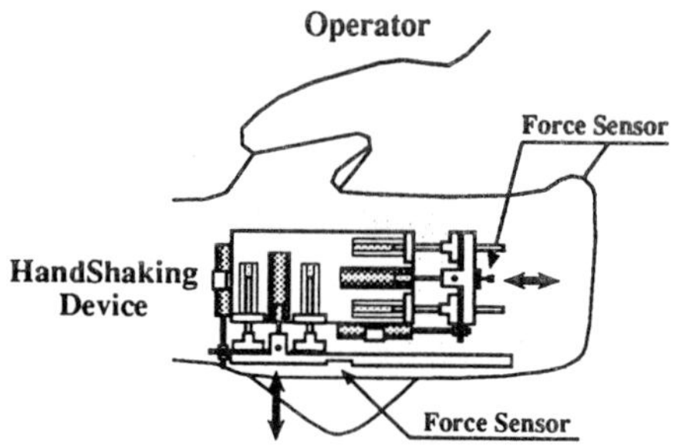

Fig.10 Handshake through HSD.

from an operator. Those information are connected to a host computer (PC) via an AD board. The linear motion motor called "AM20" is shown in Fig.11 [14]. AM20 is a three phase brushless DC servomotor with hall elements for rotor position sensing. It is composed of a gear part and a motor part. In the motor part, rolling elements (a rotor) is rotated as a synchronous motor. The power is transmitted between the screw shaft and the planetary roller by rolling elements.

HSD is composed of 2 systems: one is a master system, which is grasped by an operator, the other is a slave system, which grasps an operator's palm(Fig.10, Fig.12). The master and slave system are applied by a grasping force, $\boldsymbol{f}_m$, and a reaction force, $\boldsymbol{f}_s$. Each system has a force sensor which contain 4 strain gauges(Fig.13(a),(b)) to measure $\boldsymbol{f}_m$ and $\boldsymbol{f}_s$. These 4 strain gauges form the wheatstone bridge on each force sensor(Fig.13(c)). The force is calculated from measured v in Fig.13(c).

The experimental system structure of the HSD is shown in Fig.14. HSD system includes two kinds of computers. The motion control of HSD is realized by using GateWay2000 P5-100J(GW). This GW computer communicates to the server workstation which is a SUN Sparc Station 10(SS10). This SS10 connects HSD system with the other HSD system on a remote site with B-ISDN as an information network.

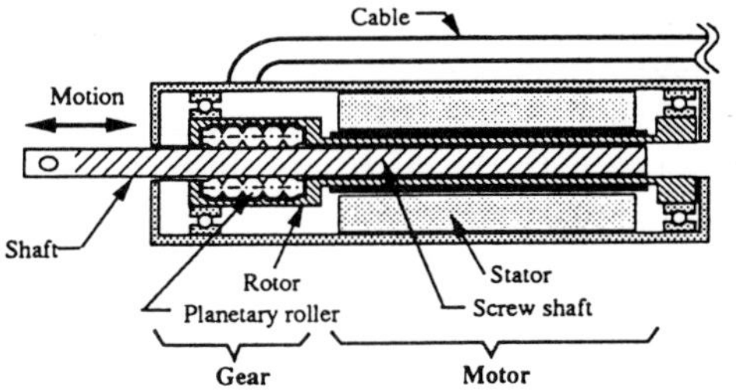

Fig.11 Linear Motion Motor : AM20

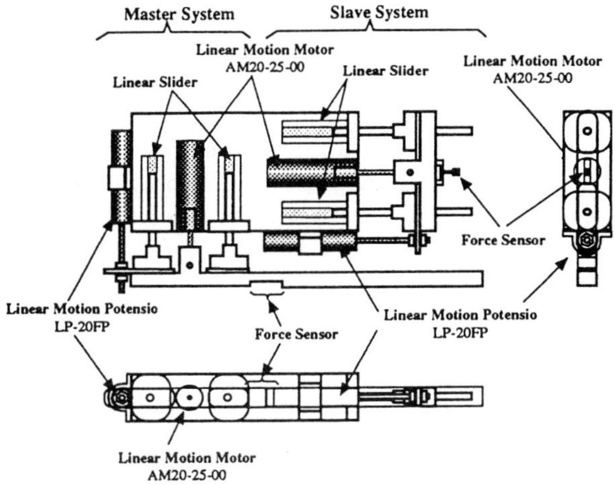

Fig.12 Schematic of HandShake Device

3.2 Virtual Dynamics Method

The human operator must feel the force information from the remote site through the system in order to perform a remote task (tele-handshake) as if he or she were in the remote site. Traditional bilateral master-slave systems, which are a position and force control, position and position control, and etc, have mechanically large impedance because the master must actuate a human operator and the slave must produce strong force like human beings to carry out some tasks in a remote site instead of the operator. Measured force or motion in master and slave system involves this large impedance of the system [15]. However this impedance reduces reality about force feeling and makes the remote task inefficient. The authors included the virtual model in order to change the dynamics of the master-slave system [16].

Dynamics of a 1DOF master and slave system is given as

$$M_m \ddot{x}_m(t) + D_m \dot{x}_m(t) = f_m(t) + \tau_m(t) \quad (14)$$

$$M_s \ddot{x}_s(t) + D_s \dot{x}_s(t) = f_s(t) + \tau_s(t) \quad (15)$$

where $x_m(t)$ and $x_s(t)$ are master and slave positions, M_m and D_m are an inertia parameter and a

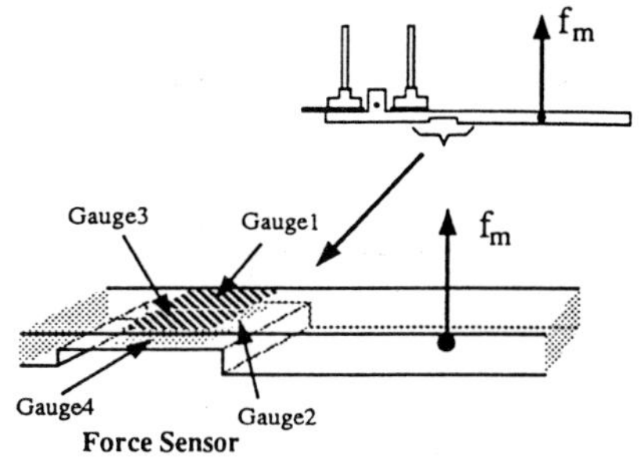

(a) Master

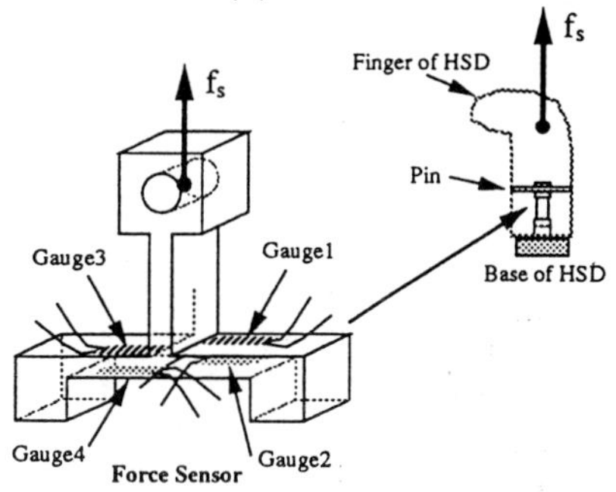

(b) Slave

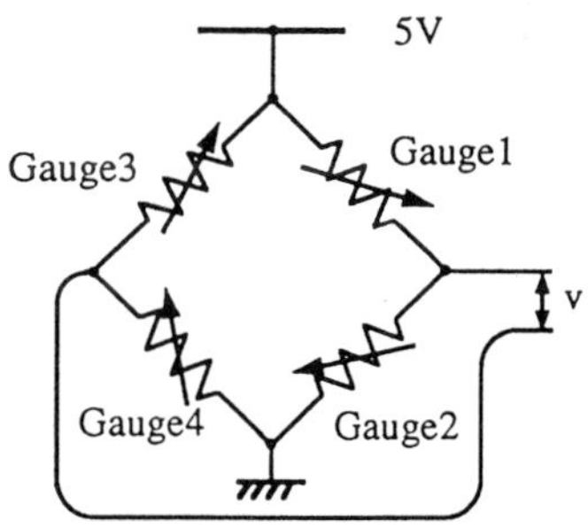

(c) Sensor Bridge Circuit

Fig.13 Schematic of Force Sensor

damping parameter, respectively. M_s, D_s are those parameters of an master and a slave. $f_m(t)$ and $f_s(t)$ are external forces, and $\tau_m(t)$ and $\tau_m(t)$ are motor torques.

A virtual model is shown in Fig.15. This model is created in a computer and this dynamics is represented as

$$M_v \ddot{x}_v + D_v \dot{x}_v + K_v x_v = f_m + f_s \quad (16)$$

where M_v, D_m and K_v correspond to the inertia, damping and stiffness parameter, respectively, and these parameters can be chosen, arbitrary, as they reform the system and keep stable. x_v is a position of a virtual model.

The virtual model connects the master and slave: it generates the reference position x_v to these sys-

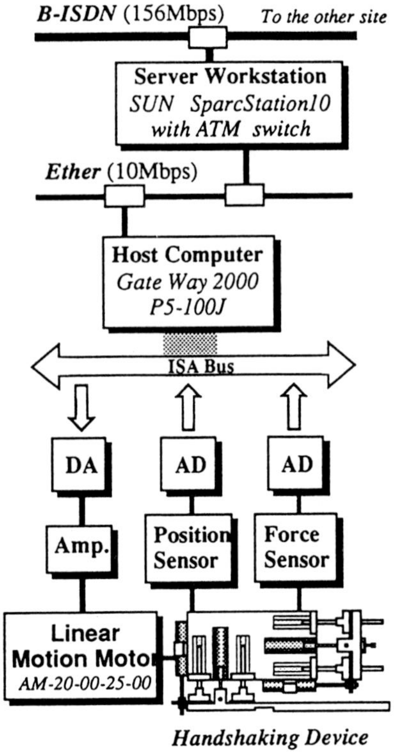

Fig.14 System Structure of HSD

tems. This reference position is calculated from the input force f_m and f_s.

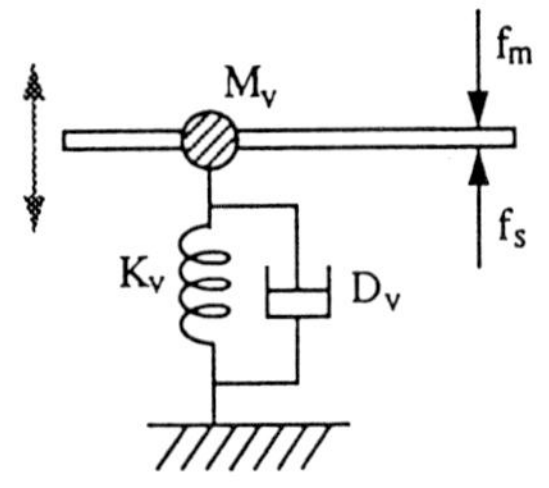

Fig.15 Virtual Dynamics Method

3.3 Tele-handshake

HSD system has two mode: one is a local mode, the other is a network mode. In the local mode, we use only one HSD system. When an operator grasps a master he or she can feel his or her own grasping-force from a slave and a reaction force from a slave is generated through a master. In this case, one virtual model is applied to make a reference signal for AM20s. The master and slave position are followed this reference.

In the network mode, 2 HDSs are used for tele-handshake. HSD is placed its own slave system at a remote site, that is, HSD system exchanges each slave system with the other site. In this mode, HSD has two virtual models on each master and slave systems(Fig.16). NNB requires realtime handshake with the operator at the other site. HSD uses only force information as data from the remote site (without position information), in order to maintain a stability for a long time-delay to communicate to the long distance. In the local site, a virtual model of a local master system is calculated with force inputs f_{m1} and f_{s2}: f_{m1} is measured on itself and f_{s2} is to the remote slave system. On the other hand, inputs of a virtual model of a local slave system are force f_{m2} and f_{s1}: f_{m2} is applied on a remote master system by the other operator and f_{s1} is generated by the operator.

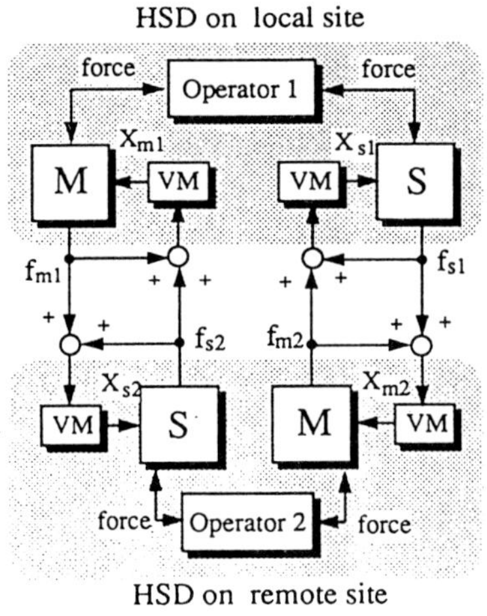

Fig.16 Schematic of Tele-handshake

3.4 Experiments

The experiments will be shown about data of those two modes of HSD system with NNB and without NNB shown Fig.17 (a) and (b). There were many operators to shake hands each other ,and those operators answered questionnaires. Many operators felt a good impression, for example, "It should be needed to feedback forces for human interface." and "It can be easy to feel the existence of Neuro-baby or a remote-site-operator through NNB system.". Its result will be shown in our presentation.

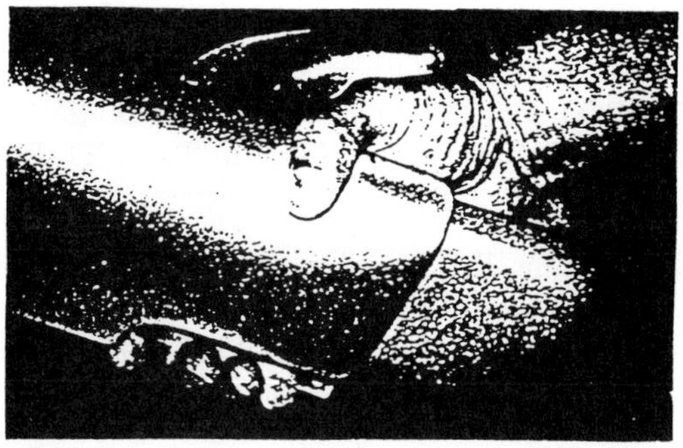

(a)

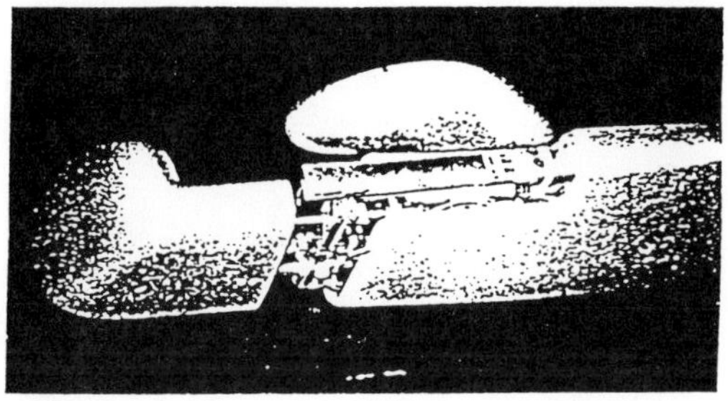

(b)

Fig.17 View of HSD

3.5 Conclusion

In this paper we proposed 1DOF tele-operation system called HandShake Device (HSD) which is a interface to a computer system with NNB. HSD system has two mode: one is the local mode which uses only one system and the other is the network mode which is connected with the other HSD located at a remote site. HSD applies the virtual model method as algorithm of tele-operation (tele-handshake). In order to keep stability of the realtime tele-handshake with a long distance, only force information (without position information) is applied to both virtual models of the master and slave. In SIGGRAPH'95 many people tried to do handshake with NNB system and another operator on the other site. Many operator felt a good impression by using HSD as human-interface of a computer or communication method of human in the other site. Detailed results of questionnaires will be shown in our presentation.

References

[1] Bejczy, A.K., Teleoperation: The Language of the Human Hand, *IEEE International Workshop on Robot and Human Communication ROMAN'92*,pp.32-43, 1992.

[2] Buss, M., Hashimoto, H., Information and Power Flow During Skill Acquisition for the Intelligent Assisting System—IAS, *Proceedings of the International Workshop on Intelligent Robots and Systems IROS'93*, pp.321-326..

[3] Iwata, H., Artificial Reality with Force-feedback: Development of Desktop Virtual Space with Compact Master Manipulator. *Computer Graphics*, Vol.24, No.4, pp.59-64, 1990.

[4] Kazerooni, H., Human-Robot Interaction via the Transfer of Power and Information Signals, *IEEE Transaction on Systems, Man and Cybernetics*, Vol.20, No.2, pp.450-463, March/April 1990.

[5] Kerr, J. and Roth, B., 1986. Analysis of Multifingered Hands. *International Journal of Robotics Research*, Vol.4, No.4, pp.3-17, Winter 1986.

[6] Mason, M.T., Salisbury, J.K., 1985. Robot Hands and the Mechanics of Manipulation. *MIT Press*, Cambridge, Massachusetts, 1985.

[7] Paul, R.P., 1982. Robot Manipulators: Mathematics, Programming and Control. *The MIT Press.*

[8] Salisbury, J.K. and Craig, J.J., 1982. Articulated Hands: Force Control and Kinematic Issues. *International Journal of Robotics Research*, Vol.1, No.1, pp.4-17, Spring 1982.

[9] Shimoga, K.B,A Survey of Perceptual Feedback Issue in Dexterous Telemanipulation:Part I. Finger Force Feedback, *Proceedings of Virtual Reality Annual International Symposium VRAIS*, pp.263-270 ,1993.

[10] Shimoga, K.B.,A Survey of Perceptual Feedback Issue in Dexterous Telemanipulation:Part II. Finger Touch Feedback, *Proceedings of Virtual Reality Annual International Symposium VRAIS*, pp.271-279 ,1993.

[11] N.Tosa, H.Hashimoto, K.Sezaki, Y.Kunii, al, "Network Neuro-Baby with robotics hand", Symbiosis of Human and Artifact, Elsevier Science B.V., pp.77-82, 1995.

[12] T.Yamada, H.Hashimoto, "Pattern recognition of emotion with Neural Network", Proceedings of International Conference on Industrial Electronics, Control, and Instrumentation, pp.–, 1995.

[13] K.Sabe, H.Hashimoto, "Real-Time Tracking Using Active Eye Sensing System", Proceedings of International Conference on Industrial Electronics, Control, and Instrumentation, pp.–, 1995.

[14] WITTENSTEIN Motion Control GmbH, "Roller screws", pp.5–7.

[15] Sheridan, Thomas B., "Supervisory Control of Anthropomorphic Teleoperators for Space, Undersea and Other Applications", TELEROBOTICS, AUTOMATION AND HUMAN SUPERVISORY CONTROL, The MIT Press, pp.99–238, 1992.

[16] S.Nojima, H.Hashimoto, "Master-Slave System With Force Feedback Based on Dynamics of Virtual Model", Proc. of ISMCR'94: Topical Workshop on Virtual Reality, pp.93–100, 1994.

Performance Measures for Haptic Interfaces

Vincent Hayward and Oliver R. Astley

McGill University
Center for Intelligent Machines
3480 University Street, Montréal, Qc, H3A 2A7, Canada

Abstract

A haptic interface is distinct from other display devices because it is bi-directional; it is capable of both reading and writing input to and from a human user. Due to both the direct human interaction and bi-directionality there has been much ambiguity in describing and evaluating these devices, making evaluation and comparison difficult. The goal of this paper is to set out requirements and guidelines for the performance measures of haptic devices and to hopefully lead towards resolving the current equivocal situation. In particular, performance measures are introduced which have so far not been pertinent in traditional robotics; these include, peak force, peak acceleration and frequency dependent measurements. Performance measures often quoted in traditional robotics are also discussed, however, the focus and relevance of these measures are different in haptic devices. Each of the suggested performance measures in this paper is discussed with respect to its importance, its measurabilty and the condition under which it should be measured.

1 Introduction

Many kinds of haptic devices and force-reflecting hand controllers have been created over the past decades. The earliest ones, devised for telemanipulation of hazardous materials, are mechanically connected to a slave manipulator. Newer ones, instrumented and powered, are connected to a slave manipulator via analog or digital control. Virtual environments, made possible by the recent availability of inexpensive computing power, replace a slave system and the task by a computational model. From this perspective, the development of hand controllers spans almost half a century.[1]

This paper is written in an attempt to progress toward collecting a set of device-specific quantitative measures. It is hoped that this work might lead to a practical means to measure progress in the art of designing and building haptic devices (defined in next section). The availability of such widely accepted set of measures would have numerous important advantages in the field of teleoperation and virtual environments.

Most industries (computer, aerospace, vehicles, electroacoustic, etc.) have already developed such commonly accepted sets of performance measures. The benefits that follow from these measures are numerous, for example:

- Device performance (and price) can be matched in an informed fashion to the tasks they are meant to address.
- Devices can be specified before they are built.
- The improvements applied to a particular device or technology can be tracked in a systematic fashion.
- Devices with different designs can be compared.

[1]Flateau, Greeb and Booker in an article dated from 1973 state in the introduction that [7]: "Teleoperator Technology (TOT) as an independent discipline was originated over 25 years ago. While this is quite ancient by today's standards of technology lifetimes, TOT still shows a considerable amount of infantilism by the degree of empiricism involved in all of its aspects not directly borrowed from the many neighbouring disciplines involved.

The advantage of this infantilicism is that activity in the field has not yet been segregated into theoretical and applied specialities whose intense activities contribute to widening the communication gap between those two groups."

These statements ring strangely up-to-date and it can be safely predicted that they will remain so for the foreseeable future.

- The importance of certain particular factors can be ascertained with respect to application areas.
- Progress in the field can be monitored.

No set of quantitative measures will replace the litmus test of actual practice (not any more than for any other technology); nevertheless, these specifications must be attempted.

The identification of single performance numbers is the key to concise reporting of characteristics. Whenever possible, one single characteristic will be associated to a single number in an attempt to clarify the associated significance. This will result in a sizeable amount of such numbers. In this spirit, we will make no attempt at specifying an ideal device given the strong dependency on applications and operating conditions.

In the paper we will make use of the concept of "ground device". This refers to the intrinsic physical characteristics of the device unchanged by control. This will not only be useful to evaluate the device before control, but also to evaluate the performance improvement and the ability of the device to modify its parameters once controlled.

This will be useful to specify the performance of the control. Conversely, it will be useful to specific the ability of the device to modify its apparent properties under control.

2 The Haptic Channel vs. Other Channels

Haptics refers to that part of physiology pertaining to the sense of touch; it has become the standard term to qualify studies and technologies concerned with this sense. Haptic feedback relates to two cognitive senses: the tactile sense that gives an awareness to stimuli on the surface of the body, and the kinesthetic sense that provides information on body position and movement. Bidirectionality is the most prominent characteristic of the haptic channel. Haptic perception always involves exchange of (mechanical) energy—and therefore of information—between the body and the world outside.

The word 'display' emphasises the unidirectional nature of information transfer, as such, it is probably not the best word to be qualified by 'haptic', as opposed to 'graphic' or 'audio'. In addition to vision and audition, vestibular perception informing humans of mechanical signals related to body motion is also uni-directional in this sense. In the case of, vision, audition and the vestibular sense, body motion is of course intimately linked to perception, which is why all these three channels can elicit vection, or sensation of motion. However, none of these channels is linked to a somatic means of modifying one's immediate physical neighbourhood. In other terms, with vision, audition or vestibular perception, there is no significant exchange of energy between the body and the world outside.

This distinction is most apparent if one considers that visual, auditory or vestibular signals (olfactory too) can be recorded and replayed (people watch movies, listen to audio recordings, or have rides in vehicle simulators). On the other hand, recording and replaying kinesthetic and tactile sensations does not make sense, except possibly for the display of textures and certain material properties.[2]

A large proportion of modifications made to one's environment involves the use of the hand. This is particularly true of the interaction with machines and the majority of haptic devices so far available.

To summarise, bi-directionality is the single most distinguishing feature of haptic devices when compared to other machine interfaces. A haptic device must be designed to "read and write" to and from the human hand (or foot, or other). As it turns out, the "read" part is relatively easy to achieve and a great many types of devices already exist (knobs, keys, joysticks, pointing devices, etc.) although many issues are still unresolved. The "write part" is comparatively much more difficult to achieve. More specifically, the function of the haptic display is to recreate constitutive properties (relationships between variables of flow and effort).[3]

These observations are at the root of a set of measures proposed here. They bear much analogy with specifications for graphic screens or electroacoustic equipment but have two major differences: they must deal with multiple dimensions and refer to both efferent and afferent channels.

In order to clarify the scope of the discussion, we will refer to any device, thereafter called a haptic device, having these two characteristics: (1) apply mechanical signals at distinct areas of the body (2) measure mechanical signals at the

[2] In effect, the perception of thermal conductivity cannot be replayed: it involves the transfer of heat from the hand to the item being touched, to measure head diffusion.

[3] That holds for the thermal conductivity example too.

same distinct areas of the body. Whether these signals refer to forces, displacements, or a combination of these and their time derivatives, is not specified in the definition. This definition also excludes tactile displays in the form of fine pitched active arrays of actuators.

The proposed performance measures are directed toward isotonic devices. This term refers to devices which in their nominal mode of operation are designed to display forces are read position variables. Thus, an ideal device nominally has neither mass nor dissipation, and is a perfect force and position transducer in the respective directions. Many of the measures about to be dicussed will suggest a 'distance' of an actual device to the ideal device. Conversely, isometric devices are designed to display displacement variables and read back forces. Thus, an ideal device nominally has infinite stiffness and is a perfect position and force transducer in the respective directions.

There has been comparatively few isometric devices developed and they would require separate performance measures to do them justice. For these reasons isometric devices are not considered, but clearly, a corresponding set of measures could easily be mirrored from those discussed for isotonic devices.

3 A Plethora of Performance Measures

Walking into the local haptic devices store, one could be forgiven for being overwhelmed by the diversity of performance specifications. What is more, performance indices which appear the same, may not be comparable with each other due to incompatible testing conditions. In this section we comment on the performance specifications of approximately twenty haptic devices, both in academia and industry.

For any robotic mechanical system, such as a hand controller, there are several essential criteria for describing the system, e.g. inertia, friction, weight and backlash. However, the duality of the hand controller to drive and to be driven causes discrepancies as to from where these measures should be taken. For example, is inertia measured as seen from the actuators, or from the output device itself? Out of the devices surveyed, only the Phantom specifications defined that inertia was taken as "Inertia at the tip." [17].

Measurements such as precision, resolution, force output, and backlash are rarely specified as to where they were taken from. For regular robots these measures have traditionally been taken at the individual joints. However for a haptic device, in which a user directly interacts with its output, the choice is not so clear. For this reason when reading performance specifications such as these, there is much ambiguity and hence confusion when evaluating the pros and cons of one device over and other. Until such time that there are recognised standards, there is a need for more specificity in such measures.

The interaction of human and machine implies that the bandwidth of the device is of great importance. This critical factor is inexplicitly overlooked in the specifications of the various haptic devices. Out of the devices we studied only Salcudean's MAG-LEV [22], and the Harwell U.K Hand Controller [23] gave any reference to position and force frequency response. Another largely overlooked factor is maximum acceleration. In emulating stiff walls, for example, the maximum acceleration is critical yet only a couple of the haptic devices we studied gave this performance measure [6].

From the lack of detail, to the miscellaneous specifications which despite their validity defy comparison because they appear in just the one data-sheet. The only reference that could be found comparing several devices by means of well defined parameters is by McAffee and Fiorini [18]. Examples of these miscellaneous measures include: interaction force resolution, effort return gain, deflection, no load accelerations, and play. This highlights the need for a common set of performance characteristics which will bring out the best and worst in individual devices and make meaningful comparison possible.

4 Gross Features

4.1 General Rule

One common difficulty with single number performance measures is their dependency on the operating point. It is proposed that for all single number measures, specifications be provided in terms of best and worst cases over the entire operating motion range of the device. In effect, best case figures are not only of limited importance in the absence of further information, but at the same can be quite misleading. A further suggestion is to specify the maximum value of the rate of change (max gradient) of the measure under scrutiny over spatial variables of displacement and orientation, whatever the case may be.

4.2 Degrees of Freedom

The most prominent feature of a haptic device or hand controller is the number and the nature of the degrees of freedom (DOF) at the active end or ends.

4.3 Device-Body Interface

Unless the active end or ends are attached to the body via some bonding method, the hand or other concerned body parts must be connected to it. Unless the interaction is at all times unilateral, there are essentially two ways to achieve this. Therefore there are three cases of device-body interface: either the hand braces (holds) the device, or device braces the body, or else the interaction is unilateral. This gross distinguishing feature specifies which of the three cases the device in question is designed for. The complete specification will also include the concerned parts of the body: finger tips, parts of the fingers, forearms, heels, ball of the foot, etc.

It is also important to specify which mechanical freedom is active (A) and which is passive (P). For each degree of freedom we specify whether it is unilateral (U), bracing (B) or held (H).

4.4 Motion Range

The motion range specification (MR) poses several problems due to the possible lack of invariance and couplings. In addition, a detailed description may rapidly become very cumbersome.

In the case of low degrees of freedom devices discussed at the end of this section, it is possible and convenient to describe the operating volume inside which all other measures are taken as simple geometrical shapes, parallepideds, spheres, encompassing the reachable locations of the device-body interface(s).

For higher degree of freedom devices, involving combinations of translations and orientations, the specification is complicated due to the lack of metric for orientations with three degrees of freedom and the presence of couplings.

4.4.1 *Independent Specification of Orientation Range*

For one degree of orientation range, one angle suffices. For two, it is accurate to specify a solid angle, the section of which being described by a simple geometrical shape. For three, this scheme breaks down.

The proposal is as follows: presumably, the device-body interface will have a specific shape defining a preferred axis. Such is the case for most handles. Even spherical handles have such an axis being attached to a driving element by a stem. It is therefore natural to specify motion range with three orientations as a combination of a solid angle, angle inside which the preferred axis may reach, with an angle specifying the amount of rotation around the preferred axis. Once the nature of the solid angle is given, the complete orientation motion range can be given in 'steradian.degrees' with little ambiguity.

4.4.2 *Dependency of Orientation Range on Position*

The orientation motion range may depend on position in two ways.

The range can assume a different origin for each position. Such is the case for example, with wrist-partitioned manipulators, which have an orientation motion range amplitude independent from position, however the origin is not. For exoskeleton type devices, this may in fact be an advantage because this dependency can be made to match the human arm [26]. For the others, this dependency must be expressed in some fashion. Recalling what was discussed in the previous subsection, it is proposed here to specify the solid angle swept by a preferred axis through the position motion range when the relative orientation is kept constant.

For systems which are not partitioned, for example, parallel driven platforms, not only the origin of the orientation motion range may vary, but also its size and shape. A concise yet unsatisfactory report of this dependency is the specification of the maximum and minimum value of the orientation range in 'steradian.degrees'.

4.5 Peak Force

Much confusion is derived from published figures regarding peak force (PKF). Most prominently, the peak force should be specified where it matters, namely at the intended device body interface. This approach lifts the confusion due to specifications in terms of forces and torques which will depend on underlying coordinates. For example, a handle with six active degrees of freedom will be intended to be grasped in a number of ways but involving a limited numbers of areas of the hand. It is at these areas that the peak force should be specified. The resulting force being a combination of forces and torques generated at some other area of the handle.

The other source of confusion arises from duration. In many cases, the actuation system has losses resulting in heat (as exemplified in electric motors or amplifiers specifications) to be dissipated (actively of passively) eventually leading to a thermal equilibrium which must be below an acceptable limit. This defines the long term peak force (LTPKF).

As discussed further down the long term peak force may not matter much in some cases, so transient peak force must be specified independently. An excessive amplitude transient force may result in two effects. Either the device saturates causing large signal distortion (by design with current limiting circuits, mechanical torque limiters, or magnetic circuit saturation) or it sustains permanent damage. This limits the nature of meaningful transients. For example a true impulse will always be limited in amplitude by saturation. Instead, following the example of electroacoustic equipment testing specification, it is proposed to define a transient as one or several square signals of an agreed upon duration or durations. The proposal here is to define a short transient as a 10 ms square pulse (short enough to approximate a pure impulse for most devices), and a persistent transient as a square signal of one second duration.

This results in three specifications for peak force: long term (smoke test) (LTPKF), short transient (STPKF), and persistent transient (PTPKF). In each case, the measured peak output force is specified which will neither saturate nor damage the device at each intended area of contact with the body. For the long term smoke test, the time taken for the actuators to overheat should also be recorded. A concise specification, rather than painstakingly going through all the possible cases, might specify the worst cases only.

The measurement of peak force is relatively easy since it involves the use of a sufficiently accurate load cell and interposing it between the active end and a stiff reference, thus under isometric conditions.

4.6 Inertia and Damping

Inertia specifications are important [20]. It is an important issue because inertia is not easily reduced by feedback. In the event it is, for the reasons outlined in the next subsection, it should be specified specifically for the ground device (GDI). Here, the same difficulties in terms of invariance arise, and the same technique should apply. Instead of specifying one or several inertia tensors, it is convenient and accurate to specify it in terms of perceived mass at the various device-body interfaces. Similarly, the specification may be simplified by reporting only the best and worst case figures over the various areas of contacts and regions of the operating volume.

If the device is indeed designed to achieve inertia reduction by control, via force or acceleration feedback for example, other dependencies are introduced. One way to solve the problem is as follows. Presumably, the device will be instrumented with adequate sensors and the placement of these sensors will determine the structure of the feedback including the non-linear coordinate transformations needed to close the loop. To decouple the ability of a device (including its sensors and control system) from the merit of a particular control method, the improvement of this figure of merit should be specified using a feedback which does not depend on frequency (i.e. fixed gains, no filters) and indicating the improvement for a specified stability margin, the most convenient being a phase margin, say 45° in isometric conditions (end clamped) (FBI).

It is clear however that direct measurement is not easy. One measurement method would consist of connecting the active end to a shaker vibrating at a known amplitude and low frequency. The same load cell as above would report the effective inertia. To handle angular motions, two motions of equal amplitude and opposite phase could be applied to specified places of the active end. The distance in between could be agreed to be 5 cm for hand held devices, a reasonable figure for most grip positions.

Damping measurements are also important and should be measured under the same conditions.

4.7 Peak Acceleration

Experience has taught us that a crucial figure of merit of a haptic device is its acceleration capability (peak acceleration, PACC). This has been reported in [8] [9] where it was mentioned that it was consistent with physiological observations. In these papers, peak acceleration was used as design guideline not only to increase its the average value but also to minimise the difference between its extrema over a target workspace. The reasoning is simple: Contacts and shocks are characterised by rapid changes in velocity, hence the need for high and uniform acceleration. The importance of peak acceleration was further confirmed in [24].

If we combine peak force with inertia, we get peak acceleration. Peak acceleration cannot be improved by feedback since it depends solely on the actuator capabilities (STPFK) combined with the inertial properties. It is therefore a fundamental figure of merit.

The measurement of peak acceleration is relatively simple given the recent availability of light weight and low cost solid-state accelerometers.

4.8 Energy Flux – Power Density

Given a device, or rather a device-body interface of given inertia and given peak acceleration, we can define its power since we can calculate its increase (or decrease) of energy by unit of time, assuming neither a change in potential energy nor dissipation. Since a device-body interface is meant to operate inside a given volume, combining all these figures we can specify its power density (per unit of distance, area, or volume whichever the case may be, PWRD). This can also be viewed as a measure of the energy flux a device is capable of. This forms a single concise figure over all the device-body interfaces. A further possibility is to include a frequency dependency leading to spectral power density.

4.9 Broad Device Classification and Examples

4.9.1 *Low DOF Devices*

A *low DOF* (degree of freedom) device does not attempt to address the literal emulation of tasks which would occur during actual performance. It has been found for example that 2 or 3 controlled mechanical freedoms could provide an operator with a task metaphor sufficiently suggestive to lead to a high level of usefulness.

An excellent example of this idea is the Phantom device which can exert forces at one point in three dimensions but is not meant to exert any torque [17]. The operator either grasps a cylindrical handle and forces are applied at the end of that handle, or one finger of the operator is braced in a single hoop gimballed to the active end. The metaphor is these cases is either to "probe" a [virtual] object or touch it with an intervening brace. Another example is the Pantograph which has only two degrees of freedom in a plane [21]. The operator uses one or several fingers to interact with a knob. In this case the metaphor is exploration of a planar world using a small hand-held object. As a further example, D. K. Pai at UBC is developing an extension to the Pantograph device with an additional rotational degree of freedom, while the task remains planar. The metaphor, in this case, is the manipulation of a small object constrained to move in a plane. Other examples could include Adelstein's spherical device [1], as well as Hannaford's miniature haptic device [4].

It has been recently been realized that the level of "usefulness" of low DOF devices is actually higher than intuition would predict, with the benefit of a great deal of construction simplification compared to a high DOF device. The combined use of several low DOF devices can lead to a richer set of metaphors as discussed in [13].

A low DOF device is labelled LDOF#, where # stands for the number.

4.9.2 *High DOF Devices*

A *high DOF* device attempts to recreate a task in its integral form. Under this category, we find all devices which are designed around a hand held manipuladum applying arbitrary forces and torques. Thus they must have at least six DOF's although the case of a 5 DOF device could arguably fall under this category too. In this case, the haptic display no longer relies on a metaphor but instead is quite literal in the attempt to recreate a task that would occur if the handle was actually used to perform a task. In this category we find many many devices developed for teleoperation such as JPL's FRHC [19] and CEA's MA23 [25]. More recently, a six (extendible to seven) DOF device is being developed at McGill for specific application to virtual environments [8]. A six DOF device has also been developed by Iwata [14]. Other examples include Salcudean and Hollis magnetic levitation devices [12] [22]. Of course there are many tasks and application which require full DOF display. An interesting case is the device developed by SRI [11] for telesurgical application which has only four degrees of freedom, but is quite literal in its use, and therefore must be classified has a high DOF device.

A high DOF device is labelled HDOF# where # stands for the number.

4.9.3 *Very High DOF Devices*

A *very high DOF* device also attempts to re-create a task in its literal form but also involves combinations of body motions: complete arm, shoulder, torso, legs, etc. In other terms, it includes more than one body-device interface and this means that, almost certainly, it must be worn, that is braced around the operator's body.

In this category we find the Sarcos exoskeleton master, Kazerooni's extenders [15], the dextrous master of Burdea *et al.* [3], and the exoskeleton system of Bergamasco *et al.* [2].

A high DOF device is labelled VHDOF# where # stands for the number.

5 Detailed Features

The questions of backdriveability, resolution, precision and repeatability consolidate many of the injurious properties associated with mechanical systems; these include in combinations, friction, backlash, and elasticity. What is more, these measures can be universally applied to the various actuator-transmission or sensory systems in terms of various defects such as hysteresis, ripple, cogging, and drift.

A haptic device being nothing but a bidirectional mechanical transducer, enables us to apply the methods used to specify more general transducers such as motors. It is possible to consolidate all these factors in terms of noise specifications.

As with any transducers, the noise on each signal will depend on the load. Consequently, isometric or isotonic measurement conditions are not strictly speaking correct, since under normal operation, the device will be in contact with fleshy tissues which have varying but finite viscoelastic properties. It is proposed that in the absence of a definition of maximum and minimum values of the viscoelastic properties of tissues, isometric (same motion) and isotonic (same force) conditions be used for now, and to be replaced later with more meaningful figures.

Similarly, measurements should not always be taken under static conditions. The mechanical noise generated by a transmission, an actuator, or sensor noise will depend dynamically on the signal. This is the case of friction, for example, seen as noise on the mechanical signal. At the very least, velocity should not identically be zero when the transmission performance is measured. The same is true for cogging (cogging causes spurious forces to appear because of cyclical mechanical or magnetic energy storage). On the other hand, ripple (periodic change of actuator gain) and digital reconstruction noise can be captured by static precision measurements.

Under specified conditions we will perform resolution and precision measurement for force, displacement and velocity signals. The repeatability figures can be found using the same measurements obtained in the precision experiments.

5.1 Resolution

It has been observed many times that resolution is the most critical detailed feature of haptic devices, while precision matters less.

The resolution is considered from the output point, rather than at the individual joints, if any. The resolution of the system represents the smallest deviation from system equilibrium which can be detected by the sensors under study. When the haptic device is under computer control the resolution may be limited by the analog to digital converter; in any case, the complete system must be considered.

To measure the position resolution of the position sensor, the device-body interface must be connected a very rigid reference to create isometric conditions. One way to achieve this is to use the bed of a milling machine which permits both static and low velocity conditions. The complete set-up should include a load-cell of appropriate characteristics interposed between the bed and and the device-body interface under study.

The measurements obtained from this are: the smallest displacement detectable in static conditions and in low velocity conditions, say $\pm.01$ m/s. Under the same isometric conditions, force readings should be taken in a wide frequency range while the device is controlled not to exert any force. The spread will provide hysteresis figures, while noise can be reported in terms of RMS form specific frequency bands, so in Newtons per root Hz. The specific bands of interest here are, low range: 1-100 Hz, and high range 100-1,000 Hz,[4] or perhaps a more finely sampled

[4]Based roughly on the response range of various skin receptors: see Reynier, F., Hayward, V. 1993. Summary of the kinesthetic and tactile function of the human upper extremity". McGill Ctr. for Int. Machines, TR CIM-93-4.

frequency scale such as 1-10-100-1000 Hz. These measurements will give a precise picture of what is commonly called "backdriveability".

The other resolution measurements will be obtained by getting the device to produce specific level of forces and repeating the same procedure as above for, say, 1 percent, 10 percent and all of its maximum force capacity (already defined). A complete test suite should include figure reports both in static and low velocity conditions. High velocity conditions (say above .1 m/s) for resolution are not necessarily important for the performance of a haptic device.

As far as isotonic conditions are concerned, they are more difficult to set-up because a pure source of force (except the null source) is not as easy to provide as a pure position source is. Similarly, instrumentation is more difficult since it will require position and velocity sensors that will not disturb measurements and that will work over significant ranges. In the event that such a test bench is available, the measurement will take place in an analogous fashion as above: forces (static and slowly ramping) are applied and the corresponding displacements measured. Fortunately, since most devices are meant to display forces and read displacements, isotonic condition are not as important as isometric ones.

Of course, as already mentioned, best and worst case figures should be provided.

5.2 Precision

The precision of the device can be described as the difference between the target coordinate and the centre of the distribution curve of the actual coordinates of the arm end point.

Here again, isometric conditions are the most useful and the milling machine will come handy again. The device-body interface is forced at various known positions of its motion range and readings of the device are noted. Conversely, the device is programmed to exert a scale of forces and readings are taken.

Repeatability will be found (independent of control) from the width of the distribution curve.

5.3 Bandwidth

This figure is particularly difficult to define because unlike noise figures, isotonic nor isometric conditions are appropriate to convey a precise meaning. In bandwidth specifications, one sometimes speaks of "small signal" and this begs the question of what signal (displacement or force) and how small. In fact, it is unclear what subjects are sensitive to; there is evidence that subjects are sensitive to all of these signals. In some cases, pressure on the skin (force per unit area) is the relevant quantity, in others the skin indentation. To compound this problem, the response of a device may depend critically on the load, or in other terms is highly sensitive to its nature [10].

There does not seem a way to escape the necessity to specify a meaningful load which would be representative of actual operating conditions. It is proposed to specify the load as a piece of defined and widely available material of normalised dimensions (say specific silicon gel), designed to approximate a typical fleshy tissue.

The frequency response and the bandwidth can then be measured with the device-body interface loaded by the sample in terms of its microdisplacements at, say 1, 10 and maximum peak force. The method for measurement must involve a small displacement transducer of appropriate characteristics. This cannot be discussed here but of number of possibilities exist in the area of optical measurement methods.

A fall back measurement is the force isometric response.

5.4 Structural Response

Similar to loudspeakers, a haptic device will 'color' the signal due to its imperfect structural properties. In fact, an experienced operator will probably, from a blind folded experiment, recognize the make of a device. So it is reasonable to also specify this in some way. The structural response measured in some specified conditions is probably the best way to do this. As before, there are two basic ways to achieve this, isometric and isotonic conditions. A large number of excitation methods are available to achieve this: impulses, sine sweep, white noise, pink noise, etc. Since presumably any device includes a number of non-linearities, these different methods will produce different results. It is suggested that the method most appropriate for application to haptic devices is the impulse excitation method. In isotonic conditions (free end), an impulse of maximum amplitude is applied at the actuators, and acceleration is measured. The spectral response is reported. In isometric conditions, the force is measured and the spectral response reported.

5.5 Dynamic Precision

5.5.1 *Cross Talk*

There is evidence that humans are capable of discriminating between the spatial properties of vibratory motions quite high in the frequency domain. One way to look at a haptic device is to consider it as multichannel transducer with one channel per DOF. Thus is is possible to specify dynamic precision in terms of 'cross talk' between these channels. Cross couplings, most probably from mechanical origin, will cause signals in one channel to spill over another one.

The way to measure this around an operating point is to constrain the device in isotonic condition along or around one dimension, (to a slide or ball bearing) causing it to apply periodic forces along or around that direction and measuring the reaction forces along or around other directions.

5.5.2 *Distortion*

Another aspect of dynamic precision is in terms of non-linear signal distortion. The device is programmed to apply a signal in some specified condition, say a force in isometric condition, and the signal distortion is reported in percent energy (RMS) for a periodic signal as it is customary in electroacoustic equipment specification. More refined distortion measurement accounting for certain types of nonlinear distortions measures the device ability to not distort transients. A high amplitude low frequency square wave is superimposed on a higher frequency sine wave. The distortion of the sine wave as the result of the square wave is reported.

5.6 Closed Loop Performance

A crude approximation of a haptic device may be a pure inertia. If position and velocity signals are fed back to the actuator, the system becomes a second order "mass-spring-damper" system. If the pure inertia approximation truly is valid, any positive values of the feedback gains, corresponding to damping and elasticity, will lead to a (possibly marginally) stable system. In practice, clearly, such is not the case. Often the device is driven by a digital system which introduces at least sampling and discretization. Thus, for some gains, the closed loop system either enters limit cycles or becomes unstable [5]. The closer the ground device is from an ideally instrumented pure inertia, the higher these gains will be.

For example, it is relatively easy to find the stability conditions for a pure inertia being driven by an ideal actuator and having ideal sensors but with the feedback closed by a sampled data control having a zero-order hold and one time delay of one sampling period.

In practice, these are only approximations and instability occurs before theoretical values are reached because of other sources of error; nevertheless, the form of these conditions remains valid. It is proposed to quantify the closed loop performance of a device including its digital control system by specifying α and β such that the system remains stable if $B < \beta\frac{M}{T}$ and $K < \alpha\frac{B}{T}$. Of course, the floor value of B is the intrinsic damping of the system.

This discussion does not consider the interaction between the haptic device and the operator although is it important [16]. The interaction will greatly influence closed loop performance and introduce many control tradeoffs the discussion of which fall outside the scope of this paper.

6 Environmental Factors

Environmental factors are extremely pertinent in haptic devices because of the direct human-machine interaction which takes place. Some of the issues associated with environmental factors can not be specified in the form of a single number; yet, as haptic devices enter the commercial market place, these environmental factors will be of equal importance to some of the measurable indices we have already described.

Environmental factors which can be measured are: weight, acoustic noise and volume. These three factors can generally exclude some designs from certain applications before the performance indices have even been considered. Non measurable, but important factors are visual intrusion and service requirements.

The human link in the haptic equation requires that safety be a critical issue. What is the maximum speed and maximum force which the device could exert on the user? In the case of electrical or mechanical failure, is the system safe? (For example, a computer crashing.) Haptic designers are now forced to address these issues more than has ever been necessary before.

7 Conclusion

This paper has set out to define a set of performance measures which are applicable to haptic devices. It has been stated that a haptic device is bi-directional, i.e. is capable of both reading and writing to the user. This bi-directionality and direct human interaction with the device, makes the emphasis of performance measures for haptic devices different to that of traditional robotics. Also many of the performance measures used to to date for haptic devices have blurred the comparison to other devices due to ambiguity in measures and different measuring conditions. By moving toward a standard, it is hoped that comparison, improvement, specification and analysis of haptic devices will become possible.

8 Acknowledgements

This paper results from discussions with and encouragements from many people here listed in alphabetical order: W. Buxton, R. Ellis, B. Hannaford, J. M. Hollerbach, R. Hui, A. Kostic, S. Lederman, J. Morrell, D. Pai, J. Payette, C. Ramstein, S. E. Salcudean, J. K. Salisbury, I. Sinclair, C. Strong.

Funding is provided by the project "Haptic Devices for Teleoperation and Virtual Environments" (AMD-5) supported by IRIS (Phase 2), the Institute for Robotics and Intelligent Systems part of Canada's National Centers of Excellence program (NCE), and an operating grant from NSERC, the National Science and Engineering Council of Canada. Initial funding provided by a research contract with the Canadian Space Agency (No. 9F009-1-1441/01-SR).

References

[1] B. D. Adelstein and M. J. Rosen. Design and implementation of a force reflective manipulandum for manual control research. *Proc. of DSC-Vol. 42, Advances in Robotics, ASME Winter Annual Meeting.*, pages 1–12, 1992. Anaheim, CA.

[2] M. Berganasco, B. Allota, L. Bisio, L. Ferretti, G. Parini, G. M. Prisco, F. Salsedo, and G. Sartini. An arm exoskeleton system for teleoperation and virtual environments applications. *IEEE Int. Conference on Robotics and Automation*, pages 1449–1454, 1992.

[3] G. C. Burdea, J. A. Zhuang, E. Rosko, D. Silver, and N. Langrama. A portable dextrous master with force feedback. *Presence: Teleoperators and Virtual Environments.*, 1:18–28, 1992.

[4] P. Buttolo and B. Hannaford. Advantages of actuation redundancy for the design of haptic-displays. *Proc. ASME Fourth Annual Symposium on Haptic Interfaces for Virtual Environment and Teleoperator Systems*, November 1995.

[5] J. E. Colgate and J. M. Brown. Factors affecting the z-width of a haptic display. *International Conference on Robotics and Automation*, pages 3205–3210, 1994. San Diego, CA.

[6] R.E. Ellis, O.M. Ismaeil, and M.G. Lipsett. Design and evaluation of a high-performance haptic interface. *Robotica*, To appear, 1996.

[7] C. R. Flateau, F. J. Greeb, and R. A Booker. Some preliminary correlations between control modes of manipulators systems and their performance indices. *Proc. First National Conference, Remotely Manned Systems, Exploration and Operation in Space*, 1973. California Institute of Technology.

[8] V. Hayward. Toward a seven-axis haptic device. *IROS '95, IEEE/RJS Int. Conference on Intelligent Robots and Systems*, 1995.

[9] V. Hayward, J. Choksi, G. Lanvin, and C. Ramstein. Design and multi-objective optimization of a linkage for a haptic interface. *Advances in Robot Kinematics*, pages 352–359, 1994. J. Lenarcic and B. Ravani (Eds.), Kluver Academic.

[10] V. Hayward and M. Cruz-Hernandez. Parameter sensitivity analysis for design and control of tendon transmissions. *Preprints of the ISER-4, International Symposium on Experimental Robotics, Stanford, CA.*, 1995.

[11] J. W. Hill, P. S. Green, J. F. Jensen, and Y. Gorfu. Telepresence surgery demonstration system. *International Conference on Robotics and Automation*, pages 2302–2307, 1994. San Diego, CA,.

[12] R. L. Hollis, S. Salcudean, and P.A. Allan. A six degree-of-freedom magnetically levitated variable compliance fine motion wrist: Design, modelling and control. *IEEE Transactions on Robotics and Automation*, 7(3):320–332, June 1991.

[13] R. Hui, A. Ouellet, A. Wang, P. Kry, S. Williams, G. Vukovich, and W. Perussini. Mechanisms for haptic feedback. *IEEE International Conference on Robotics and Automation*, pages 2138–2143, 1995.

[14] H. Iwata. Artificial reality with force feedback: Development of desktop virtual space with a compact master manipulator. *Computer Graphics*, 24(4):165–170, March 1990.

[15] H. Kazerooni. The extender technology: An example of human-machine interaction via the transfer of power and information signals. *Preprints of the Fourth International Symposium on Experimental Robotics, ISER*, June 30-July 2, 1995.

[16] D. A. Lawrence and J. D. Chapel. Performance trade-offs for hand controller design. *IEEE Int. Conference on Robotics and Automation*, pages 3211–3216, 1994.

[17] T. H. Massie and J. K. Salisbury. The phantom interface: A device for probing virtual objects. *ASME Winter Annual Meeting, Symposium on Haptic Interfaces for a virtual environment and teleoperator systems*, November 1994.

[18] D. A. McAffee and P. Fiorini. Hand controller design requirements and performance issues in telerobotics. *Proc. Fifth International Conference on Advanced Robotics*, pages 192–186, 1991.

[19] D. A. McAffee and T. Ohm. Teleoperator subsytem / telerobot demonstrator: Force reflecting hand controller equipment manual. *JPL Report D5172, JPL, California Institute of Technology, Pasadena, California, USA*, 1988.

[20] J. D. B. Paines. Design of a force reflecting hand controller for space telemanipulation studies. *33rd Congress of the International Federation*, October 10-17 1987.

[21] C. Ramstein and V. Hayward. The pantograph: a large workspace haptic device for a multi-modal human-computer interaction. *ACM/SIGCHI 1994 Conference on human factors in computing systems, Boston, MA*, April 1994.

[22] S.E. Salcudean and N.M. Wong. Course-fine motion coordination and control of a teleopertion system with magnetically levitated master and wrist. *Third international symposium on experimental robotics*, 1993. LNCIS 200. Springer Verlag.

[23] K. V. Siva, P. J. Fischer, M. H. Brown, and E. Abel. The development of a bilateral input device for use in teleoperation. Remote Handling and robotics department, AEA technology, Harwell laboratory, Oxon, OX11 0RA, U.K.

[24] L. Stocco and S. E. Salcudean. A coarse-fine aproach to force-reflecting hand controller design. *Submitted to: IEEE International Conference on Robotics and Automation*, 1996.

[25] J. Vertut, P. Marchal, G. Debrie, M. Petit, D. Francois, and P. Coiffet. The ma23 bilateral servomanipulator system. *Proceedings of the 24th conference on remote systems technology*, 1976.

[26] Y. Yokokoji and T. Yoshikawa. Design of master arms considering operator dynamics. *Proc. 1990 Japan-U.S.A. Symposium on Flexible Automation—A Pacific Rim Conference— Kyoto, Japan.*, pages 35–40, 1990.

A Measures found in the specifications of a number of devices

In this appendix we list the performance measure which we uncovered in the literature. For each category, we list in bold the performance index which we propose, followed by similar indices (if any) found in the literature.

Mechanical degrees of freedom. (P) passive or (A) active. and unilateral (U), bracing (B) or held (H): None found

Motion Range (MR): Reach, Range of Motion, Work Volume, Position, Workspace, Work Envelope, Range of Travel, (Roll, Pitch, Yaw), Volume of Operation, Angular Operation, Cubic Workspace, Angular Range, Working Areas.

Peak Force (PKF): Force, Long term peak force (LTPKF), short transient peak force (STPKF) and persistent transient peak force (PTPKF): Max force/torque, Maximum Exertable Force, Max Force Output, Long Duration Force Output, Maximum Force.

Ground Device Inertia (GDI): Imposed Inertia, Inertia (apparent mass at tip), Calculated Reflected Inertia of Motor@Hand-grip, Inertia.

Peak Acceleration (PACC): Acceleration, Maximum Acceleration, No Load Accelerations.

Energy Flux - Power Density: None found.

Broad Device Classification: LDOF, HDOF, VHDOF: Implied but never classified.

Resolution at 10, 100, and 1000Hz: Nominal Position Resolution, Positional Resolution, Backlash. Friction Breakaway, (No dependency on frequency given.)

Precision: Not found.

Structural Response Isometric and Isotonic: Not found.

Bandwidth: Mechanical Bandwidth.

Dynamic Precision: Not found.

Closed Loop Performance: Not found.

Environmental Factors: Storage Volume, Overall Mass, Weight, weight of various subassemblies, Weight: Joystick, Weight: Power and Computer Unit, Limit Stops, Power Requirements.

Operator Control of Robotic Systems

Richard P. Paul, Craig P. Sayers, Julie A. Adams
General Robotics and Active Sensory Perception Laboratory
The University of Pennsylvania
Philadelphia, PA 19104, U.S.A.
(lou,sayers,adamsj)@grip.cis.upenn.edu

Abstract - **A characteristic feature of teleoperation is that, given time, tasks are generally successfully completed. This is caused not so much by careful engineering as it is by the continuous presence of an intelligent human operator. We are interested in developing systems which are similarly as effective in performing tasks while operating under constrained communications channels or controlling multiple robotic agents. To handle the problem of constrained communications channels we have developed the teleprogramming paradigm which uses local high-bandwidth feedback at the operator and remote sites while employing relatively low-bandwidth communication between sites. To deal with multiple robotic agents a multi-level approach is required. To this end we are developing the MASC (Multiple Agent Supervisory Control) system which provides the human operator with tools to interact with all processing levels of a robotic system. In developing these implementations, the emphasis has moved from developing systems which are *correct* to systems which *may be corrected*. While the former has proven an elusive goal in robotics, the latter is well within the domain of software engineering. Our intention is not to develop hardware/software which is optimal for a particular task. Rather, the goal is to demonstrate that our approach permits successful task performance even while using imperfect equipment in a very imperfect world.**

1 INTRODUCTION

Two quite separate fields, teleoperation and automation, bear on the scope and definition of the field of robotics. Teleoperation [7, 14] has existed since the early 1950's and is representative of manual work. An operator uses tools, operated by a master/slave manipulator system, which provides bilateral feedback. In high quality teleoperation systems the teleoperator provides almost transparent control of the tool. Task feedback is provided visually with additional haptic or kinesthetic clues improving performance of some tasks.

The use of teleoperators closely resembles manual task performance: little planning is done, the task is sized-up, the appropriate tools selected and attempts to perform the task are immediately undertaken. Frequently, the tool selection is wrong, or the position of the task relative to the manipulator is bad and a task step must be retried. Watching a teleoperator perform a task is similar to watching a human perform the same task. However, even though it may follow a somewhat meandering course, task execution is usually successful. With teleoperation, as with human task execution, there is an expectation of success. While some feedback is necessary, to help perform each task step and to confirm to the operator that the task step has been successfully carried out, humans are capable of compensating for missing information, relying on experience and intuition when other senses fail. The reasoning brought to bear on a task by a human operator is, while almost subconscious, vast compared to any available computing systems. Furthermore, the reasoning brought to bear on a task is in the form of "common sense" reasoning—an elusive goal of artificial intelligence systems.

Automation is the application of external power and organization to perform a task. It is generally based on position control and the necessary maintenance of position tolerances for all components. Once a task can be performed based purely on position control then it may be performed at any rate desired by adding power and rigidity. While contact sensing might be employed to confirm completion of critical subtasks, it is rarely used during task performance since it slows down the rate of operation. While automation systems might perform the same task that an operator once performed, they do so in a quite different manner and usually orders of magnitude faster. There is no provision for error recovery, other than detecting that an error occurred and rejecting the part or stopping the machine. The occurrence of errors is an indication that the process is not under control and either the machine is incorrect or that the tolerances of the component parts are incorrect. Modifications of one or both take place to eliminate the problem. Automation is frequently associated with the development of considerable infrastructure and extensive study of all factors. Automation, like manual work, is satisfying to watch with the same expectation of success, although in the case of automation many months of debugging may be necessary to reach production rates.

Robotics was initially, seen as an intelligent activity based on sensing, planning, and execution in mimicry of an intelligent human, this initial approach soon ran into trouble as the vast intellectual resources casually brought to bear on a task by a human were not present in a computer. The substitutions, deductions from disparate clues, exploratory actions, rejections of false images, were simply not in the tool box of artificial intelligence systems. While a human might be able to use a teleoperator to perform a task, a computer was simply lost. Based on this initial experience with teleoperation the techniques of automation were instead applied to robotics. The quality of image processing was to be improved so that quality information could be extracted directly from the image without requiring the elaborate deductive process humans employed in such analysis. Manipulators were position controlled to allow for precise dispatch to given positions. In order to support this new approach geometric models were introduced. These models would make up the output of the sensing systems, allowing sensor fusion to take place and providing the input to the manipulation system. The control of such systems was, of course, by computer. Execution errors could be detected and, based on the general-purpose nature of the equipment, actions could be programmed to correct expected problems. Unfortunately, the range of possible errors was vast, hardly to be tackled by the application of the if-then-else-statement [13]. By severely limiting the task however, as in automation, robots have found application in industry as a form of flexible automation in the food, electrical machinery, and semiconductor industries.

Although advances continue to be made in the components of robotics there is little relevant work in artificial intelligence which might tie such components together to produce a device in which a task is not automated but instead simply described and then performed.

2 OPERATOR CONTROL

Teleoperation, so direct, provides no productivity improvement. It requires an operator in constant attendance while every motion made by the teleoperator is made by the operator. While some tasks might totally engage the full attention of an operator many teleoperation task are of a repetitive nature and soon lose the attention of an operator; consider the task of dismantling a nuclear reactor which might extend over years. And after years of such teleoperation no more trace of the work performed or the skills acquired will have been recorded than if the work had been directly performed manually. Is it possible to provide an operator with some tools to improve his or her productivity and to reduce boredom and consequently reduce errors?

Systems are beginning to appear which make use of many of the components of robotics in combination with an operator [5, 9, 8, 12]. The robotic components, or sub-systems: image processing, Cartesian motion control, force control, path planning, stereo, etc. are used to perform the low-level work for an operator while he or she provides the missing component of intelligence of the robotics pantheon.

Some tasks are impossible to perform using current teleoperation technology. While teleoperation implies task execution at a distance, that distance is, in reality, extremely limited. Even in tasks which do not involve any separation of master and slave system, time delay in the control loop can cause instability. If the distances become large, as in outer-space or the velocity of signal propagation becomes low, as in acoustic communication underwater, then stability becomes a problem and conventional teleoperation becomes impossible. It is in tackling this problem that teleoperation systems based on task models and robotic technology components have evolved. While it might seem that such systems were limited to the domain of time delayed teleoperation their applicability is much wider and could provide the basis for the necessary productivity improvement in time-delay-free teleoperation.

In order to work in a time-delayed environment, where both operator actions and sensory feedback are delayed by the communications time-delay, it is necessary to isolate the operator from the effects of the delay. This is usually done by providing limited task models at the master and slave stations and closing non-time-delayed feedback loops at both sites. Low-bandwidth, time-delayed, communications link both stations. In such a system it is possible to execute a task as if the time delay did not exist with execution taking place at the same rate but at a fixed time-delay with respect to the master station.

We will describe two such systems being developed at the University of Pennsylvania, teleprogramming and the MASC system. Teleprogramming is a system developed to operate a manipulator via delayed communications channels. This is our approach to the classic time-delayed teleoperation application described above. An implementation of this system was recently successfully tested outside the laboratory. An operator in Philadelphia controlled a manipulator on a submersible off the Massachusetts coast while all communication was via a combination of the internet and a simulated underwater acoustic link.

To deal with multiple robotic agents a multi-level approach is required. High level control is desirable to ease the operator burden while he or she simultaneously commands several agents. Low level control is necessary so that the operator may assume control when the system's autonomous behaviors fail. To this end we are developing the MASC (Multiple Agent Supervisory Control) system which provides the human operator with tools to interact with all processing levels of a robotic system. This latter system is nominally time-delay-free, does not attempt closed-loop operation and is representative of a much wider class of teleoperators in which productivity improvement is sought.

3 SUBSEA TELEPROGRAMMING

In applying the teleprogramming paradigm [6] to the task of subsea teleoperation our goal was to develop the means to permit an operator ashore to control a manipulator on an unmanned untethered submersible. This required that we overcome two main problems. The first is that the subsea environment is unpredictable and poorly defined. It is far removed from the safety and repeatability of the laboratory. The second difficulty is that available communications schemes are very limited. Using acoustic modems we can expect delays of as much as 10 seconds with bandwidths as low as 10 kbits/second.

When using the current implementation of the teleprogramming system, the operator's first task is to calibrate the subsea cameras (see figure 1). To do this he or she interprets images from the real cameras, indicating the correspondences between features in those images and features in the operator station model of the remote manipulator. In this task, as in other stages of the teleprogramming system, the implementation relies on cooperation between human and machine. The operator performs the difficult task of interpreting real images while the system automatically refines the operator's imprecise input to compute calibration parameters.

The operator may then use images from the subsea cameras to update the master station model of the remote environment [10] before beginning to command the remote manipulator. During this time the operator works entirely within a simulation of the remote site. The system observes this interaction and translates his or her actions into symbolic commands for transmission to the real slave robot. Sent at an average rate of only 1 or 2 per second, these commands encode a higher-level representation of the task than is common in teleoperation. They describe a desired action, expected sensory readings and additional information for the control of sensors (especially cameras) at the remote site.

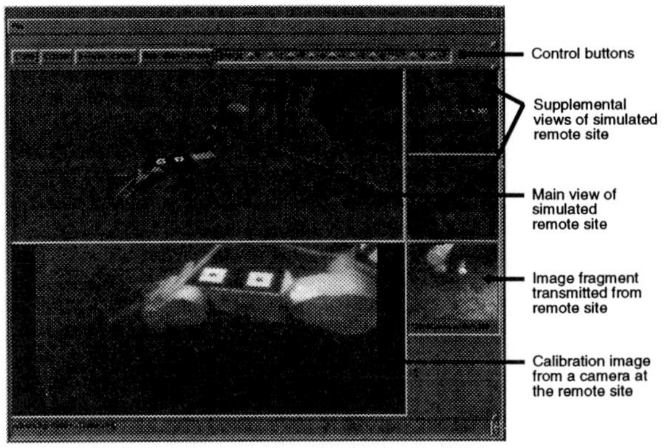

Figure 1. Teleprogramming operator interface.

As the operator works within the virtual world the system provides active assistance through the use of *synthetic fixtures* [11]. Their operation is analogous to, but more powerful than, the "snap" family of commands prevalent in computer graphics applications. The system attempts to predict the operator's future actions and generates active force and visual clues to guide his or her motions. In this way fixtures increase both the speed and precision with which operators may work. When used in the teleprogramming system, fixtures have an additional, more subtle benefit. By guiding the operator toward making deliberate unambiguous actions they simplify the task of translating those actions into commands.

At the remote site the slave system interprets and executes each command as it is received. However, due to the unpredictability of the environment and the practicality of real world interaction it is unreasonable to expect that every command will execute as planned. While some problems may be avoided by generating commands which are tolerant of remote site uncertainty there will still be cases where execution fails. These situations may be detected by having the slave compare its actual sensory readings with those predicted (and encoded within the generated commands) by the operator station. When such an error occurs the slave pauses and sends an error message stream to the operator station. This error stream contains a description of the current slave state along with a detailed record of recent slave motions.

Because of the communications time delay several seconds elapse between the time any error message is generated by the slave and the time it reaches the operator station. In the meantime the operator has continued with the task. Thus, when the master station receives the message it must alert the operator and then "wind back time" to show him or her the point where the error occurred.

At this point it is the operator's responsibility to analyze the system's preliminary error diagnosis and determine a strategy for recovery. The system can aid him or her in this task by providing simulations of expected and actual prior slave motion. The operator may step forward and back through past commands and, if necessary, correct any invalid inferences the system may have made from measured sensory data. For example, the operator may advise the system that it did not grasp an object even though the slave may "think" that it did.

In preliminary laboratory testing of the system it was found that purely symbolic replies from the slave to master station were not sufficient to enable rapid error diagnosis and recovery. The problem was that current techniques for the machine interpretation of sensory data do not cope well with the vagaries of real-world interaction. Thus, in creating the subsea system, it was considered essential to supplement the symbolic messages from the slave site with additional raw sensory data. This data, which takes the form of image fragments in the current system, provides the operator with an additional independent means of verifying the slave system's interpretation of measured data.

Providing these image fragments for the operator in real time over the narrow bandwidth acoustic link is a challenging problem [10]. We solve it by having the operator station predict (using models of the cameras, task and remote site) that fragment of the available imagery which would most help the operator if an error were to occur. That information is encoded within the transmitted commands and used by the slave to select appropriate image fragments for compression and transmission back to the operator station (see Figure 2). In addition, to further reduce bandwidth requirements, the system automatically varies both the frame rate and the rate of task execution to match that which can be supported given available bandwidth and computational constraints.

As the operator performs the task these image fragments are continually displayed and updated. When an error is detected the operator may step forward and back through the received imagery, comparing the real images with simulations of expected motion for the same point in time. This provides a powerful and natural tool for error diagnosis.

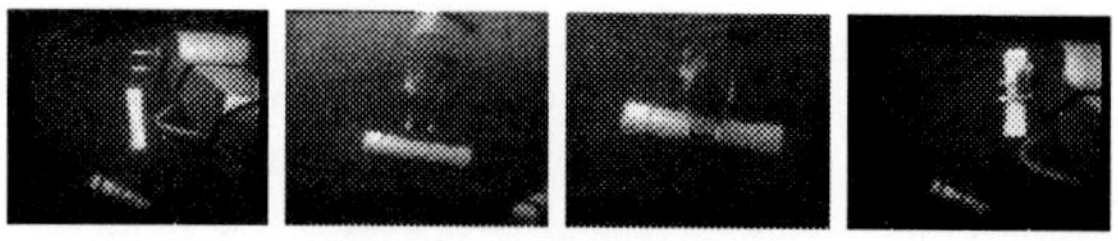

Figure 2. Examples from a sequence of image fragments displayed to the operator during subsea trials. Each image is 192x160 pixels, has been passed through contrast enhancement and JPEG compression/decompression. Note how the system automatically switches cameras while altering windowing and subsampling parameters.

The described implementation of the teleprogramming system allows tasks to be performed in the real world using very limited communication resources. In recent tests an operator in Philadelphia successfully performed retrieval operations using a manipulator mounted on the JASON subsea vehicle and submerged on the sea floor near Woods Hole, Massachusetts. Communication between sites was via a combination of the internet and simulated acoustic links. Round trip delays were between 5 and 15 seconds while bandwidths were consistent with those available from existing subsea networks.

The teleprogramming approach is well suited to cases where the environment is unpredictable or where tasks are repeated relatively infrequently. However, if operating in a more structured man-made environment where the same operations may be performed numerous times then it is appropriate to consider a greater level of autonomy. Our approach to these situations is realized in the MASC implementation.

4 MULTIPLE AGENT SUPERVISORY CONTROL SYSTEM (MASC) INTERFACE

We have developed a human-machine interface, the Multiple Agent Supervisory Control System (MASC), to interact with a multiple mobile robotic system. This interface and the interactions we have developed should permit the operator to interact with the system when undefined situations occur. The MASC interface permits the operator to assist the system such that the system should then autonomously complete the task. This section provides a brief overview of the MASC system interface and the *mediation hierarchy*.

The MASC interface has been developed for the University of Pennsylvania General Robotics and Active Perception (GRASP) Laboratory's multiagent project. The purpose of the multiagents

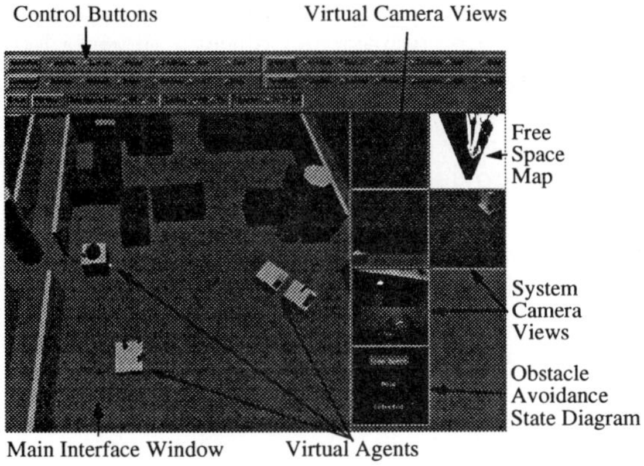

Figure 3. The MASC system interface.

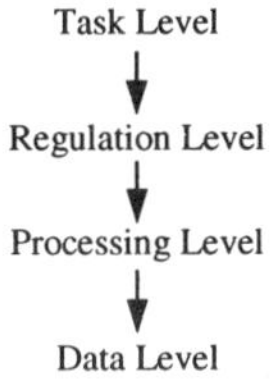

Figure 4. Hierarchical levels of human interaction.

project [3, 4] has been to investigate the coordination and monitoring of multiagent systems for intelligent material handling. The system is composed of four mobile platforms, each equipped with various sensory and manipulative capabilities, and the MASC interface. The MASC interface (shown in Figure 3) permits the individual robotic agents and their processes to be controlled by the human supervisor. Our objective is to create a semi-autonomous system which successfully completes assigned tasks in a dynamic environment.

The human's primary task is to "supervise" the agent's actions during task execution while observing sensory data and images. Each agent is composed of multiple control and processing levels, therefore MASC must permit the supervisor to interact with these levels for the successful semi-autonomous execution of tasks. This interaction will permit the supervisor to revise incorrect agent decisions and reconfigure the system after partial system failures.

The *mediation hierarchy* consisting of four levels (see Figure 4) has been formulated [1, 2]. This hierarchy defines the various intervention types which furnish the supervisor with the interaction capabilities to interact with all the robotic system levels. It should permit the supervisor to correct situations which would cause a fully autonomous system to become unstable and possibly fail its task execution. The supervisor interacts with the agents only when assistance is requested by the agents or when the supervisor detects a situation which she or he deems intervention necessary.

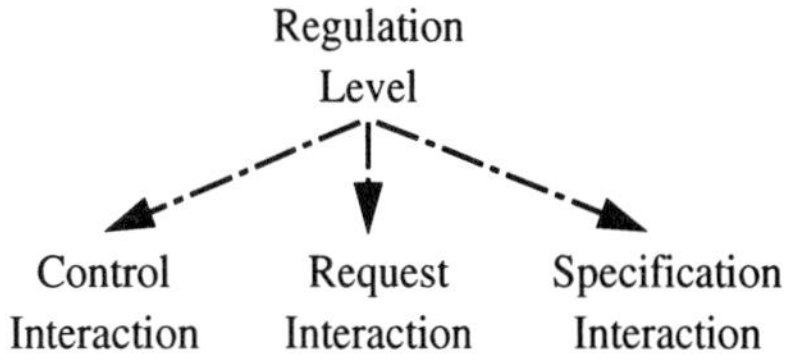

Figure 5. The interaction of the *regulation level.*

The highest level in the *mediation hierarchy* is termed the *task level.* This level permits the supervisor to specify the actions an agent, or a group of agents, are to execute to complete an assigned task. Tasks may include environmental exploration, observing another agent's task execution, carrying items such as pallets, and the navigation necessary to transport items from one location to another.

The second level, the *regulation level* is composed of three interaction types: *control interaction, request interaction* and *specification interaction. Control interaction* provides the supervisor with the capabilities to directly control the individual robotic agent's actions. This includes issuing commands such as an emergency stop or instructing them to pause their task execution until they are instructed to continue. Also included in this interaction type are situations when the supervisor may teleoperate the agent to create locomotion commands which may be employed as an alternative to the path planning methods or to assist the agent.

Systems contain various information which may be useful to the supervisor at various times throughout the task execution. The *request interaction* permits the supervisor to request the sensory data and processed information from the agent's only when it is required for error detection and/or monitoring purposes. When the information is no longer necessary, the supervisor may instruct the system to cease the information display.

Various processes require information from the supervisor prior to the commencement of processing. Such a process may be a path planning process for which the supervisor must specify the desired path's starting, intermediary and goal way points. This information may be specified through the *specification interaction* which provides the supervisor with the means to interactively specify information pertinent for a process' execution.

The third level of the *mediation hierarchy* is the *processing level.* This level of intervention may be necessary if there exist an instance when a process is incapable of determining a correct decision based upon ambiguous information and must therefore request supervisory assistance. There also exist situations when a process will formulate a correct decision in a local context but the decision will not be correct in the global scheme, thus the supervisor should either assist with the decision making process or override the decision formulated by the process. Formally, the *processing level* permits the supervisor to aid a process when it is unable arrive at a decision and to rectify incorrect decisions deduced by a process either upon the process' request for assistance or as determined by the supervisor through monitoring.

The final level of the *mediation hierarchy* is the *data level.* This level permits the supervisor to reconfigure the system when an automatic reconfiguration has failed. The *data level* also supplies the supervisor with the ability to ensure processes receive correct data for interpretations as the outcomes of the higher-level processes are dependent upon correct input data. The *data level* permits the supervisor to ensure correct data is passed up through the system for interpretations and processing. It also allows the supervisor to reconfigure the system during a hardware failure.

The MASC system interface has been designed to encompass the *mediation hierarchy.* MASC permits the human operator to control a multiple mobile robotic system in a dynamic environment. This robotic system is able to work autonomously until it either request assistance from the human operator or the human deems interaction necessary. This interaction corrects problem situations such that task assignments can be successfully completed. Thus tasks which are redundant in nature yet require a human operator for supervisory control may be successfully executed by a semi-autonomous system employing this *mediation hierarchy* theory. Our MASC interface and the multiple mobile robotics project demonstrates this capability.

5 CONCLUSIONS

We have developed both the teleprogramming and MASC implementations "from the bottom up"; starting with low-level components and moving toward higher-level representations with the goal of proposing, rather than defining, solutions. We have added components only when they made a "materially" significant contribution to improving task performance. The operator has always remained in control. He or she may interact at different levels to diagnose and correct those problems which will inevitably occur.

It has been surprising how effectively operators can compensate for missing or incorrect sensory information; relying on experience and intuition to perform tasks under difficult conditions. Most of our use of "intelligence" had been in attempting to understand what the operator intended to do so that we could help in the most useful manner. By incorporating an operator into the system we can make use of much of what is good in robotics to provide useful, productive, implementations.

6 ACKNOWLEDGMENTS

The authors wish to gratefully acknowledge the assistance of the Woods Hole Oceanographic Institution and especially Dana Yoerger, Louis Whitcomb and Josko Catipovic.

This work was funded by NSF Grant BCS-89-01352 and in part by Sandia National Laboratories Grant AN-2317, ARPA Grant N00014-92-J-1647, ARO Grant DAAL03-89-C-0031, ARPA/NSF Grant IRI94-12913, and NSF Grants CISE/CDA-88-22719 and STC-SBR8920230.

Any opinions, findings, conclusions or recommendations expressed in this publication are those of the authors and do not necessarily reflect the views of any funding agency.

REFERENCES

[1] J. A. Adams and R. Paul. Human-managed, hierarchical control of multiple mobile agents. In *Proceedings of the 33rd IEEE Conference on Decision and Control*, pages 3524 – 3529. IEEE, Dec. 1994.

[2] J. A. Adams and R. Paul. Human management of a hierarchical control system for multiple mobile robots. In *Proceedings of the 1994 IEEE International Conference on Systems, Man and Cybernetics*, pages 2780 – 2785. IEEE, Oct. 1994.

[3] J.A. Adams et al. Cooperative material handling by human and robotic agents: Module development and system synthesis. In *IEEE/RJS International Conference on Intelligent Robots and Systems.* IEEE/RJS, Aug. 1995.

[4] R. Bajcsy, V. Kumar, M. Mintz, R. Paul, and X. Yun. A small-team architecture for multiagent robotic systems. In *Workshop on Intelligent Robotic Systems: Design and Applications, SPIE's Intelligent Robotics Symposium*, Boston, MA, November 1992.

[5] A. K. Bejczy and W. S. Kim. Predictive displays and shared compliance control for time delayed telemanipulation. In *IEEE International Workshop on Intelligent Robots and Systems*, Ibaraki, Japan, July 1990.

[6] Janez Funda, Thomas S. Lindsay, and Richard P. Paul. Teleprogramming: Toward delay-invariant remote manipulation. *Presence*, 1(1):29–44, Winter 1992.

[7] Ray C. Goertz. Manipulators used for handling radioactive materials. In E. M. Bennett, editor,

Human Factors in Technology, chapter 27. McGraw Hill, 1963.

[8] S. Hirai, T. Sato, and T. Matsui. Intelligent and cooperative control of telerobot tasks. *IEEE Control Systems*, 12(1):51 – 56, Feb. 1992.

[9] G. Hirzinger, J. Heindl, and K. Landzettel. Predictive and knowledge-based telerobotic control concepts. In *Proceedings of the IEEE International Conference on Robotics and Automation*, pages 1768–1777. IEEE, May 1989.

[10] Craig Sayers, Angela Lai, and Richard Paul. Visual imagery for subsea teleprogramming. In *IEEE International Conference on Robotics and Automation*, pages 1567–1572, May 1995.

[11] Craig Sayers and Richard Paul. An operator interface for teleprogramming employing synthetic fixtures. *Presence*, 3(4):309–320, 1994.

[12] Thomas Sheridan. Space teleoperation through time delay:review and prognosis. *IEEE Trans. Robotics and Automation*, 9(5):592–606, October 1993.

[13] R. H. Taylor, P. D. Summers, and J. M. Meyer. AML: A Manufacturing Language. *The International Journal of Robotics Research*, 1(3):19–41, 1982.

[14] Jean Vertut and Philippe Coiffet. *Robot Technology*, volume 3: Teleoperations and Robotics. Prentice-Hall, 1986.

Haptic Virtual Reality: Display of Operating Feel of Dynamic Virtual Objects

Tsuneo Yoshikawa and Hitoshi Ueda
Department of Mechanical Engineering
Kyoto University
Kyoto 606, Japan
yoshi@mech.kyoto-u.ac.jp, jin@mech.kyoto-u.ac.jp

Abstract

We have been developing a haptic interface system for displaying feel of manipulating dynamic virtual objects. This system consists of several link mechanisms attached to the fingertips of an operator through which the operator can feel various dynamic force components (such as the inertial, centrifugal, Coriolis, and gravitational forces) and surface slippage of virtual objects. We have proposed the concept of impedance display and two ways of its realization, "measuring force and displaying motion" (MF/DM) and "measuring motion and displaying force" (MM/DF).

In this paper, an implementation of the second way MM/DF is presented. The contact state of the contact points between operator's fingertips and the virtual object is also explicitly defined. The validity of the proposed method is verified experimentally using a developed two-fingered display device. The operator could get the feel of manipulating dynamic virtual objects and perceive slippage on their surfaces.

1. Introduction

Virtual Reality techniques allow people to experience an illusory world without actually building the world or to obtain operating feel of a product without actually manufacturing the product. While large amounts of output information which is presented to human beings by computer is through visual display, the importance of force display has recently been recognized and studied by many researchers, e.g., [1][2][3]. Although the usefulness of the force display has been successfully demonstrated in these studies, the dynamic property of the manipulated virtual object during the operation was not taken into account rigorously. Also, no compensation for the dynamics of the force display device was considered.

Recently, we developed a haptic interface system for displaying the feel of manipulating dynamic virtual objects using a display system consisting of multiple link mechanisms attached to the fingertips of an operator [4]. We proposed the impedance display approach which aims at not just displaying the force, but displaying the relation between the force and motion of the virtual object. There are two ways for realizing the impedance display: one is "measuring force and displaying motion" (MF/DM). The other is "measuring motion and displaying force" (MM/DF). By this system the operator can feel various dynamic force components, such as the inertial, centrifugal, Coriolis, and gravitational forces, and surface slippage of virtual objects. We also stressed the indispensability of the dynamic compensation for the display device itself. We also proposed a method for considering static and dynamic frictional forces between the operator's fingertips and the object surface [5]. Both in [4] and [5], we have implemented the first way MF/DM.

In this paper, we present an implementation of the second way MM/DF. After a brief review on haptic virtual reality, the implementation of the MM/DF approach is described by dividing the whole system into three sections; the input section, the simulation section, and the output section. In the simulation section, the dynamics of the virtual objects and an elasticity and friction model are given.

Using a two-finger display device we have developed, experiments are performed. The tasks in the experiment were waving a spherical virtual object, and throwing up and catching it. The results show the validity of the proposed approach.

Table 1. Classification of Force Display Devices

(1) Interface with operator:

Type		Examples
Manipulative type	Grasp type	GROPE[1], Joy String, Sandpaper[3], Master-slave systems with grippers
	Put-on type	SPIDAR[6], Iwata[7], Master-Slave of Mitsubishi Heavy Ind.,[8] Yoshikawa et al.[4], Burdea et al.[9], Bergamasco et al.[10]
Non-manipulative type	Object mimicking type	Hirota et al.,[11] Hoshino et al.[12]

(2) Support of force:

Type	Examples
External force type	[1]-[8], [11], [12]
Internal force type	Burdea et al.[9], Bergamasco et al.[10]

(3) Generation of force:

Type	Examples
Active type	[1]-[12]
Passive type	Furusho et al.[13]

2. Related Works

Researches related to haptic virtual reality are briefly surveyed and discussed in this section.

One of the earliest force display systems is the GROPE system developed by Brooks and his group[1] in which a manipulator was used for force and torque display in molecular docking task. Agronin[2] and Sato et al.[6] proposed string-based force display devices. Minsky et al.[3] developed a *Sandpaper* system consisting of a two degree-of-freedom joystick for simulating texture. Iwata[7] developed a compact nine degree-of-freedom master manipulator as a tactile device to generate reaction force to the palm and the fingers of the operator.

Force display devices developed so far can be classified as follows (see Table 1). First, with regard to the interface device with the operator, there are two types: Manipulative type and non-manipulative type. Manipulative type devices are those which are held or put on by the operator.

Manipulative type devices can further be divided into two groups: grasp type devices and put-on type devices. Grasp type devices are those which are grasped by the operator when in use and can be released just by letting go of them when they are not needed. Put-on type devices are those which are worn by the operator or fixed to a part of the operator's body. They have to move in coordination with the operator's motion even when the operator's hand or finger is not in touch with the virtual object and hence there is no need to display any force.

Non-manipulative type devices developed so far are rather few. Those proposed in [11][12] can be called object mimicking type. They mimic the surface of the virtual object (or environment) only when the finger or hand of the operator come close to and interfere with the surface of the virtual object. The device does not need to react to the motion of the operator's hand or finger when it is far away from the virtual object.

Second, from the viewpoint of the kind of base for supporting the displayed force, the devices can be classified into two groups: External force type and internal force type. The external force type devices are those whose base is fixed to the ground or object that are not effected by the operator's motion. The internal force type devices are those whose base is mounted on some part of the operator.

Third, from the viewpoint of force generation in the devices, they can be classified into two: active type and passive type. The former uses active force generating actuators such as motors, and the latter uses passive ones such as powder clutches.

For the purpose of displaying dynamic virtual objects, display devices categorized in the external reaction type and the active type are suitable.

Application areas of haptic display system include the followings:

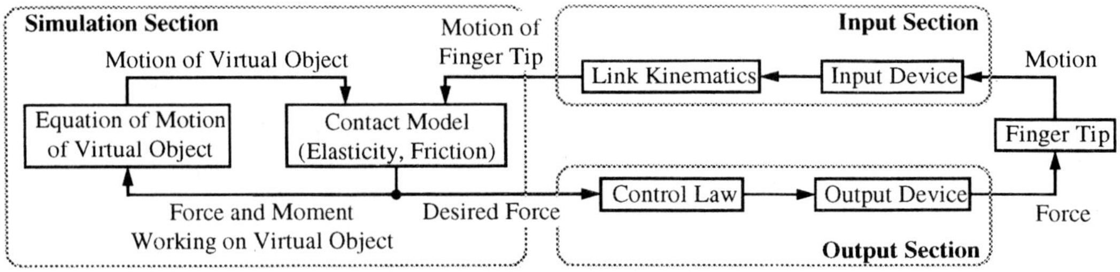

Figure 2. Algorithm for Displaying Forces

(1) Industrial design (modeling of free surface)
(2) Experience of virtual environment (game, entertainment, buildings)
(3) Experience of virtual objects (rapid prototyping)
(4) Teleoperation (Space, underwater, micro world, medicine)
(5) Training machine (rehabilitation, fitness)
(6) Interface with computer (software handling)

3. Overview of the Method

Conceptual sketch of the force display device we are developing is given in Fig.1, which consists of multiple link mechanisms (arms) with their bases fixed to the ground.

The operator inserts his/her fingers into the rings at the tips of the arms. An encoder is attached to the axis of each joint to measure the displacement of the joint. A force sensor is attached next to the finger connection device on each arm, and a force exerted by the operator's fingertip can be measured. A motor is attached to the axis of each joint, which drives the display device.

The algorithm proposed in this paper is as follows (Fig.2). In the input section, we get the motion of the fingertips by the input device. In the simulation section, we decide the forces and the moments exerted on the virtual object by the fingertips from the relative motions of the fingertips and the virtual object, and calculate the motion of the virtual object by the equation of motion.

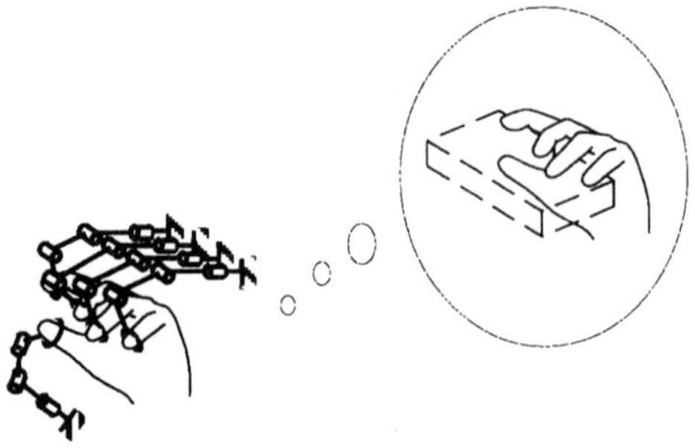
Figure 1. Force Display Device Consisting of Link Mechanisms

In the output section, we control the output device so as to display the forces calculated in the simulation section at the fingertips.

The forces working on the virtual object are determined by the elasticity and the viscosity of the surface of the virtual object.

4. Input Section

In this section, we describe the kinematics of the display devices in order to obtain the positions of the fingertips.

Consider the force display device consisting of n arms. The generalized position vector $\boldsymbol{q}_{hi}$ of the end point of the ith arm, expressed in the reference frame Σ_U, is given by

$$\boldsymbol{q}_{hi} = \boldsymbol{f}(\boldsymbol{q}_{li}) \tag{1}$$

where $\boldsymbol{q}_{li}$ is the joint vector of the ith arm and the function $\boldsymbol{f}$ is generally nonlinear.

Differentiating equation (1), we obtain

$$\dot{\boldsymbol{q}}_{hi} = \boldsymbol{J}_{hli}(\boldsymbol{q}_{li})\dot{\boldsymbol{q}}_{li} \tag{2}$$

where $\boldsymbol{J}_{hli}(\boldsymbol{q}_{li})$ is the Jacobian matrix of a hand frame Σ_{hi}, fixed to the end of the arm, with respect to $\boldsymbol{q}_{li}$ [14].

Each fingertip is represented by a point, and makes point contact with the virtual object. The position of the ith fingertip is equivalent to the position of the end of the ith arm, and is given by equation (1) using the joint vector $\boldsymbol{q}_{li}$.

5. Simulation Section

5.1. Forces Working on the Virtual Object

There are three types of contact states between a fingertip and a virtual object — non-contact state, dynamic frictional contact state, and static frictional contact state. When the ith fingertip is in each state, a force $\boldsymbol{f}_{ci}$ working on the virtual object is given as follows (Fig.3). A method to decide the contact state between the fingertip and the virtual object will be described in section 5.3.

- **Non-Contact State:**

$$\boldsymbol{f}_{ci} = 0 \tag{3}$$

- **Dynamic Frictional Contact State:**

$$\boldsymbol{f}_{ci} = \boldsymbol{f}_{cvi} + \boldsymbol{f}_{cdi} \tag{4}$$

where $\boldsymbol{f}_{cvi}$ is the contact force normal to the tangent plane, and $\boldsymbol{f}_{cdi}$ is the dynamic frictional force along the tangent plane.

- **Static Frictional Contact State:**

$$\boldsymbol{f}_{ci} = \boldsymbol{f}_{cvi} + \boldsymbol{f}_{cti} \tag{5}$$

where $\boldsymbol{f}_{cti}$ is the static frictional force along the tangent plane.

Each of $\boldsymbol{f}_{cvi}$ and $\boldsymbol{f}_{cti}$ is the sum of the elastic force and the viscous force. These forces are given by

$$\boldsymbol{f}_{cvi} = \begin{cases} K_{pv}\boldsymbol{e}_{vi} + K_{dv}\dot{\boldsymbol{e}}_{vi} & : \boldsymbol{e}_{vi}^T\dot{\boldsymbol{e}}_{vi} > 0 \\ K_{pv}\boldsymbol{e}_{vi} & : \text{otherwise} \end{cases} \tag{6}$$

$$\boldsymbol{f}_{cti} = K_{pt}\boldsymbol{e}_{ti} + K_{dt}\dot{\boldsymbol{e}}_{ti} \tag{7}$$

where K_p is the elastic coefficient, K_d the viscous coefficient, $\boldsymbol{e}_i$ the displacement between the current position and the reference position of the fingertip, and $\dot{\boldsymbol{e}}_i$ the relative velocity between these two points. The subscript v denotes the direction normal to the tangent plane, and the subscript t the direction along it. The reference position is that of the fingertip on the assumption that the virtual object is rigid.

The dynamic frictional force $\boldsymbol{f}_{cdi}$, which works in the same direction as the fingertip is moving and the value of which is proportional to that of the normal force $\boldsymbol{f}_{cvi}$, is given by

$$\boldsymbol{f}_{cdi} = \mu_d \|\boldsymbol{f}_{cvi}\| \frac{\dot{\boldsymbol{e}}_{ti}}{\|\dot{\boldsymbol{e}}_{ti}\|} \tag{8}$$

where μ_d is the dynamic frictional coefficient at the contact point.

5.2. Moments Working on the Virtual Object

We assume that the contact between the fingertip and the object surface is a soft finger contact, that is, it is a point contact with translational friction along the surface and with rotational friction around the surface normal. We call the moment caused by the rotational friction a frictional moment.

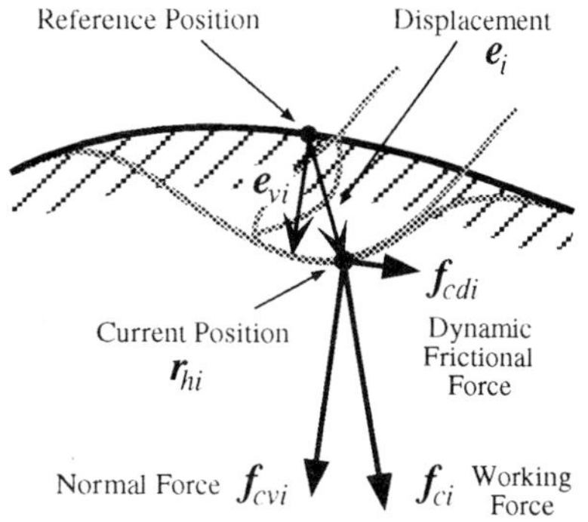

(a) Dynamic Frictional Contact State

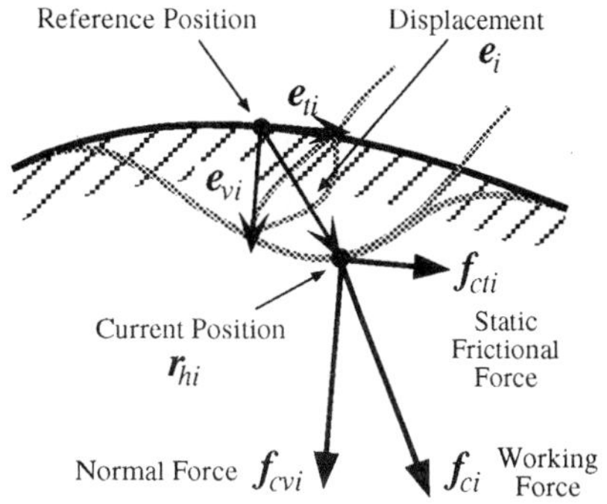

(b) Static Frictional Contact State

Figure 3. Forces Working on the Virtual Object

In this paper, for simplicity, we assume that each value of the dynamic frictional moment and the maximum static frictional moment is proportional to that of the normal force. We also assume that the elastic and the viscous moment work in static frictional contact state, the former is proportional to the relative twist angle and the latter is proportional to the relative angular velocity.

Then the frictional moment $\boldsymbol{n}_{ci}$ working on the virtual object is given as follows (Fig.4).

- **Non-Contact State:**

$$\boldsymbol{n}_{ci} = 0 \tag{9}$$

- **Dynamic Frictional Contact State:**

$$\boldsymbol{n}_{ci} = \mu_{dn} \|\boldsymbol{f}_{cvi}\| \frac{\boldsymbol{\omega}_{vi}}{\|\boldsymbol{\omega}_{vi}\|} \tag{10}$$

where μ_{dn} is the rotational dynamic frictional coefficient at the contact point, and $\boldsymbol{\omega}_{vi}$ is the normal component of the relative angular velocity between the fingertip and the virtual object.

- **Static Frictional Contact State:**

$$\boldsymbol{n}_{ci} = K_{pn}\boldsymbol{\theta}_{vi} + K_{dn}\boldsymbol{\omega}_{vi} \tag{11}$$

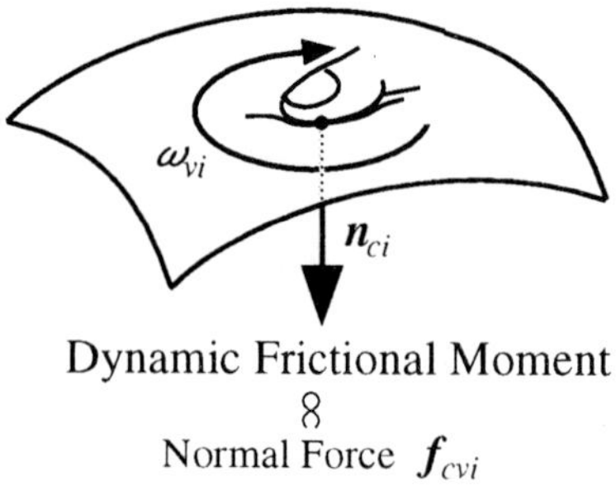

(a) Dynamic Frictional Contact State

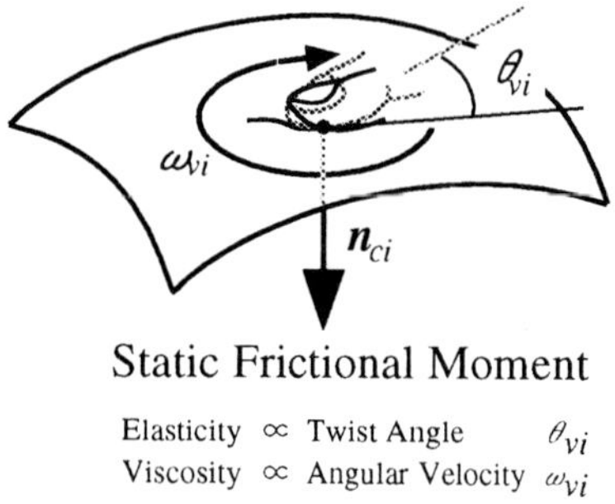

(b) Static Frictional Contact State

Figure 4. Moments Working on the Virtual Object

where K_{pn} is the rotational elastic coefficient, K_{dn} the rotational viscous coefficient, and $\boldsymbol{\theta}_{vi}$ the relative twist angle.

5.3. Transition of Contact State

Supposing that a shape of the virtual object is known, we can decide whether the ith fingertip is in contact with the virtual object.

When the fingertip and the virtual object are in contact, we can decide whether they are in dynamic frictional contact or in static frictional contact as follows (Fig.5).

- When they were in non-contact state a moment ago, or in the moment of contact, we suppose that they come in static frictional contact.

- When they were in dynamic frictional contact a moment ago, we suppose that slipping stops and they come in static frictional contact if

$$\|\dot{e}_{ti}\| = 0 \quad \text{and} \quad \|\boldsymbol{\omega}_{vi}\| = 0 \tag{12}$$

  Otherwise we suppose that they keep in dynamic frictional contact.

- When they were in static frictional contact a moment ago, we suppose that slipping starts

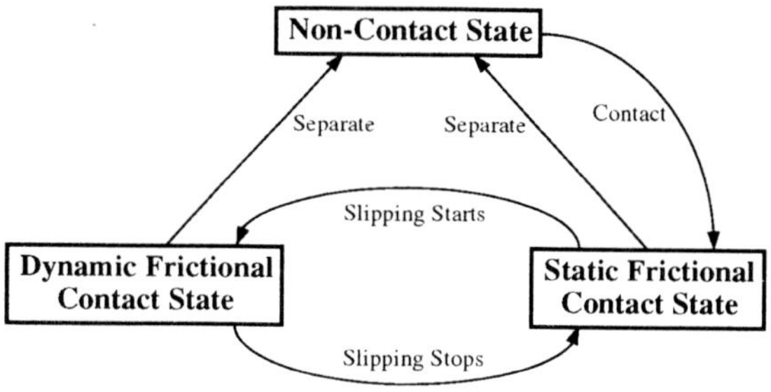

Figure 5. Transition of Contact States

and they come in dynamic frictional contact if

$$\begin{aligned} &\|\boldsymbol{f}_{cti}\| > \mu_s \|\boldsymbol{f}_{cvi}\| \\ \text{or} \quad &\|\boldsymbol{n}_{ci}\| > \mu_{sn} \|\boldsymbol{f}_{cvi}\| \end{aligned} \tag{13}$$

where μ_s is the maximum static frictional coefficient at the contact point, and μ_{sn} is the rotational maximum static frictional coefficient corresponding to the ordinary maximum static frictional coefficient.

Otherwise we suppose that they keep in static frictional contact.

5.4. Dynamics of the Virtual Object

In this paper we assume that a virtual object has an elastic surface. We use, however, the equation of motion of a rigid body, supposing that the deformation is quite little.

Consider a simple case of the virtual object as a single rigid body, on which no external forces are working except gravity and the forces exerted by the fingertips. Assuming that a force $\boldsymbol{f}_{ci}$ and a moment $\boldsymbol{n}_{ci}$ are working on the ith contact point, let $\boldsymbol{F}_{ci} \triangleq \left[\boldsymbol{f}_{ci}^T \ \boldsymbol{n}_{ci}^T \right]^T$, and also let $\boldsymbol{F}_c \triangleq col[\ \boldsymbol{F}_{c1} \cdots \boldsymbol{F}_{cn}\ ]$. Then the equation of motion of the virtual object is given by

$$\boldsymbol{M}_o(\boldsymbol{q}_o)\ddot{\boldsymbol{q}}_o + \hat{\boldsymbol{h}}_o(\boldsymbol{q}_o, \dot{\boldsymbol{q}}_o) = \boldsymbol{J}_{co}^T \boldsymbol{F}_c \tag{14}$$

where $\boldsymbol{q}_o$ is the 6-dimensional generalized vector denoting the position and the orientation of the virtual object, $\boldsymbol{M}_o(\boldsymbol{q}_o)$ the inertia matrix, and $\hat{\boldsymbol{h}}_o(\boldsymbol{q}_o, \dot{\boldsymbol{q}}_o)$ is the centrifugal, Coriolis, and gravity force term. $\boldsymbol{J}_{co}$ is defined as $\boldsymbol{J}_{co} \triangleq col[\ \boldsymbol{J}_{co1}(\boldsymbol{q}_o) \cdots \boldsymbol{J}_{con}(\boldsymbol{q}_o)\ ]$, $\boldsymbol{J}_{coi}(\boldsymbol{q}_o)$ in which is the Jacobian matrix of a contact frame Σ_{ci}, fixed to the ith contact point, with respect to $\boldsymbol{q}_o$.

6. Output Section

In this section, a control law for achieving the force $\boldsymbol{F}_{ci}$ given in the section 5.1 is described.

Let $\boldsymbol{F}_{hi}$ denote the generalized force exerted on the end of the i th arm by the operator's fingertip, which is expressed in the reference frame. Then the equation of motion of the i th arm is given by

$$\begin{aligned}\boldsymbol{\tau}_{li} = \boldsymbol{M}_{li}(\boldsymbol{q}_{li})\ddot{\boldsymbol{q}}_{li} + \widehat{\boldsymbol{h}}_{li}(\boldsymbol{q}_{li}, \dot{\boldsymbol{q}}_{li}) \\ - \boldsymbol{J}^T_{hli}(\boldsymbol{q}_{li})\boldsymbol{F}_{hi}\end{aligned} \tag{15}$$

where $\boldsymbol{\tau}_{li}$ is the joint driving torque, $\boldsymbol{M}_{li}(\boldsymbol{q}_{li})$ the inertia matrix of the arm, and $\widehat{\boldsymbol{h}}_{li}(\boldsymbol{q}_{li}, \dot{\boldsymbol{q}}_{li})$ the centrifugal, Coriolis, and gravity force term.

In order to compensate for the dynamics of the arm, let us define the joint driving torque command $\tilde{\boldsymbol{\tau}}_{li}$ as follows:

$$\begin{aligned}\tilde{\boldsymbol{\tau}}_{li} = \boldsymbol{M}_{li}(\boldsymbol{q}_{li})\ddot{\boldsymbol{q}}_{li} + \widehat{\boldsymbol{h}}_{li}(\boldsymbol{q}_{li}, \dot{\boldsymbol{q}}_{li}) \\ - \boldsymbol{J}^T_{hli}(\boldsymbol{q}_{li})\boldsymbol{u}_{fi}\end{aligned} \tag{16}$$

where $\boldsymbol{u}_{fi}$ is a new input vector. Then we obtain

$$\boldsymbol{u}_{fi} = \boldsymbol{F}_{hi} \tag{17}$$

In order to compensate for the error of the force, we define the input vector as follows using PI control law.

$$\begin{aligned}\boldsymbol{u}_{fi} = \boldsymbol{F}_{ci} - K_{fP}(\boldsymbol{F}_{hi} - \boldsymbol{F}_{ci}) \\ - K_{fI}\int(\boldsymbol{F}_{hi} - \boldsymbol{F}_{ci})dt\end{aligned} \tag{18}$$

where K_{fP} and K_{fI} are feedback gain coefficients.

7. Experiments

Based on the proposed method, a virtual world was constructed for a prototype haptic virtual reality experimental system we had built[5] for operating feel display while manipulating a virtual sphere with two fingers. Details of the system and some experimental results are given below.

7.1. Display System

The structure of the experimental display system is shown in Fig.6, and the display devices — the force display and the graphic display — are shown in Fig.7.

A pair of 3 d.o.f. miniature arms driven by geared DC motors are used for the force display devices, which are located 300 mm apart. Force sensors and finger connection devices are attached to the end of them. Since only 3 d.o.f. are available the moment that should be working on the operator's fingertip was not displayed.

Personal computers are used as a host computer (i486DX2, 66MHz) and a graphic computer (i486DX, 33MHz).

A stereoscopic display of time-sharing polarizing type is used for the graphic display.

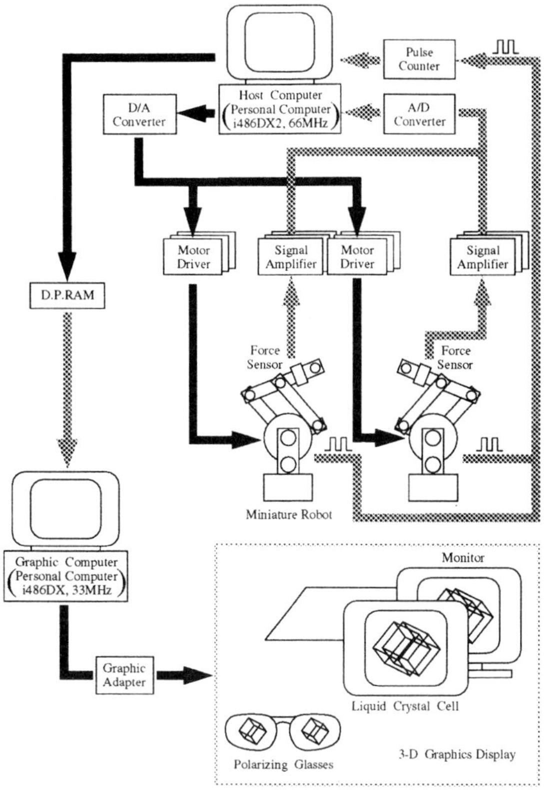

Figure 6. Structure of Display System

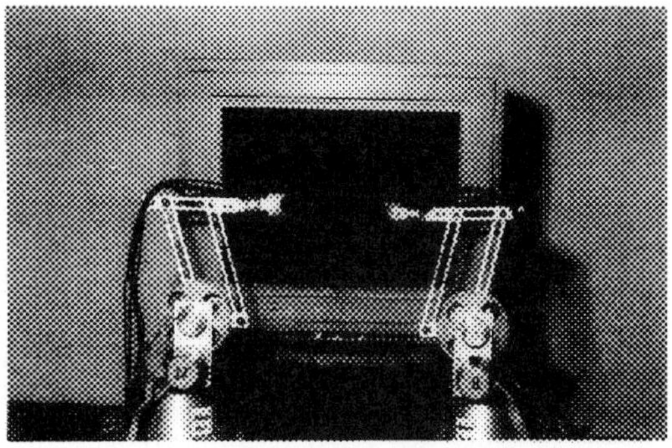

Figure 7. Display Device

7.2. Results

We will show the results of experiments in which the operator picks up a virtual object from the floor, waves it horizontally, throw it up, and catch it several times.

The virtual object is a sphere with radius 30 mm and the mass 2000 g. Since the gravitational acceleration is set $9800 \times 0.05 = 490$ mm/s^2 (one twentieth of the normal gravity on the earth), the weight of the virtual object is 0.98 N. The operator manipulates the virtual object with two fingers.

The results are shown in Fig.8. Fig.8(a) is the position of the sphere (the output of the dynamics

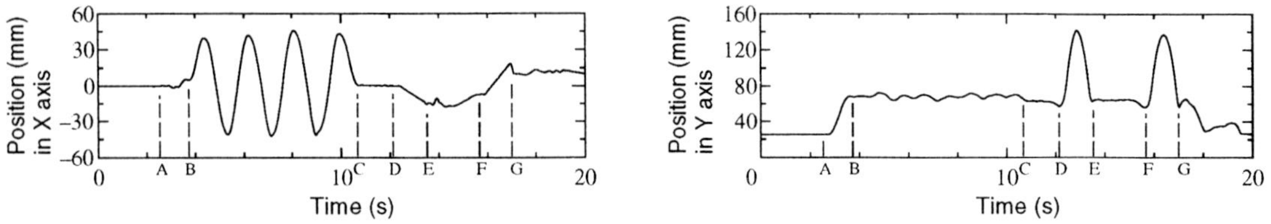

(a) Position of Virtual Object

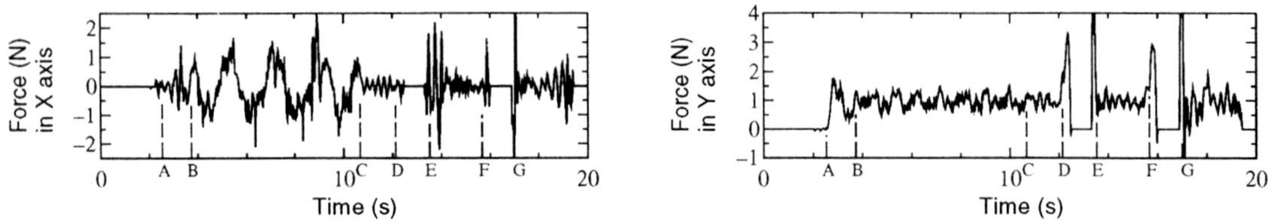

(b) Sum of Calculated Forces Working on Virtual Object

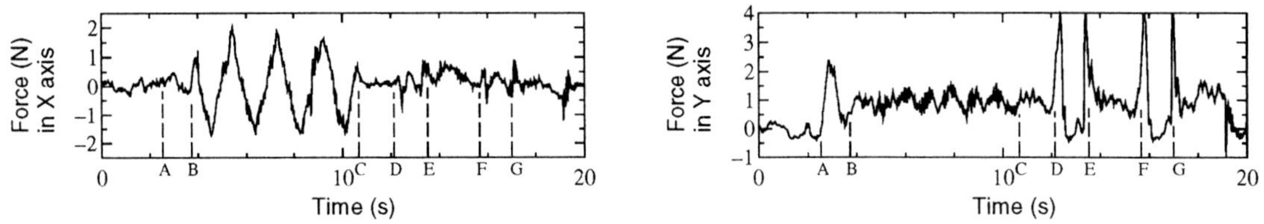

(c) Sum of Forces Measured by Force Sensors

Figure 8. Experimental Results

module for the sphere), (b) is the sum of the calculated forces working on the sphere (the outputs of the interaction modules), and (c) is the sum of the forces measured by the force sensors (approximately equal to the forces felt by the operator at the fingertips).

In the beginning the operator was not touching the sphere (~A). Next he picked it up (A~B), and waved it from side to side (B~C). Then he threw it up and caught it twice (D~E, F~G). The x, y, and z axes are in right, up, and out-from-the-screen direction as the operator faces to the display device. The graphs in the z axis are omitted, because we tried to operate the object in x-y plane.

From a good coincidence of (b) and (c), we can see a good performance of the force display device. Furthermore, from (a) and (b), we can see the following intuitively natural features of the operation.

- In the interval B~C, the phases of the position and force in the x axis are in reverse. In the y axis, the force around 0.98 N is exerted, which is the weight of the virtual object.
- In the intervals D~E and F~G, impulsive forces are observed at the moments of throwing up and catching.

Although we do not show the results concerning friction, we succeeded in displaying a natural operating feel for the case where slipping starts in grasping weakly and slipping stops in grasping strongly.

8. Conclusion

We have presented an implementation of the second method of impedance display, MM/DF, to display the operating feel of a dynamic virtual object which is in frictional contact with the operator's fingertips. It has been shown that the elasticity and friction of the surface of virtual object can be naturally modeled in this method.

We have made experiments in 3-dimensional space, using the display system consisting of two link mechanisms with 3-degrees-of-freedom. Experimental results show the validity of the proposed implementation.

References

[1] Brooks FP Jr., Ouh-Young M, Batter JJ, Kilpatrick PJ. Project GROPE — Haptic Displays for Scientific Visualization. *Comput. Graphics* 1990; 177–185

[2] Agronin ML. The Design of a Nine-String Six-Degree-of-Freedom Force-Feedback Joystick for Telemanipulation. *Proc. NASA Workshop on Space Telerobotics* 1987; 341–348

[3] Minsky M, Ouh-young M, Steele O, Brooks FP Jr., Behensky M. Feeling and Seeing: Issues in Force Display. *Comput. Graphics* 1990; 24(2):235–243

[4] Yoshikawa T, Yokokohji Y, Matsumoto T, Zheng, X-Z. Display of Feel for the Manipulation of Dynamic Virtual Objects. *Proc. Sixth Int. Conf. on Advanced Robotics* 1993; 217–222 (to appear in ASME, DSMC)

[5] Yoshikawa T, Zheng X-Z, Moriguchi T. Display of Operating Feel of Dynamic Virtual Objects with Frictional Surface. *Proc. IEEE/RSJ/GI Int. Conf. on Intelligent Robots and Systems* 1994; 731–738

[6] Sato M, Hirata Y, Kawarada H. Space Interface Device for Artificial Reality—SPIDAR— *Trans. Inst. Electron. Inf. Commun. Eng.* 1991; J74-D-II(7):887–894 (in Japanese)

[7] Iwata H. Artificial Reality with Force-Feedback: Development of Desktop Virtual Space with Compact Master Manipulator. *Comput. Graphics* 1990; 24(4):165–170

[8] Ohnishi K, Oomichi T. Manipulator System for Advanced Robot. *Journal of the Robotics Society of Japan* 1991; 9(5):628–637 (in Japanese)

[9] Burdea G, Roskos E, Gomez D, Langrana N, Richard P. Distributed Virtual Force Feedback. *Lecture Notes for Workshop on Force Display in Virtual Environments and its Application to Robotic Teleoperation* 1993; 25–44

[10] Bergamasco M, Allota B, Bosio L, et al. An Arm Exoskeleton System for Teleoperation and Virtual Environments Applications. *Proc. IEEE Int. Conf. on Robotics and Automation* 1994; 1449–1454

[11] Hirota K, Hirose M. A Curved Surface Presentation Device for Surface Display. *Proc. Tenth Sympo. on Human Interface* 1994; 193–196 (in Japanese)

[12] Hoshino H, Hirato R, Maeda T, Tachi S. Study on Virtual Haptic Space (II) —VR System Displaying the Shape of an Object—. *Proc. 33th Annual Conference of the Society of Instrument and Control Engineers in Japan* 1994; 691–692 (in Japanese)

[13] Furusho J. Development of Low-Inertia Actuator using ER Fluid and its Application to Virtual Reality. *Proc. JSME Annual Meeting* 1995; IV:265–266

[14] Yoshikawa T. Foundations of Robotics. MIT Press, Cambridge, Mass. 1990

PART 6

MOTION PLANNING AND TASK-LEVEL PROGRAMMING

SESSION SUMMARY

Jean-Claude Latombe
Department of Computer Science, Stanford University, Stanford, CA 94305, USA
email: latombe@cs.stanford.edu

The following four papers were presented in this session:

1. *Formal Verification in Robotics: Why and How?*, by B. Espiau, K. Kapellos, and M. Jourdan.
2. *Planar Manipulation on a Conveyor with a One Joint Robot*, by S. Akella, W. H. Huang, K.M. Lynch, and M.T. Mason.
3. *Robot Motion Planning in Dynamic Environments*, by P. Fiorini and Z. Shiller.
4. *A Random Sampling Scheme for Path Planning*, by J. Barraquand, L. Kavraki, J.C. Latombe, T.Y. Li, R. Motwani, and P. Raghavan.

These papers, which cover quite different topics, emphasize several important trends in robotics research, in particular the following three:

- How to do more with less.
- Reliability in robotics software.
- On-line motion planning.

How to do more with less: The first trend is best illustrated by Paper 2, which proposes effective techniques to orient a part using a robot with a single joint, a rotable fence. Actually, one could argue that the conveyor provides a second degree of freedom, but, unlike the rotable fence, it moves continuously. A remarkable outcome of the investigation is that such a very simple mechanical system is able to perform rather complex operations, at a very low cost. Since part feeding often accounts for a large portion of the cost of an assembly workcell, this type of result has great potential for industry. But the analytical analysis that comes with this work is perhaps even more important: it provides major insights in task mechanics, which can be later used to build more complicated systems, if needed. Very simple systems force clarity.

Paper 4 also illustrates the first trend, but in a very different way. While its main goal is to efficiently generate collision-free paths for robots with many joints, it proposes to solve this problem by repeating many times the same simple computation: pick a robot configuration at random and check it for collision with obstacles. Through formal analysis (backed by experimental work), the paper demonstrates the power of this general scheme.

Reliability in robotics software: The second trend is the central issue in Paper 1. As robotics systems are used to perform more complex tasks, it is increasingly important to design and implement software that meets formal requirements. Considerable research has been done on this issue in both computer science and software engineering. Surprisingly, the results of this research have still made few inroads in robotics software. Paper 1 draws from these results and adapt them to robotics software, which requires verifying both logical and temporal properties.

Since formal verification in robotics requires a prior model of the tasks (including the environments in which they will be performed), an interesting issue which arised during the discussion following the paper's presentation is whether such a model can be available. Formulated otherwise, the question is: What is an unexpected event? The notion of an unexpected event is central to the design of autonomous robots, but it seems to have different interpretations by different researchers.

The second trend is also an important aspect of Paper 4. Ideally a robot planner should be provably complete: whenever there exists a collision-free path, the planner should return one; otherwise, it should indicate that no paths exist. However, for robots with many joints (typically, more than five), provably complete planners are computationally impractical. This led the development of heuristic planners that are faster, but notoriously incomplete, hence unreliable when encapsulated in bigger systems. Paper 4 argues that it is possible to build probabilistically complete planners that come arbitrarily close to full completeness at reasonable computional cost. A question raised about this paper is whether the analysis it presents for robots in stationary environments carries over to dynamic environments. This question is still open, since the proofs of some of the results require that if a path exists between a point a (a configuration) and a point b, a path must also exist between b and a. This symmetry of path-connectedness is no longer satisfied in configuration$\times$time space.

On-line motion planning: The third trend is illustrated by Paper 3, which describes techniques for mobile robots to efficiently deal with dynamic environments. As unknown/dynamic environments are frequent in many applications of robotics, on-line planning techniques will become increasingly important. Concurrently, the reliability and actual efficiency of these techniques (second trend above) will have to be better analyzed. Competitive analysis, a general concept developed in computer science to compare on-line and off-line algorithms, will probably become a standard in future on-line robot motion planners.

Formal Verification in Robotics: Why and How?

B. Espiau, K. Kapellos, M. Jourdan
INRIA Rhône-Alpes, 46 Avenue Félix Viallet, 38031 Grenoble cedex (France)
email: Firstname.Name@inria.fr

1 Introduction

A mobile robot aimed to operate in an hazardous environment is a typical example of critical system. We mean here that, for such a system, like for a satellite, any repairing or recovery operation, even a mission reconfiguration, which would involve the intervention of a human operator is always costly, often difficult and sometimes impossible. This is why such systems should be at least provided with capacities of on-line adaption, like self replanning or sensor-based control. However, this is not sufficient and we have to be sure, as far as possible, that the system will behave correctly, *before launching*. More precisely, once a mission has been defined, we would like to verify that:

- its specifications are correct, i.e. that they correspond to the desired goals,
- its programming conforms to specifications,
- the constraints induced by real-time and implementation issues do not disturb its behavior.

Therefore, and assuming that the system hardware is given, this points out the necessity of validating all the algorithmic and software issues, from the point of view of their implementation as well as from that of their functionnalities. This mainly concerns two aspects, which should not be considered independently: the control issues, modelled through discrete- or continuous-time differential equations, and the logical ones, represented by discrete events. For example, we are led back in that last case to the need for verifying the *logical* aspects (absence of deadlocks, conformity of the results for various scenarii...), checking some *temporal* characteristics (absence of temporal deadlocks, values of lower and upper bounds on the duration of specific tasks...) and, possibly, studying *hybrid* models including other kinds of variables.

If we now adopt the user's point of view, it appears that his main interest is in the specification of complex missions or applications in an easy and safe way. For that purpose it is necessary to define properly what activities should be handled and how to *compose* them in order to meet the end user's requirements. From the verification side, the compositionnality principles must preserve the coherence between underlying mathematical models, in order to be able to perform formal computations at any level. As an example, the use of a single synchronous reactive language as a target for automatic translation is a way of preserving a logical structure whatever the complexity of the application. A consequence of this point of view is that the basic entities have to be carefully studied and also that composition operators should have a proper semantics.

The aim of this paper is to describe how and in what framework we address this class of problems, to present the used concepts and tools and to comment various examples of verification taken from robotics applications. More details may be found in [11, 12].

2 Formal Verification: From Computer Science to Robotics

2.1 Reactive systems

In 1977, Pnueli [1] proposed the use of temporal logics as a basis for proving correctness of reactive systems. Temporal logic formulas express in a declarative way the set of programs which satisfies the property. The verification process consists in testing that the program belongs to this set. Behavioral methods are also proposed as a complementary approach [2]: behavioral properties are expressed in the same formalism as the program to verify. Then, the verification process consists in comparing the program with the property by using, for instance, equivalence relations. Automation of the verification becomes possible owing to the use of finite state machines (i.e automata) to represent programs. For this reason, the definition and the efficient implementation of algorithms for automatic verification of finite state programs became an intense activity during the eighties for both methods. For temporal-logic based methods, decision procedures are known under the name of *model-checking* techniques. These algorithms are implemented through several tools: Aldebaran [3], Auto [17], The Concurrency WorkBench [5] for the behavioral approach ; EMC [16] and Xesar [6] for the logic-based one.

All the above-mentioned tools take automata as inputs. However, it is obvious that a complex system could not be described directly by an automaton which is a too low-level formalism of description. For this reason, high-level languages, the semantics of which are expressed in terms of automata, have been designed, like the *Synchronous* ones ([4]), among them ESTEREL, LUSTRE and SIGNAL are the most famous. They are used to program reactive systems, since the automaton model of a program can be translated into executable code automatically. These languages make the assumption that the outputs are simultaneous to the inputs that cause them. They offer programming environments with formal verification capabilities, as well as simulating and debugging features. Argos [8] is another synchronous language which is inspired from the graphical formalism of Statecharts [7].

2.2 Use in Robotics

Robotics, as an area integrating mechanics, computer science, automatic control and sensor technology is a domain where a natural reflex of the engineer is to *re-use* the basic techniques of the 'mother' areas before developing specific approaches. This is true also in the domain of formal verification. Indeed, it would be more exact to speak of formal verification *and* architecture design instead of verification alone, since these two issues are commonly strongly linked in robotics. In fact, considering robots as true hybrid (continuous/discrete -time) systems and exploiting this idea is a quite recent approach. This is why we will see that a review of the literature mainly exhausts the application of rather classical methods. Researchers in robotics have in fact followed two main paths:

- The first one comes more or less from questions raised by the so-called area of *Real-Time Artificial Intelligence*, and is also related to *Intelligent Control*. Here, verification aspects are usually focused on task coordination and mission reachability through reactive planning.

- A second point of view is based on the popular approach of DES (Discrete Event Systems). In the well known approach of Ramadge and Wonham ([23]), both the specification of a desired behavior and the modelling of a process have the forme of finite state machines. Then, owing to the definition of a set of controllable events, a dedicated supervisory controller can be synthetized. Very well suited to the design of controllers for flexible manufacturing systems, this theory is also the most largely used in the robotics literature.

Let us now give an overview of the existing approaches in robotics by taking the point of view of the application areas. We will distinguish three main domains.

The first is the one of flexible manufacturing systems. Of course, it is not purely robotics, and the usual approach is the synthesis of a supervisory controller *à la* RW, therefore with no major concern about verification issues. An exception is the work reported in [25, 26] in which the DES and the model checking approaches are blended in order to formally verify the logics of the synthetized supervisor. In [25], this is applied to a legged robot and in [26] to a simple workcell. Let us also mention the work of [31] where verification is done for industrial robots.

A second domain of robotics in which formal verification is clearly necessary, although only sometimes used, is the one of mobile autonomous robots. For example, in [30], coloured Petri nets are used for the design and the validation of the logical structure of a control architecture dedicated to a mobile robot with sensors. In [29], an architecture allowing to merge AI-type reasoning and real-time behavior through a graph-based representation is proposed and applied to a toy mobile robot equipped with a sonar. Again in the framework of reactive AI and planning, and even if not addressing exclusively the area of mobile robots, we find the approach of [28]. There, a structure based on sensory-motor modules and modelled as a network of concurrent communicating processes is defined. This allows to build and analyse plans including on-line reactivity. It is applied to visually-guided grasping in [27].

Another typical approach is reported in [40]: the DES/RW theory is used (mainly owing to its compositionnality properties) for synthetizing a controller handling a set of visual-based behaviors in a mobile robot. A very close application is addressed in [35], but with more verification aspects, as we will see later. Let us note that this question of compositionnality is strongly raised in the classical behavior-based models ([24]), where the need for a rich structure is obvious since it cannot exist implicitly in such approaches. That is the origin of the work reported in [41].

An important field where formal verification is required is the one of underwater vehicles or robots. A few works, like [33] or [37], verify explicitly some properties. However, the interested reader may also consult [39], where several approaches of control architecture are described, some of them allowing potentially to perform different types of verification.

Finally we cannot forget in this review the area of autonomous vehicles, which clearly demands high guarantees of safety and is therefore a critical application domain for verification methods. Let us for example mention that such requirements are expressed in [34], in the Automated Highway project. Some interesting results may also be found in automatic or aided car driving applications ([32, 38]).

3 Orccad: an Hybrid Systems-Oriented Architecture

ORCCAD ([22]) is a development environment for specification, validation by formal methods and by simulation, and implementation of robotic applications.

The formal definition of a robotic action is a key point in the ORCCAD framework. It is based on the following basic principles:

- most physical actions to be performed by robots can be stated as automatic control problems;

- the characterization of the physical action is not sufficient for fully defining a robotic action: starting and stopping times must be considered, as well as reactions to significant events observed during the task execution;

- since the overall performance of the system relies on the existence of efficient real-time mechanisms at the execution level, particular attention must be paid to their specification and their verification.

A robotic application should therefore include all these aspects coherently. Its specification must be modular, structured and accessible to users with different expertize. In ORCCAD, two entities are defined in order to fulfill these requirements: the *Robot-task* (RT), representing an elementary robotic action, where automatic control aspects are predominant although coherently merged with behavioral ones, and the *Robot-procedure* (RP), the RTs are the basic element of which, and where only behavioral aspects are considered. Therefore, in ORCCAD, a robotic application is fully specified by identifying and specifying all the RTs needed by the application, and then by composing them hierarchically in the form of RPs. The ORCCAD system provides a graphical environment, a link with a simulation system SIMPARC ([10]) and the generation of realtime code running under VXWORKS.

4 Specification Using Robot Tasks and Procedures

4.1 The Robot-task

The RT in ORCCAD is the minimal "granularity" seen by the end-user at the application level, and the maximum one considered by the control systems engineer at the control level. Formally, a RT is defined as the *parametrized specification of an elementary control law, i.e. the activation of a control scheme structurally invariant along the task duration, and of a logical behavior associated with a set of signals(events) which may occur just before, during and just after the task execution.*

The specification of the logical behavior is obtained by setting the events to be considered and their treatment. In order to make that specification easier to people from the robotics community who are not familiar with this approach, the events and the associated processings are typed. We distinguish:

• the pre-conditions: their occurence is required for starting the servoing task. Synchronization pre-conditions are related to logical conditions, while measurement ones are usually obtained through sensors. A temporal watchdog can be associated with each measurement pre-condition.

• the exceptions: they are emitted during the execution of the servoing task and indicates a failure detection. Their processing is as follows:

- type 1: the reaction to the received exception is limited to the modification of the value of at least one parameter within the control scheme,

- type 2: the exception requires the activation of new RTs. The reaction consists in killing the current one and reporting the causes of the malfunction to the adequate level. The recovering process to activate is known and specified in the RP to which the RT contributes,

- type 3: the exception is considered as fatal. The overall application is stopped and the robotic system must be driven to a safe position.

• the post-conditions: often related to the environment, they are handled as conditions for a normal termination of the RT. Watchdogs can be associated to their waiting.

Finally, a RT is completely specified by setting *temporal properties*, i.e. discretization of the time, durations of the computations, communication and synchronization between the involved processes. This is done by implementing each RT in terms of communicating real-time computing "tasks", called *Module-tasks* (MTs). Most of them are periodic and perform the calculations involved in the computation of the control algorithm. Others, called *observers*, monitor conditions and are therefore used to handle preconditions, exceptions and post-conditions. The non-periodic *reactive* behavior of the RT is handled by a special MT called the Robot-task Automaton (RTA) which may be awakened by signals coming from the RT itself through the outputs of the observers and is used to link the RT to the input/output signals associated with the application level. Modelling and analysis of the Module-tasks network are presented in [12]. Examples of RTs designed and implemented under ORCCAD may be found in [22, 21, 15, 36].

4.2 The Robot Procedure

The characterization of the interface of a RT with its environment in a clear way, using typed input/output events, allows to compose them easily in order to construct more complex actions, the so called *Robot-procedures*. The aim in designing this entity is to be able to define a representation of a robotic action that could fit any abstraction level needed by the mission specification system. In its simplest expression, a RP coincides with a RT, while the most complex one might represent an overall mission. Briefly speaking, it specifies in a structured way a logical and temporal arrangement of RTs in order to achieve an objective in a context-dependent and reliable way, providing predefined corrective actions in the case of unsuccessful execution of RTs.

More formally a RP is the full specification of

• *a main program, (nominal execution of the action), composed of RTs, RPs and conditions,*

• *a set of triplets {exception event, processing, assertion}, which specifies the processing to be applied for handling the exception, and the information to transmit to the planning level (if any),*

• *a local behavior defining the logical coordination of the previously considered items.*

The composition of RTs and RPs in the main program is obtained through operators ([13]) which express sequence, parallelism, conditions, iterations, rendez-vous and various levels of preemption. The exception events in the triplets are either type-2 exceptions detected by the participating RTs or ones specific to the RP. These elements are coordinated in the same way as in a RT : the main program is activated after satisfaction of the preconditions, and normally ends when the postconditions are satisfied. This nominal execution can be aborted for processing the exceptions. The semantics of the RP formalism is defined statically in [20]. The operational one is defined by a translation into the synchronous language Esterel [14], since its semantics is expressed in terms of automata.

4.3 Examples of Specification

4.3.1 *Automatic Vehicle Driving*

In the considered application, the PRAXITÈLE Project ([20]), we have two electric vehicles, the leader being driven and the second having to follow it like in a virtual train. Initially, the undriven car tries to catch the right signal in order to locate the first one. When done, the expected nominal execution is that the undriven car follows the driven one until the two be stopped by an operator intervention. During the execution the video signal may be lost or irrelevant. In this case a parking maneuver is started. The driven car is supposed to come back and the train to be reformed. This informal specification could be translated into a RP named FOLLOWME (see fig. 1). Its nominal execution is described in the RP main program specified as an infinite loop the body of which begins with the test of the external condition *TargetFound* which indicates that the driven car is detected by the second one. Whenever it is satisfied, within a specified elapsed time, a RP named GUARDEDFOLLOW, detailed below, is activated to control the second car. Let us emphasize that the programming is structured, in the sense that we can use a RP inside the definition of another one. Before starting the FOL-

```
Name : FollowMe
Pre-cond:OkInit[30ms],AutoMode[30ms],Start[5mn]
Main program:
        Loop
            wait TargetFound [5mn]
            start (GuardedFollow)
        EndLoop
T3 exceptions: Auto2man, MecFaill, ...
Post-cond: Stop [60mn]
```

Figure 1: Specification of FollowMe

```
Name : GuardedFollow
Nominal execution :
Parallel
        start (SensLoc)
        start (SensDir)
        Loop
           if  MoreBrake then start (BRAKE)
        EndLoop
EndParallel
T2 Exception: (TargetLost, start (PARKING))
```

Figure 2: Specification of GuardedFollow

lowMe nominal execution, a set of three preconditions must be satisfied before the indicated delays; the initialization phase (motors, sensors, ...) must have been achieved without detecting errors, the automatic mode must be activated and the "human" operator has to give the start order. The nominal execution of this main RP is stopped in two cases: either the supervisor gives a stop order or the manual mode is activated. In the first case, the RP ends normally; in the second one it is interrupted by a global type-3 exception.

In this example, six RTs are involved: two for the driving and steering motors of the car using exteroceptive sensor information, which virtually links the driven leader car to the non driven following one (named SensLoc and SensDir respectively); two allowing the undriven car to track a reference trajectory on the basis of odometry information only (respectively named CartLoc and CartDir); one (named BRAKE) using the foot-brake of the car in order to impose a desired deceleration.

The RP GuardedFollow (see fig.2) is built from the parallel composition of three RTs : SensLoc, SensDir and BRAKE . The RT BRAKE is activated every time the leading car imposes strong decelerations, indicated by *MoreBrake* event. Let us note that the first two RTs may be forced to stop if the exception of type 2 concerning the loss of the video signal between the two cars is detected (*SignalLost* event). The RP GuardedFollow handles this situation by starting a recovery program, the RP PARKING, the specification of which is analog to GuardedFollow RP; the difference is that RTs based on odometry information only are used instead of the sensor-based ones.

4.3.2 *Navigation of a Mobile Robot*

This example concerns the navigation of a mobile robot, ANIS, in an indoor environment ([35]). The mission goal is to achieve a *Reaching Target* task. Informally, it is specified as follows: the robot in a room must follow a wall while handling the presence of obstacles, reach a region of interest, find a target and go in front of it. Here, two RPs are hierarchically specified, encapsulating five RTs which include advanced control laws like a sensor-based one or a time-varying feedback; their activations are triggered by events which indicate the occurence of situations like the end of the wall, the presence of the target or of an obstacle, and failures. The abstract view of this application, in terms of RTs, as obtained by the verification process is presented fig. 8.

5 Formal Verification in the ORCCAD System

In order to perform behavorial verification, Esterel is used as a language of specification; it is interfaced with the Mauto verification tool (section 5.1). For temporal verification, we have to use Timed-Argos as a specification language and the Kronos symbolic verification tool (section 5.2). This explains why the RP and RT formalisms are both translated into Esterel and Timed-Argos according to the kind of properties one would like to check (see fig. 3).

5.1 Logical Verification

We consider here only the logical aspects of an application: time-related constraints are translated into logical events triggered by the environment. In this framework, our first goal is to formally verify the largest possible set of assertions about the logical behavior of the RTs and RPs. The systematization of the verification process requires
• an adequate systematic translation of the specification into an automaton model preserving the

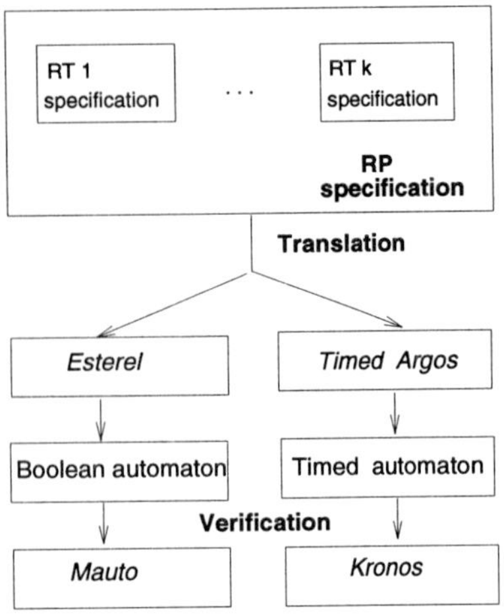

Figure 3: Specification-verification process in ORCCAD

information pertinent for the verification methods, i.e. here into an ESTEREL program;
• the classification of the properties to verify and the association of a dedicated verification method with every class. Two main categories of properties are identified: the first are related to critical working issues and the others are operational ones related to the completion of the desired objective. The generic *safety* and *liveness* properties are examples from the first category, while those concerning the *coherency of the specification with the requirements* of the application belong to the second one. In addition, verification methods are used to create adequate *abstract views* aimed to help the user during the specification phase.

5.1.1 *Robot Task Verification*

In relation with an ESTEREL program, one can generate, simulate, execute and verify automata. In particular MAUTO, an automatic verification tool, dedicated to analysis and transformation of finite automata, is directly linkable to the ESTEREL compiler. The resulted automata can be visualized before or after transformation using the MAUTO graphical interface AUTOGRAPH.

Owing to this environment the behavioral verification is performed, practically, in the following way: the global automaton is reduced with respect to the behaviors which are considered as relevant for the property to check, using the *abstraction-reduction* operations. The reduced automaton, when small, can be visually observed in order to invalidate or confirm the required property. An alternative way is to compare the reduced automaton with the specifications provided operationally in the form of an automaton. The property is verified when the reduced and the specification automata satisfy the *bisimulation* equivalence

Using that approach, we have been able to proof the *generic* correctness of any RT: it is non-blocking, and its ESTEREL implementation behaves conformly to its specification. The interest of this result is double: first, when analyzing the behavior of a RP, we can abstract the local behavior of RTs and consider it as *atomic* actions; secondly, from a software engineering point of view, the re-using of pieces of code proved correct a priori improves system reliability.

5.1.2 *Robot Procedure Verification*

The translation of a RP specification into an ESTEREL program is done in a structured way in order:
a) to enforce the reutilizability of the modules already designed. In addition, critical parts of the resulting program, independent from a particular specification are detected, programmed and verified once for ever augmenting the resulting program reliability,
b) to insert the information needed for automatizing the verification process,
c) to clearly separate the parts concerning the local behavior of the actions from the parts dealing with their sequencing.

The resulting program architecture is shown figure 4. The following modules are running in parallel:
- *action*$_i$, which contains the local behavior of the ith action involved in the RP specification. It reuses therefore either a RT program or a RP one. It is activated after reception by the *coordinator* module (see below) of a signal to start, and reports to it the way it is finished;
- *conditions*, which indicates to the *coordinator* the RP pre and post conditions satisfaction;
- *transition*, which ensures the control of the switching mechanism between two control laws applied to the same robot. The control law associated to the next RT is started after acknowledgement that the previous was stopped;
- *coordinator*, which is the principal module of the RP since it coordinates, through local signal exchange, the activation of the involved actions in accordance to the main program of the RP as given by the end-user. It implements therefore the logics of the sequencing, taking into account

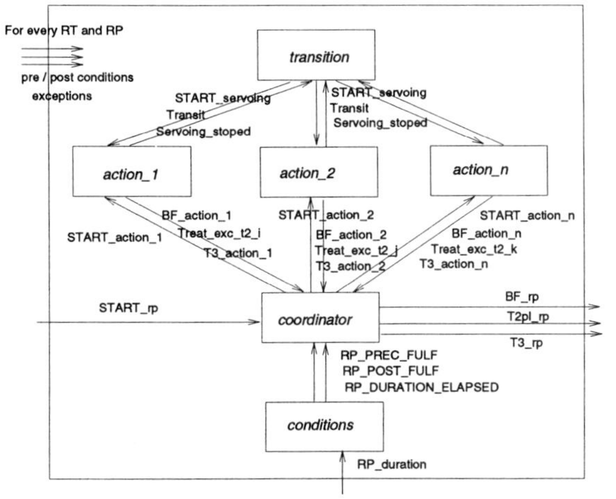

Figure 4: ESTEREL program architecture of a RP

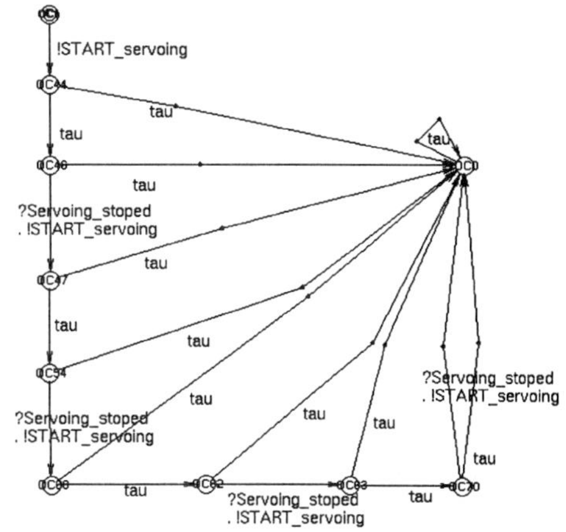

Figure 5: Abstracted automaton corresponding to starting and stopping signals of the servoing tasks

the starting and ending times of the actions only.

The RP ESTEREL program is an implementation of each of these modules. It should be emphasized that the RP definition does not give us the possibility to conclude that by construction it verifies properties analog to those validated on the RT definition since most of them are application-dependent. We describe now the classes of properties to be checked after having defined a mission in the form of a set of RPs.

5.1.3 *Crucial properties*

• Safety property

For the verification of the safety property, knowing the user's specification about fatal exceptions, abstract criteria composed by a set of abstract actions of the form $Bad_i = /Type\_3\_exception_i?\ and\ (not\ /Exc\_T3!)$ can be systematically produced. Such an abstract action indicates all the possible evolutions of the system where, although the particular exception specified as a type-3 one be an input of the system, no output asking its safe positioning is emitted. Then, the abstraction of the RP automaton with respect to this criterion is computed. The presence or absence in the resulting automaton of an action of the criterion invalidates or confirms the safety property.

• Liveness property

For proving the liveness property, we examine the equivalence by bisimulation of two automata. The first is built by abstraction of the initial automaton with respect to the criterion

$parse - criterion \quad Viv \quad = \quad tau \quad = \quad (not\ /BF\_rpr!)*,\ \ BF = /BF\_rpr!;$

which renames as *tau* (invisible action) all the sequences of actions (regular operator (*)) which do not indicate the good termination, and as BF the remaining ones. The second is specified as an automaton with one state and one transition labeled BF.

• Conflicts detection

We are interesting here to check that during the RP evolution it does not exist a time at which two different RTs are competing for the use of a particular resource of the system. For verifying that, in the case of the mobile robot ANIS, we observe the global automaton with respect to the signals *StartServoing* and *ServoingStop*. Figure 5 shows the resulting automaton. We can easily verify that the two signals appears alternatively.

5.1.4 *Coherence with the Application Requirements*

The conformity of the RP behavior to the mission constraints can also be verified. These constraints have to be expressed in a generic way as relationships between actions, events and actions or events themselves. Let us give two examples.

• Event/Action

Let us refer again to the example of autonomous vehicle driving. Mission specification requires that the brake should be activated every time the leading car imposes strong decelerations. The involved elements in the expression of this constraint are the *MoreBrake* event and the BRAKE action. The RP automaton is observed by *More-*

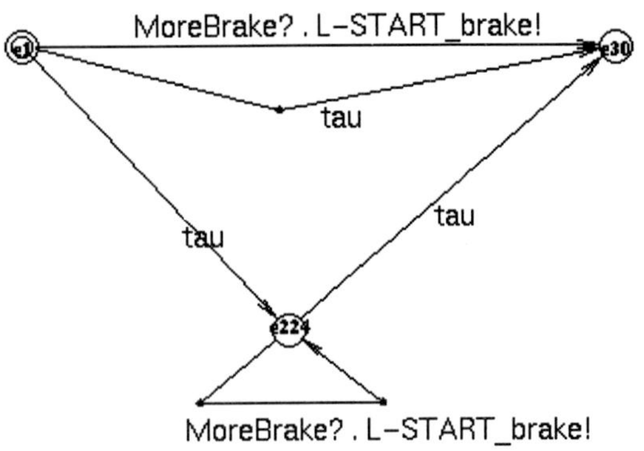

Figure 6: An abstracted automaton of the mission *Follow Me*

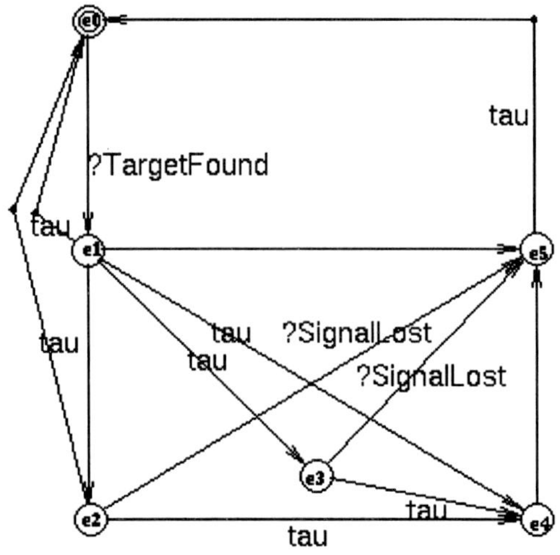

Figure 7: Abstracted automaton of the mission *Follow Me* with respect to two events

Brake and START_brake signals. The property is verified since in the resulting automaton, figure 6, there exists a loop indicating that as soon as *MoreBrake* signal is received the activation of the BRAKE action is asked for.

•Event/Event
In the same example, the specification requires that the target is searched for each time it has been lost. This is tested by specifying a behavioral property which involves the events *SignalLost* and *TargetFound*. The resulting abstracted automaton, figure 7 clearly certifies it.

An exemple of Action/Action verification in an assembly operation is also presented in [11].

5.1.5 *Abstract Views*

Finally, we consider that verification methods are also as a way of obtaining *abstract views* during the phase of RPs specification, in order to help the user. Actually, every mismatch in the behavior specification is reflected by the resulting automaton. So, by switching the body of a non-atomic behavioral activity to a single one at a relevant level of abstraction, useful views of the overall behavior can be given. Figure 8 illustrates the nominal execution of the overall mission REACHTARGET of the mobile robot ANIS at the level of abstraction of a RT; only starting and final instants of RTs are considered, making abstraction of their internal evolution.

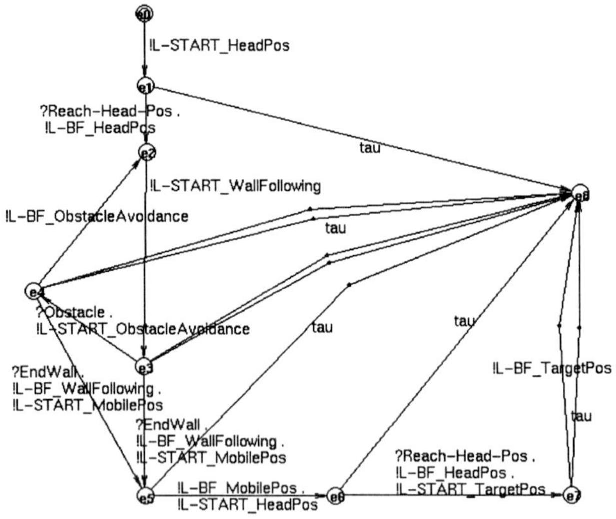

Figure 8: Abstract view of the mission *Reach Target*

5.2 Temporal Verification

5.2.1 *Timed-Argos and the Kronos tool*

The approach based on ESTEREL is well-suited for specifying and verifying the logical behavior of a robotic application. Nevertheless, it does not allow to take into account its temporal aspects. For example, it is possible to prove that an application always ends, but it is not possible to prove that its execution time is always lower than a constant T. The logical abstraction (delays are translated in logical timeout events) can also leads to inaccurate analysis if these ones are time-dependent. Proving that two execution laws are never executed simultaneously without taking into account the time parameters involved in the programs leads to approximate analysis only : the result is true if, whatever the time parameters are, the two execution laws are sequential.

We propose in this section another approach based on a synchronous language named Timed Argos, which allows to take into account the time parameters involved in a program during its formal analysis. Timed Argos [19] is an extension of the synchronous language Argos [8]. It provides the user with a set of operators that can be applied to elementary automata components to build more complex systems. These operators include parallel and hierarchic composition. The Argos semantics is expressed in terms of boolean automata.

Argos has been recently extended with a delay construction, leading therefore to Timed Argos, which allows to express watchdogs and timeouts easily. The semantics of Timed Argos is expressed in terms of timed automata. For this reason, Timed Argos is a high-level language to describe this kind of extended automata and Timed Argos programs could support quantitative analysis. Until now, Timed Argos is more specifically interfaced with the Kronos verification tool [18]. Kronos implements a symbolic model-checking algorithm for TCTL [9] (a real-time extension of the branching-time logic CTL) on Timed Automata. A property is expressed by a TCTL formula and Kronos computes the set of states of the Mealy Machines associated with the Timed Graph which satisfy it. The property is satisfied if and only if the initial state belongs to this set.

5.2.2 *A Classification of Relevant Temporal Properties*

In this section, we present a classification of temporal properties that it is interesting to check inside the ORCCAD environment. The idea is to avoid that the designer has to use TCTL, by identifying generic properties which could be automatically translated into TCTL formulas.

• **Time bound property**
This property concerns the boundedness of the maximum execution time of a Robot Procedure. The time bound property is the only one which could be checked automatically (without any designer's participation) whenever a complete specification of a Robot Procedure is built.

•**Event / Event properties**
This class of properties aims to check that in case of a complete execution (without detecting a type 3 exception) of a Robot Procedure, the maximum (resp. the minimum) elapsed time between two events of the robotic controller is lower (resp. greater) than a value T. For instance, during the specification of an automatic car parking the designer would like to check that the maximum time-lag between :

- the permission to enter given by the controller to a car and
- the parking acknowledgement given by the same car to the controller

is bounded by a value T specified by the designer.

•**Event / Action properties**
This class of properties expresses that, in case of a complete execution of a Robot Procedure, the maximum (resp. the minimum) laps of time between one event of the robotic system and the start of either a Robot Task, or a Robot Procedure or a Robot Task execution law, is lower (resp greater) than a value T. Let us take the example of the automatic vehicle driving mission to illustrate this. The designer would like to check that the maximum time-lag between the *MoreBrake* event which indicates that the engine-braking is not sufficient to stop the car, and the beginning of the Robot Task BRAKE which activates the active brake, is lower than a value T specified by the designer.

• **Action / Action properties**
This class may be splitted into two subclasses :

- Sequential Action / Action properties. The property to verify is that the maximum (resp. minimum) time-lag between two sequential actions (with the above-mentioned meaning) is lower (resp. greater) than a special value T.

- Overlapping Action / Action properties. The property to verify is that the maximum (resp. minimum) time-lag between two actions which could overlap themselves is lower (resp. greater) than a special value T.

5.2.3 *Discussion*

The experiments we conducted concern an automatic vehicle driving mission and are reported in [20]. They lead us to make the following remarks:

- Although the example we implemented was not very complex for the domain area, the size of the timed automaton which represents the controller was important: 105 states, 30000 transitions and 15 clocks. However, this timed automaton could be largely minimized for verification purposes, by developing a specific compilation process which would integrate the minimization phase, in order to improve its time performances.

- The identification of generic properties avoids that the user learns TCTL. Nevertheless, it remains to solve the "error diagnosis problem": when a property is not satisfied, the diagnosis given by Kronos is expressed in terms of timed-automaton informations. This diagnosis would be much more useful if given in terms of the source program.

6 Conclusion

We have shown in this paper how formal specification and verification methods can be used for checking the behavioral and temporal correctness in complex robotics applications. From our experiments, we may draw some conclusions which open future research directions. A first thing is that, even with a sophisticated environment like ORCCAD, it is not obvious for a non-expert user to handle verification tools. It is therefore necessary to greatly improve their interfaces. A second point is that the performances of the compilers and verification tools have to be also improved if we want to consider highly complex applications like underwater or space ones. Finally, let us emphasize that the recent approach known as *hybrid systems* seems a promising one for the analysis of complex robotics systems.

References

[1] A. Pnueli, *The temporal Logic of Programs*, 18th Annual Symp. on Foundations of computer Science, Providence, pp 46-57, 1977.

[2] R. Milner, *A Calculus of Communication Systems*, LNCS 92, Springer Verlag, 1980.

[3] J.C. Fernandez,*An Implementation of an Efficient Algorithm for Bisimulation Equivalence*, Science of Computer Programming, Vol. 13, N. 2-3, may 1990.

[4] IEEE Proc, *Another Look at Real-Time Programming*, no 9, vol 79, 1991.

[5] R. Cleaveland, J. Parrow, B. Steffen, *The Concurrency Workbench,* Workshop on Automatic Verification Methods for Finite State Systems, Vol. 407, LNCS, june 1989.

[6] J.-L. Richier, C. Rodriguez, J. Sifakis, J.Voiron , *Xesar : A Tool for Protocol Validation. User's Guide,* Technical Report, LGI-IMAG, Grenoble, France, 1987.

[7] D. Harel, *Statecharts: a visual approach to complex systems*, Weizmann Institute of Science, 1984.

[8] F. Maraninchi, *Operational and Compositional Semantics of Synchronous Automaton Compositions*, CONCUR, LNCS 630, Springer Verlag, 1992.

[9] R. Alur, C. Courcoubetis, D. Dill , *Model-checking for real-time systems* , 5th IEEE Symp. on Logics In Computer Science, IEEE Computer Society Press, 1990.

[10] C. Astraudo, J.J. Borrelly, *Simulation of Multiprocessor Robot Controllers*, Proc. IEEE Int. Conf. on Robotics and Automation, Nice, May 1992.

[11] B. Espiau, K. Kapellos, M. Jourdan, D.Simon, *On the Validation of Robotics Control Systems. Part I: High Level Specification and Formal Specification*, submitted to IEEE Trans. on Control Systems Technology, 1995.

[12] D. Simon, P. Freedman, E. Castillo, *On the Validation of Robotics Control Systems. Part II: Analysis of real-time closed-loop control tasks*, submitted to IEEE Trans. on Control Systems Technology, 1995.

[13] E. Coste-Manière, B. Espiau, E. Rutten, *A Task-Level Robot Programming Language and its Reactive Execution*, IEEE Int. Conf. on Robotics and Automation, Nice, France, May 1992, pp. 2751-2756.

[14] G. Berry, G. Gonthier, *The Synchronous Programming Language* ESTEREL*: Design, Semantics, Implementation*, Science Of Computer Programming, Vol 19 no 2, pp 87-152, 1992.

[15] E. Castillo, *Principes, techniques et outils de simulation, vérification et exécution d'actions robotiques*, PhD dissertation, INPG Grenoble, France, November 1994.

[16] E.M. Clarke, A. Emerson, A.P. Sistla, *Automatic Verification of Finite State Concurrent Systems using Temporal Logic Specifications: a practical approach*, 10th ACM Symp. on Principles of Programming Languages, pp 117-126, 1983.

[17] R. de Simone, D. Vergamini, *Aboard* AUTO, INRIA Technical Report no 111, 1989.

[18] T. Henzinger, X. Nicollin, J. Sifakis and S. Yovine, *Symbolic Model-Checking for Real-Time Systems*,LICS 92, IEEE Computer Society Press, June 1992.

[19] M. Jourdan, F. Maraninchi, A. Olivero, *Verifying quantitative real-time properties of synchronous programs*, 5th Int. Conf. on Computer-aided Verification, LNCS 697, Springer Verlag, June 1993.

[20] M. Jourdan, *Integrating formal verification methods of quantitative real-time properties into a development environment for robotic controllers*, INRIA Research Report no 2540, 1995.

[21] K . Kapellos, *Environnement de programmation des applications robotiques réactives*, PhD dissertation, Ecole des Mines de Paris, Sophia Antipolis, France, November 1994.

[22] D. Simon, B. Espiau, E. Castillo, K. Kapellos, *Computer-aided Design of a Generic Robot Controller Handling Reactivity and Real-time Control Issues*, IEEE Trans. on Control Systems Technology, vol 1, no 4, December 1993.

[23] P.J. Ramadge, W. M. Wonham,*The Control of Discrete Events Systems, Proceedings of the IEEE*, 77(1), 1989.

[24] R. Brooks: *A robust layered control system for a mobile robot*, IEEE Trans. Robotics and Automation, Vol. RA-2, No. 1, March 1986, pp 14-23.

[25] M. Antoniotti, B. Mishra, *Discrete Event Models + Temporal Logics = Supervisory Controller: Automatic Synthesis of Locomotion Controllers*, Research Report, NYU, 1995.

[26] M. Antoniotti, M. Jafari, B. Mishra, *Applying Temporal Logic Verification and Synthesis to Manufacturing Systems*, Research Report, NYU, 1995.

[27] T.G. Murphy, D.M. Lyons, A.J. Hendriks, *Visually Guided Multi-Fingered Grasping as Defined by Schemas and a Reactive System*, Workshop on Neural Architectures and Distributed AI, USC, Los Angeles, USA.

[28] D.M. Lyons, A.J. Hendriks, *Safely Adapting a Hierarchical Reactive System*, SPIE Symp. on Intelligent Robots and Computer Vision, XII, Boston, USA, 1993.

[29] D.J. Musliner, E.H. Durfee, K.G. Shin, *Reasoning about Bounded Reactivity to Achieve Real-Time Guarantees*, AAAI Spring Symp. on Selective Perception, march 1992.

[30] O. Causse, H.I. Christensen, *Hierarchical Control Design Based on Petri Net Modelling for an Autonomous Mobile Robot*, Intelligent Autonomous Systems Conf. (IAS 4), Karlsruhe, Germany, march 1995.

[31] Rahimi, Xia, *A Framework for Software Safety Verification of Industrial Robot Operations*, Computer and Industrial Engineering, vol 20 no 2 , pp 279-287, 1991.

[32] N. Lynch, H.B. Weinberg, *Proving Correctness of a Vehicle Maneuver: Deceleration* 2nd European Workshop on Real-Time and Hybrid Systems, Grenoble, France, 31-5/2-6 1995.

[33] E. Coste-Manière, M. Perrier, A. Peuch *Mission Programming: Application to Underwater Robots* 4th Int. Symp. on Experimental Robotics, Stanford, USA, June 30- July 2, 1995.

[34] A. Deshpande, P. Varaiya, *Design and Evaluation Tools for Automated Highway Systems*, 2nd European Workshop on Real-Time and Hybrid Systems, Grenoble, France, 31-5/2-6 1995.

[35] R. Pissard-Gibollet, K. Kapellos, P. Rives, J.J. Borrelly, *Real-Time Programming of Mobile Robot Actions Using Advanced Control Techniques*, 4th Int. Symp. on Experimental Robotics, Stanford, USA, June 30- July 2, 1995.

[36] P. Rives, R. Pissard-Gibollet, K. Kapellos, *Development of a Reactive Mobile Robot Using Real Time Vision*, Third International Symposium on Experimental Robotics, Kyoto, Japan, Oct 28-30, 1993.

[37] D. Simon, K. Kapellos, B. Espiau, *Formal Verification of Missions and Tasks: Application to Underwater Robotics*, Int. Conf. on Advanced Robotics, ICAR '95, Barcelona, Spain, sept. 1995.

[38] K. Kapellos, S. Abdou, M. Jourdan, B. Espiau *Specification, Formal Verification and Implementation of Tasks ans Missions for an Autonomous Vehicle*, 4th Int. Symp. on Experimental Robotics, Stanford, USA, June 30- July 2, 1995.

[39] K.P. Kalavanis and al., editors, *International Program Development in Undersea Robotics and Intelligent Control*, Proc. of the Joint US Portugal Workshop, Lisboa, Portugal, march 1995.

[40] J. Kosecka, H. Christensen, R. Bajczy, *Discret Event Modelling of Visually Guided Behaviors*, Int. J. of Computer Vision, 14, pp 179-191, 1995.

[41] J. Kosecka, H. Christensen,*Experiments in Behavior Composition*, 3rd Int. Symp. on Intelligent Robotic Systems, Pisa, Italy, July 1995.

Robot Motion Planning in Dynamic Environments

Paolo Fiorini
Jet Propulsion Laboratory,
California Institute of Technology,
Pasadena, CA 91109 USA
email: fiorini@telerobotics.jpl.nasa.gov

Zvi Shiller
Department of Mechanical, Nuclear and Aerospace Engineering,
University of California, Los Angeles,
Los Angeles, CA 90024 USA
email: shiller@seas.ucla.edu

Abstract

This paper presents a method for computing the motions of a robot in dynamic environments, subject to the robot dynamics and its actuator constraints. This method is based on the concept of Velocity Obstacle, which defines the set of feasible robot velocities that would result in a collision between the robot and an obstacle moving at a given velocity. The avoidance maneuver at a specific time is thus computed by selecting robot's velocities out of that set. A trajectory consisting of a sequence of avoidance maneuvers at discrete time intervals is generated by a search of a tree of avoidance maneuvers. An exhaustive search computes near minimum-time trajectories, whereas a heuristic search generates feasible trajectories for on-line applications. These trajectories are compared to the optimal trajectory computed by a dynamic optimization that minimizes motion time, subject to robot dynamics, its actuator limits and the state inequality constraints due to the moving obstacles. This approach is demonstrated for planning the trajectory of an automated vehicle in an Intelligent Vehicle Highway System scenario.

1 Introduction

This paper addresses the problem of motion planning in dynamic environments. Typical examples of dynamic environments include manufacturing tasks in which robot manipulators track and retrieve parts from moving conveyers, and air, sea, and land traffic, where aircraft, vessels and vehicles avoid each other while moving towards their destination.

Motion planning in dynamic environments is considerably more difficult than the widely studied static problem, since it requires the simultaneous solution of the path planning and of the velocity planning problems. Path planning involves the computation of a collision free path from start to goal without considering robot dynamics. Velocity planning, on the other hand, involves the computation of the velocity profile along a given path, satisfying system dynamics and actuator constraints. In addition, motion planning in static environments can be guaranteed to find a solution if one exists at time t_0, whereas motion planning in dynamic environments is essentially intractable, [38], [9], i.e. the solution at t_0 may not exist at a later time because of the evolution of the environment.

Previous methods consisted of a graph search in a configuration-time space of the robot [12, 13], and checking for intersection of the swept volumes of the robot and obstacles in a Cartesian-time space [5, 6, 7, 8]. Velocity constraints were considered in [37] for the solution of the asteroid problem, and acceleration constrains were satisfied in [10] and in [35] for planning for a point mass.

Approximate dynamic constraints were satisfied in [25, 26] by introducing the path-velocity decomposition method, and by limiting slope, direction and curvature of the trajectory in a position-time plane. A family of adjacent paths is developed in [17] to account for obstacles blocking the initial path, thus computing the complete trajectory by connecting different path segments. In [20] a network of possible paths is searched to compute the time-minimal trajectory. Dynamic constraints were used in [21, 22] to define the collision front, and in [18, 19] to introduce the concept of transient obstacles, also applicable to dynamic environments.

To date, on-line dynamic planning has been treated by emphasizing reasoning and decision making [38], or by creating artificial potential fields around the obstacles [27].

The time-optimal motion planning problem in static environments has been treated previously using parameter optimizations, representing the trajectory as a polynomial in time and accounting for obstacles using a penalty function [24], and representing the path by a cubic spline [40], [1], computing the motion time along the path with an efficient method developed in [1] and in [41]. Time optimal motion planning of cooperating robots moving along specified paths has been studied in [29, 41].

In this paper, we develop an efficient method for computing the trajectories of a robot moving in a time-varying environment. It utilizes the concept of Velocity Obstacle (VO), which represents the robot's velocities that would cause a collision with an obstacle at some future time. An avoidance maneuver is computed by selecting velocities that are outside of the velocity obstacle. To ensure that the maneuver is dynamically feasible, robot dynamics and actuator constraints are mapped into the robot velocity space. A trajectory consists of a sequence of such avoidance maneuvers, computed by searching over a tree of avoidance maneuvers generated at discrete time intervals. For on-line applications, the tree is pruned using a heuristic search designed to achieve a prioritized set of objectives, such as avoiding collisions, reaching the goal, maximizing speed, or computing trajectories with desirable topology.

The solutions computed with this method are conservative, since they exclude trajectories that, although feasible, include avoidance maneuvers violating the velocity obstacle. To evaluate the quality of these trajectories, they are compared to the trajectories computed using a dynamic optimization, which are not bound by the velocity obstacle approach. The optimization used here is a steepest descent [2, 11], modified to include the state-dependent inequality constraints due to the moving obstacles. The initial guess of the optimization is the trajectory generated by the search over the maneuver tree.

The advantages of this approach are multi-fold: *i*) it permits an efficient geometric representation of potential avoidance maneuvers of the moving obstacles, *ii*) any number of moving obstacles can be avoided by considering the union of their VO's, *iii*) it unifies the avoidance of moving as well as stationary obstacles, and *iv*) it allows for the simple consideration of robot dynamics and actuator constraints.

This paper is organized as follows. Section 2 defines the Velocity Obstacle, the avoidance velocities, and describes the representation used to compute the trajectories. Section 3 describes the dynamic optimization used to compute the time optimal trajectories. Examples of trajectories, and of their corresponding time optimal solutions are presented in Section 4.

2 The Velocity Obstacle

The **Velocity Obstacle** (VO) is an extension of the Configuration Space Obstacle [32] to a time-varying environment. It consists of the velocities of the robot that will cause a collision between the robot and the obstacles at some future time. Although this concept is valid for general robots and obstacles, in this paper we restrict our analysis to circular robots and obstacles in the plane.

In this section we first define the VO concept and then we combine it with the dynamic constraints of the robot to compute avoidance velocities that are also dynamically feasible. The corresponding avoidance maneuvers are used to build a tree that represents, within the given temporal resolution, all the avoidance trajectories generated to satisfy the constraints represented by the velocity obstacles.

2.1 The Velocity Obstacle

The VO is illustrated using the scenario shown in Figure 1, where two circular objects, A and B_1, are shown at time t_0 with velocities $\mathbf{v}_A$ and $\mathbf{v}_{B_1}$. Circle A represents the robot, and circle B_1 represents the obstacle. The velocities and positions of A and B_1 were chosen so that A and B_1 will collide at some time t_1, ($t_1 > t_0$), provided that $\mathbf{v}_A$ and $\mathbf{v}_{B_1}$ do not change.

To compute the VO, we first represent B_1 in the *Configuration Space* of A by reducing A to

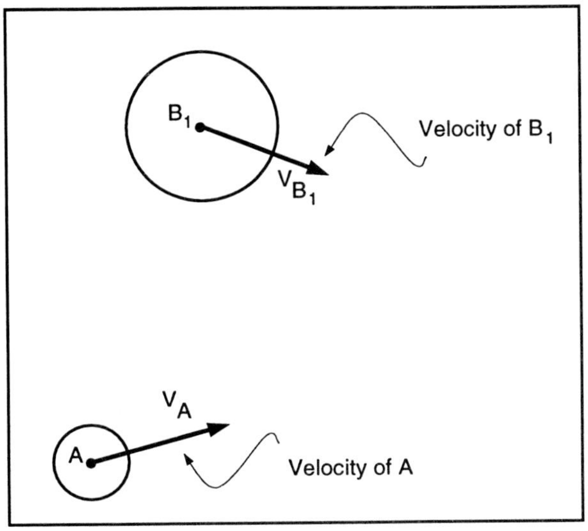

Figure 1: The robot and a moving obstacle.

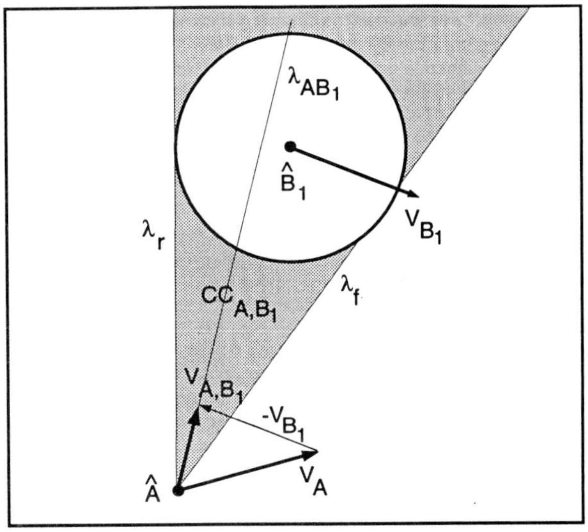

Figure 2: The Collision Cone CC_{A,B_1}.

the point $\widehat{A}$, and enlarging B_1 by the radius of A to the circle $\widehat{B_1}$ [31], and then we attach the velocity vectors to the position of $\widehat{A}$ and to the center of $\widehat{B_1}$, respectively.

By considering the relative velocity

$$\mathbf{v}_{A,B_1} = \mathbf{v}_A - \mathbf{v}_{B_1} \tag{1}$$

and by assuming that A and B_1 maintain their current velocities, a collision between $\widehat{A}$ and $\widehat{B_1}$ will occur at some future time $t_1 > t_0$ if the line λ_{A,B_1} of the relative velocity $\mathbf{v}_{A,B_1}$ intersects $\widehat{B_1}$. In fact, any relative velocity that lies between the two tangents to $\widehat{B_1}$ λ_f and λ_r will cause a collision between A and B_1. Therefore, we define the *Collision Cone*, CC_{A,B_1}, as the set of colliding relative velocities between $\widehat{A}$ and $\widehat{B_1}$ or:

$$CC_{A,B_1} = \{\mathbf{v}_{A,B_1} \mid \lambda_{A,B_1} \cap \widehat{B_1} \neq \emptyset\} \tag{2}$$

This cone is the planar sector with apex in $\widehat{A}$, bounded by the two tangents λ_f and λ_r from $\widehat{A}$ to $\widehat{B_1}$, as shown in Figure 2.

The collision cone thus partitions the space of *relative* velocities into colliding and avoiding velocities. The relative velocities, $\mathbf{v}_{A,B_1}$, lying on the boundaries of CC_{A,B_1} represent *tangent* maneuvers that would graze obstacle B_1.

The collision cone is specific to a particular pair of robot/obstacle. To consider multiple obstacles, it is useful to establish an equivalent partition of the *absolute* velocities of A. This is done simply by adding the velocity of B_1, $\mathbf{v}_{B_1}$, to each velocity in CC_{A,B_1} or, equivalently, by translating the collision cone CC_{A,B_1} by $\mathbf{v}_{B_1}$, as shown in Figure 3 [15]. The *Velocity Obstacle VO* is then defined as:

$$VO = CC_{A,B_1} \oplus \mathbf{v}_{B_1} \tag{3}$$

where $\oplus$ is the Minkowski vector sum operator.

Thus, the VO partitions the absolute velocities of A into *avoiding* and *colliding* velocities. Velocities on the boundaries of VO would result in A grazing B_1, since the corresponding relative velocities lie on the boundary of the collision cone CC_{A,B_1}. Note that the VO of a stationary obstacle is identical to its relative velocity cone, since then $\mathbf{v}_{B_1} = 0$.

To avoid multiple obstacles, the VO's of each obstacle are combined into a single velocity obstacle:

$$VO = \cup_{i=1}^{m} VO_i \tag{4}$$

where m is the number of obstacles. The VO assumes that the velocity of B_1 remains constant. To account for variable velocities, VO is recomputed at specified time intervals.

The assumption of circular robot and obstacles reduces the dimension of the configuration space, and thus greatly simplifies the computation of the VO. It also fixes the shape of the configuration space obstacles, which are generally functions of their positions in the robot's work space. For general manipulators, the VO must be periodically recomputed to account for the time varying configuration space obstacles [13].

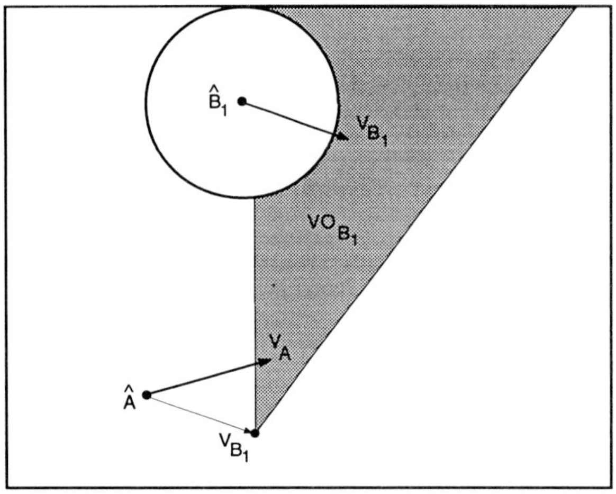

Figure 3: The velocity obstacle VO.

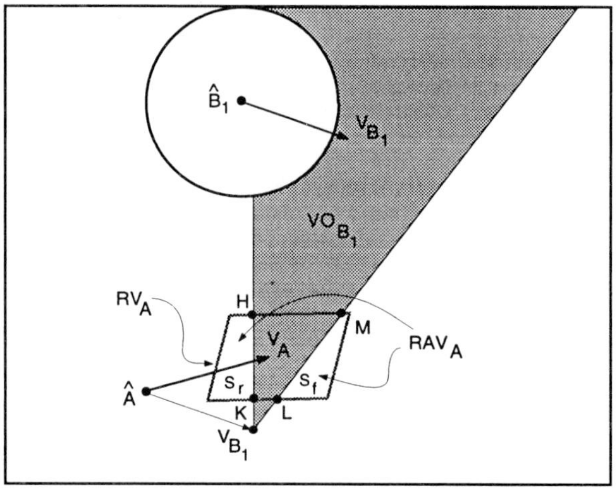

Figure 4: The reachable avoidance velocities.

2.2 The Avoidance Maneuvers

The velocities reachable by robot A at a given state over a given time interval Δt are computed by transforming the control constraints into bounds on the robot acceleration. The set of *feasible accelerations* at time t_0, $FA(t_0)$, is defined as;

$$FA(t_0) = \{\ddot{\mathbf{x}} \,|\, \ddot{\mathbf{x}} = f(\mathbf{x}, \dot{\mathbf{x}}, \mathbf{u}), \mathbf{u} \in U\} \quad (5)$$

where $f(\mathbf{x}, \dot{\mathbf{x}}, \mathbf{u})$ represents the dynamics of the robot, $\mathbf{u}$ are the actuator efforts, U is the set of admissible controls, and $\mathbf{x}$ is the position vector defined earlier. Note that the feasible acceleration range of a two degree-of-freedom system with decoupled actuator limits is a parallelogram [39].

The set of *reachable* velocities, $\mathrm{RV}(t_0 + \Delta t)$, over the time interval Δt is thus defined as:

$$\mathrm{RV}(t + \Delta t) = \{\mathbf{v} \,|\, \mathbf{v} = \mathbf{v}_A(t_0) \oplus \Delta t \cdot FA(t_0)\} \quad (6)$$

The set of *reachable avoidance* velocities, RAV, is defined as the difference between the reachable velocities and the velocity obstacle:

$$\mathrm{RAV}(t_0 + \Delta t) = \mathrm{RV}(t_0 + \Delta t) \ominus VO(t_0) \quad (7)$$

where $\ominus$ denotes the operation of set difference. A maneuver avoiding obstacle B_1 is thus computed by selecting any velocity in RAV. Figure 4 shows schematically the reachable velocity set RAV consisting of two disjoint closed sets, S_f and S_r, corresponding respectively to all feasible *front* and *rear* avoidance maneuvers. For multiple obstacles, the RAV may consist of multiple disjoint subsets.

2.3 Computing the avoidance trajectories

The trajectory that avoids static and moving obstacles, reaches the goal, and satisfies the robot's dynamic constraints is computed as a discrete sequence of elementary avoidance maneuvers, selected by a global search over the tree of all feasible maneuvers at specified time intervals. Alternatively, the global search may be reduced to a heuristic search for on-line applications, where the trajectories of the moving obstacles are not known a-priori, but are rather acquired in real-time. These two approaches are discussed next.

2.3.1 *Global Search*

We represent the state space of the robot by a tree of avoidance maneuvers at discrete time intervals. The *nodes* on this tree correspond to the positions of the robot at discrete times t_i. The *operators* expanding a node at time t_i into its successors at time $t_{i+1} = t_i + T$ are the velocities in the reachable avoidance velocity set RAV. The *edges* correspond to the avoidance maneuvers at those positions [34]. The search tree is then defined as follows:

$$\begin{aligned} n_{i,j} &= \{\mathbf{x}_{i,j} = (x_j(t_i), y_j(t_i)), & (8)\\ &\quad \mathbf{v}_{i,j} = (v_{j,x}(t_i), v_{j,y}(t_i))\} \\ o_{i,j,l} &= \{\mathbf{v}_l \,|\, \mathbf{v}_l \in \mathrm{RAV}_j(t_i)\} & (9)\\ e_{j,k} &= \{(n_{i,j}, n_{i+1,k}) \,| & (10)\\ &\quad n_{i+1,k} = n_{i,j} + (o_{i,j,l}\,\Delta t)\} \end{aligned}$$

where $n_{i,j}$ if the jth node at time t_i, $\mathrm{RAV}_j(t_i)$ is the reachable velocity set computed for node $n_{i,j}$,

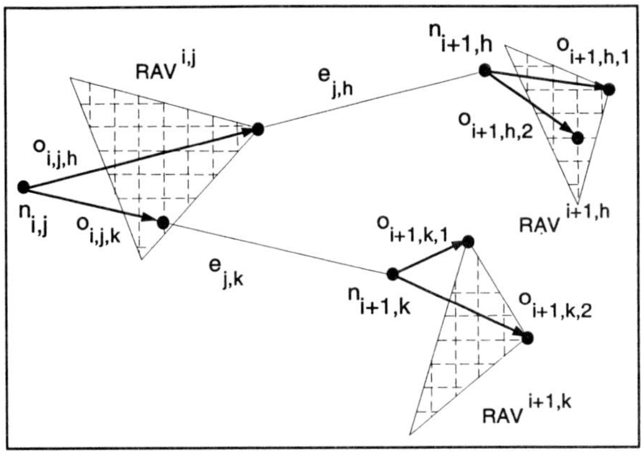

Figure 5: Tree representation for the global search.

$o_{i,j,l}$ is the lth operator on node j at time t_i, and $e_{j,k}$ is the edge between node n_j at time t_i, and node n_k at time t_{i+1}.

The tree of all feasible avoidance maneuvers is constructed as follows. At time t_i, the avoidance set RAV, corresponding to node $n_{i,j}$, is discretized by a grid. Then, the velocities corresponding to each node on the grid are used to compute the edges emanating from node n_{ij}. The positions reached by the robot at the end of each maneuver over the interval T are the successors of node $n_{i,j}$. A node $n_{i,j}$ is completely expanded when all the operators $o_{i,j,l}$ have been applied and all the edges emanating from $n_{i,j}$ have been examined. The resulting tree has a constant time interval T between nodes, a variable branch number that is a function of the shape of each RAV, and it can be searched using standard techniques [36]. Figure 5 shows schematically a subtree of some avoidance maneuvers.

Since the avoidance maneuvers are selected based on the velocity obstacles VO, they are never on a collision course with any of the considered obstacles, as discussed earlier. This excludes feasible trajectories that might, part of the time, be on a collision course with some of the obstacles. However, such trajectories may be generated by either considering only obstacles with imminent collisions, or by refining the trajectory using a dynamic optimization, as discussed later in Section 3.

2.3.2 *Heuristic Search*

For on-line applications, or when only incomplete information about the environment is available, the maneuver tree can be constructed incrementally using heuristic rules designed to satisfy the prioritized set of goals embedded in the formulation of the velocity obstacle. The survival of the robot can be guaranteed by selecting the avoidance velocities RAV; the target can be reached by selecting velocities that point towards the destination; the motion time can be minimized by choosing the highest velocity available; and the desired trajectory structure can be selected by choosing an appropriate sequence of front and rear avoidance maneuvers [14].

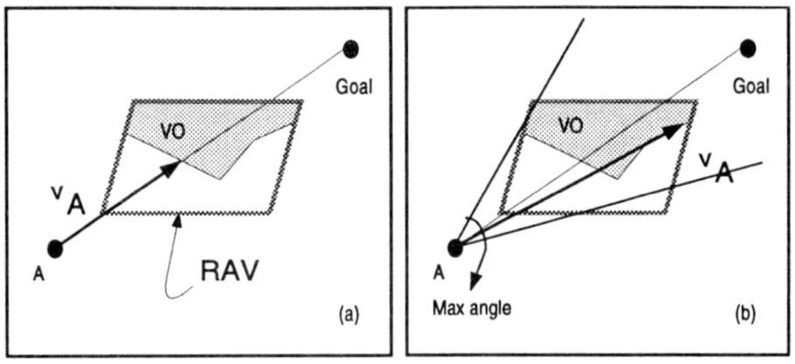

Figure 6: **a**: TG strategy. **b**: MV strategy.

We thus propose the following basic heuristics: (*a*) TG (to goal), choose the highest avoidance velocity along the line to the goal, as shown in Figure 6-a; (*b*) MV (maximum velocity), select the maximum avoidance velocity within some specified angle α from the line to the goal, as shown in Figure 6-b. Other heuristics may combine the above strategies in order to better satisfy the prioritized goals.

3 Dynamic Optimization

The trajectory computed by the heuristic search is evaluated off-line by a dynamic optimization. This optimization is formulated using Mayer's notation for the performance index J:

$$\min_{u(t)\in\mathbf{U}} J \tag{11}$$

$$J = \phi(\mathbf{x}(t_f), t_f) = t_f \tag{12}$$

subject to the following kinematic and dynamic constraints:

$$0 = \Gamma(\mathbf{x}(t_0), t_0) \tag{13}$$

$$0 = \Omega(\mathbf{x}(t_f), t_f) \tag{14}$$

$$\Psi : \bigcup_{i=1}^{n} [S_i(\mathbf{x}(t), t) = 0] \tag{15}$$

$$\dot{\mathbf{x}} = \mathcal{F}(\mathbf{x}, \mathbf{u}) = \mathbf{f}(\mathbf{x}) + \mathbf{g}(\mathbf{x})\mathbf{u} \tag{16}$$

$$\mathbf{U} = \{\mathbf{u} \mid u_i(min) \le u_i \le u_i(max)\} \quad (17)$$

where (13) is the initial manifold, (14) is the terminal manifold, (15) represents the time-varying obstacles, (16) is the robot dynamics, and (17) represents the admissible controls.

State constraints due to the presence of obstacles (15) are differentiated with respect to time until they become explicit in the controls $\mathbf{u}$, and then appended as state dependent control constraint to the Hamiltonian [4, 11]. The number of differentiations of each constraint represents the *order* p of that constraint. This approach requires an additional tangency constraint at the entry point of the constrained arc [42]. This may over-constrain the problem for constraints of order higher than two [23].

In the case of a single obstacle, the state constraint (15) is replaced by the tangency condition, denoted Ψ_1, and the control equality constraint, denoted Ψ_2:

$$\Psi_1 \; : \; \begin{pmatrix} S(\mathbf{x}(t),t)=0 \\ \dot{S}(\mathbf{x}(t),t)=0 \\ \vdots \\ S^{(p-1)}(\mathbf{x}(t),t)=0 \end{pmatrix} \quad t=t_1 \quad (18)$$

$$\Psi_2 \; : \; S^{(p)}(\mathbf{x}(t),u(t),t)=0,\; t_1 \le t \le t_2 \quad (19)$$

Then, the admissible control set $\mathcal{U}$ for the optimal control $u^*(t)$ becomes:

$$\mathcal{U}: \begin{cases} S^{(p)}(\mathbf{x}(t),u(t),t)=0 \quad \text{for} \quad S(\mathbf{x},t)=0 \\ \mathbf{U}: \mathbf{u}_{min}(\mathbf{x}) \le \mathbf{u} \le \mathbf{u}_{max}(\mathbf{x}) \end{cases} \quad (20)$$

3.1 The Necessary Optimality Conditions

The optimal control $\mathbf{u}^*(t)$ in the interval $t_0 \le t \le t_f$, that generates the optimal solution, $\mathbf{x}^*(t)$, minimizes J, and satisfies the fixed terminal manifolds Γ and Ω, is computed by satisfying the necessary conditions of Pontryagin Minimum Principle [3, 4, 30].

The Hamiltonian function $\mathcal{H}$ for this problem is defined as:

$$\mathcal{H}(\mathbf{\Lambda},\mathbf{x},\mathbf{u}) = \mathbf{\Lambda}^T(\mathbf{f}(\mathbf{x})+\mathbf{g}(\mathbf{x})\mathbf{u})+\mu^T\varphi(\mathbf{x},\mathbf{u}) \quad (21)$$

where Λ and μ are vectors of Lagrange multipliers, and $\varphi(\mathbf{x},\mathbf{u})$ is the set of active control constraints at time t, $t_0 \le t \le t_f$.

The adjoint equations for the Lagrange multipliers are given by:

$$\dot{\mathbf{\Lambda}} = -\left(\frac{\partial \mathcal{H}}{\partial \mathbf{x}}\right)^T \quad (22)$$

$$\mathbf{\Lambda}^T(t_f) = \left(\frac{\partial \phi}{\partial x} + \nu^T \frac{\partial \Omega}{\partial x}\right)_{t_f} \quad (23)$$

which exhibit the discontinuity at the entry point of the constraint given by [11]:

$$\mathbf{\Lambda}^T_{t_1^-} = \mathbf{\Lambda}^T_{t_1^+} + \eta^T \frac{\partial \Psi_1}{\partial x(t_1)} \quad (24)$$

When the constraint (19) is active, the minimization of the Hamiltonian for an autonomous system is equivalent to satisfying:

$$\mathcal{H}^*_u(\Lambda(t),\mathbf{x}^*(t),\mathbf{u}^*(t)) = 0 \qquad t_0 \le t \le t_f \quad (25)$$

and the adjoint equations for the constrained arcs become:

$$\dot{\Lambda}^T = -\Lambda^T\left[\frac{\partial \mathbf{f}(\mathbf{x})}{\partial \mathbf{x}} + \frac{\partial \mathbf{g}(\mathbf{x})\mathbf{u}}{\partial \mathbf{x}} - \mathbf{g}(\mathbf{x})\left(\frac{\partial \varphi(\mathbf{x},\mathbf{u})}{\partial \mathbf{u}}\right)^{-1}\frac{\partial \varphi(\mathbf{x},\mathbf{u})}{\partial \mathbf{x}}\right] \quad (26)$$

In summary, the trajectory minimizing the performance index $J = t_f$ is characterized by the following: *(i)* its controls $\mathbf{u}^*$ are in the admissible set $\mathcal{U}$, *(ii)* the states satisfy the terminal manifolds Γ at t_0 and Ω at t_f, *(iii)* the co-state equations are described by (22) on the free arcs and by (26) on the constrained arcs, with a discontinuity at the junction of the two given by (24), and *(iv)* the Hamiltonian is minimized over the entire interval. These conditions are satisfied by the trajectory computed by the numerical method described next [14].

3.2 Numerical Computation

The optimal trajectory is computed numerically by appending the constraints to the performance index J via appropriate arrays of Lagrange multipliers, and by computing the corrections to the controls that drive to zero the differential of the augmented performance index $d\widehat{J}$ [14]. This differential is then derived as a function of the variations in the control switches, assuming a bang-bang solution [43]. Singular arcs are approximated by a finite number of switches, generating a sub-optimal solution [33].

The differential of the augmented performance index is formed by computing the differentials of J, Ω and Ψ_1 independently, then including the effects of Ψ_1 on dJ and $d\Omega$, and finally combining all into $d\widehat{J}$. The final form of the augmented performance index is:

$$d\widehat{J} = \left(\frac{\partial\phi}{\partial t} + \nu^T \frac{\partial\Omega}{\partial t} + \mathcal{H}\right)_{t_f} dt_f + \int_{t_0}^{t_1^-} \mathcal{H}_u \delta u d\tau + \int_{t_1^+}^{t_f} \mathcal{H}_u \delta u d\tau \tag{27}$$

where the Hamiltonian is given by

$$\mathcal{H} = \Lambda^T \mathcal{F} + \mu\varphi$$

as in equation (21). The Lagrange multiplier Λ is defined as

$$\Lambda^T = \lambda_\phi^T + \nu^T \lambda_\Omega^T + \eta^T \lambda_\Psi^T$$

since the constraints Ψ_1 are not affected by the state after t_1, and η, ν are two constant Lagrange multipliers.

Bang-bang controls can only take one of two values α_M or α_m and the variation in $\mathbf{u}$ is rewritten as

$$\delta u_i = (\alpha_M - \alpha_m)\, sgn(dt_i)$$

where sgn is the signum function.

If $s_{s,r}$ indicates the number of switches in the s segment of the trajectory for the rth input, and m indicates the number of elements in $\mathbf{u}$, then $d\widehat{J}$ for bang-bang controls becomes:

$$d\widehat{J} = \sum_{i=1}^{m}\sum_{j=1}^{s_{1,i}} (\mathcal{H}_{u_i})_{t_{ij}} \delta u_i dt_{ij} + \sum_{i=1}^{m}\sum_{j=1}^{s_{2,i}} (\mathcal{H}_{u_i})_{t_{ij}} \delta u_i dt_{ij} + \sum_{i=1}^{m}\sum_{j=1}^{s_{3,i}} (\mathcal{H}_{u_i})_{t_{ij}} \delta u_i dt_{ij} + \left(\frac{\partial\phi}{\partial t} + \nu^T \frac{\partial\Omega}{\partial t} + \mathcal{H}\right)_{t_f} dt_f \tag{28}$$

This differential is minimized by choosing the steepest descent increments as [4]:

$$dt_{ij} = -\frac{(\mathcal{H}_{u_i})_{t_{ij}}}{w_{ii}\delta u_i} \tag{29}$$

$$dt_f = -\frac{1}{b}\left(\mathcal{H} + \frac{\partial\phi}{\partial t} + \nu^T \frac{\partial\Omega}{\partial t}\right)_{t_f}$$

with w_{ii} and b being suitable positive values.

The values of dt_{ij} and dt_f depend on multipliers η and ν. Their values are computed by requiring that differentials $d\Psi(t_1)$ and $d\Omega(t_f)$ be reduced by given quantities ϵ and θ. The values for the multipliers ν and η are then:

$$\eta = -I_{\Psi\Psi}^{-1}(\epsilon + I_{\Psi\Omega}\nu + I_{\Psi\phi}) \tag{30}$$

$$\nu = -\left({}^1I_{\Omega\Omega} + {}^1I_{\Omega\Psi}{}^1I_{\Psi\Psi}^{-1}{}^1I_{\Psi\Omega} + {}^2I_{\Omega\Omega} + {}^3I_{\Omega\Omega} + \frac{1}{b}\left(\frac{d\Omega}{dt}\frac{d\Omega}{dt}^T\right)_{t_f}\right)^{-1} \left(\theta + {}^1I_{\Omega\phi} - {}^1I_{\Omega\Psi}{}^1I_{\Psi\Psi}^{-1}\left(\epsilon + {}^1I_{\Psi\phi}\right) + {}^2I_{\Omega\phi} + {}^3I_{\Omega\phi} + \frac{1}{b}\left(\frac{d\Omega}{dt}\frac{d\phi}{dt}\right)_{t_f}\right) \tag{31}$$

where:

$${}^lI_{h,k} = \sum_{i=1}^{m}\sum_{j=1}^{s_{l,i}} \left(\lambda_h^T \frac{\partial\mathcal{F}}{\partial u_i} w_{ii}^{-1} \frac{\partial\mathcal{F}}{\partial u_i}^T \lambda_k\right)_{t_{ij}}$$

with

$$h = \Psi_1, \Omega, \quad k = \Psi_1, \Omega, \phi, \quad l = 1, 2, 3$$

The index l indicates the segments of the trajectory before, on, and after the state constraint. Index i on the constraint boundary represents only the independent controls.

3.3 Generating the Initial Guess

The trajectory computed by the heuristic search consists of a sequence of straight line segments and velocities assigned to each segment. Since the velocity changes abruptly between segments, we smooth it by a spline interpolation. Each trajectory segment i is divided into the two halves i_1 and i_2, and the spline interpolation is applied to the trajectory segments i_2 and $(i+1)_1$. This replaces the velocity discontinuities by a polynomial blend, chosen as a third order Hermite spline [16].

The bang-bang controls for the spline interpolation are computed as follows. First, the actuator efforts are computed from the accelerations of the spline interpolation using the inverse system dynamics. Then, the smooth actuator efforts are approximated by bang-bang controls, so that the trajectory obtained by integrating the controls has the same shape as the trajectory computed by the search. The optimal trajectory is a local minimum, in the vicinity of the initial guess,

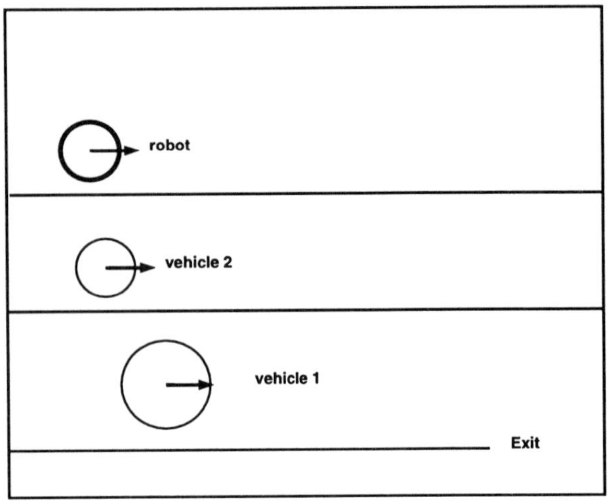

Figure 7: Example of trajectory problem.

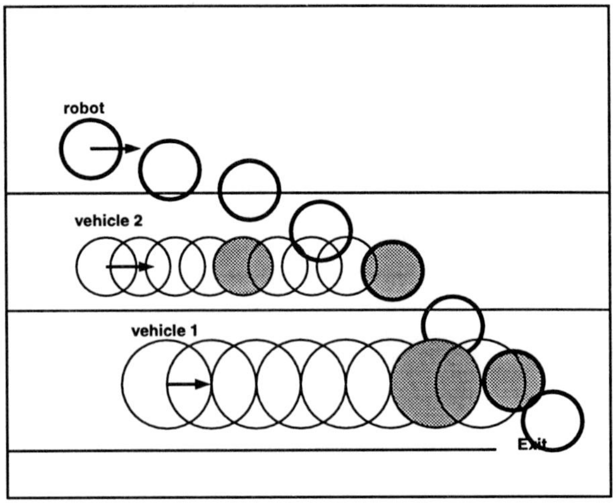

Figure 9: Trajectory computed by global search.

computed by optimizing the switching times. For the purpose of the following examples, the dynamics of the robot are modeled by two double integrators.

4 Examples

In these examples, the computation of trajectories is demonstrated for the Intelligent Vehicle Highway System (IVHS) scenario shown in Figure 7. The goal of the robotic vehicle in the leftmost lane is to reach the exit to the right without colliding with the other two vehicles. The initial velocity of the robot and of vehicle 2 is ($v_x = 30.0\ m/s$, $v_y = 0.0\ m/s$), and the initial velocity of vehicle 1 is ($v_x = 23\ m/s$, $v_y = 0.0\ m/s$). The velocities are represented as bars attached to the center of each circle.

The time interval T between nodes in the search tree is chosen arbitrarily as $T = 1s$. The trajectories are computed using the global search and the heuristic search with the TG and MV-TG strategies. The solution computed by the MV-TG strategy is then compared to the optimal solution computed by the dynamic optimization. In the following pictures, the circles representing the robot and the obstacles are displayed at .5 s intervals, and the robot is represented by the darker circle.

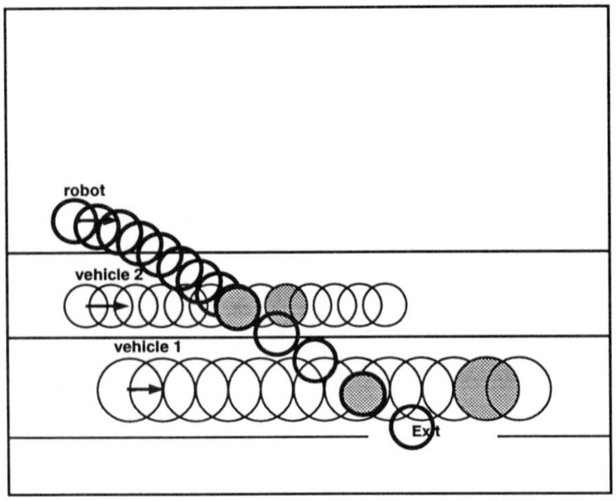

Figure 8: Trajectory computed with the TG strategy.

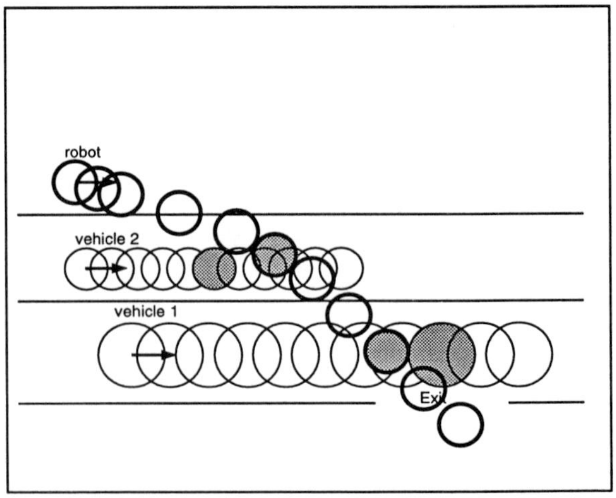

Figure 10: Trajectory computed with MV-TG strategies.

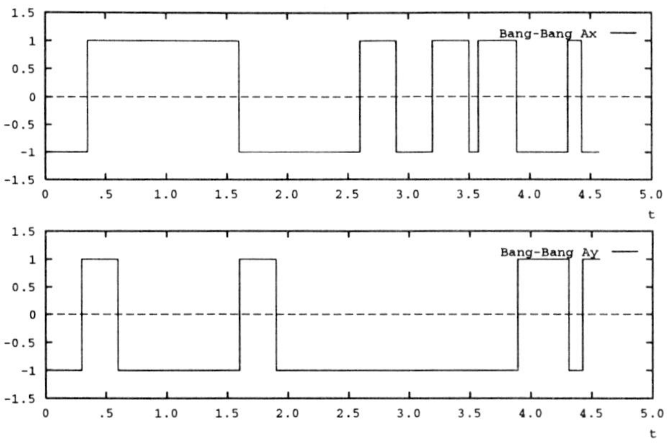

Figure 11: The Bang-Bang approximation of the controls of the initial guess.

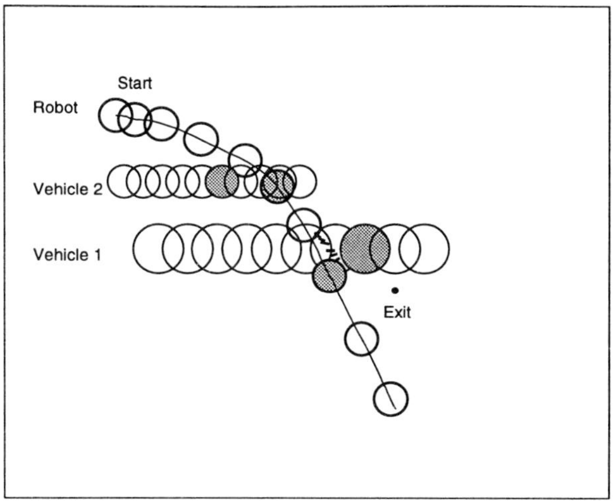

Figure 12: Initial guess of the dynamic optimization

The grey circles represent the position of the robot and of the obstacles when the robot crosses the path of the obstacles.

4.1 The Trajectories computed using the Velocity Obstacles

The trajectory shown in Figure 9 is computed using the Depth First Iterative Deepening algorithm [28]. This algorithm returns the fastest path to the target the first time it reaches the target. In the example, the RAV sets have been discretized by considering four points on each side of the boundary, and the maximum feasible velocity in the direction to the goal. The search has been expanded to 4.5 s and a depth of 5 levels . The total motion time of this solution is of 3.5 s.

Using the TG strategy resulted in the trajectory shown in Figure 8. Along this trajectory, the robot slows down and lets vehicle 2 pass, and then speeds up towards the exit, behind vehicle 1. The total motion time for this trajectory is 6.07 s.

The trajectory computed using both MV and TG strategies is shown in Figure 10. Along this trajectory, the robot first speeds up to pass vehicle 2, then slows down to let vehicle 1 pass on, and then speeds up again towards the goal. The motion time for this trajectory is 4.85 s. This trajectory was computed by using the MV heuristics for $0 \leq t \leq 2.0$ s and the TG heuristics afterwards.

4.2 The Optimal Trajectory

Figure 11 shows the bang-bang approximation of the controls for the trajectory shown in Figure 10. Figure 12 shows the initial guess of the dynamic optimization, obtained by integrating the bang-bang controls. The effects of the approximation show in the different terminal position of the robot, far from the target, and in collision with vehicle 1. This collision is represented in Figure 12 by a few bars between time $t = 2.5s$ and time $t = 3.0s$, indicating the depth of the contact between the robot and vehicle 1.

This trajectory resembles the trajectory of Figure 10, namely the type of maneuvers used in avoiding the obstacles, and the time of the avoidance of vehicle two, at approximately $t = 2.5s$. By using this trajectory as the initial guess, the dynamic optimization converges to the local minimum in the vicinity of the trajectory computed using the velocity obstacles, as shown in Figure 14. The optimal controls are shown in Figure 13.

The optimal solution of Figure 14 has the same key features of the heuristic trajectory of Figure 10, thus allowing for a meaningful comparison between the trajectory computed by the heuristic search and the corresponding optimal solution. The motion time of the optimal trajectory is 4.6 s, only slightly lower than the time of the heuristic solution of 4.85 s. The optimal trajectory violates at some points the velocity obstacle due to vehicle 1, which accounts for the shorter motion time.

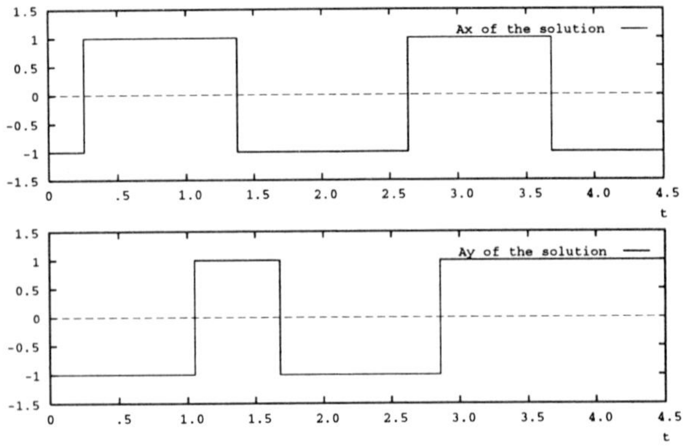

Figure 13: The optimal Bang-Bang controls.

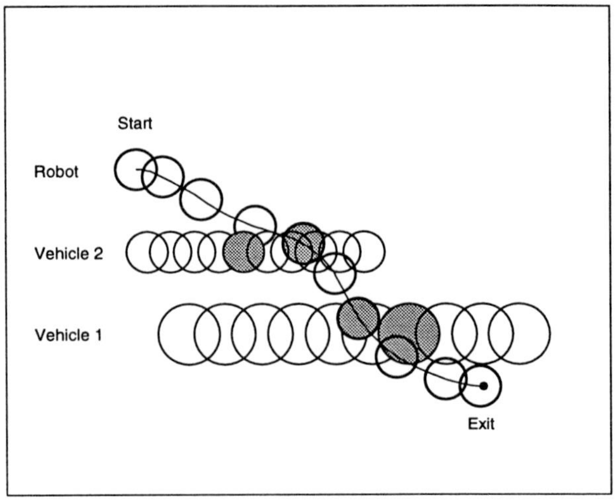

Figure 14: Optimal solution.

5 Conclusion

A novel method for planning the motion of a robot moving in a time-varying environment has been presented. It is significantly different from currently available planning algorithms, since it simultaneously computes the path and velocity profile that avoid all static and moving obstacles and satisfy the robot's dynamic constraints.

The method consists of computing, for every obstacle, its corresponding velocity obstacle, which is the set of colliding velocities between the obstacle and the robot. Then, by subtracting it from the reachable velocities of the robot, the set of reachable avoidance velocities is formed, which consists of all the velocities that avoid the obstacles and satisfy the robot's dynamic constraints. A search space is then formed by representing the state space of the robot by a tree of avoidance maneuvers. A global search over the tree yields trajectories that minimize a selected performance index, such as motion time or traveled distance. The solutions computed with this method are conservative, since each maneuver avoids all obstacles, irrespectively of their expected collision time, and they may exclude trajectories that, although feasible, violate the velocity obstacle in some interval. For on-line applications, the tree is pruned using one of several heuristic strategies, aimed at satisfying a prioritized list of goals, such as the survival of the robot, reaching the target, and minimize motion time. The quality of the heuristic trajectories is evaluated by comparing them to the optimal trajectories computed by dynamic optimization. The velocity obstacle method can be extended to three dimensional environments, by modeling the robots and the obstacles with spheres. The method is demonstrated for planning the trajectory of an automated vehicle in an Intelligent Vehicle Highway System scenario.

The main advantages of the velocity obstacle approach include the efficient geometric representation of maneuvers avoiding any number of moving and static obstacles, and the simple consideration of robot dynamics and actuator constraints. This approach, therefore, makes it possible to compute on-line safe and feasible trajectories in dynamic environments, such as Air Traffic control, Intelligent Highway Systems, and Harbor Traffic management.

6 Acknowledgment

The research described in this paper has been partially carried out at the Jet Propulsion Laboratory, California Institute of Technology, under contract with the National Aeronautic and Space Administration.

References

[1] J.E. Bobrow, S. Dubowsky, and J.S. Gibson. Time-optimal control of robotic manipulators along specified paths. *The International Journal of Robotics Research*, 4(3):3–17, Fall 1985.

[2] A.E. Bryson and W.F. Denham. A steepest-ascent method for solving optimum programming problems. *ASME Journal of Applied Mechanics*, (29):247–257, June 1962.

[3] A.E. Bryson, W.F. Denham, and S.E. Dreyfus. Optimal programming problems with inequality constraints *i*: Necessary conditions for extremal solutions. *AIAA Journal*, 1(11):2544–2550, November 1963.

[4] A.E. Bryson and Y.C. Ho. *Applied Optimal Control*. Hemisphere Publishing Corp., New York, NY, 1975.

[5] S. Cameron. A study of the clash detection problem in robotics. In *IEEE International Conference on Robotics and Automation*, pages 488–493, St. Louis, MO, March 25-28 1985.

[6] S. Cameron. Efficient intersection tests for objects defined constructively. *The International Journal of Robotics Research*, 8(1):3–25, February 1989.

[7] S. Cameron. Collision detection by four-dimensional intersection testing. *IEEE Journal of Robotics and Automation*, 6(3):291–302, June 1990.

[8] S.A. Cameron. *Modelling Solids in Motion*. PhD thesis, Edinburgh, UK, 1984.

[9] J. Canny and J. Reif. New lower bound techniques for robot motion planning problems. In *28th IEEE Symposium on Foundation of Computer Science*, 1987.

[10] J. Canny, J. Reif, B. Donald, and P. Xavier. On the complexity of kinodynamic planning. In *29th IEEEE Symposium on Foundation of Computer Science*, 1988.

[11] A.E. Denham, W.F.and Bryson. Optimal programming problems with inequality constraints *ii*: Solution by steepest ascent. *AIAA Journal*, 2(2):25–34, January 1964.

[12] M. Erdmann and T. Lozano-Perez. On multiple moving objects. In *IEEE International Conference on Robotics and Automation*, pages 1419–1424, San Francisco, CA, April 7-10 1986.

[13] M. Erdmann and T. Lozano-Perez. On multiple moving objects. *Algorithmica*, (2):477–521, 1987.

[14] P. Fiorini. *Robot Motion Planning among Moving Obstacles*. PhD thesis, University of California, Los Angeles, January 1995.

[15] P. Fiorini and Z. Shiller. Motion planning in dynamic environments using the relative velocity paradigm. In *IEEE International Conference of Automation and Robotics*, volume 1, pages 560–566, Atlanta, GA, May 1993.

[16] J.D. Foley, A. van Dam, S.K. Feiner, and J.F. Hughes. *Computer Graphics*. Addison-Wesley Publishing Company, Reading, MA, 1990.

[17] Th. Fraichard and C. Laugier. Path-velocity decomposition revisited and applied to dynamic trajectory planning. In *IEEE International Conference of Automation and Robotics*, volume 2, pages 40–45, Atlanta, GA, May 1993.

[18] K. Fujimura. On motion planning amidst transient obstacles. In *IEEE International Conference on Robotics and Automation*, pages 1488–1493, Nice, France, May 1992.

[19] K. Fujimura. Motion planning amid transient obstacles. *International Journal of Robotics Research*, 13(5):395–407, October 1994.

[20] K. Fujimura. Time-minimum routes in time-dependent networks. *IEEE Transaction on Robotics and Automation*, 11(3):343–351, June 1995.

[21] K. Fujimura and H. Samet. Time-minimal paths among moving obstacles. In *IEEE International Conference on Robotics and Automation*, pages 1110–1115, Scottsdale, AZ, May 1989.

[22] K. Fujimura and H. Samet. Motion planning in a dynamic domain. In *IEEE International Conference on Robotics and Automation*, pages 324–330, Cincinnati, OH, May 1990.

[23] D.H. Jacobson, M.M. Lele, and J.L. Speyer. New necessary conditions of optimality for control problems with state-variable inequality constraints. *Journal of Mathematical Analysis and Applications*, 35(2):255–284, 1971.

[24] D.W. Johnson and E.G. Gilbert. Minimum time robot planning in the presence of obstacles. In *IEEE Conference on Decision and Control*, pages 1748–1753, Ft. Lauderdale, FL, December 1985.

[25] K. Kant and S.W. Zucker. Towards efficient trajectory planning: the path-velocity decomposition. *The International Journal of Robotic Research*, 5(3):72–89, Fall 1986.

[26] K. Kant and S.W. Zucker. Planning collision free trajectories in time-varying environments: a two level hierarchy. In *IEEE International Conference on Robotics and Automation*, pages 1644–1649, Raleigh, NC, 1988.

[27] O. Khatib. A unified approach for motion and force control of robot manipulators: the operational space formulation. *IEEE Transaction on Robotics and Automation*, 3:43–53, 1987.

[28] R. Korf. Depth-first iterative-deepening: An optimal admissible tree search. *Artificial Intelligence*, (27):97–109, 1985.

[29] B.H. Lee and C.S.G. Lee. Collision-free motion planning of two robots. *IEEE Transactions on System Man and Cybernetics*, SMC-17(1):21–32, January-February 1987.

[30] G. Leitman. *An Introduction to Optimal Control.* McGraw-Hill Book Co., New York, NY, 1966.

[31] T. Lozano-Pérez. Spatial planning: a configuration space approach. *IEEE Transaction on Computers*, C-32(2):108–120, February 1983.

[32] T. Lozano-Pérez and M. A. Wesley. An algorithm for planning collision-free paths among polyhedral obstacles. *Communications of the ACM*, 22(10):560–570, October 1979.

[33] E.B. Meier. *An Efficient Algorithm for Bang-Bang Control Systems Applied to a Two-Link Manipulator.* PhD thesis, Stanford, December 1987.

[34] N Nilsson. *Principles of Artificial Intelligence.* Tioga, Palo Alto, CA, 1980.

[35] C. Ó'Dúnlaing. Motion planning with inertial constraints. Technical Report TR-230, NYU Robotics Laboratory, 1986.

[36] J. Pearl. *Heuristics.* Addison Wesley Publishing, Co., Reading, MA, 1985.

[37] J. Reif and M. Sharir. Motion planning in the presence of moving obstacles. In *25th IEEE Symposium on the Foundation of Computer Science*, pages 144–153, 1985.

[38] J.C. Sanborn and J.A. Hendler. A model of reaction for planning in dynamic environments. *International Journal of Artificial Intelligence in Engineering*, 3(2):95–101, April 1988.

[39] Z. Shiller and Dubowsky. Time optimal paths and acceleration lines of robotic manipulators. In *The 26th IEEE Conference on Decision and Control*, Los Angeles, CA, December 1987.

[40] Z. Shiller and S. Dubowsky. Robot path planner with obstacles, actuators, gripper and payload constraints. *The International Journal of Robotics Research*, 8(6):3–18, December 1989.

[41] K.G. Shin and N.D. McKay. Open loop minimum time control of mechanical manipulators. In *American Control Conference*, pages 1231–1236, June 1984.

[42] J. Speyer. *Optimization and Control of Nonlinar Systems with Inflight Constraints.* PhD thesis, Harvard, February 1968.

[43] A. Weinreb and A.E. Bryson. Optimal control of systems with hard control bounds. *IEEE Transactions on Automatic Control*, AC-30(d116), November 1985.

A Random Sampling Scheme for Path Planning

Jérôme Barraquand* Lydia Kavraki† Jean-Claude Latombe† Tsai-Yen Li‡
Rajeev Motwani† Prabhakar Raghavan§

Abstract

Several randomized path planners have been proposed during the last few years. Their attractiveness stems from their applicability to virtually any type of robots, and their empirically observed success. In this paper we attempt to present a unifying view of these planners and to theoretically explain their success. First, we introduce a general planning scheme that consists of randomly sampling the robot's configuration space. We then describe two previously developed planners as instances of planners based on this scheme, but applying very different sampling strategies. These planners are probabilistically complete: if a path exists, they will find one with high probability, if we let them run long enough. Next, for one of the planners, we analyze the relation between the probability of failure and the running time. Under assumptions characterizing the "goodness" of the robot's free space, we show that the running time only grows as the absolute value of the logarithm of the probability of failure that we are willing to tolerate. We also show that it increases at a reasonable rate as the space goodness degrades. In the last section we suggest directions for future research.

1 Introduction

Robot path planning has been proven a hard problem [40]. There is strong evidence that its solution requires exponential time in the number of dimensions of the configuration space, i.e., the number of degrees of freedom (dofs) of the robot. This result is remarkably stable: it still holds for rather specific robots, e.g., planar linkages consisting of links serially connected by revolute joints [17] and sets of rectangles executing axis-parallel translations in a rectangular workspace [13, 14]. Though general and complete algorithms have been proposed [6, 42], their high complexity precludes any useful application. This negative result has led some researchers to seek heuristic algorithms. While several such planners solve difficult problems, they also often fail or take prohibitive time on seemingly simpler ones. The fact that their behavior is not well characterized is a major drawback: they cannot be used as blackboxes in larger robot control systems.

The number of dofs beyond which existing complete algorithms become practically useless is low, somewhere between 3 and 5. This means that they cannot be used to compute paths for rigid objects translating and rotating in three dimensions, nor for six-dof manipulator arms, two important cases in practice. On the other hand, robot applications tend to involve more degrees of freedom than ever before. For example, an increasing number of manufacturing workcells use several cooperating robots to augment throughput and flexibility. Cells with more than twenty dofs are no longer exceptions. As costs and time for designing and deploying them become more critical, path planners integrated with CAD systems will be in higher demand to facilitate robot programming. Eventually, planners will run online to allow for non-deterministic sequences of goals and events [31]. Robots in domains other than manufacturing (e.g., medical surgery, space exploration) will also require efficient and reliable path planners. Some non-robotics domains raise a similar need as well. In computer graphics, animation of synthetic actors to produce digital movies or video games requires dealing with several dozen dofs. Here, motion planning may drastically reduce the work of human animators

*Salomon Brothers Int. Ltd., Victoria Plaza, 111 Buckingham Palace Road, London SW1W 0SB, UK. Email: jerome.barraquand@sbil.co.uk.

†Department of Computer Science, Stanford University, Stanford, CA 94305, USA. Email: {kavraki, latombe, motwani}@cs.stanford.edu.

‡Department of Computer Science, National Chengchi University, Wenshan, Taipei, Taiwan. Email: li@cs.nccu.edu.tw.

§IBM Almaden Research Center, San Jose, CA 95120, USA. Email: pragh@almaden.ibm.com.

who currently input large numbers of key frames. In molecular biology, motion planning can help compute plausible docking motions of molecules modeled as spatial linkages with many dofs.

Collision-free path planning, which assumes perfect knowledge of the world and stationary obstacles, is only the most basic motion planning problem in robotics. Clearly, we would ultimately like robot planners to also deal with issues such as uncertainties, moving obstacles, movable objects, and dynamic constraints [29, 30]. But every extension of the basic problem adds in computational complexity. For instance, allowing moving obstacles makes the problem exponential in the number of moving obstacles [6, 41]; uncertainties in control and sensing make the problem exponential in the complexity of the robot environment [6]. Before we can effectively investigate such extensions in large configuration spaces, it seems that we must better understand how to practically solve basic path planning.

Path-planning applications are so diverse that it is infeasible to design a tailor-made algorithm for every possible robot.[1] Instead, we need general path planning algorithms not bound to the specifics of any particular robot. We believe that between the two extreme types of planners suggested above – complete and heuristic – there is place for practically efficient general planners achieving a weaker form of completeness. In other words, we may perhaps trade a limited amount of completeness against a major gain in computing efficiency. Full completeness requires the planner to always answer a path-planning query correctly, in asymptotically bounded time. A weaker, but still interesting form of completeness is the following: if a solution path exists, the planner will find one in bounded time, with high probability. We call it *probabilistic completeness.* This weaker completeness becomes particularly interesting if we can show that the planner's running time grows slowly with the inverse of the failure probability that we are willing to tolerate.

With this philosophy in mind, we have designed new path planners and experimented with them in large configuration spaces. One of them, described in [3, 4, 29], is a potential-field-based planner that escapes local minima by performing random walks; in the following, we will refer to it as the *potential-field planner.* Another planner, presented in [18, 20, 21, 24], precomputes a "roadmap" (network) of simple paths connecting randomly selected configurations and tries to construct a path between any two input configurations by connecting them to this roadmap; we will refer to this planner as the *roadmap planner.* Both these planners have been successfully applied to complex problems. For example, in [26], the potential-field planner was used to automatically synthesize a video clip with graphically simulated human and robot characters entailing a 78-dimension configuration space. Both the potential-field and the roadmap planners have been used to check that parts can be removed from an aircraft engine for inspection and maintenance [8]; here, paths are generated in configuration spaces having only six dimensions, but the parts have particularly complex geometry.

These two planners achieve probabilistic completeness. For the potential-field planner, this property remains qualitative: if there exists a path, the probability that the planner finds one tends toward one as the running time increases; but the convergence speed is unknown. But, for the roadmap planner, we have proven stronger results that relate the probability that it finds a path, when one exists, to its running time [18, 19, 22, 23]. In turn, these theoretical results suggest improvements of the planner.

Other work investigating similar or related randomized planning approaches include [1, 5, 15, 16, 35, 36]. Formal attempts to predict the behavior of specific random planners are proposed in very few papers [9, 28].

This paper proposes a consistent framework to describe and study the randomized planners cited above, with the goal to eventually build more powerful planners. In the planners cited above, the robot's free space is not explicitly represented, but randomly sampled. We believe that this is the central concept underlying their success. In Section 2 we capture this concept into a general computational scheme for path planning in large configuration spaces.[2] In Section 3 we make our discussion more precise by presenting our potential-field and roadmap planners as two instances of planners using this scheme, but applying two different sampling strategies. In Section 4 we give two formal analyses of the probabilistic completeness of variants of the roadmap planner. These analyses provide a theoretical explanation for the empirically observed success of the roadmap plan-

[1] In any case, very few tailor-made planners have been successfully designed for specific robots with more than four dofs.

[2] This scheme can also be applied to configuration spaces having few dimensions; but it is less interesting in that case, since complete algorithms are then available.

ner. They are the crux of this paper. Indeed, as it has become relatively easy to design new randomized path planners that outperform previous ones on some well-selected problems, it is increasingly important to formally explain their strengths and weaknesses. This paper does not present experimental results obtained with our planners. Many such results were previously reported in [3, 8, 18, 20, 21, 24, 26, 27] and in papers by other authors [7, 11, 47]. The reader is referred to [34] for a textbook on randomized algorithms and sampling techniques.

2 General Scheme

We consider the problem of planning collision-free paths for an arbitrary n-degree-of-freedom holonomic[3] robot $\mathcal{A}$. We let $\mathcal{C}$ denote the configuration space of $\mathcal{A}$ and $\mathcal{C}_{free}$ stand for the open subset of the collision-free configurations in $\mathcal{C}$. We also refer to $\mathcal{C}_{free}$ as the *free space* and to configurations in $\mathcal{C}_{free}$ as *free configurations*. A planning problem is specified by two free configurations, $\boldsymbol{q}_{init}$ and $\boldsymbol{q}_{goal}$, called the *initial* and the *goal* configurations, respectively. Any path lying in $\mathcal{C}_{free}$ that joins these two configurations is a solution of this problem.

We assume for simplicity that $\mathcal{C}$ is the cube $[0,1]^n$, so that each configuration $\boldsymbol{q}$ is described by an n-tuple $(q_1,\ldots,q_n)$. However, it is straightforward to extend our presentation to cases where $\mathcal{C}$ is multiply connected, i.e., one or more dimensions can "wrap around". The only requirement is that $\mathcal{C}$ be measurable (in the Lebesgue sense), which is always achieved in practice.

Although there exist algorithms to construct an explicit representation of $\mathcal{C}_{free}$ given the geometry of $\mathcal{A}$ and the obstacles in semi-algebraic form, their time complexity makes them impractical for any sufficiently large value of n. On the other hand, reasonably efficient algorithms are available to compute the Euclidean distance between two objects in three-dimensional space (e.g., [10, 32, 37, 39, 45]). This leads us to assume that $\mathcal{C}_{free}$ is implicitly given by a function, CLEARANCE: $\mathcal{C} \rightarrow \mathbb{R}$, that maps any configuration $\boldsymbol{q}$ to the distance (in the workspace) between the robot placed according to $\boldsymbol{q}$ and the obstacles, or between two bodies of the robot, whichever is smaller. In the case where the robot collides with an obstacle or with itself, CLEARANCE($\boldsymbol{q}$) returns 0 or a negative number. Thus, whenever CLEARANCE($\boldsymbol{q}$) is positive, $\boldsymbol{q}$ belongs to $\mathcal{C}_{free}$. We refer to CLEARANCE($\boldsymbol{q}$) as the *clearance* of $\boldsymbol{q}$.

For any bounded robot $\mathcal{A}$, there exists a positive constant ρ such that when $\mathcal{A}$ moves along a straight path in $\mathcal{C}$ between any two configurations $\boldsymbol{q} = (q_1,\ldots,q_n)$ and $\boldsymbol{q}' = (q'_1,\ldots,q'_n)$, no point of $\mathcal{A}$ traces a curve segment longer than $\rho \max_{i\in[1,n]} |q_i - q'_i|$. We assume here that ρ is given, though for most usual robots its computation is straightforward and can be included in the planner. Rather than using a single ρ, we may also partition $\mathcal{C}$ into several regions and have a distinct constant ρ_i for each region. If the ρ_i's differ by large amounts, this could substantially reduce the running time of a planner based on the scheme presented below, but we will not go into such detail in this paper.

Definition 1 *Let* $\boldsymbol{q} = (q_1,\ldots,q_n)$ *and* $\boldsymbol{q}' = (q'_1,\ldots,q'_n)$ *be two free configurations whose respective clearances are* η *and* η'. *They are said to be adjacent if* $\rho \max_{i\in[1,n]} |q_i - q'_i| < \max\{\eta,\eta'\}/2$.

Clearly, if two configurations are adjacent the straight-line segment joining them lies entirely in $\mathcal{C}_{free}$. In fact, the coefficient $1/2$ appearing on the right-hand side of the above inequality is intended to make sure that the robot cannot collide with itself, since two bodies might then move simultaneously toward each other. This coefficient can be removed if self-collisions are known in advance to be impossible. Note also that the map:

$$(\boldsymbol{q},\boldsymbol{q}') \in \mathcal{C}^2 \mapsto \max_{i\in[1,n]} |q_i - q'_i| \in \mathbb{R}^+ \cup \{0\}$$

is the L^∞ distance in $\mathcal{C}$. We could choose other distances in $\mathcal{C}$ to define adjacency. Of course, the constant ρ depends on this choice.

Now we can state our **general planning scheme** as follows:

> Pick configurations in $\mathcal{C}$ at random; retain those configurations which lie in $\mathcal{C}_{free}$ (along with $\boldsymbol{q}_{init}$ and $\boldsymbol{q}_{goal}$) as the nodes of a graph G; and connect adjacent configurations by links of G.
>
> Return YES if $\boldsymbol{q}_{init}$ and $\boldsymbol{q}_{goal}$ belong to the same connected component of G.
>
> Return NO if no path has been found after having generated c configurations, where c is an input parameter.

The answer YES is always correct, that is, whenever the planner returns YES, there actually exists a collision-free path connecting $\boldsymbol{q}_{init}$

[3] Our presentation can be extended to nonholonomic robots. See [36, 43, 44].

to q_{goal}; moreover such a path can easily be extracted from the graph G. But the answer NO is not necessarily correct. Indeed, after a finite amount of computation (defined by the parameter c), the planner may still not have found a path between q_{init} and q_{goal}, even if one exists.

The key component of this planning scheme is the *sampling strategy* applied to generate the nodes of G. Different strategies are possible. For example, if multiple planning queries are to be made with the same robot and obstacles (*multi-query case*), then it may be suitable to invest preprocessing time to construct a network of configurations (which we call a *roadmap*). Processing each query then only requires connecting the input initial and goal configurations to the roadmap. If the queries are not known in advance, it seems reasonable to construct the roadmap by choosing configurations uniformly at random from $[0,1]^n$ since the resulting free configurations are then uniformly distributed over $\mathcal{C}_{free}$. However, we will see in Section 3.2 that, while the roadmap is being generated, it is possible to derive heuristic information from it and bias the selection of the new configurations.

Instead, if a single or very few queries are to be made with the same robot and obstacles (*single-query case*), the planner may be more successful by using a sampling strategy that picks new configurations, first in neighborhoods of the initial and goal configurations and then, iteratively, in neighborhoods of the newly generated configurations, until the two "sampling waves" meet. A similar strategy could also be used in the multi-query case to connect the initial and goal configurations to the roadmap.

Although the efficiency of any planner may be evaluated through experimentation, formal analysis is desirable to compare planners and stress their strengths and weaknesses. Ideally a planner's outcome should be YES with the highest probability whenever a free path exists and this outcome should be generated in minimal time. Hence, the analysis should relate the probability that the planner produces an incorrect answer to its running time. Given a small positive number α (the acceptable probability that the planner's output is NO while a free path exists), we would like to bound the running time of the planner by a function of α. If this function grows slowly with α^{-1}, then the planner is particularly interesting since we can get arbitrarily close to full completeness at a reasonable cost.

Let us assume for simplicity that CLEARANCE takes constant time to evaluate. Then the planner's running time mainly depends on two numbers: the number of sample configurations and the number of pairs of free configurations checked for adjacency. The number of sample configurations is at most c. But only a fraction of them, f, are in $\mathcal{C}_{free}$, hence in the constructed graph G. In practice, it is often the case that $f \ll c$. The planner should thus strive to get the greatest ratio f/c, but the distribution of the generated free configurations is also crucial to eventually find a path. The number of adjacency checks is at most proportional to f^2; however, the planner may choose not to test all pairs of free configurations for adjacency.[4]

It seems likely that no strong property can be proven for any given planner, if we do not make some assumption about $\mathcal{C}_{free}$. Moreover, no single planner is likely to be the most efficient for all possible problems. This suggests that planners should be analyzed under some well-specified assumptions. In Section 4, we will study the work carried out by a two-phase planner under two distinct assumptions. In one, the *visibility volume assumption*, $\mathcal{C}_{free}$ is such that every free configuration "sees" a subset of $\mathcal{C}_{free}$ whose volume is at least an ϵ fraction of the total volume of $\mathcal{C}_{free}$ (we then say that $\mathcal{C}_{free}$ is ϵ-good). In the second assumption, the *path clearance assumption*, there exists a collision-free path between q_{init} and q_{goal} that has some given clearance ξ. Assumptions must be carefully crafted. Indeed, if these are too specific or unrealistic, the analysis will not yield useful results; on the other hand, if they are too general, the results will be too weak.

3 Specific Planners

We now present two planners that make use of the above scheme with different sampling strategies.

3.1 Potential-Field Planner

This planner was originally described in [3, 4, 29], along with experimental results.[5] Extensions and additional experimental results were presented

[4] Checking the sample configurations for adjacency is an instance of the orthogonal range-searching problem [2, 38]. Using the range-tree technique, the pairs of adjacent configurations can be computed in $O(f \log^{n-1} f)$ time, or slightly faster. This is better than $O(f^2)$ when f is large enough.

[5] The planner is available by anonymous ftp from `flamingo.stanford.edu:/pub/li/rpp3d.tar.gz`.

in [8, 11, 26, 27]. Here, we focus on its relation to the general scheme presented above.

The planner is given a function $\boldsymbol{U} : \mathcal{C}_{free} \rightarrow \mathbb{R}^+ \cup \{0\}$, the *potential field*, with a single global minimum (0) at the goal configuration.[6] It then attempts to connect $\boldsymbol{q}_{init}$ to $\boldsymbol{q}_{goal}$ by alternating *down motions* and *escape motions*. Each down motion "descends" along $\boldsymbol{U}$ until it reaches a local minimum, while escape motions attempt to flee from local minima.

The following algorithm, in which h is an input parameter, constructs a down motion starting at configuration $\boldsymbol{q}_s$:

`Down-Motion`($\boldsymbol{q}_s$):

1. $\boldsymbol{q} \leftarrow \boldsymbol{q}_s$.
2. Until $\boldsymbol{q}$ is not labelled as a local minimum do:
   (a) Pick at random up to h configurations adjacent to $\boldsymbol{q}$, until one of them, $\boldsymbol{q}'$, satisfies $\boldsymbol{U}(\boldsymbol{q}') < \boldsymbol{U}(\boldsymbol{q})$.
   (b) If the previous step succeeded in generating $\boldsymbol{q}'$, then reset $\boldsymbol{q}$ to $\boldsymbol{q}'$; else label $\boldsymbol{q}$ as a local minimum.
3. Return $\boldsymbol{q}$.

At Step 2(a) let $\boldsymbol{q} = (q_1, \ldots, q_n)$ and $\eta = \text{CLEARANCE}(\boldsymbol{q})$. Configurations are picked at random from the volume defined by $\Pi_{i=1,\ldots,n}[q_i - \eta/\rho, q_i + \eta/\rho] \cap [0,1]^n$. According to Definition 1, all configurations in this volume are adjacent to $\boldsymbol{q}$. To guarantee that Step 2 does not loop for ever, $\boldsymbol{U}$ must have no local minimum in the boundary of $\mathcal{C}_{free}$. This can easily be obtained by including in the definition of $\boldsymbol{U}(\boldsymbol{q})$ a term proportional to $1/\text{CLEARANCE}(\boldsymbol{q})$.

The following algorithm constructs an escape motion starting at a local minimum $\boldsymbol{q}_l$:

`Escape-Motion`($\boldsymbol{q}_l$):

1. Pick at random the length, ℓ, of the motion.[7]
2. $\boldsymbol{q} \leftarrow \boldsymbol{q}_l$; $l \leftarrow 0$.
3. While $l < \ell$ do:
   (a) Pick at random a free configuration $\boldsymbol{q}'$ adjacent to $\boldsymbol{q}$.
   (b) $l \leftarrow l + \max_i |q_i - q'_i|$.
   (c) $\boldsymbol{q} \leftarrow \boldsymbol{q}'$.
4. Return $\boldsymbol{q}$.

The overall planning algorithm is the following:

`Potential-Field-Planner`($\boldsymbol{q}_{init}, \boldsymbol{q}_{goal}$):

1. $\boldsymbol{q}_l \leftarrow$ `Down-Motion`($\boldsymbol{q}_{init}$).
2. While $\boldsymbol{q}_l \neq \boldsymbol{q}_{goal}$ do:
   (a) Do:
      i. If the total number of configurations generated so far exceeds c then return NO and halt.
      ii. $\boldsymbol{q}_s \leftarrow$ `Escape-Motion`($\boldsymbol{q}_l$);
      iii. $\boldsymbol{q}'_l \leftarrow$ `Down-Motion`($\boldsymbol{q}_s$);

      until $\boldsymbol{U}(\boldsymbol{q}'_l) < \boldsymbol{U}(\boldsymbol{q}_l)$.
   (b) $\boldsymbol{q}_l \leftarrow \boldsymbol{q}'_l$.
3. Return YES.

The graph G constructed by the planner is a tree of paths rooted at $\boldsymbol{q}_{init}$. Each configuration is generated such that it is adjacent to a previously generated configuration, and is only connected to this configuration. Hence, the planner does not test pairs of configurations for adjacency, avoiding the cost of this test. During a down motion, the potential $\boldsymbol{U}$ introduces a bias in the choice of the tree path that will be followed by the planner. Although the global geometry of $\mathcal{C}_{free}$ is unknown, $\boldsymbol{U}$ can be seen as a *specialist* that gives some heuristic indications about this geometry, i.e.: which directions are promising and which aren't. The techniques in [3, 29] compute $\boldsymbol{U}$ by combining local-minima functions computed over the robot's workspace. Although the resulting $\boldsymbol{U}$ often prevents the planner from getting trapped into big obstacle cavities, it still has local minima and therefore is an imperfect characterization of the free space's geometry. On the other hand, computing local-minima-free potentials in large dimensional spaces is a difficult problem [25] that is at least as hard as path planning itself.

Using well-known properties of random motions, the potential-field planner can be shown probabilistically complete [3, 28]. Moreover, a calculation of the finite expected number of invocations of `Escape-Motion` is given in [28]. The idea underlying this calculation is the following. The basins of attraction of the local minima of $\boldsymbol{U}$ which have non-zero measure (the goal basin

[6] Techniques to automatically construct this function are proposed in [3, 29].

[7] In [3], we note that, on the average, a random walk of length ℓ ends $\sqrt{\ell}$ away from its starting point. This leads us to suggest choosing $\sqrt{\ell}$ according to a truncated Laplace distribution with mean value equal to the "radius" of the configuration space. Here, this radius is 1.

is assumed to be one of them) form a partition of the free space. For any two such basins B_i and B_j, one can define the finite transition probability p_{ij} that a random walk generated by Escape-Motion from the minimum in B_i terminates into B_j. The expected number of invocations of Escape-Motion before entering the goal basin is then expressed as a function of the p_{ij}'s. In theory, this result gives some indication of the expected complexity of the potential-field planner. But, in practice, the p_{ij}'s are unknown. In fact, a key issue in analyzing a planner based on random sampling is to translate some geometric property of the free space into probabilistic terms.

The potential-field planner has successfully solved many difficult problems (e.g., see [3, 8, 11, 26]). However, it is easy to create problems that it fails to solve in a reasonable amount of time, though they admit rather obvious solutions. Not surprisingly, most failures are due to *traps*, which cause some transition probabilities p_{ij} to be very small (in particular, the probabilities to enter the goal basin from some other basins). A trap is a basin of attraction of a local minimum of $\boldsymbol{U}$ that is almost completely surrounded by forbidden configurations (i.e., $\mathcal{C}\backslash\mathcal{C}_{free}$). Hence, a trap admits only narrow exits and the potential function inside the trap guides the robot away from these exits. If the robot's path starts within a trap or is guided into one by $\boldsymbol{U}$, each escape motion executed then has a tiny probability of escaping the local minimum. One could imagine a variant of the planner where the potential field is randomly guessed among a collection of several functions and changed several times during planning, hoping that the various potentials would entail different traps. We did some experiments with this idea, but we only got limited results.

3.2 Roadmap Planner

The problems encountered with the potential-field planner led us to develop another planner, the roadmap planner[8] described in [18, 20, 21, 24]. Several variants of this planner have been implemented; we present one below. Unlike the potential-field planner, this new planner uses no problem-specific heuristics.

The roadmap planner operates in two phases, preprocessing and query processing: The preprocessing phase consists of constructing a network R of configurations, the roadmap. The configurations in R are called *milestones*. They only form a subset of all the sample configurations generated by the planner; hence, the roadmap is not exactly the graph G mentioned in the general scheme. Every query specifies two configurations, $\boldsymbol{q}_{init}$ and $\boldsymbol{q}_{goal}$, in $\mathcal{C}_{free}$. Processing the query consists of connecting these configurations to two milestones and checking that these two milestones are in the same connected component of R.

The planner uses a simple algorithm, Connect (also called the *connector*), to construct the links of R. Given any two milestones, $\boldsymbol{m}_1$ and $\boldsymbol{m}_2$, the connector checks if the straight-line segment connecting them lies in $\mathcal{C}_{free}$. It does so by dichotomically breaking the segment into shorter segments and checking the endpoints of each segment for adjacency. It stops breaking a segment whenever its two endpoints are adjacent or one of them is not in free space. In the second case, Connect halts, without generating a link between $\boldsymbol{m}_1$ and $\boldsymbol{m}_2$ in the roadmap. Instead, if the connector eventually succeeds in partitioning the segment $\boldsymbol{m}_1\boldsymbol{m}_2$ into segments (possibly, of unequal lengths) such that the endpoints of each segment are adjacent, it establishes a link between $\boldsymbol{m}_1$ and $\boldsymbol{m}_2$ in the roadmap. The work done by the connector is part of the sampling work done by the roadmap planner, but the sample configurations it generates are not permanently stored; they are not part of the roadmap R. In other words, the planner does not attempt to find additional connections among sample configurations generated between milestones. To keep our presentation simple, we assume that Connect only tries straight paths, but it could try other canonical paths as well. We say that a configuration *sees* another configuration if the straight-line segment joining them lies entirely in $\mathcal{C}_{free}$.

In the following algorithm we could limit the total number of sample configurations to c, as in the general scheme of Section 2. However, putting the limit on the number of generated milestones instead is more convenient; it will also facilitate the analyses of Section 4. The preprocessing phase constructs an initial roadmap containing r milestones (Steps 2 and 3 in the algorithm shown below). Then it expands this initial roadmap into a final one containing $s > r$ milestones. Both r and s are input parameters. The first r milestones are chosen uniformly over $\mathcal{C}_{free}$. The remaining $s - r$ milestones are selected in small regions considered as "difficult" parts of $\mathcal{C}_{free}$. The preprocessing algorithm is the following:

[8] The planner is available by anonymous ftp from flamingo.stanford.edu:/pub/kavraki/prm.tar.gz.

`Preprocessing`:

1. $i \leftarrow 0$.
2. While $i < r$ do:
   (a) Pick a configuration $\boldsymbol{q}$ in $\mathcal{C}$ at random.
   (b) If CLEARANCE($\boldsymbol{q}$) > 0 then
      i. Store $\boldsymbol{q}$ as a milestone of the roadmap.
      ii. $i \leftarrow i + 1$.
3. For every pair of milestones $\boldsymbol{m}_1$ and $\boldsymbol{m}_2$ whose distance is less than d, do: `Connect`($\boldsymbol{m}_1, \boldsymbol{m}_2$).
4. Invoke `Resample` to expand the roadmap by $s - r$ milestones (see below).

The threshold d at Step 3 is an input parameter. It is used to limit the number of pairs of milestones that are tested by `Connect`, since in most spaces two milestones that are far apart are unlikely to see each other. Step 3 takes time at most quadratic in r, which is usually much smaller than the total number of sample configurations generated by the roadmap planner.[9]

`Resample` selects milestones with probability inversely proportional to their degrees in the roadmap (i.e., their number of connections to other milestones). Intuitively, a milestone with few or no connections lies in a difficult region of the free space. For each milestone $\boldsymbol{m}$ that it selects, `Resample` picks a number of new milestones at random in a neigborhood of $\boldsymbol{m}$ and invokes `Connect` to try to link each of these new milestones to other milestones. This expansion of the roadmap terminates when the total number of milestones is s. Experiments have been done with and without the roadmap expansion step (Step 4). Roadmaps containing the same total number of milestones have been constructed both ways. The roadmaps generated using the expansion step have been consistently better, i.e., subsequent queries were processed more quickly with less incorrect NO answers. Over a large range of problems, generating 2/3 of the milestones at Step 2 and 1/3 at Step 4 gave good results.

After preprocessing, each query is handled as follows (g is an input parameter):

`Query-Processing`($\boldsymbol{q}_{init}, \boldsymbol{q}_{goal}$):

1. For $i = \{init, goal\}$ do:
   (a) If there exists a milestone $\boldsymbol{m}$ that sees $\boldsymbol{q}_i$ then $\boldsymbol{m}_i \leftarrow \boldsymbol{m}$.
   (b) Else
      i. Repeat g times:
         Pick a configuration $\boldsymbol{q}$ at random in the neighborhood of $\boldsymbol{q}_i$
         until $\boldsymbol{q}$ sees both $\boldsymbol{q}_i$ and a milestone $\boldsymbol{m}$.
      ii. If all g trials failed then return NO and halt, else $\boldsymbol{m}_i \leftarrow \boldsymbol{m}$.
2. If $\boldsymbol{m}_{init}$ and $\boldsymbol{m}_{goal}$ are in the same component of the roadmap then return YES; else return NO.

[9]Footnote 4 applies here too.

4 Analysis

We now give formal analyses of two variants of the roadmap planner. We characterize the amount of computation that the planner must do in order to give correct answers with high probability.

The intuitive reason for the experimental success of the above planners is that there usually exist many collision-free paths joining two configurations. Hence, to bound the running time of the roadmap planner, we assume that $\mathcal{C}_{free}$ satisfies some geometric property capturing the above intuition. We propose two such properties. Our thesis is that the success of any sampling strategy will stem from a similar property.

In both analyses, we consider that the key number affecting the planner's running time is the number of milestones in the constructed roadmap. Since to generate a single milestone it might be necessary to randomly pick several (possibly, many) configurations in $\mathcal{C}$, we implicitly assume that the volume of $\mathcal{C}_{free}$ relative to the volume of $\mathcal{C}$ is not too small. If this assumption is not satisfied, any variant of the roadmap algorithm presented in Section 3.2 will behave poorly. In this case, we should probably make the additional assumption that a few free configurations are given (after all, every query will give two such configurations); then, a possible sampling strategy could be to build a roadmap by generating milestones in small regions centered at the given configurations, first, and at the newly generated milestones, next.

Note that bounding the number of milestones is not sufficient to bound the running time of the planner, because it does not account for the running time of `Connect`. If two milestones passed to `Connect` see each other, but the line segment connecting them is arbitrarily close to the free space boundary, then the connector will have to break it into arbitrarily many small segments. This problem could be eliminated by extending the general scheme of Section 2 with a primitive computing the volume swept out by the robot when it moves between two configurations along a straight path in configuration space.

Here, we only state the main results of our analyses. We give proofs of three selected theorems (1, 2, and 7) in the appendix. For more details and additional proofs, we refer the reader to [22, 23] (first analysis) and to [18, 19] (second analysis).

In the rest of this section we denote the volume of a subset $\mathcal{X}$ of $\mathcal{C}$ by $\mu(\mathcal{X})$.

4.1 Visibility Volume Assumption

For any configuration $\boldsymbol{q} \in \mathcal{C}_{free}$, let $\mathcal{S}(\boldsymbol{q})$ consist of all those configurations $\boldsymbol{q}' \in \mathcal{C}_{free}$ that $\boldsymbol{q}$ sees.

Definition 2 *Let ϵ be a positive real. A configuration $\boldsymbol{q} \in \mathcal{C}_{free}$ is ϵ-good if $\mu(\mathcal{S}(\boldsymbol{q})) \geq \epsilon\mu(\mathcal{C}_{free})$. Furthermore, $\mathcal{C}_{free}$ is ϵ-good if all the configurations it contains are ϵ-good.*

The visibility volume assumption made here is that $\mathcal{C}_{free}$ is ϵ-good, that is, each configuration in it sees a significant portion of $\mathcal{C}_{free}$. The underlying intuition is that it is then relatively easy to pick a set of milestones that, collectively, can see most of $\mathcal{C}_{free}$. But the assumption fails to prevent $\mathcal{C}_{free}$ from containing narrow passages through which it might be difficult to connect milestones. For example, consider the case where $\mathcal{C}$ is two-dimensional and $\mathcal{C}_{free}$ consists of two disks of equal size that overlap by a very small amount. Then $\mathcal{C}_{free}$ is ϵ-good for $\epsilon \approx 0.5$. But the probability that any milestone in one disk sees a milestone in the other disk is very small. For this reason, we use a variant of the roadmap algorithm that embeds a "complex planner". We assume that this complex planner is error-free in that it discovers a path between two given configurations whenever one exists, and reports failure when there is none. But of necessity such a complete planner must be expensive to run, and so we seek to use it sparingly. In particular, to keep query processing fast, we restrict the use of the complex planner to the preprocessing stage.

The preprocessing algorithm is similar to the one given in Section 3.2. The main difference is that Step 4 invokes `Permeation`, instead of `Resample`. `Permeation` uses the complex planner to improve the roadmap connectivity; unlike `Resample`, it does not add new milestones to the roadmap. The constructed roadmap contains s milestones; these are all generated at Step 2.

`Preprocessing`:

1. $i \leftarrow 0$.
2. While $i < s$ do:
   (a) Pick a configuration $\boldsymbol{q}$ in $\mathcal{C}$ at random.
   (b) If CLEARANCE($\boldsymbol{q}$) > 0 then
      i. Store $\boldsymbol{q}$ as a milestone of the roadmap.
      ii. $i \leftarrow i + 1$.
3. For every pair of milestones $\boldsymbol{m}_1$ and $\boldsymbol{m}_2$, do: `Connect`($\boldsymbol{m}_1, \boldsymbol{m}_2$).
4. Pick one representative milestone from each component of the current roadmap. Let V be the set of these representative milestones. Invoke `Permeation`(V) to improve the connectivity of the milestones (see below).

The result of this preprocessing is a roadmap such that any two nodes are in the same component if and only if the corresponding milestones are in the same component of $\mathcal{C}_{free}$. Step 3 may fail to find all possible links between the milestones due to the incompleteness of the connector. `Permeation` invoked in Step 4 fixes this problem by using the complex planner to discover additional connections between milestones. Note that `Permeation` is a last resort; hopefully, much if not all of the connectivity information should have been discovered before this step.

It is worth noticing that the algorithms of the implemented planner presented in [18, 20, 21] are similar to those described here, except that it uses the potential-field planner of Section 3.1 instead of the complex planner. The potential-field planner is not complete, but it is quite rare in practice that two milestones in the same component of the free space can be connected neither by invoking `Connect` at Step 3 of the proprocessing, nor by the potential-field planner. The planner of [18, 20, 21] seems a good approximation of the algorithms given here, and the analysis proposed below is relevant to this planner.

The query processing is handled by the following algorithm, which is similar to the query-processing algorithm of Section 3.2:

`Query-Processing`($\boldsymbol{q}_{init}, \boldsymbol{q}_{goal}$):

1. For $i = \{init, goal\}$ do:
   (a) If there exists a milestone $\boldsymbol{m}$ that sees $\boldsymbol{q}_i$ then $\boldsymbol{m}_i \leftarrow \boldsymbol{m}$.

(b) Else

i. Repeat g times:

Pick a configuration $\boldsymbol{q}$ at random in $\mathcal{S}(\boldsymbol{q}_i)$

until $\boldsymbol{q}$ sees both $\boldsymbol{q}_i$ and a milestone $\boldsymbol{m}$.

ii. If all g trials failed then return FAILURE and halt, else $\boldsymbol{m}_i \leftarrow \boldsymbol{m}$.

2. If $\boldsymbol{m}_{init}$ and $\boldsymbol{m}_{goal}$ are in the same component of the roadmap then return YES; else return NO.

Step 1(b)i differs slightly from the corresponding step in the query-processing algorithm of Section 3.2. Since in general we do not know how to compute $\mathcal{S}(\boldsymbol{q}_i)$, picking configurations in this set requires guessing configurations in configuration space and retaining only those that are seen by $\boldsymbol{q}_i$. This means that each of the g iterations performed at Step 1(b)i may require several trials to obtain a configuration visible from $\boldsymbol{q}_i$. The analysis proposed below ignores these trials and focuses only on the number g.

Another difference between the above algorithm and the one of Section 3.2 is that it returns FAILURE (instead of NO) at Step 1(b)ii. Due to the use of the `Permeation` in the preprocessing, both the answers YES and NO are now always correct. With some probability though, the query-processing algorithm may fail to give an answer.

Let us first present some performance guarantee for the preprocessing phase. Call a set of milestones *adequate* if the volume of the subset of $\mathcal{C}_{free}$ not visible from any of these milestones is at most $(\epsilon/2)\mu(\mathcal{C}_{free})$. Intuitively, if we were to place a point source of light at each milestone, we would like a fraction at least $1-\epsilon/2$ of $\mathcal{C}_{free}$ to be illuminated.

Theorem 1 *Let $\beta \in (0,1]$ be a positive real constant. Let C be a positive constant large enough that for any $x \in (0,1]$, $(1-x)^{(C/x)\log(4/x\beta)} \leq x\beta/4$. If $s \geq (C/\epsilon)(\log 1/\epsilon + \log 4/\beta)$, then preprocessing generates an adequate set of milestones with probability at least $1-\beta$.*

Note that as ϵ increases, the requirement for adequacy grows weaker, i.e., the portion of the free space that has to be visible by at least one milestone gets smaller. Intuitively, this comes from the fact that a greater ϵ will make it easier to connect query configurations to the roadmap. Naturally, the number of milestones needed becomes smaller.

Theorem 1 only says that most of $\mathcal{C}_{free}$ is likely to be visible from some milestone in the roadmap; using this property alone, we can show that queries can be answered quickly. But the adequacy of the milestones is not sufficient to imply that the roadmap is a good representation of the connectivity of $\mathcal{C}_{free}$. The use of the complex planner in the permeation algorithm is inevitable to ensure a good probability that the query processing outcomes YES or NO.

Let us choose $g = \log(2/\gamma)$ at Step 2 of `Query-Processing`, where $\gamma \in (0,1]$ is the failure probability we are willing to tolerate during a query. We can show the following performance guarantee for the query processing phase:

Theorem 2 *If the set of milestones chosen during preprocessing is adequate, then the probability that the query-processing algorithm outputs* FAILURE *is at most γ.*

In fact, our analysis (in the Appendix, Section B) implies that the expected number of executions of Step 1(b)i in the query-processing algorithm is at most 2.

Theorems 1 and 2 give performance guarantees for the preprocessing phase and the query-processing phase, respectively. This is appropriate, since many queries will be made using the same roadmap. However, we can easily blend the two theorems to bound the probability that the planner returns FAILURE for a single query by $(1-\beta)\gamma + \beta$ (this is obtained by bounding by one the probability that the planner returns FAILURE when the set of milestones is not adequate). Then, let $\alpha \in (0,1]$ be the probability of a FAILURE outcome which we are willing to tolerate. Neither the number s of milestones, nor the number g of trials at Step 2 of `Query-Processing` grow faster than $\log(\alpha^{-1})$, when $\alpha \to 0$.

We now turn to the description of `Permeation` and its analysis. Our goal is to evaluate how many times the complex planner must be invoked at Step 4 for the preprocessing algorithm.

`Permeation` must determine which milestones in V are reachable from each other, i.e., it must partition V into subsets such that all milestones in the same subset belong to the same component of $\mathcal{C}_{free}$ and no two milestones in two different subsets are in the same component of $\mathcal{C}_{free}$. If p is the size of V, it can do so with $O(p^2)$ invocations of the complex planner by trying it on every pair of milestones in V, but we show below that far fewer invocations may suffice.

We work with the following abstract version of the permeation problem. The input is a graph N with p vertices, consisting of k disjoint cliques. The goal is to determine this clique partition of N. The graph is presented as an adjacency matrix and the cost of an algorithm is measured by the number of entries it examines in the adjacency matrix of N. This is the *edge probe model* used in the study of evasive graph properties [33]. The vertices of N correspond to the milestones in V, and an edge is present between two vertices if the corresponding milestones lie in the same component of $\mathcal{C}_{free}$. The milestones from any particular component of $\mathcal{C}_{free}$ will form a clique in N, and there are no edges between two distinct cliques. A probe into the adjacency matrix corresponds to an invocation of the complex planner.

Let $N(p,k)$ denote the non-deterministic complexity of this problem. A non-deterministic algorithm is only required to verify that some partition into k cliques is the right partition. The non-deterministic complexity of the problem is clearly a lower bound on its deterministic and even randomized complexity.

Theorem 3 *For* $1 \leq k \leq p$, $N(p,k) = \Theta(p+k^2)$.

We now characterize the worst-case deterministic complexity of this problem, denoted $T(p,k)$. Consider the following deterministic algorithm: by probing all edge slots incident on an arbitrary vertex x, determine the neighborhood of x, say $\Gamma(x)$; let $C_x = \{x\} \cup \Gamma(x)$, and output C_x; then, recur on the vertex-induced subgraph $N[V \setminus C_x]$. The number of levels in the recursion is k, since one of the k cliques is removed from N prior to each recursive call. The number of probes made in the process of determining each such clique is at most p. The total number of probes is $O(pk)$. The following permeation algorithm is an iterative version of the recursive algorithm (the nodes of N are named $1, 2, \ldots, p$):

`Deterministic-Permeation`(V):

1. Mark all vertices of N as being LIVE.
2. Initialize $x \leftarrow 1$.
3. While $x \leq p$ do:
   (a) $\Gamma(x) \leftarrow \emptyset$.
   (b) For $y = x+1$ to p do:
      i. If vertex y is marked LIVE then probe the edge (x,y) in N.
      ii. If edge (x,y) is probed and found present then mark y as DEAD and add y to $\Gamma(x)$.
   (c) Output $\{x\} \cup \Gamma(x)$ as being a clique.
   (d) Mark x as being DEAD.
   (e) Set x to the smallest numbered LIVE vertex, or $p+1$ if there are no LIVE vertices left.

By the preceding discussion, we have:

Theorem 4 `Deterministic-Permeation` *correctly solves the permeation problem using* $O(pk)$ *probes.*

The following lower bound establishes that `Deterministic-Permeation` is optimal. The proof uses a non-trivial adversary argument [23].

Theorem 5 *For* $1 \leq k \leq p$, $T(p,k) = \Omega(pk)$.

We now give a randomized algorithm that beats the lower bound of Theorem 5 when the sizes of the k cliques differ significantly, which is often the case in practice (when $k > 1$). `Randomized-Permeation` labels the vertices in a random order and then invokes `Deterministic-Permeation`.

`Randomized-Permeation`(V):

1. Permute the vertices randomly. Rename the nodes by $1, \ldots, n$, in the order of the generated list.
2. Invoke `Deterministic-Permeation`(V).

Let $w_1 \geq w_2 \geq \cdots \geq w_k$ be the sizes of the cliques in an instance N arranged in a non-increasing order, where $p = \sum_{i=1}^{k} w_i$. Denote by C_i the ith largest clique in N. Define a function $u()$ on an ordered k-tuple of positive integers $n_1, n_2, \ldots, n_k$ by $u(n_1, n_2, \ldots, n_k) = \sum_{i=1}^{k} i n_i$.

Theorem 6 `Randomized-Permeation` *correctly determines the clique structure and incurs an expected cost that is at most*

$$2u(w_1, w_2, \ldots, w_k) - p - k.$$

Furthermore, with high probability, the cost is at most

$$O(u(w_1, w_2, \ldots, w_k) \log p).$$

Observe that the worst case is when all w_i are equal to p/k, in which case the expected cost is $O(pk)$. On the other hand when there is one giant clique and $k-1$ cliques of size $O(1)$ the expected cost is $\Theta(p+k^2)$, which is essentially the non-deterministic lower bound.

4.2 Path Clearance Assumption

The visibility volume assumption does not prevent the existence of narrow passages in $\mathcal{C}_{free}$. To remove the need for the "complex planner" and get closer to the roadmap planner of Section 3.2, we now consider a seemingly stronger assumption: between the two configurations given by a query, there exists a collision-free path τ that achieves some clearance ξ between the robot and the obstacles. More formally, let us parametrize τ by the arc length ℓ from the initial configuration and let L designate the path's total length, i.e., $\tau : \ell \in [0, L] \to \tau(\ell) \in \mathcal{C}_{free}$. We define $\xi(\ell)$ to be the Euclidean distance between $\tau(\ell)$ and the free space boundary, and ξ_{inf} to be $\inf_{\ell \in [0,L]} \xi(\ell)$.

We consider a variant of the roadmap planner in which the preprocessing algorithm consists of the first two steps of the algorithm given in Section 4.1. Unlike the planner of Section 3.2, it does not include the resampling step. The query-processing algorithm is also simpler than in Section 3.2, in that it only checks that the initial and goal configurations see milestones in the roadmap. If any one of these connections fails, the query-processing algorithm returns NO. Hence, this query-processing algorithm is deterministic.

Under the path clearance assumption, any NO outcome is incorrect. Let α be the probability that we are willing to tolerate for this event. The following two theorems relate the size of the roadmap to this probability, as well as to the two parameters of the assumption, that is, the length L of the hypothesized path and its clearance. They give a performance guarantee for the whole planner.

Theorem 7 *Let $\alpha \in (0,1]$ be a positive real constant. Let a be the constant $2^{-n}\mu(\mathcal{B}_1)/\mu(\mathcal{C}_{free})$ where $\mathcal{B}_1$ denotes the unit ball in $\mathbb{R}^n$. If s is chosen such that:*

$$\frac{2L}{\xi_{inf}}(1 - a\xi_{inf}^n)^s \leq \alpha, \tag{1}$$

then the planner outputs YES *with probability at least $1 - \alpha$.*

Note that, for any given L and ξ_{inf}, the quantity on the left side of the above inequality tends toward zero when $s \to \infty$. It is even more important to remark that it depends exponentially on s, so that the number of milestones s needed grows no faster than the logarithm of $1/\alpha$.

The following theorem is similar to the previous one, but makes use of the clearance distribution $\xi(\ell)$ rather than just its infimum:

Theorem 8 *Let $\alpha \in (0,1]$ be a positive real constant. Let a be the same constant as in Theorem 7. If s is chosen such that:*

$$6 \int_0^L \frac{(1 - (a/2^n)\xi^n(\ell))^s}{\xi(\ell)} d\ell \leq \alpha, \tag{2}$$

then the planner outputs YES *with probability at least $1 - \alpha$.*

Using the inequality $1 - x \leq e^{-x}$, for $x \geq 0$, we get the following easier-to-use relations:

- The bound of Theorem 7 becomes:

$$\frac{2L}{\xi_{inf}} \exp(-a\xi_{inf}^n s) \leq \alpha. \tag{3}$$

- The bound of Theorem 8 becomes:

$$6 \int_0^L \frac{\exp(-a2^{-n}\xi^n(\ell)s)}{\xi(\ell)} d\ell \leq \alpha. \tag{4}$$

Relation 3 implies that the number of milestones that the planner must generate to output YES with probability at least $1 - \alpha$ is polynomial in $1/\xi_{inf}$ and logarithmic in L.

Remark that $\xi(\ell) \geq (1/2\rho)$ CLEARANCE$(\tau(\ell))$, where ρ is the constant introduced in Section 2. Indeed, for any $\boldsymbol{q} \in \mathcal{C}_{free}$, let $\boldsymbol{q}^c$ be the configuration in the free space boundary that achieves minimum distance with $\boldsymbol{q}$ and let $\xi(\boldsymbol{q})$ be this distance. We have: $\xi(\boldsymbol{q}) \geq \max_i |q_i - q_i^c|$. By the definition of ρ, all configurations $\boldsymbol{q}'$ such that $\max_i |q_i - q_i'| \leq$ CLEARANCE$(\boldsymbol{q})/2\rho$ are in the free space. Hence, CLEARANCE$(\boldsymbol{q})/2\rho < \max_i |q_i - q_i^c| \leq \xi(\boldsymbol{q})$. Therefore, the bounds given above remain valid if we choose to define $\xi(\ell)$ as $(1/2\rho)$ CLEARANCE$(\tau(\ell))$, rather than as the distance between $\tau(\ell)$ and the free space boundary.

4.3 Example

Consider the two-dimensional problem ($n = 2$) shown in Figure 1. This problem is parametrized by ω. The walls are obstacles, but have zero thickness; hence, $\mu(\mathcal{C}_{free}) = 1$. We are interested in the order of magnitude of s when ω is small. Let $x \asymp y$ mean that $C^{-1}x \leq y \leq Cx$ for some constant $C > 0$. In particular, we have $L \asymp 1$ and $\xi_{inf} \asymp \omega$.

Estimate of s using Theorem 1: The space in Figure 1 is ϵ-good for $\epsilon = 2\omega^2$ (clearly, because of the "box" on the left). In order to bound the probability that the query-processing algorithm

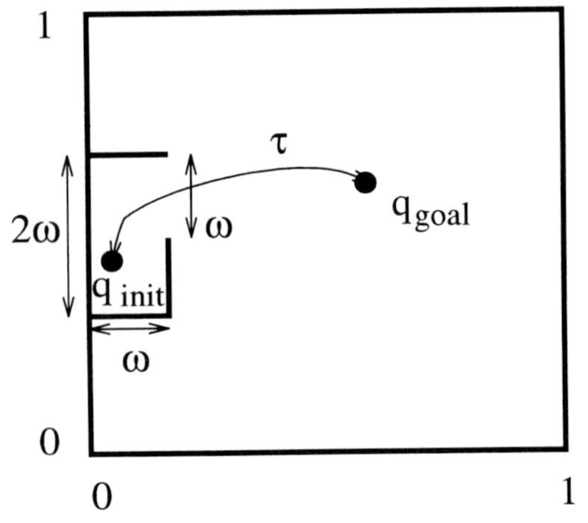

Figure 1: Illustrative example

of section 4.1 outputs FAILURE, while a collision-free path exists (Theorem 2), the roadmap must be adequate. According to Theorem 1, this is achieved with high probability if:

$$s \asymp \frac{1}{\omega^2} \log \frac{1}{\omega}. \tag{5}$$

Estimate of s using Theorem 7: Using (3) we can derive the following order of magnitude for s that will make the planning algorithm of Section 4.2 to return YES with high probability [19]:

$$s \asymp \frac{1}{\omega^2} \log \frac{1}{\omega}. \tag{6}$$

Estimate of s using Theorem 8: The bound given by (4) leads to choosing (see [19]):

$$s \asymp \frac{1}{\omega^2} \log\log \frac{1}{\omega}. \tag{7}$$

Comparison: Note that to answer queries with high probability it is necessary and sufficient in this example to pick a bounded number of milestones in the box, which happens with probability $\asymp \omega^2$. Hence, a tight estimate of s is

$$s \asymp \frac{1}{\omega^2}.$$

The estimates (5), (6), and (7) can be seen as the unavoidable quantity $1/\omega^2$ times some factor. By exploiting the fact that $\xi(\ell)$ is small only briefly, we get a smaller factor in (7) than in (6).

5 Discussion

Between complete planners, which take prohibitive time to run in large dimensional configuration spaces, and ad hoc planners, which are too unreliable, there exist planners that are reasonably efficient while achieving some form of completeness. In this paper we have discussed a class of such planners. These are based on a general computational scheme presented in Section 2, which consists of randomly sampling the free space and checking pairs of sample configurations for adjacency. By choosing an appropriate sampling strategy, this scheme can be turned into a concrete planner that achieves probabilistic completeness. In Section 3 we have presented two examples of probabilistically complete planners: the potential-field and the roadmap planners. In Section 4 we have analyzed in more detail the probabilistic completeness of variants of the roadmap planner, under two different assumptions: the visibility volume assumption and the path clearance assumption. In each case, we have established a relation between the probability that the planner finds a path, when one exists, and its running time (measured in that case by the number of milestones in the roadmap). Both analyses show that the number of milestones needed grows only as the absolute value of the logarithm of the probability of an incorrect answer that we are willing tolerate. Moreover, this number is polynomial in the inverse of the parameter used to measure the goodness of free space (ϵ in the first analysis, ξ_{inf} in the second).

How realistic and useful are our assumptions? Clearly, in most cases, there is no pratical way to verify that they are satisfied. Furthermore, it is easy to create spaces that are ϵ-good for very small values of ϵ, as well as spaces that are not ϵ-good for any positive value of ϵ. For instance, this is the case of the space comprised between two circles C_1 and C_2 tangent at some point P, with C_2 contained in C_1. On the other hand, in many applications, we are willing to discard collision-free paths if their clearance is too small, due to uncertainties in robot control and sensing. In those cases, a planner that would return a path with high probability, whenever there exists one that has enough clearance, would be quite satisfactory.[10] Remark also that, if there exists a collision-free path with a clearance greater than some given ξ_{inf}, this path must lie in a subset of $\mathcal{C}_{free}$ that is at least ϵ-good for an ϵ that can be derived[11] from ξ_{inf}. Hence, the assumption of ϵ-

[10]Note that the roadmap planner may nevertheless return a path with a smaller clearance.

[11]However, the bound derived from ξ_{inf} may be much smaller than the actual ϵ. E.g., if the free space is a rectangle having very small width ξ, then $\xi_{inf} \leq \xi$, but $\epsilon = 1$.

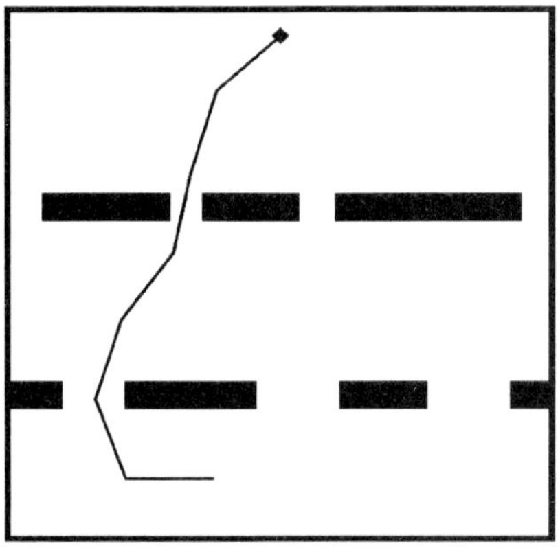

Figure 2: Robot arm example

goodness is realistic as well: for any given value of ϵ, our analysis characterizes the probability that the planner finds a path in the subset of $\mathcal{C}_{free}$ that is ϵ-good.

But we think that much more could be done. In particular, none of the two analyses we have given considers the resampling step (Step 4) in the preprocessing algorithm of Section 3.2. On the other hand, our experiments have shown that adding this step yields a much better roadmap planner. In relation to this observation, we have empirically tested the free space corresponding to the setting of Figure 2 (a seven-revolute-dof planar arm among several barriers forming multiple gates) for ϵ-goodness. To do so we have picked 9,000 milestones at random and we have computed how many other milestones each milestone can see. The milestones with the "most" visibility could only see about 0.06 (i.e., 6%) of the remaining milestones, suggesting that they are 0.06-good. As many as 3.3% of the milestones could see no other milestones, and fully 22% could see 0.001 (i.e., 0.1%) or less; in other words, only about 78% of the configuration space is 0.001-good or better. On the other hand, the implemented planner, which includes the resampling step, handles queries in this setting with quasi-perfect reliability, after having constructed a roadmap containing 5,000 to 10,000 milestones. Other experiments confirm that the use of `Resampling` improves the coverage of the free space by the milestones.

To explain the role of `Resampling` we introduce a generalization of the notion of ϵ-goodness. We say that a free configuration $\boldsymbol{q}$ is $(\epsilon, 1)$-good if $\mu(\mathcal{S}(\boldsymbol{q})) \geq \epsilon\mu(\mathcal{C}_{free})$, corresponding to our original definition of ϵ-goodness for a configuration. Next, we say a free configuration $\boldsymbol{q}$ is (ϵ, t)-good if $\mu(\{\boldsymbol{q}' \in \mathcal{S}(\boldsymbol{q}) \mid \boldsymbol{q}' \text{ is } (\epsilon, t-1)\text{-good}\}) \geq \mu(\mathcal{S}(\boldsymbol{q}))/2$. For $t > 1$, we say that $\mathcal{C}_{free}$ is (ϵ, t)-good if $\mu(\{\boldsymbol{q} \in \mathcal{C}_{free} | \boldsymbol{q} \text{ is } (\epsilon, 1)\text{-good}\}) \geq \mu(\mathcal{C}_{free})/2$ and every free configuration is (ϵ, i)-good for $i \leq t$. If $\mathcal{C}_{free}$ is (ϵ, t)-good for a small value of t, we can give a theoretical basis for the resampling strategy. The main idea is that single links discovered by the connector in the algorithms are now simulated using t-link paths found by resampling and connecting using the connector. This leads to a generalized definition of an adequate set of milestones, and eventually to a version of Theorem 2 in which the number of invocations of the connector is larger by a factor of 2^t. See [23] for more detail.

We believe that future research should develop and analyze new sampling strategies, as well as combinations of strategies, under various assumptions. In particular, it would be of particular interest to investigate strategies suitable for single-query planning problems. In this case, instead of spreading milestones all over the free space, one could expand a "cloud" of sample configurations from the initial configuration toward the goal, or in both directions concurrently. This could be achieved by iteratively generating configurations in small neighborhoods of previously generated configurations, while keeping only those configurations which are in the same free space component as the initial configuration. This strategy could use a function similar to a potential field as a probability distribution to randomly choose which configurations to expand from at every iteration. In the multi-query case, it would also be useful to consider the case where all queries are made in the same component of $\mathcal{C}_{free}$ and the preprocessing is given one configuration in that component. The case where the volume of the free space is very small relative to the total volume of the configuration space should also be investigated. We envision that ultimately randomized planners will automatically select and combine sampling strategies from a library of basic strategies to closely match the characteristics of the input problems. This sort of combination has already given good results in another area, graphic rendering [46].

We did not discuss distance calculation in this paper, but it is a key computation in the context of our sampling scheme. Indeed, our planners spend most of their time evaluating CLEARANCE. So, any improvement of the time efficiency of the algorithm implementing CLEARANCE would translate into a planning improvement of the same order of magnitude. One could also try to match the distance calculation algorithm and the sampling strategy, to get the best global result. For example, a sampling strategy generating configurations in small neighborhoods of previously generated configurations could be combined with an incremental distance calculation algorithm such as the one proposed in [32]. Such an algorithm uses the nearest points computed at the previous configuration to quickly find the nearest points at the new configuration. As another example, the distance calculation algorithm proposed in [39] (and used in our planners) precomputes data structures to later accelerate the processing of distance calculation queries. Each such data structure is a binary tree of spheres approximating a single rigid body. Given two bodies, a simple traversal of their respective trees allows us to quickly detect collision or identify the nearest points between the two bodies. A perhaps more efficient solution would be to create a single data structure for all the rigid bodies forming the robot and to efficiently update this data structure when configuration parameters vary (as proposed in [12]). Then the sampling strategy should try to generate configurations in a sequence that minimizes update costs.

The random sampling scheme for path planning proposed in this paper offers many opportunities for further research. We are confident that it can ultimately produce very fast planners able to correctly handle a wide range of problems with high probability.

Acknowledgments: The development of the potential-field and roadmap planners has been supported by ARPA contracts DAAA21-89-C0002, N00014-92-J-1809, and N00014-94-1-0721. The formal analysis of these planners and the design of the general random sampling scheme was supported by MURI grant DAAH04-96-1-0007 (Army). The experimental part of this research has also benefited from gifts by General Electric, General Motors, Renault, and Rockwell. Lydia Kavraki and Jean-Claude Latombe are partially supported by ARPA contract N00014-94-1-0721 and a MURI grant of the Army. Rajeev Motwani is supported by an Alfred P. Sloan Research Fellowship, an IBM Faculty Partnership Award, and NSF Young Investigator Award CCR-9357849, with matching funds from IBM, Schlumberger Foundation, Shell Foundation, and Xerox Corporation. This paper has benefited from useful comments by Leo Guibas, Dan Halperin, David Hsu, and Jean-Paul Laumond.

References

[1] J.M. Ahuactzin Larios, *Le Fil d'Ariane: Une Méthode de Planification Générale. Application à la Planification Automatique de Trajectoires*, Thèse de l'Institut National Polytechnique de Grenoble, Septembre 1994.

[2] S. Arya and D.M. Mount, Approximate Range Searching, *Proc. of ACM Symp. on Computational Geometry*, 172-181, 1995.

[3] J. Barraquand and J.C. Latombe, Robot Motion Planning: A Distributed Representation Approach, *The Int. J. of Robotics Research*, 10(6):628-649, 1991.

[4] J. Barraquand, B. Langlois, and J.C. Latombe, Numerical potential Field Techniques for Robot Path Planning, *IEEE Tr. on Syst., Man, and Cyb.*, 22(2):224-241, 1992.

[5] P. Bessière, E. Mazer, J.M. Ahuactzin, Planning in a Continuous Space with Forbidden Regions: The Ariadne's Clew Algorithm, *Algorithmic Foundations of Robotics*, K. Goldberg et al. (eds.), AKPeters, Wellesley, MA, 39-47, 1995.

[6] J.F. Canny, *The Complexity of Robot Motion Planning*, MIT Press, Cambridge, MA, 1988.

[7] D. Challou and M. Gini, Parallel Formulation of Informed Randomized Search for Robot Motion Planning Problems, *Proc. of IEEE Int. Conf. on Robotics and Automation*, 709-714, Nayoga, Japan, 1995.

[8] H. Chang and T.Y. Li, Assembly Maintainability Study with Motion Planning, *Proc. of IEEE Int. Conf. on Robotics and Automation*, 1012-1019, Nagoya, 1995.

[9] P.C. Chen, Adaptive Path Planning: Algorithm and Analysis, *Proc. IEEE Int. Conf. on Robotics and Automation*, 721-728, Nagoya, 1995.

[10] E.G. Gilbert, D.W. Johnson, and S.S. Keerthi, A Fast Procedure for Computing the Distance Between Complex Objects in Three-Dimensional Space, *IEEE J. of Robotics and Automation*, 4(2):193-203, 1988.

[11] L. Graux, P. Millies, P.L. Kociemba, and B. Langlois, Integration of a Path Generation Algorithm into Off-Line Programming of AIRBUS Panels. *Aerospace Automated Fastening Conf. and Exp.*, SAE Tech. Paper 922404, 1992.

[12] D. Halperin, J.C. Latombe, and R. Motwani, *Dynamic Maintenance of Kinematic Structures*, Tech. Rep. , Dept. of Computer Science, Stanford Univ., 1995.

[13] J.E. Hopcroft, J.T. Schwartz, and M. Sharir, On the Complexity of Motion Planning for Multiple Independent Objects: PSPACE-Hardness of the 'Warehouseman's Problem', *Int. J. of Robotics Research*, 3(4):76-88, 1984.

[14] J.E. Hopcroft and G.T. Wilfong, Reducing Multiple Object Motion Planning to Graph Searching. *SIAM J. on Computing*, 15(3):768-785, 1986.

[15] T. Horsch, F. Schwarz, and H. Tolle, Motion Planning for Many Degrees of Freedom - Random Reflections at C-Space Obstacles, *Proc. of IEEE Int. Conf. on Robotics and Automation*, 3318-3323, San Diego, CA, 1994.

[16] Y.K. Hwang and P.C. Chen, A Heuristic and Complete Planner for the Classical Mover's Problem, *Proc. IEEE Int. Conf. on Robotics and Automation*, 729-736, Nagoya, 1995.

[17] D.A. Joseph and W.H. Plantiga, On the Complexity of Reachability and Motion Planning Questions. *Proc. of the First ACM Symp. on Computational Geometry*, 62-66, 1985.

[18] L. Kavraki, *Random Networks in Configuration Space for Fast Path Planning*, Ph.D. Thesis, Rep. STAN-CS-TR-95-1535, Comp. Sci. Dept., Stanford Univ., January 1995.

[19] L. Kavraki, M. Kolountzakis, and J.C. Latombe, Analysis of Probabilistic Roadmaps for Path Planning, manuscript submitted to *IEEE Int. Conf. on Robotics and Automation*, 1996.

[20] L. Kavraki and J.C. Latombe, Randomized Preprocessing of Configuration Space for Fast Path Planning, *Proc. IEEE Int. Conf. on Robotics and Automation*, 2138-2145, San Diego, CA, 1994.

[21] L. Kavraki and J.C. Latombe, Randomized Preprocessing of Configuration Space for Path Planning: Articulated Robots, *Proc. IROS*, 1764-1771, München, Germany, 1994.

[22] L. Kavraki, J.C. Latombe, R. Motwani, and P. Raghavan, *Randomized Query Processing in Robot Motion Planning*, Rep. STAN-CS-TR-94-1533, Dept. of Computer Science, Stanford University, 1994.

[23] L. Kavraki, J.C. Latombe, R. Motwani, and P. Raghavan, Randomized Query Processing in Robot Path Planning, *27th Annual ACM Symp. on Theory of Computing (STOC)*, 353-362, Las Vegas, NV, 1995.

[24] L. Kavraki, P. Švestka, J.C. Latombe, and M. Overmars, Probabilistic Roadmaps for Fast Path Planning in High Dimensional Configuration Spaces, to appear in *IEEE Tr. on Robotics and Automation.*

[25] D.E. Koditschek, Exact Robot Navigation by Means of Potential Functions: Some Topological Considerations, *Proc. IEEE Int. Conf. on Robotics and Automation*, Raleigh, NC, 1-6, 1987.

[26] Y. Koga, K. Kondo, J. Kuffner, and J.C. Latombe, Planning Motions with Intentions. *Proc. of SIGGRAPH'94*, 395-408, 1994.

[27] Y. Koga and J.C. Latombe, On Multi-Arm Manipulation Planning, *Proc. IEEE Int. Conf. on Robotics and Automation*, 945-952, San Diego, CA, 1994.

[28] F. Lamiraux and J.P. Laumond, *On the Expected Complexity of Random Path Planning*, Rep. No. 95087, LAAS/CNRS, Toulouse, March 1995.

[29] J.C. Latombe, *Robot Motion Planning*, Kluwer Academic Publ., Boston, MA,1991.

[30] J.C. Latombe, Controllability, Recognizability, and Complexity Issues in Robot Motion Planning, *Proc. 36th Annual Symp. on Foundations of Computer Science (FOCS)*, 484-500, Milwaukee, Wisconsin, 1995.

[31] T.Y. Li and J.C. Latombe, On-Line Motion Planning for Two Robot Arms in a Dynamic Environment, *Proc. IEEE Int. Conf. on Robotics and Automation*, 1048-1055, Nagoya, 1995.

[32] M.C. Lin and J.F. Canny, A Fast Algorithm fir Incremental Distance Calculation, *Proc. IEEE Int. Conf. on Robotics and Automation*, 1008-1014, Sacramento, CA, 1991.

[33] L. Lovász and N. Young, *Lecture Notes on Evasiveness of Graph Properties*, Tech. Rep. CS-TR-317-91, Comp. Sci. Dept., Princeton Univ., 1991.

[34] R. Motwani and P. Raghavan, *Randomized Algorithms*, Cambridge University Press, 1995.

[35] M. Overmars, *A Random Approach to Motion Planning*, Tech. Rep. RUU-CS-92-32, Dept. of Comp. Sci., Utrecht Univ., 1992.

[36] M. Overmars and P. Švestka, A Probabilistic Learning Approach to Motion Planning, *Algorithmic Foundations of Robotics*, K. Goldberg et al. (eds.), A.K.Peters, Wellesley, MA, 19-37, 1995.

[37] M. Ponamgi, D. Manocha, and M.C. Lin, Incremental Algorithms for Collision Detection Between Solid Models, *Proc. 3rd ACM Symp. on Solid Modeling and Applications*, 293-304, Salt Lake City, UT, 1995.

[38] F.P. Preparata and M.I. Shamos, *Computational Geometry: An Introduction*, Springer-Verlag, New York, NY, 1985.

[39] S. Quinlan, Efficient Distance Computation Between Non-Convex Objects. *Proc. IEEE Int. Conf. on Robotics and Automation*, 3324-3330, San Diego, CA, 1994.

[40] J. Reif, Complexity of the Mover's Problem and Generalizations. *FOCS*, 421-4127, 1979.

[41] J.H. Reif and M. Sharir, Motion Planning in the Presence of Moving Obstacles, *proc. 25th IEEE Symp. on Foundations of Computer Science*, 144-154, 1985.

[42] J.T. Schwartz and M. Sharir, On the 'Piano Movers' Problem: II. General Techniques for Computing Topological Properties of Real Algebraic Manifolds. *Advances in Applied Mathematics*, 4:298-351, 1983.

[43] S. Sekhavat, P. Švestka, J.P. Laumond, and M. Overmars, *Probabilistic Path Planning for Tractor-Trailer Robots*, Tech. Rep., LAAS/CNRS, Toulouse, September 1995.

[44] P. Švestka, *A Probabilistic Approach to Motion Planning for Car-Like Robots*, RUU-CS-93-18, Comp. Sci. Dept., Utrecht Univ., The Netherlands, April 1993.

[45] F. Thomas and C. Torras, Interference Detection Between Non-Convex Polyhedra Revisited with a Practical Aim, *Proc. IEEE Int. Conf. on Robotics and Automation*, 587-590, San Diego, CA, 1994.
[46] E. Veach and L.J. Guibas, Optimally Combining Sampling Techniques for Monte Carlo Rendering, *Proc. of SIGGRAPH'95*, 419-428, 1995.
[47] X. Zhu and K. Gupta, *On Local Minima and Random Search in Robot Motion Planning*, Technical Report, Simon Fraser Univ., BC, Canada, 1993.

Appendix: Selected Proofs

A. Proof of Theorem 1

Let M denote the set of the s randomly chosen milestones. The volume H of configurations not visible from any of these milestones is:

$$\mu(\{q \in \mathcal{C}_{free} \mid q \notin \cup_{m \in M} \mathcal{S}(m)\}).$$

Its expected value is:

$$E[H] = \int_{q \in \mathcal{C}_{free}} \Pr[q \notin \cup_{m \in M} \mathcal{S}(m)] dq.$$

The ϵ-goodness of the free space entails that the probability that any given configuration is not visible from any of the s milestones is at most $(1-\epsilon)^s$. Thus:

$$E[H] \leq \mu(\mathcal{C}_{free})(1-\epsilon)^s \leq \mu(\mathcal{C}_{free})\epsilon\beta/4. \quad (8)$$

Given a random variable X assuming only non-negative values, the Markov inequality [34]:

$$\Pr[X \geq x] \leq E[X]/x$$

holds for all $x \in \mathbb{R}^+$. Using this inequality and the relation (8), we get:

$$\Pr[H > (\epsilon/2)\mu(\mathcal{C}_{free})] \leq \beta/2.$$

Hence, with probability $1 - \beta/2$, H is at most $(\epsilon/2)\mu(\mathcal{C}_{free})$, in which case M is adequate.

B. Proof of Theorem 2

Assume an adequate set of milestones. For any $q \in \mathcal{C}_{free}$, the volume of the subset of $\mathcal{S}(q)$ visible from some milestone is at least:

$$\mu(\mathcal{S}(q)) - (\epsilon/2)\mu(\mathcal{C}_{free}) \geq (\epsilon/2)\mu(\mathcal{C}_{free}).$$

Therefore, for either query configuration q_i ($i \in$ {init, goal}), the probability that a random configuration chosen from $\mathcal{S}(q_i)$ is not visible from any milestone is at most 1/2. The probability that `query-processing` fails to connect q_i to a milestone on $\log(2/\gamma)$ trials at Step 1(b)i is thus less than $\gamma/2$. Since Step1(b)i is performed for both query configurations, the overall failure probability is at most γ.

C. Proof of Theorem 7

We assume the existence of a path $\tau : \ell \in [0, L] \mapsto \tau(\ell) \in \mathcal{C}_{free}$ connecting q_{init} to q_{goal}, where ℓ stands for the arc length from q_{init}. For simplicity, let ξ (instead of ξ_{inf}) designate the infimum of the Euclidean distance between $\tau(\ell)$ and the free space boundary, when ℓ spans the interval $[0,1]$. Given any two configurations $q = \tau(\ell)$ and $q' = \tau(\ell')$ on τ, let $l(q, q')$ denote the path length $|\ell - \ell'|$.

Let $\mathcal{B}_r(x)$ designate the ball of radius r centered at $x \in \mathbb{R}^n$. We pick $k = \lceil 2L/\xi \rceil$ configurations on τ, denoted by $q_0 = q_{init}, q_1, \ldots, q_k = q_{goal}$, such that $l(q_i, q_{i+1}) \leq \xi/2$, for all $i \in [0, k-1]$. We have that:

$$\mathcal{B}_{\xi/2}(q_{i+1}) \subset \mathcal{B}_\xi(q_i).$$

For any two points $p_i \in \mathcal{B}_{\xi/2}(q_i)$ and $p_{i+1} \in \mathcal{B}_{\xi/2}(q_{i+1})$, the straight-line segment connecting p_i and p_{i+1} lies entirely in $\mathcal{C}_{free}$; indeed, the above relation implies that p_{i+1} also lies in $\mathcal{B}_\xi(q_i)$. So, a sufficient condition for the query-processing algorithm to find a path is that each ball $\mathcal{B}_{\xi/2}(q_i)$, $i = 1, \ldots, k-1$ contains at least one milestone. The probability that a ball of radius r lying entirely in the free space contains none of the s milestones is $(1 - \mu(\mathcal{B}_r)/\mu(\mathcal{C}_{free}))^s$. In $\mathbb{R}^n$ we have $\mu(\mathcal{B}_r) = r^n \mu(\mathcal{B}_1)$. Therefore, the probability that the planner does not find a path is at most:

$$\left(\lceil \frac{2L}{\xi} \rceil - 1\right)\left(1 - \frac{\mu(\mathcal{B}_{\xi/2})}{\mu(\mathcal{C}_{free})}\right)^s,$$

which is itself no greater than:

$$\frac{2L}{\xi}\left(1 - 2^{-n}\frac{\mu(\mathcal{B}_1)}{\mu(\mathcal{C}_{free})}\xi^n\right)^s.$$

Hence, choosing s such that the above quantity is at most $\alpha \in (0, 1]$ guarantees that the planner will find a path with probability at least $1 - \alpha$.

Planar Manipulation on a Conveyor with a One Joint Robot

Srinivas Akella   Wesley H. Huang   Kevin M. Lynch   Matthew T. Mason

The Robotics Institute
Carnegie Mellon University

Abstract

This paper explores a method of manipulating a planar rigid body on a conveyor belt using a robot with just one joint. This approach has the potential of offering a simple and flexible method for feeding parts in industrial automation applications. In this paper we outline our approach, develop some of the theoretical properties, present a planner for the robot, and describe an initial implementation.

1 Introduction

The most straightforward approach to planar manipulation is to use a rigid grasp and a robot with at least three joints, corresponding to the three motion freedoms of a planar rigid object. But three joints are not really necessary to manipulate an object in the plane. In this paper we achieve effective control of all three planar motion freedoms using a one joint robot working over a constant speed conveyor belt.

A central issue in this work is to develop a precise notion of "effective control" that is suited to the parts-feeding application. Our robot cannot impart arbitrary motions to a part on the conveyor, but it does have a set of actions sufficient to orient and position a wide class of shapes. To make this precise we define the *feeding property*:

> A system has the *feeding property* over a set of parts $\mathcal{P}$ and set of initial configurations I if, given any part in $\mathcal{P}$, there is some output configuration $\mathbf{q}$ such that the system can move the part to $\mathbf{q}$ from any location in I.

One of the goals of this paper is to demonstrate that a one joint robot can possess this property over a useful set of initial configurations and a broad class of part shapes.

The key to our approach is to use a single revolute joint to push the parts around on a constant speed conveyor belt. This approach, which we refer to as

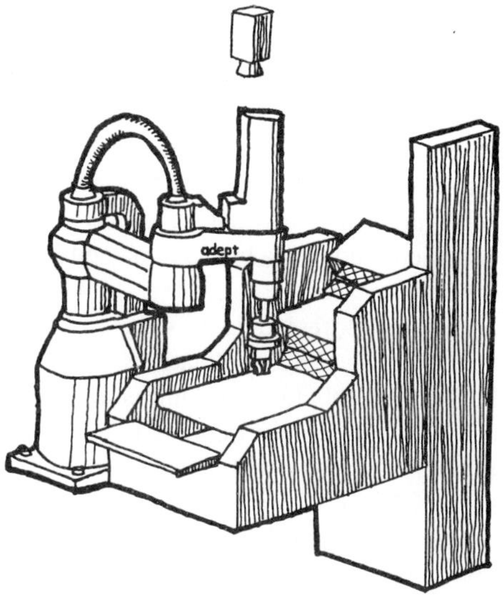

Figure 1: The Adept Flex Feeder System. A SCARA robot picks parts off the middle of three conveyors. These three conveyors, along with an elevator bucket, circulate parts; an overhead camera looks down on the back-lit middle conveyor to determine the position and orientation of parts.

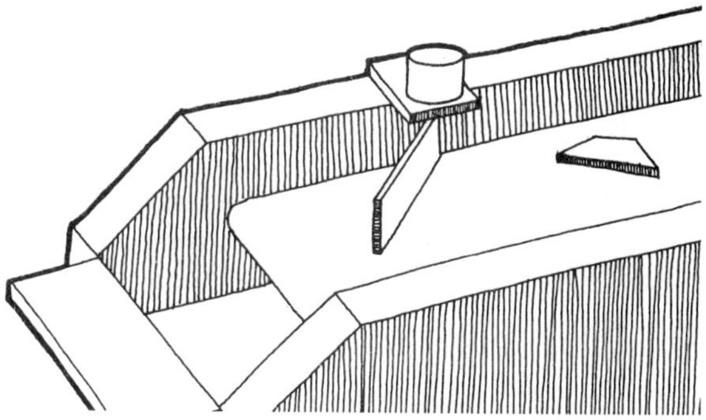

Figure 2: The Flex Feeder with a rotatable fence.

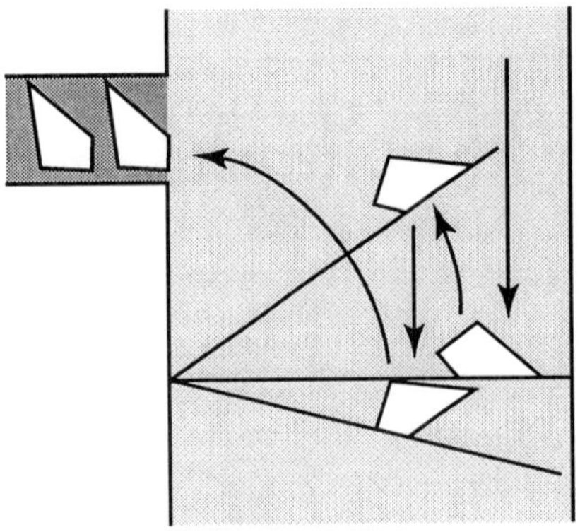

Figure 3: We can feed a part by alternately pushing it with the fence and letting it drift along the conveyor.

"1JOC" (one-joint-over-conveyor, pronounced "one jock") was initially conceived as a variation on the Adept Flex Feeder (see Figure 1). The Flex Feeder uses a system of conveyors to recirculate parts, presenting them with random orientation to a camera and robotic manipulator. Those parts that are in a graspable configuration may then be picked up by the robot and assembled into a product, placed in a pallet, or otherwise processed.

The question addressed in this paper is whether, at least in parts feeding applications, we could replace the SCARA robot with a simpler and more flexible robot, and also replace the Flex Feeder's servoed programmable conveyor with a fixed speed conveyor. Figures 2 and 3 show a possible variation on the Flex Feeder, where the SCARA robot has been replaced by a fence with a single revolute joint. By a sequence of pushing operations, punctuated by drift along the conveyor, the fence positions and orients a part and directs it into the entry point of a feeder track which carries the part to the next station.

There are many variations on this basic idea: a 2JOC, multiple 1JOCs working in parallel, curved fences, and so on. However, the purpose of this paper is an initial study of the fundamental characteristics of the idea, so we will focus on the simplest version as described above and in Figures 2 and 3. The main result is:

> It is possible to move an arbitrary polygon from a broad range of initial configurations to a specific goal; and that goal can be chosen from a broad range of possible goals. (This is a generalization of the feeding property.)

The remainder of the paper is organized as follows. After reviewing previous work and terminology from nonlinear control theory, we develop a progression of models leading to the 1JOC. We prove the feeding property for the 1JOC model and describe the implementation of a planner as well as experimental results.

1.1 Previous Work

The 1JOC approach described in this paper is an example of using a simple effector to perform nonprehensile manipulation. For a survey of work in this area, see (Erdmann [19]). Of particular relevance to this paper is previous work on parts feeding and the control of underactuated systems.

Parts Feeding Our goal is to develop parts feeders with practical use in industry. To realize this goal, we must study the parts feeding systems that have already found success in industry. Boothroyd *et al.* [8] describe parts feeding and orienting devices for automated assembly, including bowl feeders. Some recent work has focused on automating the design of bowl feeders (Boothroyd [9]; Caine [14]; Christiansen *et al.* [16]).

Another highly successful system is the SONY APOS system (Hitakawa [21]). Like a bowl feeder, the APOS system relies on vibration and shape to orient parts. Brost [13] demonstrates the use of shape alone to orient a part by dropping it on a specially designed nest.

Often, parts feeders that rely on shape are specific to a particular part or set of parts. Many researchers have studied constraint surfaces with simple shapes to handle a variety of parts. Examples of this include orienting parts by fences suspended over a conveyor (Peshkin and Sanderson [29]; Brokowski *et al.* [12]); translational pushing by a fence (Mani and Wilson [26]; Akella and Mason [1]); parallel-jaw grasping (Goldberg [20]); tray-tilting (Erdmann and Mason [18]; Christiansen [15]); vibrating a supporting plate (Böhringer *et al.* [6]; Swanson *et al.* [33]); and pushing by arrays of electromechanical manipulators (Böhringer *et al.* [7]).

Underactuated Systems The 1JOC is an underactuated system with a drift field. Such systems have been heavily studied in nonlinear control theory; see, for instance, (Brockett [11]; Crouch [17]). A good introduction to nonlinear control is given by Nijmeijer and van der Schaft [27].

Other examples of manipulation by an underactuated effector include slipping and rolling within a grasp (Bicchi and Sorrentino [5]; Brock [10]; Rao *et al.* [30]) and dynamic tasks such as snatching, rolling, and throwing (Arai and Khatib [3]; Lynch and Mason [25]).

The 1JOC can be likened to an underactuated manipulator in a gravity field, where the proximal "shoulder" (fence pivot) is directly actuated, but the distal degrees of freedom (represented by the object) are not. Research on the controllability of such serial link manipulators has been carried out by Oriolo and Nakamura [28] and Arai and Tachi [2]. Sørdalen *et al.* [31] recently developed a nonholonomic gear which allowed them to construct a controllable n-link planar arm with only two motors.

1.2 Terminology

The 1JOC has only one controlled degree-of-freedom, yet we would like to control three degrees-of-freedom of a part on the conveyor — the system is underactuated. The natural question is whether the 1JOC is "rich" enough to manipulate parts, or whether the two constraints on the motion of the fence (pivot is fixed) overly constrain the set of reachable configurations of the part.

To be precise, we borrow definitions of reachable sets from nonlinear control theory (Sussmann [32]). A part's configuration $\mathbf{q} = (x, y, \phi)^T$ is *controllable from* $\mathbf{q}$ if, starting from $\mathbf{q}$, the part can reach every configuration in the configuration space. The part is *controllable to* $\mathbf{q}$ if $\mathbf{q}$ is reachable from every configuration. The part is *accessible from* $\mathbf{q}$ if the set of configurations reachable from $\mathbf{q}$ has nonempty interior in the configuration space.

We can also define local versions of these properties. The part is *small-time accessible from* $\mathbf{q}$ if, for any neighborhood U of $\mathbf{q}$, the set of reachable configurations without leaving U has nonempty interior. The part is *small-time locally controllable from* $\mathbf{q}$ if, for any neighborhood U of $\mathbf{q}$, the set of reachable configurations without leaving U contains a neighborhood of $\mathbf{q}$.

The phrases "from $\mathbf{q}$" and "to $\mathbf{q}$" can be eliminated in these definitions if they apply to the entire configuration space.

In these terms, if the configuration of the part is accessible, we know that the 1JOC has "enough" degrees of freedom: it can transfer the part from the initial configuration to a three-dimensional subset of the configuration space. Better yet, if the part is controllable, the 1JOC can transfer the part from any configuration to any other configuration. Small-time local controllability, an even stronger condition, implies that the part can follow any path arbitrarily closely. This is a useful property for repositioning parts under time and workspace constraints.

Although controllability and small-time local controllability guarantee the ability to feed parts, they are not absolutely necessary. The minimum we require of a parts feeding system is the feeding property: the parts must be controllable to a single configuration from the set of initial configurations.

One way to reason about the controllability of a system is to write it as a system of drift and control vector fields and look at the dimension of the distribution defined by these vector fields. We can do this for an idealized model of the 1JOC (Section 2); however, this is difficult to do with the full pushing mechanics because the motion of a pushed object has no closed form. In addition, the motion of the part depends on the distribution of support forces between the part and the conveyor, which is usually unknown. The precise motion of the part is therefore unpredictable.

When we consider the full pushing mechanics in Section 3, we will limit our study to pushing motions with predictable outcomes. Despite this limited set of actions, we find that the 1JOC possesses the feeding property.

2 A Progression of Models

The 1JOC approach arises from a desire to explore the simplest mechanisms to accomplish a task. Planar objects have three degrees of freedom, suggesting the use of a robot with three or more actuated joints. However, it seems that two joints should suffice, by analogy with planar mobile robots. It is well known that a car can be arbitrarily positioned in the plane even if the steering wheel has only two settings. (For an introduction to motion planning for cars, and nonholonomic motion planning in general, see Latombe's text [22].) Each setting of the steering wheel defines a rotation center of the car. In the case of the car these two rotation centers move with the car. To see that two *fixed* rotation centers suffice to produce arbitrary motions of the car, imagine that the car is fixed and the wheels drive the plane around.

Obviously a planar object cannot be moved to an arbitrary position using a single fixed rotation center. Thus one might conclude that two controlled freedoms are necessary. However, a drift field, plus a single controlled freedom, suffices.

The problem of pushing a part across a moving conveyor is complicated by the mechanics of pushing. Some insights can be gained by first considering an idealized model: the part matches the fence's motion whenever desired, and matches the conveyor's motion otherwise. In this section we examine a progression of models leading to the 1JOC system.

One Rotation Center First we consider an idealization of a rotating fence. We envision an infinite turntable to which the part can be affixed. This turntable can give the part an arbitrary angular veloc-

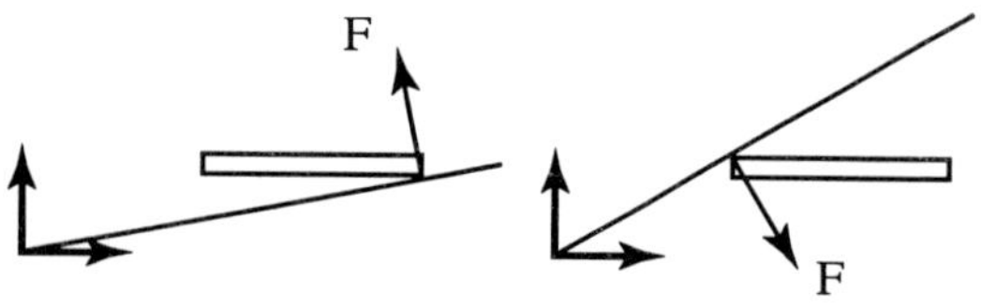

Figure 4: The thin rod is not small-time locally controllable from this configuration — it can only be rotated with a positive angular velocity.

ity about its pivot.

From any initial point in the part configuration space, the reachable set is one dimensional. If we consider the full pushing mechanics for a rotating fence, the reachable set is also just one dimensional. If the fence can swing all the way around and push on the other side, however, the part's configuration may be accessible. But we will not consider this case further.

Two Rotation Centers Now consider a pair of overlapping (yet independent) ideal turntables. At any given instant, the part is affixed to one of the two turntables and rotates about the center of that turntable. We assume that we can instantaneously switch the part from one turntable to the other.

There are two cases to consider:

- Both turntables are bidirectional. If each turntable can be driven in either forward or reverse, the situation is similar to driving a car with two different steering angles. As is well known, this system is small-time locally controllable and can approximate an arbitrary trajectory as closely as desired.

- One or both turntables are unidirectional. Now the situation is analogous to a car without reverse. The system is controllable, but not small-time locally controllable. It may be necessary to take a large excursion to accomplish a small motion.

We can model the conveyor as the limiting case of a turntable whose pivot approaches infinity. As above, the system is small-time locally controllable if both the turntable and conveyor are bidirectional, and it is simply controllable if either or both is unidirectional.

For this simple system, we can study controllability by considering the vector fields $X_1 = (0,1,0)^T$ and $X_2 = (-y,x,1)^T$ corresponding to the conveyor and the turntable, respectively. When the part tracks the conveyor, the part motion is given by $\dot{\mathbf{q}} = vX_1$, where v is the velocity of the conveyor in the y direction. When the part tracks the turntable, the part motion is given by $\dot{\mathbf{q}} = \omega X_2$, where ω is the angular velocity of the turntable and the origin is at the center of the turntable.

The Lie bracket $[X_1,X_2]$ of these two vector fields is given by:

$$[X_1,X_2] = \frac{\partial X_2}{\partial \mathbf{q}}X_1 - \frac{\partial X_1}{\partial \mathbf{q}}X_2 = (-1,0,0)^T$$

The new vector field $[X_1,X_2]$ is linearly independent of X_1 and X_2, yielding a third controlled freedom (provided both v and ω can be nonzero). The system therefore satisfies the Lie Algebra Rank Condition, and it is small-time accessible. If both the conveyor and the turntable are bidirectional (both v and ω can be positive or negative), the system is also small-time locally controllable (Nijmeijer and van der Schaft [27]). If either or both is unidirectional, the system is not small-time locally controllable, but it can be shown to be controllable by a simple constructive argument (Brockett [11]; Lynch and Mason [24]).

The 1JOC system consisting of a conveyor and a rotating fence resembles the system above with a unidirectional conveyor. One difference is that rotating the part about the pivot is only stable on a subset of the configuration space, specifically when an edge of the part is aligned with the fence.

Detailed Pushing Model We now consider pushing mechanics in the model with a conveyor and a rotating fence. We assume that the friction coefficient at the pushing contact and the distribution of support forces are known, and we place no restriction on the fence motions. Although this model makes unrealistic assumptions about the available information, any negative results will also apply to less detailed models.

Figure 4 shows that the system is not small-time locally controllable for the case of a thin rod. If the fence pushes from below, the rod's angular velocity is positive. If the fence swings around and pushes the part from above, the angular velocity is still positive. There is no maneuver combining these actions and conveyor drift that can locally achieve negative rotations.

Practical Pushing Models The detailed pushing model is difficult to use because it requires full knowledge of the pressure distribution. We can define more abstract models by restricting the allowable actions. In particular, we would like to use actions with predictable outcomes that do not impose unrealistic demands for information.

Any abstraction of the detailed pushing model will inherit its limitations, thus it is immediately clear that small-time local controllability is impossible.

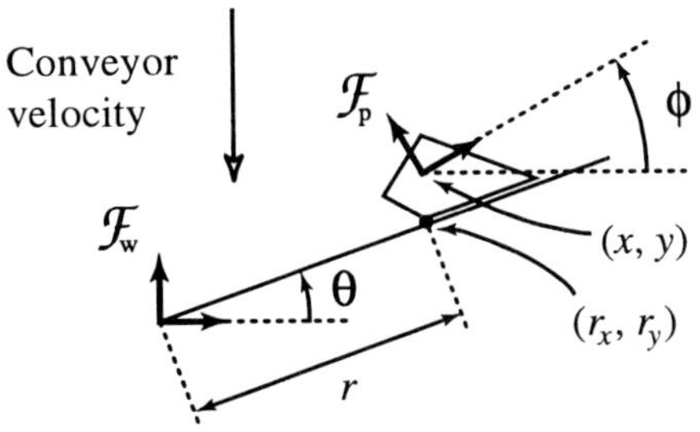

Figure 5: Notation.

3 The Feeding Property

In this section we focus on a particular model for the 1JOC and prove that it can transfer any polygonal part from any valid input configuration to a single goal configuration. To prove this property, we utilize a subset of the actions available to the 1JOC and show that they are sufficient to demonstrate the feeding property.

3.1 The 1JOC Model

The fence is a line that pivots about a fixed point on the line. The origin of a fixed world frame $\mathcal{F}_w$ coincides with the pivot point. The conveyor is the half-plane $x > 0$, with a constant drift velocity v in the $-y$ direction. The fence angle θ is measured with respect to the x axis of $\mathcal{F}_w$, and its angular velocity is given by ω. The fence angular velocity ω is our single control input.

The part can be any polygon, but because the manipulator is a line, the part can always be treated as its convex hull. The center of mass of the part lies in the interior of the convex hull and the coefficient of friction between the fence and the part is nonzero. When we refer to a "part" in the rest of the paper, we assume these properties.

We assume quasistatic mechanics. As the fence pushes the part over the conveyor, the motion of the part in the world frame $\mathcal{F}_w$ is sufficiently slow that inertial forces are negligible compared to the frictional forces. The support friction acting on the part during pushing is determined by the motion of the part relative to the conveyor, not the world frame $\mathcal{F}_w$.

It is convenient to represent the linear and angular velocity of the part relative to the conveyor in a frame $\mathcal{F}_p$ attached to the center of mass of the part. A velocity direction is simply a unit velocity vector. Velocity directions may also be represented as rotation centers in the frame $\mathcal{F}_p$.

The configuration of the part frame $\mathcal{F}_p$ in the world frame $\mathcal{F}_w$ is given by $(x, y, \phi) \in \Re^2 \times S^1$. When an edge of the part is aligned with the fence, the contact radius r defines the distance from the fence's pivot point to the closest point on the edge (the *contact vertex*), and the configuration of the part can be given in the polar coordinates (r, θ). The location of the point at r on the fence is (r_x, r_y) in the world frame. See Figure 5.

3.2 1JOC Primitives

We now describe the basic 1JOC primitives which yield the feeding property. We assume the fence is initially held at 0 degrees, perpendicular to the conveyor velocity, until the part contacts and comes to rest against the fence. See Figure 6 for an illustration of the primitives.

3.2.1 *Stable Pushes*

A stable push occurs when one edge of the part is aligned with the fence and the motion of the fence keeps the part fixed against it. The simplest example of a stable push occurs when the fence is held at 0 degrees, and a part on the conveyor drifts into contact and comes to rest on the fence. When the part is at rest, the fence is executing a stable push. The pushing direction, as seen by the part, is opposite the conveyor's velocity direction.

Of course, a stable push may also occur while the fence is rotating. In a previous paper (Lynch and Mason [24]), we described the procedure STABLE that finds a set of stable pushing directions $\mathcal{V}_{stable}$ for a given pushing edge, pushing friction coefficient, and center of mass of the part. This set of pushing directions is fixed in the part frame $\mathcal{F}_p$. The fence is guaranteed to execute a stable push if the edge is aligned with the fence and the motion of the fence and the conveyor combine to yield a pushing direction in $\mathcal{V}_{stable}$. See Figure 7 for details.

For simplicity, we will only consider the edges of the part that yield a stable push while the fence is held at 0 degrees (the pushing direction is a translation normal to the edge). The existence of these stable edges is indicated by the following lemma (Lynch and Mason [24]):

Lemma 1 *All polygonal parts have at least one edge such that the normal translational pushing direction, along with a neighborhood of this pushing direction, belongs to* $\mathcal{V}_{stable}$. *Such edges are called* stable edges.

Remark: The object may also have *metastable edges* — edges such that the center of mass lies directly above an edge endpoint when the fence is held at 0 degrees. If the fence rotates any nonzero amount in the "wrong" direction, the edge will become unstable. We will avoid these edges in this paper. If the part initially

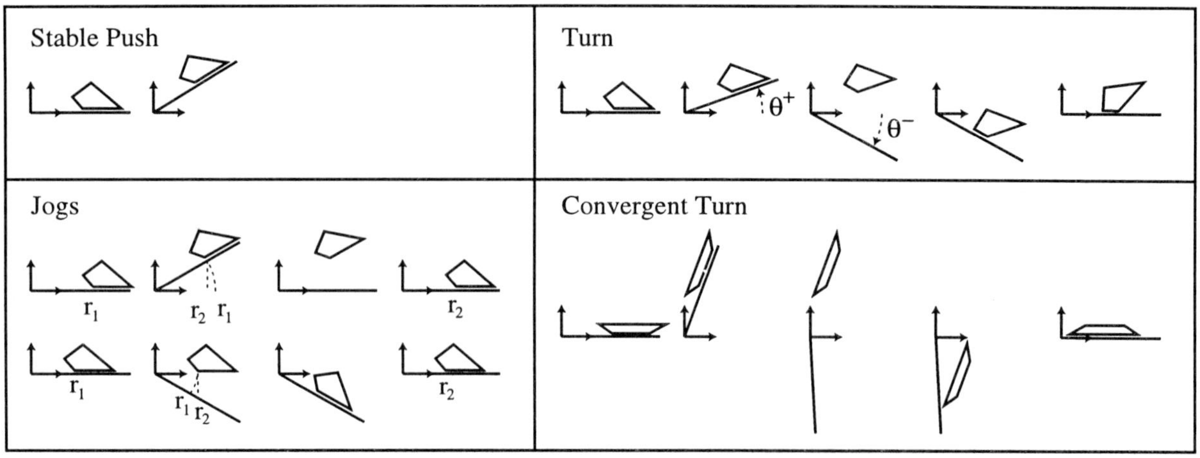

Figure 6: Step by step illustrations of the 1JOC primitives.

comes to rest on the fence on one of these edges, we can perturb the fence slightly to bring the part to a stable edge.

Lemma 1 says that a stable edge is also stable for a neighborhood of fixed fence angles around 0 degrees. In addition, if the contact radius r is sufficiently large, the fence pivot will be inside the set of stable rotation centers $\mathcal{V}_{stable}$ fixed in the part frame $\mathcal{F}_p$. If the fence angular velocity ω is large enough relative to the conveyor velocity v, the combined pushing motion is nearly a pure rotation about the pivot, and the pushing motion is stable. Therefore, it is possible to stably push the part from any θ_0 to any θ_1 in a counterclockwise (CCW) direction, where $90° > \theta_1 > \theta_0 > -90°$. Figure 8 illustrates a method for finding the *minimum stable radius*: the minimum contact radius r such that the push is stable for all fence angles in the range $[-90, 90]$ for a given fence angular velocity ω and conveyor velocity v.

For a push to be stable when the fence is rotating CCW, the contact radius r must be sufficiently large. We can change r by performing jogs, described below.

3.2.2 *Jogs*

Jogs are simple maneuvers that allow us to change the contact radius r from the fence pivot to the part without changing the stable resting edge.

Lemma 2 *A part resting on a stable edge can be moved from any initial contact radius r on the fence to any desired r in the range $(0, \infty)$ by a series of jogs.*

Proof: We use the fact that a neighborhood of pushing directions about the normal translation is stable for a stable edge. In fact, there is a range $[\theta^{min}, \theta^{max}]$ of fixed fence angles about 0 degrees that are also stable. The fence does not have to be motionless to execute a stable push, however; if the fence angular velocity ω is small enough with respect to the conveyor velocity v, then the fence can execute a stable push while moving in the angle range $[\theta^{min}, \theta^{max}]$, regardless of the contact radius r.

To decrease the contact radius r, the fence is raised to an angle θ^+ ($\theta^{max} \geq \theta^+ > 0$), keeping the part fixed to the fence by a stable push. This changes the value of r_x to $r_x \cos\theta^+$. The fence then drops to 0 degrees, immediately releasing the part. The part drifts back into contact with the fence and settles on the same stable edge. We assume there is no slip between the part and the fence. (The no-slip assumption is important for open-loop feeding plans.) The result of the jog is to decrease the contact radius from r to $r\cos\theta^+$.

To increase the contact radius r, the fence is quickly lowered to an angle θ^- ($\theta^{min} \leq \theta^- < 0$), releasing the part. The part drifts back into contact with the fence and settles on the same stable edge. The fence is then raised to 0 degrees again, pushing the part with a stable push. The contact radius of the part has changed to $r/\cos\theta^-$.

A series of jogs, either increasing or decreasing r, can bring the part to any r in the range $(0, \infty)$. □

A jog is directly analogous to the Lie bracket motion for the ideal turntable plus conveyor system analyzed in Section 2. The difference from the ideal system is that the part only tracks the fence when it is in stable contact. Nonetheless, we can now show that the configuration of the part is accessible with just stable pushes, jogs, and the conveyor drift.

Theorem 1 *The configuration of a polygonal part in stable edge contact with a fence held at 0 degrees is small-time accessible using stable pushes, jogs, and conveyor drift.*

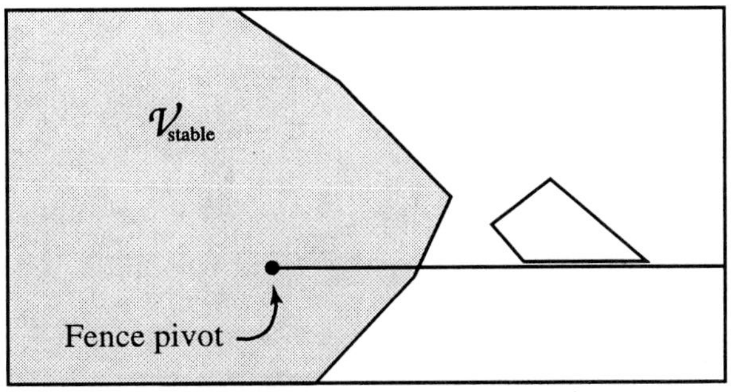

For a given edge contact, we can determine $\mathcal{V}_{stable}$, the set of pushing rotation centers that keep the object fixed to the pusher. Note that $\mathcal{V}_{stable}$ is fixed in the part frame, not the world frame.

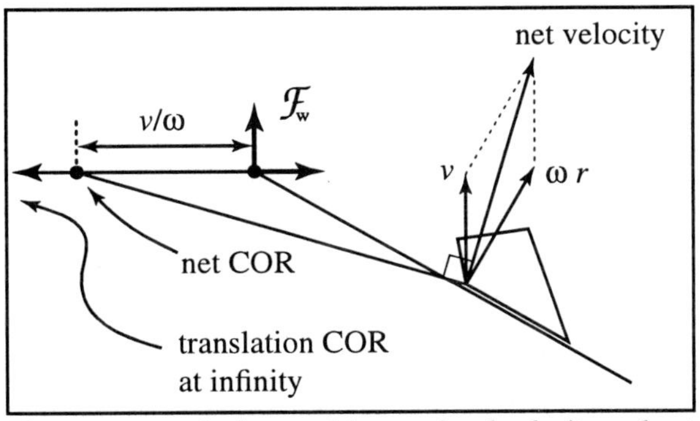

The conveyor velocity v and the rotational velocity ω about the fence pivot combine to form a net velocity at each point. This can be expressed as a net center of rotation (COR), which lies on the line through the fence pivot and perpendicular to the conveyor velocity.

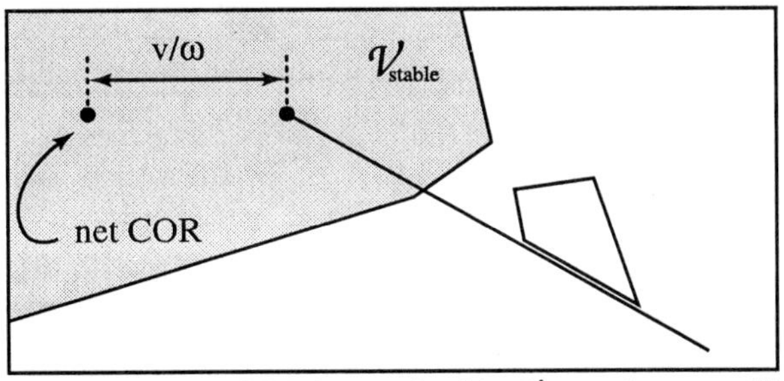

Provided the net COR is contained in $\mathcal{V}_{stable}$, the part will remain stationary with respect to the fence for that combination of conveyor and fence velocities. Note that as the fence rotates, the conveyor velocity changes direction with respect to the part, and the net COR rotates about the fence pivot at a radius of v/ω.

Figure 7

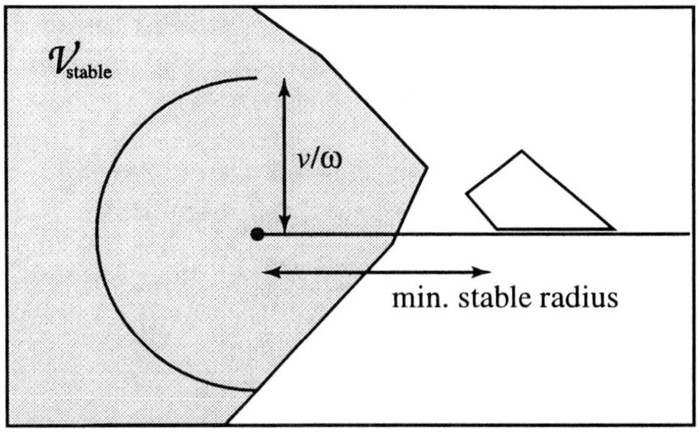

Figure 8: The locus of net COR form a semicircle about the fence pivot as the fence swings from -90 to 90 degrees. Given a fence angular velocity ω and a conveyor velocity v (which determine the radius of this semicircle), we pick a minimum stable radius so that this semicircle lies completely within $\mathcal{V}_{stable}$. Recall that $\mathcal{V}_{stable}$ is fixed with respect to the part, so increasing the contact radius moves the semicircle deeper into $\mathcal{V}_{stable}$. This conservative estimate of the minimum stable radius guarantees stable pushing in the counterclockwise direction at any fence angle in $[-90°, 90°]$.

Proof: Given any neighborhood of the part configuration while it is in stable edge contact with the fence at 0 degrees, it is possible to jog the part a nonzero distance in both directions without leaving this neighborhood. The effect of the jog is to change the x coordinate of the part configuration, exactly like the Lie bracket motion of Section 2. The other two vector fields of Section 2 can be obtained directly by conveyor drift and stable pushes. □

We call I the set of initial part configurations that drift to stable edge contact with the fence at 0 degrees. From a stable edge contact, Theorem 1 indicates that the configuration of the part is small-time accessible. Therefore, the configuration of the part is accessible from the set I.

We still have not proven the feeding property. To get this property, we need the ability to turn a part to a new stable edge.

3.2.3 *Turns*

Turns allow us to change the edge of the part aligned with the fence, subject to the constraint that the initial and final edges both be stable edges. A turn consists of raising the fence to an angle $\theta^+ \geq 0$ by executing a stable push; dropping the fence to an angle $\theta^- \leq 0$, immediately releasing the part; and reacquiring the part on the new edge with a stable push, raising the fence back to 0 degrees. The final push commences at the moment the new edge contacts the fence.

Lemma 3 *Any polygonal part has at most one stable edge from which it is impossible to turn the part to the* *next CW stable edge.* Exception: *a part has two such stable edges if they are the only stable edges and they are parallel to each other.*

Proof: The fence can execute a stable push to any angle less than 90 degrees, and it can reacquire the part with a stable push at any angle greater than -90 degrees. This indicates that it is possible to reach any final stable edge less than 180 degrees CW of the initial edge. If the part has two or more stable edges, there can only be one stable edge from which the next CW stable edge is greater than 180 degrees removed. For "exception" parts, there are two stable edges which are exactly 180 degrees removed from each other. □

There are a few important points to make about turns:

- The contact radius r, for both the initial and final edges, must be large enough that the pushing directions always remain in $\mathcal{V}_{stable}$.
- The quantity $\theta^+ - \theta^-$ is determined by the part geometry, but the actual values of θ^+ and θ^- are not. To some extent, the contact radius r after the turn can be controlled by the choice of these values. If the angle difference between the initial and final edge is 90 degrees or more, r can be changed to any value in the range $(0, \infty)$ during the turn.
- This primitive requires precise knowledge of the conveyor's motion — the final stable push must begin precisely when the new edge contacts the fence.

Consider the case where turning a part to a new stable edge requires raising the fence to $90 - \delta$ and catching the part at $-90 + \varepsilon$, where δ and ε are small positive values. It may be necessary to move the part to a large contact radius r before executing the turn to ensure that the part remains completely on the conveyor ($x > 0$) during the initial push.

3.2.4 *Convergent turns*

To perform a turn between two stable edges which are parallel, the fence must, in principle, be raised to 90 degrees, pushing the part off the conveyor. To turn such parts, we can introduce a convergent turn: raise the fence to $90 - \delta$, allow the part to drift down, and begin pushing it again at $-90+\varepsilon$. Assuming no slip at the pushing contact, the part will converge to the new stable edge as the fence is raised to 0 degrees. Convergent turns allow us to strengthen Lemma 3 to apply to all polygonal parts, with no exceptions. In this paper, the sole function of convergent turns is to handle parts with only two stable edges, which are parallel.

3.3 The Feeding Property

Using the lemmas proven above, we can now demonstrate the feeding property for the 1JOC.

Theorem 2 *The 1JOC possesses the feeding property:*

- *for all polygonal parts*
- *for the three-dimensional space I of initial configurations such that the part initially comes to rest on a stable edge, completely on the conveyor, against the fence fixed at 0 degrees, and*
- *using only stable pushes, jogs, turns, convergent turns, and conveyor drift.*

Furthermore, the goal configuration can be chosen from a three-dimensional subset of the configuration space.

Proof: If the part has a stable edge such that it cannot be turned to a new stable edge, then this edge must be chosen as the goal edge. Otherwise, any stable edge can be chosen as the goal edge. By Lemma 3 (strengthened by the convergent turn of Section 3.2.4), this goal edge can be reached using turns. Once at the goal edge, the part can be jogged to any contact radius r in the range $(0, \infty)$ (Lemma 2). By Lemma 1, the part is stable against the fence for a range of fence angles θ. The goal edge therefore allows a two-dimensional set of final stable configurations of the part in the (r, θ) space. If the fence is quickly lowered to 90 degrees, releasing the part, this set drifts to a set of reachable configurations with nonempty interior in the world configuration space. Any configuration in this set can be chosen as the goal. □

If the application of the 1JOC is to stuff parts into a feeder track, then the fence can be rotated 90 degrees, pushing the part off the conveyor into the feeder track, instead of releasing it to continue on the conveyor.

4 A Feeding Planner

The feeding property means that there exists a sequence of jogs and turns to feed any polygonal part, assuming infinite fence length and an infinite conveyor half-plane. A 1JOC with finite fence and conveyor dimensions may not be able to feed some parts. For a given part however, we can always find a fence and conveyor with finite dimensions to feed it. We describe how to find a sequence of jogs and turns to feed

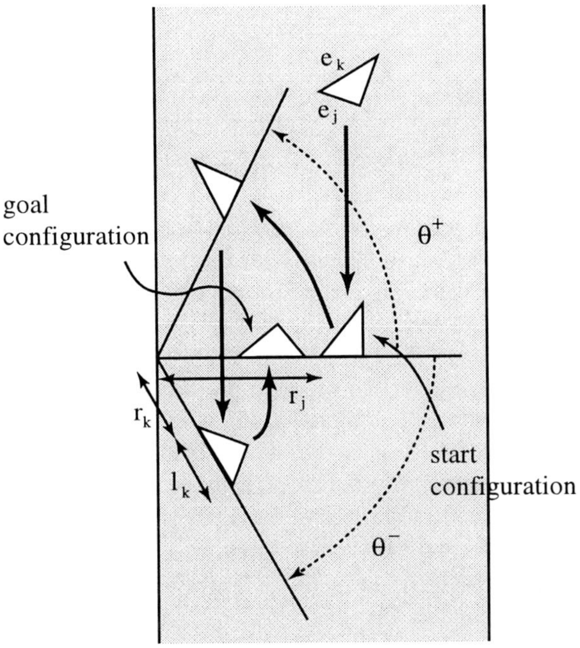

Figure 9: A feeding plan to move a triangle from start configuration (e_j, r_j) to goal configuration (e_k, r_k) by a turn with raise and drop angles of θ^+ and θ^-. Length of edge e_k is l_k.

a given part with a fence and conveyor of known dimensions in the minimum amount of time.

We assume that we know the shape and center of mass of the part, the coefficient of friction between the fence and part, the conveyor velocity, and the fence and conveyor dimensions. The coefficient of friction between the part and conveyor is assumed constant, and we pick the fence angular velocity to be one of three discrete values: ω_{stable} (for stable pushes), ω_{drop} (when lowering the fence), and zero. We assume the fence receives parts that are singulated.

A part drifts down the conveyor until it comes to rest against the fence. The feeding problem consists of getting the part from this start configuration to the goal configuration (see Figure 9), where the configurations are specified by the edge aligned with the fence, and the contact radius (distance from pivot to the closest vertex of the resting edge). We assume that the initial fence orientation is zero, and, for convenience, that the final fence orientation is zero as well.

Turns change the part orientation; they rotate the part CCW so the new edge aligned with the fence is CW to the initial edge. Turns thus enable transitions to stable edges in the CW direction. We consider all CW sequences of edges beginning at the start edge e_s and ending at the goal edge e_g which do not require rotating the part more than 360 degrees. Take an example sequence $S = \{e_s, e_j, e_k, e_g\}$. If we can find a valid sequence of jogs and turns to move the part in sequence through the edges in S and change the contact radius by the required amount, we have a feasible plan. Of all feasible plans, we take the one which requires minimum time.

4.1 A Simple Example

Consider a part resting on edge e_j (orientation ϕ_j) at radius r_j which is to be moved to the neighboring CW stable edge e_k (orientation ϕ_k) at radius r_k (see Figure 9). Here the sequence of edges is $S = \{e_j, e_k\}$. We look for a feasible solution (and ignore minimizing the time). Assume r_j and r_k are greater than the minimum stable radii for the corresponding edges, and are close enough in magnitude that no jog is required. Here, finding a plan consists of determining the fence angles θ^+ and θ^- for the turn. From the geometry, we have:

$$r_k = r_j \frac{\cos\theta^+}{\cos\theta^-} - l_k \qquad (1)$$

$$\theta^+ - \theta^- = \phi_k - \phi_j \qquad (2)$$

Solving these equations, we find:

$$\theta^- = \tan^{-1}\left(\cot(\phi_k - \phi_j) - \frac{r_k + l_k}{r_j \sin(\phi_k - \phi_j)}\right) \qquad (3)$$

$$\theta^+ = \phi_k - \phi_j + \theta^- \qquad (4)$$

This simple example has a closed form solution that provides us with a plan. (Note that when $(\phi_k - \phi_j) \geq 90$, the radius change can be arbitrarily large or small.)

4.2 Nonlinear Programming Approach

For the general case, we must find a sequence of jogs and turns that effects the desired configuration change and minimizes time. For a given sequence of edges, we must determine the parameters for each turn while ensuring that each contact edge in a turn has a contact radius that is no less than its minimum stable radius. If the contact radius is less than the minimum radius, we must translate the part prior to executing the turn. Such a *translation* may consist of a jog or a series of jogs.

A feeding plan to transfer a part from its start configuration (e_s, r_s) to its goal configuration (e_g, r_g) consists of several stages, each of which accomplishes an edge transition. Each stage consists of a translation followed by a turn; the translation for a given stage may be zero. A plan with the edge transition sequence S will take $n-1$ stages, where $n = |S|$. A final translation may also be required after the goal edge has been reached.

Consider the ith stage in a plan (see Figure 10) to accomplish the transition from edge e_j to edge e_k. First, we perform a translation of g_i, which moves the part to

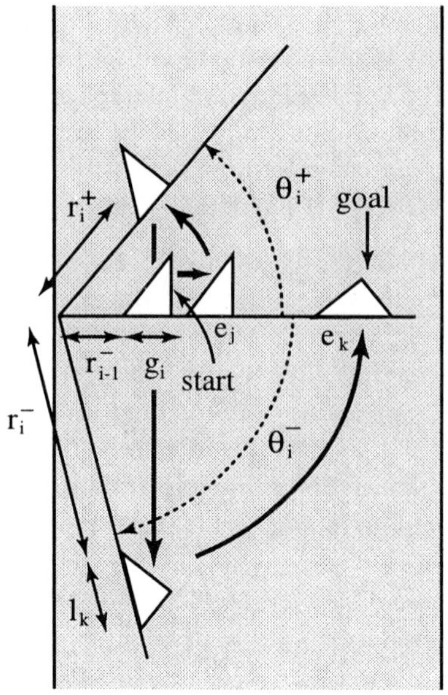

Figure 10: The ith transition from edge e_j to edge e_k with a translation of g_i, start and end turn radii r_i^+ and r_i^-, and raise and drop angles θ_i^+ and θ_i^-.

a contact radius of r_i^+. Since we are about to perform a turn, this radius must be greater than the minimum stable radius for edge e_j, r_j^{stable}. We then perform a turn with a raise angle of θ_i^+ and a drop angle of θ_i^-. The stable push at the end of the turn brings the fence to $\theta = 0$ with the new contact edge e_k. For this stable push on edge e_k, the contact radius r_i^- must be greater than the minimum stable radius for e_k, r_k^{stable}. We also have to ensure that the fence raise and drop angles are in valid ranges, that the part is within fence and conveyor bounds, and that the fence contacts the part only at the end of the drift phase.

Assume the fence pivot is on the conveyor left edge, and that the conveyor half-length exceeds its width, which exceeds the fence length. The resulting constraints are (for $i = 1, \ldots, n-1$):

$$r_i^+ = r_{i-1}^- + g_i \tag{5}$$

$$r_i^- = \frac{(r_i^+ - d_j)\cos\theta_i^+}{\cos\theta_i^-} - (l_k + d_k) \tag{6}$$

$$\theta_i^+ - \theta_i^- = \phi_k - \phi_j \tag{7}$$

$$r_j^{max} \geq r_i^+ \geq r_j^{stable} \tag{8}$$

$$r_k^{max} \geq r_i^- \geq r_k^{stable} \tag{9}$$

$$90 > \theta_i^+ \geq 0 \tag{10}$$

$$0 \geq \theta_i^- > -90 \tag{11}$$

$$(r_i^+ + a_j^v)\cos\theta_i^+ - p_j^v \sin\theta_i^+ > 0, \forall v \in V \tag{12}$$

$$\omega_{drop} r_i^+ \cos\theta_i^+ > v_c \tag{13}$$

$$\frac{(r_i^+ - d_j)\sin\theta_i^+ - (r_i^- + l_k + d_k)\sin\theta_i^-}{v_c} > \frac{\theta_i^+ - \theta_i^-}{\omega_{drop}} \tag{14}$$

where l_k is the length of edge k, ϕ_j and ϕ_k are the orientations of edges e_j and e_k, v_c is the conveyor velocity, r_j^{max} and r_k^{max} are maximum valid contact radii for edges e_j and e_k, V is the set of part vertices, and a_j^v and p_j^v are the components along and perpendicular to edge e_j of the vector from the contact vertex to a part vertex v. If e_j and e_k are neighboring edges, $d_j = d_k = 0$. Else, d_j and d_k are the distances from the (virtual) intersection vertex of (extended) edges e_j and e_k to the contact vertices of these edges.

The start and goal constraints are:

$$r_0^- = r_s \tag{15}$$

$$r_g = r_{n-1}^- + g_n \tag{16}$$

These define a set of nonlinear constraints over the variables g_i, θ_i^+, θ_i^-, r_i^+, and r_i^-. (We could also eliminate the equality constraints and solve for fewer variables.) Since the radius at the start of a turn depends on the radius at the end of the previous turn, an optimal solution must consider all variables simultaneously.

We want a solution which minimizes the total time to feed the part. The time taken for each stage of the plan is the sum of the jog time, fence raise time, drift time on the conveyor, and fence recovery time. We approximate the total time by the objective function we minimize:

$$t_n g_n^2 + \sum_{i=1}^{n-1}\left[t_i g_i^2 + \frac{\theta_i^+}{\omega_{stable}} - \frac{\theta_i^-}{\omega_{stable}} + \frac{(r_i^+ - d_j)\sin\theta_i^+ - (r_i^- + l_k + d_k)\sin\theta_i^-}{v_c}\right] \tag{17}$$

The approximate time cost coefficient of the ith jog series is $t_i = k\cos\alpha_i / v_c(r_i^+ + r_{i-1}^-)$ where k is a constant chosen to be 1000, and $\alpha_i = \min(|\theta_i^{min}|, |\theta_i^{max}|)$ is a safe jog angle for the ith jog series.

Our problem as stated above is a nonlinear programming problem whose feasible solutions yield valid feeding plans. (See Bazaraa and Shetty's text [4] for an introduction to nonlinear programming.) Once we have a solution for the g_i, θ_i^-, θ_i^+, r_i^+, and r_i^- values which satisfies the constraints (5)–(16) and minimizes our objective function, we must still determine a series of jogs to execute the translations g_i.

An inward translation distance $g_i < 0$ requires a series of decreasing jogs, such that $r_i^+ = r_{i-1}^-(\cos\gamma_i)^m \cos\rho_i$ where m is a non-negative

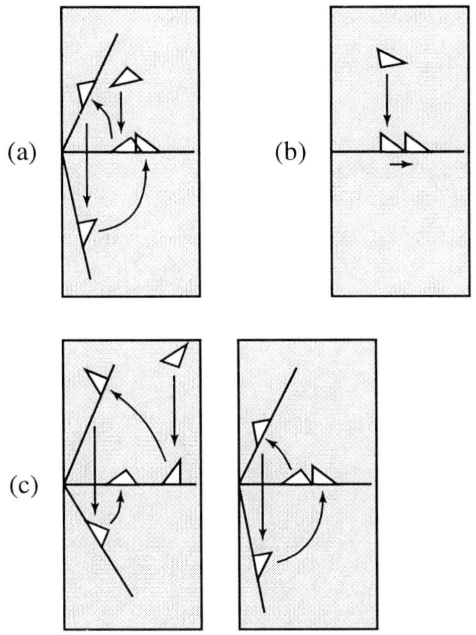

Figure 11: Example feeding plans for a triangle. The goal configuration $(\phi_g = 269.34, r_g = 150)$ is the same, while the start configurations are different: (a) $(\phi_s = 126.49, r_s = 100)$ (b) $(\phi_s = 269.34, r_s = 100)$ (c) $(\phi_s = 0, r_s = 200)$.

integer, and $\theta_i^{max} \geq \gamma_i \geq \rho_i \geq 0$. So this radius decreasing translation is accomplished by a series of m jogs of angle γ_i followed by a jog of ρ_i. Similarly, an outward translation distance $g_i > 0$ requires a series of increasing jogs, such that $r_{i-1}^- = r_i^+ (\cos\gamma_i)^m \cos\rho_i$ where m is a non-negative integer, and $0 \geq \rho_i \geq \gamma_i \geq \theta_i^{min}$.

A feasible solution to the above problem always exists if the fence and conveyor are large enough that each stable edge has a valid contact radius greater than its minimum stable radius. When this condition is satisfied, a plan with a maximum of three stages to get the part from the start to the goal configuration exists, although it may not be optimal.

4.3 Generating Feeding Plans

We now outline the process of generating feeding plans. From the part description and coefficient of friction, we compute the minimum stable radii for the stable edges and the transition angles between these edges. For the start and goal configurations, we generate candidate edge transition sequences. For each such sequence, the corresponding objective function and constraints are input to a commercial nonlinear programming package called GINO [23] to solve the problem. A feasible solution to the nonlinear programming formulation provides us with a valid feeding plan. See the example plans in Figure 11.

Currently we use minimum stable radius values empirically determined to be robust. A conservative estimate of the minimum stable radius can be determined as illustrated in Figure 8. Also, the planning procedure can be extended to use convergent turns to handle parts with only two stable edges which are parallel.

5 Implementation

We have implemented several feeding plans on a conveyor with the fence being (somewhat ironically) actuated by one joint of an Adept 550 SCARA robot. The conveyor speed and fence rotational speed have been selected to be 20 mm/s and 30 degrees/s respectively. The fence is covered with a foam material to avoid slip between the part and the fence.

Currently we place the part at the start configuration and observe the position of the part after plan execution. In a fully implemented setup, singulated parts would come down the conveyor in random configurations. The initial part orientation and contact radius would be determined by an overhead camera, and a feeding plan to get the part to the goal configuration without any further sensing would be generated.

Figure 11 illustrates three plans we ran on our setup which achieve the same goal configuration from different starting configurations. Plan (a) requires a single turn to reach the goal, plan (b) requires just a translation, and plan (c) requires two turns. These plans had position errors at the goal configuration ranging from 0 mm to 3 mm. The errors are sensitive to part shape measurements, fence turn angles, and timing in the turns. Occasional slip also causes some error.

6 Variations on a Theme

The 1JOC approach uses a fixed velocity conveyor in combination with a single servoed joint to obtain the diversity of motions required for planar manipulation. We have shown by proof and demonstration that the 1JOC is capable of useful planar manipulation: any polygon is controllable from a broad range of initial configurations to any goal chosen from a broad range of goal configurations.

For this paper we designed the system and the set of actions to simplify analysis and planning. There are many variations on the approach which may be more suitable in different contexts. Some variations on the system configuration are: a curved fence; a prismatic fence in place of the revolute fence; a rotary table in place of the conveyor; or use of gravity rather than a conveyor. Some variations on the actions are: optimization of fence rotation rates instead of using fixed values; allowing objects to slip along the fence; complete revolutions of the object to reduce jogs; and speeds high enough to require dynamic analysis instead of quasi-static. Another variation would be to use techniques similar to Bicchi and Sorrentino's to roll objects on the belt [5].

When addressing potential industrial applications, it is important to consider the whole system, including interactions with surrounding equipment. Several different scenarios have occurred to us: a 1JOC to pose objects followed by a simple pick-and-place device; a 1JOC to singulate parts for a SCARA; or two or three 1JOCs pipelined to singulate and feed parts.

Acknowledgments

Many thanks to Costa Nikou for the implementation of this work on the Adept robot. We thank Mike Erdmann, Garth Zeglin, and Nina Zumel for early discussions about this paper. We thank Adept, NSK, and SONY for helping us learn about automation and for making equipment available. This work was supported by ARPA and NSF under grant IRI-9318496.

References

[1] S. Akella and M.T. Mason. Posing polygonal objects in the plane by pushing. In *IEEE Int. Conf. on Robotics and Automation*, 1992, pp. 2255–2262.

[2] H. Arai and S. Tachi. Position Control of a Manipulator with Passive Joints Using Dynamic Coupling. *IEEE Trans. on Robotics and Automation*, vol. 7, no. 4, August 1991, pp. 528–34.

[3] H. Arai and O. Khatib. Experiments with Dynamic Skills. In *Proceedings of the 1994 Japan-USA Symposium on Flexible Automation*, pp. 81–4.

[4] M.S. Bazaraa and C.M. Shetty. *Nonlinear Programming: Theory and Algorithms*. John Wiley and Sons, 1979.

[5] A. Bicchi and R. Sorrentino. Dexterous Manipulation Through Rolling. In *IEEE Int. Conf. on Robotics and Automation*, 1995.

[6] K.F. Böhringer, V. Bhatt, and K.Y. Goldberg. Sensorless Manipulation Using Transverse Vibrations of a Plate. In *IEEE Int. Conf. on Robotics and Automation*, 1995

[7] K. Böhringer, R. Brown, B. Donald, J. Jennings, and D. Rus. Distributed Robotic Manipulation: Experiments in Minimalism. In *International Symposium on Experimental Robotics*, 1995.

[8] G. Boothroyd, C. Poli, and L.E. Murch. *Automatic Assembly*. Marcel Dekker, 1982.

[9] G. Boothroyd. *Assembly Automation and Product Design*. Marcel Dekker, 1992.

[10] D.L. Brock. Enhancing the Dexterity of a Robot Hand Using Controlled Slip In *IEEE Int. Conf. on Robotics and Automation*, 1988, pp. 249–251.

[11] R.W. Brockett. Nonlinear Systems and Differential Geometry. *Proceedings of the IEEE*, vol.64, no. 1, January 1976.

[12] M. Brokowski, M. Peshkin, and K. Goldberg. Curved Fences for Part Alignment. In *IEEE Int. Conf. on Robotics and Automation*, 1993, pp. 3:467–473.

[13] R.C. Brost. Dynamic Analysis of Planar Manipulation Tasks. In *IEEE Int. Conf. on Robotics and Automation*, 1992, pp. 2247–54.

[14] M. Caine. The Design of Shape Interactions Using Motion Constraints. In *IEEE Int. Conf. on Robotics and Automation*, 1994.

[15] A.D. Christiansen. Manipulation Planning from Empirical Backprojection In *IEEE Int. Conf. on Robotics and Automation*, 1990, pp. 762–768.

[16] A.D. Christiansen, A.D. Edwards, and C.A. Coello. Automated Design of Parts Feeders using a Genetic Algorithm. Submitted to the *1996 IEEE Int. Conf. on Robotics and Automation*.

[17] P.E. Crouch. Spacecraft Attitude Control and Stabilization: Applications of Geometric Control Theory to Rigid Body Models. *IEEE Trans. on Automatic Control*, vol. 29, no. 4, 1984, pp. 321–31.

[18] M.A. Erdmann and M.T. Mason. An exploration of sensorless manipulation. *IEEE Trans. on Robotics and Automation*, 4(4):369–379, Aug. 1988.

[19] M. Erdmann. An Exploration of Nonprehensile Two-Palm Manipulation: Planning and Execution. *The Seventh International Symposium on Robotics Research*, 1995.

[20] K.Y. Goldberg. Orienting polygonal parts without sensors. *Algorithmica*, 10:201–225, 1993.

[21] H. Hitakawa. Advanced Parts Orientation System Has Wide Application. *Assembly Automation*, vol. 8, no.3, 1988, pp. 147–50.

[22] J.C. Latombe. *Robot Motion Planning*. Kluwer Academic Publishers, 1991.

[23] J. Liebman, L. Lasdon, L. Schrage, and A. Waren. *Modeling and Optimization with GINO*. The Scientific Press, 1986.

[24] K.M. Lynch and M.T. Mason. Controllability of Pushing. In *IEEE Int. Conf. on Robotics and Automation*, 1995.

[25] K.M. Lynch and M.T. Mason. *Nonprehensile Manipulation*. Carnegie Mellon University, Robotics Institute videotape, 1995.

[26] M. Mani and W. Wilson. A Programmable Orienting System for Flat Parts. In *North American Manufacturing Research Institute Conference XIII*, 1985.

[27] H. Nijmeijer and A.J. van der Schaft. *Nonlinear Dynamical Control Systems*. Springer-Verlag, 1990.

[28] G. Oriolo and Y. Nakamura. Control of Mechanical Systems with Second-Order Nonholonomic Constraints: Underactuated Manipulators. In *Proceedings of the 30th Conference on Decision and Control*, 1991, pp. 2398–2403.

[29] M.A. Peshkin and A.C. Sanderson. Planning Robotic Manipulation Strategies for Workpieces That Slide. *IEEE Journal of Robotics and Automation*, vol. 4, no. 5, October 1988, pp. 524–31.

[30] A. Rao, D. Kriegman, and K. Goldberg. Complete Algorithms for Reorienting Polyhedral Parts using a Pivoting Gripper. In *IEEE Int. Conf. on Robotics and Automation*, 1995, pp. 2242-8.

[31] O.J. Sørdalen, Y. Nakamura, and W.J. Chung. Design of a Nonholonomic Manipulator. In *IEEE Int. Conf. on Robotics and Automation*, 1994, pp. 8–13.

[32] H.J. Sussmann. Lie brackets, real analyticity and geometric control. In *Differential Geometric Control Theory*, Brockett, Millman, and Sussman, eds. Birkhauser, 1983.

[33] P.J. Swanson, R.R. Burridge, and D.E. Koditschek. Global asymptotic stability of a passive juggler: A parts feeding strategy. In *IEEE Int. Conf. on Robotics and Automation*, 1995, pp. 1983–1988.

PART 7

LEARNING AND SKILL ACQUISITION

SESSION SUMMARY

Gerd Hirzinger
DLR Oberpfaffenhofen, Institute for Robotics and System Dynamics
82234 Wessling - Germany
email: gerd.hirzinger@dlr.de

At the 6th ISRR, Brian Carlisle, representing the American industrial robot manufacturer ADEPT, tried to formulate what he thought to be the most important tasks to be solved by robot researchers. "To teach a robot a task within typically 10 minutes so that it is able to repeat it in an at least slightly changing environment" was one of his major issues. Several years ago AI techniques seemed to be able to solve the problems of robot programming but moderate success led to a certain disappointment; apparently sensory feedback control and the adaptivity it may generate had been completely underestimated. And describing tasks by linguistic means soon had to realize its limits. In the last ten years learning by neural nets (partly based on Rumelhart's backpropagation technique, partly based on self-organizing nets like KOHONENs feature maps) celebrated quite a revival not only in robotics; and thus it does not astonish that "robot learning by showing" (be it with or without nets) has gained increasing interest in the last years. Skill transfer from man to the machine and self-improvement afterwards indeed is an old dream of robotics researchers.

To me it seems of minor importance whether the robot "passively" observes a human e.g. performing an assembly task or whether by force-reflecting teleoperation the robot is actively included in the learning process. In any of these cases the robot is supposed to become kind of an intelligent observer which is able to repeat the task it has been shown successfully and even improve itself autonomously afterwards.

The papers in the session "learning and skill acquisition" address these kind of questions. In "Behaviour Acquisition via Vision-Based Robot Learning" Minoru Asada at al. presented three topics of vision based robot learning which have in common that visual perception is tightly coupled to actuation. Vision-based reinforcement learning is demonstrated by a robot learning to shoot a ball into a goal. The concept of "motion sketch" leads to learning e.g. obstacle avoidance, and the paper also includes purposive visual control for uncalibrated camera-manipulator systems.

In answering questions M. Asada pointed out that the target has not to be found or defined by the robot but is given by the operator and that the state space implying e.g. position and size of ball or goal may be fairly large to absorb the delay and the uncertainty of image processing. He also outlined the situation when obstacles are occluding the target where the robot goes back or turns away so as to avoid the obstacles, then tries to find the target. Once he finds it, the robot tries to reach it unless other obstacles block him. It's just a result of learning, not the result of coding.

The paper "Perception and Manipulation in Robotics Neural Network Approaches" by R. Koeppe et al. gives a survey on neural net applications for the development of a new multisensory lightweight robot generation at DLR. It comprises world modelling based on sparse range data using self-organizing maps similar to those proposed by KOHONEN, discusses "shape from shading" using multilayer nets and addresses the issue of visual servoing in a tele-robotic environment based on learning elemental operations in a virtual or real environment. Learning compliant motions by task demonstration (e.g. including force feedback), learning 3 finger grasps and improving joint torque and position controlled robot control systems are the other items in the paper.

Questions referred e.g. to whether mappings to be realized by neural nets have to be smooth. Koeppe pointed out that indeed one has to work around this problem in an engineering approach, as realized by one of his colleagues using a certain angular description (as are Cardan angles) when mapping non-nominal sensory patterns into orientation errors of a robot gripper for grasping an object. He also emphasized that statistics and data approximation theory especially in the last years have helped quite a lot to shed light into the structure and topologies of neural nets. Finally he made clear that the approaches pursued at DLR try to make use of model knowledge as much as possible and to mainly learn the remaining uncertainties.

The paper "Robot Programming by Demonstration–Application of Machine Learning Techniques to Ease Robot Programming and Control" by R. Dillmann gives kind of a survey and is not restricted to neural net techniques, but focuses more on integrated symbolic and connectionist learning approaches for "Programming by demonstration". May be this is one of the most important lessons we have learned from the past that we have to widen the scope and should not restrict ourselves to just one direction. Thus in this paper the importance of functional, descriptive and methodological integration of symbolic and connectionistic components is emphasized.

In answering questions Dillmann pointed out the importance of entropy in the evaluation of learning, and the importance of scaling and projection, as well as the implication of multi-agent and multi-processing systems. He pointed out that in many cases it may be helpful to just take the exact final and endpoint of a motion and e.g. use rough visual servoing in between, as (following Hollerbach's comment) human arm motions are fairly much stereotype and may be parametrized. Dillmann also pointed out that a lot of estimation and Kalman filtering technique is necessary to derive the relevant information from noisy tracking signals.

A similar research direction was covered by the paper: "Designing skills with visual feedback for APO (assembly plan form observation)" presented by Katsushi Ikeuchi et al. The APO system uses three worlds (human, internal and robot world) and in a first step has the map to human world into an internal world, while in a second step (the issue of this paper) it maps the internal world into the robot world (task instantiation). As sensory perception and feedback is crucial for the kind of tasks considered (e.g. pick up, insert-into etc.), the paper presents a method of systematically generating sensing strategies for skills.

In the discussion Ikeuchi pointed out that concerning visual feedback it should be clear beforehand, whether e.g. the robot's configuration is the same as the human operator's configuration, as it does not make sense to observe everything visually but to concentrate on relevant features only; interpretation is absolutely necessary. In contrast to a general opinion he argued that it might not really be necessary to feed back force information for successfully observing an assembly task. In case human and robot kinematics are different (e.g. in case of articulated hands) appropriate mappings have to be used: An important problem however is to define the success of an operation combined with the fact that a human operator when repeating the demonstration of a task always does it in a (at least slightly) different way.

My impression was that the session gave an excellent overview on what is going on in learning and skill transfer; I am deeply convinced that robot learning by showing and demonstration is one of the key topics in present robotics research.

Behavior Acquisition via Vision-Based Robot Learning

Minoru Asada, Takayuki Nakamura, and Koh Hosoda
Dept. of Mechanical Eng. for Computer-Controlled Machinery, Osaka University, Suita 565 (Japan)
e-mail: asada@robotics.ccm.eng.osaka-u.ac.jp

Abstract

We introduce our approach that makes a robot learn to behave adequately to accomplish a given task at hand through the interactions with its environment with less *a priori* knowledge about the environment or the robot itself. We briefly present three research topics of vision-based robot learning in each of which visual perception is tightly coupled with actuator effects so as to learn an adequate behavior. First, a method of vision-based reinforcement learning by which a robot learns to shoot a ball into a goal is presented. Next, "motion sketch" for a one-eyed mobile robot to learn several behaviors such as obstacle avoidance and target pursuit is introduced. Finally, we show a method of purposive visual control consisting of an on-line estimator and a feedback/feedforward controller for uncalibrated camera-manipulator systems. All topics include the real robot experiments.

1 Introduction

Realization of autonomous agents that organize their own internal structure in order to take actions towards achieving their goals is the ultimate goal of AI and Robotics. That is, the autonomous agents have to learn. Recent research in artificial intelligence has developed computational approaches of agent's involvements in their environments [1]. Our final goal, in designing and building an autonomous agent with vision-based learning capabilities, is to have it perform a variety of tasks adequately in a complex environment. In order to build such an agent, we have to make clear the interaction between the agent and its environment.

In physiological psychology, Held and Hein [2] have shown that self-produced movement with its concurrent visual feedback is necessary for the development of visually-guided behaviors. Their experimental results suggest that perception and behavior are tightly coupled in autonomous agents that perform tasks. In biology, Horridge [3] similarly has suggested that motion is essential for perception in living systems such as bees.

In computer vision area, so-called "purposive active vision paradigm" [4, 5, 6] has been considered as a representative form of this coupling since Aloimonos et al. [7] proposed it as a method that converts the ill-posed vision problems into the well-posed ones. However, many researchers have been using so-called active vision systems in order to reconstruct 3-D information such as depth and shape from a sequence of 2-D images given the motion information of the observer or capability of controlling the observer motion. Furthermore, though purposive vision does not consider vision in isolation but as a part of complex system that interacts with world in specific ways [4], very few have tried to investigate the relationship between motor commands and visual information [8].

In robot learning area, the researchers have tried to make agents learn a purposive behavior to achieve a given task through agent-environment interactions. However, almost of them have only shown computer simulations, and only a few real robot applications are reported which are simple and less dynamic [9, 10]. there are very few examples of use of visual information in robot learning, probably because of the cost of visual processing.

In this paper, we introduce our approach that makes a robot learn to behave adequately to accomplish a given task at hand through the interactions with its environment with less *a priori* knowledge about the environment or the robot itself. We briefly present three research topics of vision-based robot learning in each of which visual perception is tightly coupled with actuator effects so as to learn an adequate behavior.

The remainder of this article is structured as follows: First, a method of vision-based reinforcement learning by which a robot learns to shoot a ball into a goal is presented. Next, we introduce a method to represent an interaction between the agent and its environment which is called "motion sketch" for a one-eyed mobile robot to learn several behaviors such as obstacle avoidance and target pursuit. Finally, we show a method of purposive visual control consisting of an on-line estimator and a feedback/feedforward controller for uncalibrated camera-manipulator systems. All topics include the real robot experiments.

2 Vision Based Reinforcement Learning [11]

Reinforcement learning has recently been receiving increased attention as a method for robot learning with little or no *a priori* knowledge and higher capability of reactive and adaptive behaviors [12]. In the reinforcement learning method, a robot and its environment are modeled by two synchronized finite state automatons interacting in discrete time cyclical processes. The robot senses the current state of the environment and selects an action. Based on the state and the action, the environment makes a transition to a new state and generates a reward that is passed back to the robot. Through these interactions, the robot learns a purposive behavior to achieve a given goal.

Although the role of reinforcement learning is very important to realize autonomous systems, the prominence of that role is largely dependent on the extent to which the learning can be scaled to solve larger and more complex robot learning tasks. Many researchers in the field of machine learning have been concerned with the convergence time of the learning, and have developed methods to speed it up. However, almost all of them have only shown computer simulations in which they assume ideal sensors and actuators, where they can easily construct the state and action spaces consistent with each other.

Here, we present a method of vision-based reinforcement learning by which a robot learns to shoot a ball into a goal. The robot does not need to know any parameters of the 3-D environment or its kinematics/dynamics. The image captured from a single TV camera mounted on the robot is the only source of information on the changes in an environment. Image positions and sizes of the ball and the goal are used as a state vector. We discuss several issues from a viewpoint of robot learning: a) coping with a "state-action deviation" problem which occurs in constructing the state and action spaces in accordance with outputs from the physical sensors and actuators, and b) starting with easy missions (rather than task decomposition) for rapid task learning.

2.1 Task and assumptions

The task for a mobile robot is to shoot a ball into a goal. The problem we address here is how to develop a method which automatically acquires strategies for doing this. We assume that the environment consists of a ball and a goal; the mobile robot has a single TV camera; and that the robot does not know the location/size of the goal, the size/weight of the ball, any camera parameters such as the focal length and tilt angle, or the kinematics/dynamics of itself.

2.2 Construction of State and Action Spaces

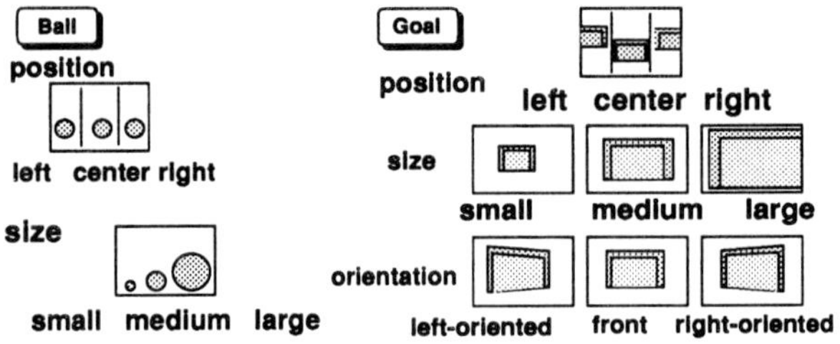

Figure 1: The ball sub-states and the goal sub-states

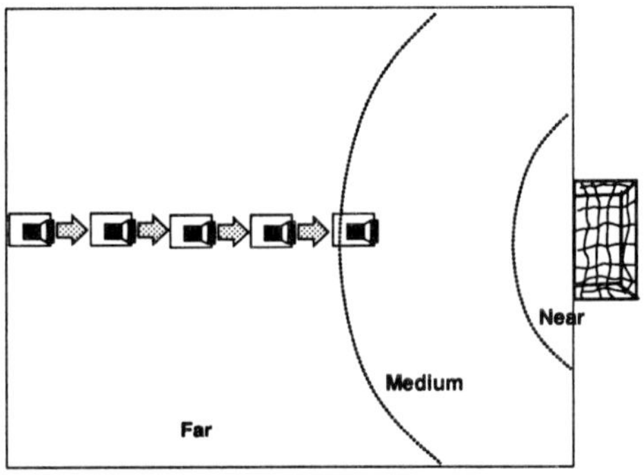

Figure 2: A state-action deviation problem

Figure 1 shows sub-states of ball and goal in which the position and the size of the ball or goal are naturally and coarsely classified into each state. Due to the peculiarity of visual information, that is, a small change near the observer results in a large change in the image and a large change far from the observer may result in a small change in the image, one action does not always correspond to one state transition. We call this the **"state-action deviation problem"**: Figure 2 indicates this problem, the area representing the state "the goal is far" is large, therefore the robot frequently returns to this state if the action is forward. This is highly undesirable because the variations in the state transitions is very large, consequently the learning does not converge correctly.

To avoid this problem, we reconstruct the action space as follows. Each action defined is regarded as an action primitive. The robot continues to take one action primitive at a time until the current state changes. This sequence of the action primitives is called an action. In the above case, the robot takes a forward motion many times until the state "the goal is far" changes into the state "the goal is medium".

2.3 Learning from Easy Missions

In order to improve the learning rate, the whole task was separated into different parts in [10]. By contrast, we do not decompose the whole task into subtasks of finding, driblling, and shooting a ball. Instead, we first used a monolithic approach. That is, we place the ball and the robot at arbitrary positions. In almost all the cases, the robot crossed over the field line without shooting the ball into the goal. This means that the learning did not converge after many trials. This is the famous *delayed reinforcement* problem due to no explicit teacher signal that indicates the correct output at each time step. To avoid this difficulty, we construct the learning schedule such that the robot can learn in easy situations at the early stages and later on learn in more difficult situations. We call this *Learning from Easy Missions* (or LEM).

2.4 Experimental results

We applied the LEM algorithm to the task in which the order of easy situations are $\mathbf{S_1}$ ("the goal is large"), $\mathbf{S_2}$ ("the goal is medium", and $\mathbf{S_3}$ ("the goal is small"). Figure 3 shows the changes of the summations of Q-values of the action-value function in the Q-learning method with and without LEM, and its temporal derivative ΔQ. The

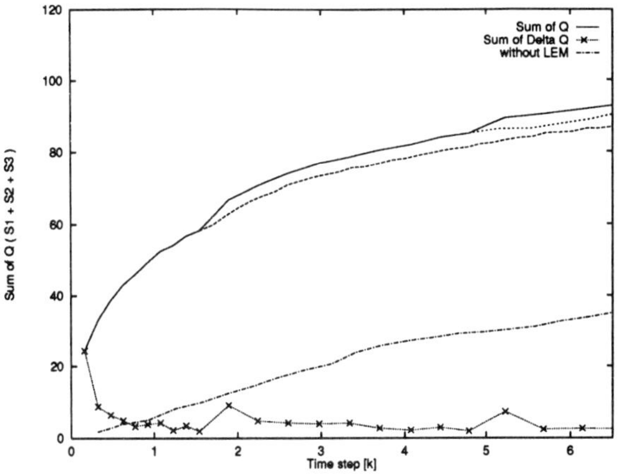

Figure 3: Change of the sum of Q-values with LEM in terms of goal size

axis of time step is scaled by M (10^6), which corresponds to about 9 hours in the real world since one time step is 33ms. The solid and broken lines indicate the summations of the maximum value of Q in terms of action in all states with and without LEM, respectively. The Q-learning without LEM was implemented by setting initial positions of the robot at completely arbitrary ones. Evidently, the Q-learning with LEM is much better than that without LEM. The broken line with empty squares indicates the change of ΔQ. Two arrows indicate the time steps (around 1.5M and 4.7M) when a set of the initial states changed from $\mathbf{S_1}$ to $\mathbf{S_2}$ and from $\mathbf{S_2}$ to $\mathbf{S_3}$, respectively. Just after these steps, ΔQ drastically increased, which means the Q-values in the inexperienced states are updated. The coarsely and finely dotted lines expanding from the time steps indicated by the two arrows show the curves when the initial positions were not changed from $\mathbf{S_1}$ to $\mathbf{S_2}$, nor from $\mathbf{S_2}$ to $\mathbf{S_3}$, respectively. This simulates the LEM with partial knowledge. If we know only the easy situations ($\mathbf{S_1}$), and nothing more, the learning curve follows the finely dotted line in Figure 3. The summation of Q-values is slightly less than that of the LEM with more knowledge, but much better than that without LEM.

We used the same experimental set up as that described in the previous section. In Figure 4 (raster order), the images are taken every second. First, the robot lost the ball due to noise, and then it turned around to find the ball, and finally it succeeded in shooting.

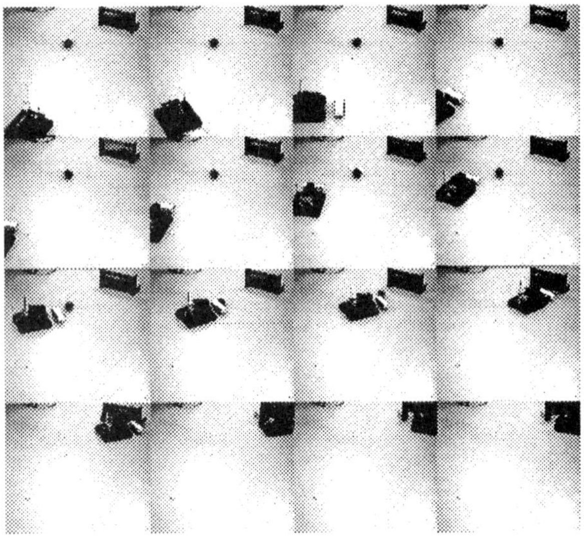

Figure 4: The robot succeeded in shooting a ball into the goal.

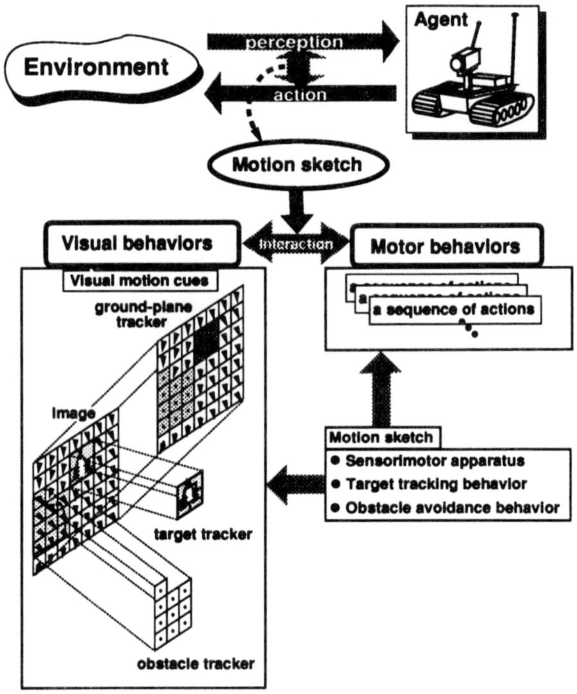

Figure 5: Motion sketch

3 Motion Sketch [13]

3.1 Basic Ideas of Motion Sketch

The interaction between the agent and its environment can be seen as a cyclical process in which the environment generates an input (perception) to the agent and the agent generates an output (action) to the environment. If such an interaction can be formalized, the agent would be expected to carry out actions that are appropriate to individual situations. "Motion sketch," we proposed here, is one of such formalizations of interactions by which a vision-based learning agent that has real-time visual tracking routines behaves adequately against its environment to accomplish a variety of tasks.

Figure 5 shows a basic idea of the motion sketch. The basic components of the motion sketch are visual motion cues and the motor behaviors.

Visual motion cues are detected by several visual tracking routines of which behaviors (called visual behavior) are determined by individual tasks. The visual tracking routines are scattered over the whole image and an optical flow due to an instantaneous robot motion is detected. In this case, the tracking routines are fixed to the image points. The image area to be covered by these tracking routines are specified or automatically determined depending on the current tasks, and the cooperative behaviors between tracking routines are performed for the task accomplishment. For the target pursuit task, the multiple templates are initialized and every template looks for the target to realize stable tracking. In the task of obstacle detection and avoidance, the candidates for obstacles are first detected by comparing the optical flow with that of non-obstacle (ground plane) region, and then the detected region is tracked by multiple templates each of which tracks the inside of the moving obstacle region.

The motor behaviors are sets of motor commands obtained by Q-learning [14], a most widely used reinforcement learning method, based on the detected motion cues and given task. The sizes and positions of the target and the detected obstacle are used as components of a state vector in the learning process.

Visual and motor behaviors work in parallel in the image and compose a layered architecture. The visual behavior for monitoring robot motion (detecting the optical flow on the ground plane on which the robot lies) is the lowest and might be subsumed in part due to occlusion by other visual and motor behaviors for obstacle detection/avoidance and target pursuits which might occlude each other.

The motion sketch does not need any calibrations nor any 3-D reconstruction so as to accomplish the given task. The visual motion cues for representing the environment does not seem dependent on scene components nor limited to the specified situations and the task. Furthermore, the interaction is quickly obtained owing to the use of real-time visual tracking routines.

The behavior acquisition scheme consists of the following four stages: i) Obtaining the fundamental relationship between visual and robot motions by correlating motion commands and flow patterns on the floor with very few obstacles. ii) Learning target pursuit behavior by tracking a target. iii) Detection of obstacles and learning an avoidance behavior. iv) Coordination of the target pursuit and obstacle avoidance behaviors. At each stage, we obtain the interaction between the agent and its environment.

3.2 Obtaining sensorimotor apparatus

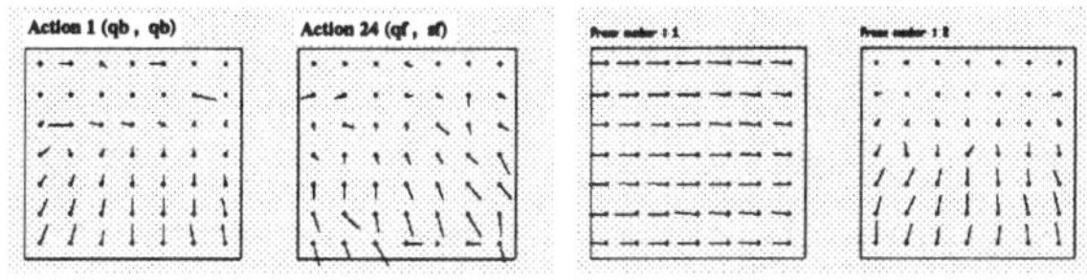

(a) examples of flow patterns

(b) obtained two principal flows

Figure 6: Acquisition of principal motion vectors

We place 49(7 × 7) visual tracking routines to detect changes in the whole image. Therefore, we obtain an optical flow composed of 49 flow vectors. In the environment without obstacles, the robot randomly selects a possible action among the action space, and executes it. While randomly wandering, the robot stores the flow patterns $\boldsymbol{p}_i$ due to its actions i. After the robot performed all possible actions, we obtain the averaged optical flows $\boldsymbol{p}_i$ removing the outliers due to noise or small obstacles based on the LMeS method. Figure 6 (a) shows examples of flows detected during random motions.

Using the averaged optical flows obtained above, we acquire principal motion patterns which characterize the space of actions. This is done by analyzing the space of averaged optical flow that robot is capable of producing. We want to find a basis for this space, i.e., a set of representative motion patterns from which all the motion patterns may be produced by their linear combinations. We can obtain representative motion patterns by using Principal Component Analysis that may be performed using a technique called Singular Value Decomposition(hereafter SVD). The first two principal components obtained in the real experiment are shown in Figure6 (b). Obviously, the left corresponds to a pure rotation and the right to a pure backward motion.

3.3 Behavior acquisition based on visual motion cues

Target tracking behavior acquisition
We use the visual tracking routines in order to pursue a target specified by a human operator and obtain the information about the target in the image such as its position and size which are used in the Q-learning algorithm [14] for acquisition of target pursuit behavior.

Obstacle avoidance behavior acquisition
We know the flow pattern $\boldsymbol{p}_i$ corresponding to the action i in the environment without any obstacles. The noise included in $\boldsymbol{p}_i$ is not so much, because this flow pattern is described as a linear combination of the two principal motion vectors. Therefore, it makes motion segmentation easy. Motion segmentation is done by comparing the flow pattern $\boldsymbol{p}_i$ with the flow pattern $\boldsymbol{p}_i^{obs}$ which is obtained in the environment with obstacles. The area in the $\boldsymbol{p}_i^{obs}$ is detected as the area for obstacle candidates if its components are different from that of $\boldsymbol{p}_i$. This information (position and size in the image) is used to obtain the obstacle tracking behavior. After obstacle detection, the visual tracking routines are set up at the positions where the obstacle candidates are detected and the regions are tracked until the region disappears from the image.

Learning to avoid obstacles consists of two stages. First, the obstacle tracking behavior is learned by the same manner as in learning the target pursuit behavior. Next, the obstacle avoidance behavior is generated by using the relation between the possible actions and the obstacle tracking behavior.

3.4 Experimental results

Figure 7 shows a configuration of the real mobile robot system. We have constructed the radio control system of the robot [11]. The image processing and the vehicle control system are operated by VxWorks OS on MVME167(MC68040 CPU) computer which are connected with host

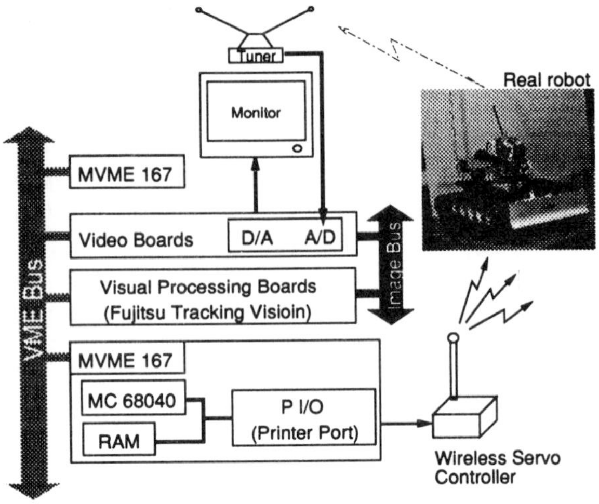

Figure 7: Configuration of the experimental system

Sun workstations via Ether net. The image taken by a TV camera mounted on the robot is transmitted to a UHF receiver and subsampled by the scan-line converter (Sony Corp.). Then, the video signal is sent to a Fujitsu tracking module. The tracking module has a function of block correlation to track some pre-memorized patterns and can detect motion vectors in real time.

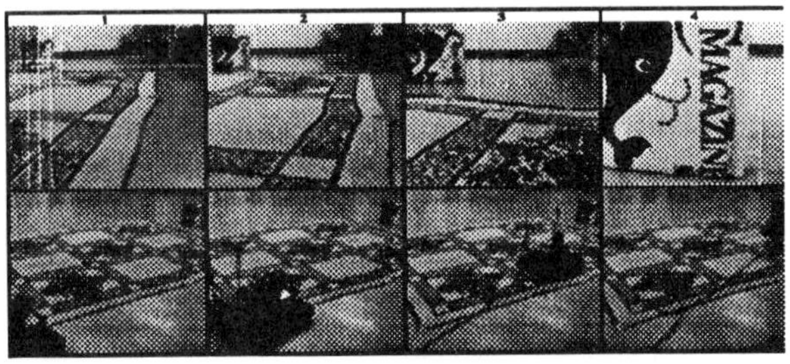

Figure 8: The robot succeeded in pursuing a target.

Figures 8 and 9 show sequences of images where the robot succeeded in target pursuit and avoiding a moving obstacle, respectively. The top shows the images taken and processed by the robot and the bottom images show how the robot behaves. In Figure 9, the rectangles indicate the obstacle candidate regions.

Figure 10 shows a sequence of images where the robot succeeded in coordinating target pursuit and obstacle avoiding behaviors. The top shows the images taken and processed by the robot and the bottom images show how the robot behaves. The rectangles indicate the obstacle candidate regions. In pursuing a target (flower pot), the robot was blocked by the foot of a student, therefore it moved backward. Then, it tracked the target again since the obstacle disappeared.

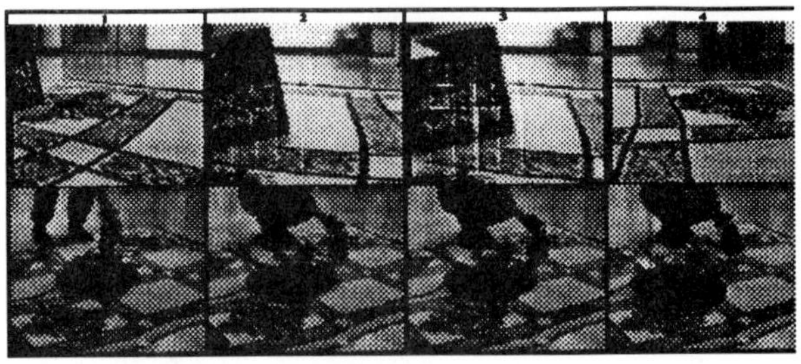

Figure 9: The robot succeeded in avoiding a moving obstacle.

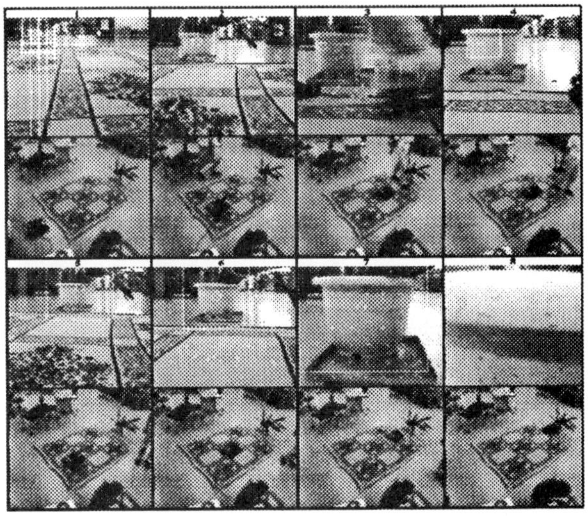

Figure 10: The robot succeeded in coordinating two behaviors.

4 Purposive Visual Control for uncalibrated camera-manipulator systems [15]

Recently, there have been several studies on visual servoing, using visual information in the dynamic feedback loop to increase robustness of the closed loop system (we can find a summary in [16]). In most of the previous works on visual servoing, they assumed that the system structure and parameters are known, or that the parameters can be identified in an off-line process or on-line parameter identification with restrictions and assumptions on the system.

On the other hand, the previous works payed attention only to feedback servoing. They sensed positions of targets and made feedback inputs subtracting the sensed positions from the desired ones. Using these controllers, the manipulator does not work until error is observed, which can be considered as *reactive* movement. For intelligent control of camera-manipulator systems, not only the reactive but also *purposive* visual movement must be realized. At the level of control, we believe that feedforward terms should play a great part in realizing the purposive movement, but no one has mentioned to the effectiveness of feedforward terms to the best of our knowledge.

Here, we propose purposive visual control consisting of an on-line estimator and a feedback/feedforward controller for uncalibrated camera-manipulator systems. It has the following features:

1. The estimator does not need any a priori knowledge on the kinematic structure nor the system parameters. We can eliminate the tedious calibration process owing to this feature.

2. There are no restrictions on the camera-manipulator system: the number of cameras, kinds of images features, structure of the system (camera-in-manipulator or camera-and-manipulator) ,the number of inputs and outputs (SISO or MIMO). The proposed method is applicable to all cases. It is strongly related with the fact that the estimator does not need any a priori knowledge on the system.

3. The aim of the estimator is not to estimate the true parameters, but to ensure asymptotical convergence of the image features to the desired values under the proposed controller. Therefore, the estimated parameters do not necessarily converge to the true values. The existing methods such as [17, 18] tried to estimate the true parameters, and therefore they need restrictions and assumptions.

4. The proposed controller can realize *purposive* movement of the system utilizing its feedforward terms. The feedforward terms of the proposed scheme are based on *estimated* parameters intending to realize visual tasks on the image planes (mentioned in 3). In this sense, this feedforward terms help realizing purposive movement at the control level.

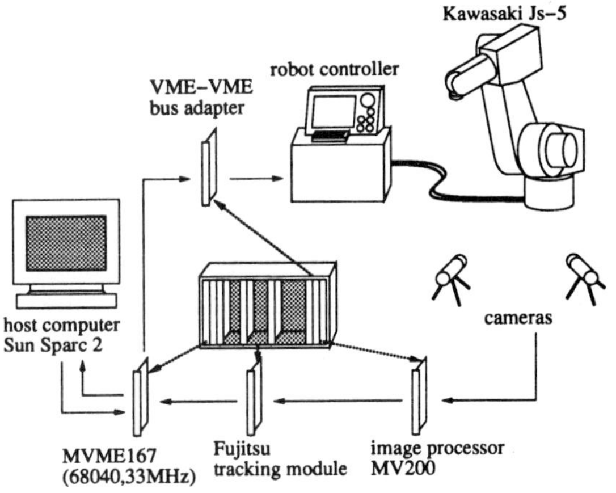

Figure 11: Experimental system

Figure 11 shows the experimental system we used. Figure 12 (a) shows an experimental set up with two cameras fixed, and (b) indicates the result of step response with and without on-line estimator, where vertical and horizontal axes indicate the error in pixels and time steps (second), respectively. Evidently, the performance without the estimator was much worse that with the estimator.

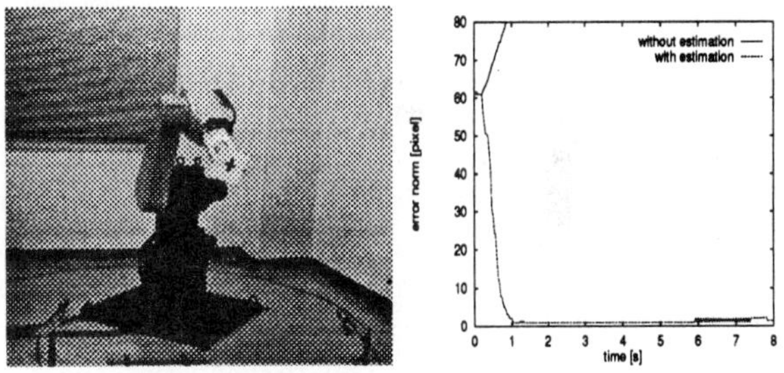

(a) experiment with cameras fixed

(b) results

Figure 12: Visual servoing with tracking vision

5 Discussion

We have shown three topics of vision based robot behavior learning. All of them do not need the knowledge about the environment or kinematics/dynamics of the robot itself because such knowledge is implicitly obtained and embodied in the internal structure that the system organizes through the learning process. As a result, the structure represents the connection between the visual behaviors and motor ones to accomplish a given task.

In the first topic, the action value function, Q(s,a) can be regarded as an internal structure the robot organizes. The task is specified as a goal state and the function reflects the relationship between the robot capability (sensing and mobility), the environment, and the task, and reduces it as a simple action value at every state. The visual behaviors of finding a ball and a goal are task specific and motor behaviors (two independent motor commands) are organized as a sequence of pairs of two motor commands to achieve the goal state.

The motion sketch can be seen as an internal structure the robot obtained during the learning processes. The basic visual behavior is visual tracking to detect an optical flow or to track obstacles and/or a target. In the first stage, the sensorimotor apparatus is obtained as a relationship between perception, action, and the environment. In the second stage, a group of tracking routines is organized so as to form an object tracking, and then obstacle avoidance and target tracking behaviors are obtained in a framework of the reinforcement learning using the motion sketch. In the third stage, coordination of the learned behaviors is managed in the motion sketch.

The image Jacobian-matrix generally represents the sensorimotor apparatus, and in the third topic, it can be regarded as an internal structure of the robot because it is estimated by the interactions with the environment in a sense of LSE (the Least Squares Estimation). The visual behavior is a simple tracking of the target, and the motor behavior, a set of torque commands to joints of the robot arm, is obtained from the control scheme consisting of both feedback and feedforward terms. The task is specified in the sensor space as image features.

To apply our approach to other kinds of tasks, we have to solve two important and difficult problems. One is how to construct the state space, in other words, how to find what is important to accomplish given tasks from sensory information through the experiences. The other is how to generalize or to abstract the learned policies to cope with a variety of similar tasks in similar environments. These are fundamental and open problems in Robotics and AI.

References

[1] Philip E. Agre. "Computational research on interaction and agency". *Artificial Intelligence*, 72:1–52, 1995.

[2] R. Held and A. Hein. "Movement-produced stimulation in the development of visually guided behaviors". *Jounal of Comparative and Physiological Psycology*, 56:5:872–876, 1963.

[3] G. A. Horridge. "The evolution of visual processing and the construction of seeing systems". In *Proc. of Royal Soc. London B230*, pages 279–292, 1987.

[4] Y. Aloimonos. "Reply: What i have learned". *CVGIP: Image Understanding*, 60:1:74–85, 1994.

[5] G. Sandini and E. Grosso. "Reply: Why purposive vision". *CVGIP: Image Understanding*, 60:1:109–112, 1994.

[6] S. Edelman. "Reply: Representatin without reconstruction". *CVGIP: Image Understanding*, 60:1:92–94, 1994.

[7] Y. Aloimonos, I. Weiss, and A. Bandyopadhyay. "Active vision". In *Proc. of first ICCV*, pages 35–54, 1987.

[8] G. Sandini. "Vision during action". In Y. Aloimonos, editor, *Active Perception*, chapter 4. Lawrence Erlbaum Associates, Publishers, 1993.

[9] P. Maes and R. A. Brooks. "Learning to coordinate behaviors". In *Proc. of AAAI-90*, pages 796–802, 1990.

[10] J. H. Connel and S. Mahadevan. "Rapid task learning for real robot". In J. H. Connel and S. Mahadevan, editors, *Robot Learning*, chapter 5. Kluwer Academic Publishers, 1993.

[11] M. Asada, S. Noda, S. Tawaratsumida, and K. Hosoda. Vision-based reinforcement learning for purposive behavior acquisition. In *Proc. of IEEE Int. Conf. on Robotics and Automation*, pages 146–153, 1995.

[12] J. H. Connel and S. Mahadevan, editors. *Robot Learning*. Kluwer Academic Publishers, 1993.

[13] T. Nakamura and M. Asada. Motion sketch: Acquisition of visual motion guided behaviors. In *Proc. of IJCAI-95*, pages 126–132, 1995.

[14] C. J. C. H. Watkins. *Learning from delayed rewards"*. PhD thesis, King's College, University of Cambridge, May 1989.

[15] K. Hosoda and M. Asada. Versatile visual servoing without knowledge of true jacobian. In *Proc. of IEEE/RSJ/GI International Conference on Intelligent Robots and Systems 1994 (IROS '94)*, pages 186–193, 1994.

[16] P. I. Corke. Visual control of robot manipulators – a review. In *Visual Servoing*, pages 1–31. World Scientific, 1993.

[17] B. Nelson, N. P. Papanikolopoulos, and P. K. Khosla. Visual servoing for robotic assembly. In *Visual Servoing*, pages 139–164. World Scientific, 1993.

[18] N. P. Papanikolopoulos, B. Nelson, and P. K. Khosla. Six degree-of-freedom hand/eye visual tracking with uncertain parameters. In *Proc. of IEEE Int. Conf. on Robotics and Automation*, pages 174–179, 1994.

Perception and Manipulation in Robotics: Neural Network Approaches

Gerhard Hirzinger, Ralf Koeppe,
Andreas Baader, Friedrich Lange, Ralph Staudte and Guo-Qing Wei
Deutsche Forschungsanstalt für Luft- und Raumfahrt (DLR)
Institute of Robotics and System Dynamics
Oberpfaffenhofen, D-82234 Wessling, Germany
email: Gerd.Hirzinger@dlr.de

Abstract

Learning and adaptation are the core paradigms of intelligent control concepts which enable to increase skilled manipulation and to achieve higher levels of autonomy. Data approximation and representation techniques like neural networks enhance or replace conventional model based approaches. We apply neural networks in two significant fields of robotics research: Perception and Manipulation. This paper provides an overview of neural network applications in robotics we developed in our lab.

1 Introduction

The Institute of Robotics and System Dynamics aims at the development of a new generation of intelligent robot systems, equipped with multisensory grippers and hands to perceive and manipulate within known and unknown environments in accessible or remote distance. Learning and adaptation are the core paradigms of the intelligent control concepts which enable to increase skilled manipulation and to achieve higher levels of autonomy.

Neural network algorithms can be used for problems, which are commonly solved by function approximation, nonparametric regression, pattern recognition and machine learning techniques. Moreover they can be placed in a common perspective to the above mentioned techniques [1]. Neural networks are used to process robot sensor and motion data, e.g. to learn controllers or to build a model of the world. They are important techniques, which enhance or replace conventional model based approaches. We apply neural networks in two significant fields of robotics research: Perception and Manipulation.

Perception is the ability to sense the state of the environment and the robot, and to interpret this data. A surface reconstruction algorithm using Kohonen's maps is used to fuse all available sensor information and build a world model therefrom (Section 2). Multilayer perceptrons are used to reconstruct the surface shape of an object based soley on the intensity values of an image (Section 3), as well as, to interpret sensor data to perceive the differences between actual and desired pose of a robot (Section 4).

Manipulation is how the robot acts in the world. This includes path planning and control. Path planning algorithms for automatic determination of a suitable path given an actual and a desired pose use a Kohonen map to represent the workspace more efficiently [2]. For skill transfer of a human operator to a robot system (Section 5) and for intelligent grasping with a 3-fingered hand (Section 6) we use neural networks to represent the mapping from sensors to motion or poses. In the field of robot control (Section 7) we use learning techniques to improve inverse dynamic models for our joint torque controlled light weight manipulator and to improve the accuracy of standard industrial robots.

2 World modeling using Kohonen's feature map

Building a geometric description of a robot's workspace is for both, fully autonomous manipulators and telerobots, of crucial importance. Task planning and execution as well as compensating long signal running times in teleoperation re-

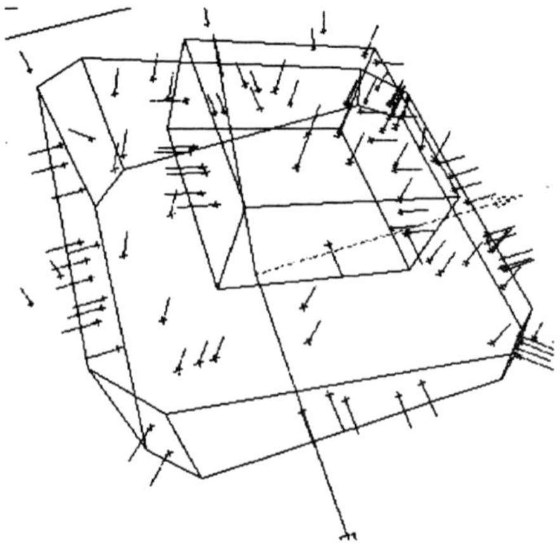

(a)

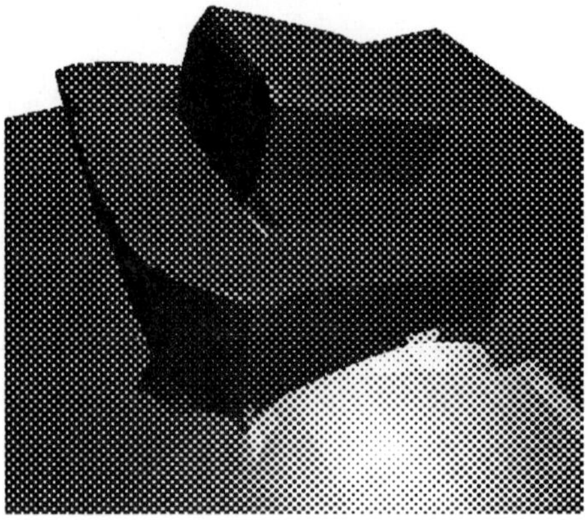

(b)

Figure 1: Reconstruction of an object in the ROTEX work cell. Part (a) shows the wireframe representation of the object and the sampled 110 point and normal vector tuples, (b) the final surface after 3000 steps and a total time of 9 seconds.

quires knowledge about the geometry of the objects to be handled.

In general the problem of world modelling is addressed by using stationary sensors of a single type. However, recent sensor development in robotics focused on integration of miniaturized components placed in a robot-gripper. Such a setup generates a new range of problems, not addressed by known perception algorithms. The following aspects characterize the world perception task for a multisensory robot-gripper: (1) The sensor data is referenced to the gripper frame. The arbitrary movements of the robot cause loss of the data order in the world reference frame; (2) Different kinds of sensors have to be integrated in the modelling process; (3) Due to the moving sensor platform, a noise robust algorithm is necessary. Not only the uncertainty in sensor reading, but the position error of the robot as well, will influence the data precision with respect to the world frame.

To address these problems a surface reconstruction algorithm, based on Kohonen's self organizing feature map was developed and successfully implemented. The importance of the surface in a perception process is presently not that widely accepted, however, the surface is the main interface to the object geometry for most robot-sensors.

The two-dimensional array of neurons in Kohonen's algorithm can be considered as a discrete parametric surface description. The parameters (s,t) denote the order of the neurons on the net and the three-dimensional weightvectors $w_{s,t}$ describe the surface's position in 3D-space. During the training process the Kohonen network establishes a mapping of a high-dimensional input space (the 3D-space of the geometric world description) on the two-dimensional neuron-grid (the parametrized surface description). This mapping preserves topology as well as data-values, regardless in which order the data is presented to the network. At the end of the training, the weightvectors $w_{s,t}$ are the measured surface data in the correct order of the surface parameters (s,t) on the neuron grid. To incorporate different types of surface information new training equations were developed for the Kohonen network. Successful surface reconstruction was performed for completely unordered 'data clouds' of different surface information (Figure 1). This method was used as a core for a complete world perception system for 'Sensor-in-Hand' robots. For a full description of the system, the experiments and the complete algorithm the reader may refer to [3].

3 Shape from shading by neural networks

The problem of shape from shading is to infer the surface shape of an object based solely on the intensity values of the image. Since the mathematical formulation and the numerical solution of the problem by Horn [4], there have been a lot of recent developments [5]. Most existing so-

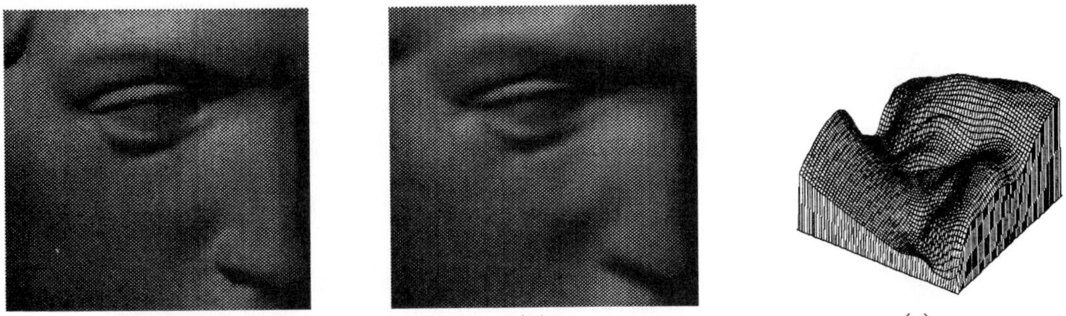

(a) (b) (c)

Figure 2: Surface recovery of the David statue. (a) The image; (b) The learned image by a network of size $2 \times 45 \times 45 \times 1$; (c) The recovered 3D surface.

lutions reformulate the problem as the calculus of variations. The solution of the corresponding Euler equations needs a smoothness term, whose weights should be gradually reduced to an appropriate value as the iteration proceeds, to avoid the walking-away of the solution from the correct one [6]. Too rapid a reduction would cause instability in the solution. Moreover, the integrability constraint has also to be imposed as a penalty term in the integrand. The reduction scheme of the smoothness weight and the choice of the appropriate integrability weight makes the results dependent on many factors.

We proposed another solution [7] of the shape from shading problem based on the optimization framework. We used a multilayer perceptron to parametrize the object surface so that both the depth and gradients can be represented by the network weights. There is no explicit use of smoothness and integrability terms in the solution. We use the gradient methods to minimize the intensity error between the given image and the generated image by the surface. After convergence, the network gives the surface reconstructed from shading. Figure 2 shows an example of our neural shape from shading approach. Figure 2 (a) is an original image of the David statue whose light source is estimated to be at tilt angle $\tau = 135°$ and slant angle $\gamma = 50°$. Figure 2 (b) is the learned image by the network. The reconstructed 3D surface is shown in Figure 2 (c). From this figure, we can see that most part of the statue has been reconstructed (the eye, the nose), except at the boundary. This is due to the lack of boundary conditions. By use of *a priori* knowledge of the surface, this ambiguity can be resolved.

4 Visual servoing by neural networks

The purpose of visual servoing is to use sensor data to position a robot such that the desired relative position with respect to an object is achieved. Contrary to conventional methods which need camera and hand-eye calibration, a neural network approach which involves no such calibrations is proposed [8]. A multilayer perceptron is used to learn the direct mapping from the sensory data to the motion commands. Here, the sensor data are of different kinds. They are stereo images and laser ranges provided by 4 laser ranging beams on the robot hand. Thus, the method also provides a practical sensor fusion method.

First, a *homing position* is defined by the sensor pattern acquired at the desired manipulation position. Since the object used in our experiment is planar, 2 feature points in each of the stereo images and 4 laser ranges of the object are used as the sensor data, which can be shown to be sufficient to uniquely determine the object pose. By making random movements of the robot, the sensor data and the amounts of movement (Cartesian) are recorded, and they are used as the training pattern for a multilayer perceptron. The standard backpropagation network is used to establish the direct mapping from sensor data to robot motions. When the homing position is changed after the network has been trained, the existing network can be modified without having to retrain the network. The geometrical relationship between the new and the old homing position need not be known *a priori*. This increases the flexibility of our neural network approach. It ought to

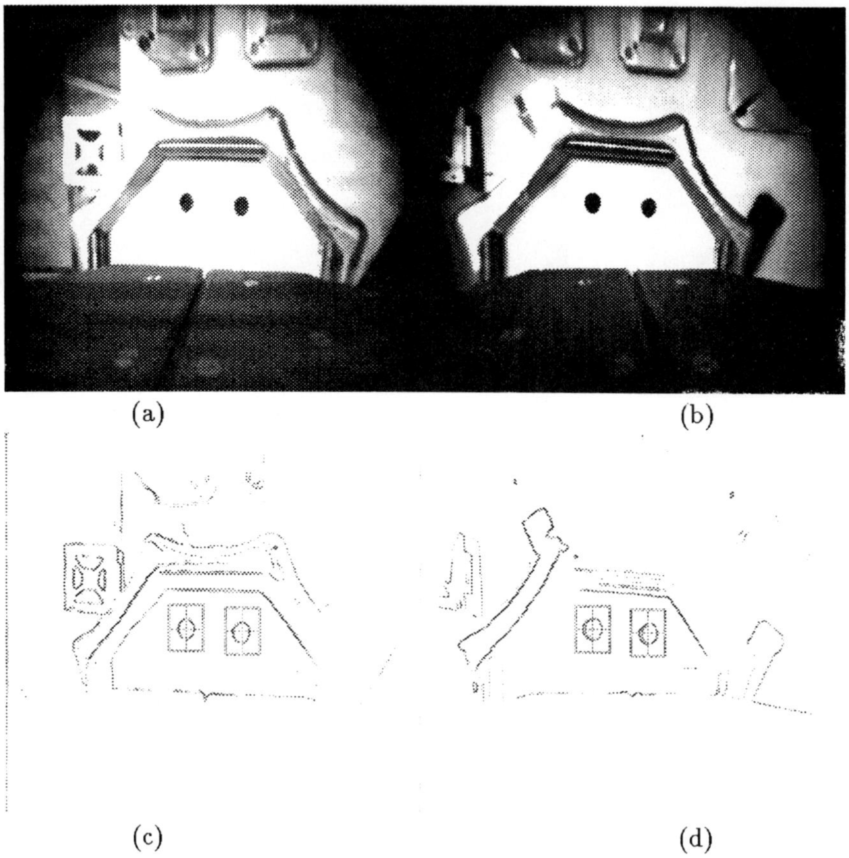

Figure 3: The images of the ORU from the hand cameras and their gradient images (used as a target position); (a),(b) the original images; (c),(d) the gradient images of (a) and (b) with the detected blobs.

be noted, however, that the new homing position cannot be arbitrary; it should be within the space where the training data were acquired.

Our method has been tested by experiments in a real robot environment. The object used is called ORU (the Orbital Replaceable Unit), and is chosen from the space robot technology experiment ROTEX in the German Spacelab-mission D2 [9]. Figure 3 (a) and (b) show the view of the ORU from the stereo hand camera in the homing position, where the two black blobs artificially marked on the object are used as the feature points. A Datacube MaxVideo 200 image processing system is used to track the feature points in real time during the robot motion. Figure 3 (c) and (d) show the corresponding Sobel gradient images superimposed by the detected blobs. Using this neural visual servoing approach, the robot can resume the homing position very accurately from any starting position. The network was size $12 \times 100 \times 6$.

5 Learning compliant motions by task demonstration

Neural networks are capable of representing nonlinear compliant motions. The training data has to consist of the desired correspondence between a measured force and the correction of motion. If sensorimotion data is directly recorded from a human demonstrating a task, the correspondence problem has to be solved, since the human executes an action $\boldsymbol{u}(t)$, due to a perception $\boldsymbol{S}(t-\tau)$.

In [10] we show that the correspondence problem is solved by approximating the function

$$\boldsymbol{u}(t) = \boldsymbol{f}[\boldsymbol{S}(t), \dot{\boldsymbol{x}}(t)]$$

with a neural network, taking the velocity $\dot{\boldsymbol{x}}(t)$ as an additional input to the network. Therefore, no reassembling of the data samples is necessary, taking into account which sensor pattern $\boldsymbol{S}$ perceived at time $(t-\tau)$ caused the operator

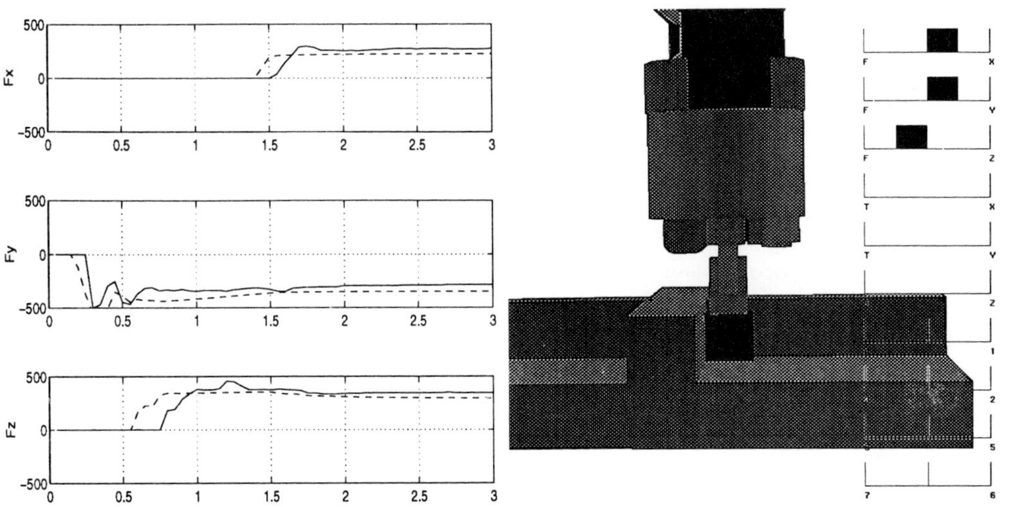

Figure 4: Experiment: Put Block in a Corner of a Box Problem. Teaching with force feedback, Task-Demonstration by human operator are represented by solid lines, Task-Execution by learning controller by dashed lines. (Signals w.r.t. world coordinate system)

to command an action $\boldsymbol{u}$ at a later time t. Nor is it necessary to determine the dead time τ of the human feedback. This is done by the network implicitly. Since humans describe tasks in a rule based way, using the concept of fuzziness, we have designed a neural network, prestructured to represent a fuzzy controller.

We investigated two types of feedback on the operator: visual feedback using a force display and a sensor ball command device, or both visual and force feedback using a haptic interface. The learning controller was verified by experiments, aligning a block in the corner of a box (Figure 4). The task needs to be demonstrated only once to be able to learn the sensorimotion mapping. The control law, determining the compliant motion, is acquired by the learning structure. By analyzing the connections of the networks, the rules by which the human operator demonstrates the task are deduced.

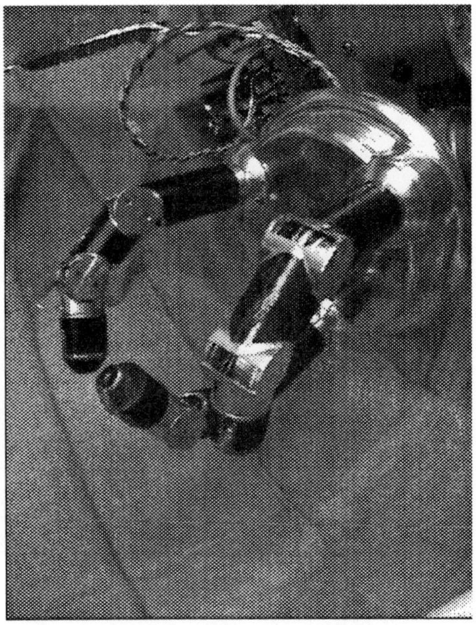

Figure 5: Prototype of the DLR-Hand

6 Grasping of objects with a 3-fingered multisensor hand

The 3-fingered anthropomorphic multisensor robot gripper (Figure 5) currently developed at our Institute is electrically driven and equipped with joint position, joint torque, tactile and force/torque sensors in the fingertips, and a miniaturized stereo camera.

The first step in grasping an object is to arrange the fingers of the robot gripper so that their tips are nearby the desired contact points after the approach phase. This preshaping will be controlled by a multilayer perceptron trained for sample objects using input patterns supplied by an image processing system and output patterns composed of the joint angles of a human operator using a data glove for teach-in. If the multisensor hand is positioned and preshaped appropriately with respect to the object a slight closing of the fingers results in a stable grasp. Maintaining the grasp is the next problem to be solved by the learning control system. The tendency of the object to slip from the robot hand is indicated by a variation of the contact forces and – if a movement already takes place – a change of the contact points. This is detectable by monitoring the data supplied by the tactile and the force/torque sensors. The deviation from the nominal values will be mapped on appropriate desired values for the joint torque control system by a second multilayer perceptron or an associative memory depending on which will appear to be more suitable for processing the sensor data of the hand [11].

7 Neural inverse dynamic for joint torque controlled robots

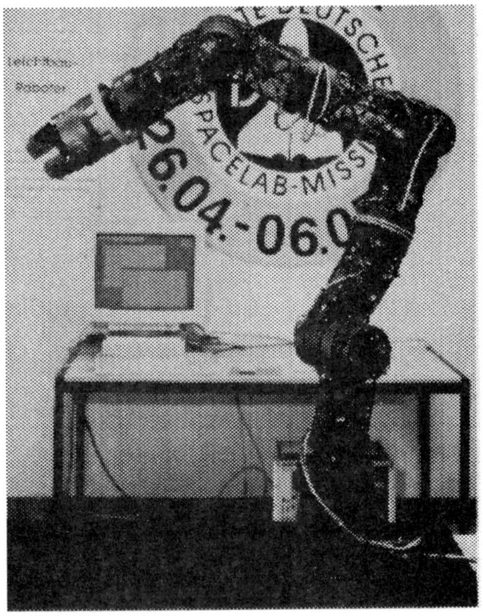

Figure 6: The DLR Light-Weight Robot

For the inverse dynamic control of the DLR light-weight robot with torque controlled joints (Figure 6), we proposed an approach using a hybrid inverse dynamic model [12]. Since an analytical inverse dynamic model requires extensive calculations and contains errors due to parameter uncertainty in the model, we use a gross analytical model, describing the most important dynamic properties. Parameter deviation of the rigid model and errors due to unmodelled structures of the robot not resolved on the joint controller level are linearized by an additional neural inverse error model. Both outputs of the analytical inverse model and the neural inverse error model are used as an input to the robot. A multilayer perceptron, trained by supervised learning, is used.

8 Improving the accuracy of standard position controlled industrial robots

This sections describes learning control for standard industrial robots. The task is to minimize the difference between the cartesian desired path w and the actual path y. The optimal control input u commands a fictive path which is deformed by the robot dynamics in such a way, that the resulting actual path coincides with the desired one. For convenience this task is splitted into subtasks which reduce the positional errors of the individual joints.

The deviations are results of statical effects as e.g. incorrect kinematic data and dynamical effects caused by acceleration, centrifugal or coriolis forces. Statical errors are not considered since the difference between actual and desired position is sensed.

For dynamical errors a method was presented in [13] to reduce delays due to acceleration forces. These delays can be minimized by independent linear feedforward controllers since they are almost decoupled and proportional to the accelerations. Neural networks are used in addition to reduce the remaining effects, mainly the nonlinear couplings.

Common approaches for learning control with neural networks require a fixed path which is trained, thus allowing repeated execution of the desired trajectory ([14] [15] [16]). On the other hand, for a fixed path no learned feedforward controller is necessary because the path memory can be modified directly as described in another method [13]. The advantage of neural networks comes into play if the path differs from time to time as it is valid for on-line planned paths, which are modified in relation to unexpected sensor data. Hence, on-line path planning of high speed movements is the adequate task for decoupling by neural networks.

The basic learning algorithm is presented in [13]. It is able to reduce existing path errors sub-

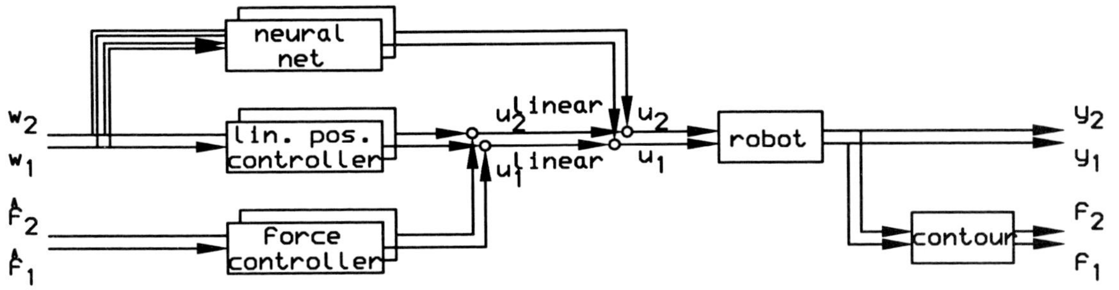

Figure 7: Structure of the controller for a robot with 2 joints: Independent linear controllers reduce path errors caused by dynamical delays, neural networks reduce path errors caused by couplings between the joints, and independent linear force controllers compensate for the influence of contact forces ($\hat{f}$ stands for the portion of the predicted joint torque which corresponds to the contact force)

stantially. Still, remaining errors exist, due to couplings. In analytical models of robots these errors can be explained by coriolis and centrifugal forces and by the non-diagonal elements of the mass matrix. Therefore we propose to extend the independent controllers by neural networks (see Figure 7 for a simple case). The hybrid approach of linear and neural controllers is adequate since for the overall system the representational error should be under 1% which can be reached best using very small output normalization values for the network.

We use one network for each joint. The inputs are the position of joints 2 to 5, and 10 additional inputs, the desired accelerations of those joints, affecting the motion of the actual joint. Multilayer perceptrons with sigmoid activation functions are used. Two hidden layers with a sufficient number of neurons each, are appropriate to solve the problem. For the experiments with a Manutec r2 robot a network structure of size 14-7-3-1 for joints 3 and 4 and of size 24-8-3-1 for joint 5 are found to perform best. The other joints are not controlled by a neural component since the are not effected by the motion of other joints.

Training is performed first for the linear and then for the neural network controllers. For training an Extended Kalman Filter algorithm is used [17]. This method is qualified for problems with up to some hundred weights since a covariance matrix is built that represents couplings between the weights and prevents overwriting of already learned information, at least for small changes of the weights. Therefore very few epochs are needed for learning. In the experiments 10 epochs are used at most before testing the controller and acquiring new training data.

The learning system was tested for high speed movements (see [18] for comprehensive experiments). It is demonstrated here for a contour following task (see Figure 8(a)) of a Manutec r2 robot with conventional positional interface. First, the linear controllers are trained. Then the neural networks are learned using training data from similar paths without contact. When applied to the case with contact, the error of the desired force is reduced by 50 % (0.6 N instead of 1.2 N) with respect to the independent controllers (see Figure 8(b)).

9 Conclusion

We presented neural network approaches for perception and manipulation. Within the last decade we increasingly used neural networks and since we have implemented conventional methods as well, (e.g. camera calibration, visual servoing, robot control) we are able to compare both methodologies for various applications. We found neural networks especially useful: for the representation and fusion of sensor data (Section 2, 3), for the representation of sensorimotion mappings (Section 5, 6), in combination with a priori knowledge (Section 5, 7, 8) and very powerful in by-passing the usage of camera calibration techniques (Section 4).

References

[1] Jerome H. Friedman. *From Statistics to Neu-*

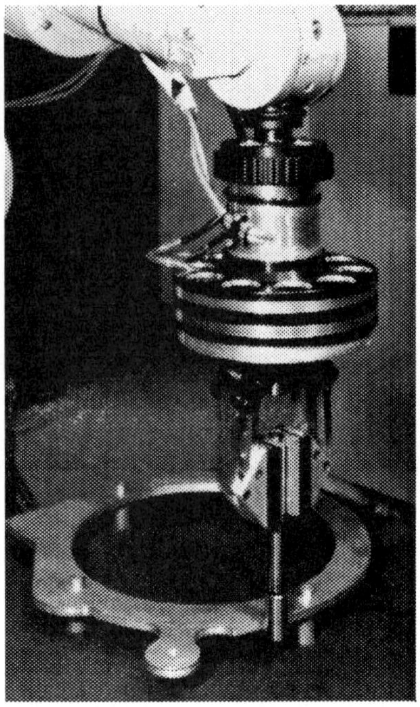

(a)

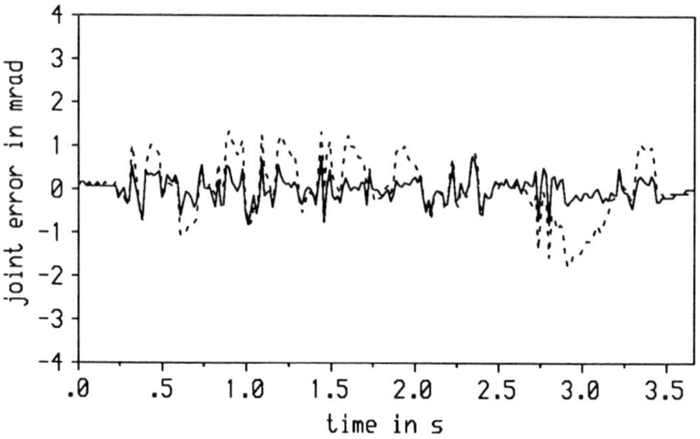

(b)

Figure 8: (a) Endeffector in starting position and the contour to be followed, (b) Error signal of joint 3 during execution of the sensed path with up to 0.7 m/s when using a controller with (solid) or without (dashed) neural networks

ral Networks - Theory and Pattern Recognition, chapter An Overview of Predictive Learning and Function Approximation, pages 1–61. NATO ASI Series F Vol. 136, Springer Verlag, 1994.

[2] E. Ralli and G. Hirzinger. Fast path planning for 6-dof robot manipulators in static or partially changing environments. In *Int. Conf. on Advanced Robotics ICAR'95*, Sant Feliu de Guixols, Spain, Sept. 20-22 1995.

[3] Andreas Baader and Gerhard Hirzinger. World modeling for a sensor-in-hand robot arm. In *IEEE Int. Conf. on Intelligent Robots and Systems, Pittsburgh*, August 1995.

[4] B.K.P. Horn. *The Psychology of Computer Vision*, chapter Obtaining shape from shading information, pages 115–155. McGraw Hill, New York, 1975.

[5] B.K.P. Horn. *Shape From Shading*. MIT Press: Cambridge, MA, 1989.

[6] B.K.P. Horn. Height and gradient from shading. *Int. Journal of Computer Vision*, 5(1):37–75, 1990.

[7] G. Hirzinger G.Q. Wei. Learning shape from shading by a multilayer network. *IEEE Trans Neural Networks*, accepted for publication.

[8] Guo-Quing Wei, Gerhard Hirzinger, and Bernhard Brunner. Sensorimotion coordination and sensor fusion by neural networks. In *Proc. IEEE Int. Conf. on Neural Networks, San Francisco*, 1993.

[9] Gerd Hirzinger, Bernhard Brunner, Johannes Dietrich, and Johann Heindl. Sensor-based space robotics - rotex and its telerobotic features. *IEEE Trans. on Robotics and Automation*, 9(5):649–663, October 1993.

[10] Ralf Koeppe and Gerd Hirzinger. Learning compliant motions by task-demonstration in virtual environments. In *Fourth Int. Symp. on Experimental Robotics, ISER, Stanford*, June 30 -July 2 1995.

[11] R. Staudte. Untersuchungen zum Einsatz neuronaler Speichersysteme innerhalb einer lernfähigen Greifersteuerung. Diplomarbeit, Inst. für Regelungstechnik, TH Darmstadt, 1992.

[12] Ralf Koeppe and Gerhard Hirzinger. Regelung von Robotern mit einem hybriden inversen Modell. In *VDI Fachtagung Intelligente Steuerung und Regelung von Robotern, Germany*, 1993.

[13] F. Lange and G. Hirzinger. Learning to improve the path accuracy of position controlled robots. In *IEEE/RSJ/GI Int. Conference on Intelligent Robots and Systems*, München, Germany, Sept. 12-16 1994.

[14] M. Höhfeld, J. Hollatz, S. Miesbach, D. Obradovic, B. Schürmann, V. Sterzing, M. Taniguchi, and V. Tresp. Advanced neural control for industrial applications. In M. van der Meer, editor, *Statusseminar Neuroinformatik*, Schloß Maurach, Germany, Oct. 1992.

[15] F. Arai, L. Rong, and T. Fukuda. Asymptotic convergence of feedback error learning method and improvement of learning speed. In *IEEE Int. Conf. on Intelligent Robots and Systems (IROS)*, Yokohama, Japan, July 26-30 1993.

[16] T. Yamada and T. Yabuta. Application of learning type feedforward feedback neural network controller to dynamic systems. In *IEEE Int. Conf. on Intelligent Robots and Systems (IROS)*, Yokohama, Japan, July 1993.

[17] F. Lange. Fast and accurate training of multilayer perceptrons using an extended Kalman filter (EKFNet). Internal paper, available by http://www.op.dlr.de/FF-DR_RS/STAFF/friedrich_lange/, Sept. 1995.

[18] F. Lange and G. Hirzinger. Learning of a controller for non-recurring fast movements. *Advanced Robotics, Special Issue on Behavior and Learning*, 1996.

Integration of Symbolic and Subsymbolic Learning to Support Robot Programming by Human Demonstration

R. Dillmann, H. Friedrich, M. Kaiser, A. Ude
Institute for Real-Time Computer Systems & Robotics,
University of Karlsruhe, D-76128 Karlsruhe, Germany.
E-Mail: dillmann@ira.uka.de

1 Introduction

One of the major cost factors related to robotic applications is the development of robot programs. Especially the use of multi sensor systems and the demand for various skills requires experienced programmers and efficient programming environments. Such sophisticated programming systems are usually available in industrial environments. However, they do not exist for robot applications in a public or personal environment. To enable new robot applications with emphasis on service tasks [40], it is necessary to develop techniques which allow untrained users to program a service robot quickly, safely, and efficiently.

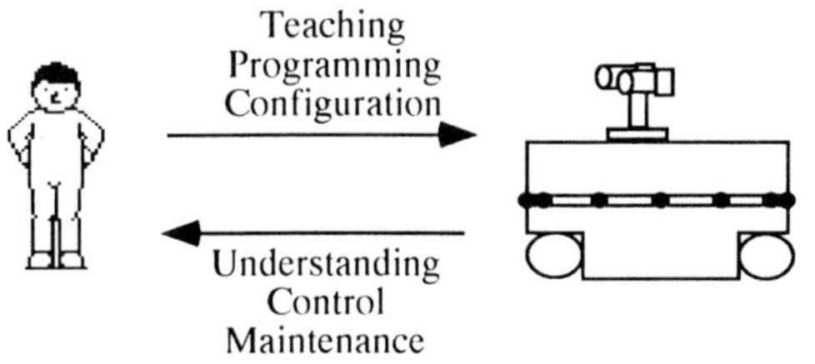

Figure 1: Aspects of Human-Robot Interaction.

Two basic aspects characterize the interaction between the robot system and the user (Fig. 1, [11]). Firstly, the user wants to configure the robot for the task and instruct it. This requires to translate the user's intention into the robot's language, i.e., to compile user intentions into correct and executable robot programs. Several techniques to perform the required task knowledge transfer and program generation do exist. On the basis of such a technique the user is acting as a teacher or programmer, and the robot as an apprentice (Fig. 2). They exchange various kinds of information via a dialogue.

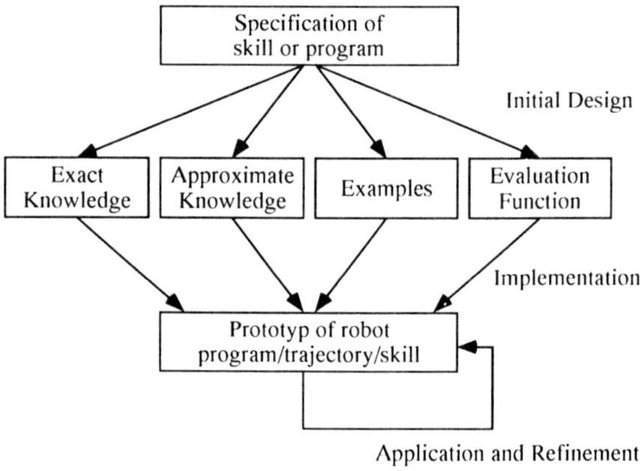

Figure 2: Approaches to the Design of Robot Programs.

Secondly, to allow the user to efficiently control and maintain the robot and its task performance, the process information from the low level control of the robot has to be translated into a representation that can be interpreted by the higher levels of system control and understood by the human operator. Hence, **symbols** have to be derived from signals. It is useful to enable the robot to perform the signal-symbol transformation task autonomously, i.e., to **learn** semantically meaningful descriptions of its own perceptions, actions, and states, and to use these descriptions both to transfer robot knowledge to the user and to interpret user demonstrations. Human skills may be derived from observing human performance. These are some of the key issues in the work related to **Robot Programming by Demonstration** [20, 31, 15] and **Robot Skill Acquisition via Human Demonstration** [5, 38, 26].

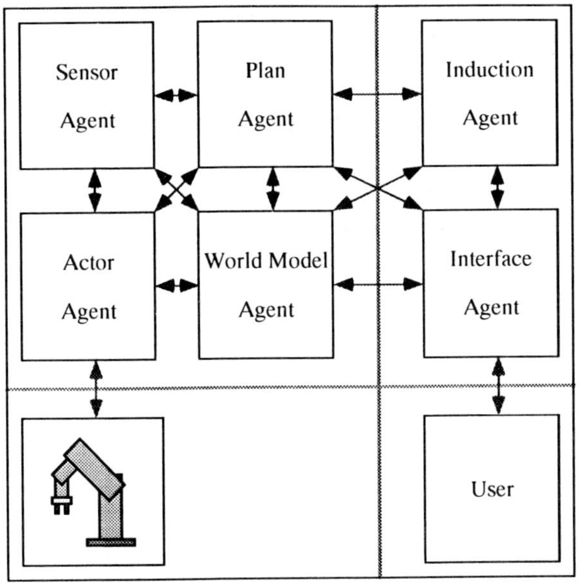

Figure 3: Agent-oriented architecture of an RPD system.

Robot skill and task learning, however, can only be processed efficiently if it is based on *information* about the robot environment, the effect of the robot's actions, and the intentions of the user. To acquire this information the user has to communicate with the robot based on the goal to be achieved and the resources the robot may use. This paper presents a method of robot programming with the help of human demonstration on different levels of robot control, considering an untrained user. A survey is given on learning techniques which make use of the human demonstration in order to derive task knowledge and robot skills. The individual components of an implemented RPD system, into which these learning techniques are integrated, are shown in Fig. 3.

2 Learning from Human Demonstration

Considering the control architecture of a typical robot system, there are numerous learning goals to be achieved. As shown in Tab. 1, they spread from the lower levels of robot control – the skill level – up to the task level. They include learning of direct sensor actor couplings on the lowest level of robot control as well as building models of the robot's environment. In addition, the continuous improvement of the robot's reasoning and planning capabilities in order to enable it to handle various complex tasks over time is an important aspect of learning. From a general point of view, two basic learning phases can be distinguished as follows [29, 30]:

- Initial acquisition of task knowledge, generation of task-level skeleton plans and generation of basic controller structures.
- Refinement of existing skills and plans and acquisition of new task plans and related controllers in order to be able to solve new problems and the related tasks.

As far as a robot can learn automatically, by using only the information being directly available (e.g., via sensor data and internal performance evaluation functions), both the acquisition of task knowledge and the refinement of plans, related controllers, and models can principally be performed at any time instant. However, if learning requires and involves information to be acquired and transformed from a human user, the learning procedure must proceed **demand-driven.** Then, two situations can be considered which require the robot to **learn**:

1. A task is specified (either internally or by the human user) that the robot cannot perform due to missing task-level knowledge, plans or skills.
2. A task is specified that involves objects that are not known or can not be identified by the robot.

2.1 Learning Tasks

The typical **learning tasks** for a robot system may be related to modelling and understanding the environment, the task to be achieved by the robot and/or the human tutor giving examples or demonstrating of how to perform a manipulation. They are hierarchically listed in table 1. The structure of the concrete learning
The structure of the concrete learning tasks depend highly on the knowledge and skills already available to the robot and on the way this knowledge is processed for robot control within the robot's control architecture. To allow a formal description of the learning tasks, we assume that the robot provides the user with the following information:

- a set of objects and related methods that allow to detect/identify and describe these objects,

Control Level	Learning Task	Representation	Learning Technique
Task	Generation of program schemata	Skeleton plans	Explanation-Based Generalisation
	Interpretation of world situations	Semantic nets	Exemplar-Based Induction
	Generation of object descriptions	Geometric models	Constructive Induction
Elementary Operation	Link skills to symbol level operations	Symbolic operators	Explanation-Based Generalisation & Exemplar-Based Induction
Skill	Transformation of perceptions into actions	Neural networks Hybrid systems	Reinforcement based Adaptation Exemplar based Induction

Table 1: Learning tasks and knowledge representation in the robot control hierarchy

- a set of elementary skills that are perception-action transformations or basic control algorithms which involve no symbolic model knowledge; these skills represent the basic capabilities of the robot,
- a reasoning mechanism which allows to generate a program to perform a specified task on the basis of the elementary skills, the task knowledge, and the available object knowledge, and
- a multi-modal user interface which supports the specification of tasks via various different input channels as well as the external supervision and control of the robot for demonstration purposes.

In the initial learning phase, no physical objects may be known to the robot. The only existing skill may be to move in ordinary point-to-point mode (without sensor feedback) to a desired location. Consequently, the **learning tasks** to be considered are the following:

1. Acquisition of a task-level program schema, which means to learn a sequence of skills including proper application conditions.
2. Acquisition of new skills, i.e., learning of procedures that transform perceptions to actions and do not involve explicit models.
3. Acquisition of a new perception skill such as learning to describe and classify a specific object.
4. Integration of the acquired skill, program schema, or object description into the knowledge base of the robot and thereby learning how to use the acquired knowledge.

Most of these tasks can not be solved by a conventional robot completely on its own. Even if an extensive knowledge base exists, finding suitable sequences of perceptions and actions is a cumbersome task that requires efficient and universal problem solvers.

2.2 User Demonstration – a Basis for Programming

Usually, the design of robot programs involves a mix of quantitative and qualitative modeling techniques as well as training procedures based on interaction with a human user (Fig. 2). Especially for users who are not deeply familiar and skilled with robot programming, **Programming by Demonstration (PbD)** [8] has a considerable potential to become a suitable programming technique. **PbD** relies on demonstrations of the considered task. The demonstrations are used as the primary input and are the basis for a learning process. **PbD** has been applied successfully in domains such as graphic editors [32], instructible software agents [33] or intelligent interfaces [34]. **Robot Programming by Demonstration (RPD)** has been realized through a number of applications and on different levels of both robot control and perception.

- Demonstrations were proven to be suitable for the acquisition of *new program schemata on task level* [41]. In [31], sequences of video images were analyzed in order to generate assembly plans. [3] presented NODDY, a system which generates generalized programs by fusing several demonstrations. Single physical demonstrations and formally defined user intentions are the basis for the robot programs generated by the system described in [15].

- On the *control level*, demonstrations can be used as basis for learning both, *open-loop and closed-loop controlled new elementary skills.* The acquisition of open-loop controlled skills is mostly focused on the reconstruction of trajectories from a sequence of demonstrated states (positions) [10, 42].
  Systems supporting the acquisition of closed-loop controlled elementary skills in general are characterized by a very task-specific control strategy design. They comprise acquisition techniques for manipulation tasks such as deburring [5] and assembly [12, 26] as well as for vehicle control [37] and autonomous robot navigation [38].

- Learning new perceptive skills for object and landmark recognition can also take place on several system control levels (Tab. 1). [1] presents an approach to learn sensor parameterizations from demonstrations. Learning active perception skills, i.e., the combination of actions and sensing for the purposes of object recognition, is the topic of work presented in [27].

Analyzing the large number of publications reporting research in the various domains related to RPD, it can be seen that it was almost never taken into account that the *crucial* role of the human teacher may influence the learning process negatively. However, especially in a service environment with demonstrations provided by unexperienced users, examples must be considered to be not at all optimal with respect to the robot or the learning system.

2.3 The Role of the Tutor

In the ideal case, the human tutor demonstrates exactly what is required by the learning system. This means, he provides – in terms of examples – an exact procedure for performing a certain action or for identifying a certain object. However, human demonstrations will only with experience result in examples of a program schema, a skill, or an object description which is optimal with respect to the robot control system or the learning system [16, 24]. Therefore, the user intention, the training context, example assessment, interpretation and transformation into a formal representation as well as on-line refinement of the initially acquired experience, and knowledge become important issues in RPD. Furthermore, for both acquired elementary sensing and action skills, the real-time availability of their operational representation (e.g., as a neural network) must be complemented by a symbolic representation. This useful in order to make the information meaningful and available for both, the robot's control system and the human tutor. The symbolic representation of skills, the **elementary operations,** allow to integrate them into task-level skeleton plans. This enables bottom-up as well as top-down acquisition of program schemata, sensing and action skills.
Thus, the process of knowledge transfer (Fig.4) from example demonstration to controller refinement becomes complex. It requires more steps than the usually considered (in robotics [18, 5] as well as in psychology [14, 2]) simple example generation, "strategy extraction" (learning), and skill application, and asks for the support of the human tutor in several phases:

1. The tutor determines the task for the robot to be learned.

2. The tutor selects the method and strategy for the generation of the examples and performs a first supervised demonstration. This demonstration may be related either to a task-level program schema or the desired, related and required skills. An example may also be related to the acquisition of world-model data like the features of an object. Alternatively, unsupervised automatic experimentation may be used. This requires the definition of permitted actions and sensing strategies and an evaluation function.

3. With the help of analysis tools the tutor, assesses the quality of the recorded example trace. Thus he's able to influence the example data preprocessing and the definition of convergence criteria. Figure 5 shows a graphical tool for trajectory analysis.

4. If a skill has been acquired, the tutor adds a symbolic interpretation. This includes the user intention and data provided by the learning robot that provides the context in which the skill or object description was generated. Thus, an elementary operation is constructed.

   In the case of program schema acquisition, the user's intention is used as the basis for generalization. The subsequent learning step results in a program schema containing branches and variables.

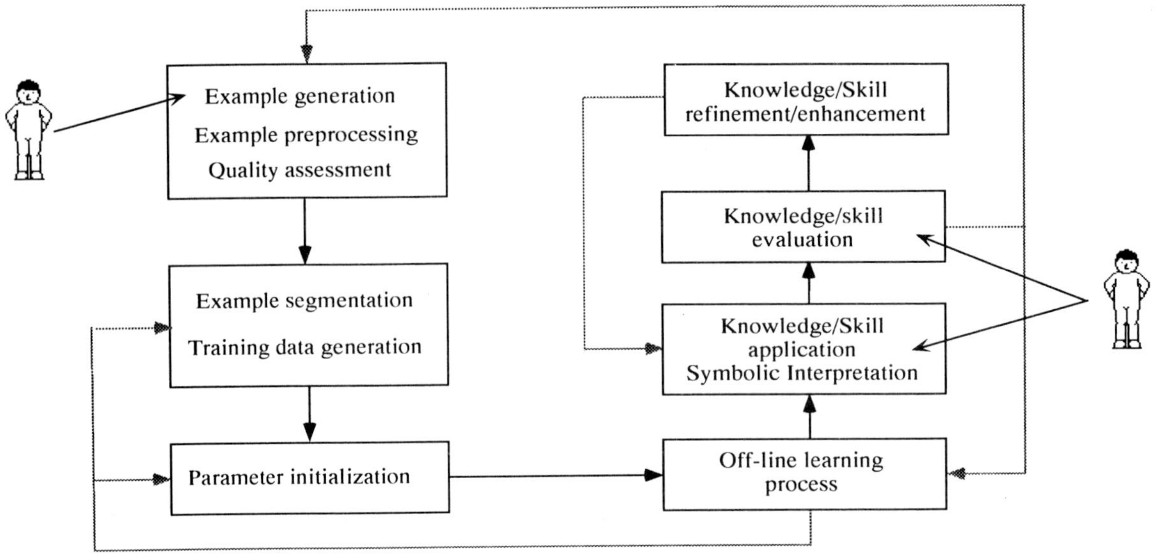

Figure 4: Different phases of robot knowledge/skill acquisition, application, and refinement. Gray arrows indicate feedback loops. Only some phases are permitted to **require** user interaction.

5. The tutor evaluates the robot's performance on-line. He may observe the system and comment on it's performance, choose a general general evaluation function (e.g., related to motion-/recognition speed), or design an evaluation function specific for the acquired skill or task-level program schema.

Thus, the human tutor is essentially involved in the robot's learning process. However, the cumbersome task to program the robot is replaced by an interactive dialogue with the robot. This dialogue can take place either on a high level of abstraction or via demonstrating task solutions instead of formally specifying them. In the following, the actual interaction sequence involved in the RPD process will be described. This interaction is related to human-robot communication on a **symbolic** level. Manipulation of numeric data, i.e., **subsymbolic** processing, which is necessary on the lower level real-time system control levels, should not be a task assigned to an unexperienced user.

3 Demonstration Techniques

An important aspect of interaction in RPD is the **identification of the task and control knowledge to be acquired.** Knowing **what to demonstrate** is a prerequisite for providing demonstrations. The second, more obvious interaction point is given by the demonstration itself. Depending on the actual demonstration technique in use, quite diverse constraints may exist that must be taken into account by the user or, alternatively, by the learning system.
In case of **unconstrained** demonstration techniques, the human user may demonstrate task-level solutions or skills without involving the robot in the demonstration. Examples of this kind of technique are the use of freely movable input devices [4, 9], that allow to specify not only motions, but also strategies relying on sensing and action, and especially **vision-based demonstration** [31, 42]. While this technique requires sophisticated monitoring and interpretation systems, it is very easy to use, since it allows the user to fully exploit his cognitive and motoric capabilities without having to be concerned with the limitations of an input device.
A crucial point in the case of a natural user demonstration is the mismatch between the human operators configuration space and the restricted configuration space of a robot. Thus, the operator needs a system feedback, whenever a trajectory which is not executable by a robot is derived from a demonstration.
At our institute, the vision based trajectory reconstruction system TRESS[1] [42] has been developed. Here, the user is being observed by a stero-vision system while manually manipulating objects. The objects motions are reconstructed from the recorded image sequence, such that they

[1] Trajectory Reconstruction System

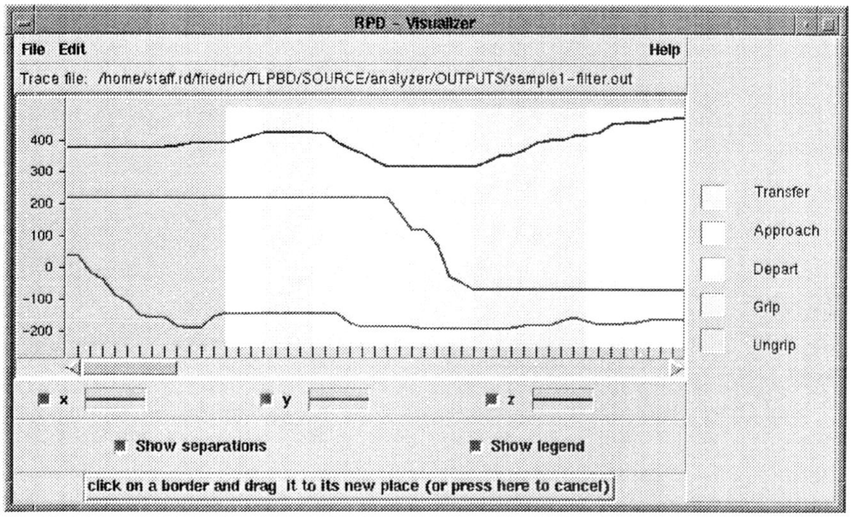

Figure 5: Translational components of a trajectory with automatically classified segments related to the different elementary operations.

can be executed by the robot later on. Fig. 6 shows a typical input sequence generated during visual demonstration of a manipulation task as well as the translational components of the reconstructed object trajectory along the x, y and z-axis.
The vision system used to monitor the demonstration consists of

(a) a stereo camera,
(b) a special real-time processor that can calculate a symbolic contour description at video rate,
(c) an object recognition system which uses pairs of line images to detect objects and calculate their poses,
(d) an object tracking system which requires a stereo sequence of line images as an input.

To allow for recognition and tracking of objects, these objects are previously modelled via a CAD system. Methods based on perceptual grouping, region-based stereo, analog representations and probability theory are used to perform the actual tracking. Special visibility tables have been developed in order to be able to handle non-convex objects. The trajectory reconstruction system is fairly resistant to occlusions, measurement noise and missing information. To reduce processing time the system focusses on the object which is identified as to be relevant for the programming system. It does not try to generate a complete geometric description of the scene. Sensing of contact relations between the manipulated object and the environment have not been considered in the TRESS system yet.

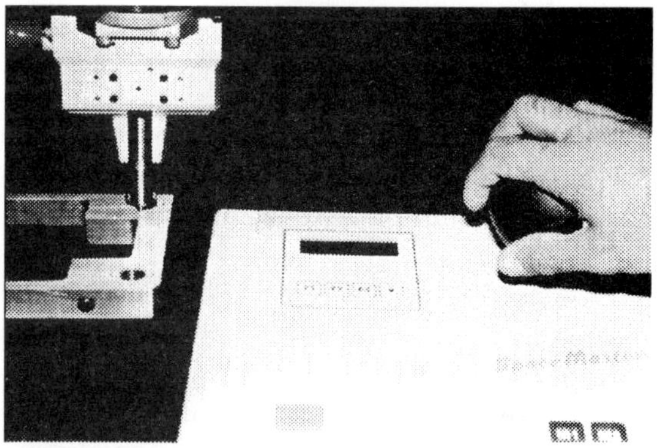

Figure 7: Demonstration of a peg-insertion skill for a Puma 260 manipulator.

A second group of robot teaching techniques involves the robot in the demonstration process. In particular, the robot is **teleoperated** by a human user with the robot's perceptions and the actions performed being sampled. In this case, input devices such as 6D mouse or joystick able to provide the human user with sensor feedback are usually being used (Figs. 7 and 8) [20, 26, 38, 17]. At our institute, a multi-modal interface environment is under development that combines different input modalities, like haptic and non-haptic input devices and speech. This system should provide the user with easy and efficient access to the robot and the learning system.
Often, demonstration via teleoperation does not

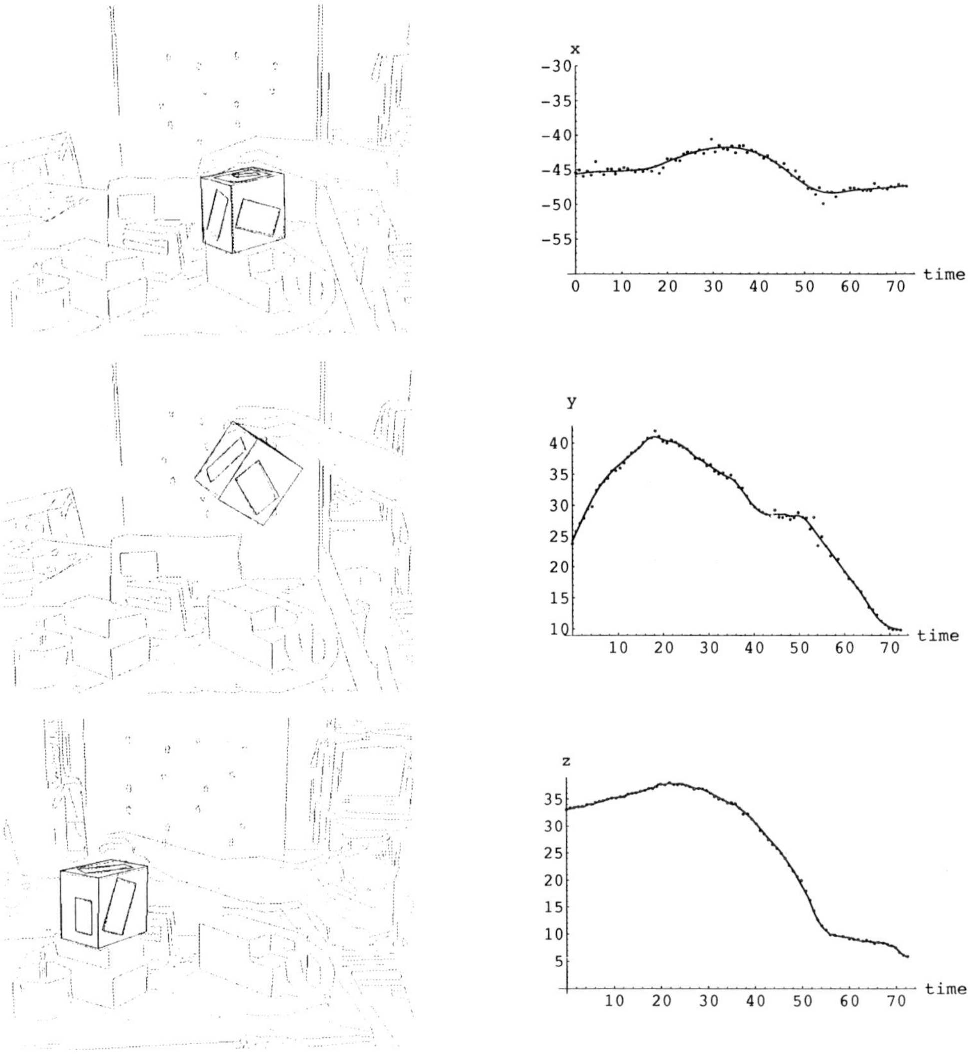

Figure 6: Left: Snapshots of a manipulation demonstration (after edge extraction). Right: Reconstructed translational components of the object trajectory (x, y, and z direction).

take place in reality but in a virtual environment [39, 28]. An obvious advantage of this kind of demonstration is the implicit consideration of the constraints defined by the robot in use. While it is, for example, quite easy to visually demonstrate trajectories that cannot be tracked by the robot to be programmed, this cannot happen if the robot itself is being used in the demonstration process. However, these constraints also make the demonstration task more difficult. From our research related to robot skill acquisition [26, 12] and task-level RPD [17], it can be seen that the demonstration data are often not nearly as smooth as in the case of unconstrained demonstration (Figs. 9 and 12). Especially if elementary skills are to be acquired, even higher demands with respect to demonstration analysis and actual learning techniques exist.

Moreover, in such cases demonstration alone will often not be sufficient to obtain a satisfactory robot performance (see also Fig. 12). Then, the RPD process must comprise an additional phase of **knowledge refinement.**

A complementary approach is the use of a direct **physical link** between the robot and the human user. These approaches include, for example, Pomerleau's ALVINN [37]. Especially if kinematic constraints play an important role and the sensor feedback available through a teleoperating device is not sufficient, the robot can also be moved directly by the human, e.g., by gripping a handle mounted at a manipulator's wrist

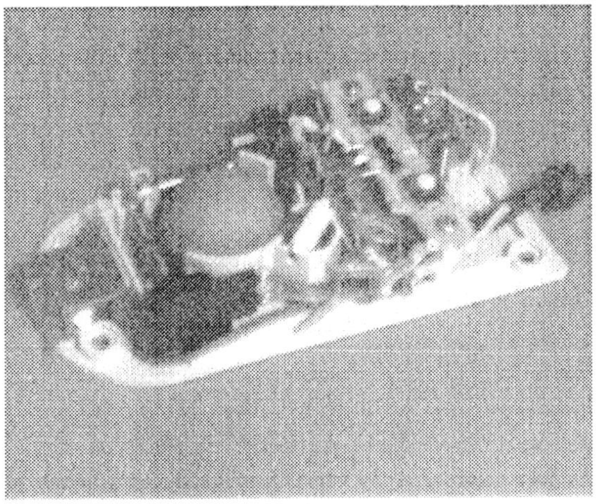

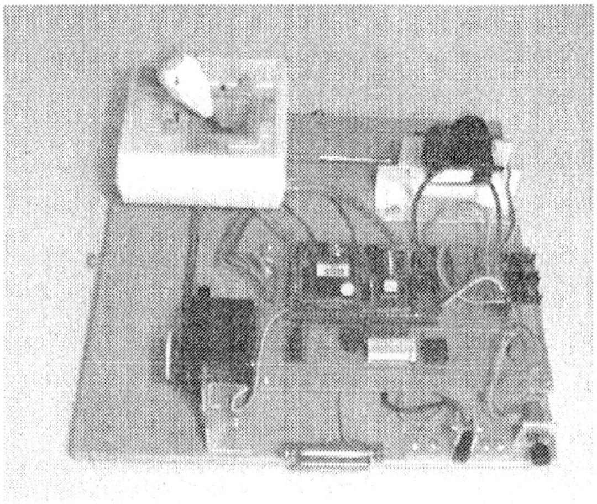

Figure 8: Haptic input devices. Left: 2D-mouse providing force-feedback. Right: 2D-joy-stick providing force-feedback.

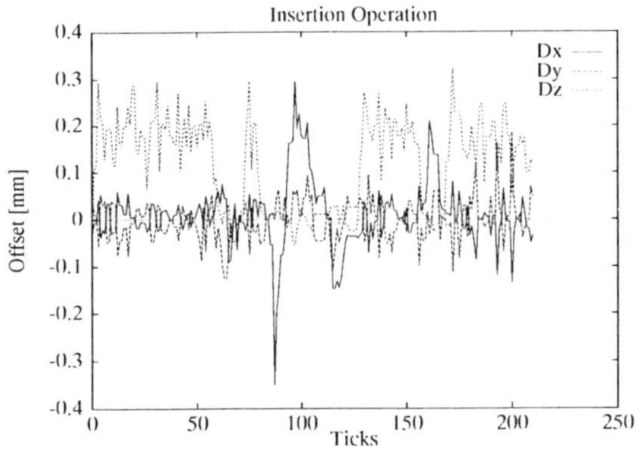

Figure 9: Actions (position offsets) (D_x, D_y, D_z) recorded from a human demonstration of the peg-into-hole skill.

[21]. In these cases, the presence of the operator may have an impact on the data recorded during the demonstration. This must be considered before these data are used for learning. Since the motions of the human user are also restricted by the robot kinematics, demonstrations can also not be expected to be optimal with respect to the given task. Therefore, such a demonstration techniques requires **preprocessing** of the demonstration data as well as **knowledge refinement.**

4 Interpretation Techniques

If the data recorded from the human demonstration is available, the task is to generate an **operational** description of the demonstrated skill, program, or trajectory from this data. Here, "operationality" means that the generated programs can be executed by the target system, i.e., by the real robot. This step of **interpretation** of human demonstrations and their **generalization** with respect to slightly changing tasks and operating conditions lies at the core of RPD and is usually treated by means of regression or learning techniques. However, to make RPD usable also for unexperienced users, the techniques supporting these steps must not require any kind of specific user knowledge. Especially, the interpretation techniques must be robust with respect to their parameterization and the characteristics of the demonstration data, such that the initial learning process can be performed automatically. The interpretation technique used in the system developed at the IPR is presented below. The methods applied in the system are represented with respect to the different levels of robot control as stated in table 1.

On the **task level**, examples given by the user (e.g., Fig. 10) consist of a sequence of symbolic operators (representing skills combined with a symbolic description and application context) that are to be combined into a robot program [15]. The relevant module in this context is the **construction unit**, which operates on a sequence of elementary operators (EOs) like **`transfer`**, **`approach`**, **`grip`** (for the case of a robot manipulator) etc. (Fig. 11). These operators are modeled following the tradition of *STRIPS'* planning operators [13] with pre- and postconditions determining their applicability as well as the changes in the environmental configuration they cause[2].

[2]These conditions as well as the add-list and the delete list are the mean to **integrate** the elementary skills de-

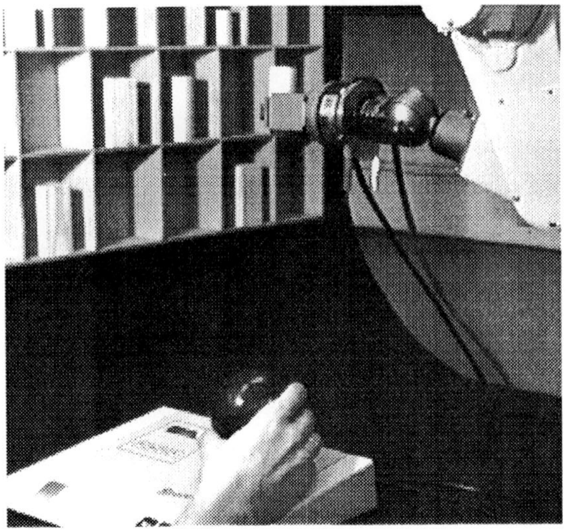

Figure 10: Task-Level Demonstration of a paletizing task.

```
move(pos~1, pos~2, cube~1)
if (on(tray, cyl) & at(pos~3, cyl)) then
  move(pos~3, pos~4, cyl)
if (at(pos~5, cube~2)) then
  move(pos~5, pos~3, cube~2)
else move(pos~6, pos~3, cube~2)
move(pos~7, pos~8, cube~3)
```

Figure 11: Generalized program schema derived from a task-level demonstration as shown in Fig. 10.

An input operator sequence is segmented into subsequences, where each subsequence (so called *Object Group*) represents the manipulation of one object. By processing the operator sequence in a straightforward way, the postcondition of the whole demonstration, i.e., the set of lasting changes resulting from the demonstration, is determined. Here, learning takes place on a **symbolic** level, such that the learning system may interact with the user. In particular, generalisations and transformations such as introducing variables and conditioned operator sequences are performed. The **user intention** is the most relevant information that is needed by the system. If this intention, i.e., the relevant subset of the established postcondition, is identified correctly, a generalized program schema can be built that is applicable as soon as its precondition is met in a given state of the world (see also Fig. 11).

In case of the demonstration of trajectories, the reconstruction of trajectories and their temporal segmentation are key operations which must be performed by a PbD system [22]. Depending on the kind of input being available, these operations can be realized in different ways. At IPR the use of a stereo image sequence vision system is emphasized with the intention to observe human demonstrations. This system can provide a sequence of object poses on the demonstrated motion trajectory. In case of low noise, the reconstruction of the trajectory can be accomplished by a simple interpolation procedure. However, the images provided by the vision system used are considerably noisy. Therefore it is necessary to carefully model the measurement noise in order to be able to reconstruct the observed motion trajectory. By propagating the estimates of measurement errors from the initial images to the calculated poses and by utilizing approximation techniques based on smoothing vector splines [42], it can be shown that it is possible to reconstruct the demonstrated trajectories with an average error of about 5mm for translational degrees of freedom and about 2 degrees in case of rotation (see also Fig. 6). The proposed methods is to be extended to take into account different types of restrictions imposed on the robot motion by the robot kinematics.

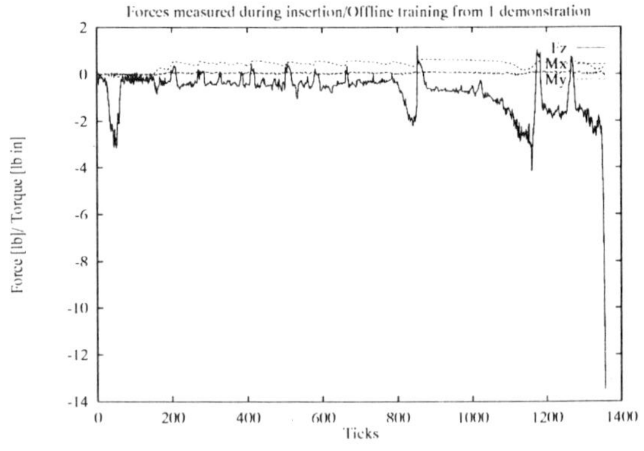

Figure 12: Forces recorded during peg-insertion using the initially learned elementary skill (no refinement).

If elementary skills are being demonstrated, a function C_s with

$$\vec{u}(t) = C_s(\vec{s}(t)) = C_s(\vec{y}(t-d), \ldots, \vec{y}(t-d-p))$$

is to be built from the demonstration data. This function represents mappings of the situations $\vec{s}$

scribed above, i.e., to make them actually usable.

defined via perceptions $\vec{y}$ of the robot to robot actions $\vec{u}$. In this case, the interpretation phase comprises the selection of the relevant sensors and the required degrees of freedom. Also, what is needed is an identification of those perceptions $\vec{y}$ that define a situation, i.e., to define the required perceptual horizon as well as the reaction time of the user. The skill acquisition system developed at the IPR [12, 26] uses both analytic and heuristic techniques to accomplish these tasks, the actual selection of techniques is based on an analysis of the demonstration quality.

After the relevant sensors and other parameters have been defined, **training data** usable for learning the function C_s are available. Then, the formalism to be used to represent C_s has to be selected. A specific constraint that must be considered during this phase is that an untrained user cannot be expected to select the structure of, e.g., a neural network to be used for function approximation. Additionally, the representation formalism must support **incremental learning,** because the initially learned skill must be refined later on in the consequent phase.

A function approximation technique whose structure can be generated from examples and which supports incremental learning of both numerical and structural parameters are neural networks based on *local receptive fields* [35], such as Radial-Basis Function Networks (RBFs) [36]. These networks calculate the output $\vec{u}$ for a given input vector $\vec{s}$ as the weighted sum of the activity of the local receptive fields. Like multi-layer perceptrons, these networks are universal approximators, and similar to MLPs, they can use time-delays for handling spatio-temporal data [7], making them applicable also for identification and control [25]. Therefore, in our system these network are used to represent the skills themselves as well as inverse models and prediction functions required for adaptation [23]. Fig. 12 shows the forces measured during the insertion of the peg while the robot was under control of an RBF network built from the preprocessed data shown in Fig. 9.

5 Refinement Techniques

Following the initial learning process, the acquired knowledge, e.g., the reconstructed trajectory, the learned elementary skill, or the generated program schema, will be applied. Since user demonstrations are not optimal with respect to the desired performance of the robot [25], the application of this initially acquired knowledge will also include a phase of **knowledge refinement.** However, in the case of open-loop operation, i.e., if the interaction of the robot with its environment is not explicitly considered, refinement can be performed **before** the acquired knowledge is actually applied [42].

During the online refinement phase, interaction with the human user is mandatory. Only the user is able to determine if the system is really performing according to his requirements. However, as discussed before, it cannot be expected that an unexperienced user is able to exactly specify the required modifications on the trajectory, skill, or schema level. Instead, he might be able to indicate if the observed performance was satisfactory or not, or, in some cases, to provide alternative actions. The refinement techniques must therefore be satisfied with a delayed scalar evaluation, but it should as well be able to use additional information in form of another demonstration, if the latter is available (an example of the latter technique can be found in [38]).

This scenario clearly asks for a multi-strategy approach that combines direct adaptation techniques as known, e.g., from the field of adaptive control and reinforcement learning [6], with indirect adaptation techniques based on either a reference model or supervised learning (see, e.g., [25] for a more detailed discussion of these issues).

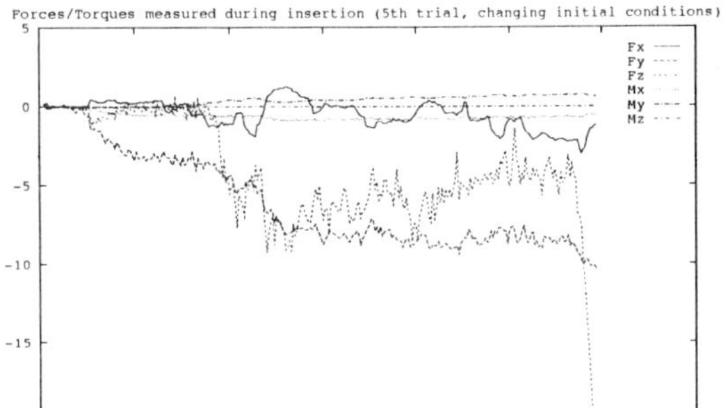

Figure 13: Forces recorded during peg-insertion under neural network control (after 5 practice runs).

In our system, refinement has to be performed using only a scalar delayed evaluation. We combine the approach originally presented by Gullapalli [19] for learning real-valued functions with an explanation-based exploration technique and exponentially discounted adaptation rate [26]. Any of the components involved in the adaptation process can be initialized from the demonstration data, such that the actual adaptation is performed sufficiently fast (see Fig. 13).

6 Conclusion

Throughout this paper, a learning robot control system has been outlined which is designed to learn from human demonstrations. Both bottom-up learning based on primitive perceptions and actions and top-down learning using symbolic operators and models of the robot's environment are supported. Currently, we are working towards the integration of the individual techniques to an all-level RPD system. To this aim, demonstration techniques that allow to obtain contact and grasp information on a symbolic level will have to be added.

On any of the levels of RPD, the techniques used to provide demonstrations and to interact with the system are of great importance. With the help of multi-modal input-/output devices allowing for gestural, natural language, and haptic communication, we expect to significantly ease the task to demonstrate skills, task-level programs, and trajectories. Since current RPD techniques are already sufficiently robust, integrated RPD systems using both multi-modal interaction and sophisticated learning techniques should become quite powerful.

Acknowledgements

This work has partially been supported by the ESPRIT Project 7274 "B-Learn II" and the German Research Foundation by the SFB 314 "Künstliche Intelligenz," Project R2. It has been performed at the Institute for Real-Time Computer Systems & Robotics, Prof. Dr.-Ing. U. Rembold and Prof. Dr.-Ing. R. Dillmann, Department of Computer Science, University of Karlsruhe, Germany. The authors would like to thank the B-Learn II partners for their support.

References

[1] M. Accame and F.G.B. De Natale. Neural tuned edge extraction in visual sensing. In M. Kaiser, editor, *Proceedings of the 3rd European Workshop on Learning Robots (EWLR-3)*, Heraklion, Crete, Greece, April 1995.

[2] J. R. Anderson. Acquisition of cognitive skill. *Psychological review*, 89(4):369 – 406, 1982.

[3] P. M. Andreae. Constraint limited generalization: Aquiring procedures from examples. In *Proceedings of the National Conference on Artificial Intelligence (AAAI)*, pages 6 – 10, 1984.

[4] H. Asada and H. Izumi. Automatic program generation from teaching data for the hybrid control of robots. *IEEE Transactions on Robotics and Automation*, 5(2), 1989.

[5] H. Asada and S. Liu. Transfer of human skills to neural net robot controllers. In *Proceedings of the 1991 IEEE International Conference on Robotics and Automation*, 1991.

[6] A. G. Barto, R. S. Sutton, and C. W. Anderson. Neuronlike elements that can solve difficult learning control problems. *IEEE Transactions on Systems, Man, and Cybernetics*, pages 835–846, 1983.

[7] M. R. Berthold. A time delay radial basis function network for phoneme recognition. In *IEEE International Conference on Neural Networks*, pages 4470 – 4472a, Orlando, Florida, 1994.

[8] A. I. Cypher. *Watch what I do – Programming by Demonstration*. MIT Press, Cambridge, Massachusetts, 1993.

[9] N. Delson and H. West. Robot programming by human demonstration: subtask compliance controller identification. In *Proceedings of the IEEE/RSJ Conference on Intelligent Robots and Systems*, Yokohama, Japan, 1993.

[10] N. Delson and H. West. The use of human inconsistency in improving 3D robot trajectories. In *Proceedings of the IEEE/RSJ Conference on Intelligent Robots and Systems*, 1994.

[11] R. Dillmann, M. Kaiser, V. Klingspor, K. Morik, and F. Wallner. Teaching and understanding intelligent service robots: A Machine Learning Approach. In *19. Deutsche Jahrestagung für Künstliche Intelligenz (KI '95 Workshop: Informationsverarbeitung in Servicerobotern)*, Bielefeld, Germany, 1995.

[12] R. Dillmann, M. Kaiser, and A. Ude. Acquisition of elementary robot skills from human demonstration. In *International Symposium on Intelligent Robotics Systems*, Pisa, Italy, 1995.

[13] R.E. Fikes, P.E. Hart, and N.J. Nilsson. Learning and executing generalized robot plans. *Artificial Intelligence*, 3(4), 1972.

[14] P. M. Fitts. Perceptual-motor skill learning. In A. W. Melton, editor, *Categories of human learning*. Academic Press, New York, 1964.

[15] H. Friedrich and R. Dillmann. Robot programming using user intentions and a single demonstration. In M. Kaiser, editor, *Proceedings of the 3rd European Workshop on Learning Robots (EWLR-3)*, Heraklion, Crete, Greece, April 1995.

[16] H. Friedrich and M. Kaiser. What can robots learn from humans? In *IFAC Workshop on Human-Oriented Design of Advanced Robotic Systems*, Vienna, Austria, 1995.

[17] H. Friedrich, S. Münch, R. Dillmann, S. Bocionek, and M. Sassin. Robot programming by demonstration: Supporting the induction by human interaction. *Machine Learning*, 1996.

[18] A. Guez and J. Selinsky. A neuromorphic controller with a human teacher. In *IEEE International Conference on Neural Networks*, pages 595 – 602, San Diego, CA, 1988.

[19] V. Gullapalli. *Reinforcement Learning and its application to control.* PhD thesis, University of Massachusetts, Department of Computer and Information Science, 1992.

[20] R. Heise. *Demonstration instead of programming: Focussing attention in robot task acquisition.* Research report no. 89/360/22, Department of Computer Science, University of Calgary, 1989.

[21] G. Hirzinger and J. Heindl. Sensor programming - a new way for teaching robot paths and forces/torques simultaneously. In *Third International Conference on Robot Vision and Sensory Controls*, Cambridge, Mass., 1983.

[22] K. Ikeuchi and T. Suehiro. Towards an assembly plan from observation. In *Proceedings of the IEEE International Conference on Robotics and Automation*, Nice, France, 1992.

[23] M. Kaiser and R. Dillmann. Hierarchical learning of efficient skill application for autonomous robots. In *International Symposium on Intelligent Robotics Systems*, Pisa, Italy, 1995.

[24] M. Kaiser, H. Friedrich, and R. Dillmann. Obtaining good performance from a bad teacher. In *International Conference on Machine Learning, Workshop on Programming by Demonstration*, Tahoe City, California, 1995.

[25] M. Kaiser, A. Retey, and R. Dillmann. Designing neural networks for adaptive control. In *IEEE International Conference on Decision and Control (34th CDC)*, 1995.

[26] M. Kaiser, A. Retey, and R. Dillmann. Robot skill acquisition via human demonstration. In *Proceedings of the International Conference on Advanced Robotics (ICAR '95)*, 1995.

[27] V. Klingspor and K. Morik. Towards concept formation grounded on perception and action of a mobile robot. In *Proceedings of the 4th Intern. Conference on Intelligent Autonomous Systems (IAS-4)*, Karlsruhe, 1995.

[28] R. Koeppe and G. Hirzinger. Learning compliant motions by task-demonstration in virtual environments. In *Fourth Int. Symp. on Experimental Robotics*, 1995.

[29] J. Kreuziger. Application of machine learning to robotics: An analysis. In *Proceedings of the Second International Conference on Automation, Robotics, and Computer Vision (ICARCV '92)*, 1992.

[30] J. Kreuziger. *An architecture for the application of symbolic learning in robotics.* PhD thesis, University of Karlsruhe, Department of Computer Science, 1994. In German.

[31] Y. Kuniyoshi, M. Inaba, and H. Inoue. Learning by watching: Extracting reusable task knowledge from visual observation of human performance. *IEEE Transactions on Robotics and Automation*, 10(6):799–822, 1994.

[32] Henry Lieberman. *Watch what I do*, chapter MONDRIAN: A teachable graphical editor. 1993.

[33] Dave Maulsby. *Instructible agents.* PhD thesis, University of Calgary, Canada, 1994.

[34] S. Minton, A. Philpot, and S. Wolfe. Specification by demonstration: The ViCCS interface. In *International Conference on Machine Learning, Workshop on Programming by Demonstration*, Tahoe City, California, 1995.

[35] J. Moody and C. Darken. Learning with localized receptive fields. In T. Sejnowski D. Touretzky, G. Hinton, editor, *Proceedings of the Connectionist Models Summer School.* Carnegie Mellon University, 1988.

[36] T. Poggio and F. Girosi. Networks for approximation and learning. *Proceedings of the IEEE*, 78(9):1481–1497, September 1990.

[37] D. A. Pomerleau. Efficient training of artificial neural networks for autonomous navigation. *Neural Computation*, 3:88 – 97, 1991.

[38] P. Reignier, V. Hansen, and J.L. Crowley. Incremental supervised learning for mobile robot reactive control. In *Intelligent Autonomous Systems 4 (IAS-4)*, pages 287 – 294. IOS Press, 1995.

[39] C. Sammut, S. Hurst, D. Kedzier, and D. Michie. Learning to fly. In *Proceedings of the 9th International Conference on Machine Learning*, pages 385 – 393, 1992.

[40] R. D. Schraft. *Serviceroboter - ein Beitrag zur Innovation im Dienstleistungswesen.* Fraunhofer -Institut für Produktionstechnik und Automatisierung (IPA), 1994.

[41] A. M. Segre. *Machine Learning of Robot Assembly Plans.* Kluwer Academic Publishers, 1989.

[42] A. Ude and R. Dillmann. Robot motion specification: A vision-based approach. *Surveys on Mathematics for Industry*, 5, 1995.

Designing Skills with Visual Feedback for APO

Katsushi Ikeuchi
Comp Sci Dept
Carnegie Mellon Univ
Pittsburgh, PA, USA
ki@cmu.edu

Jun Miura
Mech Eng Dept
Osaka Univ
Suita, Osaka, Japan
jun@cv.ccm.eng.osaka-u.ac.jp

Takashi Suehiro
RWC Project
Tsukuba, Ibaraki, Japan
suehiro@trc.rwcp.or.jp

Santiago Conanto
AI Ctr
ITESM
Monterrey, Mexico
sconant@cmu.edu

Abstract

We have been developing a novel method to program a robot from observation: the APO (assembly-plan-from-observation) method [11]. A human performs assembly operations in front of a TV camera. The APO system recognizes assembly operations and generates an assembly plan to enable a robotic arm to repeat the same assembly operations.

The APO system has three worlds: the human world to be observed, the APO internal world to be generated by the APO system, and the robot world in which a robot mimics human operations. The APO system performs two step mappings: the first step maps the human world into the internal world (task recognition), and the second step does the internal world into the robot world (task instantiation). The previously reported system mainly focused on the issues of recognition: how to obtain such an internal world from the observed human world.

This paper concentrates on the issue of instantiation. The APO system analyzes the internal world and determines an abstract command, such as pick-up or insert-into, that realizes the internal world. To each abstract command, a skill - a sequence of robot motion templates - is prepared in a skill library. By picking up an appropriate skill from the library and filling in necessary parameters, the APO system generates commands executable on a robot.

One of the important issues in designing skills is how to utilize visual information. It is generally very difficult, if not impossible, for a robot to perform fine manipulation tasks without the benefit of sensory feedback. Although the APO system cannot provide direct sensing strategy from observing human actions, useful information is obtained from analysis of the observed tasks. This paper presents a method of systematically generating sensing strategy for skills. We will present the theory for determining which skill requires visual feedback, which information is necessary, and which is the optimal configuration of a sensor. We will also report experimental results to demonstrate the validity of our method for choosing the optimal sensing strategy.

1 Introduction

Several methods to program a robot have been proposed. Such methods include: teaching-by-showing, teleoperation, textual programming, and automatic programming. In teaching-by-showing methods, an engineer uses a teach pendant to store a path along which a robot should move repeatedly. In run mode, the robot follows the path it was previously taught. This is the most common method to program a robot in industrial applications. This method is suitable for programming a robot to repeat either simple or complicated movements. However, this method requires that an engineer be in the same environment as the robot. Hence, this method is inapplicable in hazardous environments such as nuclear plants, under water, or in outer space.

To remedy this problem, the teleoperation method has been proposed. This method uses a master manipulator for teaching and a slave manipulator for execution. An engineer controls the master manipulator in a safe environment while monitoring a display from a remote TV in the hazardous environment. The slave manipulator in the hazardous environment executes real operations based on control signals from its master manipulator. Since this method does not require an operator in the execution environment, it is suitable for operation in hazardous environments. However, by using this method, we can teach a robot only trajectory information. It is difficult to program a flexible robot system to be able to use force control with error recovery capabilities. It is also true that we have to repeat the motions, even when a very minor change in the operation is desired.

Textual programming is often used in academic environments. A programmer stores a robot command sequence as a textual program in a computer.

By using a compiler or an interpreter, a command sequence in a textual program is converted into a form that the robot can execute. This method is quite flexible because one can store any kind of control program. However, this method has the disadvantage of requiring both a relatively long development period and expert programmers.

Automatic programming has been proposed to reduce the programming burden and thus speed up the programming process. This method relies on geometric reasoning systems which can use geometric information given by geometric models and task specifications to generate textual programs to control a robot. Although this direction is quite promising, there are many issues to be addressed before we have a complete automatic programming system. Such issues include: how to generate a sequence of operations; how to determine a grasp point for each operation; how to determine a global path through which to move an object while avoiding collisions with other objects. These, and other issues, make the development of a complete automatic programming system a considerable challenge.

2 Assembly Plan from Observation

We have been developing a novel programming method, the assembly-plan-from-observation (APO) method [11]. A human performs assembly operations in front of a TV camera. The APO system observes the human assembly actions, understands the actions, and generates robot programs to enable a robot arm to repeat the same assembly operations.

The APO paradigm assumes that there are three worlds: the human world to be observed, the APO internal world to be generated by the APO system, and the robot world in which a robot mimics human operations as shown in Figure 1.

The APO paradigm can be viewed as symbolic-level teleoperation. Traditional teleoperation systems connect the master and slave manipulators through a signal-level link. In the APO paradigm, perceptual information is recognized and converted into symbolic level representations, the internal world. The APO analyzes the internal world and generates abstract robot commands, such as insert-between. The APO system subsequently maps abstract robot commands to appropriate signal-level manipulator commands. Thus, the complete APO system consists of two main modules: task recognition (signal-symbolic link) to map the human world to the internal world, and task instantiation (symbolic-signal link) to map the internal world to the robot world.

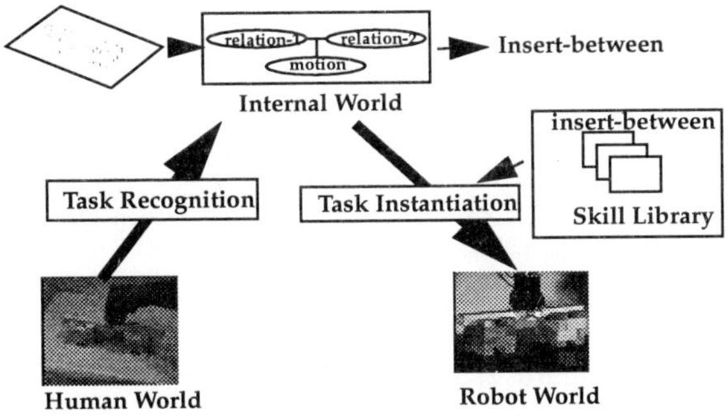

Figure 1. Assembly Plan from Observation.

2.1 Task Recognition

The primary purpose of an assembly task is to establish a particular face contact between the object manipulated by that task and other stationary environmental objects. To recognize tasks, we have created abstract task models that associate transitions of face contact relations during an assembly task and necessary assembly operations that cause such transitions. The APO system will execute the following sequence of actions in order to understand the human assembly operations:

- *Temporal Segmentation* --- Dividing the continuous sequence of images into meaningful phases, such as pre-grasp, grasp, and manipulation phases, that correspond to separate human assembly sub-actions.
- *Object Recognition* - Recognizing objects in each phase, extracting face contact relations in each phase, and determining face contact relation transitions between two consective phases.
- *Task Recognition* - Associating the extracted face contact relation transition with the abstract robot commands necessary to achieve such transition by consulting the abstract task models.

2.2 Task Instantiation

The task recognition provides an abstract command, such as pick-up or insert-into, that realizes the transition of face contact relation. Corresponding to each abstract command, a skill -- a sequence of robot motion templates -- is prepared in a skill library [25]. Each skill consists of a template of robot motions necessary to generate face contact transition. The skill also contains several empty slots to complete the template, such as grasping points or the destination of a manipulated object to complete the skill.

In run time, a skill corresponding to the actual abstract command generated by the task recognition is selected. Empty slots in a skill are filled based on the object recognition result, and actual robot motion commands are completed.

It is generally very difficult, if not impossible, for a robot to perform fine manipulation tasks without the benefit of some form of sensory feedback during actual task execution. Some tasks require visual-feed back to eliminate the uncertainty caused by the observation module. The basic questions in designing a skill library are: which skills require visual-feed back, and which do not. Certainly, it is waste of resources to employ visual-feed back in all skills.

Thus, the design issues on skills with visual feedback consist of the following four problems:

1. whether the visual feedback is necessary to perform the current skill;
2. which dimension is critical and should be observed during the skill;
3. which visual features carry such critical information; and
4. determine how to get such critical information with the sensors used.

Generally speaking, the aim of an assembly task is to reduce the degrees of freedom of a manipulated object with respect to the stationary environmental objects. Each assembly action in a task is aimed to further constrain the manipulated object against the environmental objects. Thus, when the task instantiated module is activated, the APO system knows which degrees of freedom will be reduced by realizing the corresponding robot commands. Section 3 reexamines the abstract task models and describes the method for deciding which skills need visual feedback based on the face contact relation transition during the task execution. This analysis also provides the critical dimensions to be observed. Section 4 presents a method to determine necessary features to be observed through the analysis of CAD models and the ciritical dimensions. Section 5 provides the optimal sensing strategy to observe such features, and Section 6 describes an implementation of the sensing strategies and their performance evaluations using a line laser range finder. Section 8 summarizess the report.

3 Which Skills Require Feedback

This paper assumes that a robot has the capability to perform passive compliant motions [19]. Under this assumption, this section examines which skills really require visual feedback, and which skills can be performed only with passive compliant motions. The analysis is based on face contact relation transitions between manipulated and environmental objects. First, we analyze polyhedral objects with three translational motions; then, we extend the analysis to include cylindrical objects with four dimensional, three translational and one rotational motions.

3.1 Face Contact Relations

We assume that each assembly operation involves one manipulated object and several stationary environmental objects that have face contacts with the manipulated object. Face contact relations are defined between the manipulated and the environmental objects.

Let us suppose a face of the manipulated object has a face contact to a face of an environmental object (See Figure 2). Each face contact pair constrains the possible translation motions of the manipulated object: $\delta T \bullet N \geq 0$ where δT denotes possible translational vectors of the manipulated object and N denotes the normal direction of an environmental face patch.

We use points on the Gaussian sphere to specify both a constraint vector and all possible motion vectors. Each motion unit vector is translated so that its start point is located at the center of the Gaussian sphere and its end point exists at some point on the surface of the Gaussian sphere. We use this end point to denote the vector.

The constraint given by a contact face pair defines several regions in the Gaussian sphere. We refer to the plane perpendicular to the normal, N, as the *constraint plane*; this plane divides the Gaussian sphere into two hemispheres. Assuming that the normal points to the north pole of the Gaussian sphere without loss of generality, the northern hemisphere corresponds to possible translational motion directions; the southern hemisphere corresponds to prohibited motion directions toward which the manipulated object cannot move.

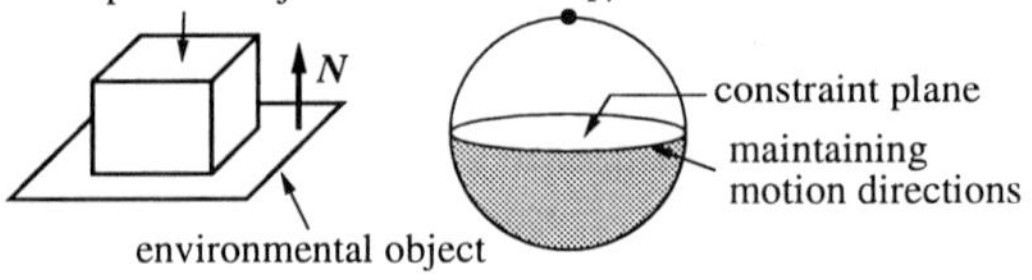

Figure 2. Constraint inequality on the Gaussian Sphere.

In Figure 2, motion directions corresponding to the boundary between the hemispheres (the equator) maintain the current face contact between the manipulated and environmental objects. Those directions and those degrees of freedom (DOF) are referred to as the *maintaining directions* and the *maintaining*

DOF, respectively. In this example, maintaining DOFs are two. Motions of the directions corresponding to the inside of the detaching hemisphere break the face contact, and is referred to as the *detaching* motion. In this example, a pure detaching motion, which does not contain any maintaining motion component, is toward the north pole; it has one degree of freedom (*detaching DOF*).

When several surface patches of different orientations make contact, possible motion directions are constrained through simultaneous linear inequalities. These constraints are represented as a combined region in the Gaussian sphere.

Figure 3 shows the case where two normal vectors of environmental objects have the opposite directions. The possible motion directions of the manipulated object can be represented as *the entire great circle* perpendicular to the axis connecting the two poles. These motions are maintaining motion, and those maintaining DOFs of these motions are two degrees. There are no detaching motions; the detaching DOF is zero. The direction along the surface normal is completely constrained; the degrees of freedom of the constraint directions (*constraining DOF*) is one.

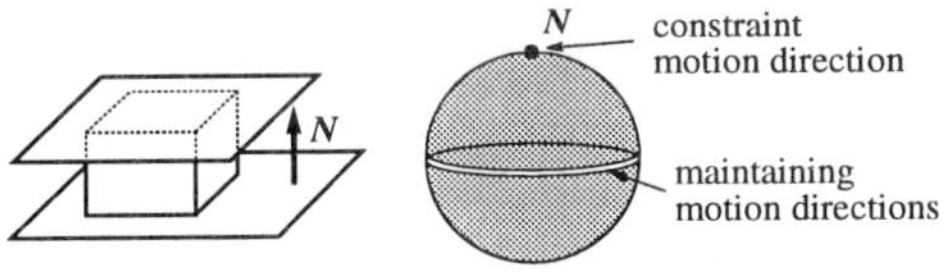

Figure 3. A bidirectional constraint.

We can specify face contact relations by using a triplet of maintaining, detaching, and constraining DOFs. Using this triplet, for example, the relations of Figure 2 and Figure 3 are represented as (2,1,0) and (2,0,1), respectively.

In general, possible contact relations are classified into ten contact relations as shown in Figure 4 [11]. The triplet of DOFs for each relation is also indicated in the figure. Note that the sum of the maintaining DOF, the detaching DOF and the constraining DOF is three, the entire DOF of three-dimensional translational motions.

Each assembly task causes a *transition* from one face contact relation to another. We extract *admissible transitions* between the ten contact states based on the following criteria.

Selection criteria for admissible transitions

- If a *direct detach* motion (a motion that immediately breaks a face-contact) exists, choose it.
- If a *lateral* motion (a motion that maintains the same contact state) that would break face-contacts by crossing a certain boundary exists, choose it.

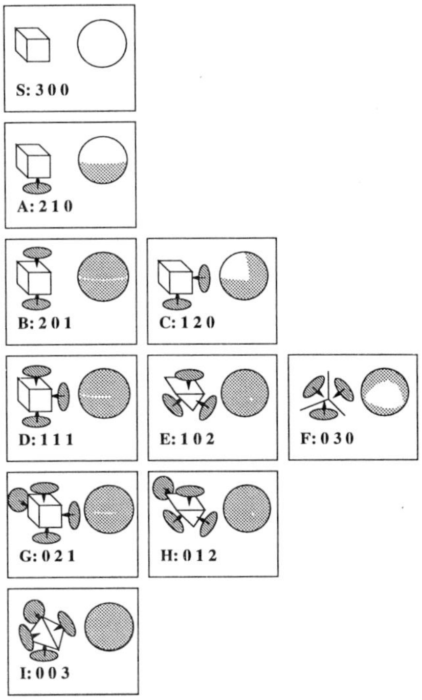

Figure 4. Ten contact states. The white areas in the Gaussian sphere denote possible motion vectors. The three digits denote maintaining DOF, detaching DOF, and constraining DOF.

- If several candidate motions satisfy criterion 1 or criterion 2, choose the motion that least reduces the number of face contacts.

The application of these criteria to the analysis of face contact relations results in the contact state transition graph as shown in Figure 5.

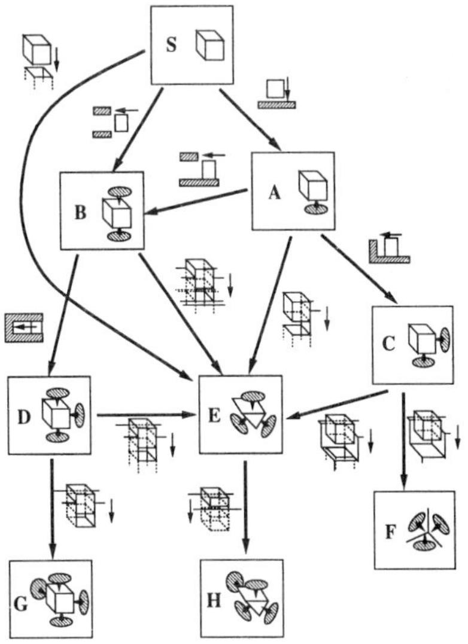

Figure 5. Contact relation transitions as a directional graph.

3.2 DOF Transitions

An assembly operation (i.e., a transition of a face

contact state) always increases constraints on some degrees of freedom of the manipulated object. This increment is classified into three classes: from maintaining DOF to detaching DOF; from detaching DOF to constraining DOF; and from maintaining DOF to constraining DOF. Figure 6 shows typical examples of the three classes.

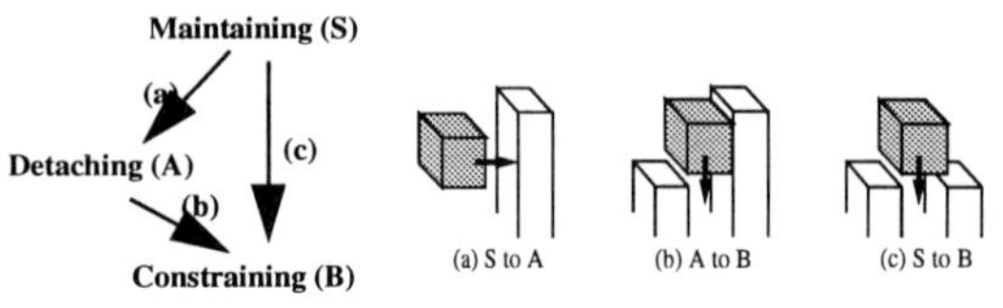

Figure 6. Three cases of DOF increment.

Let us examine how the type of the DOF transit in examples in Figure 6. In example (a), the horizontal operation belongs to class (a) operations that translate the maintaining DOF (horizontal freedom) into the detaching DOF. The approaching direction of the block is parallel to the direction of the pure detaching motion at the goal relation A, i.e., the normal vector against the wall. This class of operations can be performed by a compliant motion without visual feedback such as a *move-to-contact* operation.

In example (b), the vertical operation translates the detaching DOF of the horizontal direction to the constraining DOF. A compliant motion maintains the contact between the block and the right wall and achieves the desired horizontal position. The final constraining DOF is automatically achieved by this compliant motion, e.g., *move-with-maintaining-contact,* without visual feedback.

In example (c), the vertical operation translates the maintaining DOF of the horizontal direction to the constraining DOF. The horizontal position of the block needs to be adjusted before mating so that both the left and the right face contacts are achieved simultaneously. Before this operation, along the horizontal maintaining degree, a manipulated object has no physical contact; after this operation, the constraining degree from both walls occur simultaneously. Thus, simple compliant motion cannot achieve such translation. This operation needs visual feedback; this degree should be observed by visual feedback[1].

We can summarize the above arguments into the following criteria:

Selection criteria for visual feedback

- maintaining DOF to detaching DOF: no visual feedback is necessary
- detaching DOF to constraining DOF: no visual feedback is necessary
- maintaining DOF to constraining DOF: visual feedback is necessary

Applying these criteria to thirteen admissible transitions in Figure 5 provides four transitions that require visual feedback (S-to-B, S-to-E, A-to-E and B-to-E) indicated with bold lines in Figure 7.

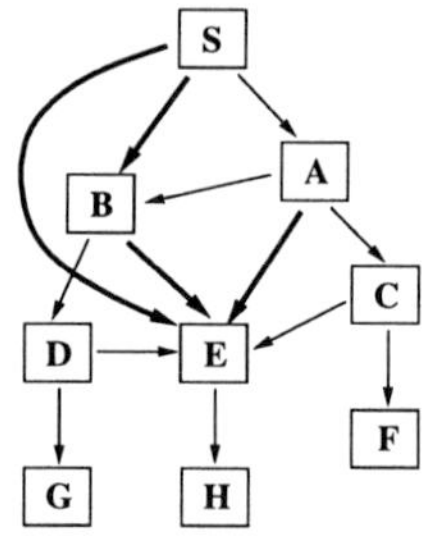

Figure 7. Classification of state transitions. Bold lines indicate the transitions that require visual information.

3.3 Extension of Face Contact State

So far the analysis covers only three-dimensional translation motions with planar face contacts. We will extend the analysis to include one additional rotational motion about the Z axis. This extended analysis denotes face contact relations with a sextuplet of DOFs, one triplet for translational DOFs and another triplet for rotational DOFs.

We also include cylindrical objects for the analysis. We define the three classes of contact relations between cylindrical surfaces as shown in Figure 8. No contact, half contact, and full contact in Figure 8 correspond, respectively, to maintaining DOF, detaching DOF, and constraining DOF with respect to the translational motion perpendicular to the axis of rotation.

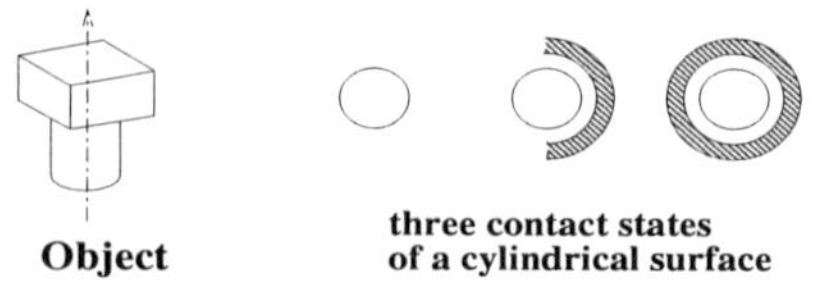

Figure 8. Three contact states for cylindrical surfaces.

This extension allows the system to handle a large class of realistic operations with industrial parts, including gears mating and volt-nuts operations. It is also true that our Robotworld testbed has these four degrees of freedom.

1. A sophisticated force control-based manipulation strategy may be employed to perform this kind of assembly operation without visual feedback. Even in such a case, reducing errors by visual information would be useful. For example, with the help of visual information, the number of motion steps could be reduced.

Figure 9 summarizes the analysis result. Here, the possible face contact relations are arranged according to their sextuplets of DOFs.

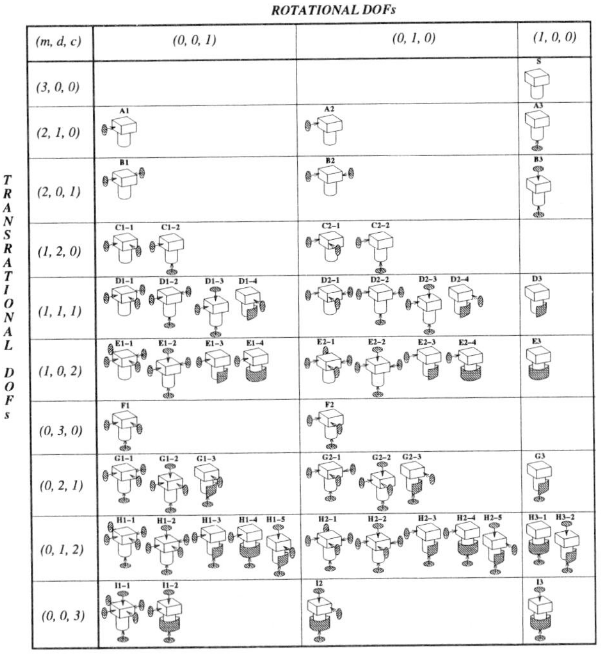

Figure 9. Summary of extended contact state analysis.

We enumerate possible transitions between contact relations. Eighty five admissible transitions are extracted, as shown in Figure 11, by applying the selection criteria for admissible transitions to all possible transitions.

We consider an infinitesimal rotation. Such an infinitesimal rotational motion is equivalent to an infinitesimal translational motion in terms of face-contact relation transition. We can use the same criteria for selecting skills that require visual feedback in this four-dimensional case.

By examining the sextuplet transitions in eighty five cases, nineteen transitions are selected as those that require visual feedback. These selected transitions are as shown with bold solid and bold broken arcs in Figure 11. We can group these nineteen cases into six groups with respect to the degrees of freedom necessary to be observed. Figure 10 shows the reprentative examples of these six groups.

4 Feature Selection

This section describes how to select a set of features that carry sufficient dimensional information for executing the operation with visual feedback resulting in the transition from maintaining DOF to constraint DOF.

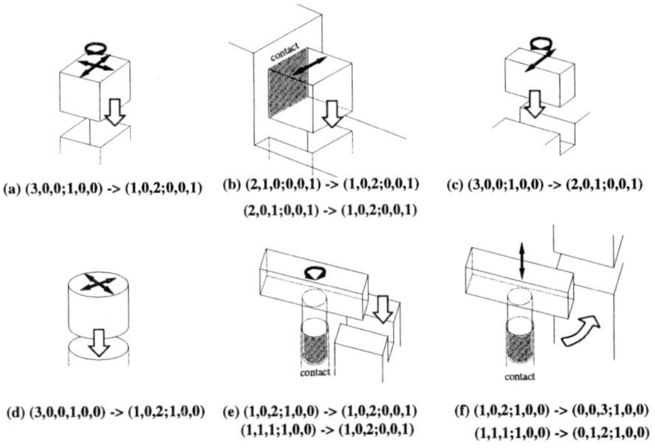

Figure 10. Transition groups which need visual information. Thick arrows indicate the direction of movement. Thin arrows indicate degrees of freedom to be monitered.

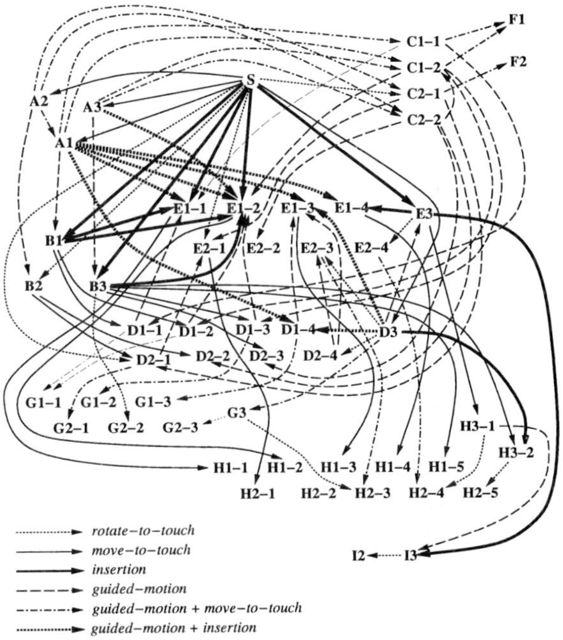

Figure 11. Transition graph for extended analysis.

4.1 Sensing Primitive

A sensing primitive describes the relationship between an observable feature, such as edges or planar faces, and its degrees of freedom obtainable by observing it. Examples of sensor primitives are shown in Figure 12. For example, by observing a straight line in 3D space, all the translation freedoms except those along the straight line are constrained. This constrained area, A_i^T , given by this primitive feature is denoted as the shaded area on the Gaussian sphere. All the rotations except that about the line are also constrained; the entire sphere except for a pair of points is constrained.

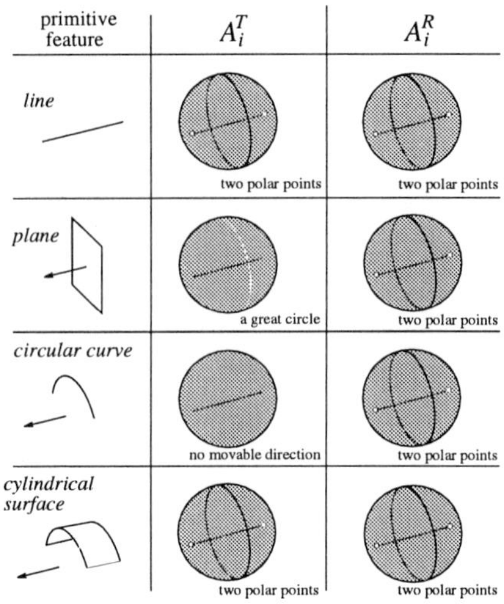

Figure 12. Example sensing primitives. Constrained area are represented as the shaded area on the Gaussian spheres.

4.2 Feature Selection Process

The analysis of face-contact relation transition by this current task provides constraining DOF added by this task. A set of observable features is obtained by examining current CAD-based internal-world model[2]. By consulting prepared sensing primitives with the observable features, a set of features required to be observed is obtained.

As an example of this process, let us consider case (a) peg-in-hole operation shown in Figure 13. For the sake of simplicity, we will consider the case (a)'s translation operations. Before the insertion operation, the manipulated object has a full three dimensional maintaining DOF, M_{before} . After this operation, the manipulated object has a equator of constraining DOF, C_{after} . The (a) insertion operation translates two dimensional maintaining DOF perpendicular to the insertion direction to two dimensional constraining DOF, $A_{curr} = M_{before} \cap C_{after}$ referred to as critical DOF. The DOF to be monitored with visual feedback is the two dimensional plane perpendicular to the insertion direction. Figure 13 (a) depicts the full equator on the Gaussian sphere to depict these DOF. This dimension requires visual feedback.

Before case (b) operation, the manipulated object has a two-dimensional maintaining DOF as a equator perpendicular to the contact face. The constraining DOF is the same as is the (a) operation. The resulting, $A_{curr} = M_{before} \cap C_{after}$, provides a pair of points on the equator as depicted in Figure 13 (b). The visual feedback is necessary along the X axis, but is not necessary along the Y axis.

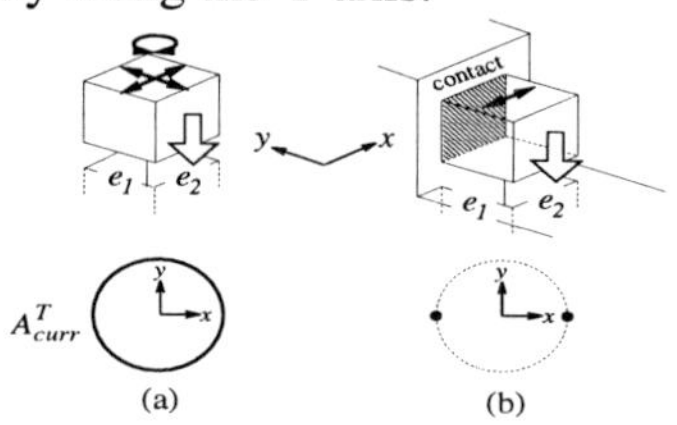

Figure 13. Two insertion operations.

Figure 14 shows two sensing primitives. By observing edge e_1 , all the translation motions, except those along the y axis are constrained. The intersection of constraint directions by this edge e_1 , $A_{e_1}{}^T$ with the previous equator is a full circle except a pair of points along the Y-axis as depicted Figure 14 (a). Edge e_2 also provide similar constraints. In order to eliminate the full circle of the critical DOF in case (a), both e_1 and e_2 are necessary to be observed. On the other hand, only e_1 is sufficient for case (b) to eliminate the pair of points of the critical DOF.

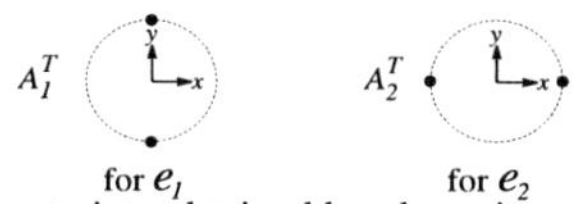

Figure 14. Constraints obtained by observing two edges: A_1 for edge e_1 and A_2 for edge e_2 .

This procedure can be formalized as follows: Since constraints for translational operations and that for rotational operations can be considered separately, we take two Gaussian spheres, G^T and G^R , and use points on each sphere to represent *movable directions* (on G^T) or *possible rotation axes* (on G^R) under a certain set of constraints. Let us denote:

- $A_{curr}{}^T\left(A_{curr}{}^R\right)$ as a set of points on $G^T\left(G^R\right)$ to be translated from maintaining DOF to constraining DOF (critical DOF).
- $A_i{}^T\left(A_i{}^R\right)$ as a set of points on $G^T\left(G^R\right)$ which represents constraints given by the observation of the i-th sensing primitive.

By observing n different sensing primitives, the following two resultant point sets, $A_{obsd}{}^T$ and $A_{obsd}{}^R$, are obtained:

$$A_{obsd}{}^T = A_{curr}{}^T \cap \left(\bigcap_{i=1}^{n} A_i{}^T \right)$$

In order that this set of features provides sufficient constraints, the resultant area should be null set. Namely, the following condition must be satisfied:

$$\Phi = A_{obsd}{}^T \quad \text{and} \quad \Phi = A_{obsd}{}^R$$

5 Evaluation of Sensing Strategies

Once a set of features to be observed is selected, a set

2. The object recognizer determines each object configuration in the real world and generates a CAD-based world model.

of possible sensor positions, from which all of the selected features are observable, is enumerated. We determine the optimum sensing position among those possible sensing positions by evaluating them with respect to accuracy of the estimated object position. Evaluation methods based on covariance matrix are relatively common in unconstrained environments such as navigation application [2], [28]. Such covariance methods are no so not satisfactory to reflect the relative importance of sensing dimensions given by the relationship between faces of manipulated and environmental objects. Thus, we decided to develop our novel method based on an operation's success probability.

5.1 Predicted Success Probability

A poor positional estimation of a manipulated object provides poor operational performance; an accurate estimation provides high performance. Thus, the predicted success probability of a current operation provides a measure of accuracy of the sensing strategy.

The success probability is determined by three steps. The first step calculates a success region, the sub-space of the positional parameters in critical DOF such that if the uncertainty is inside the space, the current operation succeeds. The rectangular area in Figure 15 is an example of the success region. The relative tolerance of the manipulated and environmental objects along the critical dimensions, given by the analysis of CAD models, provides the area. The second step calculates the uncertainty distribution of the measurements given by the current sensing strategy (the ellipse area in Figure 15). The final step obtains the predicted success probability by the intersection between the success region and the uncertainty distribution of the measurements.

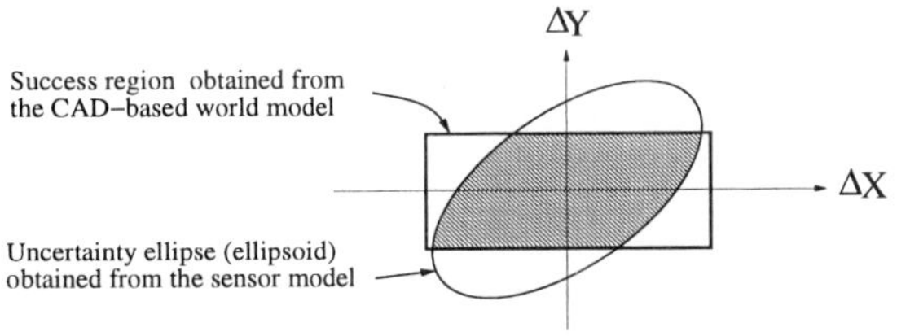

Figure 15. Calculation of the predicted success probability. This figure shows the case where the uncertainty in the position parameter of the object is two-dimensional, $(\Delta X, \Delta Y)$.

After calculating the success probabilities for all feasible sensing strategies, the one with the highest probability is selected as the optimal sensing strategy.

5.2 Success Region

A success region is formed in a space of critical DOF, those translated from maintaining to constraining DOF, in each operation. For example, the (a) peg-in-hole operation in Figure 13 has three degrees of freedom of critical DOF, two translational on the plane perpendicular to the insertion direction and one rotational around the insertion direction.

A success region is a representation of the clearance of the operation in the critical DOF from the CAD models of objects. Since the current system consists of the assembly operations with planar or cylindrical surfaces, the following three cases are sufficient for consideration:

- Case 1: insertion of a polygonal cross-section peg with a hole (cases (a) and (b) in Figure 9). In this case, for each pair of vertices of a peg and a hole, two inequalities are generated; if the peg vertex is convex, inequalities are generated between the peg vertex and the two edges crossing at the hole vertex; if the peg vertex is concave, inequalities are generated between the hole vertex and the two edges crossing at the peg vertex. Figure 16 shows the two possibilities; V , L_1 and L_2 indicate the vertex and the edges used for generating inequalities.

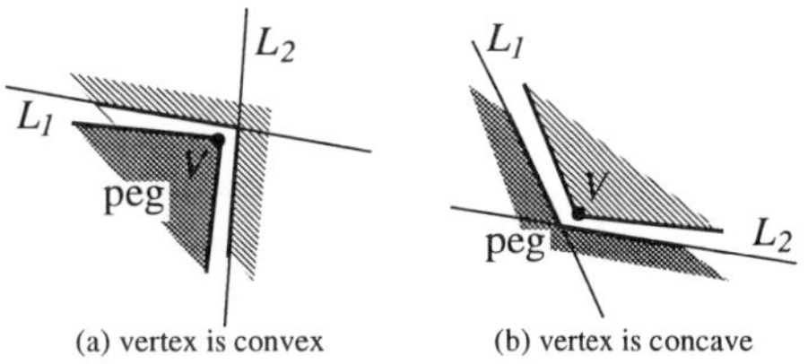

Figure 16. Vertex (V) and edges (L_1 and L_2 used for generating inequalities in the case of insertion of a peg with a polygonal cross-section into a hole.

- Case 2: insertion of a peg with a polygonal cross-section into a parallel gap (cases (c), (e) and (f) in Figure 9). In this case, for each pair of a vertex and an edge, one inequality is generated. Figure 17 shows two possibilities; V and L indicate the vertex and the edge for generating an inequality.

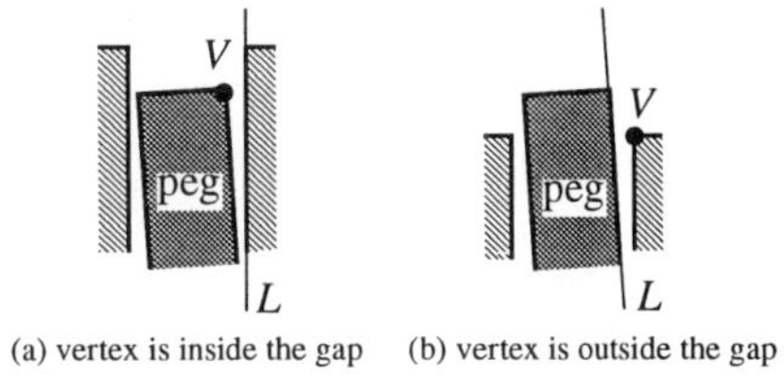

Figure 17. Vertex (V) and edge (L) used for generating an inequality in the case of insertion of a peg with a polygonal cross-section into a parallel gap.

- Case 3: insertion of a peg with a circular cross-section into a hole (case (d) in Figure 9). In this case, the success region is represented as a circle in a two-dimensional space composed of two translational DOFs; the radius of the circle is the difference between the radius of the hole and that of the peg.

The above calculations of the success region deal with a general case, that is, the case where there are no constraints on the peg before the operation. If there are some constraints before the operation e.g., case (b) in Figure 9, the actual success region is given as a cross-section of the general success region cut by fixed parameter values under the constraints.

6 Experiments

6.1 Laser Range Finder

The proposed sensing method has been implemented using a Toyota line laser range finder (LRF) as the sensor (see Figure 18) [21]. The LRF emits a slit laser, detects the highlighted portion of the object with a TV camera, and obtains a line of 3D measurements. The LRF is attached to one of the arms in the Robotworld. All the arms have four degrees of freedom: three degrees of freedom for translation and one degrees of freedome for rotation about the vertical axis.

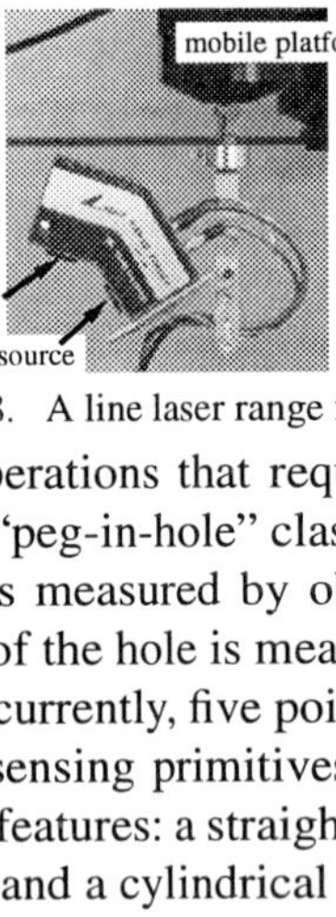

Figure 18. A line laser range finder.

All assembly operations that require visual feedback belong to the "peg-in-hole" class operation. The location of a peg is measured by observing its side faces; the location of the hole is measured by observing several points (currently, five points) on its edges. Thus, we prepare sensing primitives for the following four geometric features: a straight edge, a circular edge, a planar face and a cylindrical surface.

We use a general sensing strategy that measures by moving the LRF in parallel with the insertion direction; the relative displacement on the plane perpendicular to the insertion direction is important for the operation. We also control the position of the range finger so that each measured point is within a certain area of the slit laser; the uncertainty of the measurement with the LRF is considered to be constant in this area. Thus, the only parameter that specifies the position of the range finder is the angle φ between the direction of the laser and some axis of the plane perpendicular to the insertion direction.

The actual skills are implemented to performed a "stop and sense" mode operation. First, a peg is moved by a manipulator to the position just before a hole. Then, the LRF is placed in the planned position, and measures the position of the hole and the peg. If the error in the relative position between the peg and the hole is within the success region, the peg is inserted. Otherwise, the peg position is adjusted and the peg is observed again. This final step is repeated until the relative position becomes satisfactory, and then the peg is inserted.

For evaluation of the skills, we conducted three operations: a peg-in-hole operation, an operation that consisted of putting the tip of a screwdriver into the slot of a bolt head, and a gear-mating operation. We then compared the success probability predicted by the object models and sensor models, with the actual success ratio, obtained through fifty trials of the operation by the robot arm and the sensor.

6.2 Uncertainty Model

Our laser range finder provides quite accurate measurements: less than 0.1 mm in depth and less than 0.3 mm in the horizontal position. The purpose of this report is, however, not to construct an uncertainty model of our laser range finder, but to demonstrate that our method can generate the optimal sensing strategy if the uncertainty model of the sensor is given. Thus, we artificially added a relatively large Gaussian noise to the measurement; we added a Gaussian noise with a standard deviation 0.12 mm to the depth measurement, and a Gaussian noise with a standard deviation 0.30 mm to the horizontal position measurement; These two Gaussian noises are set to be independent of each other.

6.3 A Peg-in-Hole Operation

6.3.1 *Face Contact Analysis and Actual Operation*

The peg-in-hole skill inserts a rectangular-cross-sectional peg into a hole of the same shape. This operation belongs to group (a) in Figure 9; the state transition is from (3,0,0;1,0,0) to (1,0,2;0,0,1). The critical DOF consists of two translational and one rotational degree. These dimensions will be monitored through visual feedback.

Let us consider the case in Figure 19. This insertion operation establishes the following face contact: $f_1-f'_1$, $f_2-f'_2$, $f_3-f'_3$ and $f_4-f'_4$. The candidate features for observation include f_1, f_2, f_3 and f_4 for the peg, and e'_1, e'_2, e'_3 and e'_4 for the hole. The sensing primitive analysis indicates that observing two neighboring faces, such as f_1 and f_2, and edges, such as e'_1 and e'_2, provides sufficient information.

Considering the conditions that five points are completely observed on an edge, and that the LRF does not collide with the robot manipulating the peg, the possible position of LRF is on the circle. The circle's center is located at the vertex at the intersection of the two neighboring edges. The LRF is pointing toward the vertex. The position of the sensor is measured as the angle (φ) from the x axis in the Robotworld testbed.

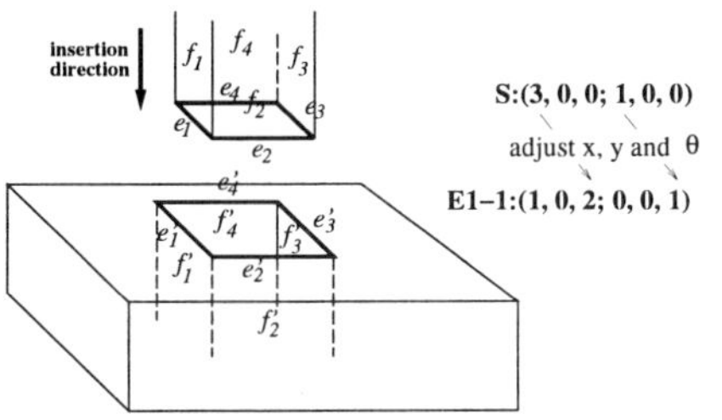

Figure 19. Face contact analysis of the rectangular peg-in-hole operation. The triplet o DOF's changes from (3,0,0) to (1,0,2).

6.3.2 *Evaluation of Predicted Success Probability*

We verify the accuracy of the predicted success probability using the actual success ratio in the following two sets of the objects:

- Case (a): The cross-section of the peg is a square of 19.05 mm X 19.05 mm. The clearance ratio of the hole is 0.043.
- Case (b): The cross-section of the peg is a rectangle of 38.1 mm X 19.05 mm. The clearance ratio of the hole is 0.043.

In each case, the observation angle, φ, is sampled every 10 degrees; at each angle, the insertion operations were performed 50 times to obtain the ratio between success and failure[3].

Figure 20 shows the verification results; in both cases, the results given by actual performances (dots) coincide with the predicted success probability (solid curves).

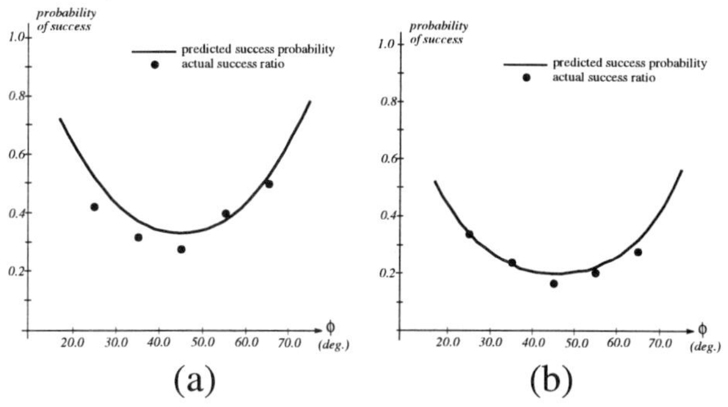

Figure 20. Comparison of the predicted success probability with the actual success ratio. The success ratio for each viewing angle ϕ was obtained by repeating the same operation 50 times by the RobotWorld, and by counting the numbers of success and failure of the operation.

6.4 Putting Screwdriver on Bolt

This example considers the insertion of a screwdriver into the slot on a bolt head as shown in Figure 21. This operation belongs to group (c) in Figure 9. During this operation, the critical DOF are one DOF for translation and one DOF for rotation. The face contacts to be achieved are $f_1 - f'_1$, and $f_2 - f'_2$. The candidates for observed features are f_1, f_2, f_3 and f_4 for the screwdriver, and e'_1, and e'_2 for the hole.

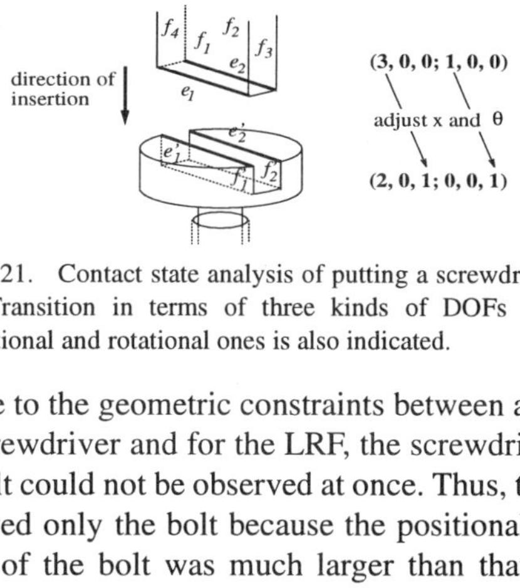

Figure 21. Contact state analysis of putting a screwdriver on a bolt. Transition in terms of three kinds of DOFs for both translational and rotational ones is also indicated.

Due to the geometric constraints between arms for the screwdriver and for the LRF, the screwdriver and the bolt could not be observed at once. Thus, the LRF observed only the bolt because the positional uncertainty of the bolt was much larger than that of the screwdriver. Edges e'_1, and e'_2 of the bolt were observed. Figure 22 shows a successful operation of putting a screwdriver on a bolt.

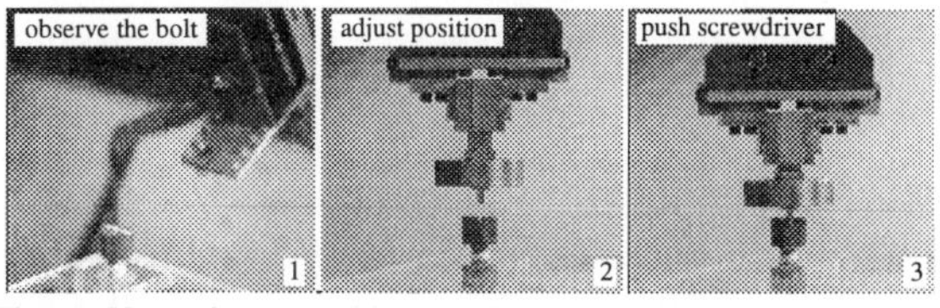

Figure 22. The screwdriver was successfully inserted into the slot of the bolt head.

3. Since the number of the data points for the peg is much larger than that for the hole, the uncertainty of the peg position is small (about 100 times more accurate) compared to that of the hole position. Thus, we neglected the uncertainty of the peg position and considered only the uncertainty of the hole position in calculation of predicted success probabilities.

Figure 23 shows the verification result of the predicted success probability with the actual success ratio. The predicted success probability coincides with the actual success ratio.

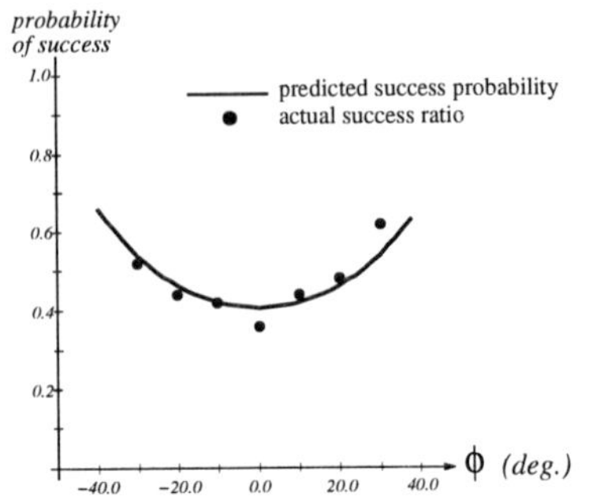

Figure 23. Comparison of the two success probabilities in the screwdriver-bolt operation. The angle ϕ indicates the relative angle between the direction of the slot and the viewing direction of the range finder. In each viewing angle, the same operation was repeated 50 times to obtain the success ratio.

6.5 Gear Mating

A gear-mating operation, shown in Figure 24, belongs to group (e) in Figure 9. In this operation, *a priori* knowledge about how gears are mated is necessary because there are many potential matches between gear teeth. First, two virtual edges e_1 and e'_1 are generated; one edge is placed on the center of the nearest tooth to the line connecting two gear centers; another edge is placed on the center of the nearest gap to the line. Then, the orientation of the inserted gear is adjusted so that these two virtual edges are aligned.

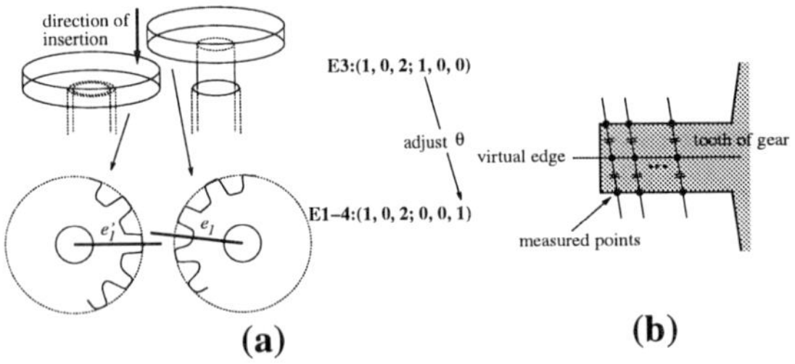

Figure 24. Gear mating. (a) Contact state analysis for gear mating; (b) Measuring the tooth position from edge positions.

The position of a virtual edge is calculated from the position of the edges on the tooth (or gap) on which the virtual edge is set (see Figure 24). Assuming that the shape of the tooth is almost rectangular, the virtual edge is obtained by fitting a line of the center points of pairs of edge points.

Figure 25 shows a successful gear-mating operation. Figure 26 shows the comparison of the predicted success probability with the actual success ratio. The predicted success probability is consistent with the actual success ratio.

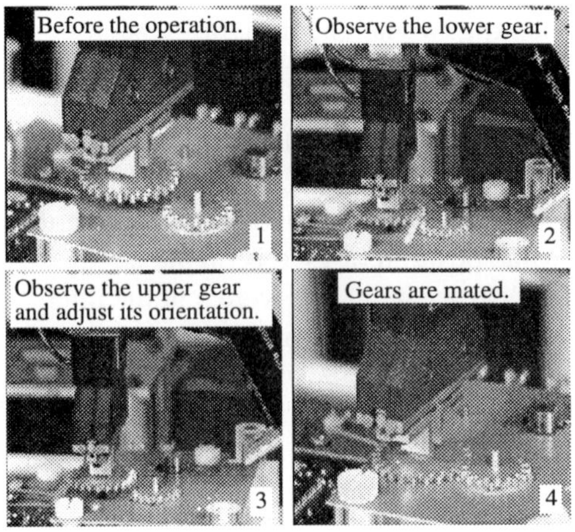

Figure 25. The gears were successfully mated.

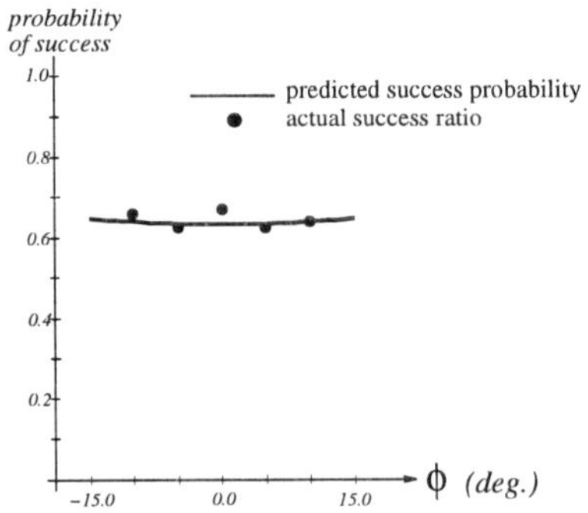

Figure 26. Comparison of the predicted success probability and the actual success ratio in the gear-mating.

7 Summary

7.1 Conclusions

We have described a method of systematically generating visual sensing strategies for a skill library. The analysis of face contact transitions during an observed assembly operation decides 1) whether a skill to realize such an assembly operation requires visual feedback, and 2) in which, if any, degrees of freedom (critical DOF) a skill is required to monitor the operation through visual feedback, if any. Then, sensing primitives convert the critical DOF and the CAD models into a set of visual features to be observed. The final sensing strategy selected is the one with the highest predicted probability among possible sensing strategies that observe the set of visual features. The proposed skills are implemented using a line laser range finder as the sensor. The experiments are conducted to verify our evaluation method for choice of the optimal sensing strategy.

7.2 Related Work

7.2.1 *CMU APO Systems*

Ikeuchi and Suehiro have developed a system called the Assembly Plan from Observation (APO) system [11]. The system observes a human performing an assembly task and recognizes object relations and relation transitions. Then, it maps relation transitions to assembly tasks that will initiate such relations. The relations are defined with respect to face contacts between the just manipulated object and environmental objects [17], [20].

The original system could analyze only translation operations of polyhedral objects. Ikeuchi, Kanade, and Suehiro extended the system to include curved surfaces such as spherical and cylindrical surfaces [12]. Paul and Ikeuchi proposed the extension of the translational analysis to rotational analysis using the screw theory [22]. Hirukawa, Suehiro and Ikeuchi are working on the full six dimensional analysis.

Visual information obtained by observation sometimes contain noise. The APO system collects two different type of information from observation: face contact relation and motion parameters. Even when the observation contains noise, face contact relations are usually identified correctly, because the system devides all possible relations into ten representative relations. On the other hand, the motion parameters are direct conversions of the observed date; they are noise-contaminated; the robot, if it utilizes such noise-contaminated motion parameters, will fail the performance of the observed operation. Suehiro and Ikeuchi have developed a method to clean up noise-contaminated motion parameters from the correctly obtained face contact relations [26]. This demonstrates one of the strengths of symbolic level teleoperation.

The original system stores predetermined grasping strategies of an object with the geometric model of the object. Kang and Ikeuchi have been analyzing image sequences obtained during human assembly operations for obtaining the grasping strategy from observation. The research issues include: how to segment a sequence of images into pre-grasp, grasp, and manipulation phases [13], how to recognize human grasping strategies [14], and how to map such recognized grasping strategies into a robot grasping strategy [15].

7.2.2 *Other APO Related Systems*

The APO paradigm is related to several bodies of research. Kuniyoshi, Inoue and Inaba developed a system that tracks movements of a human hand for program generation [16]. Since this system does not employ the knowledge obtained by analyzing geometric models of objects, it is restricted to simple pick-and-place operations.

Asada and Hirai proposed a system to recognize the state of a series of assembly operations at a given point in time based on face contact relations between geometric models [1]. They derive a taxonomy of contact states from all the possible face contact relations. In order to estimate the state of the assembly process, force and displacement information acquired in the process is interpreted using the taxonomy.

Hirai and Sato described a system which is capable of recognizing manipulator motions in order to maintain the consistency between an internal world model and the real world. Their primary goal, however, is to monitor assembly processes [8].

7.3 Visual Sensing Strategy

The work reported in this paper is directly related to the research on determining visual sensing strategies [6], [7],[16], [27].

Hutchinson and Kak dealt with the problem of resolving the ambiguity of sensor information. They used the Dempster-Shafer theory to represent the uncertainties of hypotheses in object identification. An entropy of a set of hypotheses was used as a utility function; a sensor placement which minimizes the entropy [9], [10] was selected.

Rimey presented a framework of task-oriented vision which can solve high-level vision problems such as determining the succession of objects to be search in order answer a query [23]. The knowledge of the task is represented by a Bayesian network, and the sensing action which has the highest expected utility is selected. The utility function is defined as the combination of the predicted information value and the sensing cost.

Birnbaum et al. presented a vision system which can explain a scene of a block world in terms of the stability of block structures. Using the rules derived from causal knowledge of naive physics, the focus of attention is changed to look for evidence that explains the situation [5].

Acknowledgment

The authors would like to thank Marie Elm and Sing Bing Kang for helpful comments on the draft of this report and Masato Kawade for the experimental setup. This research was conducted at the Task-Oriented Vision Laboratory, The Robotics Institute, Carnegie Mellon University.

Reference

[1] Asada H, Hirai S. Toward a symbolic-level force feedback. In: Miura H, Arimoto S (eds) Robotics Research, vol 5, MIT Press, 1990, pp341-346

[2] Ayache N, Faugeras OD. Maintaining representations of the environment of a mobiel robot. IEEE R&A 1989; RA-5(6): 804-819

[3] Bajcsy R. Active perception. Proceedings of IEEE 1988; 76(8)

[4] Ballard DH. Reference frames for animate vision In: Proc. IJCAI, 1989, pp1635-1641

[5] Birnbaum L, Brand M, and Cooper P. Looking for trouble: using causal semantics to direct focus of attention. In: Proc. ICCV, 1993, pp49-56

[6] Cowan CK. Automatic camera and light-source placement using CAD models In: Proc. IEEE Workshop on Direction in Automated CAD-Based Vision, 1991, pp22-31

[7] Gremban KD, Ikeuchi K. Planning multiple observations for object recognition. Intern. Journal of Computer Vision 1994; 12(2-3): 137-172

[8] Hirai S, Sato T. Motion understanding for world model management of telerobot. In: Proc IROS, 1989, pp124-131.

[9] Hutchinson SA, and Kak AC. Planning sensing strategies in a robot work cell with multi-sensor capabilities. IEEE Trans R&A 1989; RA-5(6): 756-783

[10] Hutchinson SA, Kak AC. SPAR: a planner that satisfies operational and geometric goals in uncertainty environments. AI Magazine 1990; 11(1): 31-61

[11] Ikeuchi K, Suehiro T. Task recognition with polyhedral objects. IEEE Trans. R&A 1994; 10(3): 368-385

[12] Ikeuchi K, Kawade M, Suehiro T. Assembly task recognition with planar, curved, and mechanical contacts. In: Proc. ICRA, vol 2, 1993, pp 688-694

[13] Kang SB, Ikeuchi K. Determination of motion breakpoints in a task sequence from human hand motion In: Proc. ICRA, 1994, pp 551-556

[14] Kang SB, Ikeuchi K. Toward automatic robot instruction from perception - recognizing a grasp from observation. IEEE Trans. R&A 1993; 9(4): 432-443

[15] Kang SB, Ikeuchi K. Mapping human grasps to manipulator grasps. In: Proc IROS, 1994, pp 97-14

[16] Kuniyoshi Y, Inoue H. Qualitative understanding of ongoing human action sequences. In: Proc. IJCAI, 1993, pp 1600-1609

[17] Laugier C. Planning fine motion strategies by reasoning in the contact space. In: Proc ICRA, 1992, pp 653-659

[18] Lozano-Perez T, Mason MT, Taylor RH. Automatic synthesis of fine-motion strategies for robots. In: Brady and Paul (eds) Robotics Research, vol 1, MIT press, 1984, pp 65-96

[19] Mason MT. Compliance and force control for computer controlled manipulation. IEEE Trans SMC 1981; SMC-11(6): 418-432

[20] Mello LSH, Sanderson AC, A correct and complete algorithm for the generation of mechanical assembly sequences. In: Proc ICRA, 1989, pp 56-61

[21] Ozeki O, Higuchi K, Yamamoto S. Automated dimension inspection system for automotive plastic parts with a Laser probe. In: Proc ROBOTS 12, vol 5, 1988, pp51-60

[22] Paul GV, Ikeuchi K. Partitioning contact-state space using the theory of polyhedral convex cones In: Proc ICRA, 1995, pp 421-426

[23] Rimey RD. Control of selective perception using bayes nets and decision theory. TechReport 468, U. Rochester, 1993

[24] Sakane S, Sato T, Automatic planning of light source and camera placement for an active photometric stereo system In: Proc ICRA, 1991, pp 1080-1087

[25] Suehiro T, Takase K, (1989): "Representation and control of motion in contact and its application to assembly tasks," In: Miura H, Arimoto S (eds) Robotics Research, vol 5, MIT press, 1989, pp 367-374

[26] Suehiro T, and Ikeuchi K, Fine localization based on face contact relations In: Proc IROS, 1992, pp 2095-2102

[27] Tarabanis KA, Allen PK, Tsai RY. A survey of sensor planning in computer vision. IEEE Trans R&A 1995; 11(1): 86-103

[28] Whaite P, Ferrie FP. From uncertainty to visual exploration. IEEE Trans PAMI; 13(10): 1038-1049

PART 8

MULTI-ROBOT INTERACTING SYSTEMS

SESSION SUMMARY

Georges Giralt
L.A.A.S./C.N.R.S., 7 Avenue du Colonel Roche, 31077 Toulouse Cedex, France
email: giralt@laas.fr

Multi-robot systems attract a growing interest both for the promising and difficult research issues they arise and for their real world application impact.

The most developed and mature aspect of the subject is undoubtedly the coordinated control of two arms to handle heavy, flexible or fragile objects.

The session discarded this important domain and was set in the perspective of the very open and broad spectrum of issues involved when several robots, typically a group of robots including platform mobility capacities, interact and cooperate to carry out a task or a set of tasks.

Let us assume that the multi-robot system is endowed with some appropriate level of *decisional and operational autonomy*. A first categorization relates then to the way tasks are dealt with:

i) Tasks can be *assigned* at a given time by a *higher level agent*: Human Operator, Central Station...

ii) They can be precisely *defined* at *system design* level and be automatically handled by it as asynchronous events.

iii) They are in part or entirely *implicitly* defined and thus only *emerge* throughout an *experimentation and learning* process.

According to the role played by the individual agent, i.e. the elementary machine, several research streams have to be considered:

i) Every robot corresponds to a full breadth, stand alone machine. Related research work is quite often targeted towards real-world applications.

ii) Simple machines or agents possessing limited capacities that have to be integrated into a complex multi-agent system to achieve meaningful tasks. Three, quite different in fact, research orientations can be identified:

- Individual agents are machine modules whose integration into a complex multi-robot system is engineered at design level. Related work has been done by P. Khosla and its coworkers at CMU, and very recently, by the proponents community of the RISC concept which is, clearly, manufacturing applications oriented.

- Every single element is considered as a cell and as such it has to be mechanically and informationally part of a *Cellular Robotic System* to become fully operative. A central objective in this stream of research is to divise the individual cell mechanisms and functionalities so that they can *self-organize* to dynamically match the requirements of a given task. Related exploring research work is being carried on by T. Fukuda and his group in the project CEBOT at Nagoya University.

- Mobile multi-robot systems in the framework of *Swarm Robotics*. Here too, every individual agent, i.e. machine, has very limited perceptual, decisional, and operational capacities. The objective is, in order to carry efficiently on – or just be able tõ– a meaningful task, to have a *collective behaviour* emerging from individual behaviours and dynamic interaction. This research domain is covered by a large community mostly of which concentrates in exploratory work at the interplay of biological and technological systems.

The three papers presented in this session do not cover, of course, all the scope and the research topics of the domain. Yet, they provide, in their diversity, very interesting insights on basic technical and scientific issues definitely situated at the very frontier of the field and, beyond, of Advanced Robotics.

Although several interesting questions related to other subjects were asked and discussed, only issues and matters directly concerned with the session central theme will be briefly reported.

The first paper, "A Multi-Robot Cooperation Scheme based on Incremental Plan-Merging" by R. Alami, stems directly from a large-scale real-world application. Rooted in the context of a set of cargo transhipment tasks assigned by a central station to a fleet of autonomous mobile robots, the paper presents and discusses a generic decisional scheme for multi-robot cooperation based on an incremental and distributed plan merging process.

Within the framework of the Plan-Merging Paradigm, each robot, autonomously and incrementally builds, refines and executes its own plans , to carry on an assigned task possibly including temporal or other constraints, taking into account the interaction with the other agents.

Individual agents possess, besides decisional and operational autonomy, communication capacities both with a Central Station agent and, locally, with other neighbourhood agents.

In the discussion, several central questions were addressed, related to planning and execution coordination in the proposed incremental and interaction driven mode. It can be summarized as follows:

First, regarding the possibility to predict and to handle future deadlocks as well as to use backtracking when plan merging fails, the answer was to recall the key features of the approach which purposefully limit coordination within time and space bounds, thus, furthermore providing for deadlock detection and handling.

A similar comment was made regarding the possibility to back-out to a safer place from where to go forward which can be proceeded with either as nominal re-planning or in the worse case, as exception handling.

Finally, the critical issue is certainly to define on which conditions the paradigm will work in general. It can be contended that it will work, as indeed it does in the real-world case presented, for well designed and well operated application cases where shared resources, such as space, as well as constraints, such as time constraints, allow for some flexibility and that no artificial overall optimality criterium is introduced. One can see that this characterizes a system in which local on-line decisions are reasonably well enough decoupled.

Future work will attempt to a more general and certainly more formal characterization aiming to several real-world extensions.

The second paper, "Behaviour Matching by Observation for Multi-Robot Cooperation" by Y. Kuniyoshi, presents a conceptual and exploratory research work supported with laboratory experimentations. Set in the research stream of decentralized tacit cooperative multi-robot systems, usually developed in the framework of a few somewhat stereotyped and very simple tasks, it aims to reach a higher level of generality and to include richer and more complex patterns of cooperative behaviours. The proposed approach for action selection, i.e. task oriented cooperative behaviours, which integrates some level of mainly static shared knowledge and explicit communication, relies on the central concepts of robot observed behaviour and resource identification.

The discussion covered some technical perception related points and focused on such key aspects of the approach as behaviour identification and behaviour initialization, or shifting, to carry on a complex cooperative task.

It can be argued that for many meaningful tasks intention or purpose, possibly represented by some behaviour sequence, must also be known to a Robot to interfere positively, i.e. to engage in a cooperative task. Identification by observation only, already quite demanding for an isolated behaviour, calls necessarily then for repetitive task patterns.

Y. Kuniyoshi further commented on the key issues central to his current research work of general behaviour identification as well as on the need and role of more structured control, i.e. explicit commands and information exchange for complex tasks.

The third paper, "Force Strategies for Cooperative Tasks in Multiple Mobile Manipulation Systems" by O. Khatib et al. described an ongoing effort to develop mobile manipulation capabilities oriented to satisfy the needs of new non-manufacturing application, a generic example being given in the paper, related to Draywall construction.

Two or more holonomic mobile platforms are equiped with robot manipulator arms ($\geq$ 6 d.o.f.) and will interact by physical contact in cooperative tasks such as large objects handling/transportation.

The contribution builds on previous generic results obtained by O. Khatib and his group related to task oriented robot motion and force control. Two salient features characterize the development which is proposed:

i) A unique system integrates and combines the high bandwidth and low range motion of the manipulator arm with the low dynamics and large range motions of the mobile platform.

ii) A decentralized control structure based on the *augmented object* and *virtual linkage* models.

The discussion besides some technical comments related to the operational space approach on which the work presented is grounded attempted to clarify the issues of:

i) Safety, since such a system is meant to operate among – or to cooperate with – people. This remains an important and open issue.

ii) Saturation of some of the high bandwidth d.o.f. Here it was argued that the movements of a multiple mobile-platform system in a dynamic environment may enter into this kind of behaviour for one or more of the arms – for instance when platforms attempt to avoid a collision – thus possibly leading to a breakdown of the decentralized control. O. Khatib countered that the set of methods implemented built a coordinated system intrinsically robust for all regular operational cases. But, of course, exception states need to be further investigated.

In summary, the three contributions presented, stemming from very different research streams, cover a broad, yet incomplete, set of issues related to the session theme. Thus, they offer a quite satisfactory multiple viewpoint assessment of leading-edge research directions in multi-robot cooperative systems.

In fact, beyond the interest itself of some of the results that the speakers have obtained, it can be contended that what really matters the most are the questionable aspects that were and still remain, object of debate and, therefore, of future developments.

A Multi-Robot Cooperation Scheme Based on Incremental Plan-Merging

Rachid ALAMI
LAAS / CNRS
7, Avenue du Colonel Roche
31077 Toulouse - France
e-mail: rachid@laas.fr

1 Introduction

We present and discuss a generic cooperative scheme for multi-robot cooperation. It is based on an incremental and distributed plan-merging process.

Each robot, autonomously and incrementally builds and executes its own plans taking into account the multi-robot context.

The robots are assumed to be able to collect the other robots current plans and goals and to produce plans which satisfy a set of constraints that will be discussed.

We discuss the properties of this cooperative scheme (coherence, detection of dead-lock situations) as well as the class of applications for which it is well suited.

We show how this paradigm can be used in a hierarchical manner, and in contexts where planning is performed in parallel with plan execution.

We also discuss how this paradigm "fills the gap" between (centralized or distributed) goal / task allocation and distributed task achievement[1]

We finally illustrate this scheme through an implemented system which allows a fleet of more than ten autonomous mobile robots to perform load transfer tasks in a route network environment with a very limited centralized activity, and important gains in system flexibility and robustness to execution contingencies.

[1]We use here the term "task achievement" instead of "task execution" in order to emphasize the fact that task achievement involves further (context-dependent) task refinement and sensor-based plan execution.

2 Two multi-agent cooperation issues

In the field of multi-agent cooperation we distinguish two main issues:

- **C1:** the first issue involves goal/task decomposition and allocation to various agents
- **C2:** the second issue involves the simultaneous operation of several autonomous agents, each one seeking to achieve its own task or goal.

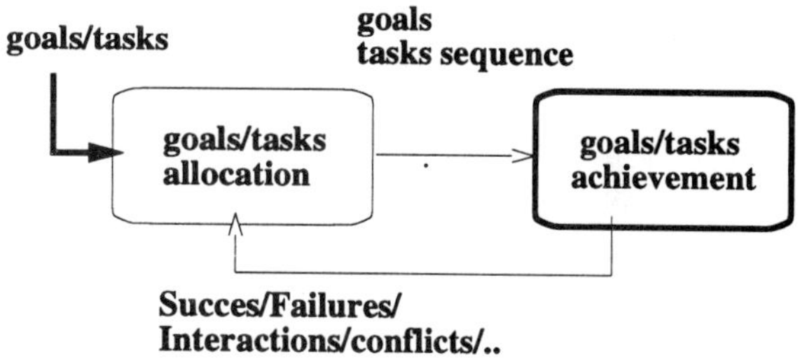

Figure 1: The two main issues in multi-agent cooperation

While several contributions have concentrated more particularly on one issue or the other, we claim that in numerous multi-agent applications, both issues appear and even "invoke" one another "recursively".

This is particularly true for autonomous multi-robot applications and, more generally, when the allocated tasks or goals cannot be directly "executed" but require further refinement.

The case of multi-robot applications is indeed directly concerned because the robots act in a same physical environment and because of the multiplicity of uncertainties.

Let us assume a set of autonomous robots, which have been given, through a centralized system or after a distributed process, a set of (partially ordered) tasks or goals. One can reasonably assume that this process is finished when the obtained tasks or goals have a sufficient range and are sufficiently independent to cause a substantial "selfish" robot activity. However, each robot, while seeking to achieve its task will have to compete for resources, to comply with other robots activities. Hence, several robots may find themselves in situations where they need to solve a new goal/task interaction leading to a new goal/task allocation scheme.

Indeed, these two issues are somewhat different in nature, and should call for different resolution schemes. While the first (**C1**) is more oriented towards the collective search for a solution to a problem and calls for a purely deliberative activity, the second (**C2**) involves a more "compliant" behavior of the agents and integrates a closer interaction between deliberation and action.

While several generic approaches have been proposed in the literature concerning task or goal decomposition and allocation (Contract Nets[25], Partial Global Planning[9], distributed search[10], negotiation [14, 5, 12, 23, 6], motivational behaviors [21, 11]), cooperation for achieving independent goals have been mostly treated using task-specific or application-specific techniques [16, 20, 22, 27]

We argue that there is also a need for generic approaches to **C2**. One can of course make the agents respect a set of rules (e.g. traffic rules), or more generally "social behaviors"[24], which are specially devised to avoid as much as possible conflicts and to provide pre-defined solutions to various situations. However, this cannot be a general answer applicable to various domains.

We would like to devise a scheme which guarantees a coherent behavior of the agents in all situations (including the avalanche of situations which may occur after an execution failure) and a reliable detection of situations which call for a new task distribution process.

In the following, we propose a paradigm[3, 4] which, we believe, provides a generic framework for **C2** issues and clearly establishes a link with **C1** issues.

3 The *Plan-Merging* Paradigm

Let us assume that we have a set of autonomous robots equipped with a reliable inter-robot communication device which allows to broadcast a message to all robots or to send a message to a given robot.

Let us assume that each robot processes sequentially the goals it receives, taking as initial state the final state of its current plan. Doing so, it incrementally appends new sequences of actions to its current plan.

However, before executing any action, a robot has to ensure that it is valid in the current multi-robot context, i.e. that it is compatible with all the plans currently under execution by the other robots. This will be done by collecting all the other robot plans and by "merging" its own plan with them. This operation is "protected" by a mutual exclusion mechanism and is performed *without modifying* the other robots plans or inserting an action which may render one them invalid in order to allow the other robots to continue execution.

We call this operation, a *Plan-Merging Operation* (PMO) and its result a *Coordination Plan* (i.e. a plan valid in the current multi-robot context). Besides new actions for the robot such a plan will specify the necessary synchronization between these new actions and the other robots coordination plans.

Note that such an operation involves only communication and computation and concerns future (near term) robot actions. It can run in parallel with the execution of the current coordination plan.

3.1 The "global plan" and its properties

Everything works as if there was a *global plan* produced and maintained by the set of robots. In fact, no robot elaborates, stores or maintains a complete representation of such a *global plan*.

At any moment, a robot has its own *coordination plan* under execution. Such a *coordination plan* consists of a sequence of actions and events to be signaled to other robots as well as events which are planned to be signaled by other robots. Such events correspond to state changes in the multi-robot context and represent temporal constraints (precedence) between actions involved in different individual coordination plans.

At any moment, the "global plan" is the graph

representing the union of all current robot coordination plans. Such a global plan is valid (i.e. it does not contain inconsistent temporal constraint) if it can be represented by a *directed acyclic graph* (*dag*).

The key point here is how to devise a system composed of a set of robots which should, as much as possible, plan independently to achieve their tasks while maintaining such property of the global plan.

3.2 The *Plan-Merging* Protocol

Let us assume here that:

1. there exists a mean which allows a robot to get the right to perform a PMO while having the guarantee that it is the only robot doing so. This right should be thought of as a resource allocation[2]
2. there is a mean allowing a robot (which has obtained the right to perform a PMO) to ask for and obtain all the other robots coordination plans.
3. there exists a mean allowing robots to ask or inform one another about the occurrence of an event.

Robots are assumed to plan from time to time (whenever it is necessary).

When a robot is not planning and even when it is waiting to obtain the right to perform a PMO, it must be able to send its current coordination plan to another robot (which currently has the right to perform a PMO).

When a robot has to plan, it uses the following protocol which we call the *Plan Merging Protocol* (see Figure 2):

1. It asks for the right to perform a PMO and waits until it obtains it together with the coordination plans of all the other robots.
2. It then builds the *dag* corresponding to the union of all coordination plans (including its own coordination plan)
3. It then tries to produce a new plan which can be inserted in the *dag*, *after* its current coordination plan. The new plan insertion may only add temporal constraints which impose that some of its actions must be executed after some time-points from other robots coordination plans.

   Besides, the insertion must maintain the fact that the obtained global plan is still a *dag*.
4. If it succeeds in producing the desired plan, the robot appends it to its current coordination plan.
5. And finally, it releases the right to perform a PMO.

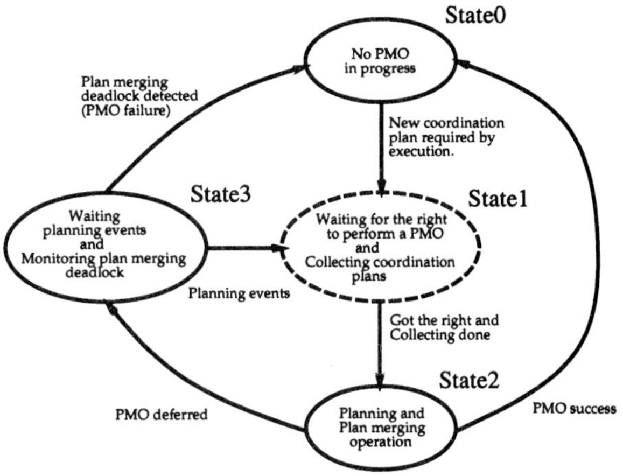

Figure 2: The general protocol state graph

When a robot executes its coordination plan, if it reaches a step with a temporal constraint linked to another robot time-point, it asks that robot if it has passed that time-point or not. Depending on the answer, the robot will wait until the other robot informs it or will immediately proceed.

3.3 Situations where PMO is deferred or where deadlock is detected

When a robot tries to perform a PMO, it may fail to produce a plan which satisfies the properties discussed earlier.

This may happen in two situations:

1. the goal can never be achieved. This can be detected if the robot cannot produce a plan even if it was alone in the environment.
2. the robot can generate a plan but this plan cannot be inserted in the global plan. This means that the final state of another robot forbids it to insert its own plan.

[2] A simple way to do it is, for instance, to maintain a "token" through communication

In such situation, the robot can simply abandon the PMO and decide to wait until the robots, that it has identified, have performed a new PMO which may possibly make them change the states preventing it to insert its plan.

Hence, we have introduced two types of events:

1. *execution events:* i.e. events which occur during plan execution and which allow robots to synchronize their execution.
2. *planning events:* i.e. events which occur whenever a robot performs a new PMO. These events can also be awaited for.

Note that, even when a robot fails in its PMO, it leaves the global plan in a correct state (it is still a *dag* and its execution can continue).

In order to detect deadlocks, a robot which finds itself in a situation where it has to wait for a *planning event* from a particular robot, must inform it. Then, it becomes possible for a robot to monitor and detect *deadlock situations* by propagating and updating, the list of robots waiting (directly or by transitivity) for *planning events* from itself. Indeed, a deadlock is detected when a robot finds itself in the list of robots waiting for itself.

When a deadlock occurs, it is necessary to take explicitly into account, in a *unique planning operation*, a conjunction of goals (which have been given separately to several robots).

This simply means that the global mission was too constrained to be solved using the Plan-Merging Paradigm. Here we must recall that we do not claim that the Plan-Merging paradigm can solve or help to solve multi-robot planning problems. The main point here is that the Plan-Merging paradigm is *safe* as it includes the detection of the deadlocks i.e. situations where a cooperation scheme of type **C1** should take place.

Note also that, in the case where only a small number of robots are involved in a deadlock, one can decide to allow the robot, which detected the deadlock, to plan for all the concerned robots . The Plan-Merging paradigm remains then applicable: the inserted plan will then concern several robots at a time.

A detailed discussion on the properties of the Plan-merging paradigm as well as on its ability to cope with execution failures can be found in [3].

The paradigm and the protocol presented so far is generic. We believe that it can be used in numerous applications. Several instances of the general paradigm can be derived, based on different planners: action planners in the stream of STRIPS, as well as more specific task planners or motion planners.

One class of applications which seems particularly well suited is the control of a large number of autonomous mobile robots.

We present in the sequel an application in the case of a fleet of autonomous mobile robots.

4 A fleet of autonomous mobile robots

We have applied the Plan-Merging Paradigm in the framework of the MARTHA project[3] which deals with the control of a large fleet of autonomous mobile robots for the transportation of containers in harbors, airports and railway environments.

In such context, the dynamics of the environment, the impossibility to correctly estimate the duration of actions (the robots may be slowed down due to obstacle avoidance, and delays in load and un-load operations, etc..) prevent a central system from elaborating efficient and reliable detailed robot plans.

The use of the Plan-Merging paradigm allowed us to deal with several types of conflicts in a general and systematic way, and to limit the role of the central system to the assignment of tasks and routes to the robots (without specifying any trajectory or any synchronization between robots) taking only into account global traffic constraints.

4.1 A Plan-Merging Protocol for Multi-Robot Navigation

For the case of a number of mobile robots in a route network environment, we have devised a specific *Plan-Merging Protocol* (PMO) based on spatial resource allocation (see [4]). It is an instance of the general protocol described above, but in this context, *Plan-Merging Operation* is done for a limited list of required spatial resources: a set of cells which will be traversed during the plan to merge. The robot broadcasts the set or required cells, receives back the set of coordination plans from other robots which have already planned to use some of the mentioned cells,

[3]MARTHA: European ESPRIT Project No 6668. "Multiple Autonomous Robots for Transport and Handling Applications"

and then tries to perform a plan insertion which ensures that the union of the considered plans is a directed acyclic graph.

One of the most interesting attributes of this protocol is that it allows several PMOs to be performed simultaneously if they involve disjunctive resource sets. This is particularly useful when there are several local conflicts at the same time.

4.1.1 *Plan-Merging for cell occupation:*

In most situations, robot navigation and the associated Plan-merging procedure are performed by trying to maintain each cell of the environment occupied by at most one robot. This allows the robots to plan their trajectories independently, to compute the set of cells they will cross and to perform Plan-Merging at cell allocation level.

In order not to constrain unnecessarily the other robots, the allocation strategy makes a robot allocate one cell ahead when it moves along lanes, while it allocates all the cells necessary to traverse and leave crossings.

4.1.2 *When reasoning about cells is not sufficient*

While, most of the time, the robots may restrict their cooperation to cells allocation, there are situations where this is not enough. This happens when they have to cross large (non-structured) areas or when an unexpected obstacle, encountered in a lane or in a crossing, forces a set of robots to maneuver simultaneously in a set of cells. In such situations, a more detailed cooperation (using the same protocol but a different planner: the motion planner) takes place allowing robots to coordinate their actions at trajectory level. Figure 3 illustrates the results of a *PMO* at trajectory level leading to trajectory synchronizations: R_i-W_j stands for a position where robot r_i should stop and wait that robot r_j has passed position R_j-S_i.

Thus, we have a hierarchy of PMOs

- first, at the cell level, based on resource (cells) allocation
- then, depending on the context, at trajectory level: motion planning in a set of common cells determined by the first level

This hierarchy authorizes a "light" cooperation, when possible, and a more detailed one, when necessary.

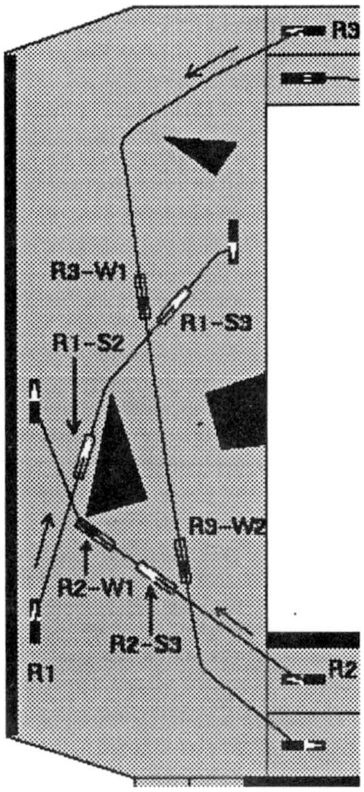

Figure 3: The result of a PMO at trajectory level (in an area)

4.2 Examples

We shall now illustrate the plan-merging paradigm and its capabilities with some sequences from our experimentation with simulated robots in a route network with open areas. The first example presents a PMO at a crossing, the second, a PMO in an open area.

An example of PMOs at a crossing

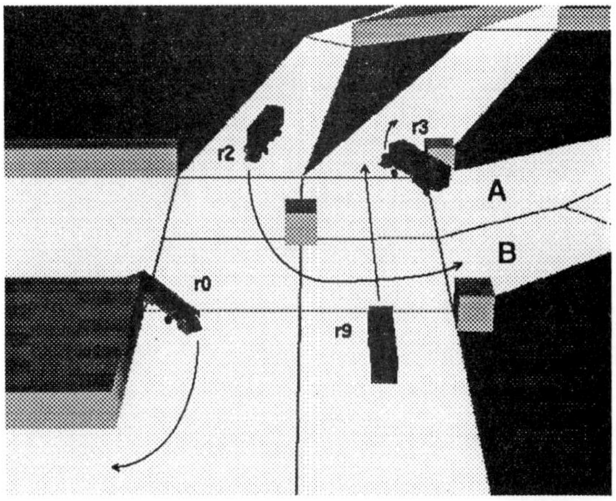

Figure 4: Crossing (Step 1)

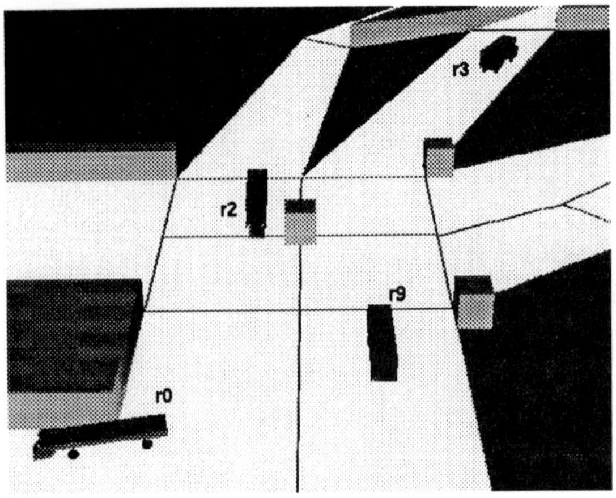

Figure 5: Crossing (Step 2)

Note that as presented in section 4.1.1, PMO performed at crossing allocates all the cells necessary to traverse and leave the crossing.

Crossing, Step 1 (Figure 4):
This snapshot has been edited to exhibit the routes that the robots must follow.

• Robot $r3$, coming from position A, and $r0$ have disjunctive list of resources. Therefore, they can perform PMOs in parallel and traverse the crossing without any synchronization.

• Robot $r9$ follows $r3$, but its PMO fails (because $r3$ has not yet planned an action to free the cell that $r9$ must allocate to exit the crossing). As a result $r9$ must wait a *planning event* from $r3$ (i.e. a new PMO).

• Robot $r2$, which wants to go to position B, finds itself in a situation where it can merge its plan with the other robots plans.

Crossing, Step 2 (Figure 5):

• Robot $r2$ traverses the crossing, after an execution synchronization with $r0$ (it must wait until $r0$ leaves the lower left cell of the crossing before entering it).

• Robot $r9$ has received the awaited *planning event* from $r3$ (which has now planned an action to exit its current cell) and its PMO succeeds, but it must synchronize with $r2$ and $r3$.

Crossing, Step 3:

• Robot $r2$ frees the crossing cells and sends the corresponding *execution event* to $r9$.

• Robot $r9$ can now traverse the crossing.

An example of PMOs in an Open Area

As mentioned earlier, open areas are not exclusive resources. In our example, $r9$ goes to position B, $r4$ moves backward from a station in the area to position A and $r5$ enters the area to dock at a station.

Open Area, Step 1 (Figure 6):
The three robots have perform theirs PMOs, first at "cell level", then at "trajectory level", in the following order: $r4$, $r9$ and $r5$. $r9$ and $r5$ (which has entered the area) are both waiting until $r4$ passes a computed curvilinear abscissa. Each of them waits at a particular position (*R9-W4* and *R5-W4*) that $r4$ passes respectively *R4-S9* and *R4-S5*.

Open Area, Step 2 (Figure 7):
Robot $r5$ goes from *R5-W4* to *R5-W9* and waits for an *execution event* from $r9$ (which will be sent as soon as $r9$ passes *R9-W5*).

Open Area, Step 3:
$r9$ has sent the execution even and all the robot can now continue their mission.

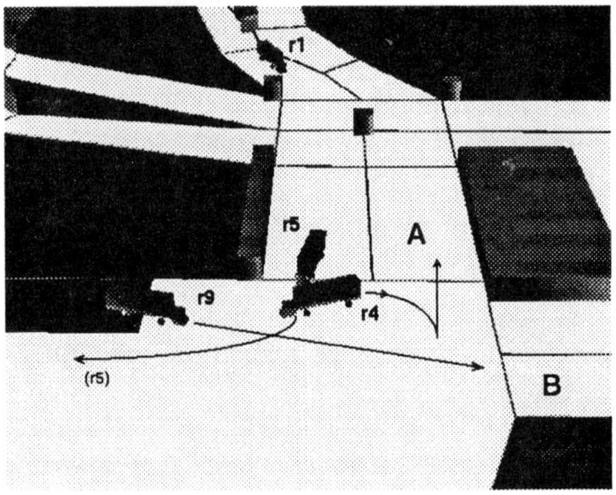

Figure 6: Open Area (Step 1)

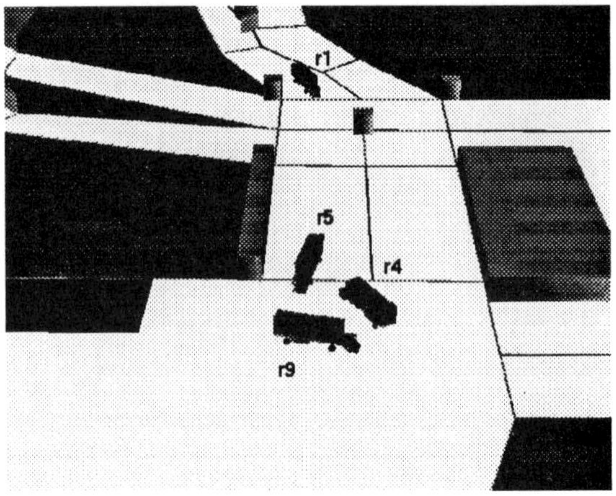

Figure 7: Open Area (Step 2)

The two examples, presented above, exhibit the

following properties of the PMO:

- Planning and execution is done in parallel.
- Several robots may use the crossing simultaneously (r0 and r3).
- The example exhibits the two types of synchronization: *execution events* ($r4$, $r5$ and $r9$ in the area example), and *planning events* ($r9$ and $r3$ in the crossing example).
- Each robot produces and merges its plans iteratively, and the global plan for the use of the crossing is incrementally built through several PMOs performed by various robots.
- It is not a first arrived first served execution. In the crossing example $r9$ arrived second, was blocked by $r3$, but did not block the crossing, letting $r2$ enter the crossing before itself.

4.3 Implementation and Results

We have developed a complete robot control system which includes all the features described. Its architecture is based on a generic control architecture for autonomous mobile robots developed at LAAS [7, 2]. It is instantiated in this case by adding an intermediate layer for performing Plan-Merging operations.

For testing and demonstration purposes, it has been linked to a robot simulator.

Experiments have been run successfully on a dozen of workstations (each workstation running a complete robot simulator) communicating through Ethernet. A 3-d graphic server has been built in order to visualize the motions and the load operations performed by all the robots in a route network (Figure 6) or in-door (Figure 8) environments. The simulated robots where able to achieve navigation missions, performing hundreds of PMOs and solving local conflicts. Motion planning and PMOs were sufficiently efficient to allow most often the robots to elaborate and merge their plans without stopping unless necessary.

Extensive experiments have also been performed using three lab robots (Fig 9).

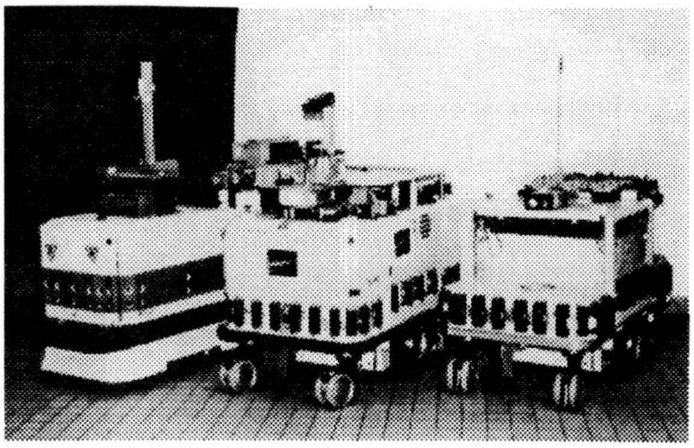

Figure 9: The Hilare family

5 Related work

There are numerous contributions dealing with multi-robot cooperation. However, the term "cooperation" has been used in several contexts with different meanings.

We will not consider here contributions to cooperation schemes at servo level (e.g. [17]) nor contributions which aim at building an "intelligent group" of simple robots (e.g. [18]). We will limit our analysis to contributions which involve an effective cooperation (at plan or program level) between several robots.

Several approaches have been proposed, such as generation of trajectories without collision (e.g. [8, 26]), traffic rules [13, 15], negotiation for dynamic task allocation [16, 5], and synchronization by programming [19, 22].

Inter-robot communication allows to exchange various information, positions, current status, future actions, etc [5, 22, 27] and to devise effective cooperation schemes.

Traffic rules have been proposed as a way to allow several robots to avoid collision and to synchronize their motion (with limited or even without communication). However, many aspects should be taken into account in order to build the set of traffic rules: the tasks, the environment, the robot features, and so on. This entails that the generated rules are valid only under the considered assumptions. If some of them are changed, the rules have to be modified or sometimes be regenerated completely. Besides, these systems are generally built heuristically and do not provide any guarantee such as deadlock detection.

Negotiation have been used for dynamic task [5] or resource [16] allocation to several robots on a situation-dependent basis.

Most contributions which make use of synchronization through communication are based on a pre-defined set of situations or on task dependent properties.

Indeed, most of the methods listed here, deal

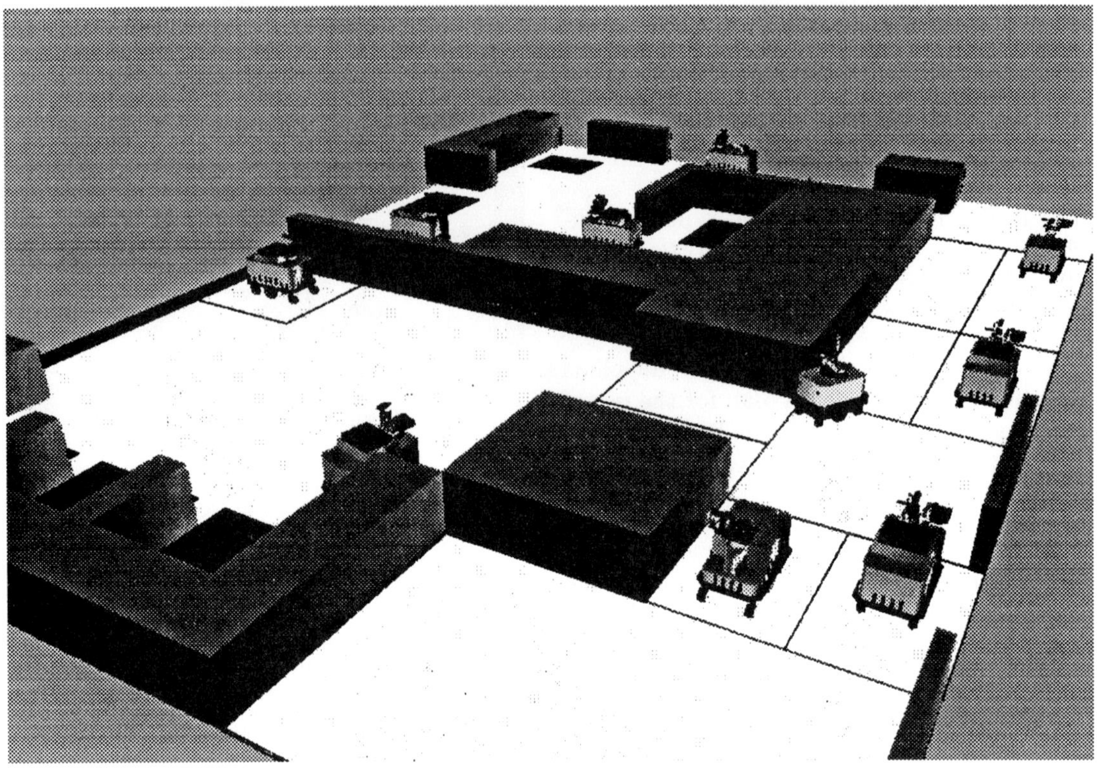

Figure 8: Several (10) Hilare using the Plan-Merging Paradigm (simulation)

essentially with collision avoidance or motion coordination and cannot be directly applied to other contexts or tasks.

We claim that our Plan-Merging paradigm is a generic framework which can be applied in different contexts, using different planners (action planners as well as motion planners). It has some clean properties (and clear limitations) which should allow, depending on the application context, to provide a coherent behavior of the global system without having to encode explicitly all situations that may encoded.

6 Conclusion

We have argued that, in the field of multi-robot (and more generally multi-agent) cooperation, it is useful to distinguish between two main issues: **C1** goal/task decomposition and allocation, and **C2** cooperation while seeking to achieve loosely coupled goals.

We have even claimed that in numerous multi-agent applications, both issues appear and even "invoke" one another "recursively".

We have then proposed a "generic" approach called *Plan-Merging Paradigm* which deals with **C2** issues and clearly establishes a link with **C1** issues.

The Plan-Merging paradigm has the following properties;

1. It makes possible for each robot to produce a coordination plan which is compatible with all plans executed by other robots.

2. No system is required to maintain the global state and the global plan permanently. Instead, each robot updates it from time to time by executing a PMO.

3. The PMO is safe, because it is robust to plan execution failures and allows to detect deadlocks.

We believe that it can be applied to a large variety of contexts and with different planners (from action planners to task or motion planners), and at different granularities.

Such a multi-robot cooperation scheme "fills the gap" between very high level planning (be it centralized or distributed) and distributed execution by a set of autonomous robots in a dynamic environment.

Indeed, it appears to be particularly well suited to the control of a large number of robots navigating in a route network. The application that we have implemented clearly exhibits its main features. It allowed us to make a large number of autonomous robots behave coherently and efficiently without creating a huge activity at the central system.

Besides the investigation of other classes of applications and the work on a more formal description of the proposed approach, our future work will concentrate on developing new cooperation schemes by embedding a multi-robot planning activity.

References

[1] L. Aguilar, R. Alami, S. Fleury, M. Herrb, F. Ingrand, F. Robert *Ten Autonomous Mobile Robots (and even more) in a Route Network Like Environment* In, *IEEE International Conference on Intelligent Robots and Systems*, IROS'95, Pittsburgh (USA), July 1995.

[2] R. Alami, R. Chatila, and B. Espiau. Designing an Intelligent Control Architecture for Autonomous Robots. in *International Conference on Advanced Robotics*, Tokyo (Japan), October 1993.

[3] R. Alami, F. Robert, F. Ingrand, and S. Suzuki. A paradigm for plan-merging and its use for multi-robot cooperation. In *IEEE International Conference on Systems, Man, and Cybernetics*, San Antonio, Texas (USA), October 1994.

[4] R. Alami, F. Robert, F. Ingrand, S. Suzuki. Multi-robot Cooperation through Incremental Plan-Merging In *IEEE International Conference on on Robotics and Automation*, Nagoya (Japan), May 1995.

[5] H. Asama, K. Ozaki, et al. Negotiation between multiple mobile robots and an environment manager. In *International Conference on Advanced Robotic*, Pisa (Italy), June 1991.

[6] Ronen I. Brafman, Yoav Shoham. Knowledge considerations in robotics and distribution of robotics tasks. *IJCAI 95*, Montréal (Canada), August 1995.

[7] R. Chatila, R. Alami, B. Degallaix, and H. Laruelle. Integrated planning and execution control of autonomous robot actions. In *IEEE International Conference on Robotics and Automation*, Nice, (France), May 1992.

[8] H. Chu and H.A. EiMaraghy. Real-time multi-robot path planner based on a heuristic approach. In *IEEE International Conference on Robotics and Automation*, Nice, (France), May 1992.

[9] E. H. Durfee, V. Lesser Partial Global Planning: A Coordination Framework for Distributed Hypothesis Formation In *IEEE Transactions on Systems, Man and Cybernetics*, Vol 21 (5), October 1991.

[10] E. H. Durfee, T. A. Montgomery Coordination as Distributed Search in a Hierarchical Behavior Space In *IEEE Transactions on Systems, Man and Cybernetics*, Vol 21 (6), December 1991.

[11] Eithan Ephrati, Motty Perry, Jeffrey S. Rosenschein. Plan execution motivation in multi-agent systems. *AIPS'94*, Chicago, June 1994.

[12] George Ferguson, James F. Allen. Arguing about plans: plan representation and reasoning for mixed-initiative planning. *AIPS'94*, Chicago, June 1994.

[13] D.D. Grossman. Traffic control of multiple robot vehicles. *IEEE Journal of Robotics and Automation*, 4(5):491–497, October 1988.

[14] N.R. Jennings. Controlling cooperative problem solving in industrial multi-agent systems using joint intentions *Artificial Intelligence*, Vol 73 (1995) 195-240.

[15] S. Kato, S. Nishiyama, and J. Takeno. Coordinating mobile robots by applying traffic rules. In *IEEE International Conference on Intelligent Robots and Systems*, IROS'92, Raleigh, (USA), July 1992.

[16] C. Le Pape. A combination of centralized and distributed methods for multi-agent planning and scheduling. *IEEE International Conference on Robotics and Automation*, Cincinnati (USA), May 1990.

[17] Z.W. Luo, K. Ito, and M Ito. Multiple robot manipulators cooperative compliant

manipulation on dynamical environments. In *IEEE International Conference on Intelligent Robots and Systems*, IROS'93, Yokohama, (Japan), July 1993.

[18] M. Mataric. Minimizing complexity in controlling a mobile robot population. In *IEEE International Conference on Robotics and Automation*, Nice, (France), May 1992.

[19] F.R. Noreils. Integrating multi-robot coordination in a mobile-robot control system. In *IEEE International Conference on Intelligent Robots and Systems*, IROS'90, Tsuchiura (Japan), July 1990.

[20] K. Ozaki, H. Asama, et al. Synchronized motion by multiple mobile robots using communication. *IEEE International Conference on Intelligent Robots and Systems*, IROS'93, Yokohama (Japan), July 1993.

[21] Lynne E. Parker. Heterogeneous multi-robot cooperation. *MIT Technical Report AITR-1465* , February 1994.

[22] S. Yuta, S. Premvuti. Coordination autonomous and centralized decision making to achieve cooperative behaviors between multiple mobile robots. *IEEE International Conference on Intelligent Robots and Systems*, IROS'92, Raleigh (USA), July 1992.

[23] Jeffrey S. Rosenschein, Gilad Zlotkin. Designing conventions for automated negotiation. *AI MAGAZINE, Volume 15, (1994)*

[24] Yoav Shoham, Moshe Tennenholtz. On social laws for artificial societies: Off-Line design. *Artificial Intelligence, 73 (1995) 231-252*

[25] Reid G. Smith. The Contract Net Protocol: High-level communication and control in a distributed problem solver. *IEEE Transactions on Computers*, Vol C-29 (12), December 1980.

[26] T. Tsubouchi and S.Arimoto. Behavior of a mobile robot navigated by an "iterated forecast and planning" scheme in the presence of multiple moving obstacles. In *IEEE International Conference on Robotics and Automation*, San Diego, (USA), May 1994.

[27] J. Wang. On sign-board based inter-robot communication in distributed robotic systems. In *IEEE International Conference on Robotics and Automation*, San Diego, (USA), May 1994.

Acknowledgments: This work was partially supported by the MARTHA (ESPRIT III) Project, the CNRS, the Région Midi-Pyrénées and the ECLA ROCOMI Project.

It is the fruit of a very intensive collaboration between numerous researchers: F. Robert, F. Ingrand, S. Fleury, M. Herrb, L. Aguilar. I would like also to acknowledge the help and the effective involvement of M. Khatib, H. Bullata, S. Suzuki, J. Perret, T. Siméon, B. Dacre-Wright, M. Devy.

Force Strategies for Cooperative Tasks in Multiple Mobile Manipulation Systems

O. Khatib, K. Yokoi,* K. Chang, D. Ruspini, R. Holmberg, A. Casal, A. Baader†
Robotics Laboratory, Department of Computer Science
Stanford University, Stanford, California 94305

Abstract

Mobile manipulation capabilities are key to many new applications of robotics in space, underwater, construction, and service environments. This article discusses the ongoing effort at Stanford University for the development of multiple mobile manipulation systems and presents the basic models and methodologies for their analysis and control. This work builds on four methodologies we have previously developed for fixed-base manipulation: the Operational Space Formulation for task-oriented robot motion and force control; the Dextrous Dynamic Coordination of Macro/Mini structures for increased mechanical bandwidth of robot systems; the Augmented Object Model for the manipulation of objects in a robot system with multiple arms; and the Virtual Linkage Model for the characterization and control of internal forces in a multi-arm system. We present the extension of these methodologies to mobile manipulation systems and propose a new decentralized control structure for cooperative tasks. The article also discusses experimental results obtained with two holonomic mobile manipulation platforms we have designed and constructed at Stanford University.

1 Introduction

A central issue in the development of mobile manipulation systems is vehicle/arm coordination. This area of research is relatively new. There is, however, a large body of work that has been devoted to the study of motion coordination in the context of kinematic redundancy. In recent years, these two areas have begun to merge, and algorithms developed for redundant manipulators are being extended to mobile manipulation systems [1] [2] [3] [13].

Typical approaches to motion coordination of redundant systems involve the use of pseudo- or generalized inverses to solve an under-constrained or degenerate system of linear equations, while optimizing some given criterion. These algorithms are essentially driven by kinematic considerations and the dynamic interaction between the end effector and the manipulator's internal motions are ignored.

Our approach to controlling redundant systems is based on two models: an *end-effector dynamic model* [10] obtained by projecting the mechanism dynamics into the operational space, and a *dynamically consistent force/torque relationship* [11] that provides decoupled control of joint motions in the null space associated with the redundant mechanism. These models form the basis for the dynamic coordination strategy we are implementing on the mobile manipulation platforms. With this strategy, the vehicle/arm system can be viewed as the mechanism resulting from the serial combination of two sub-systems: a "macro" structure with coarse, slow, dynamic responses (the mobile base), and a relatively fast and accurate "mini" device (the manipulator).

Another important issue in mobile manipulation concerns cooperative operations between multiple vehicle/arm systems [4]–[8]. An example of cooperative operations involving multiple vehicle/arm systems in construction tasks, is illustrated in Figure 1. Our research in cooperative manipulation has produced a number of results which provide the basis for the control strategies we are developing for mobile manipulation platforms. Our approach is based on the integration of two basic concepts: The *augmented object* [14] and the *virtual linkage* [15]. The *virtual linkage* characterizes internal forces, while the *augmented object* describes the system's closed-chain dynamics. These models have been successfully used in cooperative ma-

*Dr. Yokoi is a visiting scholar from MEL, Japan
†Dr. Baader is a visiting scholar from DLR, Germany

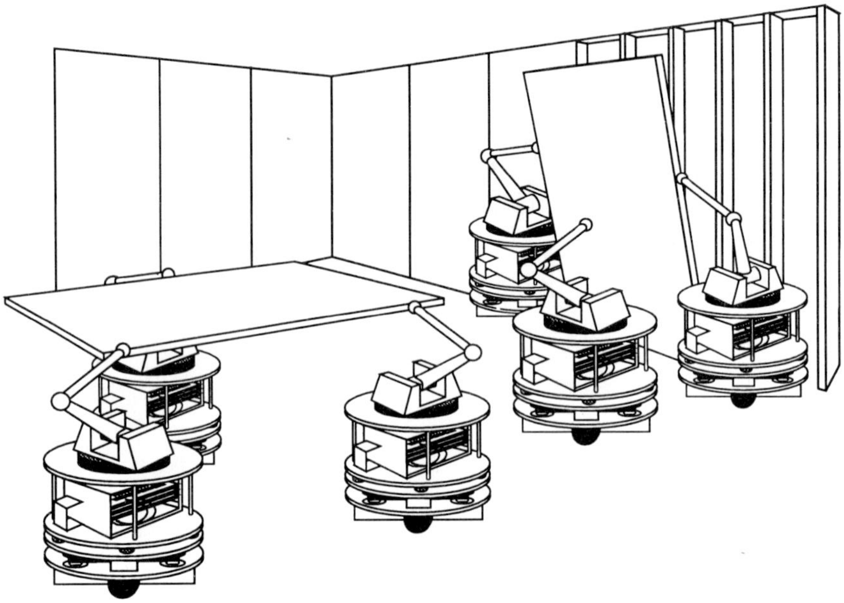

Figure 1: Robotics in Construction: Drywall

nipulation for various compliant motion tasks performed by two and three PUMA 560 manipulators [16].

The *augmented object* model we developed for fixed base multi-arm robots has been extended to the multiple vehicle/arm systems. The dynamics at the operational point of a multiple vehicle/arm system are described by an *augmented object* model which is obtained by combining the dynamics of the individual mobile manipulators and the object. Control of this highly redundant system relies on both the *dynamically consistent* relationships between joint forces and end-effector forces for each mobile arm.

While providing accurate description of cooperative manipulation, the *augmented object* and *virtual linkage* models have been implemented in an architecture that requires some level of centralized control, which is not quite suited for autonomous mobile manipulation platforms. The article presents a new strategy based on the *augmented object* and *virtual linkage* models for decentralized cooperative operations between multiple mobile manipulation platforms.

2 Vehicle/Arm Coordination

Dynamics

The joint space dynamics of a manipulator are described by

$$A(\mathbf{q})\ddot{\mathbf{q}} + \mathbf{b}(\mathbf{q}, \dot{\mathbf{q}}) + \mathbf{g}(\mathbf{q}) = \mathbf{\Gamma}; \qquad (1)$$

where $\mathbf{q}$ is the n joint coordinates and $A(\mathbf{q})$ is the $n \times n$ kinetic energy matrix. $\mathbf{b}(\mathbf{q}, \dot{\mathbf{q}})$ is the vector of centrifugal and Coriolis joint-forces and $\mathbf{g}(\mathbf{q})$ is the gravity joint-force vector. $\mathbf{\Gamma}$ is the vector of generalized joint-forces.

The operational space equations of motion of a manipulator are [10]

$$\Lambda(\mathbf{x})\ddot{\mathbf{x}} + \mu(\mathbf{x}, \dot{\mathbf{x}}) + \mathbf{p}(\mathbf{x}) = \mathbf{F}; \qquad (2)$$

where $\mathbf{x}$, is the vector of the m operational coordinates describing the position and orientation of the effector, $\Lambda(\mathbf{x})$ is the $m \times m$ kinetic energy matrix associated with the operational space. $\mu(\mathbf{x}, \dot{\mathbf{x}})$, $\mathbf{p}(\mathbf{x})$, and $\mathbf{F}$ are respectively the centrifugal and Coriolis force vector, gravity force vector, and generalized force vector acting in operational space.

Redundancy

The operational space equations of motion describe the dynamic response of a manipulator to the application of an operational force $\mathbf{F}$ at the end effector. For non-redundant manipulators, the relationship between operational forces, $\mathbf{F}$, and joint forces, $\mathbf{\Gamma}$ is

$$\mathbf{\Gamma} = J^T(\mathbf{q})\mathbf{F}; \qquad (3)$$

where $J(\mathbf{q})$ is the Jacobian matrix.

However, this relationship becomes incomplete for redundant systems. We have shown that the relationship between joint torques and operational forces is

$$\mathbf{\Gamma} = J^T(\mathbf{q})\mathbf{F} + \left[I - J^T(\mathbf{q})\overline{J}^T(\mathbf{q})\right]\mathbf{\Gamma}_0; \qquad (4)$$

with

$$\overline{J}(\mathbf{q}) = A^{-1}(\mathbf{q})J^T(\mathbf{q})\Lambda(\mathbf{q}); \qquad (5)$$

where $\overline{J}(\mathbf{q})$ is the *dynamically consistent generalized inverse* [11]. This relationship provides a decomposition of joint forces into two dynamically decoupled control vectors: joint forces corresponding to forces acting at the end effector ($J^T\mathbf{F}$); and joint forces that only affect internal motions, $\left([I - J^T(\mathbf{q})\overline{J}^T(\mathbf{q})]\mathbf{\Gamma}_0\right)$.

Using this decomposition, the end effector can be controlled by operational forces, whereas internal motions can be independently controlled by joint forces that are guaranteed not to alter the end effector's dynamic behavior. This relationship is the basis for implementing the dynamic coordination strategy for a vehicle/arm system.

The end-effector equations of motion for a redundant manipulator are obtained by the projection of the joint-space equations of motion (1), by the *dynamically consistent* generalized inverse $\overline{J}^T(\mathbf{q})$,

$$\begin{aligned} \overline{J}^T(\mathbf{q})\,[A(\mathbf{q})\ddot{\mathbf{q}} + \mathbf{b}(\mathbf{q},\dot{\mathbf{q}}) + \mathbf{g}(\mathbf{q}) = \Gamma] \\ \Longrightarrow \quad \Lambda(\mathbf{q})\ddot{\mathbf{x}} + \mu(\mathbf{q},\dot{\mathbf{q}}) + \mathbf{p}(\mathbf{q}) = \mathbf{F}; \quad (6) \end{aligned}$$

The above property also applies to non-redundant manipulators, where the matrix $\overline{J}^T(\mathbf{q})$ reduces to $J^{-T}(\mathbf{q})$.

Inertial Property

A mobile manipulator system can be viewed as the mechanism resulting from the serial combination of two sub-systems: a "macro" mechanism with coarse, slow, dynamic responses (the mobile base), and a relatively fast and accurate "mini" device (the manipulator).

The mobile base referred to as the *macro structure* is assumed to be holonomic. Let Λ be the *pseudo kinetic energy matrix* associated with the combined macro/mini structures and Λ_m the operational space *kinetic energy matrix* associated with the mini structure alone.

The magnitude of the inertial properties of macro/mini structure in a direction represented by a unit vector $\mathbf{w}$ in the m-dimensional space can be described by the scalar [11]

$$\sigma_{\mathrm{w}}(\Lambda) = \frac{1}{(\mathbf{w}^T\Lambda^{-1}\mathbf{w})};$$

which represents the effective inertial properties in the direction $\mathbf{w}$.

Our study has shown [11] that, *in any direction* $\mathbf{w}$, *the inertial properties of a macro/mini-manipulator system are smaller than or equal to the inertial properties associated with the mini-manipulator in that direction:*

$$\sigma_{\mathrm{w}}(\Lambda) \le \sigma_{\mathrm{w}}(\Lambda_m). \qquad (7)$$

A more general statement of this *reduced effective inertial* property is that the inertial properties of a redundant system are bounded above by the inertial properties of the structure formed by the smallest distal set of degrees of freedom that span the operational space.

Coordination Strategy

The reduced effective inertial property shows that the dynamic performance of a combined macro/mini system can be made comparable to (and, in some cases, better than) that of the lightweight mini manipulator. The idea behind our approach for the coordination of macro and mini structures is to treat them as a single redundant system. High dynamic performance for the manipulated object task (motion and contact forces) can be achieved with an operational space control system using essentially the fast dynamic response of the mini structure [13]. However, given the mechanical limits on the mini structure's joint motions, this would rapidly lead to joint saturation of the mini-structure degrees of freedom.

The *dynamic coordination* we propose is based on combining the operational space control with a minimization of deviation from the midrange joint positions of the mini-manipulator. This minimization must be implemented with joint force control vectors selected from the *dynamically consistent* null space of equation (4). This will eliminate any effect of the additional control forces on the end-effector task.

Let $\overline{q}_i$ and $\underline{q}_i$ be the upper and lower bounds on the ith joint position q_i. We construct the potential function

$$V_{\mathrm{Coordination}}(\mathbf{q}) = k_c \sum_{i=n_M+1}^{n} \left(q_i - \frac{\overline{q}_i + \underline{q}_i}{2}\right)^2; \qquad (8)$$

where k_c is a constant gain and n_M is the macro structure's number of degrees of freedom. The gradient of this function

$$\mathbf{\Gamma}_{\mathrm{Coordination}} = -\nabla V_{\mathrm{Coordination}}; \qquad (9)$$

provides the required attraction to the mid-range joint positions of the mini-manipulator. The interference of these additional forces with the end-effector dynamics is avoided by projecting them into the null space of $J^T(\mathbf{q})$. This is

$$\mathbf{\Gamma}_{\text{Null-Space}} = \left[I - J^T(\mathbf{q})\overline{J}^T(\mathbf{q}) \right] \mathbf{\Gamma}_{\text{Coordination}}; \tag{10}$$

where $\mathbf{\Gamma}_0$ is the joint force in the null space. In addition, joint limit avoidance can be achieved using an "artificial potential" function [12].

3 Cooperative Manipulation

Augmented Object

The *augmented object* model provides a description of the dynamics at the operational point for a multi-arm robot system. The simplicity of these equations is the result of an additive property that allows us to obtain the system equations of motion from the equations of motion of the individual mobile manipulators.

The *augmented object* model is

$$\Lambda_\oplus(\mathbf{x})\ddot{\mathbf{x}} + \mu_\oplus(\mathbf{x}, \dot{\mathbf{x}}) + \mathbf{p}_\oplus(\mathbf{x}) = \mathbf{F}_\oplus; \tag{11}$$

with

$$\Lambda_\oplus(\mathbf{x}) = \Lambda_\mathcal{L}(\mathbf{x}) + \sum_{i=1}^{N} \Lambda_i(\mathbf{x}); \tag{12}$$

where $\Lambda_\mathcal{L}(\mathbf{x})$ and $\Lambda_i(\mathbf{x})$ are the kinetic energy matrices associated with the object and the ith effector, respectively. The vector, $\mu_\oplus(\mathbf{x}, \dot{\mathbf{x}})$, of centrifugal and Coriolis forces also has the additive property

$$\mu_\oplus(\mathbf{x}, \dot{\mathbf{x}}) = \mu_\mathcal{L}(\mathbf{x}, \dot{\mathbf{x}}) + \sum_{i=1}^{N} \mu_i(\mathbf{x}, \dot{\mathbf{x}}); \tag{13}$$

where $\mu_\mathcal{L}(\mathbf{x}, \dot{\mathbf{x}})$ and $\mu_i(\mathbf{x}, \dot{\mathbf{x}})$ are the vectors of centrifugal and Coriolis forces associated with the object and the ith effector, respectively. Similarly, the gravity vector is

$$\mathbf{p}_\oplus(\mathbf{x}) = \mathbf{p}_\mathcal{L}(\mathbf{x}) + \sum_{i=1}^{N} \mathbf{p}_i(\mathbf{x}); \tag{14}$$

where $\mathbf{p}_\mathcal{L}(\mathbf{x})$ and $\mathbf{p}_i(\mathbf{x})$ are the gravity vectors associated with the object and the ith effector. The generalized operational forces $\mathbf{F}_\oplus$ are the resultant of the forces produced by each of the N effectors at the operational point.

$$\mathbf{F}_\oplus = \sum_{i=1}^{N} \mathbf{F}_i. \tag{15}$$

Virtual Linkage

Object manipulation requires accurate control of internal forces. Recently, we have proposed the *virtual linkage* [15], as a model of internal forces associated with multi-grasp manipulation. In this model, grasp points are connected by a closed, non-intersecting set of virtual links (Figure 2.)

In the case of an N-grasp manipulation task, a *virtual linkage* model is a $6(N-1)$ degree of freedom mechanism that has $3(N-2)$ linearly actuated members and N spherically actuated joints. Forces and moments applied at the grasp points of this linkage will cause forces and torques at its joints. We can independently specify internal forces in the $3(N-2)$ members, along with $3N$ internal moments at the spherical joints. Internal forces in the object are then characterized by these forces and torques in a physically meaningful way.

The relationship between applied forces, their resultant and internal forces is

$$\begin{bmatrix} \mathbf{F}_{res} \\ \mathbf{F}_{int} \end{bmatrix} = \mathbf{G} \begin{bmatrix} \mathbf{f}_1 \\ \vdots \\ \mathbf{f}_N \end{bmatrix}; \tag{16}$$

where $\mathbf{F}_{res}$ represents the resultant forces at the operational point, $\mathbf{F}_{int}$ the internal forces and $\mathbf{f}_i$ the forces applied at the grasp point i. $\mathbf{G}$ is called the grasp description matrix, and relates forces applied at each grasp to the resultant and internal forces in the object. Furthermore, $\mathbf{G}$ can be written as

$$\mathbf{G} = [\ \mathbf{G}_1 \mathbf{G}_2 \ldots \mathbf{G}_N\ ];$$

where each $\mathbf{G}_i$ represents the contribution of the ith grasp to the resultant and internal forces felt by the object. Also, $\mathbf{G}_i$ can be further decomposed

$$\mathbf{G}_i = \begin{bmatrix} \mathbf{G}_{res,i} \\ \mathbf{G}_{int,i} \end{bmatrix};$$

where $\mathbf{G}_{res,i}$ is the contribution of $\mathbf{G}_i$ to the resultant forces in the object and $\mathbf{G}_{int,i}$ to the internal ones.

The inverse $\mathbf{G}^{-1}$ provides the forces required at the grasp points to produce the resultant and internal forces acting at the object.

$$\begin{bmatrix} \mathbf{f}_1 \\ \vdots \\ \mathbf{f}_N \end{bmatrix} = \mathbf{G}^{-1} \begin{bmatrix} \mathbf{F}_{res} \\ \mathbf{F}_{int} \end{bmatrix}. \tag{17}$$

Similarly, $\mathbf{G}^{-1}$ can be written as

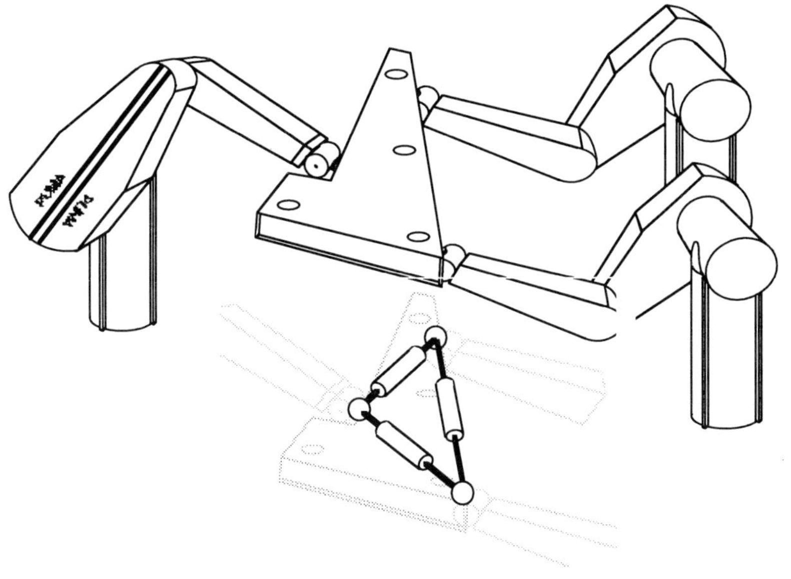

Figure 2: The Virtual Linkage

$$\mathbf{G}^{-1} = \begin{bmatrix} \overline{\mathbf{G}}_1 \\ \vdots \\ \overline{\mathbf{G}}_N \end{bmatrix};$$

with

$$\overline{\mathbf{G}}_i = [\ \overline{\mathbf{G}}_{res,i}\ \ \overline{\mathbf{G}}_{int,i}\ ];$$

where $\overline{\mathbf{G}}_{res,i}$ represents the part of $\overline{\mathbf{G}}_i$ that correspond to the resultant forces at the object; and the matrix $\overline{\mathbf{G}}_{int,i}$ represents the part corresponding to the internal forces.

Centralized Control Structure

For fixed base manipulation, the *augmented object* and *virtual linkage* have been implemented in multiprocessor system using a centralized control structure. This type of control is not suited for autonomous mobile manipulation platforms. Before presenting the decentralized implementation, we begin with a brief summary of the centralized control structure.

The overall structure of the centralized implementation is shown in Figure 3. The force sensed at the grasp point of each robot, $\mathbf{f}_{s,i}$, is transformed, via $\mathbf{G}$, to sensed resultant forces, $\mathbf{F}_{res,s}$, and sensed internal forces, $\mathbf{F}_{int,s}$, at the operational point, using equation (16)

$$\begin{bmatrix} \mathbf{F}_{res,s} \\ \mathbf{F}_{int,s} \end{bmatrix} = \mathbf{G} \begin{bmatrix} \mathbf{f}_{s,1} \\ \vdots \\ \mathbf{f}_{s,N} \end{bmatrix}.$$

The centralized control strategy consists of (i) a unified motion and force control structure for the *augmented object* corresponding to the resultant force vector, $\mathbf{F}_{res}$; and (ii) a force control vector, $\mathbf{F}_{int}$, corresponding to the internal forces acting on the *virtual linkage*. These are

$$\mathbf{F}_{res} = \mathbf{F}_{motion} + \mathbf{F}_{contact}; \tag{18}$$

where

$$\mathbf{F}_{motion} = \hat{\Lambda}_{\oplus}\Omega\mathbf{F}^*_{motion} + \hat{\mu}_{\oplus} + \hat{p}_{\oplus}; \tag{19}$$

$$\mathbf{F}_{contact} = \hat{\Lambda}_{\oplus}\overline{\Omega}\mathbf{F}^*_{contact} + \mathbf{F}_{contact,s}. \tag{20}$$

$\hat{\Lambda}_{\oplus}$, $\hat{\mu}_{\oplus}$, and $\hat{p}_{\oplus}$ represent the estimates of $\Lambda_{\oplus}$, $\mu_{\oplus}$, and $p_{\oplus}$. The vector $\mathbf{F}^*_{motion}$ and $\mathbf{F}^*_{contact}$ represent the inputs to the decoupled system. Ω is the *generalized selection matrix* associated with motion control and $\overline{\Omega}$, its complement, is associated with force control.

The control structure for internal forces is

$$\mathbf{F}_{int} = \hat{\Lambda}_{\oplus}\mathbf{F}^*_{int} + \mathbf{F}_{int,s}; \tag{21}$$

where the vector $\mathbf{F}^*_{int}$ represents the inputs to the decoupled system. A suitable control law can be selected to obtain $\mathbf{F}^*_{motion}$, $\mathbf{F}^*_{contact}$ and $\mathbf{F}^*_{int}$.

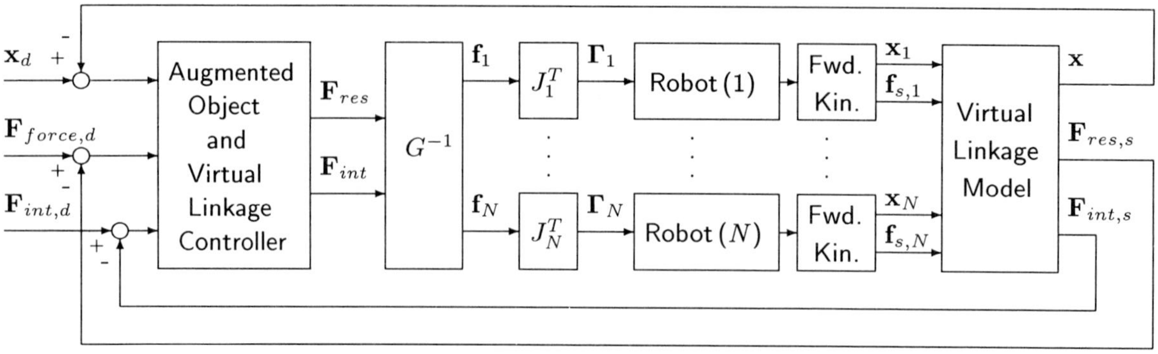

Figure 3: Centralized Control

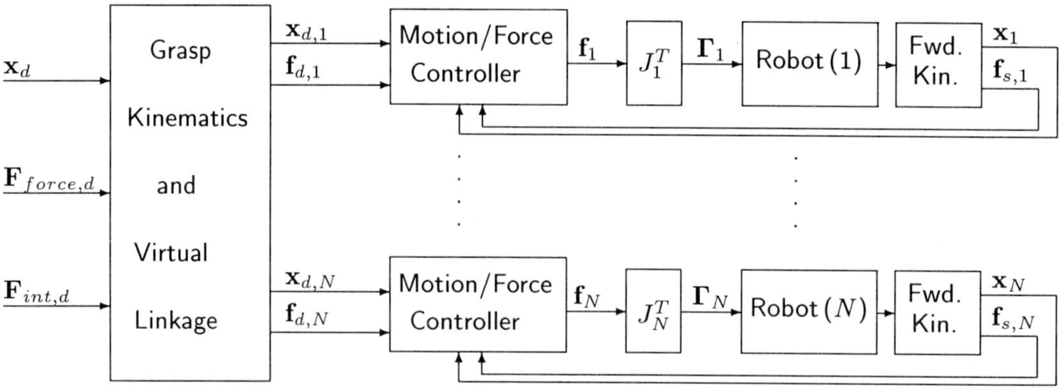

Figure 4: Decentralized Control Structure

The control forces of the individual mobile manipulator, $\mathbf{f}_i$, are given by using equation (17),

$$\begin{bmatrix} \mathbf{f}_1 \\ \vdots \\ \mathbf{f}_N \end{bmatrix} = \mathbf{G}^{-1} \begin{bmatrix} \mathbf{F}_{res} \\ \mathbf{F}_{int} \end{bmatrix}.$$

The above strategy has been successfully implemented for two and three PUMA 560 arms [16].

Decentralized Control Structure

In a multiple mobile robot system, each robot has real-time access only to its own state information and can only infer information about the other robots' grasp forces through their combined action on the object. In the decentralized control structure we propose, the object level specifications of the task are transformed into individual tasks for each of the cooperative robots. Local feedback control loops are then developed at each grasp point. The task transformation and the design of the local controllers are accomplished in consistency with the *augmented object* and *virtual linkage* models. The overall structure of the proposed decentralized control structure is shown in Figure 4.

The local control structure at the ith grasp point is

$$\mathbf{f}_i = \mathbf{f}_{motion,i} + \mathbf{f}_{force,i}. \tag{22}$$

The control vectors, $\mathbf{f}_{motion,i}$, are designed so that the combined motion of the various i^{th} grasp points results in the desired motion at the object operational point. On the other hand, the vectors $\mathbf{f}_{force,i}$ create forces at the grasp points, whose combined action produces the desired contact and internal forces on the object.

The motion control at the ith grasp point is

$$\mathbf{f}_{motion,i} = \hat{\Lambda}_{\oslash,i}\Omega\mathbf{f}^*_{motion,i} + \hat{\mu}_{\oslash,i} + \hat{p}_{\oslash,i}; \tag{23}$$

with

$$\hat{\Lambda}_{\oslash,i} = \hat{\Lambda}_{g,i} + \overline{\mathbf{G}}_{res,i}\hat{\Lambda}_{\mathcal{L}}\overline{\mathbf{G}}^T_{res,i}; \quad (24)$$

where $\hat{\Lambda}_{g,i}$ is the kinetic energy matrix associated with the ith effector at the grasp point. The second term of equation (24) represents the part of $\hat{\Lambda}_{\mathcal{L}}$ assigned to the ith robot and described at its grasp point.

The vector, $\hat{\mu}_{\oslash,i}$, of centrifugal and Coriolis forces associated with the ith effector is

$$\hat{\mu}_{\oslash,i} = \hat{\mu}_{g,i} + \overline{\mathbf{G}}_{res,i}\hat{\mu}_{\mathcal{L}}; \quad (25)$$

where $\hat{\mu}_{g,i}$ is the centrifugal and Coriolis vector of the ith robot alone at the grasp point. $\overline{\mathbf{G}}_{res,i}\hat{\mu}_{\mathcal{L}}$ represents the part of $\hat{\mu}_{\mathcal{L}}$ assigned to the ith robot and described at its grasp point. Similarly, the gravity vector is

$$\hat{p}_{\oslash,i} = \hat{p}_{g,i} + \overline{\mathbf{G}}_{res,i}\hat{p}_{\mathcal{L}}; \quad (26)$$

where $\hat{p}_{g,i}$ is the gravity vector associated with the ith end effector at the grasp point. $\overline{\mathbf{G}}_{res,i}\hat{p}_{\mathcal{L}}$ represents the part of $\hat{p}_{\mathcal{L}}$ assigned to the ith robot and described at its grasp point.

The sensed forces at the ith grasp point, $\mathbf{f}_{s,i}$, combine the contact and internal forces felt at the ith grasp point, together with the acceleration force acting at the object. The sensed forces associated with the contact and internal forces alone, $\mathbf{f}_{\bar{s},i}$, are therefore obtained by subtracting the acceleration effect from the total sensed forces

$$\mathbf{f}_{\bar{s},i} = \mathbf{f}_{s,i} - \overline{\mathbf{G}}_{res,i}\left(\hat{\Lambda}_{\mathcal{L}}\ddot{\mathbf{x}}_d + \hat{\mu}_{\mathcal{L}} + \hat{p}_{\mathcal{L}}\right). \quad (27)$$

Here, the object desired acceleration has been used instead of the actual acceleration, which would be difficult to evaluate.

The force control part of equation (22) is

$$\mathbf{f}_{force,i} = \hat{\Lambda}_{\oslash i}\mathbf{f}^*_{force,i} + \mathbf{f}_{\bar{s},i}; \quad (28)$$

where $\mathbf{f}^*_{force,i}$ represents the input to the decoupled system associated with the contact forces and internal forces. $\mathbf{f}^*_{force,i}$ can be achieved by selecting

$$\mathbf{f}^*_{force,i} = -\mathbf{K}_f(\mathbf{f}_{\bar{s},i} - \mathbf{f}_{d,i}) - \mathbf{K}_{vf}\dot{\mathbf{f}}_{\bar{s},i}. \quad (29)$$

The vector $\mathbf{f}_{d,i}$ is the desired force assigned to the ith mobile manipulator. Using equation (17), this vector is

$$\begin{bmatrix} \mathbf{f}_{d,1} \\ \vdots \\ \mathbf{f}_{d,N} \end{bmatrix} = \mathbf{G}^{-1}\begin{bmatrix} \mathbf{F}_{res,d} \\ \mathbf{F}_{int,d} \end{bmatrix}; \quad (30)$$

where the desired resultant forces are

$$\mathbf{F}_{res,d} = \overline{\Omega}\mathbf{F}_{contact,d}; \quad (31)$$

where $\mathbf{F}_{contact,d}$ is the desired contact force vector.

The assumptions in the above control structure is that the object is rigid and that there is no slippage at the grasp points. Gripper slip in the real system will result in errors in the grasp kinematic computation and inconsistencies with the *virtual linkage* model. To compensate for these effects, some level of communication between the different platforms will be needed for updating the robot state and modifying the task specifications. The rate at which this communication is required is much slower than the local servo control rate. Such communication can be achieved over a radio Ethernet link (at 10-20 Hz).

4 Experimental Platforms

Recently, we have designed and built two autonomous mobile manipulation platforms (Figure 5). The Stanford Assistant Mobile Manipulator(SAMM) platforms have been developed in collaboration with Nomadic Technologies and Oak Ridge National Laboratories.

Each platform consists of a PUMA 560 arm mounted on a holonomic mobile base (Figure 6). The PUMA manipulator is equipped with a 6-axis force sensor on the wrist, and an electric two fingered gripper. The base consists of three "lateral" orthogonal universal-wheel assemblies which allow the base to translate and rotate holonomically in relatively flat office-like environments [9].

The base houses two Personal Computers, the motor amplifiers for both the base and arm, and batteries for approximately two hours of autonomous operation. In addition, a 1500 Watt AC power supply is included for tethered operations. Selection between on-board and tethered operations can be made while the robot is in operation.

Environmental sensing is currently provided by 48 sonars divided into two rings encircling the base. Sonars in the upper and lower rings fire simultaneously to allow through cross-echoes. This allows the detection of objects even if they are located in the middle of the two sonar rings. We are currently working on adding vision capabilities to allow for more complex tasks.

The two Personal Computers with 90MHz Pentium processors were selected over other competing technologies because of their relatively low cost and power consumption and the large selection of

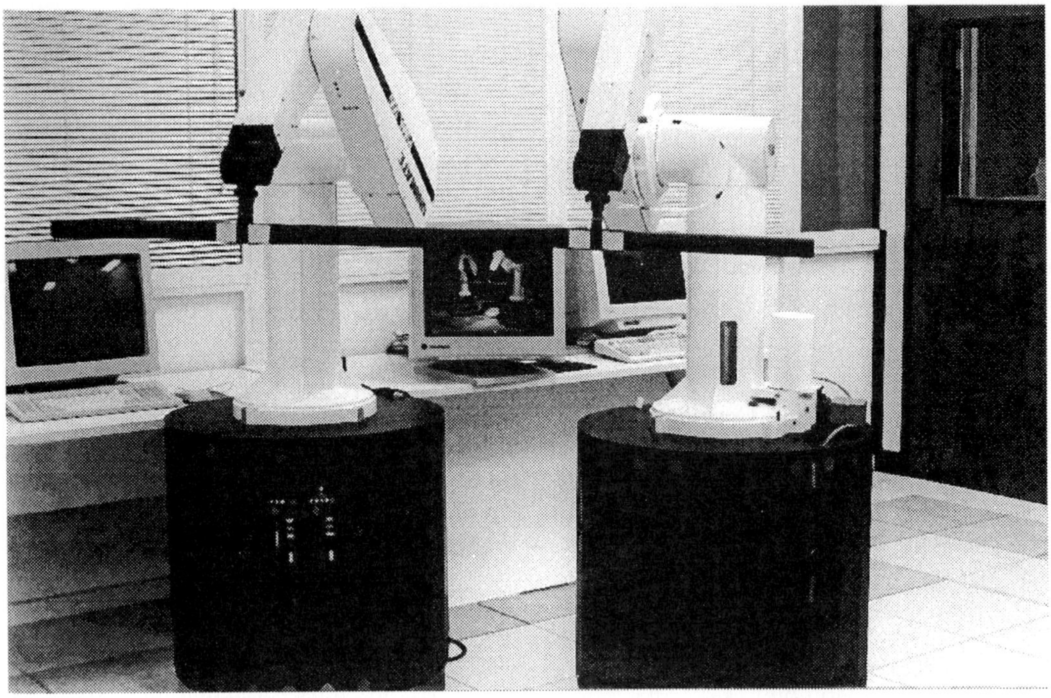

Figure 5: Stanford Mobile Platforms

add-on cards. Also, this architecture will continue to be easily upgraded as computer technology advances.

The two computers in each robot are the heart of the multi-processing architecture. On each platform, one computer is devoted to robot control and other real-time tasks. The second computer is responsible for navigation, path planning, communication and other less time-critical functions. Within the robot, the two computers communicate with each other through a wire Ethernet backbone. Communication between robots or with other workstations is accomplished via a radio Ethernet link. This architecture is illustrated in Figure 7.

The control computer, running the real-time operating system *VxWorks*, interfaces with the robot via an off-the-shelf, compact, modular multi-axis controller which provides 12 encoders channels, 16 D/A channels and 8 A/D channels on a single EISA backplane slot. This is used to read the joint positions of the arm and base and send the motor torques to the amplifiers. Information from the force sensor is received via a fiber optic link through a dedicated interface board at speeds of up to 1000Hz. In addition, the control computer communicates via memory-mapped, dual-ported RAM with a Motorola 6811 microcontroller board. This board provides basic watchdog functions such as enabling and disabling amplifiers and motor brakes, and processing assorted error signals.

The navigation computer running the *Linux* operating system is responsible for tasks that do not require real-time performance. These include path planning and modification, environmental sensing and communication with the other robots and the outside world. The sonar firing and timing is accomplished with a Motorola 6811 microprocessor board identical to the one used in the control computer. This frees almost all of this computer's processing power for planning, navigation and self localization. In the future, this computer will also perform some vision processing. A hard disk with 1 GB capacity is provided for completely autonomous startup and development.

The control strategies discussed above have been successfully implemented in a system of two platforms. The dynamic coordination strategy has allowed full use of the relatively high bandwidth of the PUMA. Object motion and force control performance with these mobile platforms are comparable with the results obtained with fixed base PUMA manipulators.

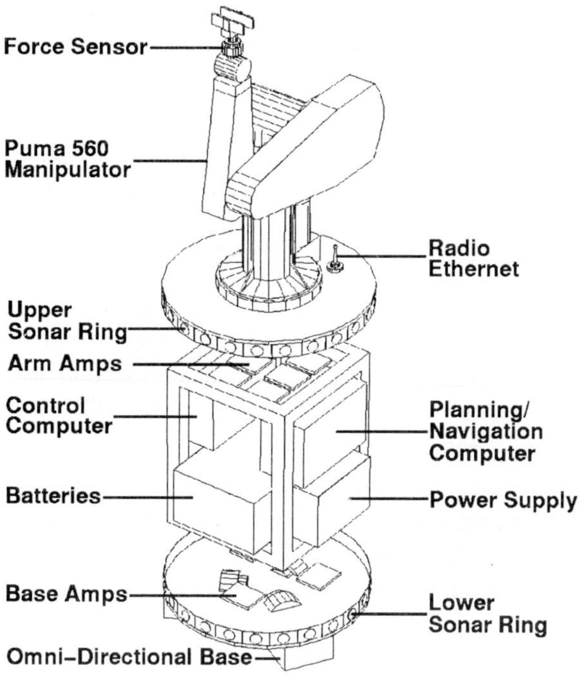

Figure 6: Platform Components

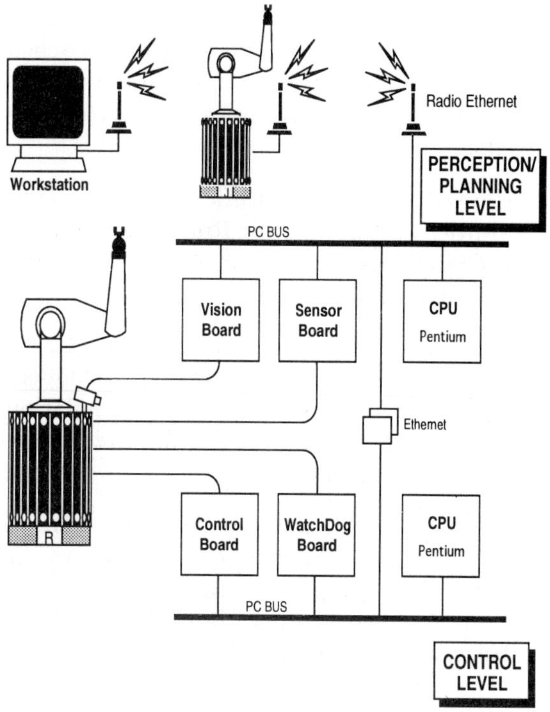

Figure 7: SAMM Architecture

5 Conclusion

We have presented extensions of various operational space methodologies for fixed-base manipulators to mobile manipulation systems. A vehicle/arm platform is treated as a macro/mini structure. This redundant system is controlled using a dynamic coordination strategy, which allows the mini structure's high bandwidth to be fully utilized. For cooperative operations, we have developed a new decentralized control structure based on the *augmented object* and *virtual linkage* models that is better suited for mobile manipulator systems.

Vehicle/arm coordination and cooperative operations have been successfully implemented on two mobile manipulator platforms developed at Stanford University.

Acknowledgments

The financial support of Boeing, General Motors, Hitachi Construction Machinery, and NSF (grants IRI-9320017 and CAD-9320419) is gratefully acknowledged. Many thanks to Alan Bowling, Oliver Brock, Francois Pin, James Slater, John Slater, Stef Sonck and Dave Williams who have made significant contributions to the design and construction of the Stanford robotic manipulation platforms.

References

[1] Ullman, M., Cannon, R., "Experiments in Global Navigation and Control of a Free-Flying Space Robot," *Proc. Winter Annual Meeting*, Vol. 15, 1989, pp. 37-43.

[2] Umetani, Y., and Yoshida, K., "Experimental Study on Two-Dimensional Free-Flying Robot Satellite Model," *Proc. NASA Conf.* Space Telerobotics, 1989.

[3] Papadopoulos, E., Dubowsky, S., "Coordinated Manipulator/Spacecraft Motion Control for Space Robotic Systems," *Proc. IEEE Int. Conf. Robotics and Automation*, 1991, pp. 1696-1701.

[4] Zheng, Y.F. and Luh, J.Y.S., "Joint Torques for Control of Two Coordinated Moving Robots," *Proc. IEEE Int. Conf. Robotics and Automation*, 1986, pp. 1375-1380.

[5] Uchiyama, M. and and Dauchez, P., "A symmetric Hybrid Position/Force Control Scheme for the coordination of Two Robots," *Proc. IEEE Int. Conf. Robotics and Automation*, 1988, pp. 350-356.

[6] Hayati, S., "Hybrid Position/Force Control of Multi-Arm Cooperating Robots", *Proc. IEEE Int. Conf. Robotics and Automation*, 1986, pp. 1375-1380.

[7] Tarn, T. J., Bejczy, A. K., and Yun, X., "Design of Dynamic Control of Two Cooperating Robot Arms: Closed Chain Formulation." *Proc. IEEE Int. Conf. Robotics and Automation*, 1987, pp. 7-13.

[8] Adams, J. A., Bajcsy, R., Kosecka, J., Kumar, V., Mandelbaum, R., Mintz, M., Paul, R., Wang, C., Yamamoto, Y., and Yun, X.,"Cooperative Material Handling by Human and Robotic Agents: Module Development and System Synthesis," *Proc. IEEE/RSJ Int. Conf. Intelligent Robotics and Systems*, 1995, pp. 200-205.

[9] Pin, F. G. and Killough S. M., "A New Family of Omnidirectional and Holonomic Wheeled Platforms for Mobile Robots," *Oak Ridge National Laboratory Technical Reports*, 1994.

[10] Khatib, O., "A Unified Approach to Motion and Force Control of Robot Manipulators: The Operational Space Formulation," *IEEE J. Robotics and Automation*, vol. 3, no. 1, 1987, pp. 43-53.

[11] Khatib, O., "Inertial Properties in Robotics Manipulation: An Object-Level Framework," *Int. J. Robotics Research,* vol. 14, no. 1, February 1995. pp. 19-36.

[12] Khatib, O., "Real-Time Obstacle Avoidance for Manipulators and Mobile Robots," *Int. J. of Robotic Research*, vol. 5, no. 1, 1986, pp. 90-98.

[13] Russakow, J. and Khatib, O., "A New Control Structure for Free-Flying Space Robots," *Int. Symposium on Artificial Intelligence, Robotics and Automation in Space*, 1992, pp. 395-403.

[14] Khatib, O., "Object Manipulation in a Multi-Effector Robot System", *Robotics Research 4*, R. Bolles and B. Roth, eds., pp. 137-144, Cambridge: MIT Press 1988.

[15] Williams, D. and Khatib, O., "The Virtual Linkage: A Model for Internal Forces in Multi-Grasp Manipulation," *Proc. IEEE Int. Conf. Robotics and Automation*, 1993, pp. 1025-1030.

[16] Williams, D. and Khatib, O., "Multi-Grasp Manipulation," *IEEE Int. Conf. Robotics and Automation Video Proceedings*, 1995.

Behavior Matching by Observation for Multi-Robot Cooperation

Yasuo Kuniyoshi
Electrotechnical Laboratory
1-1-4 Umezono, Tsukuba, Ibaraki 305 JAPAN
email: kuniyosh@etl.go.jp

Abstract

Based on formal analysis of qualitative robot behaviors, a criterion for classifying multi-robot tasks to either static or dynamic is presented. It suggests that *common resource recognition* is essential in dynamic cooperative tasks. In the presented framework *Cooperation by Observation*, a robot finds a common resource and chooses an appropriate helpful behavior by observing other agents' behaviors. Demonstrative experiments with real mobile robots are presented. The robots have actively controlled binocular vision and controlled by an extended behavior based architecture. The vital part of the architecture consists of *attentional buffers*, which handles behavior coordination by temporarily remembering the common resource and initializing new behaviors using it.

1 Introduction

Over the last decade, various cooperative frameworks for decentralized tacit multiple robot systems have been developed, which have made significant academic contributions. However, if we step back and view the whole variety of cooperative tasks realized so far, we cannot suppress an impression that most of them (with a few exceptions) fall into rather monotonous patterns, such as collecting/accumulating objects, marching/formation keeping and coordinated object carrying/manipulation. A naive but important intuition tells us that these patterns cover only a narrow band of a possibly rich spectrum of cooperative behaviors.

I would propose here that this is the time to investigate the underlying common structure of a much wider spectrum of cooperative behaviors, thereby extracting guidelines for developing new task patterns and generalizing/extending existing frameworks.

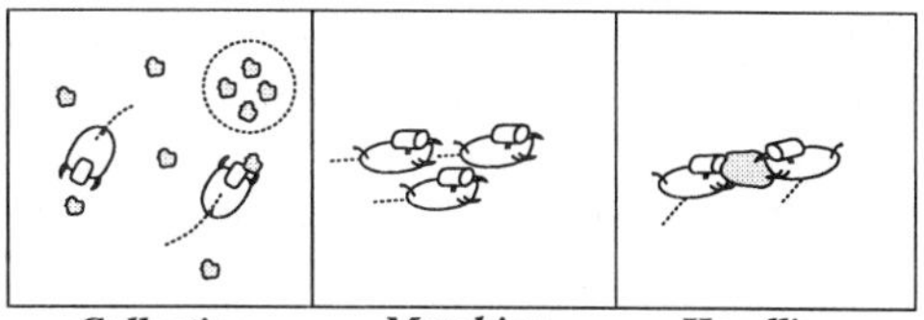

Figure 1: *Any other task patterns? Can robots switch among them?*

Very recently, a frontier attempt has been made by Brafman and Shoham [1] towards a mathematical formulation of a task distribution process, i.e. a provably correct procedure for deriving distributed agent programs from a global task specification. Although this is an important step towards a general theory of multi-agent coordination, it is still in its infancy, dealing with a simplistic artificial problem with strong assumptions. Presumably, a mathematical formulation and solution of general multi-robot tasks will require another decade or more of continuous efforts.

In this paper, I will first propose one axis of generalization, i.e. viewing cooperation as dynamic matching of *behavior resources* and *behavior types*. This aspect has not been focused by previous works, which lead to the rather static task patterns in most cases. Then I will present a recently developed cooperative framework called "Cooperation by Observation", which treats the dynamic matching as deictic, in other words, grounded in robot behaviors. It proposes novel directions of extending the variety of task patterns.

2 Nature of Dynamic Cooperative Behavior

In this section I will explain why each of the typical cooperative behavior patterns as in Fig. 1 are considered *static* and what is considered important for a cooperative behavior pattern to be *dynamic*, in my sense.

2.1 Definition of an interactive task

Intuitively, any cooperative task should involve at least two robots *interacting* through a physical environment. So, let us consider this weak form for a while, an *interactive* task. Here I adopt a simplified notion of *interaction* for a two-robot case as follows:

> Robot R_i and R_j are *interacting* through the behaviors $Beh_l(R_i)$ and $Beh_m(R_j)$ if the two behaviors *involve* a common *resource.*

Here, a *resource* can be a set of objects (including robot bodies) or spatial regions, or both. For instance, in the case of coordinated handling, the held object and the arms are the resource, and in the case of marching, the robot bodies are the resource. A *behavior* can be defined as achieving and maintaining a particular state defined over the *resource*O_p (or *involves* the resource.). Thus, $Beh_l(R_i) = Achieve(R_i, O_p, S_l)$. A state can include relative positions of objects, qualitative movement of objects such as rotating or translating, or characteristic of spatial regions such as empty or occupied, and so forth. Function $Res()$ extracts the resource of a behavior: $Res(Beh_l(R_i)) = O_p$. Using the above notions, an interactive task is expressed as follows:

> Two robots R_i and R_j are doing interactive task (Beh_l, Beh_m) (the task formed by a combination of the two behaviors) if and only if there exists such tasks $Beh_l(R_i)$ and $Beh_m(R_j)$ with non-empty common resource: $CRes_{lm} = Res(Beh_l(R_i)) \cap Res(Beh_m(R_j)) \neq \phi$

This can be extended to cases with more than two robots by allowing both chaining and intersecting the above definition.

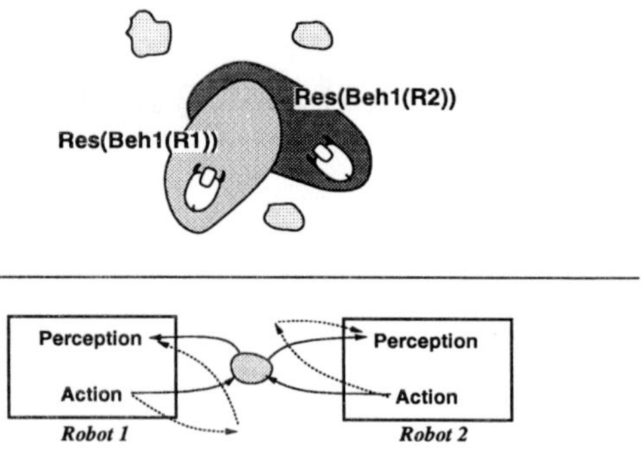

Figure 2: *Cooperation require merged behaviors.*

2.2 Static interactive task patterns

An interactive task pattern is called *static* if the common resource $CRes_{lm}$ is temporally invariant, and *dynamic* if it changes over time.

Let us consider a leader-follower task. In this case, a navigational behavior Beh_1 of the leader robot R_1 involves at least the body of the R_1. And a chasing behavior $Beh_2(R_2)$ involves $\{R_1, R_2\}$ so that it maintains a constant distance between the two. (If R_1 is aware of the follower R_2, $Res(Beh_1(R_1)) \supset \{R_1, R_2\}$.) Therefore, this is a static interactive task with a temporally invariant common resource $CRes_{12} = \{R_1\}$.

Similarly, a coordinated manipulation is also a *static* task because the common resource is always the held object.

In a typical cooperative object collecting task, every robot collects samples scattered around the workspace, and accumulates at a particular place P_{acc}. Thus the common resource $CRes_{1..N} = P_{acc}$ which is invariant over time, so the task is static. [1]

2.3 Dynamic interaction and common resource matching

In a dynamic interactive task, the common resource changes over time. This change usually

[1] This case is subtle because if we have a closer look, we might find some local interactions which are dynamic. Indeed, some of such cases are considered later as good examples of "cooperation by observation". I will elaborate on this point later. But we should note here that at the level of overall cooperative collecting task, it is considered static. As shown by this example, the notion of static/dynamic can vary depending on which level of the task to describe.

occurs at initial and final time point of an interactive behavior pattern. For example, when a team of robots creates a formation, or when they initially find and grasp a common object. Although the process of this transition has not been carefully studied, it plays an important role when we want the team of robots to exhibit complex task patterns realized by combining basic interactive behaviors.

Here arises an important question: "How can a team of robots match their behavior resources to establish an interactive task pattern?" or "How can they initialize their attention so that they match?" (See Fig. 3). From a robot's point of view, the resource of another robot's behavior can only be deduced by recognizing its behavior and extracting the objects or places relative to the behavior. In most cases, this process takes place before the robots physically touches a common object and vision becomes an important sensing method. The "cooperation by observation" is an approach to solving with these problems.

3 Cooperation by Observation

In this section I will attempt to answer the following questions: "How can a team of robots match their behavior resources? And how this leads to a cooperative task patterns?" The answers are integrated as a cooperative framework called "Cooperation by Observation". I will also present an experimental validation of the framework.

> "Cooperation by Observation" is defined as "Observing other agent's action, and Choosing appropriate actions regarding the observed action and the current task situation."

It has a potential to realize a high level of task coordination by decentralized autonomous robots allowing minimum explicit communication. Its source of power lies in an advanced capability given to each robot for recognizing other agent's actions by (primarily visual) observation. This provides rich information about the current task situation around each robot which facilitates highly-structured task coordination.

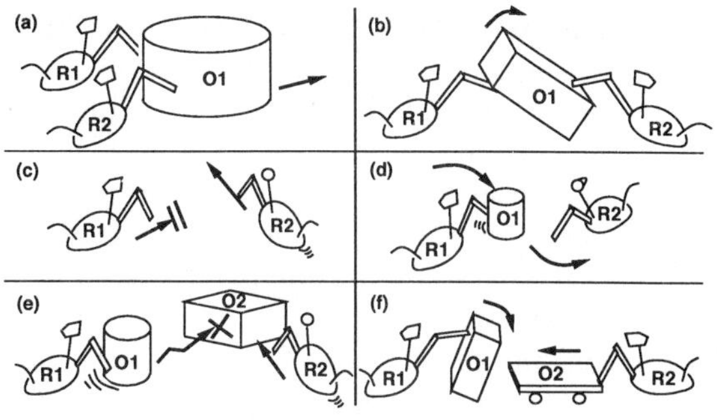

Figure 3: Cooperative task patterns: (a) Pushing. (b) Tumbling. (c) Giving way. (d) Passing. (e) Unblocking. (f)Cart loading. *How can the robots match their attention?*

3.1 Behavior matching by observation

Let us consider the examples given in Fig. 3. How can the pairs of robots initialize their behaviors in each situation? The procedure consists of two steps:

1. Find the behavior resource of the other robot by observation.
2. Choose and invoke an appropriate action regarding the identified resource and the observed behavior of the other robot.

In patterns (a) and (b), the robots should find one common object. This could be done by chance while wandering around, or more efficiently, finding and approaching an object which the other robot is heading. In patterns (c), R_1 can find the common resource $\{R_2\}$ (or $\{R_1, R_2\}$, if R_2 is aware of R_1) by always looking ahead. In pattern (d), R_2 finds O_1 by observing R_1 releasing an object. In pattern (e), R_2 finds O_2 by predicting the trajectory of R_1 and checking for possible future collisions. In pattern (f), each robot finds the object handled by the other by searching from the other's body towards its arm-tip.

In summary, a common resource is found by observing the other robot's behavior and extracting an object (or a place) relevant to the observed behavior. This is equivalent to an existing method called "deictic representation"[2].

It is extremely important to note that the *detected common resource is immediately used for establishing self behavior.* So the observation process is essential to ensure the common resource is directly perceived by the observer and can be included in a newly established perception-action cycle. Otherwise, if the robot is given an ungrounded data such as "grasp the object 001" via a communication link, it is not much help because the receiver of the message must locate the object in the real environment and relate it to its appropriate perception for the current behavior.

Another important question remains: *How can a robot choose an appropriate action?* Presumably, there is much work to be done before arriving at an ultimate solution. My current analysis of the problem is as follows: "Action selection relies on (1) observed other robot's behavior, (2) static shared knowledge, and (3) explicit communication." The weights of these three components varies depending the task. Also there is a trade of between them. My current framework is limited to "routine cooperative behaviors". It assumes that there is a one to one, static correspondence between the observed behavior and an appropriate self behavior. It corresponds to the level of reactive behaviors. A good example would be such situation as when we see somebody with handsful of luggages approach a door, we would open the door for the person.

4 Basic Cooperative Behavior Patterns

I have chosen and implemented three basic patterns of *Cooperation by Observation* [3], as shown in Fig. 4.

In this paper, I concentrate on the behaviors of the observer/helper robot "R2". Although these are the simplest cases, they include no physical coupling between "R1" and "R2". Thus, they are good examples where visual observation is inevitable.

In pattern (A), "Posing", the observer robot is always oriented toward the target robot and keeps a constant distance from it by maneuvering its mobile base. I assume a complex dynamic environment, so the observer robot must not be distracted by other static or moving objects. Posing serves as a base behavior for vision-based cooperation, since it keeps the helper robot attending to the target robot.

Pattern (B), "Unblocking", is analogous to a common event in our everyday life, that is, if someone with a handful of baggage approaches a closed door, someone should open the door for

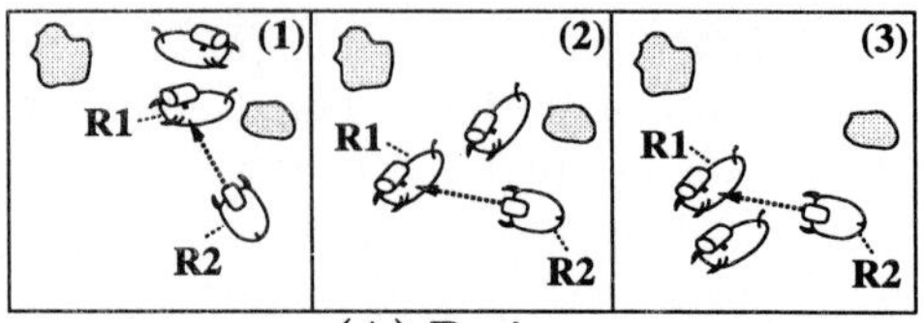

(A) Posing.

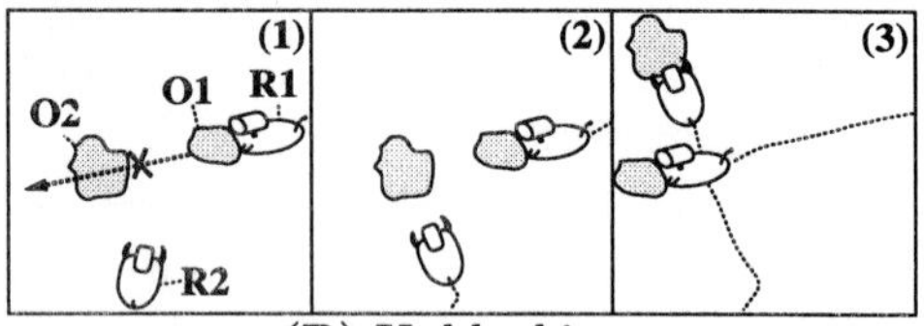

(B) Unblocking.

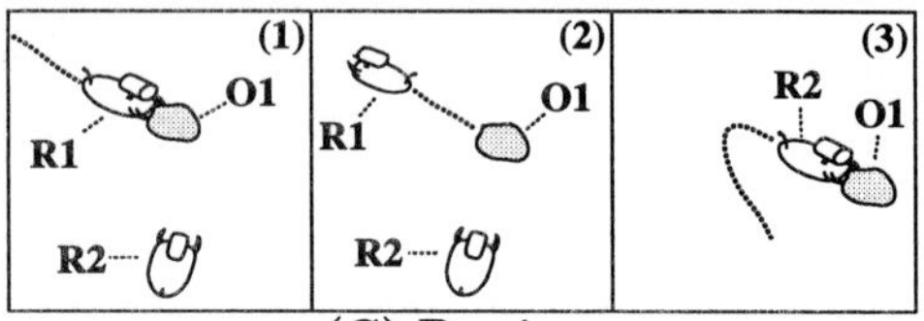

(C) Passing.

Figure 4: Cooperative Behavior Patterns. The three basic cases are chosen and realized. (A) Chasing the target robot among others keeping a constant distance. (B) Helping the other transporter robot by removing the obstacle. (C) Passing a cargo from one robot to another.

the person. The robot "R1" is pushing "O1" and on its way ahead, there is an obstacle "O2". The helper robot "R2" anticipates the possible collision between "O1" and "O2", and helps "R1" by pushing "O2" aside from the path.

Pattern (C), "Passing", is an example of "bucket brigading" type of cooperation. The robot "R1" brings a cargo "O1" and leaves it in front of another robot "R2". The relay robot, "R2" waits until "R1" releases "O1", and then approaches and pushes the "O1" to carry it further.

5 Basic Visuo-Motor Functions

Currently, we assume a mobile robot with vergence controlled stereo cameras. Since the details of vision algorithms are out of the scope of this paper, we will describe them only briefly.

(1)Find: This redirects an observer's attention to a new target, e.g. a moving robot and obstacles. First, an optical flow field is continuously computed by an incremental version of the Horn&Schunck algorithm [4]. Our

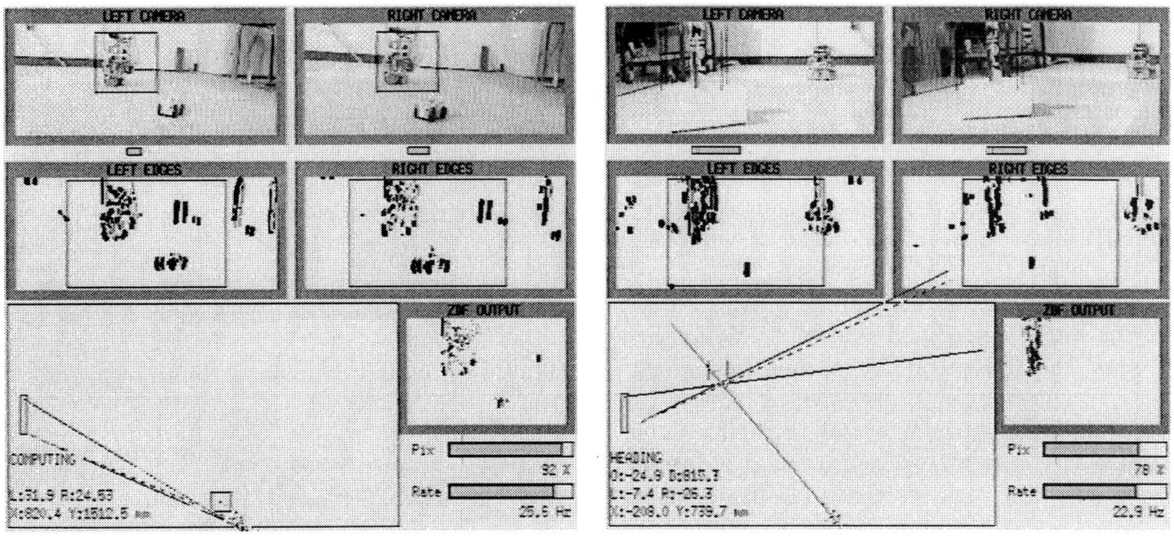

Figure 5: Tracking a mobile robot (left). Trajectory estimation and obstacle detection (right). In both figures, raw stereo images are shown at the top, extracted edges in the middle, ZDF outputs at bottom-right, and results of computation are shown at bottom-left. The whole processing is done at near frame rate (20 – 30Hz).

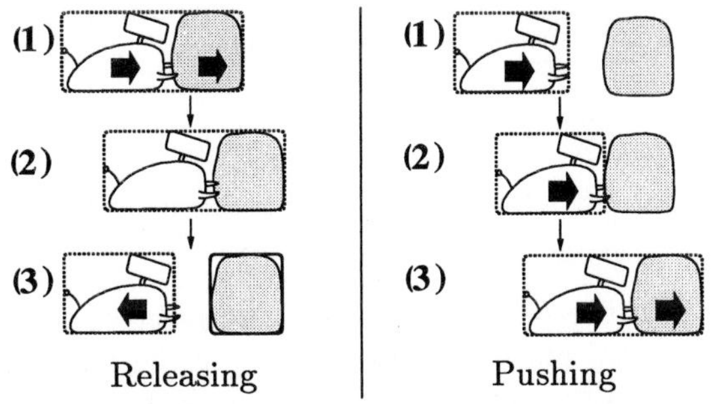

Figure 6: Detecting Release and Pushing Actions. The key features are sudden changes of the size of the bounding boxes of the segmented optical flow fields.

current implementation executes 1 or 2 iterations per each frame acquisition within about 100 [msec]. Then, noise removal by local averaging is applied. Next, the flow field is compared with a theoretical background flow to cancel self motion and to detect interesting targets. In order to deal with the significant noise in the flow field, we currently assume pure translation or pure rotation as the self motion and do the comparison with coarsely quantized intensity and orientation of the flow vectors. The background flow for a rotation is trivial. For a translation, we compute a theoretical flow produced by a virtual vertical wall in front of the robot at a preset distance (which is actually a time to collision), using self velocity. Using this *virtual wall* field as a threshold, we can ignore distant targets and extract only close obstacles and quickly approaching robots. The result field is then segmented by clustering the residual vectors. The above method can detect moving/static objects from a moving robot.

(2)Track: Strong spatial selectivity is required for tracking in a complex environment. We adopt vergence controlled stereo with zero disparity filtering (ZDF) [5], which has strong depth selectivity. The ZDF extracts zero disparity image features by first extracting vertical edges from the left and right edge images and simply taking "AND" of the results. The result contains only those features lying on a "horopter", a circle defined by the two camera centers and the current fixation point. In combination with the classical 2D region of interest method, the system can actively select a 3D region to extract visual information. The ZDF output edges are tracked by controlling the vergence angle of stereo cameras. In order to follow the target motion in depth axis, we extended the original ZDF by incorporating a limited search around near-zero disparities to find the best match, and feedback the result to vergence control [6].

(3)Anticipate: In the unblock situation, the helper, R2, need to anticipate a possible collision of O1(R1) with O2. When the target motion is tracked over a constant distance, a 3D line is fit to the recorded trajectory by the least mean square method. Then the system quickly scans the gaze point along this line by vergence control. During the scan, the ZDF output is continuously monitored to detect an object. An experimental result is shown in Fig. 5(right). Near-zero disparity search is turned off in this phase, because this time we have a top-down specification of the target depth.

(4)Event Detection: For a passing type of task, the receiving robot must synchronize its action by recognizing a release action of another robot. This is done by monitoring the temporal change in the segmented optical flow field, as shown in Fig. 6. The event is signaled when the area of segmented flow suddenly decreases and the average velocity of the area is reversed (Tistarelli and Sandini [7] presented a similar method for recognizing a push action). In order to simplify the computation and to cope with the noise, we assume that the observer is not moving while detecting the release action, and approximate the segmented region by its surrounding rectangle in our current implementation.

(5)Navigation: This guides a robot to the target position, e.g. O2 in the unblocking situation, and the released object in the passing situation. When there is no other constraints and the robot can take a straight line path to the visible target (as in the unblocking situation), the robot uses visual guidance. The steering angle is selected from {-5, 0, 5}[deg] according to the sign of the displacement angle of the current fixation point of the cameras from the current forward direction of the mobile base. Of course, this crude control does not converge well, but is sufficient for preliminary experiments since the cycle time is short (about 100[msec])compared to the speed of the mobile base. In some cases, a robot cannot directly approach the target but must follow a precomputed trajectory. For example, in the passing situation, the receiver robot must push the target object in the same direction as the first robot did. In such cases, the robot must compute an approach path in order to cope with the non-holonomic nature and/or preserve enough approach distance so that the visual guidance routine can correct the significant odometric error in self orientation imposed by turning itself to the desired pushing direction.

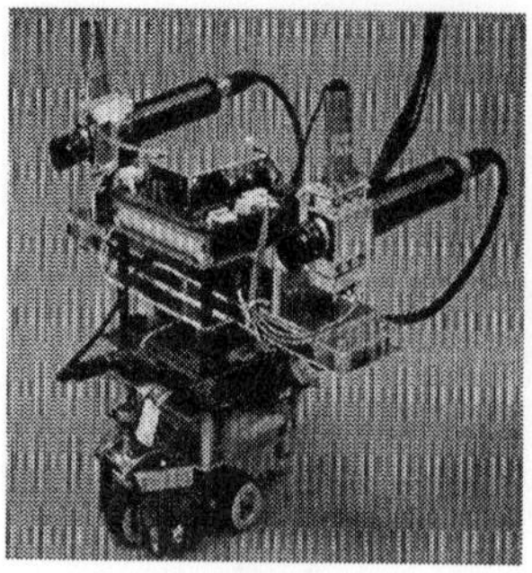

Total spec:	
length	: 0.20[m]
width	: 0.30[m]
height	: 0.28[m]
weight	: 2.5[kg]
max.speed:	0.20[m/s]
Head	:
DOF	: 2(verge)
baseline	: 0.25[m]
range	: ±30[deg]
precision	: ±0.4[deg]
max.spd.	: 50[deg/s]

Figure 7: A mobile robot with a stereo gaze platform.

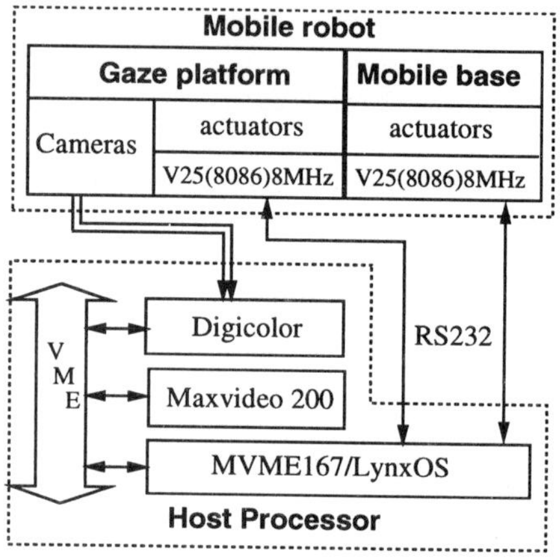

Figure 8: Overview of the prototype system.

6 Experiments

In order to verify the effectiveness of the visuomotor routines and to present a concrete examples of cooperation by observation, we have realized three task patterns; posing, passing and unblocking.

6.1 The prototype system

Fig. 7 shows one of our mobile robot equipped with a stereo gaze platform. The gaze platform has 2 DOF for vergence control. The robot has two onboard processors, one for controlling the mobile base and another for the gaze platform. Image processing is currently done on a remote host processor which consists of pipeline image processors (Datacube Maxvideo system), and a CPU board running a real time operating system. The system configuration is given in Fig. 8.

6.2 Posing

The mission of the chaser robot is to keep a fixed relative position (distance and orientation) with regard to the target robot. There is a distractor robot moving around, that sometimes comes in between the chaser and the target. The chaser must lock onto the target. The result is shown in Fig. 9. The experiment proved that the behavior is quite stable and robust.

6.3 Unblocking

Fig. 10 shows a result. The helper robot successfully found the anticipated obstacle and helped the transporter robot by pushing away the obstacle just in time for the transporter to pass.

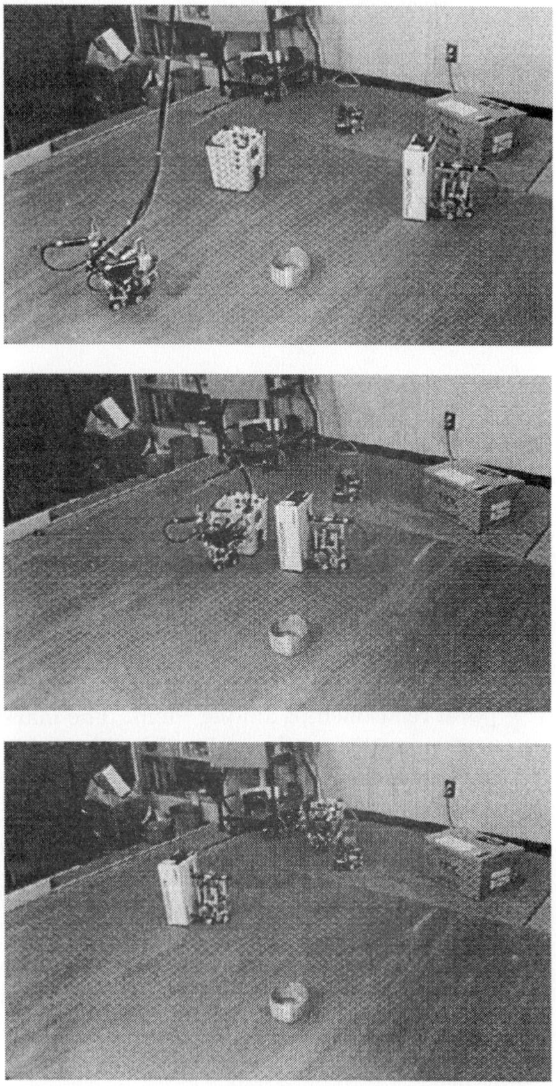

Figure 10: Unblocking behavior: The binocular robot (helper) on the left watched the monocular transporter robot pushing a block from right to left, estimated the path and found the possible obstacle on the path (top figure), then quickly approached the obstacle using visual servo (middle figure), and successfully pushed it away to help the transporter (bottom figure).

6.4 Passing

The realized "Pass" behavior pattern is shown in Fig. 11- 12. Figure 11(a)-(c) shows the result of motion segmentation and release detection. The released can is detected and the receiver robot starts to follow the computed approach trajectory (d). This is done in an open-loop manner using the mobile space buffer because the can goes out of sight temporarily. After a "Pose" behavior, the can becomes visible again (e), then the robot approaches it by visual servoing (f) - (g). Figure 12 shows a bird-eye view of the entire situation.

7 Integrated Architecture

Figure 13 shows the abstract organization of an agent architecture which integrates the basic behaviors presented before.

The basic structure consists of four stages, sensor data preprocessing, intermediate buffers, behavior modules, and arbitration networks. It is partly inspired by the architectures proposed by Brooks [8] and Chapman [9], but with a significant extension using two types of buffers.

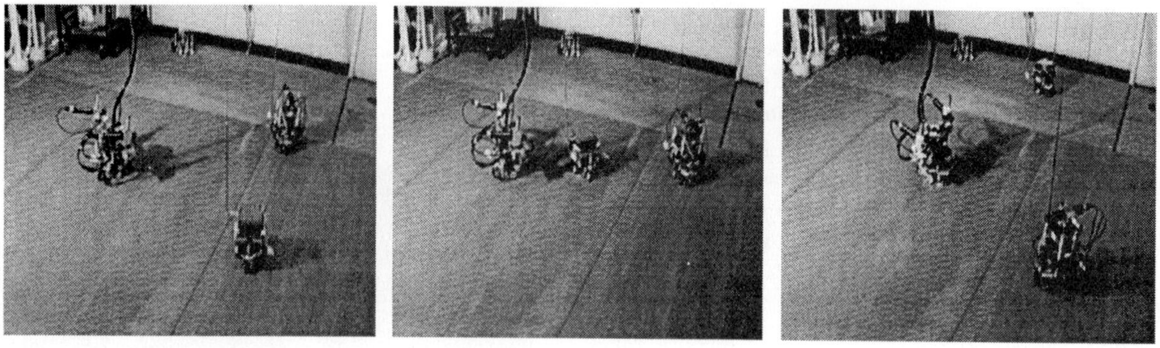

Figure 9: Posing behavior. The chaser (left) follows the target (moving downward). Even when the distractor (moving upward) comes in between the chaser and the target, the chaser is not disturbed and locks onto to the target.

The *buffers* mediate perceptual data flow. They contain *markers* which realize the following dynamic functions.

- Intermediate representation: MSB (mobile space buffer) represents 2D egocentric limited range map. ISB (image space buffer) represents 2D image space.
  The markers "mark" particular perceptual features and represent geometric and temporal relationships among them. The markers are continuously updated by the sensor data. And the relationships reflect the current state of the world. In this sense, markers can be regarded as "floating around" in the perceptual data stream and "mark" particular structure over selected substreams.

- Filters, or selective attention: The behavior modules take inputs from the markers. Thus the behavior modules reacts to the selected features.

- Short term memory: The markers reside in the buffer for some duration of time, until it is removed by some behavior module or fades away by auto temporal degradation (forgetting). When a marker has a direct perceptual input, it is continuously updated by the data. But when the input is lost for a short while, the marker tries to predict the current status. For example, when a marker is associated with a visible object and later the object falls off the view angle of the cameras, the marker predicts the current position by motion prediction and self motion cancellation. This is only valid for a short term because of uncertainty. Another aspect of short term memory is to serve as a mechanism for behavior coordination and chaining. This is discussed separately.

- Behavior coordination: Since the markers serve as short term memory, and they also feed data to behavior modules, the markers coordinate the behaviors. There are mainly two types of coordination: (1) Simultaneous combination and (2) Behavior chaining. In (1), if more than two behavior modules get inputs from the same marker, these modules react to the same data, producing combined effects. In (2), if a marker triggers one behavior, and that behavior places another marker which triggers another behavior, a systematic chaining of many behaviors can be realized. This is quite similar to a production system. The major difference is that the contents of the short term memory is always updated by perceptual inputs. This assures that the behavior chaining is always grounded in the real situation.

The details of the architecture is somewhat out of scope of this paper (see [10]). The important point is that the behavior chaining function of the buffers realizes the *chaining* of *common resource* detected in one behavior context to the next one, thereby giving the robot a combinatorial power.

For example, in one demonstration, the robot first does "posing", marking the target robot in the MSB. Then, the target position is accumulated and a module predicts the target trajectory, placing a line marker in the MSB. This invokes a visual searching behavior, which controls the fixation point along the line. When a sig-

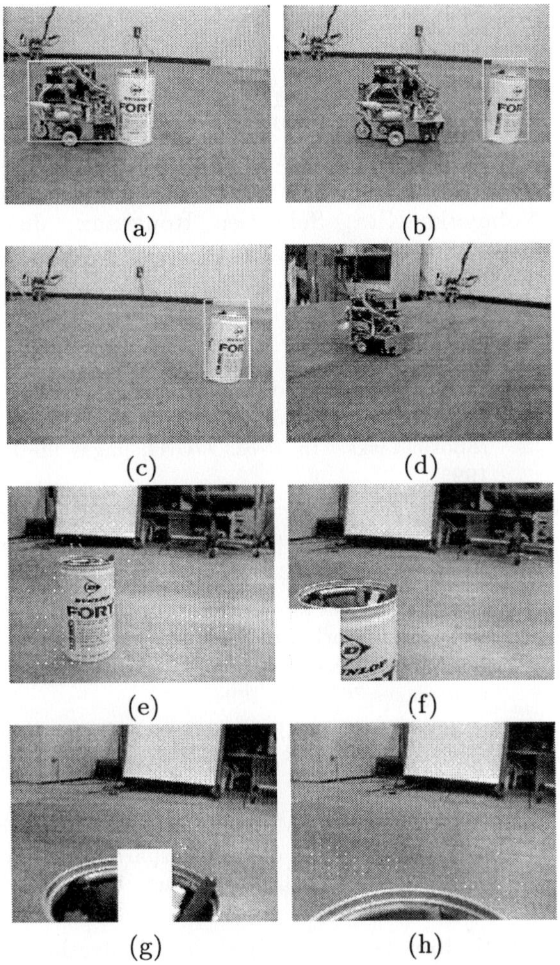

Figure 11: Vision Processing during a Receiving Action; A view from the receiver robot. (a) A segmented optical flow region including the robot and the can. (b) A release action is detected. (c) The released can is segmented. (d) Starting an open loop MoveTo behavior. (e) MoveTo behavior completed and recaptured the can. (f),(g),(h) VisServo behavior based on optical flow. The white box indicates the HOME marker.

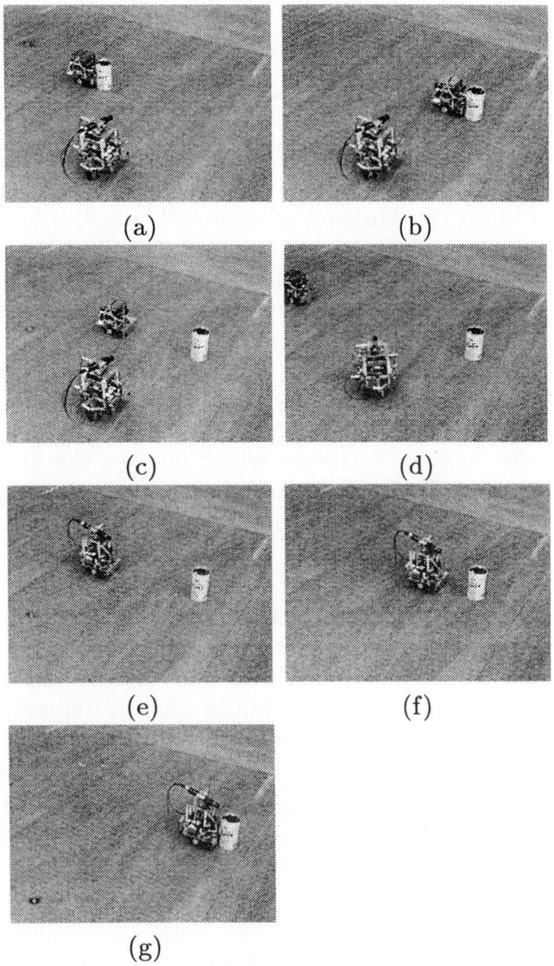

Figure 12: Passing Task. (a)(b) A robot is pushing a can from left to right. (c) The first robot releases the can and returns to fetch another can, and the helper robot recognizes the situation. (d) The helper robot determines the pushing direction, and starts to follow a computed approach trajectory. (e)(f) The helper robot approaches the can using visual servo. (g) Pushing the can in the same direction as the first robot did.

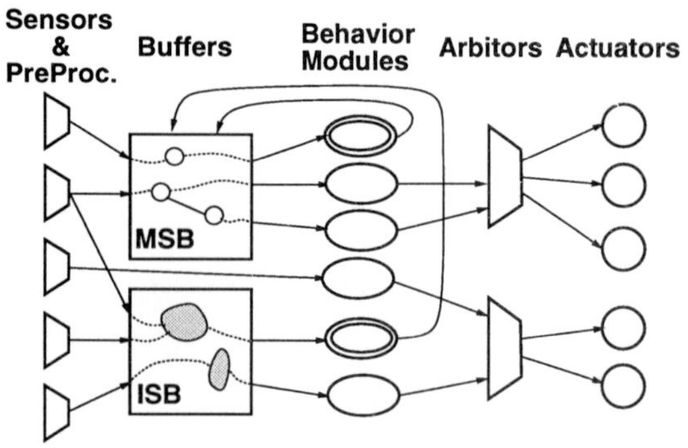

Figure 13: Abstract structure of behavior architecture. MSB: Mobile Space Buffer. ISB: Image Space Buffer. Double ellipses denote abstract behaviors which affect buffers.

nificant feature is detected through ZDF during the search, a goal marker is placed to represent the detected obstacle, which invokes the visual servo navigation toward the obstacle to realize the "unblock" behavior.

8 Discussions

Behavior switching is quite important for realizing a combinatorial power in cooperating multi-robot systems. Recently, a few frameworks have been proposed, which basically adopts common methods. For example, ALLIANCE[11] introduces "motivations" or activation values of behavior modules. These values are dynamically changed by each behavior module based on the environmental state. The work also considers the effect of knowing other agents' task state through explicit communication.

This paper introduced a different aspect of dynamic nature of cooperative tasks and proposed that common resource recognition is essential in dynamic cooperative task patterns. This is quite a novel concept and has not been recognized in previous works on cooperative robots.

The common resource must be searched within the current context of each robot, and once such objects or places are found, they must be immediately used in the next selected action. This way, a robot can temporally combine different behaviors while matching them with other robot's behaviors to realize complex cooperative behavior patterns.

As an integrated framework of such notions, "cooperation by observation" has been presented. Concrete implementation of basic task patterns was presented as first examples of the framework.

Many important problems remain to be solved, such as dealing with ambiguous situations, and adding flexibility to behavior type selections.

Acknowledgements

The author is grateful to the following contributors to the implemented system: Makoto Ishii, Nobuyuki Kita, Sebastien Rougeaux, Jukka Riekki and Kenji Konaka.

References

[1] R. I. Brafman and Y. Shoham. Knowledge considerations in robotics and distributions of robotic tasks. In *Proc. IJCAI*, pages 96–102, 1995.

[2] P. E. Agre and D. Chapman. Pengi: An implementation of a theory of activity. In *Proc. AAAI*, pages 268–272, 1987.

[3] Y. Kuniyoshi et al. Cooperation by observation – the framework and basic task patterns –. In *Proc. IEEE Int. Conf. Robotics and Automation*, pages 767–774, 1994.

[4] B. K. P. Horn and B. G. Schunck. Determining optical flow. *Artificial Intelligence*, 1:185–203, 1981.

[5] P. von Kaenel, C. M. Brown, and D. J. Coombs. Detecting regions of zero disparity in binocular images. Technical report, University of Rochester, 1991.

[6] S. Rougeaux, N. Kita, Y. Kuniyoshi, and S. Sakane. Tracking a moving object with a stereo camera head. In *Proc. 11th Annual Conf. of Robotics Society of Japan*, 1993.

[7] M. Tistarelli and G. Sandini. Dynamic aspects in active vision. *CVGIP: Image Understanding*, 56(1):108–129, 1992.

[8] R. A. Brooks. A robust layered control system for a mobile robot. *IEEE Trans. Robotics and Automation*, RA2:14–23, 1986.

[9] D. Chapman. *Vision, Instruction and Action.* MIT Press, 1991. ISBN 0-262-03181-7.

[10] Y. Kuniyoshi et al. Vision-based behaviors for multi-robot cooperation. In *Proc. IEEE Int. Conf. Intelligent Robots and Systems*, 1994.

[11] Lynne E. Parker. ALLIANCE: An architecture for fault tolerant, cooperative control of heterogeneous mobile robots. In *Proc. IROS*, pages 776–783, 1994.

PART 9

3D VISION

SESSION

Chair: Takeo Kanade

The Robotics Institute, Carnegie-Mellon University

5000 Forbes Avenue, Pittsburgh, PA 15213, USA

3D Vision - Shape and Colour

Ray Jarvis
Intelligent Robotics Research Centre
Monash University, Clayton, Victoria 3168
Australia
Ph: 61 3 9905 3470
Fax: 61 3 9905 3454
e-mail: Ray.Jarvis@eng.monash.edu.au

ABSTRACT

The potential complementarity of surface shape and colour in support of 3D Vision is often neglected, as is evidenced by the relative paucity of such studies in the technical journal literature. This paper argues for greater attention to be paid to such combinations and presents three case studies and two future projects to emphasise this stance.

1. INTRODUCTION

In an earlier publication [1] the author conjured up a 'magic powder' which could provide a rich data source for 3D Vision:

> *It would be ideal if the exposed surfaces of objects could be sprayed with some magic powder, which, when gathered up, could, particle by particle, be interrogated as to where it had been in space and what the properties of the surface it touched were.*

This 'ideal' has yet to be realised but its potential value has not diminished. The properties of the surface local to points on it that were referred to could include shading, specularity, colour and visual texture as well as thermal conductivity, smell, feel, temperature, chemistry, ultrasonic absorption, etc. but, for 3D Vision, one is primarily interested in the first four of these. For Lambertian surfaces only the first three need be considered.

It is interesting to consider that, whilst surface properties such as colour, texture and specularity have been used extensively in 2D image analysis, including segmentation, parameter extraction and recognition, such information is often ignored in the domain of 3D Vision, where shape and shading (brightness) have ruled supreme. One explanation may be that recent rangefinding methodologies have largely emphasised triangulation geometries for both active and passive approaches and, in addition, time-of-flight notions for active. Spectral surface attributes are usually not considered of sufficient importance to justify the building of special apparatus to capture them along with range data. However, careful consideration of how colour (and perhaps also texture and specularity) can enhance shape and shading in supporting the 3D scene analysis process, including segmentation, object recognition, pose determination and multi-object structure, can reveal a richness worth exploiting.

Three examples and two proposals involving the exploitation of combined shape and colour information are presented in the following sections.

2. RANGE ENHANCED 3D ROBOT COLOUR VISION [2]

The aim of this work was to provide the scene analysis basis for simple robotic manipulator operations of primitive objects in cluttered environments. In this example, effective colour based scene segmentation was considered to be restricted to situations where individual untextured objects could be visually isolated through distinct intensity and colour variations at boundaries. Shadow effects which could cause false boundaries (non object boundaries) were largely illiminated by using a simple computation called a 'hue transform' on each pixel in turn:

For each pixel calculate
$I = \Sigma Ij$, j = red, green, blue
$Hj = K(Ij/I)$, j = red, green, blue

where the I's are intensity values and K is a scaling factor. The alternative of using $I = \sqrt{(I\ red)^2+(I\ green)^2+(I\ blue)^2}$ would be computationally more expensive. In fact, a table look-up version was used on low resolution data.

The Rosenfeld [3] binary connectivity algorithm was used as the essential segmentation tool, this being applied a multitude of times over binary slices through the hue transformed, colour image data. Each slice was defined by a 'key' of red, green and blue values and a tolerance band. The 'keys' were selected in raster order from the image data itself. Each binary slice was segmented using the connectivity procedure and the next as yet un-segmented pixel in raster order used as the next key, and so on until exhaustion. A simple merge procedure allowed regions with small variations across their borders to be combined. This two phase simple procedure was surprisingly fast and effective. Details are to be found in [2]. The application of this procedure to a simple scene is shown in Figure 1.

Where visual juxtapositions of like coloured objects occur, the above procedure would fail, particularly since only pure hue information remains after the hue transform. The addition of range data from a time-of-

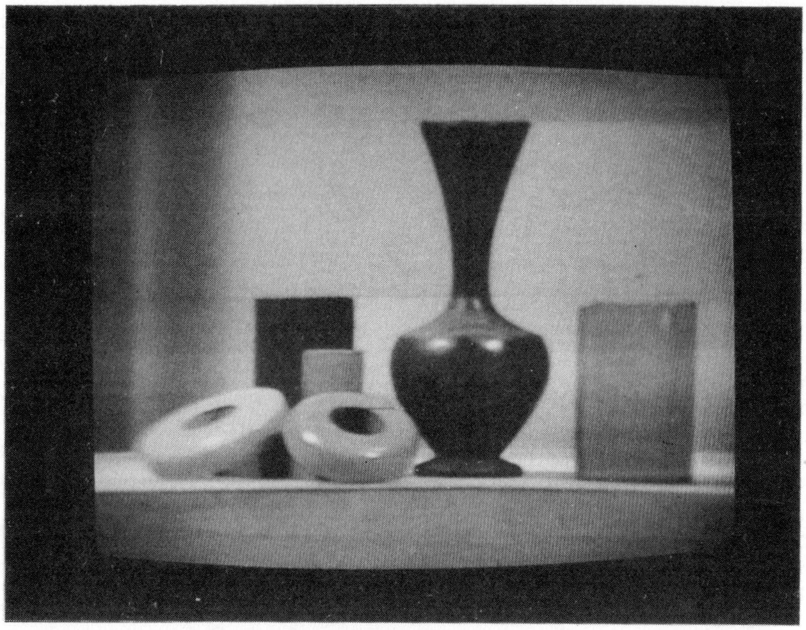

(a) Simple Colour Scene

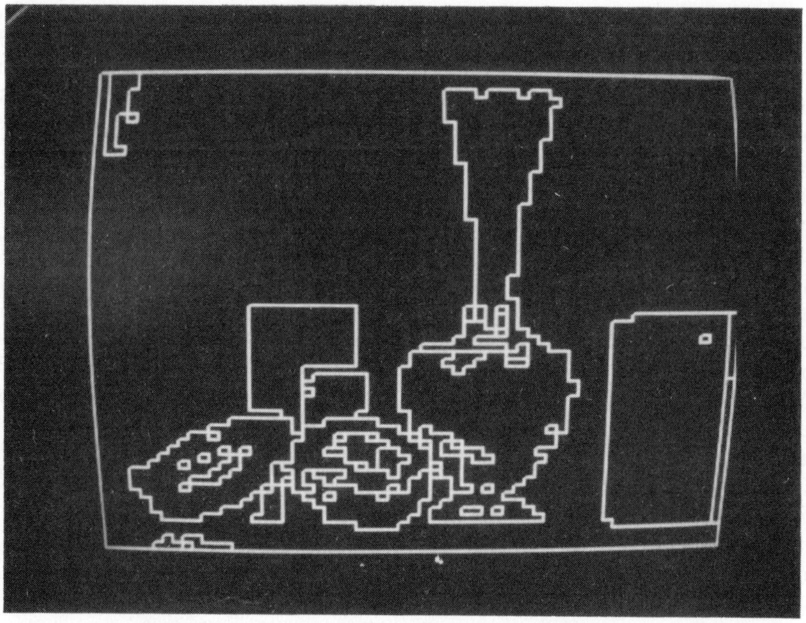

(b) Colour Based Segmentation of (a)

Figure 1. Colour Based Scene Segmentation

flight laser rangefinder [4] was seen as a solution to this problem. Each pixel/rangel was represented as a four element vector comprised of red, green, blue and range components. The same process of binary slice connectivity analysis segmentation and simple merge procedure was used as for the pure colour based segmentation. A typical result for a scene which could not be adequately segmented before the addition of the range data is shown in Figure 2.

In this colour plus range segmentation example, it is clear that a basic methodology designed for image data has been extended to accommodate range data. That the objects in the scenes used to illustrate the approach were brightly coloured and quite primitive in shape with mostly flat surfaces facing forward (towards the sensors) contributed to the acceptable results of Figure 2. Obviously, more complex shaped objects would not have been adequately dealt with using the very simple approach described above. In such cases it is likely that dealing essentially with the shape (range data) as the primary mode of analysis and adding the colour data to enhance the result would be the preferred approach.

3. 3D Shape and Surface Colour Sensor Fusion [5]

Many rangefinder methodologies have been developed to provide 3D shape data using active or passive instrumentation [6]. One of the most successful and affordable active approaches is based on using stripe lighting and a displaced camera whose images are analysed to extract range from the known triangulation geometry of the equipment set-up. Whilst the analysis of images of single stripes can obviate the possible stripe identity ambiguity that can occur in multiple object scenes, this approach is slow when using commercially available video cameras which have been developed for analog television compatibility. An alternative approach is to use a binary coding scheme [7] which limits the number of images to be analysed to $\log_2 n+1$, where n is the number of light stripes; this scheme resolves the potential ambiguity problem and yet is certainly less tedious than analysing one image for each stripe. Doubling the resolution requires only one more image to be analysed. Normally, a monochrome video camera is used for this type of instrument. However, the use of a colour camera, whilst likely to reduce the accuracy somewhat (since colour sensitive triads of active elements are used), can guarantee good registration of range and colour image data and thus provide a rich and convenient source of information for 3D scene analysis. Since the range data is extracted from image arrays which are uniform, the simple immediate neighbourhood of pixels allows meshes of range data to be constructed; deriving such meshes from unordered range data (x,y,z triplets in cartesian space) is computationally very expensive. From the range meshes it is easy to generate the local surface unit normals at each range point by calculating the normalised cross product between two vectors from a range point to its neighbours which are easily identified in the mesh. Figure 3 shows a mesh and a set of unit surface normals.

Now the whole data set can be considered to be a set of feature vectors, each of which has nine attributes:

x,y,z the cartesian coordinates of the surface point
r,g,b the colour intensity components of that point
n_x, n_y, n_z the unit surface normal components

Natural structure within this nine element vector data set can be discovered using cluster analysis. Sub-spaces involving only particular components as well as the whole space can be subjected to such analysis.

The Shared Near Neighbour clustering procedure devised by the author [8] was used to see just what could be achieved in this way. A particular advantage of the Shared Near Neighbour clustering procedure is that it auto-scales the feature vector volume domain of influence in proportion to the sparcity of samples. It is also computationally efficient and can handle a considerable range of difficult clustering problems with tuning being restricted to adjustments of two integer parameters, one being the neighbourhood size in sample number terms and the other a threshold to determine whether a cluster link is made or discarded. It is based on the simple concept that two individual points in Euclidean space should be in the same cluster provided that their respective specified neighbourhoods share a sufficient number of common samples, provided one or both of the pair of points is not an outlier. The outlier problem is easily accommodated by requiring each point to contain the other in its specified neighbourhood. Some typical results are shown in Figure 4. This clustering method is ideal for the shape/colour segmentation method presented here as it is very tolerant of a variety of cluster types and can be tuned to be essentially 'centre of gravity' based or 'continuity' based or anywhere in between. Its degree of segmentation resolution is also variable according to the needs and expectations of the particular applications. The results can be summarised as follows:

(i) Clustering in x,y,z space can segment the scene into objects which are geometrically separated.

(ii) Clustering in unit surface normal component space can group points according to surface normal directionality; plain faces of objects are easily discovered this way.

(iii) Clustering in colour component space allows homogeneously coloured pixels to be grouped.

(iv) Clustering in the whole nine dimensional space can group pixels which are geometrically connected and of similar colour and surface normal directionality.

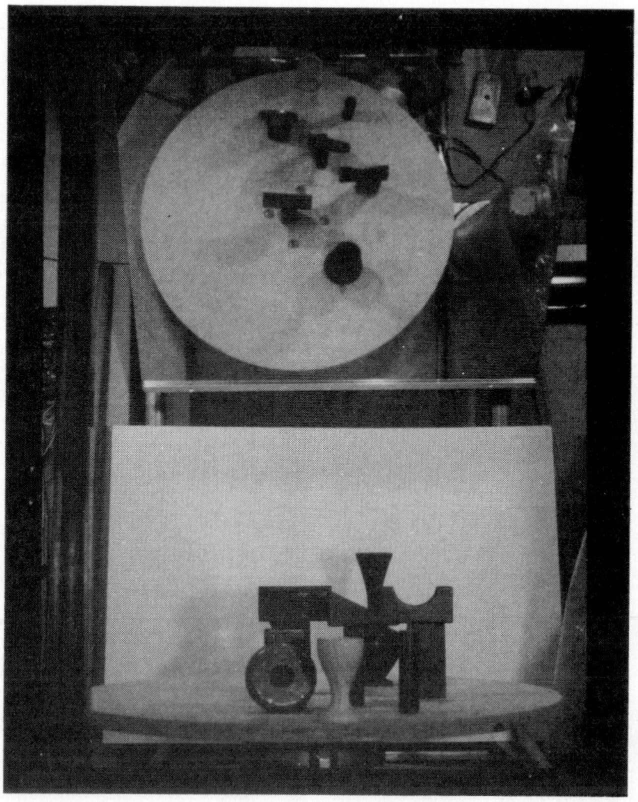

(a) Complex Colour Scene

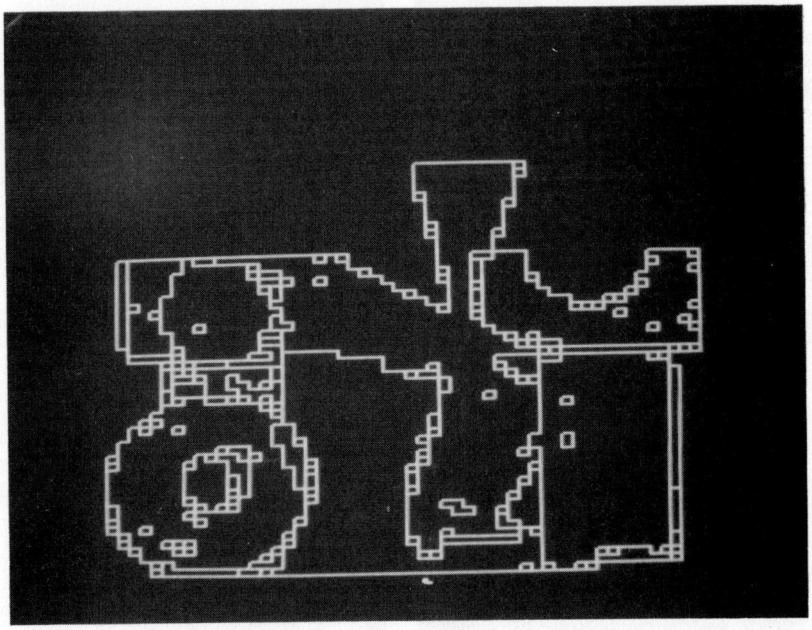

(b) Colour Based Segmentation of (a)

Figure 2. Range Enhanced Colour Based Scene Segmentation

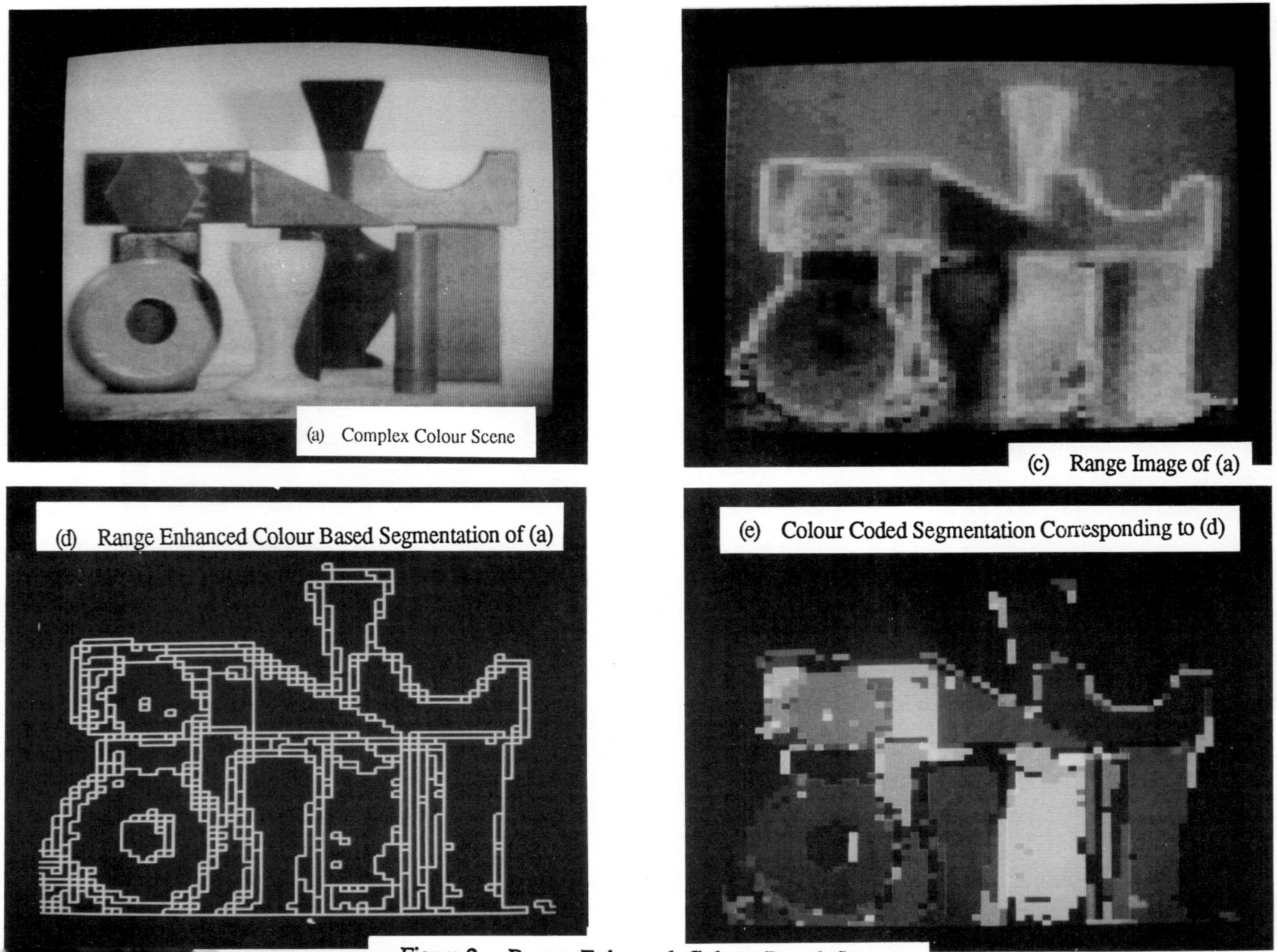

(a) Complex Colour Scene

(c) Range Image of (a)

(d) Range Enhanced Colour Based Segmentation of (a)

(e) Colour Coded Segmentation Corresponding to (d)

Range Enhanced Colour Based Scene

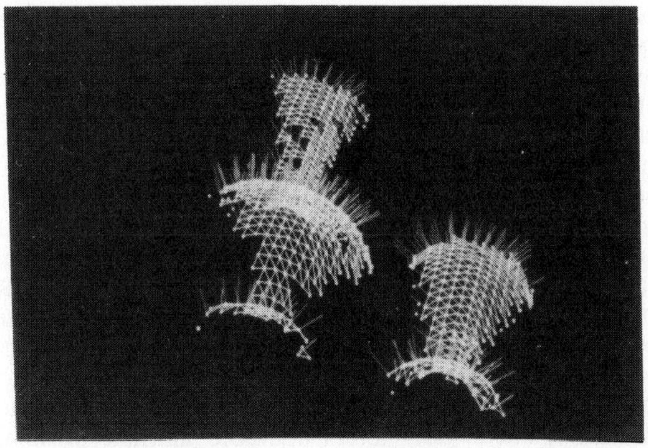

Figure 3. A 3D Mesh and Surface Normal Needle Diagram

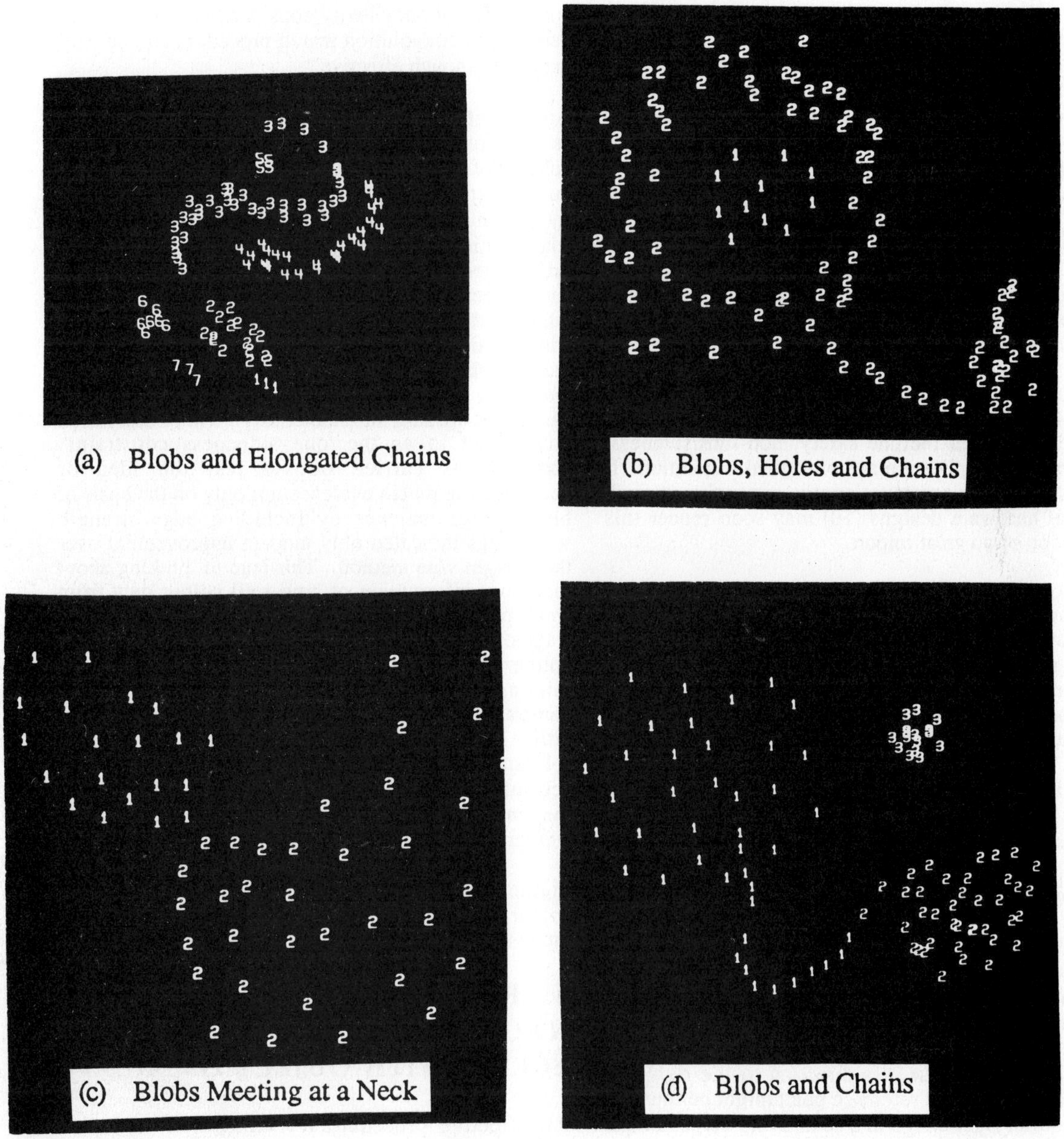

(a) Blobs and Elongated Chains

(b) Blobs, Holes and Chains

(c) Blobs Meeting at a Neck

(d) Blobs and Chains

Figure 4. Some Typical Shared Near Neighbour Clustering Results

Figure 5 illustrates some of the results of such analysis. Clearly this approach is entirely general and low level and has not taken advantage of any domain specific knowledge which might be available and ought to be exploited at some stage. The type of grouping structures which can be discovered by the clustering approach in coordinate/colour/surface normal component space are very supportive of subsequent scene analysis procedures. For example, geometrically separated objects can be analysed in terms of colour and surface normal distributions for recognition purposes and principal axes of the geometric coordinate data extracted for pose determination. The 'centres of mass' of the objects also allow for positional inter-relationships amongst the objects to be derived. Multiple view rangefinding can provide a fairly complete set of surface property feature vectors for a complex scene.

In this example range data and derived surface normal data play dominant roles but the addition of registered colour component data has been achieved with little extra effort, yet enriches this shape data considerably.

4. Quad Camera Stereopsis Ranging with Colour Match Verification [9]

This example uses colour data to help extract reliable range data from passive stereopsis by assisting in the disparity match process.

There have been many attempts [9,10,11,12] to exploit stereopsis for range extraction. The advantages of this approach include safety, non-intrusiveness, generality of application and biological inspiration. It is also usually computationally expensive though special hardware designs [10] may soon render this objection of no great import.

If more than the minimum number of two cameras are used, then each pair can contribute towards improving the reliability and accuracy of the result. With n cameras there are n(n-1)/2 such pairs so the advantage is square law related with respect to the number of cameras. It is well known that disparity/baseline is an invariant with respect to a point in space whose image is matched between cameras. After all, disparity/baseline is inversely range related and the range to a fixed point in space is constant. This property can be exploited [9,10].

If four cameras are arranged with their axes parallel and at the corners of a square with their scan lines parallel with the top and bottom sides of that square, a simple way of exploiting the above invariance exists. For example, if the top right most image is taken as the reference, an edge point in that image may match a point in the top left image along a search line parallel to the scan line and in a direction to the right. A match for that same reference point might be found in the bottom right image in a search line perpendicular to the camera scan line and in an upward direction. A third match could be found in the bottom left image along a diagonal search line in a upward and to the right direction. If all three matches are correctly found, the disparity between the top two images should be the same as that between the two right images and the diagonal disparity should be $\sqrt{2}$ times this. Thus a synchronised (locked step) search from the coordinate position of the reference edge point transferred to the other images with a pixel index incrementation to the right in the top left image, upwards in the bottom right image and both to the right and upwards in the bottom left image should be able to accumulate up to three votes for supporting an edge match hypothesis. Three votes would be accepted with confidence and fewer with more suspicion. But if only edge position is used to contribute to such votes, a considerable amount of image data useful in strengthening the confidence of or discarding a hypothesised match will have been wasted. However, using only edge data for matching can reduce the computational complexity considerably. A full large kernel area convolution match procedure can be very expensive though effective.

An elegant compromise is to use edge matching votes as cues for further hypothesis testing concerning the validity of matches. Of course, not all low votes are due to suspicious matches but could have been caused by the 'missing parts problem' which plagues all triangulation based ranging methods. This problem arises when not all that can be seen by any one camera can be seen by all the others. But camera separation, which is the source of this problem, is the basis of the triangulation approach. The effect of the missing parts problem is most critical at what are called 'obscuring edges'. However, both sides of such edges cannot be missing simultaneously. This observation is referred to as the 'one sidedness constraint'. Experiments carried out to test the idea of accumulating match evidence not only on the basis of binary edge matches by including edge strength weightings indicated only modest improvement over the straight vote method. This lead to thinking about including information of how well colour data from small patches around the hypothesised matched edge fragments matched amongst the images. But a poor colour match on one side may have been due to the missing parts problem. Thus the one sidedness constraint was accommodated by accumulating votes if either side patches matched within a specified tolerance. Thus each hypothesised match can accumulate one vote for a particular disparity if both the binary edge position and the one sidedness colour similarity tests are successful.

Figure 6 illustrates the effectiveness of the above colour match verification in improving the quality of the result.

5. Free-Form Object Recognition in Cluttered Scenes with Object Overlap

An effective procedure for recognition and pose determination of free-form objects with respect to a library of possible objects was recently devised in the Intelligent Robotics Research Centre at Monash

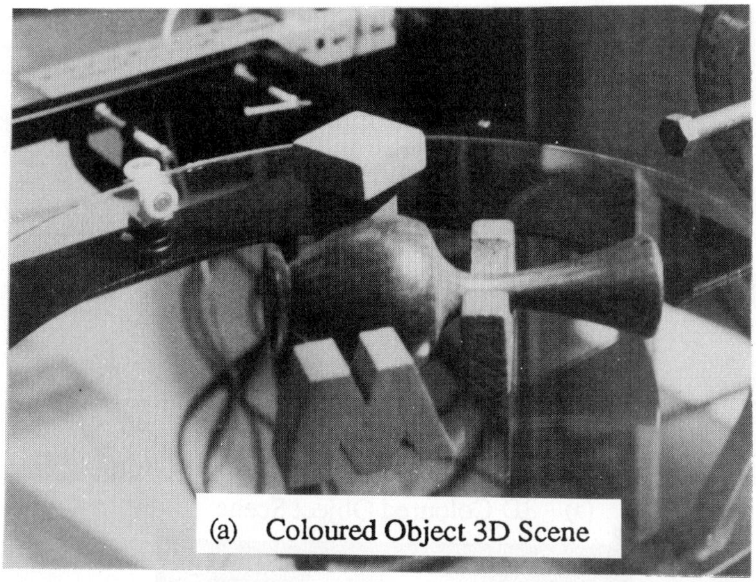

(a) Coloured Object 3D Scene

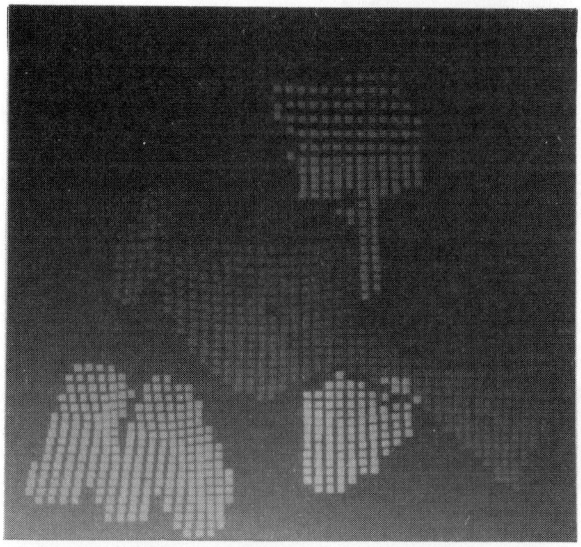

(b) Object Segmentation by Clustering in Spatial Coordinate Space

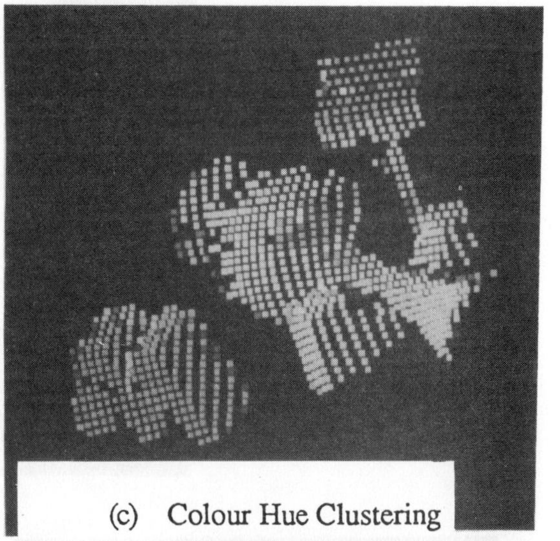

(c) Colour Hue Clustering Based Segmentation

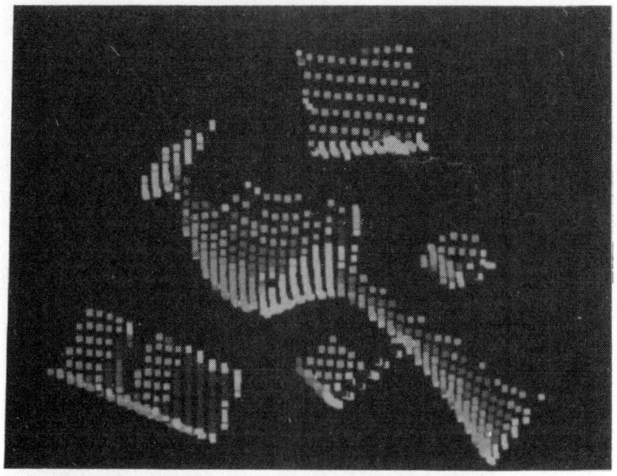

(d) Surface Normal Component Clustering Based Segmentation

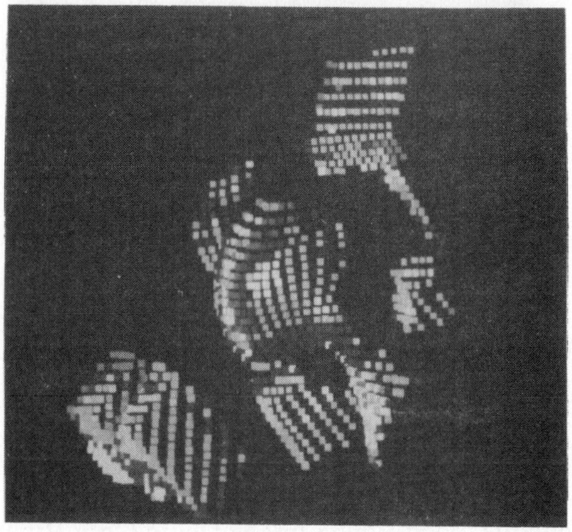

(e) Colour/Normals/Coordinate Space Clustering Based Segmentation

Figure 5. Cluster Analysis in Colour, Spatial Coordinate and Surface Normal Component Space

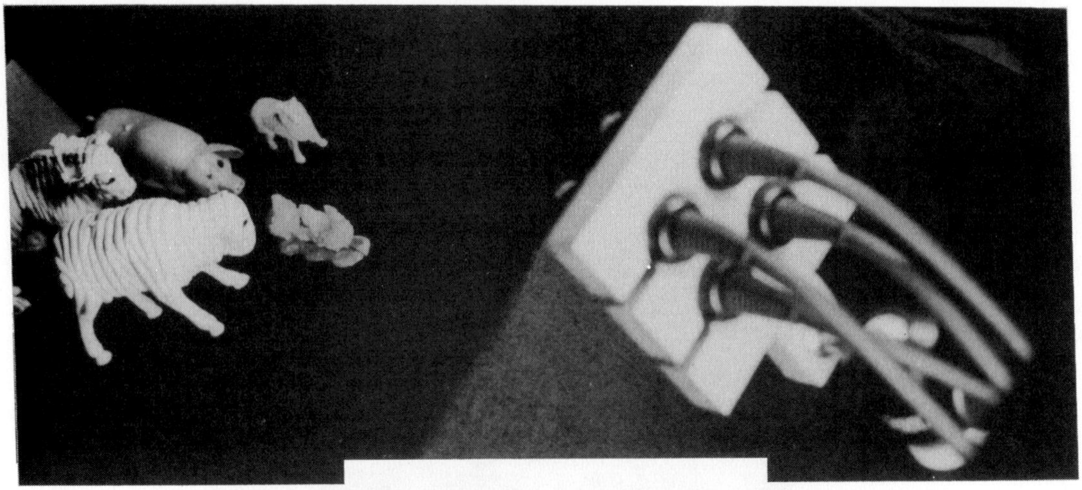

(a) 3D Coloured Object Scene

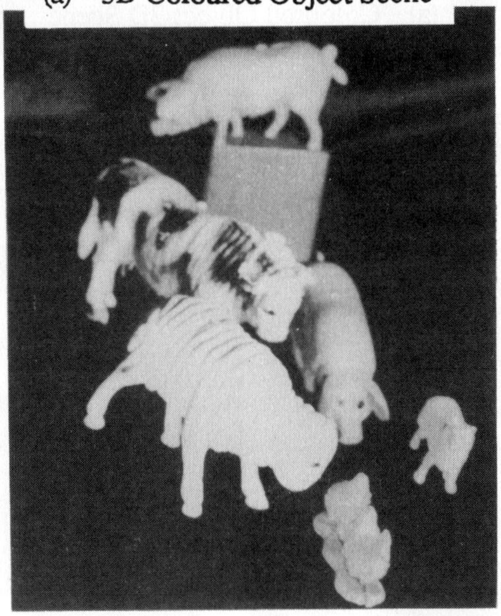

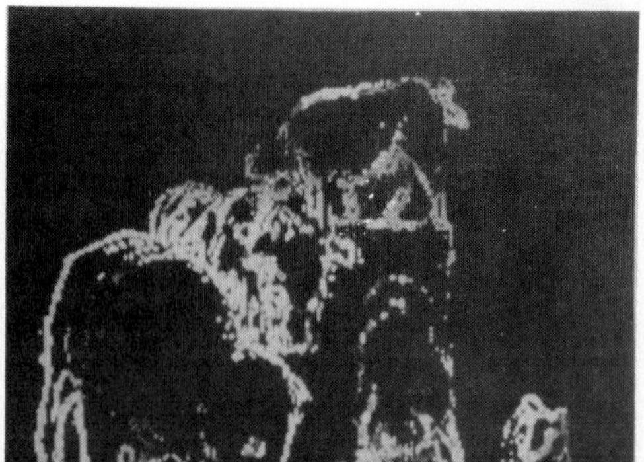

(b) Intensity Coded Range Image Using Only Binary Edge Match Disparities

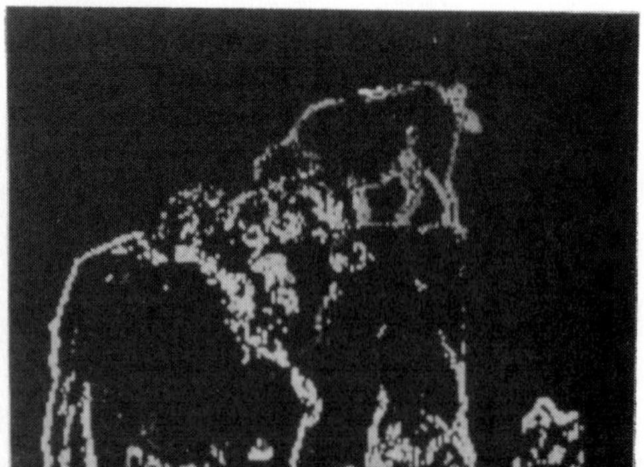

(c) Intensity Coded Range Image Using Binary E and Colour Verification Matching

Figure 6. Quad Camera Stereopsis with Colour Match Verification

University [13,14]. Some quite complex scenes could be correctly analysed using a combination of what were termed 'point signatures', three point matching searches and hypothesis verifications based on transforming whole sets of data points using parameters derived from matching triplets of surface points. The three point matching procedure was based on:

(i) geometric separation and local shape within a coordinate framework based on surface normal and principal curvature directions,
(ii) principle curvature value correspondence and
(iii) an elegant pruning method applied at the full point set transformation verification stage.

The free-form objects are considered to be rigid and no scaling was considered. Selecting which models from the library to test was based on 'point signature' matching and vote accumulation. A 'point signature' is a cyclic set of distances from points at the intersection of a sphere of specified radius centred on a surface point and the plane constructed by fitting a plane in a least means squared sense to this set of intersection points and translating it along its normal to the original surface point whose signature is being determined. In fact, these distances are determined at quatized angles around the 'circle' of intersection as a vector, $d(\theta)$, $\theta=0$ to 360^0 at intervals of K degrees. K is typically 5 to 20^0.

Such 'signatures' are translation and rotation invariant, periodic and not restricted to smooth surfaces. They also obviate the need to extract derivatures which are often used and can be very unreliable when the spatial data is noisy. The maximum and minimum element values of each signature extracted from the candidate object are used, with tolerances, to index into a library of signatures previously extracted from the models. Signature matches are used as votes for model selection. False selections are eventually discarded by an extensive verification procedure, but an ordering of hypothesised models on the basis of vote counts can usefully reduce the computational burden by at least discarding unlikely candidates at an early stage.

Figure 7 gives a very convincing illustration of what can be achieved using this methodology. For the scene of Figure 7(a) with its range data shown in Figure 7(b), it is possible to extract individual objects [Figure 7(c)] which came from the model set of Figure 7(d).

In this work to date all objects were deliberately made to have uniform, light coloured surfaces to enhance the quality of the range data. Not only is this unrealistic for many practical applications of the methodology, but a very rich source of recognition supporting data has been ignored.

If surface colour were available this could have been used to improve both major phases of the analysis. In the early phase the 'point signatures' could have included colour information and be described as $d,r,g,b(\theta)$, $\theta=0$ to 360 at K^0 intervals. That is, each component of the signature around the 'circle' is itself a four element vector. The average colour components could be used along with the maximum and minimum d values as index elements for access to the model signature database. So as to avoid lighting variations problems, hue transforms could be used instead of the raw colour component values.

In this way the indexing would be more precise and full matching of signatures more reliable. This in turn would lead to faster and surer extraction of the most likely models to be tested in the second phase of analysis. Whilst more data is available and this might seem to increase the computational burden, increased reliability would allow less tedious search procedures through safer pruning.

In the second phase the principle curvatures used can be augmented by colour hue components. This would allow more efficient triple point matching to be carried out, since all of point distribution geometry, curvature and colour evidence would be available.

Overall, the very effective existing methodology for free-form shape analysis could be improved by the addition of surface colour data. The above ideas may well be tested in a future research project to extend the work already carried out.

6. Shape and Colour Extraction for Landmark Based Robot Navigation

The use of fixed path robot navigation supported by localisation based on under-floor signal wires or on-floor markings has already given way to free-ranging navigation based on various beacon localisation methodologies. However, natural landmark based mobile robot navigation has many advantages over the previous approaches. These include generality of application both indoor and out, not needing to set up the environment beforehand and the elegance of combining environmental modelling, path planning and localisation in a unified way with potentially high reliability. Importantly, in general, cluttered obstacle fields can provide the basis of accurate localisation (at least relative to the immediate surroundings) just when it is most needed, whilst sparse obstacle fields may provide less reliable localisation in formation but this is less critical since collisions are less likely. With beacon based localisation schemes, obstacle clutter might make a sufficient number of the beacons inaccessible and thus render the method useless just when it is needed most. Ideally, natural landmarks should be distinct, unique, spatially compact, geometrically well distributed, temporally stable and relatively viewpoint invariant.

Both ultrasonics and vision have been extensively used for both beacon localisation and environmental modelling. Even though research progress in the former has come a long way from the simple but intrinsically unreliable first return time-of-flight pulse

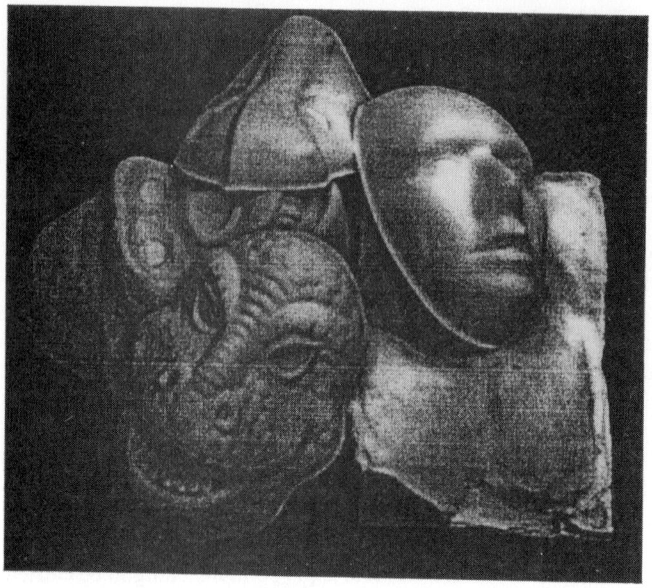

(a) A Complex 3D Scene with Overlapping Objects

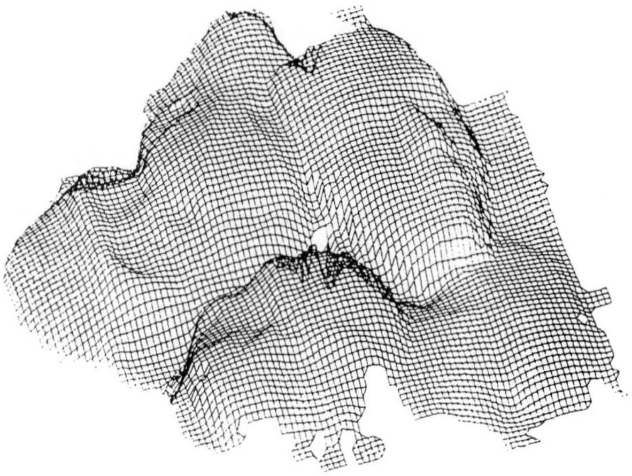

(b) Range Data for (a)

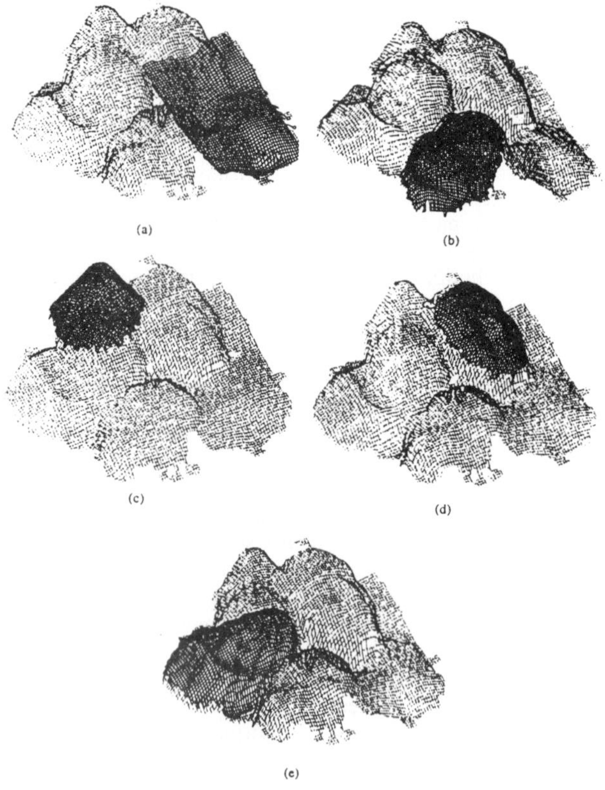

(c) Object Recognition and Pose Determination

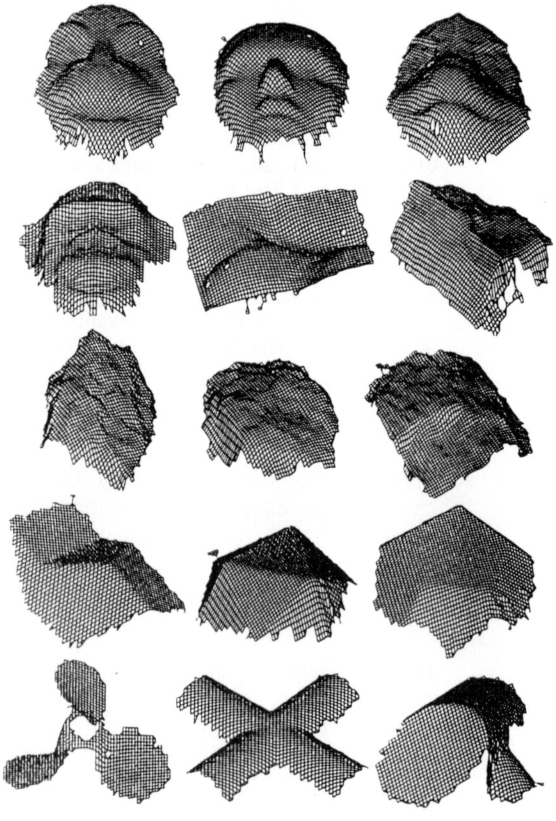

(d) Model Reference Set

Figure 7. Model Based Free-Form 3D Object Recognition and Pose Determination methods to those which use multiple sensors, physical scanning and the entire envelope analysis of the returned signal, they are still of little value beyond a range of 15 metres in air due to severe attenuation in that medium. Vision will emerge as the only way to go.

Nowadays, vision can include both image based ranging and scene analysis as well as active devices such as stripe light systems and laser time-of-flight sensors. Having quality (accurate and dense) range data with a high maximum range (say 50 metres) is of great advantage in combining environmental modelling with natural landmark based localisation. Even one horizontal slice of range data at the appropriate height above the ground can be very useful, at least indoors, where many vertical planes touching the ground are usually found.

Range data of this kind obtained over a large sweep of the working environment (say 180^0), with the robot at differing positions and orientations, needs to be matched to localise the robot. Fortunately, odometry and motion continuity can simplify the matching search over the three degrees of freedom required. What is sought is the affine transformation that moves robot centred range data to match the global model of the environment so far obtained.

Odometry can provide a good starting point for the search, especially if only small incremental movements are carried out between fixes. The main problem which arises in carrying out the match search is that obscuration can render the range data severely incomplete and reduce the uniqueness of the correct match due to missing components. For example, in Figure 8(b) which shows a single horizontal (at waist height) 180^0 range scan of the scene shown in Figure 8(a), the gap between the boxes in the centre of the range scan but slightly left in the photograph is clearly evident as one of many distinct range changes around the room. Shifting the viewpoint of the range scanner would cause many of those range 'jumps' to shift and also some to disappear (e.g. if the gap between the boxes was obscured), making the matching search somewhat difficult even if a reasonable starting guess were provided by odometry. Clearly, small moves not only provide accurate incremental odometry based fixes but reduces the variations between range scans. Thus, localising often, if possible, is a good idea, in general; this effectively means the problem is more one of tracking than matching *ab initio*.

If it were possible to capture colour data along the range scan line, perhaps by simply placing a wide angle colour camera in a position which is effectively coaxial with the centre of the range scanner sweep and choosing an appropriate image scan line that corresponds to the ranging beam position, the matching process could be improved, since each x,y range data point would also be associated with an r,g,b triple. Both search and verification could be improved in this manner, the former especially so if there happens to be one or two distinctly coloured element sequences along the colour scan line, as this would provide good starting alignment data. One or two distinctive range features as well would practically clinch the process. Effectively, the two sources of data can reduce interpretation ambiguities in a very satisfying way. It is hoped that this approach will be tested in the laboratory in the not too distant future on a new robot system currently being constructed.

7. DISCUSSION

Each of the case studies and proposals presented above illustrates how colour and shape can work together to produce results in 3D scene analysis better than either could produce on its own. Whilst these examples are distinct they all echo the same theme concerning the advantages of combining modalities of object surface data in an intelligent way. One critical factor shared by the five examples is that the range and colour data in all cases are registered and non-contradictory. In many sensor fusion research projects in robotics, say solving the problem of combining vision and ultrasonic modes of environmental modelling, one does not have the luxury of simple registration and non-contradiction. Extracting the full advantages of such mixed modes can be very difficult. Thus it might be worthwhile to consider the problems of registration and consistency in the design of the 3D Vision instruments we build. It may be entirely possible to build the one instrument to collect colour, texture, intensity and range in a registered way which also allows simple representation of sensor data in ordered arrays of feature vectors so that uniform computational processes might be easily applied, perhaps using massively parallel architectures which are becoming more and more affordable. Such an instrument would be much sought after, initially by the research community but eventually in the industrial marketplace.

8. CONCLUSION

This paper has presented three case studies and two proposals in which shape and colour sensor data have been combined to improve some aspect of 3D scene analysis. Considerably more research should be aimed at strengthening this worthwhile liaison and perhaps instrumentation should be built to capture both types of data simultaneously and in a registered way. Only time will tell whether these hopes will be realised.

9. BIBLIOGRAPHY

1. Jarvis, R.A., A Semantic-Free Approach to 3D Robot Colour Vision, invited chapter for book edited by Prof. Takeo Kanade, published by Kluwer Academic Publishers, 1987, pp.565-609.

2. Jarvis, R.A., Expedient Range Enhanced 3-D Robot Colour Vision, Robotica, Vol. 1, 1983, pp.25-31.

3. Rosenfeld, A. and Pfaltz, J.L., Sequential Operations in Digital Picture Processing, J. Assn. for Comp. Mach. 13, No.4, 471-494, Oct. 1966.

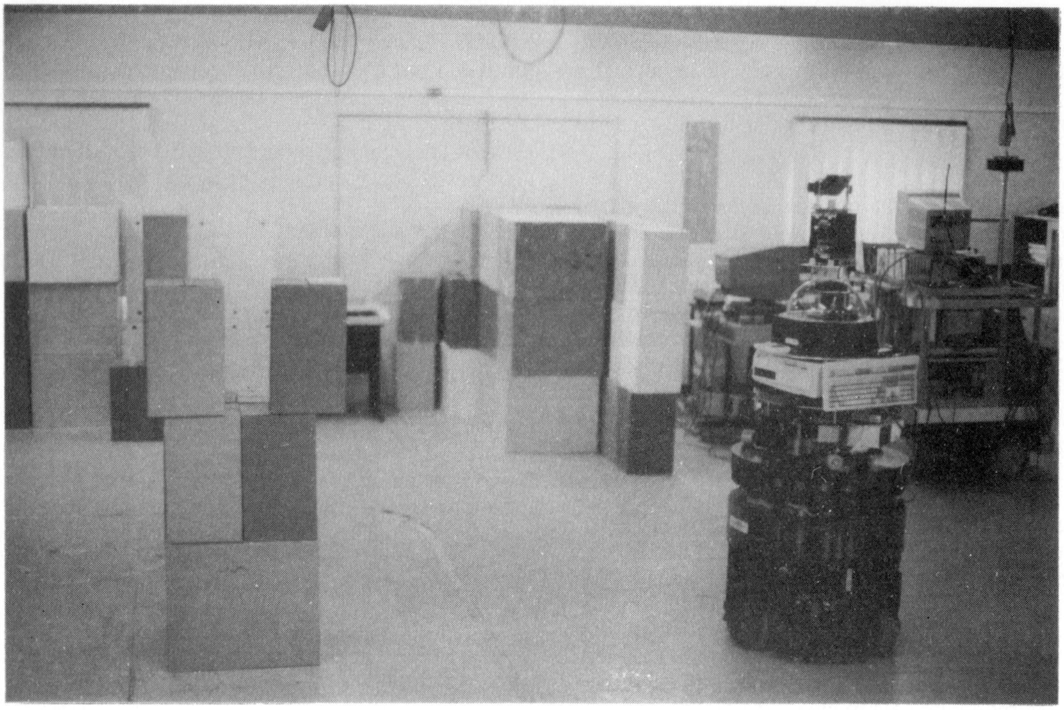

(a) Robot Navigation Laboratory Scene

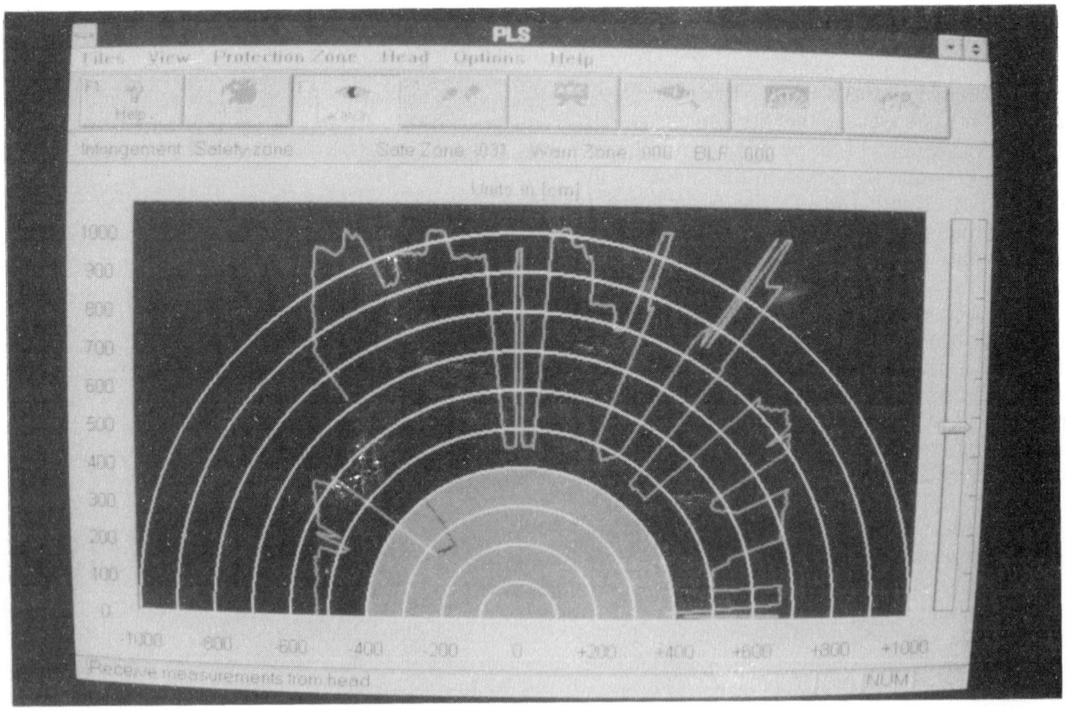

(b) 180^0 Range Scan of (a)

Figure 8. Range Based Environmental Mapping

4. Jarvis, R.A., A Laser Time-of-Flight Range Scanner for Robotic Vision, IEEE Transactions on Pattern Analysis and Machine Intelligence, PAMI - 5, NO. 5, Sept. 1983, pp.505-512.

5. Jarvis, R.A., 3D Shape and Surface Colour Sensor Fusion for Robot Vision, accepted by Robotica, 1992, Vol.10, pp.389-396.

6. Jarvis, R.A. Range Sensing for Computer Vision, chapter in book entitled 3D Object Recognition, Jain, A. and Flynn, P. (Editors) Elsevier Science Publishers, 1993.

7. Alexander, B.F., High Accuracy Non Contact Three Dimensional Shape Measurement, PhD thesis (Department of Electrical and Computer Systems Engineering, Monash University, September 1989).

8. Jarvis, R.A. and Patrick, E.A., Clustering Using a Similarity Measure Based on Shared Near Neighbours, IEEE Transactions on Electronic Computers, Vol.C-22, No.11, November 1973.

9. Jarvis, R.A., Quad-Vision Ranging for Robotic Application, Proc. 4th Australian Joint Conference on Artificial Intelligence, November 21-23, 1990, Hyatt Regency, Perth, Western Australia, pp.682-698.

10. Kanade, T., Very Fast 3-D Sensing Hardware, Proceedings, The Sixth International Symposium on Robotics Research, Pittsburgh, Pennsylvania, U.S.A., Oct 2-5, 1993. Published by the International Foundation for Robotics Research, pp185-198.

11. Bolles, R.C., Spatiotemporal Consistency Checking of Passive Range Data, Proceedings, The Sixth International Symposium on Robotics Research, Pittsburgh, Pennsylvania, U.S.A., Oct 2-5, 1993. Published by the International Foundation for Robotics Research, pp165-183.

12. Moravec, H.P., Visual Mapping by a Robot Rover, Proc. Int'l Joint Conf. on Artificial Intelligence, Tokyo, Japan, Aug. 1979, pp598-600.

13. Chua, C.S. and Jarvis, R.A. 3D Free-Form Surface Registration and Object Recognition, accepted by International Journal of Computer Vision.

14. Chua, C.S. and Jarvis, R.A. Point Signatures : A New Prepresentation for 3D Object Recognition, submitted to the International Journal of Computer Vision.

A Multi-body Factorization Method for Motion Analysis*

João Costeira **Takeo Kanade**
School of Computer Science, Carnegie Mellon University
5000 Forbes Avenue, Pittsburgh, PA 15213
http://www.cs.cmu.edu/~tk

Abstract

The structure-from-motion problem has been extensively studied in the field of computer vision. Yet, the bulk of the existing work assumes that the scene contains only a single moving object. The more realistic case where an unknown number of objects move in the scene has received little attention, especially for its theoretical treatment. In this paper we present a new method for separating and recovering the motion and shape of multiple independently moving objects in a sequence of images.

1 Introduction

A motion image sequence allows for the recovery of the three-dimensional structure of a scene. While a large amount of literature exists about this structure-from-motion problem, most previous theoretical work is based on the assumption that only a single motion is included in the image sequence; either the environment is static and the observer moves, or the observer is static and only one object in the scene is moving. More difficult and less studied is the general case of an unknown number of objects moving independently. Suppose that a set of features has been extracted and tracked in an image sequence, but it is not known which feature belongs to which object. Given a set of such feature trajectories, the question is whether we can segment and recover the motion and shape of multiple objects contained in the image sequence.

The previous approaches to the structure-from-motion problem for multiple objects can be grouped into two classes: image motion-based (2D) and three-dimensional (3D) modeling. The image-motion based approach relies mostly on spatio-temporal properties of an image sequence. For example, regions corresponding to different velocity fields are extracted by using Fourier domain analysis [Watson and Ahumada, 1983][Adelson and Bergen, 1985] or scale-space and space-time filters [Bergen *et al.*, 1990, Burt *et al.*, 1991, Irani *et al.*, 1994, Jasinschi *et al.*, 1992]. These image-based methods have limited applicability either because object motions are restricted to a certain type, such as translation only, or because image-level properties, such as locality, need to be used for segmentation without assuring consistent segmentation into 3D objects.

To overcome these limitations, models of motion and scene can be introduced which provide more constraints. Representative constraints include rigidity of an object [Ullman, 1983] and smoothness (or similarity) of motion [Sinclair, 1993, Navab and Zhang, 1992, Adiv, 1985, Boult and Brown, 1991]. Then the problem becomes segmenting image events, such as feature trajectories, into objects so that the recovered motion and shape satisfy those constraints. It is now a clustering problem with constraints derived from a physical model. Though sound in theory, the practical difficulty is the cyclic dilemma: to check the constraints it is necessary to segment features and to segment it is necessary to compute constraints. So, developed methods tend to be of a "generate-and-test" nature, or require prior knowledge of the number of objects (clusters). Ullman [Ullman, 1983] describes a computational scheme to recursively recover shape from the tracks of image features. A model of the object's shape is matched to the current position of the features, and a new model that maximizes rigidity is computed to update the shape. He suggests that this scheme could be used to segment multi-body scenes by local application of the rigidity principle. Since a

*This research is partially sponsored by the Advanced Research Projects Agency under the Department of the Army, Army Research Office under grant number DAAH04-94-G-0006. The views and conclusions contained in this document are thos e of the authors and should not be interpreted as representing the official poli cies or endorsements, either expressed or implied, of the Department of the Army , or the U.S. government.

single rigid body model does not fit the whole data, collections of points that could be explained by a rigid transformation would be searched and grouped into an object. Under the framework of the factorization method [Tomasi and Kanade, 1992], this view of the problem is followed by Boult and Brown [Boult and Brown, 1991] and Gear [Gear, 1994], where the role of rigidity is replaced by linear dependence between feature tracks. Since the factorization produces a matrix that is related with shape, segmentation is obtained by recursively clustering columns of feature trajectories into linearly dependent groups.

This paper presents a new method for segmenting and recovering the motion and shape of multiple independently moving objects from a set of feature trajectories tracked in a sequence of images. Developed by using the framework of the factorization by Tomasi and Kanade [Tomasi and Kanade, 1992], the method does not require any grouping of features into an object at the image level or prior knowledge of the number of objects. It directly computes shape information and allows segmentation into objects. This has been made possible by introducing a linear-algebraic construct of object shapes, called the shape interaction matrix. The entries of this matrix are invariant to individual object motions and yet is computable only from tracked feature trajectories without knowing their object identities (ie, segmentation). Once the matrix is computed, transforming it into the canonical form results in segmenting features as well as recovering the shape and motion of each object. We will present our theory by using the orthographic camera model. It is, however, easily seen that the theory, and thus the method, works under a broader projection model including weak perspective (scaled orthography) and paraperspective [Poelman and Kanade, 1994] up to an affine camera [Koendering and van Doorn, 1992]

In the next section, we will first re-derive the factorization method in homogeneous coordinates including the translational motion component in order to facilitate the development of the multi-body factorization method in section 3. Finally in sections 4 and 5 we present experimental results, and discuss implications of our method.

2 Factorization Method: A New Formulation Including Translation

The factorization method was originally introduced by Tomasi and Kanade[Tomasi and Kanade, 1990, Tomasi and Kanade, 1992] for the case of single object motion.

The original formulation addressed the case of a moving camera observing a static scene. In this section we will reformulate the method in such a way that a static camera observes a scene with a moving object. Also, whereas the translation component of motion is first eliminated in the Tomasi-Kanade formulation, we will retain that component in our formulation. Though equivalent for the single object case, the new formulation with these two changes simplifies some of the representations and allows for easy extension to the case of multiple moving objects.

2.1 World and Observations

Let us assume for the moment that a static camera observes a single moving object. To represent the situation we need two coordinate systems: a moving system **O** attached to the object, and a static system **C** attached to the camera as shown in figure 1.

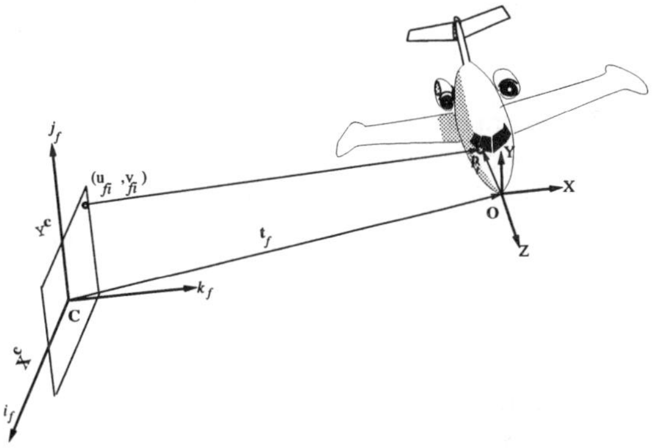

Figure 1: This figure shows the camera and the object with its coordinate system. The (unit) vectors **i**, **j** define the image plane and **k** its normal.

Consider a point $\mathbf{p}_i$ on the object. Its position represented in the camera coordinate system at instant f is given by the transformation,

$$\mathbf{p}_{fi}^{C} = \mathbf{R}_f\, \mathbf{p}_i + \mathbf{t}_f. \tag{1}$$

Here

$$\mathbf{R}_f \equiv \begin{bmatrix} \mathbf{i}_f^T \\ \mathbf{j}_f^T \\ \mathbf{k}_f^T \end{bmatrix} \tag{2}$$

is the rotation matrix whose rows $\mathbf{i}_f^T = \begin{bmatrix} i_{x_f} & i_{y_f} & i_{z_f} \end{bmatrix}$, $\mathbf{j}_f^T = \begin{bmatrix} j_{x_f} & j_{y_f} & j_z \end{bmatrix}$ and $\mathbf{k}^T = \begin{bmatrix} k_{x_f} & k_{y_f} & k_{z_f} \end{bmatrix}$ are the axes of the camera coordinate frame **C** expressed in the object's frame. The vector

$$\mathbf{t}_f \equiv \begin{bmatrix} t_{x_f} \\ t_{y_f} \\ t_{z_f} \end{bmatrix} \tag{3}$$

represents the position of the object's coordinate frame at instant f in the camera frame. The representation (1) can be simplified if we use homogeneous coordinates,

$$\mathbf{s} \equiv \begin{bmatrix} \mathbf{p} \\ 1 \end{bmatrix} \equiv \begin{bmatrix} X \\ Y \\ Z \\ 1 \end{bmatrix} \tag{4}$$

for the object's point. In the homogeneous coordinates, equation (1) can be expressed as

$$\mathbf{s}_{fi}^C \equiv \begin{bmatrix} \mathbf{p}_{fi}^C \\ 1 \end{bmatrix} = \begin{bmatrix} \mathbf{R}_f & \mathbf{t}_f \\ 0_{1\times 3} & 1 \end{bmatrix} \begin{bmatrix} \mathbf{p}_i \\ 1 \end{bmatrix} \tag{5}$$

$$= \begin{bmatrix} \mathbf{R}_f & \mathbf{t}_f \\ 0_{1\times 3} & 1 \end{bmatrix} \mathbf{s}_i \tag{6}$$

The camera is modeled as an orthographic projection. It produces the image by projecting a world point parallel to the optical axis onto the image plane. The image of point $\mathbf{p}_i$ at time f is then given by the first two elements of $\mathbf{p}_{fi}^C$:

$$\begin{bmatrix} u_{fi} \\ v_{fi} \end{bmatrix} = \left[\begin{array}{ccc|c} i_{x_f} & i_{y_f} & i_{z_f} & t_{x_f} \\ j_{x_f} & j_{y_f} & j_{z_f} & t_{y_f} \end{array}\right] \mathbf{s}_i. \tag{7}$$

The object moves relative to the camera which acquires images. In the sequence we track feature points from frame to frame. Suppose that we track N feature points over F frames, and that we collect all these measurements into a single matrix

$$\mathbf{W} \equiv \begin{bmatrix} u_{11} & u_{12} & \dots & u_{1N} \\ & \vdots & \vdots & \\ u_{F1} & u_{F2} & \dots & u_{FN} \\ v_{11} & v_{12} & \dots & v_{1N} \\ & \vdots & \vdots & \\ v_{F1} & v_{F2} & \dots & v_{FN} \end{bmatrix} \tag{8}$$

Each row of $\mathbf{W}$ lists the image coordinates u or v of all the feature points in each frame, and each column represents the image trajectory of one feature over the whole image sequence.

Using (7) we can represent $\mathbf{W}$ as the matrix product,

$$\begin{bmatrix} u_{11} & u_{12} & \dots & u_{1N} \\ & \vdots & \vdots & \\ u_{F1} & u_{F2} & \dots & u_{FN} \\ v_{11} & v_{12} & \dots & v_{1N} \\ & \vdots & \vdots & \\ v_{F1} & v_{F2} & \dots & v_{FN} \end{bmatrix} = \left[\begin{array}{ccc|c} i_{x_1} & i_{y_1} & i_{z_1} & t_{x_1} \\ & \vdots & & \vdots \\ i_{x_F} & i_{y_F} & i_{z_F} & t_{x_F} \\ j_{x_1} & j_{y_1} & j_{z_1} & t_{y_1} \\ & \vdots & & \vdots \\ j_{x_F} & j_{y_F} & j_{z_F} & t_{y_F} \end{array}\right] \begin{bmatrix} s_{x_1} & & s_{x_N} \\ s_{y_1} & \dots & s_{y_N} \\ s_{z_1} & & s_{z_N} \\ 1 & & 1 \end{bmatrix} \tag{9}$$

If we denote

$$\mathbf{M} \equiv \left[\begin{array}{ccc|c} i_{x_1} & i_{y_1} & i_{z_1} & t_{x_1} \\ & \vdots & & \vdots \\ i_{x_F} & i_{y_F} & i_{z_F} & t_{x_F} \\ j_{x_1} & j_{y_1} & j_{z_1} & t_{y_1} \\ & \vdots & & \vdots \\ j_{x_F} & j_{y_F} & j_{z_F} & t_{y_F} \end{array}\right] \tag{10}$$

$$\mathbf{S} \equiv \begin{bmatrix} s_{x_1} & & s_{x_N} \\ s_{y_1} & \dots & s_{y_N} \\ s_{z_1} & & s_{z_N} \\ 1 & & 1 \end{bmatrix} \tag{11}$$

as the motion and shape matrices respectively, we have the compact representation

$$\mathbf{W} = \mathbf{MS}. \tag{12}$$

Compared with the original formulation in [Tomasi and Kanade, 1992], note that the motion matrix contains translational components t_{x_f} and t_{y_f}. In the original formulation they were eliminated from the bilinear equation corresponding to (9) by subtracting the means of the measurements beforehand. That procedure reflected the fact that under orthography the translation components does not provide any information about shape. In the case of a scene with multiple independent moving objects, the situation is not the same: on one hand the translation component of each object cannot be computed without segmentation, and on the other hand the translation component does provide information for segmenting the multi-body scene. For these reasons, we include the translation component in our formulation of the factorization method.

2.2 Solution for Shape and Motion by Factorization

We have derived the bilinear relationship (9) by modeling the imaging process. The problem of recovering the shape and motion is to start with a given matrix $\mathbf{W}$ and obtain a factorization into motion matrix $\mathbf{M}$ and shape matrix $\mathbf{S}$. By simple inspection of (9) we can see that since $\mathbf{M}$ and $\mathbf{S}$ can be at most rank 4, $\mathbf{W}$ will be at most rank 4. In real situations $\mathbf{W}$ is constructed from noisy measurements, so the rank of $\mathbf{W}$ can be higher due to noise in the feature tracking. Among all the possible matrix decompositions, singular value decomposition (SVD) is the most robust and the best rank revealing of all [Stewart, 1992] to approximate $\mathbf{W}$ by a rank-4 matrix. With SVD, $\mathbf{W}$ is decomposed and approximated as:

$$\mathbf{W} \quad = \quad \mathbf{U}\mathbf{\Sigma}\mathbf{V}^T \tag{13}$$

Matrix $\mathbf{\Sigma} = diag(\sigma_1, \sigma_2, \sigma_3, \sigma_4)$ is a diagonal matrix made of the four biggest singular values which reveal the most important components in the data. Matrices $\mathbf{U} \in R^{2F\times 4}$ and $\mathbf{V} \in R^{N\times 4}$ are the left and right singular matrices respectively, such that $\mathbf{U}^T\mathbf{U} = \mathbf{V}^T\mathbf{V} = \mathcal{I}$ (the 4×4 identity matrix).

By defining,

$$\hat{\mathbf{M}} \equiv \mathbf{U}\mathbf{\Sigma}^{\frac{1}{2}} \tag{14}$$

$$\hat{\mathbf{S}} \equiv \mathbf{\Sigma}^{\frac{1}{2}}\mathbf{V}^T \tag{15}$$

we have the two matrices whose product can represent the bilinear system $\mathbf{W}$. However, this factorization is not unique, since for any invertible 4×4 matrix $\mathbf{A}$, $\mathbf{M} = \hat{\mathbf{M}}\mathbf{A}$ and $\mathbf{S} = \mathbf{A}^{-1}\hat{\mathbf{S}}$ are also a possible solution because

$$\mathbf{MS} = \left(\hat{\mathbf{M}}\mathbf{A}\right)\left(\mathbf{A}^{-1}\hat{\mathbf{S}}\right) \quad = \quad \hat{\mathbf{M}}\hat{\mathbf{S}} = \mathbf{W}. \tag{16}$$

In other words, the singular value decomposition (13) provides a solution both for shape and motion up to an affine transformation.

The exact solution can be computed, using the fact that $\mathbf{M}$ must have certain properties. Let us denote the 4×4 matrix $\mathbf{A}$ as the concatenation of two blocks,

$$\mathbf{A} \equiv \left[\mathbf{A}_R | \mathbf{a}_t\right], \tag{17}$$

The first block $\mathbf{A}_R$ is the first 4×3 submatrix related to the rotational component and the second block $\mathbf{a}_t$ is a 4×1 vector related to translation. Now, since

$$\mathbf{M} \quad = \quad \hat{\mathbf{M}}\mathbf{A} = \left[\hat{\mathbf{M}}\mathbf{A}_R | \hat{\mathbf{M}}\mathbf{a}_t\right], \tag{18}$$

we can impose motion constraints, one on rotation and the other on translation, in order to solve for $\mathbf{A}$.

2.2.1 Rotation Constraints

Block $\mathbf{A}_R$ of $\mathbf{A}$, which is related to rotational motion, is constrained by the orthonormality of axes vectors $\mathbf{i}_f^T$ and $\mathbf{j}_f^T$: each of the $2F$ rows entries of matrix $\hat{M}A_{\mathbf{R}}$ is a unit norm vector and the first and second set of F rows are pairwise orthogonal. This yields a set of constraints:

$$\hat{\mathbf{m}}_i \mathbf{A}_R \mathbf{A}_R^T \hat{\mathbf{m}}_i^T \quad = \quad 1 \tag{19}$$

$$\hat{\mathbf{m}}_j \mathbf{A}_R \mathbf{A}_R^T \hat{\mathbf{m}}_j^T \quad = \quad 1 \tag{20}$$

$$\hat{\mathbf{m}}_i \mathbf{A}_R \mathbf{A}_R^T \hat{\mathbf{m}}_j^T \quad = \quad 0 \tag{21}$$

where $\hat{\mathbf{m}}_i, \hat{\mathbf{m}}_j$ are rows i and j of matrix $\hat{\mathbf{M}}$ for $i = 1 \ldots F$ and $j = F + 1 \ldots 2F$. This is an overconstrained system which can be solved for the entries of $\mathbf{A}_R\mathbf{A}_R^T$ by using least squares techniques, and subsequently solving for $\mathbf{A}_R$. See [Tomasi and Kanade, 1992] for a detailed solution procedure.

2.2.2 Translation Constraints

In orthography, the projection of the 3D centroid of an object features into the image plane is the centroid of the feature points. The X and Y position of the centroid of the feature points is the average of each row of $\mathbf{W}$:

$$\overline{\mathbf{w}} \equiv \begin{bmatrix} \frac{1}{N}\sum u_{1,i} \\ \vdots \\ \frac{1}{N}\sum v_{F,i} \end{bmatrix} \tag{22}$$

$$= \mathbf{M}\overline{\mathbf{s}} = \left[\hat{\mathbf{M}}\mathbf{A}_R | \hat{\mathbf{M}}\mathbf{a}_t\right] \begin{bmatrix} \overline{\mathbf{p}} \\ 1 \end{bmatrix}, \tag{23}$$

where

$$\overline{\mathbf{p}} \equiv \frac{1}{N}\sum \mathbf{p}_i \tag{24}$$

is the centroid of the object.

The origin of the object's coordinate system is arbitrary, but we can choose to place it at the centroid of the object, that is $\overline{\mathbf{p}} = 0$. Then it follows immediately from (23) that

$$\overline{\mathbf{w}} = \hat{\mathbf{M}}\mathbf{a}_t \tag{25}$$

This expression is also an overconstrained system of equations, which can be solved for the entries of $\mathbf{a}_t$ in the least square sense. The best estimate will be given by

$$\mathbf{a}_t \quad = \quad (\hat{\mathbf{M}}^T\hat{\mathbf{M}})^{-1}\hat{\mathbf{M}}^T\overline{\mathbf{w}} \tag{26}$$

$$= \quad \mathbf{\Sigma}^{-1/2}\mathbf{U}^T\overline{\mathbf{w}}, \tag{27}$$

which completes the computation of all the elements of matrix $\mathbf{A}$.

3 The Multi-body Factorization Method

So far we have assumed that the scene contains a single moving object. If there is more than one moving object, the measurement matrix $\mathbf{W}$ will contain features (columns) which originate from different motions. One may think that solving the problem requires first sorting the columns of the measurements matrix $\mathbf{W}$ into submatrices, each of which contains features solely from one object, so that the factorization technique of the previous sections can be applied individually. We will show in this section that the multi-body problem can be solved without prior segmentation. For the sake of simplicity in presentation we will present the theory and method for the case of two bodies, but it will be clear that the method is applicable to the general case of an arbitrary unknown number of objects.

3.1 Multi-body Motion Recovery Problem: Its Difficulty

Suppose we have a scene in which two objects are moving and we take an image sequence of F frames. The relevant coordinate systems in this case are depicted in figure 2. Suppose also that the set of features that we have observed and tracked in the image sequence actually consists of N_1 feature points from object 1 and N_2 from object 2 which are observed in an image sequence of F frames.

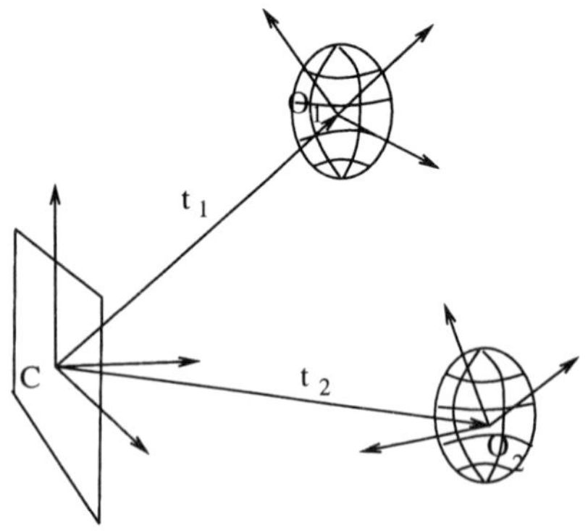

Figure 2: Two bodies: The coordinate systems

Imagine for the moment that somehow we know the classification of features and thus could permute the columns of $\mathbf{W}$ in such a way that the first N_1 columns belong to object 1 followed by the N_2 columns from object 2. Matrix $\mathbf{W}$ would have the canonical form:

$$\mathbf{W}^* \equiv \left[\mathbf{W}_1 \mid \mathbf{W}_2 \right]. \tag{28}$$

Each measurement submatrix can be factorized as

$$\begin{aligned} \mathbf{W}_l &= \mathbf{U}_l \mathbf{\Sigma}_l \mathbf{V}_l^T & (29) \\ &= \mathbf{M}_l \mathbf{S}_l = (\hat{\mathbf{M}}_l \mathbf{A}_l)(\mathbf{A}_l^{-1} \hat{\mathbf{S}}_l) & (30) \end{aligned}$$

with $l = 1$ and 2 for object 1 and 2 respectively. Equation (28) now has the canonical factorization:

$$\begin{aligned} \mathbf{W}^* &= [\mathbf{M}_1 | \mathbf{M}_2] \begin{bmatrix} \mathbf{S}_1 & 0 \\ 0 & \mathbf{S}_2 \end{bmatrix} & (31) \\ &= [\mathbf{U}_1 | \mathbf{U}_2] \begin{bmatrix} \mathbf{\Sigma}_1^{\frac{1}{2}} & 0 \\ 0 & \mathbf{\Sigma}_2^{\frac{1}{2}} \end{bmatrix} \\ & \quad \begin{bmatrix} \mathbf{A}_1 & 0 \\ 0 & \mathbf{A}_2 \end{bmatrix} \begin{bmatrix} \mathbf{A}_1^{-1} & 0 \\ 0 & \mathbf{A}_2^{-1} \end{bmatrix} \\ & \quad \begin{bmatrix} \mathbf{\Sigma}_1^{\frac{-1}{2}} & 0 \\ 0 & \mathbf{\Sigma}_2^{\frac{-1}{2}} \end{bmatrix} \begin{bmatrix} \mathbf{V}_1^T & 0 \\ 0 & \mathbf{V}_2^T \end{bmatrix} & (32) \end{aligned}$$

By denoting

$$\begin{aligned} \mathbf{M}^* &\equiv [\mathbf{M}_1 | \mathbf{M}_2], \\ \mathbf{S}^* &\equiv \begin{bmatrix} \mathbf{S}_1 & 0 \\ 0 & \mathbf{S}_2 \end{bmatrix}, \\ \mathbf{A}^* &\equiv \begin{bmatrix} \mathbf{A}_1 & 0 \\ 0 & \mathbf{A}_2 \end{bmatrix} & (33) \end{aligned}$$

$$\begin{aligned} \mathbf{U}^* &\equiv [\mathbf{U}_1 | \mathbf{U}_2], \\ \mathbf{\Sigma}^* &\equiv \begin{bmatrix} \mathbf{\Sigma}_1 & 0 \\ 0 & \mathbf{\Sigma}_2 \end{bmatrix}, \\ \mathbf{V}^{*T} &\equiv \begin{bmatrix} \mathbf{V}_1^T & 0 \\ 0 & \mathbf{V}_2^T \end{bmatrix} & (34) \end{aligned}$$

we express a factorization in a similar way to a single object case, where the canonical measurement matrix relates to shape and motion by:

$$\begin{aligned} \mathbf{W}^* &= \mathbf{M}^* \mathbf{S}^* & (35) \\ \mathbf{S}^* &= \mathbf{A}^{*-1} \mathbf{\Sigma}^{*\frac{1}{2}} \mathbf{V}^{*T} & (36) \\ \mathbf{M}^* &= \mathbf{U}^* \mathbf{\Sigma}^{*\frac{1}{2}} \mathbf{A}^* & (37) \end{aligned}$$

From equation (31), we see that $\mathbf{W}^*$ (and therefore $\mathbf{W}$) will have at most rank 8, since $\mathbf{W}_1$ and $\mathbf{W}_2$ are at most rank 4. Let us consider for the remainder of this paper the non-degenerate case where the rank of $\mathbf{W}$ is in fact equal to 8; that is, the object shape is actually three-dimensional (not planar or line) and the motion vectors span 3D for both objects.

In reality, we do not know which features belong to which object, and thus the columns of the given measurement matrix $\mathbf{W}$ are a mixture of features from object 1 and 2. We can still apply singular value decomposition (SVD) to the measurement matrix, and obtain

$$\mathbf{W} = \mathbf{U\Sigma V}^T. \tag{38}$$

Then it may appear that the remaining task is to find the linear canonical transformation $\mathbf{A}^*$ in (33) such that shape and motion will have the block structure of equations (36) and (37).

There is, however, a fundamental difficulty in doing this. The metric (rotation and translation) constraints (eq.(19)-(20) and (25)-(27)) were obtained in section 2.2 by considering that the motion matrix for one object, that is, by assuming that the measurement matrix consists of features from a single object. Those constraints are therefore applicable only after knowing the segmentation. This is exactly the mathematical manifestation of the cyclic dilemma mentioned earlier.

Faced with this difficulty, a usual approach would be to group features bit by bit so that we segment $\mathbf{W}$ into two rank-4 matrices and obtain the factorization of the form (31). For example, a most simplistic procedure would be like the following. Pick the first four columns of $\mathbf{W}$ and form a rank-4 subspace. If the fifth column belongs to the subspace (ie. is linear dependent on the first four, or "almost" linear dependent in the case of noisy measurement), then classify it to the same object as the first four columns and update the subspace representation. Otherwise, it belongs to a new object. Apply this procedure recursively to all the remaining columns. This approach is in fact essentially the one used by [Boult and Brown, 1991] and [Gear, 1994] to split matrix $\mathbf{W}$, and similar to what was suggested by Ullman [Ullman, 1983], whose criteria for merging was local rigidity.

There are a few disadvantages in this cluster-and-test approach. First, there is no guarantee that the first four columns, which always form a rank-4 subspace, are from the same object. Second, if we use a sequential procedure like the one above or its variation, the final result is dependent on where we start the procedure, and alternatively, the search for the globally optimal segmentation most likely will be computational very expensive. Finally, the prior knowledge of the number of objects becomes very critical, since depending on the decision criteria of subspace belongingness the final number of objects may vary arbitrarily.[1]

[1]While this is beyond the scope of the assumption in this section, this cluster-and-test approach also requires the prior knowledge of the ranks of objects as well. Since for example a rank-8 measurement matrix might have been generated by two line (rank-2) objects and one full 3D (rank 4) object instead of two full 3D objects, and therefore committing to find two rank-4 subspaces might be wrong.

3.2 Mathematical Construct of Shapes Invariant to Motions

The main difficulty in the multi-body structure-from-motion problem revealed above is that shape and motion interact. Mathematically, the equation (31) indicates that the rank-8 measurement space is originally generated by the two subspaces of rank 4 each, represented by the block-diagonal shape matrix $\mathbf{S}^*$. However, the recovered shape space $\mathbf{V}^T$, obtained by the singular value decomposition of the non-canonical $\mathbf{W}$, is in general a linear combination of the two subspaces and has lost the block-diagonal structure.

There is however a mathematical construct that preserves the original subspace structure. Let us define $\mathbf{Q}$ as $(N_1 + N_2) \times (N_1 + N_2)$ square matrix

$$\mathbf{Q} \equiv \mathbf{V}\mathbf{V}^T. \tag{39}$$

We will call this matrix the *shape interaction matrix*. Mathematically, it is the orthogonal operator that projects $N = (N_1 + N_2)$ dimensional vectors to the subspace spanned by the columns of $\mathbf{V}$. This matrix $\mathbf{Q}$ has several interesting and useful properties. First, by definition it is uniquely computable only from the measurements $\mathbf{W}$ without knowing the segmentation, since $\mathbf{V}$ is uniquely obtained by the singular value decomposition of $\mathbf{W}$.

Secondly, permuting columns of $\mathbf{W}$ does not change the set of values $\{Q_{ij}\}$ that appear in $\mathbf{Q}$ though their arrangement in $\mathbf{Q}$ does; swapping columns l and m of $\mathbf{W}$ results in swapping columns l and m of $\mathbf{V}^T$. Therefore it results in simultaneously swapping columns l and m and rows l and m in $\mathbf{Q}$, but not their entry values.

Thirdly, each element of $\mathbf{Q}$ provides important information about whether a pair of features belong to the same object. Since the set of values do not change, let us compute $\mathbf{Q}^*$, the shape interaction matrix for the canonical measurement matrix $\mathbf{W}^*$. By substituting (36) into (39), we obtain

$$\begin{aligned}
\mathbf{Q}^* &= \mathbf{V}^*\mathbf{V}^{*T} && (40)\\
&= \mathbf{S}^{*T}\mathbf{A}^{*T}\mathbf{\Sigma}^*\,\mathbf{A}^*\,\mathbf{S}^* && (41)\\
&= \mathbf{S}^{*T}(\mathbf{A}^{*-1}\mathbf{\Sigma}^{*-1}\mathbf{A}^{*-T})^{-1}\mathbf{S}^* && (42)\\
&= \mathbf{S}^{*T}\left[(\mathbf{A}^{*-1}\mathbf{\Sigma}^{*-1/2}\mathbf{V}^{*T})\right.\\
&\qquad \left.(\mathbf{V}^*\mathbf{\Sigma}^{*-1/2}\mathbf{A}^{*-T})\right]^{-1}\mathbf{S}^*
\end{aligned}$$

$$= \mathbf{S}^{*T}(\mathbf{S}^{*}\mathbf{S}^{*T})^{-1}\mathbf{S}^{*} \tag{43}$$

$$= \begin{bmatrix} \mathbf{S_1}^T & 0 \\ 0 & \mathbf{S_2}^T \end{bmatrix} \begin{bmatrix} \mathbf{\Lambda_1}^{-1} & 0 \\ 0 & \mathbf{\Lambda_2}^{-1} \end{bmatrix} \begin{bmatrix} \mathbf{S_1} & 0 \\ 0 & \mathbf{S_2} \end{bmatrix} \tag{44}$$

$$= \begin{bmatrix} \mathbf{S_1}^T\mathbf{\Lambda_1}^{-1}\mathbf{S_1} & 0 \\ 0 & \mathbf{S_2}^T\mathbf{\Lambda_2}^{-1}\mathbf{S_2} \end{bmatrix} \tag{45}$$

where $\mathbf{\Lambda_1}$ and $\mathbf{\Lambda_2}$ are the 4×4 matrices of the moments of inertia of each object. This means that the canonical $\mathbf{Q}^*$ matrix for the sorted $\mathbf{W}^*$ has a very defined block-diagonal structure. Moreover, each entry has the value

$$Q^*_{ij} = \begin{cases} \mathbf{s}_{1_i}^T \mathbf{\Lambda_1}^{-1} \mathbf{s}_{1_j} & \text{if feature trajectory } i \text{ and } j \text{ belong to object 1} \\ \mathbf{s}_{2_i}^T \mathbf{\Lambda_2}^{-1} \mathbf{s}_{2_j} & \text{if feature trajectory } i \text{ and } j \text{ belong to object 2} \\ 0 & \text{if feature trajectory } i \text{ and } j \text{ belong to different objects.} \end{cases} \tag{46}$$

Finally and most importantly, the set of values $\{Q^*_{ij}\}$, which is the same as $\{Q_{ij}\}$ are invariant to motion. This is true since equations (46) include only S's, and not M. In other words, in whatever way the objects move they will produce the same set of entries in matrix $\mathbf{Q}$.

In summary, we have shown that without knowing the segmentation of features we can compute matrix $\mathbf{Q}$ whose element Q_{ij} can be interpreted as a measure of the interaction between feature i and j: if the value is non zero, they belong to the same object, and if they don't belong to the same object, the value is zero. Also, if the features are sorted correctly into the canonical form of the measurement matrix $\mathbf{W}^*$, then the corresponding canonical shape interaction matrix $\mathbf{Q}^*$ must be block diagonal.

3.3 Sorting Matrix Q into Canonical Form

The problem of segmenting and recovering motion of multiple objects now has reduced to sorting the entries of matrix $\mathbf{Q}$ by swapping pairs of rows and columns until it becomes block diagonal. Once achieved, the corresponding permutations of columns of $\mathbf{W}$ will transform it its canonical form where features from one object are grouped into contiguous columns. This relationship between sorting $\mathbf{Q}$ and permuting $\mathbf{W}$ is illustrated in figure 3.

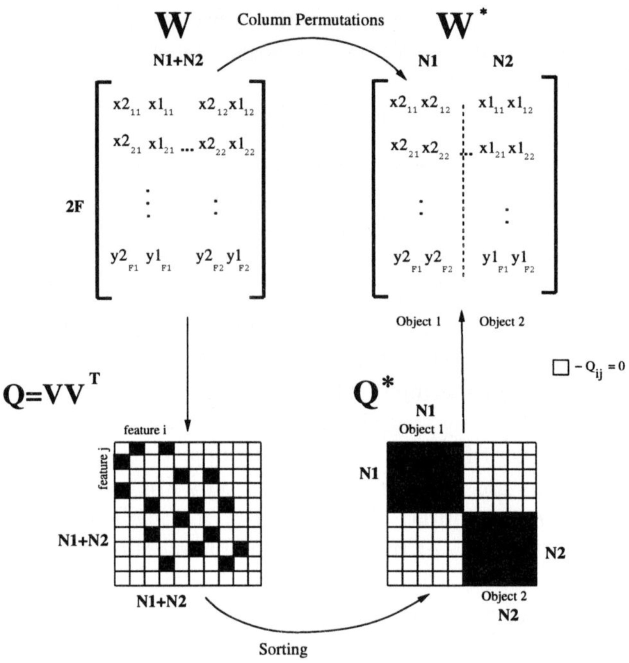

Figure 3: Segmentation process

With noisy measurements, a pair of features from different objects may exhibit a small non-zero entry in $\mathbf{Q}$. We can regard Q^2_{ij} as representing the energy of the shape interaction, and the block diagonalization of $\mathbf{Q}$ can be achieved by minimizing the total energy of all possible off-diagonal blocks over all set of permutations of rows and columns of $\mathbf{Q}$. We found that a simple iterative minimization procedure suffices for our purpose. Alternatively, we can regard matrix $\{Q^2_{ij}\}$ as defining a graph of $N_1 + N_2$ nodes, where the Q^2_{ij} indicates the weight of the link (i, j). We also found that graph-theoretical algorithms, such as the minimum spanning tree, can be used to achieve the block diagonalization more efficiently than the energy minimization.

3.4 Summary of Algorithm

While we have presented the theory for the case of two full-3D objects, it is easy to see tht its essential part holds for more general cases. First the matrix $\mathbf{Q}^*$ has the block diagonal structure for an arbitrary number of moving objects, that is, an entry $\mathbf{Q}_{ij}$ of the $\mathbf{Q}$ matrix equals to zero if features i and j belong to different objects. Furthermore, this property holds even when the shape matrix of the objects has rank less than 4 (planes and lines). The computation of $\mathbf{Q}$

by (39) requires only the knowledge of the total rank of $\mathbf{W}$, which we can determine by SVD. Finally once $\mathbf{Q}^*$ is obtained, instead of permuting columns of $\mathbf{W}$ we can use the equivalent permutation of $\mathbf{V}^T$, since it is more computationally efficient.

The whole algorithm of the multi-body factorization method is now summarized as:

1. Extract and track features in the input image sequence and create matrix $\mathbf{W}$
2. Compute $r = rank(\mathbf{W})$
3. Decompose matrix $\mathbf{W}$ using SVD
4. Compute shape interaction matrix $\mathbf{Q}$ using the first r rows of $\mathbf{V}^T$
5. Block-diagonalize $\mathbf{Q}$
6. Permute matrix $\mathbf{V}^T$ into submatrices, each corresponding to a single object
7. Compute $\mathbf{A}_i$ for each object, and thus its shape and motion.

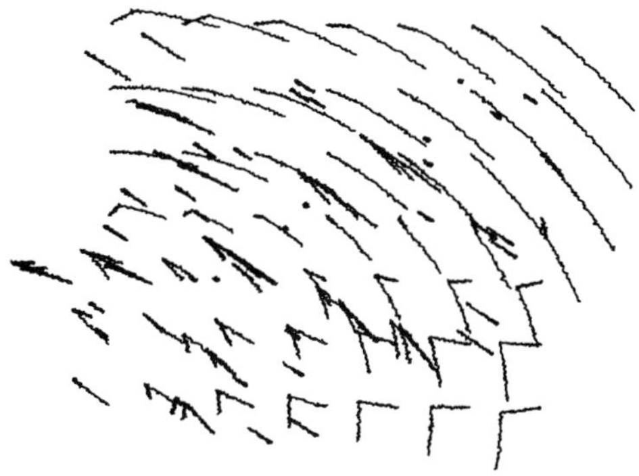

Figure 5: Trajectories of image 118 feature points

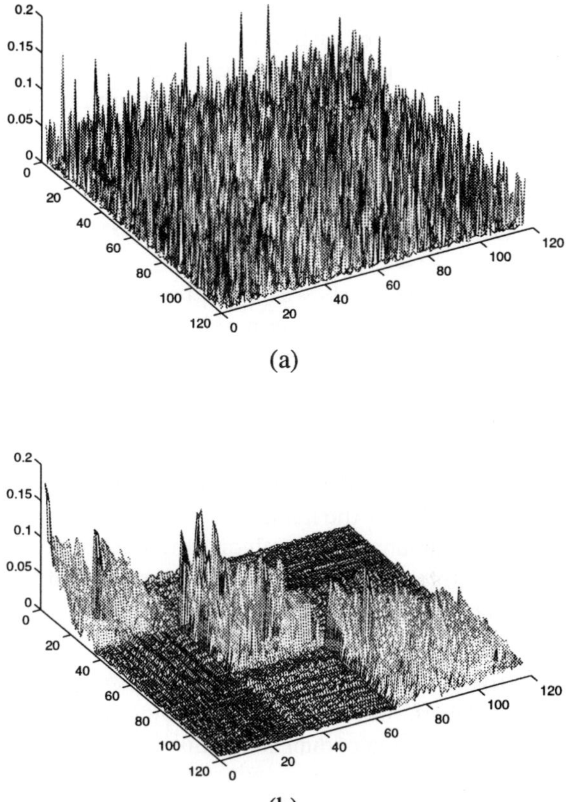

(a)

(b)

Figure 6: The shape interaction matrix for the synthetic scene with three transparent objects: (a) Unsorted matrix $\mathbf{Q}$, and (b) sorted matrix $\mathbf{Q}^*$.

4 Experiment

We will present an experiment result with synthetically generated feature trajectories to demonstrate how the algorithm works.

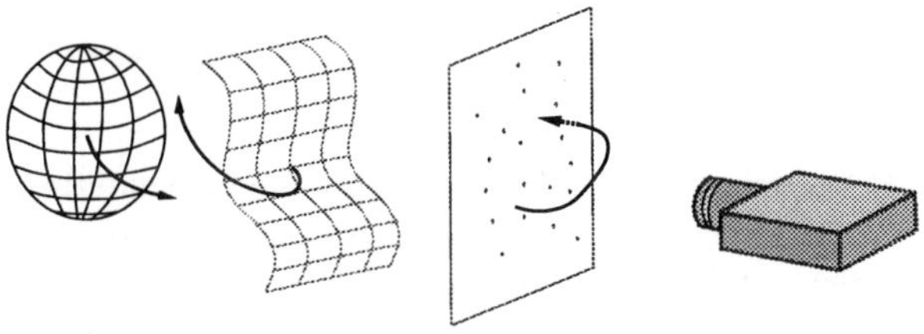

Figure 4: Synthetic scene. Three objects move transparently with arbitrary motion

Figure 4 shows the 3D synthetic scene. It contains three transparent objects in front of each other moving independently. A static camera takes 100 images during the motion. The closest object to the camera is planar (rank 3) and the other two are full 3D objects (rank 4). So this is in fact a shape-degenerate case. Each object translates slightly and rotates over its own centroid in such a way that the features of all objects are completely intermingled in the image plane. This complication is intentionally introduced in order to demonstrate the fact that our motion segmentation and recovery method does not use any local information in the images. One hundred and eighteen (118) points in total on three objects are chosen: 33 features from the first object, 49 from the second, and

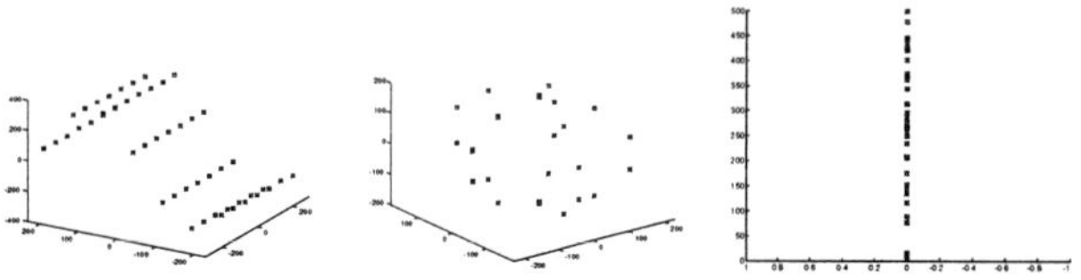

(a) (b) (c)

Figure 7: Recovered shape of the objects

36 from the third. Figure 5 shows the projections of 118 scene points onto the image plane during the motion, that is, the simulated trajectories of tracked image features. Independently distributed Gaussian noise with one pixel of variance was added to the image feature positions for simulating errors in feature tracking. Of course, the identities of the features are assumed unknown, so the measurement matrix created by randomly ordering the features was given to the algorithm.

Figure 6(a) shows the shape interaction matrix **Q**: the height is the square of the entry value. The result of sorting the matrix into a bloackdiagonal form is shown in figure 6(b). We can observe the three blocks corresponding to objects 3, 2 and 1: all of the 118 features are correctly classified. Figures 7(a) (b) and (c) show one view of each of the recovered shapes of the three objects in the same order as figure 4. Figure 7(c) showing the planar object viewed from its edge indicates the correct recovery of its shape.

5 Discussion and Conclusion

In this paper we have shown that the problem of multibody structure-from-motion problem can be solved systematically by using the shape interaction matrix. The striking fact is that the method allows for segmenting or grouping image features into separate objects *based on* their shape properties *without* explicitly computing the individual shapes themselves. Also, no prior knowledge of the number of moving objects in the scene is assumed in the algorithm. This is due to the interesting and useful invariant properties of the shape-interaction matrix **Q**.

Acknowledgments: The authors wish to thank Martial Hebert, José Moura and José Bioucas for useful suggestions and comments to this work.

References

[Adelson and Bergen, 1985] Edward. Adelson and James Bergen. Spatiotemporal energy models for the perception of motion. *Journal of the Optical Society of America*, 2(2):284–299, 1985.

[Adiv, 1985] G. Adiv. Determining three dimensional motion and structure from optical flow generated by several moving objects. *IEEE Transactions on Pattern Analysis and Machine Intelligence*, 7(4):384–401, July 1985.

[Bergen *et al.*, 1990] J. Bergen, P. Burt, R. Hingorani, and S. Peleg. Computing two motions from three frames. In *Proceedings of the IEEE International Conference on Computer Vision*, December 1990.

[Boult and Brown, 1991] Terrance Boult and Lisa Brown. Factorization-based segmentation of motions. In *Proceedings of the IEEE Workshop on Visual Motion*, October 1991.

[Burt *et al.*, 1991] Peter Burt, Rajesh Hingorani, and Raymond Kolczynski. Mechanisms for isolating component patterns in the sequential analysis of multiple motion. In *IEEE Workshop on visual motion*, pages 187–193, October 1991.

[Gear, 1994] C. W. Gear. Feature groupin in moving objects. In *Proceedings of the workshop on motion of non-rigid and articulated objects*, Austin, Texas, 1994.

[Irani *et al.*, 1994] Michal Irani, Rousso Benny, and Shmuel Peleg. Computing occluding and transparent motions. *International Journal of Computer Vision*, 12(1):5–16, February 1994.

[Jasinschi *et al.*, 1992] Radu S. Jasinschi, A. Rosenfeld, and K. Sumi. Perceptual motion transparency: the role of geometrical information. *Journal of the Optical Society of America*, 9(11):1–15, November 1992.

[Koendering and van Doorn, 1992] Jan Koendering and Andrea van Doorn. Affine structure from motion. *Journal of the Optical Society of America*, 12(1):5–16, February 1992.

[Navab and Zhang, 1992] N. Navab and Z. Zhang. From multiple objects motion analysis to behavior-based object recognition. In *Proc. 10th European Conference on Artificial Intelligence, EC AI 92*, 1992.

[Poelman and Kanade, 1994] Conrad. Poelman and Takeo. Kanade. A paraperspective factorization method for shape and motion recovery. In *Proc. Third European Conference on Computer Vision*, pages 97–108. Stockholm, Sweden, May 1994. Part of this was originally published as CMU Technical Report TR-CMU-CS-93-219, December 1993.

[Sinclair, 1993] D Sinclair. Motion segmentation and local structure. In *Proceedings of the 4th International Conference on Computer Vision*, 1993.

[Stewart, 1992] G.W. Stewart. On the early history of the singular value decomposition. TR-92-31, University of Maryland,Institute for Advanced Computer Studies, March 1992.

[Tomasi and Kanade, 1990] Carlo Tomasi and Takeo Kanade. Shape and motion without depth. In *International Conference on Computer Vision*, pages 91–95. IEEE Computer Society, 1990.

[Tomasi and Kanade, 1992] Carlo Tomasi and Takeo Kanade. Shape from motion from image streams under orthography: A factorization method. *International Journal of Computer Vision*, 9(2):137–154, November 1992. Originally published as CMU Technical Report CMU-CS-90-166, September 1990.

[Ullman, 1983] Shimon Ullman. Maximizing rigidity: The incremental recovery of 3-d structure from rigid and rubbery motion. A.I. Memo No. 721, MIT, June 1983.

[Watson and Ahumada, 1983] A.B. Watson and A.J. Ahumada. A look at motion in the frequency domain. TM-84352, NASA, 1983.

Focus Range Sensors

Shree K. Nayar
Department of Computer Science, Columbia University New York, NY 10027, U.S.A.

Minori Noguchi, Masahiro Watanabe, Yasuo Nakagawa
Productional Engineering Research Laboratory, Hitachi Ltd. Yokohama, 244 Japan

Abstract

Structures of dynamic scenes can only be recovered using a real-time range sensor. Focus analysis offers a direct solution to fast and dense range estimation. It is computational efficient as it circumvents the correspondence problem faced by stereo and feature tracking in structure from motion. However, accurate depth estimation requires theoretical and practical solutions to a variety of problems including recovery of textureless surfaces, precise blur estimation, and magnification variations caused by defocusing. Both textured and textureless surfaces are recovered using an illumination pattern that is projected via the same optical path used to acquire images. The illumination pattern is optimized to ensure maximum accuracy and spatial resolution in computed depth. A prototype focus range sensor has been developed that produces up to 512x480 depth estimates at 30 Hz with an accuracy better than 0.3%. In addition, a microscopic shape from focus sensor is described that uses the derived illumination pattern and a sequence of images to recover depth with an accuracy of 1 micron. Several experimental results are included to demonstrate the performances of both sensors. We conclude with a brief summary of our recent results on passive focus analysis.

1 Introduction

Of all problems studied in computational vision, recovery of three-dimensional scene structure has by far attracted the most attention. This has resulted in a panoply of sensors and algorithms [15] [2] that can be broadly classified into two categories; passive and active. Passive techniques such as shape from shading and shape from texture attempt to extract structure from a single image. These algorithms are still under investigation and, given the assumptions they are forced to invoke, they are expected to prove complementary to other techniques but not serve as standalone strategies. Other passive methods such as stereo and structure from motion use multiple views to resolve shape ambiguities inherent in a single image. The primary bottleneck for these methods has proved to be correspondence and feature tracking. Recently, it was demonstrated that stereo could achieve real-time performance, but only with the use of significant customized hardware [16]. Further, passive algorithms have yet to demonstrate the accuracy and robustness required for high-level perception tasks such as object recognition and pose estimation.

Hitherto, high quality depth maps have resulted only from the use of active sensors based on time of flight or light striping [15]. From a practical perspective, light stripe range finding has emerged as a clear winner. In structured environments, where active radiation of a scene is feasible, it offers a robust yet inexpensive solution to a variety of problems. However, it has suffered from one inherent drawback, namely, speed. To achieve depth maps with sufficient spatial resolution, a large number (say, N) of closely spaced stripes are used. If all stripes are projected simultaneously it is impossible to associate a unique stripe with any given image point, a process that is necessary to compute depth by triangulation. The classical approach is to obtain N images, one for each stripe. If T_f is the time required to sense and digitize an image, the scanning of N stripes takes at least $N \,.\, T_f$. Substantial improvements can be made by assigning gray codes to the stripes and scanning the entire collection of stripes in sets [14]. All the information needed is then acquired in $log_2(N)\, T_f$, a significant improvement. An alternative approach uses color-coded stripe patterns [5]; this however is practical only in a gray-world that reflects all wavelengths of light. New hope for light stripe range finding has been instilled by advances in VLSI. Based on the notion of cell parallelism [17], a computational sensor is developed where each sensor element records a stripe detection time-stamp as a single laser stripe sweeps the scene at high speed. Depth maps are produced in as little as 1 msec, though present day silicon packaging limits the total number of cells, and hence spatial depth resolution, to 28x32 [12].

Here, we summarize our work on a class of range sensors that are based on focus analysis. In particular, we describe a real-time range sensor that produces high-resolution (512x480) depth maps at 30 Hz (video rate) [25] [32]. Focus analysis has a major advantage over stereo and structure from motion. Two or more images of a scene are taken under different optical settings but from the same viewpoint, as initially demonstrated by [27][29] and subsequently by others[1]. This circumvents the need for correspondence or feature tracking. The real-time sensor mentioned above here uses only two scene images. These images correspond to different levels of focus, and local frequency analysis implemented typically via linear operators yields depth estimates. However, differences between the two images tend to be very subtle and we believe that previous solutions to depth from defocus have met with limited practical success as they are based on rough approximations to the optical and sensing mechanisms involved in focus analysis. In contrast, our approach is based on a careful physical modeling of all the optical, sensing, and computational elements at work; the optical transfer function, defocus, image sensing and sampling, and focus measure operators.

Depth from defocus shares one inherent weakness with stereo and motion, in that, it requires that the scene have high frequency textures. A textureless surface appears the same focused or defocused and the resulting images do not contain information necessary for depth computation. This has prompted us to develop a range sensor that uses active illumination. The key idea is to force a texture on the scene and then analyze the relative defocus of the texture in two images. Illumination projection has been suggested in the past [9][28] for both depth from defocus and depth from pattern size distortion under perspective projection. However, these projected patterns were selected in a more or less arbitrary fashion and do not guarantee desired precision in computed depth. A critical problem therefore is determining an illumination pattern that would maximize the accuracy and robustness of depth from defocus. We arrive at a solution to this problem through a detailed Fourier analysis of the entire depth from defocus system. First, theoretical models developed for each of the optical and computational elements of the system are expressed in spatial and Fourier domains. The derivation of the illumination pattern (or filter) is then posed as an optimization problem in Fourier domain. The optimal pattern is one that maximizes sensitivity of the focus measure to depth variations while minimizing the size of the focus operator to achieve high spatial resolution in computed depth.

An implementational problem that has repeatedly surfaced in previous work is the variation in image magnification that occurs when images are taken under different focus settings [33]. This manifests into a correspondence-like problem. It has forced investigators to resort to techniques varying from image registration and warping [7] to the use of precise lens calibration for correcting magnification variations [33] [7]. We have a simple but effective optical solution to this problem [31]. By appending an additional aperture to the optics, we show that the focus setting of an imaging system can be varied substantially without altering magnification.

A prototype real-time focus range sensor has been developed. It uses two CCD image detectors that view the scene through the same optical elements [25][32]. The derived illumination pattern is fabricated using micro-lithography and incorporated into the sensor. The illumination pattern is projected onto the scene via the same optical path used to image the scene. This results in several advantages. It enables precise registration of the illumination pattern with the sampling grid of the image sensors. Light rays projected out through the imaging optics are subjected to similar geometric distortions as rays reflected back to the sensors. Therefore, despite ever-present lens distortions, the illumination pattern and the sensing grid of the detector are well registered. The coaxial illumination and imaging also results in

[1] All work in focus based depth computation can be broadly classified into depth from focus and depth from defocus. The former relies on a large number of images taken by displacing the sensor in small increments and uses a focus operator to detect the image of maximum focus for each scene point (see [20, 7, 22, 21, 19, 1, 34, 26]). In contrast, depth from defocus typically uses two images and estimates relative blurring to get depth (see [27, 29, 11, 28, 4, 8, 34, 10]).

a shadowless image; all surface regions that are visible to the sensor are also illuminated. Furthermore, since both images are acquired from the same viewing direction, the missing part or occlusion problem in stereo is avoided. Several experiments have been conducted to evaluate the accuracy and real-time capability of the sensor. In addition to a quantitative error analysis, real-time depth map sequences of moving objects are shown.

Finally, we describe a second system we have developed that is based on focus analysis [26]. This system is complementary to the first one, in that, it is based on depth from focus (rather than defocus) and is capable of reocovering microscopic objects with an accuracy as good as 1 micron. The starting point for this system is an algorithm described in [22], where a shape from focus method was developed for microscopic objects. The high magnification of a microscope results in images that capture brightness variations caused by the micro-structure of the surface. Most surfaces that appear smooth and non-textured to the naked eye produce highly textured images under a microscope. Examples of such surfaces are paper, plastics, ceramics, etc. Microscopic shape from focus [22] was therefore demonstrated to be an effective approach, offering solutions to a variety to challenging shape inspection problems. However, there exist surfaces that are smooth at the micro-structure level and consequently do not produce sufficient texture even under a microscope.

The illumination pattern we used in the real-time range sensor was initially incorporated into the microscopic shape from focus system [26]. The illumination pattern is projected onto the sample via the path of the bright field illumination of the microscope. As with the depth from defocus sensor, this enables very precise registration of the illumination pattern with the sampling grid of the image sensor; light rays projected out through the imaging optics are subjected to the same geometric distortions as rays reflected back to the image sensor. The motorized stage of the microscope is used to automatically acquire a set of images (typically 10-20) by moving the sample towards the objective lens. The depth from focus algorithm [22] is then applied to the image set to obtain a complete depth map of the sample. As examples, we show accurate and detailed depth maps of structures on silicon substrates and solder joints. These are samples of significant industrial import that have been found hard to recover using other vision techniques.

We conclude with a brief discussion on our most recent result on focus analysis that include a passive bifocal vision sensor [23] and a novel algorithm for passive depth from defocus [30] that uses a minimal operator set to recover scenes with unknown and complex textures.

2 Depth from Defocus

Fundamental to depth from defocus is the relationship between focused and defocused images [3]. Figure 1 shows the basic image formation geometry. All light rays that are radiated by object point P and pass the aperture A are refracted by the lens to converge at point Q on the image plane. For a thin lens, the relationship between the object distance d, focal length of the lens f, and the image distance d_i is given by the Gaussian lens law:

$$\frac{1}{d} + \frac{1}{d_i} = \frac{1}{f}. \tag{1}$$

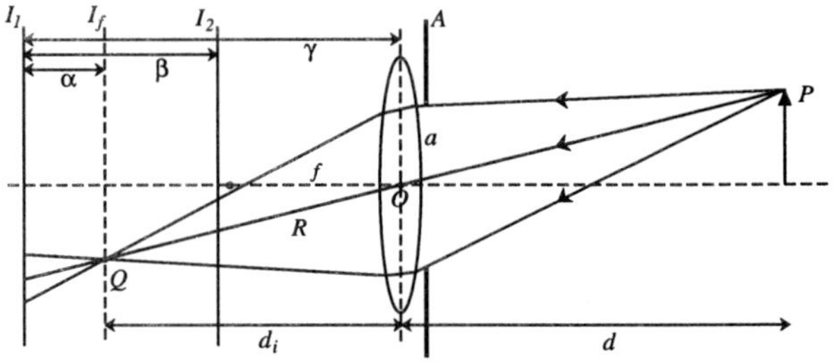

Figure 1: Image formation and depth from defocus.

Each point on the object plane is projected onto a single point on the image plane, causing a clear or *focused* image I_f to be formed. If, however, the sensor plane does not coincide with the image plane and is displaced from it, the energy received from P by the lens is distributed over a patch on the sensor plane. The result is a blurred image of P. It is clear that a single image does not include sufficient information for depth estimation as two scenes defocused to different degrees can produce identical images. A solution to depth is achieved by using two images, I_1 and I_2, separated by a known physical distance β. The problem is reduced to analyzing the relative blurring of each scene point in the two images and computing the distance α of its focused image. Then, using $d_i = \gamma - \alpha$, the lens law (1) yields depth d of the scene point. Simple as this procedure may appear, several technical problems emerge when implementing an algorithm of practical value. These include (a) accurate estimation of relative defocus in the two images, (b) recovery of textured **and** textureless surfaces, and (c) achieving constant magnification that is invariant to the degree of defocus.

3 Constant-Magnification Defocus

We begin with the last of the problems mentioned above. In the imaging system shown in Figure 1, the effective image location of point P moves along ray R as the sensor plane is displaced. This causes a shift in image coordinates of P that in turn depends on the unknown scene coordinates of P. This variation in image magnification with defocus manifests as a correspondence-like problem in depth from defocus as the right set of points in images I_1 and I_2 are needed to estimate blurring. We approach this problem from an optical perspective rather a computational one. Consider the image formation model shown in Figure 2. The only modification made with respect to the model in Figure 1 is the use of the external aperture A'. The aperture is placed at the *front-focal plane*, i.e. a focal length in front of the *principal point* O of the lens. This simple addition solves the prevalent problem of magnification variation with distance α of the sensor plane from the lens. Simple geometrical analysis reveals that a ray of light R' from any scene point that passes through the center O' of aperture A' emerges parallel to the optical axis on the image side of the lens [18]. Furthermore, this parallel ray is the axis of a cone that includes all light rays radiated by the scene point, passed through by A' and intercepted by the lens. As a result, despite blurring, the effective image coordinates of point P in both images I_1 and I_2 are the same, namely, the coordinate of its focused image Q on I_f.

This invariance of magnification to defocus holds true for any depth from defocus configuration (all values of α and β). It can also be shown that the constant-magnification property is unaffected by the aperture radius a'. Furthermore, the lens law of (1) remains valid. This modification is realizable not only in single lens systems but any compound lens system. Given an off-the-shelf lens, such an aperture is easily appended to the casing of the lens. The resulting optical system is called a *telecentric lens*. While the nominal and effective *F-numbers* of the classical optics in Figure 1 are f/a and d_i/a, respectively, they are both equal to f/a' in the telecentric case.

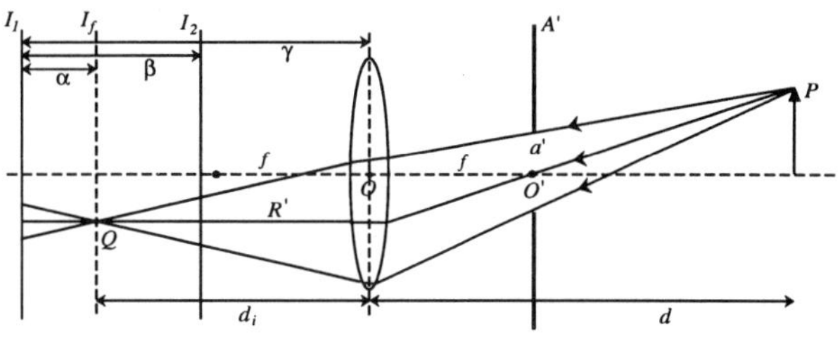

Figure 2: A constant-magnification imaging system for depth from defocus is achieved by simply placing an aperture at the front-focal plane of the optics.

4 Modeling

Effective solutions to both illumination projection and depth estimation require careful modeling and analysis of all physical phenomena involved in depth from defocus. There are five different elements, or components, that play a critical role, namely, the illumination pattern, optical transfer function, defocusing, image sensing, and the focus operator. All of these together determine the the relation between the depth of a scene point and its two focus measures. Since we have used the telecentric lens (Figure 2) in our implementation, it's parameters are used in developing each model. However, all of the following expressions can be made valid for the classical lens system (Figure 1) by simply replacing the factor $\frac{f}{a'}$ by $\frac{d_i}{a}$. Though we use both spatial and Fourier (frequency) models of the above components, for brevity we will present Fourier models only when they are needed to make pertinent observations.

4.1 Illumination Pattern

Before the parameters of the illumination pattern can be determined, an illumination model must be defined. Such a model must be flexible in that it must subsume a large enough variety of possible illumination patterns. As we will describe shortly, the image sensor used has rectangular pixels arranged on a rectangular spatial grid. Hence, the basic building block of the model is a rectangular illuminated patch, or cell, with uniform intensity:

$$i_c(x,y;b_x,b_y) = {}^2\Pi(\frac{1}{b_x}x, \frac{1}{b_y}y) \quad (2)$$

where, ${}^2\Pi()$ is the two-dimensional *Rectangular* function [6]. The *unknown* parameters of this illumination cell are b_x and b_y, the length and width of the cell. This cell is assumed to be repeated on a two-dimensional grid to obtain a periodic pattern. This periodicity is essential since our goal is to achieve spatial invariance in depth accuracy. The periodic grid is defined as:

$$i_g(x,y;t_x,t_y) = {}^2\mathrm{III}(\tfrac{1}{2}(\tfrac{1}{t_x}x + \tfrac{1}{t_y}y),\ \tfrac{1}{2}(\tfrac{1}{t_x}x - \tfrac{1}{t_y}y)) \quad (3)$$

where, ${}^2\mathrm{III}()$ is the 2-dimensional *Shah* function [6], and $2t_x$ and $2t_y$ determine the periods of the grid in the x and y directions. The final illumination pattern is obtained by convolving the cell $i_c(x,y)$ with the grid $i_g(x,y)$, i.e. $i(x,y) = i_c(x,y) * i_g(x,y)$. The exact pattern is therefore determined by four parameters, namely, b_x, b_y, t_x and t_y. The above illumination grid is not as restrictive as it may appear upon initial inspection. For instance, the parameters b_x, b_y, $2t_x$ and $2t_y$ can each be stretched to obtain repeated illumination and non-illumination stripes in the horizontal and vertical directions, respectively. Alternatively, they can also be adjusted to obtain a checkerboard illumination pattern with large or small illuminated patches. The exact values for b_x, b_y, t_x and t_y will be evaluated by the optimization procedure described later.

The Fourier transforms of the illumination cell, grid, and pattern are denoted as $I_c(u,v)$, $I_g(u,v)$, and $I(u,v)$, respectively, and are related as:

$$I(u,v;b_x,b_y,t_x,t_y) = I_c(u,v) \cdot I_g(u,v) \quad (4)$$

4.2 Optical Transfer Function

Adjacent points on the illuminated surface reflect light waves that interfere with each other to produce diffraction effects. The angle of diffraction increases with the spatial frequency of surface texture. Since the lens aperture of the imaging system (Figure 2) is of finite radius a', it does not capture the higher order diffractions radiated by the surface (see [3] for details). This effect places a limit on the optical resolution of the imaging system characterized by the optical transfer function (OTF):

$$O(u,v;a',f) \quad (5)$$

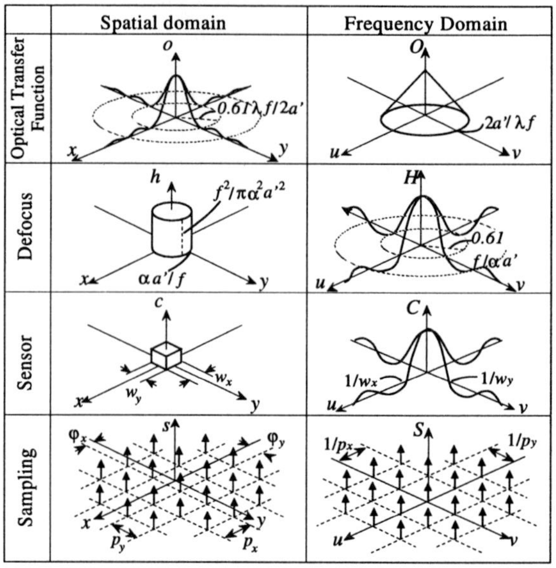

Figure 3: Spatial and frequency models for the optical and sensing elements of depth from defocus.

$$= \begin{cases} (\frac{a'}{f})^2(\gamma - \sin\gamma), & \sqrt{u^2+v^2} \le \frac{2a'}{\lambda f} \\ 0, & \sqrt{u^2+v^2} > \frac{2a'}{\lambda f} \end{cases}$$

$$\text{where } \gamma = 2\cos^{-1}(\frac{\lambda f}{a'}\frac{\sqrt{u^2+v^2}}{2}).$$

where, (u,v) is the spatial frequency of the two-dimensional surface texture as seen from the image side of the lens, f is the focal length of the lens, and λ is the wavelength of incident light. It is clear from the above expression that only spatial frequencies below the limit $\frac{2a'}{\lambda f}$ will be imaged by the optical system (Figure 3). This in turn places restrictions on the frequency of the illumination pattern.

4.3 Defocusing

The defocus function is described in detail in previous work (see [3][13] for example). As in Figure 2, let α be the distance between the focused image of a surface point and its defocused image formed on the sensor plane. The light energy radiated by the surface point and collected by the imaging optics is uniformly distributed over a circular patch on the sensor plane. This patch, also called the *pillbox*, is the defocus function (Figure 3):

$$h(x,y;\alpha,a',f) = \frac{f^2}{2\pi a'^2\alpha^2}\Pi(\frac{d}{2a\alpha}\sqrt{x^2+y^2}) \quad (6)$$

where, a' is the radius of the telecentric lens aperture. The Fourier transform of the defocus function is:

$$H(u,v;\alpha,a',f) \quad (7)$$
$$= \frac{f}{2\pi a'\alpha\sqrt{u^2+v^2}} J_1\left(\frac{2\pi a'\alpha}{f}\sqrt{u^2+v^2}\right)$$

where J_1 is the first-order Bessel function [3]. As is evident from the above expression, defocus serves as a low-pass filter. The bandwidth of the filter increases as α decreases, i.e. as the sensor plane gets closer to the plane of focus. Note that in a defocused image, all frequencies are attenuated at the same time. In the case of passive depth from focus or defocus, this poses a serious problem; different frequencies in an unknown scene are bound to have different (and unknown) magnitudes and phases. It is difficult therefore to estimate the degree of defocus of an image region without the use of a large set of narrow-band focus operators that analyze each frequency in isolation. Hence, it would be desirable to have an illumination pattern that has a single dominant frequency, enabling robust estimation of defocus and hence depth.

4.4 Image Sensing

We assume the image sensor to be a typical CCD TV camera that can be modeled as a rectangular array of rectangular sensing elements (pixels). The quantum efficiency [13] of each pixel is assumed to be uniform over the area of the pixel. Let $m(x,y)$ be the continuous image formed on the sensor plane. The finite pixel area has the effect of averaging the continuous image $m(x,y)$. In spatial domain, the averaging function is the rectangular cell:

$$s_c(x,y;w_x,w_y) = {}^2\Pi\left(\frac{1}{w_x}x, \frac{1}{w_y}y\right) \quad (8)$$

where, w_x and w_y are the length and width of the pixel, respectively. The discrete image is obtained by sampling the convolution of $m(x,y)$ with $s_c(x,y)$. This sampling function is a rectangular grid:

$$s_g(x,y;p_x,p_y,\varphi_x,\varphi_y) \quad (9)$$
$$= \frac{1}{p_x p_y}{}^2\mathrm{III}\left(\tfrac{1}{p_x}(x-\varphi_x), \tfrac{1}{p_y}(y-\varphi_y)\right)$$

where, p_x and p_y are spacings between discrete samples in the two spatial dimensions, and (φ_x,φ_y) is phase shift of the grid. The final discrete image is therefore:

$$m_d(x,y) = (s_c(x,y) * m(x,y)) \cdot s_g(x,y) \quad (10)$$

The parameters w_x, w_y, p_x, and p_y are all determined by the particular image sensor used. These parameters are therefore known and their values are substituted after the optimization is done. On the other hand, the phases (φ_x,φ_y) of the sampling function is with respect to the illumination pattern and are also viewed as parameters to be optimized. In Fourier domain, the final discrete image is:

$$M_d(u,v) = (S_c(u,v)\cdot M(u,v)) * S_g(u,v) \quad (11)$$

4.5 Focus Operator

Since defocusing has the effect of suppressing high-frequency components in the focused image, it is desirable that the focus operator respond to high frequencies in the image. For the purpose of illumination optimization, we use the Laplacian. However, the derived pattern will remain optimal for a large class of symmetric focus operators. In spatial domain, the 3x3 discrete Laplacian is:

$$l(x,y;q_x,q_y)$$
$$= 4\delta(x)\cdot\delta(y) - [\delta(x)\cdot\delta(y-q_y) + \delta(x) \quad (12)$$
$$\cdot\delta(y+q_y) + \delta(x-q_x)\cdot\delta(y) + \delta(x+q_x)\cdot\delta(y)]$$

Here, q_x and q_y are the spacings between neighboring elements of the discrete Laplacian kernel and are given by the image sensor. The Fourier transform of the Laplacian is:

$$L(u,v;q_x,q_y) = 4 - 2\cos(2\pi q_x u) - 2\cos(2\pi q_y v) \quad (13)$$

The required discrete nature of the focus operator comes with a price. It tends to broaden the bandwidth of the operator. Once the illumination pattern has been determined, the above filter will be tuned to maximize sensitivity to the fundamental illumination frequency while minimizing the effects of spurious frequencies caused either by the scene's inherent texture or image noise.

4.6 Focus Measure

The focus measure is simply the output of the focus operator. It is related to defocus α (and hence depth d) via all of the components modeled above. Note that the illumination pattern $(i_c * i_g)$ is projected through optics that is similar to that used for image formation. Consequently, the pattern is also subjected to the limits imposed by the optical transfer function o and the defocus function h. Therefore, the texture projected on the scene is:

$$i(x,y;b_x,b_y,t_x,t_y)*o(x,y;a',f)*h'(x,y;\alpha',a',f) \quad (14)$$

where, α' represents defocus of the illumination itself that depends on the depth of the illuminated point. However, the illumination pattern, once incident on a surface patch, plays the role of surface texture and hence defocus α' of illumination does not have any significant effect on depth estimation. The projected texture is reflected by the scene and projected by the optics back onto the image plane to produce the discrete image:

$$\{i(x,y;b_x,b_y,t_x,t_y)*o(x,y;a',f)^{*2} *h'(x,y;\alpha',a',f)*h(x,y;\alpha,a',f) *s_c(x,y;w_x,w_y)\}\cdot s_g(x,y;p_x,p_y,\varphi_x,\varphi_y) \quad (15)$$

where, $o^{*2}=o*o$. The final focus measure function $g(x,y)$ is the result of applying the discrete Laplacian to the above discrete image:

$$g(x,y)=\{(i*o^{*2}*h^{*2}*s_c)\cdot s_g\}*l \quad (16)$$

Since the distance between adjacent weights of the Laplacian kernel must be integer multiples of the period of the image sampling function s_g, (16) can be rearranged as:

$$g(x,y)=(i*o^{*2}*h'*h*s_c*l)\cdot s_g=g_0\cdot s_g \quad (17)$$

where, $g_0=i*o^{*2}*h'*h*s_c*l$. Alternately, in Fourier domain we have:

$$G(u,v)=(I\cdot O^2\cdot H'\cdot H\cdot S_c\cdot L)*S_g=G_0*S_g \quad (18)$$

The above expression gives us the final output of the focus operator for any value of the defocus parameter α.

5 Optimization

The illumination optimization problem is formulated as follows: Establish closed-form relationships between the illumination parameters (b_x,b_y,t_x,t_y), sensor parameters $(w_x,w_y,p_x,p_y,\varphi_x,\varphi_y)$, and discrete Laplacian parameters (q_x,q_y) so as to maximize the sensitivity, robustness, and spatial resolution of the focus measure $g(x,y)$. High sensitivity implies that a small variation in the degree of focus results in a large variation in $g(x,y)$. By robustness we mean that all pixels with the same degree of defocus produce the same focus measure independent of their location on the image plane. This ensures that depth estimation accuracy is invariant to location on the image plane. Lastly, high spatial resolution is achieved by minimizing the size of the focus operator.

The details of the optimization process are given in [24] and will be omitted in the interest of space. Here, we briefly outline the arguments we have used to arrive at the optimal pattern. In order to minimize smoothing effects and maximize spatial resolution of computed depth, the support (or span) of the discrete Laplacian must be as small as possible. This in turn requires the frequency of the illumination pattern be as high as possible. However, the optical transfer function described in section 4.2 imposes limits on the highest frequency that can be imaged by the optical system. This maximum allowable frequency is $\frac{2a'}{\lambda f}$, determined by the numerical aperture of the telecentric lens. Our objective then is to maximize the fundamental spatial frequency $(1/t_x,1/t_y)$ of the illumination. In order to maximize this frequency while maintaining high detectability, we must have $\sqrt{(1/t_x)^2+(1/t_y)^2}$ close to the optical limit $\frac{2a'}{\lambda f}$. This in turn pushes all higher harmonics in the illumination pattern outside the optical limit. What we are left with is a surface texture whose image has only the quadrapole fundamental frequencies $(\pm 1/t_x, \pm 1/t_y)$. Using this observation, the illumination pattern parameters $(b_x,b_y,t_x,t_y,)$ and the illumination phase shift (φ_x,φ_y) that maximize $\| G(\frac{1}{t_x},\frac{1}{t_y}) \|$ are determined in [24]. Two optimal patterns were found and are shown in Figure 4. Exactly how such high resolution patterns can be projected and perfectly registered with the image detector will be described in the experimental section.

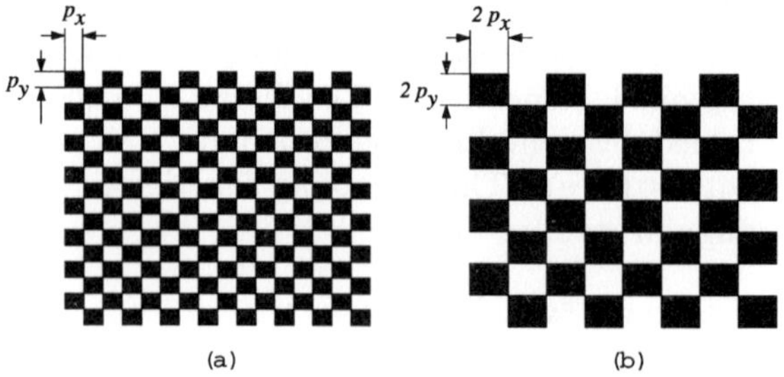

Figure 4: Optimal illumination filter patterns: (a) $t_x=2p_x, t_y=2p_y, \varphi_x=0, \varphi_y=0$; and (b) $t_x=4p_x, t_y=4p_y, \varphi_x=1/8t_x, \varphi_y=1/8t_y$. Here, (t_x,t_y) is the illumination period, (p_x,p_y) is the pixel size, and (φ_x,φ_y) is the illumination phase shift with respect to the sensing grid.

6 Tuned Focus Operator

For the purpose of illumination optimization, we used the Laplacian. The resulting illumination pattern has only a single dominant absolute frequency, $(1/t_x, 1/t_y)$. Given this, we are in a position to further refine our focus operator so as to minimize the effects of all other frequencies caused either by the physical texture of the scene or image noise.

Given that the operator must eventually be discrete and of finite support, there is a limit to the extent to which it can be tuned. To constrain the problem, we impose the following conditions. (a) To maximize spatial resolution in computed depth, we force the operator kernel to be 3x3. (b) Since the fundamental frequency of the illumination pattern has a symmetric quadrapole arrangement, the focus operator must be rotationally symmetric. (c) The operator must not respond to any DC component in image brightness. The above conditions, yield a set of equations with the operator elements as variables [32]. These equations were solved to find that the operator with its symmetric structure has only one variable. This variable was optimize so as to yield a frequency response with sharpest peaks, i.e. power spectrum with the lowest second moment around the illumination frequency $(\pm 1/t_x, \pm 1/t_y)$. This tuned focus operator was found to have substantially sharper peaks than the discrete Laplacian.

7 Depth from Two Images

Depth estimation uses two images of the scene, $I_1(x,y)$ and $I_2(x,y)$, that correspond to different effective focal lengths as shown in Figure 2. Depth of each scene point is determined by estimating the displacement α of the focused plane I_f for the scene point. The tuned focus operator is applied to both images to get focus measure images $g_1(x,y)$ and $g_2(x,y,)$. Since the image now has a single dominant frequency, namely $(\pm 1/t_x, \pm 1/t_y)$, a relation between the focus measures and defocus α can be derived using (18):

$$q = \frac{g_1 - g_2}{g_1 + g_2} = \frac{H(\frac{1}{t_x}, \frac{1}{t_y}; \alpha) - H(\frac{1}{t_x}, \frac{1}{t_y}; \alpha - \beta)}{H(\frac{1}{t_x}, \frac{1}{t_y}; \alpha) + H(\frac{1}{t_x}, \frac{1}{t_y}; \alpha - \beta)} \tag{19}$$

As shown in Figure 5, q is a monotonic function of α such that $-p \le q \le p$, $p \le 1$. In practice, the above relation can be pre-computed and stored as a look-up table that maps q to a unique α. Since α represents the position of the focused image, the lens law (1) yields the depth d of the corresponding scene point. Note that the tuned focus operator designed in the previous section is a linear filter, making it feasible to compute depth maps of scenes in real-time using simple image processing hardware.

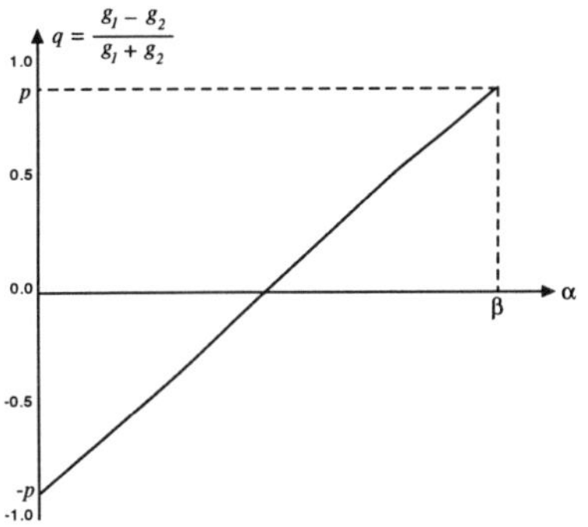

Figure 5: Relation between focus measures g_1 and g_2 and the defocus parameter α.

8 Real Time Range Sensor

Based on the above results, we have implemented [25][32] the real-time focus range sensor shown in Figure 6. The scene is imaged using a standard 12.5 mm Fujinon lens with an additional aperture added to convert it to telecentric. Light rays passing through the lens are split in two directions using a beam-splitting prism. This produces two images that are simultaneously detected using two Sony XC-77RR 8-bit CCD cameras. The positions of the two cameras are precisely fixed such that one obtains a near-focus image while the other a far-focus image. In this setup, a physical displacement of 0.25mm between the effective focal lengths of the two CCD cameras translates to a sensor depth of field of approximately 30 cms. This detectable range of the sensor can be varied either by changing the sensor displacement or the focal length of the imaging optics.

The illumination pattern shown in Figure 4(b) was etched on a glass plate using microlithography, a process widely used in VLSI. The filter was then placed in the path of a 300 W Xenon arc lamp. The illumination pattern generated is projected using a telecentric lens identical to the one used for image formation. A half-mirror is used to ensure that the illumination pattern projects onto the scene via the same optical path used to acquire images. As a result, the pattern is almost perfectly registered with respect to the pix-

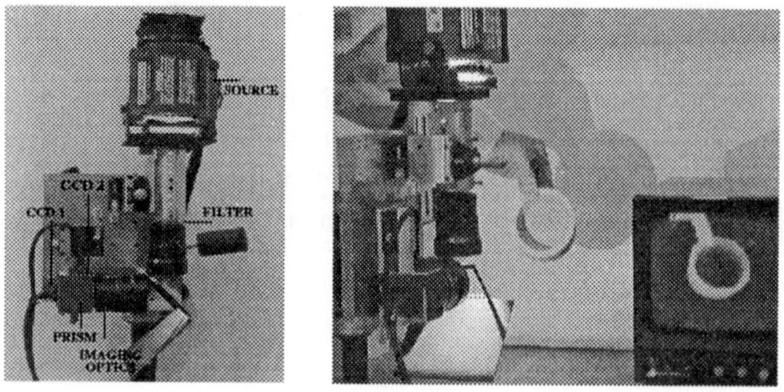

Figure 6: (a) The real-time focus range sensor and its key components. (b) The sensor can produce depth maps up to 512x480 in resolution at 30 Hz.

els of the two CCD cameras. If objects in the scene have a strong specular reflection component, cross-polarized filters can be attached to the illumination and imaging lens to filter out specularities.

Images from the two CCD cameras are digitized and processed using MV200 Datacube image processing hardware. The present configuration includes the equivalent of two 8-bit digitizers, two A/D convertors, and one 12-bit convolver. This hardware enables simultaneous digitization of the two images, convolution of both images with the tuned focus operator, and the computation of a depth map, all within a single frametime of 33 msec with a lag of 33 msec. A look-up table is used to map each pair of focus measures to a unique depth estimate (see [32] for details).

Several experiments were conducted on both textured and textureless surfaces to test the performance of the sensor [25] [32]. The performance evaluation results are summarized in Table 1 and discussed in detail in [32].

	Simulatneous Image Grab	Successive Image Grab
Depth Accuracy (rms)	0.24 %	0.34 %
Repeatability (rms)	0.23 %	0.29 %
Spatial Resolution	256 x 240	512 x 480
Speed	30 Hz	30 Hz
Delay	33 msec	33 msec

Table 1: Performance characteristics of the sensor.

As stated earlier, structures of dynamic scenes can only be recovered using a real-time sensor such as the one we have developed. Figure 7 illustrates the power of such a high-speed, high-resolution sensor. The figure shows two brightness images and the computed depth map of a cup with milk flowing out of it.

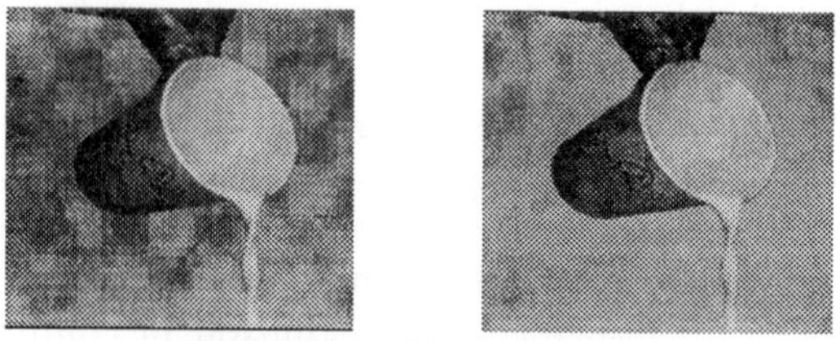

(a)

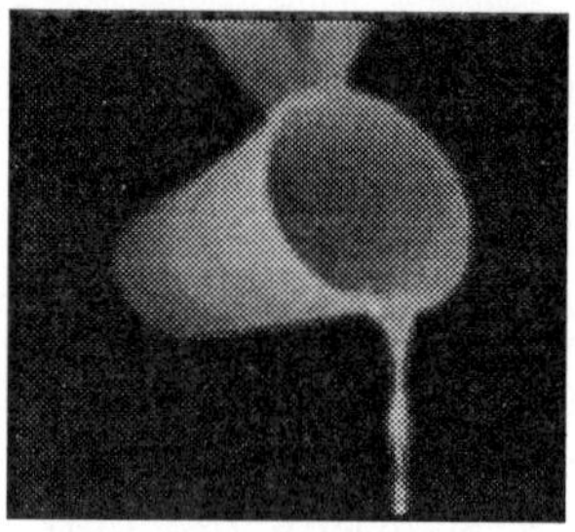

(b)

Figure 7: (a) Two images of a scene taken using different focus settings. (b) A depth map of the scene computed in 33 msec by the focus range sensor.

Figure 8 shows a scene with polyhedral objects. The computed depth map in Figure 8(b) is fairly accurate despite the complex textural properties of the objects. All surface discontinuities and orientation discontinuities are well preserved. Figure 9 shows an object's depth map computed as it rotates on a motorized turntable. Such depth map sequences are valuable for automatic CAD model generation from sample objects.

9 Microscopic Shape from Focus

The optimal illumination filter shown in Figure 4(b) was also incorporated into the shape from focus system developed in [22]. A set of sample images are obtained by automatically moving the microscope stage in increments of $\Delta z = z_i - z_{i-1}$. The Laplacian focus operator is applied to each image to obtain a set of focus measure values at each image point (x, y); the number of focus measures equals the number of images taken. The discrete stage position z_j that yields the maximum focus measure value at an image point, can be used as an approximation of the depth of the corresponding surface point. A more accurate depth estimate z is obtained by applying Gaussian interpolation [22] to the three focus measures corresponding to z_{j-1}, z_j, and z_{j+1}.

It was shown in [22] that at least 3 focus measures are needed for depth estimation by Gaussian interpolation. In the active illumination system proposed here, the fundamental frequency of the surface texture is determined by the illumination filter used and is simply $\sqrt{(\frac{1}{t_x})^2 + (\frac{1}{t_y})^2}$. For this frequency the defocus measure has a zero-crossing at defocus value α' such that $\frac{2\pi a\alpha'}{d}\sqrt{(\frac{1}{t_x})^2 + (\frac{1}{t_y})^2} = 3.83$. This gives us an upper limit on the usable defocus range for any point on the surface, which is, $-\alpha' \leq \alpha \leq \alpha'$. Therefore, on the image side of the microscope optics, we need to obtain (for any surface point) at least 3 images within the above defocus range. This gives us the following maximum distance between consecutive focused images on the image side of the optics:

$$\Delta z' \leq \frac{2\alpha'}{4} = \frac{1}{2} \cdot 3.83 \cdot \frac{d}{2\pi a\sqrt{(\frac{1}{t_x})^2 + (\frac{1}{t_y})^2}} \qquad (20)$$

This distance on the image side is related to the maximum allowable microscope stage displacement (between consecutive images) on the sample side of the optics by the magnification M of the objective lens used:

$$\Delta z \simeq \Delta z' \frac{1}{M^2} \qquad (21)$$

In our experiments, the magnification of the objective lens is $M = 20$, the ratio $a/d = 0.025$, and illumination parameters are $t_x = 44\ \mu m$, $t_y = 52\ \mu m$. Using these values in eqs.(20) and (21) we get the maximum allowable stage displacement $\Delta z \leq 0.5\ \mu m$. Experiments reported in [26] illustrate that $\Delta z = 0.5\ \mu m$ does in fact produce the best results. Further decreasing Δz does not significantly improve the accuracy of the depth maps.

The sample in Figure 10 has rectangular structures fabricated on a smooth silicon wafer. Silicon wafer inspection is of great relevance in a variety of chip manufacturing processes. The surface of the wafer is very smooth, resulting in images (see Figure 10(a)) that are more or less textureless. This renders the original shape from focus system

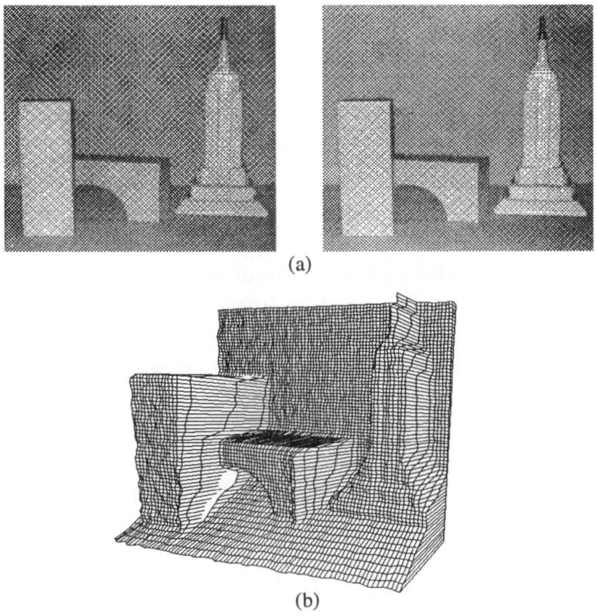

Figure 8: (a) Near and far focused images of a set of polyhedral objects. (b) Computed depth map.

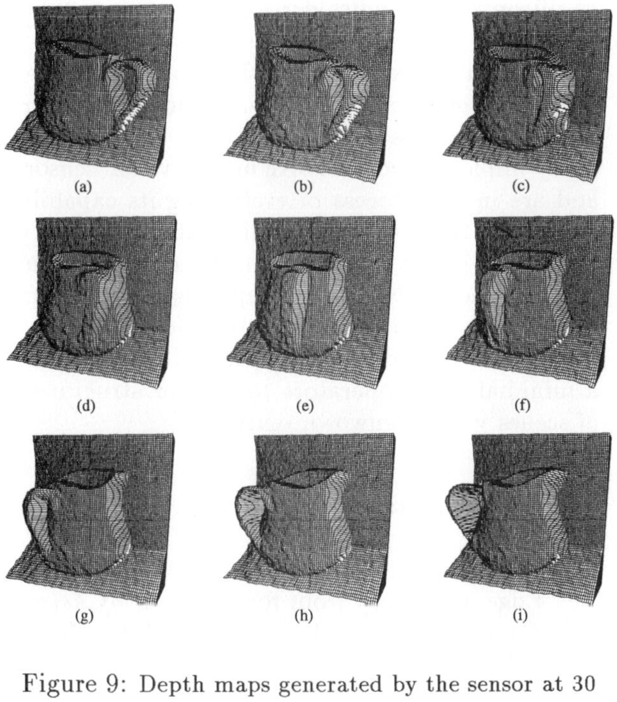

Figure 9: Depth maps generated by the sensor at 30 Hz while an object rotates on a motorized turntable.

[22] ineffective. A total of 16 images were taken for this sample. The derived illumination pattern produces very accurate shape information that is far superior to that produced by the bright field illumination of the microscope [26]. Similar results were obtained for the solder joint sample shown in Figure 11. For this sample, an objective lens with $M = 10$ was used and a total of 23 sample images were taken. This sample exhibits noticeable texture under a microscope. However, the texture is not consistent over the entire surface. As a result, the illumination pattern is necessary to get an accurate depth map. Solder shape inspection has remained a challenging and unresolved industrial problem. These results indicate that microscopic shape from focus may provide an effective solution to this important problem.

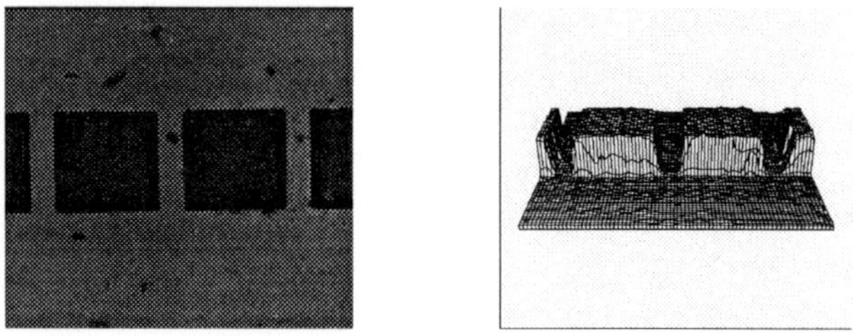

(a) Camera image (b) Depth Map

Figure 10: Image and depth map of rectangular structures on a smooth silicon substrate. The structures are approximately $13\mu m$ tall.

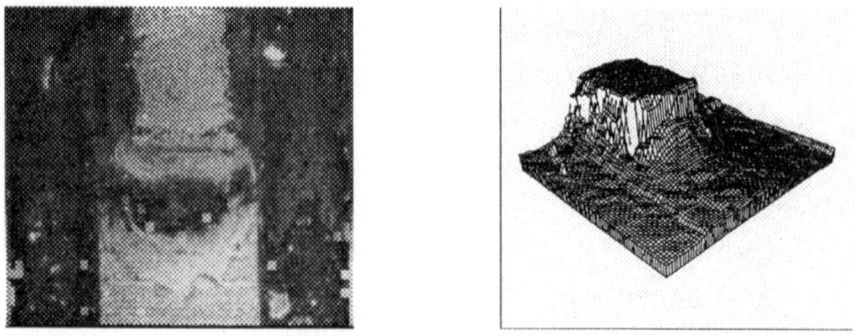

(a) Camera image (b) Depth Map

Figure 11: Image and depth map of a solder joint on a circuit board. The solder joint is approximately $150\mu m$ high and $100\mu m$ wide.

10 Summary

We have summarized our results on a variety of issues related to depth estimation by focus analysis. Accurate modeling of optics and sensing were shown to be essential for precise depth estimation. Both textured and textureless surfaces are recovered by using an optimized illumination pattern that is registered with the image sensor. We also presented an optical solution to constant magnification defocusing, a problem that has limited the precision of depth from defocus algorithms. All of these results were used to implement a real-time focus range sensor that produces high resolution depth maps at frame rate. This sensor is unique in its ability to produce fast, dense, and precise depth information at a very low cost. With time we expect the sensor to find applications ranging from visual recognition and robot control to automatic CAD model generation for visualization and virtual reality.

The second system we described targets a different class of objects, namely, microscopic structures. Using the derived illumination pattern, we have demonstrated a fully automated microscopic shape from focus system that can recover depth with an accuracy within 1 micron. This system has well-defined applications in the industrial arena, where a depth sensor for samples such as silicon wafers and solder joints is much sought after.

The obvious extension to this work is the development of passive focus range finders for both indoor as well as outdoor scenes. We have already implemented a passive bifocal vision sensor and are in the process of evaluating its capabilities [23]. Such a sensor cannot afford the luxury of projected illumination. It must rely on complex scene textures for depth estimation. In this regard, we have recently developed an efficient depth from defocus algorithm [30] that uses a minimal set of operators to recover structures of scenes with unknwown textures.

References

[1] N. Asada, H. Fujiwara, and T. Matsuyama. Edge and depth from focus. *Proc. of Asian Conf. on Comp. Vis.*, pages 83–86, Nov. 1993.

[2] P. J. Besl. Range imaging sensors. Technical Report GMR-6090, General Motors Research Laboratories, March 1988.

[3] M. Born and E. Wolf. *Principles of Optics.* London:Permagon, 1965.

[4] V. M. Bove, Jr. Entropy-based depth from focus. *Jrn. of Opt. Soc. of Am. A*, 10:561–566, Apr. 1993.

[5] K. L. Boyer and A. C. Kak. Color-encoded structured light for rapid active sensing. *IEEE Trans. on Pattern Analysis and Machine Intelligence*, 9(1):14–28, January 1987.

[6] R. N. Bracewell. *The Fourier Transform and Its Applications.* McGraw Hill, 1965.

[7] T. Darrell and K. Wohn. Pyramid based depth from focus. *Proc. of IEEE Conf. on Comp. Vis. and Patt. Rec.*, pages 504–509, June 1988.

[8] J. Ens and P. Lawrence. A matrix based method for determining depth from focus. *Proc. of IEEE Conf. on Comp. Vis. and Patt. Rec.*, pages 600–609, June 1991.

[9] B. Girod and S. Scherock. Depth from focus of structured light. *Proc. of SPIE: Optics, Illum., and Image Sng for Mach. Vis. IV*, 1194, Nov. 1989.

[10] M. Gokstorp. Computing depth from out-of-focus blur using a local frequency representation. *Proc. on Intl. Conf. on Patt. Recog.*, October 1994.

[11] P. Grossman. Depth from focus. *Patt. Recog.*, 9(1):63–69, 1987.

[12] A. Gruss, S. Tada, and T. Kanade. A vlsi smart sensor for fast range imaging. *Proc. of ARPA Image Understanding Workshop*, pages 977–986, April 1993.

[13] B.K.P. Horn. *Robot Vision.* MIT Press, 1986.

[14] S. Inokuchi, K. Sato, and F. Matsuda. Range imaging system for 3-d object recognition. *Proc. of 7th Intl. Conf. on Pattern Recognition*, pages 806–808, July 1984.

[15] R. A. Jarvis. A perspective on range finding techniques for computer vision. *IEEE Trans. on Pattern Analysis and Machine Intelligence*, 5(2):122–139, March 1983.

[16] T. Kanade. Development of a Video-Rate stereo machine. *Proc. of ARPA Image Understanding Workshop*, pages 549–557, Nov. 1994.

[17] T. Kanade, A. Gruss, and L. R. Carley. A very fast vlsi rangefinder. *Proc. of Intl. Conf. on Robotics and Automation*, pages 1322–1329, April 1991.

[18] R. Kingslake. *Optical System Design.* Academic Press, 1983.

[19] A. Krishnan and N. Ahuja. Range est. from focus using a non-frontal imaging camera. *Proc. of AAAI Conf.*, pages 830–835, July 1993.

[20] E. Krotkov. Focusing. *Intl. Jrnl. of Comp. Vis.*, 1:223–237, 1987.

[21] H. N. Nair and C. V. Stewart. Robust focus ranging. *Proc. of IEEE Conf. on Comp. Vis. and Patt. Rec.*, pages 309–314, June 1991.

[22] S. K. Nayar and Y. Nakagawa. Shape from focus. *IEEE Trans. on Patt. Anal. and Mach. Intell.*, 16(8):824–831, Aug. 1994.

[23] S. K. Nayar and M. Watanabe. Passive bifocal vision sensor. Technical Report (in preparation), Dept. of Computer Science, Columbia University, New York, NY, USA, Dec 1995.

[24] S. K. Nayar, M. Watanabe, and M. Noguchi. Real-time focus range sensor. Technical Report CUCS-028-94, Dept. of Computer Science, Columbia University, New York, NY, USA, November 1994.

[25] S. K. Nayar, M. Watanabe, and M. Noguchi. Real-time focus range sensor. *Proc. of Intl. Conf. on Computer Vision*, pages 995–1001, June 1995.

[26] M. Noguchi and S. K. Nayar. Microscopic shape from focus using active illumination. *Proc. of Intl. Conf. on Patt. Recog.*, Oct. 1994.

[27] A. Pentland. A new sense for depth of field. *IEEE Trans. on Patt. Anal. and Mach. Intell.*, 9(4):523–531, July 1987.

[28] A. Pentland, S. Scherock, T. Darrell, and B. Girod. Simple range cameras based on focal error. *Jrnl. of Opt. Soc. of Am. A*, 11(11):2925–2935, Nov. 1994.

[29] M. Subbarao. Parallel depth recovery by changing camera parameters. *Proc. of Intl. Conf. on Comp. Vis.*, pages 149–155, Dec. 1988.

[30] M. Watanabe and S. K. Nayar. Rational filters for passive depth from defocus. Technical Report CUCS-035-95, Dept. of Computer Science, Columbia University, New York, NY, USA, September 1995.

[31] M. Watanabe and S. K. Nayar. Telecentric optics for constant magnification imaging. Technical Report CUCS-026-95, Dept. of Computer Science, Columbia University, New York, NY, USA, September 1995.

[32] M. Watanabe, S. K. Nayar, and M. Noguchi. Real-time implementation of depth from defocus. *Proceedings of SPIE Conference*, October 1995.

[33] R. G. Willson and S. A. Shafer. Modeling and calibration of automated zoom lenses. Technical Report CMU-RI-TR-94-03, The Robotics Institute, Carnegie Mellon University, Jan. 1994.

[34] Y. Xiong and S. A. Shafer. Variable window gabor filters and their use in focus and correspondence. *Proc. of IEEE Conf. on Computer Vision and Pattern Recognition*, pages 668–671, June 1994.

Is Structure-From-Motion Worth Pursuing?

Carlo Tomasi and John Zhang
Computer Science Department, Stanford University
Stanford, CA 94305
tomasi@cs.stanford.edu, zhang@sccm.stanford.edu

1 Introduction

Structure-from-motion is the problem of recovering the motion of a camera and the shape of objects in front of it from the images that the camera produces. If vision is to compute a representation of the world from images, structure-from-motion is undeniably a central problem. Yet decades of research have shown that this computation is very sensitive to errors in image measurements, both random and systematic. The time has come to assess this sensitivity quantitatively. Is the problem so bad that only extremely controlled experiments with carefully calibrated, high-quality equipment can yield acceptable reconstructions of the world's geometry? Or can we hope to devise methods that extract useful three-dimensional information from the images produced by an off-the-shelf camera in a wide array of imaging scenarios? It may seem surprising that no clear and complete answer to this question is known yet. After all, if structure-from-motion is a hopeless endeavor, hundred of man-years are being wasted. Conversely, if solutions are practically feasible, only a good understanding of what can and cannot be computed reliably is likely to lead to success.

Presumably, there are two main reasons why a systematic sensitivity analysis of structure-from-motion has not been done yet. One is that the function to be analyzed, which takes image measurements as inputs and produces structure and camera motion as outputs, is not explicitly available. Reconstruction methods exist, but they differ in important ways, so selecting a particular method severely limits the generality of any conclusion that is drawn from its analysis. The other reason is that if sensitivity is expressed as a Jacobian of the ouput versus the input of the reconstruction process, this very large matrix of numbers is by itself uninspiring and needs interpretation, a less than straightforward task.

The main contribution of this paper is a framework for the sensitivity analysis of structure-from-motion that addresses both these difficulties. Although not available directly, the reconstruction function is completely specified by the formulation of structure-from-motion as a least squares problem. Implicit differentiation then leads to the desired Jacobian. The interpretation of the latter is made possible by the Singular-Value Decomposition (SVD) which exposes the structure of the matrix in a way that lends itself directly to geometric interpretation.

Of course, a sensitivity analysis is not the whole answer. Algorithms can fail because of outliers in the data or because of convergence to local minima, whether inherent in the function to be minimized or produced by a particular reconstruction method. Consequently, only an actual working reconstruction system will eventually prove feasibility. While sensitivity analysis gives, in a sense, an expected lower bound on the structure and motion errors that one can hope to attain for a given image quality, actual experiments with specific algorithms provide upper bounds. However, experiments with image sequences can be expensive in terms of setup, computation, and analysis. In fact, processing a whole sequence of images provides but one data point in the large space of experimental parameters. To address this difficulty, we have developed a method that reconstructs structure and motion in Flatland, that is, in a two-dimensional slice of the world, using single image scanlines as its input. The input is small enough that the method works in real time on a workstation. Yet the sensitivity issues are not trivialized relative to the three-dimensional case. In fact, a simple equation counting argument shows that the data constrain the solution better in three than in two dimensions. If we understand the sensitivity issues in Flatland, we

[0]This research was supported under NSF Grants IRI-9201751 and IRI-9496205.

have a good handle on their three-dimensional counterpart as well.

The remainder of this paper describes this two-prong approach to understanding the sensitivity of structure-from-motion. In section 2, we describe our real-time system and what we have learned from our experiments. Then, in section 3, we show how structure-from-motion can be analyzed without explicit reference to a specific reconstruction method. Section 4 discusses implications and future work.

2 A Real-Time System for Experimentation

The theory and full implementation details of our real-time strcture-from-motion system in Flatland are explained in [9]. Here we first sketch the main steps of the computation. We then show how we have been using our system, and we summarize some of the lessons we have learned from our experiments in terms of sensitivity.

2.1 Image Measurements

Suppose that the camera and the world points all live in a two-dimensional world. Point 0 serves as the origin of the global reference system. For every frame $f = 1, \ldots, F$ the camera records the tangents t_{fp} of the angles formed by the projection ray of the feature points $1, \ldots, P$ with that of feature point 0, so with $P+1$ feature points there are P tangents per frame. The tangent t_{fp} can be found by simple geometry to be (see also [7])

$$t_{fp} = \frac{u_f z_p - w_f x_p}{1 - u_f x_p - w_f z_p} \qquad (1)$$

where $\mathbf{s}_p = (x_p, z_p)^T$ is the position of feature number p in the world and

$$\mathbf{k}_f = \begin{bmatrix} u_f \\ w_f \end{bmatrix} = \mathbf{m}_f / |\mathbf{m}_f|^2 \qquad (2)$$

is the vector obtained by reflecting the camera coordinates $\mathbf{m}_f$ across the unit circle. This reflection is introduced to make equation (1) bilinear in motion and shape.

The FP measurements $t_{11}, \ldots, t_{FP}$ can be collected into an $F \times P$ matrix T. Each row represents one snapshot, and each column is the evolution of one tangent over time. If the reflected camera coordinates u_f, w_f and the shape coordinates x_p, z_p are collected in a $F \times 2$ reflected-motion matrix and a $2 \times P$ shape matrix, equation (1) can be rewritten in matrix form for the entire sequence as follows:

$$T = \pi(K, S) \qquad (3)$$

where the projection function π operates on the f-th row of K and on the p-th column of S to produce entry t_{fp} of T according to equation (1). The reconstruction method described in [9] solves the matrix equation (3) for shape S and reflected motion K in a series of steps, each of which either solves a linear system or takes a ratio of scalars. Initial estimates of shape are refined and the new camera coordinates are computed every time a new image becomes available. Here are, in summary the steps of the computation:

1. Find shape $\hat{S}$ up to an affine transformation for each quadruple of points with subscripts $(0, 1, 2, p)$ where p ranges from 3 to P. Points $0, 1, 2$ establish a common affine reference system for all the quadruples.

2. Compute Euclidean shape S by determining a 2×2 matrix A such that
$$S = A\hat{S} \ . \qquad (4)$$

3. Compute the matrix K of reflected camera positions from equation (3).

4. Determine the matrix M of camera positions by reflecting the rows of K back across the unit circle through the inverse of transformation (2),
$$\mathbf{m}_f = \mathbf{k}_f / |\mathbf{k}_f|^2 \ . \qquad (5)$$

It turns out that the critical step for the understanding of the sensitivity of the computation is the first one. In this step, the scalar projection equation (1) is repeated three times for points $1, 2, p$, so the reflected camera coordinates u_f, w_f can be eliminated to yield the following homogeneous linear equation (see [7] for details)

$$\begin{aligned} & a_1^{(p)} t_{f1}(t_{f2} - t_{fp}) + a_2^{(p)} t_{fp}(t_{f2} - t_{f1}) \\ + \; & a_3^{(p)} t_{f1}(1 + t_{f2} t_{fp}) + a_4^{(p)} t_{f2}(1 + t_{f1} t_{fp}) \\ + \; & a_5^{(p)} t_{fp}(1 + t_{f1} t_{f2}) = 0 \ . \end{aligned}$$

where

$$\begin{aligned} a_1^{(p)} &= -x_p(x_1 - x_2) - z_p(z_1 - z_2) \\ a_2^{(p)} &= -x_1(x_2 - x_p) - z_1(z_2 - z_p) \\ a_3^{(p)} &= -x_2 z_p + z_2 x_p \\ a_4^{(p)} &= x_1 z_p - z_1 x_p \\ a_5^{(p)} &= -x_1 z_2 + z_1 x_2 \ . \end{aligned} \qquad (6)$$

Writing this equation once for every frame $f = 1, \ldots, F$ yields an $F \times 5$ homogeneous linear system in $a_1^{(p)}, \ldots, a_5^{(p)}$:

$$H\mathbf{a}^{(p)} = 0 \; . \tag{7}$$

An $F \times 5$ homogeneous system of the form (7) is solved for every point $p = 3, \ldots, P$. In section 2.3 below we show that matrix H is the key to understanding the sensitivity of this particular reconstruction method.

2.2 Experimental Setup

In our experiments, we use a Pulnix CCD camera with a 6.6×8.8 mm sensor and a high quality Schneider Cinegon f1.8/4.8mm lens. Because of the wide field of view (105 degrees along the diagonal), distortion is unavoidable. We calibrate it away by the procedure described in [1]. Frames are acquired by a Digital J300 frame grabber that interfaces directly with the TurboChannel bus of a Digital Alpha 600 workstation. Features are tracked by a one-dimensional version of the system described in [5] at a rate of about one feature per frame per millisecond. Feature selection at this point requires user interaction and the camera is required to remain still until the selection is completed. In our experiments, either the camera or the object is moved by sliding it on a table.

Although the tracker updates image feature coordinates at every frame, shape computation waits until the changes in these coordinates are large enough to warrant incorporating a new set of input data. To check for this event, we first monitor the RMS displacement of all the features in the scanline. Once this measure has exceeded one pixel, a new row of the matrix T of angle tangents (equation (3)) is computed, and its RMS variation with respect to the previous row used for reconstruction is checked against another threshold (0.005 radians in our implementation). Only when this threshold is exceeded is the most recent frame passed to the reconstruction algorithm. Consequently, the more expensive part of the computation is performed only rather occasionally for a slowly moving camera. In summary, all the frames produced by the camera are tracked, but only a few of them, called *significant frames*, are used for reconstruction. These significant frames are displayed on the computer screen as the object or the camera are moved. The shape and motion results are also redisplayed whenever they are updated.

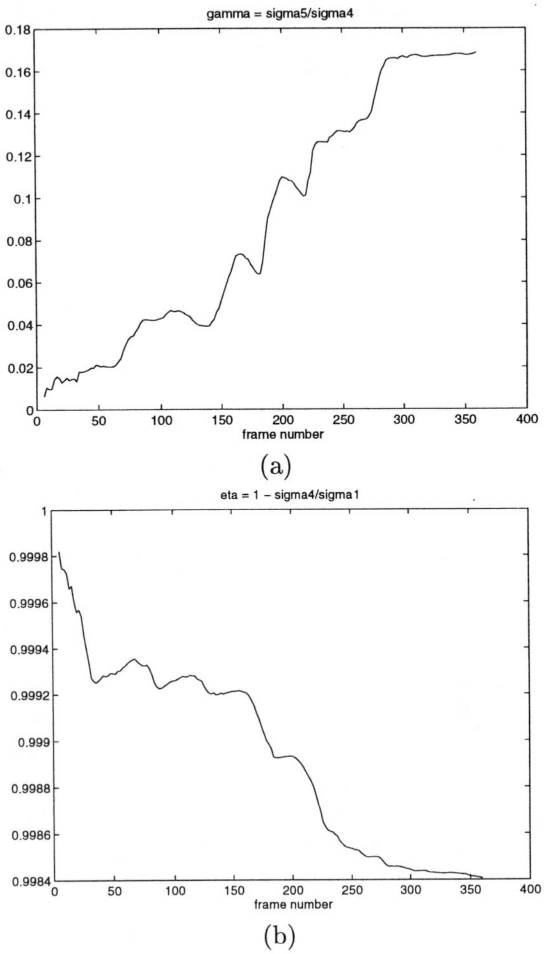

Figure 1: (a) Noise factor γ and (b) sensitivity η as a function of "significant frame" number for a forward moving camera.

2.3 Sensitivity Considerations

Sensitivity to noise is the dominant issue in the reconstruction problem. For meaningful results, the field of view of the camera must be wide enough [4], [3], motion must be sufficiently extended around the object [8], and image measurements must be both sufficiently accurate (low bias) and precise (low standard deviation).

From a numerical standpoint, since the homogeneous $F \times 5$ system (7) is expected to have exactly one affine shape solution with perfect data and nontrivial motion, its matrix H should be of rank 4. Of its singular values $\sigma_1 \geq \ldots \geq \sigma_5$, σ_4 should be nonzero and σ_5 should be zero. In reality, noise increases σ_5, and closeness to degenerate shape or motion decreases σ_4. For instance, if the camera does not move, all the rows of H are ideally equal, and H is rank 1: $\sigma_2 = \ldots = \sigma_5 = 0$.

Also, when the camera's field of view approaches zero width (telephoto lens) the rank of H tends to 3: $\sigma_4 = \sigma_5 = 0$ [7]. Yet reconstruction under perspective requires a stable, substantial gap between the last two singular values, leading to a low *noise factor*

$$\gamma = \frac{\sigma_5}{\sigma_4} \ll 1$$

and a good conditioning of the rank-4 part of the system, leading to a low *sensitivity factor*

$$\eta = 1 - \frac{\sigma_4}{\sigma_1} \ll 1$$

(the condition number is usually defined as σ_1/σ_4; we prefer η because it remains between 0 and 1; $\eta = 0$ is ideal). These two conditions can be optimized by proper setup (η) and good image measurements (γ). Figures 1 (a) and (b) show the two parameters γ and η for a typical point quadruple seen by a forward-moving camera. The object is initially 25 cm away, and the four points span about 15 degrees of the field of view. These plots show the difficulty of the problem. While noise is relatively under control ($\gamma < 0.2$), the sensitivity factor η is dangerously close to 1 throughout. Furthermore, the sensitivity η declines very slowly when more frames are added (the camera moved forward by about 8 cm during the 360 frames), and the noise factor γ increases. The increase of γ may be counterintuitive at first, but is due to the fact that new frames often add more noise than new shape information.

In summary, our system for reconstructing shape and motion from image sequences in real time makes it possible for us to run many experiments with little effort, and at the same time forces us to consider the real difficulties of the problem. Our experiments suggest that reconstruction is indeed possible with sufficient accuracy at least for navigation and as a guidance to manipulation. Sensitivity to image noise is by far the dominant problem in reconstruction, and can be understood by looking at the singular values of the matrix H that appears in the system that solves for affine shape. For good results, the field of view must be wide enough and the camera must move by a sufficiently large amount; image measurements must be accurate (good calibration) and precise (low noise); the formulation of the problem must be in terms of a minimal number of parameters (camera and feature positions, but no camera rotation); and the algorithms must be numerically sound.

The extension of reconstruction to three dimensions is mathematically far from straightforward, and the computation requires more time or resources than in two dimensions. However, the sensitivity of the problem should, if anything, improve, because the ratio of unknowns to measurements is reduced by a factor of 3/4 from approximately 2(P+F)/PF to 3(P+F)/2PF, where P is the number of points and F is the number of frames.

3 Sensitivity Analysis of Structure-From-Motion

The successful experiments reported in the previous section and in [9] imply that structure-from-motion is not out of reach. However, the quality of the results is still far from optimal. When objects are moved away from the camera, performance degrades quickly because image parallax is reduced and becomes comparable to image noise and the residual lens distortion. When they are moved too close to the camera, performance degrades because of unmodeled near-field lens distortion effects. In brief, the system is still rather sensitive to errors in the image measurements. In this section, we show how this sensitivity can be understood without making reference to a particular reconstruction system.

The most important point is that *sensitivity is a property of the problem.* Sensitivity does not depend on the solution algorithm, although a poor algorithm may add its own problems, and does not depend on noise, although sensitivity *to noise* is what is being measured. In fact, sensitivity expresses a *ratio* of output noise per amount of input noise, assuming that input noise is small. Consequently, sensitivity analysis techniques should exist that depend neither on a particular algorithm nor on a noise generator. Doing away with these two factors has important advantages. First, sensitivity is a local property of a problem, but an actual algorithm operates in the global space of all solutions. If an actual algorithm is used, the suspicion lingers that some of the poorer results may depend on a failure of that algorithm to converge to the correct minimum. Second, using a noise generator makes it necessary to compute averages of performance, leading to potentially long running times, and to the question of how small noise really should be in the simulations, given that sensitivity is a local property. Furthermore, since in these noise-based studies sensitivity is measured by collecting scatter statistics, the results do depend on the statistics of noise which, in principle, they should not do.

In this section, the main issues involved in sensitivity analysis are identified, and a general methodology is developed. In structure-from-motion, the projection relation from unknowns to measurements is simple, but the reverse reconstruction relation from measurements to unknowns is hard to determine and sensitive to noise. Furthermore, reconstruction problems share with many other problems in vision and physics the property that they depend on many parameters, thereby bringing forth important issues of data visualization and summarization. In fact, the latter are perhaps the most important challenges to address if sensitivity analysis is to be of any use.

In the following, the basic structure of a reconstruction problem is identified. Then, function minimization is introduced as the prototypical method for inverting the projection equations. Minimization is, in a sense, the conceptual "solution method" for any reconstruction problem. After that, sensitivity is defined and related to the parameters of the minimization method, and a general sensitivity analysis method is outlined.

3.1 Reconstruction

In all reconstruction problems, the set of image measurements is written as a function of motion and structure parameters. Image measurements can be point feature coordinates, their velocities, a vector field, sometimes raw image intensities as in [3], and can come from one, two, or more images. Motion, on the other hand, can be represented by translation and rotation, or by the corresponding velocities. Structure can be represented as depth or shape, in viewer-centered or object-centered coordinates. Whatever the particular case, however, the structure of the problem is the same, and can be represented by the following generic notation:

$$u = p(m, s)$$

where u represents image measurements, m and s represent motion and shape, respectively, and p is the projection function. In discrete problems, u, m, s are vectors of real numbers. In continuous problems, u and s are functions of two variables and m is a function of one variable (essentially time).

Because reconstruction is sensitive to image measurement errors, it should be posed as a minimization problem. To this end, the residue

$$r(u, m, s) = u - p(m, s) \tag{8}$$

is defined and either the sum or the integral of $\|r\|^2$ is minimized over all possible choices of m and s, depending on whether a discrete or a continuous formulation of the problem is being used. Here, we restrict our approach to a discrete, least squares formulation, so the double vertical bars represent the standard Euclidean norm. The minimization may be constrained by some requirement on the solution. Because absolute size cannot be inferred from images alone, camera translation is often normalized in some way.

3.2 Sensitivity

For the time being, it is not necessary to draw a distinction between motion and shape variables. We will therefore introduce the *world* vector

$$\mathbf{w} \triangleq [\mathbf{m}, \mathbf{s}] \tag{9}$$

of motion and shape combined. Let q be the number of entries of $\mathbf{w}$. The projection equations are now represented by

$$\mathbf{u} = \mathbf{p}(\mathbf{w}) \tag{10}$$

and we look for the[1]

$$\begin{aligned} \min_{\mathbf{w}} e(\mathbf{u}, \mathbf{w}) &= \frac{1}{n} \sum_{i=1}^{n} \|\mathbf{r}_i(\mathbf{u}, \mathbf{w})\|^2 \\ &= \frac{1}{n} \mathbf{r}^T(\mathbf{u}, \mathbf{w}) \mathbf{r}(\mathbf{u}, \mathbf{w}) . \end{aligned}$$

If there are h measurements per image point (typically $h = 1$ or 2), the residue vector $\mathbf{r}_i$ has h components, which are stacked into the hn-component vector $\mathbf{r}$.

A necessary condition for $\mathbf{w}$ to be a minimum is that the partial derivatives of e with respect to $\mathbf{w}$ vanish:

$$\mathbf{g}(\mathbf{u}, \mathbf{w}) \triangleq \frac{\partial e}{\partial \mathbf{w}} = 0 . \tag{11}$$

For sensitivity analysis, the distinction between a generic stationary point and the global minimum is irrelevant: since we know the solution, we need not search for it. The system of equations $\mathbf{g}(\mathbf{u}, \mathbf{w}) = 0$ is called the system of *normal equations* for the reconstruction problem.

[1] Parentheses denote functional dependency.

Sensitivity analysis addresses the following question: How does the solution $\mathbf{w}$ change as a result of small variations in the image measurements $\mathbf{u}$? In other words, we consider the normal equations as implicitly defining a transformation

$$\mathbf{w} = \mathbf{f}(\mathbf{u}) \tag{12}$$

and seek to characterize the derivatives of $\mathbf{w}$ with respect to $\mathbf{u}$, that is, we look for the Jacobian of $\mathbf{f}$. Notice that equation (12) is, conceptually, the inverse of the projection equation (10). For this reason, we call equation (12) the *reconstruction* equation. The Jacobian of the reconstruction function $\mathbf{f}$ can be found from the partial derivatives of $\mathbf{g}$ through the implicit function theorem [6]. The derivatives of $\mathbf{g}$ can then be written in terms of those of the projection function $\mathbf{p}$ thanks to equation (11). These two steps are carried out in the Appendix. The final result is that the Jacobian $J_{\mathbf{f}} = \frac{\partial \mathbf{f}}{\partial \mathbf{u}^T}$ of $\mathbf{f}$ is the solution of the linear system

$$\left(J_{\mathbf{p}}^T J_{\mathbf{p}}\right) J_{\mathbf{f}} = J_{\mathbf{p}}^T \tag{13}$$

where $J_{\mathbf{p}} = \frac{\partial \mathbf{p}}{\partial \mathbf{w}^T}$ is the Jacobian of the projection function.

The term in parentheses turns out to be the Hessian matrix of the total error function $e(\mathbf{u}, \mathbf{w})$ with respect to $\mathbf{w}$ at the solution point of the reconstruction problem. If this Hessian happens to be invertible, that is, if the Jacobian $J_{\mathbf{p}}$ of the projection function $\mathbf{p}$ has full rank, then equation (13) can be multiplied by $\left(J_{\mathbf{p}}^T J_{\mathbf{p}}\right)^{-1}$ from the left and by $J_{\mathbf{p}}$ from the right to obtain the new equation

$$J_{\mathbf{f}} J_{\mathbf{p}} = I_q \tag{14}$$

where I_q is the $q \times q$ identity matrix. Therefore, if the Hessian is invertible, the Jacobian of the reconstruction function $\mathbf{f}$ is the left pseudoinverse of the Jacobian of the projection function $\mathbf{p}$.

3.3 Interpretation of the Jacobian

In this section, we show an example of the sensitivity analysis outlined above. The main point here is to show how the Jacobian can be interpreted. Interpretation is not a trivial matter, as there are important issues related to choosing a reference system in which comparing different variables becomes meaningful. This problem also includes choosing an appropriate scaling of all the variables.

The most obvious difficulty stems from the fact that some variables are incommensurable. Most notably, rotations on one hand and translations and point coordinates on the other cannot be compared directly. From an abstract point of view, two approaches to this issue appear possible. We either find a way to scale variables so they become somehow comparable, or we avoid comparisons altogether.

The first option, scaling variables into a common frame, could be realized by considering relative sensitivities instead of absolute ones. In this approach, both image and world variables are divided by their typical values, represented by suitable individual or cumulative averages. Comparison are now possible, because all quantities are dimensionless. However, relative sensitivities are not too useful. Often, a rotation error of, say, one degree is equally significant whether it affects a zero-degree rotation or a sixty-degree rotation. Similarly, an uncertainty of one pixel on an image coordinate can be achieved with similar effort (after lens calibration) at the center and at the periphery of the field of view. In brief, absolute sensitivities are really what we are after.

Another method for scaling variables is to multiply them by coefficients that make the columns of the projection Jacobian equal in norm (this is called column scaling, see for instance [2]). This, however, merely makes all absolute sensitivities similar to each other, without altering the nature of the problem. One would still have to determine whether a sensitivity of, say, one rotation unit is more or less than a sensitivity of one translation unit: quantities are still incommensurable.

A third scaling scheme is to compare camera rotations and translations by their effects on the image. A given camera rotation moves image points of a given scene by a certain amount, measured, say, by the average motion of all the image points. The average motion of image points caused by a camera translation can now be compared with that produced by the rotation in homogeneous units. In other words, the existence of a viewer, and the fact that we are interested in the relation between images and the world, induces an inherent metric for the comparison of rotations and translations: rotations and translations are compared in terms of their perceptual importance.

Whether this comparison is useful, however, depends on the application. For a general analysis, the less committal approach of avoiding comparisons when they are not possible is both conceptually simpler and more general. We propose

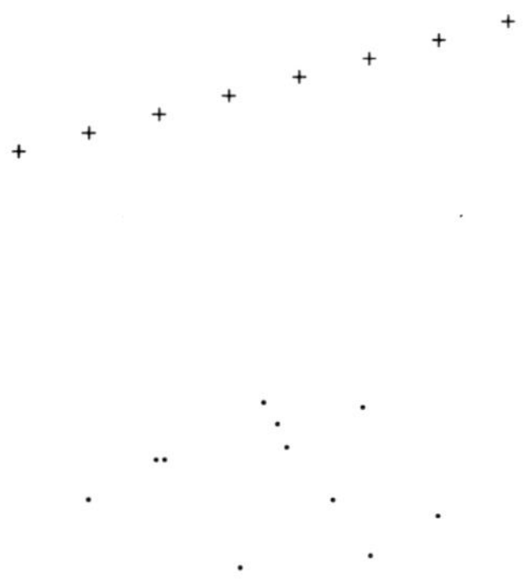

Figure 2: Eight camera positions (crosses) and eleven points in the world (dots). As the camera moves laterally, it fixates one of the points.

to analyze each of shape, translation, and rotation independently, and in terms of their absolute sensitivities. Comparisons between different types of quantities can be made *a posteriori*, once the accuracy requirements of a given application are known. To illustrate, consider a camera moving laterally while observing a cloud of points. Figure 2 shows this scenario. There are eight camera positions and eleven feature points.

The z axis connects the centroid of the world feature points with the centroid of the camera positions. This choice makes z essentially equal to the camera-to-scene distance. The x axis is along the camera motion, and x, y, z are an orthonormal reference system. Projection equations are written for this scenario, and the derivative of each image coordinate is computed analytically with respect to each world variable. Eight images of eleven points yield $11 \times 8 \times 2 = 176$ image coordinates. Of the world parameters, $11 \times 3 = 33$ coordinates specify the positions of the points in space, while $8 \times 6 = 48$ rotation and translation parameters specify the camera positions, for a total of $33 + 48 = 81$ world parameters. This leads to a 176×81 projection Jacobian $J_{\mathbf{p}}$. This Jacobian, however, is singular because the image measurements determine the world parameters only up to an overall scale factor, a rotation, and a translation. To remove this degeneracy, one of the eleven points in space is made to coincide with the origin, and one with point $[1, 0, 0]^T$. A third point is forced to have a y coordinate of zero. The columns of $J_{\mathbf{p}}$ corresponding to these seven fixed components are then removed, leaving a full-rank 176×74 Jacobian. Image coordinates are listed in the 176-dimensional vector

$$\begin{aligned} \mathbf{u} &= [u_{11}, v_{11} \ldots u_{1,11}, v_{1,11} \ldots \\ &\quad \ldots u_{81}, v_{81} \ldots u_{8,11}, v_{8,11}]^T \end{aligned}$$

where the first subscript denotes the frame number and the second the feature number. The world parameters are in the 74-dimensional vector

$$\mathbf{w} = [\mathbf{s}, \mathbf{t}, \mathbf{r}]^T$$

where shape is represented by the $33 - 7 = 26$ components of

$$\mathbf{s} = [x_3, z_3, x_4, y_4, z_4 \ldots x_{11}, y_{11}, z_{11}]^T ,$$

translation is the 24-dimensional vector

$$\mathbf{t} = [t_{x1}, t_{y1}, t_{z1} \ldots t_{x8}, t_{y8}, t_{z8}] ,$$

and rotation is represented by the 24 entries of

$$\mathbf{r} = [r_{\alpha 1}, r_{\beta 1}, r_{\gamma 1} \ldots r_{\alpha 8}, r_{\beta 8}, r_{\gamma 8}] ,$$

which represent the eight camera rotations with quantities similar to Euler angles. Correspondingly, the Jacobian $J_{\mathbf{p}}$ of the projection function can be partitioned into three Jacobians

$$J_{\mathbf{p}} = [\ J_s \ \mid \ J_t \ \mid \ J_r \ ] .$$

Inverting these Jacobians would not lead to reliable results. In fact, these Jacobians are very poorly conditioned, a direct consequence of the high sensitivity of the reconstruction problem to image perturbations. The part of each Jacobian that can be inverted reliably is revealed by taking its Singular Value Decomposition (SVD). Figure 3 shows the singular values of J_s, J_t, and J_r.

In each plot, there are several large singular values, a clear gap, and several singular values that are much smaller. The magnitudes of the entries of the corresponding right singular vector matrices, shown in figure 4, show which components correspond to the large and to the small singular values, respectively. Each column of a right singular vector matrix corresponds to a singular value, and each row corresponds to an entry of $\mathbf{s}$, $\mathbf{t}$, or $\mathbf{r}$. For instance, figure 4 (a) shows that the 18 larger singular values correspond to the x and y components of shape, while the last eight columns of the right singular vector matrix have nonzero entries essentially in correspondence with the z components. Since in the inverse Jacobians the sizes of the singular values are inverted, this means that the z component of shape (approximately along

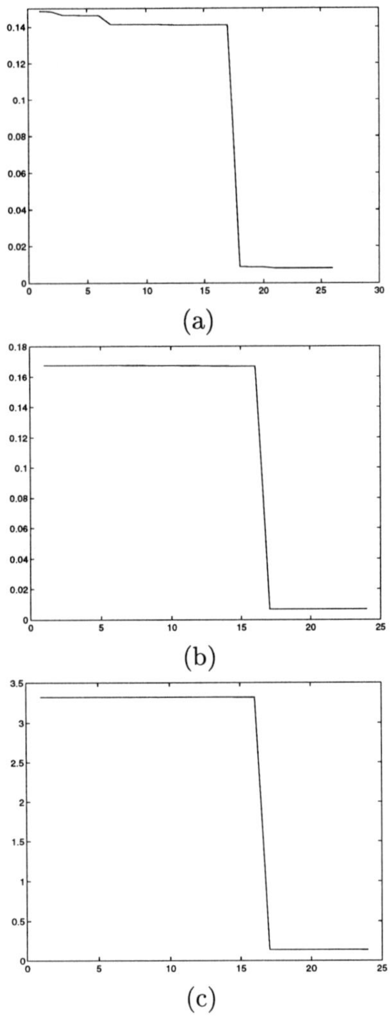

Figure 3: The singular values of (a) the shape Jacobian J_s, (b) the translation Jacobian J_t, and (c) the rotation Jacobian J_r of the projection function.

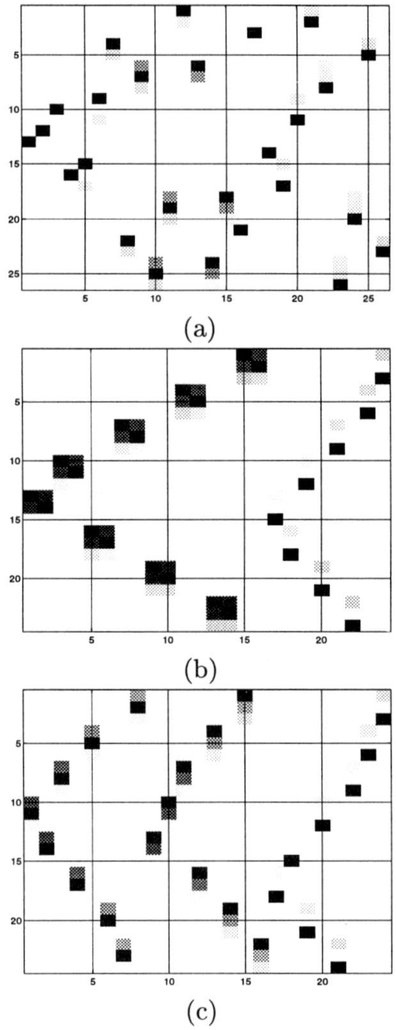

Figure 4: The right singular vectors of (a) the shape Jacobian J_s, (b) the translation Jacobian J_t, and (c) the rotation Jacobian J_r of the projection function. Singular vectors are columns, and large entries are dark.

the optical axes) can be reconstructed with an accuracy that is about 17 times worse than for the x and y components. Similar conclusions can be drawn for figures 4 (b) and 4 (c).

These results stand to reason. In fact, they show that in this scenario depth and translation along the optical axis can be recovered only with much less accuracy than the other parameters, because of the relatively large distance between scene and cameras. Also, and for the same reason, a rotation around the optical axis can be determined with accuracy about 24 times worse than a rotation around axes parallel to the image plane.

4 Conclusion

Successful experiments are always the definitive proof that structure-from-motion is feasible. Our preliminary experiments in Flatland show that reconstruction is indeed possible. In addition, they show how explicit attention to sensitivity from the very beginning of the design pays off in the end. The matrix formulation of the input to our reconstruction method provides an interesting handle for understanding sensitivity issues, as we have demonstrated by our arguments regarding singular values of the matrix H in equation (7). However, the performance of our system depends critically on geometric factors like the shape to be recovered and its distance from the camera; on camera calibration; and on the quality of image measurements. Consequently, we do not propose this system as a tool to use directly in the applications, but as a vehicle for investigation of a particular approach to structure-from-motion.

In order to solve structure-from-motion well we must first understand the sensitivity of the problem itself. In this paper, we have presented a framework for this task which we claim is superior to the traditional approach of simulation of noisy images. In fact, our approch is not tied to any particular solution algorithm nor to a particular noise model. Computing derivatives of the reconstruction function, which is only defined implicitly by the optimization criterion, was possible thanks to the implicit function theorem, a mathematical result of deep importance in the analysis of the sensitivity of inverse problems. Again, a matrix proved to be the central element of our analysis. Also, the SVD turned out once more to be the ideal mathematical tool for understanding the structure of the matrix. Our framework needs to be fleshed out, and scaling issues must be addressed in more detail. Most importantly, the theoretical results from our analysis must be compared with results from experiments. All the tools, however, are now in place. We hope that our investigation will soon produce a satisfactory and useful analysis of the sensitivity of structure from motion, and that this analysis will in turn generate useful and reliable algorithms.

A The Jacobian of the Reconstruction Function

In this Appendix, we derive the fundamental equation (13) for the Jacobian of the reconstruction function f. To write the computations compactly, we introduce the following matrix notation for derivatives. Only the following types of derivatives are allowed: derivatives of a scalar with respect to a vector or a matrix, or vice versa; and derivatives of a row vector with respect to a column vector, or vice versa. The meaning of these derivatives is that of either a vector or a matrix, with the following rules. Let $A(r,s)$ be an $r \times s$ matrix (possibly with $r=1$ or $s=1$, or both). Then,

$$\frac{\partial A(r,s)}{\partial A(1,1)} = \left[\ \frac{\partial A_{ij}}{\partial A_{11}}\ \right]$$

$$\frac{\partial A(1,1)}{\partial A(r,s)} = \left[\ \frac{\partial A_{11}}{\partial A_{ij}}\ \right]$$

$$\frac{\partial A(1,s)}{\partial B(r,1)} = \left[\ \frac{\partial A_{j}}{\partial B_{i}}\ \right]$$

$$\frac{\partial A(r,1)}{\partial B(1,s)} = \left[\ \frac{\partial A_{i}}{\partial B_{j}}\ \right]$$

where i is the row index and j is the column index.

By replacing the reconstruction function (12) into the normal equations (11), we obtain the new equation

$$\gamma(\mathbf{u}) = \mathbf{g}(\mathbf{u}, \mathbf{f}(\mathbf{u})) = 0 \qquad (15)$$

where only image measurements appear as variables. From the implicit function theorem (see for instance [6]) we obtain

$$\frac{\partial \mathbf{g}}{\partial \mathbf{u}^T} + \frac{\partial \mathbf{g}}{\partial \mathbf{w}^T}\frac{\partial \mathbf{f}}{\partial \mathbf{u}^T} = 0 \, . \qquad (16)$$

Because all quantities involved are vectors, this is a matrix equation. The function $\mathbf{g}$ has q components, because it gathers the partial derivatives of the error e with respect to the q components of the world vector $\mathbf{w}$ (see equation (11)). Since there are hn scalar image measurements in the measurement vector $\mathbf{u}$, the matrix $\frac{\partial \mathbf{g}}{\partial \mathbf{u}^T}$ is $q \times hn$. For similar reasons, $\frac{\partial \mathbf{g}}{\partial \mathbf{w}^T}$ is $q \times q$. Finally, the reconstruction function $\mathbf{f}$ has one component for each of the q entries in the world vector (see equation (12)), so that $\frac{\partial \mathbf{f}}{\partial \mathbf{u}^T}$ is a $q \times hn$ matrix.

Equation (16) expresses the Jacobian $\frac{\partial \mathbf{f}}{\partial \mathbf{u}^T}$ of the implicit function $\mathbf{f}$ as the solution to a linear equation whose coefficients are derivatives of $\mathbf{g}$. We now compute the sensitivity matrix, that is, the Jacobian of the reconstruction function $\mathbf{f}$, by writing the derivatives of the normal function $\mathbf{g}$ in terms of those of the projection function $\mathbf{p}$. This manipulation pays off, since it leads to simplifications. In fact, although the first derivatives of $\mathbf{g}$ are combinations of first and second order derivatives of $\mathbf{p}$, the terms that contain second order derivatives vanish at the solution of the reconstruction problem, leaving only first order derivatives to compute.

To see this, consider the total error

$$e(\mathbf{u}, \mathbf{w}) = \frac{1}{n}\mathbf{r}^T(\mathbf{u}, \mathbf{w})\mathbf{r}(\mathbf{u}, \mathbf{w})$$

where the residue vector $\mathbf{r}$ is (see equation (8))

$$\mathbf{r}(\mathbf{u}, \mathbf{w}) = \mathbf{u} - \mathbf{p}(\mathbf{w}) \, .$$

The derivatives of $e(\mathbf{u}, \mathbf{w})$ with respect to the world vector $\mathbf{w}$ are

$$\frac{\partial e}{\partial \mathbf{w}^T} = -\frac{2}{n}\mathbf{r}^T\frac{\partial \mathbf{p}}{\partial \mathbf{w}^T}$$

and the second derivatives can be collected in the $q \times q$ *Hessian matrix* of the residue e:

$$\frac{\partial^2 e}{\partial \mathbf{w} \partial \mathbf{w}^T} = \frac{2}{n} \left(\frac{\partial \mathbf{p}^T}{\partial \mathbf{w}} \frac{\partial \mathbf{p}}{\partial \mathbf{w}^T} - \mathbf{r}^T \frac{\partial^2 \mathbf{p}}{\partial \mathbf{w} \partial \mathbf{w}^T} \right) . \tag{17}$$

Although the general expression of the Hessian of e is given by equation (17), we are only interested in the Hessian *at the solution points*, that is, where $\mathbf{w} = \mathbf{f}(\mathbf{u})$. At these points, the residue $\mathbf{r}$ vanishes, and

$$\frac{\partial^2 e}{\partial \mathbf{w} \partial \mathbf{w}^T} = \frac{2}{n} \frac{\partial \mathbf{p}^T}{\partial \mathbf{w}} \frac{\partial \mathbf{p}}{\partial \mathbf{w}^T} , \tag{18}$$

which simplifies the computation of derivatives considerably. A similar argument holds for the derivatives $\frac{\partial \mathbf{g}}{\partial \mathbf{u}^T}$ that appear in equation (16), for which we obtain

$$\begin{aligned} \frac{\partial \mathbf{g}}{\partial \mathbf{u}^T} &= \frac{\partial^2 e}{\partial \mathbf{u}^T \partial \mathbf{w}} = \frac{\partial}{\partial \mathbf{u}^T} \left[-\frac{2}{n} \frac{\partial \mathbf{p}^T}{\partial \mathbf{w}} \mathbf{r} \right] \\ &= -\frac{2}{n} \left[\frac{\partial^2 \mathbf{p}^T}{\partial \mathbf{u}^T \partial \mathbf{w}} \mathbf{r} - \frac{\partial \mathbf{p}^T}{\partial \mathbf{w}} \right] \end{aligned}$$

since

$$\frac{\partial \mathbf{r}}{\partial \mathbf{u}^T} = \frac{\partial \mathbf{u}}{\partial \mathbf{u}^T}$$

is the identity matrix. At solution points, this general expression simplifies to the following, since $\mathbf{r}$ vanishes there:

$$\frac{\partial \mathbf{g}}{\partial \mathbf{u}^T} = \frac{\partial^2 e}{\partial \mathbf{u}^T \partial \mathbf{w}} = -\frac{2}{n} \frac{\partial \mathbf{p}^T}{\partial \mathbf{w}} . \tag{19}$$

By replacing the expressions (18) and (19) into equation (16) with $J_{\mathbf{f}} = \frac{\partial \mathbf{f}}{\partial \mathbf{u}^T}$ and $J_{\mathbf{p}} = \frac{\partial \mathbf{p}}{\partial \mathbf{w}^T}$ we obtain the equation (13) for the Jacobian of the reconstruction function $\mathbf{f}$.

References

[1] Fleck M. Shape and the wide-angle image. Technical Report 04, University of Iowa, 1994.

[2] Gill PE, Murray W, Wright MH. *Practical Optimization.* Academic Press, London San Diego New York, 1981.

[3] Horn BKP, Weldon Jr. EJ. Direct methods for recovering motion. *Int'l J Computer Vision*, 1988; 2:51–76.

[4] Koenderink JJ, van Doorn AJ. Facts on optic flow. *Biological Cybernetics*, 1987; 56:247–255.

[5] Shi J, Tomasi C. Good features to track. In: *Proc IEEE Conf Computer Vision and Pattern Recognition*, 1994; 593–600.

[6] Taylor AE, Mann WR. *Advanced Calculus.* John Wiley and Sons, New York, NY, 1983.

[7] Tomasi C. Pictures and trails: a new framework for the computation of shape and motion from perspective image sequences. In: *Proc IEEE Conf Computer Vision and Pattern Recognition*, 1994; 913–918.

[8] Tomasi C, Kanade T. Shape and motion from image streams – a factorization method. In: *Proc Nat'l Acad Sci USA*, 1993; 90(21):9795–9802.

[9] Tomasi C, Zhang J, Redkey D. Experiments with a real-time structure-from-motion system. In: *Proc IV Int'l Symp Experimental Robotics*, 1995; in press.

PART 10

ACTIVE/REAL-TIME VISION

SESSION SUMMARY

Yoshiaki Shirai
Dept. of Computer-Controlled Machinery, Faculty of Engineering, Osaka University
Yamadaoka, Suita, Osaka, 565, Japan

1 Introduction

Recently, active vision and real time vision are popular research area. They are closely related because real time vision is undispensable for an active vision system to work effectively in the real world.

One of the earliest active vision research appeared in 1971, which attempted to extract edges by adjusting the gain of a camera if the contrast of a particular part is not suitable [1]. In those days, however, vision played a passive role: from a given image, obtain useful information as much as possible. It is assumed that the position of a camera is set at the best position and the viewing condition is also adjusted.

The name "active vision" was first used as the title of a paper by Aloimonos in 1987 [2]. The paper showed that a problem of shape recovery can be simplified by moving a camera. Later the concept of active vision has been extended to acquisition of necessary or interesting information actively by moving a camera.

The purpose of active vision illustrated above are adjust the position or parameters of a camera such that the target object may well recognized. This also means a closed loop of sensing and action. If the sensor da is not enough to make a decision, more useful information can be acquired based on the result.

Another purpose is to recognize useful parts of the environment at necessary times. This happens when a robot execute a task which requires a sequence of hand eye coordinations.

An interesting application is a mobile robot with active vision, where the robot has to look at the environment to move and at the same time it has to move to recognize the environment.

Real time vision is indispensable for practical use in industry (e.g. AGV) or in agriculture. We should demonstrate that vision is quick enough to be used in the real world.

2 Active vision for executing a given task

In many industrial applications, a task is farely well defined because the environment of a robot is constrained. In such cases, a plan can be generated not only for a manipulator but also the position of a camera.

At least, the camera should be placed at the position from which the object of interest should be seen. Given the geometric model of the environment, the best viewing position can be obtained by using a geodesic dome [3]. Similarly, the position of the light source can be determined which provide the best illumination condition for feature extraction [4].

The first talk by Laugier deals with more complex environments. The deterministic approach is not easy because the search space is too large. A method of finding the suboptimal solution is proposed.

A question to this presentation is how to extend to multiple cameras. A simple answer is to apply the evaluation method to each camera. However, this can not be applied to the case where the effect of joint observation from multiple cameras should be considered. This problem is to be solved in the future.

Note that one of the related works is camera position control for laparoscopic surgery [5], where the geometric model of the environment can not easily be made in this case.

3 Object Tracking

Object tracking is one of the most popular tasks for real time active vision because only a real time system can track a moving object. Most systems controls the direction of a camera head which consists of a monocular camera or stereo pair of cameras. Many institutions took much effort to make the computer controlled camera head.

Early works assumed that the brightness of an object is different from that of the background in order to release the computation load. Recently an object is assumed to move differently from the background. Initially a camera is stable and tries to detect a moving object, and then starts to track the object.

The motion is detected either by the difference of images taken at consecutive instances, correlation, or optical flows. The first one is advantageous in a computational view point although the movement of the camera head should be precisely known for compensating the background motion. The second is more reliable although it requires much computation. Better VLSI chip may be available for correlation because such chips are also useful for motion image compression. Optical flow approach is in the middle: several DSP (digital signal processor)s can compute the optical flow in the entire image and the flow is not accurate at the boundary of objects.

Object tracking becomes more reliable if more information is available such as depth data, color, or edges. Color and motion information is used for human motion analysis [6]. They used Data Cube for low level image processing and a microprocessor for high level processing which limit the processing cycle of half a second. Several systems integrate motion and depth data in real time. An important problem is how to simplify the computation for easy implementation.

Real time image processing systems use Transputors, Data Cube, DSP, correlation VLSI, FPGA(field programmable gate array), or combination of them. Unfortunately, initial efforts is currently not negligible. We are planning to develop a DSP board with DSP, memories, a controller, and data buses. By employing multiple DSPs a flexible system can be constructed easliy. Many members of a robotics group and computer vision group orderd about hundred boards.

4 Discussions

Current active vision uses only vision sensors. De we need more sensors? In fact, many mobile robots use sonar sensors or inertia sensors. They are useful because information processing is usually much easier than vision. This is one of the reasons why vision people deal with rather difficult problems of integration of various vision data. An interesting research theme is how to integrate different kind of information: some qualitative and others quantitative. Also a problem is how to balance the trade off between reliability of sensing and the cost of using many sensors.

Another issue is the importance of recognition, which has long been a central problem in computer vision. Is recognition essential for actions? Sometimes it is not. An example is tracking an object presented in this session. Another important remark is that recognition becomes easy by action such as a gaze control for capturing a high quality image.

In spite of the importance of the action, recognition is still important depending on applications. Image database handling, for example, requires recognition rather than action. I would also like to point out that current tracking systems locate an object from the difference of motion, depth, or color, not from the recognition result. If reliable recognition is possible in real time, more flexible tracking can be realized.

As pointed out in another session, a cat can recognize the environment better with active motion than passive motion. Although we do not know the similarity of vision between computer and animals, there seem to be many interesting research themes in integration of action and recognition.

References

[1] Pingle, K.K. and Tenenbaum, J.M: An Accommodation Edge Follower, Proc. 2nd IJCAI, pp.1-7 (1971)

[2] Aloimonos, J. et al: Active Vision, Proc. ICCV'87, pp.35-54 (1987)

[3] Sakane, S. et al: Occlusion Avoidance of Visual Sensors Based on a Hand-Eye Action Simulator System, Proc 3rd Int. Conf. on Advanced Robotics, pp163-174 (1987)

[4] Niepold, R. et al: Vision Sensor Set-Up Planning for a Hand Eye System Using Environmental Model, Proc. SICE'87, pp.1037-1040 (1987)

[5] Casals, A. Amat, J. and Laporte, E.: Robotic Aids for Laparoscopic Surgery Problems, in this book

[6] Etoh, M. and Arakawa, H.: An Integration Algorithm for Stereo, Motion and Color in Real-Time Applications, IAPR Workshop on Machine Vision Applications, pp.186-189 (1994)

Visual Guidance of an AGV

F. Li, Michael Brady and H. Hu
Department of Engineering Science
University of Oxford
Oxford OX1 3PJ, U.K.
email:{fuxing,jmb,huosheng}@robots.ox.ac.uk

Abstract

Image processing plays an important role in the visual navigation of a mobile robot. A real-time visual tracking system, using an active stereo vision head mounted on a mobile robot, is described in this paper to demonstrate the real-time implementation of a classic stereo method, featuring algorithm simplification for a real application. Disparity information is used directly to guide visual tracking by integrating tracking tightly to its final purpose, to provide control signals for obstacle avoidance by the mobile robot. Using an approach proposed here, the stereo correspondence algorithm can be applied to compute disparity information in real time from stereo images captured from active stereo system with two cameras verging dynamically.

Key words: Image Processing, Vision, Tracking.

1 Introduction

Robot vehicles have been routinely deployed in manufacturing industry for over twenty years, primarily for material parts transfer. Their advantages over conveyor systems include the size of the loads that can be carried and the distance over which they can be carried. The vast majority of fielded systems are wire-guided, not, primarily, because of the simplicity of the controller, but because of the end-point accuracy that can be achieved. To date, non-wire-guided vehicles have substantially less endpoint accuracy, though laser positioning systems are challenging strongly [1]. This is significant because it has begun to open up application of *free-ranging* vehicles, which, freed from the fixed path of the wire, can plan paths to chosen endpoints, and can automatically detect and avoid obstacles. Significant and real possibilities now exist in: stockyarding; civil construction; in-shore marine; truck loading; and servicing the input bay of a factory/mail office. Characteristics of these environments are: (1) an environmental model may be available, but will inevitably be partial, and subject to change; (2) there are limited classes of objects in the environment; (3) the environment imposes severe constraints on manoeuverability (obstacles in a stockyard, marked locations in civil construction); (4) tight clearances to pass through (typically less than 10cm); (5) predominantly metallic and noisy environments; and (6) outdoor working, subject to the vagaries of weather.

The key problems are sensing and real-time sensor-guided control. We have explored the use of a variety of sensors, including: laser triangulation, (lateral-effect Photodiode) proximity, sonar (both Polaroid and FM), and infrared time-of-flight range. None are wholly satisfactory, all have major weaknesses. We have been forced, based on the limitations of other sensors, our own experience, and the demands of our industrial collaborators, to include (real-time, of course) vision.

The advantages of vision are several: (i) it has high angular resolution; (ii) it is increasingly inexpensive, has high reliability, low weight, and low power sensors; (iii) it is passive, and the environments that arise in the applications listed above have few (if any) specularities; (iv) vision can operate over a substantial range, easily out to tens of metres, and down to a meter (with the same lens); and (v) it has high bandwidth. The disadvantages of vision, relative to other sensors, are:

- it necessitates substantial computing power and memory, though this is already essentially not a problem.

[1] Current systems attain $\pm 2mm$ over $100m$

- visual processing has (relatively) long latency, compromising control stability, though again (as we shall see) this is less and less an issue.

- the vast majority of published algorithms do not work with sufficient reliability to be used for extended periods in unpredictable lighting in environments that are subject to change. This is the *real* issue to be addressed!

In the rest of the paper we discuss some of the issues that arise in developing active vision systems for mobile robot navigation, aimed at: (i) obstacle detection and avoidance; (ii) object tracking; and (iii) traversing narrow gaps. We begin with a discussion of active vision and the way it can address the key unreliability issue. Then we sketch the latiest features of the robot vehicle architecture and stereo 'head' we have developed.

2 Active Vision

Until quite recently the number of images that could be processed to yield information likely to be of use to a robot was quite small. For example, as recently as five years ago, it typically took an hour to process a stereo pair of images to produce a depth map. Now, with the assistance of any of a number of parallel distributed processors, the same computation may be completed 10 times per second. Initially, this vastly increased processing power was used simply to execute previously slow algorithms more quickly. Aside from the enhanced understanding resulting from better visualisation, the technology had little effect *per se* on the capabilities of mobile robots.

But it was soon realised that the increased speed of processing enabled the loop to be closed between sensing and action, so that a robot's behaviour could be modified according to what it had seen. Even though the design of visually-guided control loops is difficult, and still poorly understood, there have already been sigificant gains from active or purposive vision. Simply stated, this means that on the basis of previous visual processing the robot can act not only to achieve a goal but also to elicit additional, useful, visual information. We illustrate this idea with four examples.

An important part of the early processing of an image by the human brain aims to isolate significant intensity changes. Intensity changes arise in several very different ways, including: a reflectance change, a lighting change (say, a shadow boundary), a sharp change in surface orientation, and a depth discontinuity. (see [11] for more details). It has proved remarkably difficult to distinguish these different physical events on the basis of the intensity distribution across an intensity change. Recently, Cipolla and Blake [4] showed that some differentiation is possible, and reliably, if the viewer makes a deliberate motion. More precisely, some intensity changes correspond to a physical contour that is fixed in space (eg a reflectance boundary or a sharp physical edge), while at others the surface normal turns smoothly away from the viewer. Cipolla and Blake showed that if a deliberate motion is made orthogonal to the viewing direction, these two cases can be distinguished reliably and, in the latter case, the surface curvature normal to the bounding contour can be estimated. However prosaic this may seem, it has supported the construction of a path planner by which a robot can navigate amongst curved obstacles [2].

Building on the work referred to in the previous paragraph, Cipolla and Blake have also shown that a sequence of deliberate motions can enable a robot to determine its time-to-contact a surface without explicitly computing the distance to it. More precisely, if a planar contour is tracked, and Cipolla and Blake show that if a deliberate move is made orthogonal to the view direction, then the surface orientation of the patch can be determined. If a subsequent deliberate move is made along the line of sight, the time-to-contact can be computed from the divergence of the boundary of the patch (after correcting for the surface slant). Recently, Merron [12] observed that similar considerations apply to vertical edges, so that the time-to-contact can be determined knowing only one's forward speed, and ignoring lateral and rotational motions. This is ideally suited for enabling a mobile robot to navigate about its environment.

A further implication of such work is to change in some cases the information that it is required to compute. Stereo vision is a case in point. Conventionally, stereo algorithms operate in two stages. The first establishes correspondences between feature points in the two images, by determining that they arise from the same point in space. (A number of algorithms have been proposed for this stage, notably PMF [16][15], which is fast and reliable.) The correspondences established in the first stage each have an associated disparity, a measure of the change in image position from the left to the right view. The second stage converts disparity to depth. Roughly, depth is inversely proportional to disparity: but the exact relationship depends on the careful calibration of the two cameras, a delicate, non linear task. Nevertheless, it has been assumed that the goal of stereo is to produce explicit measures of depth, despite the complexity and error introduced by the second stage. It now seems, as a result of recent work in active vision, that the second stage is unnecessary: the changing disparity field encodes most of what is needed to navigate about the environment. Working in a similar vein, Shapiro, Zisserman and Brady [17] have shown that one can compute the image of the instantaneous axis of rotation of a moving object, under certain quite weak assumptions.

Our final example concerns gaze (or attentional) control: where to look? it is well known that the human eye is an exquisite device that comprises two very different subsystems. A remarkably small fovea is dedicated to detailed shape analysis and colour processing, whereas the majority of the retina, though quite coarse in its spatial sampling, detects temporal changes very quickly. Such changes serve to direct gaze to parts of the scene that may pose a threat. The complex mechanism of attentional control is poorly understood (though see [3]). Recently, robotics researchers have developed steerable camera platforms [5] [13] [14], not only to support experimentation, but to exploit the information that gaze control potentially provides. Threat detection aside, a system that can be made to track a target smoothly can build up a representation of its shape, perhaps even identify it, without being sidetracked by the complex problem of compensating for image motion (optic flow, retinal slip). Similarly, if a (stereo) pair of cameras both track an object, stereo matching can be made much easier because disparities can be kept small.

3 The Experimental System

As noted above, over the past six years, we have developed a variety of sensors, including a vision system, a laser scanner, a sonar array, a lateral effect photodiode (LEP) range sensor and a stereo vision "head". As a result, a distributed control architecture is required to integrate these sensors and provide a unified action in real time. In such an architecture, the information flow from

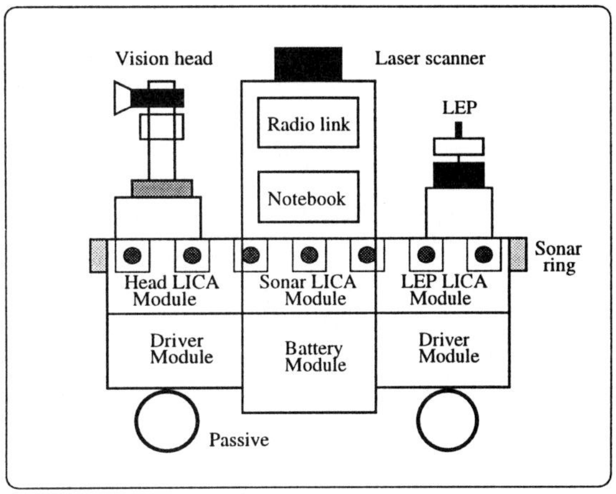

Figure 1: *The Modular structure of our mobile robot.*

sensing to action should be maximised to provide minimal delay in responding to the dynamically changing environment. Our mobile robot is a modular structure, consisting of drive modules, battery modules, different sensor modules and control modules, as shown in Figure 1.

A standard approach to the design and construction of a complex control system is to maximise modularity to support the step-by-step development of individual modules, and a evolutionary improvement of the system as a whole. To this end, we have fixed on a processing module that we call a *Locally Intelligent Control Agent* (LICA). As a concept, the LICA is independent of any particular processor realization. Central to the LICA design is the concept of *Communication Sequential Processes* (CSP) which was developed by Hoare [6]. We have already implemented the LICA concept using transputers which have onchip FPU, 4k RAM, 4 20MBit/s serial links, and a micro-coded scheduler for parallelism.

It is important to realise that the LICA architecture is *not* restricted to T800 technology, which is already no longer state of the art. Rather, the LICA architecture is a *virtual* design which uses the communicating sequential processing primitives for communication; but which uses more sequential processing as compute servers at each node. We are currently experimenting for example with C40s.

Figure 2 shows the current control system based on LICAs. There are 6 main function blocks for distributed sensing, planning, and control tasks. The communication between LICAs

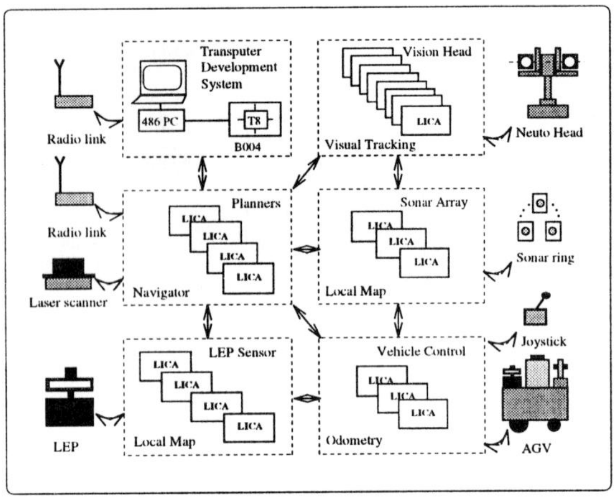

Figure 2: *A LICA-based architecture for our mobile robot.*

uses high speed serial links (20MB/s). The LICA modularity offers flexibility in changing the system configuration between different applications by exploiting the communications topology. It provides a degree of robustness and reliability, as well as a cost-effective solution to problems of designing complex AGV systems. Note that the number of LICAs for each function module is flexible, and depends upon the particular sensor being used and the system time constraints.

Further descriptions of the system architecture can be found in [9] [8].

4 Object Tracking using Disparity Information

To explore the potential of using an active vision system, the LICA-based active vision system has been incorporated onto the vehicle. Using such a robot head, the vision system is able to actively change its geometry to adapt to the task requirements and to plan and execute accurate camera motions. We have demonstrated spot light tracking using this method based on parallel implementation in [7]. But it is extremely difficult to track any other object in real time using that method. In this paper, an approach is adopted which integrates tracking tightly with its final purpose, ie. to provide a control signal for obstacle avoidance, similar to [1]. First of all, the nearest object on the path, ie., the one with largest disparity, is detected. Then a limited tracking is applied to keep in view the object to be tracked until it is passed by the mobile robot on which the head is mounted. Subsequently, a new nearest object on the path will be tracked.

Image processing is the main part of the tracking system. The image processing results are used to guide tracking. Not only do the reliability and accuracy of the image processing affect tracking quality, but also its delays affect tracking speed and quality. A classic stereo method is used to compute the disparity for each object. To realise the tracking using this method in real time, some measures have been taken to reduce the image processing delay. These amount to algorithm simplifications sufficient for the application and intensive parallel implementation in hardware. We next describe the system software and hardware in detail.

4.1 Algorithm Simplification

As shown in Figure 3, the visual tracking system consists of two parts:

1. image processing, including:
   - image grabbing, which yields images at frame rate. The images we use have resolution 240 × 240;
   - edge detection;
   - stereo matching;
   - head motion planning according to the image information;
2. vision head control, including:
   - adaptation which converts a planned head motion into a head control input;
   - vision head controller;

As classical stereo methods try to obtain depth information and positional accuracy, this always entails considerable computational expense and makes it difficult to apply in real-time systems. In this section, we describe the image processing methods used for different stages and the respective simplifications.

As mentioned above, we propose to use disparity information directly for tracking and visual navigation of the mobile robot. Since indoor environments contain many vertical contours, and the obstacles to be avoided by a mobile robot are often standing on the ground, we use vertical edges to represent them. The idea is that after capturing a stereo image pair, we perform vertical edge

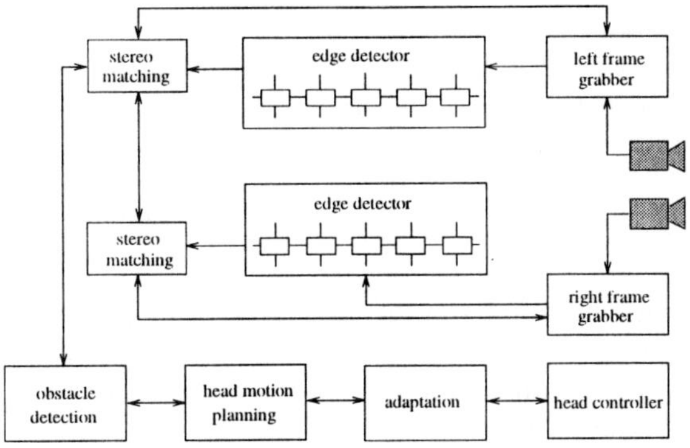

Figure 3: *The control structure of the visual tracking system.*

detection then stereo matching to find the vertical edges in space. We need to distinguish which vertical edges represent an object standing on the ground and which are lying on the ground, so are not an obstacle. This is done by using the ground plane disparity map constructed a priori.

4.1.1 *Edge Detector*

A Sobel vertical edge detector is used for edge detection, which reduces the time that would have been needed for horizontal edge detection. Two thresholds of edge contrast value are adopted. Detected vertical edges whose contrast values exceed a low threshold are connected in different edge strings according to their position and direction. Hysteresis is applied to delete those strings whose lengths are below a length threshold and those strings none of whose contrast values exceed the high contrast threshold. The remaining strings are used for stereo matching. This greatly reduces the computational expense of stereo matching.

4.1.2 *Stereo Matching*

The PMF[16][15] stereo matching algorithm is used because of its reliability and speed, especially for vertical edges. To simplify matching, parallel camera geometry has been adopted so far. Since we use edge strings for matching, only selected seed edge points from strings are used to compute matches and then matching is relaxed to the whole string. This technique greatly reduces the computational expense. Matching is performed from the left image to the right and from the right image to the left in parallel. If two edge points choose each other as their matching point, these two points form a match pair;

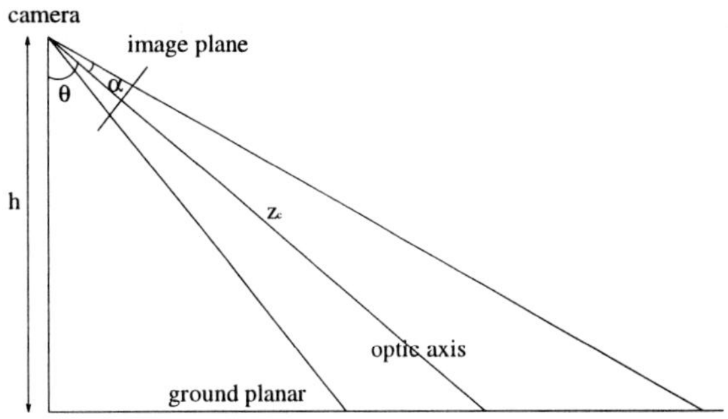

Figure 4: *Analytic computation of ground disparity.*

4.1.3 *Obstacle Detection*

The easiest way to detect an obstacle standing on the ground is to use dense prediction of ground plane disparity to distinguish whether the point is on the ground or over the ground. In this section, we first investigate whether the ground disparity can be represented as a plane in the image coordinates x and y. Then, we use the ground images with special line features on the ground and without obstacles to train the system and get the parameters of the predicted ground disparity plane. This can then be used to detect obstacles standing on the ground.

1. Ground disparity plane

   From Figure 4, when the head elevation angle θ has a certain constant value, the ground disparity plane obeys:

$$\begin{aligned}\delta(y) &= \frac{f*d}{h}*(cos(\theta)-\frac{y}{f}*sin(\theta)) \\ &= \frac{f*d*cos(\theta)}{h}-\frac{y*d}{h}*sin(\theta) \\ &= K_1+K_2*y. \qquad (1)\end{aligned}$$

   Here $\delta(y)$ is the ground disparity as a function of y; y is the image coordinate; f, d, h are vision geometry parameters: d represents the distance between the two camera optic centres, and f is the camera focal length.

2. Parameter Fitting

   The problem is: given a set of observation data x, y (image coordinates) and z (here representing disparity), to fit parameters a, b, c to obey the equation:

$$z = a * x + b * y + c. \qquad (2)$$

This can be considered as a least squares problem, ie., to minimise the function

$$E = \sum(\tilde{a} * x_i + \tilde{b} * y_i + \tilde{d} * z_i + \tilde{c})^2.$$

Straightforward computation yields

$$\tilde{c} = -(\tilde{a} * \overline{x} + \tilde{b} * \overline{y} + \tilde{d} * \overline{z}), \qquad (3)$$

$$\begin{aligned} E &= \sum(\tilde{a} * \Delta x_i + \tilde{b} * \Delta y_i + \tilde{d} * \Delta z_i)^2 \\ &= (\tilde{a}\ \tilde{b}\ \tilde{d})W(\tilde{a}\ \tilde{b}\ \tilde{d})^T, \qquad (4) \end{aligned}$$

where W is the real symmetric matrix

$$W = \begin{pmatrix} \sum \Delta x_i^2 & \sum \Delta x_i * \Delta y_i & \sum \Delta x_i * \Delta z_i \\ \sum \Delta x_i * \Delta y_i & \sum \Delta y_i^2 & \sum \Delta y_i * \Delta z_i \\ \sum \Delta x_i * \Delta z_i & \sum \Delta y_i * \Delta z_i & \sum \Delta z_i^2 \end{pmatrix}$$

, It is known that when $(\tilde{a}\ \tilde{b}\ \tilde{d})$ is chosen to be the smallest eigenvector of the matrix W, E takes its minimum value so by simply doing an eigenvalue decomposition of W, we can compute $\tilde{a}$, $\tilde{b}$, $\tilde{d}$. From equation (3) , we can also estimate $\tilde{c}$.

There may be some observations which are not correct. In order to get the best fit of the predicted ground disparity map, we discard incorrect observations according to its covariance as follows:

After obtaining $\tilde{a}$, $\tilde{b}$, $\tilde{c}$, $\tilde{d}$ through eigenvalue decomposition, we compute the covariance for each observation

$$EZ_i = (\tilde{a} * x_i + \tilde{b} * y_i + \tilde{d} * z_i + \tilde{c})^2. \qquad (5)$$

If $EZ_i > T$, we discard this observation, here T is the threshold.

A new set of $\tilde{a}$, $\tilde{b}$, $\tilde{c}$, and $\tilde{d}$ is computed using the remaining observations and this process is repeated until no observation can be discarded.

Then a, b, c can be computed as:

$$a = -\frac{\tilde{a}}{\tilde{d}}, \qquad (6)$$

$$b = -\frac{\tilde{b}}{\tilde{d}}, \qquad (7)$$

$$c = -\frac{\tilde{c}}{\tilde{d}}. \qquad (8)$$

Alternatively, we could use least median squares, random sampling, or M-estimation. In pratice, there are few outliers, so least squares suffices.

3. Ground disparity computation when the head elevates

   From the analysis of the ground disparity plane, the coefficients of the plane equation form sine and cosine relations with the head elevation angle. In order to check the feasibility of using this kind of plane, we captured a set of image pairs with different head elevation angles and fit plane parameters according to their matches. The results shown in Figure 5 verify this.

   A look-up table is estabilished for three parameters of the ground disparity plane as functions of the head elevation angle according to these experimental results using spline interpolation.

After obtaining stereo matches, we can compute the average x-coordinate in the cyclopean image and the average disparity for each vertical edge string. From the head state we know the elevation angle, and so the ground disparity plane parameters can be found from the look-up table. Comparing the disparity of the vertical edge and the ground disparity at that position, if the difference is larger than a threshold, then the vertical edge is from an object standing on the ground: it is an obstacle.

4.1.4 *Object Tracking*

Since we use parallel camera geometry, we can control head pan and elevation angle to fixate the tracked vertical edge in the middle of the cyclopean image. After we have computed the correspondence between the left and right images, we can compute the disparity and its coordinate in the cyclopean image. Then, from projective geometry, the difference between this coordinate and the centre of the image can be taken as the input signal of the head motion controller. A simple proportional controller has been used. From Figure 6, the pan angle that the head needs to move through according to the image position of the tracked vertical edge is:

$$\delta\theta = -tan^{-1}(\frac{x}{f}), \qquad (9)$$

$$\theta_d(k) = \theta(k-1) + \delta\theta, \qquad (10)$$

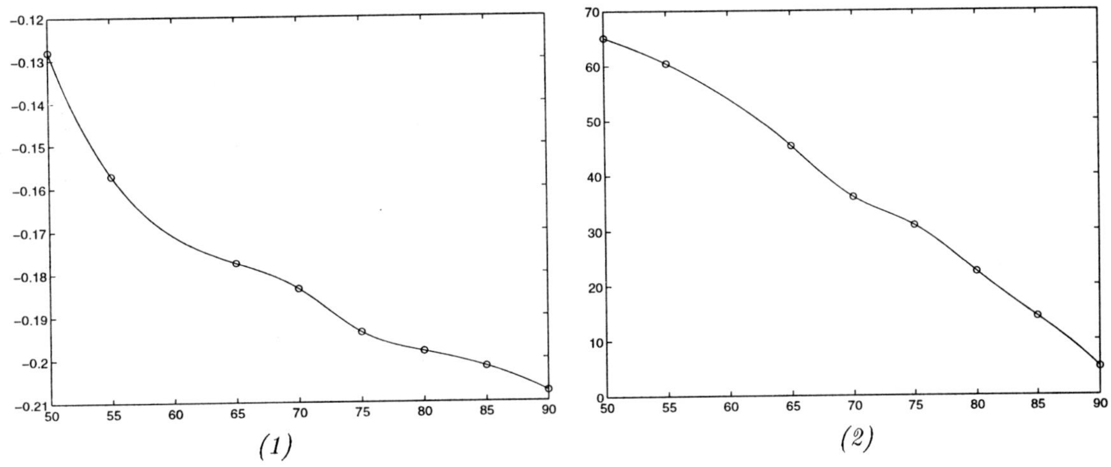

Figure 5: *Parameters of ground disparity plane z=ax+by+c; (1) parameter b; (2) parameter c.*

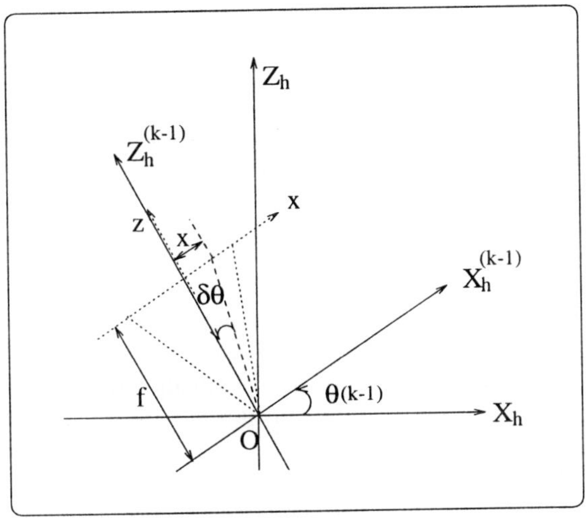

Figure 6: *The head motion planning.*

where x is the relative pixel distance of the vertical edge to the middle of the cyclopean image, f is the focal length in pixel units of the camera. $\theta(k-1)$ is the head pan angle feedback at the last sample time. $\theta_d(k)$ is the demand head pan angle for this sample time.

In order to include the target speed into the controller, we use $\theta(k) - \theta(k-1)$ as a prediction of the target speed of the next sample time, so the whole controller becomes:

$$\theta_d(k) = \theta(k) + \delta\theta. \tag{11}$$

To maintain the head tracking an object standing on the ground, two approaches can be used:

1. move the tracking window down and up as the object becomes closer or further away;

2. move the head down and up, but keep the tracking window fixed as the object becomes closer or further away;

We consider that the second method is better than the first since the image size is fixed, so that if we keep the head elevation angle fixed, the object might go out of view. Since we suppose that the object is standing on the ground, we can fixate the head at a position on the object slightly above the ground. This can be computed according to its disparity and the ground disparity map. Suppose we have detected a vertical edge and its average position in the cyclopean image, then from the feedback head state, we know the head elevation angle, so can find the parameters of the ground disparity plane hence the intersection of the vertical edge and the ground disparity plane can be computed. This is used to control the head elevation angle.

4.2 Parallel Implementation

Three kinds of parallelism can be adopted to reduce the image processing time. They are:

1. *pipeline different stages:* In theory this can reduce the processing time from the sum of the three stages (image grabbing, edge detection and PMF matching) to the largest of them. Although this can not reduce the latency of the whole system, it reduces the

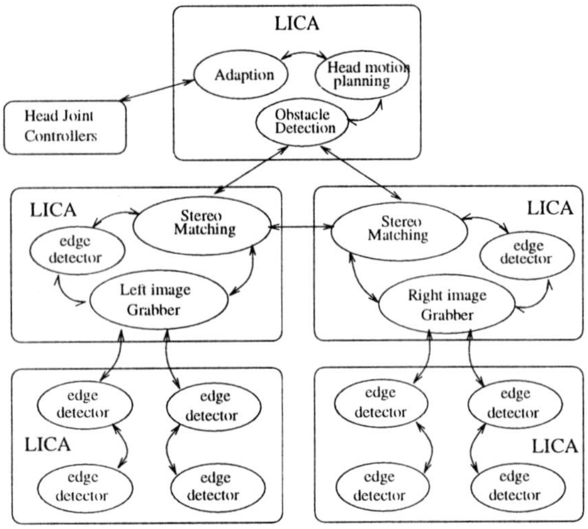

Figure 7: *The hardware configuration and the software mapping of the vertical edge tracking system.*

system sample time and raises the efficiency of individual processors;

2. *parallel processing of left and right images:* In theory, this can save almost half of the whole processing time especially when the processing of the two sides is independent, such as image grabbing and edge detection;

3. *parallel edge detection of different parts of the same image:* In theory, this will reduce the processing time proportional to the number of the processors used;

Based on the above analysis, the tracking system is implemented in the on-board transputer network, based on 5 LICA [8] boards. Figure 7 presents the hardware configuration and software mapping of the different processors.

4.3 Experimental Results

Experimental results show that these parallel processing methods are successful in reducing the processing time. Table 1 shows the processing times for edge detection when different numbers of transputers are used. Two window sizes have been used in experiments: a large one for initialisation of tracking, and a small one for tracking. Note that when the tracking process is performed in parallel in a small size window, the system sample rate is limited by the image grabbing rate

Table 1: Time spent for edge detection with different windows

Window Size	200 × 15 pixels	50 × 5 pixels
1 T800	520ms	70ms
5 T800s	130ms	40ms

which is 25Hz. The disparity and the head position are used as references to keep tracking the same vertical edges. The PMF stereo match is done by one T800 transputer currently in 600ms during initialisation, and reaches real time during tracking.

Figure 8 shows the view direction of the target in space, the head pan angle and the tracking error when the target rotates on the turn table at different speed. Figure 10 presents the left image of an image sequence. The head follows the vertical edge motion to keep it on the middle of the cyclopean image horizontally.

5 Fast Computation of the Fundamental Matrix

To solve the feature-based correspondence problem in real time, a fast method must be employed to compute the epipolar geometry, hence to limit the search from two dimensions to one for correspondence. Furthermore, during active verging, all camera-related parameters are unchanged apart from the camera verge angles, which are available from the head state feedback.

We have investigated specific forms of the fundamental matrix, the algebraic representation of the epipolar geometry, for common elevation platform stereo systems. In a common elevation stereo system, the two cameras have independent vergence joints and use a common elevation platform. A pan joint is used to increase the field of view of the system. There is a platform for each of the cameras so that their optical centres can be adjusted to a position near the intersection of the elevation axis and the vergence axes to ensure that the elevation and vergence of the head approximately causes a pure rotation of the cameras. The corresponding coordinate system is shown in Figure 9 and P is the fixation point of two cameras, θ_e is the elevation angle of the system, and θ_l and θ_r are the two camera angles in the X-Z plane. Clearly, the epipolar geometry depends only on the two vergence joint angles. When the cameras verge, the epipolar ge-

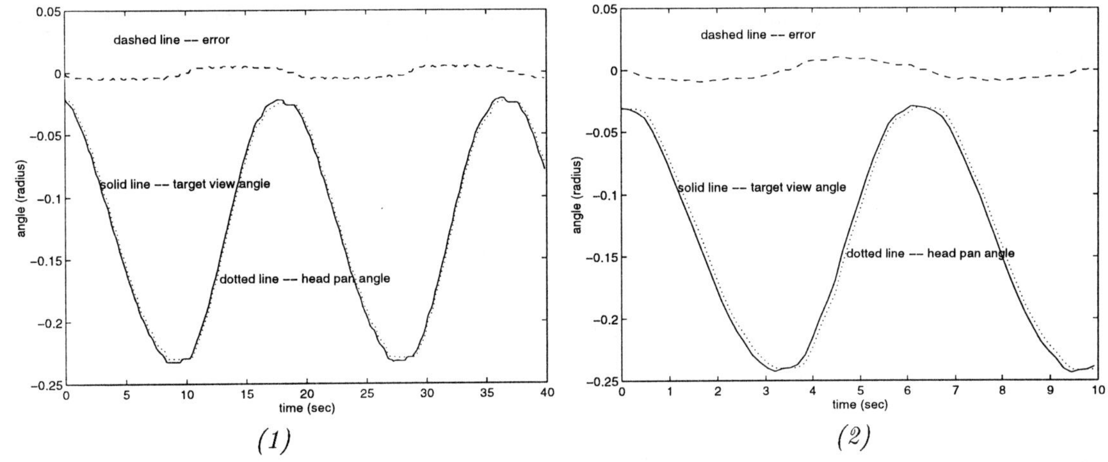

Figure 8: *Vertical edge tracking performance of the head when the target rotates at: (1) 0.056Hz; (2) 0.167Hz.*

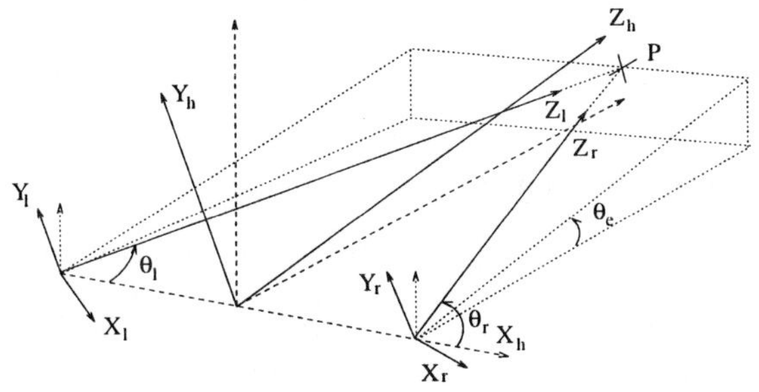

Figure 9: *The coordinate system for an active stereo.*

ometry changes dynamically. As a result, we have developed a number of methods to compute the fundamental matrix quickly and reliably. From experimental results of applying these methods, the epipolar geometry related camera intrinsic parameters can be calibrated approximately.

The image pair shown in Figure 11 was captured by our active stereo head with camera angles $\theta_l = 75^o$ and $\theta_r = 105^o$. The camera vergence angle is 30° which is about the maximum value that arises in real applications.

On the basis of extensive experiments, we have developed the calibration algorithm, shown in Figure 12. Note that it shows how vision can compensate for inevitable manufacturing errors which, if not take into account, make calibration ineffective.

The real fundamental matrix of each image pairs computed directly from the intrinsic parameters, the camera angles and the manufacturing error compensation is listed in table 2, where $d_t < 1$. If we take $d_t < 1.5$, then all match points are inliers. The number of inliers among the point pairs used for the computation is used as one of the criteria to evaluate the quality of the computed fundamental matrix. If $d_i = \frac{d(p'_i, Fp_i) + d(p_i, F^\top p'_i)}{2} < d_t$, we say that the computed fundamental matrix fits for the point pair (p_i, p'_i), and point pair (p_i, p'_i) is a inlier. Here $(p_i, p'_i), i = 1, \cdots, n$, are the n correspondence point pairs and $d(\cdot, \cdot)$ is the point-to-line Euclidean distance expressed in pixels. d_t is the threshold value which reflects the quality of the computed fundamental matrix as another criterion.

It is clear that this method is practical and efficient without sacrificing accuracy. The epipolar geometry computed from the approximately calibrated situation for a sequence of four image pairs is shown in figure 13.

For real time tasks, a process of calibrating the camera intrinsic parameters has been developed. We have shown that it is possible to attain the required efficiency, only slightly decreasing accuracy as a trade-off.

Table 2: *The results of the epipolar geometry for different image pairs from the direct computation using the calibration parameters and camera angles.*

Vergence angle, θ_v	Fundamental matrix, $\mathbf{F}$	Epipole $\mathbf{e}$ and $\mathbf{e}'$	inlier/total point
Parallel image pair	$\mathbf{F} = \begin{bmatrix} -0.000000 & -0.000000 & 0.024713 \\ -0.000000 & 0.000000 & 1.000000 \\ 0.001243 & -1.000000 & -4.720147 \end{bmatrix}$	$\mathbf{e} = \begin{bmatrix} 8.24435e+07 \\ 7.16923e+07 \end{bmatrix}$ $\mathbf{e}' = \begin{bmatrix} -2.90098e+09 \\ 102441 \end{bmatrix}$	19/20
10°	$\mathbf{F} = \begin{bmatrix} -0.000006 & -0.000245 & 0.052390 \\ -0.000245 & 0.000006 & 1.000000 \\ 0.030384 & -0.938134 & -13.125493 \end{bmatrix}$	$\mathbf{e} = \begin{bmatrix} 4080.28 \\ 213.844 \end{bmatrix}$ $\mathbf{e}' = \begin{bmatrix} -3829.27 \\ 123.975 \end{bmatrix}$	18/20
20°	$\mathbf{F} = \begin{bmatrix} -0.000011 & -0.000479 & 0.078777 \\ -0.000479 & 0.000011 & 1.000000 \\ 0.058886 & -0.879146 & -21.207220 \end{bmatrix}$	$\mathbf{e} = \begin{bmatrix} 2088.92 \\ 164.534 \end{bmatrix}$ $\mathbf{e}' = \begin{bmatrix} -1836.19 \\ 123.009 \end{bmatrix}$	20/20
30°	$\mathbf{F} = \begin{bmatrix} -0.000016 & -0.000705 & 0.104344 \\ -0.000705 & 0.000017 & 1.000000 \\ 0.087203 & -0.821995 & -29.082062 \end{bmatrix}$	$\mathbf{e} = \begin{bmatrix} 1418.68 \\ 147.98 \end{bmatrix}$ $\mathbf{e}' = \begin{bmatrix} -1165.74 \\ 123.712 \end{bmatrix}$	20/20

6 Conclusion

To use active vision for vehicle navigation, real-time processing of visual information is required. Classical stereo methods entail considerable computational expense to obtain depth information and positional accuracy. Accurate camera calibration is crucial to reduce errors. However, it is meaningless for an active vision system to perform such calibration since the extrinsic parameters change dynamically. We have shown that a simplified classical stereo method can still be used to keep an active head tracking a detected obstacle in real-time, when implemented in parallel. Only disparity information from stereo matching is used. This avoids accurate camera calibration.

By proposing an algorithm for fast computation of the fundamental matrix for a common elevation stereo system, we demonstrated the possibility that the classical stereo correspondence method can be extened to do stereo matching in real time for stereo images captured form stereo system verged dynamically.

Based on our research described in this paper, it is possible to use the active vision head to do real-time navigation of the mobile vehicle using stereo. Current work investigates the navigability of narrow gaps.

References

[1] P.A. Beardsley, A. P. Zisserman, and D.W. Murray. Navigation using affine structure and motion. In *Proccedings of 3rd European Conference on Computer Vision*, pages 85–96, 1994.

[2] A. Blake, J.M. Brady, R. Cipolla, Z. Xie, and A. Zisserman. Visual navigation around curved objects. In *Proc. IEEE International Conference on Robotics and Automation*, pages 2490–2495, 1991.

[3] R. H. S. Carpenter. *Movements of the Eyes.* Pion, London, 1988.

[4] R. Cipolla and A. Blake. Surface orientation and time to contact from deformation of apparent contours. In *Proc. European Conf. Computer Vision 1992*, pages 187–202, 1992.

[5] F. Du and J.M. Brady. An active computer vision system for vehicle navigation. *International Journal of Pattern Recognition and Artificial Intelligence*, 1994.

[6] C.A.R. Hoare. *Communication Sequential Process.* Prentice Hall, London, England, 1985.

[7] H. Hu, J.M. Brady, F. Du, and P.J. Probert. A transputer network for real-time control of an intelligent mobile robot. In *IEEE Int. Conf. on Control Applications*, pages 1049–1054, 1994.

[8] H. Hu, M. Brady, J. Grothusen, F. Li, and P. Probert. Licas: A modular architecture for intelligent control of mobile robots. In *Procced-*

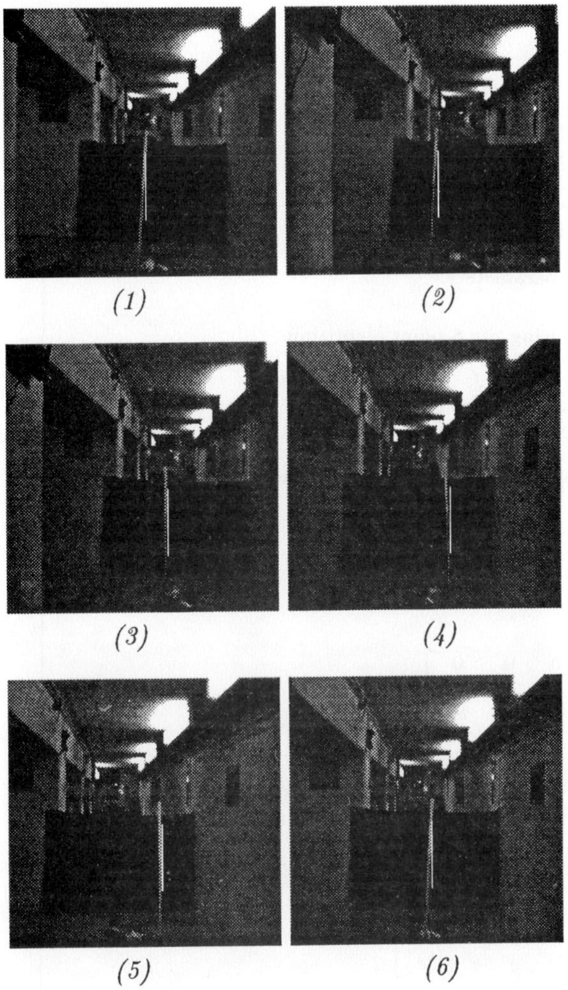

Figure 10: *Vertical edge tracking by the active vision head.*

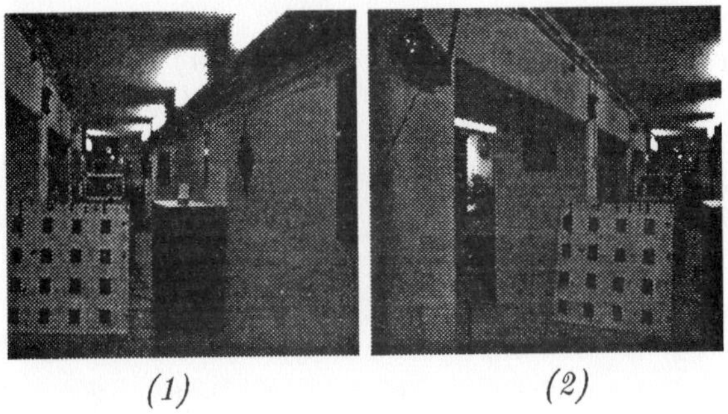

Figure 11: *A stereo image pair with vergence angle* 30°*: (1) the left image; (2) the right image.*

ing of IROS'95 International Conference on Intelligent Robots and Systems, 1995.

[9] H. Hu and Michael Brady. A parallel processing architecture for sensor-based control of intelligent mobile robots. *Int. Journal of Robotics & Autonomous Systems*, 1995.

[10] F. Li, J.M. Brady, and C. Wiles. Calibrating the camera intrinsic parameters for epipolar geometry computation. Technical report, OUEL 2075/1995, 1995.

[11] D. Marr. *Vision*. W. H. Freeman, 1985.

[12] J. Merron. Alternative depth representation for road-based vision. PROMETHEUS report 5, Department of Engineering Science, Oxford University, 1994.

[13] D.W. Murray, F. Du, P.F. McLauchlan, I.D. Reid, P.M. Sharkey, and M. Brady. Design of stereo heads. In A. Blake and A. Yuille, editors, *Active Vision*, pages 155–172. MIT, 1992.

[14] K. Pahlavan and J. O. Eklundh. A head-eye system for active, purposive computer vision. TRITA-NA-P9031, CVAP 80, Royal Institute of Technology, Stockholm, Sweden, 1990.

[15] Stephen B Pollard, John E W Mayhew, and John P Frisby. Implementation details of the pmf stereo algorithm. In John E W Mayhew and John P Frisby, editors, *3D Model Recognition From Stereoscopic Cues*, pages 33–39. MIT, 1991.

[16] Stephen B Pollard, John Porrill, John E W Mayhew, and John P Frisby. Disparity gradient, lipschitz continuity, and computing binouclar correspondences. In John E W Mayhew and John P Frisby, editors, *3D Model Recognition From Stereoscopic Cues*, pages 25–32. MIT, 1991.

[17] L.S. Shapiro, A. Zisserman, and J.M. Brady. Motion from point matches using affine epipolar geometry. In *Proc. 3rd European Conference on Computer Vision*, volume II, pages 73–84, May 1994.

Input $I_l(0), I_r(0)$: Image pair with parallel geometry;
$I_l(30), I_r(30)$: Image pair verged 30°;
Output fk_u, u_0, v_0 : Camera intrinsic parameters.
Algorithm

1. Obtain the stereo match pairs for each image pair using the 8-point algorithm.

$$\mathbf{F} = \begin{bmatrix} f_1 & f_2 & f_3 \\ f_4 & f_5 & f_6 \\ f_7 & f_8 & f_9 \end{bmatrix};$$

2. Identify and correct small manufacturing errors[10] so that both fundamental matrices have the form:

$$\mathbf{F} = \begin{bmatrix} 0 & a & ab \\ c & 0 & 1 \\ b'c & e & b' + be \end{bmatrix};$$

3. Compute δb from

$$\mathbf{F(0)} = \begin{bmatrix} 0 & 0 & 0 \\ 0 & 0 & 1 \\ 0 & -1 & \delta b \end{bmatrix},$$

ie.

$$\delta b = \frac{\sum_{i=1}^{n}(y_i - y_i')}{n};$$

4. Compute $v_0 = -b$ and $v_0' = -b' = -(b + \delta b)$ using the 5-point algorithm from:

$$\mathbf{F(30)} = \begin{bmatrix} 0 & a & ab \\ c & 0 & 1 \\ b'c & e & b' + be \end{bmatrix};$$

5. Compute fk_u and u_0 using SVD from:

$$\begin{bmatrix} sin(\theta_l)(y' - v_0') - sin(\theta_r)(y - v_0) & cos(\theta_l)(y' - v_0') - cos(\theta_r)(y - v0) \end{bmatrix} \begin{bmatrix} fk_u \\ u_0 \end{bmatrix}$$
$$= (y' - v_0')x * cos(\theta_l) - (y - v_0)x' cos(\theta_r),$$

here $\theta_l = 75°$, $\theta_r = 105°$. v_0 and v_0' are computed from above step.

Figure 12: *The algorithm of calibrating the camera intrinsic parameters for computing the epipolar geometry.*

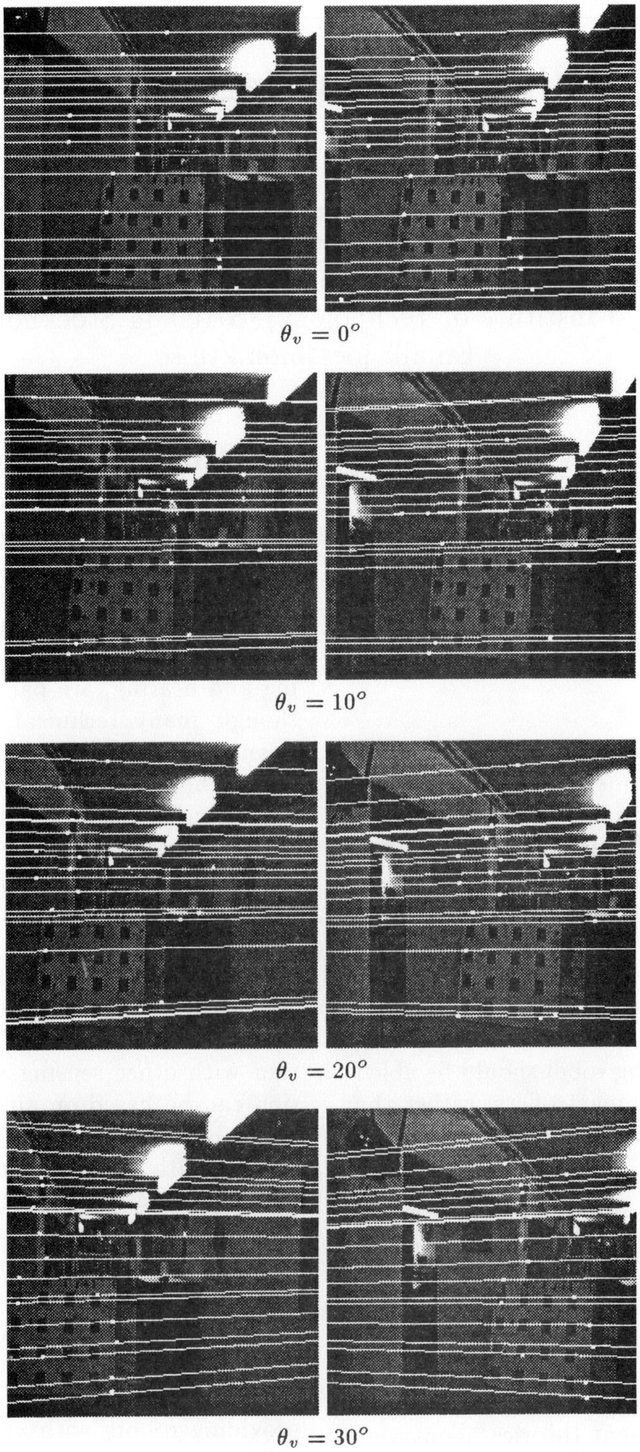

Figure 13: *Epipolar lines computed directory from the calibration parameters and camera angles for different image pairs taken at different vergence angles,* θ_v.

Active Vision and Seeing Robots

Jan-Olof Eklundh, Tomas Uhlin, Peter Nordlund, and Atsuto Maki

Computational Vision and Active Perception Laboratory (CVAP)
Department of Numerical Analysis and Computing Science
KTH (Royal Institute of Technology), S-100 44 Stockholm, Sweden
Email: joe@bion.kth.se

Abstract

In this paper we argue that the study of seeing systems or robots should be performed from a systems perspective. This implies that the importance of a computational mechanism should be established in view of the tasks that the system can perform using it.

Since such robots usually are aimed at functioning in real environments the figure-ground problem becomes essential. Computational approaches based on minimal information, commonly studied in the field, could hence be applied first after such problems are solved. We argue that for a seeing robot, capable of actively fixating and holding gaze on object in three dimensions, this problem is manageable, in particular if multiple cues can be used. An important point here is that a seeing robot should be able to utilize what the environment offers, rather than relying on a predetermined set of features.

Furthermore, if there are early processes for figure-ground segementation the local statistics of an observed (yet unknown) objects should be simpler than that of the entire scene. That suggest that processing to derive further properties, of e.g the shape or motion of that object, or even recognizing it, could be based on other features than those used to segment out the object. Such a view fits well with recent theories about view-based recognition. It also allows efficient implementations in terms of a visual-front-end, since both the target selection and the target analysis can be based on the output from such a layer, even though different features are used.

We support these ideas with an implementation of a mobile visual observer, functioning in realtime in a cluttered environment.

1 Introduction

Autonomous robots, like living creatures, can only act in a complex world by guidance from their senses. For humans our far senses, seeing and hearing, are particularly important. Although many technical sensing modalities are possible there are several reasons why vision should be essential also to attain sophisticated and flexible robots capable of performing tasks in an unknown and uncontrollable environment. Three such reasons are worth mentioning. The first one is that vision not only tells us *where* something is, or happens, but also *what* it is. Furthermore, and from a technical point of view, vision provides high resolution over a large field of view, a feature that is difficult, or costly, to obtain with other sensing modalities. Finally, vision can be based on ambient light, which so to speak is for free and doesn't depend on, or interferes with other activities. So, although robot behaviors might well be guided by a multiplicity of sensors, the capability of vision would be most desirable, if it could be indeed be supplied.

Computer vision research is commonly considered to aim at understanding and developing computer based seeing systems and hence also at providing robots with vision. However, despite certain success in limited contexts, such as for vehicle guidance and for pick-and-place, highly competent seeing robots hardly exist. In this paper we will pinpoint some of the reasons for this, stemming from the way computer vision has considered the problem. We will also propose a research approach that can be furthered towards the realization of seeing systems and illustrate this with some experimental evidence.

2 Computer Vision and Seeing Systems

Until the past few years the prevailing view was that the recovery of scene structure in terms of surfaces and volumes and their properties was a key step towards general purpose vision. In reaction to this so-called reconstructionist approach, Aloimonos (1990), stressed the purposiveness of perception, while Ballard (1991) and others put the behaviors in focus, arguing that general purpose capabilities could be obtained by combining such behaviors, e.g. in a layered architecture in the spirit of Brooks (1991), or by adding components of learning. Others still have explicitly focused on seeing systems that combine the flexibility and performance of the human visual system. It is important to note that unless we specify the capabilities and behaviors of the system we aim at (note, there exists no such system), it is unclear what sensor structure it should have and what activities it should be engaged in. If human vision is the yardstick by which performance is measured then anthropomorphic features become essential. On the other hand a system capable of, say, catching moving prey aided by vision, e.g. mimicking a frog catching flies, can also be of interest, see e.g. Arbib (1987). In robotics the desired repertoire of behaviors may also be limited. However, an important reason for using vision in robotics is that it can be used in many different situations. Hence, the implementation of each such behavior is useful only if it can be applied together with other desirable behaviors.

Vision, as we observe it in nature, is trivially active, both because it is tied to action and in the sense that it has to respond to dynamical changes in the world surrounding the visual agent. However, as argued by Pahlavan *et al.* (1993a), this does not by itself exclude the interest or use of "passive" or "reconstructionist" approaches. It is true that for most of the tasks we want to solve using vision all but a small fraction of the potentially available visual information is irrelevant. Nevertheless, this does not imply that active vision is the only way to address these problems. At the same time the importance of the active vision approach follows from a methodological argument: if we want to study computer-based seeing systems it is natural to study them in the context of active and purposive behaviors, because this is how the only seeing systems we know about, the biological ones, function.

As explained in the cited paper, such a perspective puts constraints on what information is available, on the possible algorithms and on what should be computed. A more elaborate analysis of the same type of problems, although from a different viewpoint, has been presented by Tsotsos (1995). Notwithstanding that, many problems remain at the level of computational theory, using the term of Marr.

To summarize one can say that active or animate vision is the appropriate methodology for the study of computer vision and hence also robot vision, but nevertheless more traditional approaches are needed to address competence problems about what information is available and how it can be computed.

This paradigm puts the emphasis on issues different from those of more traditional approaches to computer vision. In this paper we will consider some of these aspects in detail, arguing that seeing robots may indeed be less elusive than progress up to date may indicate.

3 Animate Vision in a Rich Environment.

Minimal information vs. salient characteristics. Current research in computer vision is largely concerned with how information about the world can be derived from single cues, like motion, stereo or texture. The approaches are generally *minimalistic*[1]: problems are addressed on how properties of the environment, like motion and structure, can be determined from *as little information* as possible. Such considerations may be well motivated scientifically and adapt to the standards in psychophysics. To understand the importance of a specific piece of information influences from other sources must be eliminated or controlled. Hence, there is a wealth of results on e.g. how many points or lines are needed to compute scene or object structure from a set of image frames. We contend that however important it may be to understand such limit cases, it hardly brings us closer to achieving a seeing system. More precisely such a system should function in a real environment, which inevitably will contain a rich amount of information and therefore finding and maintaining a stable perception of these features will often be more difficult than computing the scene structure from them.

This is well understood by computer vision re-

[1] Ballard (1991) used this term referring to the assumed generality of the models.

searchers, who approach the problem by performing grouping operations, including uncertainties and the like. However, if this is done on the basis of prespecified features, as is customary, the available visual information is not used to its advantage. What makes it possible to detect an object or observe a feature, and even what constitutes them, depends very much on the situation and the surrounding environment. A grey lump of materia may be observed as being an object by its color if it rests on or moves over a white background. However, if e.g. due to the relative displacement or a change of illumination the background also becomes grey, then this feature is no longer appropriate. The optic flow pattern or the binocular disparities may on the other hand then be used to select out the lump as an "object". The key point is that it is the world that determines both what is useful for the computations and what constitutes an object. So, while the complexity of the real world poses problems for some feature detectors, it may well make the job for others quite easy. To utilize this, we must be prepared to compute a number of different features simultaneously and in the particular situation use the ones that stand out sufficiently well against their surroundings[2].

It is also important to note that this lump of materia by standing out from its environment becomes an object and thereby gets an identity, whether we have a memorized category for it or not. We can hence ascribe properties to this entity, based on what we can observe, rather than on some model-based features given a priori. Such considerations are certainly well established in psychology and are elaborated e.g. by Gibson (1979), but computer vision approaches have hitherto not been based on them.

Attention and expectations. What we have discussed so far in this section can be regarded as an attentional step, where objects or features pop up from the environment by being different from nearby structure. Such mechanisms have been considered in computer vision by e.g. Culhane and Tsotsos (1992), who argues for the need for attention to overcome the complexity offered by visual tasks in the real world. Here, we want to stress that attention offers a way to utilize what can be observed from the world and also a mechanism that tells us that we see "something" and at what location this "something" is at the moment. We will later return to how these computations can be performed. Let us now just point out that they critically upon the use of multiple cues.

Of course, the process described could be mediated by various expectations even if the saliency of the computational features nevertheless is essential. Notably there are two cases when expectations play a central role. One, mentioned by Ballard, is when we have specific knowledge about a particular object, like the color of "my own coffee cup". Another case is when we keep track of something that we just noticed. The expectation then comes from our observations rather than from memory. However, in both cases we can do with very simple computations: we can either establish some coarse global feature, e.g. if the object has some dominating color, texture or shape feature, or adapt our computations to finding some quite characteristic local feature that we either know since before or just observed. Again, it is the situation at hand that tunes our computations, not some prespecified agglomeration of a set of bottom-up structures.

4 The systems approach

As is obvious from our previous discussion a vision system capable of providing robust behaviors in a complex and dynamically changing environment must be considered in its entirety. Such systems oriented approaches have recently been suggested by various researchers, see e.g. Crowley and Christensen (1995), but their impact on computer vision research is yet limited. Several important questions arise in this context. The most central one is of course how we model the seeing agent displaying the behaviors guided by its vision. This is a very far-reaching question that actually touches upon our understanding of intelligence as such. Fortunately, this problem can be finessed in robotics applications. Then the scope often can be limited to a visual observer who is able to look at, identify, and to some extent recognize objects and structures, and on the basis of that capable of reactive behaviors like obstacle avoidance during navigation, and certain limited active tasks, like searching for more or less well-specified known features and categories. However, even in such a limited scenario two major questions arise: which algorithms and features should be used for acceptable system performance (i.e. for solving the tasks) and how can the sys-

[2] It is worth stressing that this is true also for "objects" like the ground plane or the sky, as long as there is a way of characterizing them. We will use the word "object" in a somewhat fuzzy sense to denote any configuration of interest in the environment.

tem choose between and integrate the different features?

Rather than trying to provide any definite answers to these questions we want to point out three important aspects of the systems approach to machine vision.

Feature selection. First, the selection of features and algorithms should be made and evaluated on the basis of systems performance. Hence, considering algorithms in isolation, as is common in computer vision research, becomes less informative. In particular, an open system design, say, using a layered architecture, is necessary to try alternative approaches.

Control. Secondly, the control issue becomes central, also at the level of visual processing. How can it decide that something is currently a good descriptor of what it is looking at? How does this depend on the expected action of the system? These are questions that will go unanswered in this paper, but are essential in the continuation of the approach outlined here. Individual features can be computed in parallel, but conjunctions of features are sometimes derived in a sequential step, if they are complex enough. Such interactions between parallel and sequential computations are well-known in the literature on human attention, and they should therefore hardly be unexpected in computer vision systems.

Continuous operation. A final general point to make on the systems approach, obvious from the focus on behaviors, is that the time aspect is crucial. A seeing system naturally functions over time. Processing is, in principle, continuous and the system must respond to events and changes in the environment as they occur. This puts an emphasis on real-time processing that goes beyond the mere goal of fast algorithms. The important point is rather that the system continuously receives input which it uses to solve certain tasks, that in turn also may vary with time. These thoughts may seem obvious in robotics, but are important to stress, since computer vision generally has considered problems of pictorial vision and aimed at one single interpretation.

5 The Visual Front-End

Current research on early processes in computer vision to increasing extent seems to converge on certain basic principles for early vision which fits well in with our previous considerations. Generally speaking most approaches rely on the computation of directional derivatives of low orders and combinations or functions thereof, computations performed at several different scales. The manner in which this is done varies: some approaches use Gaussian or Gabor filters, others tunable filters of a more general nature, wavelets or more sophisticated anisotropic models. It is beyond the scope of this paper to discuss these different techniques in detail, we refer the reader to the literature on the topic. However, we note that such methods in general are indeed very appropriate for developing systems of the type we are aiming at. By implementing a first layer of retinotopic processing, a Visual Front-End in the terminology as proposed by Koenderink and van Doorn (1987), we obtain a highly efficient implementation at the same time as we can base our low level computations on state-of-the-art techniques. We also get a scalable design, which allows the extensions we foresee, without cumbersome additions.

More precisely, by computing a set of low order derivatives at multiple scales in a VFE layer, we obtain output that can be shared by all our subsequent modules to derive monocular, binocular and motion cues at later stages. See Jones and Malik (1992) for an example. Without arguing for or against whether such a model provides the most reasonable architecture for a computer vision system, we observe that at least it maps well onto already existing hardware. Existing pipeline and signal processors are well suited for such retinotopic computations, but less so for the more general and less image oriented ensuing computations, which usually are performed in coarse grained parallelism. In the end and in our general spirit we may want to develop other pathways for more direct computations of, say time-to-collision or certain scene characteristics, but currently we favor the VFE structure. The details of the approach we have used in our experiments is described elsewhere, e.g. in Gårding and Lindeberg (1995).

6 System Description

We shall now describe an experimental system by which we are able to demonstrate some of these ideas, with the aim to design and implement of an autonomous mobile observer. This observer should while moving be able to detect and look at a possibly moving object in the world on the basis of its standing out in some way from its back-

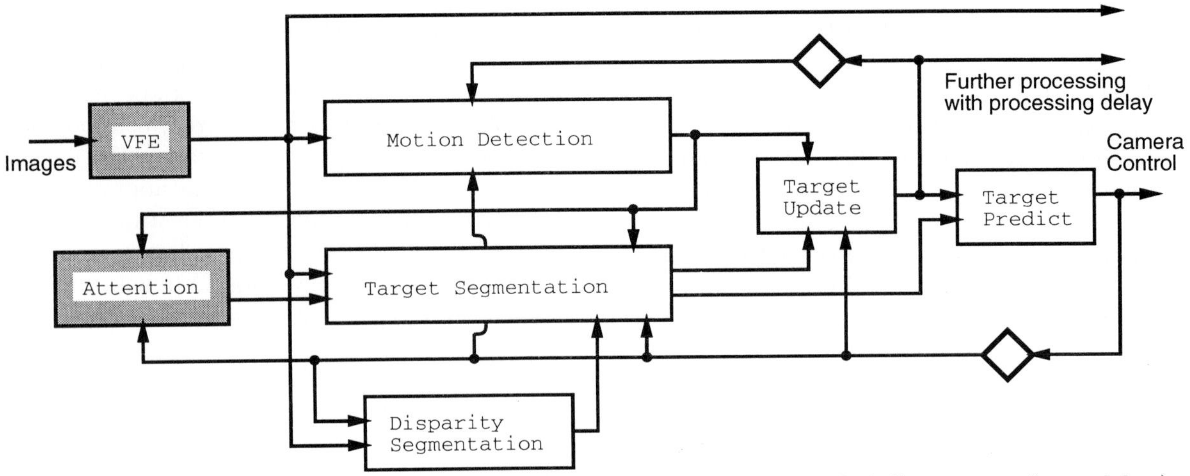

Figure 1: The system implementation is shown schematically. (The diamonds indicate a one frame delay.)

ground. Currently, motion and binocular disparities are used in the figure-ground separation step, but the architecture is open for additional cues. So far fundamental skills in terms of fixation, target pursuit and target discrimination, have been implemented. The basic system is at the moment implemented partly in real-time on an existing mobile platform, and partly as a post-processing stage working on images captured in real-time during the execution of the former processes.

The full system includes the integration of three cues for target selection and target discrimination. These are used by the moving observer to smoothly pursue moving or stationary targets binocularly while maintaining vergence. Mechanisms for discovering moving targets also form integral and vital parts of the system, since that provides means of attention, without which cues to changing the state of the system do not exist. Any active vision system must have two components, one to maintain attention, and another one to find and select new locations to attend to. Moreover, these must function in parallel, because the system would otherwise be without choices. The pursuit and motion detection mechanisms here provide this basic behavioral level for the active observer (Figure 1). (Currently, motion provides the only cue to changing attention, but other cues can easily be incorporated.)

Terminology and Comments. Throughout this section we will use the term *target* and with this we mean the portion in the image which is currently attended and supposed to represent a scene-object. A *target model* is used, which is coded as an image mask, including pixels that are considered as belonging to the target. All segmentation modules produce similar masks after a thresholding step. These masks are used for the integration.

Attention and Smooth Pursuit

A key feature of the system lies in its ability to smoothly pursue an arbitrary target that is described by its location, extent and visual appearance. No a priori knowledge about the target's visual appearance, i.e. texture or shape, is built into the system, although we indeed leave open that additional constraints or knowledge can be included dynamically as information becomes available, and similarly that such information can be excluded as it becomes obsolete. The features that are seen, are assumed by the pursuit module to change smoothly. Non smooth changes will trigger attentional mechanisms.

As the basis for this implementation we have chosen a coarse to fine correlation scheme similar to the one reported in Pahlavan *et al.* (1993b), see Section 7. This technique works very well when no occlusions are present. We will here present an extension handling occlusions as well.

In order to take care of occluding objects and distracting things in the background, motion detection is integrated to filter out parts of the scene that can be parts of the target, namely those that are moving. This is provided in the system by a connection from `Motion Detection` to `Target Segmentation`, see Figure 1. We will see later in the experiments that this is not enough in many cases when there are occluding objects that are themselves moving, since they will also be detected as moving. Disparity, which is an essential component in our binocular system can in these

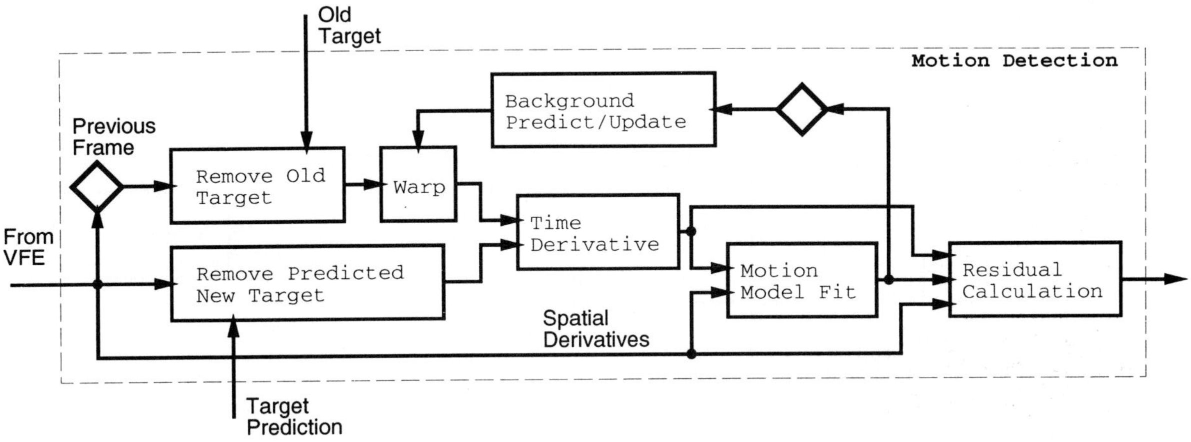

Figure 2: The motion detection is shown schematically. This is a detailed description of the `Motion Detection` box in Figure 1. (The diamonds indicate a one frame delay.)

situations aid in depth discrimination by also providing a clue to where occluding objects may lie.

Motion Detection. Motion detection provides the system with a recovering mechanism if the rest of the system for some reason should fail to pursue a target. (In real-world situations this will happen sooner or later.) It functions continuously during pursuit and is used as an integral part in selecting points where the target is lying (Figure 2).

Affine Background Motion Model. Using the brightness constancy constraint a global affine image velocity model is fit to the gradient and time derivative information (see `Motion Model Fit` in Figure 2). Two steps involving feedback are included to account for object motion and large background motion.

- The predicted and previous position and extent of the target are used to mask out parts of the image which likely belong to the object, so that they do not affect the calculation of the affine parameters for the background. See feedback into `Motion Detection` in Figures 1 and 2. This is a crucial step to avoid iterations that otherwise would have been necessary to get reasonable performance even if targets are large. We could say that the iterations are performed in time instead of in a single frame.
- The accumulated affine parameters are used over time to cancel out the majority of the time difference, see feedback into `Warp` in Figure 2. Thereby we can avoid iterations in fitting the affine model, and instead iteratively refine our motion model in time. This is very important to increase the computational speed and decrease the subsequent lag of the pursuit mechanism. We have here an example of how one can tradeoff the performance of a single algorithm for better overall performance.

Background Segmentation. Once the fit is made to the background, and the affine parameters are available, the residual normal image velocity is calculated in the entire image, `Residual Calculation` in Figure 2. The obtained residual is thresholded and high residuals are considered as possible independently moving areas in the scene. The threshold on the residual is adaptive and is set relative to the difference between target and background motion. The result of this can be seen in Figure 8.

Target Segmentation. The aim of target segmentation is to determine which parts of the scene are moving consistently with what is currently believed to be the target. This is the typical situation in connection with occlusions, when the target gradually disappears, to later emerge on the other side of the occlusion. It also supplies the system with the ability to discriminate between several moving targets, found by the motion detection, and distinguish parts of the scene that lie at the same depth, found by the disparity segmentation (Figure 3).

Affine Target Motion Model. The calculations on the target are performed in analogy with what

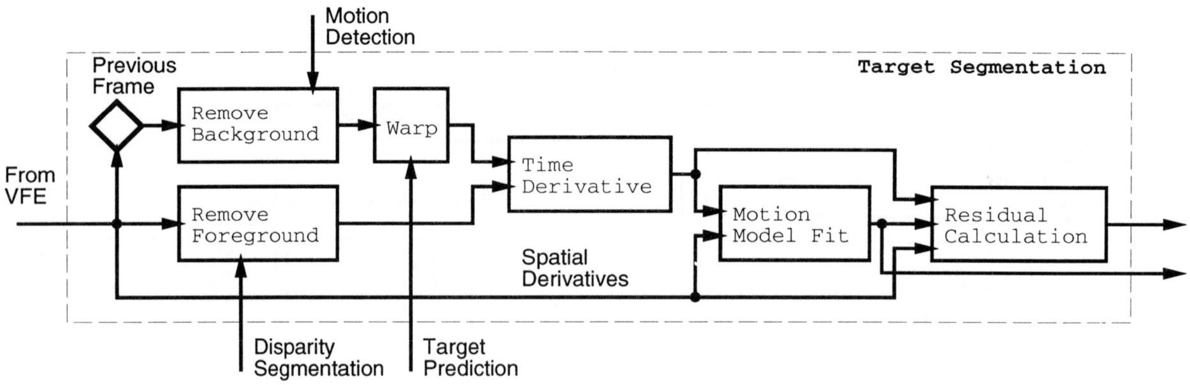

Figure 3: The target segmentation is shown schematically. This is a description of the `Target Segmentation` box in Figure 1. (The diamonds indicate a one frame delay.)

is done for the background motion, with the advantage that most of the computations spent on the background can be used also here (spatial derivatives and smoothed images from the VFE). By sharing these data we greatly increase the performance.

Target Segmentation. Analogous to the motion detection in Section 6, the residual normal image velocity is calculated and an adaptive thresholding is applied. A sample result of this can be seen in Figure 8 in the presentation of the experiments.

Disparity Segmentation. To increase performance of the system it has also proven useful to have relative depth information, if areas are closer than the target or behind the target (Figure 4). Of course having a disparity measure is necessary to obtain vergence.

Disparity Computation. The disparity computation is carried out with a phase based method. The phase based method has been chosen because of its low complexity, and because it fits well into the VFE framework, which means that we can use the same data for vergence and pursuit movements. It also performs well in a real-time implementation for vergence on the KTH-Head-Eye system (see Section 7).

Disparity Selection. The object of disparity selection is to select the disparities that belong to the target in the presence of disparities that arise from other locations in the scene. The method is described in Maki *et al.* (1994) and has been further developed to take advantage of the feedback obtained when the system is run in a closed loop together with other cues to follow a single target, to ensure that version and vergence components in smooth pursuit remain consistent. The basis for selection is the predicted location and extent of the target. Also other researchers have presented work on disparity selection, e.g. Taylor *et al.* (1994).

Integration

The integration is performed mainly in **`Target Update`** (Figure 1), but in the **`Target Segmentation`** (Figures 1,3), motion detection information and disparities are also integrated. The target extent is kept in a mask. The integration produces an updated target mask using masks produced in Section 6 (see Figure 8). There are two parts needed to update the target mask, one to exclude pixels and and one to include pixels.

Target Exclusion. Target areas that do not get support from either motion detection or target segmentation are excluded from the target model. Also the disparity module detects areas in the scene that lie in front of the pursued target, which are then excluded from the target model.

Target Inclusion. Areas in the scene that are detected as both moving independently, detected by motion detection, and are moving consistently with the target image velocity model from the target segmentation, are added to the target model. That is, if pixels arise in both in the top and bottom of Figure 8, they are considered as target pixels.

7 Real-Time Implementation

Presently we have implemented parts of what has been described above to perform real-time visual

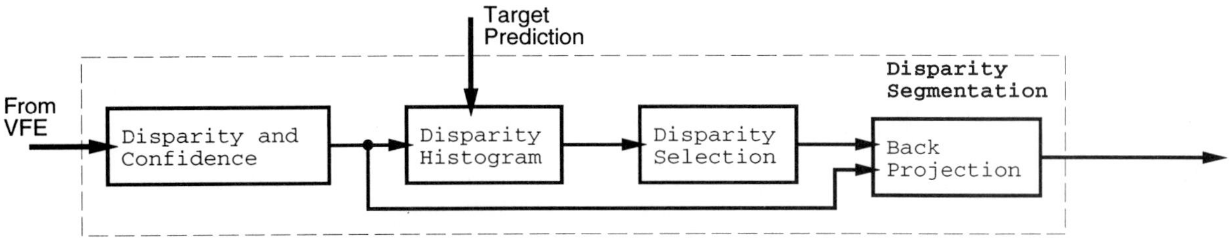

Figure 4: The disparity segmentation is shown schematically. This is a detailed description of the **Disparity Segmentation** box in Figure 1.

processing coupled to motor control of the KTH Head-Eye system. These are briefly described here, and visual results are presented in the experimental section.

Real-Time Pursuit

An implemementation of a pursuit system, running in real-time, includes not only parts from above, motion detection, target velocity and disparity estimation, but also motor control of the KTH-Head-Eye system, thus implementing smooth pursuit, vergence and saccades in a closed loop. The implementation is made on a T800 Transputer network, where both the motor control and visual processing takes place.

Image Velocity Estimation for Smooth Pursuit. The velocity estimation is done in the center of a cyclopean image using a binomial pyramid with four layers, in which the velocity estimation is derived coarse to fine (see Uhlin *et al.* (1995) for details).

The maximum measured motion is 15 pixels between frames in both the horizontal and vertical direction, and it operates in 25 Hz with about 50 ms delay from the occurrence in the world.

Disparity Estimation. In the most integrated version of the pursuit system, a log-polar mapping has been used to give disparity estimation for vergence control. The disparity estimates are computed at 5 Hz, while the pursuit signal computed from image velocity is measured, and the pursuit controlled, 25 times per second. For sample experiments see Section 8.

Maki *et al.* (1993) have demonstrated the use of phase based disparity estimation for vergence control on our Head-Eye system. Vergence errors were measured with this phase based disparity estimator at 25 Hz to control the cameras to continuously and smoothly verge correctly on an approaching moving target.

Frame Differencing for Target Selection. For detecting targets to attend to, the real-time system uses frame differencing. This process runs at 10 Hz, and a saccade is triggered with about a 150 ms delay.

Real-Time Motion Detection

A real-time implementation of a more complex algorithm for detecting moving targets in the presence of camera and observer ego-motion, similar to what is presented in Section 6, has also been made.

This system runs continuously, capturing image data directly from one camera on the robot head, and in a closed loop controls the camera motion to track the moving object. The loop runs in 25Hz. The image processing takes place on a MaxVideo200. The robot head is controlled by a T800 Transputer network.

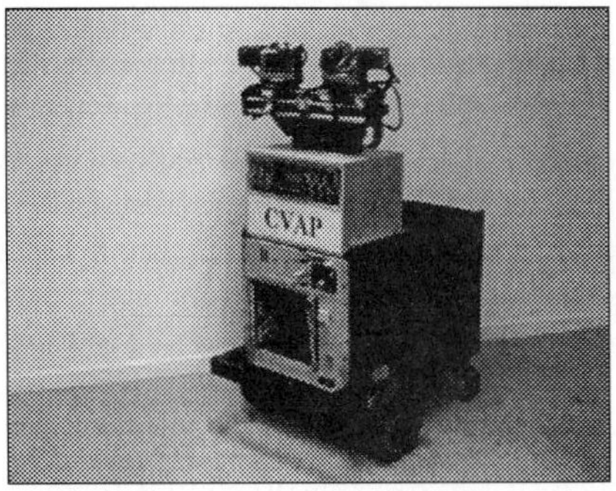

Figure 5: The experimental platform is a Head-Eye-System mounted on a mobile platform.

8 Experiments and Results

We will here show how the system performs in a number of different situations, both in the presence of all cues and when some cues are missing. The target will in most examples undergo complex movement, including rotations and movements towards and from the camera, while the observer undergoes ego-motion. No knowledge about the motion is assumed.

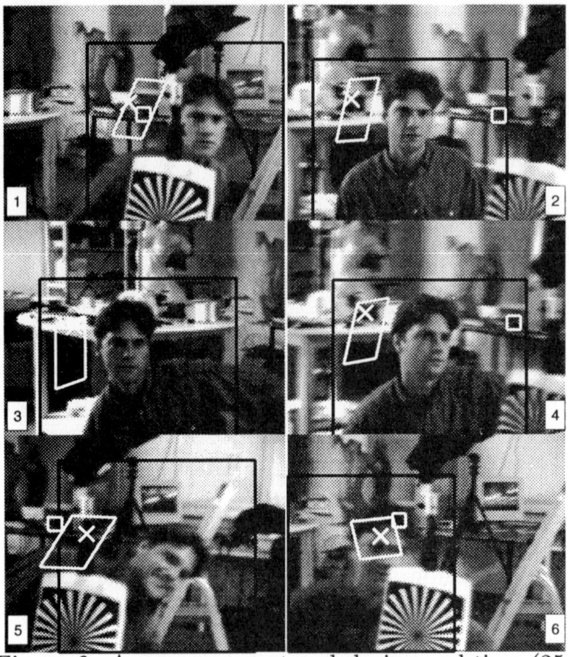

Figure 6: A sequence captured during real-time (25 Hz) pursuit. Shown are every 24th frame.

During all the experiments (except the experiment with the rotating umbrella) the images used were recorded during real-time pursuit performed by the already existing pursuit mechanism on the Head-Eye system (see Section 7). Therefore the experiments are performed on quite realistic data since all the noise due to inaccurate control of the head, motion blur, out of focus blur and vergence errors are present. All image sequences were captured at 25 Hz, during camera motion which is purely image driven with no human interference (except that there are humans walking in front of the cameras, being pursued). This facility enables us to experiment on a variety of sequences under different conditions exemplifying the performance of the system presented in Section 6 in many different situations. For performance figures see Uhlin *et al.* (1995).

Real-Time Pursuit with Motor Control. The performance of the real-time pursuit mechanism is shown in Figures 6, 9 and 11. For an explanation of the overlay we refer to the next section. We can see how the pursuit mechanism centers a target visually while the target is moving across the room. The visual processing and control are performed at 25 Hz and these sequences are used in the subsequent three experiments to

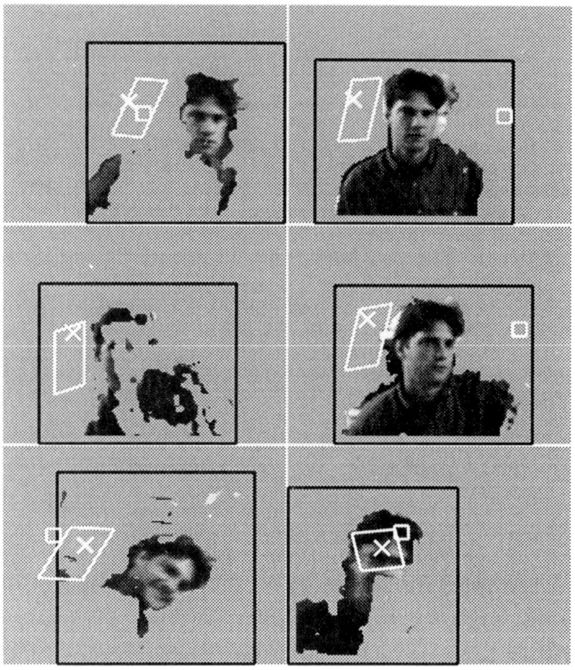
Figure 7: Resulting target pixels as extracted by the system (after integration and update) from the sequence shown in Figure 6.

show the increased performance achieved with the system described in Section 6, especially in conjuntion with occlusions. In Figure 6, when the head moves behind the occluding object in the last frame, it is not pursued across to the other side, but stays on the occluding pole.

Handling Occlusions. With figure ground segmentation provided by the system descibed in Section 6, the system can handle partial occlusions and segment out the target, see Figure 7. We can see how the occluding pole is excluded from the target. The target was lost by the simplified real-time pursuit process, but is now correctly pursued past the occluding object.

The result of the affine fit to the target is shown in Figure 7 as a rectangle which is allowed to distort accordingly. The absolute location and shape of this distorted rectangle is of no relevance. The initial location is determined when a moving target is detected and its subsequent movement is driven by the affine fit to the target image velocity. The rectangle's relative movement between frames, reflects target movement, lateral, vertical, rotational and expansion, all of which are important descriptors in the pursuit of a target. The small rectangle shows a fixed point on the background as calculated during background cancellation in motion detection. The white cross shows

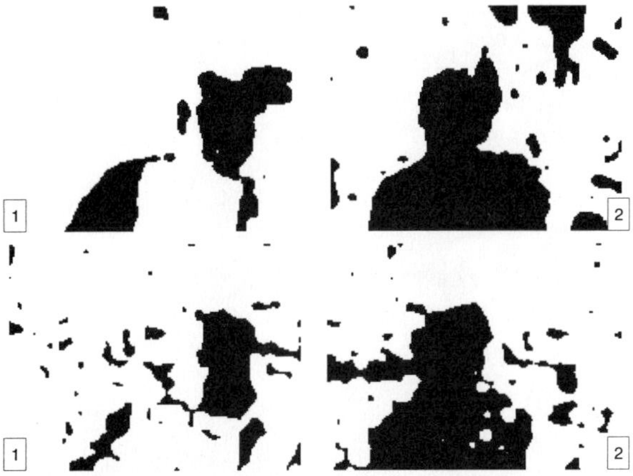

Figure 8: Motion detection (1st row) and Target segmentation (2nd row) return masks (before integration is performed) used to produce the final target masks in Figure 7 (compare with the 1st row).

the result of coarse-to-fine correlation when performed with the target masks produced by the system. The large black rectangle shows a window that is automatically placed around the centroid of the target pixels which are flagged as belonging to the target in each frame. Only pixels inside this rectangle are kept to the next frame. To achieve this result the system uses two masks (see Sections 6 and 6, and Figure 8).

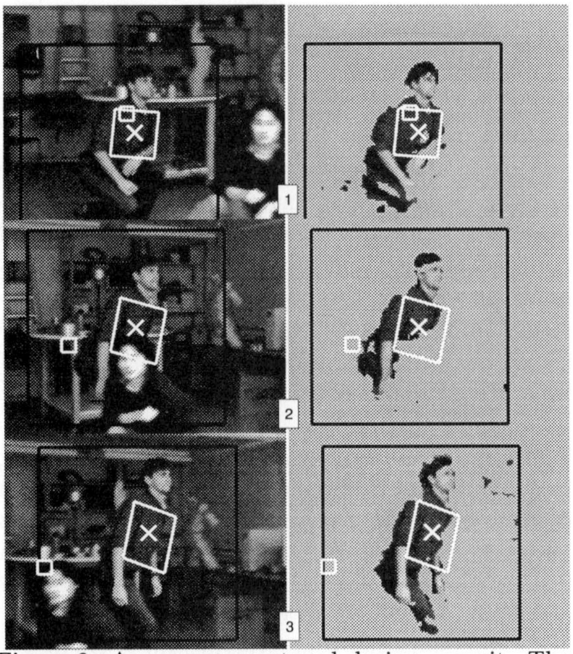

Figure 9: A sequence captured during pursuit. The pursuit is performed at 25 Hz, and shown are every 12th frame.

Occlusion and Other Moving Objects. To show also how the system performs in the presence of other moving targets, we have performed experiments when another object not only moves in the scene, but also partially occludes the pursued target as it moves across the scene, see Figure 9.

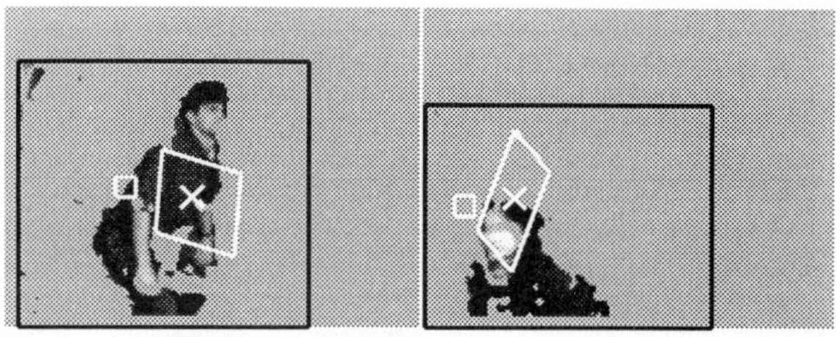

Figure 10: Target pixels as extracted by the system from the sequence shown in Figure 9, but without the disparity cue. The attention shifts to the other moving person. Every 12th frame is shown here.

When an occluding object is present, and it also is moving, it will be included by motion detection as a possible target location. If this object in some way dominates, it may well take over the attention of the system if it relies only on motion. To illustrate that, we have removed the disparity detection of areas in front of the target with the result shown in Figure 10. The attention is shifted to the person moving in front, although the attention was initially on the person moving behind. Even though this may be an unwanted behavior of the system, it anyway shows that the system can stably change its attention.

We see that without the disparity cue, the system is easily distracted by moving occluding targets, while with the disparity cue, successful pursuit is achieved even in the presence of distractors.

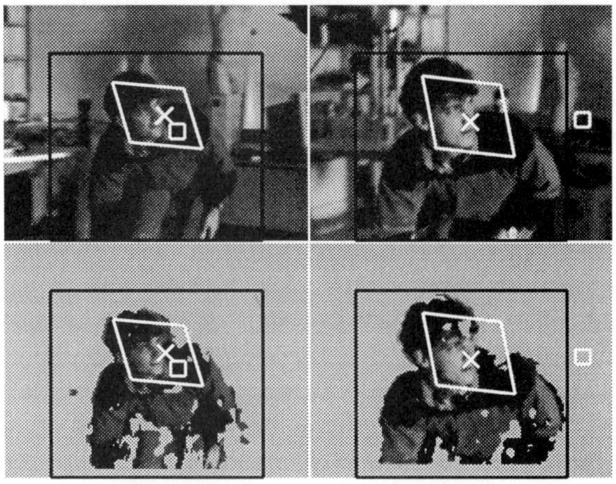

Figure 11: Pursuit during target expansion. On the top is shown the original sequence recorded during real-time pursuit, where every 12th frame is shown. In the bottom row is shown the result of the corresponding segmentation.

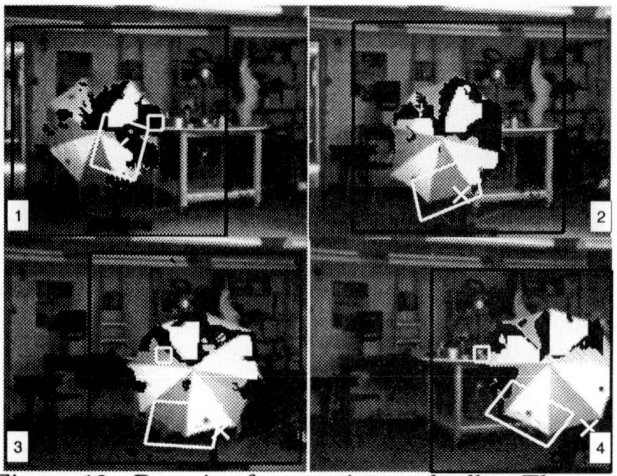

Figure 12: Pursuit of a rotating umbrella. The sequence shows every 18th frame in a 25 Hz sequence. The black areas show which areas are used for the affine fit depicted by the distorted white rectangle.

Expanding and Rotating Targets. The ability to detect and pursue an object that moves toward the observer is of great importance since such an object presents a potential collision. We then need a measure which is directly linked to the approach of the target so that an avoidance maneuver can be deployed in time. Although just qualitatively shown here, the affine model provides such a measure, which is depicted by the distorted white rectangle in Figure 11, together with the result of the target pursuit and segmentation. This measure is however not used explicitly here, but we see that pursuit is functioning also when the target is approaching.

The performance during target rotation is shown in Figure 12, where the segmentation is shown with black overlay. It is seen that the affine model does pick up the rotation of the target and that a correct pursuit is performed.

Real-Time Motion Detection with Motor Control. Figure 13 shows the result of an experiment with the algorithm described in Section 7. The system is tracking the white paper-box moving diagonally through the room from left to right, seen from the camera. While the camera tracks the box, the platform moves approximately straight ahead, the platform slows down, stops and moves straight backwards. Notice that the system is capable of tracking the object both in the presence of a rather smooth untextured background and a quite cluttered background.

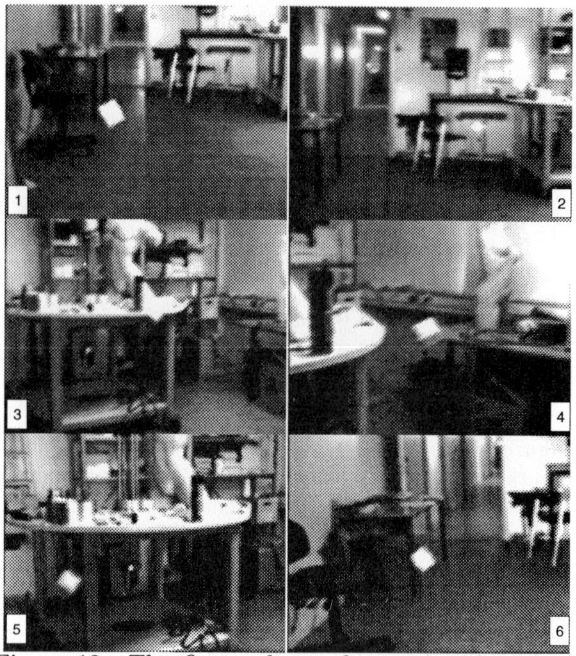

Figure 13: The figure shows the motion detection mechanism with real-time motor-control feedback, tracking the white box (see Section 8). (8 seconds between frames)

9 Discussion

We have argued that the study of seeing systems or robots should be performed from a systems perspective. This implies that the importance of a computational mechanism should be established in view of the tasks that the system can perform using it.

A particular aspect of this view follows from the fact that seeing robots usually are aimed at functioning in real environments. The computational approaches based on minimal information, commonly studied in the field, assume that the necessary (minimal) information can be robustly derived. Such segmentation or figure-ground problems are generally considered to be very difficult in image understanding. We argue that for a seeing robot, capable of actively fixating and holding gaze on objects in three dimensions, this problem is manageable, in particular if multiple cues can be used, as is well-known from studies fo human vision. An important point here is that a seeing robot should be able to utilize what the environment offers, rather than relying on a predetermined set of features.

Furthermore, if there are early processes that segment out a target in a cluttered world, one can assume that the local statistics of that (yet unknown) object will be simpler than that of the entire scene. That suggest that processing to derive further properties, of e.g the shape or motion of that object, or even recognizing, could be based on other features than those used to segment the object from its background. Such a view fits well with recent theories about view-based recognition, see e.g. Bültoff and Edelman (1992). Our approach also allows early inclusion of three dimensional information as favored by Nakayama and his co-workers, see e.g. Nakayama and Silverman (1988), and Shimojo *et al.* (1988). Finally, such a technique can be efficiently implemented in terms of a visual-front-end, since both the target selection and the target analysis can be based on the output for such a layer, even though different features are used.

We support these ideas with an implementation of a mobile visual observer, functioning in realtime in a cluttered environment.

10 Acknowledgements

The authors want to thank Kourosh Pahlavan for valuable discussions on the topic of the article. The reported work has been supported by a grant from the Swedish Research Council for Engineering Sciences. This support is gratefully acknowledged.

References

[1] Aloimonos, Y.: Purposive and Qualitative Active Vision. Proc. DARPA Image Understanding Workshop. (1990) 816-828

[2] Arbib, M.A.: Levels of Modeling Visually Guided Behavior (with peer commentary and author's response). Behavioral and Brain Science, **10** (1987) 407–415

[3] Ballard, D.H.: Animate vision. Journal of Artificial Intelligence. **48** (1991) 57–86.

[4] Brooks, R.A. (1991). Intelligence without representation. Journal of Artificial Intelligence, **47** (1991) 137–160.

[5] Bültoff, H. and Edelman, S. (1992). Psycophysical support for a 2-d view interpolation theory of object recognition. In *Proc. Nat. Acad. of Sci.*, volume 89, pp. 60–64.

[6] Crowley, J.L. and Christensen, H.I. (eds): Vision as Process. Springer Verlag, Berlin (1995).

[7] Culhane, S. and Tsotsos, J. (1992). An attentional prototype for early vision. In *Proc. 2nd European Conference on Computer Vision*, pp. 551–562, Santa Margherita Ligure, Italy.

[8] Gårding, J. and Lindeberg, T.: Direct Computation of Shape cues Based on Scale-Adapted Spatial Derivative Operators. International Journal of Computer Vision. (1995) To appear.

[9] Gibson, J. J. (1979). *The echological approach to visual perception.* Houghton and Mifflin.

[10] Jones, D.G. and Malik, J.: A Computatiopnal Framework for Determining Stereo Correspondences from a Set of Linear Spatial Filters. Proceeding of the 2nd ECCV. Santa Margherita Ligure, Italy. Springer Verlag, Berlin (1992) 395–410.

[11] Koenderink, J. J. and van Doorn, A. J. (1987). Representation of local geometry in visual system. *Biological Cybernetics*, *55*, 367–375.

[12] Maki, A., Uhlin, T., and Eklundh, J.-O. (1993). Phase-based disparity estimation in binocular tracking. In K. A. Høgdra, B. Braathen, K. Heia, editor, *Proc. 8th Scandinavian Conference on Image Analysis*, pp. 1145–1152, Tromsø, Norway. Norwegian Society for Image Processing and Pattern Recognition.

[13] Maki, A., Uhlin, T., and Eklundh, J.-O. (1994). Disparity selection in binocular pursuit. In *Proc. 4th IAPR (International Association for Pattern Recognition) Workshop on Machine Vision Applications*, pp. 182–185.

[14] Nakayama, K. and Silverman, G. H. (1988). Serial and parallel processing of visual feature conjunctions. *Nature*, , no. 320, 264–265.

[15] Pahlavan, K. Uhlin, T. and Eklundh, J-O.: Active vision as Methodology. Active Perception. Aloimonos, Y. (ed.): Lawrence Erlbaum Associates, Hillsdale, NJ. (1993a)

[16] Pahlavan, K., Uhlin, T., and Eklundh, J.-O. (1993b). Dynamic fixation. In *Proc. 4th International Conference on Computer Vision*, pp. 412–419, Berlin, Germany. IEEE Computer Society Press.

[17] Shimojo, S., Silverman, G. H., and Nakayama, K. (1988). An occlusion-related mechanism of depth perception based on motion and interocular sequence. *Nature*, , no. 222, 265–268.

[18] Taylor, J., Olson, T., and Martin, W. N. (1994). Accurate vergence control in complex scenes. In *Proc. IEEE Comp. Soc. Conf. on Computer Vision and Pattern Recognition, 1994*, pp. 540–545, Seattle, Washington.

[19] Tsotsos, J.K.: Behaviorist intelligence and the scaling problem. Journal of Artificial Intelligence, **75** (1995) 135–160.

[20] T. Uhlin, P. Nordlund, A. Maki, and J.-O. Eklundh, "Towards an active visual observer", Tech. Rep. ISRN KTH/NA/P--95/08--SE, Royal Inst. Tech., 1995.

Automatic Task Planning for Robot Vision

Bill Triggs & Christian Laugier
LIFIA, INRIA Rhône-Alpes,
46, avenue Félix Viallet, 38031 Grenoble, France.
Bill.Triggs@imag.fr, Christian.Laugier@imag.fr

Abstract

This paper describes an automated planner for visual inspection and monitoring tasks using a robot mounted CCD camera. It is capable of finding heuristically near-optimal viewing positions for a series of visual tasks, and of ordering them and planning robot paths between them to produce a near minimal cost overall task plan for the sequence. It has been optimized for intervention-style tasks with relatively large, cluttered workspaces, where it is difficult for a human operator to take account of all the constraints involved. The guiding principle of the work has been to aim for reliability by using quantitative, physically-based measures of quality wherever possible, and by using global search to ensure that possible solutions are not overlooked. We discuss a novel global function optimization technique based on decision theory and subjective models of function behaviour, that allows global viewpoint searches to run in the time usually required for local ones.

Keywords: Robot Task Planning, Machine Vision, Image Prediction, Global Optimization.

1 Introduction

This paper describes an automated task planner designed for visual inspection and monitoring using a robot mounted CCD camera. The system is capable of finding heuristically near-optimal viewing positions for a series of visual tasks, and of ordering the tasks and planning robot paths between them to produce a near minimal cost overall task plan for the sequence.

The design has been optimized for intervention-style applications such as the maintenance and decommissioning of nuclear power plants and deep sea oil rigs. The environments and tasks in these applications are often extremely complex. They can usually be modelled only imperfectly at a coarse scale. Reliability is a key issue, so active sensing is required to reduce uncertainty and ensure robustness. In practice the sensor needs to be mobile (robot-mounted) to avoid occlusion and direct attention towards the points where it is most needed, but this brings up the problem of sensor management. When the complexities of robot kinematics, environmental collision avoidance and visual occlusion are combined with task constraints and more traditional visual indices such as field of view, resolution and focus, it is seldom easy for a human operator to guess where the sensor should best be placed to execute its task effectively, so automatic placement planning is required.

To get some idea of the kind of task the planner was designed for, consider fig. 1. The problem is loosely based on a routine task in nuclear power plant maintenance. The area around the base of the columns must be inspected for damage or detritus. Here we have planned viewpoints only for the difficult-to-see regions *behind* the columns. The sequence of views selected by the system is shown in the lower panel. Note that the robot is at the limit of its reach for several of the views.

Our system is currently in its second generation, being a significantly extended rewrite of the single viewpoint planner described in [21]. It can be configured for a variety of tasks. For example the first version of the planner was integrated into a visually guided grasping system developed under the European Esprit collaboration SECOND [4, 11], where it was used to choose viewpoints for workpiece verification and pose correction and for visual servoing of the grasp approach.

The central activity of the system is the selection of viewing positions and the associated camera robot poses on the basis of task, camera, robot and environmental preferences and constraints. In practice the constraints are often very restrictive and it is necessary to reach a compromise solution. To support this we have adopted an approach based on the global optimization of a heuristically weighted sum of quality metrics, over a search region defined by the true hard constraints such as kinematic reachability. Each metric measures a single distinct aspect of viewpoint quality with respect to the relevant constraints. As far as possible, the assessments are based on objective physical and optical

This work was supported by the European Community through Esprit programs HCM and SECOND.

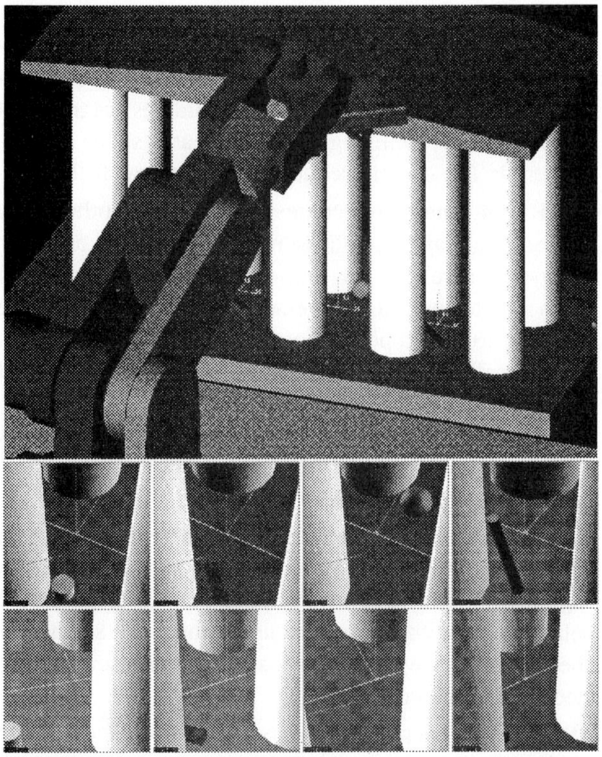

Figure 1: A routine inspection task. The areas behind the columns must be inspected for damage or detritus. The sequence of views selected by the system is shown in the lower panel, reading anticlockwise from top right.

quantities. For image-based metrics, we predict the expected image and perform direct measurements on it.

At present the planner is entirely model-based. There is no feedback from the images taken, although ultimately this would be desirable. As input, it requires a camera calibration and a list of specifications of visual tasks to be executed. These include the location and geometry of the object or region to be viewed and constraints on the viewing direction, field of view, resolution and focus required to complete the task.

The system has three principal components, which are relatively loosely coupled in the present implementation: the viewpoint assessor, the viewpoint search routine, and the tour planner. Each is discussed more fully below, but in brief they work as follows:

- The viewpoint assessor estimates the quality of a single prospective camera placement for a given visual task, as above.
- For each task, the search routine calculates a space of potentially feasible camera placements and searches it to find a globally quasi-optimal viewpoint, using the assessor to evaluate the trial viewpoints it generates. To speed this step we have developed a novel and highly efficient global search technique based on decision theory and subjective models of likely function behaviour. This will be described in detail.
- Once the search routine has found viable viewpoints (camera robot poses) for each of the visual tasks, the tour planner is called. This takes the set of planned viewpoints and converts it into a complete task plan by choosing an efficient ordering for the tasks and planning manoeuvres to move the camera robot safely through the sequence of viewing positions.

2 Previous Work

There have been relatively few previous studies of camera placement planning, mostly concentrated on visual and task constraints. Cowan [6, 7] developed geometric constraint based models of resolution, focus, aperture and occlusion, and also considered simultaneous camera and light source placement. Yi, Haralick and Shapiro [22] studied the camera/light source problem too. But perhaps the most complete previous system is that of Tarabanis and Tsai [18, 19], who also took a constraint-based approach and developed a fairly sophisticated CAD based model of workpiece self-occlusion for their MVP (Machine Vision Planner) system.

We feel that the above work has several major limitations that make it unsuitable for our application:

- *Workspace constraints* on the camera-carrying robot are not considered. This is unrealistic because in practice these are often the most restrictive constraints in the whole problem, especially for fixed-base robots in large workspace intervention-style applications. As far as we are aware, the only previous study to consider workspace constraints is that of Al-Chami [1, 2], which can be viewed as a precursor of the present work.
- *Image-based quantities* are avoided. For example, occlusion is evaluated by back-projecting the 'occlusion shadows' of obstacles as seen from the task and excluding every shadowed viewpoint. This sort of 'backwards' approach does not generalize easily, *e.g.* to quantitative measures of occlusion. Ultimately, the most satisfactory way to judge expected image quality is to actually predict the image and make direct measurements on it.
- The approach is *constraint based* rather than *decision based*. The emphasis is on passively defining regions of viewpoints with 'nominally satisfactory' focus, resolution, occlusion and so forth, rather than on actively searching for a single 'optimal' solution. But real applications need to make *decisions*: a single viewpoint must be chosen, and in practice the choice must often be a compromise between several conflicting constraints. A heuristic combination of continuous quality metrics is needed for this: binary classifications do not provide sufficient discrimination.

• *Search issues* are largely ignored. The viewpoint assessment function often turns out to be highly nonconvex, with several distinct feasible regions where occlusion or kinematic constraints cut the workspace in two. The use of a global optimization technique is essential for reliability. Otherwise it is impossible to guarantee that a tentative solution is the *global* extremum and not just a relatively poor local one.

In summary, the approach we advocate is based on the global optimization of a heuristically weighted viewpoint evaluation function, that takes robot workspace constraints into account as well as the more traditional optical and task constraints. As far as possible, the evaluations should be based on objective physical measurements, with image prediction and measurement being used for image-derived quantities.

A program such as this must draw on a rather diverse range of sources. Apart from physical optics (focus, resolution, lighting...), computational geometry (task & environment modelling) and computer graphics (image prediction), we will mention just two: work on general decision-theoretic methods of sensor management (for example by Durrant-Whyte and co-workers, who take a Bayesian decision-theoretic stance [5, 13]) and work on quantitative task-oriented measures of visual quality (for example the control-theoretic framework for visual servoing described by Espiau, Chaumette and Rives [8]).

3 Viewpoint Assessment

The heart of the camera placement planner is the viewpoint assessment routine. This takes a partially specified hypothetical camera placement, completes the specification, and then runs a series of independent tests on the placement and its associated robot pose. Each test considers a single distinct aspect of the placement and returns a scalar assessment of its prospective quality from that point of view. The final overall assessment is a heuristically weighted sum of the individual test results. When possible, the individual assessments are based on objective, physically-based quality metrics, although in practice a significant heuristic component is unavoidable.

As an aid to weighting, test values are normalized to be "χ^2-like" in the sense that zero indicates an ideal situation and values much greater than one a marginal one. Hard constraint violations are encoded as essentially infinite test results, but where possible we also include a term proportional to the degree of constraint violation to guide search methods towards the feasible region.

For speed, the tests are arranged roughly in order of computational cost and selectivity, and candidates whose accrued assessment becomes unacceptably large are discarded immediately. This is particularly important in the initial phases of search, when many inaccessible or otherwise unacceptable candidates are tried before the more promising regions of search space are discovered. The current sequence of tests is as follows:

1) Robot kinematics comes first as many other tests depend on it and can not be run if it fails. The assessor supports search in task, end-effector or joint space and provides forward and inverse kinematics to fill in the missing joint and pose information[1]. For inverse kinematics, all of the possible solutions are assessed and the best is chosen. Most of the tests do not depend on the kinematics solution in any case. Note that our 3R spherically-wristed SCEMI robots have relatively simple closed-form kinematics.

2) The camera position is assessed with respect to the environment. This is currently just a quick check against a list of rough workspace bounds to eliminate placements with the camera under the work surface or behind a wall, without the cost of full interference detection (which comes later).

3) The camera position is assessed with respect to the task. A task is characterized by a frame defining a task origin and coordinate system, a polyhedral volume that must be visible, and a set of viewing position and angle bounds and preferences. This test assesses the viewing angle and position of the camera. Simple types of task self-occlusion constraint can be encoded, but a more specialized assessor would be required for tasks with complicated self-occlusion geometry (*c.f.* [18]).

4) The task is assessed with respect to the camera. This is by far the most complex test and the only one to use camera parameters. The image that would be seen by the camera is predicted and assessed. To save the effort of writing a 3D hidden surface graphics library, the current prediction mechanism is heuristic rather than exact. We only project approximate bounding hulls of objects[2], and we use heuristic depth sorting rather than full visible surface prediction when evaluating occlusion and clutter.

4.1) After a quick test to verify that the camera is pointed in roughly the right direction, the task region is projected and its silhouette is clipped against the image borders. A penalty is charged for the fractional area that is clipped.

4.2)

The closest and farthest points of the task region are determined and used for focus and resolution assessment. Lack of focus is penalized according to

[1] In fact, this flexibility is not currently needed as all of the existing search routines use inverse task space kinematics.

[2] Most of the objects in our environment models are assembled from convex primitives in any case.

the area of the blurred image of a point[3], and lack of resolution according to the squared resolution deficit $\left(\frac{\text{desired resolution}}{\text{actual resolution}} - 1\right)^2$.

4.3) For each obstacle in the environment that may occlude or clutter the task, its image is predicted and the degree of its impingement on the task image is assessed. Obstacles lying in front of the task cause occlusion, penalized according to the percentage of the task image area that is occluded. Obstacles lying behind or close to the task image cause clutter. Although it does not actually obstruct the view of the task, clutter also needs to be penalized because it makes visual routines such as segmentation much more difficult. Underlying clutter is penalized according to the underlayed percentage of the task area, and nearby clutter according to the inverse image distance between it and the task region image.

5) The mobility of the camera robot in the chosen pose is assessed, and poses very close to kinematic singularities or joint limits are penalized. Without this softening of the constraints the optimal viewpoint would very often be hard against a kinematic or joint limit. This would make life difficult for local viewpoint search techniques, for our local path planner, and not least for real robot controllers.

6) Full robot-environment interference detection comes last owing to its very high cost (at least 80–90% of the run time of the assessor). Our current system does not allow us to evaluate the degree of penetration so we simply return a yes/no answer.

In practice the assessor performs well. Most importantly, the viewpoints it selects seem to correspond very well with the ones a human might choose, although the absolute values of its assessments are probably not an infallible index to "absolute view quality" as judged by a human. The basic approach of assessing view quality by image prediction followed by quantitative predicted-image-based quality metrics seems to be effective, and has no doubt contributed to the robustness of the evaluations: despite its many parameters the current assessor has never needed any significant parameter tuning.

However, several aspects of the implementation could be improved. A more precise image prediction mechanism would allow much more refined assessments of probable image quality, as well as providing support for important visual routines such as tracking, model-image matching and so forth. The model of task geometry should be extended to consider more complex types of task self-occlusion. It would also be useful to include a simple model of lighting (although exact light levels are notoriously hard to predict).

[3]The diameter of this 'defocus disk' for a point at depth d is $\frac{a}{p}\left|\frac{d^{-1}-D^{-1}}{f^{-1}-D^{-1}}\right|$ pixels in the geometric optics limit, where a is the camera aperture, p is the pixel size, f is the focal length, and D is the in-focus depth.

4 Viewpoint Search

Given the viewpoint assessment function, we need to optimize it to find the best available viewpoint for the task. Several features of the assessment function combine to make it difficult to optimize:

- It is relatively expensive owing to the large amount of geometric computation required for kinematics, image prediction, and particularly collision detection. The search method must make the most of each function evaluation, even if this involves a substantial amount of subsidiary computation.
- It is highly nonlinear with sudden 'cliffs' where hard constraints switch on and relatively shallow local minima lying hard up against constraint boundaries. The search method must be robust as this type of behaviour will almost certainly cause problems for any method that makes strong smoothness assumptions.
- There are often several distinct feasible regions because occlusion and accessibility constraints frequently cut the workspace in two. For reliability, the search method must be as global as possible: local viewpoint searches *do* in practice get caught in local minima.
- The search dimension is potentially quite high (*e.g.*, six for a search over the full space of 3D camera poses or robot configurations). This is sufficiently high to preclude simple ground covering methods such as grid search. Any search method needs to be well focused, and if a global method is used the search dimension probably needs to be reduced.

On balance we feel that a well-directed global search over a reduced-dimension space is most appropriate. In fact, even for local methods we have preferred to restrict the search to the 3D space of accessible upright camera poses directed towards the task centre, in order to guarantee upright, centred images. Global searches of this 3D space turn out to be quite feasible, and as far as we can tell relatively few good viewpoints are lost by restricting attention to it.

A further potential advantage of global methods is that if there are several minima of similar quality, they can output all of them and leave the choice to the user. Such backtracking might substantially improve the robustness of a system based on the viewpoint planner.

In our implementation we have hedged our bets by providing a choice of three different optimization routines, which can be used singly or in combination: (*i*) maximum expected gain (a quasi-optimal global region-based search technique); (*ii*) simulated anneal-

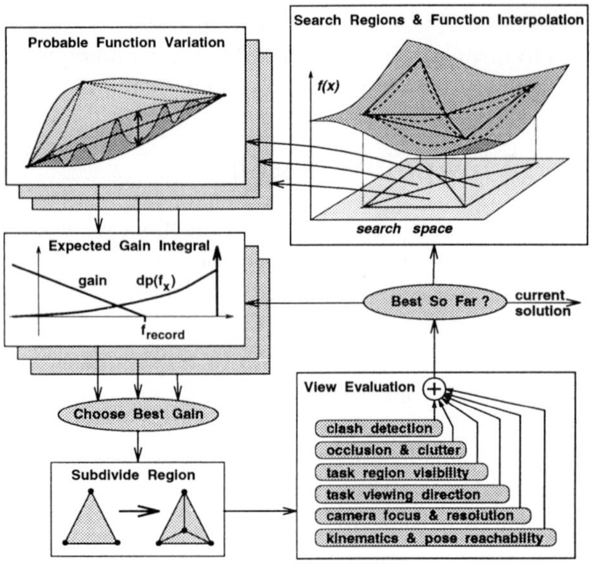

Figure 2: The maximum expected gain search loop.

ing; and (*iii*) Powell's method (a classical smooth function minimization technique). We will concentrate on the global method as we feel that it has the best overall performance, with brief notes on the others for comparison.

4.1 Maximum Expected Gain Search

To meet the above requirements we have developed a new approach to global function optimization called **maximum expected gain** search. This is a region-based recursive decomposition technique that uses decision theory and subjective-probabilistic models of function behaviour to choose a quasi-optimal location for each function sample. Its key advantage is that it allows knowledge or expectations about likely function behaviour to be encoded in a form directly adapted to search control, so that search effort can be focused very precisely on the regions where it is likely to prove most profitable. The basic search loop is illustrated in fig. 2. More details will appear in [20].

4.1.1 General Approach

To understand the method, imagine that you're trying to minimize a function $f(\mathbf{x})$ over some domain, and that you've already evaluated it at several points $\mathbf{x}_1, \ldots, \mathbf{x}_n$. What's the best place to try next? — In general it's impossible to be sure and one has to guess. The only information relevant to this guessing is the values one thinks the function is *likely* to have at different points. These are constrained by general prior information about function behaviour such as continuity, smoothness, and boundedness, and (via continuity) by the known function values at existing nearby sample points.

Taking a Bayesian approach, the most direct way to encode such information is as a **subjective probability distribution** for the function value $f_{\mathbf{x}}$ at $\mathbf{x}$ given the samples $f_{\mathbf{x}_1}, \ldots, f_{\mathbf{x}_n}$ and the prior information:

$$\mathrm{dp}(f_{\mathbf{x}}) \;=\; \mathrm{dp}(f_{\mathbf{x}} \,|\, \mathbf{x}, f_{\mathbf{x}_1}, \ldots, f_{\mathbf{x}_n}, \text{prior-information})$$

This is a powerful technique because *any* information relevant to optimization can be encoded. Bounds on the function and its derivatives limit the support of the distributions, while softer estimates of typical behaviour control their shape. As a general rule, near an existing sample point $\mathbf{x}_i$ the subjective probabilities will be sharply peaked near $f_{\mathbf{x}_i}$, while further away they will grow increasingly diffuse as the function value becomes more and more uncertain. The amount of spread will depend principally on the distance moved and the expected smoothness of the function.

Of course, such subjective distributions have no existence in objective reality: they simply reflect our ignorance or laziness in refusing to evaluate the function exactly. For search guidance, they are useful only to the extent that they are both (*i*) tractable and (*ii*) accurate estimates of the true function behaviour. Their exact forms are entirely subjective and can only be specified heuristically according to intuition or computational convenience.

Now we can answer our question about where to look next. In general, points that have a high probability of having a good function value will be good places to look. But we can be much more specific. At any given point in the search there will be some *best currently known function value* or **current record** f_{record}. The aim of optimization is to find the best possible record in the fewest possible function evaluations. Evaluations that do not improve the current record are in some sense 'wasted'. This suggests that the best measure of the prospective value of a trial point is the **expected improvement in the current record** anticipated from evaluating the function at that point. We will call this improvement the **gain** and advocate a search régime that greedily selects the point with the **maximum expected gain** for evaluation at each step.

The expected gain is given by an expectation value integral:

$$\langle \text{gain}_{\mathbf{x}} \rangle \;=\; \int_{-\infty}^{f_{\text{record}}} (f_{\text{record}} - f)\; \mathrm{dp}(f_{\mathbf{x}} \,|\, \mathbf{x}, \ldots)$$

Note that the result depends only on the part of the distribution below the current record. As the record improves, the integral gets pushed further and further out into the tail of the distribution. For this reason, we will concentrate mainly on capturing the form of the asymptotic tails of the subjective function value distributions.

The current record and the subjective value distributions both evolve as samples are added, so the implementation will need to keep track of the changes in the expected gains. Although this can represent a significant overhead, in many practical cases it is amply justified by a dramatic increase in search efficiency because the *absolutely most promising point* (within the limits of the current information and the subjective models) is investigated at each step. As the record improves, the expected gain of any given point decreases and relatively unpromising points become less and less likely to contribute: hence the search is well focused. On the other hand, once the most promising regions have been thoroughly investigated, the method turns increasingly towards verifying that apparently less promising but less well investigated regions do not contain large unexpected fluctuations that create islands of good function values: hence the search is complete.

As it stands, the above method is not practical: (*i*) it is hard to tie down the forms of the subjective distributions, which grow increasingly complex as more samples are added; (*ii*) it is difficult to optimize the gain integral globally over the continuum of values $\mathbf{x}$ (after all, this is the sort of problem we were trying to solve in the first place!).

Both of these problems can be tackled by making a major simplifying assumption. We will suppose that: (*i*) at each search step the search domain can be subdivided into a set of non-overlapping **regions of influence**, each 'under the influence of' an associated set of **control points** chosen from among the existing function samples; and (*ii*) the function value distributions in each region can be approximated by ones depending *only* on the region's controlling samples. In other words, for all $\mathbf{x}$ in the region controlled by samples $\mathbf{y}, \ldots, \mathbf{z}$:

$$\mathrm{dp}(f_{\mathbf{x}}) \approx \mathrm{dp}(f_{\mathbf{x}} | \mathbf{x}, f_{\mathbf{y}}, \ldots, f_{\mathbf{z}}, \text{prior-information})$$

This should be a reasonable subjective model provided the sets of control points can be chosen to be sufficiently simple to make the model tractable, while still containing the samples that have subjectively significant influences on the function values in each region. Usually, the control points will be (a subset of) the samples nearest to the region.

This assumption simplifies the problem enormously because the forms of the function value distributions can be specified parametrically once and for all for a generic influence region, and then instantiated for particular regions on the fly. Even more significantly, given a sufficiently simple form for the value distributions, the location and gain of the most promising point in the generic region can be evaluated (or more commonly estimated) once and for all as a function of the region parameters and the current record. This reduces the difficult problem of ranking an infinite number of potential sample points to the much simpler one of ranking a finite number of regions according to the *best potential sample point* each contains.

4.1.2 Overview of Implementation

At this point we can sketch the implementation. The central data structure is a priority queue of records describing influence regions. The user must supply a region evaluation routine that determines a (quasi-) optimal sample location in the region and its expected gain with respect to the current record, presumably by using the subjective function value model outlined above.

For efficiency it is essential to postpone sorting and evaluation work for as long as possible, so an optimal, lazy queue implementation is required. We have used a version of Fredman and Sedgewick's elegant pairing heap queue [10]. This is ideal for our application as it is efficient (amortized $\mathcal{O}(\log n)$ updates), extremely simple to implement, and naturally lazy in that the minimum possible amount of sorting is always performed at the last possible moment.

To reduce the evaluation workload, regions are ranked in the queue according to *possibly out-of-date* expected gains. At each search step, regions are repeatedly pulled off the head of the queue, re-evaluated with respect to the current record, and reinserted, until a region with an up-to-date expected gain is found. This region has the best expected gain in the queue even with respect to the *current* record because gains can only decrease as the record improves. The function is evaluated at the chosen location(s) for the selected region and the region decomposition is updated, with deleted regions being deleted from the queue and newly created ones evaluated and inserted. Often, the update will simply be a fixed subdivision of the selected region.

The process continues until a sufficiently good sample has been found or a sufficiently thorough search has been made. The running gain provides a rough but quantitative guide to the improvement further search is likely to yield. Expected gains can also be replaced by 'expected profits' (*i.e.* expected gain less the cost of a further search step), in which case search can continue until the expected profit is zero.

4.1.3 Region Models

To implement the method, forms must be chosen for the influence regions and the subjective distributions. We have investigated two main types of region decomposition: 2^d-trees and Delaunay triangulations. The

Delaunay model allows slightly better search focusing and is perhaps more refined, but the 2^d-tree model is less complicated, numerically more robust and has a much lower overhead in geometric computation. Both models are strictly 'local', with function samples at each region vertex acting as the controlling vertices of the region. It would also be interesting to investigate slightly less local influence models such as k-nearest-neighbour regions.

In the 2^d-tree model the search arena is subdivided into a set of non-overlapping hypercubes that form the leaves of a 2^d-tree (*e.g.* an oct-tree for our 3D search problem)[4]. There is a function sample at every cube vertex. At each search step the most promising region is removed from the queue and bisected along each coordinate to produce 2^d child cubes, which replace the parent in the queue. The model does not quite fit the above general paradigm in that several function samples may be created at each step, but the only change needed is to hypothesize a subjective distribution for the *best* value among all the new function samples when estimating the expected gain integrals. The fact that a single subdivision may create as many as $3^d - 2^d$ new vertices (and hence function samples) is a potential source of inefficiency, however in practice the amount of oversampling is usually small: away from the boundary there are never more than $(3/2)^d - 1$ samples per region, and if the subdivision is locally fairly uniform (as is often the case) the true number is usually much closer to 1.

In simplex-based models like the Delaunay one, the regions form a simplicial complex with a function sample at each vertex. At each step, the most promising region is selected, a single new function sample is created at a chosen location inside it, and the triangulation is updated to include the new vertex. To ensure that the simplices remain fairly uniform (and hence provide good control over the function values in their interior), the update process must have a strong tendency to break up large facets, especially edges. Delaunay triangulations [3] are interesting in this respect because their geometric properties ensure that the simplices are locally as well-shaped as possible. To preserve the Delaunay property, the amortized $\mathcal{O}(\log n)$ incremental Delaunay vertex insertion process must retriangulate the 'conflict neighbourhood' of each newly inserted vertex.

The principal advantages of the Delaunay model over the 2^d-tree one are:

- Only one new sample is created at each step, so it is potentially very economical in function evaluations.
- The sample can be located anywhere in the simplex, so there is maximal scope for effective sample placement and the method can be used with pre-existing samples, ones created by local search strategies, *etc.*
- The initial search region can be an arbitrary convex polytope rather than simply a hypercube, so less time is spent locating the feasible region during the initial phases of search.

The disadvantages of the Delaunay model are:

- It is geometrically more complex and significantly harder to implement.
- The computational cost of the retriangulation process can easily negate any extra efficiency the technique may have, so it is probably only suitable for fairly expensive evaluation functions.
- Round-off error tends to make retriangulation rather delicate, particularly in high dimensions. In practice, the sample placement heuristic must be chosen quite carefully to avoid numerical problems. It is also essential to make the code robust to vertex insertions on simplex facets (including external ones).

In the current viewpoint planner we have preferred the 2^d-tree for its simplicity and robustness, but we have experimented extensively with both and find their performance to be comparable for viewpoint assessment. With a more expensive evaluation function the Delaunay model would probably have the edge.

4.1.4 Models of Function Variation

The final component needed for the implementation is a plausible model of probable function behaviour for each region. The precise form this takes is necessarily very subjective, but to be useful for search focusing it must reflect the likely influence of existing nearby function samples, region size and shape, and prior constraints such as positivity or smoothness on function values in the region. It must also be sufficiently tractable to allow the most promising sample point to be estimated, and the expected gain integral to be evaluated.

The models we have used are based on "probabilistic Lipschitz bounds" on function variation. Classical Lipschitz bounds limit the maximum change in a function $f(\cdot)$ between any two points of a region to the form $|f(\mathbf{x}) - f(\mathbf{y})| < K\|\mathbf{x} - \mathbf{y}\|^\nu$, where K and ν are positive constants and $\|\cdot\|$ is some positive metric. Here, we will view this formula as a heuristic subjective estimate of the *probable* amount of variation in a region of a given size rather than as a strict bound. Specifically, we will suppose that the function variation Δf across a region of size $\Delta\mathbf{x}$ has a subjective probability distribution of the form

$$\mathrm{dp}(\Delta f) \;\approx\; \rho(\mu\, \Delta f / \|\Delta\mathbf{x}\|^\nu)$$

[4] To reduce the amount of pointer chasing, we do not not actually store the tree structure. Vertices are accessed by hashing their coordinates.

where μ and ν are constants characteristic of $f(\cdot)$ and $\rho(\cdot)$ is some fixed probability distribution. This is nothing more than a simple heuristic model aimed at capturing the intuition that large variations become less likely as the region gets smaller.

Without further information, the choice of $\rho(\cdot)$ is necessarily rather arbitrary and must be made experimentally. This is unfortunate as the key parameters for search focusing are the exponent ν and the asymptotic form of the tail of $\rho(\cdot)$, since it is these that govern how quickly small regions with relatively poor function samples will be discounted from the search. For simplicity we have based $\rho(\cdot)$ on an exponential distribution in most of our experiments, although a power law or a Gaussian might have been equally plausible choices.

However none of the above distributions can be used without modification. They all have tails stretching to $-\infty$, whereas we know that our viewpoint assessment function is bounded below by zero. In fact, across particular regions it is often possible to find a strictly positive lower bound for the function. When available, such bound information can be extremely valuable: it puts strong limits on the subjective function value distributions in the region, and hence allows the prospective gain to be estimated much more tightly, and it allows regions whose lower bound is greater than the current record to be rejected outright. In our case, lower bounds turn out to be especially important for rejecting regions that contain no feasible viewpoints, but they are also used to discount feasible regions in relatively poor areas of the search space. Again, the method used to incorporate the bound information into the subjective distributions is essentially arbitrary. At present we simply truncate the distributions at the bound value, although we have also experimented with smoother types of cut-off.

Putting all of the above together, we can outline the models adopted for sample point choice and expected gain evaluation in the viewpoint planner. Instead of applying the Lipschitz model directly to the viewpoint assessment function, we apply it to the *deviation* of the function from a fiducial interpolation passing through the existing samples at the region vertices. In the Delaunay case, this is simply the unique linear interpolant of the function through the simplex vertices, while in the 2^d-tree case it is the corresponding d^{th}-order tensor-product spline[5].

In the Lipschitz model, the expected deviation scale $|\Delta f|$ increases monotonically and isotropically as the point moves away from any existing sample vertex. Hence, the point of the region with the maximum scope for deviation is always the point maximally distant from all the vertices. In the case of the 2^d-tree model this is the centre of the cube, while in the Delaunay case it is the circumcentre of the simplex (the centre of the unique sphere passing through the vertices) — provided this lies within the simplex.

For finite records, this central point is usually *not* the point with the maximum expected gain: some point further 'downhill' (towards the best sample vertex) is usually better. However in general the best point turns out to be quite hard to estimate. Since the asymptotic search complexity is governed mainly by the asymptotic rate at which small regions far above the current record are discounted, we have concentrated on getting the model right in this limit. And in this limit the central point *is* optimal, because the large deviation required to better the record completely swamps any gradient term in the fiducial interpolation.

Hence, for simplicity and (at least) asymptotic efficiency, we always place the next function sample at the centre of a region. The expected gain of the region is then estimated as the expected gain at this point, based on the Lipschitz deviation model with an exponential characteristic distribution truncated at the best lower bound we have for the function on the region.

4.1.5 Discussion

In practice the maximum expected gain viewpoint search method works very well. It runs as quickly as the local methods we have implemented and seems to converge reliably to the globally optimal viewpoint. Despite the very heuristic nature of the probabilistic models used to derive it, it seems robust and easy to tune. (For example it performs well with any Lipschitz exponent ν between about 0.5 and 1.0 and any scale factor μ within a factor of 2–4 of the optimum). Of course, it *can* be badly tuned, but in such cases it either runs slowly but reliably (like a global grid search), or quickly converges to a (possibly local) minimum like a local search technique. The distinction between global and local search may be muddied, but at least the failure modes are relatively benign.

Perhaps the main disadvantage of the method is its relative complexity. Although the underlying ideas are simple there are quite a few pieces to implement. Another problem is termination: as with other multidimensional search methods it is difficult to know when to stop searching. In theory the expected gains can be used as a guide, but in practice they decay quite slowly and irregularly[6], so that simple thresholding is not a reliable stopping criterion. Currently we just search for

[5] If the cube coordinates are normalized to $[0,1]^d$ and $\phi(x,q)$ is x for $q=1$ and $1-x$ for $q=0$, this is $\sum f(\mathbf{q}) \cdot \prod_{j=1}^{d} \phi(x_j, q_j)$, where the sum is over the 2^d cube vertices $q_i = \{0,1\}, i=1,\ldots,d$.

[6] For any one region the expected gain decays monotonically, but new regions are created and destroyed all the time.

a fixed number of samples — hardly an optimal policy. The Lipschitz scaling model for expected function variation could also be improved to allow slightly more control over the search profile, so that (for example) the search could be practically global down to a certain resolution, and then become rapidly local after that.

4.2 Local Search Methods

For comparison with the above global method we have implemented two local search techniques: Powell's direction set method and simplex-based simulated annealing, both derived from (but not identical to) routines from Numerical Recipes [16].

Powell's method (as implemented in [16]) is a quasi-quadratically convergent method designed for smooth functions, that does not require function derivatives. It works by repeated line searches along an evolving set of approximately conjugate directions. We had hoped to use it to 'polish' coarse solutions found by the global search, but its performance has proved disappointing in our application: it was not designed for functions with sudden steps and tends to 'stick' on constraint walls rather than sliding along them. By softening the constraints slightly (*e.g.* by penalizing poses close to the robot joint limits) it can be made to converge, but it is neither so reliable nor so fast as the global technique.

By contrast, the simplex-based annealing method makes very few assumptions about function smoothness. It works by applying the simulated annealing acceptance criterion to trials, each of which flips, stretches or shrinks a simplex of trial solutions. Annealing methods are often considered to be quasi-global, but this depends entirely on the annealing schedule. In our case, for realistic schedules, the method is perforce rather local. Certainly, if it is started in a suboptimal feasible region there is no significant chance of its tunnelling through an infeasible barrier to a better feasible one.

In practice the annealing method performs very respectably for the most part, although it *does* sometimes get caught in local minima. Apart from these cases, it can be tuned to give similar results to the global method for similar amounts of computation, so the simplex-based trial generator seems to be making quite effective use of function evaluations (a traditional weak point of annealing methods). On the whole we tend to prefer the global method because of its greater reliability, but the performance and relative simplicity of the annealing method make it an attractive second choice.

5 Tour Planning

The tour planner is responsible for creating a feasible overall plan for a sequence of visual tasks, starting from an unordered set of viewpoints (camera robot poses) produced by the viewpoint planner. It must select an efficient ordering of the tasks and ensure that collision-free paths can be found to move the camera robot through the sequence of viewing positions. Nothing about it is specific to the viewpoint problem: it simply plans a low-cost tour visiting an arbitrary list of n 'sites' (robot poses).

An underlying path planner is needed to plan segments of the tour. We are currently using a potential field based method [9, 14] that is fast but relatively unreliable. Planning is intrinsically costly and is likely to dominate the run time relative to simpler operations such as graph manipulation in any practical implementation, so we will measure the cost of tour planning by counting planner calls. The minimum possible cost is $\mathcal{O}(n)$ as the tour visits n sites.

The classical tour planning problem is the Travelling Salesman problem: find a minimal cost tour that visits each vertex of a completely connected graph exactly once. This is NP hard as an exact problem, and hence intractable even for moderate n. However for many graphs (notably those with costs based on geometric proximity) good approximate solutions can be found reliably in $\mathcal{O}(n^2)$ using simulated annealing [12, 15, 16].

On the other hand, a direct travelling salesman attack on our problem would require a complete matrix of edge (robot path) costs between each pair of sites in the tour-graph. These would have to be produced by $\mathcal{O}(n^2)$ calls to the underlying path planner, which would make the method intolerably slow relative to the $\mathcal{O}(n)$ output. Although it is not possible to avoid $\mathcal{O}(n^2)$ calls in the worst case, we would like to find a method closer to '$\mathcal{O}(n)$ in favourable cases', even if this involves a considerable cost in graph manipulations.

One simple way to achieve this is to construct candidate tours using *estimated* path costs, and then to attempt to verify them using the path planner. If the planner fails an alternative tour has to be planned with improved cost estimates, but 'in favourable cases' the method will hopefully converge within a few cycles.

Another point is that although the tour must visit each site at least once, we do not really want to insist on its visiting each site *exactly* once. In fact, given the unreliable nature of our path planner this would often make the problem insoluble: a site connected to only one other could have no tour passing through it. The tour planner should fail only when it is impossible to connect the site-graph using the local path planner, although in difficult cases it may generate long tours with many detours around missing graph edges.

This suggests that we should base our candidate tours on estimated inter-site *route* costs for travel via any route through the graph, rather than simply on estimated *edge* costs for direct (path planner) paths between sites. Such route costs can be evaluated in $\mathcal{O}(n^2 \log n)$ using Dijkstra's algorithm [17], and updated in the same worst case time (but in practice often much more quickly) using an incremental version of this algorithm. The use of route costs also allows us to plan with respect to a background route map (perhaps built up over previous calls to the planner) and hence to speed up planning and enhance reliability.

Putting all of this together, the complete tour planning algorithm goes as follows:

0) Add the tour sites to the route map graph.

1) For each pair of sites in the graph, estimate the edge (path planner path) cost between them as either: (*i*) the true path cost if a path has been planned (∞ if the planner failed); or (*ii*) an estimate of the probable path cost based on 'straight line distance in C-space' or some similar lower bound on path cost, inflated by some small factor k.

2) Find the estimated minimal route cost between each pair of sites in the graph using the Dijkstra algorithm (or the incremental one if only a few edge costs have changed).

3) Find a trial tour of the sites based on the estimated route costs, using the simulated annealing travelling salesman algorithm.

4) If the tour cost is ∞, fail. (The tour contains some edge for which the path planner failed, which indicates that the corresponding state is unreachable).

5) For each edge of the tour for which a plan has not yet been attempted, plan a path. If any plan fails or its cost is higher than the estimate, abandon the trial tour and go back to step 1.

6) Output the current tour and exit.

An important enhancement to step 5 is that if the graph has not yet been connected we only attempt to plan edges leading to currently unconnected sites: tour feasibility (graph connection) is initially a higher priority than tour optimality.

The only heuristic parameter in the method is the cost estimate inflation factor k used in step 1. It trades route optimality for planning effort. In essence, routes (or at least those selected for tours) are replanned until the method *proves* that the ones it has are within a factor k of the shortest possible ones in the completely path-planned graph. With $k = 1$ the behaviour is akin to that of A^* and guaranteed shortest routes are found, while with larger k there is a tendency to reuse existing routes, even if some untried ones may be slightly shorter. The final tour is within a factor $k\delta$ of the true minimal one, where δ is the cost overrun in the travelling salesman solution (difficult to estimate, but often no more than a few percent).

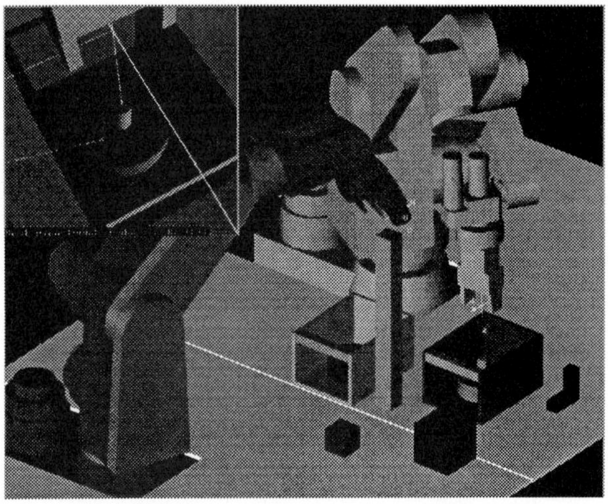

Figure 3: A planned camera placement, and (inset) the corresponding predicted image. The task was to get a clear view of the central peg in the right hand box.

In practice the algorithm works well and usually succeeds or proves failure within a few n planner calls, despite the unreliability of the local path planner. The tours it generates seem subjectively good, although they often have to make detours around edges whose path plans have failed. If there are a few unreachable sites this is usually detected within a few n planner calls, although the worst case run time is certainly $\mathcal{O}(n^2)$, for example when failing exhaustively to connect two disconnected $\mathcal{O}(n)$-site cliques.

It must also be emphasized that the '$\mathcal{O}(n)$ favourable case' run time has been bought at a high worst-case cost in graph operations: at each of the $\mathcal{O}(n^2)$ worst-case iterations there can be an $\mathcal{O}(n^2 \log n)$ route cost update and an $\mathcal{O}(n^2)$ travelling salesman problem. However order of magnitude run time estimates suggest that in our case the trade-off is amply justified out to $n \approx 10^3$, and probably far beyond.

One potential weakness of the current implementation is that there is no feedback from the tour planner to the choice of viewpoints. If a self-consistent but inaccessible viewpoint is chosen, the entire task plan will fail. Feedback would presumably help to reduce the problem, but we have not included any as it was not clear what form it should take.

6 Implementation & Experiments

The current version of our system contains about 10,000 lines of fairly modular C code and runs on Silicon Graphics hardware under the robot modeller ACT [14]. A typical application is shown in fig. 1. The problem is loosely based on a routine inspection task in the maintenance of steam generators in nuclear power plants. The area around the bases of the columns must be examined for damage or detritus. The task is formulated as a request to view a series of overlapping disk-shaped areas covering the region. (For clarity we have only included areas hidden behind columns, since these are the hardest to get a clear view of). The current robot can not quite reach the ideal viewpoints for the most difficult areas, so a compromise solution is found. Some areas can be partially seen from either side of their column, so the search space has several local minima. The planned sequence of views is shown in the lower panel. In each case the viewpoint was found by a global search over the entire reachable space of upright camera poses directed towards the centre of the area to be viewed. The tour planner has reordered the views so that left-looking ones precede right-looking ones, as the left-right reorientation is relatively expensive for the robot. On an SGI R4000 machine each viewpoint search takes about 4 seconds. Planning a robot path between viewpoints takes 1–2 seconds, so the total run time is about 5–10 seconds per viewpoint.

Another problem is illustrated in fig. 3. The task was simply to get a clear view of the central peg in the right hand box. Again occlusion cuts the search space in two so the problem has several deep local minima. The final pose is very near the limit of the robot's reach. Fig. 4 shows a grasp approach being visually servoed from a viewpoint chosen by the first version of the planner.

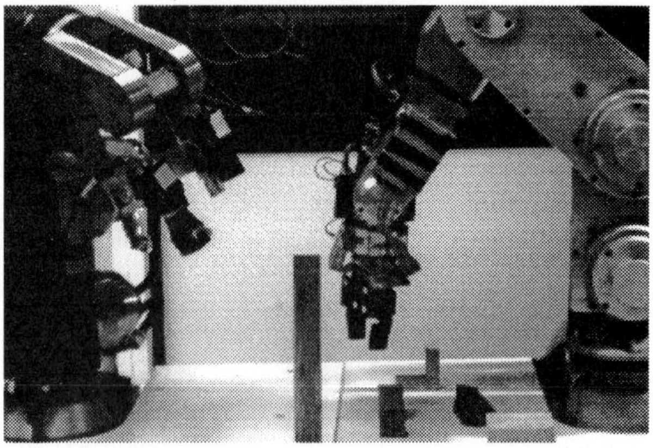

Figure 4: Visually servoing a grasp approach from a planned viewpoint.

7 Discussion

Overall we are pleased with the viewpoint planner's performance. Despite the large number of parameters involved, the viewpoint assessment and search routines have proved to run reliably with very little tuning. The global search seems robust and surprisingly efficient. Most importantly, the viewpoints selected really do seem to correspond to ones a person might choose as the subjectively "best available camera placements".

The tour planner also performs reasonably well, within the limits of the underlying path planner. For the simple problems we have tried, either a tour is found or one of the sites is shown to be unreachable (according to the path planner), within a few times n (the number of sites) planner calls. However even for 'simple' problems the current local path planner fails fairly often, so that tours frequently need to make long detours around the gaps left by its failures.

As with any large system, there are many minor improvements that could be made, and several major limitations. As it stands, the whole system is rather too static and model based. Real images are never used. Plans are based entirely on internal geometric models that may not be an accurate reflection of the true geometric environment, let alone the true visual one. No attempt is made to model important visual effects such as shadowing and specularities. Any real system is likely to need the ability to compensate for such unmodelled effects by making on-line adjustments to a planned viewpoint, so some means of feeding visual information back into the viewpoint selection process would be desirable.

Similarly, there is currently no feedback from the tour planner to the viewpoint selection process: a viewpoint is chosen once and for all for each task, and then the path planner either succeeds or fails to find paths connecting these viewpoints. The whole process may fail when there was a perfectly acceptable alternative viewpoint for which it would have succeeded.

Another major omission is the ability to deal with time constraints and moving targets. This would be useful for visual servoing and process monitoring applications. It probably lies more in the domain of active vision than visual planning, however there is a real (and difficult) motion planning problem too: generate a task plan that optimizes a complex continuous criterion (sustained view quality), in a workspace encumbered with hard constraints (joint limits, collisions, time constraints…). The difficulties are: (*i*) the size of the search space; (*ii*) the fact that there may be points at which the camera has to lose sight of the task in order to pass an obstacle or reorient itself, so that the search has a mixed continuous/discrete character.

Finally, the current user interface needs improvement. In particular, it would be useful to have a more efficient means of specifying repetitive visual tasks such as seam or surface inspection. At present these must be discretized and programmed by hand.

As far as future work is concerned, our immediate priorities for the system are improving the user interface (especially the task-specification aspects), the integration with visual servoing, and the reliability of the path planner. In the longer term we are interested in incorporating visual feedback into the plans and in planning sustained visual tasks such as visual tracking in cluttered environments.

8 Summary

We have described an automated planner for visual inspection and monitoring tasks using a robot mounted CCD camera. It is optimized for intervention-style applications with large potentially cluttered workspaces, and therefore needs to take account of workspace constraints on the camera-carrying robot as well as task constraints and more traditional visual indices such as field of view, focus, resolution and occlusion. It is capable of finding heuristically near-optimal viewing positions for a series of visual tasks, and of ordering them and planning robot paths between them to produce a complete, near-minimal-cost overall task plan for the sequence.

The guiding principle of the design was to aim for reliability by using quantitative, physically-based measures of quality wherever possible, and by using global search to ensure that possible solutions are not overlooked. As a means to this, we have developed a novel global function optimization technique based on decision theory and subjective models of function behaviour, that allows global viewpoint searches to run in the time usually required for local ones.

References

[1] O. Al-Chami. *Contribution à l'intégration Robotic-Vision en Manipulation Automatisée: Modélisation de la Tâche, Placement d'une Caméra Mobile et Localisation Fine d'Objet*. PhD thesis, LIFIA, Grenoble, 1994.

[2] O. Al-Chami and C. Laugier. Stratégie perceptive pour positionner une caméra. In *Proc. AFCET Reconnaisance des Formes et Intélligence Artificiel*, volume 1, pages 617–22, 1994.

[3] F. Aurenhammer. Voronoi diagrams — a survey of a fundamental geometric data structure. *ACM Computing Surveys*, 23(3):345–405, 1991.

[4] C. Bard, C. Bellier, J. Troccaz, C. Laugier, B. Triggs, and G. Vercelli. Achieving dextrous grasping by integrating planning and vision based sensing. *Int. J. Robotics Research*. Accepted, to appear in 1995.

[5] A. Cameron and H. Durrant-Whyte. Optimal sensor placement. *Int. J. Robotics Research*, 9(3), 1990.

[6] C. K. Cowan and A. Bergman. Determining the camera and light source location for a visual task. In *IEEE Int. Conf. Robotics & Automation*, pages 509–14, 1989.

[7] C. K. Cowan and P. D. Kovesi. Automatic sensor placement from vision task requirements. *IEEE Trans. Pattern Analysis & Machine Intelligence*, 10(3):407–16, May 1988.

[8] B. Espiau, P. Chaumette, and P. Rives. A new approach to visual servoing in robotics. *IEEE Trans. Roboics & Automation*, 8(3):313–26, June 1992.

[9] B. Faverjon and P. Tournassaud. A local based approach for path planning manipulators with a high number of degrees of freedom. In *IEEE Int. Conf. Robotics & Automation*, Raleigh, March 1987.

[10] D. Fredman and R. Sedgewick *et.al.* The pairing heap: A new form of self-adjusting heap. *Algorithmica*, 1:111–29, 1986.

[11] R. Horaud, F. Dornaika, C. Bard, and C. Laugier. Integrating grasp planning and visual servoing for automatic grasping. In 4th *Int. Symposium on Experimental Robotics*, Stanford, 1995.

[12] S. Kirkpatrick, C. D. Gelatt, and M. P. Vecchi. *Science*, 220:671–80, 1983.

[13] J. M. Manyika and H. F. Durrant-Whyte. *Data Fusion and Sensor Management*. Ellis Horwood, 1994.

[14] E. Mazer, J. Troccaz, and *et.al.* ACT: A robot programming environment. In *IEEE Int. Conf. Robotics & Automation*, pages 1427–32, Sacramento, California, April 1991.

[15] R. H. J. M. Otten and L. P. P. P. van Ginneken. *The Annealing Algorithm*. Kluwer, Boston, 1989.

[16] W. H. Press, S. A. Teukolsky, W. T. Vetterling, and B. P. Flannery. *Numerical Recipes in C: The Art of Scientific Computing*. Cambridge University Press, 1992. 2nd edition.

[17] R. Sedgewick. *Algorithms*. Addison-Wesley, Reading, MA, 1988.

[18] K. Tarabanis and R. Y. Tsai. Sensor planning for robotic vision. In O. Khatib, J. Craig, and T. Lozano-Pérez, editors, *Robotics Research 2*. MIT Press, 1992.

[19] K. Tarabanis, R. Y. Tsai, and P. K. Allen. Automated sensor planning for robotic vision tasks. In *IEEE Int. Conf. Robotics & Automation*, pages 76–82, 1991.

[20] B. Triggs. Global optimization using subjective models of function behaviour. Technical report, INRIA Rhône-Alpes, Grenoble, France, 1995. (To appear).

[21] B. Triggs and C. Laugier. Automatic camera placement for robot vision tasks. In *IEEE Int. Conf. Robotics & Automation*, pages 1732–7, Nagoya, 1995.

[22] S. Yi, R. M. Haralick, and L. G. Shapiro. Automatic sensor and light source positioning for machine vision. In *10th Int. Conf. Pattern Recognition*, pages 55–9, June 1990.

Object Tracking by Using Optical Flows and Edges

Yoshiaki Shirai, Yasushi Mae, and Shin'ya Yamamoto
Dept. of Computer-Controlled Machinery, Faculty of Engineering, Osaka University
Yamadaoka, Suita, Osaka, 565, Japan
email:shirai@ccm.osaka-u.ac.jp

Abstract

This paper describes real time object tracking in a complex background by extracting optical flows, and integration of optical flows and contrast. In real time tracking, optical flows are calculated based on the generalized gradient model. The system is a linear array processor capable of processing directly TV images in real time. Our system detects a moving object, and tracks it from the optical flow data. The system controls a camera head to follow the object. Although the background moves, the movement of the object is discriminated from the background.

If two objects should be tracked simultaneously, each object is tracked simultaneously. When two objects overlap, foreground object is recognized, and each object is tracked without confusion.

If an object stops or overlaps with another object with a similar motion, edged extracted from an intensity image is used together with optical flows. While the object is moving, the contour is extracted using optical flows and the contrast. When the object can not be tracked only with optical flows, the edge information is used to determine the the object contour.

1 Introduction

Object tracking has been a popular research theme. Early research tried to extract objects in a static image using the difference of properties of the objects from those of the background. Later, motion itself is used to extract objects.

Calculation of the optical flow from image sequences is essential for the latter approach. Most methods try to eliminate errors caused by noises, disregarding the computational cost. Horn and Schunck's method [1], for example, requires iterations for regularization.

For a limited number of specified points in a image the optical flow can be extracted by correlation. Tracking of road boundaries [2], for example, is actually extracting the motion of the boundaries by correlation using multiple transputers. Special hardware is developed for computing optical flows by correlation and for segmenting the flows [3]. In Oxford University, a reactive camera head is developed Murray 93 which extract a moving object and track it while controlling pan and tilt. One problem with the correlation method is the computation cost, which make it difficult to extract dense flows. A powerful chip has been employed to compute correlations of many points (about a hundred) [5]. However, it is not enough because we need about 10,000 flow vectors to achieve motion segmentation.

We are developing a real-time optical flow processor to calculate the optical flow based on a generalized gradient model [6] ,[7], which measures spatial and temporal intensity gradients. The algorithm is implemented easily on a pipeline processor.

Real-time extraction of a moving object has been realized for a scene with only one moving object. Extraction of multiple moving objects in a stationary background is also performed by clustering optical flows [3]. It is not easy to extract multiple moving objects in a moving background because much computation is required [8], [9].

Section two describes real time object tracking, assuming that the movement of an object is different from the other object (including the background) . Section three describes tracking of two objects simultaneously by assigning each object to a different processor to track the object independently. When two objects overlap, the two processors communicate to determine which is the foreground object, and to track each object without confusion.

Section four describes a more difficult case where an object stops in the stationary background or it overlaps with another object with a similar motion. Since the optical flow alone is not enough to segment the object, the contrast of light intensity is also used to find the object contour.

There are some works for integrating motion and other properties: optical flows and color [10], or optical flows and edges [11] An important problem is that an object is hardly detected even with integrated properties at some instances. To solve this problem the contour is extracted, while the object is moving, using optical flows and the edges. When the object can not be tracked only with optical flows, the stored contour and the extracted edges are used to determine the object contour.

2 Extraction of Optical Flow

The optical flow extraction is based on the generalized gradient model. In a typical gradient method the relation between a sequence of two-dimensional image, $f(x, y, t)$, and the velocity u and v at (x, y, t) is represented by the following equation

$$u \cdot f_x + v \cdot f_y + f_t = 0$$

where subscripts denote partial differentiation.

Orientation-selective spatial Gaussian filters are used to obtain the constraint equations. We apply two orientation-selective spatial filters to the original image: one is sensitive to vertical edges, the other to horizontal edges.

Let $g(x, y)$ and $h(x, y)$ denote the two filters and $g \otimes f$ denote the convolution of $f(x, y, t)$ and g. The following constraint equations are given.

$$u \cdot g \otimes f_x + v \cdot g \otimes f_y + g \otimes f_t = 0$$
$$u \cdot h \otimes f_x + v \cdot h \otimes f_y + h \otimes f_t = 0$$

In order to decrease the effect of noise, a temporal filter $p(t)$ is applied for smoothing. The equations above are modified by convolving the temporal filter.

$$u \cdot g_x \otimes f \otimes p + v \cdot g_y \otimes f \otimes p + g \otimes f \otimes p_t = 0$$
$$u \cdot h_x \otimes f \otimes p + v \cdot h_y \otimes f \otimes p + h \otimes f \otimes p_t = 0$$

where p_t is the derivative of temporal filter p. From the two filtered images we derive two independent constraint equations at each pixel and by solving the equations we obtain the optical flow vector.

The reliability of the velocity measurement at each pixel location is low unless the following two conditions are satisfied.

$$(g_x \otimes f \otimes p)^2 + (g_y \otimes f \otimes p)^2 + (h_x \otimes f \otimes p)^2 + (h_y \otimes f \otimes p)^2 > n_1$$

where n_1 is a predetermined gradient threshold value. This means that the spatial gradient of the image must be large.

$$|(g_x \otimes f \otimes p)(h_y \otimes f \otimes p) - (g_y \otimes f \otimes p)(h_x \otimes f \otimes p)| > n_2$$

where n_2 is a denominator threshold value. This means that the two equation must be independent.

The advantage of this method is that it requires only three independent operations: spatial filtering, temporal filtering, and velocity calculation which integrates the filtering results.

3 Object Tracking

We propose an efficient object tracking method which distinguishes a target object flow vectors from the others [12]. An object is tracked by updating a rectangular window which circumscribe an object region in the image sequences. The neighborhood of the window is searched for the next object region. This method is not subject to the influence of the flow vectors which are far from the target object.

Before tracking, we have to determine a moving object to track in a scene. In our experiment we track the first object which moves in the image. In order to find a moving object, the camera direction is fixed and the window size is initialized to the image size. When enough flow vectors are obtained, we consider them as belonging to a moving object.

Once a moving object is obtained, it is tracked by the following procedure.

(1) Calculate optical flows in the image (Figure 1 (a) shows an example of optical flows).

(2) If the number of pixels with a flow vector in the window is more than a threshold, go to (3). Otherwise, go to (1) to detect a moving object again.

(3) Calculate the mean flow vector in the window.

(4) Search the window and the neighborhood of the window for the pixels whose flow vector are similar to the mean flow. A set of these pixels is the object region in the next frame (Figure 1 (b) shows an example).

(5) Update the window to circumscribe the object region. If the number of pixels of the region is less than a threshold, the window is not updated.

(6) Based on the position of the object region and the flow vector, the camera direction is controlled to capture the object in the image center.

(7) return to (1).

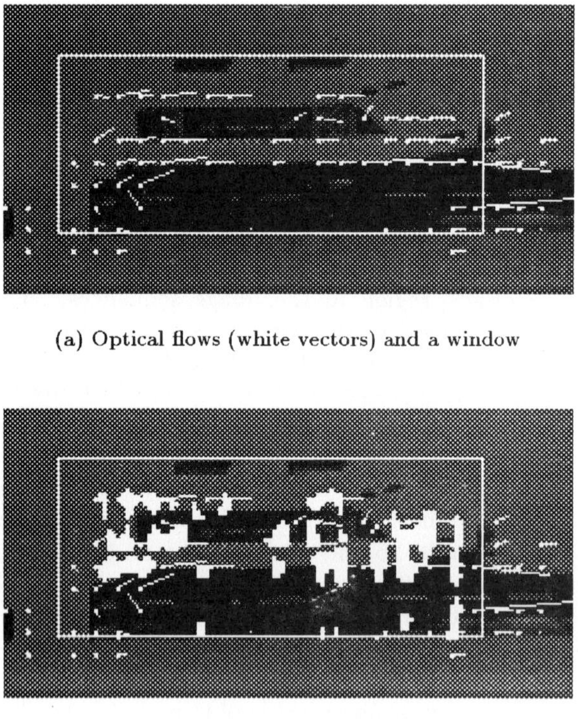

(a) Optical flows (white vectors) and a window

(b) Object region.

Figure 1: Detected object region (white regions)

Figure 2 shows an example of tracking of a person walking from left to right in the image. The direction of the camera is controlled from left to right. The black lines at the image center represent the vertical and the horizontal velocities of the target object.

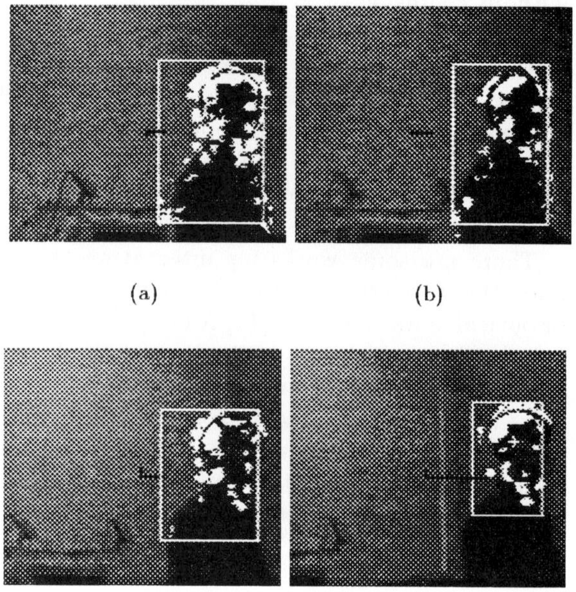

(a) (b)

(c) (d)

Figure 2: Tracking an object

The object tracking is realized by an active vision system as shown in Figure 3 which consists of an active camera, a real time image processor, and a personal computer.

The active camera has two cameras with pan, tilt, and vergence mechanisms which are actuated by D.C. servo motors. We use only one camera in the current experiment.

As shown in Figure 4, the real time image processor includes four DSP boards, each of which has two DSP chips (TMS320c40). The personal computer receives the position and velocity data of the moving object, and controls the direction of the camera.

4 Multiple Object Tracking

The tracking method described above can be extended to multiple objects tracking [13]. Real time tracking of two objects is realized as follows. At first, each window size is initialized to image size and assigned to a DSP. If a DSP finds a moving object, it begins to track it with the window. Meanwhile another DSP is searching a

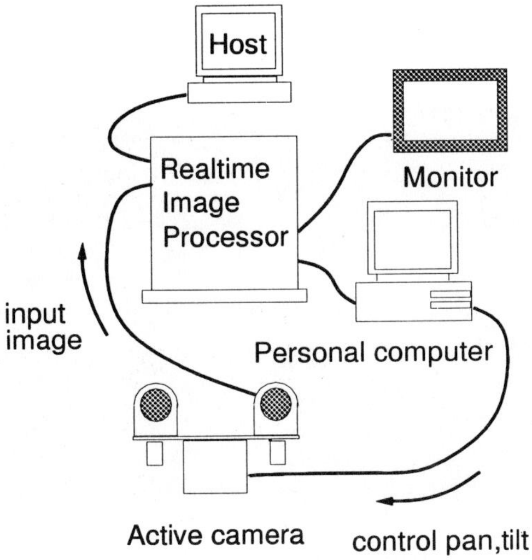

Figure 3: System organization

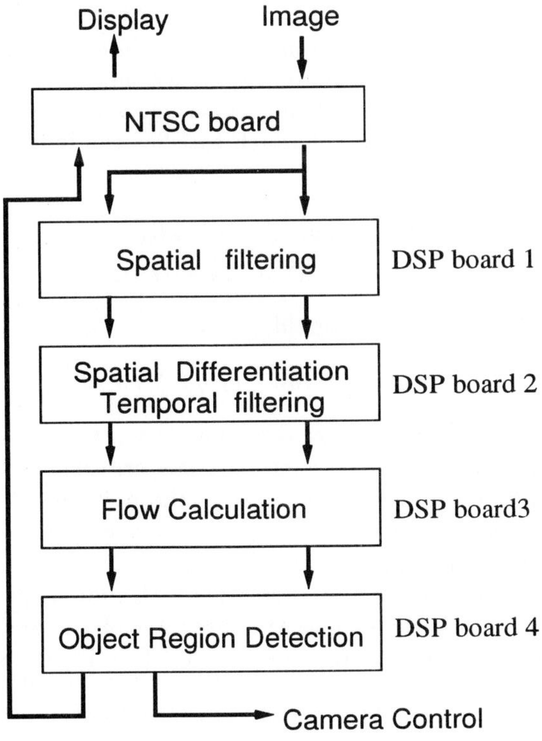

Figure 4: Realtime image processor

new object. When another object is found, the DSP begins to track it with another window.

If two objects overlap, it must be determined which is foreground so that windows are correctly updated. Figure 4 shows a situation where window-A and window-B overlap. Each overlapping window is divided into two regions: "overlapping region" and "non overlapping region". Let us suppose the following cases according to the relation of the two windows.

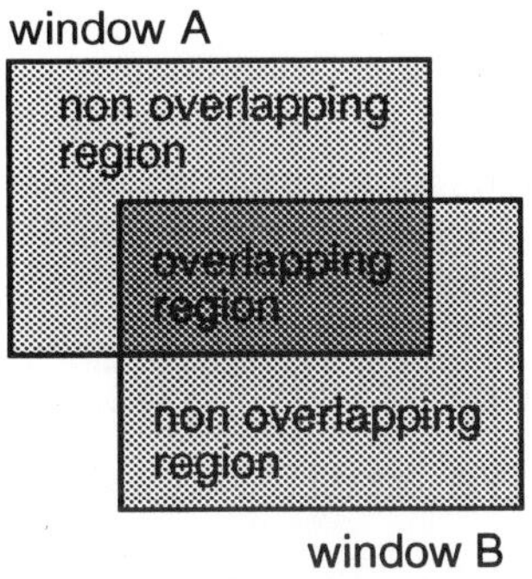

Figure 5: Overlapping windows

case 1: A part of window-A and a part of window-B are considered to be overlapping.

case 2: window-A (window-B) includes window-B (window-A).

case 3: window-A and window-B are considered to be fully overlapping.

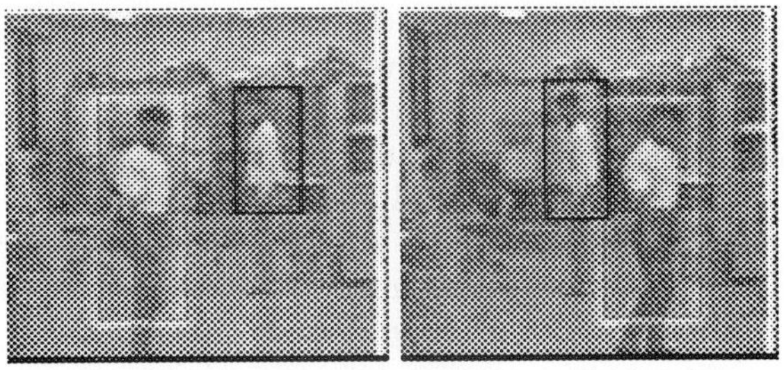

Figure 6: tracking two men

For each case, a decision is made which window is foreground as follows. Let $\boldsymbol{v}_o$ denote the mean flow vector of the overlapping region, and $\boldsymbol{v}_{n_A}$ and $\boldsymbol{v}_{n_B}$ denote the mean flow vector of the non overlapping region of window-A and window-B respectively.

case 1: If $||v_o - v_{n_A}|| < ||v_o - v_{n_B}||$, window-A is the foreground. Otherwise, window-B is the foreground.

case 2: When large window is window-A, if $||v_o - v_{n_B}|| < v_t$(a threshold), window-A is foreground. Otherwise, window-B is foreground.

case 3: The state of each window is the same as the previous frame because the decision can not be made.

For each case, a decision is made which window is foreground as number of reliable optical flows.. Figure 4 shows an example of an experiment for two persons crossing in the image.

5 Integration of Optical Flows and Edges

This section describe a case where the motion of an object gets similar with the background. Then, optical flows are not enough to segment the object. Humans can easily find the object if the brightness is different from the background. In a textured background, however, object detection from a static image is difficult even for humans. If the object moves, it is more easily detected.

In our method, while an object is moving the object region is determined from optical flows as described in the previous sections. Then edges are extracted near the boundary of the object region, and the object contour is determined by modifying the object region with the edges. This contour is used for processing the following images. The contour is refined gradually with newly extracted edges.

If the motion boundary is not obtained, the stored contour is used to determine the current object contour. Edges are detected near the current contour to determine whether the contour really exists or occluded.

5.1 Extraction of Object Contour

The object region obtained from the optical flow is not precise because the the optical flow is extracted with spatial filters assuming the similar flow in the filtered region. Therefore it is not accurate near the boundary of different optical flows Figure 7(a) shows an example of a synthesized image where an object is moving from left to right, and Figure 7(b) shows extracted optical flows. We can see that the optical flow is not accurate near the contour of the object.

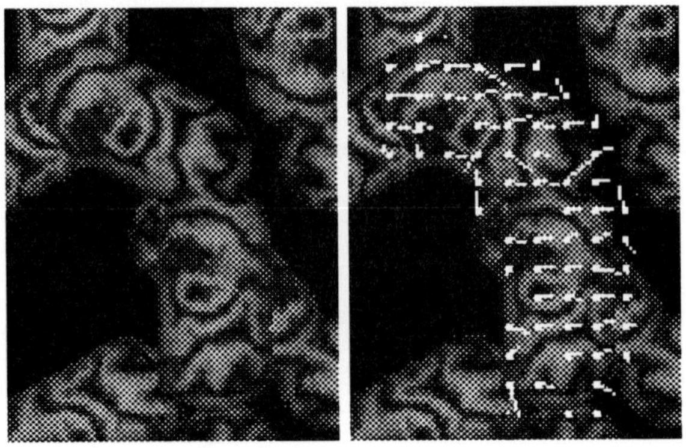

(a) synthesized image (b) extracted flows

Figure 7: Extracted optical flows

In order to obtain a more precise contour, edges are extracted near the boundary of the object boundary. The neighbor of the boundary depends on the variance of the Gaussian filter σ^2 for optical flow detection. Here the neighbor is defined to be within 3σ distance from the boundary.

Let $f(x, y)$ denote the image intensity and $G(x, y)$ denote the Gauss function. The edges are defined as the pixel which satisfies the following equations.

$$\nabla^2 G(x,y) \otimes f(x,y) = 0$$
$$|\nabla G \otimes f| > T_i$$

where T_i is a threshold.

$$\theta_e(x, y) = tan^{-1}(f_y/f_x) + \pi/2$$

Each extracted edge is examined whether or not it is a part of the contour. The criterion is the difference of the direction of the edge and that of the contour. The edge direction is defined as perpendicular to vector $\nabla G \otimes f$.

The contour direction of the s-th contour point at $\mathbf{c}(s)$ is defined as the direction of the vector from the $\mathbf{c}(s + \Delta s)$ to $\mathbf{c}(s - \Delta s)$ (see Figure 8).

$$\theta_c(s) = tan^{-1}(\Delta y/\Delta x)$$

Now the whole contour position is matched to the detected edges such that the reliable edges may overlap with the contour. Given a candidate position of the contour, the matching candidate

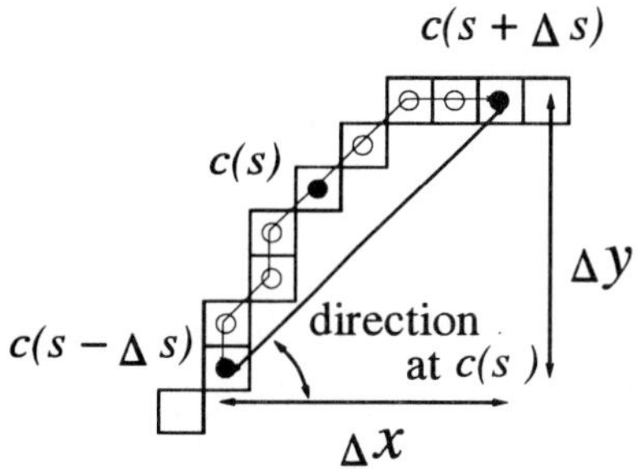

Figure 8: direction of motion boundary

of each point on the contour is the edge which is nearest along the perpendicular direction of the contour. If there is no edge, the candidate is called a null point. Let $C = \{\mathbf{c}_i; i = 1, \cdots, n\}$ denote the sequence of the contour position, and $E = \{\mathbf{e}_i; i = 1, \cdots, m\}$ denote the corresponding edge sequence (E contains null points). The evaluation function for the similarity of C and E is defined in terms of the difference of the position and that of the direction.

$$S'(C, E) = \sum_{i=1}^{n} cos(\frac{|\mathbf{c}_i - \mathbf{e}_i|}{3\sigma}\frac{\pi}{2})cos(\theta_d(\mathbf{c}_i, \mathbf{e}_i)) \quad (1)$$

where $\theta_d(\mathbf{c}_i, \mathbf{e}_i)$ denotes the difference of direction, and the distance and the difference of the direction of the null point are defined to be the allowable maximum values (3σ and $\pi/2$, respectively).

Now let $S(C, E, \mathbf{p}_k)$ denote the the value of $S'(C, E)$ when C is at $\mathbf{p}$. The best position is given by the following equation.

$$\mathbf{p}_m = arg \max_{\mathbf{p}_k} S(C, E, \mathbf{p}_k)$$

Finally the object contour is modified by the edges which is successfully matched to the contour. Since these edges are more reliable then the stored contour points, they replace the contour points. Note that the more edges are stored as the contour points as the more images are analyzed.

5.2 Determination of Occlusion

When an object overlaps with another object, we would like to know which object is foreground. Parts of contours of foreground object may be detected in the predicted regions As shown in Figure

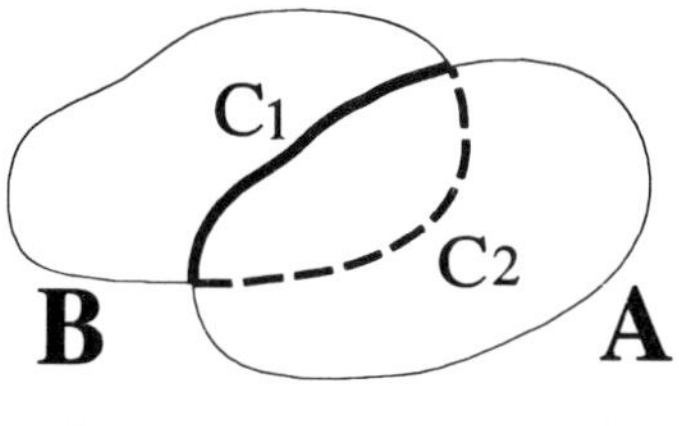

Figure 9: Occluding contour

9, the occlusion is determined by finding edges of objects near the predicted contours $C_1 and C_2$.

If the number of contour points n_k of C_k is large for $k = 1, 2$, the occlusion is examined. First edges are detected near the contour of each object, and then determined which contour they belong. The decision is based on the following evaluation function which is similar to equation (1).

$$S_k(C_k, E) = \frac{1}{n}\sum_{i=1}^{n_k} cos(\theta_d(\mathbf{c}_i, \mathbf{e}_i)) \quad (2)$$

The value of the evaluation function is computed for each contour. Contour C_1 is regarded to be foreground if the following conditions are satisfied:

1. $S_1 > S_t$, where S_t is a threshold.
2. $S_1 - S_2 > S_d$, where S_d is a threshold.

Note that no decision is made in ambiguous cases.

Once the occlusion is determined, this relation hold so long as the two objects have large overlapping regions.

5.3 Experiments and Discussion

A tracking experiment is performed for a synthesized image sequence, one of which is shown in Figure 7(a). Figure 10shows the results of sample frames where the white lines represent the stored contours. Since not enough edges have been extracted from the figure (a), the contour includes the motion boundary. In figure (b) where the object is not moving, only the stored boundary is used to determine the contour. Note that the stored boundary is more precise because it includes more edges.

Next an experiment of multiple object tracking in a real image sequence is performed. The first image is shown in Figure 11. The task is to track two toy cars which move from the right to the left of the image. The result is shown in Figure 12, where the white line shows the predicted contour. The right car (A) gradually approaches the left car (B) and overlaps in the image. Because the optical flows of the two cars are almost the same, the occlusion relation is not determined only from the flow. In frame 24, the system determined that car A occludes car B. The value of evaluation function is shown in Figure 13.

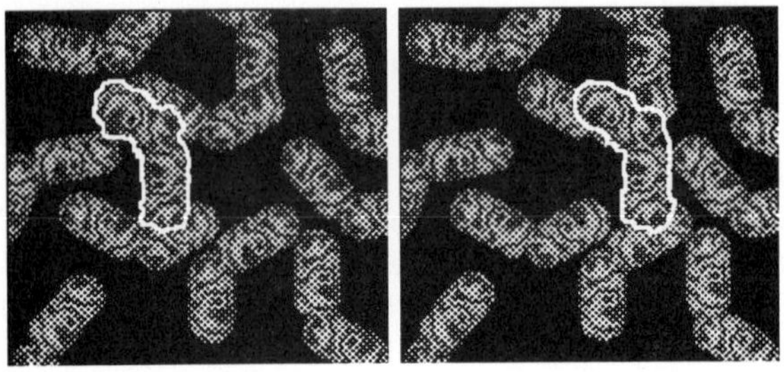

(a) frame 10 (b) frame 40

Figure 10: Extracted contour

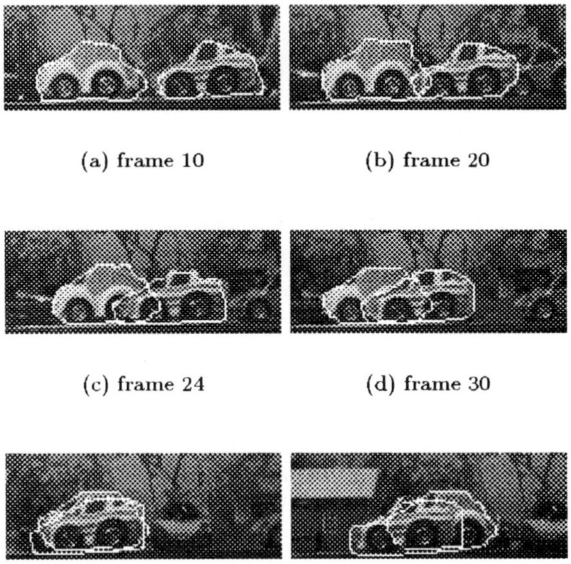

(a) frame 10 (b) frame 20

(c) frame 24 (d) frame 30

(e) frame 40 (f) frame 50

Figure 12: Tracking two objects

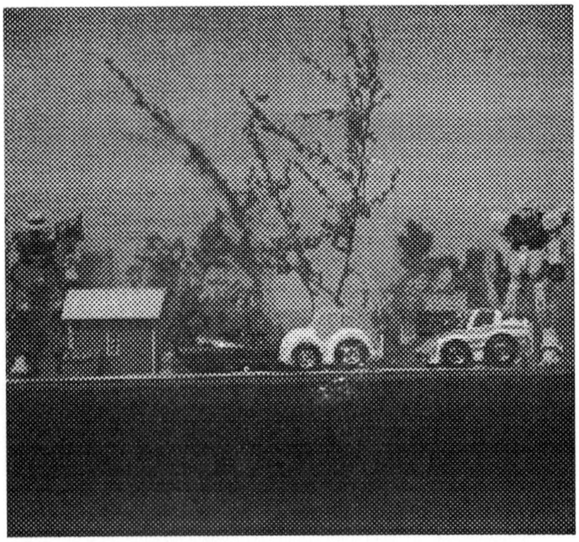

Figure 11: example of tracking toy cars.

From the evaluation value of frame 24, the occlusion relation is clearly determined. The overlap ratio is defined here as the ratio of the overlapping length along the motion direction and the length of a car. The value become similar when both cars almost overlap (around frame 40). In this case, however, the correct decision is obtained from the continuity constraint. Figure 13 also shows the validity of the evaluation function.

Although the effectiveness of integration of optical flows and edges has been proved, this method requires too much computation for real time processing. We are now preparing to realize faster processing by using many standard image processing DSP boards. This board contains a single DSP with input memory for double buffering and local memory for storing intermediate results or outputs. By organizing the boards for multiple pipe line processing, the system may be

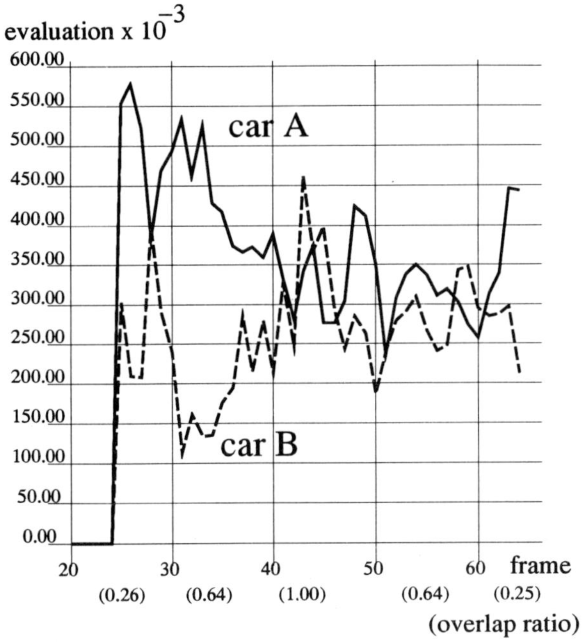

Figure 13: change of evaluation function.

able to execute general image processing faster than conventional computers. Because of generality, the board is expected to be used by many robotics and computer vision people.

References

[1] Horn B. K. P. and Schunck, B. G.: Determining optical flow, Artif. Intell., Vol.17, pp.185-203, (1981)

[2] Dickmanns, E.D., Behringer, R., Brudigam, C., Dickmanns, D., Thomanek, F., and Holt, V.v.: An All-Transputer Visual Autobahn-Autopilot/Copilot, Proc. IEEE ICCV'93, pp.608-615 (1993)

[3] Brandenbusch, M., Sung, Chen-Ko, and Wirtkopp, T.: A vision system for real-time motion analysis in image sequences, FhG IITB Mitteilungen pp.31-35 (1992)

[4] Murray, D.W. and McLauchlan, and P.F. Reid, I.D. and Sharkey, P.M.: Reactions to Peripheral Image Motion using a Head/Eye Platform, Proc. ICCV'93, pp.403-411 (1993)

[5] Inoue, H. and Tachikawa, T. and Inaba, M.: Robot Vision System with a Correlation Clip for Real-time Tracking, Optical Flow and Depth Map Generation, Proc. IEEE Int. Conf. R&A, Vol.2, pp.1621-1626 (1992)

[6] Srinivasan, M.V.: Generalized gradient schemes for the measurement of two-dimensional image motion, Biol. Cybern., Vol.63, pp.421-431 (1990)

[7] Sobey, P. and Srinivasan, M.V.: Measurement of optical flow using a generalized gradient scheme, Journal of the Optical Society of America, pp.1488-1498 (1991).

[8] Adiv, G.: Determining 3-D Motion and Structure from Optical Flow Generated by Several Moving Objects, IEEE Trans. Pattern Anal.& Mach. Intell., Vol.PAMI-7, pp.384-401 (1985)

[9] Chen, H. -J. and Shirai, Y.: Detecting Multiple Image Motions by Exploiting Temporal Coherence of Apparent Motion, Proc.IEEE Conf. on Computer Vision and Pattern Recognition, pp.899-902 (1994)

[10] Etoh, M. and Shirai, Y.: Segmentation and 2D motion estimate by region fragments, Proc. 4th Int. Conf. Comput. Vision, pp.192-199 (1993)

[11] Thompson, W.B.: Combining Motion and Contrast for Segmentation, IEEE Trans. Pattern Anal. Mach. Intell., Vol.PAMI-2, No.6 pp.543-549 (1980)

[12] Mae, Y. Yamamoto, S. Shirai, Y. and Miura, J.: Optical Flow Based Object Tracking by Active Camera, Proc. 2nd Japan-France Congress on Mechatronics, pp.545-548 (1994).

[13] Yamamoto, S., Mae, Y., Shirai Y. and Miura, J.: Realtime Multiple Object Tracking Based on Optical Flows, Proc. IEEE International Conference on Robotics and Automation, pp.2328-2333 (1995).

PART 11

HOW DO WE BUILD RELIABLE AND EFECTIVE MOBILE ROBOTS?

SESSION SUMMARY

Robert Bolles
AI Center, M/S EK290, SRI International
333 Ravenswood Avenue, Menlo Park, CA 94025, USA
email: bolles@ai.sri.com

In this session, four very impressive robotic vehicles are described. One vehicle deftly navigates the halls of an office building, dynamically constructing maps of landmarks and obstacles. Two of the vehicles operate outdoors in natural environments, driving across fields and around rocks and trees. And the fourth vehicle drives on the autobahn at 130 kilometers per hour, automatically selecting times to pass slower vehicles – and if you've ever driven on the autobahn, you know how nerve-racking that can be.

These vehicles, although impressive, also illustrate the relatively poor state of the art with respect to reliability and robustness. For the most part, these research vehicles are single-thread systems, which are inherently fragile. They operate properly only as long as all their sensors work and the vehicles stay within their areas of competence. For example, if the Siemens indoor robot encounters a chair with legs that meet the floor at a 75-degree angle, instead of 90 degrees, it may bump into them because the energy from its sonar sensors is not reflected back to its receivers. Or, if the vehicle on the autobahn drives toward the setting sun, it may not be able to detect the necessary road markings or the nearby cars.

Future users of this technology, on the other hand, require safe, effective, reliable, and inexpensive vehicles – they don't care *how* a vehicle performs its task, they just want it to work. Users expect a robot vehicle to continue functioning appropriately if one of its sensors fails or if a sensor intermittently returns strange measurements. The vehicle should even work if the primary location determination algorithm gets lost once in awhile.

The robotics community is just starting to take reliability seriously – they've developed a wide range of effective sensing and behavior modules, as convincingly demonstrated in these papers, and now they are ready to tackle the next level of challenges. So, how do we build reliable and effective mobile robots? The papers include the following suggestions:

1. incorporate redundant sensors,
2. develop smart sensor subsystems that monitor their own behavior and annotate their results,
3. develop a central control architecture that monitors each subsystem and dynamically re-plans actions as needed,
4. include a person in the loop to handle situations outside the competence of the automatic system.

All these alternatives, implicitly or explicitly, require the ability of the system to recognize when a sensor or subsystem is having trouble. When has a sensor stopped working? When is a sensor producing incorrect measurements? Or, when has a sensor drifted out of calibration?

At the heart of this type of evaluation is a characterization of what to expect from a subsystem. What's normal? And what's an anomaly? The robotics community is beginning to explore representations for describing expectations and techniques for computing the factors that form these descriptions. For example, Chatila and Lacroix, in their paper in this section, describe a characterization of the ground's roughness, its effect on mobility, and a method for learning the association between roughness and navigation strategies. As another example, Dickmanns discusses a combination of differential and integral representations to support the spatio-temporal control required for complex high-speed systems. Characterizations such as these will form the basis for the next generation of self-aware vehicles that will be safe, effective, and reliable.

Adaptive Navigation for Autonomous Mobile Robots

Raja Chatila and Simon Lacroix

LAAS-CNRS
7, Ave. du Colonel Roche – 31077 Toulouse Cedex, France
email: {raja,simon}@laas.fr

Abstract

The effectiveness of a mobile robot is related to its capacity to *adapt* to its environment. This adaptivity is only possible if the robot possesses a) the processes which enable it to select the more appropriate way to perform its actions, and b) the representations that are adequate for these processes. The navigation of an autonomous mobile robot in a natural environment will be used as an illustrative example to present these ideas. This task involves several levels of reasoning, several environment representations, and several action modes. It has to select sub-goals, navigation modes, and perception actions according to the situation. In order to achieve the task reliably, a measure of the confidence in the models used is necessary. Experimental results illustrate the approach.

1 Introduction

Effective robots must be able to *adapt* to their environment. Otherwise they will not be able to cope with a variety of situations and to use their hardware, i.e., sensors, effectors, computing power, etc. efficiently to accomplish their tasks. Therefore they have to possess the ability to select their actions, and the manner in which they are accomplished, according to the actual situation and an anticipation of its evolution.

The reliability of the robot depends on several factors, and is related to the *provability* and predictibility of their behavior. This in turn is directly related to the quality of their representations, and we shall focus on this last issue only.

We consider the example of a planetary rover to illustrate these issues. Navigation is the basic task that has to be solved by this autonomous mobile robot, and in which effectiveness is central. Navigation is in general an incremental process that can be summarized in four main steps:

1. Environment perception and modelling: any motion requires a representation of the local environment at least, and often a more global knowledge. The navigation process has to decide where, when and what to perceive.
2. Localization: the robot needs to know where it is with respect to its environment and goal.
3. Motion decision and planning: the robot has to decide where or which way to go, locally or at the longer term, and if possible compute a trajectory;
4. Motion execution: the commands corresponding to the motion decisions are executed by control processes - possibly sensor-based and using environment features.

The complexity of the navigation processes depends on the general context in which this task is to be executed (nature of the environment, efficiency constraints,...) and should be adapted to it.

2 Navigation

The problem of long range navigation in unknown outdoors environments is not very frequently addressed. Systems that were actually experimented were demonstrated for road following [1], or motion in rather limited environment conditions [2, 3]. Two important achievements are Ambler [4] and the navigation of the UGV [5].

A canonical navigation task is reaching a goal: ``Go To [goal]'', where the argument goal is a distant target to reach autonomously. Any more complex robotic mission (exploration, sample collection...) will include one or more instances of this task. Given the variety of terrains the robot wil have to traverse, this task involves in our approach several levels of reasoning, several environment representations, and three different motion modes. It raises a need for a specific decisional level (the *navigation* level), that is in charge of deciding which environment representation to update, which sub-goal to reach, and which motion mode to apply. The paper is especially devoted to this level, which is a key component of the system, since it controls the perception and motion activities of the robot for this task.

The paper is organised as follows: the next section presents the adaptive approach to autonomous navigation in unknown outdoor environments. Section 4 then presents how the terrain representations required by the navigation decisional level is incrementally built on the basis of 3D data, produced either by a Laser Range Finder (LRF) or by stereo-vision. The algorithms that perform the selection of sub-goals and perception tasks (*i.e.* that compose the navigation level) are described in section 5. Finally, we present experimental results and conclude the paper.

3 A General Strategy for Navigation in Outdoor Unknown Environments

3.1 An adaptive approach

Using its own sensors, effectors, memory and computing power efficiently is certainly a feature that we would like to implement in a robot. This becomes even more a necessity for a planetary rover which has important limitations on its processing capacities, memory and energy. To achieve an efficient behavior, the robot must adapt the manner in which it executes the navigation task to the nature of the terrain and the quality of its knowledge on it [12, 7]. Hence, three motion modes are considered:

- A **reflex** mode: on large flat and lightly cluttered zones, it is sufficient to determine robot locomotion commands on the basis of a goal (heading or position) and informations provided by "obstacle detector" sensors. The terrain representation required by this mode is just the description of the borders of the region within which it can be applied;
- A **2D planned** mode: when the terrain is mainly flat, but cluttered with obstacles, it becomes necessary for efficiency reasons to plan a trajectory. The trajectory planner reasons on a binary description of the environment, which is described in terms of empty/obstacle areas.
- A **3D planned** mode: when the terrain is highly constrained (uneven), collision and stability constraints have to be checked to determine the robot locomotion commands. This is done thanks to a 3D trajectory planner [8], that reasons on a fine 3D description of the terrain (an elevation map or numerical terrain model [9]);

The existence of different motion modes enables more adpated and efficient behavior, at the price of complicating the system since it must be able to deal with several different terrain representations and motion planning processes. It must especially have the ability to determine which motion mode to apply: this is performed thanks to a specific planning level, the *navigation planner* which requires its own representations.

3.2 The Navigation Planner

We assume the terrain on which the robot must navigate is initially unknown, or mapped with a very low resolution. In this last case, it is possible for a user to specify a graph of possible *routes*, *i.e.* corridors that avoid largedifficult areas, and within which the robot has to move autonomously. In this context, the navigation task ``Go To'' is achieved at three layers of planning (figure 1):

- *route planning* which selects long-term paths to the goal on the basis of the initial informations when available (the route map that may cover several kilometers). The route planner selects a sub-goal for the navigation planning level;
- *navigation planning* which reasons on a global qualitative representation of the terrain, built from the data acquired by the

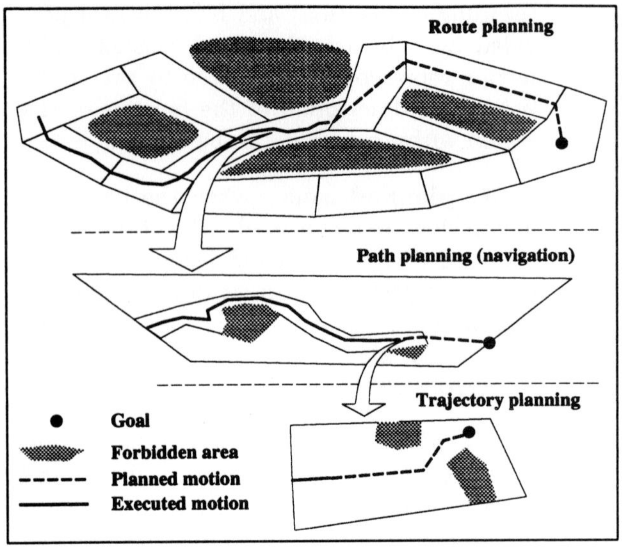

Figure 1: *Three levels of planning*

robot's sensors. The navigation planner selects the next perception task to perform, the sub-goal to reach and the motion mode to apply: it selects and controls the trajectory planner;

- Finally *trajectory planning* which determines the trajectory to execute (in one of the above-mentioned three motion modes) to reach the goal defined by the navigation planning level.

Splitting the decisional process into three layers of planning has the advantage of structuring the problem: each planning layer controls the one that is directly below by specifying its goal and its working domain. It has also the great advantage of helping to analyze failures: when a planner fails to reach its goal, it means that the environment representation of the immediatly higher layer is erroneous and therefore that it has to be revised.

The navigation planner (layer 2) is *systematically* activated at each step of the incremental execution of the task: each time 3D data are acquired, they are analyzed to provide a description of the perceived zone in terms of navigation classes. This description is fused with the model acquired so far to maintain a *global qualitative representation* of the environment, (the *region map*), on which the navigation planner reasons to select a sub-goal, a motion mode, and the next perception task to execute.

3.3 Several terrain representations

Each of the three different motion modes requires a particular terrain representation. The navigation planner also requires a specific terrain representation, and during navigation, an exteroceptive localisation process has to be activated frequently, which requires an other terrain representation. Aiming at building a "universal" terrain model that contains all the necessary informations for these various processes is extremely difficult, inefficient, and moreover not really useful. It is more direct and easier to build different representations adapted to their use: the environment model is then *multi-layered and heterogeneous.* Several perception processes coexist in the system, each dedicated to the extraction of specific representations: perception is *multi-purpose.* Figure 2 presents the various terrain representations required during navigation: one can distinguish the numerical terrain model [9] necessary to the 3D trajectory planner, the region map dedicated to the navigation planner, and three different ways to build a localisation model: *(i)* by modelling objects (rocks) with ellipsoids or superquadrics [10], *(ii)* by detecting interesting zones in the elevation map represented by a B-spline based model [11], or *(iii)* by detecting poles in the 3D raw data. Coherence relationships between these various representations are to be maintained when necessary.

4 Building the region map

For the purpose of navigation planning, a global representation that describes the terrain in terms of navigation classes (flat, uneven, obstacle, unknown) is required. This representation enables to select the adequate motion mode. We focus in this section on the algorithms developed to build such a model from 3D data (produced either by a laser range finder or a correlation stereo-vision algorithm).

4.1 3D data classification

Applied each time 3D data are acquired, the classification process produces a description of the perceived areas in term in *terrain classes.* It relies on a specific discretization of the perceived area respecting the sensor resolution (figure 3), that defines "cells" on which different characteristics (attributes) are determined: density (number of 3D points in a cell compared with a nominal

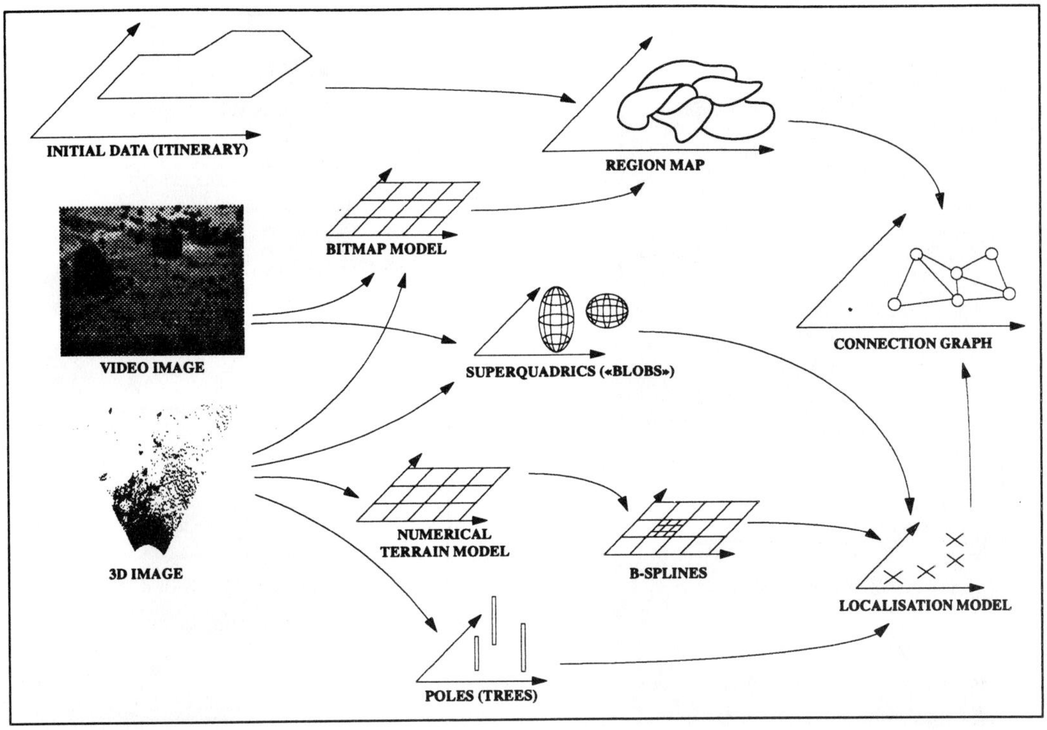

Figure 2: *The various representations used in the system. Arrows represent the constructive dependencies between them*

density defined by the discretization rates), mean altitude, variance on the altitude, mean normal vector and corresponding variances...

A non-parametric Bayesian classification procedure is used to label each cell: a learning phase, based on prototypes classified by a human, leads to the determination of probability density functions, and the classical Bayesian approach is applied, which provides an estimate of the probability for each possible label. A decision function that prefers false alarms (*i.e.* labelling a flat area as obstacle or uneven) instead of the non-detections (*i.e.* the opposite: labelling an obstacle as a flat area) is used (figure 4). A simpler but faster technique based on thresholds on the cell attributes has also been implemented.

This technique proved its efficiency and robustness on several hundreds of 3D images. Its main interest is that it provides an estimate of the confidence of its results: this information is given by the *entropy* of a cell. Moreover, a statistical analysis of the cell labelling confidence as a function of its label and distance to the sensor (directly related to the measure uncertainty) defines a predictive model of the classification process.

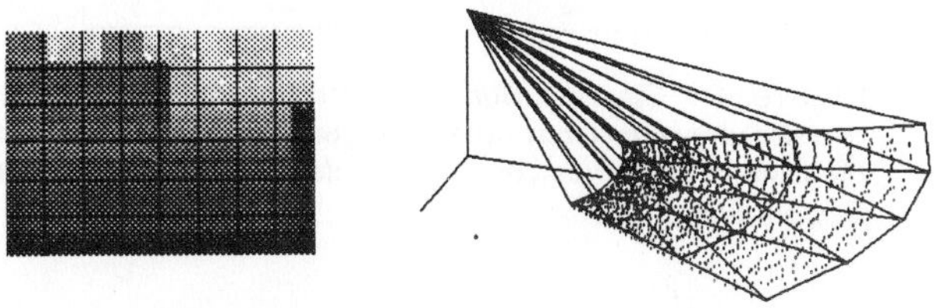

Figure 3: *Discretization used for the classification procedure: a regular discretisation in the sensor frame (left: a 3D image is represented as a video image, where the gray levels corresponds to point depth) defines a discretization of the perceived zone that respects the sensor resolution (right)*

4.2 Incremental fusion

The partial probabilities of a cell to belong to a terrain class and the variance on its elevation allow to perform a fusion procedure of several classified images. The fusion procedure is performed on a bitmap, in the pixels of which are encoded cell attributes determined by the classification procedure (label, label confidence, elevation and variance on the elevation).

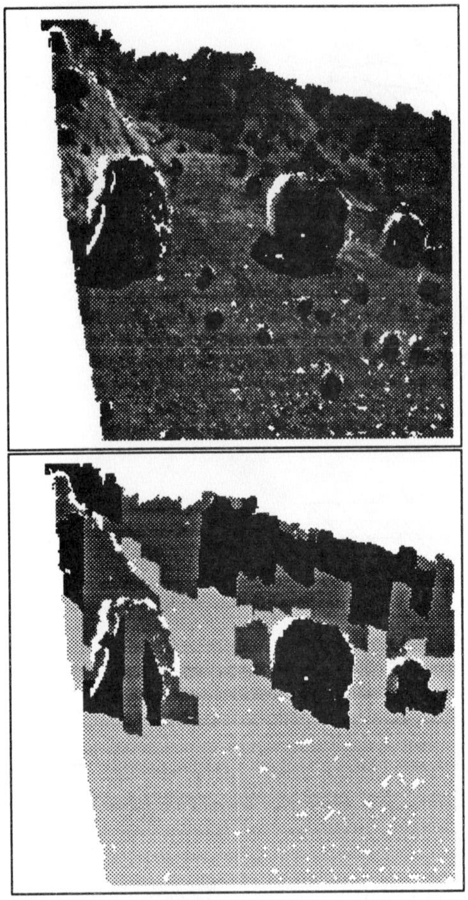

Figure 4: *Classification of a correlated stereo image: correlated pixels (top) and reprojection of the result in the camera frame (bottom - from clear to dark: unknown, flat, uneven and obstacle)*

4.3 Model structure and management

For the purpose of navigation planning, the global bitmap model is structured into a region map, that defines a connection graph. Planning a *path* (as opposed to planning a *trajectory*) does not require a precise evaluation of the static and kinematic constraints on the robot: we simply consider a robot point model, and therefore perform an obstacle growing in the bitmap before segmenting it into regions. The regions define a connection graph, whose nodes are on their borders, and whose arcs correspond to a region crossing (figure 6).

In order to satisfy memory constraints, the global model is explicited as a bitmap only in the surroundings of the robot's current position, and the region model (much more compact) is kept in memory during the whole mission (figure 7).

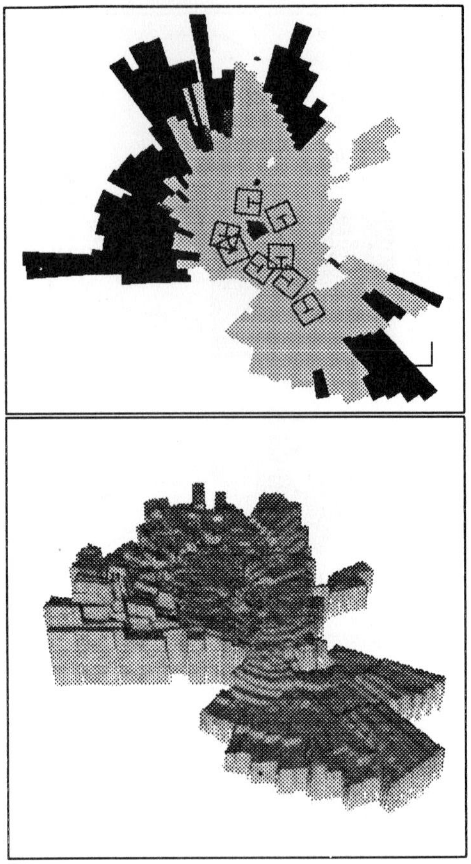

Figure 5: *Fusion of 8 different classified laser images: terrain classes (top) and elevation (bottom)*

5 Navigation planning

Each time 3D data are acquired, classified and fused in the global model, the robot has to answer the following questions:

- Where to go? (sub-goal selection)
- How to go there? (motion mode selection)
- Where to perceive? (data acquisition control)
- What to do with the acquired data? (perception task selection)

For that purpose, the navigation planner reasons on the robot capabilities (action models for perception and motion tasks) and the global terrain representation.

5.1 Planning motion versus planning perception

A straightforward fact is that motion and perception tasks are strongly interdependent: planning

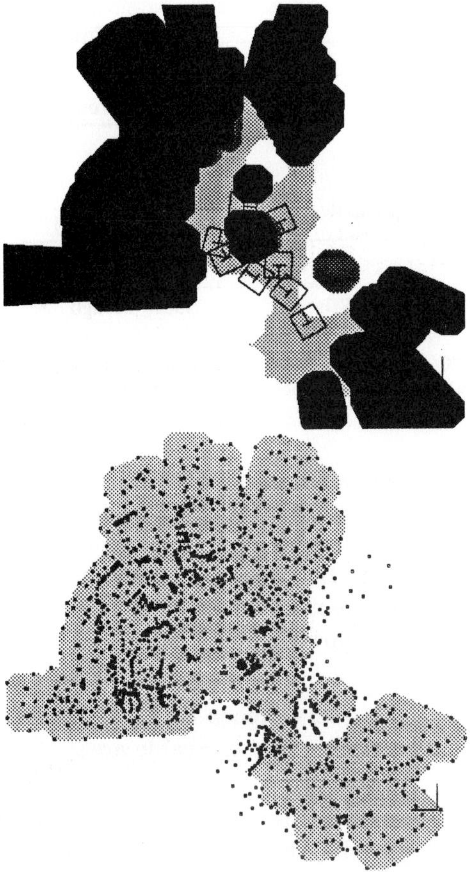

Figure 6: *The model of figure 5 after obstacle growing (top) and the nodes defined by the region segmentation (bottom)*

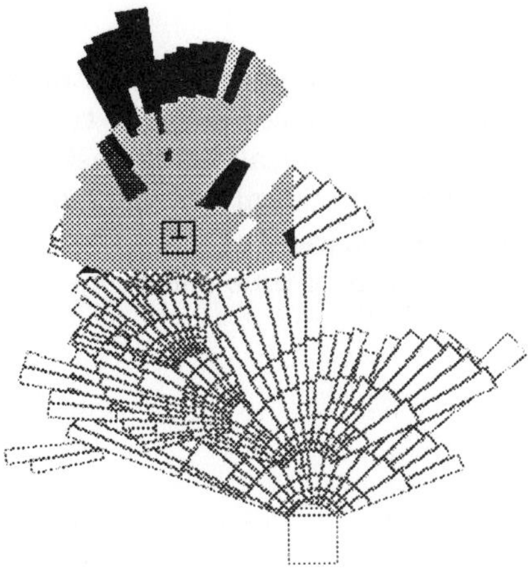

Figure 7: *Robot surroundings explicited as a bitmap*

and executing a motion requires to have formerly modelled the environment, and to acquire some specific data, a motion is often necessary to reach the adequate observation position.

Planning motion tasks in an environment modelled as a connection graph is quite straightforward: it consists in finding paths in the graph that minimise some criteria (time and energy, that are respectively related to the terrain classes and elevation variations). This is easily solved by classical graph search techniques, using cost functions that express these criteria.

Planning perception tasks is a much more difficult issue: one must be able to predict the results of such tasks (which requires a model of the perception processes), and the *utility* of these results to the mission:

- Localization processes can be modelled by a simple function that expresses the gain on the robot position accuracy, as a function of the number and distances of perceivable landmarks - assuming all landmarks are intrinsically equally informative (figure 8);

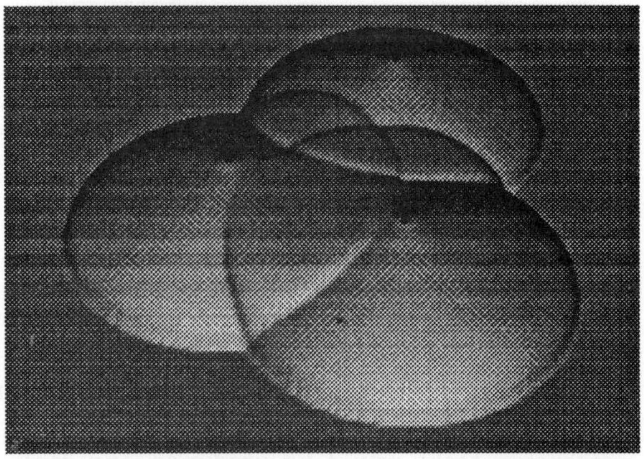

Figure 8: *A simple localisation model: the gain on the robot position precision increases as the distance of the perceived landmarks decreases, and is better as the number of perceivable landmarks increases. This figure presents this gain in an environment with three landmarks and no occlusions*

- With the confidence model of the 3D data classification process, one can predict the amount of information a classification task can provide. But it is much more difficult to express the utility of a classification task to reach the goal: the model of the classification task cannot predict *what* will be effectively perceived. It is then difficult to estimate the interest of these tasks (figure 9).

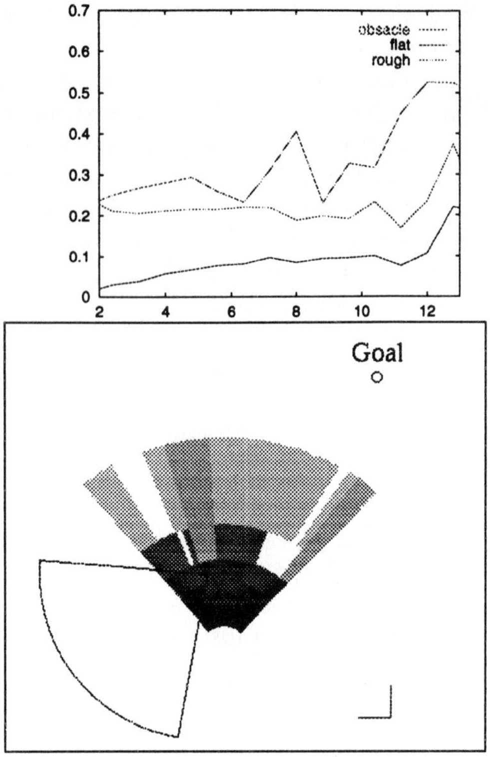

Figure 9: ***The confidence model of the classification procedure (top) and the labelling confidence of the terrain model (represented as grey levels in the bottom image) allow to determine the classification task that maximises the gain in confidence from a given view point. But the result of this task is of a poor interest to reach the goal***

5.2 Approach

A direct and brute force approach to answer the former questions would be to perform a search in the connection graph, in which *all* the possible perception tasks would be predicted and evaluated at *each* node encountered during the search. Besides its drastic algorithmic complexity, this approach appeared unrealistic because of the difficulty to express the utility of a predicted classification task to reach the goal.

We therefore choose a different approach to tackle the problem: the perception task selection is *subordinated* to the motion task. A search algorithm provides an *optimal* path, that is analyzed afterwards to deduce the perceptions tasks to perform along it. The "optimality" criterion takes here a crucial importance: it is a linear combination of time and energy consumed, weighted by the terrain class to cross and the confidence of the terrain labelling (figure 10).

Introducing the labelling confidence in the crossing cost of an arc comes to consider *implicitly* the

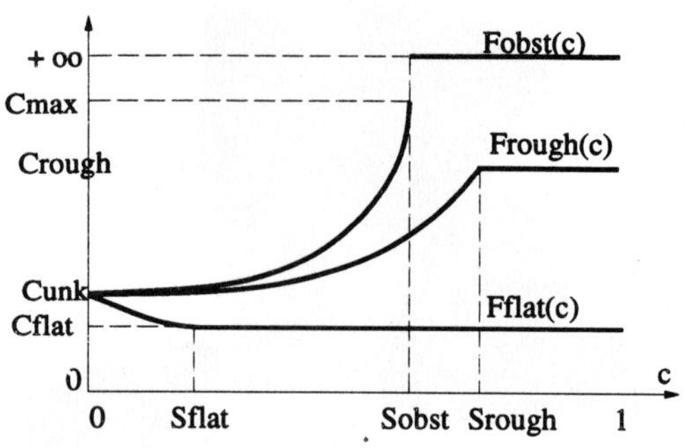

Figure 10: ***Weighting functions of an arc cost, as a function of the arc label and confidence***

modelling capabilities of the robot: tolerating to cross obstacle areas labelled with a low confidence means that the robot is able to acquire easily informations on this area. Off course, the returned path is not executed directly, it is analysed according the following procedure:

1. The sub-goal to reach is the last node of the path that lies in a crossable area;

2. The labels of the regions crossed to reach this sub-goal determine the motion modes to apply;

3. And finally the rest of the path that reaches the global goal determines the aiming angle of the sensor.

Controlling localization: the introduction of the robot position uncertainty in the cost function allows to plan localization tasks along the path. The cost to minimise is the integral of the robot position accuracy as a function of the cost expressed in terms of time and energy (figure 11)

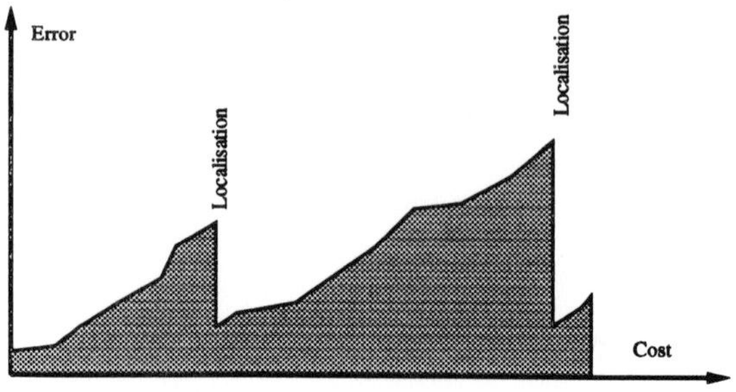

Figure 11: ***Surface to minimise to control localisation tasks***

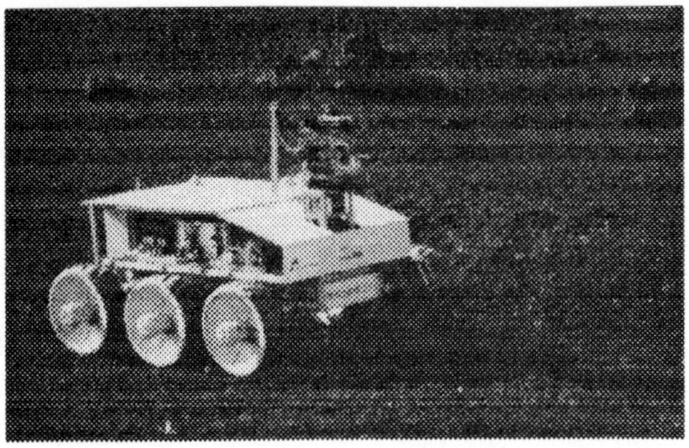

Figure 12: *ADAM in the Geroms test site*

6 Results and discussion

The terrain modelling procedures and navigation planning algorithm have been intensively tested with the mobile robot Adam[1]. We performed experiments on the Geroms test site in the French space agency CNES, where Adam achieved several ``Go To [goal]'' missions, travelling over 80 meters, avoiding obstacles and getting out of dead-ends (for more details concerning Adam and the experimental setup, refer to [12]).

Figure 13 presents two typical behaviours of the navigation algorithm in a dead-end, and figure 14 shows the trajectory followed by the robot to avoid this dead-end, on the terrain model built after 10 data acquisitions.

The navigation planner proved its efficiency on most of our experiments. The adaptation of the perception and motion tasks to the terrain and the situation enabled the robot to achieve its navigation task efficiently. By possessing several representations and planning functions, the robot was able to take the adequate decisions. However, some problems raised when the planned classification task did not bring any new information: this happened in some very particular cases where the laser range finder could not return any measure, because of a very small incidence angle with the terrain. In these cases, the terrain model is not modified by the new perception, and the navigation planner re-planned the same perception task. This shows clearly the need for an explicit sensor model to plan a relevant perception task. And this generalizes to all the actions of the robot: the robot control system should possess a model of the motion or perception actions in order to select them adequately.

[1]ADAM is property of Framatome and Matra Marconi Space currently lent to LAAS

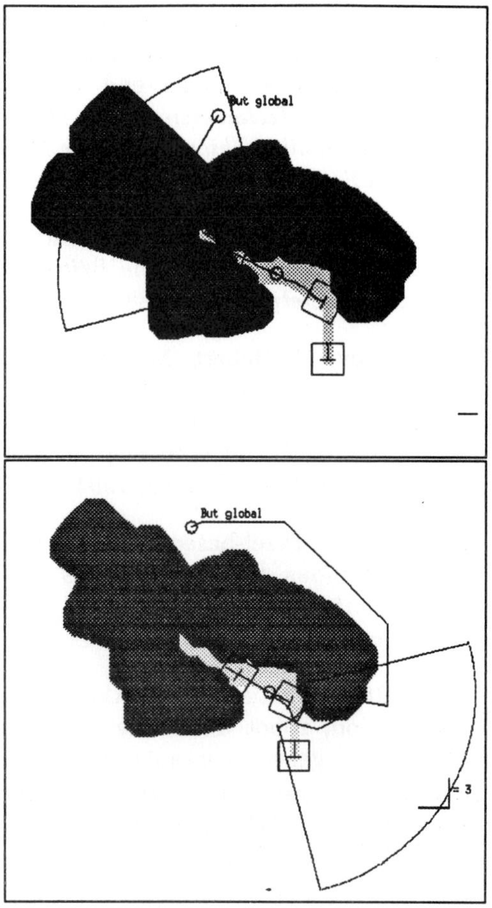

Figure 13: *The navigation planner explores a dead-end: it first tries to go through the bottom of the dead-end, which is modelled as an obstacle region, but with a low confidence level (top); after having perceived this region and confirmed that is must be labelled as obstacle, the planner decides to go back (bottom)*

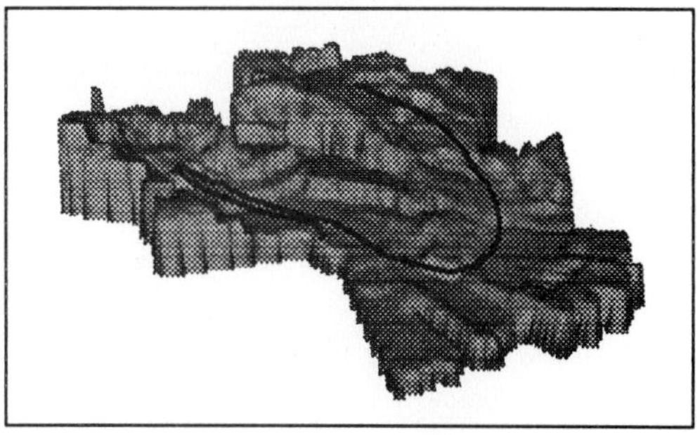

Figure 14: *A trajectory that avoids a dead-end (80 meters - 10 perceptions)*

References

[1] C. Thorpe, M. Hebert, T. Kanade, and S. Shafer. Toward autonomous driving : the cmu navlab. part i : Perception. *IEEE Expert*, 6(4), August 1991.

[2] C.R. Weisbin, M. Montenerlo, and W. Whittaker. Evolving directions in nasa's planetary rover requirements end technology. In *Missions, Technologies and Design of Planetary Mobile Vehicules. Centre National d'Etudes Spatiales, France*, Sept 1992.

[3] B. Wilcox and D. Gennery. A Mars Rover for the 1990's. *Journal of the British Interplanetary Society*, 40:484–488, 1987.

[4] E. Krotkov, M. Hebert, M. Buffa, F. Cozman, and L. Robert. Stereo driving and position estimation for autonomous planetary rovers. In *IARP 2nd Workshop on Robotics in Space, Montreal, Canada*, 1994.

[5] M. Hebert. Pixel-based range processing for autonomous driving. In *IEEE International Conference on Robotics and Automation, San Diego, California*, 1994.

[6] G. Giralt, R. Chatila, and R. Alami. Remote intervention, robot autonomy and teleprogramming : generic concepts and real-world application cases. In *IEEE International Conference on Intelligent Robots and Systems, Yokohama (Japan)*, July 1993.

[7] S. Lacroix, R. Chatila, S. Fleury, M. Herrb, and T. Simeon. Autonomous navigation in outdoor environment : Adaptative approach and experiment. In *IEEE International Conference on Robotics and Automation, San Diego, California*, 1994.

[8] T. Simeon and B. Dacre-Wright. A practical motion planner for all-terrain mobile robots. In *IEEE International Conference on Intelligent Robots and Systems, Yokohama (Japan)*, July 1993.

[9] F. Nashashibi, P. Fillatreau, B. Dacre-Wright, and T. Simeon. 3d autonomous navigation in a natural environment. In *IEEE International Conference on Robotics and Automation, San Diego (USA)*, May 1994.

[10] S. Betge-Brezetz, R. Chatila, and M. Devy. Natural scene understanding for mobile robot navigation. In *IEEE International Conference on Robotics and Automation, San Diego (USA)*, May 1994.

[11] P. Fillatreau, M. Devy, and R. Prajoux. Modelling of unstructured terrain and feature extraction using b-spline surfaces. In *International Conference on Advanced Robotics, Tokyo (Japan)*, November 1993.

[12] R. Chatila, S. Fleury, M. Herrb, S. Lacroix, and C. Proust. Autonmous navigation in natural environment. In *Third International Symposium on Experimental Robotics, Kyoto (Japan)*, October 1993.

Parallel Use of Differential and Integral Representations for Realising Efficient Mobile Robots

E.D. DICKMANNS
Universität der Bundeswehr München
Werner-Heisenberg Weg 39, D-85577 Neubiberg, Germany
e-mail: Ernst.Dickmanns@rz.unibw-muenchen.de

Abstract

In the 4-D approach to dynamic machine vision, representations of objects, both with respect to 3-D shape and to motion behavior over time, play a central role. Both differential and integral representations in both space and time have been used from the beginning; however, these specific representations for different subtasks have not been discussed explicitly in depth. Apparently, this has caused some lack of understanding of the general power of this method especially for solving the symbol grounding problem.

Therefore, a more extended treatment is given here and demonstrated at the task of road vehicle guidance. Performance levels achieved in high-speed driving with the new autonomous passenger car **VaMP** on freeways are dicussed.

1. Introduction

Road vehicle guidance has become a rather intensively studied field of research since the launch of the European EUREKA-project Prometheus in the second half of the last decade. 'Autonomous Land Vehicle' (ALV) activities in the US considered driving on roads at low speeds as a first step towards general cross-country type driving. In Japan, the Personal Vehicle System (PVS) has been a first government supported step before most of the car manufacturing companies started their own developments at the turn of the decade.

A good survey may be obtained by tracking the proceedings of such annual conferences as IEEE 'Robotics and Automation', SPIE 'Mobile Robots', the bi- (tri-) annual international conference on 'Intelligent Autonomous Systems' (all starting at about 1986) and, especially, those of the annual International Symposium on 'Intelligent Vehicles' since 1992 (see list of references [1,2,3,4]). 1986 may be considered as the year when real-world experiments with these systems started at a larger scale.

At that time sufficiently powerful microprocessors (16- and 32-bit word length) had become available for embedded systems in dense packaging for on-board use at reasonable prices and with affordable power consumption. Special architectures for machine vision were in the development phase, several of them considering single-bit processors for each picture element (pixel or pel) as the basic image processing layer (Image Understanding Architecture (IUA), Connection Machine (CM), Massively Parallel Processors (MPP)). Quasi-steady approaches predominated which evaluate each image (or temporal as well as binocular image pair) in a sequence separate from the temporal process as a whole; this requires solving the initialisation task for recognizing a continuous process in the time domain over and over again.

Extending recursive estimation techniques, well-known in the field of system dynamics, to perspective projection and image sequence processing with the use of physics-based dynamical models in 3-D space and time, the '4-D approach' developed at UBM in the mid-80ies lifted the performance level achievable with general-purpose microprocessors at hand by about two orders of magnitude; it was, however, confined to well structured environments with an appropriate set of well visible edge-features for achieving real-time performance. This real-time capability has been defined as requiring cycle times of less than about 150 ms (preferably less than 100, optimally less than or equal

40 ms). As compared to human performance levels, such systems may be considered to have the same dynamical bandwidth.

In 1987, **VaMoRs** was capable of performing 'road running' on a stretch of Autobahn free of other traffic participants at maximum speed (96 km/h) [5], controlling both lateral and longitudinal degrees of freedom based on road curvature recognition.

2. Spatial and temporal embeddings

This performance level with the relatively small computing power on board the vehicle has only been possible through an efficient parallel use of differential and integral representations which will be detailed here. Table 1 shows the systematic subdivisions chosen of the independent variables 3-D space (vertical) and time (horizontal).

All interaction with the real world happens at the point here and now (upper left field (1,1) of matrix). Inertial sensors pick up its accelerations, while an imaging sensor measures the intensity values over some spectral range at that point.

In the 4-D approach to dynamic vision it is deliberately avoided to take temporal differenes of image data because of the huge amount of data and of the noise problem involved. Time derivatives of state variables of objects are computed by well known observer techniques exploiting temporal (socalled 'dynamical') models of motion processes by prediction error feedback. This will be discussed in column 3 for local temporal integrals.

At each point in time, however, due to remote sensing with the speed of light, spatial neighborhoods may be explored. The immediate neighborhood yields spatial intensity gradients in the image, edge directions, edge curvatures and their change rate; all of this is subsumed under the term visual 'features'. They constitute spatial differentials as bottom-up information (matrix element (2,1)). These differentials may be integrated in somewhat larger environments in order to characterize pieces of the world by invariants. Rigid bodies, for example, are typical local integrals with invariant shape in 3-D space; note that this shape invariance is lost through perspective projection in the image plane. Therefore, all spatial representations are done directly in 3-D space. Knowledge about the real world is mainly affixed to 3-D objects and their motion behavior again in 3-D space and time. None-the-less, 2-D

	point in time	temporal differentials	local temporal integrals	→	global temporal integrals
point in space	intensity value *accelerations* • *translatory* • *rotatory* } *'states'*	temporal intensity gradients	local intensity range, short term intensity distribution	multiple	long term intensity and color ranges, temporal distributions
spatial differentials	edge direction edge curvature spatial intensity gradients (higher moments) 'features'	optical flow feature motion *(incremental states)*	feature trajectories (short term)	time	long term feature trajectories
local spatial integrals	object shape segmented regions of an object (color, texture) *rotational speeds, angular positions*	object state incl. speeds, shape changes *incremental object states (vel., pos., angles)*	short term feature-, object trajectory integration of dynamical models *short term state ranges (transl., rot.) of objects*	scales →	long term object trajectory *[long term history of objects (drift problems)]*
↓	multiple space scales			→	
global spatial integrals	arrangement of objects, mission space; actual situation (as seen 'at a glance')	actual situation change rate *relative object states*	extended presence, actual situation *short term development of situation*		long term development of situation mission performance *long term development of situation (drift problems)*
	based on visual measurements		*based on inertial measurements*		

Table 1: Different scale regions for representation in the 4-D approach

spatial regions in the image may be considered also to be local spatial integrals characterized by some visual property.

The transition to larger environments or to global space may be performed at different space scales depending on the task at hand and the mission context; the state of an arrangement of objects in the mission context is called a situation. Situations play an important role in decision taking since behaviors are defined in connection with mission elements to be performed, usually characterized by the situation. Several space scales may be used in parallel at different organizational levels of the overall system. However, no fixed scheme of spatial scales is considered to be useful independent of the task at hand.

The third column in table 1 is the one containing the aspects most characteristic for the 4-D approach to dynamic vision. Especially, element (3,3) constitutes the central hub for dynamic scene understanding. As will be discussed in more detail below, it serves for temporal embedding of the understanding of motion processes in three functions:

1. single step predictions for recursive state estimation for objects with known (or assumed) dynamics,
2. control computation for the own vehicle with the assignment of eigenvalues (dynamical behavioral properties) including delay time compensation, and
3. multiple prediction steps for the generation of maneuver elements (both feedforward control and reference state time histories) for the own vehicle, and of likely trajectories of other objects (e.g. for collision avoidance).

This latter property is an input to the last entry in this column characterizing the actual situation in the sense of 'extended presence' as the basis for short term decision taking.

This orientation towards physical objects (row 3) and temporal expectations (column 3) is the core of the 4-D approach. Maneuver elements of these objects of relatively short duration with well defined initial and final conditions are the units for an easy transition from the control engineering methods at this level to the quasi-static 'Artificial Intelligence'- (AI-) methods usually preferred for handling knowledge based overall mission performance (to the right and below, in table 1).

At this level, also, the integration (sometimes called data fusion) of inertial (lower italic entries) and visual sensor data (upper entries in the relevant fields of table 1) is performed. These data have nice complementary properties: inertial data are available immediately and are so at high rates; they may be integrated numerically to velocity components and position changes. Good dynamic properties are available at low costs; however, longer term drift problems exist. On the contrary, results from image data, usually, become available only after some longer delay time; higher angular rates cause motion blur due to light intensity integration times in the sensor elements and may destroy the image content. However, if the image sensors are short-term stabilized inertially, elements of the stationary environment can be recognized and longterm inertial stabilization is easily realized by visual feedback.

As opposed to [19] and previous publications to the 'RCS'-scheme there is no hierarchy of sensory or actuatory elements prescribed in this approach. Also, the transition from local to global scales is left open for optimal adaptation to the special task at hand. Here, the central elements are objects and subjects [20] (defined as objects with the capability of control actuation and mode switching depending on their internal 'mental state', i.e. the totality of internal representations of the recursively estimated state relative to other objects which constitute the situation). Control time histories, feedback control laws, and their effects upon motion behavior are the key knowledge elements in the 4-D approach.

This knowledge is represented in symbolic form on the mission control level allowing for task decomposition according to these behavioral capabilities in the planning phase. In the mission performance phase, switching between control modes is done by a monitoring process. For reliable mission performance under perturbations it has to be guaranteed that the radius of convergence of the ensuing control mode is larger than the outer bound of the final state errors of the previous mode at the actual time of switching.

This general scheme will be discussed in the sequel for the example of lane changing on a freeway.

3. Differential Representations

Road curvature in a plane (reduced to that of its skeletal line) is the most simple differential geometry description for representing road heading as a function of arc length: Curvature as the inverse of the radius of curvature R ($C_R = 1/R$) is the change of the road heading angle χ_R over arc length l

$$d(\chi_R) = C_R \cdot dl. \quad (1)$$

Driving at speed $V = dl/dt$ along the road, the temporal change of curvature at the local point of the vehicle centre of gravity (cg) is, by applying the chain rule of differentiation,

$$dC_R/dt = dC_R/dl \cdot dl/dt = C_{R_1} \cdot V. \qquad (2)$$

Assuming the rate of change of curvature along the arc length C_{R_1} as constant, as has been specified as international standard for the construction of high speed roads, a simple ordinary differential equation results.

This road building standard has been adopted on the ground that curvatures of trajectories driven by cars with front wheel steering directly correspond, to first order approximation, to the actual steer angle chosen, as may be seen from fig.1.

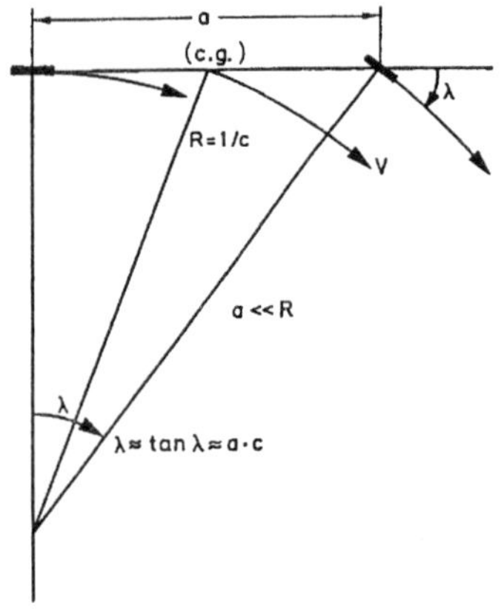

Figure 1. Horizontal curvature of a trajectory driven by a vehicle with front wheel steer angle λ

From kinematic relationships it can be seen immediately that the curvature of the trajectory C_t depends on the axle distance a and the steer angle λ as the control variable for lateral maneuvering:

$$\sin(\lambda) = a/R_f = a \cdot C_{tf}. \qquad (3)$$

Since, usually, the axle distance is much smaller than the radius of curvature of the front wheel trajectory, and since the horizontal radii of curvature of the front wheel and the trajectory of the c.g. are approximately the same, eq. (3) may be reduced to the form

$$C_t = \lambda/a. \qquad (4)$$

With this simple model, staying on the road by lateral control means to select the steer angle such that $C_t = C_R$ once the vehicle has been aligned to the road; this is called the anticipatory part of lateral control.

Since there will be perturbations and dynamical deviations from this simple kinematic model, a feedback component for coping with these perturbations has to be added on top.

Resorting to dynamical models of the road vehicle for lateral guidance this can easily be achieved by conventional control engineering methods; state vector feedback is known to solve this problem in an optimal way. Well acceptable single track (bicycle) models for normal guidance tasks have been discussed in [6,7,8]. The following one is taken from [9]); it has five state variables: For automatic control the steer angle is the output of an electric torque motor modeled as a first order integrator; therefore, the steer rate $\dot{\lambda}$ is the control variable. β is the vehicle slip angle (difference between trajectory and vehicle heading), $\dot{\psi}_{abs}$ is the inertial yaw rate of the vehicle, ψ_{rel} the heading direction of the vehicle relative to the road tangent and y the lateral offset from lane center.

$$x^T = [\dot{\psi}_{abs}, \beta, \psi_{rel}, y, \lambda]; \quad \dot{x} = Ax + bu + g\, C_R,$$

with

$$A = \begin{bmatrix} -1/T_\psi & 0 & 0 & 0 & V/(a\,T_\psi) \\ -1 & -1/T_\beta & 0 & 0 & 1/(2\,T_\beta) \\ 1 & 0 & 0 & 0 & 0 \\ 0 & V & V & 0 & 0 \\ 0 & 0 & 0 & 0 & 0 \end{bmatrix}$$

$$b^T = [0 \;\; 0 \;\; 0 \;\; 0 \;\; 1]$$

$$g^T = [0 \;\; 0 \;\; -V \;\; 0 \;\; 0] \qquad (5)$$

$u = \dot{\lambda}$ (control); $\qquad C_R$ = horizontal curvature;

$T_\beta = mV/(2b)$; $\qquad T_\psi = 2V\, I_z/(b\, a^2)$

m = vehicle mass, V = vehicle speed, b = cornering coeff. of tires, I_z = inertia around vertical axis, a = axle distance

This is the set of differential equations representing knowledge about the temporal development of the motion process including control applications; it will be used in order to predict future states and to intelligently control the extraction of features from the vast data stream coming from several video cameras. Note, that this representation does not contain any initial values for the state variables, and that the control terms are yet unspecified. It constitutes knowledge about the invariance properties of the plant irrespective of actual control inputs $u(t)$. By specifying $u(t)$, the trajectory can be determined by integration (see below).

4. Local Integrals

4.1 Spatially local integrals for visual perception in look-ahead range:

The parameters in the differential geometry model of the curved road are determined by visually measuring the lateral positions of multiple boundary line elements simultaneously. Therefore, the first and second integrals along the skeletal line have to be evaluated in order to obtain the cartesian coordinates favorable for performing perspective projection.

The first integral of eq.(1) with the clothoid model

$$C_R = C_{R_0} + C_{R_1} \cdot l \qquad (6)$$

yields

$$\chi - \chi_0 = C_{R_0} \cdot l + C_{R_1} \cdot l^2/2. \qquad (7)$$

Within a coordinate system locally tangent to the road skeletal line ($y_C = 0$) at the longitudinal position of the projection center ($\chi_0 = 0$) the following relations hold

$$dx = dl \cdot \cos(\chi)\,; \quad dy = dl \cdot \sin(\chi).$$

For small angles χ (< 15°) the cosine may be approximated by 1 and the sine by its argument, so that follows:

$$x - x_0 = l \text{ and } y_C = C_{R_0} \cdot l^2/2 + C_{R_1} \cdot l^3/6\,. \quad (8)$$

This integral representation has been exploited in order to estimate the curvature parameters by machine vision [10,11].

In general shape recognition problems, tracking contour lines with a tangent operator may lead to similar piecewise differential representations of 2-D shapes which are position-, orientation- and scale invariant [12].

4.2 Temporally Local Integrals

Single step for state estimation: Single step integration over one cycle time is needed for prediction in the context of recursive estimation. These integrals are determined analytically beforehand with one of the standard methods available (e.g. algebraically in the frequency domain after Laplace transformation) and evaluated numerically in each time-step after innovation by the recursive estimation process and with the actual control input.

Local integrals for control determination: Two aspects are of importance here:

1. Derive feedforward control components taking care of the spatial structure recognized in the look-ahead range (e.g. curvatures), and
2. determine feedback control components in order to cope with perturbations and with model errors.

In quasi-steady curve driving (index 'c') the following relations hold:

$$\dot{\psi}_{abs,c} = V \cdot C_R; \qquad \beta_c = C_R \cdot (a - mV^2/b)/2$$

$$\psi_{rel,c} = C_R\,(mV^2/b - a)/2 = -\beta_c\,; \quad \lambda_c = a\,C_R\,. \ (9)$$

The lateral offset should be $y_c = 0$; in order to enforce this, feedback control of the integral $y_I = \int y(t)dt$ is added to the state vector feedback with $y_{I,C} = 0$,

$$x_{fb} = x_c - x\,. \qquad (10)$$

Differentiating the last term in eq. (9) with respect to time yields

$$\dot{\lambda}_c = a\, d(C_R)/dt = a \frac{d\,C_R}{dl} \cdot \frac{dl}{dt} = a\,C_{R_1}\,V. \qquad (11)$$

According to eq. (4) and the last line of eq. (5) this directly yields the feedforward control component

$$u_{ff} = a\,C_{R_1}\,V. \qquad (11a)$$

In order to achieve fast and robust overall road following properties, superimposed feed-forward (u_{ff}) and feedback control

$$u_{fb} = -K^T x_{fb}, \quad u = u_{uff} + u_{fb} \qquad (12a,b)$$

is applied according to figure 2.

For determining the feedback gain vector K the dynamical model is locally linearized and used as a frozen state model so that Laplace transformation yields an algebraic set of equations in the frequency domain. In this transformed space, integration can easily be done just by division by the independent complex variable (s). With all (the estimated) physical state variables available the eigenvalues can be assigned to desirable locations (pole placement); this yields the feedback vector K (for details see [6,8]).

Since several terms in the dynamical model depend on speed V, the gain computation is done at every cycle repeatedly.

Multiple Prediction Steps for Maneuver Elements: For specific, frequently needed maneuver elements like lane changes with stereotypical control time history inputs over the maneuver time span $T_s = t_f - t_0$,

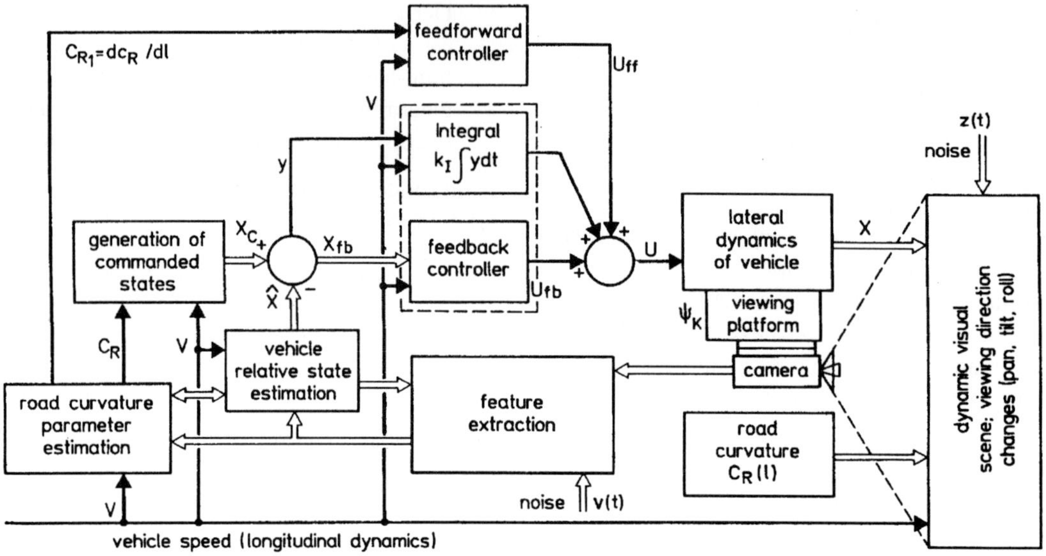

Figure 2. Parallel use of differential and integral representations for state and parameter estimation as well as for control computation

integrals may be computed in advance (analytically or numerically). The complete transition of state variables from t_0 to t_f may be represented on higher system levels by a symbol like 'lane change'; on this level, just the maneuver time and the final lateral offset are of interest, maybe, in addition, some extremal values during the maneuver, like maximal lateral acceleration or maximal heading angle change with their times of occurrence after the trigger point.

On the execution level, the control time sequence u(t) has to be stored either point by point or coded in some easily computable form. Figure 3 shows a typical steer control time history for lane change.

Parameters are the magnitude of $\dot{\lambda}$, the duration of each segment, the relative duration of the central null-input section s (in percent of T_c), and the total control input maneuver time T_c.

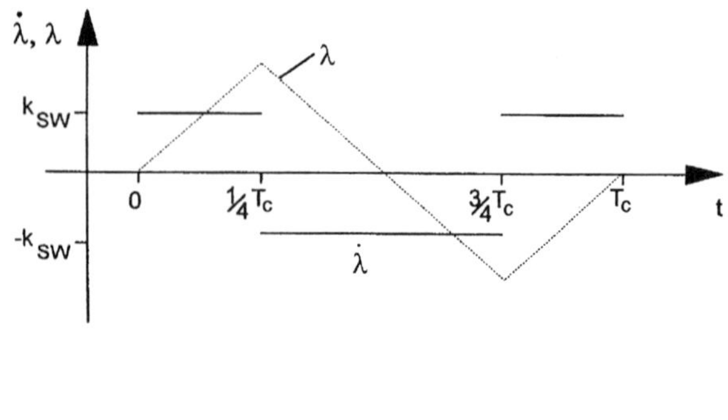

Figure 3. Control ($\dot{\lambda}$) and steer angle time history for lane change

Figure 4 shows some state variable time histories resulting from the control law of figure 3 for a speed of 100 km/h (27.8 m/s) and different parameters T_c = 6 and 2 sec, and s = 0 respectively 20 % (0.2). The parameters for a lane change of 3.75 m width, some maximal values of the maneuver, as well as total maneuver time T_s and distance are given in table 2 for speeds of 40 and 100 km/h and some more parameter combinations; the trajectories of fig. 4 may be found in rows 4, 6 and 9.

It is seen that both speed V and control maneuver time T_c do have a strong influence on the trajectories and their extremal values. The shorter the maneuver times (column 2), the larger the control inputs (column 3) have to be for the same lane width, and the higher are the extremal values of the state variables (columns 4 to 8) and the lateral acceleration (column 7) encountered. Column 5 gives the maximum heading change, column 6 the maximal amplitude of the slip angle, and column 8 the maximal lateral speed.

In column 9 the integral of the absolute value of the lateral acceleration as a measure of comfort is shown; in this simple maneuver this is, of course, twice the value of the maximal lateral speed. Column 10 gives the total time T_s for the state variables to settle to the steady state after the lane change maneuver; it can be seen from rows 6 and 9 that at higher speeds for

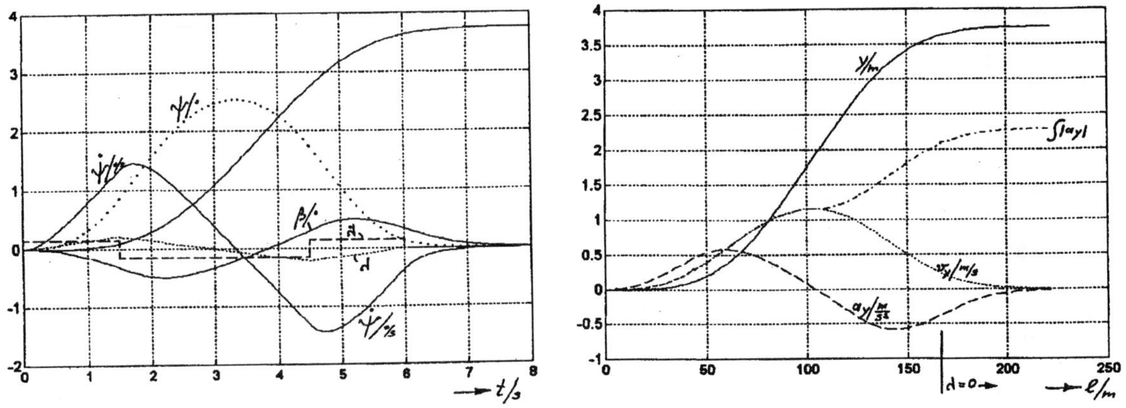

Fig. 4.1. V = 100 km/h; T_c = 6s; no central 0-input time span (s = 0)

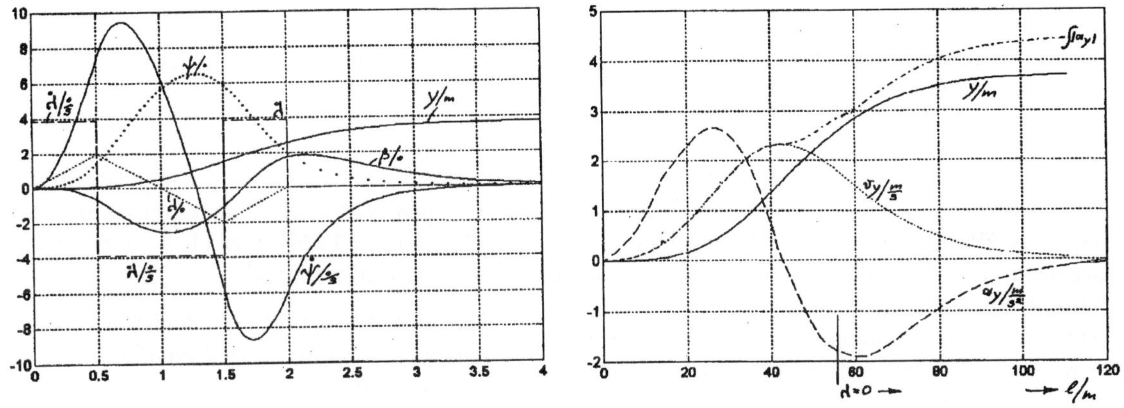

Fig. 4.2. V = 100 km/h; T_c = 2s; no central 0-input time span (s = 0)

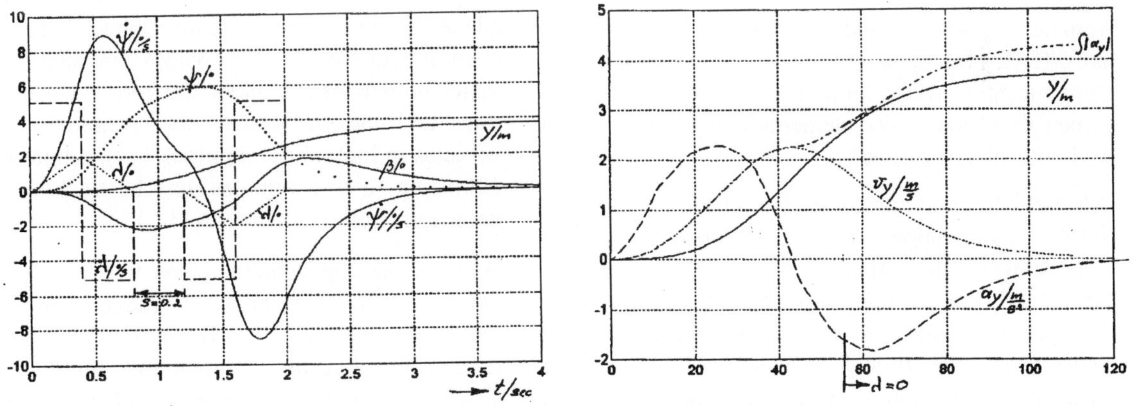

Fig. 4.3. V = 100 km/h; T_c = 2s; s = 20 % central 0-input time span

a) steer-(λ), slip-(β) and yaw angle (ψ) [in degrees] ($^{\bullet}$) = time derivative

b) lateral acceleration a_y [m/s^2] lateral speed v_y [m/s] lateral position y [m]

Figure 4. 'Lane change' control ($\dot{\lambda}$) and state variable time histories as parameter dependent knowledge elements; parameters: driving speed V, control actuation time T_c

Column	①	②	③ T_c	④	⑤	⑥	⑦	⑧	⑨	⑩ T_s	⑪	
	Speed V km/h	control maneuver time / s	$\dot{\lambda}$ deg/s	λ_{max} / deg	ψ_{max} / °	$\lvert\beta\rvert_{max}$ / °	$a_{y_{max}}$ / m/s^2	$v_{y_{max}}$ / m/s	$\int\lvert a_y\rvert$ / m/s	state maneuver time / s	maneuver distance / s	remark
s = 0	40	6	0.9	1.4	6.4	0.16	0.7	1.2	2.4	6.5	75	
	40	4	3.1	3.1	9.6	0.36	1.5	1.8	3.6	4.5	54	
	40	2	24	12	18.5	1.7	$5.2_{-4.8}$	3.2	6.3	2.7	34	
s = 0	100	6	0.16	0.22	2.5	0.5	0.6	1.2	2.3	7.9	220	fig 3.1
	100	4	0.5	0.5	3.7	0.95	$1.1_{-1.05}$	1.6	3.2	6.0	168	
	100	2	3.9	2	6.5	2.6	$2.7_{-1.9}$	2.35	4.7	4.4	125	fig 3.2
s = 0.2	100	6	0.18	0.23	2.2	0.5	0.6	1	2	7.9	~220	
	100	4	0.64	0.51	3.2	0.9	1_{-1}	1.45	2.8	6	168	
	100	2	5.1	2	6	2.2	$2.3_{-1.8}$	2.26	4.3	4.4	125	fig 3.3

Table 2. Expectations for lane change maneuvers $\Delta y = 3.75m$ according to the 5th-order bicycle model with piecewise constant steer rates $\dot{\lambda}$

dynamic maneuvers (small T_c) this total time may be more than twice the feedforward control input time. Figures 4.2b and 4.3b show that the maximal negative accelerations occur after finishing the feedforward control input at zero steer angle when also the slip angle attains its maximum. Comparing rows 3, 6 and 9 in column 7 it is recognized that peak lateral acceleration decreases strongly with speed and slightly with a central zero steer angle control arc dividing the lane change maneuver into two impulse-like maneuver elements. Note that, because of the eigendynamics of the vehicle, the positive and negative acceleration peaks are different for the same control input for these highly dynamic maneuvers; for larger separation times between these maneuver elements they become equal and opposite, of course.

All these different maneuvers are subsumed under the symbol 'lane change'; it achieves its specific meaning depending on the situation in which it is invoked. Speed and lane width are taken from the actual measurements (tachometer resp. vision) while the selected control maneuver time T_C is specified by the calling module 'behavior decision'; this time, however, is not equal to the actual state transition time T_S. From these examples it should have become clear that interfacing the control engineering methods with AI-methods has to be done with care, taking dynamical and situational aspects into account.

If perturbations are to be expected, the state variable time histories computed serve as command input to a superimposed feedback loop counteracting the perturbations.

Figure 5 shows control and state time histories of a slow real lane change maneuver of the new autonomous vehicle **VaMP** (see [13]) containing the superimposed feedback mode.

It is seen that appreciable deviations from the ideal maneuver occur in practical situations with continuous corrective control action.

Another application area of local integrals for situation understanding is the prediction of trajectories of passing vehicles in order to arrive at the decision whether a lane change can be made safely in two-lane traffic on a unidirectional highway. **VaMP** has active bifocal vision both to the rear and to the front. It can detect and track and perform relative state estimation to other vehicles in look-ahead and look-back ranges of up to 120m; up to five vehicles may be tracked simultaneously in both hemispheres [14,15].

When a passing vehicle leaves the rearward wide-angle field of view its reappearance in the forward wide-angle field of view is predicted, based on the relative speed vector and the known blind area to the side of the vehicle. Keeping 'mentally' track of all vehicles leaving the rearward viewing cone, and checking these against those reappearing in the frontal viewing cone it can be concluded when the neighboring lane is free for a lane change. Of course, the time to contact of another vehicle approaching from the rear with high excess speed with a virtual vehicle in the neighboring lane has also to be checked.

All these computations (approximate integrals) are performed in a simple form under the assumption of

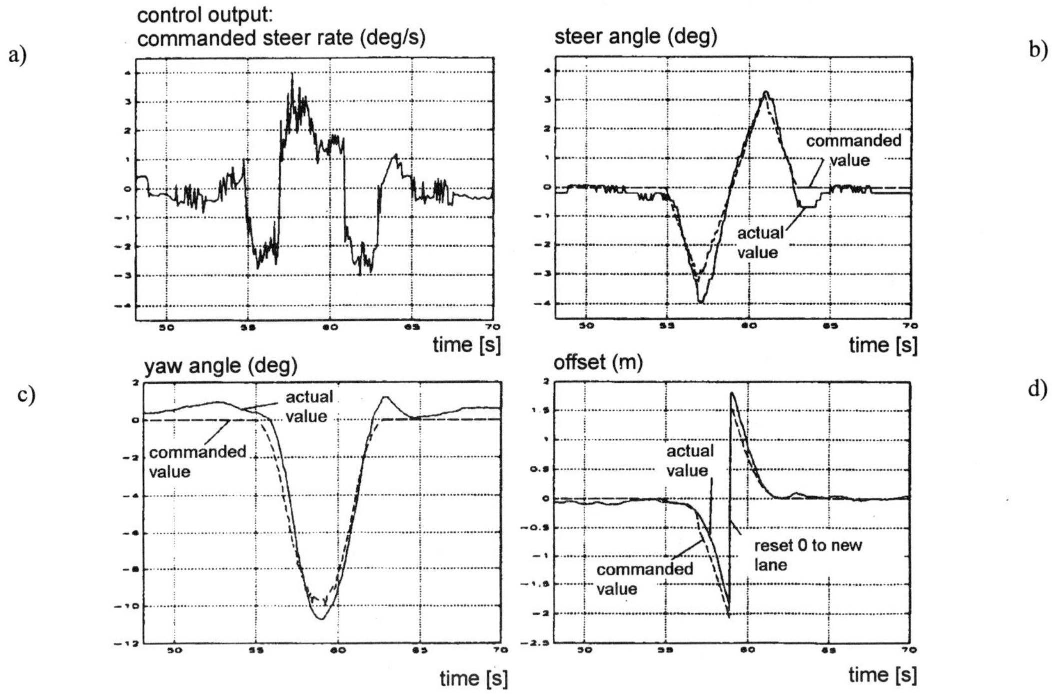

Figure 5. Time history of steer rate (a), steer angle (b), yaw angle (c) and lateral offset (d) commanded for a lane change maneuver; superimposed feedforward and feedback control (after [8]).

unchanged driving strategies of the other partners. In October 1994 **VaMP** was the first vehicle to make the decision for a lane change completely on its own in normal French Autoroute traffic on the A1 from Paris to Senlis [16]. The safety driver, of course, checked the actual situation independently before allowing the lane change maneuver to be performed automatically. Fully autonomous driving at speeds up to 130 km/h has been demonstrated.

5. Global Integrals for Mission Performance

On freeways, global maneuvering is being performed by proper lane selection and lane following.

On state roads with same level intersections, the detection of crossroads, junctions and road forks is another important perceptual capability. The behavioral capability of turning-off onto cross-roads and of picking the proper road fork is the additional performance element necessary for achieving global mission performance capabilities.

Proper sequencing of all these behavioral modes then allows to solve general transportation tasks in topologically known road networks.

The combined perceptual and behavioral capability of detecting crossroads, recognizing the intersection parameters by fixation-type viewing direction control, and of performing the turn onto the cross-road by combined feedforward and feedback components has been developed at UBM over the last years [17,18]. Exact knowledge about the intersection parameters is not needed beforehand; this is determined while approaching the intersection by adapting parameters in generic cross-road models using visual feedback.

With this additional maneuver element (local integrals), global mission performance can be done on general road networks too. Note that because of the local behavioral capabilities no geographically exact map is needed in order to perform the mission; topological map information in conjunction with the capability of recognizing proper landmarks is sufficient for guiding the vehicle to its destination.

6. Conclusions

Exploiting differential and integral representations in parallel, very efficient algorithms for perception and control can be obtained by taking advantage of both control engineering and AI methods. Figure 6 shows

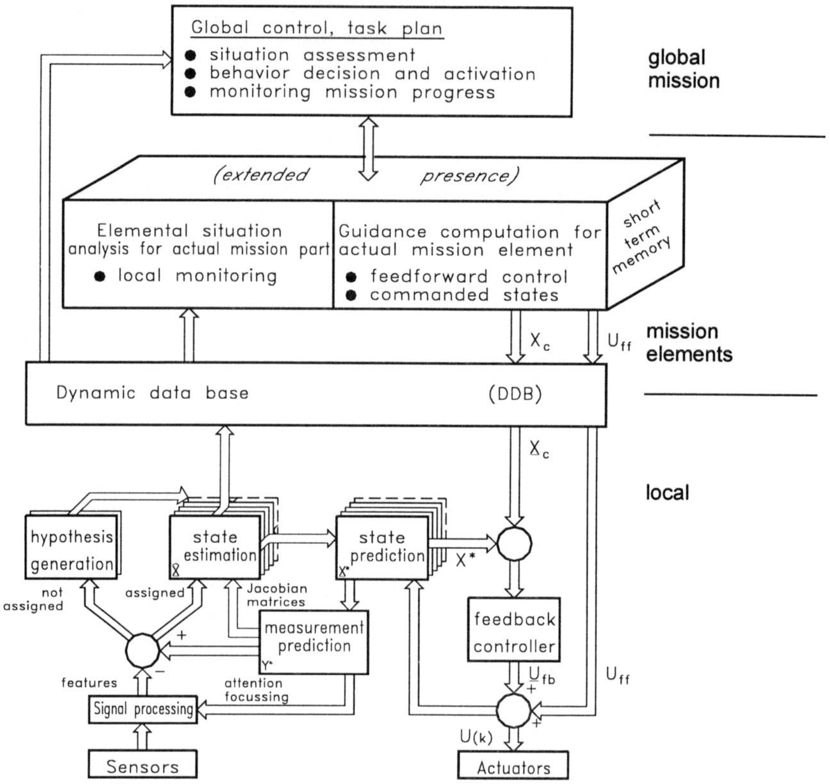

Figure 6. General block diagram for hierarchical control of complex mission using differential and integral representations in parallel in both space and time

in a general block diagram how complex tasks can be performed working with differential (lower level) and both local (medium level) and global integrals (higher level) in parallel. Switching between modes has to be done according to the situations encountered.

7. Literature

[1] Masaki I. Proceedings of Internat. Symposium on 'Intelligent Vehicles '92', Detroit, 1992.

[2] Masaki I. Proceedings of Internat. Symposium on 'Intelligent Vehicles '93', Tokyo, 1993.

[3] Masaki I. Proceedings of Internat. Symposium on 'Intelligent Vehicles '94', Paris, 1994.

[4] Masaki I. Proceedings of Internat. Symposium on 'Intelligent Vehicles '95', Detroit, 1995.

[5] Dickmanns ED, Zapp A. Autonomous High Speed Road Vehicle Guidance by Computer Vision. 10th IFAC World Congress Munich, Preprint 4, 1987, pp 232-237.

[6] Zapp A. Automatische Straßenfahrzeugführung durch Rechnersehen. Dissertation, Universität der Bundeswehr München, Fakultät für LRT, 1988.

[7] Mysliwetz B. Parallelrechner-basierte Bildfolgen-Interpretation zur autonomen Fahrzeugsteuerung. Dissertation, Universität der Bundeswehr München, Fakultät für LRT, 1990.

[8] Brüdigam C. Intelligente Fahrmanöver sehender autonomer Fahrzeuge in autobahnähnlicher Umgebung. Dissertation, Universität der Bundeswehr München, Fakultät für LRT, 1994.

[9] Brüdigam C. Das Fahrzeugführungsmodul VC. UniBwM/LRT/WE 13/IB/93-3, 1993.

[10] Dickmanns ED, Zapp A. A Curvature-based Scheme for Improving Road Vehicle Guidance by Computer Vision. In: 'Mobile Robots', SPIE Proc. Vol. 727, Cambridge, Mass., Oct. 1986, pp 161-168.

[11] Dickmanns ED, Mysliwetz B. "Recursive 3D Road and Relative Ego-State Recognition".

IEEE-Trans. PAMI, Vol.14, No.2, Special Issue on 'Interpretation of 3D Scenes', February 1992, pp. 199-213.

[12] Dickmanns ED. 2D-Object Recognition and Representation using Normalized Curvature Functions. In M.H. Hamza (Ed): Proc. IASTED Int. Symposium on Robotics and Automation, Acta Press, 1985, pp 9-13.

[13] Maurer M, Behringer R, Dickmanns D, Hildebrandt T, Schiehlen J, Dickmanns ED. VaMoRs-P An Advanced Platform for Visual Autonomous Road Vehicle Guidance. SPIE Conference 'Mobile Robots IX', Oct. 31 - Nov. 4, 1994.

[14] Thomanek F, Dickmanns ED, Dickmanns D. Multiple Object Recognition and Scene Interpretation for Autonomous Road Vehicle Guidance. Int. Symposium in Intelligent Vehicles '94, Paris, Oct. 1994.

[15] Thomanek F. Visuelle Erkennung und Zustandsschätzung von mehreren Straßenfahrzeugen zur autonomen Fahrzeugführung. Dissertation, Universität der Bundeswehr München, Fakultät für LRT, 1995.

[16] Kujawski C. Deciding the Behaviour of an Autonomous Mobile Road Vehicle in Complex Traffic Situations. 2nd IFAC Conf. on Intelligent Autonomous Vehicles-95, Helsinki, University of Technology, Otaniemi (Finland), June 1995.

[17] Müller N. Feedforward Control for Curve Steering for an Autonomous Road Vehicle. IEEE Int. Conf. Robotics and Automation, Nice, 1992, pp 200-205.

[18] Müller N, Baten S. Image Processing Based Navigation with an Autonomous Car. Int. Conf. on Intelligent Autonomous Systems (IAS-4), Karlsruhe, März 1995.

[19] Albus. Outline for a theory of intelligence. IEEE Trans. Systems, Man, and Cybernetics, Vol. 21, No. 3, May/June 1991

[20] Dickmanns ED. Subject-Object Discrimination in 4-D Dynamic Scene Interpretation Machine Vision. Proc. IEEE-Workshop on Visual Motion, Newport Beach, 1989, pp 298-304

Robust Autonomous Low-Cost Mobility

Wendelin Feiten and Gisbert Lawitzky

Siemens AG, Zentrale Forschung und Entwicklung, ZFE T SN 4
D 81730 München,
e-mail: [wendelin.feiten | gisbert.lawitzky]@zfe.siemens.de

Abstract

To become a widely useable technology, mobile robots have to be very robust and exhibit a high degree of autonomy. On the other hand, they have to meet severe cost limits. This makes us think that the first generation of mobile service robots will be mainly based on ultrasonic sensors. In this paper we discuss various approaches how to implement both robust manoeuvring and obstacle avoidance as well as navigation on this basis.

1 Introduction

The first applications for autonomous mobile robots are transport and contract cleaning. Both types of robots are there, but have not yet reached the necessary level of autonomy. In fact it seems that mobile service robots need a very high degree of autonomy to be effecient enough to earn their money. Part of this autonomy is that the robot does not require prestructuring like the installation of beacons or time consuming preparation of maps.

In order to substitute human labour, autonomous robots have to operate reliably without user interaction for most of the time, e. g. more than 80% of the time in contract cleaning. Even if the degree of autonomy would generally be high enough, this still requires a high degree of robustness including selfdiagnosis and reconfiguration of the robot, error recovery and exception handling. The costs for the existing robots are not much too high, maybe less than a factor of two. Prices have come down in other areas on a larger scale, given a certain piece count.

The requirements concerning robustness and autonomy, however, have not yet been met.

In the following chapters, we give an overview of promising approaches to the various problems along the way to this goal, together with some results of the related work in our group.

2 Hardware Design Issues

Low cost means to restrict oneself to components that will predictably be low cost when the product is supposed to be marketed.

The main hardware components of mobile robots are sensors, computers, chassis, motors and batteries and communication.

The last four issues are mainly defined by the task at hand. Hence the main design decisions where we can influence the cost of autonomous mobility concern the sensors and the algorithms, i.e. computers as far as hardware is concerned.

The tasks of the sensors, with respect to mobility, are obstacle avoidance and localisation. The trivial sensor for obstacle avoidance is the bumper, and the trivial sensor for localisation is odometry. Every robot has them.

Among the sensors with a wider range, the competition is mainly between ultrasonic sensors, video cameras, optical distance switches and optical range finders (in general laser range finders), scanning 1 or 2 dimensions.

Laser range finders can provide very valuable information. Given the price limit for our first robots, however, we didn't include them in that generation.

For navigation and obstacle avoidance, optical distance switches offer no sufficient resolution of length measurement.

Therefore our decision was between ultrasonic sensors and vision. Ultrasonic sensors are well understood, simple and low-cost. Their drawbacks are low angular resolution, large share of wrong readings and low data rate. However, these drawbacks can be overcome by the use of appropriate information processing techniques.

Video cameras give high angular resolution, but distance measurement poses a problem. It also is a problem to watch the entire volume around the robot. This will probably require a pan-tilt-vergence head, with all the complexity that comes with it. Nevertheless, with prices coming down significantly and advances in algorithms, computer vision is a technology that will become very important in the future.

From these observations we conclude that the first products will be based on ultrasonic sensors and odometry for indoor applications, possibly with additional gyros and laser scanners for navigation and optical sensors for collision avoidance, depending on the application.

In difficult environments the navigation can be improved by the use of inertial sensor systems, esp. gyroscopes [1]. In office environments however, according to our results they might be dispensible.

Consequently, we try to achieve as much functionality as possible based on very simple sensor hardware. In our first robot aimed at indoor applications [Fig. 1], we use a conventional ultrasonic sensor system with 24 transducers and odometry on the driven wheels.

Which computer to use in a laboratory model is more a question of ease of development than of cost. It is important to note, however, that computer power is becoming ever cheaper. In the four years of development in our group, we upgraded from a 386, 25 Mhz to a 486, 66 Mhz, and the last board did not cost much more than the first one. Now a Pentium is available for the same price, and this will probably go on for a long time. For a product, we expect PC-based embedded control or corresponding high-end controllers to be available.

3 Software Design

With our simple sensor hardware, a lot of processing is required to achieve the desired behaviour. The robot should

- robustly avoid obstacles at a reasonable speed
- efficiently manoeuver around obstacles

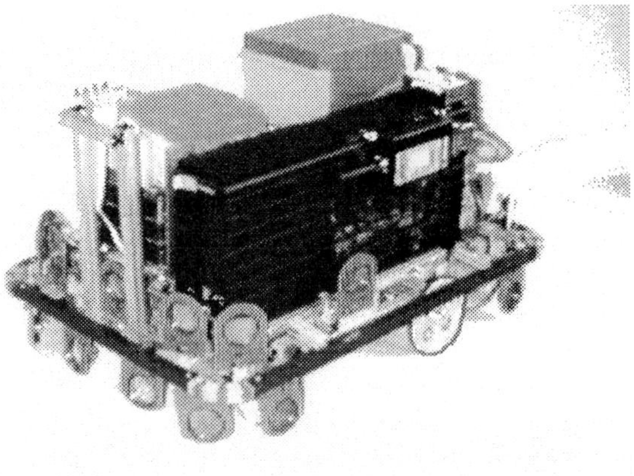

Fig. 1: Experimental platform ROAMER 1

- know its own position and orientation during the entire period of operation
- be able to use given map information
- be able to autonomously acquire map information
- plan geometric paths
- plan sequences of actions.

3.1 Robust Obstacle Avoidance

Robust obstacle avoidance depends mainly on an up to date, complete image of the environment. Due to the well known limitations of ultrasonic sensors (i.e. poor lateral resolution, low firing rate, many unuseful readings), such an image can not be expected to be acquired in one shot and not even in one complete scan. Therefore a sensor data integration over the time needs to be done. We use a variant of Borenstein's grid map [2]. In this approach, for each measurement below a preset time-out distance the occupancy value of one randomly selected grid map on the arc segment corresponding to this distance and to the sensors beam width is incremented, and along the line from the sensor to this cell, the occupancy values are decreased [3].

Our experience of some years of frequent operations in unprepared everyday home or office environments is that this grid map is reactive and precise enough to provide a basis for robust obstacle avoidance. With a firing rate of about five complete scans per second, it allows the robot to drive at up to 50 cm per second depending on the clutteredness of the environment.

3.2 Efficient Reactive Manoeuvring

The aim of efficient reactive manoeuvring (the pilot) is to take the robot from its current position to a given intermediate target position while robustly avoiding the obstacles as given in the local grid map. The manoeuvres should start as early as possible, the robot should not slow down more than necessary, and they should be efficient in approaching the target point. Last but not least the implementation should be fast. Below we give a sketch of our approach; the details can be found in [4], [5].

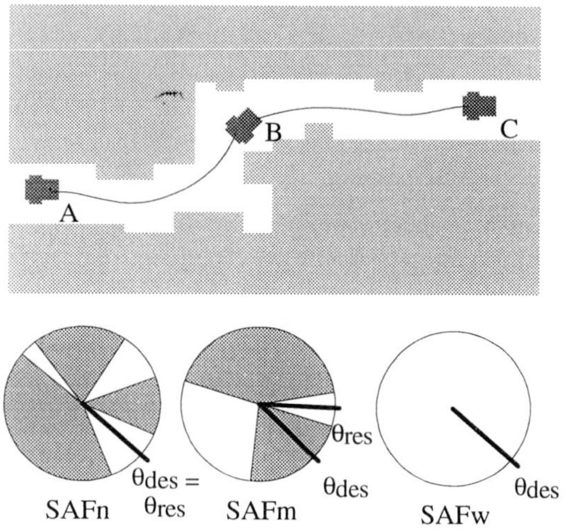

Fig. 2: Steer angle fields

In order to make the robot as agile as possible, we take geometry and kinematics of the robot into account right from the beginning. For every grid cell in the local map, those steer angles of the robot are precalculated and stored at which the robot would not hit this particular grid cell when travelling a corresponding arc for a given distance. Finding the free steer angles is then reduced to a simple intersection of the corresponding steer angles for all occupied grid cells.

Given the free steer angles, a steer angle for the robot is selected which approaches the target point as closely as possible. These operations are fast enough to be performed every 50 ms.

The robot is able to travel in most cases reliably and efficiently from one point to another point, if these points are not too far apart and if there is a more or less direct passage [Fig. 2]. Of course, these two condition need not necessarily be true for the starting point and the target point in a more complex user defined task.

If the target point can't be reached, a planner [see Section 3.5], [6] is invoked to generate a new intermediate target point.

3.3 Localisation

To be able to perform coherent operations, extended over a longer time and over wider distances, the robot should be able to know where it is at any time during operation.

Internal sensors like odometry and inertial navigation systems will help in this task, but they will not suffice in the long run. The robot has to establish a reference to some known feature in its environment.

Two different approaches to localisation can be distinguished

- symbolic navigation and
- cartesian navigation.

In symbolic navigation, the stations that the robot needs to reach or to traverse are marked by semantic labels like 'door', 'crossroad', 'T-joint' or 'X's office' and so on, and a graph is constructed joining these stations [7]. The difficult part in this approch is to reliably state whether or not the robot has reached a certain station.

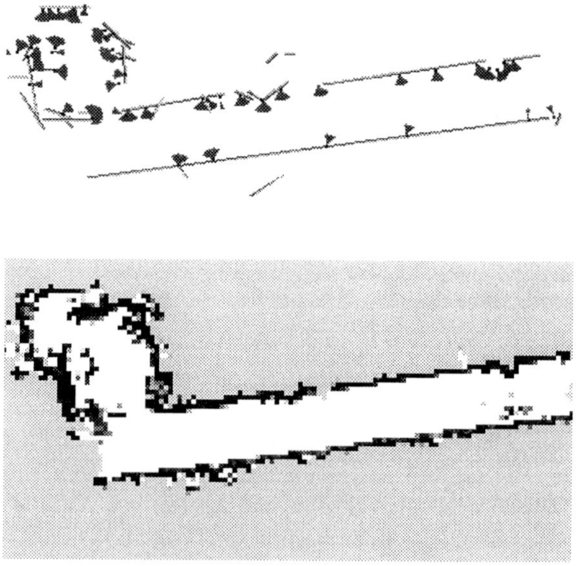

Fig. 3: Cartesian landmark based navigation and global grid map

Although ultrasonic sensors can in principle be used for this type of navigation, we didn't want to rely exclusively on this. Instead, as a basis which can be complemented by symbolic navigation, we implemented cartesian navigation using natural landmarks, in our case vertical walls end edges [8], [9]. The landmarks are given in absolute cartesian coordinates.

For this moment, let's assume that enough landmarks are known. Later on we will discuss how these landmarks are acquired under operation.

The procedure of localisation is based on a model of the robot's motion and on the entries in the landmark map. The robot's position and orientation is estimated using an Extended Kalman Filter. Based on this estimate the measurements of the ultrasonic sensors are predicted and then compared to the real measurements. Those measurements are rejected that fall outside a validation gate depending on the covariance matrix of the robot's position and orientation estimate. This takes care of the case that landmarks are obscured by people or by things placed in front of them. It also gets rid of errouneous sensor readings. The remaining measurements are used to update the robot's position and orientation estimate and recalculate the corresponding covariance matrices in the Kalman filter. We experimented a lot with this localisation module and found it stable for at least several hundred meters.

In the case that the localisation process should become unstable, there is a module for complete relocalisation. For this purpose, a sufficiently large local landmark map is extracted [see Section 3.4] and fitted into the original map.

3.4 Landmark Extraction

If not sufficiently many landmarks are known in advance, the robot is able to extract the landmarks on its own. For this purpose, it starts out with the raw ultrasonic measurements rather than using the information in the grid map to avoid the error due to rasterization.

The sensor readings are clustered according to two different hypotheses:

- that they originate from a wall or
- that they originate from a vertical edge.

Sensor readings that originate from a wall are characterized by the fact that there is a common tangent to the arc segments describing the possible positions of the origin of the echo, sensor readings originating from corner exhibit a common intersection point.

These clusters are candidates for landmarks. If such a cluster can be confirmed often enough, i.e. enough measurements are consistent with the corresponding hypothesis, a landmark is generated. The landmarks are given by random variables which are updated with new measurements. The decisive point is to assign the right amount of correction to the landmark position estimate on the one hand and to the robot position estimate on the other hand. This amount of correction is calculated based on the respective covariance matrices for those two estimates.

3.5 Path Planning

Path planning is part of the autonomy of a mobile robot. What can be planned depends on the available information about the environment and the task.

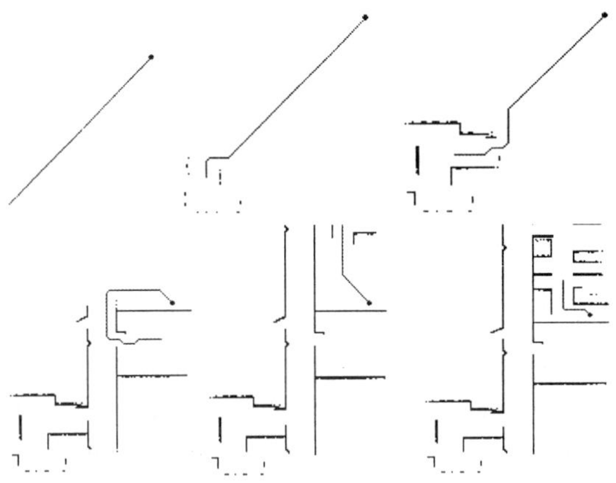

Fig. 4: Global path planning

In some tasks a basic set of landmarks can be derived from CAD maps of the building which the robot is supposed to operate in. In these maps it is natural to inscribe annotated graphs with suitable intermediate target points. Path planning is then reduced to finding a suitable sequence of subtargets from this graph.

If no a-priori knowledge is available, the robot is able to extract, along with the feature based map, a global grid based map of the environment using the navigation described above. In general the mapping (or exploration) process described above has to be supported, since it needs sensor measurements from different positions and the position estimate itself can be corrupted by the poor quality of the odometry. Therefore, the robustness of the landmark extraction process depends very much on the speed of the robot and on the trajectories travelled. An integrated planner, based on the grid based global map and the A*-algorithm, is used to find a path which mediates between position estimation, landmark extraction and the goal of reaching some target position [10], [11]. A simulation run [Fig. 4] illustrates the alternating sequence of intermediate target point generation and reactive manoeuvring, where the emphasis is on reaching a target position.

Those positions at which the planned path in the grid map takes turns are used as intermediate target points. Plan execution is done by the pilot by successively steering the robot to these intermediate target points. If the map of the environment used for planning was complete and correct, the robot will in general be able to travel this path.

If the pilot is not able to travel the planned path, that is the next intermediate target point can't be approached sufficiently closely, the planner starts over.

4 Discussion

In this paper we have summarized the work in our group on how to implement robust low-cost autonous mobility. The result of this investigation is that with conventional ultrasonic sensing and odometry alone a robust mobile robot with a high degree of autonomy can be built.

Conventional ultrasonic sensing is the most reliable, mature low-cost technology for obstacle avoidance at this time. However, using advanced signal processing techniques, the performance of ultrasonic sensing can be significantly enhanced [12]. Although good results have been obtained with ultrasonic sensors alone, there are applications where they will obviously fail, e.g. in wide open areas like very large hotel lobbies or airports.

Another sensor with large future potential both in obstacle avoidance and navigation is computer vision. It is closing in quickly as far as hardware cost is concerned, partly driven by the consumer electronics and has a larger overall potential than ultrasonic sensing. We expect it to play a very important role in the future. It might prevent laser scanners from becoming a high-piece-count, low-cost technology.

Obviously, to find the right mix of sensors and the appropriate algorithms for sensor data integration will still require many experiments in many different environments.

5 Literature

[1] Barshan, B., Leonard. J. Inertial Navigation Systems for Mobile Robots. IEEE Trans. on Robotics and Automation 1995, Vol. 11 No. 3: 328 - 342

[2] Borenstein, J., Koren, Y. Real-time obstacle avoidance for fast mobile robots in cluttered environments. In: Proc. 1990 IEEE Conf. on Robotics and Automation: 572 - 577

[3] Lawitzky, G., Feiten, W., Möller, M. Sonar sensing for low-cost indoor mobility. Robotics and Autonomous Systems 1995, No. 14: 149 - 157

[4] Bauer, R., Feiten, W., Lawitzky, G. Steer Angle Field - An Approach to Robust Manoeuvring in Cluttered Unknown Environment. In: Proc. Int. Workshop on Intelligent Robotic Systems '93, Warschau, Grenoble, 1993, pp. 67-70.

[5] Feiten, W., R. Bauer, R., Lawitzky, G. Robust Obstacle Avoidance in Unknown and Cramped Environments. In: Proc. 1994 IEEE Int. Conf. on Robotics and Automation, May 1994: 2412-2417.

[6] U. Wienkop. G. Lawitzky, W. Feiten: Intelligent Low-cost Mobility. In: Proc. IROS 94: 1708-1715

[7] Hügli, H., Facchinetti, C., Tiche, F. et al. Architecture of an Autonomous System: Application to Mobile Robot Navigation. In: PNR-23 Symposium 1994, Switzerland

[8] Rencken, W. D. Concurrent Localization and Map Building for Mobile Robots. In: Proc. IROS 93 Conference: 2198-2203.

[9] Rencken, W. D. Autonomous Sonar Navigation in Indoor, Unknown and Unstructured Environments. In: Proc. IROS 94: 431-438.

[10] Bauer, R. Dynamic Path Planning Integrating Self-Localisation and Landmark Extraction. In: Proc. Intelligent Autonomous Systems IAS-4, Karlsruhe, March 27-30, 1995: 420-426.

[11] Bauer, R., Rencken, W.D. Sonar Feature Based Exploration. In: Proc. IROS 95: 148 - 153.

[12] Rencken, W. D., Peremans, H., Möller, M. Triaural versus Conventional Sonar Localisation and Map Building. In: Proc. Intelligent Autonomous Systems IAS-4, Karlsruhe, March 27-30, 1995, pp. 398-407.

Obstacle Detection for Unmanned Ground Vehicles: A Progress Report*

Larry Matthies, Alonzo Kelly, Todd Litwin, and Greg Tharp
Jet Propulsion Laboratory, 4800 Oak Grove Drive, Pasadena, CA, USA 91109
email: lhm@robotics.jpl.nasa.gov

Abstract

To detect obstacles during off-road autonomous navigation, unmanned ground vehicles (UGV's) must sense terrain geometry and composition (ie. terrain type) under day, night, and low-visibility conditions. To sense terrain geometry, we have developed a real-time stereo vision system that uses a Datacube MV-200 and a 68040 CPU board to produce 256×45-pixel range images in about 0.6 seconds/frame. To sense terrain type, we are using the same computing hardware with red and near-infrared imagery to classify 256×240-pixel images as a rate of 10 frames/second. This paper reviews the rationale behind the choice of these sensors, describes their recent evolution and on-going development, and summarizes their use in demonstrations of autonomous UGV navigation over the past five years. This work has been the first to show that stereo vision can be practical for autonomous UGV navigation, and is now the first to show a real-time terrain classification system with very low computing requirements.

1 Introduction

Unmanned ground vehicles (UGV's) are being developed for a variety of applications in the military, in hazardous waste remediation, and in planetary exploration. Such applications often involve limitations on communications that require UGV's to navigate autonomously for extended distances and extended periods of time. Under these circumstances, UGV's must be equipped with sensors for detecting obstacles in their path. This paper provides a progress report on work being done at the Jet Propulsion Laboratory (JPL) on obstacle detection sensors for the "Demo II" UGV Program, which is sponsored by the Advanced Research Projects Agency (ARPA) and the Office of the Secretary of Defense (OSD).

The goal of the Demo II Program is to develop technology enabling UGV's to perform autonomous scouting missions. In its full generality, this application will require operating during the day or night, in clear or obscured weather conditions, and over terrain that will include a variety of natural and man-made obstacles. Obstacle detection sensors for the full problem must be able to perceive the terrain geometry, perceive the material type of any ground cover (ie. terrain type), and do so at night and through haze.

We have been addressing the problem of perceiving terrain geometry (by day and by night) and we are beginning to address the problem of perceiving terrain type (by day). For sensing geometry, we are using stereo vision, because its properties of being non-emissive, non-scanning, and non-mechanical make it attractive for military vehicles that require a low signature sensor and well-registered range data while jostling over rough terrain. In collaboration with other labs, we have recently begun experimenting with stereo vision on thermal imagery to address night operations. For sensing terrain type, we have done preliminary investigations of discriminating soil, vegetation, and water using visible, near infrared, and polarization imagery.

Section 2 reviews the current status of the real-time stereo vision system we have developed for UGV applications. The system currently produces range imagery from a 256×45 pixel window of attention in about 0.6 second/frame, using a Datacube MV-200 image processing board and a 68040 CPU board as the sole computing engines. This vision system is installed on a roboticized HMMWV[1] that serves as a testbed UGV. Section 3 describes a number of enhancements currently in progress on this system, including recent tests with thermal imagery, simple algorithms for real-time obstacle detection, methods to support focus of attention, and approaches to terrain classification. Section 4 reviews how the vision

*The work described in this paper was sponsored by the Advanced Research Projects Agency.

[1]High Mobility Multipurpose Wheeled Vehicle – the modern military jeep.

system has evolved through four major demonstrations over the last five years, and shows results from an autonomous navigation trial conducted with our HMMWV on a dirt road near the laboratory. These demonstrations have driven HMMWV's at speeds in the neighborhood of 5 to 10 kph over gentle, but not barren cross-country terrain.

This work has been the first to show that stereo vision can provide range data of sufficient quality, at sufficient speeds, and with sufficiently small computing resources to be practical for UGV navigation. Future work will attempt to increase the quality of the range data, miniaturize the computing system, and integrate terrain classification with range imaging.

2 The JPL Stereo Vision System

Previous versions of JPL's real-time stereo vision system have been described in [1, 2]. Here, we will outline the current version of the algorithm, then discuss how and why it has changed. Current steps in the algorithm are:

1. Digitize fields of the stereo image pairs.
2. Rectify the fields.
3. Compute image pyramids by a difference-of-Gaussian image pyramid transformation.
4. Measure image similarity by computing the sum-squared-difference (SSD) for 7×7 windows over a fixed disparity search range.
5. Estimate disparity by finding the SSD minimum independently for each pixel.
6. Filter out bad matches using the left-right-line-of-sight (LRLOS) consistency check [3, 4].
7. Estimate sub-pixel disparity by fitting parabolas to the three SSD values surrounding the SSD minimum and taking the disparity estimate to be the minimum of the parabola.
8. Smooth the disparity map with a 3×3 low-pass filter to reduce noise and artifacts from the sub-pixel estimation process.
9. Filter out small regions (likely bad matches) by applying a blob filter that uses a threshold on the disparity gradient as the connectivity criterion.
10. Triangulate to produce the X-Y-Z coordinates at each pixel and transform to the vehicle coordinate frame.
11. Detect "positive" obstacles[2] by thresholding the output of a simple slope operator applied to the range image.

Since the first version, this algorithm has evolved as follows. The original version digitized full frames instead of fields (step 1), because it was first used on a Mars rover prototype vehicle that stopped to acquire imagery. At the time, the remainder of the algorithm consisted of steps 3, 4, 5, 7, 8, and 10, plus a Bayesian posterior probability measure that was applied in place of step 6 above to filter out bad matches. Matching was done at only at the 64×60-pixel level of resolution in the image pyramid. Changes to the system and the rationale behind them are described below.

Digitizing fields. Since the vehicle now drives continuously, fields are digitized instead of frames to avoid temporal misregistration from the field-interlaced cameras used on the vehicle.

Rectification. With the 64×60 resolution and 30 degree field of view (FOV) of the original system, it was possible to align the cameras well enough for stereo matching by mounting the cameras on mechanically adjustable brackets, although the alignment procedure was painstaking. Since then, we have moved to matching at the 256×240 level of resolution and have used FOV's as wide as 85 degrees. These changes produce sufficiently high angular resolution and radial distortion to make mechanical alignment impractical. Rectification currently works well enough that the residual vertical disparity at 256×240 resolution is less than half a pixel. Since the adjustable camera mounts had potential to go out of alignment, this approach is more robust, as well as much easier to use.

LRLOS consistency check. This procedure ensures that disparities obtained by choosing best matches along left camera lines of sight agree with those obtained by choosing matches along right camera lines of sight. We have found that this procedure runs faster than the posterior probability measure and works better, especially at occluding boundaries.

Blob filter. Small islands of bad matches occasionally survive the LRLOS test. These can be contiguous with larger regions of good matches or completely disjoint from good regions. A simple blob coloring algorithm [5] can be used to reject small disjoint regions. Furthermore, it is possible to eliminate small regions of bad matches that are contiguous to larger,

[2] Positive obstacles are those that extend upward from the nominal ground plane, like rocks, bushes, and fence posts. "Negative" obstacles extend downward, like potholes, man-made ditches, and natural ravines.

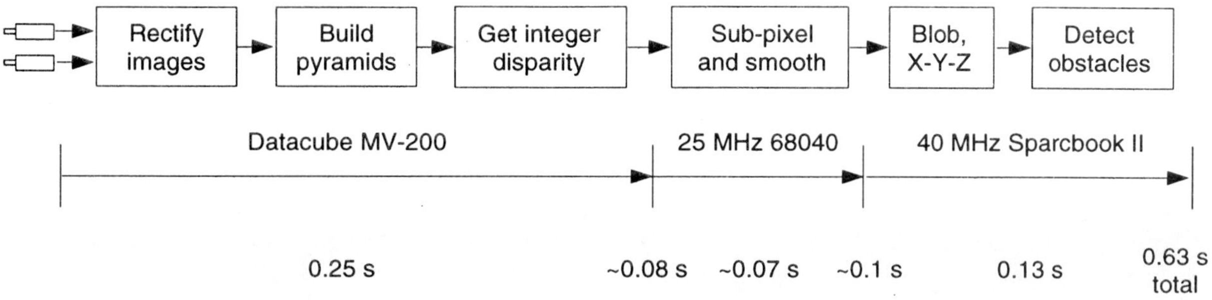

Figure 1: Major functional units of the stereo algorithm, their mapping onto hardware, and approximate run-times for each processor and I/O between the processors. Run-times are for a 256 × 45 window processed at the 256 × 240 level of resolution, with a disparity search range of 32 pixels.

good regions by making the connectivity criterion in the blob coloring algorithm be a threshold on the disparity gradient.

Positive obstacle detection. As discussed elsewhere [6], obstacles can be difficult to define and expensive to compute. To enable real-time operation, we initially implemented an algorithm that detected only positive obstacles using a simple slope estimation algorithm applied to columns of the range image [2]. This algorithm assumes that obstacles are vertical step displacements of minimum height H on an otherwise flat ground plane. On each scanline, we project a line of sight out to intersect the ground plane, assume an obstacle of height H at that point, and determine on what scanline the top of the obstacle would appear in the image. A table of vertical scanline offsets is determined in this way for every pixel in the image. To detect obstacles, for every pixel p_1 in the range image, we use it and its higher partner p_2 in the same column to compute the change in height at that point in the image; if the change in height exceeds a threshold, the entire interval of the range image from p_1 to p_2 is declared to contain an obstacle. Although this algorithm is very simple, it is reasonably robust and very fast.

On our HMMWV, these algorithms are implemented in a computing system comprised of a Datacube MV-200 image processing board, a 68040 CPU board that controls the MV-200 over the VME bus, and a Sparcbook II notebook computer that communicates with the 68040 over ethernet. Figure 1 shows how the major units of the algorithm map onto each processor, together with approximate run-times for processing a 45 scanline window at the 256 × 240 level of resolution.

Figure 2 shows range data produced by the system. There are a few gaps in the disparity map in the foreground, but no gross errors are evident. The rock in the middle of the road was 43 cm high and was placed 10 m from the cameras; it is perceived as approximately 42 cm high and 9 m from the cameras. The road is seen to be quite smooth up to the rock; bumps behind the rock may correspond to bushes on the side of the road. The elevation map shows a relatively smooth area corresponding to the roadway, high areas corresponding to the bushes beside the road, and a small bump corresponding to the rock on the road. Overall, the results are quite good.

3 Ongoing Algorithm Extensions

A number of extensions to the basic stereo system are in progress. These have the goals of enabling night driving by using forward looking infrared (FLIR) cameras, detecting negative obstacles, increasing speed by using windows of attention and dual FOV camera systems, providing software exposure control, increasing the resolution by moving to full frame cameras, and augmenting obstacle detection by doing terrain classification as well as range imaging. These are listed in order from the most simple to the most complex. Initial implementations and results have been obtained for several of these functions; full frame imaging and terrain classification are not yet integrated into the real-time system. The current status for each issue is summarized below.

FLIR imagery. UGV's must be able to perform missions at night. Stereo vision with thermal imagery obtained from FLIR cameras is a possible way to detect obstacles at night. The Demo II Program has acquired several FLIR cameras operating in the 3 to 5 μm wavelength range with 256 × 256 detectors. In collaboration with SRI International, a pair of these cameras were mounted on the JPL HMMWV and used for a preliminary evaluation of real-time range imaging. The cameras were mounted with a 30 cm baseline and had a field of view of approximately 30 degrees;

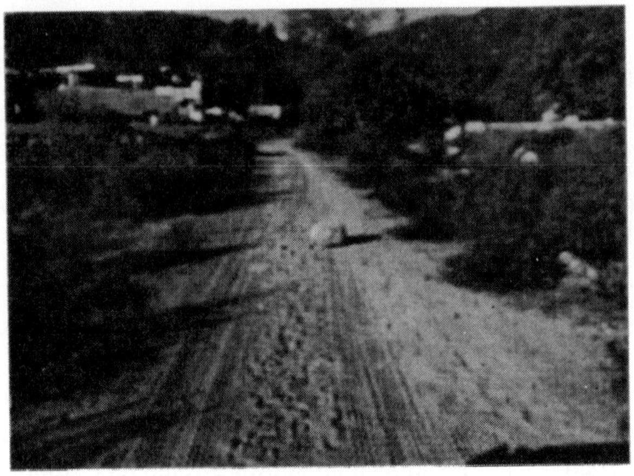

(a) Intensity image

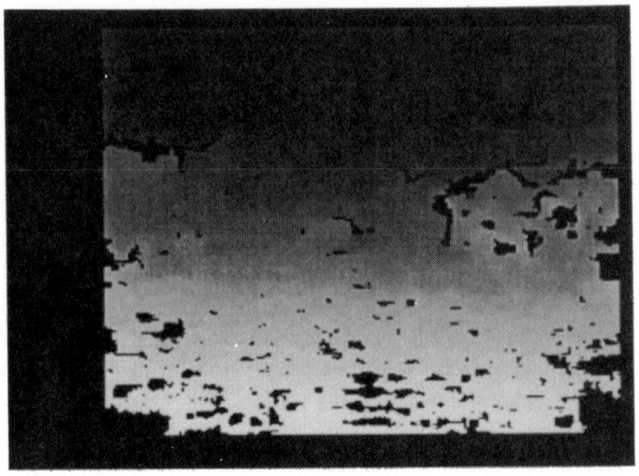

(b) Disparity map

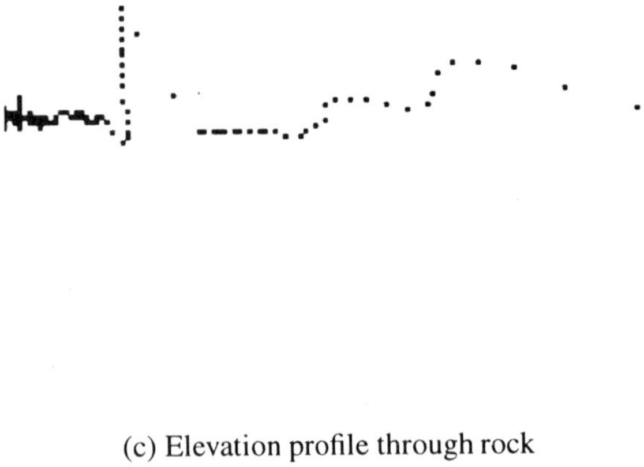

(c) Elevation profile through rock

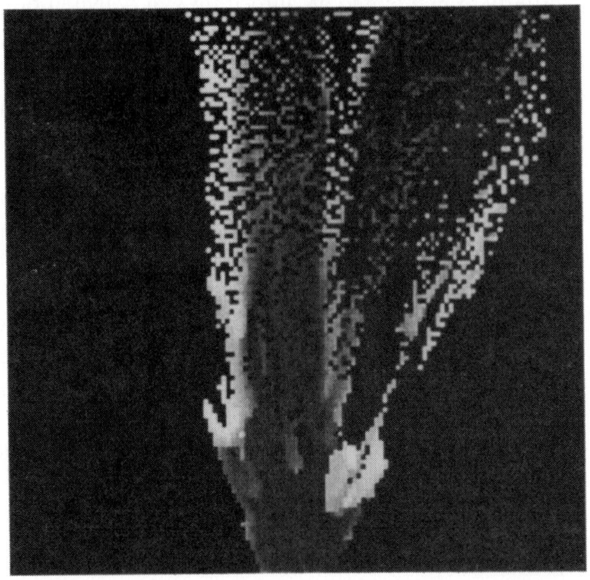

(d) Elevation map

Figure 2: Range data for a dirt road in the Arroyo Seco, next to JPL. The disparity map was generated at 256 $\times$ 240 resolution; bright is close, dark is far. The elevation profile in (c) shows range pixels for a column through the rock in the middle of the road; the scale is 4 to 35 m horizontally and roughly -1 to +1.5 m vertically. In the elevation map, bright is high, dark is low; the scale is 4 to 35 m downrange and roughly -15 to +15 m crossrange.

this compares with 35 cm and 57 degrees for the visible imagery shown in figure 2. Figure 3 shows results obtained with FLIR imagery of the same scene as figure 2, processed at 128 $\times$ 120 resolution. Note that this data was processed with the *identical* algorithms used for the visible imagery. Qualitatively, the results compare quite well to the visible imagery. Results at 256 $\times$ 240 resolution are much sparser, for reasons yet to be determined. These initial results are encouraging. An important next step will be to characterize the quality of the range data quantitatively, in terms of the thermal sensitivity of the cameras and temperature variations of the scene, at various times of the day and night.

Negative obstacles. For cross country navigation, UGV's must detect many types of obstacles. To extend the capability of the basic system while staying within the power of the computing system, we are developing a simple algorithm for detecting negative obstacles. This algorithm is an extension of the column-oriented algorithm for positive obstacles. For each pixel in the range image, the algorithm projects a line of sight out to the ground plane and computes what would be the distance to the next range pixel above it in the same column, if that pixel were on the same ground plane. If the actual distance to the second pixel is greater than expected (with some allowance for noise), then by the geometry of the camera it must also be lower. This indicates a gap in the range data, a negative slope of the terrain, and a possible negative obstacle (eg. ravine) between the two pixels; in this case, the first pixel is labelled as a potential negative obstacle. Figure 4 illustrates results for both positive and negative obstacle detection for a cross-country scene with moderately tall grass in the foreground, a ravine about 3.5 m wide in the upper right, and a tree in the ravine. The disparity map is fairly good, though there is some missing data on the tall grass. Positive obstacles are detected around the tree and negative obstacles are detected consistently at the near edge of the ravine. The elevation maps span 3 to 30 m downrange, -13 to +13 m crossrange, and roughly -1 to +2 m in elevation. The white area on the far side of the ravine in figure 4e is the tree. The additional white areas in figure 4f show the locations of both the positive and negative obstacles, around the tree and the near edge of the ravine, respectively. The results suggest that very simple algorithms do a sufficiently good job of detecting these classes of obstacles to enable credible UGV demonstrations in these environments.

Window of attention and dual FOV. It is widely appreciated that focus of attention mechanisms may increase the performance of an autonomous system by reducing the amount of computation that must be done. UGV navigation makes this issue concrete, because the task, the computing limitations, and the criteria for focusing attention all can be well defined. Vehicle speed and reaction times determine the part of the image that must be examined next [7]; essentially, the faster the vehicle goes, the higher up the image it must look and the higher the level of resolution it must use. Here, we only note that two mechanisms are being developed to allow the stereo system to focus attention. The first is the ability to select a rectangular subregion of the image for processing; the second is that ability to switch between alternate stereo camera pairs for input ("dual FOV"). Because of restrictions in the Datacube architecture, windows of attention cannot be changed arbitrarily at run-time. However, a moderate number of fixed windows can be defined at initialization time and selected at frame rates at run-time; these can be at any level of resolution and more than one window can be processed per input image pair. The dual FOV capability involves relatively straightforward switching of inputs and calibration data at frame rate; for example, this enables changing from a wide FOV camera pair to a narrow FOV camera pair as vehicle speed increases.

Software exposure control. Even with auto-iris lenses, obtaining satisfactory dynamic range on the desired part of the image is always difficult, due to changing cloud cover, changing albedo of the terrain, and touchy adjustments on the lenses. We have implemented functions in the MV-200 that will be used in the near future to attempt to control exposure in software. Computational elements within the MV-200 memory modules do most of the work needed to compute the mean, min, max, and standard deviation of intensity over a pre-defined window of attention. We expect to filter these quantities over time and use them to drive lens aperture and/or image exposure time. Experiments with software exposure control will be conducted in the fall of 1995.

Full frame, non-interlaced range imaging. One of the least satisfactory characteristics of the stereo system is that it must discard half the available vertical resolution, due to the interlaced scanning of the cameras. However, cameras with full frame, non-interlaced scanning are now available (eg. the Pulnix TM-9700). The stereo software can be configured easily to process this imagery; however, three potential problems lurk:

- there will be more demanding requirements for image alignment because of the higher angular resolution of the images;

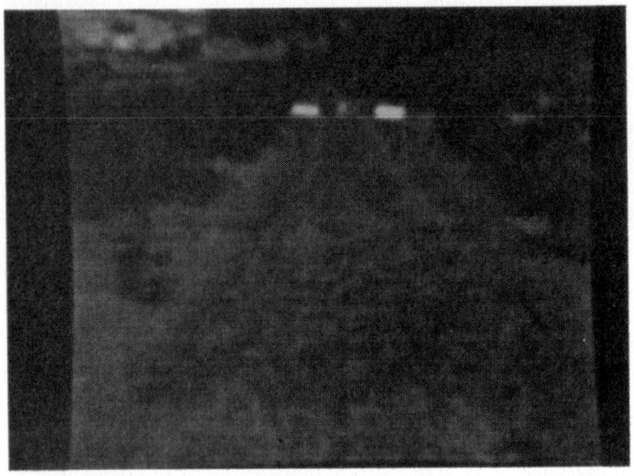

(a) Left image

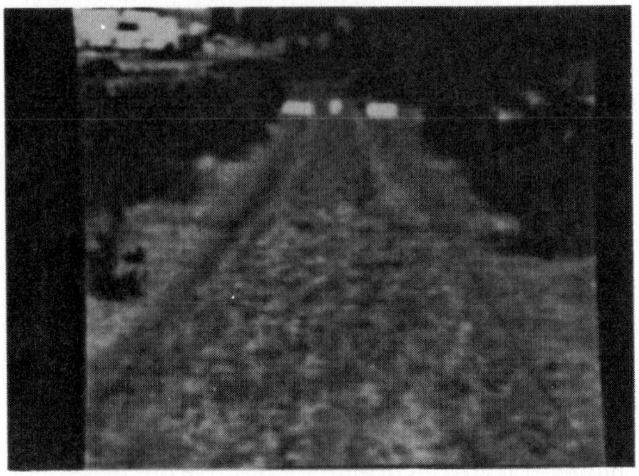

(b) Right image

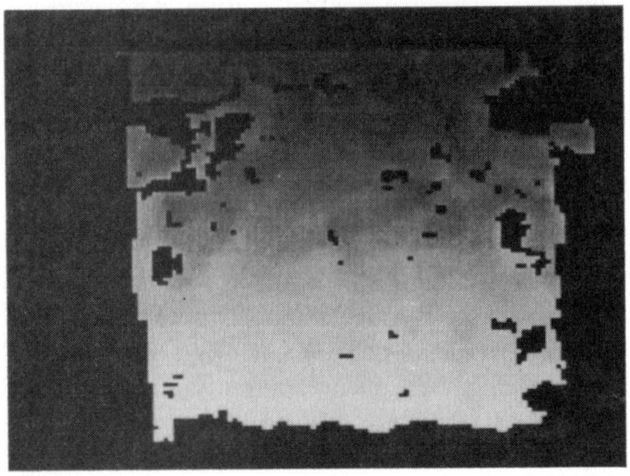

(c) Disparity map

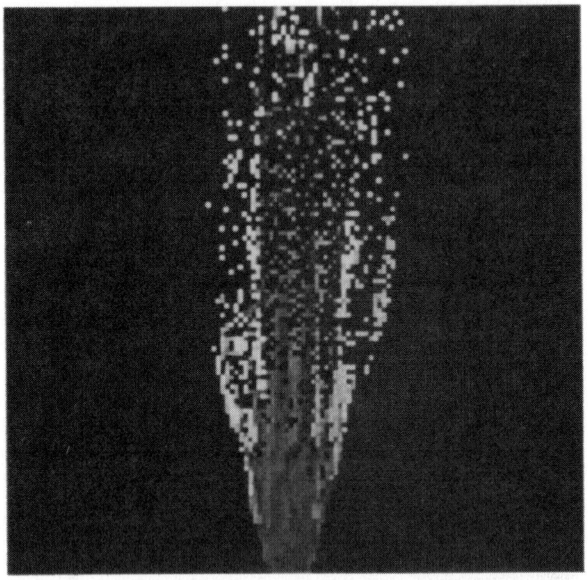

(d) Elevation map

Figure 3: Range data obtained with stereo FLIR imagery of the same road as figure 2. The disparity map was generated at 128×120 resolution. The scale of the elevation map matches figure 2.

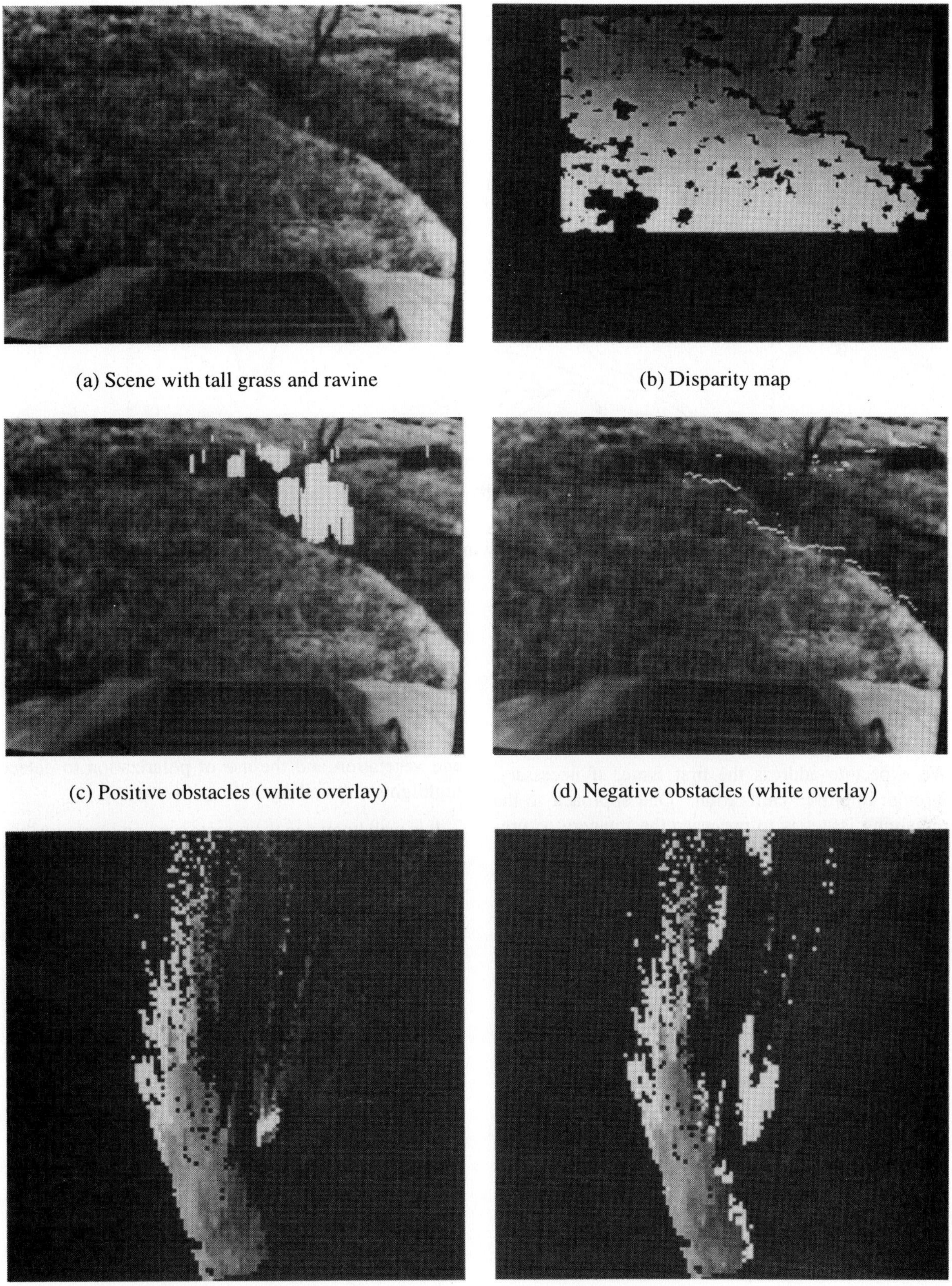

(a) Scene with tall grass and ravine

(b) Disparity map

(c) Positive obstacles (white overlay)

(d) Negative obstacles (white overlay)

(e) Elevation map

(f) Elevation map with obstacles (in white)

Figure 4: Obstacle detection results. See text for discussion.

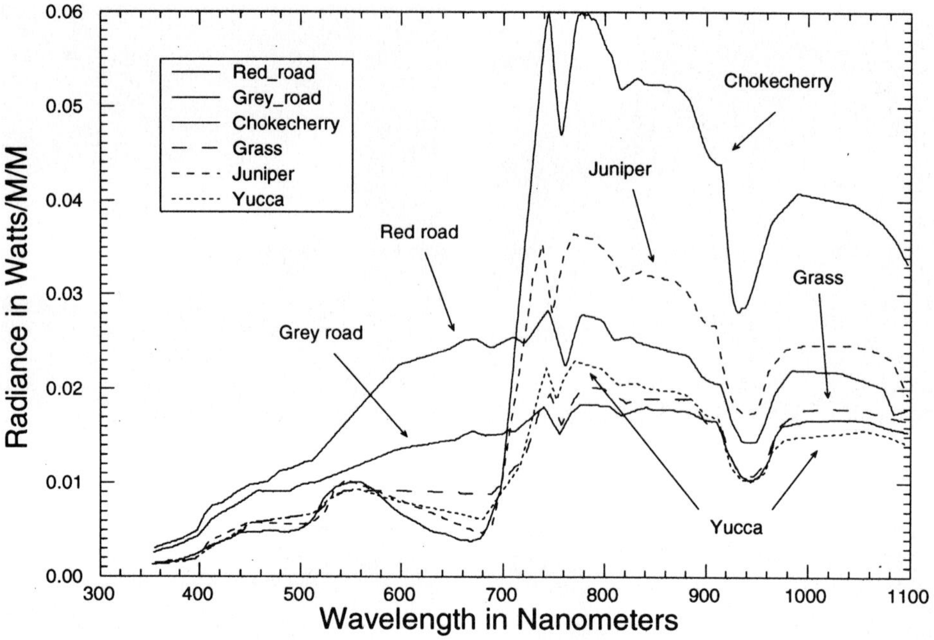

Figure 5: Spectral radiance for six terrain types, showing the generally brighter response of vegetation in the near infrared.

- there may be more matching ambiguity because of wider disparity ranges;
- there may be slower run-times because of the higher resolution.

We expect to address the first issue, if necessary, through improved calibration. One approach to the remaining issues is to pre-warp the imagery to produce zero disparity for the nominal ground plane; provided that no part of the scene deviates substantially from that plane, this technique reduces ambiguity and run-time by reducing the necessary disparity search range. Unfortunately, this is a fragile assumption, which will often be violated in cross-country navigation. However, by also matching at a lower resolution over a wider search range, we expect that a multiresolution algorithm can be developed that will achieve the strengths of pre-warping with only slightly greater cost. Thus, the real-time constraints of UGV navigation are likely to lead to new multiresolution stereo algorithms, because they will address different requirements than the off-line algorithms developed in the past.

Terrain classification. Clearly, geometry is not the only scene property important for obstacle detection; material or terrain type is also important. A great deal of research has been done on classifying and segmenting images based on color, texture, and other features. As with stereo, however, real-time performance is essential for terrain classification. Therefore, we looked for non-traditional image features that might improve the robustness of classification with very simple, low-level algorithms. Two such possibilities are the use of near infrared imagery to distinguish between soil and vegetation and the use of polarization to detect highlights on standing bodies of water.

It is well known in the remote sensing literature that live vegetation is very reflective in the near infrared [8]. During the ARPA Autonomous Land Vehicle (ALV) Program in the mid-1980's, the U.S. Army Engineer Topographic Laboratory took spectral reflectance measurements, at ground level, from several different types of soil and ground cover. Figure 5 shows such spectra for roads of grey and red dirt and for four types of vegetation. In general, we see that soil gets gradually brighter from blue through near infrared. Vegetation is dark in blue and red, somewhat brighter in green, and anywhere from a little brighter to dramatically brighter in near infrared. CCD cameras are sensitive out to about 1000 nm, so in principle this entire spectral range can be sensed by using appropriate filters with existing cameras.

Figure 6 makes this point further by showing ground level imagery taken with a CCD camera and blue, green, red, and near infrared filters. The scene contains a pond in the foreground and various soils, bushes, trees, and buildings in the background. Although there are differences between the visible bands,

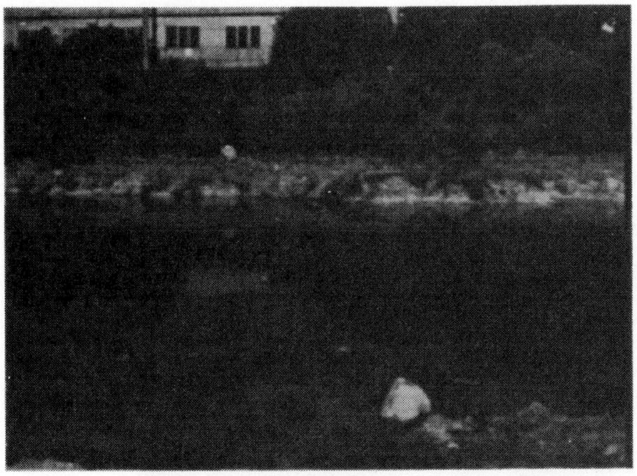

(a) Blue band (450 nm)

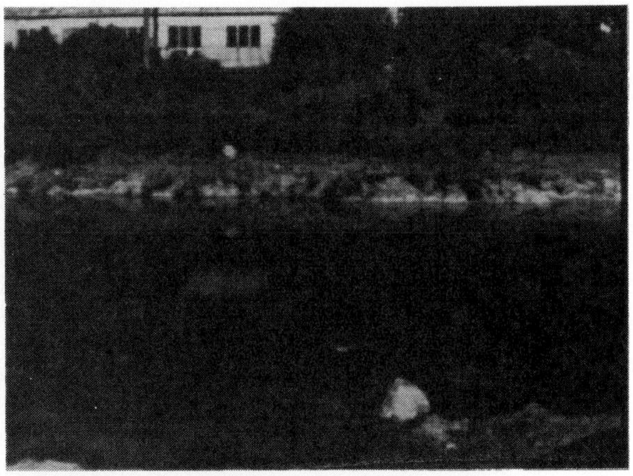

(b) Green band (550 nm)

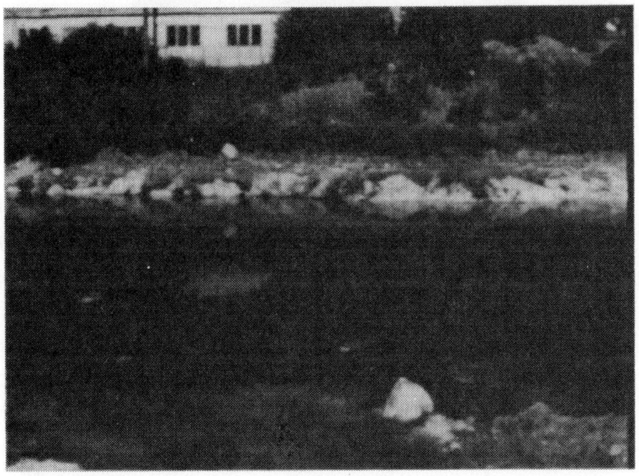

(c) Red band (650 nm)

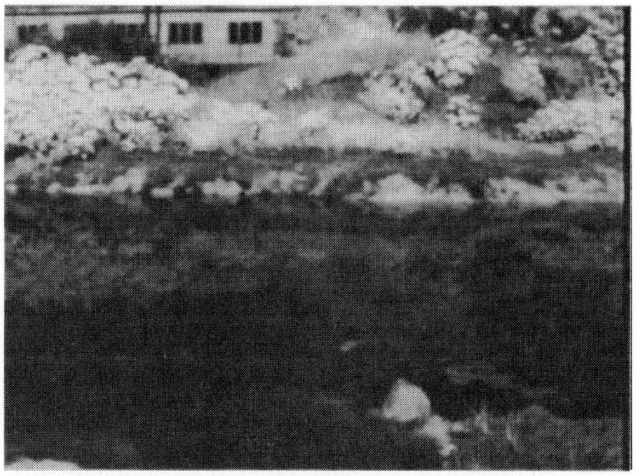

(d) Near infrared band (800 nm)

Figure 6: Multispectral images

they are much less dramatic than the contrast reversal shown by the soil and vegetation in the near infrared band. We have used red, near infrared, and polarization imagery of this scene to obtain good discrimination between vegetation, water, and "other" pixels, using very simple, pixel-wise ratios of the bands. The results are dependent on the scene, however, and even more dependent on the season; in winter, with dead vegetation, the near infrared cue is unavailable. With other summertime imagery that has been segmented manually to estimate ground truth, we have seen improvements of up to 15% in correct classification rates using red and near infrared imagery, as compared to using red and green imagery. We have developed a real-time implementation of red/near infrared classification using a filter wheel camera that obtains six spectral bands in 1/10 of second (one field per filter); with the Datacube hardware, we achieve classification 10 fields per second.

4 System Integration and Demonstrations

The stereo vision system described in section 2 has been integrated into robotic vehicles and used in the following configurations and demonstrations:

- On the Mars rover testbed vehicle "Robby", running at 64×60 resolution in ≈ 6 seconds per frame, for autonomous navigation over 100 m of sandy terrain interspersed with bushes and mounds of dirt (Sept'90) [1]. In this demonstration, the vehicle stopped at each frame.
- On the JPL HMMWV (figure 7), running at 64×60 resolution in ≈ 1.5 seconds/frame, for autonomous obstacle detection and halting during the U.S. Army "Demo I" exercise at Aberdeen Proving Ground (April'92). The HMMWV drove continuously at ≈ 1 to 3 m/sec (≈ 5 to

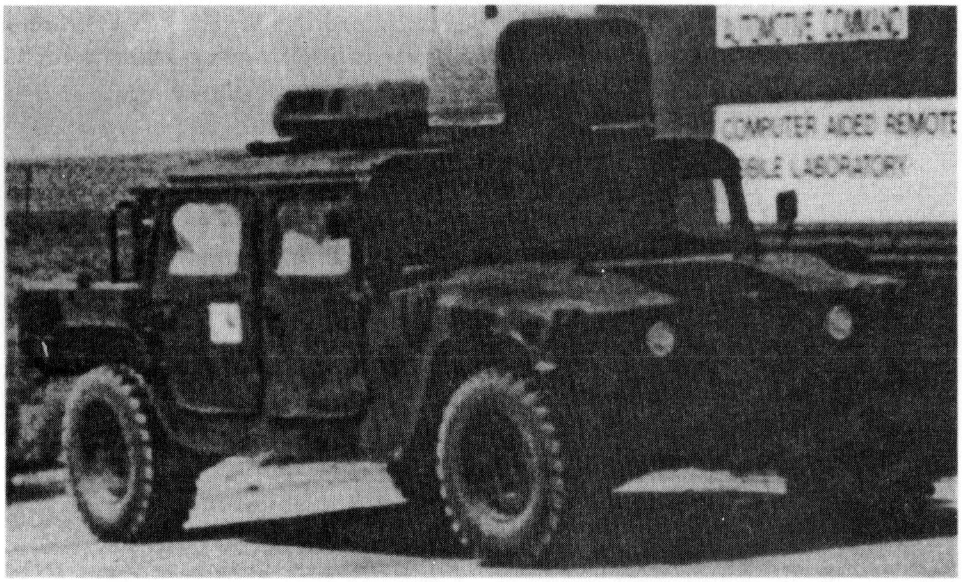

Figure 7: The JPL HMMWV testbed vehicle

10 kph) while detecting half-meter obstacles in time to stop.

- On a Lockheed Martin HMMWV, integrated with the "Ganesha" and "Ranger" obstacle mapping and path planning systems developed at CMU [9, 10], running at 128×120 resolution in under one second per frame, for autonomous navigation on flat terrain with moderately tall grass and positive obstacles $\geq$ 1.0 m high ("Demo B", June'94). The vehicle velocity in this case was 2 m/sec (7.2 kph).

- On a Lockheed Martin HMMWV, integrated with the "Smarty" obstacle mapping and path planning system developed at CMU [10], running at 256×240 resolution, with a 45-scanline window of attention, in about 0.75 second per frame, for autonomous navigation on grassy terrain with positive obstacles $\geq$ 1.0 m high ("Demo C", June'95). The vehicle velocity in this case was 2 to 3 m/sec (up to 10.8 kph).

In addition, the stereo vision system was tested with the Ranger path planning system on the JPL HMMWV in December'94. Figure 8 shows results from a test run that travelled about 200 m down the road shown in figure 2. The left side of the figure shows 256×45 windows of attention from eight positions along the run; the right side of the figure shows the corresponding range images. A composite elevation map of the entire run is shown in the middle; figure 9 shows a wireframe rendering of the composite map. The vehicle was instructed to follow its current heading whenever it could, but to swerve to a new heading as necessary to avoid obstacles. Two significant swerves are seen in this run. The first was to avoid a tree on the right side of the road; the second was to avoid a bush on the left side of the road, just beyond the tree. This was a very successful run, and typical of results obtained in this kind of terrain. Due to long computing latencies experienced with the existing computing system (figure 1), the vehicle speed was only 0.75 m/sec. The computing system has since been upgraded by putting a Sparc-5CE CPU board in the same card cage as the 68040. This increases the speed of the stereo vision system to 0.4 second/frame for the 256×45-pixel window of attention.

5 Summary and Discussion

In this paper, we reviewed the current state of development of JPL's real-time stereo vision system, described ongoing development of new capabilities, and summarized significant field demonstrations that have used this system to perform autonomous navigation. The system currently produces range data from 256×45 windows of attention in approximately 0.6 seconds/frame. Ongoing development will extend the system to process imagery at 512×480 resolution using full frame, non-interlaced cameras and a variety of techniques to mitigate the computational burden. Recent experiments with FLIR cameras have shown good results with imagery processed at 128×120 resolution; the results shown here were with thermal

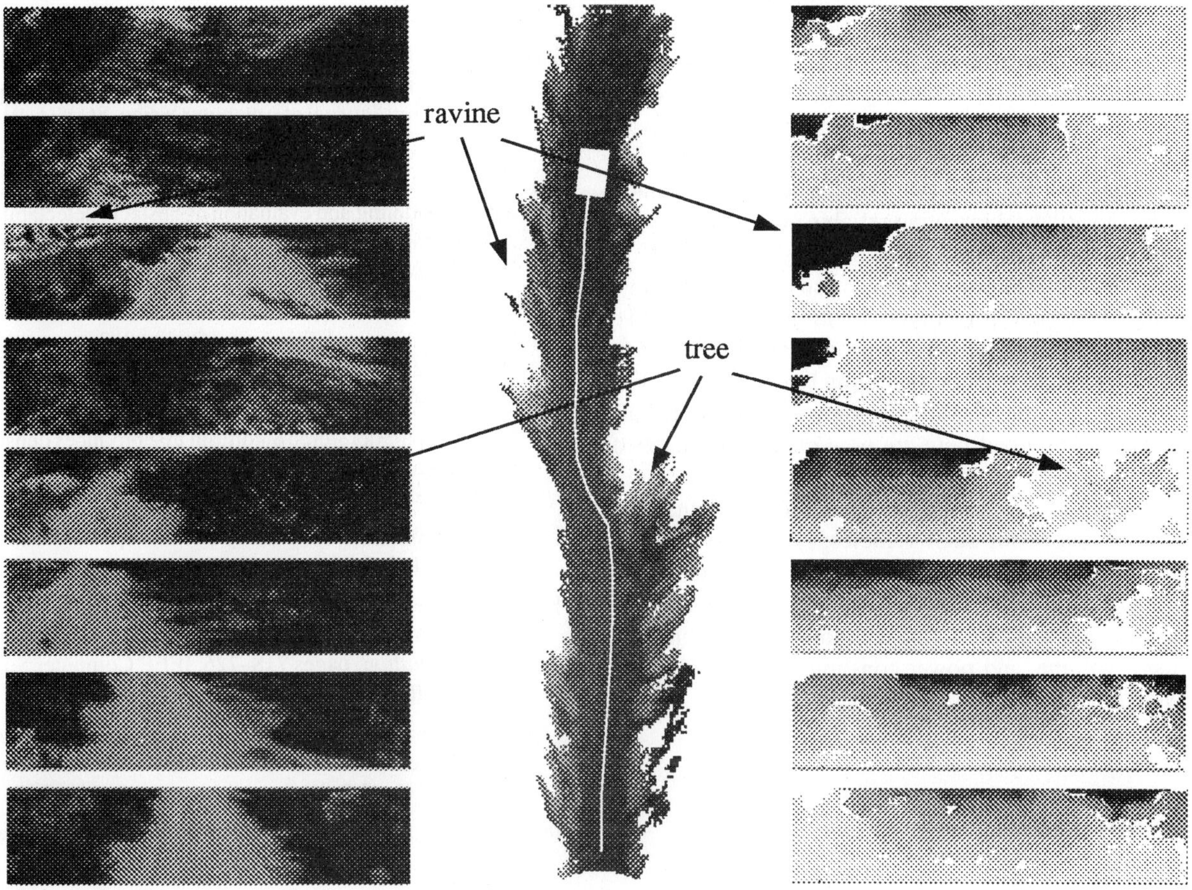

Figure 8: Road run with Ranger, covering about 200 meters. Shown are windows of attention from representative intensity images (left side), corresponding range images (right side), and a composite elevation map from the whole run (center). The white curve and rectangle on the elevation map represent the vehicle path and the vehicle itself at the end of the run.

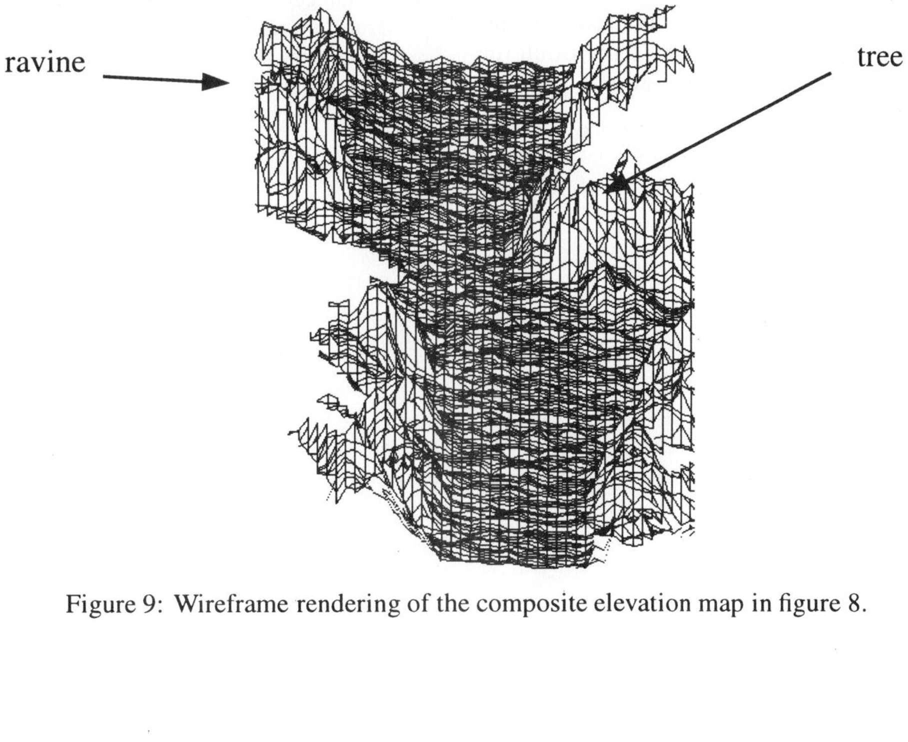

Figure 9: Wireframe rendering of the composite elevation map in figure 8.

imagery collected by day, but qualitatively similar results have been achieved at night. Field demonstrations have driven HMMWV's at speeds on the order of 5 to 10 kph over relatively flat, cross-country terrain covered with calf-high grass and sprinkled with positive obstacles 0.5 m high and larger.

This work has been undertaken with a pragmatic view toward demonstrating reliable performance with simple, fast algorithms. This strategy has paid off; it has shown that stereo vision can provide usable range data for UGV's in a timely, relatively cost-effective manner and that stereo is a promising alternative to emissive range sensors, including laser scanners. For example, optimized, software-only versions of the algorithms described here are now planned to be used on prototype microrovers for Mars exploration; until recently, stereo was still considered too computationally expensive for such vehicles. Future work will improve the quality of the range data, integrate terrain classification capabilities with range imaging, and use new imager and computing technologies to reduce the mass, volume, and power consumption of the vision system.

Acknowledgements

Thanks to Bob Bolles and Harlyn Baker of SRI International for bringing the FLIR cameras to JPL, as well as assisting in mounting them on the HMMWV, interfacing them to the vision system, calibrating them, and evaluating the results. Thanks to Terry Boult for helping to collect the multispectral imagery.

References

[1] L. H. Matthies. Stereo vision for planetary rovers: stochastic modeling to near real-time implementation. *International Journal of Computer Vision*, 8(1):71–91, July 1992.

[2] L. H. Matthies and P. Grandjean. Stochastic performance modeling and evaluation of obstacle detectability with imaging range sensors. *IEEE Transactions on Robotics and Automation*, 10(6):783–791, December 1994.

[3] M. Drumheller and T. Poggio. On parallel stereo. In *Proc. IEEE Conf. on Robotics and Automation*, pages 1439–1448. IEEE, 1986.

[4] P. Fua. A parallel stereo algorithm that produces dense depth maps and preserves image features. Technical Report No. 1369, INRIA, January 1991.

[5] D. H. Ballard and C. M. Brown. *Computer Vision*. Prentice-Hall, Englewood Cliffs, NJ, 1982.

[6] M. Daily et al. Autonomous cross-country navigation with the ALV. In *Proc. IEEE Int'l Conf on Robotics and Automation*, pages 718–726. IEEE Computer Society, April 1988.

[7] P. Grandjean and L. Matthies. Perception control for obstacle detection by a cross-country rover. In *Proc. IEEE Int'l Conf. Robotics and Automation*, May 1993.

[8] C. Elachi. *Introduction to the Physics and Techniques of Remote Sensing*. John Wiley and Sons, New York, NY, 1987.

[9] A. Kelly, A. Stentz, and M. Hebert. Terrain map building for fast navigation on rugged outdoor terrain. In *Proceedings SPIE Conference on Mobile Robots*. SPIE, November 1992.

[10] D. Langer, J. K. Rosenblatt, and M. Hebert. An integrated system for autonomous off-road navigation. In *Proc. 1994 IEEE Int'l Conf. on Robotics and Automation*, pages 414–419, May 1994.

PART 12

VISUAL SERVOING

SESSION SUMMARY

Michael Brady
Department of Engineering Science, Oxford University
19 Parks Road, Oxford OX1 3PJ, UK

The two papers in this section, and several others elsewhere in the volume, concern visual servoing. High performance visual control of robotic devices, whether for object tracking or acquisition, obstacle avoidance, or manipulation, presents a number of problems that are at once familiar, but arise in more severe forms, than control based on force or low performance sensors such as sonar.

First, the amount of information that is generated in real time by a camera, or stereo pair of cameras, is typically three orders of magnitude greater than by sonar/force. This problem is becoming less and less important as the power of processors increases. Still, it has motivated work of lasting value on multiscale image analysis and control of focus of interest, a subregion of the image. Second, the latency of visual processing, typified by a stereo algorithm, can reduce sharply the stability margin of a system. Work on "active vision" reported elsewhere in this volume, investigates predictive control and steerable camera platforms ("heads"). Third, in the current state of the art, vision algorithms, particularly early visual processing such as edge detection, are not reliable. Even when they perform well, the information produced is deficient: the projective loss of depth and the aperture problem of motion are examples. The response has been to develop systems that combine vision with other sensors, and which actively control the visual process. An important finding in the latter case has been that a number of problems previously considered hard, become easy. Finally, in view of the uncertain information produced at variable rates by different sensors, the design of the control system has a considerable effect on the performance of the system. Todate, most work has concentrated on the Kalman filter.

The first paper in this section is by Peter Yorke, who offers a reminder that effective control requires taking account of system dynamics. He stresses the need for quantifiable performance metrics. The second paper, by Masa Inaba, describes the sensor control architecture that supports a number of remarkable demonstrations: legged robots that can stand from a prone posture, and can perform gymnastics. The key idea is to regard the body as a sensing unit as well as an actuating unit, and to map contact sensory information into an image representation, at which point it is fused with vision. This is an elegant dual to the steerable camera platforms discussed elsewhere.

Dynamic Issues in Robot Visual-Servo Systems

Peter I. Corke
CSIRO Division of Manufacturing Technology
P.O. Box 883, Kenmore, 4069
AUSTRALIA.
pic@brb.dmt.csiro.au

Abstract

This paper poses a number of questions related to the performance and structure of closed-loop visual control, or visual servo, systems. While the fundamentals of visual servo control are well known and systems have been demonstrated for many years, the achieved performance is far less than could be expected. In particular the questions discussed relate to fundamental control system structure, performance metrics, compensator design and the choice of feedback versus feedforward control.

1 Introduction

The use of visual sensors with robot manipulators has a history of over two decades, dating back to early work with block worlds. However the performance of modern closed-loop visual control, or visual-servo, systems is significantly poorer than would be expected from a control systems point of view. In this paper it is argued that the reason for such poor performance is that the dynamics of the system, vision and robot, have been ignored. To this end, a distinction is introduced between *visual servo kinematics* and *visual servo dynamics.* The former, essentially a "pure computer-vision" problem, is well addressed by the literature, while the latter has been largely ignored. Investigating dynamics, rather than just the kinematics, of visual servo systems leads to advantages beyond the obvious ones of tracking objects whose pose is unknown and possibly time varying. Visual servoing admits the possibility of "closing the loop" on the dynamics of the actuator, transmission and structural compliance, allowing superior control of end-point dynamics than can be achieved using axis sensing alone.

Related research [1] introduces the concept of *visual dynamic control* which is concerned with dynamic effects due to the manipulator and machine vision sensor which limit performance and must be explicitly addressed in order to achieve high-performance control. That research takes a very different approach to that done by others in the field, in emphasising temporal and dynamic issues over image processing. "Simplistic computer vision" techniques and a specialised hardware architecture combine to provide a high sample-rate and minimum latency vision system. This is an ideal testbed with which to bring closed-loop dynamic problems to the fore, conduct modelling and experiment with various control strategies.

One problem that dogs any discussion of visual servo performance is the lack of agreed, quantitative, performance metrics. It is actually relatively simple to build a visual closed-loop robot system which will converge on a target, if the controller is sufficiently detuned to ensure stability. Most researchers propose the use of simple proportional controllers but it can be shown that these have two undesirable properties. Firstly, feedback-only controllers have a loop-gain limit due to the open-loop delay, and this limits the achievable closed-loop bandwidth. Secondly, simple feedback controllers have significant phase lag characteristics which lead to poor tracking. Much research has focussed on solving the computer-vision part of the problem, but the control or dynamics problem is generic but not adequately investigated.

The remainder of this section discusses existing research on visual servoing from a control systems perspective, and then describes the experimental platform on which the research described in this paper was conducted. Section 2 then poses a number of questions which seem to have been overlooked by the research community, and offers some observations and partial answers. Finally Section 3 provides some conclusions.

1.1 Research to date

Research in visual servoing has, to date, focussed largely on the visual kinematic issues such as estimating object pose from n points, or analysing the relationship between Cartesian motion and the corresponding image plane motion as encapsulated in an image Jacobian matrix. These computer vision problems underpin, respectively, the *position-based* and *image-based* approaches to visual servoing [2]. Both approaches are well known and have been demonstrated experimentally under laboratory conditions. It could perhaps be argued that the problem is solved for all practical purposes, and the computational costs of these two approaches are comparable and readily achieved. Control of a 6-DOF robot was demonstrated by Ganapathy [3] over a decade ago using a closed-form solution that executed in only 4 ms on a 68000 processor. An iterative approach has even been patented [4]. The 3-D pose estimation problem can be significantly simplified by using non-anthropomorphic direct 3-D sensors — those based on structured lighting are now compact and fast enough to use for visual servoing [5].

Recently, however, there has been a considerable growth in the literature which has been fuelled by personal computing power crossing the threshold which allows analysis of scenes at a sufficient rate to "servo" a robot manipulator. Prior to this, researchers required specialised and expensive pipelined pixel processing hardware.

An obvious characteristic of many reported visual servo systems is that they appear slow and "shaky", and exhibit significant lag with respect to the motion of the tracked object. This can be seen in video tape presentations and in the time axis scaling for many published results. Reported results and comments such as "*slight instability*" are clear indications of closed-loop systems close to their stability limit. Another common comment is of "*noticeable lag*" in the observed performance, which is indicative of a poorly designed control system, inappropriate control architecture or both. To the casual observer there are obvious similarities between the performance of such systems and that of early force controlled robot systems. The force control community learned, long ago, the importance of dynamic modelling and how the dynamics of the manipulator, sensor and environment impact on closed-loop stability [6]. In this paper it is argued that the visual-servo community has yet to learn this lesson.

Vision is admittedly a "difficult" sensor to use for closed-loop control due to its low sample rate, high latency (often one or more sample intervals) and low spatial resolution. However most reports demonstrate a closed-loop bandwidth which is only a fraction of that which would be expected. Evidence of stability problems, or tracking lag, has tended to be ignored or blamed on the low sample rate or low-bandwidth communications link between the vision system and robot. Unfortunately, rectifying these "deficiencies" will not solve a problem whose roots are in the fundamentals of the system dynamics. The evidence points to an underlying lack of control systems knowledge in this research community. This "dynamics problem" is generic, that is, its existence is independent of the type of image processing or scene interpretation performed. Its existence is due to the temporal characteristics of the camera and vision system, and the robot and its controller.

Problems of delay in visual servo systems were first noted over 15 years ago [7] but are not always taken into consideration. The sources of delay include the charge integration time within the camera, serial pixel transport from the camera to the vision system, and the computation time for feature extraction. If the target motion is constant then prediction can be used to compensate for the latency, but combined with low sample rate leads to poor disturbance rejection and long reaction time to target "maneuvers". Grasping objects on a conveyor belt is an ideal, though non-challenging, application for prediction since the target velocity is constant.

1.2 Experimental system

The research described [1] takes a different approach from others in the field, by emphasising temporal and dynamic issues over image processing. "Simplistic" computer vision techniques and a specialised hardware architecture combine to provide a high sample-rate and minimum latency vision system. This is an ideal testbed with which to bring closed-loop dynamic problems to the fore, conduct modelling and experiment with various control strategies. A block diagram of a 1-DOF system is given in Figure 1. The camera lens is modelled as a simple gain, K_{lens}, which, due to perspective, is a function of target distance. The system is multi-rate due to the different sample intervals for the robot's servo system and the camera's field rate. Manipulator structural dynamic effects are represented by $G_{struct}(s)$, which

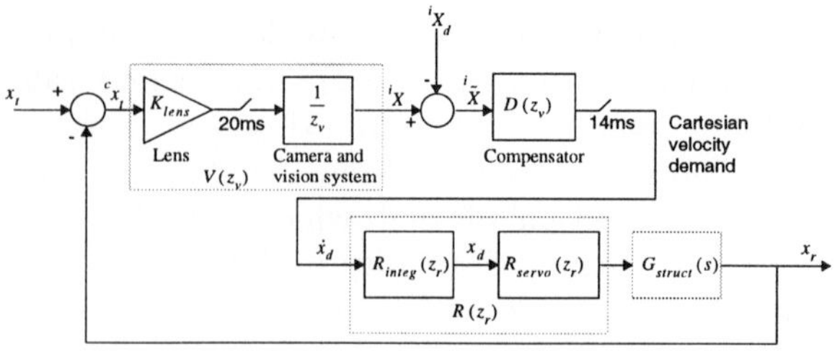

Figure 1: Block diagram of the 1-DOF visual feedback system. x_t is the world coordinate location of the target, and ${}^iX_d$ is the desired location of the target on the image plane. Note the two samplers in this system.

is assumed, for now, to be unity.

The performance of this system, with proportional control $D = K_p$, has been measured and Figure 2 shows a step response to a visual square wave excitation. The 'early' and 'late' response are related to when the visual stimulus occurs with respect to the system's samplers. The closed-loop frequency response is given in Figure 3 and shows evidence of a low frequency pole at approximately 2 Hz, and time delay. The phase is linear with respect to frequency and indicates a delay of 58.0 ms.

To simplify analysis, the system may be approximated by a single-rate model operating at the visual sampling interval of 20 ms. The open-loop transfer function may be written as

$$V(z)D(z)R(z) = \frac{K}{z^2(z-1)} \tag{1}$$

where

$$K = \frac{T_v}{T_r} K_p K_{lens} \tag{2}$$

and T_v and T_r are the vision and robot servo sample intervals respectively. The overall delay is Ord(denominator) − Ord(numerator), in this case three sample periods or 60 ms which agrees with previous observations of latency and the step response of this model.

The single-rate open-loop model (1) can be used for stability analysis as well as state-estimator and controller design using classical digital control theory.

The root-locus diagram in Figure 4 shows that for the known gain setting, $K = 0.132$, the poles are located at $z = 0.789$, 0.528 and -0.316. The dominant pole corresponds to a frequency of 1.9 Hz which is in good agreement with the estimate of 2 Hz made earlier. This pole is very sensitive to changes in loop gain such as those

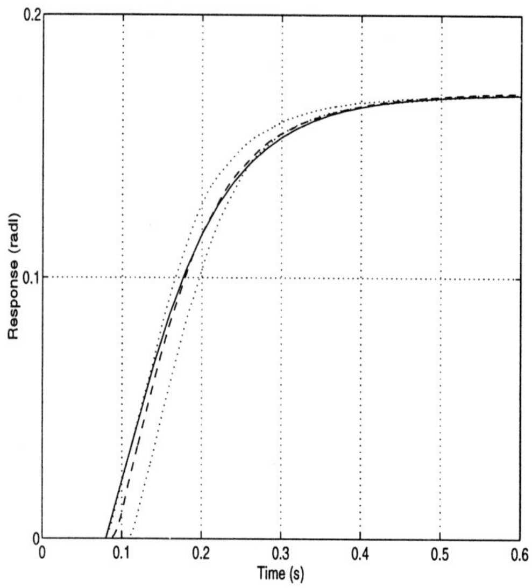

Figure 2: Comparison of measured and simulated step response. Measured early and late step responses (dotted), single-rate model (solid), and the multi-rate model (dashed).

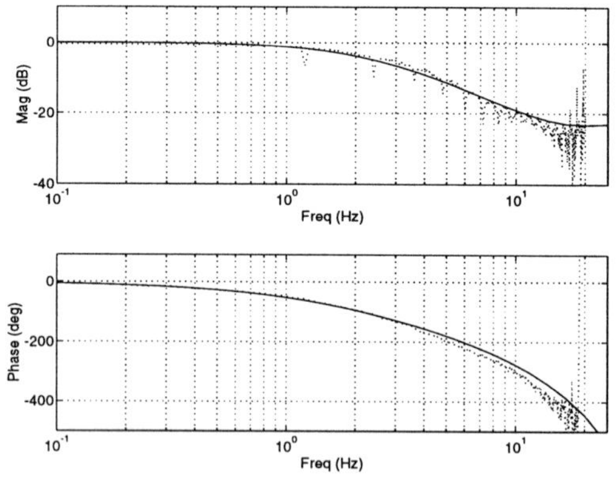

Figure 3: Bode plot of single-rate model closed-loop transfer function (solid) with measured frequency response (dotted) overlaid. Note the considerable phase error for moderate frequency demands, eg. 1 Hz. ($K_p = 1.3 \times 10^{-4}$, $K_{lens} = 708$)

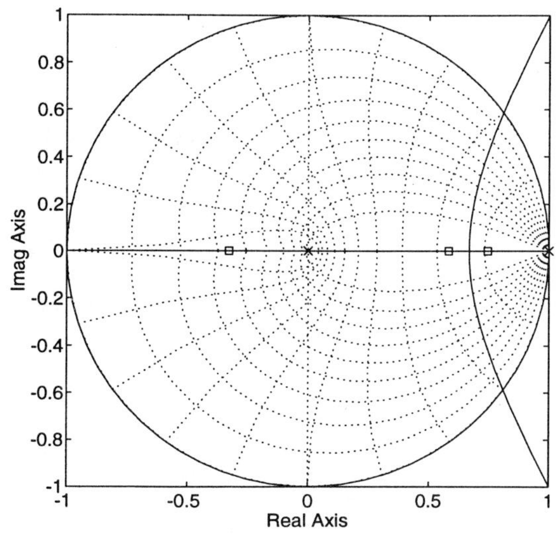

Figure 4: Root-locus of single-rate model. Closed-loop poles for gain $K_p = 1.3 \times 10^{-4}$ are marked ($K_{lens} = 708$).

caused by variation in target distance. Figure 3 shows the Bode plot of the single rate model overlaid on the experimentally determined frequency response function from above.

2 Some questions

This section poses a number of questions which need to be addressed in order to achieve high performance visual closed-loop control.

2.1 Is image feature trajectory generation necessary?

Most robot controllers accept absolute position setpoint commands but an end-effector mounted vision sensor provides relative position information. In general, due to limited torque capabilities, the robot will be unable to reach an arbitrary target position within one setpoint interval and thus requires some "planned" motion over many intervals. In robotics this is the well-known trajectory generation problem, but that approach has a number of disadvantages in this situation, since a moving target requires continual path adjustment [8] while also maintaining continuity of position and its time derivatives.

A far simpler approach is to consider the positioning task as a control problem:

> Perceived error generates a velocity demand which moves the robot toward the target.

This results automatically in target-following behaviour, but without the complexities of an explicit trajectory generator. This behaviour is analogous to the way we control cars and boats within an unstructured environment: we do not compute a sequence of precise spatial locations, but rather a velocity which is continually corrected on the basis of sensory inputs until the goal is achieved. Stated another way, we consider visual control as a *steering problem.*

2.2 What is the most appropriate axis control method to employ in a visual servo system?

At a very fundamental level a robot link is moved by an actuator that applies a torque or force. However it is common to provide local axis-level control loops which create the abstraction of a link whose position or velocity may be specified. Most industrial robot applications involve positioning tasks, bringing the tool to a known point in the workspace, and this requires position controlled axes. When the robot is in contact with a surface some positional degrees of freedom are lost and the robot must perform compliant motion requiring some, or all, joints to be torque (or force) controlled. In some teleoperation applications it is desirable to consider axes as being velocity controlled. Robot joints can thus be considered as torque, velocity or position "sources", and the choice is dependant upon the robot's application.

The argument in Section 2.1 would indicate that axis velocity control was the "natural" choice for closed-loop visual control. Surprisingly the majority of reported experimental systems implement a visual feedback loop "on top of" the robot's existing position-control loops. This has occurred, probably, for the pragmatic reason that most robot controllers present an abstraction of the robot as a position-controlled device. However some researchers in visual servoing who have "built their own" robot controllers at the axis level have all chosen to implement axis-position control [9–11]. The redundant levels of control add to system complexity and may impact on closed-loop performance [12]. Exceptions to this approach are the citrus-picking robot [13] which closes the vision loop about a hydraulic actuator which is a "natural" velocity source, and also [14, 15]. Weiss [2] proposed closing a visual-feature control loop around a torque servo but conducted only simulation studies.

Sending incremental motion commands to a robot position controller is essentially velocity control — the controller is integrating the increments to create an absolute coordinate as a setpoint for the position servos. From a control systems perspective, however, there are some subtle differences between a velocity controlled axis and an incremental position controlled axis. In particular the inner position loop will generally be very "tight", that is high-gain, in order to minimize position error. For an ideal robot this is not a problem, but it will exacerbate resonance problems due to structural and/or drive compliance. It has been shown [16] that for the case of significant structural compliance it is preferable to use a "soft" velocity loop in order to increase system damping. This is particularly desirable in the non-colocated control situation of an eye-in-hand camera.

In [1] the three axis control modes were compared for a simple linear discrete-time visual servo simulation model with proportional control. Torque control resulted in the slowest closed-loop poles, and position control the fastest, but with a steady state error to a step or higher order target motion. The addition of an extra integrator remedies this, but as discussed above, the structure is now essentially velocity controlled. However when a full non-linear motor model was simulated, the torque-mode controller did not behave as expected — simulation resulted in very lightly damped oscillatory responses. This was found to be due to a combination of non-linear friction, integral action and the relatively low sample rate. The unmodeled dynamics, particularly stick/slip friction, are complex and have time constants that are shorter than the visual sample interval. It is a "rule of thumb" that the sample rate should be 4 to 20 times the natural frequency of the modes to be controlled. Improved torque-mode control would require one or more of the following: a higher sample rate, friction measurement and feedforward, adaptive, or robust control. A more straightforward option is the use of axis velocity control. Velocity feedback is effective in linearising the axis dynamics and eliminating much of the parameter variation. Parameter sensitivity studies showed robustness with changes of up to ±25%.

2.3 How should visual servo performance be evaluated?

Performance of the visual servo system can be expressed in terms of time or frequency domain metrics. In the frequency domain, performance is related to the closed-loop bandwidth. While "high closed-loop bandwidth" is desirable to reduce error it can also lead to a system that is sensitive to noise and unmodeled dynamics. Additionally, the notion of "bandwidth" is not straightforward since it implies both magnitude (3 dB down) and phase (45° lag) characteristics. Time delay introduces a linear increase in phase with frequency, and the Bode plot of Figure 3 indicates 45° lag occurs at less than 1 Hz. For tracking applications phase is a particularly meaningful performance measure — for instance the citrus picking robot's tracking specification [13] was in terms of phase for a particular fruit swinging frequency.

Common time domain metrics include settling time, overshoot and polynomial signal following errors. However many published visual servo results should be subject to close scrutiny about what they really demonstrate. Any closed-loop system of Type 1 or greater will have zero steady state error to a step demand. Since, as described above, most visual servo systems incorporate at least one integrator (by virtue of incremental positioning), zero settling error is to be expected. In fact settling error is the *least challenging* test of control performance. For example, the feedback system introduced above has a fast square wave response (200 ms) but simulated tracking of sinusoidal motion in Figure 5 shows a robot lag of approximately 30° and the centroid error peaking at over 80 pixels.

A better metric for tracking performance would be related to tracking error such as peak-to-peak error, or

$$\eta = \sqrt{\frac{\int_{t_1}^{t_2} \left({}^iX(t) - {}^iX_d(t)\right)^2 dt}{t_2 - t_1}} \tag{3}$$

which is the RMS pixel error value over the time interval $[t_1, t_2]$. The choice of appropriate target motion with which to evaluate (3) depends upon the application. For instance, if tracking a falling or flying object it would be appropriate to use a parabolic input. A sinusoid is particularly challenging since it is a non-polynomial with significant and persistent acceleration that can be created readily in the laboratory using a pendulum or turntable.

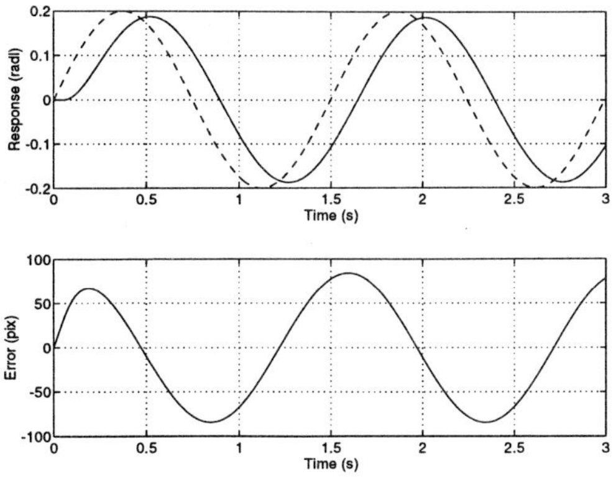

Figure 5: Simulated tracking performance of visual feedback controller, where the target is moving sinusoidally with a period of 1.5 s (0.67 Hz). Note the considerable lag of the response (solid) to the demand (dashed).

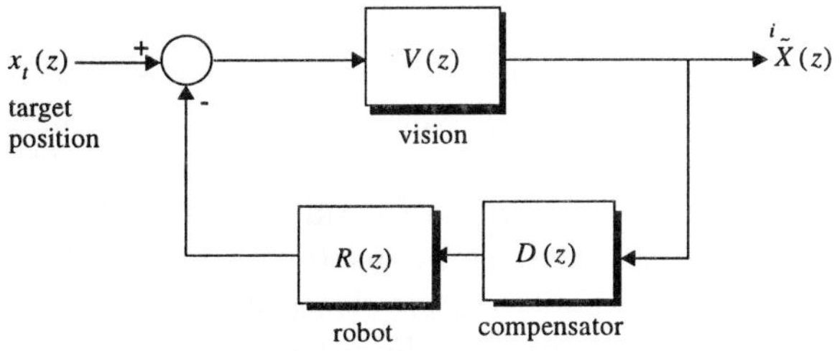

Figure 6: Block diagram showing target position as input and image-plane error as output.

2.4 What performance should be expected?

The camera frame rate is the sample rate of a visual servo system and this is a function of the standard video formats typically employed (CCIR or RS170). Most high-performance systems reported achieve field-rate processing at best, giving a sample rate of 50 or 60 Hz, which is a major performance limiting factor. Franklin [17] suggests that the sample rate of a digital control system be between 4 and 20 times the desired closed-loop bandwidth. Such a rule implies a closed-loop bandwidth of between 2.5 Hz and 12 Hz. Corresponding response times would be in the range 80 to 400 ms, which the experimental system achieves.

2.5 What are appropriate control laws to achieve high closed-loop performance?

In control system terms the visual servo system is a regulator (trying to maintain a constant image plane target centroid) and the target motion is a non-measurable disturbance input. Tracking performance can be expressed as a transfer function for image plane error response to target input, as shown in Figure 6

$$\frac{^i\tilde{X}(z)}{x_t(z)} = \frac{V(z)}{1+V(z)R(z)D(z)} \quad (4)$$

$$= \frac{K_{lens}z(z-1)D_D(z)}{z^2(z-1)D_D(z)+K_{lens}TD_N(z)} \quad (5)$$

where $^i\tilde{X} = {}^iX - {}^iX_d$ is the image plane error and ^iX_d is assumed, for convenience, to be zero.

The steady-state value of the image plane error for a given target motion, $x_t(z)$, can be found by the final-value theorem. For proportional control, and substituting previously identified models for $V(z)$ and $R(z)$, the steady-state error is

$$^i\tilde{X}(\infty) = \lim_{z\to1}(z-1)\frac{K_{lens}z(z-1)}{z^2(z-1)+K_pK_{lens}}x_t(z) \quad (6)$$

which may be evaluated with the target motion chosen as one of

$$x_t(z) = \begin{cases} \frac{z}{z-1} & \text{for a step input} \\ \frac{Tz}{(z-1)^2} & \text{for a ramp input} \\ \frac{T^2}{2}\frac{z(z+1)}{(z-1)^3} & \text{for a parabolic input} \end{cases} \quad (7)$$

For the steady-state tracking error to be zero, the numerator of (6) must cancel the poles at $z=1$ of $x_t(z)$ and retain a factor of $(z-1)$. This numerator comprises the poles of the robot and compensator transfer functions — the latter may be selected to give the desired steady-state error response by inclusion of an appropriate number of integrator terms, or $(z-1)$ factors. The number of such open-loop integrators is referred to as the *Type* of the system in classical control literature. Equation (6) has one open-loop integrator and is thus of Type 1 giving zero steady-state error to a step input. A ramp input however will result in a finite error of T/K_p pixels which can be seen as the substantial lag in Figure 5. Also note that the numerator of the closed-loop transfer function (5) has a zero at $z=1$ which acts as a differentiator. From (5) compensator poles will appear as additional closed-loop zeros.

Common approaches to achieving improved tracking performance are:

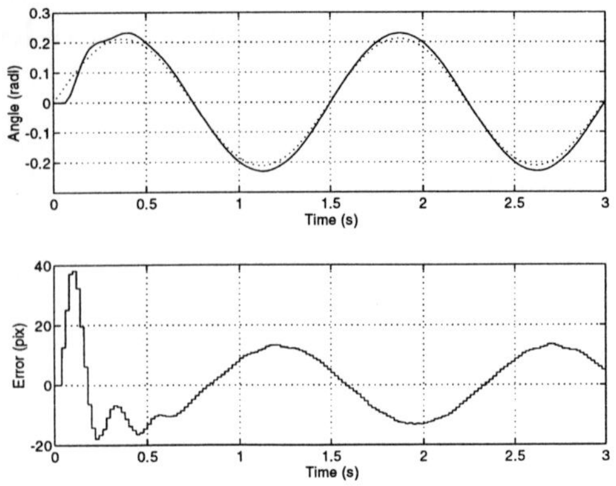

Figure 7: Simulation of pole-placement fixation controller, showing response (solid) and demand (dotted). Closed-loop poles at $z = 0.6 \pm 0.4j, 0.4$.

1. Increasing the loop gain, K_p, which minimises the magnitude of ramp-following error. However for this system, as shown in the root-locus plot of Figure 4, this approach is limited, since the poles leave the unit circle as gain is increased. It is not possible to achieve the stable high-gain control required for accurate tracking.
2. Increasing the Type of the system, by adding additional open-loop integrators.
3. Introducing feedforward of the signal to be tracked. This is an important issue in machine-tool control and has been discussed, for instance, by Tomizuka [18]. This approach is discussed in Section 2.7.

A number of control laws including PID, Smith's predictor, pole-placement, and LQG, were synthesised and evaluated in both simulation and experiment. In order to achieve good sinusoid tracking performance, it was found necessary to increase the Type of the open-loop system to 2. However the location of the closed-loop poles is fundamentally constrained if a stable compensator is required. As closed-loop bandwidth is increased the robustness with respect to plant parameter variation, in particular saturation, is greatly reduced.

Figure 7 shows the simulated closed-loop response for sinusoidal target motion, a detailed non-linear system model, and a pole-placement controller. The peak-to-peak error in pixel space is approximately 28 pixels — clearly a great improvement over simple proportional control — and the best performance achieved in this work. However a number of characteristics of the response deserve comment:

- The error is greatest at the peaks of the sinusoid, which correspond to greatest acceleration. The Type 2 controller has zero error following a constant velocity target (which the slopes of the sinusoid approximate), but will have finite error for an accelerating target. That performance is dependent on the closed-loop pole locations which cannot be made arbitrarily fast.
- There is considerable delay at $t = 0$ before the robot commences moving, due to delay in the vision system and controller.
- Once again this is a Type 2 system and the closed-loop response will include a double differentiator. This effect is evident in the initial transient shown in the simulated closed-loop response of Figure 7.

The closed-loop poles cannot be made arbitrarily fast and are constrained by the practical requirement for compensator stability. Linear discrete-time simulations indicate no difficulty with an unstable compensator, but non-linear simulations show that the closed-loop system will be unstable. This is believed to be due to non-linearities in the plant which change the apparent dynamics at saturation which is exacerbated by high controller gains. The addition of integrators, while raising the system Type, places stringent constraints on closed-loop and estimator pole locations.

The Smith's predictor also warrants comment since it is commonly cited by the active vision community. The essential characteristic of this predictor is cancellation of open-loop poles. In linear simulation this approach works well, but in non-linear simulation or experiment it lacks robustness. Stability can be achieved but only by choosing "slow" closed-loop poles.

2.6 What are the limitations to closed-loop performance?

For a feedback system such as described so far the limiting factor is the low loop gain necessary to ensure stability in the presence of open-loop delay. The architecture of the system, and in particular communications between the vision and robot control subsystems should be designed in

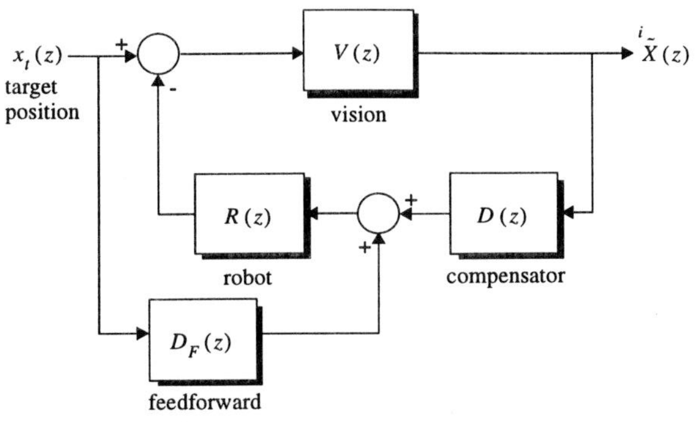

Figure 8: Block diagram of visual servo with feedback and feedforward compensation.

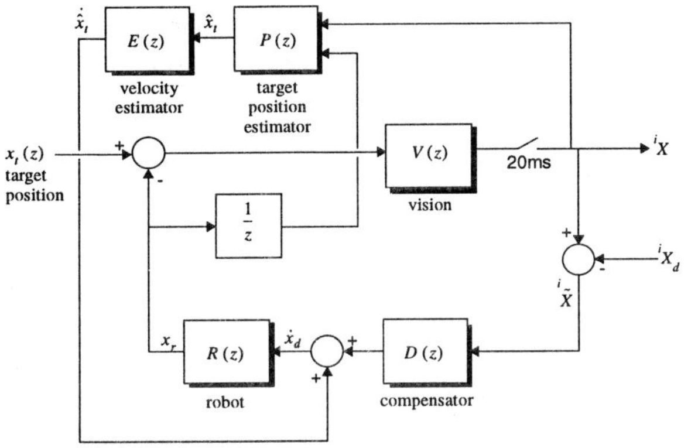

Figure 9: Block diagram of velocity feedforward and feedback control structure as implemented. $P(z)$ estimates target bearing.

order to minimize this delay. Prediction can be used to "cover" system delay, but at the expense of poor disturbance rejection with respect to unmodeled target motions. More sophisticated compensators can be employed, but they remain fundamentally constrained by the requirements for compensator stability and robustness.

2.7 Why are most visual servo systems based only on feedback control?

The term "visual servo" is almost synonymous with "visual feedback" but this need not be so. The design constraints inherent in feedback-only control lead naturally to the consideration of feedforward control strategies, since additional design degrees of freedom are available by manipulating the system zeros as well as the poles. The feedforward quantity, target position or velocity in this case, is not directly measurable but can be estimated. The feedforward term is shown in Figure

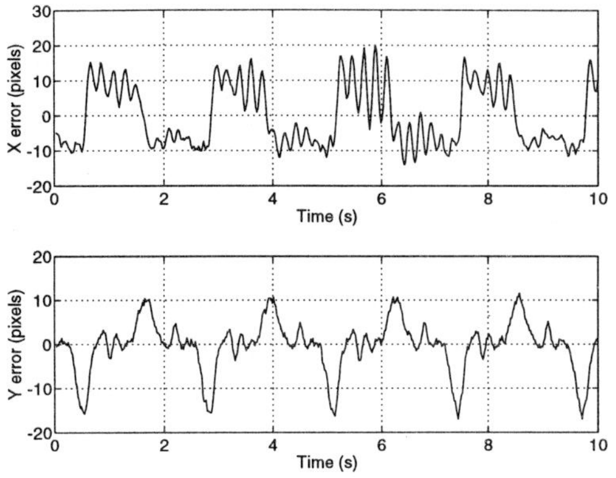

Figure 10: Measured tracking performance, centroid error, for target on turntable revolving at 4.2 rad/s.

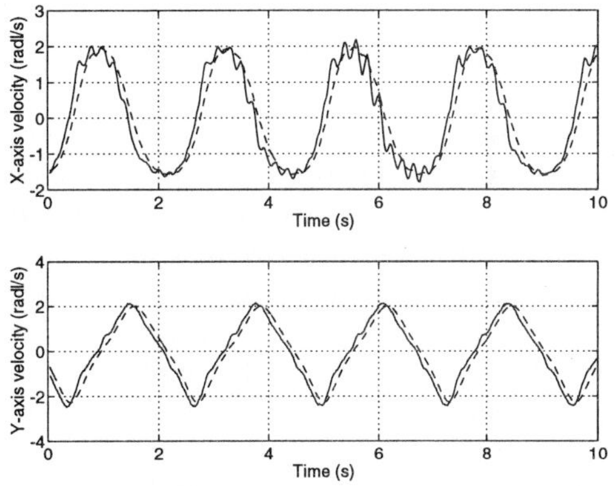

Figure 11: Measured tracking v velocity for target on turntable revolving at 4.2 rad/s, showing axis velocity (solid) and estimated target velocity (dashed).

8, and the closed-loop transfer function becomes

$$\frac{{}^i\tilde{X}}{x_t} = \frac{V(z)\,(1 - R(z)D_F(z))}{1 + V(z)R(z)D(z)} \tag{8}$$

which has the same denominator as (5) but an extra factor in the numerator. Clearly if $D_F(z) \equiv R^{-1}(z)$ the tracking error would be zero, but such a control strategy is not realizable since it requires (possibly future) knowledge of the target position which is not directly measurable.

Such a controller has been implemented, see Figure 9, based on estimated target velocity feedforward and with high-performance digital axis velocity-control loops. In the experiment the camera has 2-DOF rotational motion and is fixated on an object revolving on a turntable. The resulting centroid tracking error is shown in Figure 10. Figure 11 compares the measured and estimated target velocity. Since the majority of the control effort is due to the feedforward path, lower feedback gains can be used thus minimising problems due to latency. The use of a feedforward term has been proposed previously by De Schutter [19] in the context of robotic force control.

2.8 What other advantages are offered by "visual dynamic" control?

The analysis of visual servo dynamics leads to advantages beyond the obvious ones of improved tracking performance. Visual servoing admits the possibility of "closing the loop" on the dynamics of the actuator, transmission and structural compliance, allowing superior control of end-point dynamics than can be achieved using axis sensing alone.

Robot speed, or cycle time, is a critical factor in the economic justification of a robot. Although machines capable of extremely high tool-tip accelerations now exist, the overall cycle time is dependent upon other factors such as settling time and overshoot. High speed and acceleration are often achieved at considerable cost since effects such as rigid-body dynamics, non-linear friction and link and transmission flexibility [20] become significant. To achieve precise end-point control using joint position sensors the robot must be engineered to minimize these effects by employing gearless transmissions and massive links to minimise friction and compliance. End-point relative position measurement admits direct end-point, rather than joint-angle, control. With appropriate control strategies this may result in reduced overshoot and thus lower cycle time.

Figure 12 shows the measured end-effector motion of a Puma robot (under native position controller with arm fully outstretched and horizontal) after a high velocity motion. The end-effector has overshot by around 12 mm. Although not visible in the figure, the camera acceleration peaks at 4 g and oscillates at around 19 Hz. This mode is due to torsion in the vertical drive shaft.

A computed torque controller, with friction feedforward, runs at 500 Hz

$$\tau_d = \mathbf{M}\left\{\ddot{\theta}_d + K_p(\theta_d - \theta) + K_v(\dot{\theta}_d - \hat{\dot{\theta}})\right\} + \mathbf{F}(\hat{\dot{\theta}}) \qquad (9)$$

This is overlaid with a variable structure control strategy which moves the arm under open-loop trajectory generator control to the vicinity of the target, and once found incorporates visual control for fine positioning

$$\dot{\theta}_d = {}^iK_p({}^iX_d - {}^iX) + {}^iK_v{}^i\hat{\dot{X}} \qquad (10)$$

Experimental results with this controller are shown in Figure 13. The end-effector has followed an almost critically damped path to the target which was displaced approximately 20 mm from the destination of the trajectory generator. The target came into view at 0.32 s and the transition in control strategies occured four field times later at approximately 0.40 s. The end-effector has settled to within 1 pixel or 0.78 mm of the target whose position was unknown in advance.

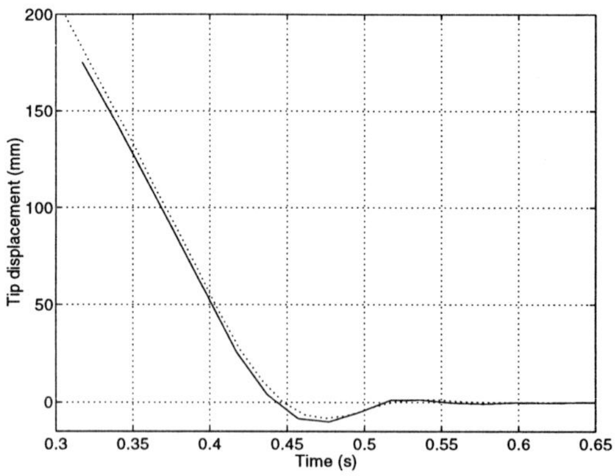

Figure 12: Measured tip displacement determined from axis sensing (dotted) and end-effector mounted camera (solid).

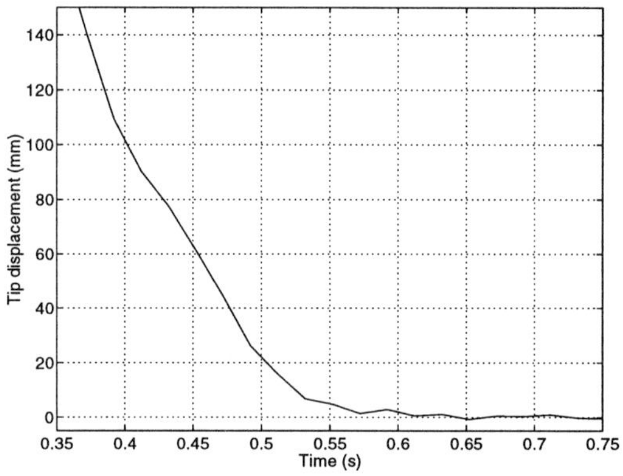

Figure 13: Measured tip displacement determined from end-effector mounted camera for hybrid visual control strategy.

3 Conclusions

This paper has discussed several important aspects of visual closed-loop systems that receive little attention in the literature. It has been argued that ignoring the temporal and dynamic characteristics of the system being controlled leads to the poor closed-loop performance demonstrated in many papers. It is also argued that axis velocity control has many advantages for use in visual servo systems. These include a 'natural' interpretation of visual servoing as a steering problem, and improved dynamic performance for manipulators with structural compliance.

The key conclusions are that in order to achieve high-performance visual servoing it is necessary to minimize open-loop latency, have an accurate dynamic model of the system and to employ a feedforward type control strategy. Prediction can be used to overcome latency but at the expense of reduced high frequency disturbance rejection. Open-loop latency is reduced by choice of a suitable control architecture.

An accurate dynamic model is required for control synthesis. Feedback-only controllers have a loop gain limit due to the significant delay in pixel transport and processing. Simple feedback controllers have significant phase lag characteristics which lead to poor tracking. More sophisticated feedback controllers can overcome this but the solution space becomes very constrained and the controllers are not robust with respect to plant parameter variation. Feedforward control results in a robust controller with excellent tracking capability. In addition dynamic visual servoing can be used to "close the loop" on the dynamics of the actuator, transmission and structural compliance, allowing superior control of end-point dynamics than can be achieved using axis sensing alone.

References

[1] P. I. Corke, *High-Performance Visual Closed-Loop Robot Control.* PhD thesis, University of Melbourne, Dept. Mechanical and Manufacturing Engineering, July 1994.

[2] L. Weiss, *Dynamic Visual Servo Control of Robots: an Adaptive Image-Based Approach.* PhD thesis, Carnegie-Mellon University, 1984.

[3] S. Ganapathy, "Real-time motion tracking using a single camera," Technical Memorandum 11358-841105-21-TM, AT&T Bell Laboratories, Nov. 1984.

[4] J.-C. Yuan, F. Keung, and R. MacDonald, "Telerobotic tracker," Patent EP 0 323 681 A1, European Patent Office, Filed 1988.

[5] S. Venkatesan and C. Archibald, "Realtime tracking in five degrees of freedom using two wrist-mounted laser range finders," in *Proc. IEEE Int. Conf. Robotics and Automation*, pp. 2004–2010, 1990.

[6] S. D. Eppinger and W. P. Seering, "Introduction to dynamic models for robotic force control," *IEEE Control Systems Magazine*, vol. 7, no. 2, pp. 48–52, 1987.

[7] J. Hill and W. T. Park, "Real time control of a robot with a mobile camera," in *Proc. 9th ISIR*, (Washington, DC), pp. 233–246, Mar. 1979.

[8] J. Feddema, *Real Time Visual Feedback Control for Hand-Eye Coordinated Robotic Systems.* PhD thesis, Purdue University, 1989.

[9] N. Papanikolopoulos, P. Khosla, and T. Kanade, "Vision and control techniques for robotic visual tracking," in *Proc. IEEE Int. Conf. Robotics and Automation*, pp. 857–864, 1991.

[10] P. Sharkey, I. Reid, P. McLauchlan, and D. Murray, "Real-time control of a reactive stereo head/eye platform," *Proc. 29th CDC*, pp. CO.1.2.1–CO.1.2.5, 1992.

[11] N. Houshangi, "Control of a robotic manipulator to grasp a moving target using vision," in *Proc. IEEE Int. Conf. Robotics and Automation*, pp. 604–609, 1990.

[12] P. Corke and M. Good, "Dynamic effects in high-performance visual servoing," in *Proc. IEEE Int. Conf. Robotics and Automation*, (Nice), pp. 1838–1843, May 1992.

[13] T. Pool, *Motion Control of a Citrus-Picking Robot.* PhD thesis, University of Florida, 1989.

[14] K. Hashimoto, T. Kimoto, T. Ebine, and H. Kimura, "Manipulator control with image-based visual servo," in *Proc. IEEE Int. Conf. Robotics and Automation*, pp. 2267–2272, 1991.

[15] P. Corke and M. Good, "Controller design for high-performance visual servoing," in *Proc. IFAC 12th World Congress*, (Sydney), pp. 9–395 to 9–398, 1993.

[16] M. Roberts, "Control of resonant robotic systems," Master's thesis, University of Newcastle, Australia, Mar. 1991.

[17] G. Franklin and J. Powell, *Digital Control of dynamic systems.* Addison-Wesley, 1980.

[18] M. Tomizuka, "Zero phase error tracking algorithm for digital control," *Journal of Dynamic Systems, Measurement and Control*, vol. 109, pp. 65–68, Mar. 1987.

[19] J. D. Schutter, "Improved force control laws for advanced tracking applications," in *procra*, pp. 1497–1502, 1988.

[20] M. C. Good, L. M. Sweet, and K. L. Strobel, "Dynamic models for control system design of integrated robot and drive systems," *ASME Journal of Dynamic Systems, Measurement and Control*, vol. 107, p. 53, 1985.

Extended Vision with the Robot Sensor Suit: A Primary Sensor Image Approach to Interfacing Body to Brain

Masayuki Inaba

Department of Mechano-Informatics
The University of Tokyo
7-3-1 Hongo, Bunkyo-ku, Tokyo, JAPAN

Abstract

We present an approach to interface body to brain through building primary sensor image integrating visual and somatic sensing systems. The fundamental idea of this approach is that a robot body is a sensor to observe the world. The primary sensor image gives the robot brain a projection of the world. We have developed methods to build sensor images. One method is to use sensor display of somatic sensors so that the vision can observe their state. Another method is to multiplex the somatic sensors onto the video image electrically. In order to attach large number of tactile sensors distributed on a robot body, we have built a robot sensor suit by using electrically conductive fabric. A humanoid wearing the sensor suit is introduced.

1. Introduction

The body of a robot rules the ability of the brain. The methodology of how to build the real body with rich sensors and actuators is crucial to success in development of brains.

The robots developed so far usually have limitation in variety of sensors. Recently it has become popular to use vision since the performance has become high. However, it is few to use both vision and other sensors effectively. We need a good methodology to put rich sensors on a robot body.

Although vision is important function for a robot, somatic sensors such as tactile sensors distributed on the entire body are indispensable in order to accommodate itself to a real environment. When we wish to give a lot of sensors to a robot body, the technology of hardware implementation becomes important. In such a case the main problems are how to build (1) sensor systems, (2) wiring methods, (3) an interface to brain, and (4) an environment for development.

If we think about behaviors of the entire body, it is more important to cover the body with a lot of tactile sensors. The distribution of the sensors is key in such situations.

The more sensors there are, the harder the wiring problem becomes because the bunch of the wires becomes thicker and thicker. Wiring usually requires many connectors which have considerable volume. A robot with a computer and many sensors has to select a good connection technology[1].

The events of the dynamic world occur in parallel. The brain has to have the capability to receive and process such events as rapidly as it can, which requires the brain to have multiple processors observe the events.

The environment for building brains for bodies should provide a capability to show what happens to bodies during interaction with the environment. Visualization of the signals from sensors is a key to such capability.

In the following sections we introduce our approach and concrete modules to build real bodies to develop brains.

2. Sensor Image Approach

2.1. Sensing Through Body

The basic idea in our approach is that a robot body is not only an active unit but also a sensing unit to observe the world. The primary sensor image is defined as a projected image of the world through the entire body.

Figure 1 is shown to explain the concept. (a) is the usual scheme of vision which looks the real world through on eye. The visual image is an optical projection of the outer world through the eye. The sensor image shown in (b) is like a visual image but is given through a body. The outer world is projected to the sensor image through physical interactions by the body.

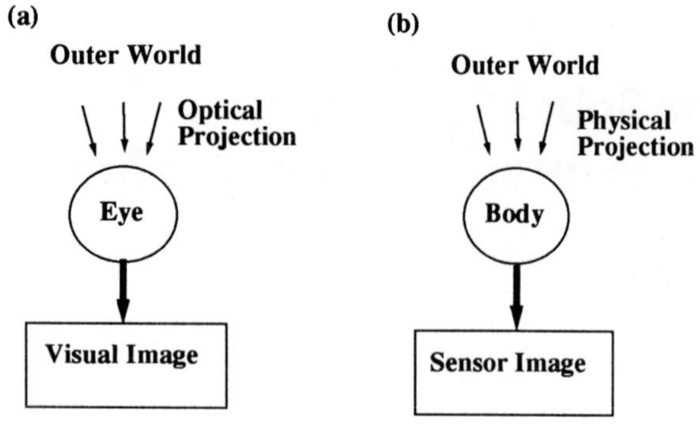

Figure 1. Basic concept in extended vision

2.2. Parallel Processing Through Sensor Image Bus

Extended vision looks the real world through the whole sensor of body and gives the sensory image projected from the outer world to the robot brain. The image contains multi modal information consisting of somatic information as well as visual information. The robot brain observes this and manages the body with actions.

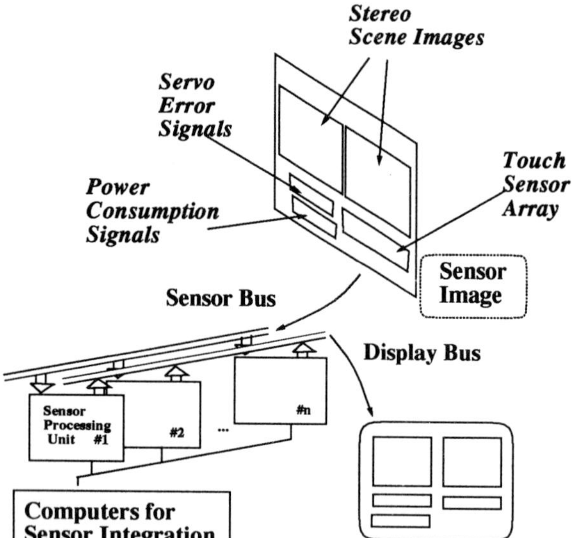

Figure 2. Basic configuration for vision-based multisensor integration

Figure 2 shows the design concept for the system based on the sensor image approach[3]. The sensor image contains primary information from sensors such as eyes, touches, joint torques, etc. The sensor image is processed by multiple sensor processors which receive the image through the sensor image bus. In this configuration, each processor can access all the sensor information and generate secondary information by integrating the primary information. The sensor processors can transmit the processed information to robot brain systems.

2.3. Methods in Building Sensor Image

The sensor image containing the multiple sensor information can be generated by two methods: (a) 'electrical multiplexer' and (b) 'optical multiplexer', as shown in Figure 3. The electrical multiplexer has a circuit to superimpose sensor signals on a video signal. The optical multiplexer is the method of a camera looking at a sensor display which shows the sensor signals optically.

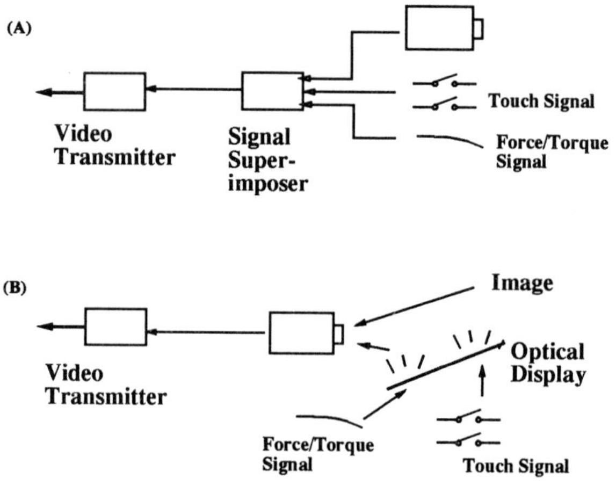

Figure 3. Two approaches to building the sensor image

2.4. Uniform Treatment in Behavior Development

The advantage of the sensor image approach is that we can define all the input data to a brain from the environment and operate them in an uniform way. So far we have developed a series of human-formed robots to investigate brain development in the remote-brained design[4]. Figure 5 shows the models of the robots we have developed. They can share brain software in the remote-brained environment and could effect knuckle walking, bipedal walking[5], carrying an object, and some desk work. Their sensing systems can be extended though designing the sensor images according to desired behaviors.

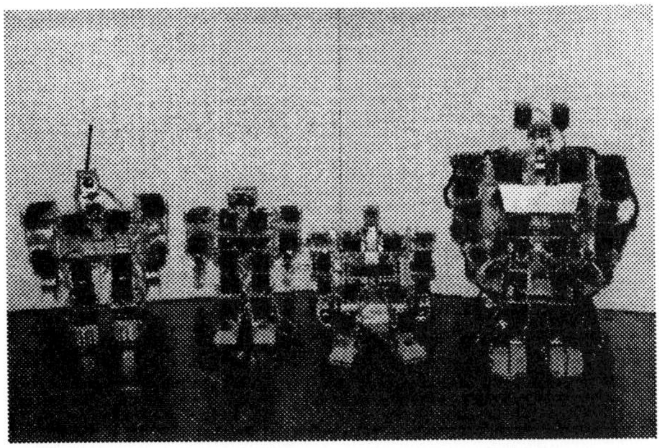

Figure 4. Developed human-form robots

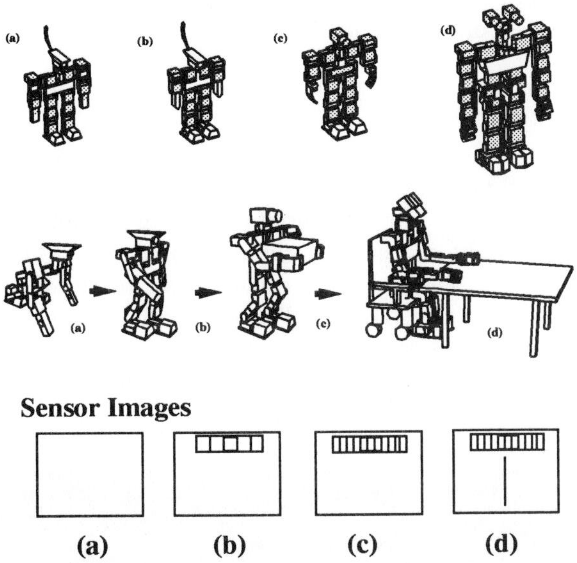

Figure 5. Behavior development with sensor images

3. Modules for Image Based Sensor Multiplexer

We have designed and implemented several tools to advance the sensor image approach. The following sections explain some of them. We can design several kinds of extended vision systems using these modules.

3.1. Stereo Viewer

In order to multiplex stereo views into an image, we have developed a stereo viewer using four mirrors[6]. It gives an binocular image to a single camera. Figure 6 shows the stereo adapter attached to a video camera.

The internal structure of the stereo viewer is

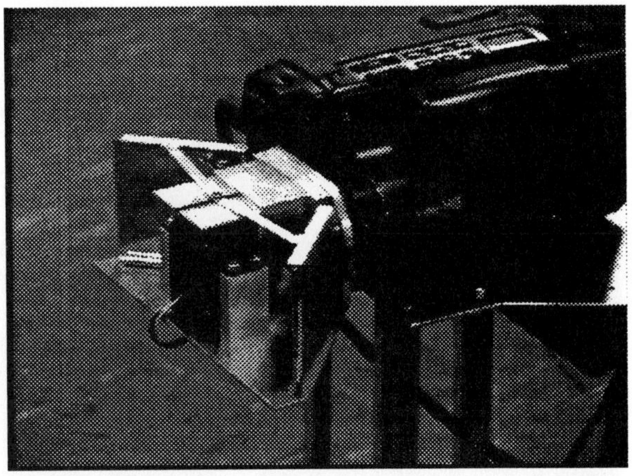

Figure 6. The stereo viewer with a hand video camera

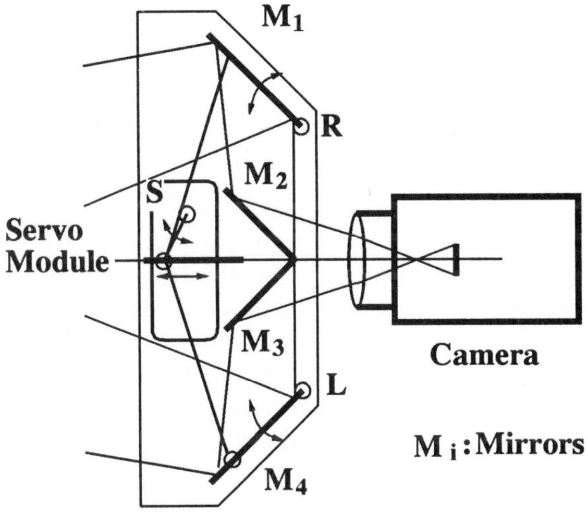

Figure 7. Top view of the internal structure of the stereo adapter

shown in Figure 7. The stereo adapter guides the outside image to the camera screen from two different viewing points through four mirrors (M_1, M_2, M_3, M_4). The outside mirrors (M_1, M_4) can rotate around the axes of L and R. The inside ones (M_2, M_3) are fixed, so that they are perpendicular. The box S is a servo module which moves the outside mirrors using two links.

Figure 8 shows an example of the image viewed through the stereo viewer. The input image is split into left and right views. The squares in the left and right views are regions that correspond each other, as which are derived by the block image matching method. This matching process searches for an image region that has the

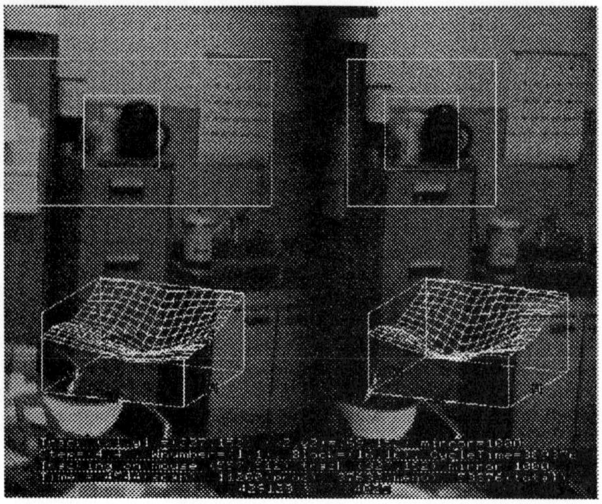

Figure 8. A view through the stereo viewer

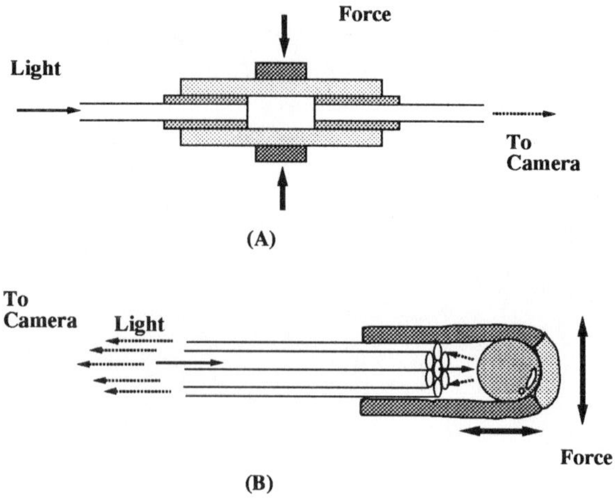

Figure 9. Mechanism of touch and force sensors based on optical fibers

minimum value in terms of sum of absolute difference (SAD) between square images in the left view and the right view. The 3D graphs displayed on the images show the SAD distribution around the minimum position.

The stereo viewer works like a lens filter to get stereo images. With a single camera and image processor set, it is possible to get not only gray-level images but also range information. The viewer does not depend on a special camera platform. A vision system with the viewer can grab two images at the same time. The system can thus track the 3D trajectory of an object in real time.

The viewer controls vergence with M_1 and M_4. As the mirrors are light and small, it is possible to obtain fast vergence control. Vergence control needs fast stereo tracking function. In our system, it has been shown that the block matching process, using a special processor to calculate the sum of absolute difference between two gray-level images, can perform both motion tracking and stereo matching in real-time[7].

3.2. Optical Fiber Based Sensors

If it is possible to build optical transducer for touch and force sensing, the optical multiplexer can be done by simply showing the optical intensity to a camera. We have designed and developed several touch and force sensors using optical fibers. Figure 9 shows examples. The sensor (a) is built with two pieces of fiber aligned on a line where one piece is connected to a light source and the other is connected a light sensor. The mechanism to measure pressure force is based on light interference.

The sensor (b) is based on light reflection to sense haptic motion in touch. It consists of a bunch of fibers. The central fiber of the bunch is connected to a light source. The light through the central fiber is reflected on the metal ball to the outer fibers of the bunch. The outer fibers are connected to a camera. The direction of the motion is measured by the pattern of the light projected through the outer fibers[8].

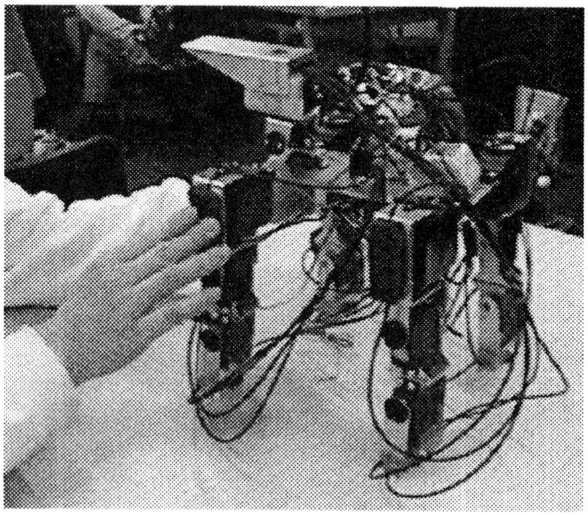

Figure 10. The quadraped robot with optical fiber based tactile sensors

We have applied the optical sensors in a quadraped robot as shown in Figure 10. The robot is designed for research on sensor-based interactive behaviors using vision and touch. The system configuration of the robot is shown in Figure 11. The robot has tactile sensors on shoulders, knees, and feet. The ends of the fibers are connected to the top of the camera. The camera observes the ends of the fibers through a mirror placed at the upper front of the camera.

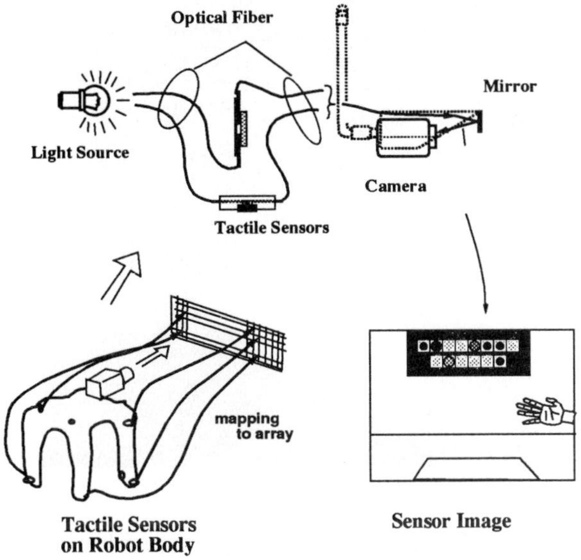

Figure 11. A quadraped robot with fiber-based tactile sensors

The view seen by the robot is shown in Figure 12. The robot can observe not only the visual scene in the lower area in the image but also the states of the fiber-based sensors in the upper area of the view.

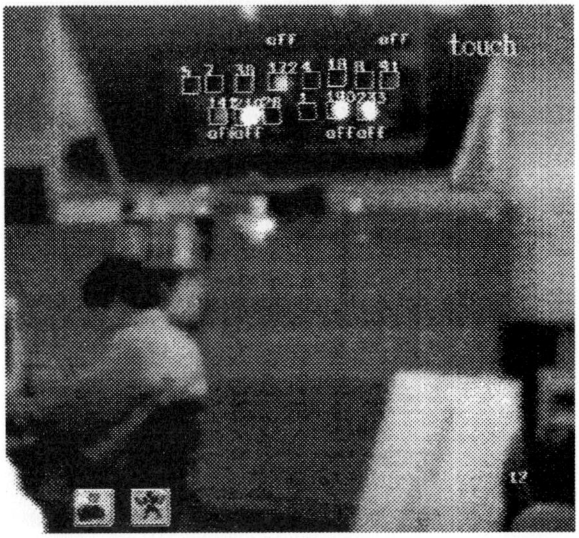

Figure 12. Sensor image of the quadraped robot

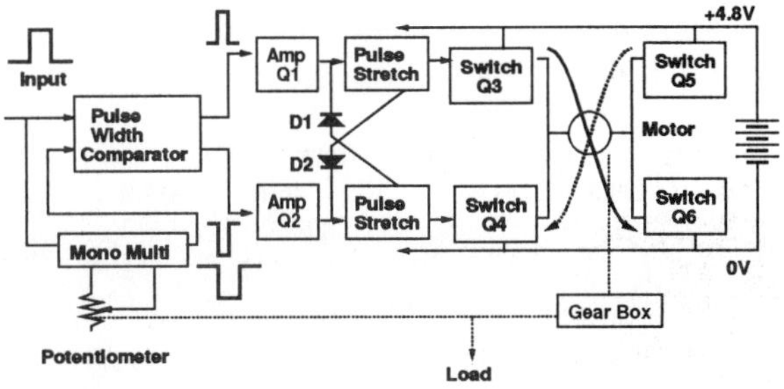

Figure 13. Circuit diagram of servo module

3.3. Sensor Display using Electro-Optic Transformation

If a robot can detect reactive joint torque, it can infer obstacles, balance, grasping force, configurations, and so on. The reactive joint torque can be estimated from the servo error. Our robots use servo modules designed for radio control model toys. The servo module contains a motor, a potentiometer, gears, and a position feedback controller in a box. Figure 13 shows the block diagram of the circuit. It is possible to show the servo error signal from this circuit to know the torque value and its direction relative to the joint.

We can build a sensor display which contains LED array connected to servo error signals and other tactile sensors. Figure 14 shows the configuration as applied to a two-armed bipedal robot as shown in Figure 15. The sensor image of the robot is shown in Figure 16. The LED array has 12 LEDs in this case.

The torque of a joint is shown by two LEDs corresponding to the rotational directions. The higher

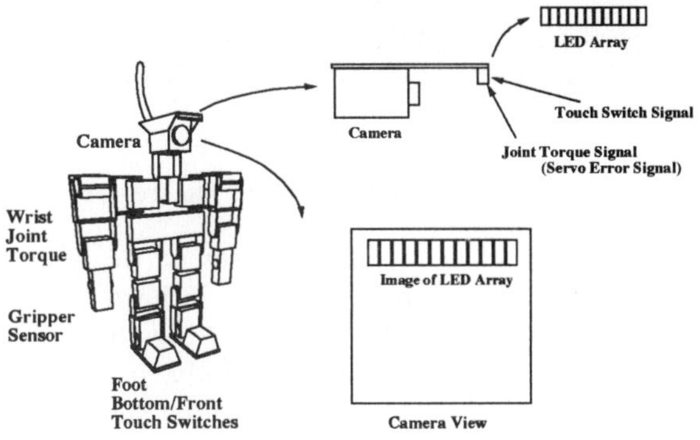

Figure 14. An application in two-armed bipedal robot

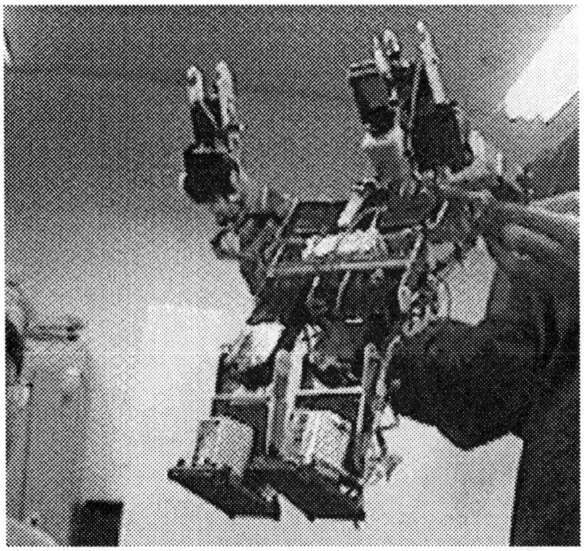

Figure 15. The two-armed bipedal robot

the servo error is, the longer the LED is switched on.

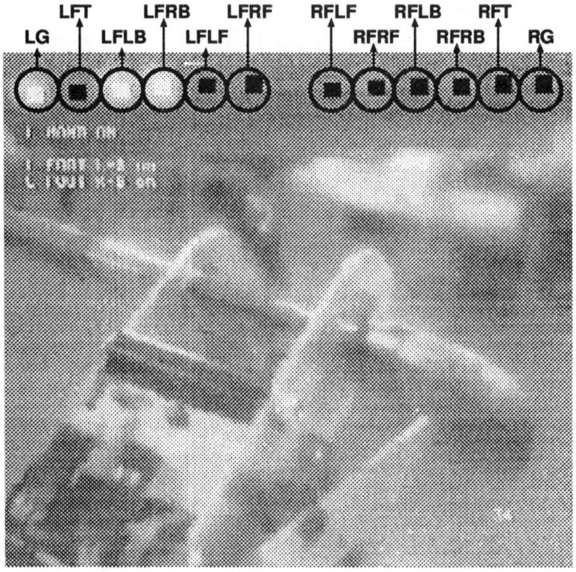

Figure 16. Sensor image of the robot

This method of using a LED array as the sensor display is good in cases where it is hard to put extra circuits on a robot body. If the robot can afford to have more circuits on body, it is possible to multiplex the signal into video signal of a camera by an electrical superimposing circuit.

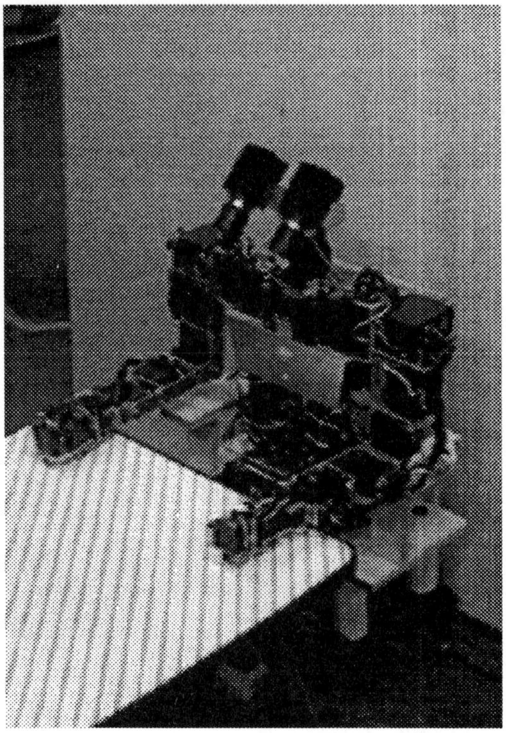

Figure 17. Overview of the humanoid with 35 joints

4. The Robot Sensor Suit

Our approach is based on two methods: sewing with ECS and the video multiplexer to pass on all signals.

4.1. A Humanoid with 35 joints

We have developed several human-form robots. The biggest one has 35 joints and could grasp an object on a table sitting on a chair as shown in Figure 17. It has two 6DOF arms, two 4DOF hands, two 6DOF legs and a head on a 3DOF neck. Figure 18 shows the structure of the robot.

4.2. Actions Coordinating Legs and Arms

The humanoid can perform actions coordinating the legs and arms. When the robot sit on a chair and handles objects on a table, it can grasp an object even if the object is outside of the reach of the robot arms. If the robot can not reach an object, it tries to reach it by changing the upper body posture or standing to get a longer reach. The system can perform such actions based on environment models and visual information.

Figure 19 shows a case where the robot grasps an object on a table by standing up. In this system,

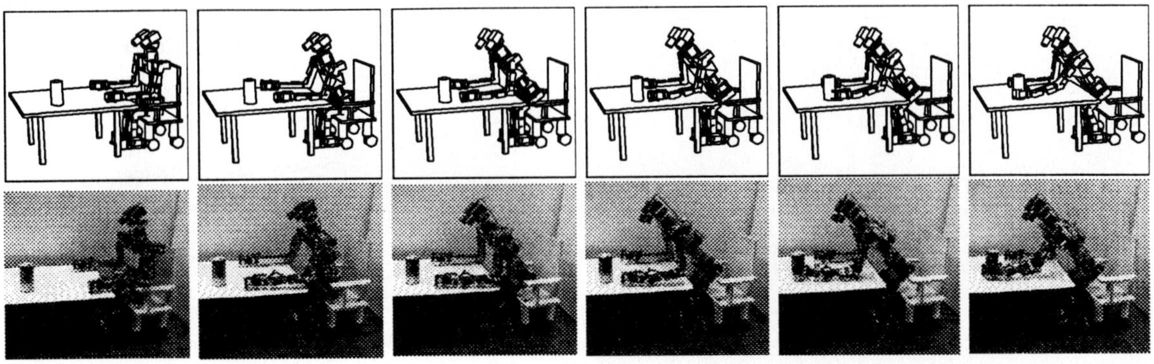

Figure 19. Experiment in grasping an object with legs and arms cooperation

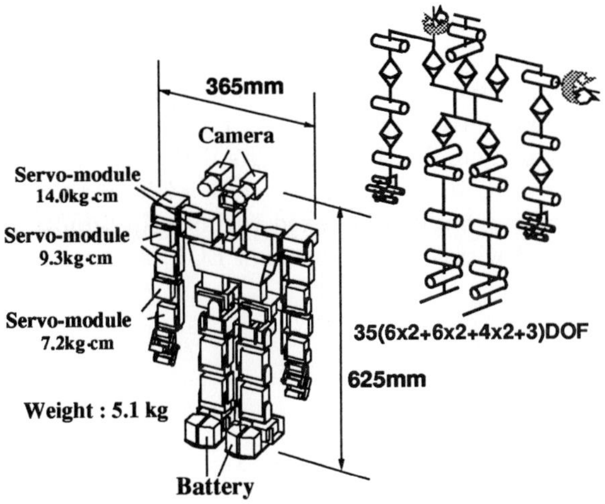

Figure 18. Structure of the humanoid with 35 joints

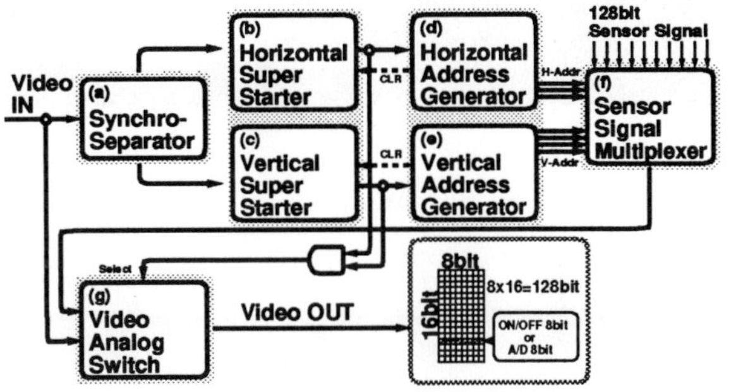

Figure 20. Circuit for video multiplexer

when the robot grasps an object, it finds it by vision and measures the location of the table. The robot changes the posture of the body according to the distance. The reachability is decided by checking the movable range of all the joints by calculating inverse kinematics of the arm. When the solution of joint angles is over the range, the robot tries to move the shoulders forward while sitting. When the object is out of the range when the body is sitting down, the robot generates a plan to stand up by placing a hand on the table. This arm placing motion helps the legs to stand up.

4.3. The Video Multiplexer

We have built a circuit to produce sensor images by electrically multiplexing sensor signals into the video signal. The configuration of an example circuit is shown in Figure 20. The circuit can multiplex 128 sensor signals on a video signal.

In order to study robot behaviors based on sensor image approach with this robot, we adopted new robot head which provides a binocular visual image with robot sensor suit. The binocular image is given by the stereo viewer. The signals from the robot sensor suit are superimposed on the binocular image through the video multiplexer.

Figure 21 shows the robot which uses the new robot head. The image shown in Figure 22 is a view seen when the robot looks at its own left hand. The upper middle region is the region for displaying the other sensor signals.

4.4. Sensor Suit Using Electrically Conductive Fabric

The experiment shown in the Figure 19 is done by model and vision based system. The robot assumes that the model describes the real world and the interaction between the body and the table/chair does not break the model. In model and vision based robots, it is hard for the robot to know whether the robot interacts with the objects. The robot sitting on a chair actually does

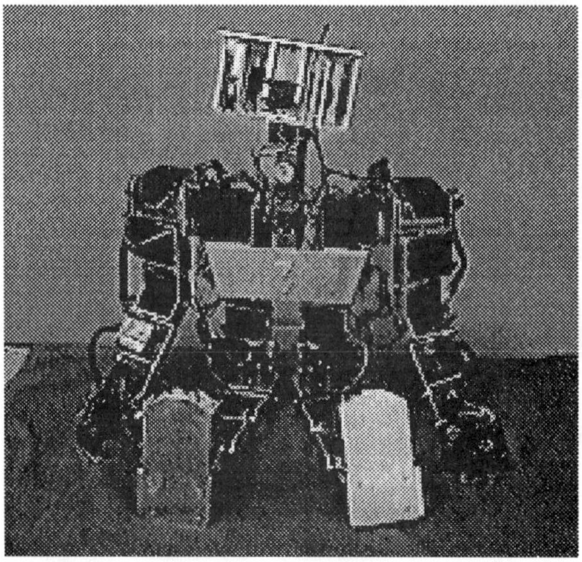

Figure 21. The humanoid robot with the stereo viewing head

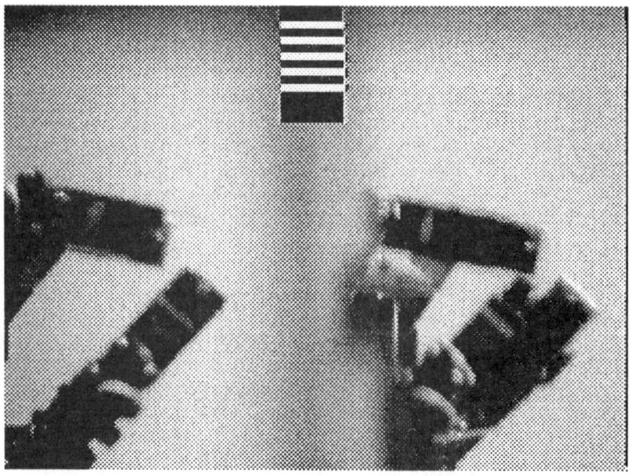

Figure 22. The image of the stereo viewer with multiplexed signals

not know whether the robot body is touching the chair or not. It should have tactile sensors distributed on the body to know this at least.

In order to give large number of distributed tactile sensors to a robot, we have developed a sensor suit from an electrically conductive fabric. The electrically conductive fabric (ECF) is electrically plated cloth.

The structure of the tactile sensor is shown in Figure 23. It has six layers. The top and bottom layers covers the sensing element and the wiring layer. The sensing element is a switch made of

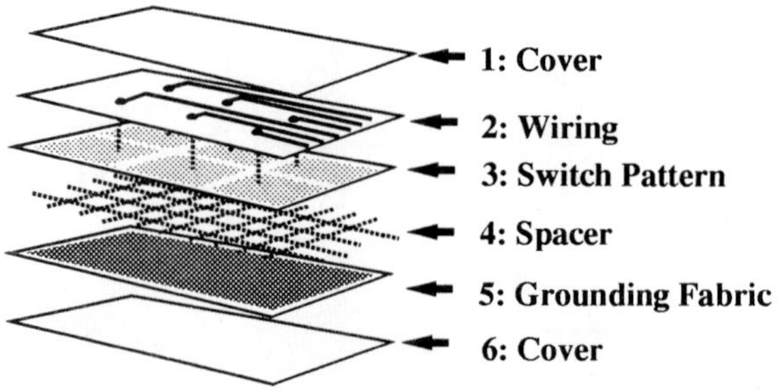

Figure 23. Structure of the tactile sensor made of electrically conductive fabric

three layers. The layer 5 is a cloth of ECF. The layer 4 has reticulate structure and works as a spacer between the layer 4 and 6. The layer 4 has distributed segments of ECF. Each ECF segment is connected to the video multiplexer by electrical conductive string (ECS). The layer 2 is non-electrical conductive cloth. As wiring is achieved by sewing on the non ECF cloth with an ECS, it is easy to build complexed wiring using multiple layered wiring with other non ECF cloths.

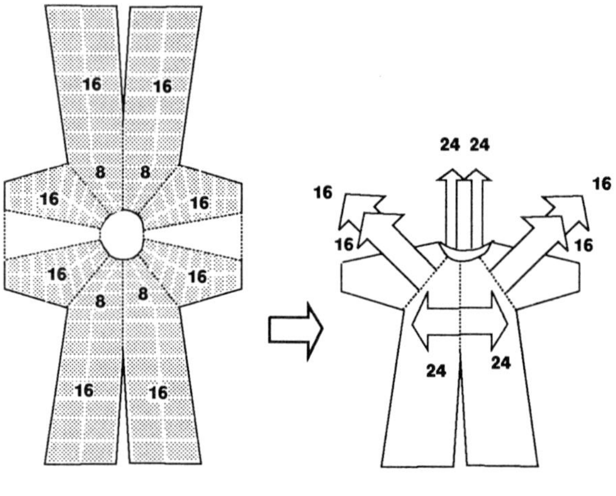

Figure 24. Structure of the sensor suit for a 35 axes humanoid

The structure of the entire suit is shown in Figure 24. The suit has 160 sensing regions. Each region is a tactile sensor made of ECF. The sensing segments have different sizes depending on the position of the suit. The regions at the bottom of the leg are about 10cm by 5cm. The upper arm has sensing segments of about 5cm by 5cm.

The key issure in building a suit with a large number of sensitive regions is how to wireing the cable from the sensing segments to the processing subsystem.

As wiring is done by electrical conductive string(ECS), it keeps the suit flexible enough to allow movement of the body. It is possible to have more segments than the current version.

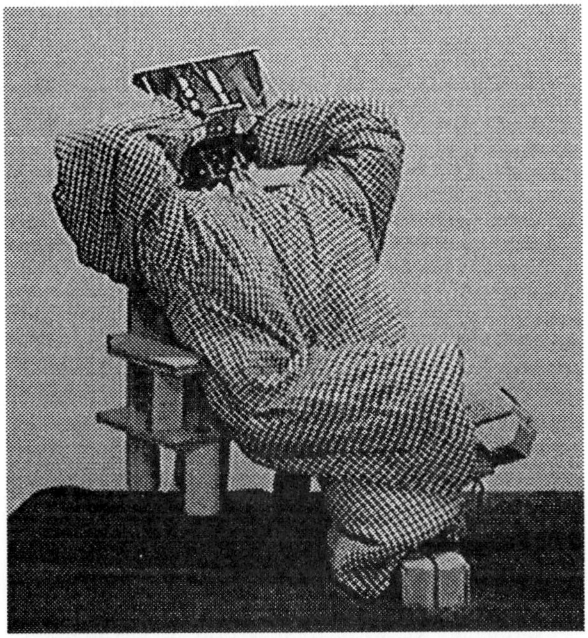

Figure 25. The humanoid in the sensor suit

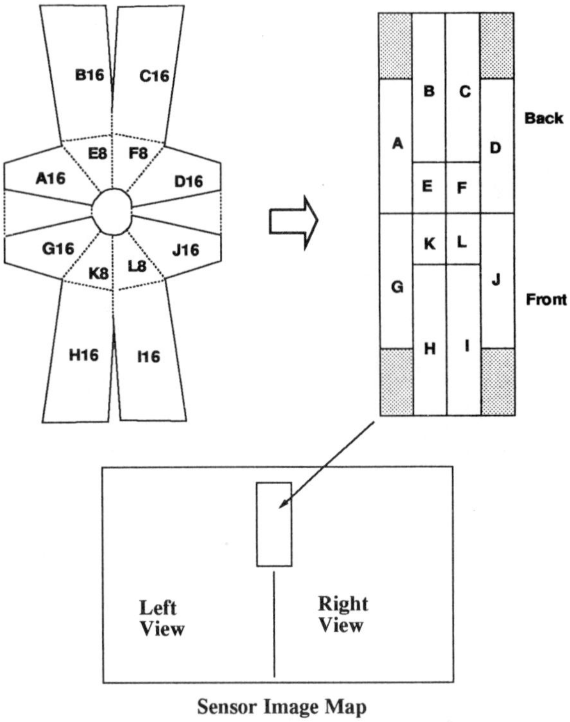

Figure 26. Sensor image map

Figure 25 shows a humanoid wearing the robot sensor suit.

4.5. Sensor Image Map and Experiment

In order to map all the sensing regions of the sensor suit, the super impose circuit is extended to accept 192 (8 x 24) sensing points. The sensing regions of the sensor suit are devided as shown in figure 26. The hashed region is reserved for other sensors. Figure 27 shows a real sensor image seen from the robot. Figure 28 shows an experiment of interactions with humans. In this experiment, if a human touch a left or a right arm, the robot look to the corresponding direction. The program for this behavior monitors the sensing regions (G, J) and detects pattern changes in these regions. If the change is larger than a threshold, it decides whether to move its head to look the human.

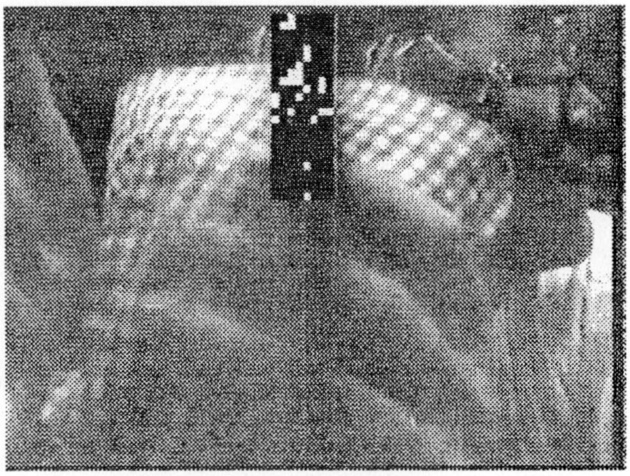

Figure 27. A sensor image of the humanoid

5. Concluding Remarks

We presented an approach to interface body to brain through building a primary sensor image integrating visual and somatic sensing systems. The basic idea of this approach is that a robot body is not only a mobile unit but also a sensing unit to observe the world. The primary sensor image is a projected image of the world through a body.

When we study development process of robot behaviors by increasing the complexity of the robot body, complexity in the sensing system should be handled in a uniform way. The primary sensor image approach provides a platform to reduce the complexity of sensing system by defining the

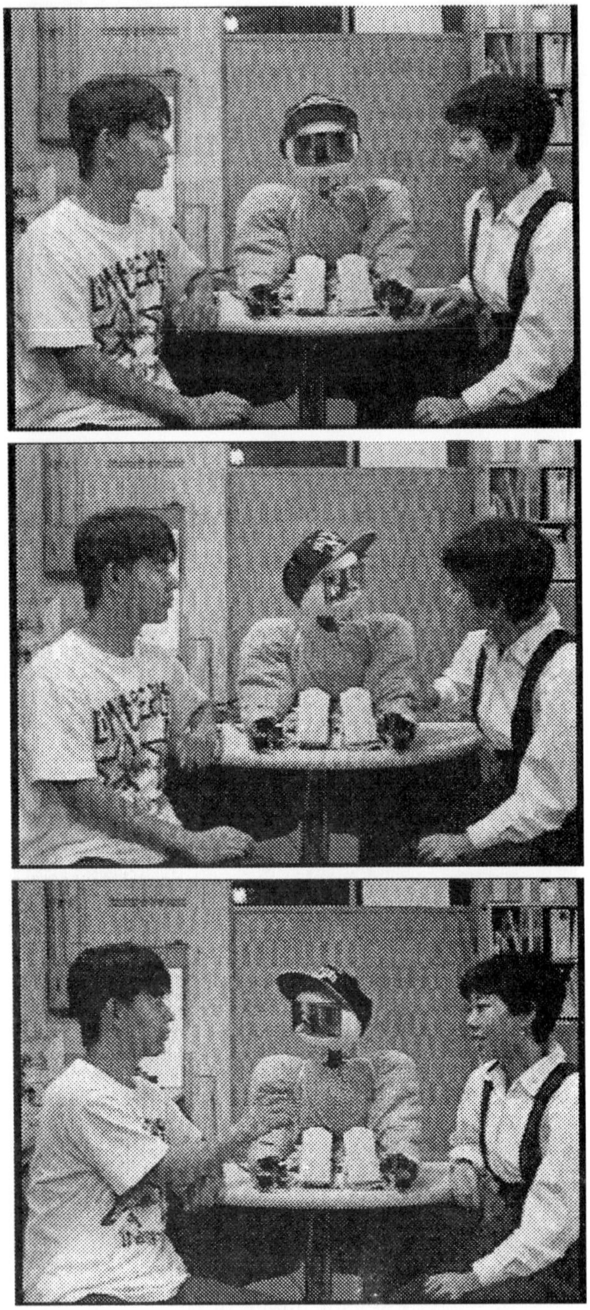

Figure 28. Experiment of interactions

interface from body to brain. The robot sensor suit shown here is a promising research tool to advance behavior based control through body interactions.

Acknowledgement

The author would like to note his great gratitude to Professor H. Inoue for fruitful discussions on everything. The students in our laboratory helped me to advance this approach. The author appreciates them all, especially S. Kagami for his great efforts building the remote-brained environment, F. Kanehiro for the develpment of humanformed robots, T. Igarashi for building the 35 DOF robot, T. Ninomiya for experiments with the 35 DOF robot, K. Nagasaka for building the optical sensor display and video multiplexer, and Y. Hoshino for building the robot sensor suit. My sincere thanks to them all.

References

[1] Anita M.Flynn and Rodney A. Brooks. Battling Reality. Technical Report AIM-1148, Massachusetts Institute of Technology, Artificial Intelligence Laboratory, 1989.

[2] Rodney A. Brooks. Building Brains for Bodies. Technical Report AIM-1439, Massachusetts Institute of Technology, Artificial Intelligence Laboratory, 1993.

[3] M. Inaba, S. Kagami, K. Sakaki, F. Kanehiro, and H. Inoue. Vision-Based Multisensor Integration in Remote-Brained Robots. In *1994 IEEE International Conference on Multisensor Fusion and Integration for Intelligent Systems*, pp. 747–754, 1994.

[4] Masayuki Inaba. Remote-Brained Robotics: Interfacing AI with Real World Behaviors. In *Robotics Research: The Sixth International Symposium*, pp. 335–344. International Foundation for Robotics Research, 1993.

[5] M. Inaba, F. Kanehiro, S. Kagami, and H. Inoue. Two-armed bipedal robot that can walk, roll-over and stand up. In *Proceedings of the IEEE/RSJ International Conference on Intelligent Robots and Systems*, Vol. 3, pp. 297–302, 1995.

[6] Masayuki Inaba, Takeyori Hara, and Hirochika Inoue. A stereo viewer based on a single camera with view-control mechanisms. In *Proceedings of IEEE/RSJ International Workshop on Intelligent Robots and Systems IROS '93*, pp. 1857–1864, 1993.

[7] H. Inoue, T. Tachikawa, and M. Inaba. Robot vision system with a correlation chip for realtime tracking, optical flow and depth map generation. In *Proceedings of the 1992 IEEE International Conference on Robotics and Automation*, pp. 1621–1626, 1992.

[8] Ken Sasaki. Tactile Sensor Using Optical Fiber (in Japanese). In *Preprint of Jpn. Soc. of Precision Eng.*, pp. 975–976, 1993.

PART 13

SYMBOLIC GEOMETRY AND EFFECTIVE ALGEBRA FOR ROBOTICS AND VISION

SESSION SUMMARY

Olivier Faugeras
INRIA, 2004 route des Lucioles
B.P. 93, 06902 Sophia-Antipolis, France
faugeras@sophia.inria.fr

The discussion after the two session papers focussed on the following three issues:

Similarity between vision and robotics problems

It was pointed out that there is an intriguing algebraic similarity between problems in inverse kinematics such as the Stewart platform problem and problems in vision such as the structure from motion problem. The question of whether this similarity can be understood better and perhaps shed new light on some of other problems encountered in robotics and vision was thought to be important.

Systems for symbolic algebraic manipulation

The use of such systems is spreading quickly in robotics and vision but people are worried by the often quite large execution times, the difficulty of interacting in a meaningful way with those systems while they are executing and by the usual impossibility of interpreting the intermediate results.

Numerical stability

It was also pointed out during the discussion that many of the algebraic methods were often quite unstable numerically and there was a consensus on the need for the development of more robust algorithms.

Symbolic Geometry and Effective Algebra for Solving Problems in Computer Vision: Where do we Stand, Where may we Go *

Olivier Faugeras
INRIA, 2004 route des Lucioles
B.P. 93, 06902 Sophia-Antipolis, France
faugeras@sophia.inria.fr

1 Introduction

In this article we describe a few ideas that come from geometry and algebra and have proven to be quite successful in handling a variety of problems in computer vision. We also speculate that some of these ideas will be useful in solving applications involving geometric reasoning. The guiding light in writing this paper has been to attempt to convey the basic ideas without going too deep into the mathematical details which the interested reader can find in the references. Moreover we have attempted each time it was possible to illustrate the ideas with real examples.

The paper is organized as follows. In section 2 we introduce the basic classical geometric tools which are needed in order to describe the geometry of a system of cameras. The main thrust is that such a geometry is best described in the framework of projective geometry and that when affine or Euclidean ideas are necessary they are best thought in that framework as well. In sections 3 and 4 we connect these geometric tools to the geometry of a system of cameras and introduce some newer ideas from algebra, the Grassmann-Cayley or double algebra, which shed a different light on the same problem and provide a unified framework to describe the geometry of the cameras and the geometry of the scene. The important point is that this algebra is the right tool to manipulate incidence relations and intersections which can be used as high level descriptors of the environment. In section 5 we present a number of applications of these ideas to robotics problems and conclude in section 6.

2 Notations and definitions

We use projective geometry in all this article. An elementary introduction can be found in [32, 10]. Capital letters indicate 3-D entities, lower case letters 2-D or image entities. Bold letters indicate vectors or matrixes. For example, M represents a 3-D projective point and $\mathbf{M}$ one of its 4×1 coordinate vectors, m_i is the image of M in the i-th camera, $\mathbf{m}_i$ one of its 3×1 coordinate vectors of coordinates x_i, y_i, z_i. The i-th perspective projection matrix is noted $\mathbf{P}_i$, a 3×4 matrix whose 3 row vectors will be denoted by the digits $3*(i-1)+j$, $i = 1, \cdots, N$, $j = 1, 2, 3$. The perspective projection equation of the first camera can thus be written as:

$$\mathbf{m}_1 = \mathbf{P}_1\mathbf{M} = \begin{bmatrix} \mathbf{1} \\ \mathbf{2} \\ \mathbf{3} \end{bmatrix} \mathbf{M},$$

or

$$x_1 : y_1 : z_1 = \mathbf{1} \cdot \mathbf{M} : \mathbf{2} \cdot \mathbf{M} : \mathbf{3} \cdot \mathbf{M} \quad (1)$$

where for example $\mathbf{1} \cdot \mathbf{M}$ represents the usual inner product of the row vector $\mathbf{1}$ with the column vector $\mathbf{M}$. This equation is equivalent to the three (linearly dependent) equations

$$\begin{aligned} x_1 \mathbf{2} \cdot \mathbf{M} - y_1 \mathbf{1} \cdot \mathbf{M} = 0 \\ y_1 \mathbf{3} \cdot \mathbf{M} - z_1 \mathbf{2} \cdot \mathbf{M} = 0 \\ z_1 \mathbf{1} \cdot \mathbf{M} - x_1 \mathbf{3} \cdot \mathbf{M} = 0 \end{aligned} \quad (2)$$

Finally, given four row or column vectors $\mathbf{U}_i, i = 1, \cdots, 4$ of size four, the notation

*This work was partially supported by the EEC under ESPRIT projects 6448, Viva and 8878, Realise.

$[\mathbf{U}_1, \mathbf{U}_2, \mathbf{U}_3, \mathbf{U}_4]$ will denote the 4×4 determinant of these four vectors, their bracket.

Let us denote by $\mathbf{E}_n$ the vector space of dimension n on the real numbers. We will be using both the cross-product of two vectors $\mathbf{x}$ and $\mathbf{y}$ of $\mathbf{E}_3 \equiv \mathbf{R}^3$, noted $\mathbf{x} \times \mathbf{y}$, a vector of $\mathbf{E}_3$ and the exterior product of two vectors of $\mathbf{E}_n$, noted $\mathbf{x} \wedge \mathbf{y}$, a vector of $\overset{2}{\wedge} \mathbf{E}_n$ which is a vector space of dimension $\binom{n}{2}$, i.e. also 3 for $n = 3$. The projective space associated to $\mathbf{E}_{n+1}$ will be denoted by $\mathcal{P}^n$. We assume that the reader is familiar with the basic results of the exterior calculus and the Grassmann-Cayley algebra. A good introduction, targeted at computer vision researchers, can be found in [4], more details can be found in the quite accessible book by Sturmfels [33], and in the more advanced article by Barnabei et al. [2].

3 One camera

Let us now turn to the sensor that we use to measure the world: the camera. We model classically a camera as a pinhole. This has proven to be an excellent approximation for most practical purposes. Even though it is important to keep in mind that the pinhole model is only an approximation, albeit usually a very good one, of a real physical camera, we hope to convince the reader in what follows of the usefulness of forgetting for some time the actual physical device and of thinking of the camera as a projective geometric engine.

3.1 The perspective projection model

This projective engine maps the two-dimensional projective space $\mathcal{P}^3$ onto the two-dimensional projective plane $\mathcal{P}^2$ by perspective projection from a center of projection C (the optical center of the camera) onto a plane $\mathcal{R}$ not containing C (the retinal plane of the camera).

This projection operation is projective linear in the sense that if we choose a projective basis of $\mathcal{P}^3$ and a projective basis of $\mathcal{P}^2$, the correspondence between a point M of $\mathcal{P}^3$ represented by $\mathbf{M}$ and its image m of $\mathcal{P}^2$ represented by $\mathbf{m}$ can be written in vector form

$$\mathbf{m} = \mathbf{P}\mathbf{M} \tag{3}$$

where $\mathbf{P}$ is a 3×4 matrix of rank 3 defined up to the multiplication with a non zero scalar. This matrix depends therefore upon 11 parameters and is called the *perspective projection* matrix of the camera.

Note that if we change projective basis in the world by $\mathbf{M}' = \mathbf{K}\mathbf{M}$ ($\mathbf{K}$ a 4×4 matrix of rank 4) and in the retinal plane by $\mathbf{m}' = \mathbf{H}\mathbf{m}$ ($\mathbf{H}$ a 3×3 matrix of rank 3), then the perspective projection matrix becomes $\mathbf{P}' = \mathbf{H}\mathbf{P}\mathbf{K}^{-1}$.

Given the perspective projection matrix $\mathbf{P}$ and without any further assumption about the world, we can recover the coordinates of the optical center in the projective basis of the world. Indeed, the optical center is the point for which the perspective projection is not defined, it has no image and must therefore satisfy

$$\mathbf{P}\mathbf{C} = \mathbf{0}$$

which shows that C is represented by any non zero vector of the nullspace of the matrix $\mathbf{P}$ which is by definition of dimension 1 since $rank(\mathbf{P}) = 3$.

3.2 One camera and the absolute conic: Measuring the angle between two optical rays

If the image of the absolute conic is known in a camera, it then becomes a metric measurement device that can compute angles between optical rays [10]. This can be readily seen by using Laguerre's formula as follows. Let m and n be two image points and consider the two optical rays $\langle C, m\rangle$ and $\langle C, n\rangle$. Let us call α the angle (between 0 and π) that they form, let M and N be their intersections with the plane at infinity, and let A and B be the two intersections of the line $\langle M, N\rangle$ with the absolute conic Ω. The angle α between $\langle C, m\rangle$ and $\langle C, n\rangle$ is given by Laguerre 's formula $\frac{1}{2i}\log(\{M, N; A, B\})$. The reason for this is that the line at infinity of the plane defined by the three points C, m, n is the intersection of that plane with the plane at infinity, i.e., the line $\langle M, N\rangle$. The absolute points of that plane are the intersections A and B of that line with the absolute conic Ω.

The cross-ratio $\{M, N; A, B\}$ is preserved under the projection to the retinal plane, and thus the angle between $\langle C, m\rangle$ and $\langle C, n\rangle$ is given by $\frac{1}{2i}\log(\{m, n; a, b\})$, where the points a and b are the "images" of the points A and B.

Since a and b are the two intersections of the line $\langle m, n\rangle$ with ω, this shows that the angle can be computed only from the image ω of the absolute conic. The situation is depicted in figure 1.

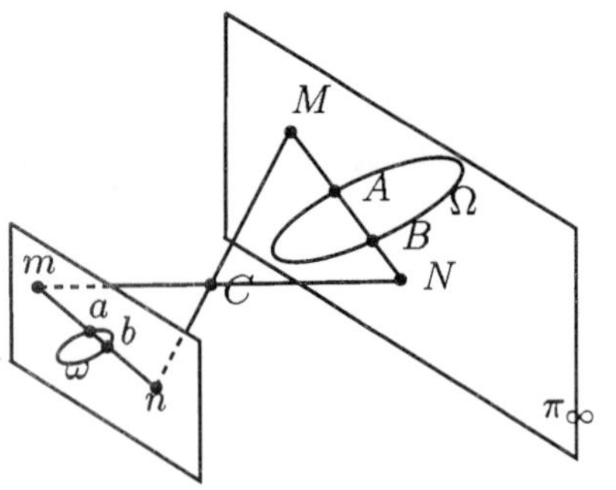

Figure 1: How to compute the angle between the optical rays $\langle C, m\rangle$ and $\langle C, n\rangle$ using the image of the absolute conic.

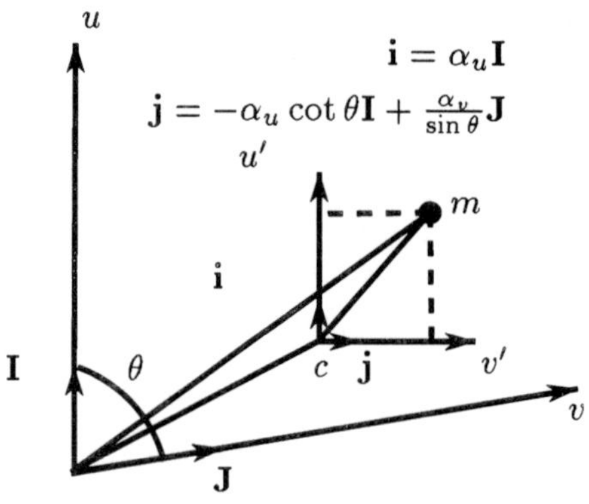

Figure 2: From pixel-coordinates (u, v) to normalized coordinates (u', v').

In detail, the line $\langle m, n\rangle$ is represented by $\mathbf{m} + \theta\mathbf{n}$. The variable θ is a projective parameter of that line. Point m has projective parameter 0, and point n has projective parameter equal to ∞. The reader should not worry about this, since the magic of the cross-ratio will take care of it!

In order to compute the projective parameters of a and b we write, S being the equation of ω, the following quadratic equation:

$$S(\mathbf{m}) + 2\theta S(\mathbf{m}, \mathbf{n}) + S(\mathbf{n})\theta^2 = 0$$

Let θ_0 and $\overline{\theta_0}$ be the two roots, which are complex conjuguate. According to the definition of the cross-ratio, we have

$$\{m, n; a, b\} = \frac{0 - \theta_0}{0 - \overline{\theta_0}} : \frac{\infty - \theta_0}{\infty - \overline{\theta_0}}$$

The ratio containing ∞ is equal to 1 (that is the magic!), and therefore

$$\{m, n; a, b\} = \frac{\theta_0}{\overline{\theta_0}} = e^{2iArg(\theta_0)}$$

where $Arg(\theta_0)$ is the argument of the complex number θ_0. In particular, we have $\alpha = Arg(\theta_0)\ (\pi)$. A straightforward computation shows that the two roots are equal to

$$\frac{-S(\mathbf{m}, \mathbf{n}) \pm i\sqrt{S(\mathbf{m})S(\mathbf{n}) - S(\mathbf{m}, \mathbf{n})^2}}{S(\mathbf{n})} \tag{4}$$

Simple considerations show that

$$\cos\alpha = -\frac{S(\mathbf{m}, \mathbf{n})}{\sqrt{S(\mathbf{m})S(\mathbf{n})}} \tag{5}$$

an equation which uniquely defines α between 0 and π. The sine is therefore positive and given by $\sqrt{1 - \cos^2\alpha}$.

3.3 The absolute conic and the intrinsic parameters

We have seen previously that the image ω of the absolute conic did not change when we moved the camera in space. This, together with the fact that ω contains only complex points, has some strong implications on the coefficients of the equation defining ω. We now examine these consequences. Let $\mathbf{A}$ be the symmetric matrix defining the equation of ω in the retinal plane:

$$S(\mathbf{m}) = \mathbf{m}^T\mathbf{A}\mathbf{m}$$

Since ω does not contain any real point, this means that $S(\mathbf{m})$ is either strictly positive or strictly negative for all points m with real coordinates. Let us assume that it is strictly positive. The quadratic form defined by matrix $\mathbf{A}$ is accordingly positive definite and we can use a theorem which says that a necessary and sufficient condition for a quadratic form to be positive definite is that its matrix can be written as $\mathbf{W}\mathbf{W}^T$, where $\mathbf{W}$ is a lower-triangular matrix with positive diagonal elements [15]. This decomposition is called the Cholesky decomposition of matrix $\mathbf{A}$ and is unique.

If we define $\mathbf{p} = \mathbf{W}^T\mathbf{m}$, the equation of ω can be written $S(\mathbf{p}) = \mathbf{p}^T\mathbf{p}$. The matrix $\mathbf{W}$ can be interpreted as defining a change of projective coordinates in the retinal plane. For reasons which will become clear later, we are more interested in the matrix $\mathbf{W}^{-T}$ which is upper triangular. Let us write this matrix

$$\mathbf{W}^{-T} = \begin{bmatrix} a & b & c \\ 0 & d & e \\ 0 & 0 & f \end{bmatrix} \tag{6}$$

It is easy to see that $\mathbf{W}^{-T}$, like $\mathbf{W}$, has positive diagonal elements and since $\mathbf{W}$ is, like $\mathbf{A}$, defined up to a scale factor, we can assume that $f = 1$ (f cannot be equal to 0 because otherwise the rank of $\mathbf{A}$ would be less than 3). We also have $a > 0$ and $d > 0$. Changing notations, we write

$$\begin{array}{ll} a = \alpha_u & b = -\alpha_u \cot\theta \\ d = \frac{\alpha_v}{\sin\theta} & c = u_0 \\ e = v_0 & \end{array} \tag{7}$$

These equations uniquely define the five parameters α_u, α_v, u_0, v_0, and θ. This is clear for α_u, u_0, v_0. For θ and α_v, we see that the equation $b = -\alpha_u \cot\theta$ defines θ between 0 and π. Thus the sine is positive and since $d > 0$ this uniquely defines α_v as a positive number.

These parameters have been introduced by several authors from physical and heuristic considerations in the past [10, 23] and are called the intrinsic parameters of the camera. Here they appear without such considerations, as a consequence of the fact that the image of the absolute conic is an imaginary curve.

4 Geometry of systems of two cameras

4.1 Two cameras and the fundamental matrix: the projective stratum

If we now consider a binocular stereo rig, we can bring in some more geometric information which has profound implications for computer vision problems. Let us call C' the optical center of the second camera and $\mathcal{R}'$ its retinal plane. The line $\langle C, C'\rangle$ intersects $\mathcal{R}$ (resp. $\mathcal{R}'$) in a point that we denote by e (resp. e'). These two points are called the epipoles of the stereo rig. By construction, any plane containing the line $\langle C, C'\rangle$, called an epipolar plane, intersects $\mathcal{R}$ (resp. $\mathcal{R}'$) along a line going through the epipole e (resp. along a line going through the epipole e'), see figure 3 . Two such lines are called corresponding epipolar lines and have an immense importance for stereo algorithms. We can rephrase the situation in projective terms by saying that the pencil of epipolar planes induces in each retinal plane a pencil of epipolar lines. These three pencils are projective lines, i.e. one-dimensional projective spaces $\mathcal{P}^1$.

The fundamental property of this geometric construction is that the "natural" correspondence between the two pencils of epipolar lines is projective linear, it is a homography between the two pencils considered as projective lines. The "natural" correspondence consists in associating with each epipolar line of the first pencil the corresponding epipolar line of the second, i.e. the intersection of the epipolar plane defined by the first one and the two optical centers with the second retinal plane. The reason why it is an homography is because it is one to one and preserves cross-ratios: The cross-ratio of four lines of the first pencil is equal to the cross-ratio of the four corresponding epipolar planes which is equal to the cross-ratio of the four corresponding epipolar lines in the second retinal plane. This homography is at the heart of many of the ideas which will be presented in the next sections (see figure 4). We call it the *epipolar* homography.

Having presented the geometric viewpoint, let us now present its algebraic face. In order to do this, we will adopt a slightly different view, namely we will characterize the relationship between a point m in the first retinal plane and its epipolar line l'_m in the second. This correspondence is also clearly projective linear (it is a projective linear mapping between the first retinal plane considered as a $\mathcal{P}^2$ and the dual of the second retinal plane, also considered as a $\mathcal{P}^2$) and therefore there exists a 3×3 matrix $\mathbf{F}$, defined up to a scale factor, such that

$$\mathbf{l}'_m = \mathbf{Fm}$$

The matrix $\mathbf{F}$ is *not* of rank 3 since if m coincides with the epipole e its epipolar line is undefined and therefore

$$\mathbf{Fe} = \mathbf{0}$$

Let us now further consider a point m' on the epipolar line l'_m of m. This point satisfies the relation

$$\mathbf{m}'^T\mathbf{Fm} = 0 \tag{8}$$

which shows that the epipolar line $l_{m'}$ in the first retinal plane of m' is represented by $\mathbf{F}^T\mathbf{m}'$:

$$\mathbf{l}_{m'} = \mathbf{F}^T\mathbf{m}'$$

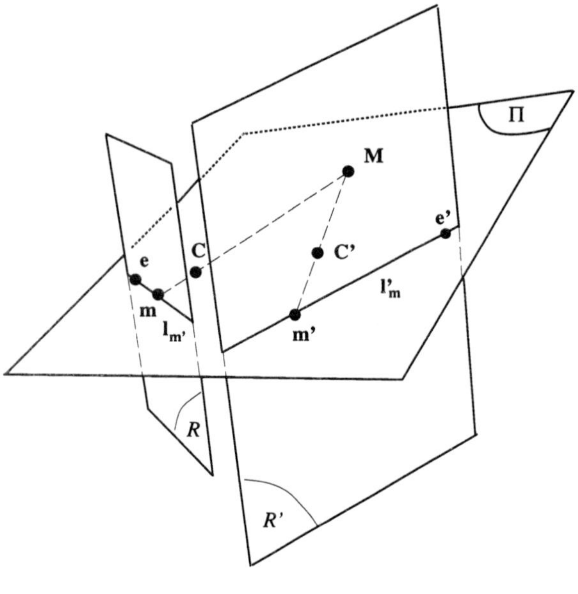

Figure 3: The epipolar geometry.

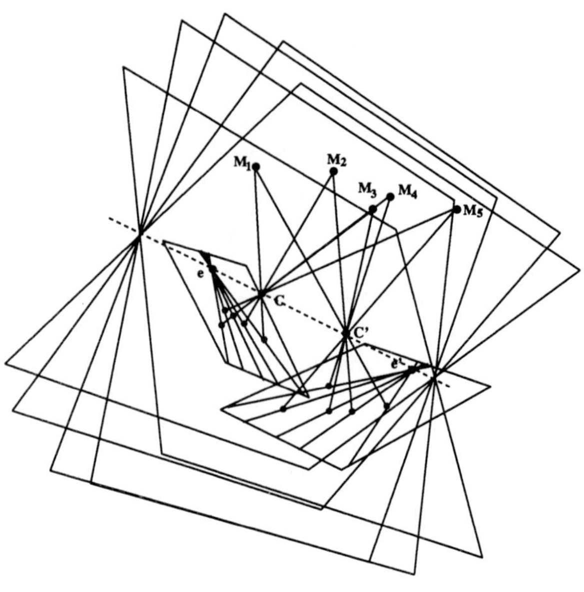

Figure 4: The epipolar pencils.

In particular we have

$$\mathbf{F}^T \mathbf{e}' = \mathbf{0}$$

The matrix $\mathbf{F}$ expresses algebraically the epipolar correspondence between the two retinal planes. It is called the fundamental matrix [23, 12, 24]. Its rank is in general equal to 2 and it therefore depends upon seven free parameters. A set of such parameters which have a neat geometric interpretation are the four ratios of projective coordinates of the two epipoles and the three ratios of the coefficients of the homography between the two pencils of epipolar lines.

Just as in the one-camera case where we related the optical center to the perspective projection $\mathbf{P}$, in the two-cameras case, we can also relate the fundamental matrix $\mathbf{F}$ to the two perspective projection matrices $\mathbf{P}$ and $\mathbf{P}'$. The interested reader is referred to, for example [10].

4.2 A complementary viewpoint, the Grassmann-Cayley algebra [11]

4.2.1 *Mathematical background*

We denote by $\mathbf{E} = \mathbf{E}^4$ the vector spaces of dimension 4 over the reals $\mathbf{R}$. The rows of the projection matrices belong to this space. They represent planes in $\mathcal{P}^3$. If $\mathbf{U} = (U_1, U_2, U_3, U_4)$ is such a vector, the plane associated to it, has the equation $U_1\, M_1 + U_2\, M_2 + U_3\, M_3 + U_4\, M_4 = 0$. A non-zero multiple of $\mathbf{U}$ represents the same plane so that the set of planes $\hat{\mathcal{P}}^3$ has naturally a structure of projective space of dimension 3. It is called the dual space of $\mathcal{P}^3$.

In order to be able to manipulate subspaces of $\mathcal{P}^3$, we need to introduce the **exterior algebra** $\wedge\mathbf{E}$ **of** $\mathbf{E}$. This algebra has an anti-commutative product (noted $\wedge$) which satisfies $e_i \wedge (\lambda\, e_j + \mu\, e_k) = \lambda\, e_i \wedge e_j + \mu\, e_i \wedge e_k$ for all $\lambda, \mu \in \mathbf{R}$. Let $e_1, \ldots, e_4$ be the canonical basis of $\mathbf{E}$. Then $\wedge\mathbf{E}$ is a vector space of basis $(e_{i_1} \wedge \cdots \wedge e_{i_k})$ with $1 \le i_1 < \cdots < i_k \le 4$.

For any vectors $u_1 = (u_{1,1}, \ldots, u_{4,1})$, $\ldots$, $u_m = (u_{1,m}, \ldots, u_{4,m})$ of $\mathbf{E}$, the coordinates of $u_1 \wedge \cdots \wedge u_m$ in the basis $(e_{i_1} \wedge \cdots \wedge e_{i_m})_{1 \le i_1 < \cdots < i_m \le 4}$ of $\wedge^m\mathbf{E}$ are the corresponding determinants $|u_{i_k,j}|_{1 \le j \le m, 1 \le i_k \le 4}$. So $u_1 \wedge \cdots \wedge u_m = 0$ if and only if the vectors $\{u_1, \ldots, u_m\}$ are linearly dependent.

We associate naturally to an element $u_1 \wedge \cdots \wedge u_m$ the subspace of $\mathcal{P}^3$ corresponding to the intersection of the hyperplane $u_1, \ldots u_m$. For instance $u_1 \wedge u_2$ is the line of intersection of the two planes u_1, u_2 and $u_1 \wedge u_2 \wedge u_3$ represents a point of $\mathcal{P}^3$. Conversely any linear subspace L of $\mathcal{P}^3$ can be represented by a unique element $u_1 \wedge \cdots \wedge u_m$ (up to a non-zero scalar). We just take any basis $u_1, \ldots u_m$ of L. The elements which represent linear subspace of $\mathcal{P}^3$ form a subvariety of $\mathcal{P}(\wedge\mathbf{E})$, the projective space defined from the vector space $\mathbf{E}$, called the *Grassmannian*.

We illustrate these notions in our context:

- $x_1\,\mathbf{2} - y_1\,\mathbf{1}, y_1\,\mathbf{3} - z_1\,\mathbf{2}$ *are planes which contain the optical center* O_1 *(because* $\mathbf{P}_1\mathbf{C}_1 = 0$*) and the point* $\mathbf{M}$ *by construction.*

- $(x_1\,\mathbf{2} - y_1\,\mathbf{1}) \wedge (y_1\,\mathbf{3} - z_1\,\mathbf{2}) = y_1(z_1\mathbf{1} \wedge \mathbf{2} - y_1\mathbf{1} \wedge \mathbf{3} + x_1\mathbf{2} \wedge \mathbf{3})$, *is the* **optical ray** $\langle C_1, M\rangle$.

- $\mathbf{1} \wedge \mathbf{2} \wedge \mathbf{3}$ *is the intersection of the planes defined by the rows of* $\mathbf{P}_1$. *It is the optical center* C_1 *of the first camera.*

We check immediately that for any vectors $\mathbf{U}_1, \ldots, \mathbf{U}_4$ of $\mathbf{E}$, we have the properties $(\mathbf{U}_1 \cdot \mathbf{U}_2 \wedge \mathbf{U}_3 \wedge \mathbf{U}_4) = [\mathbf{U}_1, \ldots, \mathbf{U}_4]$, where $[\mathbf{U}_1, \ldots, \mathbf{U}_4]$ denotes the 4×4 determinant of these vectors. In particular two lines $(\mathbf{a} \wedge \mathbf{b})$, $(\mathbf{c} \wedge \mathbf{d})$ intersect if and only if $[\mathbf{a}, \mathbf{b}, \mathbf{c}, \mathbf{d}] = 0$ or equivalently $(\mathbf{a}\wedge\mathbf{b})\wedge(\mathbf{c}\wedge\mathbf{d}) = 0$. As $\mathbf{a}\wedge\mathbf{b}\wedge\mathbf{c}\wedge\mathbf{d} = [\mathbf{a}, \mathbf{b}, \mathbf{c}, \mathbf{d}]\, e_1 \wedge e_2 \wedge e_3 \wedge e_4$, with a slight abuse of notations we may identify $\mathbf{a} \wedge \mathbf{b} \wedge \mathbf{c} \wedge \mathbf{d}$ with $[\mathbf{a}, \mathbf{b}, \mathbf{c}, \mathbf{d}]$.

A natural approach, when dealing with intrinsic quantities, is to consider the determinants of points as variables $[j_1, \ldots, j_4]$. Therefore we have to manipulate polynomials in these variables. Especially, we want to be able to check when a polynomial is zero or not. This is not straightforward, because there exist relations between these variables. A fundamental result in invariant theory says that these relations are generated by the **Plücker relations**. This algebra of determinants falls in the the category of **algebra with straightening laws**, for which there is an algorithmic way to normalize any element. Moreover, the monomials which are normalized can be described in terms of a partial order on these variables (see for instance [8], [9], or [33]). In this paper, the computations and tests of equality to zero are based on this normalization procedure that we have implemented in Maple.

More details about invariant theory, exterior calculus and the Grassmann-Cayley algebra can be found in the classical book [34], or in the quite accessible book by Sturmfels [33], and in the more advanced article by Barnabei et al. [2]. A good introduction, targeted at computer vision researchers, can also be found in [4].

4.2.2 *The fundamental matrix in the Grassmann-Cayley formalism*

Let us start with the case of two cameras. Even though this case is now well understood, it is worthwhile going through its analysis because it turns out that the way we approach it is rather different from the usual way and can be extended in a straightforward fashion to the trilinear and quadrilinear cases. Let us choose the first two equations among the three equations (2) for each of the two cameras. These four equations can be written in matrix form as:

$$\begin{bmatrix} x_1\mathbf{2} - y_1\mathbf{1} \\ y_1\mathbf{3} - z_1\mathbf{2} \\ x_2\mathbf{5} - y_2\mathbf{4} \\ y_2\mathbf{6} - z_2\mathbf{5} \end{bmatrix} \mathbf{M} = 0 \qquad (9)$$

The 4×4 matrix which appears in this equation is a linear function of the two points m_1 and m_2. The question is, what is the necessary and sufficient condition that the two points m_1 and m_2 must satisfy so that they can be considered as the images of the same point M. The answer is well known and is that they must satisfy the epipolar constraint [1, 18, 10, 16].

Looking at equations (9) we see that the condition we seek is that the 4×4 matrix must be of rank less than or equal to 3 which is equivalent to writing that its determinant must be equal to 0. Expanding this determinant, we find

$$y_1 y_2 \left(\mathbf{m}_2^T \mathbf{F}_{1,2} \mathbf{m}_1\right)$$

where $\mathbf{F}_{1,2}$ is a 3×3 matrix which we recognize as the fundamental matrix of the first two images [23, 12, 35].

Its elements are 4×4 determinants built from the row vectors of the perspective projection matrices $\mathbf{P}_1$ and $\mathbf{P}_2$. In detail

$$\mathbf{F}_{1,2} = \begin{bmatrix} [2,3,5,6] & -[1,3,5,6] & [1,2,5,6] \\ -[2,3,4,6] & [1,3,4,6] & -[1,2,4,6] \\ [2,3,4,5] & -[1,3,4,5] & [1,2,4,5] \end{bmatrix}. \qquad (10)$$

the factor $y_1 y_2$ comes from the choice of the specific equations among (2) to build the determinant. Another choice of equation will produce another factor but the matrix $\mathbf{F}_{1,2}$ will remain unchanged. Note that this constraint can also be rewritten in a symmetric way, using some standard tools of the Grassmann-Cayley algebra.

$$(x_1\mathbf{2} \wedge \mathbf{3} + y_1\mathbf{3} \wedge \mathbf{1} + z_1\mathbf{1} \wedge \mathbf{2}) \wedge (x_2\mathbf{5} \wedge \mathbf{6} + y_2\mathbf{6} \wedge \mathbf{4} + z_2\mathbf{4} \wedge \mathbf{5}) = 0 \qquad (11)$$

which also explains why it is independent of the choice of the equations. This constraint can be interpreted geometrically as follows. The vectors $\mathbf{1}, \mathbf{2}, \mathbf{3}$ represent planes through the optical center of the first camera. So $\mathbf{2} \wedge \mathbf{3}, \mathbf{1} \wedge \mathbf{3}, \mathbf{1} \wedge \mathbf{2}$ are lines through this center and $(x_1\mathbf{2} \wedge \mathbf{3} + y_1\mathbf{3} \wedge \mathbf{1} + z_1\mathbf{1} \wedge \mathbf{2})$ is just the optical ray from the center to the point M or m_1. The equation (11) corresponds to the fact that the rays from the center of the two cameras have a common point M in the space $\mathcal{P}^3$.

The fact that the fundamental matrix $\mathbf{F}_{1,2}$ is of rank less than or equal to two can be found in several ways. The first way is to expand its determinant as a function of the brackets $[i, j, k, l]$ and use a straightening algorithm to rewrite it in standard form (see [33, 27]). The result is found to be the null polynomial. The second way is to show that there are scalars $\lambda_i, i = 1, 2, 3$ such that $\lambda_1 \mathbf{f}_1 + \lambda_2 \mathbf{f}_2 + \lambda_3 \mathbf{f}_3 = \mathbf{0}$ where the $\mathbf{f}_i$'s are the column vectors of $\mathbf{F}_{1,2}$. This is done explicitly in the next paragraph.

The epipoles can also be found quite easily. One way is to find values of x_1, y_1, z_1 such that the line $(x_1 \mathbf{2} \wedge \mathbf{3} + y_1 \mathbf{3} \wedge \mathbf{1} + z_1 \mathbf{1} \wedge \mathbf{2})$ contains the center of the second camera which is the intersection $\mathbf{4} \wedge \mathbf{5} \wedge \mathbf{6}$ of the three planes. Using the standard operators of intersection and sum in the Cayley Algebra (see [2, 26]), this yields

$$\begin{gathered}(x_1[2,4,5,6] - y_1[1,4,5,6])\mathbf{3} \\ -(x_1[3,4,5,6] - z_1[1,4,5,6])\mathbf{2} \\ +(y_1[3,4,5,6] - z_1[2,4,5,6])\mathbf{1} = 0\end{gathered}$$

As the vectors $\mathbf{1}, \mathbf{2}, \mathbf{3}$ are independent, we obtain a system of 3 (dependent) equations, which determines the epipole $e_{1,2}$.

Another way to compute the coordinates of the epipole $e_{1,2}$ in the first image is to use the cross-product of any two row vectors of $\mathbf{F}_{1,2}$. Computing the cross-product of the first two rows, we find a vector whose coordinates are quadratic polynomials of the brackets. Rewriting these polynomials in standard form, we find that $\mathbf{e}_{1,2}$ is proportional to the vector

$$[[1,4,5,6], [2,4,5,6], [3,4,5,6]]^T$$

Similarly, the epipole $e_{2,1}$ in the second image is found to be represented by the vector

$$[[1,2,3,4], [1,2,3,5], [1,2,3,6]]^T$$

These vectors can be also be interpreted as follows: the coordinates of the image m_1 of a point M in the first camera are given by $[\mathbf{1} \cdot \mathbf{M}, \mathbf{2} \cdot \mathbf{M}, \mathbf{3} \cdot \mathbf{M}]^T$. As the epipole $e_{1,2}$ is the image of the center of the second camera defined by $\mathbf{4} \wedge \mathbf{5} \wedge \mathbf{6}$, the coordinates of $e_{1,2}$ are

$$\begin{gathered}[\mathbf{1} \cdot \mathbf{4} \wedge \mathbf{5} \wedge \mathbf{6}, \mathbf{2} \cdot \mathbf{4} \wedge \mathbf{5} \wedge \mathbf{6}, \mathbf{3} \cdot \mathbf{4} \wedge \mathbf{5} \wedge \mathbf{6}]^T \\ = [[1,4,5,6], [2,4,5,6], [3,4,5,6]]^T .\end{gathered}$$

One can check that $\mathbf{F}_{1,2} \mathbf{e}_{1,2} = 0$:

```
> evalm(F12 &* E12);
```

$$\begin{bmatrix} [2,3,5,6][1,4,5,6] - [1,3,5,6][2,4,5,6] + [1,2,5,6][3,4,5,6] \\ -[2,3,4,6][1,4,5,6] + [1,3,4,6][2,4,5,6] - [1,2,4,6][3,4,5,6] \\ [2,3,4,5][1,4,5,6] - [1,3,4,5][2,4,5,6] + [1,2,4,5][3,4,5,6] \end{bmatrix}$$

```
> map(stgt,");
```

$$\begin{bmatrix} 0 & 0 & 0 \end{bmatrix}$$

These computations have been done in MAPLE. The function **stgt** that we have implemented (see [27]) gives a normal form to any polynomial in the determinants of 4 points. This normal form is zero if and only if the polynomial is a consequence of the Plücker relations. This evaluation shows that

$$[1,4,5,6]\mathbf{f}_1 + [2,4,5,6]\mathbf{f}_2 + [3,4,5,6]\mathbf{f}_3 = \mathbf{0}$$

4.3 Two cameras looking at planes: the affine stratum

It is known that in order to go from a projective representation of the world to an affine representation, we have to identify the plane at infinity. Before explaining how this can be achieved, we will start with a brief description of the relationship between planes in the world and a pair of cameras.

Indeed, planes in the world have very interesting properties with respect to our stereo rig. In effect, a plane π induces in general a projective linear correspondence, a collineation, between the two retinal planes.

This can be readily seen by noting that the perspective projection from π to $\mathcal{R}$ (resp. from π to $\mathcal{R}'$) is one to one if π does not contain the optical center C (resp. the optical center C') and preserves cross-ratios and is therefore a collineation. Composing the inverse of the first collineation with the second defines a collineation from $\mathcal{R}$ to $\mathcal{R}'$ called the collineation induced by π that we note H_π and represent by the 3×3 matrix $\mathbf{H}_\pi$. Even though a collineation of $\mathcal{P}^2$ depends upon 8 parameters, there is no contradiction with the fact that a plane depends upon 3 parameters. Indeed, the collineation is related to the fundamental matrix [23] in the following manner. Let m be a point of $\mathcal{R}$. The point m' of $\mathcal{R}'$ represented by $\mathbf{H}_\pi \mathbf{m}$ is the image in the second camera of the intersection of the optical ray $\langle C, m \rangle$ with π. Therefore it belongs to the epipolar line of m and we have

$$(\mathbf{H}_\pi \mathbf{m})^T \mathbf{F} \mathbf{m} = 0$$

for all points m. This implies that the matrix $\mathbf{H}_\pi^T\mathbf{F}$ is antisymmetric:

$$\mathbf{H}_\pi^T\mathbf{F} + \mathbf{F}^T\mathbf{H}_\pi = \mathbf{0} \tag{12}$$

This imposes six homogeneous constraints on the collineation.

The interaction between the geometry of the stereo rig and H_π can also be seen as follows. Let us consider the point of intersection P of the line $\langle C, C'\rangle$ with π. The images of P in the two retinal planes are the two epipoles e and e' which therefore correspond to each other through H_Π. A consequence of this is that if the fundamental matrix is known (and thus the epipoles), three pairs of corresponding points are sufficient to determine the collineation since a fourth pair (e, e') is already available.

4.4 Affine calibration

To compute the homography of the plane at infinity $\mathbf{H}_\infty$, we must know correspondences of points at infinity $(\mathbf{m}'_i, \mathbf{m}_i)$ and compute $\mathbf{H}_\infty$.

The only way to obtain correspondences of points at infinity is to assume some additional knowledge.

First, we can assume that we have some additional knowledge of the observed scene that allows to identify, in the images, some projections of points at infinity, like, for instance, the vanishing points of parallel lines of the scene, or the images of some points on the horizon, which provide sufficiently good approximations to points at infinity.

Another way to proceed is to assume that we have an additional pair of views. More precisely, if this second pair differs from the first only by a translation of the rig, any pair (M, N) of stationary object points (see figure 5), seen in the first views as $(\mathbf{m}_1, \mathbf{m}'_1)$ and $(\mathbf{n}_1, \mathbf{n}'_1)$, and in the second as $(\mathbf{m}_2, \mathbf{m}'_2)$ and $(\mathbf{n}_2, \mathbf{n}'_2)$, gives us the images $(\mathbf{i}_1, \mathbf{i}'_1)$ and $(\mathbf{i}_2, \mathbf{i}'_2)$ in the four images of the intersection I of the line $\langle M, N\rangle$ with the plane at infinity.

Because the motion of the rig is a pure translation, $i_1 = i_2 = i$ and $i'_1 = i'_2 = i'$. On the other hand, as I lies on $\langle M, N\rangle$, i_1 lies on $\langle m_1, n_1\rangle$, i_2, on $\langle m_2, n_2\rangle$, i'_1, on $\langle m'_1, n'_1\rangle$ and i'_2, on $\langle m'_2, n'_2\rangle$. Consequently, i and i' are obtained as the intersections of $\langle m_1, n_1\rangle$ with $\langle m_2, n_2\rangle$ and of $\langle m'_1, n'_1\rangle$ with $\langle m'_2, n'_2\rangle$, respectively:

$$\mathbf{i} = (\mathbf{m}_1 \times \mathbf{n}_1) \times (\mathbf{m}_2 \times \mathbf{n}_2)$$

and

$$\mathbf{i}' = (\mathbf{m}'_1 \times \mathbf{n}'_1) \times (\mathbf{m}'_2 \times \mathbf{n}'_2)$$

Once $\mathbf{H}_\infty$ has been obtained, the ratio of the lengths of any two aligned segments of the scene can be computed directly in the images. Indeed, given three points M_1, M_2 and M_3 on a line, as in figure 6, from their images (m_1, m'_1), (m_2, m'_2) and (m_3, m'_3), we can compute the images (m, m') of the intersection of this line with the plane at infinity, using $\mathbf{H}_\infty$. We can compute in each image the cross-ratio of those four points. As a projective invariant, this cross-ratio is then exactly equal to the ratio of M_1, M_2 and M_3. More precisely:

$$\frac{\overline{M_1M_3}}{\overline{M_2M_3}} = \frac{\overline{M_1M_3}}{\overline{M_1M_\infty}} : \frac{\overline{M_2M_3}}{\overline{M_2M_\infty}} = \frac{\overline{m_1m_3}}{\overline{m_1m}} : \frac{\overline{m_2m_3}}{\overline{m_2m}}$$

4.5 Two cameras and the absolute conic

Let us now consider the pair of images of the absolute conic in the two retinal planes of a stereo rig and the two epipolar planes which are tangent to the absolute conic (they are complex conjuguate, intersecting along the real line $\langle C, C'\rangle$). These two planes intersect the two retinal planes along two pairs of corresponding epipolar lines, by definition, and these epipolar lines are tangent to the images of the absolute conic in the two retinal planes. This property will be used in section 4.6.1.2 to derive the Kruppa equations.

4.6 Metric calibration

In this section we want to push our ideas to their final stage and show that a stereo rig for which the fundamental matrix, the collineation induced by the plane at infinity and the two images of the absolute conic have been estimated allows to recover the third stratum of the world, its euclidean structure or, more precisely, its structure up to a similitude. It is in fact redundant to know the collineation at infinity and the two images of the absolute conic.

4.6.1 *Estimating the image of the absolute conic*

We now describe several ways of estimating the image of the absolute conic. First, in some cases, some similitude invariants of the scene may be available. For example, we may know the angle

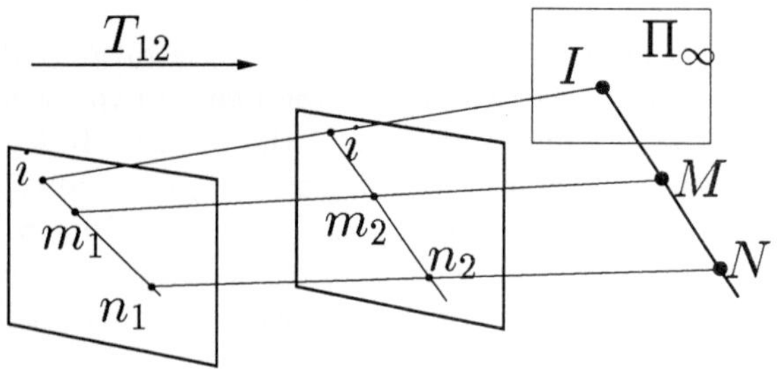

Figure 5: Determining the images of points at infinity.

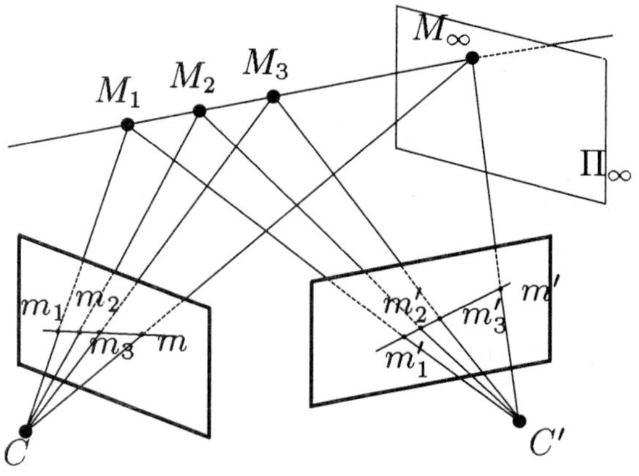

Figure 6: Determining ratios of lengths in affine calibration (see text).

between two lines, or the ratio of the lengths of two non parallel segments. Each such bit of information yields a constraint on the image of the absolute conic, an idea that is used, at least in the case of angles, in [3].

4.6.1.1 A priori information about the scene

If we know the angle α between two lines in the world, according to the properties of the absolute conic, and to equation (5), this yields the following constraint on the coefficients of the equation of ω

$$S(\mathbf{m}, \mathbf{n})^2 = S(\mathbf{m})S(\mathbf{n})\cos^2\alpha \tag{13}$$

This equation is seen to be a quadratic constraint on the coefficients of the equation of ω.

If we have two images of the scene for which we know the plane at infinity, then we can obtain the vanishing point v_{pq} of any line $\langle p, q\rangle$. Now if we know the ratio of the lengths of two non coplanar segments AB and CD, we can use equation (18) which will be derived in the section 4.6.2.2 to derive another constraint on the coefficients of the equation of ω. Let us call r the (known) ratio $\frac{AB}{CD}$ and define

$$D(\mathbf{v}_{pq}, \mathbf{v}_{st}) = S(\mathbf{v}_{pq}, \mathbf{v}_{st})^2 - S(\mathbf{v}_{pq})S(\mathbf{v}_{st})$$

Using equations (13) and (18) we obtain the following constraint on the coefficients of the equation of ω

$$\begin{aligned} D(\mathbf{v}_{ac}, \mathbf{v}_{bc})D(\mathbf{v}_{cd}, \mathbf{v}_{bd})S(\mathbf{v}_{ab}) = \\ r^2 D(\mathbf{v}_{ac}, \mathbf{v}_{ab})D(\mathbf{v}_{bc}, \mathbf{v}_{bd})S(\mathbf{v}_{cd}) \end{aligned} \tag{14}$$

which is seen to be a polynomial of degree 5 in the coefficients of the equation of ω.

A similar constraint on the image of the absolute conic in the second image can be written. If that image has been obtained with the same camera without changing the internal parameters, then we obtain a second constraint on ω.

4.6.1.2 Moving the camera and using Kruppa's equations

If no a priori information about the scene is available, we can still estimate the image of the absolute conic by using motions of the cameras. Note that the camera motions do not have to be known and can be anything as long as they are not pure translations or pure rotations. This observation was made in [25] and turned into an algorithm and a working method in [23, 12].

We now show that each such motion yields two quadratic polynomial equations in the coefficients of the equation of the dual of the image of the absolute conic. In order to do this, we note that if we move a camera from position 1 to position 2 without changing its internal parameters, the image of the absolute conic remains the same [10]. Also, as we noticed in section 4.5, the two tangents from the epipoles to this image correspond to each other in the epipolar homography. Expressing these two facts algebraically yields the two equations.

Let $\mathbf{B}$ be the matrix of the dual of the image of the absolute conic. Let m be a point in the retinal plane, e the epipole. The line $\langle e, m\rangle$ is represented by the cross-product $\mathbf{e} \times \mathbf{m}$ which we write in matrix form $[\mathbf{e}]_\times \mathbf{m}$ with $[\mathbf{e}]_\times$ being the antisymmetric matrix representing the cross-product with the vector $\mathbf{e}$. To say that this line is tangent to ω is equivalent to saying that the point represented by $[\mathbf{e}]_\times \mathbf{m}$ is on the dual conic ω^*. Hence we write the algebraic equation

$$([\mathbf{e}]_\times \mathbf{m})^T \mathbf{B} [\mathbf{e}]_\times \mathbf{m} = 0$$

or

$$\mathbf{m}^T [\mathbf{e}]_\times \mathbf{B} [\mathbf{e}]_\times \mathbf{m} = 0 \qquad (15)$$

This quadratic equation in the coordinates of $\mathbf{m}$ is the equation of the two tangents from e to ω.

From the previous considerations, for each point m on either one of these two tangents, its epipolar line must also be tangent to the image of the absolute conic in the second image. But we have seen that the image of the absolute conic in the second image is identical to its image in the first. The same is of course true of the dual conics. Therefore, introducing the fundamental matrix $\mathbf{F}$, we can write that

$$\mathbf{m}^T \mathbf{F}^T \mathbf{B} \mathbf{F} \mathbf{m} = 0 \qquad (16)$$

if and only if the point m is on one of the previous two tangents. This second quadratic equation in the coordinates of $\mathbf{m}$ therefore also defines the two tangents from e to ω and, thus, The two equations (15) and (16) are equivalent.

This yields a priori five quadratic equations in the coefficients of $\mathbf{B}$. But in fact, because equations (15) and (16) represent a pair of lines and not a general conic, only two of these five equations are independent since it is sufficient to look at the intersection of the tangents with another line not going through e. In [13, 14, 25, 23], the line was chosen to be the line at infinity but in principle any line not going through e will do.

The two equations are called the Kruppa equations in recognition of the work of this Austrian mathematician [21] who worked on a variant of a problem posed by Chasles [6]. For details about the implementation of these ideas and experimental results, see [12, 24].

4.6.2 *From a pair of images of the absolute conic, the plane at infinity to similitude invariants*

4.6.2.1 Angles

We know from a previous section that the knowledge of the image of the absolute conic in one camera allows to compute angles between optical rays. Knowing the images of the absolute conic in two cameras allows to compute angles between any lines. For example, given three points A, B, C in the world with images a, b, c (resp. a', b', c') in the first (resp. second) retinal plane, how do we compute the angle between the lines $\langle A, B\rangle$ and $\langle A, C\rangle$? Let P (resp. Q) be the points at infinity of $\langle A, B\rangle$ (resp. $\langle A, C\rangle$). The angle θ is obtained by a straightforward application of Laguerre's formula. In order to be able to compute the cross-ratio that appears into it, we must be able to compute the images of P and Q, i.e. the vanishing points of the image lines $\langle a, b\rangle$ (resp. $\langle a', b'\rangle$) and $\langle a, c\rangle$ (resp. $\langle a', c'\rangle$). This is possible since, according to the previous section, we know the plane at infinity. We can thus call upon our **Line-Plane** or **Vanishing-Point** procedures. More precisely, and according to equation (5), we have

$$\cos(\langle A, B\rangle, \langle A, C\rangle) = -\frac{S(\mathbf{p}, \mathbf{q})}{\sqrt{S(\mathbf{p})S(\mathbf{q})}} = -\frac{S'(\mathbf{p}', \mathbf{q}')}{\sqrt{S'(\mathbf{p}')S'(\mathbf{q}')}} \qquad (17)$$

and of course the sine is obtained as $\sqrt{1 - \cos^2(\langle A, B\rangle, \langle A, C\rangle)}$.

More generally, the angle between two general lines $\langle A, B\rangle$ and $\langle C, D\rangle$, not necessarily coplanar is obtained by considering the point at infinity P of $\langle A, B\rangle$ and the point at infinity Q of $\langle C, D\rangle$ (i.e. the directions of the two lines) and computing the cross-ratio of P and Q and the points of intersection of the line $\langle P, Q\rangle$ with the absolute conic, this computation being of course performed in the images, i.e. using equation (17).

4.6.2.2 Ratios of lengths

Let us now describe how we can compute the other type of similitude invariants, the ratios of lengths. Using the fact that angles can be computed, we show how to use them to compute ratios of lengths. Let us consider four points A, B, C and D and suppose we want to compute the ratio $\frac{AB}{CD}$. Considering the two triangles ABC and BCD we can write, for the first triangle

$$\frac{AB}{\sin\gamma} = \frac{BC}{\sin\alpha}$$

and, for the second

$$\frac{BC}{\sin\delta} = \frac{CD}{\sin\beta}$$

from which we obtain the ratio $\frac{AB}{CD}$ as a function of the four angles α, β, γ and δ

$$\frac{AB}{CD} = \frac{\sin\gamma \cdot \sin\delta}{\sin\alpha \cdot \sin\beta} \qquad (18)$$

Using the procedure **Line-Plane** or **Vanishing-Point**, we construct the four vanishing points v_{ab}, v_{ac}, v_{bc}, v_{bd}, and v_{cd} of the image lines $\langle a, b\rangle$, $\langle a, c\rangle$, $\langle b, c\rangle$, $\langle b, d\rangle$ and $\langle c, d\rangle$. The sines which appear in equation (18) are then obtained through equation (5). More specifically, we obtain the neat formula for the ratio $\frac{AB}{CD}$ computed in the first image in which appears only the equation of ω:

$$\frac{AB}{CD} = \sqrt{\frac{S(\mathbf{v}_{ab})}{S(\mathbf{v}_{cd})} \cdot \frac{S(\mathbf{v}_{ac})S(\mathbf{v}_{bc})-S(\mathbf{v}_{ac},\mathbf{v}_{bc})^2}{S(\mathbf{v}_{ab})S(\mathbf{v}_{ac})-S(\mathbf{v}_{ab},\mathbf{v}_{ac})^2}} \cdot \sqrt{\frac{S(\mathbf{v}_{bd})S(\mathbf{v}_{cd})-S(\mathbf{v}_{bd},\mathbf{v}_{cd})^2}{S(\mathbf{v}_{bc})S(\mathbf{v}_{bd})-S(\mathbf{v}_{bc},\mathbf{v}_{bd})^2}} \quad (19)$$

5 Application to navigation [28, 29]

In this section we show three robotic applications of the geometric properties presented in the above sections. In the first one, stereo is used to detect close obstacles. In the second one, the robot uses affine geometry to follow the middle of a corridor. In the third one, relative heights and orientations with respect to the ground plane are used for trajectory planning.

5.1 Detecting obstacles

This section describes how to use the previous results to provide a robot with the ability to detect obstacles. The only requirement is that the robot is equipped with a stereo rig which can be very simply calibrated, as explained next.

Let us imagine the following calibration steps:

- some correspondences between two views taken by the cameras are found;
- these correspondences are used to compute the fundamental matrix;
- three particular correspondences are given to the system; they correspond to three object points defining a virtual plane Π in front of the robot;
- the H-matrix of Π is computed.

The fundamental matrix, as well as the plane H-matrix, remain the same for any other pair of views taken by the system, as long as the intrinsic parameters of both cameras and the attitude of one camera with respect to the other do not change.

By repeatedly performing rectifications with respect to Π, the robot knows whether there are points in front between itself and Π by looking at the sign of their disparity and can act in consequence. If the distance d_0 of the robot to Π is known, the robot may, for example, move forward of a distance d_0. Furthermore, if Π and the focal plane, Π'_f, of the second camera intersect sufficiently far away from the cameras, it can detect whether the points are moving away or towards itself. Indeed, Π and Π'_f may then be considered as parallel around the images, so that, for the points considered, $d(M,\Pi)$ is proportional to $d(M,\Pi'_f) - d(\Pi,\Pi'_f)$, where $d(\Pi,\Pi'_f) = d(M_\Pi,\Pi'_f)$ for any point $M_\Pi \in \Pi$. Writing $\mathbf{H}_\Pi \mathbf{m} = (\hat{x}', \hat{y}', \hat{z}')$ and $\mathcal{D} = \hat{x}' - x'$, it is possible to show [28] that $\hat{z}'\mathcal{D}$ is proportional to $\frac{d(M,\Pi)}{d(M,\Pi'_f)}$ and hence approximatively proportional to

$$1 - \frac{d(\Pi,\Pi'_f)}{d(M,\Pi'_f)}$$

thus, is a monotonic function of the distance of M to Π'_f.

At last, since we are only interested in the points near the plane, which have a disparity close to zero, we can limit the search along the epipolar line of the correspondent point $\hat{m}'$ of any point $\hat{m}$ to an interval around $\hat{m}$, which significantly reduces the computation time.

An example is given in Figures 7 and 8. Figure 7 shows as dark square boxes the points used to define a plane and the image of a fist taken by the left camera and the left and right images once rectified with respect to this plane. Figure 8 shows the disparity map obtained by correlation and the segmentation of the disparity map in two parts. On the left side, points with negative disparities, that is points in front of the reference plane, are shown. The intensity encodes closeness to the camera. Similarly, the right side of the figure shows the points with positive disparities, that is the points which are beyond the reference plane.

5.2 Navigating along the middle of a corridor

If we add to the computation of the fundamental matrix during the calibration stage, the computation of the homography of the plane at infinity, the robot becomes able to compute ratios of three aligned points, thus, for example, the middle of a corridor and do visual servoing.

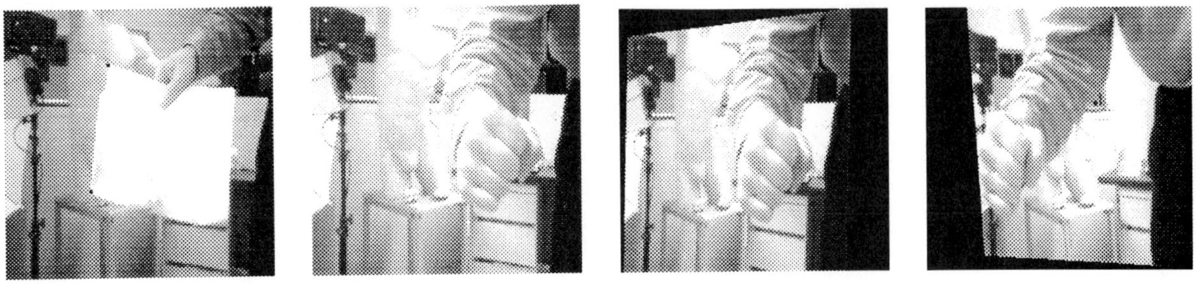

Figure 7: The paper used to define the plane and the left image of the fist taken as an example.The left and right rectified images of the fist.

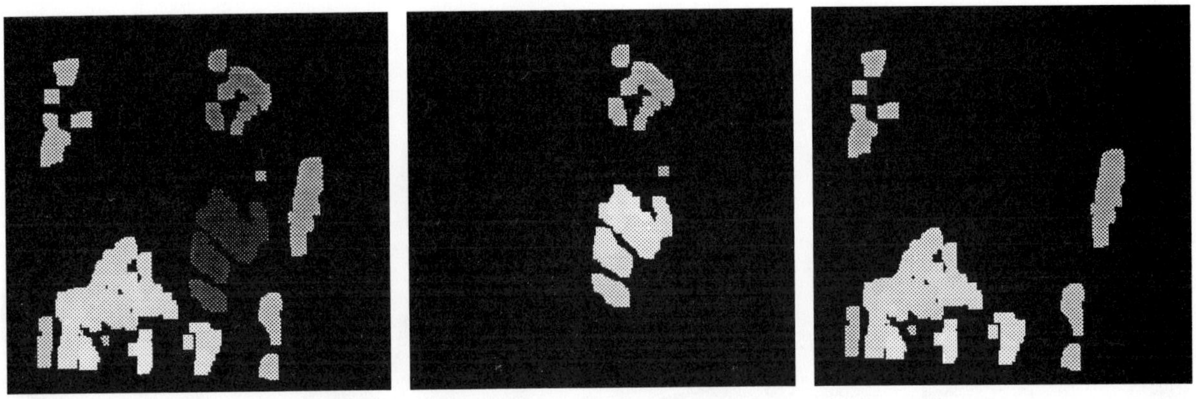

Figure 8: The disparity map obtained from the rectified images of Figure 7. The absolute value of the negative disparities on the left, showing that the fist and a portion of the arm are between the robot and the plane of rectification, and the positive disparities on the right, corresponding to the points located beyond the plane.

Indeed, if we represent the projections of the sides S_1 and S_2 of the corridor by s_1 and s_2 in the first image and s'_1 and s'_2 in the second image (see Figure 9) and choose any point a of s_1 and any point b of s_2, projections in the first image of a point A of S_1 and a point B of S_2, then the corresponding points a' and b' in the second image are computed as the intersections, respectively, of the epipolar line e_a of a with s'_1 and the epipolar line e_b of b with s'_2. Having (a, a') and (b, b') allows to compute the projections (m, m') of the midpoint M of A and B. If we consider S_1 and S_2 as locally parallel, then M lies on the local middle line of the corridor and computing the projections of another point of this line the same way as M allows to have the projections of this line in the two images.

Figure 10 shows some real sequences used to perform the affine calibration of a stereoscopic system. Six strong correspondences between the four images have been extracted, from which fifteen correspondences of points at infinity have been computed to finally get the homography of the plane at infinity. Figure 11 shows some midpoints obtained once the system calibrated: the endpoints are represented as black squares and the midpoints as black crosses and the midline of a corridor obtained from another affinely calibrated system: the endpoints are represented as numbered oblique dark crosses, the midpoints as black crosses and the midline as a black line.

5.3 Trajectory evaluation using relative elevation

A limitation of the conventional approach to stereo driving is that it relies on precise metric calibration with respect to an external calibration target in order to convert matches to 3-D points. From a practical standpoint, this is a serious limitation in scenarios in which the sensing hardware cannot be physically accessed, such as in the case of planetary exploration. In particular, this limitation implies that the vision system must remain perfectly calibrated over the course of an entire mission. Nevertheless, navigation should not re-

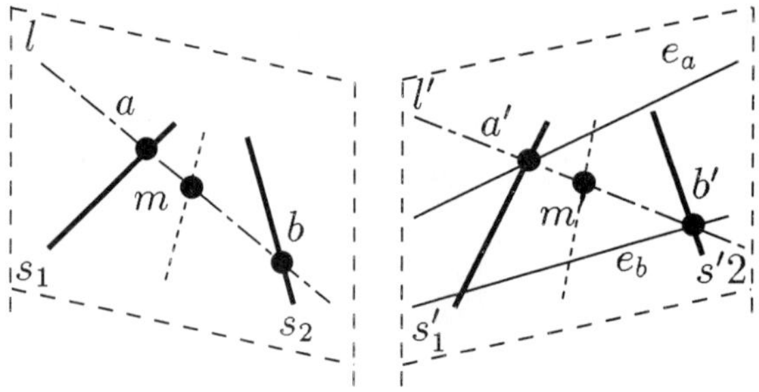

Figure 9: Determining a point at the middle of a corridor (see Section 5.2).

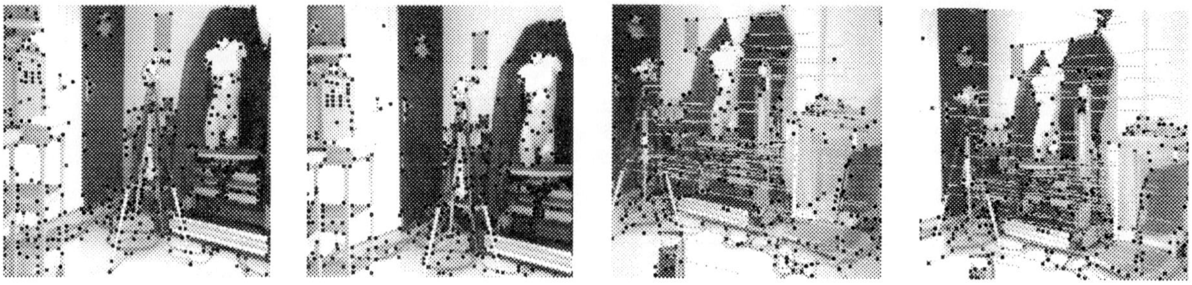

Figure 10: The two left images correspond to a first pair of views taken by a stereoscopic system and the two right images to a second pair taken by exactly the same system after a translation. Among the 297 detected corners of the top left image and the 276 of the top right image, 157 points correspondences have been found by stereo points matching, among which 7 outliers have been rejected when computing the fundamental matrix. The top to bottom correspondences matching has been obtained by tracking.

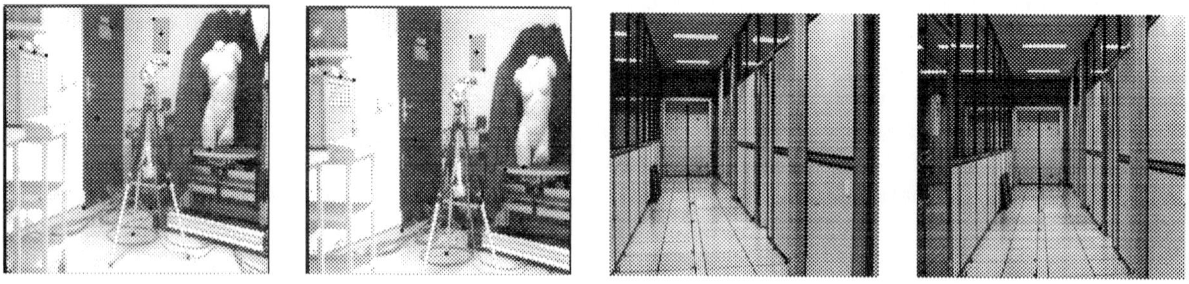

Figure 11: Midpoints obtained after affine calibration (left). Midline of a corridor obtained after affine calibration (right).

quire the precise knowledge of the 3-D position of points in the scene: What is important is how much a point deviates from the reference ground plane, not its exact position. Based on these observations, we have developed an approach which relies on the measure of relative height with respect to a ground plane.

5.3.1 *The driving approach*

We give only an overview of the approach since a detailed description of the driving system is outside the scope of this book. A detailed description of the stereo driving system can be found in [20]. The autonomous navigation architecture is described in [22] and [17].

In autonomous driving, the problem is to use the data in the stereo images for computing the best steering command for the vehicle, and to update the steering command every time a new image is taken. Our basic approach is to evaluate a set of possible steering directions based on the relative heights computed at a set of points which project onto a regular grid in the image. Specifically, a given steering radius corresponds to an arc which can be projected into a curve in the image. This curve traces the trajectory that the vehicle would follow in the image if it used this steering radius. Given the points of the measurement grid and the set of steering radii, we compute a vote for every arc and every point of the grid which reflects how drivable the arc is. The computed value lies between -1 (high obstacle) and +1 (no obstacle) (Figure 12).

For a given steering radius, the votes from all the grid points are aggregated into a single vote by taking the minimum of the votes computed from the individual grid points. The output is, therefore, a distribution of votes between -1 and +1, -1 being inhibitory, for the set of possible steering arcs. This distribution is then sent to an external module which combines the distribution of votes from stereo with distributions from other modules in order to generate the steering command corresponding to the highest combined vote. Figure 13 shows examples of vote distributions computed in front of visible obstacles.

5.3.1.1 Characterization of the reference ground plane:

The homography of the reference ground plane is estimated from a number of point correspondences related to scene points on this plane. These point correspondences are obtained by selecting points in the first image. Their corresponding points in the second image are computed automatically through the process of rectification and correlation along the epipolar lines.

5.3.1.2 Measuring obstacle heights

Let us we consider one point of the grid in the first image, for which a corresponding point in the second image has been found by the stereo process. Based on the previous ideas, we can compute its height with respect to the ground. The unit height is defined by a point of the scene, selected manually in one of the two images at the beginning of the experiment and matched automatically in the other image.

This measurement is not sufficient, since we aim at measuring heights along trajectories which are estimated in the ground plane. So, after determining the elevation of a point selected in one image, we determine the point of the ground plane to which this elevation has to be assigned, by projecting the measured 3D point on the (horizontal) ground plane, along the vertical direction. This means computing the intersection between the ground plane and the vertical line passing through the 3D point. We need to compute the images of the vertical the line passing through the observed point. For this, we compute the images of the point at infinity in the vertical direction (also called vertical vanishing point) in both images. First, we select manually four points representing two vertical lines in the left image. Matching two of these points we obtain one of the two corresponding lines in the right image. The left vertical vanishing point is obtained by intersecting the two lines in the left image. Computing the intersection of its epipolar line with the line in the right image, we obtain the right vertical vanishing point.

5.3.1.3 Computing images trajectories

This approach assumes that a transformation is available for projecting the steering arcs onto the image plane. Such a transformation can be computed from features in sequences of images using an approach related to the estimation algorithm described above for computing the homography induced by the ground plane.

We first introduce a system of coordinates in the ground plane, attached to the rover, which we call "rover coordinates". At each time, we know in rover coordinates the trajectory which will be followed by the rover for each potential steering command. Further more, for a given motion/steering command sent to the robot, we know from the mechanical design the expected change of rover coordinates from the final position to the initial one. We can even estimate the actual motion using dead-reckoning. Since the transformation is a change of coordinates in the plane, it can be represented by a 3×3 matrix $\mathbf{T}^r$ operating on homogeneous coordinates.

The transformation which we compute is the homography $\mathbf{H}^{ir}$ which maps pixel coordinates in the left image onto rover coordinates. The inverse of this matrix then allows us to map potential rover trajectories onto the left image plane.

Computation of $\mathbf{H}^{ir}$ is done by tracking points across the left images taken at various rover positions. Let us consider two images acquired at positions 1 and 2, with a known rover motion $\mathbf{T}^r_{12}$. Given a point correspondence (p_1, p_2) we have the following equation (up to a scale factor):

$$\mathbf{H}^{ir}\mathbf{p}_1 = \mathbf{T}^r_{12}\mathbf{H}^{ir}\mathbf{p}_2$$

where the only unknown is the matrix $\mathbf{H}^{ir}$. This can be also written

$$\mathbf{H}^{ir}\mathbf{p}_1 \times (\mathbf{T}^r_{12}\mathbf{H}^{ir}\mathbf{p}_2) = 0$$

This yields a system of two independent quadratic equations in the coefficients of $\mathbf{H}^{ir}$. Given a set of displacements and point coordinates, we can write a large system of such equations, which we solve in the least-squares sense using the Levenberg-Marquardt technique.

5.3.1.4 Using heights to speed up stereo matching

The relative height is also used for limiting the search in the stereo matching. More precisely, we define an interval $[h_{min}, h_{max}]$ of heights which we anticipate in a typical terrain. This interval is converted at each pixel to a disparity range $[d_{min}, d_{max}]$. This is an effective way of limiting the search by searching only for disparities that are physically meaningful at each pixel.

5.3.2 *Experimental results*

This algorithm has been successfully used for arc evaluation in initial experiments on the CMU HMMWV [17], a converted truck for autonomous navigation. In this case, a 400 points grid was used. The combination of stereo computation and arc evaluation was done at an average of 0.5s on a Sparc-10 workstations. New steering directions were issued to the vehicle at that rate. This update rate is comparable to what can be achieved using a laser range finder [22].

An important aspect of the system is that we are able to navigate even though a relatively small number of points is processed in the stereo images. This is in contrast with the more conventional approach in which a dense elevation map is necessary, thus dramatically increasing the computation time. Using such a small number of points is justified because it has been shown that the set of points needed for driving is a small fraction of the entire data set independent of the sensor used [19], and because we have designed our stereo matcher to compute matches at specific points.

6 Conclusion

We have shown in this article how the use of geometry in computer vision is a key factor to both a comprehensive theoretical understanding of the geometry of systems of multiple cameras and to the development of sophisticated applications.

The projective framework is now recognized as the natural one for describing computer vision problems because of its closeness to the actual geometrical operation that is performed by a camera, i.e. perspective projection. We have also shown here and elsewhere how the projective framework could also accomodate gracefully the affine and Euclidean frameworks which are useful in many applications.

The use of the Grassmann-Cayley formalism has been shown to be the natural one for studying the geometry of the systems of cameras and of the scene. In this article we have mostly focussed on the geometry of the cameras but the same tools can also be used to describe the geometry of the scene [5, 7, 30, 31]. It is clear that much remains to be done in order to fully appreciate the power of this approach for computer vision and robotics in general. For example, the extension of this formalism to the affine and Euclidean cases remains to be worked out, the extension to the instantaneous case where in fact one is dealing with camera and scene motions also has not been worked out just as the case of more complicated primitives than points and lines such as general curves and surfaces.

Apart from these problems, I foresee the use of these methods at a higher level than that of what may be called the first-order geometry of

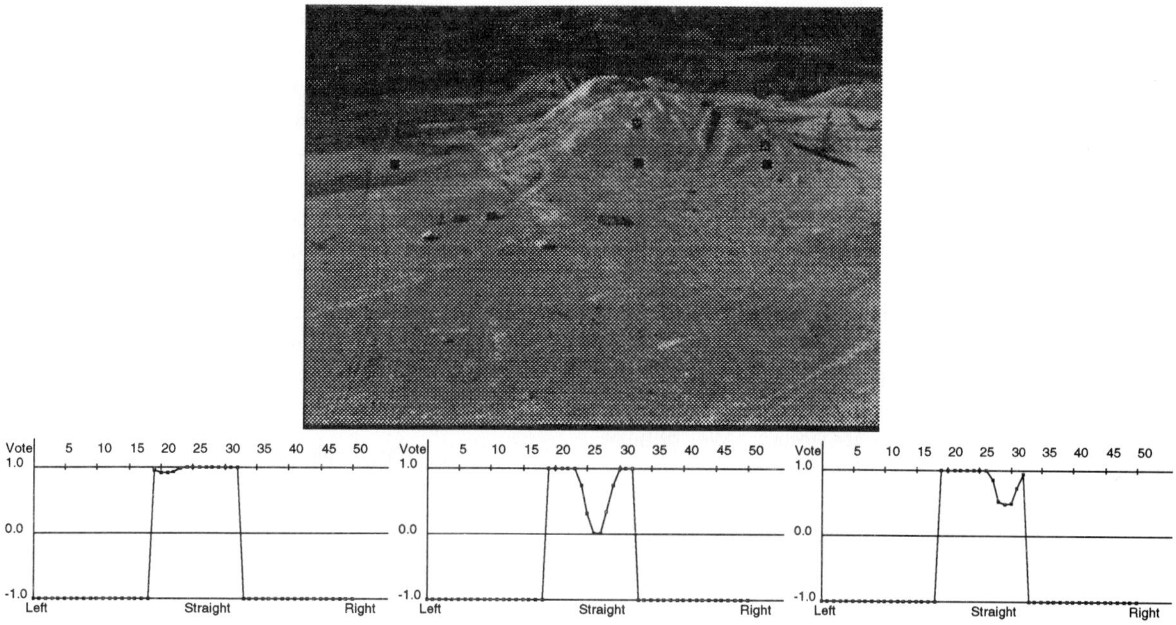

Figure 12: Evaluating steering at individual points; (top) Three points selected in a stereo pair and their projections; (bottom) Corresponding votes for all steering directions.

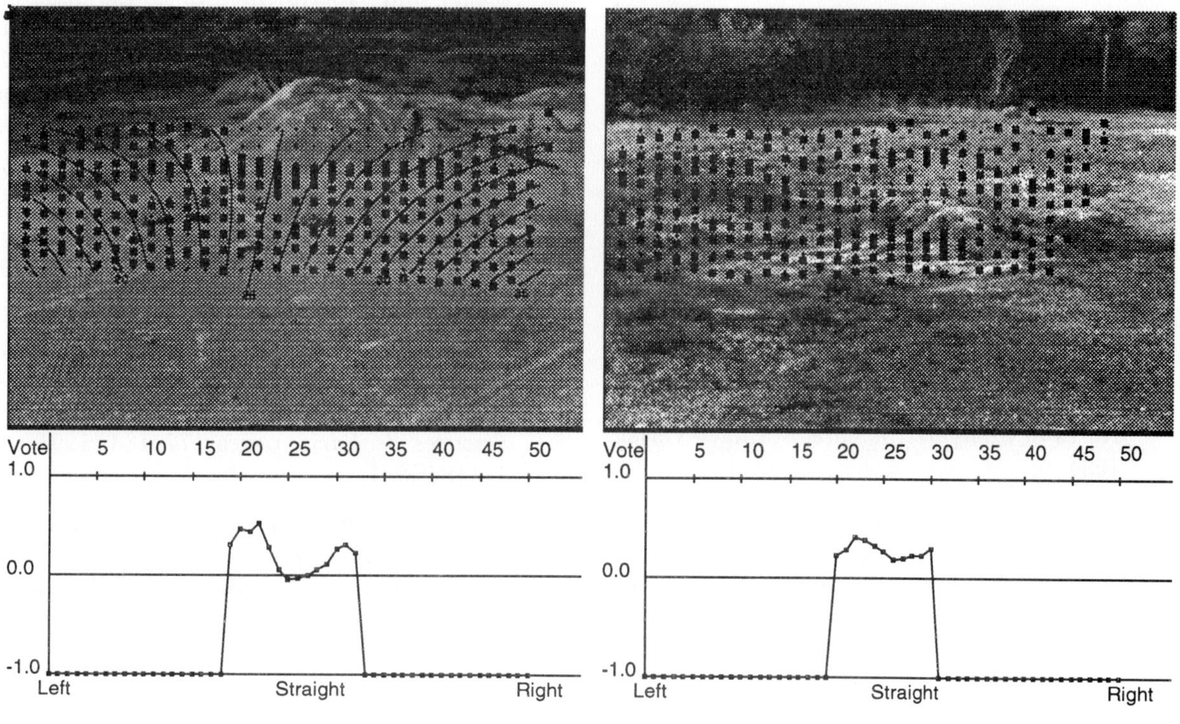

Figure 13: Evaluating steering directions in the case of a large obstacle (left) and a small obstacle (right). (Top): Regular grid of points (dots) and corresponding projections (squares); (Bottom): Distribution of votes (see text).

the scene, namely at the level of describing the relations between the geometric elements. Simple examples of such relations are the well-known incidence properties such as coplanarity, collinearity, parallelism. It should be clear that they are bound to play a crucial role in describing the geometry of scenes and objects in future systems. The nice thing about the formalism that has been presented here is that it allows to express these properties very compactly and symbollically, i.e. without using coordinates. But even more than that, the same formalism can be used to discover new geometric properties by formal manipulation of these symbolic representations. We believe that the use of geometric theorem proving methods in computer vision and robotics will turn out to be one of the major features of future systems.

References

[1] D.C. Ballard and C.M. Brown. *Computer Vision.* Prentice-Hall, Englewood Cliffs, New Jersey, 1982.

[2] M. Barnabei, A. Brini, and G.C. Rota. On the exterior calculus of invariant theory. *J. of Algebra*, 96:p 120–160, 1985.

[3] Boubakeur Boufama, Roger Mohr, and Franccoise Veillon. Euclidean Constraints for Uncalibrated Reconstruction. Technical Report RT96 IMAG 17 LIFIA, LIFIA, INSTITUT IMAG, March 1993.

[4] Stefan Carlsson. Multiple image invariance using the double algebra. In Joseph L. Mundy, Andrew Zissermann, and David Forsyth, editors, *Applications of Invariance in Computer Vision*, volume 825 of *Lecture Notes in Computer Science.* Springer-Verlag, 1994.

[5] Stefan Carlsson. Multiple image invariance using the double algebra. In Joseph L. Mundy, Andrew Zissermann, and David Forsyth, editors, *Applications of Invariance in Computer Vision*, volume 825 of *Lecture Notes in Computer Science*, pages 145–164. Springer-Verlag, 1994.

[6] M. Chasles. Question No. 296. *Nouv. Ann. Math.*, 14:50, 1855.

[7] Gabriella Csurka and Olivier Faugeras. Computing three-dimensional projective invariants from a pair of images using the Grassmann-Cayley algebra. In R. Mohr and C. Wu, editors, *Proceedings of Europe-China Workshop on Geometrical Modeling and Invariants for Computer Vision*, pages 150–157, Xi'an, China, April 1995. Xidian University Press.

[8] C. DeConcini, D. Eisenbud, and C. Procesi. Hodge algebras. *Astérisque*, 91, 1982.

[9] P. Doubilet, G. C. Rota, and J. Stein. Foundations of Combinatorics IX: Combinatorial methods in Invariant Theory. *Studies in Appl. Math.*, 53:185–216, 1974.

[10] Olivier Faugeras. *Three-Dimensional Computer Vision: a Geometric Viewpoint.* The MIT Press, 1993.

[11] Olivier Faugeras and Bernard Mourrain. About the correspondences of points between n images. In *Proceedings of the Workshop on the representation of visual scenes*, Cambridge, Massachusetts, USA, June 1995.

[12] Olivier D. Faugeras, Tuan Luong, and Steven Maybank. Camera self-calibration: theory and experiments. In Giulio Sandini, editor, *Proceedings of the 2nd European Conference on Computer Vision*, pages 321–334. Springer-Verlag, Lecture Notes in Computer Science 588, 1992.

[13] Olivier D. Faugeras and Steven Maybank. Motion from point matches: multiplicity of solutions. *The International Journal of Computer Vision*, 4(3):225–246, June 1990. also INRIA Tech. Report 1157.

[14] Olivier D. Faugeras and Steven Maybank. Mouvement à partir de points : nombre de solutions. *Comptes rendus de l'Académie des Sciences de Paris*, pages 177–183, 1991. t. 312, Série II.

[15] Gene H. Golub and Charles F. Van Loan. *Matrix computations.* The John Hopkins University Press, Baltimore, Maryland, 1983.

[16] Robert M. Haralick and Linda G. Shapiro. *Computer and Robot Vision.* Addison-Wesley, 1993.

[17] M. Hebert, D. Pomerleau, A. Stentz, and C. Thorpe. A behavior-based approach to autonomous navigation systems: The cmu ugv project. In *To Appear in IEEE Expert, 1994*, 1994.

[18] Berthold Klaus Paul Horn. *Robot Vision.* MIT Press, 1986.

[19] A. Kelly. A partial analysis of the high speed autonomous navigation problem. Technical Report CMU-RI-TR-94-16, The Robotics Institute, Carnegie Mellon, 1994.

[20] E. Krotkov, M. Hebert, M. Buffa, F.G. Cozman, and L. Robert. Stereo driving and position estimation for autonomous planetary rovers. pages 320–328.

[21] E. Kruppa. Zur Ermittlung eines Objektes aus zwei Perspektiven mit innerer Orientierung. *Sitz.-Ber. Akad. Wiss., Wien, math. naturw. Kl., Abt. IIa.*, 122:1939–1948, 1913.

[22] D. Langer, J. Rosenblatt, and M. Hebert. A reactive system for autonomous navigation in unstructured environments. In *Proc. International Conference on Robotics and Automation*, San Diego, 1994.

[23] Quang-Tuan Luong. *Matrice Fondamentale et Calibration Visuelle sur l'Environnement-Vers une plus grande autonomie des systèmes robotiques.* PhD thesis, Université de Paris-Sud, Centre d'Orsay, December 1992.

[24] Quang-Tuan Luong and Olivier Faugeras. The Fundamental matrix: theory, algorithms, and stability analysis. *The International Journal of Computer Vision*, 1994. To appear.

[25] S.J. Maybank and O.D. Faugeras. A Theory of Self-Calibration of a Moving Camera. *The International Journal of Computer Vision*, 8(2):123–152, August 1992.

[26] B. Mourrain. New aspects of geometrical calculus with invariants. *Advances in Mathematics*, 1994. to appear.

[27] B. Mourrain and N. Stolfi. *Invariants methods in Discrete and Computational Geometry*, chapter Computational Symbolic Geometry. Kluwer acad. pub., 1994. (to appear).

[28] L. Robert, C. Zeller, O. Faugeras, and M. Hebert. Applications of non-metric vision to some visually-guided robotics tasks. Technical Report 2584, INRIA, June 1995.

[29] L. Robert, C. Zeller, O. Faugeras, and M. Hébert. Applications of non-metric vision to some visually-guided robotics tasks. In Y. Aloimonos, editor, *Visual Navigation: From Biological Systems to Unmanned Ground Vehicles*, chapter ? Lawrence Erlbaum Associates, 1996. to appear, also INRIA Technical Report 2584.

[30] C. Rothwell, G. Csurka, and O. Faugeras. A comparison of projective reconstruction methods for pairs of views. In *Proceedings of the 5th Proc. International Conference on Computer Vision*, pages 932–937, Boston, MA, June 1995. IEEE Computer Society Press.

[31] C. Rothwell, G. Csurka, and O. Faugeras. A comparison of projective reconstruction methods for pairs of views. Technical Report 2538, INRIA, Sophia Antipolis, April 1995.

[32] J.G. Semple and G.T. Kneebone. *Algebraic Projective Geometry.* Oxford: Clarendon Press, 1952. Reprinted 1979.

[33] Bernd Sturmfels. *Algorithms in Invariant Theory.* Springer-Verlag Wien New york, 1993.

[34] H. Weyl. *The Classical Groups, their invariants and representations.* Princeton University Press, 1939.

[35] Z. Zhang, R. Deriche, O. Faugeras, and Q.-T. Luong. A robust technique for matching two uncalibrated images through the recovery of the unknown epipolar geometry. Research Report 2273, INRIA Sophia-Antipolis, France, May 1994. accepted to *Artificial Intelligence Journal.*

Sine-Cosine Polynomial Systems

J. Nielsen and B. Roth
Department of Mechanical Engineering
Stanford University
Stanford, CA 94305-4021
email: nielsen@leland.stanford.edu, roth@flamingo.stanford.edu

1 Introduction

Frequently, in the design and analysis of robotic mechanisms, sets of nonlinear equations appear and must be solved. The nonlinearities often arise due to the kinematic couplings between members, which lead to mathematical descriptions in terms of products of trigonometric functions. The sets of equations which must be solved, then, are often polynomials in the sine and cosine of a group of variables. Because these sine-cosine polynomial systems occur so frequently in mechanical analysis, it is highly desirable to study such systems, and the solutions to which they lead.

Recently a lot of progress has been made in solving several long-standing problems of fundamental importance. References to these papers are given in [2]. Here, we will briefly list the basic results: In regard to the inverse kinematics of a general six-degree-of-freedom open chain manipulator, we know that chains of general dimensions have sixteen different configurations for each end-effector pose if the chain is composed of either six revolute joints or five revolute joints and one prismatic joint. For four revolute and two prismatic jointed chains, the number of configurations drops to eight, while for three revolute and three prismatic joints there are only two configurations for each end-effector pose. In all these cases the number of real configurations may be less than the cited number by any multiple of two.

Combinations of structural parameters which produce special geometry (such as parallel or intersecting joint axes) can cause the number of configurations to be reduced from the above cited numbers. Open chain six-degree-of-freedom manipulators have been found which will yield 2, 4, 6, 8, 10, 12, 14 or 16 configurations at a given end-effector pose. Similar results can be found for one-degree-of-freedom single loop closed linkage chains.

For in-parallel systems, the direct kinematics of a platform coupled to a ground link by six independent telescoping legs paired with spherical joints to the ground and moving platform (arranged as in the so-called generalized Stewart platform) is known to have, in general, forty solutions. That means for a structure with general dimensions, if the six leg lengths are specified there are forty possibly platform poses in to which the system can be assembled. Not all forty poses need be real, and the real poses can be less than forty in multiples of two. Furthermore it is known that specific special geometries will reduced the number from forty to various lower even numbers.

Even though much has been learned in obtaining these powerful recent results, much is still unknown about the basic structure of sets of sine-cosine polynomials. To gain some insight into the behavior of sine-cosine polynomial systems, small but representative systems, for which solution methods are readily available, may be analyzed. In this paper both two- and three-equation systems will be studied. Beginning with generic sine-cosine polynomials of a certain degree and with a specified set of monomials, we will see how the specialization of coefficients in the system affects the total number of solutions, the number of solutions at infinity, and the possible existence of spurious roots in the univariate polynomial resulting from elimination of variables. These spurious roots have been previously examined in [1].

The behavior of the sine-cosine system must be viewed in the context of the particular solution method used. There are a variety of such methods, including graphical, numerical, and analytical approaches [2]. We are concerned in this paper with the use of resultants to solve the nonlinear system, and the various equation manipulation techniques used to simplify the creation of a resultant. These manipulation techniques have proven very important in previous research, including the solution of the inverse kinematics problem for general 6R serial manipulators [3].

2 Two-Equation System

The basic system which will be analyzed is a sine-cosine polynomial system with two equations and two unknown angles. The two equations are as follows:

$$k_{11}c_1c_2 + k_{12}c_1s_2 + k_{13}s_1c_2 + k_{14}s_1s_2 + k_{15}c_1 + k_{16}s_1 + k_{17}c_2 + k_{18}s_2 = k_{19} \quad (1)$$

$$k_{21}c_1c_2 + k_{22}c_1s_2 + k_{23}s_1c_2 + k_{24}s_1s_2 + k_{25}c_1 + k_{26}s_1 + k_{27}c_2 + k_{28}s_2 = k_{29} \quad (2)$$

where c_i and s_i represent $\cos\theta_i$ and $\sin\theta_i$, respectively, the k_{ij} are coefficients in the complex field, and θ_1, θ_2 are unknowns. Note that there are no squared terms in either variable, although there are second-order cross terms between θ_1 and θ_2. This system was chosen because it could arise in planar kinematics, where squared terms are rarely encountered.

We begin by considering equations (1,2) above with general coefficients. The first step towards a solution is the introduction of the well-known tangent-of-the-half-angle substitution; the subtleties of this substitution will be encountered later. We set:

$$\cos\theta_i = \frac{1-x_i^2}{1+x_i^2} \quad \sin\theta_i = \frac{2x_i}{1+x_i^2} \quad \text{where} \quad x_i = \tan\frac{\theta_i}{2} \quad (3)$$

then clear the denominators by multiplying (1, 2) above by $(1+x_1^2)(1+x_2^2)$. This leads to a system of two polynomial equations in x_1 and x_2, which may be written as:

$$\sum_{p=0,1,2;q=0,1,2} d_{pqj}x_1^p x_2^q = 0 \qquad j = 1,2 \quad (4)$$

Since we have only two equations in two unknowns, Sylvester's resultant may be used to find the solutions to (4) above. Resultants will be used throughout this summary to obtain a univariate polynomial (in either x_1 or x_2), whose roots correspond to solutions of the entire system.

Our first step is to get a bound on the number of solutions which might exist. A Bezout count gives 16 possible solutions. A tighter bound can be achieved by calculating the BKK bound [4, 5]. Calculating this bound for (4) we find that there will be a maximum of 8 solutions, corresponding to the case where all coefficients are general. Using randomly generated coefficients for the sine-cosine system, it can be shown that the resulting polynomial system does indeed have 8 numerical solutions.

Beginning from this point, where we have a maximum number of solutions corresponding to a generic coefficient set, we will proceed to specialize the coefficients systematically, showing how this specialization affects the solutions to the system.

2.1 Rotation Matrices

In kinematics, sines and cosines often appear in equations due to the application of rotation matrix multiplications. For instance, equations (1,2) above might have been generated by the following matrix equation:

$$A_1R_{\theta_1}R_{\theta_2}v_1 + A_2R_{\theta_1}v_2 + A_3R_{\theta_2}v_3 = v_4$$
$$\text{where} \quad R_{\theta_i} = \begin{bmatrix} \cos\theta_i & -\sin\theta_i \\ \sin\theta_i & \cos\theta_i \end{bmatrix} \quad (5)$$

The A_i and v_i are 2×2 and 2×1 matrices, respectively, with arbitrary elements from the complex field. We will find that equations created by the matrix multiplication above will always have a solution at $x_1 = i$, $x_2 = -i$ and one at $x_1 = -i$, $x_2 = i$ after the introduction of the tangent-of-the-half-angle substitution. The total number of solutions to the tangent-of-the-half-angle system remains at eight, but now the locations of two of those solutions are known. These two solutions do not transfer to the sine-cosine system, which now has six solutions.

To see why the two roots at $\pm i$ must exist, we write equation (5) above as it would appear after the introduction of the tangent-of-the-half-angle substitution, and clearing of denominators. Equation (5) above then becomes

$$A_1T_{x_1}T_{x_2}v_1 + \left(1+x_2^2\right)A_2T_{x_1}v_2 + \left(1+x_1^2\right)A_3T_{x_2}v_3 = \left(1+x_1^2\right)\left(1+x_2^2\right)v_4$$
$$\text{where} \quad T_{x_i} = \begin{bmatrix} 1-x_i^2 & -2x_i \\ 2x_i & 1-x_i^2 \end{bmatrix} \quad (6)$$

Using this representation, it is easy to see that at the points $(i,-i)$ and $(-i,i)$, all terms which are multiplied by either $\left(1+x_1^2\right)$ or $\left(1+x_2^2\right)$ will go to zero. Only the first term in (6) above remains;

setting $x_1 = i, x_2 = -i$, this term becomes:

$$A_1 \begin{bmatrix} 2 & -2i \\ 2i & 2 \end{bmatrix} \begin{bmatrix} 2 & 2i \\ -2i & 2 \end{bmatrix} v_1 \qquad (7)$$

and we can see that:

$$\begin{bmatrix} 2 & -2i \\ 2i & 2 \end{bmatrix} \begin{bmatrix} 2 & 2i \\ -2i & 2 \end{bmatrix} = \begin{bmatrix} 0 & 0 \\ 0 & 0 \end{bmatrix} \qquad (8)$$

so that all terms in (6) go to zero, and that equation is satisfied. The behavior of the system is similar at the point $x_1 = -i, x_2 = i$. So when the equations are obtained through matrix multiplication as in (5), the location of two solutions of the corresponding tangent-of-the-half-angle system are automatically known: $(i, -i)$ and $(-i, i)$.

Here something should be said about the tangent half-angle substitution and its consequences. When the denominators in the equations are cleared, if all terms (after cancelation) are multiplied by a term of the form $(1 + x_i^2)$, then clearly there will be solutions at $x_i = \pm i$. The requirement for this condition to be avoided is simple to state: in each equation, the highest power sine-cosine term for each variable must occur in at least one term. If such a term exists, then it will not be multiplied by a term like $(1 + x_i^2)$, and thus will not necessarily go to zero at $x_i = \pm i$.

In the equations (5) above, there does in fact exist a term containing the highest power sine-cosine term for each variable. Since the highest power of sine-cosine terms is linear, the cross terms like $\cos\theta_1 \sin\theta_2$ meet this requirement. So from the theory above, we do not necessarily expect solutions at $\pm i$ after introducing the tangent-of-the-half-angle substitution. In this case, though, these solutions exist because of some additional structure in the equations. It is important to recognize that structure, rather than dismissing the solutions at $\pm i$ as simple artifacts of the clearing of denominators.

2.2 Rotation Coefficient Matrices

We will now consider one possible specialization for the elements of the A_i matrices in (5) above. Consider the case where coefficient matrices A_1 and A_2 (that is, the coefficient matrices for the cross terms and the θ_1 rotation matrix) both have orthogonal rows of equal length. Then those two coefficient matrices may be written as:

$$A_1 = \begin{bmatrix} p & -q \\ q & p \end{bmatrix} \qquad A_2 = \begin{bmatrix} r & -s \\ s & r \end{bmatrix} \qquad (9)$$

Note that all rotation matrices have orthogonal rows with equal length, and they are frequently encountered in kinematics calculations.

At $x_1 = i$, (6) then becomes:

$$\begin{bmatrix} p & -q \\ q & p \end{bmatrix} \begin{bmatrix} 2 & -2i \\ 2i & 2 \end{bmatrix} \begin{bmatrix} 1 - x_2^2 & -2x_2 \\ 2x_2 & 1 - x_2^2 \end{bmatrix} v_1 + \left(1 + x_2^2\right) \begin{bmatrix} r & -s \\ s & r \end{bmatrix} \begin{bmatrix} 2 & -2i \\ 2i & 2 \end{bmatrix} v_2 = \begin{bmatrix} 0 \\ 0 \end{bmatrix} \qquad (10)$$

What we are left with, then, is two quadratic equations in x_2. Because the equations result from multiplication of rotation matrices, we know that $(i, -i)$ must be a solution; therefore $x_2 = -i$ must satisfy both quadratic equations in (10). With arbitrary coefficient matrices, the second root would not necessarily be the same for each quadratic equation. In the case of orthogonal-row coefficient matrices as in (10), though, we will show that the second root is identical for each quadratic, and therefore represents a solution to the tangent-of-the-half-angle system.

If we write (10) as:

$$\begin{bmatrix} f_1(x_2) \\ f_2(x_2) \end{bmatrix} = \begin{bmatrix} 0 \\ 0 \end{bmatrix} \qquad (11)$$

then simple algebra shows that $f_1(x_2) + i f_2(x_2)$ is identically 0. That is, the two quadratic equations are found to be linearly dependent. Thus, they will share both roots, and those roots will correspond to solutions to the tangent-of-the-half-angle system. So corresponding to $x_1 = i$, there will be two solutions for x_2: $-i$ and some other root in the complex field. This second root can prove particularly insidious, because the univariate polynomial in x_2 gives no indication that the second root is in fact at infinity ($x_1 = i$) [1], and thus does not represent a solution to the sine-cosine system.

The situation is exactly analogous at $x_1 = -i$. Also, we can switch the roles of x_1 and x_2 in the above arguments and find a similar situation when coefficient matrices A_1 and A_3 meet the above conditions.

If the second term in (10) happens to be zero (as would be the case if $r = s = 0$), then the above analysis still holds, but it can be shown in addition that the resulting quadratic equations in x_2 contain $(x_2 + i)^2$ as a factor. So $x_1 = i, x_2 = -i$ becomes a double point, and there will be a corresponding double point at $x_1 = -i, x_2 = i$.

2.3 Reductions in Number of Solutions

All of the coefficient specializations described thus far have resulted in solutions appearing at infinity, with either x_1 or x_2 (or both) equal to $\pm i$. The total number of solutions to the tangent-of-the-half-angle system has always remained at eight, despite the appearance of these solutions at infinity. We will now discuss specializations which cause the number of solutions to drop.

The BKK bound will prove to be vital in our discussion of these specializations. It has been our experience that the BKK bound provides a remarkably accurate root count for a variety of kinematics problems. Looking at equation (5), the specializations which lead to reductions in number of solutions involve equalities among coefficient matrices and coefficient vectors.

We begin by examining the behavior of the system when the coefficient matrices and vectors for the cross terms (A_1 and v_1) are the same as the coefficient matrices and vectors for one of the uncoupled terms (such as A_2 and v_2). Using $A_1 = A_2 = A$ and $v_1 = v_2 = v$, we rewrite (6) after combining some terms and ignoring lower order terms:

$$\begin{aligned} & A\begin{bmatrix} -x_1^2 - x_2^2 + x_1^2x_2^2 & 2x_1^2x_2 + 2x_1x_2^2 \\ -2x_1x_2^2 - 2x_1^2x_2 & -x_1^2 - x_2^2 + x_1^2x_2^2 \end{bmatrix} v \\ + & A\begin{bmatrix} -x_1^2 + x_2^2 - x_1^2x_2^2 & -2x_1x_2^2 \\ 2x_1x_2^2 & -x_1^2 + x_2^2 - x_1^2x_2^2 \end{bmatrix} v \\ + & A_3\begin{bmatrix} x_1^2 - x_2^2 - x_1^2x_2^2 & -2x_1^2x_2 \\ 2x_1^2x_2 & x_1^2 - x_2^2 - x_1^2x_2^2 \end{bmatrix} v_3 \\ & = \left(1 + x_1^2 + x_2^2 + x_1^2x_2^2\right) v_4 \end{aligned} \tag{12}$$

Because of the equality of coefficients, it is easy to see that

$$A\begin{bmatrix} -x_2^2 + x_1^2x_2^2 & 2x_1x_2^2 \\ -2x_1x_2^2 & -x_2^2 + x_1^2x_2^2 \end{bmatrix} v \tag{13}$$

from the first term in (12) will be canceled by

$$A\begin{bmatrix} x_2^2 - x_1^2x_2^2 & -2x_1x_2^2 \\ 2x_1x_2^2 & x_2^2 - x_1^2x_2^2 \end{bmatrix} v \tag{14}$$

in the second term. This cancelation means that there will be no $x_1x_2^2$ term in either polynomial equation. Moreover, when we examine the remaining terms, it becomes clear that the x_2^2 and $x_1^2x_2^2$ terms always occur together, and with the same sign. This means that these two terms will have the same coefficient within each of the two equations; these terms may thus be eliminated in *one* of the two polynomial equations.

Calculating the BKK bound given our new set of non-zero coefficients, we find that the maximum number of solutions to the polynomial system has dropped to six. Numerical examples confirm that there are indeed six solutions to the system. The roles of x_1 and x_2 could be reversed in the above argument, so that there are also six solutions if $A_1 = A_3$, and $v_1 = v_3$. If all coefficient matrices are the same, that is $A_1 = A_2 = A_3$ and $v_1 = v_2 = v_3$, then a similar analysis leads to a BKK bound of four. Numerical examples also confirm this result.

When there is equality among coefficient matrices, the analysis of section 2.2 no longer holds, because the two quadratic equations described at the end of that section become linear equations.

Simplifications, such as the ones encountered in the case of equality among coefficients, point out an important point when solving equations by suppressing one variable then calculating the resultant. If the x_2^2 and $x_1^2x_2^2$ terms were not eliminated from one of the two polynomial equations, the calculation of the resultant would have lead to extraneous solutions. This difficulty can be easily seen in the following problem: if we wanted to solve the polynomial system

$$\begin{aligned} x^2y + x + y &= 1 \\ x^2y - 3x + 4y &= 2 \end{aligned} \tag{15}$$

then resultants would be a natural choice to find all solutions. Without any modification of the system, a resultant calculation gives the following univariate polynomial in y:

$$9y^4 - 6y^3 + 29y^2 - 20y = 0 \tag{16}$$

If, however, we recognize that the term x^2y appears with the same coefficient in both equations, and we eliminate that term from *one* of the equations, then a resultant calculation (leading to the determinant of a smaller matrix than before) gives the univariate polynomial:

$$9y^3 - 6y^2 + 29y - 20 = 0 \tag{17}$$

and we now see that the original resultant calculation led to an extraneous $y = 0$ factor. It is vital, then, that we recognize all such possible simplifications.

2.4 Two-Equation Example

The results of the previous sections may be used to predict the behavior of certain nonlinear sys-

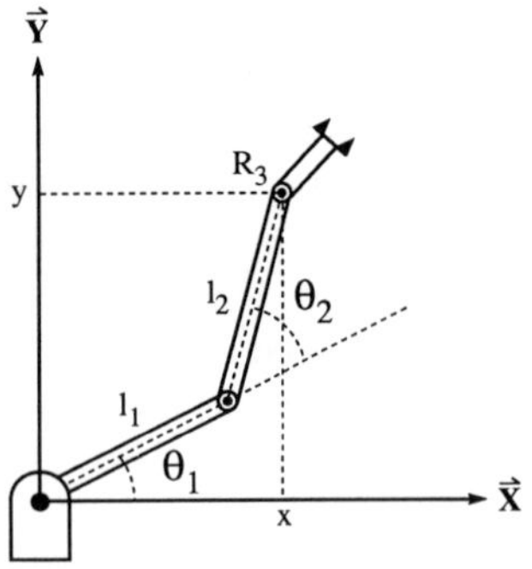

Figure 1: Three-link planar mechanism

tems in kinematics. As an example, we will look at the inverse kinematics problem for a three-link planar mechanism with three revolute joints, as shown in Figure 1. With a specified position and orientation of the end effector, we wish to determine the angles θ_1 and θ_2. Given the end effector's position and its orientation, we may easily calculate the position of the revolute joint R_3. The inverse kinematics problem, then, may be expressed as:

$$\begin{bmatrix} c_1 & -s_1 \\ s_1 & c_1 \end{bmatrix} \begin{bmatrix} c_2 & -s_2 \\ s_2 & c_2 \end{bmatrix} \begin{bmatrix} l_2 \\ 0 \end{bmatrix} + \begin{bmatrix} c_1 & -s_1 \\ s_1 & c_1 \end{bmatrix} \begin{bmatrix} l_1 \\ 0 \end{bmatrix} = \begin{bmatrix} x \\ y \end{bmatrix} \quad (18)$$

Comparing this nonlinear system to our generic-coefficient example in (2), the terms are the same except that (18) does not include linear terms in θ_2. We know, then, that (18) has a maximum of 8 solutions. Moreover, this system was created by rotation matrix multiplication, so that we know *a priori* of the existence of one set of roots to the tangent-of-the-half-angle system at $\pm i$. Finally, using the notation of (5), the A_i or "coefficient" matrices in (18), which are actually identity and zero matrices, satisfy the row-orthogonality requirements of section 2.2.

We can now make certain statements about the behavior of the sine-cosine polynomial system corresponding to the inverse kinematics problem for this manipulator. Due to the matrix multiplication, there is one set of roots to the tangent-of-the-half-angle system at $\pm i$. There is another set at $\pm i$, making those double points, because A_1 and A_3 meet the row-orthogonality condition and A_3 is a zero matrix. Lastly, there will be a pair of solutions with $x_1 = \pm i$ and x_2 in the complex field, because A_1 and A_2 meet the row-orthogonality condition. We have accounted for 6 of the 8 possible solutions to the tangent-of-the-half-angle system, leaving just 2 which may exist and be finite. These 2 may then be transferred to solutions to the sine-cosine polynomial system.

Solving the system by resultant methods confirms that these 2 finite solutions exist; they correspond, of course, to the elbow-up and elbow-down positions for the manipulator.

2.5 Solution Methods

There is a much simpler way to solve the two-equation example described above. To facilitate the creation of smaller resultants, methods have been developed in the past to create new, linearly independent equations from the original set [2]. If we square both equations in (18) and add them together, then simplify the result, we may obtain the following equation:

$$l_1^2 + l_2^2 + 2 \begin{bmatrix} l_1 & 0 \end{bmatrix} \begin{bmatrix} c_2 & -s_2 \\ s_2 & c_2 \end{bmatrix} \begin{bmatrix} l_2 \\ 0 \end{bmatrix} = x^2 + y^2 \quad (19)$$

which is linear in both $\cos\theta_2$ and $\sin\theta_2$.

After the tangent-of-the-half-angle substitution, the above equation becomes a quadratic in the variable x_2. For each root of this quadratic equation, the corresponding value of x_1 may be found to solve the original system. We see, then, that there are only 2 solutions to the new system composed of the original equations augmented by the new, linearly independent equation. By creating this new equation and adding it to the original system, we have eliminated the 6 solutions at infinity described in section 2.4.

In general, the solutions at infinity which appear after the introduction of the tangent-of-the-half-angle substitution are of very little interest. Their elimination through the introduction of modified equations, then, is highly desirable. Not only do newly introduced equations actually eliminate certain roots at infinity (by changing the ideal of the polynomial system), they also facilitate the construction of a small resultant. The elimination of solutions at infinity may prove particularly important when the univariate polynomial gives no indication that certain roots are actually at infinity.

3 Three-Equation Systems

Beyond two-equation, two-unknown systems, standardized solution methods become more difficult to apply, and the number of solutions to the problem (both finite and infinite) becomes difficult to predict. Nevertheless, it is still possible to see why certain solutions, particularly those at infinity, exist for certain three-equation, three-unknown systems. The ability to spot the conditions which lead to solutions at infinity could prove highly valuable for much larger systems.

3.1 Euler-angle rotations

In spatial kinematics, one commonly-used representation for the arbitrary rotation of a reference frame is Euler angles. Starting with a frame $\{B\}$ coincident with a reference frame $\{A\}$, we may obtain $\{B\}$ with an arbitrary orientation by first rotating $\{B\}$ about $\hat{Z}_B$ by θ_1, then rotating about $\hat{Y}_B$ by θ_2, then rotating about $\hat{X}_B$ by θ_3. The transformation matrix between $\{B\}$ and $\{A\}$ could then be calculated as follows:

$$^A_BR = \begin{bmatrix} c_1 & -s_1 & 0 \\ s_1 & c_1 & 0 \\ 0 & 0 & 1 \end{bmatrix} \begin{bmatrix} c_2 & 0 & s_2 \\ 0 & 1 & 0 \\ -s_2 & 0 & c_2 \end{bmatrix} \begin{bmatrix} 1 & 0 & 0 \\ 0 & c_3 & -s_3 \\ 0 & s_3 & c_3 \end{bmatrix} \tag{20}$$

Using this transformation, we may express the equality between two arbitrary vector in the two reference frames as follows:

$$^A_BR \; ^Bv_2 = {}^Av_1 \tag{21}$$

where $^Bv_2$ and $^Av_1$ are two arbitrary 3×1 vectors in frames $\{B\}$ and $\{A\}$ respectively. Equation (21) above represents three equations in the three unknown variables θ_1, θ_2, and θ_3. Clearly, there may only be a solution to this sine-cosine polynomial system if the lengths of vectors $^Av_1$ and $^Bv_2$ are the same, since length is invariant under rotation.

After the introduction of the tangent-of-the-half-angle substitution, though, we can show that the resulting polynomial system in fact has an infinite number of solutions, all at infinity. To see why, we again write the equations in matrix form as they would appear after the clearing of the denominators from the tangent-of-the-half-angle substitution, with the substitution $x_2 = i$:

$$\begin{bmatrix} 0 \\ 0 \\ 0 \end{bmatrix} = T_{x_1} \begin{bmatrix} 2 & 0 & 2i \\ 0 & 0 & 0 \\ -2i & 0 & 2 \end{bmatrix} \begin{bmatrix} 1+x3^2 & 0 & 0 \\ 0 & 1-x_3^2 & -2x_3 \\ 0 & 2x_3 & 1-x_3^2 \end{bmatrix} {}^Bv_2$$

$$\text{where} \quad T_{x_1} = \begin{bmatrix} 1-x_1^2 & -2x_1 & 0 \\ 2x_1 & 1-x_1^2 & 0 \\ 0 & 0 & 1+x_1^2 \end{bmatrix} \tag{22}$$

Multiplying out the final three terms on the right side of equation (22), we have:

$$\begin{bmatrix} 2+2x_3^2 & 4ix_3 & 2i-2ix_3^2 \\ 0 & 0 & 0 \\ -2i-2ix_3^2 & 4x_3 & 2-2x_3^2 \end{bmatrix} {}^Bv_2 \tag{23}$$

and we can see once again that we have two linearly dependent quadratic equations (in x_3). Provided that this quadratic equation is satisfied, then the final three terms in equation (22) go to zero and the full system may be satisfied by choosing any value for x_1. Thus, we have an infinity of solutions at infinity ($x_2 = i$). There is a corresponding infinity at $x_2 = -i$.

Once again, equation manipulation techniques prove useful in eliminating the unwanted solutions at infinity. Calculating the dot product of the vector equation in (21) with itself, we obtain one scalar equation which equates the lengths of the vectors $^Av_1$ and $^Bv_2$. If those lengths differ, then the new equation is contradictory and provides the desirable result that there are no solutions. If those lengths are the same, then the new equation is uniquely satisfied, indicating an infinity of solutions. This infinity of solutions corresponds to the free rotation of frame $\{B\}$ about the vectors $^Av_1$ and $^Bv_2$, once they have been aligned.

3.2 Three-link Spatial Manipulator

Another three-equation, three-unknown system can be found by studying the inverse kinematics of a three-link, spatial robot. Such a manipulator may achieve any position in its workspace (with unspecified orientation). The inverse kinematics problem may be expressed as follows:

$$R_z(\theta_1)\, R_x(\alpha_1)\, R_z(\theta_2)\, R_x(\alpha_2)\, R_z(\theta_3) \begin{bmatrix} l_3 \\ 0 \\ 0 \end{bmatrix} + R_z(\theta_1)\, R_x(\alpha_1)\, R_z(\theta_2) \begin{bmatrix} l_2 \\ 0 \\ 0 \end{bmatrix} + R_z(\theta_1) \begin{bmatrix} l_1 \\ 0 \\ 0 \end{bmatrix} = \begin{bmatrix} x \\ y \\ z \end{bmatrix} \tag{24}$$

where l_1, l_2, and l_3 are the link lengths, α_1, α_2, and α_3 are the link twists (in the Denevit-Hartenberg sense), and R_x and R_z are rotation matrices about the x and z axes respectively. Equation (24) above represents three equations in the three unknowns θ_1, θ_2, and θ_3.

As always, the first step in the solution process is the introduction of the tangent-of-the-half-angle substitution. Then, based on our previous results, we look for an infinity of solutions when the inner variable in the highest-power term (the first term above) is at $\pm i$. As we expect, there is in fact an infinity of solutions corresponding to $x_2 = i$; linear dependency appears between the equations, leaving x_1 free to take on any value.

In this case, however, the introduction of a new, linearly independent equation does not serve to eliminate these excess solutions at infinity. Nevertheless, it does simplify the solution to the problem. The identification of the excess solutions at infinity also tells us to pursue a univariate polynomial in either x_2 or x_3. Any resultant in x_1 would have to be identically 0, since this particular set of equations can be satisfied by any value of x_1.

4 Conclusions

Sine-cosine polynomial systems have proven to be very important in many robotics calculations. To analyze the behavior of such systems, particularly with regards to the often-used tangent-of-the-half-angle substitution, small systems for which solution methods are readily available were studied. Conditions between coefficients which reduce the number of solutions or the number of finite solutions were identified. The tangent-of-the-half-angle substitution was shown to sometimes introduce an infinity of solutions, and under certain conditions it introduced into the univariate solution polynomial deceptive spurious roots.

Solution methods which involved the creation of new equations from the original set were shown to sometimes alleviate the problems caused by the tangent-of-the-half-angle substitution. These methods changed the problem being solved and they eliminated solutions at infinity. Moreover, the new equations simplified the creation of a resultant.

5 Acknowledgment

The financial support of the National Science Foundation is gratefully acknowledged.

References

[1] P. Kovács and G. Hommel. On the tangent-half-angle substitution. In J. Angeles, G. Hommel, and P. Kovacs, editors, *Computational Kinematics*, pages 27–39. Kluwer Academic Publishers, 1993.

[2] M. Raghavan and B. Roth. Solving polynomial systems for the kinematic analysis and synthesis of mechanisms and robot manipulators. *Special 50th Anniversary Design Issue, Trans. ASME*, 117:71–79, June 1995.

[3] M. Raghavan and B. Roth. Kinematic analysis of the 6R manipulator of general geometry. In H. Miura and S. Arimoto, editors, *Robotics Research, The Fifth International Symposium*, pages 263–270. MIT Press, 1990.

[4] D. N. Bernstein. The number of roots of a system of equations. *Funct. Anal. and Appl.*, 9(2):183–185, 1975.

[5] I. Emiris. *Sparse Elimination and Applications in Kinematics.* PhD thesis, University of California at Berkeley, 1994.

PART 14

MECHANISMS AND DESIGN

SESSION SUMMARY

Marc H. Raibert
EECS Dept, Room 765, MIT AI Lab.
545 Technology Square, Cambridge, MA 02139, USA

I had the pleasure of chairing the session entitled "Mechanisms and Design". John Hollerbach, Shigeo Hirose, and Joel Burdick made presentations that both painted a broad backdrop for their areas and reported their recent findings.

John Hollerbach started the session by introducing the concept of the "Calibration Index". The calibration index is an information tool for understanding the many ways one might calibrate a given robot system. It allows one to consider diverse mechanical structures, diverse instrumentation, and diverse calibration methods, all in a common framework. The calibration index collapses the contributions made by parallel and serial linkages, externally applied constraints and fixtures, and sensors attached to the joints of the mechanical structure as well as to the endpoints, all into a single implicit equation. A key idea is that the mobility of the system, the numbers of sensors, and the calibration poses must work together to provide adequate information for calibration to succeed. The calibration index also opens the door for a more rational accommodation of what is known about relative errors in the various sources of measurement. Hollerbach and his co-author, Charles Wampler, used the calibration index to understand the natural relationships that exist between the myriad calibration methods that have been developed for real devices. Perhaps their approach will help robotics practitioners determine more easily the adequacy of proposed calibration approaches for their particular system.

Shigeo Hirose reviewed the topic of hyper-redundant robot arms, including the many innovative designs he has considered and implemented over the years. This work ranges from seven kinds of snake to trunk-like articulated structures. He then described a new kind of hyper-redundant manipulator that "defies gravity". It uses pantograph mechanisms in each link in such a way that the gravity loading seen at each joint is independent of the positions of the more peripheral links. That structure, combined with anti-gravity springs, result in an exciting new kind of robot arm. There was an animated discussion of this mechanism, its practicality, and the details. I would summarize by saying that Hirose has pulled a remarkable mechanical rabbit out of his apparently bottomless hat.

Joel Burdick presented work on the behavior of a broad class of non-holonomic systems, including locomotion systems that mix kinematic constraints and dynamic constraints. I think of non-holonomic systems as being those with fewer degrees of freedom at any instant of time than they exhibit at longer scales of time. Parallel parking of an automobile is the classic example. Instantaneously, the car can only move in the direction of the wheels, not sideways. However, if we wait a minute between observations (and there is a skilled driver parking the car), we can see the car travel sideways. The apparent number of freedoms is determined by the interval at which we observe and its relationship to the fundamental rates of the mechanism freedoms.

Burdick's paper, co-authored with Jim Ostrowski, takes this a step further. They use recent results from the applied mechanics literature on nonholonomy to develop robotic systems that take advantage of both kinematic and dynamic constraints to control their behavior. They demonstrate the power of the approach by applying it to the "Snakeboard", a mechanism with two sets of steerable wheels and a dynamic rotor. The interplay of the steering oscillation and rotor oscillation can be seen in their exceptional videotape, which shows a working snakeboard. It travels fore and aft, sideways, and turns in the plane, all depending on oscillatory accelerations of the rotor and their frequency relationship to changes in steering angle. The patterns of interacting oscillation that are so obvious in the videotape reminded me of solutions to the tower of Hanoi stacking problem, a non-holonomic manipulation task. Burdick and Ostrowski's work is a truly outstanding step in applying abstract mechanics theory to robot motion planning and control.

Geometric Perspectives on the Mechanics and Control of Robotic Locomotion

Jim Ostrowski (jpo@robby.caltech.edu) and Joel Burdick (jwb@robby.caltech.edu)
Dept. of Mechanical Engineering, California Institute of Technology, Pasadena, CA 91125

Abstract: *This paper uses geometric methods to study basic problems in locomotion. We consider in detail the case of "undulatory locomotion," which is generated by a coupling of internal shape changes to external nonholonomic constraints. Such locomotion problems can be modeled as a connection on a principal fiber bundle. The properties of connections lead to simplified results for both the dynamics and controllability of locomotion systems. We demonstrate the utility of this approach on a novel "Snakeboard" and a multi-segmented serpentine robot which is modeled after Hirose's ACM.*

1 Introduction and Motivation

A large body of research has developed in the area of robotic locomotion, since mobility is an important capability for autonomous systems. Most mobile robots are wheeled vehicles, since wheels provide the simplest means for mobility. The assumption that these wheels do not slip provides nonholonomic kinematic constraints on a vehicle's motion, and these kinematic nonholonomic systems have been extensively studied [18, 21, 25, 31].

Robotics researchers have also considered legs and/or tracks to generate robot movement. Multi-legged quasi-static locomotion has been the most extensively studied means of legged locomotion, and several quadrupedal and hexapedal robots have been successfully developed and demonstrated [8, 33]. Beginning with Raibert [30], hopping robots have received considerable attention [16, 22, 34]. Bipedal walking and running has also been an active area of study [9, 14, 20, 23]. Other researchers have investigated various forms of "snake-like" locomotion [4, 7, 12], as snake-like robots can potentially access environments, such as tunnels and pipes, that are inaccessible to legged or wheeled vehicles.

In an attempt to derive strong results for specific examples, prior locomotion studies have focused either on a particular set of locomotion assumptions (such as the quasi-static assumption) or on a particular robot morphology (such as a biped). However, results derived for one morphology may not extend to other morphologies.

This paper introduces a unifying approach for the study of *undulatory locomotion.*

> **Definition:** Undulatory locomotion *is the process of generating net displacements of a robotic mechanism via a coupling of internal deformations to an interaction between the robot and its environment.*

Common biological examples of undulatory locomotion include worms, snakes, amoeba, and fish. In this paper, we limit these interactions to those modeled by nonholonomic kinematic constraints. This restriction allows us to model a rich class of systems and results in enough structure to make the problem tractable. We believe that this framework can ultimately be extended to a very large class of locomotory systems.

This paper shows that many locomotion problems can naturally be described as a *connection* on a *principal fiber bundle.* Second, we introduce a specialized form of the dynamical equations for mechanical systems with Lagrangian symmetries and nonholonomic constraints. Third, we show how these results lead to a simple and appealing insight into undulatory locomotion. Finally, we show that the framework is a superset of prior work on the mechanics of wheeled nonholonomic vehicles and free-floating satellites. The ideas are illustrated by two examples: a "snakeboard" and a serpentine robot which is roughly modeled on Hirose's Active Cord Mechanism.

2 Relation to Previous Work

While there is a vast literature on robotic locomotion, very few works have attempted to uncover the underlying mathematical structure that is common to all locomotion systems.

One of the earliest relevant works is that of Shapere and Wilczek [32] who studied the movement of small organisms through a highly viscous fluid. Using gauge theory (which is the language used by physicists for a connection), they have shown how the organism's net displacement results from cyclical changes in the shape of the body.

More recently, Chirikjian and Burdick coined the term *hyper-redundant* to describe robots with a very large number of degrees of freedom, such as snake-like robots. Chirikjian and Burdick considered locomotion schemes that were reminiscent of the sidewinding [4] and creeping [7] gaits of snakes. They also considered gaits that are analogous to those of inchworms and earthworms.

Kelly and Murray [15] have modeled a number of locomotive systems, such as idealized inchworms and sidewinding snakes, using kinematic constraints. They show how these systems can be modeled using a *principal kinematic connection* (which is described below). They also provide results on controllability, as well as an interpretation of movement in terms of geometric phases.

Tsakiris and Krishnaprasad [17] have investigated the movement of Variable Geometry Truss (VGT) mechanisms which employ no-slip wheel constraints. They term these models "G-snakes" since each mechanism segment must move within a subset of a Lie group, G. Hence, the mechanisms' configuration space is a principal fiber bundle, and the kinematic constraints define a principal kinematic connection.

Many locomoting mechanisms use wheels to provide nonholonomic kinematic constraints on the robot's motion. These nonholonomic systems have largely been treated as purely *kinematic* systems, i.e. their dynamics are constrained in a manner such that only configuration *velocities* need be considered. This assumption has led to some excellent progress in areas such as controllability [2] stabilization [6] and trajectory generation [26]. The framework outlined below includes all of these systems as a special case. Dynamic effects are *essential* to the motion of some systems. In one class of systems, momentum conservation laws are treated as a type of *internal* nonholonomic constraint. These problems primarily arise in rigid body reorientation, such as the spinning satellite [5, 27, 35] and the falling cat [24]. There are a number of systems, such as the snakeboard example that is considered below [19], where *both* kinematic constraints and symmetry (dynamic) constraints come into play. These *mixed* nonholonomic systems have rarely been treated in the literature. Of notable exception is Ref. [2], where control results were established for the restrictive assumption that the unconstrained directions are fully actuated, and Ref. [3] which has been a valuable foundation for many of the results presented here.

One of the earliest studies of snake-like robots was conducted by Hirose [10, 12]. Hirose formulated a *serpenoid* curve—a curve representing the path that a snake would trace out as it slithers forward. Hirose showed that a snake-like vehicle could generate a net forward force by applying torques along the length of its body in the manner dictated by the serpenoid curve. To demonstrate these results, he built a snake-like robot capable of propelling itself forward using only internal torques. In the ensuing sections, we develop a framework for understanding and controlling such systems.

3 Background

It is always possible to divide a locomoting robot's configuration variables into two classes. The first set of variables describes the *position* of the robot, which is the displacement of a coordinate frame attached to the moving robot with respect to a fixed reference frame. The set of frame displacements is $SE(m), m \leq 3$, or one of its subgroups—i.e., a Lie group. The second class of variables defines the internal configuration, or *shape*, of the mechanism. We only require that the set of all possible shapes (the "shape space") be described by a manifold, M. The total configuration space is $Q = G \times M$. The shape and position variables are coupled by the constraints acting on the robot. Hence, by making changes in the shape variables, it is possible to effect changes in the position variables through the constraints. The relationship between shape changes and position changes can be described by a *connection*.

Let's first consider the conventional Lagrangian approach for studying the dynamics of locomotion systems. We assume the existence of a Lagrangian function, $L(q, \dot{q})$, on TQ and k constraints which are linear in velocities:

$$\omega^i_j(q)\dot{q}^j = 0, \qquad \text{for} \quad i = 1 \dots k, \quad j = 1, \cdots, n. \tag{1}$$

This class of constraints includes most commonly investigated nonholonomic constraints. In conventional engineering mechanics, the constraints are incorporated into Lagrange's equations through the use of Lagrange multipliers, λ:

$$\frac{d}{dt}\left(\frac{\partial L}{\partial \dot{q}^i}\right) - \frac{\partial L}{\partial q^i} + \lambda_j \omega^j_i - \tau_i = 0. \quad \text{for } i = 1, \cdots, n; \; j = 1, \cdots, k \tag{2}$$

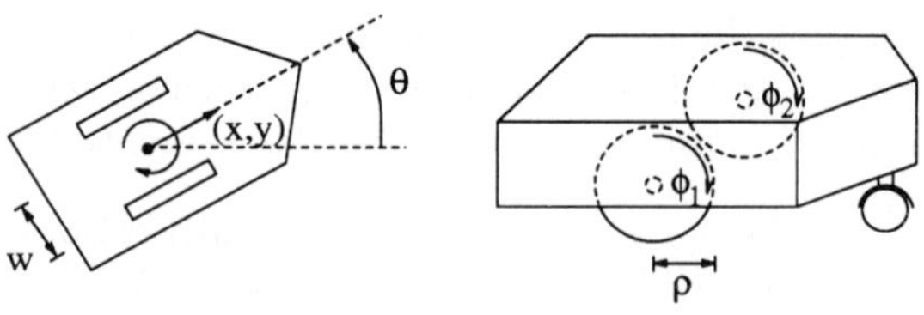

Figure 1: Two wheeled planar mobile robot.

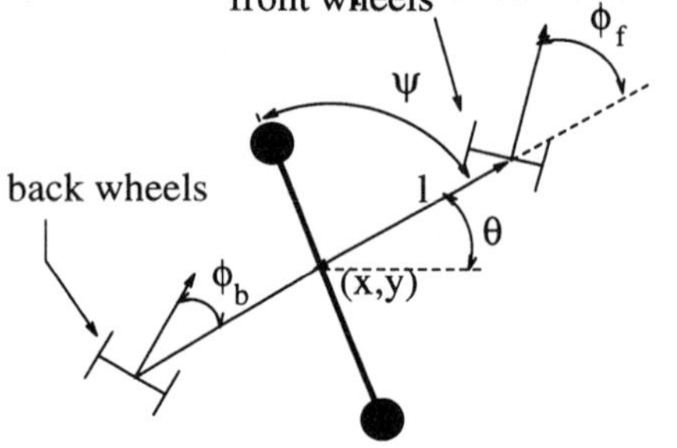

Figure 2: The simplified model of the Snakeboard

where τ is a forcing function.

Example #1: Consider the two wheeled planar mobile robot shown in Fig. 1. The robot's position, $(x, y, \theta) \in SE(2)$, is measured via a frame located at the center of the wheel base. The wheels' angles, (ϕ_1, ϕ_2), are measured relative to vertical. Each wheel is assumed to rotate independently and without slipping. The configuration space is $Q = SE(2) \times (\mathcal{S}^1 \times \mathcal{S}^1)$. The Lagrangian for this problem is: $L = \frac{1}{2}m(\dot{x}^2+\dot{y}^2)+\frac{1}{2}J\dot{\theta}^2+\frac{1}{2}J_w(\dot{\phi}_1^2+\dot{\phi}_2^2)$, where m is the mass of the robot, J is its inertia, and J_w is the inertia of each of the wheels. The constraints defining the no-slip condition can be written as in Eq. (1). However, we will rearrange the constraints into the more revealing form:

$$\begin{bmatrix} \cos\theta & \sin\theta & 0 \\ -\sin\theta & \cos\theta & 0 \\ 0 & 0 & 1 \end{bmatrix} \begin{bmatrix} \dot{x} \\ \dot{y} \\ \dot{\theta} \end{bmatrix} = \begin{bmatrix} \frac{\rho}{2}(\dot{\phi}_1 + \dot{\phi}_2) \\ 0 \\ \frac{\rho}{2w}(\dot{\phi}_1 - \dot{\phi}_2) \end{bmatrix} \tag{3}$$

The motion in the group variables, $(\dot{x}, \dot{y}, \dot{\theta})$, is strictly a function of the motion in the internal shape variables, $(\dot{\phi}_1,\ \dot{\phi}_2)$, since there are three independent kinematic constraints on the 3-dimensional set of body displacements. If we assume that the base variables are controllable, then given the time evolution of (ϕ_1, ϕ_2), we can completely solve for the robot's motion using Eq. (3). Hence, Eq. (3) describes the relationship between the robot's "internal" motions and its net movement.

Example #2: The *Snakeboard* [19, 28] is a variant of the skateboard in which the passive wheel assemblies can pivot freely about a vertical axis. By coupling a twisting of the human torso with a turning of the wheels, a rider can generate a snake-like locomotion pattern *without* having to kick off the ground. A simplified model that captures the essential behavior is shown in Fig. 2. It consists of a rigid body connecting the two sets of wheels. The angle of the wheel axles can be independently specified. A momentum wheel rotates about a vertical axis through the center of mass. By a proper choice of the relative frequency and phasing of the main rotor rotation with the wheel axles' rotations, the snakeboard can move in any direction. Fig. 3 shows a robotic snakeboard prototype.

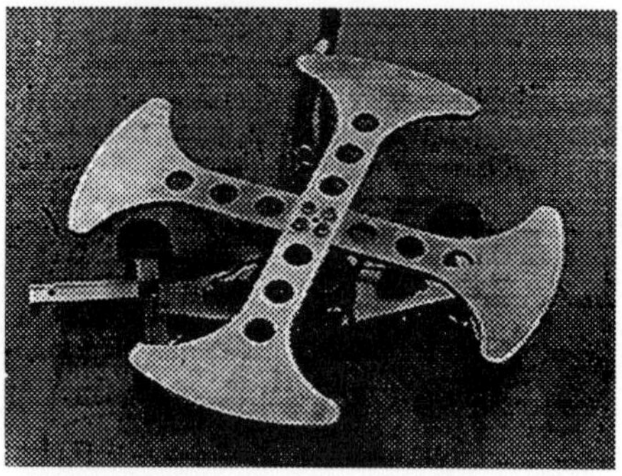

Figure 3: A prototype robotic snakeboard

The snakeboard's position variables, $(x, y, \theta) \in SE(2)$, are determined by a frame affixed to its center of mass. The shape variables are (ψ, ϕ_b, ϕ_f), and so $Q = G \times M = SE(2) \times \mathcal{S}^1 \times \mathcal{S}^1 \times \mathcal{S}^1$. The Lagrangian is

$$\begin{aligned} L &= \tfrac{1}{2}m(\dot{x}^2 + \dot{y}^2) + \tfrac{1}{2}\hat{J}\dot{\theta}^2 + \tfrac{1}{2}J_r(\dot{\psi} + \dot{\theta})^2 \\ &\quad + \tfrac{1}{2}J_w((\dot{\phi}_b + \dot{\theta})^2 + (\dot{\phi}_f + \dot{\theta})^2) \end{aligned} \tag{4}$$

where J is the snakeboard body inertia, J_r is the rotor inertia, J_w is the inertia of the wheels about a vertical axis, and $\hat{J} = J + J_r + 2J_w$. Control torques at the rotor and wheel axles are assumed, so $\tau = (0, 0, 0, \tau_\psi, \tau_b, \tau_f)$. The assumption that the wheels do not slip in the direction of the wheel axles determines two constraints of the form of Eq. (1):

$$\begin{aligned} -\sin(\phi_f + \theta)\dot{x} + \cos(\phi_f + \theta)\dot{y} + l\cos(\phi_f)\dot{\theta} &= 0 \\ -\sin(\phi_b + \theta)\dot{x} + \cos(\phi_b + \theta)\dot{y} - l\cos(\phi_b)\dot{\theta} &= 0. \end{aligned} \tag{5}$$

There are not enough kinematic constraints to uniquely define the Snakeboard's motion. The technique that was employed in Example #1 to use the kinematic constraints to solve for the robot's motion as a function of shape changes is therefore no longer viable. The snakeboard's dynamics must come into play. To determine the robot's dynamics, engineers have traditionally been relegated to substituting the Lagrangian and the constraints into Eq. (2) and explicitly solving for the Lagrange multipliers. There are many drawbacks to this approach. First, the system is equivalent to $2n + k$

first order differential equations—i.e., 14 first order coupled equations in the case of the Snakeboard. Second, when eliminating the Lagrange multipliers, we do not have a relationship, such as Eq. (3), which readily shows the effect of shape changes on robot motion. We now present an alternate approach that makes use of the inherent structure found in problems of locomotion.

4 The Mechanics of Locomotion

The concept of conservation of linear and angular momentum is well known. We revisit the basis for this theory in a geometric setting, as this will lead to a new result of great use for locomotion analysis.

If the robot's initial body fixed frame position is denoted by h (e.g. h is a homogeneous transformation matrix), and it is displaced by an amount g, then its final position is gh. This displacement can be thought of as a map $L_g : G \to G$ given by $L_g(h) = gh$, and is termed a *left translation*. The left translation induces a *left action* of G on Q.

Definition: A *left action* of a Lie group G on a manifold Q is a map $\Phi : G \times Q \to Q$ such that: (1) $\Phi(e, q) = q$ for all $q \in Q$, and e the identity element of G; and (2) $\Phi(h, \Phi(g, q)) = \Phi(hg, q)$ for every $g, h \in G$ and $q \in Q$. The left action can be viewed as a map from Q into Q, with the element $g \in G$ held fixed. Notationally, $\Phi_g : Q \to Q$ is given by $(s, r) \mapsto (\Phi(g, s), r)$. The *lifted action*, which describes the effect of Φ_g on velocity vectors in TQ, is the linear map, $D_q\Phi_g : T_qQ \to T_qQ$.

Recall the division of the configuration space, $Q = G \times M$. Such a configuration space is termed a *trivial principal fiber bundle.*

Definition: Let M be a manifold and G a Lie group. A *trivial principal fiber bundle* with *base* M and *structure group* G consists of the manifold $Q = G \times M$ together with the free left action of G on Q given by left translation: $\Phi_g(s, r) = (gs, r)$ for $r \in M$ and $g, s \in G$. Given a point $(s, r) \in G \times M = Q$, we define the natural projections $\pi_1 : Q \to G : (s, r) \mapsto s$ and $\pi_2 : Q \to M : (s, r) \mapsto r$.

We say that Q is a *fiber bundle* with *fibers* G and *base space* M. Q is "trivial" because the product structure is global, and it is a "principal" bundle because the fiber is a Lie group. The additional structure arising from the Lie group component (which is common to *all* locomotion systems) is very important for the ensuing developments.

Example #2 *(cont'd)*. The snakeboard's configuration space is $Q = SE(2) \times (\mathcal{S}^1 \times \mathcal{S}^1 \times \mathcal{S}^1)$, and a configuration is denoted $q = (x, y, \theta, \psi, \phi_b, \phi_f)$. The action is the left action of $SE(2)$ on itself. Given $g = (a_1, a_2, \alpha) \in SE(2)$, the action and lifted action are:

$$\Phi_g(q) = \begin{pmatrix} x\cos\alpha - y\sin\alpha + a_1 \\ x\sin\alpha + y\cos\alpha + a_2 \\ \theta + \alpha \\ \psi \\ \phi_b \\ \phi_f \end{pmatrix} \tag{6}$$

$$D_q\Phi_g(v_q) = \begin{pmatrix} v_x\cos\alpha - v_y\sin\alpha \\ v_x\sin\alpha + v_y\cos\alpha \\ v_\theta \\ v_\psi \\ v_b \\ v_f \end{pmatrix} \tag{7}$$

where $v_q = (v_x, v_y, v_\theta, v_\psi, v_b, v_f)$.

Associated with a Lie group, G, is its Lie algebra, denoted $\mathfrak{g}$. The Lie algebra can be identified with T_eG and locally generates G via the exponential mapping, $\exp : \mathfrak{g} \to G$ (see [1]). The exponential mapping also associates with each $\xi \in \mathfrak{g}$ a vector field on G, and by extension on $Q = G \times M$, called the *infinitesimal generator*, ξ_Q, given by

$$\xi_Q(q) = \frac{d}{ds}(\Phi(\exp(s\xi), q))|_{s=0}. \tag{8}$$

Each infinitesimal generator is tangent to the fiber, and the set of all such vectors at $q \in Q$ forms the *vertical* subspace of T_qQ,

$$V_qQ = \{\xi_Q(q) \in T_qQ |\ \xi \in \mathfrak{g}\}. \tag{9}$$

That is, a vector $v_q \in V_qQ$ represents a net velocity of the body with respect to an inertial frame, such that $v_q = (v_g, 0)$ with $v_g \in T_gG$.

Example #2 *(cont'd)*: The Lie algebra for $SE(2)$ is denoted $se(2)$, and the relationship between an element, $\xi = (a_1, a_2, \alpha) \in se(2)$, and the corresponding infinitesimal generator is

$$\begin{aligned} &\xi_Q(x, y, \theta, \psi, \phi_b, \phi_f) \\ &\quad = (a_1 - y\alpha, a_2 + x\alpha, \alpha, 0, 0, 0). \end{aligned} \tag{10}$$

The vertical subspace is given trivially by $TG \times 0$:

$$\begin{aligned} V_qQ &= \{(v_q, w_q) \in T_{\pi_1(q)}G \times T_{\pi_2(q)}M | w_q = 0\} \\ &= \text{span}\{\tfrac{\partial}{\partial x}, \tfrac{\partial}{\partial y}, \tfrac{\partial}{\partial \theta}\}. \end{aligned} \tag{11}$$

4.1 Mechanics with Symmetries

Conservation laws naturally arise when a Lagrangian remains invariant under the action of a Lie group, as stated in Noether's theorem [1]:

Theorem: Let $L(q,\dot{q})$ be a Lagrangian which is invariant under the action of a Lie group, G, (i.e., $L(\Phi_g(q), D_q\Phi_g v_q) = L(q, v_q)\ \forall g \in G$, for all $v_q \in T_qQ$). Then, for all curves $c(t) : [a,b] \to Q$ satisfying Lagrange's equations:

$$\frac{d}{dt}\left\langle \frac{\partial L}{\partial \dot{q}}(\dot{c}(t)); \xi_Q(c(t))\right\rangle = 0 \tag{12}$$

for all $\xi \in \mathfrak{g}$. Equivalently, $\dot{p} = 0$, where $p =< \frac{\partial L}{\partial \dot{q}}; \xi_Q >$ is the *generalized momentum.*

The invariance of the Lagrangian with respect to a Lie group is termed a *symmetry.* When G is $SE(2)$ or $SE(3)$, Noether's theorem is equivalent to conservation of linear and angular momentum. However, undulatory locomotion relies upon an interaction with the environment. Unfortunately, conservation laws *may not be preserved* in the presence of constraints. The next section describes a recent extension to the classical theory that combines symmetries with constraints.

4.2 Symmetries and Constraints

Let the *constraint distribution* be the set of all velocities that satisfy the kinematic constraints):

$$\mathcal{D}_q = \{v_q \in T_qQ \mid \omega^i_j v^j_q = 0,\ i = 1,\ldots,k\}. \tag{13}$$

Let $\mathcal{S}$ denote the *constrained fiber distribution,* which is given at each q by the intersection of the constraint distribution, $\mathcal{D}_q$, with the vertical subspace, V_qQ:

$$\mathcal{S}_q = \mathcal{D}_q \cap V_qQ. \tag{14}$$

If the constraints act *vertically* (i.e. $\mathcal{S}$ is nonempty), then we have the following [3]:

Proposition #1: Let L and $\mathcal{D}$ define a constrained system on $Q = G \times M$ whose Lagrangian is G-invariant. If $c(t)$ is a curve which satisfies the Lagrange-d'Alembert equations (Eq. 2) for a system with nonholonomic constraints (Eq. 1), then the following *generalized momentum equation* holds for all vector fields, $\xi^c_Q \in \mathcal{S}$:

$$\frac{d}{dt}p^c = \frac{\partial L}{\partial \dot{q}^i}\left(\frac{d}{dt}[\xi^c(c(t))]\right)^i_Q + \tau_i\left(\xi^c(c(t))\right)^i_Q \tag{15}$$

where

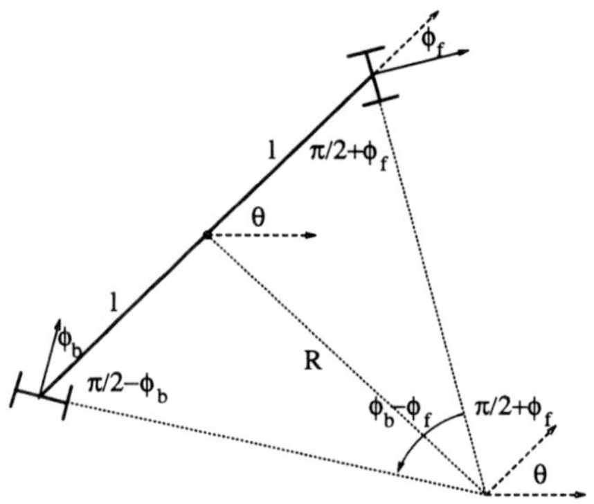

Figure 4: The snakeboard's instantaneous center of rotation

$$p^c = \frac{\partial L}{\partial \dot{q}^i}(\xi^c(c(t))^i_Q \tag{16}$$

is the *constrained,* or *nonholonomic,* momentum. The generalized momentum equation can always be factored into the quadratic form [29]: $\dot{p} = \frac{1}{2}\dot{r}^T\sigma_{\dot{r}\dot{r}}(r)\dot{r} + p^T\sigma_{p\dot{r}}(r)\dot{r} + \frac{1}{2}p^T\sigma_{pp}(r)p$.

Prop. #1 states that in the presence of constraints, momentum-like quantities exist, but may not be conserved. Eq. (15), determines how these momentum-like quantities evolve. *The nonconservation of these momentum-like quantities is the key to dynamic undulatory locomotion* (and many other forms of locomotion). It describes why the snakeboard can build up momentum, even though no external forces act on the system.

Example #2 *(cont'd).* The snakeboard Lagrangian, Eq. (4), is invariant with respect to an $SE(2)$ group action. The wheel constraints of Eq. (5) can be expressed as a constraint distribution:

$$\mathcal{D}_q = \text{span}\{a\frac{\partial}{\partial x} + b\frac{\partial}{\partial y} + c\frac{\partial}{\partial \theta}, \frac{\partial}{\partial \psi}, \frac{\partial}{\partial \phi_b}, \frac{\partial}{\partial \phi_f}\}, \tag{17}$$

where

$$\begin{aligned} a &= -l[\cos\phi_b\cos(\phi_f+\theta) + \cos\phi_f\cos(\phi_b+\theta)] \\ b &= -l[\cos\phi_b\sin(\phi_f+\theta) + \cos\phi_f\sin(\phi_b+\theta)] \\ c &= \sin(\phi_b - \phi_f). \end{aligned} \tag{18}$$

The vertical distribution was defined in Eq. (11), and the constrained fiber distribution is: $\mathcal{S}_q = \mathcal{D}_q \cap V_qQ = \text{span}\{a\frac{\partial}{\partial x} + b\frac{\partial}{\partial y} + c\frac{\partial}{\partial \theta}\}$. $\mathcal{S}_q$ corresponds to instantaneous rotations of the snakeboard about the point where the two wheel axes intersect (Fig. 4). The *nonholonomic momentum* of Eq. (16) is:

$$\begin{aligned} p^c &=< \tfrac{\partial L}{\partial \dot{q}}; (\xi^c)_Q(q) > \\ &= (mR^2 + \hat{J}c)\dot{\theta} + J_r c\dot{\psi} + J_w c(\dot{\phi}_b + \dot{\phi}_f), \end{aligned} \tag{19}$$

R is the radius from the instantaneous center of rotation to the snakeboard's center of mass. Thus, *p^c corresponds to the snakeboard's angular momentum about the instantaneous center of rotation.* If the front and back wheels were fixed, this momentum would be conserved, as the fixed wheels provide a holonomic constraint of rotation about the fixed center. However, since the constraints are variable, the snakeboard's momentum can be altered by internal forces.

4.3 The Connection and Locomotion

We now introduce a key concept in the theory of principal fiber bundles which also plays an important role in locomotion analysis.

Definition 1 *A* connection *is an assignment of a* horizontal *subspace, $H_qQ \subset T_qQ$, for each point $q \in Q$ such that H_qQ depends smoothly on q and:*

(1) $T_qQ = V_qQ \oplus H_qQ$;

(2) $D_q\Phi_g H_qQ = H_{g\cdot q}Q$.

Condition (1) implies that T_qQ can everywhere be divided into a vertical subspace, V_qQ, and a complementary (but not necessarily orthogonal) subspace, H_qQ, which is termed the horizontal space. Connections are useful because of the following fact. The horizontal subspace defined by the connection is everywhere isomorphic to the tangent space of the base: $H_qQ \simeq T_{\pi_2(q)}M$. The *horizontal lift* is the isomorphism which maps vectors in $T_{\pi_2(q)}M$ to the corresponding lifted vectors in $H_qQ \subset T_qQ$. The horizontal lift determines the relationship between motion in the base space (i.e., tangent vectors in $T_{\pi_2(q)}M$) and motion in the total space, Q, where locomotion is effected. We will choose H_qQ to encode the action of the constraints. Hence, *the connection will determine how internal shape changes create net robot motion.*

A connection on Q is alternatively described by a Lie algebra-valued one-form termed the *connection form.* That is, the connection form is a mapping $\mathcal{A}(q) : T_qQ \to \mathfrak{g}$, where $\mathfrak{g}$ is the Lie algebra of G. The connection form has the properties

(1) $\mathcal{A}(q) \cdot \xi_Q = \xi$ for $\xi \in \mathfrak{g}$,

(2) $\mathcal{A}(\Phi_g q) \cdot D_q\Phi_g \dot{q} = Ad_g \mathcal{A}(g) \cdot \dot{q}$.

Condition (1) implies that $\mathcal{A}(q)$ takes infinitesimal generators to their associated Lie algebra elements. The connection form vanishes on horizontal vectors: $\mathcal{A}(q)\dot{q} = 0$ for all $\dot{q} \in H_qQ$. Hence, H_qQ can alternatively be defined as the set of tangent vectors upon which the connection form vanishes.

For the case of mixed constraints, we will define H_qQ as $H_qQ = \{v_q \in T_qQ |\ v_q \in \mathcal{D}_q$ and $\ll v_q, w_q \gg = 0\ \forall w_q \in \mathcal{S}\}$, where $\ll \cdot,\cdot \gg$ is the inner product with respect to the kinetic energy metric. In other words, horizontal vectors satisfy both the kinematic constraints and the symmetry constraints. If $\dot{q}$ is a horizontal velocity vector, with $\dot{q} = (\dot{g}, \dot{r})$, then it can be shown that the connection form can be expressed as:

$$g^{-1}\dot{g} = -A(r)\dot{r} + I(r)^{-1}p, \tag{20}$$

where $A(r)$ is the "local form" of the connection, and $I(r)$ is the *locked inertia tensor.* $I(r)$ describes the total inertia of the system when all joints are frozen at configuration r. The connection plays the most central role in the mechanics of locomotion, since it determine the robot's motion (note that $g^{-1}\dot{g}$ is the velocity of the robot's reference frame) as the combination of built-up momentum, p and internal shape changes, $\dot{r}$.

In summary, we have reduced the system of n second order ODE's with k first order constraints (Eqs. (1), (2)) to a system of $\sigma = dim(G)$ first order constraints (the connection form), $\sigma - k$ first order generalized momentum equations, and $dim(M)$ second order equations on the base space (termed the "reduced" dynamics):

$$\begin{aligned} g^{-1}\dot{g} &= -A(r)\dot{r} + I(r)^{-1}p, \\ \dot{p} &= \tfrac{1}{2}\dot{r}^T\sigma_{\dot{r}\dot{r}}(r)\dot{r} + p^T\sigma_{p\dot{r}}(r)\dot{r} + \tfrac{1}{2}p^T\sigma_{pp}(r)p, \\ M(r)\ddot{r} &= -C(r,\dot{r}) + N(r,\dot{r},p) + \tau. \end{aligned} \tag{21}$$

The reduced dynamics are not important for understanding locomotion, though they may be important in the synthesis of a controller, as they describe the dynamics of the mechanism's internal shape deformations. Thus, the equations which are important for understanding locomotion are reduced to two first order equations that make explicit how shape changes lead to robot motion.

Example *(concl):* The snakeboard's connection and generalized momentum equation take the form of Eq. (21), where:

$$A(r) = I^{-1}(r)f(r) \tag{22}$$

$$I^{-1}(r) = \frac{1}{W(r)} \cdot \begin{bmatrix} 2l\cos\phi_b\cos\phi_f \\ l\sin(\phi_b + \phi_f) \\ -\sin(\phi_b - \phi_f) \end{bmatrix} \tag{23}$$

$$\dot{p}^c = \frac{1}{2}\frac{d}{dt}\left(\log\left((mR^2 + \hat{J})c\right)\right) \cdot (p^c - f(r)\dot{r}) + \dot{f}(r)\dot{r}, \tag{24}$$

where $W(r) = \hat{J}\sin^2(\phi_b - \phi_f) - ml^2\sin^2(\phi_b + \phi_f) - 4ml^2\cos^2\phi_b\cos^2\phi_f$, and $f(r) = \sin(\phi_b - \phi_f)(J_r, J_w, J_w)$. This approach leads to a very concise formulation of the important dynamical relationships that govern the snakeboard's movement.

We now briefly consider special cases of these equations in order to show that Eq. (21) includes many previously studied problems as a special case.

Kinematic Constraints. With a sufficient number of kinematic constraints, the group symmetries are annihilated, and the system's motion is fully determined by the constraints. In this *principal kinematic case*, the connection takes the form:

$$g^{-1}\dot{g} = A(r)\dot{r}. \tag{25}$$

which we term a *kinematic connection*. Many wheeled vehicles, such as Example #1, fall into this category, and have been extensively studied in the literature on nonholonomic systems.

Symmetry Constraints. There are no kinematic constraints in the cases of falling cats or platform divers. Yet, these systems have inherent Lagrangian symmetries, and therefore conserved momenta. In these cases, the connection and generalized momentum equations take the form:

$$\begin{aligned} g^{-1}\dot{g} &= A(r)\dot{r} + I^{-1}(r)\mu \\ \dot{\mu} &= 0. \end{aligned} \tag{26}$$

μ is the generalized momenta, and is constant. If $\mu = 0$, these equations take the form of Eq. (25).

5 Controllability

A locomotion mechanism is *controllable* if given any initial configuration q_i and final configuration q_f, there exists an admissible control law which drives the system from q_i to q_f. Controllability is an important issue for several reasons. First, when designing a locomoting robot, we would like to know if it can reach all possible positions in its environment. Second, Section 6 will briefly discuss the relationship between "gaits" and certain controllability calculations. Surprisingly, *controllability can be determined via the connection and the generalized momentum equation.* This result further illustrates the usefulness of these equations as tools for locomotion analysis.

To express the system of equations in the standard control form of $\dot{z} = f(z) + h_1(z)u_1 + \cdots + h_m(z)u_m$, we must dynamically extend the control inputs by redefining them as higher derivatives of the input variables. Equivalently, the accelerations are the control inputs. Letting $v = \dot{u}$, and using $z = (g, p, r, \dot{r})$, the connection and generalized momentum equations can be put in the standard control form with:

$$f(z) = \begin{bmatrix} g(-A\dot{r} + I(r)^{-1}p) \\ \frac{1}{2}\dot{r}^T\sigma_{\dot{r}\dot{r}}\dot{r} + p^T\sigma_{p\dot{r}}\dot{r} + \frac{1}{2}p^T\sigma_{pp}p \\ \dot{r} \\ 0 \end{bmatrix} \tag{27}$$

$$h_i(z) = [0 \quad 0 \quad 0 \quad e_i^T]^T \tag{28}$$

where e_i is the m-vector ($m = \dim M$) with a "1" in the ith row and "0" otherwise.

There are two commonly used notions of control—*accessibility* and *controllability* (see [13] for descriptions of these concepts). For nonlinear systems, *small-time local controllability* is the closest that we can come to our intuitive notion of controllability. Accessibility is determined by computing the *accessibility distribution*. Let $\Delta_0 = span\{f, h_1, \ldots, h_m\}$, and iteratively define the distributions

$$\Delta_k = \Delta_{k-1} + span\{[X, Y] \mid X, Y \in \Delta_{k-1}\}. \tag{29}$$

This nondecreasing sequence of distributions terminates at some k_f under certain regularity conditions. Δ_{k_f} is the accessibility distribution. A system is *accessible* if $\dim \Delta_{k_f} = \dim T_zN$ for all $z \in N$. It can be shown [29] that the system in Eq. (27) is small-time locally controllable from equilibrium points if: (1) the accessibility distribution is full rank; and (2) $\sigma_{\dot{r}\dot{r}}$ is an onto map; and (3) $(\sigma_{\dot{r}\dot{r}})_{ii} = 0$ for $i = 1, \ldots, m$.

While these controllability results are useful for the analysis of locomotion systems, the next section shows that there is a more fundamental reason to undertake this controllability analysis.

6 Geometric Phases and Gaits

The snakeboard moves by coupling periodic motions of the rotor and wheel axles. For example, Fig. 5 shows the position of the snakeboard's center of mass versus time for the case in which the rotor and wheel axles oscillate with the same frequency (which we term the "drive gait"), the case in which the axles oscillate at twice the frequency of the rotor (the "rotate gait," as the robot essentially rotates in place), and the case in which the

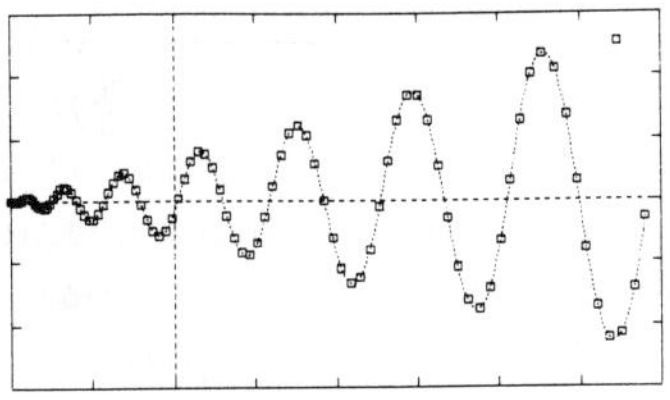

Figure 5: Center of mass position (drive gait)

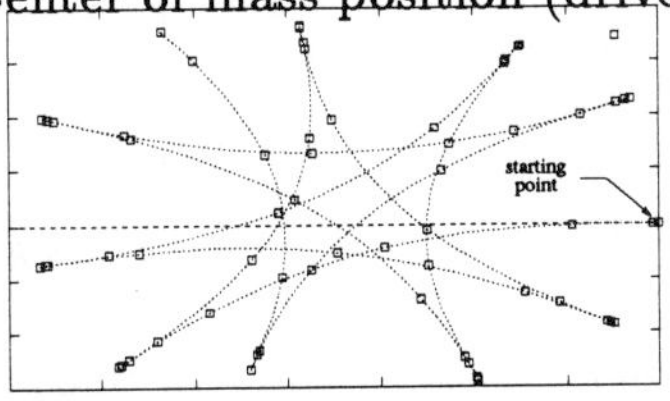

Figure 6: Center of mass position (rotate gait)

axles oscillates three times for every two oscillations of the rotor (the "parallel parking gait"). Intuitively, a gait is a cyclic pattern of internal shape changes that results in a net robot displacement. Different gaits correspond to different cyclic input patterns. In the principal kinematic case, the net displacement of the mechanism that arises from periodic inputs is the *geometric phase*, or holonomy, associated with the connection. In the mixed case, the net displacement is a combination of the geometric phase and the *dynamic phase.*

We do not yet have a complete geometric understanding of the notion of gaits. However, we have found that different gaits can be associated with the brackets that are computed for the controllability test in Section 5. For the snakeboard, let $\alpha_\phi = [h_\phi, f]$—the bracket between the control vector field associated with the turning of the wheel axles and the drift vector field. Similarly, let $\alpha_\psi = [h_\psi, f]$. These two brackets loosely correspond to an integration of the control torques to produce velocity controls. For the drive gait, which employs a 1-to-1 frequency ratio of rotor and axle oscillations, the bracket $[\alpha_\phi, \alpha_\psi]$ has non-zero terms only in the $\dot{x}$ component, which corresponds to forward motion of the vehicle. Similarly, the rotate gait employs a 2-to-1 frequency ratio of the axle and rotor oscillations. The two-level bracket $[\alpha_\phi, [\alpha_\phi, \alpha_\psi]]$, where α_ϕ appears two times, has non-zero terms only in the $\dot{\theta}$ direction, and thus corresponds to the rotation gait. Thus, the controllability analysis of Section 5 has been useful in identifying candidate gaits.

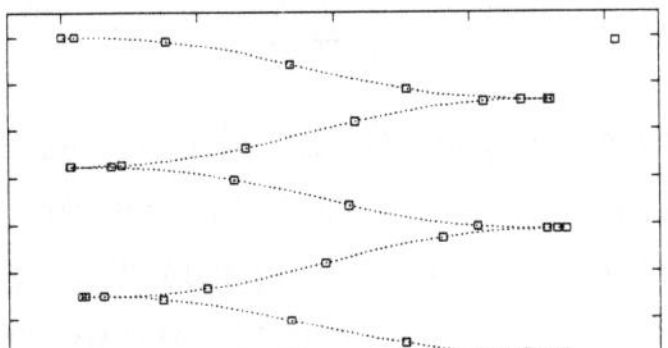

Figure 7: Position of the center of mass for the "parallel parking gait"

7 The Kinematic Snake Model

To demonstrate the utility of this approach, let's consider a "snake-like" system which is modeled after Hirose's Active Cord Mechanism (ACM-III) [11]. Fig. 8 shows a schematic of three segments of the mechanism. Each segment is connected to adjacent segments through an actively controlled revolute joint. The axles of the passive wheel assemblies can be rotated about a vertical axis. Let's first consider a mechanism comprised of two segments. There are two wheel constraints which constrain the 3-dimensional Lie group of robot motion. The kinematic constraints do not fully specify the system's motion in $SE(2)$. This is a system with *mixed* constraints, where the third constraint is based on a generalized momentum.

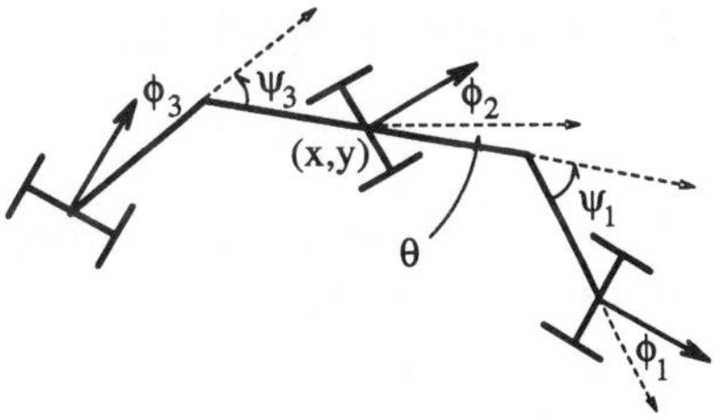

Figure 8: A model for the kinematic snake

With three segments, there are the same number of constraints as there are dimensions of $SE(2)$, and so the kinematic constraints define a *principal kinematic connection.* In the case of additional body segments, the first three segments define the robot's motion in $SE(2)$, and the wheel constraints of the additional segments are used as governing equations for the motion of these segments. This system will have a "following" behavior, in which the lead segments define the path to be traced, and the additional segments follow this lead. In real snakes, additional segments can provide greater stability or can be used for more complicated maneuvers, such as crossing over gaps.

For the three segment device in Fig. 8, let $(x, y, \theta) \in SE(2)$ denote the center of the middle segment, while the wheel angles of segments 1, 2, and 3 are denoted $(\phi_1, \phi_2, \phi_3) \in S^1 \times S^1 \times S^1$. The relative orientation of segment 1 with respect to segment 2, and segment 2 with respect to segment 3, is denoted by $(\psi_1, \psi_3) \in S^1 \times S^1$. Each no-slip wheel constraint takes the form:

$$(-\sin\tilde{\phi}_i, \cos\tilde{\phi}_i) \cdot \begin{bmatrix} \dot{\bar{x}}_i \\ \dot{\bar{y}}_i \end{bmatrix} = 0, \tag{30}$$

where $\tilde{\phi}_i$ is the absolute angle of the ith wheel, and $(\bar{x}_i, \bar{y}_i)$ is the Cartesian positioning of the center of rotation for the ith wheel. The kinematic constraint equations can be expressed in the form:

$$W(r)T_g L_{g^{-1}}\dot{g} = \Gamma(r)\dot{r}. \tag{31}$$

Since the constraints are G-invariant, the kernel of these constraints defines a connection on $Q = SE(2) \times S^1 \times S^1 \times S^1 \times S^1 \times S^1$. We can invert the constraint equations directly to write the local form of the connection one-form as

$$\xi = g^{-1}\dot{g} = -\mathcal{A}(r)\dot{r}, \tag{32}$$

where $\xi \in \mathfrak{g}$ and:

$$\begin{aligned} \xi^1 &= -\tfrac{l^2\cos\phi_2}{\det W}\left[\dot{\psi}_1\cos\phi_1\left(\cos(\phi_3-\psi_3)+\cos\phi_3\right)\right. \\ &\quad \left.+\dot{\psi}_3\cos\phi_3\left(\cos(\phi_1+\psi_1)+\cos\phi_1\right)\right] \\ \xi^2 &= \xi^1\tan\phi_2 \\ \xi^3 &= \tfrac{l}{\det W}\left[\dot{\psi}_1\cos\phi_1\sin(\phi_3+\psi_3-\phi_2)\right. \\ &\quad \left.-\dot{\psi}_3\cos\phi_3\sin(\phi_1-\psi_1-\phi_2)\right], \end{aligned} \tag{33}$$

with

$$\begin{aligned} \det\ W &= l[\sin(\phi_1-\psi_1-\phi_2)(\cos(\phi_3+\psi_3) \\ &\quad +\cos\phi_3)+\sin(\phi_3+\psi_3-\phi_2) \\ &\quad (\cos(\phi_1-\psi_1)+\cos\phi_1)]. \end{aligned} \tag{34}$$

Suppose that we add a fourth segment. Let ϕ_4 and ψ_4 denote the angles of the wheels and the body segment, respectively. Letting $\tilde{\phi}_4 = \phi_4 + \psi_3 + \psi_4 + \theta$, the no-slip kinematic wheel constraint is:

$$\begin{aligned} &-\sin\tilde{\phi}_4\,\dot{x} + \cos\tilde{\phi}_4\,\dot{y} - l(\cos(\phi_4+\psi_3+\psi_4) \\ &+2\cos(\phi_4+\psi_4)+\cos\phi_4)\dot{\theta} \\ &-l(2\cos(\phi_4+\psi_4)+\cos\phi_4)\dot{\psi}_3 = l\cos\phi_4\,\dot{\psi}_4. \end{aligned} \tag{35}$$

We have added two degrees of freedom (ϕ_4 and ψ_4) and one kinematic constraint (Eq. (35)). Both ϕ_4 and ψ_4 are controlled, but we must also satisfy the constraint. This is done by choosing to control the wheel angle, while inverting Eq. (35) to establish a governing equation for ψ_4:

$$\begin{aligned} \dot{\psi}_4 &= \tfrac{1}{\cos\phi_4}\Big(\tfrac{1}{l}\big(-\sin\tilde{\phi}_4^e\,\xi^1 + \cos\tilde{\phi}_4^e\,\xi^2\big) \\ &\quad -(\cos\tilde{\phi}_4^e + 2\cos(\phi_4+\psi_4)+\cos\phi_4)\xi^3 \\ &\quad -(2\cos(\phi_4+\psi_4)+\cos\phi_4)\dot{\psi}_3\Big), \end{aligned}$$

where $\tilde{\phi}_4^e = \phi_4 + \psi_3 + \psi_4$, i.e., $\tilde{\phi}_4\big|_{g=e} = \tilde{\phi}_4\big|_{\theta=0}$. Notice that $\dot{\psi}_4$ depends upon a term with $\cos\phi_4$ in the denominator. We assume that the wheels cannot pivot to an angle of $\pm\frac{\pi}{2}$). Repeating this process, we can add additional segments so as to guarantee that each following segment satisfies all of the constraints. It is always possible to arrange the additional constraint equations in the form:

$$\begin{aligned} &B(\phi_4,\ldots,\phi_k)\begin{bmatrix}\dot{\phi}_4 \\ \vdots \\ \dot{\phi}_k\end{bmatrix} \\ &= C(\phi_4,\ldots,\phi_k,\psi_3,\ldots,\psi_k)\begin{bmatrix}\xi \\ \dot{\psi}_3\end{bmatrix} \end{aligned}$$

where k is the total number of body segments. The matrix B is lower triangular, with determinant

$$\det(B) = \prod_{j=4}^{k} l\cos\phi_j,$$

and is always invertible.

7.1 Locomotive Gaits

We now consider the undulatory gait employed by many snakes. Hirose described this gait as being closest to a "serpenoid" curve, and we show below that one of the gaits generated by our approach is strikingly similar to the serpenoid curve. We also present two other gaits not normally seen in nature, but which arise for the particular model we are using. Like the snakeboard, the gaits are based on integrally related frequencies of the shape inputs. The ratios relate the frequency of bending of the inter-segment angle, ψ_i, with the frequency of the wheel rotations, ϕ_i.

The "serpentine" gait: Forward motion can be created by using a 1:1 frequency ratio. For this, we use periodic control inputs of the form:

$$\phi_i = a_i^\phi \sin(\omega_i^\phi t + p_i^\phi), \tag{36}$$

where similar values for ψ_i will be superscripted with a ψ. The serpentine gait can be demonstrated using the values

$$\begin{aligned} a_1^\phi = a_3^\phi = 0.2 = -a_2^\phi, &\qquad a_1^\psi = a_3^\psi = 0.6, \\ \omega_1^\phi = \omega_2^\phi = \omega_3^\phi = 1, &\qquad \omega_1^\psi = \omega_3^\psi = 1, \end{aligned} \tag{37}$$

where the length from the wheel base to inter-segment pivot point is 0.1m (hence a segment has length 0.2m). For the phasing, we use a traveling wave down the length of the snake (by incrementing the phase of the wheels at each segment), while forcing the inter-segment angle, ψ_i, to move

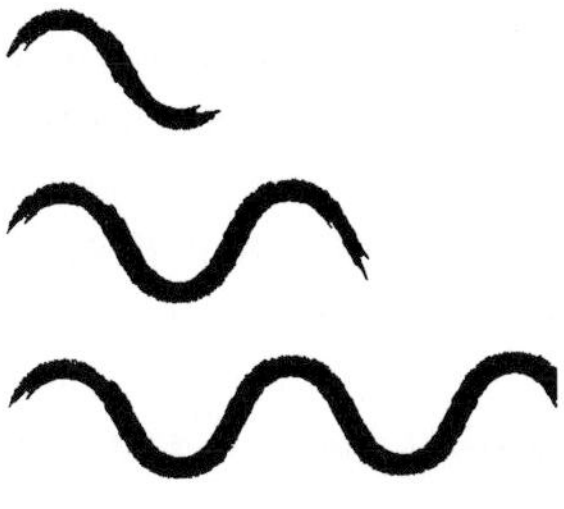

Figure 9: A trace of the serpentine mode

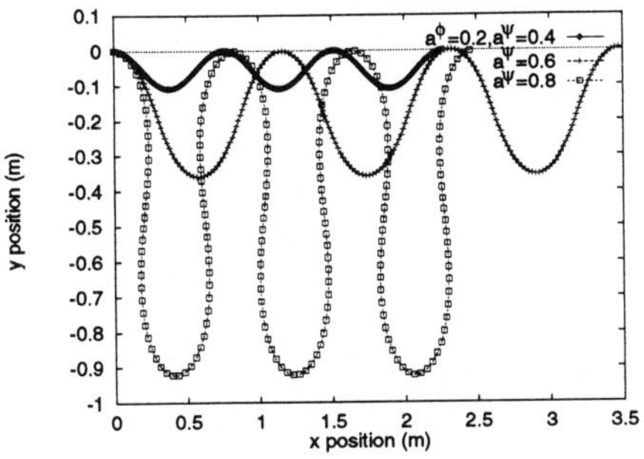

Figure 10: Three different serpentine gait shapes

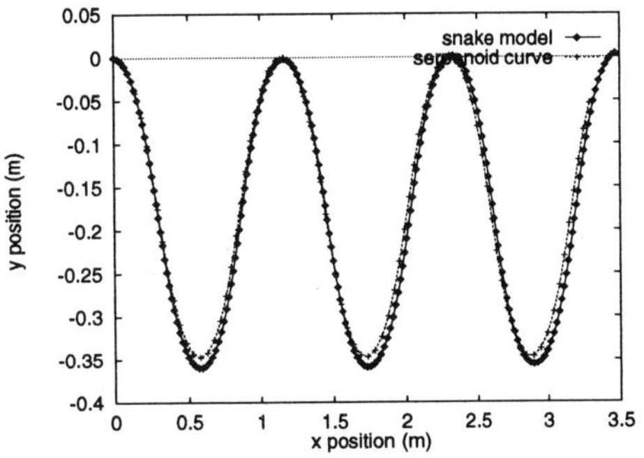

Figure 11: A comparison of the kinematic snake model versus the serpenoid curve

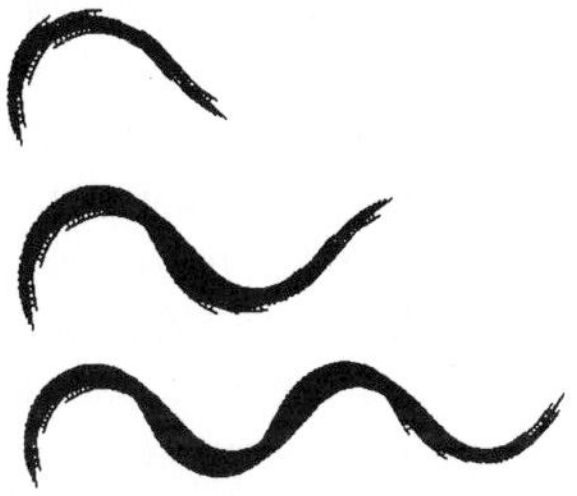

Figure 12: Traces of the 5-link kinematic snake

90° out of phase with their corresponding wheels. Thus, for the simulation shown in Fig. 9, the phases are given by

$$\begin{array}{lll} p_1^\phi = -\frac{\pi}{16}, & p_2^\phi = 0, & p_3^\phi = \frac{\pi}{16}, \\ p_1^\psi = \frac{7\pi}{16}, & & p_3^\psi = \frac{9\pi}{16}. \end{array} \tag{38}$$

Each of the wheel angles differs by $\frac{\pi}{16}$, while the joint angles are $\frac{\pi}{2}$ out of phase with their respective wheel angles. A trace of the motion appears in Fig. 9.

By varying the magnitudes of the wheel angles (or the inter-segment joint angles), slightly different patterns of locomotion are found to occur. Fig. 10 shows the resultant gaits for three different values of $a_1^\psi = a_3^\psi = a^\psi$. Each of these simulated gaits is run for the same length of time, which indicates that certain parameter values will result in a greater distance being traveled.

Let us compare this motion with the serpenoid curve proposed by Hirose [10]. The serpenoid curve has the parameterization:

$$\begin{aligned} \dot{x} &= \cos\left(\alpha \sin(\beta s)\right) \\ \dot{y} &= -\sin\left(\alpha \sin(\beta s)\right). \end{aligned} \tag{39}$$

The parameters α and β can be chosen such that the serpenoid curve will closely match any of the serpentine patterns generated using the 1:1 gait. An example is shown in Fig. 11, with $\alpha = 0.85$ and $\beta = \frac{23\pi}{16} \simeq 4.52$rad.

The serpentine gait seems to work well when additional segments are added using the methods described in above. Fig. 12 shows the trace of a 5-segment snake robot model.

The "rotate" gait: There are at least two gaits that lead to a net rotation. The first gait is reminiscent of a "three-point turn." For the gait shown in Fig. 13, the frequency ratio is 2:1, with the parameters used being the same as above, except that $\omega_1^\psi = \omega_2^\phi = 2$ and all joint magnitudes, $a_1^\phi, \ldots, a_2^\psi$ are 0.4 and 0.5, respectively. The 1:2 rotate gait, with the parameters $a^\phi = \frac{\pi}{4}$ and $a^\psi = \frac{3\pi}{8}$, is shown in Fig. 14, which plots the variation of θ with respect to time. The (x, y) position of this segment moves only very insignificantly during this gait.

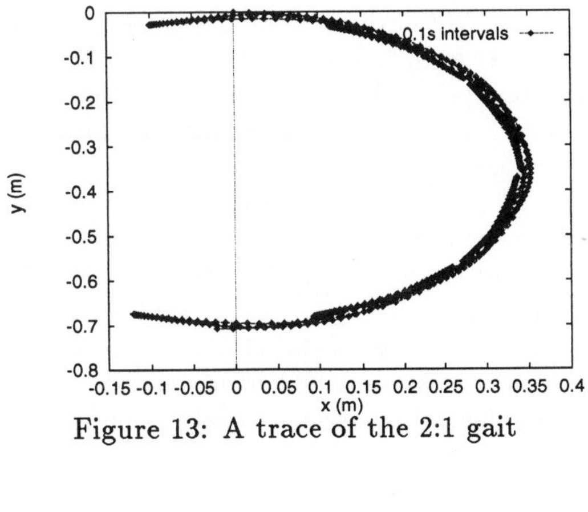

Figure 13: A trace of the 2:1 gait

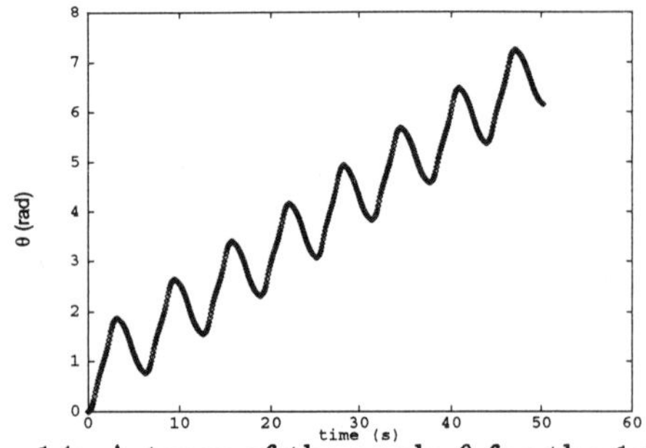

Figure 14: A trace of the angle θ for the 1:2 gait

8 Extension of the Approach

Preliminary work has shown that many of the basic concepts presented in this paper can be extended to a large class of locomotory systems, since virtually all forms of locomotion use some type of constraint to generate propulsive forces. There are two basic efforts required to extend this theory to other classes of locomotors. First, not all constraints that arise in different locomotion systems will take the form of Eq. (1). Hence, the derivations that lead to Proposition #1 must be reworked for different types of constraints. Similarly, the form of the connection will also be modified. Second, the theory must be extended to handle the intermittent constraints that arise in legged locomotion. This extension might proceed as follows. During the motion of a legged robot, a leg will repeatedly make and break contact with the supporting terrain. This nonsmooth interaction with the environment can be modeled as a system that evolves on a configuration manifold, Q, with a boundary, ∂Q. The boundary corresponds to those states where the leg contacts the terrain. There will be two sets of dynamical equations, and hence two connections. One set will describe the system when it is in contact with the terrain, while one will describe the system when the leg is not in contact. While our dynamics methodology extends easily to this case, the issue of controllability becomes more complex. We have recently developed a controllability test for driftless systems on a manifold with boundary. Omitting a few technicalities, it can be shown that if $\overline{\Delta}^c_Q$ is the involutive closure of the accessibility distribution of the control system operating in Q, and if $\overline{\Delta}^c_{\partial Q}$ is the involutive closure of the accessibility distribution of the control system operating in ∂Q, then the system is locally controllable if $\overline{\Delta}^c_Q(q) + \overline{\Delta}^c_Q(q) = T_qQ$ for all $q \in U$, where U is a neighborhood of $q \in \partial Q$. In general, a legged mechanism will have more than one leg. The configuration space of such a system will be a *stratified set*. For multi-legged robots, the individual strata physically correspond to states in which different numbers of legs are in contact with the ground. This is the generalization of the problem of control on a manifold with boundary.

9 Discussion and Conclusions

Undulatory locomotors have no thrusters, tracks, or legs to generate motion. Instead, motion is generated by a coupling of internal shape changes to external constraints. This paper focused on a class of systems with nonholonomic kinematic constraints, which includes not only the snakeboard and Hirose's ACM, but also any undulatory robotic system that uses wheels to provide motion constraints. Also, many of the snake and worm-like systems discussed in [7] can be analyzed using these techniques. A key observation is that the constraints inherent in undulatory systems provide the means to determine motion as a function of internal shape change. When kinematic constraints do not uniquely determine a robot's motion, dynamic symmetries provide the additional constraints. We used the formal language of connections on principle bundles because the connection encompasses much of the information which is essential to locomotion. Using these tools, we can parameterize the dynamics in terms of physically meaningful variables of momenta, internal shape, and robot reference frame motion. The connection also leads directly to the analysis of locomotor controllability. We believe that this framework will ultimately form the basis of a unifying theory for robotic locomotion.

References

[1] Ralph Abraham and Jerrold Marsden. *Foundations of Mechanics.* Addison-Wesley, Reading, MA, second edition, 1978.

[2] A. M. Bloch, M. Reyhanoglu, and N. H. McClamroch. Control and stabilization of nonholonomic dynamic systems. *IEEE Trans. on Aut. Control*, 37(11):1746–1757, 1992.

[3] A.M. Bloch, P. S. Krishnaprasad, J.E. Marsden, and R.M. Murray. Nonholonomic mechanical systems and symmetry. Technical report, California Institute of Technology, Pasadena, CA, 1994. Submitted to the *Archive for Rational Mechanics and Analysis.* Available via http://avalon.caltech.edu /cds/reports/cds94-013.ps.

[4] J.W. Burdick, J. Radford, and G.S. Chirikjian. A 'sidewinding' locomotion gait for hyper-redundant robots. *Advanced Robotics*, 9(3):195–216, 1995.

[5] C. I. Byrnes and A. Isidori. On the attitude stabilization of rigid spacecraft. *Automatica*, 27:87–95, 1991.

[6] C. Canudas de Wit and O. J. Sørdalen. Exponential stabilization of mobile robots with nonholonomic constraints. *IEEE Trans. on Aut. Control*, 37(11):1791–1797, Nov 1992.

[7] G.S. Chirikjian and J.W. Burdick. The kinematics of hyper-redundant locomotion. *IEEE Trans. on Robotics and Automation*, 11(6):781–793, 1995.

[8] S. Hirose et. al. Titan III: A quadruped walking vehicle. In 2nd *Int. Symp. on Robotics Research*, Tokyo, Japan, 1985.

[9] J. Furusho and A. Sano. Sensor-based control of a nine-link biped. *Int. J. Robotics Research*, 9(2):83–98, April 1990.

[10] S. Hirose. *Biologically Inspired Robots: Snake-like Locomotors and Manipulators*. Oxford Univ. Press, Oxford, 1993.

[11] S. Hirose and A. Morishima. Design and control of a mobile robot with an articulated body. *The Int. J. of Robotics Research*, 9(2):99–114, 1990.

[12] S. Hirose and Y. Umetani. Kinematic control of active cord mechanism with tactile sensors. In *Proc.* 2nd *Int. CISM-IFT Symp. on Theory and Practice of Robots and Manipulators*, pages 241–252, 1976.

[13] A. Isidori. *Nonlinear Control Systems: an Introduction*. Springer-Verlag, Berlin, 1989.

[14] Shuuji Kajita and Kazuo Tani. Study of dynamic biped locomotion on rugged terrain. In *Proc. IEEE Int. Conf. Robotics and Automation*, pages 1405–1411, Sacramento, CA, April 1991.

[15] S. Kelly and R. Murray. Geometric phases and robotic locomotion. *J. Robotic Systems*, 12(6):417–431, June 1995.

[16] D. Koditschek and M. Buhler. Analysis of a simplified hopping robot. *Int. J. of Robotics Research*, 10(6):587–605, Dec. 1991.

[17] P.S. Krishnaprasad and D.P. Tsakiris. *G*-snakes: Nonholonomic kinematic chains on Lie groups. In *Proc.* 33rd *IEEE Conf. on Decision and Control*, Lake Buena Vista, FL, Dec 1994.

[18] J-P. Laumond, P. Jacobs, M. Taix, and R. M. Murray. A motion planner for nonholonomic mobile robots. *IEEE Trans. on Robotics and Automation*, 10(5):577–593, 1994.

[19] A. Lewis, J. Ostrowski, R. Murray, and J. Burdick. Nonholonomic mechanics and locomotion: the snakeboard example. In *Proc. IEEE Int. Conf. on Robotics and Automation*, pages 2391–2397, San Diego, CA, 1994.

[20] T. McGeer. Passive dynamic walking. *Int. J. of Robotics Research*, 9(2):62–82, April. 1990.

[21] R. T. M'Closkey and R. M. Murray. Exponential stabilization of driftless nonlinear control systems using homogeneous feedback. Technical Report CIT/CDS 95-012, California Inst. of Technology, 1995.

[22] R.T. M'Closkey and J.W. Burdick. On the periodic motions of a hopping robot with vertical and forward motion. *Int. J. of Robotics Research*, 12(3):197–218, June 1993.

[23] H. Miura and I. Shimoyama. Dynamic walk of a biped. *Int. J. of Robotics Research*, 3(2):60–74, 1984.

[24] R. Montgomery. Isoholonomic problems and some applications. *Comm. in Math. Physics*, 128(3):565–592, 1990.

[25] R. M. Murray and S. S. Sastry. Nonholonomic motion planning: Steering using sinusoids. *IEEE Trans. on Automatic Control*, 38(5):700–716, 1993.

[26] R.M. Murray and S.S. Sastry. Steering nonholonomic systems in chained form. In *Proc. 31st IEEE Conf. on Decision and Control*, pages 1121–1126, Brighton, England, December 1991.

[27] Y. Nakamura and R. Mukherjee. Exploiting nonholonomic redundancy of free-flying space robots. *IEEE Trans. on Robotics and Automation*, 9(4):499–506, August 1993.

[28] J. Ostrowski, J. Burdick, A. Lewis, and R. Murray. The mechanics of undulatory locomotion: the mixed dynamic and kinematic case. In *Proc. IEEE Int. Conf. on Robotics and Automation*, Nagoya, Japan, May 1995.

[29] J.P. Ostrowski. *The Mechanics and Control of Undulatory Robotic Locomotion*. PhD thesis, California Inst. of Technology, Pasadena, CA, Sept. 1995.

[30] M.H. Raibert. *Legged Robots that Balance*. MIT Press, 1986.

[31] P. Rouchon, M. Fliess, J. Levine, and P. Martin. Flatness and motion planning: the car with *n* trailers. In *Proc. European Control Conference*, pages 1518–1522, 1992.

[32] Alfred Shapere and Frank Wilczek. Geometry of self-propulsion at low Reynolds number. *J. Fluid Mechanics*, 198:557–585, January 1989.

[33] S.M. Song. and K.J. Waldron. *Machines that walk: the Adaptive Suspension Vehicle*. MIT Press, 1989.

[34] A.F. Vakakis, J. W. Burdick, and T.K. Caughey. An interesting attractor in the dynamics of a hopping robot. *Int. J. of Robotics Research*, 10(6):606–618, Dec. 1991.

[35] Gregory C. Walsh and S. Shankar Sastry. On reorienting linked rigid bodies using internal motions. *IEEE Trans. on Robotics and Automation*, pages 139–146, 1995.

Design of Hyper-Redundant Arm

Shigeo Hirose

Tokyo Institute of Technology
Department of Mechano-Aerospace Engineering
Ookayama Meguro-ku Tokyo 152, Japan
hirose@mes.titech.ac.jp

Abstract

It has long been one of the dreams of robotic researchers to realize a hyper-redundant multi-joint arm which, just like elephant trunk, can enter into a rugged and narrow environment while actively avoiding obstacles. Thus far, there have been many simulation studies on the redundant control problems encountered in controlling hyper-redundant arms, but it has proved very difficult to design a real arm mechanism, and only a few practical models have been made and used so far. We have been studying this problem for more than twenty years. This paper reviews our past experiences in designing hyper-redundant arms which include a mechanism with oblique swivel joints, a coupled tendon driven arm, and a "Moray" drive. The paper also proposes a new mechanism that we have coined the "Float Arm", which is composed of a series of parallel links on which are installed spring compensation mechanisms to cancel the weight of arbitrarily shaped segments of the hyper-redundant arm. The experimental results of a mechanical model of the Float Arm are also included.

1. Introduction

One of the dreams of robotic researchers is to realize a hyper-redundant multi-joint arm which can enter into a complicated environment while avoiding obstacles. There are many studies that have been conducted to date which focus on the realization of a hyper-redundant arm for this purpose, and which examine the control problems involved. In particular, the control problem of how to control a hyper-redundant arm in an environment full of obstacles is now regarded as one of the typical field of robotic research.

However, it is equally true that, regrettably, under the current circumstances there have been hardly any examples thus far of having actually created a prototype and realized a hyper-redundant multi-joint arm. Why is it that there has been so little sound research attempting to prototype an actual mechanical model proportionate to the great expectations held for making a practical hyper-redundant arm? The author believes that the difficulty of configuring a joint drive system should be cited as the greatest reason for this. A hyper-redundant arm must operate while supporting a long arm in resistance to gravity. For this reason, it is necessary for the drive system installed in each joint to be sufficiently light weight and to have high output power at the same time. Also, that drive system must have large load resistance characteristics which can also resist the unexpected contact force working around the arm. We do not yet have a drive system that is notably light weight, powerful and highly solid which can respond to these kinds of use conditions. For these reasons, it is incredibly difficult to realize a hyper-redundant arm designed by ~~a~~ conventional drive system.

In this research, we will introduce three attempts that the author has made in the past to develop a mechanical design method to overcome the difficulties of configuring this kind of drive system which is a bottle neck to the making of a practical hyper-redundant arm. At the same time, we will propose a new design method and experimentally verify its validity. In this way, by studying the methods of realizing a hyper-redundant arm mechanism in an integrated manner, we will offer a specific basis for realizing a practical hyper-redundant arm in the future.

2. Hyper-redundant Arm Mechanisms

The authors have made several attempts thus far to break through the restrictions to the drive system for a hyper-redundant multi-joint arm. These can be classified into the following 3 kinds.

1) Use of light weight joint driving mechanism.
2) Use of the of coupled tendon-drive effect.
3) Positive use of a root drive mechanism.

Furthermore, in this paper we will offer a new fourth attempt.

4) Introduction of a complete gravity compensation mechanism.

The authors believe that the hyper-redundant arms that will be developed in the future will probably be realized by several of these methods or by combining them. The essential features of the design principles for these methods will be described hereafter in order.

3. Light Weight Joint Driving Mechanism

If a mechanism which arranges an actuator in each joint is adopted, it must be possible to configure the mechanism in a light weight small scale actuator utilizing the actuator output as efficiently as possible. Thus far there have been many proposals for arm configuration methods that utilize universal joint mechanisms [1,2,3]. Here, we will introduce an oblique swivel mechanism which the author has proposed as a mechanism that can be made particularly compact and that can yield a great output. This is a mechanism characterized by having oblique swivel joints which rotate around slanted axes. Figure 1 is the ACM IV which is configured by alternately connecting oblique swivel joints and coaxial swivel joints [4,5].

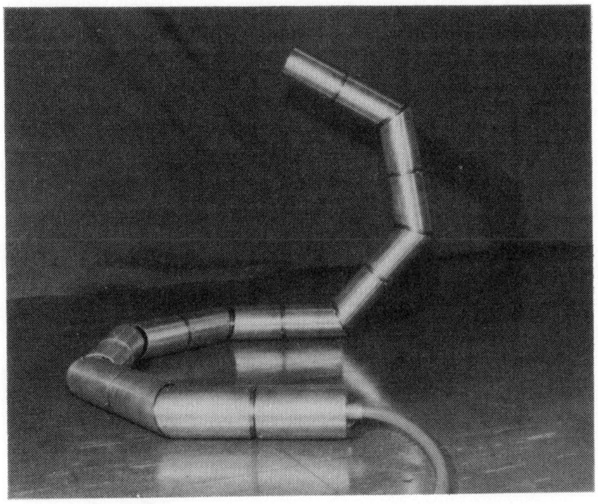

Fig.1 ACM VI (Active Cord Mechanism) [4,5]

This kind of oblique swivel mechanism has the movement restriction that the motion range of the joints is limited to $\pm 2\theta$ when the motion range of the oblique swivel axis is $\pm\theta$. However, the angle of curvature of each joint does not need to be that great in an arm which connects multiple joints, and this characteristic is not that great an

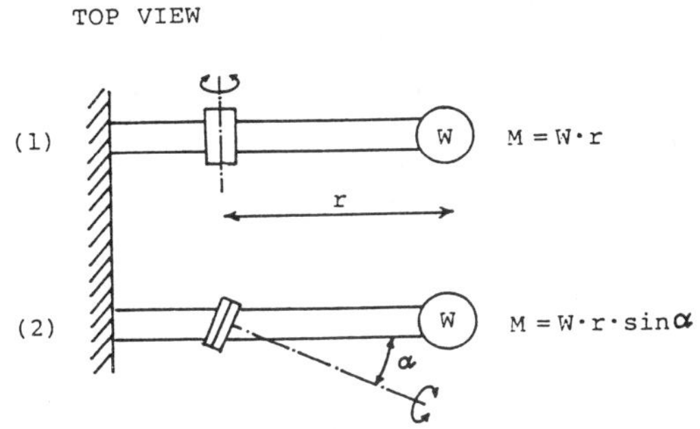

Fig.2 Weight bearing characteristics of Oblique Swivel Joint

impediment. Conversely, if the swivel axis is perpendicular to the arm as in Fig. 2(a), the actuator must support all of the load, but if the swivel axis is arranged to be at an angle along the arm as in Fig. 2(b), even if the arm is supported by being cantilevered as indicated in Fig. 2, the majority of the weight of the arm can be supported by the large diameter ball bearings of the joint part, and this gains the desirable characteristic that the output torque of the drive system can at least produce a large support force.

In addition to this kind of characteristic, oblique swivel mechanisms also have the following kinds of properties.

1) The oblique swivel axis can be designed compactly together with coaxial swivel axes, and a smooth external shape can be configured with few protrusions.

2) Because the segments can be configured like an exoskeleton, a light weight and highly rigid frame can be configured.

3) Wiring can be stored inside the segments, and the wiring will not interfere with the movement of the segments.

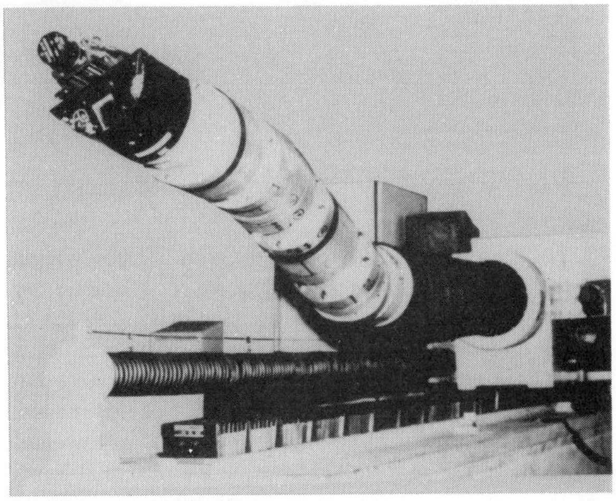

Fig.3 Measuring/Grinding System for Water Turbine Runner [6]

Fig. 3 is a robotic arm which utilizes this oblique swivel mechanism, and which is inserted into the water turbine runners for hydroelectric power generator and measures the internal wall conditions and conducts grinding [6]. This is a configuration which alternately arranges 2 oblique swivel joints and 3 coaxial swivel joints with a total arm length of 2.5 m and a diameter of 200 mm, and this mechanism is capable of heavy operation by bending and extending the arm into the narrow water turbine runners.

4. Coupled Tendon-drive Mechanism

There have thus far been several mechanisms in which, in order to lighten the segments, the base part rather than each joint is equipped with the drive system, and this drive force is transmitted to the joints with wires or hydraulics. Among these, Fig. 4 indicates the ACMVII, prototypes of which have been made the authors [5,7]. This is a device which comprises joints with coil springs, and realizes bending movement by transmitting the tractive force of the actuator installed in the base to each joint using 3 conduit wires. Because it consists of 8 segments, this is an arm which has a total of 24 degrees of freedom.

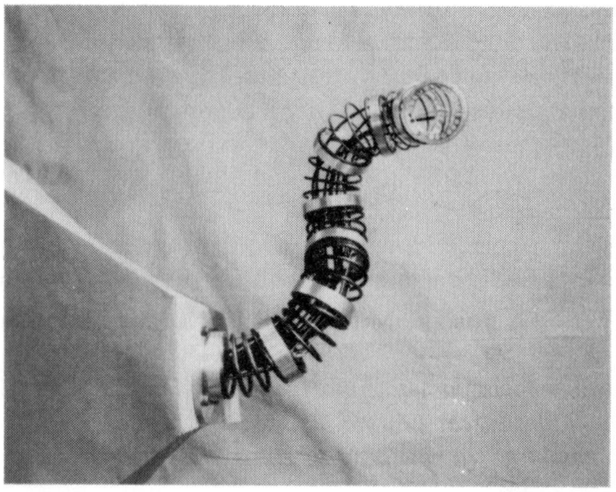

Fig.4 ACM VII; Conduit wire controlled arm [7]

The authors afterward foundthat even lighter weight and higher output arm could bc configured by positively using the effect of coupled driving forces of traction tendons [8]. Normally, to support an arm like a cantilever beam, a larger moment must be produced at the joints of proximal end of the arm. For this reason, when designing a hyper-redundant arm, a powerful actuators must be installed to drive the proximal joints of the arm even though it is driven by conduit wires. However, the mechanism that the authors have proposed can support a hyper-redundant arm which forms a cantilever posture with same power of joint actuators from the tip to the root.

This mechanism is utilized in 2 mechanism design principles which we proposed. They are named as "connected differential mechanism" [9] and "coupled drive" [10]. A differential mechanism is a mechanism which has 3 input and output ports. Then, a coupled differential mechanism is one in which a pair of ports of this kind of differential mechanism are connected in a series, and the energy which is supplied from one port can be output uniformly to the remaining one port in each mechanism. The connected differential mechanism for a hyper-redundant arm is provided with links and pulleys with free rotational movement around the axes of each segment, the joint units around which the wire is stretched are connected, and one wire stretches from the distal to the proximal part as shown in Fig.5(a). The authors have called this mechanism the "soft gripper mechanism" because it has already been used in

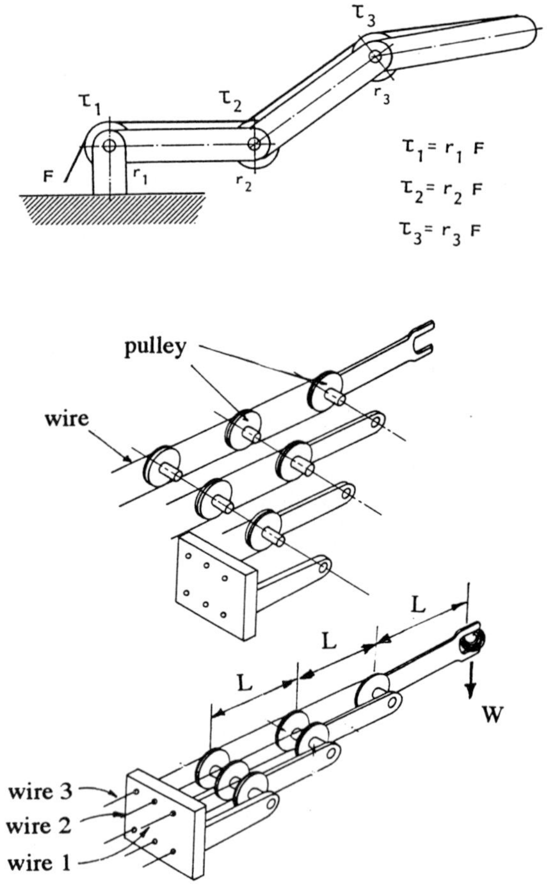

Fig.5 Basic concept of Soft Gripper mechanism and Coupled Tendon-driven arm

a mechanism which the author named the Soft Gripper [11]. The soft gripper mechanism is configured such that one wire connects one pair of ports for each joint, and the one remaining output port of the joint produces curved movement as the output. Because it has this kind of configuration, if friction of the moveable parts can be ignored, the soft gripper mechanism is arranged such that by pulling the wires, a moment proportional to the radius of the pulleys is produced on each joint.

Let us consider a device which introduces this soft gripper mechanism into every series which drives each joint as in Fig. 5(b)(c). It is assumed that the length of the segments is L and is equal for all, that the weight of the segments can be ignored, and that an object of weight W is being grasped at the tip of the arm. Moreover, it is assumed that the radius of the pulleys is r, and that the friction can be ignored. At this time, the torque required to support the third segment of Fig. 5(c) is WL, and if wire 3 is pulled with a force of $(L/r)W$ in order to produce this torque, the same torque Wr is produced also on the first and second segments. The torque required to support segment 2 is $2Wr$. However, torque Wr is already being produced by the tensile force of wire 3, the wire 2 to drive segment 2 can support the arm with the tensile force of $(L/r)W$. Furthermore, with the first segment, because a torque of $2Wr$ is already being produced by the tensile force of wire 3 and wire 2, the tensile force of the wire 1 should be $(L/r)W$.

In this way, this mechanism is able to drive the arm which is supported in a cantilever with the same traction force from the tip to the root based on the fact that the coupling effect of the wire drive forces is utilized. Fig. 6 indicates the CT arm (Coupled Tendon-driven Arm) which utilizes this kind of wire coupled drive effect [8].

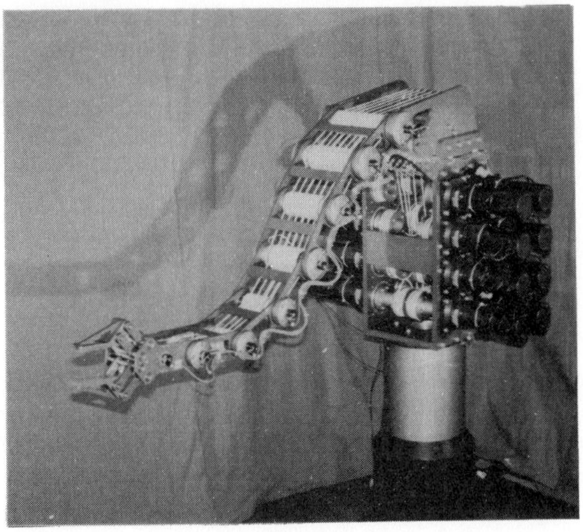

Fig.6 CT-I; Coupled Tendon-driven arm

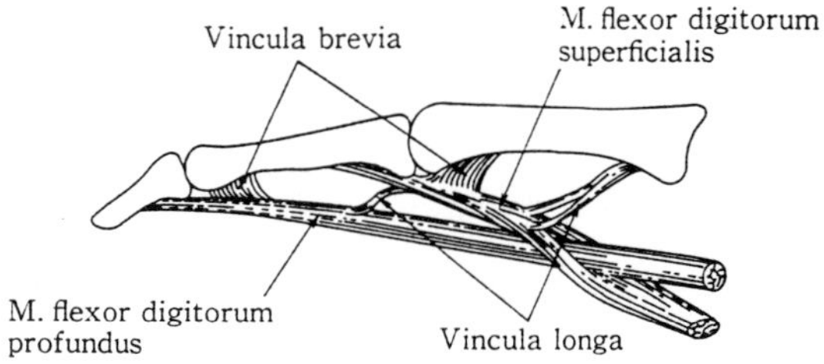

Fig.7 Coupled drive observed in finger tendon drive mechanism [8]

This kind of coupled tendon-drive mechanism is already widely used in the natural world. Fig. 7 is the tendon drive system of a human finger, and not only does the tendon which drives the finger tip segment produce torque on the tip joint, but it also produces torque in the same way on the proximal joints. For this reason, we can understand that the same coupling drive effect as in Fig. 6 is produced with this mechanism, and that a powerful grasping force can be produced by the finger.

5. Powerful Root Drive Mechanism

With hyper-redundant arms, the operational performance of the arm is severely restricted by the fact that drive actuators in each joint increase the weight. However, the root-most segment which is mounted on the base has little weight restrictions, and the operational performance of the entire arm is not

Fig.8 Basic concept of Moray arm [12,13]

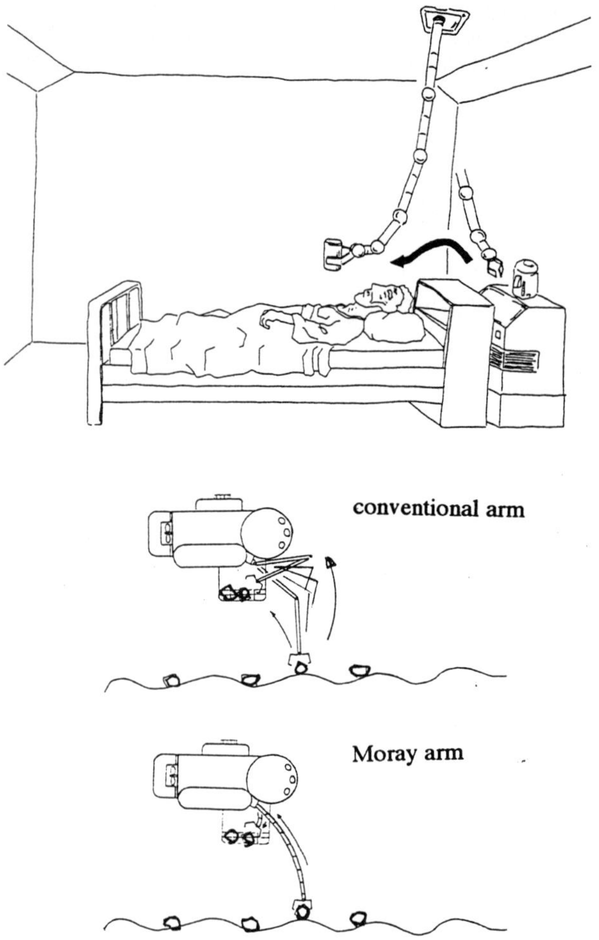

Fig.9 Expected applications of Moray arm

lowered that much by providing that segment with a heavy and powerful actuator. For this reason, if, as in Fig. 8, prismatic movement is produced by providing a powerful sliding mechansim on the base of the hyper-redundant arm, and if dynamic movement is produced that synchronizes to the extending motion of the base part sliding mechanism and the bending motion of the arm joint parts, even if the actuators equipped on the arm joints are small scale and have restricted output power, there is the possibility of being able to realize agile arm movement as a whole. Because this kind of arm movement resembles the movement of a Moray eel which jumps out from holes in the rocks of the ocean floor, the authors have called this the "Moray arm" [12,13].

In addition to the characteristics described above, the Moray arm has the following characteristics.

i) Because only the length of arm required is extended and utilized, normal operation can be conducted with the necessary minimum arm length, and it is rigid and limited in waste of energy.
ii) Because it is based on trajectory following movement, the mechanism performs very well in avoiding obstacles. Also, there is little effect of fluid resistance on the arm even when operating under water, and it is possible to have high speed movement even if there are limits to the capacity of the joint actuator.

An example applying the Moray arm is indicated in Fig.9. Fig. 9(a) is a robot arm that efficiently collects manganese nodules which are scattered on the ocean floor. Because the Moray arm collects the manganese nodules into the basket while extending and retracting without swinging the arm around, it is possible to have little fluid resistance and good energy efficiency. Moreover, there will be few reaction force effecting the submarine. Fig. 9(b) depicts a Moray arm embedded in the ceiling which is nursing a patient. When not necessary, the arm is kept out of the way by storage in the ceiling, and also, because it is based on an extending and retracting movement, it easily avoids obstacles.

The basic posture of a Moray arm extending from a storage vessel in free space is a Clothoid curve. We also studied control methods to produce the kind of movement which places as much as possible the load on the output of the base part actuator when swinging the arm from an option initial posture to an optional final posture [13].

6 Gravity Compensation Mechanism

The method of installing a spring on a robot arm in order to compensate the force of supporting its own weight is well known as a method of lightening the load on the arm drive system. As in Fig. 10, nearly all weight compensation mechanisms introduced thus far

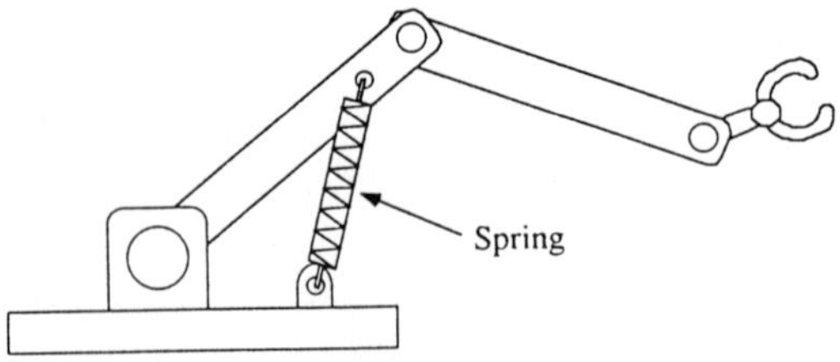

Fig.10 Conventional weight compensation mechansim for robot manipulator

are devices which use a spring extended from the stand to the first arm to compensate the weights of the first arm which is installed on the base joint of the stand, and the weight of the tip arm from that. With this compensation mechanism, the weight compensation torque to be produced on the base joint will vary depending on the postures of the arms which are equipped on the tip from the first arm, and weight compensation will not always be complete. However, in a normal industrial robot arm, the deviation of weight support torque is comparatively small in accordance with the posture, and in this situation, the conventional approach is rather effective.

However, if it is the kind of hyper-redundant arm that is the subject of this paper, it is meaningless unless weight compensation springs are installed not just on the first arm of the root joint, but also on every joint. Moreover, the weight compensation torque to be produced at every joint will vary greatly depending on the posture of the arm. For this reason, it was believed that, with a hyper-redundant arm, it was completely impossible to use springs to compensate the weight support force to be produced at each joint. In fact, until now there have been no reports of attempts to compensate the weight of hyper-redundant arms.

As a result of studying this problem, the author has discovered that a new mechanism which can completely compensate the arm weight when a hyper-redundant arm assumes any kind of posture can be realized by combining the two mechanisms of:

1) An arm stacking mechanism which is composed by parallel links which can lighten the weight supporting task by eliminating the torque component which changes in accordance with the posture;

2) A weight compensation spring mechanism for each parallel link.

In the remainder of the paper we will call this kind of gravity compensation hyper-redundant arm we are proposing, the "Float arm", and we will explain its basic principles.

6.1 Parallel Link Arm Mechanism

Let us consider a model of the kind of hyper-redundant arm in Fig. 11 which links the segments such that the weight of the segment is concentrated on the segment tip. In this case, the torque τ_i that the first i joints of a hyper-redundant arm is to produce in the posture indicated in Fig. 11 is

$$\tau_i = \sum_{j=i}^{n} (m_i g L_i \cos\theta_i) \qquad (1)$$

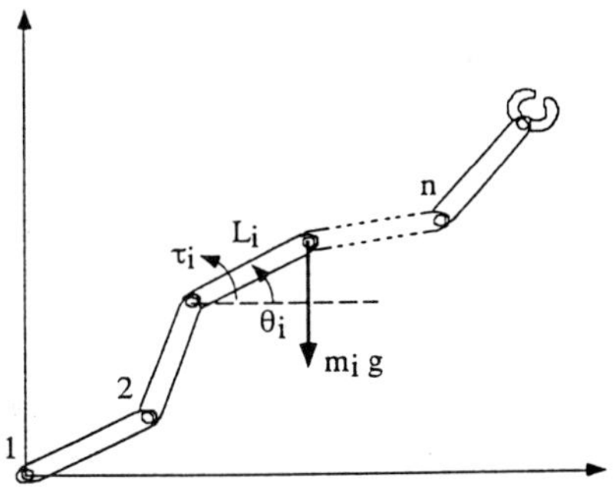

Fig.11 Static joint torque τ_i required to support the weight of conventional serial joint arm

From Equation (1) it is understood that the weight support torque to be produced by each joint of the hyper-redundant arm is a function of the posture of the distal segments, and it is necessary to adjust the compensation torque corresponding to that posture.

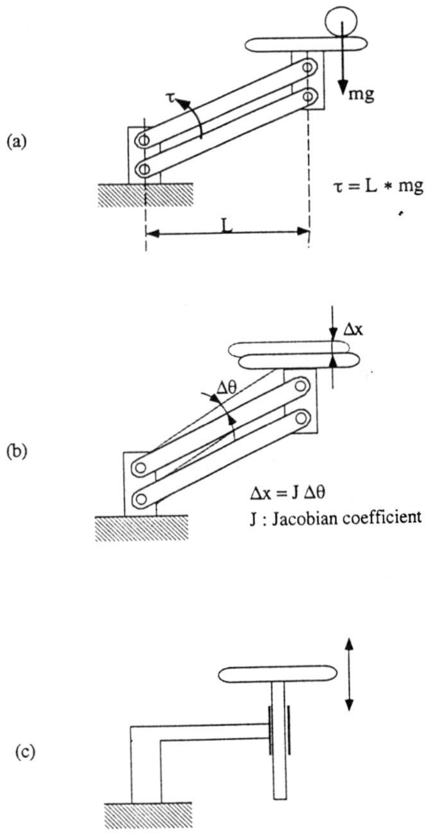

Fig.12 Characteristic of parallel link mechanism (a)(b) and equivalent mechansim (c)

However, if parallel limkage mechansim is introduced into the indivisual segments, the conditions change altogether. With the kind of parallel links indicated in Fig. 12(a), the support torque to be produced at the root part joint does not change even if a load is placed anywhere on a flat plate attached to the tip. This can be understood from the fact that, when rotating a parallel link mechanism a small amount as indicated in Fig. 12(b), a uniform small displacement is produced everywhere on the flat plate. The reason is understand easily by the fact that the Jacobian coefficient, which indicates the function of load displacement in relation to the joint displacement of the root part, is constant everywhere on the flat plate. Specifically, this parallel link mechanism is equivalent to the kind of drive mechanism of prismatic joints as in Fig. 12(c).

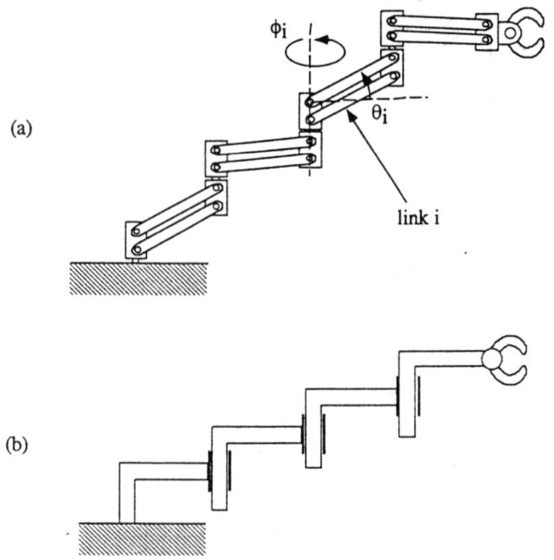

Fig.13 Basic mechansim of "Float arm" (a) and equivalent mechansim (b)

Now then, the Float arm we are proposing is the kind of hyper-redundant arm with stacked parallel links indicated in Fig. 13(a). Because it has this kind of configuration, the torque τ_i which swivel axis of joint i of the Float arm is to produce is

$$\tau_i = \left(\sum_{j=i}^{n} m_i g\right) \cdot L_i cos\theta_i \qquad (2)$$

Furthermore, with the Float arm, the swivel axis Φ_I of the root part of parallel link segment i is constantly maintained perpendicularly, and segment i may swivel around a perpendicular axis. From Equation (2), it can be understood that the Float arm yields the remarkable characteristic property that the torque which balances segment i of the Float arm in regard to swivel τ_i is determined only by the weight applied to the tip of that segment i and the posture of segment i, and does not depend on the posture of segments $i+1 \sim n$ of the tips thereafter. This kind of effect of the Float arm can be easily understood by replacing this arm mechanism with an equivalent mechanism composed of movement that swivels around a perpendicular axis and that moves prismatically in the direction of the perpendicular axis as indicated in Fig. 13(b).

6.2 Weight Compensation Spring Mechanism

By introducing the parallel link mechanism, the problem of the self-weight compensation of segment i of the float arm which assumes optional postures returns to the problem of compensating the arm which takes the weights of segment i and segments $i+1 \sim n$ of the distal side from there to be the load, and it was argued in the previous section that this relationship does not depend on the postures of segments $i+1 \sim n$. Because there is this kind of characteristic, the weight of the Float arm as a whole can be perfectly

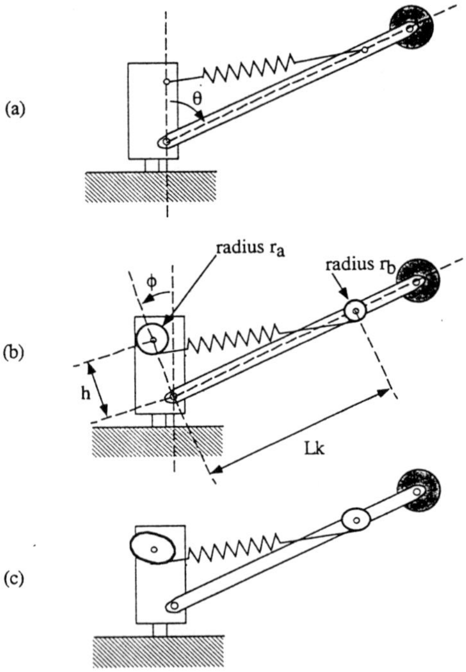

Fig.14 Spring balancing mechansim
(a) basic structure
(b) introduced mechansim
(c) complete balancing mechanism with non-circular pulleys

compensated irrespective of the posture if each links equipped the compensation springs to compensate the weight of own segment as well as the weight of distal segments. For this reason, in this section we will study a spring compensation mechanism to be used for indivisual link which is supporting a load as indicated in Fig. 14.

The basic compensating mechanism of the weight of the arm will be a structure which extends a spring as in Fig. 14(a). When configured in this way, as the arm is tilted from the perpendicular posture and θ increases, the moment around swivel axis Θ goes on increasing. As this happens, the spring is simultaneously stretched, and the moment in the opposite direction increases. The weight of the arm can be expected to be suitably balanced in this way. However, the balance cannot be completely made with this mechanism as is.

For this reason, in our research for realizing better compensation, we have assumed that wires which stretch the spring have been wrapped around pulleys on both ends, and position of the supporting pulley is assumed to inclined. Introduction of pulleys on both end of the wires is important from mechanical viewpoint, because the pulleys ease the wire-end fixation and it can be used as wire-length adjusting mechanism. As a result, in the mechanism of Fig.14(b), there are in total seven adjustable variables, namely: radii r_a and r_b of these pulleys A and B, distance variable h and angle variable ψ of the attached pulley A, variable L_k of the position at which pulley B is attached, and spring constant k and initial pulling force of that spring installed.

In addition, it is assumed that to implement the optimum design the permissible range of the arm is $0° \leqq \theta \leqq 120°$. Also, the evaluation function J to seek the optimum variables is set under minimized conditions such that the load torques τ_{mg} between the permissible motion range approach being the spring compensation torque τ_{sp}, specifically,

$$J = \int_{\theta=0}^{\theta=120} \{\tau_{mg}(\theta) - \tau_{sp}(\theta)\}^2 d\theta \tag{3}$$

Furthermore, although it was not studied in this research, there is also the possibility that if a pulley for support of the aforementioned spring can be suitably designed as in Fig. 14(c), there can be complete torque compensation. We have been studying a design method for this kind of pulley shape in the design of Shape Memory Alloy actuator system as reported in reference [14], and such an approach can be applied to this as well.

6.3 Optimized Compensation Mechanisms for Passive Mechanical Model

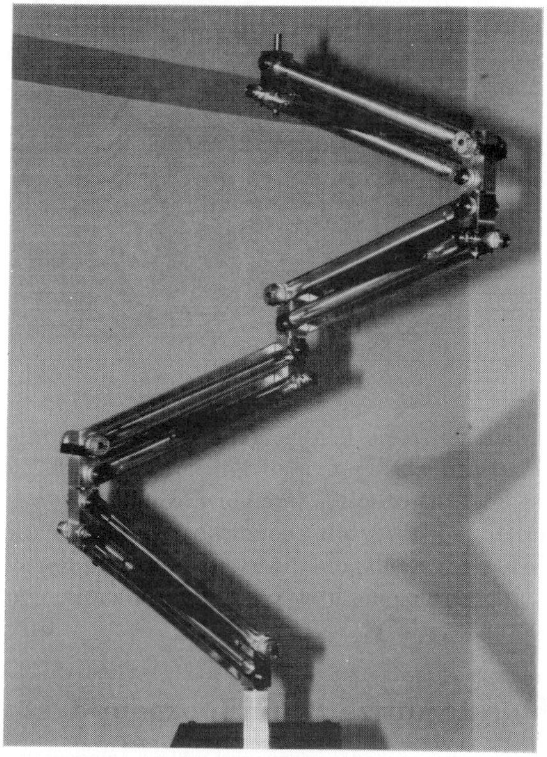

Fig.15 Mechanical model of the "Float Arm"
(Only the spring balancing mechanism is installed)

We designed a compensation spring mechanism for the hyper-redundant arm composed of six segments as indicated in Fig. 15. The specifications of the arm obtained in the design results are indicated in Table 1.
It is assumed that as the load increases from the tip segment to the root segment, only the spring constant will be varied corresponding to this. The spring

Table 1 Optimized parameters of the Float arm

Joint length	L (mm)	300
Length	L_k (mm)	300
Pulley radius A	r_a (mm)	12
Pulley radius B	r_b (mm)	-6
Spring length	d_{init} (mm)	225
Height	h (mm)	36
Angle	ψ (deg)	12
Stretch length (θ=0°)	d_{min} (mm)	39.824
Stretch length (θ=120°)	d_{max} (mm)	74.653

Table 2 Spring constants of the springs for Float arm

Number of joint	N	1	2	3	4	5	6
Equivalent load	M (kg)	0.45	1.35	2.25	3.15	4.05	4.95
Spring constant	k (kgf/mm)	0.037	0.113	0.118	0.263	0.338	0.413
Spring force θ=0°	fk_{min} (N)	14.64	43.91	78.08	102.5	131.7	161.4
Spring force θ=120°	fk_{max} (N)	27.44	82.31	137.2	192.1	246.9	301.5

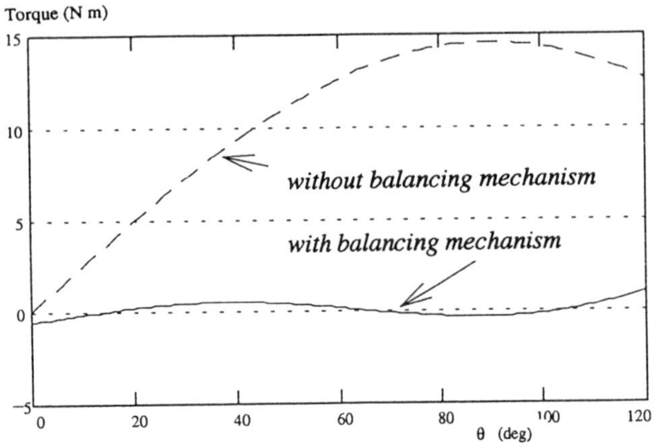

Fig.16 Simulation estimation of balancing mechansim

constant of each segment was taken to be that in Table 2. In addition, the design value of the support torque when there was no first segment support spring, the design value of the torque support when a compensation spring was installed, and the actually measured data are indicated in Fig. 16. It was found that the support torque was reduced to approximately 1/30 by installing the compensation spring, and that the compensation spring produced a remarkably high compensation effect. In addition, although there was still slight friction in the joints without the actuator being installed in the mechanical model that was prototyped in Fig. 15, it moved easily in the range of $0° \leqq \theta \leqq 120°$, and it could maintain optional postures.

7 Conclusion

In this research we focussed on mechanism design for the purpose of realizing a hyper-redundant arm, and in addition to reporting on the methods and theories already attempted by the authors such as: 1) devising a method to configure the drive system itself; 2) utilization of the coupled tendon-drive effect; and 3) positive utilization of the root drive mechanism, the authors have proposed the introduction of a complete gravity compensation mechanism, and they have verified its effectiveness using a mechanical model of a hyper-redundant arm composed of six segments. In the future these ideas will be developed further, and we would like to continue development of a practical hyper redundant arm.

References

[1] K. Asano, Control System for a Multi-joint Inspection Robot, Proc. 1984 National Topical Meeting on Robotics and Remote Handling in Hostile Environments, pp.375-382 (1984)

[2] P.K.J. Smith, The "WARRIOR" Welding Manipulator, Proc. Int. Advanced Robotics Programme, 1st Workshop on Manipulators, Sensors and steps towards Mobility, pp.107-113 (1987)

[3] D. Smith, C. Blume, M. Wadle, A Multi-Link Multi-Purpose Advanced Manipulator with a Large Handling Hemisphere for Out-door Applications, Proc. 5Th Int. Symp. On Robotics in Construction, pp.625-634 (1988)

[4] S. Hirose, Y. Umetani; An Active Cord Mechanism with Oblique Swivel Joints and Its Control, Proc. 4th RoManSy Symp., Zaborow,Poland, PWN-Polish Scientific Publishers , pp.327-340 (1981)

[5] S. Hirose, Biologically Inspired Robots (Snake-like Locomotor and Manipulator), Oxford University Press (1993)

[6] H.Shimada, S.Hirose, Y.Umetani, Measuring /Grinding System for Water Turbine Runner, Proc. 2nd Int. Conf. on Advanced Research, pp.199-206 (1985)

[7] S. Hirose, T. Kado, Y. Umetani, Tensor Actuated Elastic Manipulator, Proc. 6th IFToMM World Congress, New Delhi ,2, pp.978-981 (1983)

[8] S. Ma, S. Hirose, H. Yoshinada, CT ARM-I: Coupled Tendon-Driven Manipulator Model I -Design and Basic Experiments-, Proc. IEEE Int. Conf. on Robotics and Automation, pp.2094-2100 (1992)

[9] S. Hirose, Connected Differential Mechanism and Its Applications, Proc. 2nd Int. Conf. on Advanced Research, pp.319-326 (1985)

[10] S. Hirose, M. Sato, Coupled Drive of the Multi-DOF Robot, Proc. IEEE Int. Conf. on Robotics and Automation, Scottsdale, Arizona, pp.1610-1616 (1989)

[11] S. Hirose, Y. Umetani, The Development of Soft Gripper for the Versatile Robot Hand, Mechanism and Machine Theory, Pergamon Press ,13, pp.351-359 (1978)

[12] S.Hirose, S. Ma; Moray Drive for Multijoint

Manipulators, Proc. 5th Int. Conf. Advanced Robotic, Pisa, Italy, pp.521-526 (1991)
[13] S. Hirose, K. Yokoshima, S. Ma, 2 DOF Moray Drive for Hyper Redundant Manipulator, Proc. 1992 IEEE/RSJ Int. Conf. on Intelligent Robots and Systems, Raleigh, NC, pp.1735-1740B(1992)
[14] S. Hirose, K. Ikuta, K. Sato, Development of Shape Memory Alloy Actuator. Improvement of Output Performance by the Introduction of a σ-Mechanism, Advanced Robotics ,3, 2, pp.89-108 (1989)

The Calibration Index and the Role of Input Noise in Robot Calibration

†John M. Hollerbach and ‡Charles W. Wampler
†Dept. Computer Science, Univ. Utah, Salt Lake City, UT 84112
‡Mathematics Dept., General Motors Research and Development, Box 9055, Warren, MI 48090

1 Introduction

This paper addresses two topics in robot calibration: the formulation of a calibration index and the presence of input noise. The calibration index is introduced to succinctly capture the essence of a kinematic calibration method and allow insightful comparison between that method and other methods. Input noise, e.g., joint angle errors, is a potential source of error that is usually ignored, but is known to lead to bias errors in the estimates (Norton 1986). The two topics also will be seen to have a relationship.

Many kinematic calibration methods have been proposed by now, and more are continually introduced. These methods differ in terms of whether an external metrology system is required (open-loop methods) or not (closed-loop methods), how many components of pose are measured and by what means, how many components of pose are passively constrained and how, whether there are unsensed joints, and whether a serial or parallel mechanism is being calibrated (Hollerbach 1993). It can be very difficult to understand how one method is related to another method, or what the tradeoffs are in selecting a calibration method. This statement is particularly true for parallel linkages, with different mixtures of sensed and unsensed joints, and passive and active joints. It is not always immediately obvious how easy a parallel linkage is to calibrate, especially by a closed-loop method.

The literature on camera calibration is usually distinct from that on manipulator calibration. By modeling cameras and light beams as prismatic legs, it is possible to fold camera calibration into manipulator calibration. An advantage is that mixed manipulator/camera calibration problems can be handled uniformly.

2 Closed-Loop Viewpoint

An enabling step to clarify this picture was suggested in (Wampler and Arai, 1992; Hollerbach 1993; Wampler et al. 1995): nearly all kinematic calibration approaches can be viewed as closed-loop approaches, where the end effector measurement in open-loop methods is considered to form a joint. All measurements, whether from joints or metrology systems, are put on an equal footing, as are all unsensed joints, whether from unsensed components of pose, passive environmental constraints, or joints in the chain without sensors. In the case of parallel linkages, sufficient numbers of loop closure equations are formulated to characterize the kinematics and are combined at each pose.

Figure 1 illustrates this idea for a calibrated stereo camera metrology system that measures the 3D coordinates of a distinguished point on the end effector of an uncalibrated robot. On the right this camera system has been replaced by a prismatic leg, which stylistically represents a 6-DOF joint that provides equivalent 3D coordinate measurements. The result is a closed-loop mechanism.

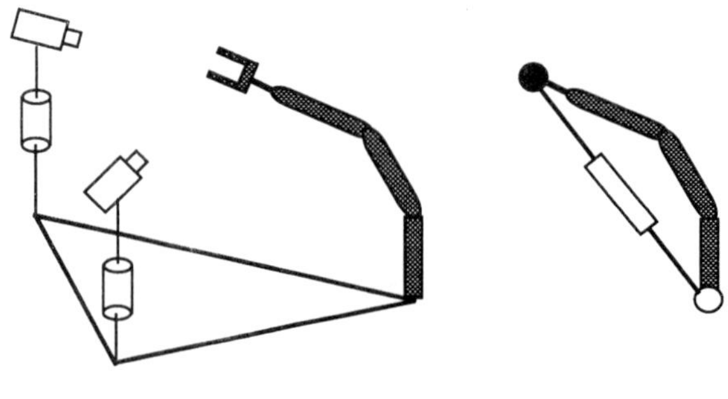

Figure 1: External metrology system modeled as a 6-DOF joint.

2.1 Uncalibrated Visual Sensors

As suggested in the Introduction, visual sensors can be modeled as prismatic legs. In the case of a laser interferometer, the laser beam is typically scanned by pan-tilt mirrors to a retroreflector on the endpoint, and can be modeled as a Hooke joint (pan-tilt mirror), followed by a sensed prismatic joint (the beam length measured interferometrically), and ending with a spherical joint (reflection from the retroreflector). Thus a scanning laser interferometer can be viewed very much as a 6-DOF leg on a Stewart platform or other parallel-drive platform (Figure 2).

Figure 2: A camera and light beam modeled as prismatic leg (box), with hook joint (open circle) and spherical joint (solid circle) at either end.

In the case of a camera, the detector tracks a fixed distinguished point in the environment. The light ray from the camera to the tracked point is also modeled as a prismatic leg. The fixation point on the end effector is modeled as forming a spherical joint with the light ray. The detector measures in terms of linear image coordinates, and can be stylized also as a Hooke joint located at the lens focal point. The scenario of Figure 1 can now be replaced with Figure 3, where the stereo camera system as well as the manipulator are uncalibrated, and both can be calibrated together (Bennett et al. 1991).

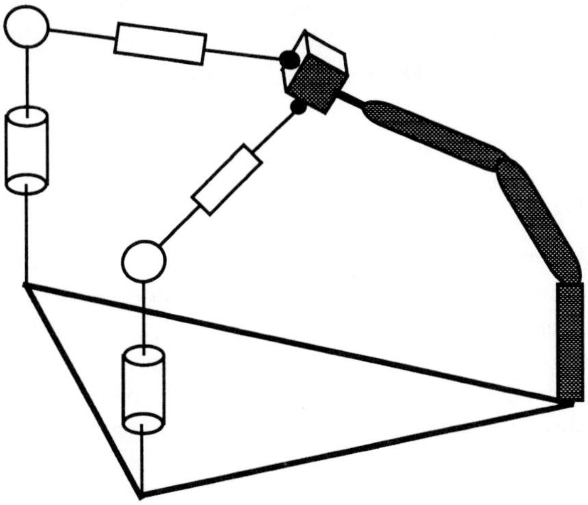

Figure 3: Mechanical equivalent of a stereo camera system on single-DOF tripods tracking a point on a 6-DOF arm.

In manipulator calibration, we distinguish between geometric parameters, such as the Denavit-Hartenberg parameters, and non-geometric parameters, which typically are confined to a joint and express some functional relation between the true joint angles and the joint angle measurements. Nongeometric parameters include joint angle gains, joint angle offsets, and gear eccentricities (Whitney et al., 1986). Similarly, lens and detector parameters for a camera, such as focal length, offsets between the image plane readings and the optical center (Bennett et al. 1991), and radial lens distortion (Zhuang et al. 1995) are analogous camera nongeometric parameters that can be calibrated.

2.2 Loop Closure Equations

Assume the robot is placed into p poses. Following the formulation in (Hollerbach 1993; Wampler et al. 1995), all kinematic calibration methods are considered as closed-loop methods, wherein any endpoint measurement system is considered to form a joint. Consequently, form the kinematic loop closure equations $\mathbf{f}^i$ for the ith pose ($i = 1, ..., p$):

$$\mathbf{f}^i(\mathbf{x}^i, \boldsymbol{\phi}) = \mathbf{0} \tag{1}$$

where $\boldsymbol{\phi}$ is a vector of robot parameters to be calibrated, and $\mathbf{x}^i$ is a vector of joint sensor readings and possibly external sensor readings. Combine (1) for the p poses into a single matrix equation:

$$\mathbf{f}(\boldsymbol{\phi}) = [\mathbf{f}^{1^T} \ldots \mathbf{f}^{p^T}]^T = \mathbf{0} \tag{2}$$

In (2), we treat the sensor readings $\mathbf{x}^i$ as constants for each pose i. Linearize (2) around the nominal values of the parameters:

$$\Delta\mathbf{f} = \frac{\partial \mathbf{f}}{\partial \boldsymbol{\phi}} \Delta\boldsymbol{\phi} = \mathbf{J}\Delta\boldsymbol{\phi} \tag{3}$$

where $\Delta\mathbf{f}$ is the deviation of the computed loop closure equations from zero, $\mathbf{J}$ is the identification Jacobian, and $\boldsymbol{\Delta\phi}$ is the correction to be applied to the current parameter estimate. The calibration problem is then solved by minimizing $\Delta\mathbf{f}$ via iterative least squares.

3 The Calibration Index

3.1 The Mobility Index

With all calibration methods viewed as closed-loop methods, we apply the mobility equation to characterize the degrees of freedom in any calibration setup. The mobility index is (McCarthy 1990):

$$M = 6(L-1) - \sum_{i=1}^{J} D_i \tag{4}$$

where $M =$ mobility index,
$L =$ number of links,
$J =$ number of joints, and
$D_i =$ number of constraints at joint i.

L includes the base link and any extra links attached to the robot to constrain or measure its motion. J includes joints of any additional linkages added for calibration. For a rotational or prismatic joint $D_i = 5$, while for a ball or spherical joint $D_i = 3$. For an external measurement system for a freely moving end effector $D_J = 0$, while for rigid attachment of the endpoint $D_J = 6$.

While (4) is generally correct, there are exceptions due to special or to degenerate mechanisms. The mobility index for planar mechanisms and spherical mechanisms is discussed in (McCarthy 1990). Immobile joints may arise in special cases, such as the immobile elbow joint in anthropomorphic manipulators with fixed end link (Bennett and Hollerbach, 1991). The analysis of special or degenerate mechanisms has to take place on a case-by-case basis, and is best done by looking for linear dependence in the columns of the Jacobian matrix (3). This would happen automatically when checking observability, which reveals any ill conditioning of the Jacobian matrix (Mooring et al., 1991).

3.2 The Sensor Index

The total number of sensors S in the joints can be written as:

$$S = \sum_{i=1}^{J} S_i \tag{5}$$

where $S_i =$ the number of sensed degrees in joint i. We impose the restriction $S_i \leq 6 - D_i$, i.e., there cannot be more sensors than degrees of freedom at a joint. Usually $S_i = 1$ for the lower-order pairs typical of actuated joints, while $S_J = 6$ for full pose measurement of the end effector "joint" J. For an unsensed joint, such as in passive environment kinematics, $S_i = 0$.

3.3 The Calibration Index

Calibration can proceed if $S > M$, that is to say, there is redundant sensing with regard to positioning of the chain. The excess of sensed joints over mobility represents the number of equations per pose that can be used for calibration:

$$C = S - M \tag{6}$$

where we term C the *calibration index.* If P is the number of poses, then CP is the total number of equations for the calibration procedure. Clearly a larger C means less poses are required, other things being equal.

For the single-loop case consisting of a series of sensed lower-order robot joints ($S_i = 1, D_i = 5, \quad i = 1, \ldots, J-1$) with a final joint (S_J, D_J) connecting the end-effector to ground, one has from Eqs. (4-6):

$$C = S_J + D_J \tag{7}$$

According to the calibration index, using full endpoint constraints is an equivalent kinematic calibration method to using full pose measurements. There is potentially a problem with smaller pose selection available with endpoint constrained, but otherwise the mathematics are the same.

4 Legs and Limbs

We are now in a position to compare and contrast calibration methods that have appeared in the literature. We can also readily conceive new calibration methods, by picking a calibration index and posing appropriate endpoint measurements or constraints. We now give examples for different values of the calibration index C.

4.1 Metal Limbs

4.1.1 *Serial-Link Manipulators*

All manners of pose measurement and endpoint constraints have been proposed that result in different calibration indices. Existing methods almost never mix measurements and constraints and choose one or the other exclusively. Examples are given for each value of calibration index.

C=6: $D_J = 0$ and $S_J = 6$ corresponds to full pose measurement (e.g., Vincze et al. 1994).

$D_J = 6$ and $S_J = 0$ corresponds to rigid endpoint attachment (Bennett and Hollerbach 1991).

C=5: $D_J = 0$ and $S_J = 5$ corresponds to 5-DOF pose measurement (e.g., Lau et al. 1985).

$D_J = 5$ and $S_J = 0$ corresponds to 5-DOF endpoint constraints, such as manipulation of an unsensed passive hinge joint (Giugovaz and Hollerbach 1994).

C=4: There is a glaring hole here: no published methods exist for $C = 4$ for serial link manipulators! We will fix this lapse of the field. Let's choose $S_J = 0$ and $D_J = 4$, i.e., a closed-loop method with 4 endpoint constraints. Motion along a line with orientation constraint leaves 4 equations per pose: we can rotate about the axis of the line, but not about the other directions. An example is placing a round peg into a round hole. Another example is sliding a door along its hinges.

C=3: $D_J = 0$ and $S_J = 3$ corresponds to 3-DOF pose measurement, which is provided by basic triangulation systems or CMMs (coordinate measuring machines) (Renders et al., 1991).

$D_J = 3$ and $S_J = 0$ corresponds to 3-DOF endpoint constraints, such as fixed point contact (Bennett and Hollerbach 1991; Craig 1993).

C=2: $D_J = 0$ and $S_J = 2$ corresponds to 2-DOF pose measurements, such as is provided by a single theodolite (Whitney et al., 1986).

$D_J = 2$ and $S_J = 0$ corresponds to 2-DOF endpoint constraints. Motion along a line provides two constraints (Newman and Osborn, 1993). This is contrasted to the case for $C = 4$ in that now there is no orientation constraint. Tracking an edge with a stylus is one example; this case is considered more fully in Section 4.2. Motion along a line is just one example of a path constraint on a plane. Actually, tracing any known curve will do, such as a circle.

C=1: $D_J = 0$ and $S_J = 1$ corresponds to measurement of just 1-DOF pose, provided by a linear transducer such as an LVDT or wire potentiometer (Driels 1993). This is like using a single measuring tape.

$D_J = 1$ and $S_J = 0$ corresponds to plane constraint (Zhong and Lewis 1995). Touching a probe anywhere on a plane yields a viable closed-loop approach.

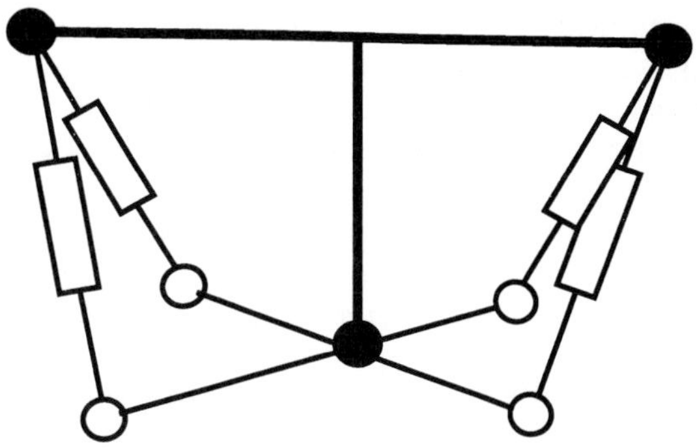

Figure 4: Redundant parallel drive shoulder joint.

4.1.2 *Parallel-Link Manipulators*

Calibration procedures have been increasingly applied to parallel-drive linkages. Often these linkages are platforms activated by prismatic legs (Figure 2), which have a sensed prismatic joint in the middle. Six-DOF rotary legs (i.e., manipulator arms) are also used. The number of legs varies.

Wampler et al. (1995) calibrated the 6-legged Stewart platform ($M = 6$) via a closed-loop procedure. In addition to leg length measurements, all angles on one of the legs were measured ($S = 11$). Hence $C = 5$, and Wampler et al. (1995) were able to forgo an external measurement system. For a regular Stewart platform without the instrumented leg, $S = 6$ and hence $C = 0$. External pose measurement is required; for example, with full pose measurement $S = 12$ and $C = 6$ (Masory et al., 1993).

Nahvi et al. (1994) calibrated a spherical shoulder joint (first introduced by Kurtz and Hayward 1992), driven redundantly by four prismatic legs (Figure 4). For this system, $M = 3$ and $S = 4$, so that $C = 1$. Hence self-calibration is possible. This is one advantage of the redundant design (another is increased workspace); without the extra leg and the sensing it provides, one would have $C = 0$. Nahvi et al. (1994) formulated four equations per pose by comparing one sensor reading against the three others; it follows that three of these equations must have been redundant.

Hollerbach and Lokhorst (1995) calibrated a commercial hand controller, which can be considered a platform driven by three rotary legs. In each arm, there is an unsensed spherical wrist and sensed upper arm joints ($S = 9$). Given that $M = 6$, therefore $C = 3$.

Some papers calibrate serial-link arms, but employ external linkages which can be considered as instrumented prismatic legs. Hence the resulting systems can be considered as two-legged platforms. Goswami et al. (1993) employed a ball bar with linear extension measured by an LVDT. This ball bar can be viewed the same as a leg in a Stewart platform. Here $S_J = 1$ and $D_J = 0$, and so $C = 1$. An equivalent setup was the single-wire potentiometer employed by Driels and Swayze (1994).

Driels (1993) employed a ball bar with fixed length. The length constraint on the ball bar yields one calibration equation for each pose; here $S_J = 0$ and $D_J = 1$, and so $C = 1$. This method is equivalent to the use of the measured extendable ball bar of Goswami et al. (1993) above in terms of yielding one equation per pose, and again illustrates the equivalence of endpoint measurements and constraints. There may be other reasons for preferring one method over another. An advantage of the fixed length ball bar is the lack of instrumentation; an advantage of the variable length ball bar is the automatic adjustment to particular endpoint positions and a larger set of reachable poses.

4.2 Laser Limbs

Laser interferometers have been successfully used as external metrology systems, or have themselves been the subjects for kinematic calibration. When multiple beams are employed, a parallel linkage can be considered to be formed if the laser beams are considered as kinematic linkages (Figure 2). Depending on which joints of these "legs" are measured, different calibration procedures result.

$S_J = 3$: Sometimes the pan-tilt mirror angles are sensed, as well as the linear displacement, as part of the calibration. The laser tracker is then serving as a 3D point measurement source, like many other metrology systems such as camera-based triangulation systems.

$S_J = 5$: Additionally, a measured steerable pan-tilt mirror on the end effector can be employed instead of a retroreflector (Lau et al. 1985).

$S_J = 1$: If a retroreflector is employed and pan-tilt measurements are not provided, then only linear displacement is sensed. Something roughly equivalent was done by Tang and Liu (1993), who employed a single laser displacement meter. Their method required the laser to be positioned roughly perpendicular to the flat surfaces of a stack of blocks and that manipulator displacements be made roughly perpendicularly. The requirement for perpendicularity obviates the need for a scanning head or retroflector.

A problem with laser interferometers is that they provide only incremental distance measurements; an extra calibration procedure is required to establish absolute lengths. The Chesapeake Bay Laser System is a 3-beam laser tracking system that provides 3D point measurement of a retroreflector (Mooring et al. 1991); as in the case just discussed, pan-tilt angles of the scanning mirrors are not provided. Zhuang et al. (1992) proposed to calibrate autonomously for absolute lengths by adding a fourth scanning laser interferometer; at the same time relative locations of each tracker are determined.

We now analyze how addition of a fourth tracker permits auto-calibration, through the equivalent kinematic linkage (Figure 5). Three trackers are modeled as 6-DOF prismatic legs as in Figure 2. The fourth tracker is a 3-DOF leg with a solid block at the end instead of a spherical joint; the other trackers attach with spherical joints to this solid block. The mobility $M = 3$ while $S = 4$, hence $C = S - M = 1$. Thus one equation per pose results, and closed-loop calibration is possible. If only 3 trackers were used, then again $M = 3$ but $S = 3$ and $C = 0$, i.e., self-calibration is not possible. In their paper, Zhuang et al. (1992) actually formulated 16 equations per pose, by comparing each sensor reading against combinations of the other 3 sensor readings; our analysis indicates that 15 of these equations are redundant.

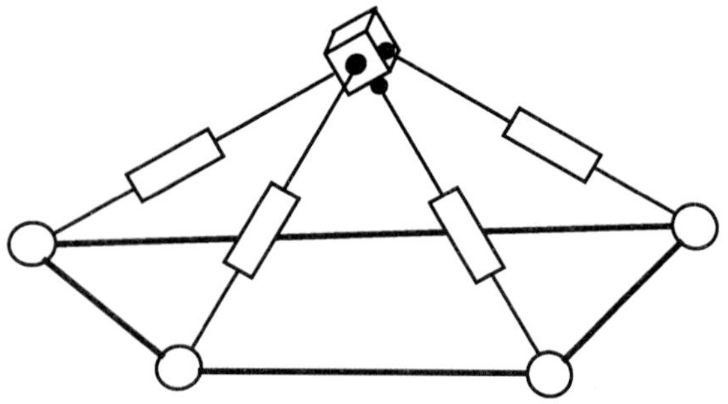

Figure 5: Equivalent linkage for 4-beam laser scanning system.

Newman and Osborn (1993) used a laser to define a line in space, rather than to measure length interferometrically. A retroreflector mounted on the end effector reflects the laser light back to a 4-quadrant detector, whose output is used to servo the manipulator to track the line. For a serial link manipulator, then $D_J = 2$ because orthogonal distance from the line is constrained to be zero, the retroreflector does not constrain orientation, and motion along the line is unconstrained. Since length along the line is not measured, then $S_J = 0$ yielding $C = 2$. In their procedure, Newman and Osborne minimize based on predicted 3D coordinates, which yields one redundant equation.

4.3 Light Limbs

Zhuang et al. (1993) developed a closed-loop procedure for calibration of a robot with a single camera mounted on the end effector. Unlike interferometers, length is not measured, whereas the camera's detector is considered the equivalent of measuring the Hooke joint. A single loop is formed with $S_J = 2$ and $D_J = 0$; hence $C = 2$. In the lens model, radial distortion was parameterized through a quadratic function and identified.

Bennett et al. (1991) showed that an uncalibrated stereo camera system could be simultaneously calibrated with an uncalibrated manipulator. Consider the mechanical linkage equivalent of their system (Figure 3). Two cameras on tripods track a distinguished point on a 6-DOF arm. The rays from each camera to the tracked point are modeled as prismatic legs. Each tripod is presumed to have a sensed rotation about the azimuth axis. For this system, it turns out that $M = 8$ and $S = 12$; hence $C = 4$ and self-calibration is possible. It follows that two of the six equations used in Bennett et al. (1991) must have been redundant.

5 The Role of Input Noise

The closed-loop viewpoint of calibration methods leads to an implicit kinematic equation (1), in which there is no longer a notion of input (joint angles) or output (endpoint pose) variables. Moreover, joint angle measurement error or noise is often significant relative to endpoint measurement error or noise. The approach of ordinary least squares, applied iteratively to the linearized kinematic equations by practically all previous investigators, is therefore problematical. Ordinary least squares minimizes the error in the output variables while assuming there is no noise in the input variables. It is known that the presence of input noise in ordinary least squares leads to bias errors in the parameter estimates (Norton 1986).

For kinematic calibration methods employing passive endpoint constraints, such as rigid attachment, then there can be virtually no endpoint measurement error (except for any attachment backlash) and hence error sources are dominated by joint angle measurement errors. Hence the use of ordinary least squares is inappropriate. Similar comments can be made about other calibration problems, such as inertial parameter estimation, where the error in joint velocities or accelerations is usually significant relative to joint torque error and where the use of ordinary least squares is dominant.

5.1 Orthogonal Distance Regression

A very few papers in robotics have recently begun to address input noise, which we survey. We first examine related developments in the statistics and numerical analysis communities, which have proposed methodologies to handle input noise in linear or nonlinear parameter estimation problems. In the statistics community, the approach is termed by some as orthogonal distance regression, while in the numerical analysis community the approach is called total least squares.

As a simple example, consider the linear equation $y = ax$, where only the slope a is to be determined. If the uncertainties in the input/output variables are σ_x and σ_y, then the weighted normal distances from points (x_i, y_i) to the line are minimized:

$$\chi^2 = \sum_{i=1}^{P} \frac{(y_i - ax_i)^2}{\sigma_y^2 + a^2\sigma_x^2} \tag{8}$$

which is a χ^2 statistic (Bevington 1992; Press et al. 1992). To solve this equation, Press et al. (1992) proposed an iterative nonlinear optimization procedure.

The formulation and solution of equations like (8) have appeared under various names and guises over the years, including total least squares (Van Huffel and Vandewalle 1991), orthogonal distance regression (Boggs et al. 1987), or errors-in-variables regression (Fuller 1987). One common application is to fit a plane to a set of measured 3D points; all 3 components of measured position have noise. For example, this problem arises when measuring multiple points on an end effector and establishing a coordinate system (e.g., An et al. 1988). Another example is circle point analysis, where a plane is to be fit to the circle of points generated by moving a single joint (Mooring et al. 1991). Total least squares solves such equations analytically using singular value decomposition, making unnecessary the iterative solution proposed in Press et al. (1992).

Even more pertinent to kinematic calibration, nonlinear versions of the total least squares problem have been developed. As another simple example, suppose we are now trying to fit a scalar function $y = f(x, a)$ by adjusting some parameters a (Boggs et al. 1987). Assuming again equivalent error in x and y measurements, we should minimize the Euclidean distances from the measurements to the fitted curve f (Figure 6).

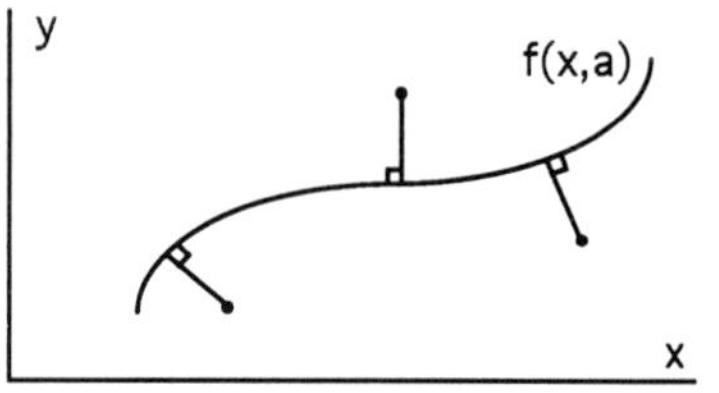

Figure 6: Fitting a curve $y = f(x, a)$ with both input and output errors.

Consider open-loop calibration of serial manipulators, where an explicit kinematic equation results:

$$\mathbf{y}^i = \mathbf{f}^i(\mathbf{x}^i, \boldsymbol{\phi}) \tag{9}$$

where we now separate the endpoint pose measurements $\mathbf{y}^i$ from the joint angle measurements $\mathbf{x}^i$. This forward kinematics function is again stacked for all poses:

$$\mathbf{y} = \mathbf{f}(\mathbf{x}, \boldsymbol{\phi}) \tag{10}$$

where $\mathbf{y} = (\mathbf{y}^1, \ldots, \mathbf{y}^P)$, etc. This equation is rewritten to reflect both the input $\Delta\mathbf{x}$ and output $\Delta\mathbf{y}$ measurement errors:

$$\mathbf{y} - \Delta\mathbf{y} = \mathbf{f}(\mathbf{x} - \Delta\mathbf{x}, \boldsymbol{\phi}) \tag{11}$$

Let the covariances of the output and input variables be $\mathbf{R} = \mathbf{E}(\Delta\mathbf{y}\Delta\mathbf{y}^T)$ and $\mathbf{P} = \mathbf{E}(\Delta\mathbf{x}\Delta\mathbf{x}^T)$, where $\mathbf{E}$ is the expectation operator. Then minimize the weighted orthogonal squared distances from the measurements $(\mathbf{y}, \mathbf{x})$ to the multidimensional surface represented by $\mathbf{y} = \boldsymbol{f}(\mathbf{x}, \boldsymbol{\phi})$:

$$\min_{\boldsymbol{\phi}, \Delta\mathbf{y}, \Delta\mathbf{x}} \chi^2 = \Delta\mathbf{y}^T\mathbf{R}^{-1}\Delta\mathbf{y} + \Delta\mathbf{x}^T\mathbf{P}^{-1}\Delta\mathbf{x} \tag{12}$$

subject to (11). The solution proceeds by substituting (11) into (12) to eliminate $\Delta\mathbf{y}$, and an unconstrained optimization problem results which is solved as usual by linearization and iteration (Schwetlick and Tiller 1985).

Such an approach for robot calibration was taken by Zak et al. (1994). Consider the linearization of (9):

$$\Delta\mathbf{y}^i = \mathbf{J}^i\boldsymbol{\phi} \tag{13}$$

where $\mathbf{J}^i$ is the Jacobians of $\mathbf{f}^i$ with regard to the joint angle sensors $\mathbf{x}$ at pose i. Zak et al. augmented the output variance in this linearized equation with the transformed input variance as follows:

$$\tilde{\mathbf{R}} = \mathbf{R} + \mathbf{J}\mathbf{P}\mathbf{J}^T \tag{14}$$

where $\mathbf{J} = \mathrm{diag}(\mathbf{J}^1, \ldots, \mathbf{J}^P)$ is a block-diagonal matrix. This weighting equation is the generalization of the weighting in (8). Zak et al. (1994) then apply weighted least squares iteratively, at each step treating $\mathbf{J}$ (which contains the unknown parameters) as a constant matrix based on the current estimates, and obtaining the weighted least squares estimate with the modified $\tilde{\mathbf{R}}$.

5.2 Implicit Loop Method

Renders et al. (1991) and Wampler et al. (1995) proposed methods for kinematic calibration equivalent to nonlinear total least squares, with the addition of *a priori* parameter estimates as in damped least squares. Following the development in (Wampler et al. 1995), an implicit equation results from the unifying view of open-loop kinematic calibration methods as closed-loop methods (2). Again, all measurements, whether from joint angle sensors or endpoint sensors, are grouped together.

Let us suppose that the true measurements are $\mathbf{x} = \mathbf{x}_0 + \mathbf{x}_e$, where $\mathbf{x}_0$ is the measurement and $\mathbf{x}_e$ is the measurement error. Similarly, factor the true parameter values into $\boldsymbol{\phi} = \boldsymbol{\phi}_0 + \boldsymbol{\phi}_e$, where $\boldsymbol{\phi}_0$ is our initial estimate and $\boldsymbol{\phi}_e$ is the estimation error. Because $\mathbf{x}_0$ and $\boldsymbol{\phi}_0$ are known, they are incorporated as constants into the loop closure equation (2), which is then a function

$$\mathbf{f}(\mathbf{x}_e, \boldsymbol{\phi}_e) = \mathbf{0} \tag{15}$$

The maximum likelihood estimate minimizes

$$\chi^2 = \mathbf{x}_e^T\mathbf{R}^{-1}\mathbf{x}_e + \boldsymbol{\phi}_e^T\mathbf{M}^{-1}\boldsymbol{\phi}_e \tag{16}$$

subject to (15), where $E(\mathbf{x}_e\mathbf{x}_e^T) = \mathbf{R}$ and $E(\boldsymbol{\phi}_e\boldsymbol{\phi}_e^T) = \mathbf{M}$.

This minimization problem is solved by linearization of the constraints and iteration. Let the scaled variables $\mathbf{y}$ and $\mathbf{q}$ be defined via $\mathbf{x}_e = \mathbf{R}^{1/2}\mathbf{y}$ and $\boldsymbol{\phi}_e = \mathbf{M}^{1/2}\mathbf{q}$, where the superscript 1/2 means the symmetric square root. Iterate from an initial guess $\mathbf{y} = \mathbf{0}$ and $\mathbf{q} = \mathbf{0}$ to find corrections $\Delta\mathbf{y}$ and $\Delta\mathbf{q}$ to minimize

$$\chi^2 = (\mathbf{y} + \Delta\mathbf{y})^T(\mathbf{y} + \Delta\mathbf{y}) + (\mathbf{q} + \Delta\mathbf{q})^T(\mathbf{q} + \Delta\mathbf{q}) \tag{17}$$

subject to the linearized constraints

$$\begin{aligned} -\mathbf{f}'(\mathbf{y},\mathbf{q}) &= \frac{\partial \mathbf{f}'}{\partial \mathbf{y}}(\mathbf{y},\mathbf{q})\Delta\mathbf{y} + \frac{\partial \mathbf{f}'}{\partial \mathbf{q}}(\mathbf{y},\mathbf{q})\Delta\mathbf{q} \\ &\equiv \mathbf{J}_y\Delta\mathbf{y} + \mathbf{J}_q\Delta\mathbf{q} \end{aligned} \tag{18}$$

Compute the QR-decomposition $\mathbf{Q}\tilde{\mathbf{R}} = \mathbf{J}_y^T$ (this is actually done individually for each pose) and define

$$\mathbf{D} = \tilde{\mathbf{R}}^{-T}\mathbf{J}_y, \quad \mathbf{E} = \mathbf{Q}^T\mathbf{y} - \tilde{\mathbf{R}}^{-T}\mathbf{f}(\mathbf{y},\mathbf{q}) \tag{19}$$

The variables $\Delta\mathbf{y}$ are eliminated and the step in $\mathbf{q}$ found from:

$$\begin{bmatrix} \mathbf{D} \\ \mathbf{I} \end{bmatrix} \Delta\mathbf{q} = \begin{bmatrix} \mathbf{E} \\ -\mathbf{q} \end{bmatrix} \tag{20}$$

and the updated error estimates are

$$\mathbf{y} + \Delta\mathbf{y} = \mathbf{Q}(\mathbf{E} - \mathbf{D}\mathbf{q}) \tag{21}$$

Covariance matrices are scaled *post facto* based on the value of χ^2.

Because the loop closure equation (15) is an implicit function of all measurements, Wampler et al. (1995) coined the term *Implicit Loop Method.* Whereas in nonlinear total least squares (12) the output errors could be directly eliminated by using the explicit constraint equation, for the implicit loop method the linearization of the implicit constraint equation was required for this elimination. The earlier approach of Renders et al. (1991) deals only with open-loop calibration of serial arms; the resulting explicit constraint equation is incorporated using Lagrange multipliers rather than using variable elimination as in (12). Like Zak et al. (1992), linearization is at nominal values, whereas in the Implicit Loop Method the unknown parameters from the linearization also come into play.

6 Discussion

This paper has proposed the calibration index, which indicates how many equations per pose result for a calibration method. Use of this index can help one to understand how a particular calibration method relates to others, and what the tradeoffs are in fashioning a calibration method. To get enough equations for calibration, one needs sufficient numbers of poses. With a higher calibration index, a particular calibration method would require fewer poses. On the other hand, more complex metrology systems could be required to get more measured components of pose if open-loop calibration is used.

It should be be emphasized that the calibration index is only a first-order characterization of calibration methods. Even if two putative methods have the same calibration index, other considerations might dictate a preference of one over the other. Perhaps an external metrology system is not available, in which closed-loop methods would be perferred. On the other hand, closed-loop methods typically constrain pose set selection more than open-loop calibration, so one has to analyze observability of parameters carefully.

With the view that kinematic calibration methods are all closed-loop methods, an implicit loop closure equation results that doesn't distinguish input from output variables. The issue then arises of handling input noise as well as output noise, which is scarcely recognized in the robotics calibration literature. Orthogonal distance regression has been developed to handle such problems in the statistics literature. A related development is the implicit loop method of Wampler et al. (1995), which also incorporates *a priori* parameter estimates.

Acknowledgments. Support for this research was provided by the NSERC NCE IRIS II Project AMD-5, and by NSF Grant MIP-9508588.

References

An, C.H., Atkeson, C.G., and Hollerbach, J.M. 1988. *Model-Based Control of a Robot Manipulator.* Cambridge, MA: MIT Press.

Bennett, D.J., and Hollerbach, J.M. 1991. Autonomous calibration of single-loop closed kinematic chains formed by manipulators with passive endpoint constraints. *IEEE Trans. Robotics and Automation.* 7: 597-606.

Bennett, D.J., Hollerbach, J.M., and Geiger, D. 1991. Autonomous robot calibration for hand-eye coordination. *Int. J. Robotics Research.* 10: 550-559.

Bevington, P.R., and Robinson, D.K. 1992. *Data Reduction and Error Analysis for the Physical Sciences.* N.Y.: McGraw-Hill.

Boggs, P.T., Byrd, R.H., and Schnabel, R.B. 1987. A stable and efficient algorithm for nonlinear orthogonal distance regression. *SIAM J. Sci. Stat. Comput.*. 8: 1052-1078.

Craig, J.J. 1993 (November). Calibration of industrial robots. *Proc. 24th Intl. Symp. on Industrial Robots.* Tokyo: pp. 889-893.

Driels, M.R. 1993. Using passive end-point motion constraints to calibrate robot manipulators. *J. Dynamic Systems, Meas., Control.* 115: 560-565.

Driels, M.R., and Swayze, W.E. 1994. Automated partial pose measurement system for manipulator calibration experiments. *IEEE Trans. Robotics and Automation.* 10: 430-440.

Fuller, W.A. 1987. *Measurement Error Models.* NY: John Wiley & Sons.

Giugovaz, L., and Hollerbach, J.M. 1994 (Sept. 12-16). Closed-loop kinematic calibration of the Sarcos Dextrous Arm. *Proc. IEEE/RSJ Intl. Conf. on Intelligent Robots and Systems.* Munich, Germany: pp. 329-334.

Goswami, A., Quaid, A., and Peshkin, M. 1993. Identifying robot parameters using partial pose information. *IEEE Control Systems.* 13(5): 6-14.

Hollerbach, J.M. 1993 (Oct. 2-5). Advances in robot calibration. *Robotics Research: The Sixth International Symposium.* Hidden Valley, PA: pp. 319-326.

Hollerbach, J.M., and Lokhorst, D. 1995. Closed-loop kinematic calibration of the RSI 6-DOF hand controler. *IEEE Trans. Robotics and Automation.* 11: 352-359.

Kurtz, R., and Hayward, V. 1992. Multiple-goal kinematic optimization of a parallel spherical mechanism with actuator redundancy. *IEEE Trans. Robotics and Automation.* 8: 644-651.

Lau, K., Hocken, R., and Haynes, L. 1985. Robot performance measurements using automatic laser tracking techniques. *Robotics & Computer-Integrated Manufacturing.* 2: 227-236.

Masory, O., Wang, J., and Zhuang, H. 1993 (May 2-7). On the accuracy of a Stewart platform - part II Kinematic calibration and compensation. *IEEE Intl. Conf. Robotics and Automation.* Atlanta: pp. 1:725-731.

McCarthy, J.M. 1990. *Introduction to Theoretical Kinematics.* Cambridge, MA: The MIT Press.

Mooring, B.W., Roth, Z.S., and Driels, M.R. 1991. *Fundamentals of Manipulator Calibration.* NY: Wiley Interscience.

Nahvi, A., Hollerbach, J.M., and Hayward, V. 1994 (May 8-13). Closed-loop kinematic calibration of a parallel-drive shoulder joint. *Proc. IEEE Intl. Conf. Robotics and Automation.* San Diego: pp. 407-412.

Newman, W.S., and Osborn, D.W. 1993 (May 2-7). A new method for kinematic parameter calibration via laser line tracking. *IEEE Intl. Conf. Robotics and Automation.* Atlanta: pp. 2:160-165.

Norton, J.P. 1986. *An Introduction to Identification.* London: Academic Press.

Press, W.H., Teukolsky, S.A., Vetterling, W.T., and Flannery, B.P. 1992. *Numerical Recipes in C.* Cambridge, UK: Cambridge Univ. Press.

Renders, J.-M., Rossignol, E., Becquet, M., and Hanus, R. 1991. Kinematic calibration and geometrical parameter identification for robots. *IEEE Trans. Robotics and Automation.* 7: 721-732.

Schwetlick, H., and Tiller, V. 1985. Numerical methods for estimating parameters in nonlinear models with errors in variables. *Technometrics.* 27: 17-24.

Tang, G.-R., and Liu, L.-S. 1993. Robot calibration using a single laser displacement meter. *Mechatronics.* 3: 503-516.

Van Huffel, S., and Vandewalle, J. 1991. *The Total Least Squares Problem: Computational Aspects and Analysis.* Philadelphia: SIAM.

Vincze, M., Prenninger, J.P., and Gander, H. 1994. A laser tracking system to measure position and orientation of robot end effectors under motion. *Intl. J. Robotics Research.* 13: 305-314.

Wampler, C., and Arai, T. 1992 (Sept. 24-26). Calibration of robots having kinematic closed loops using non-linear least-squares estimation. *Proc. IFTOMM Symp.*. Nagoya, Japan: pp. 1:153-158.

Wampler, C.W., Hollerbach, J.M., and Arai, T. 1995. An Implicit Loop Method for kinematic calibration and its application to closed-chain mechanisms. *IEEE Trans. Robotics and Automation.* 11: 710-724.

Whitney, D.E., Lozinski, C.A., and Rourke, J.M. 1986. Industrial robot forward calibration method and results. *ASME J. Dynamic Systems, Meas., Control.* 108: 1-8.

Zak, G., Benhabib, B., Fenton, R.G., and Saban, I. 1994. Application of the weighted least squares parameter estimation method for robot calibration. *J. Mechanical Design.* 116: 890-893.

Zhong, X.-L., and Lewis, J.M. 1995 (May 21-26). A new method for autonomous robot calibration. *IEEE Intl. Conf. Robotics and Automation.* Nagoya: pp. 1790-1795.

Zhuang, H., Li, B., Roth, Z.S., and Xie, X. 1992. Self-calibration and mirror center offset elimination of a multi-beam laser tracking system. *Robotics and Autonomous Systems.* 9: 255-269.

Zhuang, H., Wang, K., and Roth, Z.S. 1995. Simultaneous calibration of a robot and a hand-mounted camera. *IEEE Trans. Robotics and Automation.* 11: 649-660.

Zhuang, H., Wang, L., and Roth, Z.S. 1993 (May 2-7). Simultaneous calibration of a robot and a hand-mounted camera. *IEEE Intl. Conf. Robotics and Automation.* Atlanta: pp. 2:149-154.

PART 15

MICROROBOTICS

SESSION SUMMARY

Hirofumi Miura
Dept of Mechano-Informatics, The University of Tokyo
7-3-1 Hongo, Bunkyo-ku, Tokyo 113, Japan
e-mail: miura@leopard.t.u-tokyo.ac.jp

An NSF report "Small machines, Large opportunities, a report on the emerging field of microdynamics (Report of the NSF-Workshop on Microelectromechanical Systems [MEMS] Research, 1988, by K. Gabriel, J. Jarvis and W. Trimmer)" has spurred research activities on microtechnology at many institutes around the world. This report focused on the results of three workshop held in 1987 and 1988. These workshops covered the microfabrication of micromechanisms using silicon process originaly developed for manufacturing of semiconductors and integrated circuits. Following the above report, many examples of micromechanical fabrication were reported.

Although possible applications of micromechanisms are suggested by the examples reported, these examples are only trials and cannot be put to practical use without improvements. Microsensors based on microfabrication are now in practical use in some fields, but these are exceptions. These micromechanisms only possess passively deformative parts. In these several years, many national research projects on micromechanical systems have started in many countries. Many researchers on robotics also have started researches on microrobots. In the Sixth International Symposium on Robotics Research(6th ISRR) held in Pittsburgh, Pennsylvania, 1993, the session "*Micro/Miniature Mechanisms*" was introduced first, and four papers were presented – (1) Microrobotics, (2) Simple Microflight Mechanism on Silicon Wafer, (3) Lorentz Levitation Technology: A New Approach to Fine Motion Robotics, Teleoperation, Haptic Interfaces and Vibration Isolation, (4) Miniature Direct-Drive Rotary Actuators. Micromechanism technology is a very new field and has not been defined well. Some persons call cm-sized mechanisms micromechanisms and other persons are much interested in handling molecule-sized structures. Both of these researches are called "microtechnology" in these days. In this Seventh International Symposium on Robotics Research, the session of *Microrobotics* includes different kinds of microtechnology. In this session, four papers were presented. Two of them, Ronald S. Fearing's paper and Jehuda Ish-Shalom's paper, deal with "multi-robot planar motion system". The size of robots is the same size to the parts to be manipulated. For assembly and testing purpose, for example, assembly or testing of hybrid circuit modules, millimeter and smaller size objects need to be transported over distances on the order of several object diameters. In order to handle these sub-millimeter size mechanical or electronic components, a miniature manipulator system must be developed. Many kinds of microfabrication technology must be applied for developing these kinds of system. For example, electromagnet microcoil array with sub-mm sized diameter must be developed on the planar work space to manipulate magnetized miniature robots. Some sucessful example are shown in these papers.

Tomomasa Sato is trying to develop the robot with three dimensional motion which assemble the sub-milli sized mechanical structure under the microscope. In this case, the robot itself is nomal-size, not microsize. Techniques for microteleoperation are also reported. Isao Shimoyama is developing very interest robots. They are named hybrid robots. They are consisted of mechanical parts and living organisms. An example shown in the paper is a mobile robot with the antennae of a silkworm moth. This robot follows the pheromone stream of a female silkworm moth. His idea is that the insect may be a good model for the microrobot. In the insect we can find many good structures to get the design principle of the microsized structure.

A Planar Milli-Robot System on an Air Bearing*

Ronald S. Fearing
Department of EE&CS
University of California
Berkeley, CA 94720
email: ronf@eecs.berkeley.edu

1 Introduction

For assembly and testing purposes, for example, assembly or testing of hybrid circuit modules, millimeter and smaller size objects need to be transported over distances on the order of several object diameters. In order to handle these submillimeter size mechanical or electronic components, a miniature manipulator system has been developed. There are many advantages to shrinking robots and mechanical actuators to the same size as the parts to be manipulated. Extremely delicate forces can be applied, robots can be readily parallelizable, and the relative accuracy required can be markedly reduced. One of the major difficulties in building millimeter scale microrobots is overcoming forces due to friction and wiring. Friction forces can be reduced by using levitation or using fluid lubrication, such as an air-bearing. This paper considers some initial steps towards implementing a miniature robotic system using magnetically driven platforms about 7 mm square on a 35 mm square workspace. The position of the platform is sensed using an array-type capacitive proximity sensor. The system should be scalable to one tenth the present size. The mobile platforms can be used in cooperation to grab and position small objects.

When parts to be handled are less than one millimeter in size, adhesive forces between gripper and object can be significant compared to gravitational or inertial forces. These adhesive forces arise primarily from surface tension, Van der Waals, and electrostatic attractions and can be a fundamental limitation to part handling in a gas environment. While it is possible to fabricate miniature versions of conventional robot grippers, for example from polysilicon, it appears that it will be difficult to overcome adhesion effects for

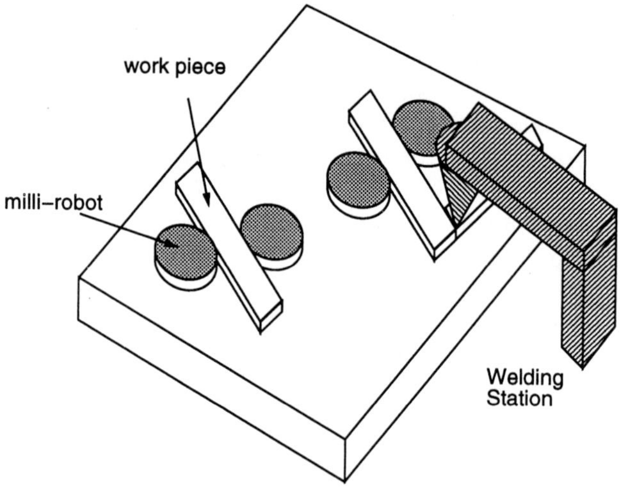

Figure 1: Generic model of miniature planar robotic workcell. The goal is to build a system on which miniature parts can be transported, tested, and assembled.

the smallest parts. Thus, manipulation of parts on the order of 10 micron or smaller may best be done in a fluid medium. Figure 1 shows a typical concept for a planar milli-manipulator system. It can be thought of as a miniature version of the RobotWorld system (see for example [Ish-Shalom, 1994] for a description of some current RobotWorld research). I believe Prof. H. Fujita is one of the main originators of the minia-

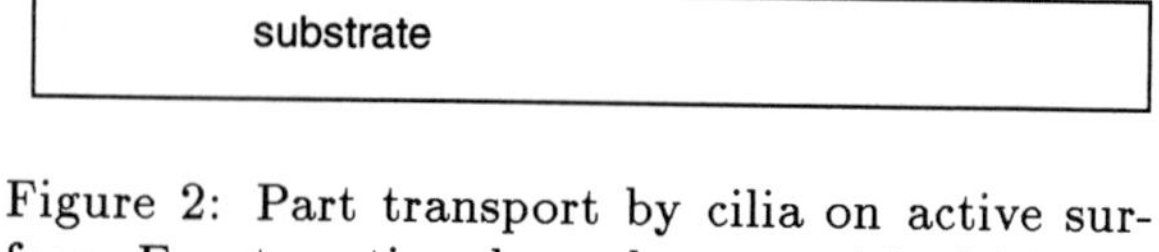

Figure 2: Part transport by cilia on active surface. Exact motion depends on model of friction properties.

*This work was funded in part by: NSF-PYI grant IRI-9157051.

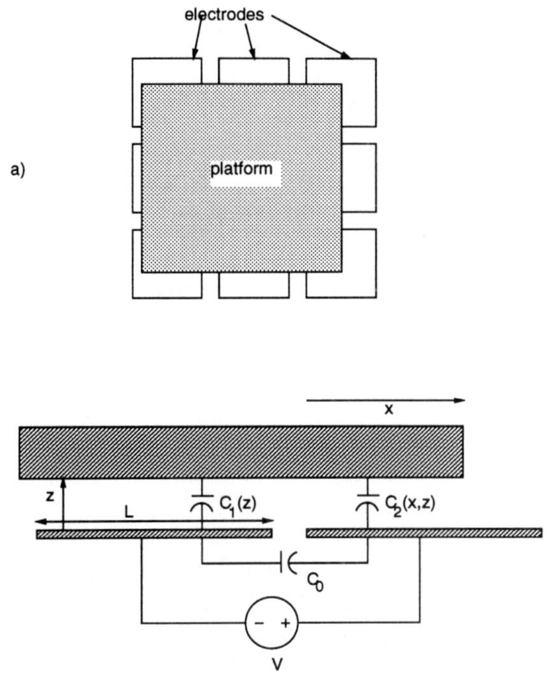

Figure 3: Planar electrostatic drive with levitated platform. a) Electrode configuration. b) Equivalent circuit.

ture planar manipulator system concept [Fujita and Omodaka, 1987; Fujita 1989]. There are two basic types of systems-the active surface where motion of the surface applies mechanical forces to any objects on the surface, and the electromagnetic type where parts and pushing elements can be driven by magnetic or electrostatic forces.

A typical example of an active surface is the "cilia" type drive of Figure 2. By coordinated wave gaits of cilia, parts can be transported in any direction on the surface [Fujita and Gabriel, 1991]. It is possible that sticking effects between cilia and parts in a non-liquid environment could cause indeterminate behavior [3, 2, 19, 20]. There have been a wide variety of actuation schemes for cilia, including thermal [Ataka et al 1993], ultrasonic [Furuhata Hirano Fujita 1991], electrostatic [Boehringer et al 1994], and magnetic [Liu and Will, 1995] using magnetic flap actuators developed by [Liu, Tsao and Tai, 1995].

Due to large friction forces on small dry parts, a fluid medium and lubrication (either gas or liquid) is commonly used. For example, whole silicon wafers on a gas bearing can be driven electrostatically [Tokisue et al 1991]. Another type of active surface uses controlled jets of air to push parts around. This method is used in semiconductor wafer transport and has been prototyped at the micro scale [Konishi and Fujita 1993]. Alternative methods to avoid problems with friction include levitation due to the Meissner effect [Fujita, 1989], [Kim, Katsurai, Fujita 1990] [Maeda, Aihara, Fujita 1993] and a partially levitated motion phase for a miniature magnetic actuator [Busch-Vishniac et al 1990], [Wang et al, 1991]. Other possibilities are active magnetic levitation [Robichaux and Ahmed, 1992], or self-pressurized air bearings [Fearing, 1992]. With some magnetic actuators, the forces are strong enough to overcome friction, and lubrication is not essential [Wagner et al 1993], [Inoue et al 1995].

Biological cells in a liquid medium can be moved by electrostatic attraction. Integrated systems using arrays of electrodes to apply electrostatic fields have been used to transport and fuse cells [Fuhr, et al 1991; Sato et al, 1990; Washizu, 1990]. Sensing systems can also be integrated to determine cell position. Dry particles can also be moved by AC potentials [Moesner and Higuchi, 1995]. As some objects may be damaged by high electric fields, mechanically driven transport systems are also needed.

An integrated system for manipulating dry parts in the plane using multiple mobile manipulation units was proposed by Pister et al [1990]. This device consists of a 1 cm^2 substrate with an air bearing to support individual $1mm^2$ platforms. The individual platforms are driven in the plane by electrostatic forces (see Figure 3), and could carry grippers, probes for sensing, or tools for processing. (To be useful, silicon micro-robots will need tools to interact above the plane of the wafer, not just in it. One promising approach, called "silicon origami" [Suzuki et al, 1992] or micro-hinges [Pister, et al 1992] allows planar fabrication, followed by 3 dimensional assembly of structures.) By incorporating capacitive position sensing of the platforms, an integrated micro system for parts handling could be made on a single chip.

1.1 Actuation Methods

One scaling law that is helpful for micro-actuators is the increased break-down field strength of very small gaps due to the Paschen effect. Thus for small sizes, the obtainable electrostatic forces can be stronger than electromagnetic forces [Trimmer and Jebens, 1989], [Price et al, 1989].

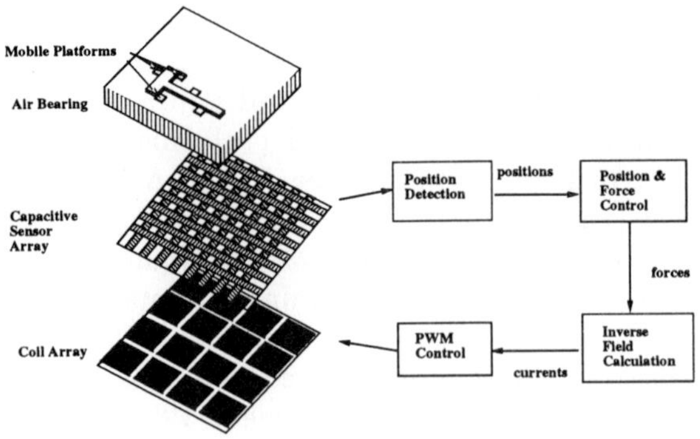

Figure 4: Planar milli-robot system block diagram. Positions of robots and parts are detected using capacitive sensing, and robots are driven magnetically. The servo rate is 120 Hz, limited by sensor electronics scan time. The sensor is an 8 by 8 array, and the electromagnets are a 6 by 6 array.

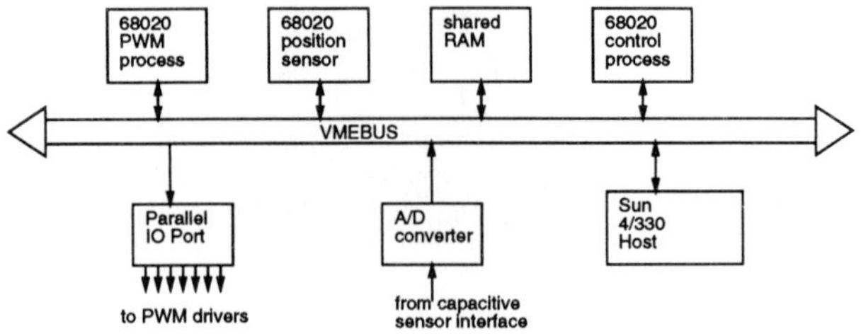

Figure 5: Multiprocessor system for milli-robot control. Two processors are dedicated solely for sensor and PWM processing, and a third processor implements the control algorithm.

Unfortunately, electrostatic actuation may have problems with charge accumulation in the dielectric, see for example [Anderson and Colgate, 1991]. Another problem is that an electrostatic drive won't work in a conductive fluid medium such as water. Magnetic actuation may thus be an attractive alternative, see for example [Busch-Vishniac, 1991]. Recently, magnetic actuators have been made at the micro-scale using extensions of micromachining techniques, see for example [Ahn et al 1993], [Guckel et al 1993] [Wagner and Benecke, 1991], [Wagner et al 1991]. Comprehensive surveys of available actuator technology and scaling effects have been published [Dario et al 1992], [Fujita and Gabriel, 1991], [Shimoyama, 1995], [Trimmer and Jebens, 1989].

2 System Description

As parts become small, the inertial forces become negligible while the friction forces and electrostatic forces can become very large. An air bearing works very well at the macro and micro scales to reduce friction. To avoid constructing a miniature air bearing with thousands of air holes, a self-pressurizing squeeze-film air bearing is used. Since the bearing gap is less than about $5\mu m$, surfaces must be very clean and very flat for the bearing to function properly.

The design considered in this paper uses a mobile platform made from high permeability or magnetic material, with embedded coils in a stator. The forces required for translation of the platform are very low, so even low strength magnetic fields may work well. Magnetics may also be attractive when using an internal high-current low-voltage source like a single cell battery, since the coil will usually be fairly low impedance.

It is useful to be able to sense the position of both driven platforms and passive parts floating on the air bearing. For example, magnetically driven platforms could be used to push and grasp passive pieces of silicon which are not controlled by the magnetic fields. Capacitive sensing is used to detect the presence and position of dielectrics and conductive materials. Of course a vision system could be used for platform position detection, [Inoue et al 1995] but the frame rate is typically fairly low at 60 Hz, and optics can limit the workspace for vertical assembly operations. Figure 4 shows the overall view of the milli-robot system, with the closed loop control system shown in Figure 5. Conventional printed circuit techniques, with a resolution of $75\mu m$ are used to fabricate the sensor and electromagnet arrays. The sensor and magnet layers are glued together, and a thin piece of cover glass is used for the bottom surface of the air bearing. The circuit board layers are then glued to a rigid backing plate and a piezoelectric driver to form the complete mechanical system (Figure 6). (The vertical axis in the figure is not to scale). It is interesting to note that there is no precision assembly required other than ensuring that the cover glass is flat to within $2\mu m$. The mobile platforms are simply square or round pieces of silicon or glass with a magnet glued on.

The 7.5 mm diameter platforms float on a self-pressurized air bearing created by the squeeze-film effect between the platform and glass covering the capacitive sensor array. The capacitive sensor array and magnetic coil array are fabricated on $250\mu m$ thick printed circuit board mate-

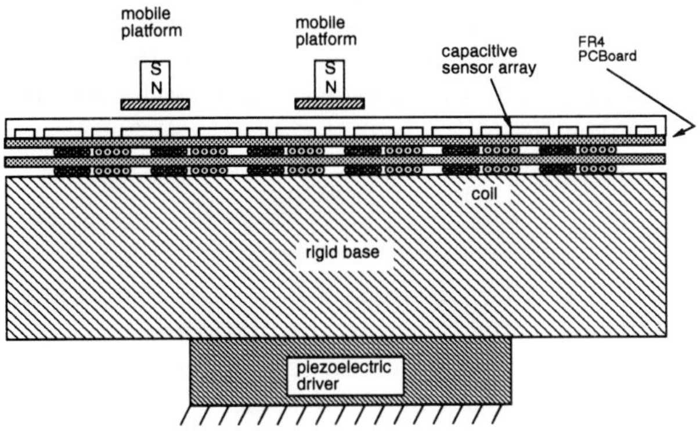

Figure 6: Overall construction of milli-robot system.

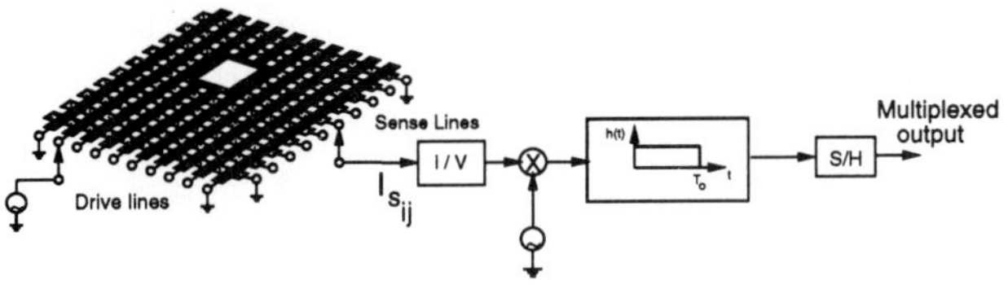

Figure 7: Position sensor array and interface electronics. Sensitivity of the electronics is about $5 \times 10^{-16}F$ at 120 Hz bandwidth.

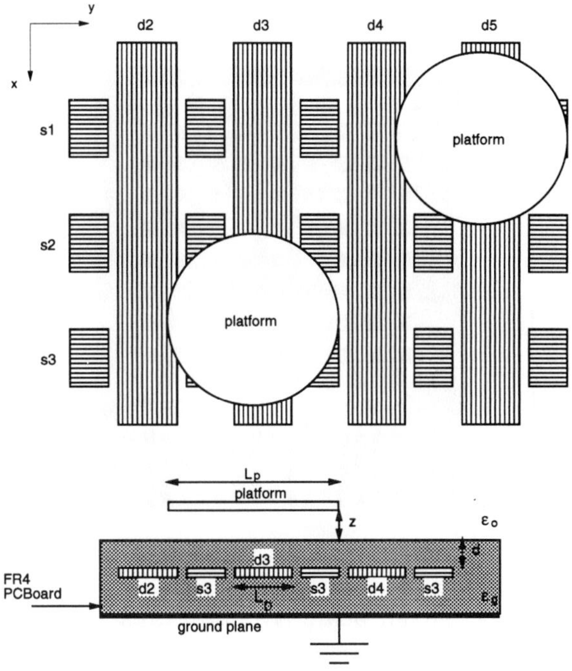

Figure 8: Geometry for calculating sensor capacitances. Platform is 7.5 mm wide, drive and sense line width is approximately 2.5 mm.

rial chosen to reduce the gap distance for the magnetic field. A rigid aluminum base, approximately 1 cm thick, is used to reduce flexible bending modes. A piezo-electric driver moves the whole assembly vertically at 20 KHz with an amplitude of approximately 10 nm, giving a calculated air bearing thickness of about $5\mu m$.

3 Capacitive Position Sensing Array

The capacitive position sensor consists of an array formed by 8 driven electrodes running perpendicular over 8 sense electrodes, as depicted in Figure 7. The array is covered with approximately $1.5 \times 10^{-4}m$ of glass to give a smooth surface for the air bearing. A 250 KHz sine wave is injected into one of 8 drive lines, and the capacitive coupling current is measured on one of 8 sense lines. The array is scanned at a 120 Hz frame rate, with unaddressed rows and columns being grounded. Stray capacitance to ground on the sensor output, for example cable capacitance, is suppressed using a current-to-voltage converter. The current to voltage converter ideally has zero input impedance. A synchronous demodulator is used to measure the magnitude of the coupling current. The sensor easily measures a capacitance change of 1 part in 1000. The capacitive sensor has rows and columns spaced at $2.5 \times 10^{-3}m$.

The coupling currents from the sensor will be a function of the conductive and dielectric properties and position of objects resting above the sensor surface. The sensing array can be used to detect the position of a platform as shown in Figure 8. The mobile platform consists of a 7.5 mm diameter conductive plate. (The coupling current is maximized by using a small conductive platform). The actual top layer circuit board used is shown in Figure 9. Other approaches to planar capacitive sensors can be found in [Bonse et al 1994] and [Miller, 1990].

3.1 Height Measurement

With the platform directly centered above one sense line and one drive line (that is, no overlap with other sense or drive lines), the float height can be found as a function of the measured currents. Let I_o' be the sensed current when the platform is on the substrate, and let I_f' be the current when the platform is floating, with air gap permittivity ϵ_o, and the permittivity of the

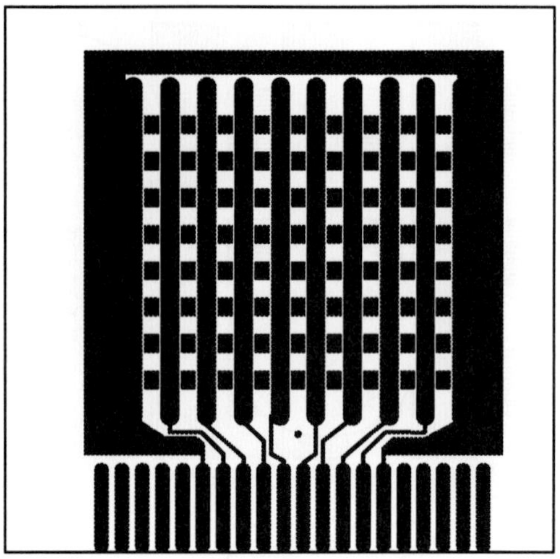

Figure 9: Circuit board layout for capacitive sensor array. Element spacing is 2.5 mm.

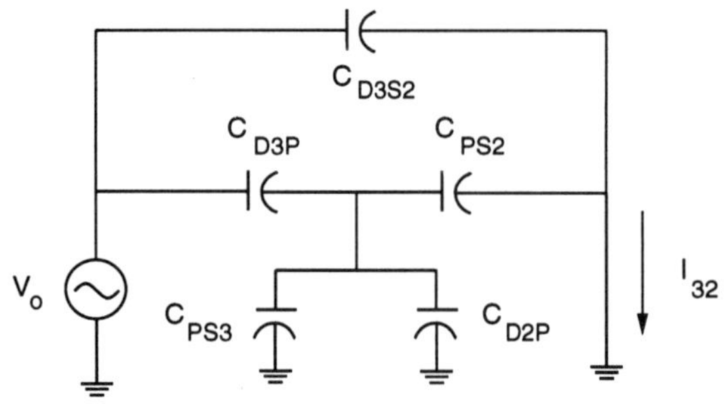

Figure 10: Equivalent circuit for capacitive sensor array. Fixed coupling capacitance C_{D3S2} is subtracted as an offset. Capacitance is measured with a current amplifier at virtual ground, so shielding capacitance can be neglected.

covering layer of glass $\epsilon_g \approx 4.8\epsilon_o$. The height of a square platform can be determined from

$$z = \frac{\epsilon_o}{\epsilon_g(2L_P - L_D)}\left(\frac{1}{I_f'} - \frac{1}{I_o'}\right), \qquad (1)$$

where L_P and L_D are the width of the platform and drive electrode respectively.

3.2 Position Measurement

With the platform as shown in the figure, the equivalent electrical circuit is shown in Figure 10 when column 3 ("D3") is driven, and row 2 ("S2") is sensed. Let I_{Sij} be the sensed current when drive i is driven, and sense j is read. Note that for this configuration, the unused grounded lines D2 and S3 drain off coupling current from I_{S32}. Scanning all 4 combinations, the sensed currents are given by

$$I_{Sij} = \frac{C_{DiP}C_{PSj}}{C_\Sigma} j\omega V_o \qquad (2)$$

where $C_\Sigma = C_{Di} + C_{Di+1} + C_{Si} + C_{Si+1}$ plus any additional capacitance to the platform.

An earlier version of this paper [Fearing 1992] considered measuring the position of square platforms. However, the orientation of a platform is not directly controlled in this system, so a circular platform is preferred as there are only 2 degrees of freedom to sense. The capacitance between the platform and drive and sense lines can be approximated as a parallel plate capacitor. Thus the capacitance from the platform to drive strip 1 is given by

$$C_{D1P} = \frac{\epsilon_o \epsilon_g g(y)}{\epsilon_o d + \epsilon_g z}, \qquad (3)$$

where $g(y)$ is the area of intersection between the drive line and the circular platform for the platform at relative position y, and d is the glass layer thickness, about $150\mu m$. Similarly, the capacitance from platform to drive strip 2 is

$$C_{D2P} = \frac{\epsilon_o \epsilon_g g(y - 2L_D)}{\epsilon_o d + \epsilon_g z}, \qquad (4)$$

where L_D is the drive line width. After some algebra, it can be shown that the current ratio

$$\frac{I_{D1S1} + I_{D1S2}}{I_{D2S1} + I_{D2S2}} = \frac{g(y)}{g(y-2)}. \qquad (5)$$

This ratio is approximately a parabola, thus the relative position can be obtained through the square root of the current ratio.

Extracting the x position is slightly more complicated, as there is coupled dependence on x and y. The capacitance from platform to sense lines is proportional to the intersection of a circle with two squares:

$$C_{PS1} = \frac{\epsilon_o \epsilon_g h(x, y)}{\epsilon_o d + \epsilon_g z}, \qquad (6)$$

where $h(x, y)$ is the area of intersection of the sense line with the circular platform. Similarly for sense line 2,

$$C_{PS2} = \frac{\epsilon_o \epsilon_g h(x - 2L_D, y)}{\epsilon_o d + \epsilon_g z}. \qquad (7)$$

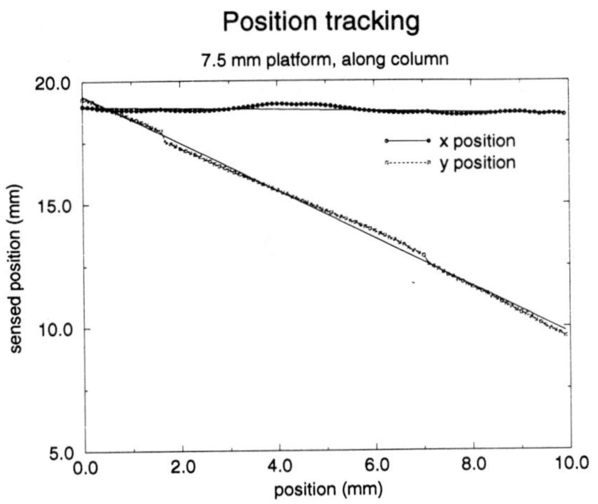

Figure 11: Position estimation for motion in y direction.

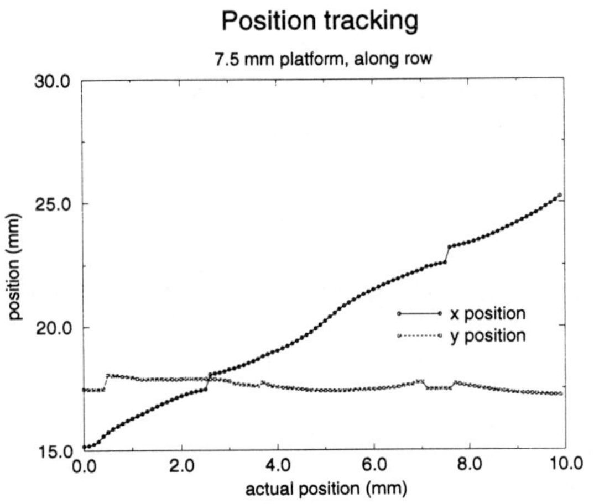

Figure 12: Position estimation for motion in x direction.

After some algegra, it can be shown that the current ratio

$$\frac{I_{D1S1} + I_{D2S1}}{I_{D1S2} + I_{D2S2}} \tag{8}$$

is equivalent to

$$\frac{h(x,y)[g(y) + g(y - 2L_D) + h(x - 2L_D, y)]}{h(x - 2L_D, y)[g(y) + g(y - 2L_D) + h(x,y)]}. \tag{9}$$

This current ratio can be approximated as a separable function of x and y, and solved for x given the y position obtained above. This is the basic form of the equations for determining relative platform displacement. There are actually four different configurations of the platform, depending on how it is centered with respect to the array.

The current ratio is generally a well-behaved measure, which has the advantage of eliminating dependencies on dielectric constants and the float height. (Note that the float height is not constant, and is a function of the magnetic field strength). Before the current ratio method can be used, the offset current is subtracted from each sensed current. Preliminary experiments were performed to evaluate the position sensing algorithm. In the experiment, a micrometer stage with $10\mu m$ accuracy was moved in $100\mu m$ steps while a 7.5 mm diameter platform was levitated on the air bearing to remove friction effects, and maintain constant height. The platform was moved using a permanent magnet attached to the micrometer stage. Results are shown in Figure 11 and Figure 12. Several imperfections are apparent, particularly the discontinuities at 5.1 mm spacing which corresponds to the sensor element spacing. A detailed examination of the discontinuities has shown that they result from coupling of the fringing fields which have been ignored in the parallel plate model. A slightly larger platform could reduce this problem. The sensed x position is more problematic; the separable approximation used for inverting equation 8 is a likely source of the problem. While the accuracy of the sensing algorithms needs improving, the resolution of the sensor is reasonable, with standard deviation better than $10\mu m$ with the coils off.

Further refinements, such as embedding a more complicated conductor pattern in the platform could be used to determine rotation angle as well. More general purpose capacitive imaging algorithms are also needed which can accurately determine the position and orientation of arbitrary polygons in the plane.

4 Fluid Bearings for Platform Levitation

A stable and relatively stiff fluid bearing is essential to ensure smooth operation of the mobile platform. If the rotational stiffness is too low, the platform could tilt and touch down, probably sticking to the surface. With some actuation methods, for example, a linear variable-capacitance motor, the normal force into the substrate may be 100 times greater than the tangential drive force. The more common approach to an air bearing uses an external air supply. However, for small dimensions and narrow air gaps, a form of self-pressurization can be obtained by us-

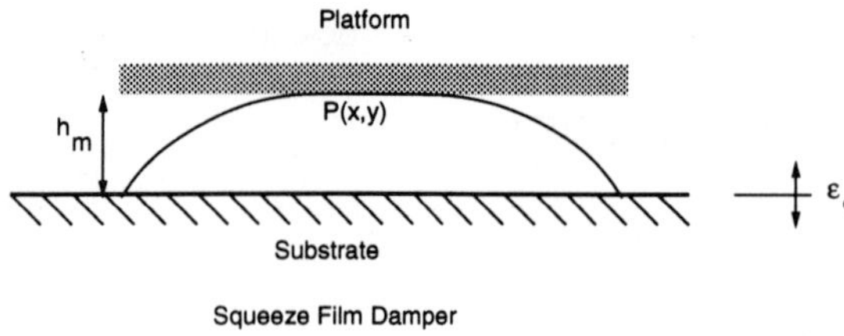

Figure 13: Squeeze film bearing. High frequency vibration of the substrate compresses air in the narrow gap between the platform and substrate.

ing a squeezed film of gas. This section examines the feasibility of using the squeeze film to support the platform.

4.1 Externally Pressurized Bearing

An air bearing can be constructed using surface micro-machining as described by [Pister et al, 1990]. That bearing had an array of 10 μm air nozzles spaced at 100 μm intervals. Each nozzle was made by a plasma etch through a 1 μm membrane, and ran with a supply pressure of approximately 500 pascals. Several platforms up to 750 μm thick were supported on this bearing, with float heights of about 20 μm. The low nozzle resistance and low supply pressure tend to encourage this high float height. For electrostatic drive, the equilibrium height should ideally be less than 10μm to increase the drive force on the platform.

4.2 Squeeze Film Bearing

A squeeze film bearing action is defined by [Pan and Broussard, 1967]:

- *In a squeeze film bearing, high frequency transverse oscillations of one of the bearing surfaces provides a pumping action. The oscillatory squeeze motion results in a time-averaged pressurization effect primarily due to the compressibility of the gas film, and the degree of this pressurization increases monotonically with the amplitude of the oscillation relative to the average gap.*

The intuition behind a squeeze film bearing can be improved by the following example. Consider two square flat plates with a very thin layer of air in between them. If these two plates are brought together slowly, the air between them will escape out the sides of the plates, and the two plates will touch. If the two plates are brought rapidly together, the gas will not all have time to escape, and will be compressed in the central region, increasing the fluid pressure there. Now consider a floating platform at average height h_m above a rapidly vibrating substrate as shown in Figure 13, driven by $z(t) = e_o cos\omega t$. The acceleration of the substrate is thus $-e_o\omega^2 cos\omega t$. Suppose the pressure under the platform is slightly less than the normal load on the platform, then the platform will start to decrease in height. However, when the substrate moves up on the next cycle, the fluid will be more compressed since the gap is now smaller, increasing the pressure under the platform, and forcing it back up to its equilibrium height.

Assuming constant fluid density, Reynold's equation for compressible gas film lubrication is [Sherman, 1990]:

$$\vec{\nabla} \cdot [(\frac{h^3}{\mu})\vec{\nabla} p] = 12\frac{\partial(h)}{\partial t} + 6\vec{\nabla} \cdot (hU) \qquad (10)$$

where h is the height, p is pressure, and U is velocity. If the variations in pressure δp are small compared to P_a, the ambient pressure, then in linearized form [Blech, 1983]:

$$\frac{h_m^2 P_a}{12\mu}\nabla^2\frac{\delta p}{P_a} - \frac{\partial}{\partial t}\frac{\delta p}{P_a} = \frac{\partial}{\partial t}\frac{e_o \cos\omega t}{h_m}. \qquad (11)$$

This linearized version can be solved for sinusoidal excitation in the steady state using a series solution. The first term of the series solution gives the fundamental mode response. The in-phase response corresponds to a spring force, and the out-of-phase response corresponds to a damping force.

The force response can be characterized by the non-dimensional *squeeze number* σ. For a square plate of area $A = a^2$, driven sinusoidally by $e_o cos\omega t$, the squeeze number is given by

$$\sigma = \frac{12\mu a^2}{P_a h_m^2}\omega, \qquad (12)$$

where mean film thickness is h_m. For a 5 μm thick air bearing, $10^{-6}m^2$ area platform, and drive frequency of $\omega = 10^5 s^{-1}$, the squeeze number is approximately $\sigma \approx 10$. For large squeeze numbers, the fluid layer has increased resistance to flowing out between the plates, increasing the effective stiffness of the air bearing.

The damping force f_d is given by

$$f_d = \frac{128\sigma}{4\pi^6 + \sigma^2\pi^2}\frac{e_o}{h_m}P_a A \qquad (13)$$

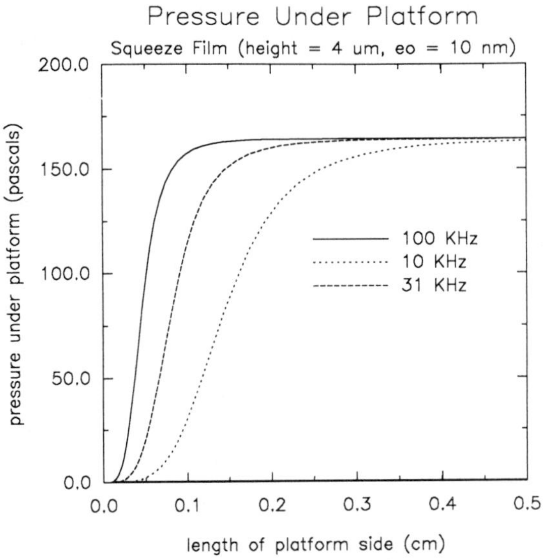

Figure 14: Theoretical pressure for various platform sizes, assuming 10 nm amplitude vibration. Note that a pressure of 100 pascal can support a piece of silicon about 4 mm thick. Parts less than 1 mm in size may drag on the substrate.

The spring force f_k is given by

$$f_K = \frac{64\sigma^2}{4\pi^8 + \sigma^2\pi^4} \frac{e_o}{h_m} P_a A \qquad (14)$$

The cutoff squeeze number $\sigma_c = 2\pi^2$ occurs when the spring and damping force are equal. At the cutoff squeeze number, the spring force is proportional to the inverse 5th power of the plate separation:

$$f_k \approx \frac{64\sigma^2}{8\pi^8} \frac{e_o}{h_m} P_a A = \frac{8}{\pi^8} \left(\frac{12\mu a^2}{P_a} \omega \right)^2 \frac{e_o}{h_m^5} P_a A \qquad (15)$$

while the damping force is proportional to the inverse cube of the float height. The squeeze-film bearing will resist the attraction of the platform to the substrate from gravitational and magnetic forces. Since f_k is inversely proportional to the fifth power of the mean platform height, this bearing will be very stiff, but well damped due to the damping force f_d. With ω in the high audio range, a piezo-electric actuator can easily generate the required oscillation. See Figure 14 for trade-offs between platform size, pressure, and vibration frequency. See [Zhang et al 1992] for comparisons of theoretical and experimental results for squeeze film damping.

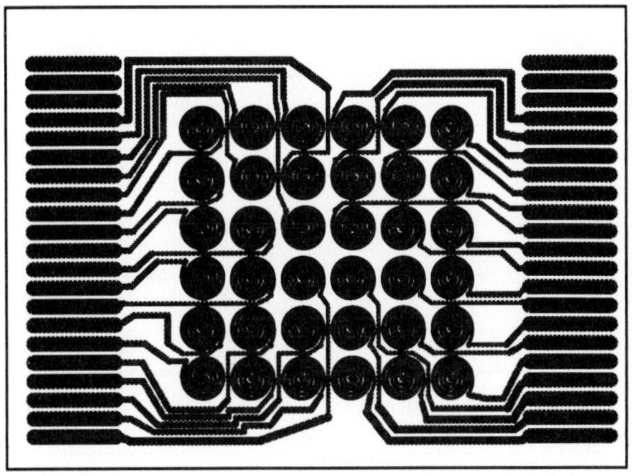

Figure 15: Electromagnet coil array. Coils are spaced 6.6 mm center-to-center. Copper traces are about $400\mu m$ wide and $30\mu m$ thick. Coil can handle 2 amps continuous duty without excessive temperature rise. Switching diodes and array connections are mounted on main control circuit board.

4.3 Squeeze Film Bearing Implementation

A piezo-electric transducer is used to drive the substrate in vertical oscillation, as shown in Figure 6. For stiff support of the weight of the platforms, the squeeze number σ should be greater than the cutoff squeeze number σ_c. For small platforms, this implies that the drive frequency may need to be relatively high. The drive frequency should also be high enough to be inaudible. One limitation on the drive frequency is the excitation of bending modes in the substrate. These bending modes could easily have greater amplitude than the air bearing thickness. A relatively thick aluminum plate was chosen for its relatively high stiffness and low mass, with dimensions of approximately 5 by 3.5 cm. This size was chosen to provide complete support for the active sensor area of the substrate. The lowest bending mode [Timoshenko, Young, and Weaver, 1974] occurs at:

$$f_1 = \frac{\pi h_s}{a_s^2} \sqrt{\frac{E}{12\rho}} , \qquad (16)$$

where the density of aluminum is $\rho_{Al} = 2.7 \times 10^3 Kgm^{-3}$, the area a_s^2 is approximately $1.5 \times 10^{-3} m^2$, Young's Modulus is $E = 7 \times 10^{10} Nm^{-2}$,

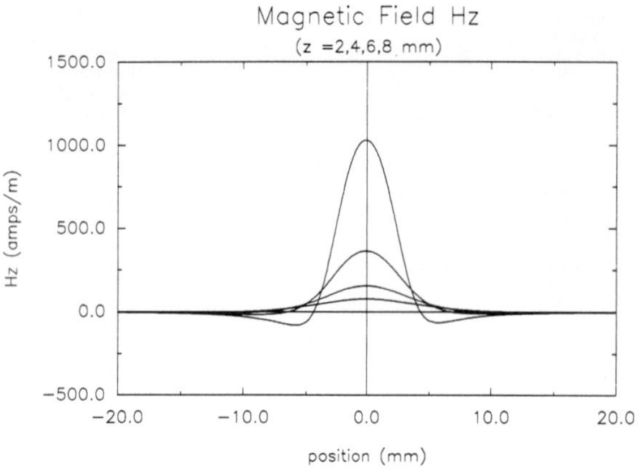

Figure 16: Relative vertical component of magnetic field for various heights above magnet. Because of the capacitive sensor array, the platform is approximately 2 mm above the electromagnets.

and the thickness of the substrate is $h_s \approx 1.2 \times 10^{-2}m$ which includes a glass covering layer over the capacitive array sensor. The first resonant frequency is then at about 40 KHz. We choose a drive frequency of 20KHz to reduce heat dissipation in the piezo drive, as the load of the platform, about 50 grams, is significant at such high frequencies.

5 Magnetic Actuator Design

The electro-magnet actuator consists of an array of planar coils fabricated on a printed circuit board as shown in Fig. 15. The coils are made of 2 layers of 4 turns of copper cladding, approximately 400 μm in width and 30μm thick. The coils are spaced 6.6 mm apart. (For comparison, Wagner and Benecke [1991] use a single coil of 17 turns in a 5.8 mm square.) A permanent magnet is attached to a 7.5 mm diameter glass or silicon platform, which is driven by the fields from the coil array. For a coil with current I amps of negligible cross-section, the magnetic field is obtained from

$$\vec{H}(\vec{r}) = \frac{I}{4\pi}\int_l \frac{\vec{dl} \times \vec{r}}{|\vec{r}|^3}, \qquad (17)$$

where μ_o is the permeability of free space, ($\mu_o = 4\pi \times 10^{-7}Hm^{-1}$), and $\vec{r}$ is the vector from the current element $I\vec{dl}$ to the calculation point. Using numerical integration, the z component of the magnetic field is plotted in Figure 16, assuming 1.0 amp of current for heights of 2, 4, 6, and 8 $\times$ 10^{-3}m above the magnet surface. These heights are significant compared to the conductor size, thus ignoring the cross-section of conductors is reasonable. Note that the height of the platform above the magnetic layer includes the thickness of the capacitive array sensor.

For a permanent magnet with constant magnetization $\vec{M} = M_o\hat{z}$, the force per unit volume can be obtained (e.g. Jackson, [1975]) from

$$d\vec{F} = \nabla(\vec{M}\cdot\vec{B}) = M_o\nabla B_z = \mu_o M_o \nabla H_z , \qquad (18)$$

where B_z is the z component of the magnetic flux density, and H_z is the z component of the magnetic field. Using a permanent magnet with a magnetic flux density of 1.0 Tesla, $M_o = 1.0T/\mu_o = 8 \times 10^5 ampm^{-1}$. Since the applied field strength is an order of magnitude less than the magnet's internal field, it is reasonable to assume a constant magnetization.

The total force is obtained from

$$\vec{F} = \int_{V_o} d\vec{F}dV . \qquad (19)$$

The vertical force F_z is supported by the air bearing, but may decrease the float height for large currents in the coil. Typically, the vertical forces are 10 times the horizontal force.

The torque per unit volume is obtained from

$$\vec{\tau} = \vec{M}\times\vec{B} = M_o\hat{z}\times\vec{B} = -M_oB_y\hat{x} + M_oB_x\hat{y} . \qquad (20)$$

Note that with uniform vertical magnetization, there is no component of torque about the vertical axis. The torques about the horizontal axes may cause tilting of the platform, but the stiffness of the air bearing should prevent the platform from contacting the substrate.

The horizontal acceleration of the platform is limited by the gradient of the magnetic field and the magnetization of the permanent magnet. The horizontal force is given by

$$F_x = \int_{V_o} \mu_o M_o \frac{\partial H_z}{\partial x}\, dV . \qquad (21)$$

A rough estimate of the force on the magnet can be made assuming an approximately constant field gradient at some position. With a current of 0.4 amp, the peak field gradient $\frac{\{\partial H_z\}}{\{\partial x\}}$ is approximately $1.2 \times 10^4 ampsm^{-2}$. Using $\mu_o M_o$ equal to

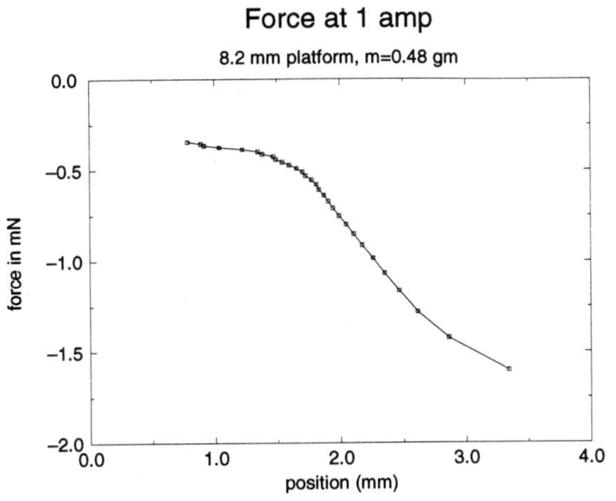

Figure 17: Force profile for platform on electromagnet with 1 amp.

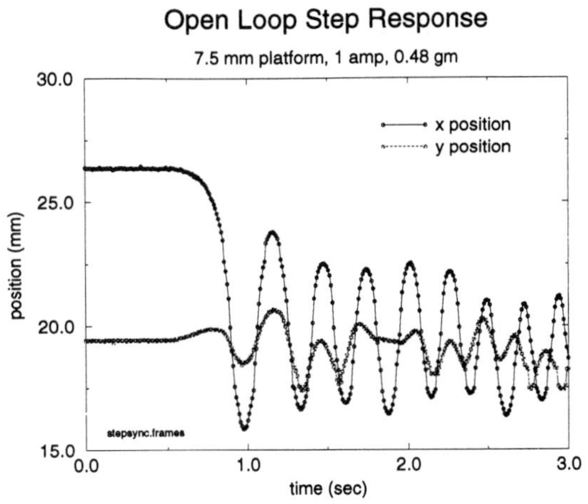

Figure 18: Open loop step response with 1 amp. Peak velocity $.15ms^{-1}$, and peak acceleration $\approx 3ms^{-2}$.

1.0 Tesla, the force per volume is $1.2 \times 10^4 Nm^{-3}$. A typical permanent magnet used is cylindrical, with diameter 6.3 mm and 5.0 mm length, for a total volume of approximately $1.6 \times 10^{-7} m^3$. The net force is then about $1.9 \times 10^{-3} N$. With a platform mass of $1.0 \times 10^{-3} Kg$, this gives an acceleration of about $2ms^{-2}$ which is quite reasonable for a workspace of less than $10cm^2$.

The static electromagnetic force on the platform can be measured as follows. We assume first that the magnetic field is circularly symmetric. The minimum force occurs when the platform is centered exactly above a magnet as can be seen by considering the gradient of Figure 16. The maximum force f_{max} occurs at some distance say $r_{max} > 0$. When a constant force less than f_{max} is applied to the platform, a stable equilibrium position will be reached which is a function of the current. Thus by applying a constant force and measuring the equilibrium position as a function of current, one can map out the force reponse of the system. A platform of mass 0.48 gram is placed on the air bearing at a tilt angle of 7.8 degrees. Thus the tangential force on the platform is a constant 0.64 mN. Figure 17 shows the measured force profile for this platform. The peak force occurs at a distance of about 3.3 mm from the coil center. This method can not be used to estimate the force profile at distances greater than r_{max} since the system is unstable. With 1 amp, peak accelerations of $3ms^{-2}$ are predicted. (The coils can withstand 3 amps peak, and 2 amps continuous without overheating). Note that the force profile will be a function of the magnet dimensions.

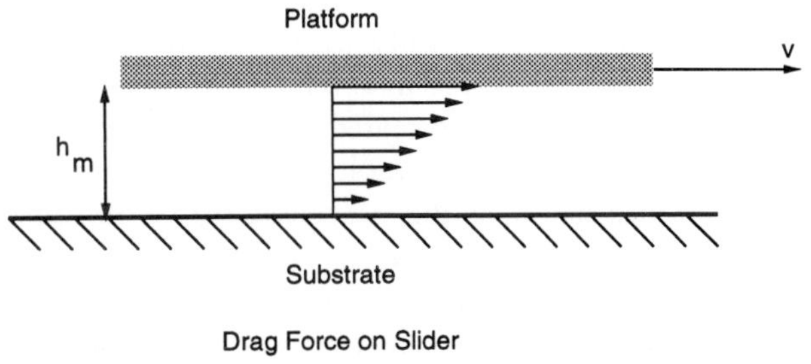

Figure 19: Drag force on slider.

The predicted accelerations are in line with what can be seen from measured step responses. A short portion of a 1 amp step response (open-loop) is shown in Figure 18. The natural frequency is on the order of 5 Hz, and damping ratio approximately 0.005. From Figure 17, the slope is approximately $0.5Nm^{-1}$ so the predicted natural frequency $\frac{1}{2\pi}\sqrt{\frac{k}{m}}$ of 5.1 Hz is in close agreement.

5.1 Drag Force Due to Air Bearing

Although the air bearing has no static friction component, it does contribute a velocity proportional damping term. Consider a platform moving at velocity $v = \dot{x}$ horizontally as in Figure 19. The relative velocity of the fluid at the fixed and moving surfaces should be zero, thus there is a shearing force on the platform due to the viscosity of air. This drag force is found from

$$F_{drag} \approx \frac{\mu A}{h_m}\dot{x} = b\dot{x} \qquad (22)$$

where viscosity for air is assumed to be 2×10^{-5} Nsm^{-2}, the platform area $A = 4.4 \times 10^{-5} m^2$. The damping ratio $\zeta = \frac{b}{2\sqrt{km}} \approx 0.006$ is in close agreement with the open loop step response. This result indicates that the air bearing is preventing any significant coulomb friction. Note that the power required to overcome this viscous drag is much less than $10^{-6} watt$. Another damping term to consider is the induced voltage in adjacent conductors due to the changing magnetic flux density.

6 Conclusions and Future Work

This paper has shown a relatively simple method to implement a miniature mobile platform system. With inherently low friction, position resolution is limited mainly by sensor noise. The almost pure-mass system is relatively easy to control. The coils can be scaled down somewhat using printed circuit technology, but lithographic techniques will probably be necessary to obtain a factor of 10 reduction in coil size.

The eventual goal of this work is to implement a micro-robot system consisting of a work surface and multiple planar micro-robots, which can autonomously move in the plane with three degrees-of-freedom. The micro-robots can work cooperatively to grasp and manipulate objects which may be larger than themselves. Eventually, a system with practical applications, such as assembly of micro-mechanical parts could be built. In the system concept, welding and inspection systems could be added at the periphery of the work surface, and parts could be shuttled between stations by the micro-robots, thus implementing a micro-assembly cell. Each micro-robot could be 1000 μm × 1000 μm in area, and only 100 μm thick. The micro-robot would rest on a fluid film (an air bearing) and never move while in frictional contact.

Acknowledgments

I thank Steve Albrecht for layout and design of the printed circuit coil array and sensor array, and Emile Sahouria for implementing the multi-processor closed-loop control system. I also thank Hiroaki Furuichi and Michael Cohn for stimulating discussions.

References

[1] C.H. Ahn, Y.J. Kim, and M.G. Allen, "A Planar Variable Reluctance Magnetic Micromotor with Fully Integrated Stator and Wrapped Coils" *Proc. IEEE Micro Electro Mechanical Systems*, pp. 1-6, Fort Lauderdale, FL Feb. 7-10, 1993.

[2] F. Arai, D. Ando, T. Fukuda, Y. Nonoda, and T. Oota, "Micro Manipulation Based on Micro Physics-strategy based on attractive force reduction and stress measurement", *Proc. IEEE/RSJ Intelligent Robots and Systems*, pp. 236-241, Pittsburgh, PA August 3-5, 1995.

[3] K. M. Anderson and J. Edward Colgate, "A model of the attachment/detachment cycle of electrostatic micro actuators", *ASME Micromechanical Sensors, Actuators, and Systems*, DSC-vol. 32, pp. 255-268, Atlanta, GA Dec. 1-6, 1991.

[4] M. Ataka, A. Omodaka, N. Takeshima, and H. Fujita, "Fabrication and Operation of a Polyimide Bimorph Actuators for a Ciliary Motion System", *Jnl. of Microelectromechanical Systems* , vol. 2, no. 4, pp. 146-150, Dec. 1993.

[5] J.J. Blech, "On Isothermal Squeeze Films", *ASME Jnl. of Lubrication Technology*, vol. 105, pp. 615-620, 1983.

[6] K. Boehringer, B.R. Donald, R. Milhailovich, N.C. MacDonald, "Sensorless Manipulation Using Massively Parallel Microfabricated Actuator Arrays", *IEEE Int. Conf. on Robotics and Automation* , pp. 826-833, San Diego: May 8-13, 1994.

[7] M.H.W. Bonse, F. Zhu, and J.W. Spronck, "A New Two Dimensional Capacitive Position Transducer", *Sensors and Actuators* , vol. 41-42, pp. 29-32, 1994.

[8] I.J. Busch-Vishniac, S. Chen, M. Jeong, S. Li, and I. Wang, "Magnetic Levitation-Based Micro-Automation of Mechanical Processes in Semiconductor Fabrication", *Proc. IEEE Micro Electro Mechanical Systems*, pp. 142-146, Napa Valley, CA, Feb. 1990.

[9] I.J. Busch-Vishniac, "The Case for Magnetically Driven Micro-Actuators", *ASME Micromechanical Sensors, Actuators, and Systems*,DSC-vol. 32, pp. 287-302, Atlanta, GA Dec. 1-6, 1991.

[10] P. Dario, R. Valleggi, M.C. Carrozza, M.C. Montesi, and M. Cocco, "Microactuators for microrobots: a critical survey", *J. Micromech. Microeng.* vol. 2, pp. 141-157, 1992.

[11] R.S. Fearing, "A Miniature Mobile Platform on an Air Bearing", *Third Int. Symp. on Micro Machine and Human Sciences*, Oct. 14-16, 1992, Nagoya, Japan.

[12] G. Fuhr, R. Hagedorn, T. Muller, B. Wegner, and W. Benecke, "Linear Motion of Dielectric Particles and Living Cells in Microfabricated structures induced by traveling electric fields", *IEEE Micro Electro Mechanical Systems"*,Nara, Japan, Feb. 1991, pp. 259-264.

[13] H. Fujita, "Studies of Micro-Actuators in Japan", *IEEE Int. Conf. on Robotics and Automation*, Scottsdale, AZ, May 1989, pp. 1559-1564.

[14] H. Fujita and A. Omodaka, "Electrostatic Actuators for Micromechatronics", *IEEE MicroRobots and Teleoperators Workshop,* Hyannis, MA Nov. 9-11, 1987.

[15] H. Fujita and Kaigham J. Gabriel, "New Opportunities for MicroActuators", 1991 Int. Conf. on Solid-State Sensors and Actuators (Transducers '91) June 1991, San Francisco, CA, pp. 14-20.

[16] T. Fukuda and T. Tanaka, "Micro-Electro Static Actuator with Three Degrees of Freedom", *Proc. of the IEEE Workshop on Micro-Electro Mechanical Systems* , Napa Valley, CA Feb. 11-14, 1990.

[17] T. Furuhata, T. Hirano, and H. Fujita, "Array-Driven Ultrasonic Microactuators- arrayed microactuator modules that have swing pins", 1991 Int. Conf. on Solid-State Sensors and Actuators (Transducers '91), June 1991, San Francisco, CA, pp. 1056-1059.

[18] H. Guckel, T.R. Christenson, K.J. Skrobis, T.S. Jung, J. Klein, K.V. Hartojo, and I. Widjaja, "A First Functional Current Excited Planar Rotational Magnetic Micromotor", *Proc. IEEE Micro Electro Mechanical Systems*, pp. 7-11, Fort Lauderdale, FL Feb. 7-10, 1993.

[19] W.R. Harper, *Contact and frictional electrification*, Oxford: Clarendon Press, 1967.

[20] R.G. Horn, D.T. Smith, "Contact electrification and adhesion between dissimilar materials." *Science*, 17 April 1992, vol.256, (no.5055):362-4.

[21] T. Inoue, K. Iwatani, I. Shimoyama and H. Miura, "Micromanipulation Using Magnetic Field", *IEEE Int. Conf. Robotics and Automation* , pp. 679-684, Nagoya, Japan May 1995.

[22] J. Ish-Shalom, "Sawyer Sensor for Planar Motion Systems", *IEEE Int. Conf. on Robotics and Automation* , pp. 2652-2658, San Diego, CA, May 8-13, 1994

[23] J.D. Jackson, *Classical Electrodynamics*, 2nd ed., Wiley:New York, 1975.

[24] Y. Kim, M. Katsurai, and H. Fujita, "Fabrication and Testing of a Micro Superconducting Actuator using the Meissner Effect", *Proc. IEEE Micro Electro Mechanical Systems*, pp. 61-66, Napa Valley, CA, Feb. 1990.

[25] S. Konishi and H. Fujita, "A Conveyance System Using air flow based on the concept of distributed micro motion systems", *7th Int. Conf on Solid-State Sensors and Actuators (Transducers '93)* , pp. 28-31, Yokohama, Japan, June 1993.

[26] C. Liu, T. Tsao, and Y.-C. Tai, "Out of Plane Permalloy Magnetic Actuators for Delta-Wing Control", *Proc. IEEE Micro Electro Mechanical Systems* , pp. 7-12, Amsterdam Jan. 29-Feb 2, 1995.

[27] W. Liu and P. Will, "Parts Manipulation on an Intelligent Motion Surface", *Proc. IEEE-RSJ Intelligent Robots and Systems*, pp. 399-404, Pittsburgh, PA August 3-5, 1995.

[28] Y. Maeda, K. Aihara, and H. Fujita, "A Vacuum-compatible X-Y micro-actuator using the levitational effect of a superconductor" *Robotics, Mechatronics and Manufacturing Systems*, ed. by T. Takamori and K. Tsuchiya, North-Holland Elsevier 1993, pp. 509-514.

[29] G.L. Miller, "Capacitive Incremental Position Measurement and Motion Control", United States Patent 4893071, Jan. 9, 1990.

[30] F.M. Moesner and T. Higuchi, "Devices for Particle Handling by an AC Electric Field" *Proc. IEEE Micro Electro Mechanical Systems* , pp. 66-71, Amsterdam Jan. 29-Feb 2, 1995.

[31] C.H.T. Pan and P.H. Broussard, "Squeeze-Film Lubrication", Proc. Gas Bearing Symposium on Design Methods and Applications, pp. 12/1-12/35, Univ. of Southampton, April 1967.

[32] K.S.J. Pister, M.W. Judy, S.R. Burgett, and R.S. Fearing, "Microfabricated Hinges", *Sensors and Actuators A*,vol. 33, pp. 249-256, 1992.

[33] K.S.J. Pister, R.S. Fearing, and R.T. Howe, "A Planar Air Levitated Electrostatic Actuator System", *IEEE Workshop on Micro Electro Mechanical Systems*, Napa Valley, CA Feb. 12-14 1990.

[34] R.H. Price, J.E. Wood, and S.C. Jacobsen, "Modelling Considerations for Electrostatic Forces in Electrostatic Microactuators", *Sensors and Actuators* , vol. 20, pp. 107-114, 1989.

[35] J. Robichaux and S. Ahmed, "Magnetically Levitated Micro-Robots: Design, Control, Experimentation and Applications", *Robotics and Manufacturing: recent trends in research, education, and applications, Proc. of the 4th Int. Symp. on Robotics and Manufacturing* , Nov. 11-13, 1992, Sante Fe NM pp. 9-14.

[36] K. Sato, Y. Kawamura, S. Tanaka, K. Uchida, and H. Kohida, "Individual and Mass Operation of Biological Cells using Micromechanical Silicon Devices", *Sensors and Actuators*, vol. A21-A23, pp. 948-953, 1990.

[37] F.S. Sherman, *Viscous Flow*, McGraw-Hill, New York: 1990.

[38] I. Shimoyama, "Scaling in Microrobots", *Proc. IEEE/RSJ Intelligent Robots and Systems* , pp. 208-211, Pittsburgh, PA August 3-5, 1995.

[39] K. Suzuki, H. Miura, I. Shimoyama and Y. Ezura, "Creation of an Insect-based Microrobot with an External Skeleton and Elastic Joints", *Proc. IEEE Micro Electro Mechanical Systems Workshop,* Travemunde, Germany, February 4-7, 1992, pp. 190-195.

[40] H. Tokisue, N. Tsumaki, H. Inoue, and M. Tsukagoshi, "Development of Wafer Handling Systems in Gas, Liquid and Vacuum Environments for ULSI Manufacturing", *Proc. 69th JSME Fall Annual Meeting,* pp. 354-356, Nagoya, Japan, Oct. 16-17, 1991.

[41] W. Trimmer and R. Jebens, "Actuators for Micro Robots", IEEE Intern. Conf. on Robotics and Automation, Scottsdale, AZ, May 1989, pp. 1547-1552.

[42] B. Wagner, M. Kreutzer, and W. Benecke, "Permanent Magnet Micromotors on Silicon Substrates", *Jnl. of Microelectromechanical Systems* , vol. 2, no. 1, pp. 23-29, 1993.

[43] B. Wagner and W. Benecke, "Microfabricated Actuator with Moving Permanent Magnet", *Proc. IEEE Micro Electro Mechanical Systems* , pp. 27-32, Nara, Japan, Feb. 1991.

[44] B. Wagner, M. Kreutzer, and W. Benecke, "Electromagnetic MicroActuators with Multiple Degrees of Freedom" 1991 Int. Conf. on Solid-State Sensors and Actuators (Transducers '91), June 1991, San Francisco, CA, pp. 614-617.

[45] I. A. Wang, S. Li, and I. Busch-Vishniac, "A Magnetic Levitation Transport Path" *IEEE Trans. on Semiconductor Manufacturing*, vol. 4, no.2, May 1991.

[46] M. Washizu, "Electrostatic Manipulation of Biological Objects in Microfabricated Structures", *Integrated Micro-Motion Systems: Micromachining, Control, and Applications*, Proc. Third Toyota Conference, Aichi, Japan, 22-25 Oct. 1989, pp. 417-431.

[47] L. Zhang, D. Cho, H.Shirashi, and W. Trimmer, "Squeeze film damping in microelectromechanicl systems", in *Micromechanical Systems* , Winter Annual Meeting ASME DSC-Vol. 40, Anaheim, CA Nov. 8-13, 1992 pp. 149-160.

Sub-Micron Large Motion Multi-Robot Planar Motion System
or
ROBOTWORLD – The Next Generation

Jehuda Ish-Shalom
Columbia University
6 Oak Place, Croton-on-Hudson, NY 10520-9771, USA
e-mail: jehuda@photon.poly.edu

Abstract

A table-top sub-micron large motion high performance modular manufacturing system is discussed using several improvements to commercial Sawyer planar motor technology. Commercial planar Sawyer motors provide uniform performance over their entire workspace, offering fairly high speeds, accelerations, accuracy and resolution. They also offer mechanical and kinematic simplicity, modularity, large motion ranges, and a short path to mechanical ground. On the other hand, primarily because they are operated as open-loop step motors, they can miss steps, have long settling times, and fall short of their potential capabilities. These short comings are overcome and the system is improved by: (i) A novel very high resolution (0.4 μm) Sawyer planar sensor which can be used for electronic commutation and closed-loop control. (ii) Improved accuracy by thermal deformation compensation. (iii) Motor speed and force improvement by a new composite structure. (iv) Motion extension to 6 degrees-of-freedom using the same planar motor technology with a simple modular parallel manipulator configuration.

1. Introduction

The goal of this work is to advance and refine the base capabilities and framework of ROBOTWORLD™, a Sawyer planar multi-robot system, presented by Scheinman in 1987 ISRR [1] (similar commercial systems [2-4]) Fig. 1. There are two parts to this goal: (1) Extend the framework of ROBOTWORLD™ to be an even more modular table-top manufacturing system, with stronger support for Canny & Golberg RISC approach to robotics [5]. (2) Extend the basic capabilities of the multiple "robots" in the system to better support objective (1). "Robots" is in quotations here to signify that they provide programmable motion, but each one individually may provide less than 6 degrees-of-freedom (DOF) of motion. Additional DOF, if needed, would come from motion devices at each workstation.

Partitioning the required DOF provides possibly lower cost and higher performance. A familiar example of this decomposition is a common X-Y pen plotter. The X motion is provided by the pen and the Y motion by the paper.

Electronic product sizes are shrinking so manufacturing using a dozen work-station steps can now be done using a single table top planar motor motion system, replacing a large manufacturing hall. Furthermore higher resolution and accuracies are required. Some products are getting so small, and require such high resolution motion, that a manual solution is no longer even an option. A HP LED array assembly requires 1 μm resolution placement.

ROBOTWORLD - the next generation goal is to be "The **P**C for **M**anufacturing" **(PM)**. In this analogy the platen (motors' "table" in Fig. 1) would be the base "PC-Motherboard" which supports multiple workstations' locations on a single table-top. Integrated with the system would be planar motors that function as both the gross planar motion conveyors between workstations and fine motion planar X-Y-Yaw (small rotation in the plane) positioning. Besides positioning, the motors should provide other programmable task level defined planar motion functions. A simple task level function is applying force along a task specified planar direction. The purpose of programmable task level defined motion is to reduce the required special tooling at each workstation. The PM would provide the common base conveyor and planar motion requirements. The personalization of the system to achieve different process steps would be in: (i) the modular planar robots used; (ii) the software and computing hardware controlling each planar motor motion (supporting task level defined motions) and (iii) in each workstation minimal additional motion and processing or assembly hardware (*e.g* Z motion). The notion is to allow several layers of "personalization" of the system to support as many possible different types of system construction and upgrade objectives. Such objectives range from fast system construction for low volume and prototyping, to high speed low cost production. One would like the production system to support a consistent simple evolution path along the progression between the above objectives with the option of incremental system upgrades. This paper outlines some technical solutions to achieve these goals.

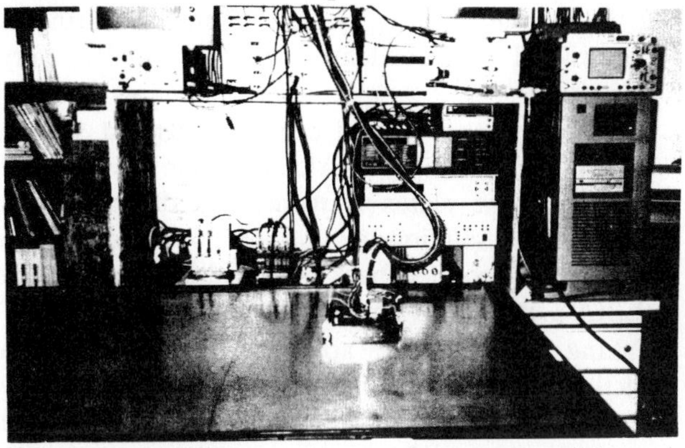

Fig. 1: Platen with one Sawyer planar motor with test Sawyer sensors.

In the proposed PM the required motion is partitioned between: (i) the basic motion provided by multiple planar motion platforms operating over a common, well defined, accurate plane, and (ii) additional motion, if needed, provided separately at each workstation.

The paper includes the following sections: Section 2 discusses the basic requirements for PM. Following we review current planar Sawyer motor technology in Section 3 and objectives for sensing planar position in Section 4. Section 5 describes our implemented Sawyer planar position sensor. To overcome the "large" thermal deformation error from the platen reference grid Section 6 describes a real-time compensation method we developed. On the basic motor performance side, Section 7 describes what improvements a composite magnetic structure can offer. Finally, Section 8 reviews novel kinematic structures that provide six DOF motions using the same planar motor technology.

2 Basic Requirements for PM

The following built-in properties are desired for a basic multi-robot planar manufacturing/processing system technology: 1. Large motion dynamic range. 2. Integrated position sensing. 3. Use of planar motors. 4. Offer extensions beyond planar motion.

2.1 Large Motion Dynamic Range

To increase the range of uses of a motion technology one would like to be able to move both large and very-small distances. To provide both scales of motion by the same system a large dynamic range motion system is required, *i.e.* the ratio of largest motion possible to smallest resolution motion achievable should be as large as possible. In general it is very difficult to achieve sub-micron resolution motion within say one meter reach. This range spans over six orders of magnitude. For example, Si wafers are currently 0.2 meter (8 inch) in diameter and have features of less than 0.4 μm. Currently large dynamic range motion systems are built with a fine motion system piggy-backed on a course motion system [6] (CD players currently use this approach). In these cases, such an approach may be the only possibility, but comes with increased complexity and additional cost (and this approach does not even always work). An actuator that directly provides the whole required dynamic range of motion at low cost is advantageous. Furthermore, if one can achieve a built-in large dynamic range motion at low cost, one may afford to use the motion system for both the conveyors between workstations and course/fine motion apparatus at each station.

The obstacle in achieving a large dynamic range is the difficulty in dealing with a large dynamic range for analog physical variables. This problem is not new and is overcome in most cases by transforming the problem into a counting problem (digital computers use this principle). For position measurements, incremental encoders do exactly that. If the count's resolution is very fine, one gets into a counting speed problem when motion is fast. To overcome this problem and increase resolution, coarser counting grids are used with analog interpolation between counts. This is somewhat like a built-in course/fine system.

For actuators, the corresponding solution to achieve large dynamic range motion, is a walking actuator (step counting) with the ability to interpolate motion between steps. A step motor, with μ-step control of its motion, does exactly that. The problem with magnetic motors is the limited resolution of μ-stepping (analog world) due to operation in a highly nonlinear region. If one could control the motor with a higher resolution sensor feedback, then higher resolution motion would be possible (assuming other factors, such as friction, are not a limiting factor).

2.2 Integrated Position Sensing

Counting/Walking motors, like a μ-step controlled Sawyer motor, that operate without position monitoring or feedback are very good as long as they obey the motion commands given to them. Disturbances to motions or exceeding actuator motion specifications can cause an actuator not to follow the commands exactly. A more severe problem can occur if the actuator has a cyclic control (like a step motor) and it misses (losses) multiple whole cycles, a condition termed **loss-of-sync**. The **loss-of-sync** condition is a major system failure. The failure is not even known and therefore cannot be recovered from. To recover, system shut down and manual initialization is required (in most cases).

Another possible error is from accumulation of errors from the analog variation of the size of each count. This accumulation causes actual position of the actuator to drift away from the expected position. This error is characteristic of a counting/walking actuator that walks over a surface with no grid. If there is a built-in surface grid, like in a planar Sawyer motor, this type of error does not occur.

The tradeoff involves the additional cost of having the grid built into the base surface. If the grid is already built-into the system, one can use it for independent position sensing as developed later in this paper. Such a grid can act as a common physical reference for position measurement of all objects in the system.

If integrated position sensing with sufficient resolution and accuracy is built-into the system then one can overcome (within physical limits) by feedback control the three types of positioning errors (discussed before): 1) Recoverable, command following errors. 2) Unrecoverable, loss-of-sync error. 3) Step-size accumulation of error.

Even just motor position monitoring at step size resolution can detect the loss-of-sync condition. Detection can be used to initiate corrective actions before damage occurs. This element is **critical** to obtain a reliable manufacturing system in most cases and does not exist in current ROBOTWORLD™ systems (see subsection 3.3 for partial exception for Mega 1 system). In a conservative system design and where the motion environment is carefully controlled loss-of-sync can be a nonexistent problem (e.g. very successful use of open-loop planar Sawyer motor wafer prober since 1973).

Having a high resolution and accuracy sensor can improve respectively the motor resolution and accuracy. Without high resolution feedback, submicron resolution motions are currently not possible with 1mm pitch planar Sawyer motor technology. Position feedback can also improve positioning accuracy when there are some uncertainties and disturbances to the motion, which are not possible with open-loop μ-step control (*e.g.* disturbance from the motor umbilical cord).

2.3 Planar Motors

Many planar motor designs can provide the required base planar X-Y motion with multiple planar robots operating over a large area. A planar motor offers kinematic simplicity, modularity and a large motion range. A planar motor uses a surface to constrain the motor to the 3 DOF in a plane. A reference plane is a simple shape and as such is relatively simple to construct. Thus it can be made accurately at relatively low cost.

A low friction bearing system for the motor relative to the plane is not as simple. A very good bearing is a fluid or air-bearing system. If an air-bearing is constructed independently, it is usually very expensive.

A planar Sawyer motor comes integrated with a built-in air-bearing. The friction of the moving part relative to the platen is reduced to the point that ultra-high resolution motion of 20 nm resolution with closed-loop feedback is available commercially (LEIA 2 using laser position feedback [7]).

Taking the opposite approach to dealing with friction is to use it for motion. NanoMotion walking actuator does that. It can achieve 1 nm resolution motion at up to 0.3 m/sec speed using control with multiple step sizes [8]. The main innovation here compared to previous ultra-sound or surface-wave motors is the multiple step size construction and control, which allows the motor to achieve both ultra-high resolution stepping and high speed with the same motor. A big advantage of this motor is that it does not required any built-in grid in its base surface, which can be just a flat sheet of glass which is a common item and very low cost.

In both above examples the plane is "completely" passive. If magnets are embedded into the surface, a DC type planar motor [9] can be constructed such as the Anorad Microglide X-Y Planar Stage [10].

The magnet grid is too coarse, so laser feedback is used to achieve 20 nm linear and 0.01 arc sec rotation resolution. Unfortunately the laser position sensing limits the travel and does not allow in general operation of multiple motors over a common platen (same for LEIA system [7]).

If one makes the platen active, the forcers (moving part of planar motor) can be passive with no umbilical cord connecting them. This simplifies motion planning, but comes at an inherently more complex and costly platen. This method becomes "economical" with a large number of moving robots over the common platen.

One may want to shrink the scale of the whole system down to the micro world. In this case the umbilical cord becomes an even more difficult problem. Smaller scale (few cm on the side) magnetic actuation active platen systems were constructed [11], [12]. Fearing [11] uses both a cleaver squeeze-film air-bearing system and a built-in common grid capacitive position sensing system. Currently the achieved motion dynamic range of these small scale planar motor systems is small.

2.4 Beyond Planar Motion

Although it is considered here that X-Y motion is the basic motion required, it is not always sufficient. Even the additional small Yaw rotation motion (up to a few degrees) that can be provided may not be sufficient. It is not even sensible in all cases to decompose the required relative motion into two parts, as done in the pen plotter example. An example is the Sawyer planar motor used as a wafer prober (the current largest "user" of planar Sawyer motors). It has Z and Yaw motions built on top of each planar motor not at the workstation (probe-card station). Here it is much simpler, more accurate, and faster, to move the light wafer under test then the probe-card with all its numerous connection wires.

Another example is current commercial ROBOTWORLD™ like systems that provide each robot with additional Z and Yaw motions built on top of each planar motor. The additional two DOF use a different motion technology. The notion is that the "robots" should provide all the motions in most cases and four DOF is considered the minimal useful robot in many cases.

Thus the simple notion of motion decomposition is not always the best or even a possible solution and the system should provide motion extensions for even six DOF motions (*i.e.* a general robot). It is advantageous to provide the motion extensions within the same motion technology framework. Using the same technology supports the modularity of the system and would simplify its use. Generally, adding DOF comes at a sacrifice of performance and additional cost but cannot be avoided in all cases. So a built-in option should be provided to reduce additional burden.

2.5 Areas for Research

There are many areas of research in a few engineering and science disciplines related to PM: 1) Planar sensor. 2) Real-time thermal deformation compensation. 3) Fast and task driven motion control. 4) Alternative planar motors. 5) Alternative magnetic and composite structures. 6) Actuator scaling theory. 7) Beyond planar motion. 8) Multi-robot "umbilical" path planning [13]. 9) Multi-robot motion programming.

In this paper we discuses only a few of these areas: 1), 2), 5) and 7). The general requirements from planar motors, area 4), was discussed in some detail in Subsection 2.3 before.

3 Planar Sawyer Motor Systems

A planar Sawyer motor consists of two parts. A thin flat passive platen (stator) and a forcer (rotor). In ROBOTWORLD™ (also in most other uses) the platen is stationary. Multiple forcers can be placed over a common platen forming a multi robot system. An air-bearing prevents frictional contact between the forcer and the platen and is maintained by injecting high-pressure air from the motor (about 4 bar (60 psi) supply pressure). The thickness of the air-gap is typically 10 μm to 25 μm (our system had about 20 μm).

A limited travel multi DOF Sawyer motor was patented by Hinds in 1974 [14]. A planar version with a moving platen and stationary forcer was constructed by Hollis in 1992 [15]. The device provides limited travel of ±1 mm in X and Y, and Yaw rotation in the plane of ±1.75 degree. The moving platen and limited travel enables this design to achieve higher peak acceleration, 12 g compared to common planar Sawyer motors [16] that achieve only 5 g. Hollis fine positioner also demonstrates use of internal sensing for closed-loop feedback to achieve very high resolution motion of 0.2 μm linear and 0.0001 degree in rotation.

3.1 Sawyer Motor Platen

The platen is a thin (1.8 mm typical) low carbon steel plate glued on top of a honey-comb structure. The top thin steel plate has groves cut in it at a fixed pitch, p, along both axes, to form a waffle pattern. The groves are cut conventionally by two methods: (i) photo chemical etching (most common Fig. 2); (ii) mechanical cutting. After the groves are cut they are refilled with a non-ferromagnetic material (epoxy) and polished to be an air-bearing surface. Fig. 2 shows rounding, shape variability and scratches on the platen's teeth.

The honey-comb structure is used to increase the transverse stiffness of the platen structure (i.e. keep the platen flat). A stiff steel frame surrounds and supports the platen and the honey-comb structure. A thin steel sheet covers the back side of the honey-comb adding addition rigidity to the structure.

Thicker and liquid filled platens are used to distribute the motor heat better. Other materials can also be used to make the platen. LEIA uses a corrosion-resistant alloy. New plastic coated powder iron [17] can be used to reduce eddy currents and cost by molding the teeth. Using these materials compromises the maximum motor force that can be obtained.

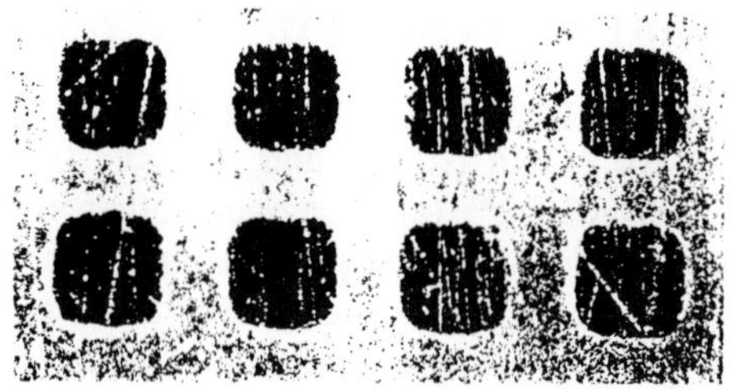

Fig. 2: Top view of commercial 0.040 inch (1.016 mm) pitch platen.

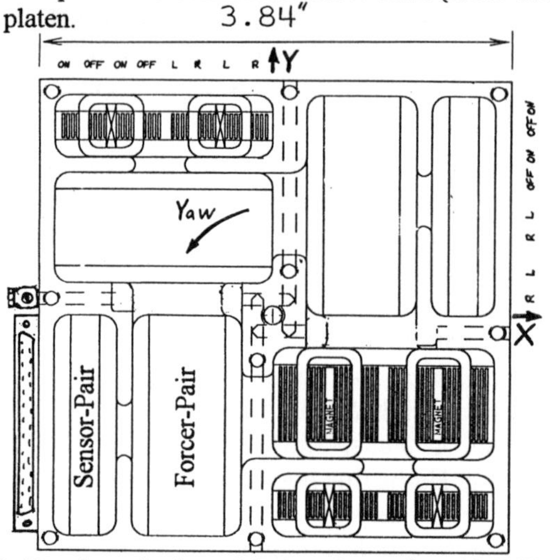

Fig. 3: Top through view of designed two phase X-Y-Yaw Sawyer planar integrated motor/sensor. Motor and sensor teeth are aligned to each other.

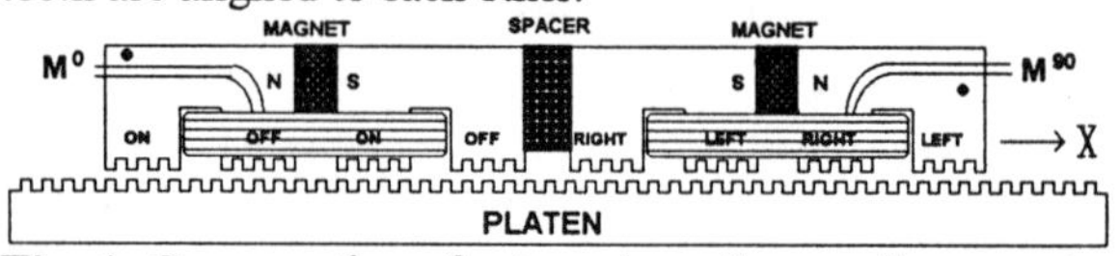

Fig. 4: Cross-section of a two phase Sawyer linear motor forcer-pair.

3.2 Sawyer Motor Operation

A Sawyer planar motor is composed of at least two perpendicular linear motor forcer-pairs to obtain XY planar motion. To obtain planar rotational control and higher stiffness, a third forcer-pair is added to produce rotational torque in the plane. Most often a planar motor includes four symmetrically arranged forcer-pairs, two along each axis (Fig. 3). The laminated Si iron forcer-pairs Fig. 4 are mounted inside an aluminum housing Fig. 3.

Fig. 4 shows the principle of operation of a two phase motor forcer-pair [16] (other possible magnetic paths configuration exist and are used). Each pair is composed of two identical phases M^0 and M^{90} offset by (1/4+n) tooth pitch, n=integer. An intuitive explanation of motor's operation is given here, a detailed magnetic model is given in [18, 28].

We start with the motor at the position shown in Fig. 4, with no currents. The M^0 phase magnet produces magnetic flux that closes through the aligned poles labeled ON since the poles' teeth are aligned with the platen teeth. In contrast, the OFF poles' teeth are miss-aligned and thus a smaller amount of flux closes through them. In the M^{90} phase all four poles are "half" miss-aligned and the flux distributes evenly between the RIGHT and LEFT poles (ignoring edge effects). At this position the net lateral X force is zero since M^0 has zero net force from each pole and in M^{90} the RIGHT and LEFT forces balanced.

Putting current through a motor winding adds magnetic flux closing between each pair of adjacent poles on each side of the MAGNET (the MAGNET is equivalent to a large air-gap isolating the two sides). With appropriate current polarity delivered to M^{90}, additional flux is added to each LEFT pole and the same amount subtracted from each RIGHT pole. Assuming that the current induced flux does not exceed the MAGNET flux, then there is no longer a balance of forces on M^{90} and there is a net force that would pull the motor in the positive X direction until the LEFT poles' teeth are aligned to the platen's teeth, becoming ON poles (RIGHT=>OFF and for M^0 ON=>RIGHT and OFF=>LEFT). The amount of movement in this STEP is 1/4 cycle in the +X direction. To continue to move, another 1/4 cycle +X, one would apply the same principle, *i.e.*, increase the current in M^0 while decreasing the current in M^{90} to zero. The poles' alignment at the end of the motion would be (left to right, M^0 and M^{90}):

OFF ON OFF ON LEFT RIGHT LEFT RIGHT.

With two more steps, applying current in the negative polarity, the motor would move one whole tooth pitch distance and return to the same pole arrangement started with in Fig. 4. With a 1,016 μm tooth pitch motor, this full step resolution is 254 μm. Finer position resolution is achieved by "μ-stepping," *i.e.*, interpolating the currents so that the net balance of forces between the two motor phases M^0 and M^{90} gives a very large set of equilibrium points (beyond the four discussed before). Using open-loop μ-stepping "control," the 1,016 μm pitch motor above can achieve 2 μm (0.08 mil) resolution. The "magnetic internal position closed-loop" can achieve high stiffness, 1,500 lb/inch, leading to good position performance, even when using just μ-stepping with "external" open-loop control. This is why such motors are very useful even as they are now.

3.3 Closed-Loop Control in Current Products

In current Sawyer planar motor products three types of feedback sensor elements are used:

1) Accelerometers. 2) Back EMF (induced voltage in motor coils due to motor motion). 3) Laser position measurement.

Three linear accelerometers are used to obtain motor X, Y and Yaw rotation velocities (after integration). These velocities are used in a closed-loop controller to reduce the tracking error and damp motor oscillations particularly at move-end. Accelerometers are used in: Xynetics plotters, LEIA system and Megamation Mega 1 system [19]. Megamation also uses the measured acceleration to detect motor loss-of-sync by detecting an acceleration that deviates from the motor commanded acceleration. This method is only effective in some cases.

Probers use a four-phase planar Sawyer motor and require only fast settle-time (no trajectory tracking). To achieve this a special closed-loop control is engaged a short time after the commanded move ends. Two phases of the motor are disconnected from the power amplifier and their back EMF is used as velocity feedback to the other two phases driving the motor. This method is very effective and reduces settling time by a factor of 10 or more [19].

To improve planar motor resolution and accuracy laser feedback is used. The LEIA 2 planar Sawyer motor X-Y table is specified to be able to achieve 20 nm resolution and an accuracy of ±1.5 μm over an area of 0.23 m by 0.23 m [7] (with just μ-step control ElectroGlas prober accuracy is ±5 μm over 0.2 m square area [19]).

4. Planar Position Sensing for Sawyer Motor

A planar motor has the six DOF of a rigid body. The sensor's primary function is to measure the motor's three controlled planar DOF positions and velocities: two large XY and a small YAW rotation. The other three motor DOF have high stiffness and change very little relative to the first three DOF. The motor's six DOF can be relative to a **fixed reference** frame, or relative to a **local frame** aligned with the platen surface. Both measurements are important.

The resolution and accuracy objectives for measuring the motor's linear controlled motions in the local platen surface frame are classified into four ranges (for 0.040 inch = 1.016 mm pitch platen):

1. **Coarse motor phase**, 1/4 tooth pitch (254 μm).
   - Detect and correct linear motor loss of synchrony.
2. **Fine motor phase,** 1% tooth pitch (10 μm).
   - Good electronic commutation => higher force can be utilized.
   - Detect and correct motor rotation in the plane to avoid loss of synchrony.
   - Course velocity computation.
   - Monitor motor performance.
3. **Very fine resolution position**, 0.2% tooth pitch (2 μm, μ-step resolution)
   For closed-loop control to achieve:
   - Better disturbance rejection.
   - Reduce motor settling time by active damping.
   - Correct motor inaccuracies (requires also accuracy in local frame).
   - More accurate motor positioning (requires also positional accuracy in fixed frame).
4. **Ultra fine resolution position**, 0.02% tooth pitch (0.2 μm) or better:
   - Improved closed-loop control in areas of 3. above.
   - Finer motor resolution compared to μ-step.

For collocated actuator control, the sensor measurements are relative to the local platen surface reference frame. But for the control to achieve **accurate positioning** relative to an external point, sensor measurements relative to a fixed reference frame are also required. Such position measurements are typically done with a laser interferometer system. One would like to use a **single position sensor** to achieve both frames of reference position measurements.

The platen can be used as a reference grid to achieve position measurements, relative to both frames, but one has to be careful. **Accurate positioning** (60 μm or better), relative to a given fixed reference frame, requires consideration of the accuracy and the thermal changes to the platen reference grid. For even higher accuracy and more accurate motor control, the other uncontrolled three DOF of the motor (fly height and, pitch and roll attitude), should be measured and considered, as well as the platen flatness.

An important objective for the sensor is to be able to obtain the above measurements for multiple independent motors on a single platen without mutual interference. This can be achieved using the platen grid as a reference. Using the platen as a reference has other advantages too: it is already there and it is a common reference to all motors.

Several means were proposed to sense position using the platen as a reference [24]: (a) Magnetic [20-25] (b) Capacitive [26] and (c) Optical [20], [27-29].

In this paper we describe a magnetic method which measures the most direct property needed for accurate motor commutation and has extremely high resolution (recall magnetic disk recording technology). The magnetic method described here uses a separate structure while the capacitive method [26] has the advantage (and disadvantage) of using the motor's teeth.

The optical method [28] is insensitive to coupling from the motor's field but is sensitive to scratches on the platen surface, imperfect platen fill and surface finish. These optical properties vary with time, cannot be easily controlled and are not critical for accurate motor operation. Some of these problems are overcome by a cleaver wave length shift method proposed by Brennemann [29].

5. Sawyer Sensor

The Sawyer sensor [24] described here uses the dual effect of flux steering used by the Sawyer motor and thus was named in honor of the motor's inventor.

A magnetic sensor uses the platen's teeth as a reference and measures the changes of reluctance as the sensor moves relative to the platen. To achieve minimal cross axes coupling the sensor's relative magnetic reluctance should be designed to change as much as possible along the measurement axis and change as little as possible along the orthogonal axis. Correspondingly, the sensor teeth along the measurement axis have the same tooth pitch as the platen, p, and along the orthogonal direction, width of the teeth w=n•p, n=integer (Fig. 5 w=5•p).

A single axis platen with groves along one axis only, as used with variable reluctance motors, can have a maximal reluctance change of 50% between the perfectly aligned and miss-aligned teeth positions. In contrast, the two axis waffle platen has a maximal relative reluctance change of only half of that, 20% measured. Thus only ±10% change is available to measure position and the 90% base-line reluctance needs to be subtracted out. To aggravate the situation, the motor/sensor fly height can vary, changing the 90% base-line reluctance considerably. This base-line change is the largest error contributing factor for measuring position from the relative reluctance changes as a function of position. To overcome this problem a differential magnetic field measurement can be used between (i) aligned and (ii) miss-aligned, teeth's structures, that are in as close proximity as possible.

The sensor configuration chosen in Fig. 5 has a single coil that does the "differential" part rather then two coils that can only be matched up to a point. Thus the "differencing" to cancel the large 90% base-line is done at a "lower level" (*B* field) and closer to perfection. This is the KEY important design consideration of the Sawyer sensor in Fig. 5. For further detail see [24].

To simplify usage an LVDT (Linear Variable Differential Transformer) design is chosen to measure the sensor-teeth's reluctance changes. In an LVDT arrangement the transformation ratio between excitation and sense coils changes as a function of the sensor position.

An integrated motor/sensor was designed, Fig. 3 with four pairs of sensors. The teeth of each motor/sensor pair are parallel and are aligned to each other for simple construction. Three sensor pairs provide measurement of the planar three DOF of the motor. The extra fourth sensor pair allows to overcome problems when crossing from one platen section to another.

The sensor was mounted externally on a 149 mm square two-phase Sawyer motor and the motor position x was measured by a laser interferometer system. The sensor's synchronously demodulated output voltage as a function of x is shown in Fig. 6 (for a fixed excitation voltage at a frequency of 22.5 kHz). Fig. 6 shows a $2{\bullet}p$ window of measurements going from x=0 to $x=8{\bullet}p$ and back to x=0. Comparison to a sine function (thin line) and the excellent repeatability of the sensor can be seen.

5.1 Sensor Characteristics

The motor's position, x, was piece-wise interpreted from the two quadrature sensor-pair outputs (other methods are used by [25], [28]). First by counting quarter sensor's cycles and secondly by interpolating position in between them. Two 3rd order polynomial fits for the center 70% range of each sensor's output (one for positive slop and one for negative slop) were used to interpret the sensor's voltage within each 1/4 cycle. The maximum error for the 4 fits (2 for each sensor) were less than ±1.5 µm. The sensor's equivalent position noise level measured in another experiment with a sensor's band-width of 4 kHz was σ=0.12 µm, *i.e.* sensor's resolution of 3σ=0.36 µm. As can be seen from Fig. 7, the actual sensor's noise may be even smaller. The externally induced vibrations of the motor relative to the platen, can be seen to damp out to about 0.1 µm. By using two sensor-pairs to measure planar rotation a resolution of 0.0006 degree can be achieved for the geometry in Fig. 3.

A critical problem for a magnetic sensor is the coupling from the motor. At low speed, less than ±2 µm equivalent was measured and at higher speed, to 2 m/sec, the coupling increased, with a maximum at about 1 m/sec, of ±6 µm. Modeling of the coupling can be used to reduce the interference, but not carried out so far.

The sensor has a planar rotational angular sensitivity. Fig. 5 sensor with w=5.08 mm (0.2 inch 5•p) should have zero output at 11.4 degree rotation.

The sensor's maximum output changes also as a function of motor air-bearing fly height, that changes dynamically as well. The changes are on the order of a few percent and need to be evaluated more precisely. These changes lead to errors in our position interpretation from the sensor.

The sensor real-time interpretation of the quadrature Sawyer sensor output (1/4 count + 3rd order polynomial interpolation) was performed by a SPARTA system [30] at 10 kHz sample rate, using a an AT&T DSP32C 25 MFLOPS processor.

For a very small 10 μm step move using the μ-step controller the damping time was over 100 msec and the motor natural frequency is 78 Hz. The motor's maximum velocity of 5 mm/sec during this move is obtained from the sensor time derivative. At a bandwidth of 3 kHz the velocity noise level was σ =0.2 mm/sec, i.e. a velocity dynamic range of 76 dB (6,700, 3σ= 0.6 mm/sec out of a ±2 m/sec range).

5.2 Results of Sensor Usage

Using the sensor in a PD (Proportional Derivative) control loop around an electronically commutated planar Sawyer motor showed that much better motor damping can be achieved. The motor can settle within 20 msec for a small 8 μm step move, (Fig. 8) compared to over 100 msec for the open-loop μ-step controlled motor. Judging from the higher natural frequency the motor's stiffness was improved 7 fold to 2.5 10^6 Nt/m (14,000 lb/inch). Using closed-loop control the motor's resolution was improved to less than 0.25 μm, a factor of 8 over the μ-step control. These results are preliminary, and are attributed to the low-noise position and velocity signals obtained from the sensor.

Considering absolute accuracy for the large motion dynamic range (10^7) of this system, evaluating the sensor is a major challenge and a laser interferometer system was used for these tests. The largest error is thermal expansion of the reference platen which can be as high as 60 μm Fig. 10. A real-time compensation method for the platen expansion is shown in Section 6. Other errors are from platen scale factor inaccuracy, platen teeth variations, motor rotations and from the sensor interpretation method used.

These results suggest the advantages of **sensor uses** listed below. The disadvantages of using a sensor are: added system complexity and somewhat lower motor base performance (additional area and weight required by sensor).

Sensor uses

Detection of motor's loss of synchrony. Velocity feedback to damp the motor's oscillations to obtain much faster settling time. Position feedback for electronic commutation to obtain higher usable force (30% to 100%) and speed. Electronic commutation also allows to use the motor as a force source. Better angular position control. Reduce requirements for motor's sections alignment. Help cross-over between platen sections by dynamic motor and platen sections' alignment. Local position sensing for disturbance rejection (e.g. from the motor umbilical cord). Local position sensing to improve motor's resolution. Lower heating from motor with appropriate electronic commutation method. Thus more accurate positioning could be obtained due to lower thermal expansion.

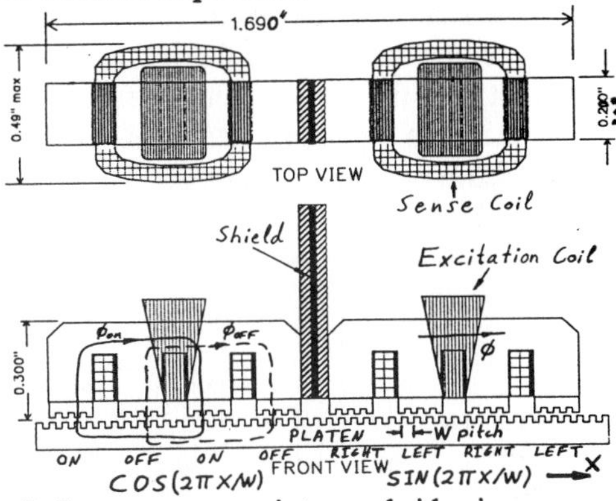

Fig. 5: Sawyer sensor pair top and side view.

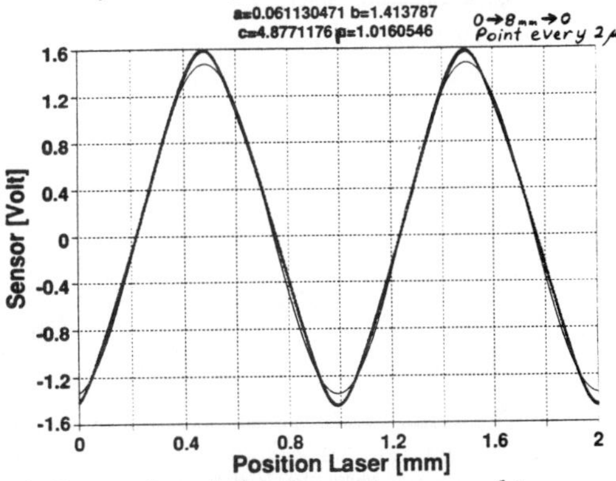

Fig. 6: Sensor demodulated output compared to: $a+b \sin(2\pi\, x/p+c)$. x from laser interferometer.

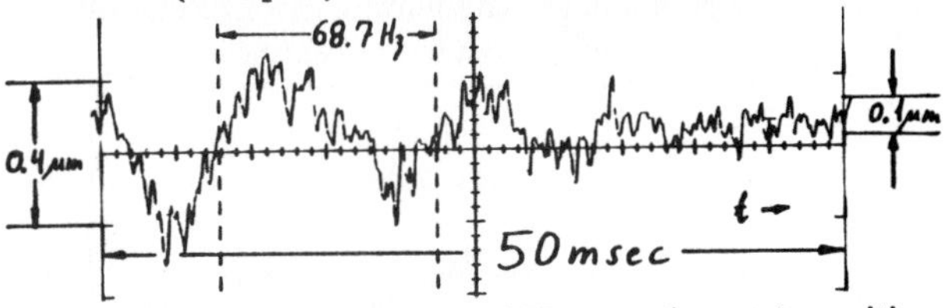

Fig. 7: Sensor real-time (10 kHz sample rate) position with no commanded motion, a damped vibration transmitted from the building can be seen.

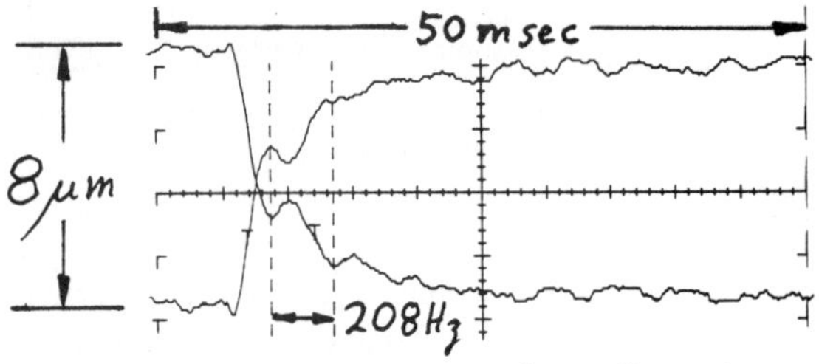

Fig. 8: Closed-loop response to pos. & neg. 8 μm step.

6. Real-Time Compensation of Platen Thermal Deformation

The position accuracy that can be achieved relative to a fixed reference frame depends on the accuracy of the reference grid, in this case the platen grid. This affects both the accuracy of µ-step controlled motor position and Sawyer sensor position measurement. The platen grid has manufacturing inaccuracy of the positions of the teeth on the platen, their size and shape. These fixed inaccuracies are mapped and compensated for in the motor control to a large extent. As the platen temperature changes additional variable platen deformation occurs causing the teeth's position to change.

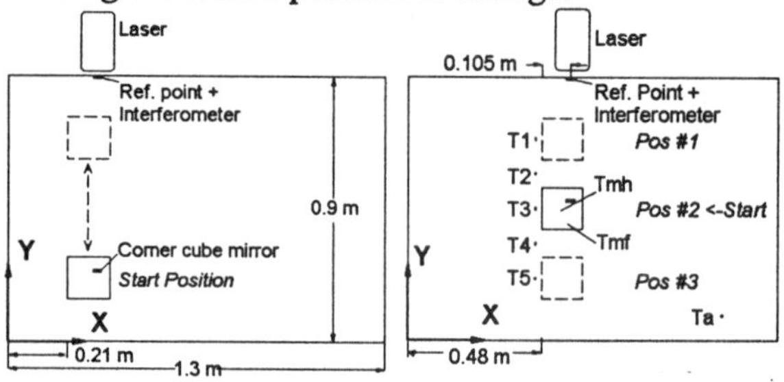

Fig.'s 10-11 Fig.'s 12-15

Fig. 9: Configuration of the experiments.

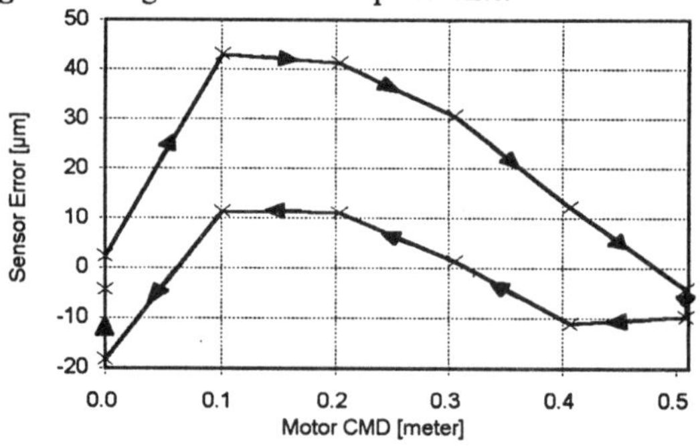

Fig. 10: Sensor error relative to laser interferometer with integer number of tooth pitch's moved. Error mostly due to platen thermal expansion from motor heating.

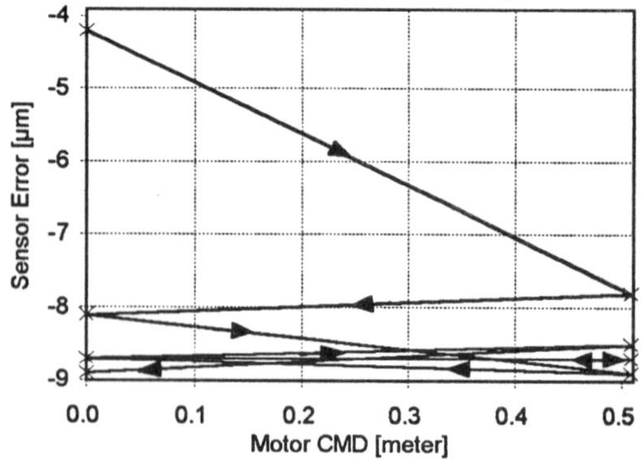

Fig. 11: Sensor repeatability error relative to laser interferometer. Note convergence to thermal equilibrium.

To evaluate these variable changes the experiments Fig. 10-11 were performed. The motor was commanded to move over the width of the platen, y, by µ-step control Fig. 9(left). The motor moved to the dashed position 0.5 m away and returned stopping at 13 points along the way (experiment took about 20 minutes). The arrows show the order of positions visited. The stopping points were at integer number of tooth pitch cycles so as to eliminate any sensor interpolation interpretation errors from this measurement. The reference motor position was measured by a laser interferometer system. The reference measurement point (and interferometer location) was at the edge of the platen, Fig. 9(left). Thus the reference point moves relative to inertial space as the platen deforms, Fig. 12. The motor had a corner cube mirror on it at an off-center location of 30 mm, Fig. 9.

One can see in Fig. 10 that there is a sensor position "measurement hystereysis" error of 30 µm and a peak to peak error of 60 µm. Even when the motor is at a fixed position (0 or 0.5 m) repeated sensor measurement error drifts 14 µm within minutes. These errors limit the accuracy of the system to 60 µm from this effect alone (other errors are smaller). To evaluate the source of this error the continuation of the experiment in Fig. 10 is shown in Fig. 11 where the motor was moved back and forth 0.5 m four times at even pace (within about 4 minutes).

After convergence to thermal equilibrium the sensor's repeatability is better than 1 µm. Thus if thermal effects can be accounted for—as in the special case of reaching (or close to it) thermal equilibrium—the error can be reduced by a factor of 100 to less than 1 µm over a motion distance of 0.5 m.

Electroglas prober Sawyer planar motor position accuracy with just µ-step control is specified to be ±5 µm over a 0.2 m diameter wafer (smaller travel). Even higher accuracy is desired by the manufacturer and thermal deformation is considered to be a major contributor to this error.

Our objective is to compute the deformation (grid error) in the unsteady heat flow condition resulting from arbitrary motor motion and to use the computation results to compensate for the actual grid deformation error.

The source of the heating of the platen is from one or more planar motors moving over its surface with a very small 10 µm to 25 µm air-gap. Inevitably, some of the electric energy supplied to the motor is converted to heat (about 75 Watt at v=0) and part of this heat is transferred to the platen (total of 720 Watt at v=2 m/sec). Due to this heat, and the heat generated directly in the platen from eddy currents and hysteresis losses, the temperature of the part of the platen under the motor increases causing the platen along with the grid to expand. Thus, the actual position, $\vec{x}_{actual}$, of a point on the platen is equal to the nominal position, $\vec{x}_{nominal}$, plus some thermal deformation $\Delta\vec{x}$:

$$\vec{x}_{actual} = \vec{x}_{nominal} + \Delta\vec{x}. \tag{6.1}$$

Furthermore, when the motor moves to a new position, the new platen portion under the motor heats up and expands. At the same time, the portion of platen which was under the motor cools down and contracts, due to heat convection to the surroundings and heat conduction through the platen. Thus, the motion of the motor creates complex temperature and deformation fields in the platen.

6.1 Thermal Compensation Method

Real-time inputs at every time step to the deformation computation are:

1) Measurement of motor temperature, T_m.
2) Motor position in the plane, x, y.
3) Motor velocity, v.
4) Ambient temperature, T_a.

The steps followed to compute the thermal deformation compensation at each time-step in real-time are:

1) Compute the evolution of platen temperature field using a unsteady heat transfer model with the real-time inputs specified above.
2) Using a platen structural model, compute the deformations resulting from the material stress caused by the uneven temperature of the platen.
3) Use the computed deformation $\Delta\vec{x}$ to correct the measured sensor position. Or for a μ-step controlled motor, use the deformation to correct the commanded location of the motor.

To make the computation achievable in real-time a planar 2D lumped Finite Element (FE) model was used. The detailed thermal, structural and FE model derivation are given in [31].

6.2 Experimental Results

The experimental setup Fig. 9(right) included the planar Sawyer-motor system, a laser interferometer, three thermocouples and a 486 PC computer to sample their measurements every 5 seconds. The motor was commanded to move over the platen according to Fig. 13 and the resulting error in position was measured by the laser interferometer, Fig. 14. The temperature inputs for the model real-time compensation computation where: ambient T_a, center of the inside housing of the motor T_{mh} and the top of one of the motor's forcer elements T_{mf}. The latter two temperatures were used to estimate the effective motor temperature T_m.

The time required to compute both the temperature and displacement fields using the 15×15 node FE model used for one time-step was 0.6 sec on a 486-33MHz PC. This time is sufficiently small compared to the thermal time constants which can be inferred from Fig. 14, assuming the averaging of the heat transfer from the motor during motion is adequate. Thus, the model can be used for real-time thermal deformation compensation.

In Fig. 14, the range of error in the motor's position is 53 μm. Fig. 15 shows the difference between the actual position error and the error predicted by the model in Fig. 14. Fig. 15 shows that the range of error of the model prediction is about 12 μm which is an improvement of factor of 4.5 in positional accuracy.

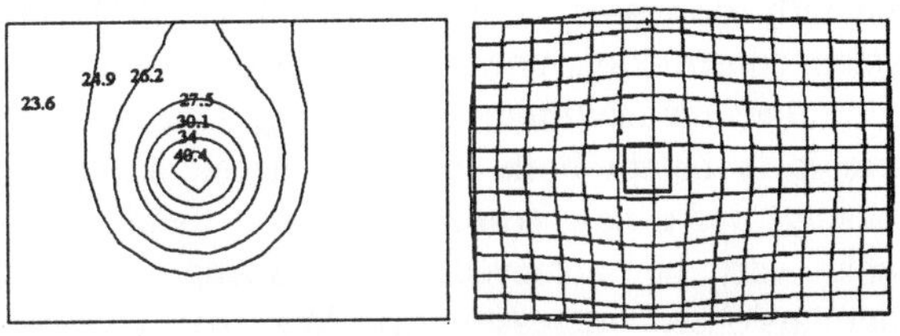

Fig. 12: Temperature field (^oC) on surface of platen (left). Displacement field magnified x2,000 (right). Both at t=100 min with motor at pos #2 after being at pos #1.

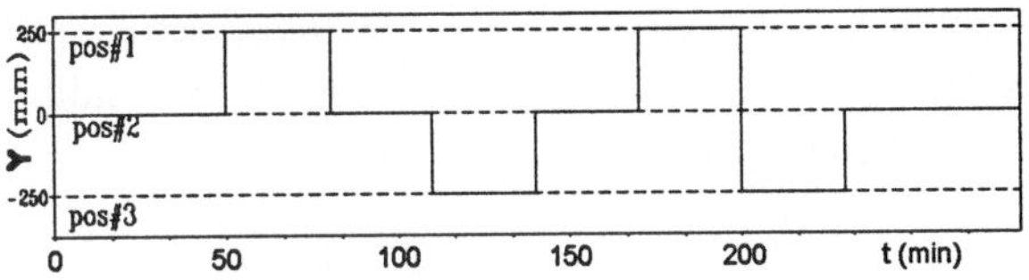

Fig. 13: Center of motor position, y, versus time, t.

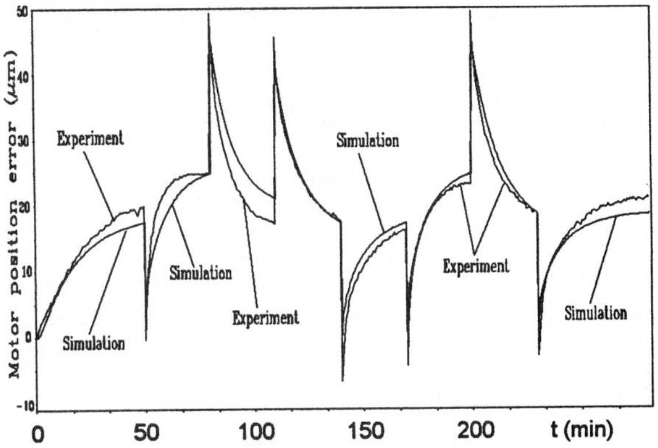

Fig. 14: Comparison between the motor position error and error predicted by the FE model.

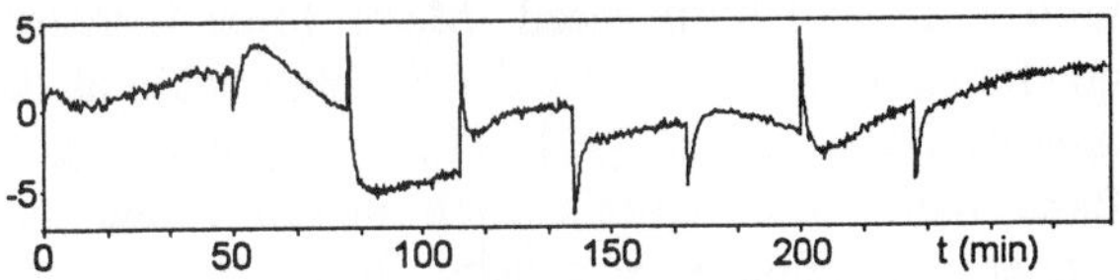

Fig. 15: Motor pos. error μm after thermal compensation.

7. Composite Magnetic Structure

To obtain higher motor force and speed both system and basic motor improvements can be used. On the system side, using electronic commutation (possible with a Sawyer sensor), the motor maximum available force can be utilized—which can provide 30% to 100% higher force compared to μ-step control. Beyond that, basic motor improvements are required to obtain higher force for the same size and weight motor, *i.e.* higher possible acceleration. Among other things higher acceleration would simplify operation of the system with the platen in a vertical orientation which increases the flexibility of the system usage.

The maximum available motor acceleration is a function, $a_m(v)$, of the motor's velocity, v. $a_m(v)$ generally decreases as v increases [32], [33]. The system construction, power amplifier used and $a_m(v)$ determines the maximum velocity, v_m. Our objective here is to obtain **both** increased acceleration and velocity, **not a new tradeoff** between them. These goals can be achieved by showing how to simultaneously increase: (a) low speed motor acceleration and (b) reduce the drop of $a_m(v)$ with increased velocity, leading to higher v_m. These improvements also lead to faster response time, which may be the only concern in some cases. A relationship of improved motors to improved overall system performance is given by "Actuator Scaling Theory" [34], [35].

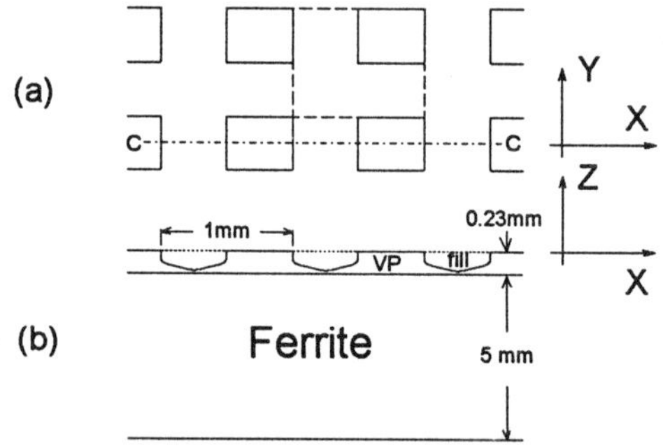

Fig. 16: Composite magnetic structure for platen of Sawyer motor. (a) Top view showing 4:1 area ratio. (b) Cross-section of Vanadium Permendur (VP) and Ferrite.

7.1 Composite Structure

Use of a single material for the magnetic flux path requires a compromise of the material properties [36]. In Sawyer motor design the choice is between high force and high speed. Motor force increases quadratically with the flux density obtainable at the motor's teeth surface [33]. Thus to obtain higher force requires a motor's teeth material that has higher maximum flux density, B_{sat}. High motor speed corresponds to high frequency flux oscillations in the motor's soft magnetic material. Thus to obtain low drop in force as motor speed increases requires low losses at high frequency, which correspond to high resistivity, ρ, of the magnetic material and small hysteresis losses. No known material has both these properties (see end Section 3.1 regarding new powder iron material).

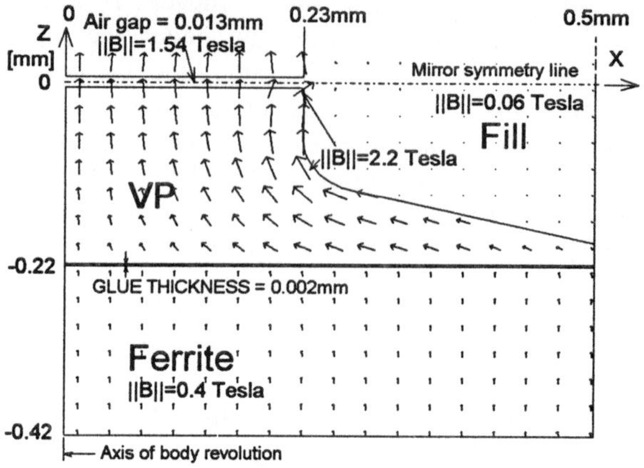

Fig. 17: Solution of magnetic field of revolute body model of composite platen.

For example, use of Vanadium Permendur, **VP** (a Co-Fe alloy) can achieve very high motor force due to its high B_{sat} of 2.4 Tesla. But high motor speed cannot be achieved without a very fine laminated structure, required to compensate for the material's low resistivity (ρ=40 10^{-6} Ohm•cm). On the other hand, use of a ferrite material offers the ability to achieve, without lamination, much higher speeds, since it has a million times higher resistivity, (*e.g.*, MnZn ferrite ρ=100 Ohm•cm). The problem with ferrite is that only much lower motor force can be obtained, because MnZn ferrite has much lower B_{sat} by a factor of four to six (0.6 to 0.4 Tesla). Thus a ferrite motor would produce sixteen to thirty-six times lower force, which makes a ferrite motor impractical to use. Achieving high motor speed is most problematic when considering the solid platen for a planar Sawyer motor [16] (cannot be laminated in two directions).

Fig. 16 shows a composite layered magnetic platen for a planar Sawyer motor. The VP teeth at the surface occupy only 1/4 of the area (Fig. 16a top view) but carry the same total vertical magnetic flux as the supporting ferrite substrate (neglecting the "stray flux" around the VP tooth). Thus the required flux density, B_{sat}, of the Ferrite is only 1/4 of that of the VP teeth (Ferrite B_{sat}=0.6 Tesla supports teeth VP B_{sat}=2.4 Tesla). Fig. 16b cross section shows a generic flux concentration shape required to quadruple B from Ferrite to VP teeth surface.

To validate the flux distribution a 3D symmetric revolute model Fig. 17 was used. Boundary conditions were set to obtain a constant total flux along the axis of body revolution. The highest B is at the inner corner of the VP flux concentrator and tooth tip, **not** at the tooth surface. The model shows that a 10% smaller tooth width should be used (0.45 mm Fig. 17) to achieve a 3.8 ratio of tooth surface B to bulk ferrite B. The lower ratio results from 22% "stray flux" around the tooth neglected before. The required ferrite B_{sat} is 5.5 times smaller than in VP.

7.2 Results

A test composite platen Fig. 16 [36] was compared to a commercial low carbon iron platen by measuring the series inductance and resistance $L(f)+R(f)$ as a function of frequency of a commercial laminated SiFe Sawyer motor section. Inductance function $L(f)$ is related to the drop of motor force as a function of increased speed (frequency, f). $R(f)$ is the motor losses (heating of motor and platen) as a function of motor speed (frequency).

The small signal $L(f)$ measurements [36] suggest that the composite platen should allow an order of magnitude higher motor speed. $R(f)$ combined losses of motor and platen are lower for the composite platen for f<10kHz (10 m/sec). For 10kHz<f<100kHz we measured higher R (losses) of the composite platen which result from higher B possible (higher force). At high frequency the losses become dominated by the laminated motor section. Tests at nominal motor current show that the composite can achieve the required higher B field at the teeth's surface compared to bulk ferrite [37].

The composite platen takes advantage of each material's desirable properties. Using a similar composite structure for the motor should give 40% to 100% higher motor force and even higher speed. The suggested composite platen construction has other advantages including lower cost of manufacturing as demonstrated by a similar construction used in a HP plotter [38].

8. Six Degrees-of-Freedom Motion

Planar Sawyer motors have many advantages but also have a major limitation of not being able to provide an end-effector with six DOF. A very simple novel extension of the planar motor technology to provide six DOF motion, by use of three planar motors was patented by Tahmasebi & Tsai [39] and independently developed by Shoham & Ben-Horin [40]. They suggest a parallel robot structure based on a closed spatial kinematic chain, Fig. 18. A major advantage of this P6DOF (Planar 6 DOF) robot is that it has a very small number of parts.

The P6DOF robot's end-effector plate is connected to the three planar motors by three simple inextensible rods. Several joint arrangements are possible. In Fig. 18 prototype, built by Shoham and Ben-Horin [40], there are 3 rotary joints one on top of each planar motor and 3 ball joints connecting the rods to the end-effector-plate. The planar motors have relatively low ability to withstand torques applied to them in the plane. The air-bearing also operates better without torque loading. These conditions can be obtained by exchanging the ball joint and rotary joint locations (ball joints on motor's top). This symmetric configuration seems to be preferable for use with planar Sawyer motors. Mobility analysis using Kutzback equation [40] shows that a third configuration is feasible using two universal joints (2 DOF each) on both ends of each rod [39]. In some medical applications a passive movable clamp may be desired. Such clamping of the end-effector position at a desired location can be obtained by shutting off the air supply to the air-bearing and utilizing the very strong magnetic force that clamps the planar motors to the platen.

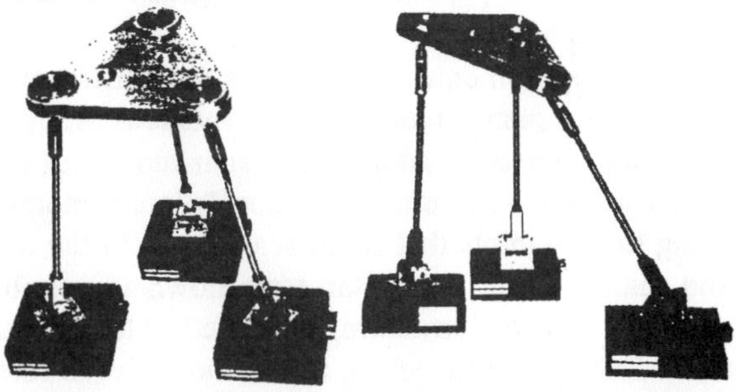

Fig. 18: Six DOF parallel robot structure using three planar motors for actuation from [40].

Since the robot can move along the whole platen-base-plane the work volume of the end-effector is much larger then any known 6 DOF parallel structure, such as the Stewart platform [41]. The range of vertical motion (Z axis perpendicular to the planar motors plane) can be tailor for a given application by simply changing the length of the 3 inextensible robs. The length of the rods and size of the end effector plate can be used to tailor the dimensionally-uniform decoupled central stiffness matrix [39].

Multiple P6DOF robots can operate on a common base platen. These P6DOF robots are very modular in the planar motor world. They can be assembled simply from 3 planar motors and a light weight modular 6 DOF kinematic extension structure that has very few parts. This robot can obtain very high dynamic performance since all the motors are at the base, the extension structure is light weight and it has a closed-kinematic structure providing a stiff structure. The additional weight of the extension structure of the robot prototype [40] was only 50%

(0.67 Kg) of the total weight of the small 3 base planar motors used. The achievable robot end-effector resolution is very good resulting from the good performance of the planar motors. Unfortunately, both the end-effector resolution and dynamic performance are inherently lower then that of the planar motors by themselves due to the additional joints and structure length involved. Nevertheless, one can expect that this "poorer performance" is still much better then that of conventional serial robots. Two P6DOF robots can even be used as an input/out two hand haptic interface with very large dynamic range.

The inverse kinematics solution found [40] is very simple. Analysis of the forward kinematics shows that there are at most 8 configurations for any given set of 3 planar motor positions. This is the lower bound on the number of configurations for any parallel robot [40].

Analysis of a dimensionally-uniform stiffness matrix shows that this parallel robot has greater stiffness than that of a Stewart platform [39].

Another six DOF configuration was demonstrated by [12] using 4 planar motors with 5 inextensible rods. This method uses an additional planar motor but was demonstrated with much smaller "planar motors" using flexure joints that could scale down to the mili and micro world. A similar scale down is possible with the 3 motor structure, discussed before, using two 2 DOF universal joints, at the ends of each fixed length link, assuming ball joints are harder to make for small structures.

9. Conclusions

In this paper we looked at planar motion Sawyer motor technology as a foundation for an advanced modular manufacturing system with the platen as a common base to build upon. Multiple planar motors can operate over the common platen providing the basis for a compact multiple step integrated manufacturing system, PM. Sawyer motors offer high performance rapid motion with a very large dynamic range of motion of 7 orders of magnitude: meters travel with sub-micron resolution. This extended range is expect to cover many current and future manufacturing systems needs.

To arrive at these capabilities and further extend them several advances to current commercial Sawyer planar motor technologies were shown:

1) Sub-micron, $3\sigma=0.4$ μm, planar position sensing method that can operate with multiple motors on a common platen. Position sensing is critical to yield a reliable manufacturing system to avoid motor loss-of-sync (not available in commercial systems). The integrated Sawyer sensor can increase both system performance and functionality using closed-loop control.

2) A real-time thermal deformation compensation method can improve the accuracy that can be achieved from the reference platen. Currently an improvement by a factor of 4.5 to 12 μm over 0.5 m was demonstrated. One can expect that another order of magnitude of improvement may be possible with further research.

3) To improve dynamic response (increased force, acceleration and, speed of motion and response) and reduce platen heating a composite magnetic structure was developed. So far, measurements show that the composite structure can be expected to improve motor speed by a factor of 5 at least. The composite structure also offers lower cost construction method. Full composite platen construction was not done yet.

4) A very simple modular extension to achieve 6 degrees-of-freedom (DOF) motion with the same planar motor technology was shown. The extension uses a novel P6DOF parallel manipulator kinematic design, [39-40], with 3 planar motors and was experimentally demonstrated by [40]. The parallel manipulator structure has high stiffness and low weight and thus should provide very good dynamic response with good 6 DOF motion performance. Multiple 6 DOF robots can move over the very large area of the platen.

Acknowledgments

Thanks to the many people that gave their generous help, encouragement and advise during this work: A. Brennemann, J.W. Chang, E.I. Cooper, M. Datta, B. Hammer, D.A. Herman Jr., M. Ish-Shalom, W. Jecusco, P. Kazanzides, F. Khorrami, Y.Y. Lin, J.R. Marino, S. Ostrander, E.J. O'Sullivan, J. Pawletko, B. Petek, L.T. Romankiw, R.V. Shenoy, T. Wasfy, Ed Yarmchuk., J. Yeh. Partial support provided by NSF DMI-9413543 grant and IBM donation of equipment.

References

[1] V. Scheinman, "Robot World - A Multiple Robot Vision Guided Assembly System," Robotics Research, The 4th Int. Symp., Ed. R.C. Bolles & B. Roth, 1988.

[2] P.F. Lilienthal et. al., "A Flexible Manufacturing Workstation," AT&T Technical Journal pp. 5-14, March 1988.

[3] R. Tella et. al., "Generic Multi-Robot Workcell and its Application to Printed-Circuit-Board Assembly," Surface Mount Int. Conference. & Expo, San Jose, pp. 677-685, Aug 1993.

[4] "Speed and Precision from Novel Assembly Robot," Assembly Automation, IFS Pub., pp. 85-87, May 1989.

[5] J. Canny and K. Goldberg, "A RISC Approach to Robotics," IEEE Robotics and Automation Magazine, pp. 26-28, March 1994.

[6] A. Sharon and D. Hardt, "Enhancement of Robot Accuracy Using Endpoint Feedback and A Macro-Micro Manipulator System," American Control Conference Proceedings, San Diego, pp. 1836-1842, June 1984.

[7] LEIA Air-cushioned single-level stage with integral drive for rapid positioning with submicron precision. Jenoptik Technology GmbH Micro-fabrication Division, sales information, April 1994.

[8] LN2000 Ultra-Precision Ceramic Linear Motors. NanoMotion Ltd., Gutwirth Park, Technion, Haifa 32000, Israel. Sales information July 1994.

[9] W. Hinds, "Single Plane Orthogonally Movable Drive System," US Patent 4,654,571, March 1987.

[10] "Microglide" X-Y Planar Stage, Anonews, Anorad Corporation, Vol. 13, No. 4, December 1994.

[11] R.S. Fearing, "A Miniature Mobile Platform on an Air Bearing," 3rd International Symposium on Micro Machine and Human Sciences, Nagoya, October 1992.

[12] T. Inoue, K. Iwatani, I. Shimoyama, and H. Miura, "Micromanipulation Using Magnetic Field," IEEE International Conference on Robotics and Automation, Nagoya, Vol. 1, pp. 679-684, May 1995.

[13] S. Hert and V. Lumelsky, "The Ties That Bind: Motion Planning for Multiple Tethered Robots," IEEE International Conference on Robotics and Automation, San Diego, pp. 2734-2741, May 1994.

[14] W. Hinds, "Plural Axis Linear Motor Structure," US Patent 3,851,196, November 1974.

[15] R. Hammer, R.L. Hollis, C.H. An and F. Hendriks, "Design and Control of an Air-bearing Supported Three Degree-of-Freedom Fine Positioner, Proceedings IEEE International Conference on Robotics and Automation, Nice, pp. 677-684, May 1992.

[16] B.A. Sawyer, "Magnetic Positioning Device," US patent 3,457,482, July 1969.

[17] C.G. Oliver, "Advances in Powder Metallurgy of Soft Magnetic Materials," *IEEE Transactions on Magnetics,* Vol. 31, No. 6, pp. 3982-3984, November 1995.

[18] W.E. Hinds & B. Nocito, "The Sawyer Linear Motor," in **Theory & App. of Step Motors**, B.C. Kuo Editor, West Pub. Co., New York, pp. 327-340, 1973.

[19] E.R. Pelta, "Two-Axis Sawyer Motor for Motion Systems," IEEE Control Systems Magazine, pp. 20-24, October 1987.

[20] B.A. Sawyer, "Linear Magnetic Drive System," US patent 3,735,231, May 1973.

[21] B.A. Sawyer, "Multi Axes Linear Movement Positioning System," US patent 3,836,835, Sep. 1974.

[22] W. Hinds, "Hall Effect Encoder for LSM," Northern Magnetics Inc., Personal Communication, Dec. 1988.

[23] A. Brennemann et al, "Magnetic Sensor for 2-D Linear Stepping Motor, IBM TDB, June 1992.

[24] J. Ish-Shalom, "Sawyer Sensor for Planar Motion System," IEEE International Conference on Robotics and Automation, San Diego, pp. 2652-2658, May 1994.

[25] D. Crawford, F.Y. Wong, and K. Youcef-Toumi, "Modeling and Design of a Sensor for Two Dimensional Linear Motor," IEEE International Conf. Robotics and Automation, Nagoya, Vol. 3, pp. 2367-2372, May 1995.

[26] G.L. Miller, "Capacitively Commutated Brushless DC Servomotors," US patent 4,958,115, Sep. 1990.

[27] B.D. Hoffman and S.H. Pollack., "Method and Apparatus for Precision Control of Linear Motors and the Like Employing a Closed Loop Technique," US. patent 4,823,062, April 1989.

[28] E. Nicolson et. al., "Optical Position Sensing for Closed-Loop Control of Linear Stepper Motors," Int. Conf. Adv. Mechatronics, Tokyo, August 1993.

[29] A. Brennemann et al, "An Optical Means of Sensing Position of a Sawyer Motor on a Magnetic Grid Surface," IBM TDB, 1994.

[30] J. Ish-Shalom and P. Kazanzides, "SPARTA: Multiple Signal Processors for High-Performance Robotic Control," IEEE Transactions on Robotics and Automation, pp. 628-640, October 1989.

[31] J. Ish-Shalom and T. Wasfy, "A Finite Element Model for Real-Time Compensation of the Thermal Deformation of the Platen of a Planar Step Motor," IEEE International Conference on Robotics and Automation, San Diego, pp. 1490-1495, May 1994.

[32] E.R. Pelta, M. Powell and P. Smit, "Relative Performance of a Profiled Force Applied to Linear (and Rotary) Stepper - Driven Motion Systems," Incremental Motion Control Symposium, Champaign, pp. 165-174, June 1987.

[33] T. Kenjo, *Stepping Motors and their Micro-processor Controls,* Oxford Univ. Press, NY, 1990.

[34] S. Grotzinger and J. Ish-Shalom, "Task Independent Relationship of System to Actuator Performance via Scaling," IEEE International Conference on Robotics and Automation, Atlanta, pp. 697-704, May 1993.

[35] J. Ish-Shalom, "Task Independent Bounds on System Performance with Improved Actuators that Do Not Scale," IEEE International Conference on Robotics and Automation, Nagoya, Vol. 2, pp. 1757-1762, May 1995.

[36] J. Ish-Shalom, "Composite Magnetic Structure for Planar Motors," *IEEE Transactions on Magnetics,* Vol. 31, No. 6, pp. 4077-4079, November 1995.

[37] J. Ish-Shalom, "Large-Signal Measurements of a Composite Magnetic Structure for Planar Motors," *The Electrochemical Society Extended Abstracts,* Vol. 95-2, pp. 509-510, October 1995.

[38] L.-W. Tsai and R. Ciardella, "Linear Step Motor Design Provides High Plotter Performance at Low Cost," pp. 7-14, Hewlett-Packard Journal, Feb. 1979.

[39] F. Tahmasebi and L-W Tsai, "On The Stiffness of A Novel Six-DOF Parallel Minimanipulator," First World Automation Congress (WAC'94), Maui, TSI Press, Vol. 2, pp. 189-194, August 1994.

[40] R. Ben-Horin, "A Six-Degrees-of-Freedom Parallel Manipulator with Three Planarly Actuated Links," M.Sc. Thesis, Department of Mechanical Engineering, Technion - Israel Institute of Technology, August 1994.

[41] D. Stewart, "A Platform with Six Degrees of Freedom," Proc. Institute of Mechanical Engr., London, Vol. 180, pp. 371-386, 1965.

Intelligent Manipulation Robot in Micro and Nano meter World

Tomomasa SATO, Hideki MIYAZAKI
RCAST, The University of Tokyo 4-6-1 Komaba, Meguro-ku, Tokyo, 153 (Japan)
email: tomo@lssl.rcast.u-tokyo.ac.jp, miyazaki@lssl.rcast.u-tokyo.ac.jp

1 Introduction

Miniaturization of target objects is progressing in many fields such as electronics, mechanical engineering and medical care in consequence of the improvements of microscope, precision measurement, precision positioning, and fine machining technologies. There are growing demands to reveal the nature of a much smaller world based on manipulating smaller objects in such scientific fields as biology and physics.

Figure 1 shows the present state of the imaging, fabrication, and manipulation technologies in the world ranging from 1m to 1Å. While the microscope technologies have been established in the whole range, the manipulation technologies remain immature even though we have the following technologies.

Manipulation as precise as from 1 μm up to 1mm (refered to Micro manipultion hereafter) can be executed under optical microscopes. In order to handle cells, commercial micro manipulators are utilized. Recently, micro manipulators specially designed for assembly of micromachines or micro-surgery have been developed [1][2][3], and techniques for micro-teleoperation have also been reported [4][5][6]. Additionally, manipulation technologies utilizing electric field [7] or laser beam [8] have been proposed.

In the world of atom manipulaton, scanning tunneling microscopes (STM's), which were originally developed to image ultra micro objects, are becoming regarded as the manipulation tool of a single atom[9]. STM can be considered the ultimate manipulation technology, however, we can neither handle micro parts as large as 100nm nor correct micro defects of a LSI mask by STM, because a STM can only be utlized to handle objects of atomic scale.

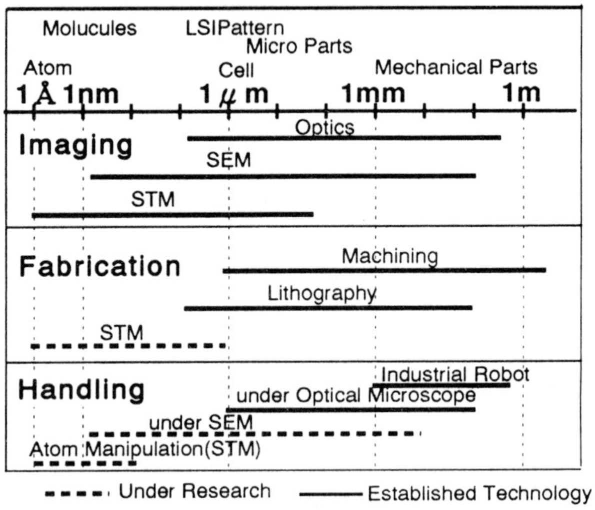

Fig. 1 Coverage of Technology

In the intermediate region of the precision ranging from 1nm to 1μm (mesoscopic range), where neither STM nor manipulation under optical microscope can contribute, manipulation technologies such as handling (i.e. pick up, transfer, and place an object), assembling, and machining are in early stage. In this paper, we call the manipulation with precision ranging from 1nm to 10 μm, Nano Manipulation. In order to realize nano manipulation, micro manipulators under scanning electron microscope (SEM) have been developed [10][11][12], and a few tasks, such as assembly of micro parts around several hundred μm in size and cutting of LSI wire several μm in width, were executed by teleoperation.

Micro/nano manipulation stated above remains at an early stage because we do not know much about the micro manipulation world in terms of microphysics and skills for micro/nano object handling. The differences between the micro and the macro worlds, and the difficulties of micro/nano manipulation originating from that gap have been reported, however, any approach toward reliable micro/nano manipulation overcoming the difficulties has not reported so far. We need a whole set of new robotics to cope with the difficulties of manipulations in micro and nano world, i.e. we need new manipulator mechanism, control, and intelligence for micro and nano maipulation.

In this paper, we firstly discuss the features of the micro/nano manipulation world and clarify the required functions for such a manipulation system. Then the paper proposes 1) a micro manipulation system named 'A Concentrated Manipulator' suitable for micro manipulation, 2) a micro teleoperation system featuring 'Manual Task Execution Posture', 3) an automatic manipulation system based on visual feedback control named 'Nano Hand-eye System', and finally 4) 'A Pick&Place Forming Method' to realize optical quantum devices as a typical product of nano manipulation. In the last portion, the paper touches upon the scope of the application areas of micro/nano manipulation as well as the scope of the research field.

2 Features and required functions of micro/nano manipulation

As the result of miniaturization of object feature size, we have to become conscious of many remarkable characteristics of the object properties, the physical environment and human-machine interaction, which are different from our macro world robotics. In this section, we point out the features of micro/nano manipulation world and make clear the functional requirements for such manipulation system.

2.1 Micro object

2.1.1 *Invisible by Human eyes*

The most essential feature of micro/nano manipulation lies in the fact that the target object is small. Therefore the manipulation system must be equipped with a microscope with high resolution because the object cannot be seen by naked eyes.

However with high magnification, the field of view is restricted to a small area that it is very difficult to grasp the global information of each object. Therefore, variable magnification zooming up and down functions in a wide range is also indispensable.

In an optical microscope with very high magnification, the depth of a focus is often very short compared with the lateral size of the field of view and the working distance (the distance between the objective lens and the object) also tends to be short. To realize three-dimensional manipulation, a long depth of focus is required. Further, to monitor the interaction between the object and the manipulator, the microscope should overlook them from above the manipulator. A long working distance is another demand, i.e., a microscope with long working distance enables the sytem to monitor and manipulate the same position simultaneously in order to realize effective manipulation.

To obtain the information of three-dimensional location, multi-directional imaging (multiple-views) is very effective. One more important function essential to the visual feedback control is the realtime image processing capability such as displacement and/or deformation measurement, detection of the object status, and visual tracking of the object.

2.1.2 *Light and Weak Structure*

As the object geometries become smaller, the objects become lighter, more elastic, and fragile. This is because the object's mass decreases in proportion to the third power of the length reduction, its structural rigidity falls proportionally, and its cross-sectional area, to the second power. Therefore, we tend to apply excessive force, which makes the micro object 1) be suddenly flipped away, 2) be elastically deformed to a nonnegligible degree, 3) sometimes make a plastic deformation, and 4) even be easily destroyed.

To cope with the problem that the object is suddenly flipped away, we need some means to prevent it. Also the function to search the flipped-away object is necessary. Since all of these accidents of sudden movement, deformation, or destroyal of the object are caused by improper force, we can prevent them by monitoring the small force acting on the object and extremely precise manipulation accurately based on the force information. Therefore, sensitive force detection is a requisite function. A precise force sensor is an essential device.

Force produced by physical contact in nano manipulation is estimated to range from 1nN to

1mN, however, a multi-axial force sensor with high sensitivity of 1mN or less is very difficult to realize. On the other hand, we can visually recognize and estimate much smaller force by the slight deformation of an elastic end effector. Consequently, visual information can be utilized for the force monitoring. Further, in case of accidental motion or deformation by excessive force, visual information can contribute to recognize the status and even continue the task adaptively.

2.1.3 *Complexity of Shape*

The complexity of the shape is another feature of the micro object. Since the accuracy of the manufacturing process of the object is not small enough compared with the size of the object itself, its surface is remarkably rough and never agrees with the designed ideal outline when it is observed with a microscope. Even the grain shape of each crystal can be recognized. Consequently, we cannot rely on nominal designed shape, i.e., since the manipulation based on the given model is impractical, visual feedback control has to be adopted. It is also requisite to ensure the repeatability of the task.

2.2 Micro world

2.2.1 *Smallness in Size*

The most essential feature of micro/nano world lies in the fact that the size of the world is small.

Let's consider the size reduction of a triangle. While the length of edges become shorter as the triangle is scaled down, the inner angles do not change. This implies that the micro/nano manipulation indeed requires translational motion with high accuracy, but does not need higher accuracy for rotational motion. However, rotational movement requires wide rotational range. Needless to say, it is preferable to have a large number of Degrees Of Freedom (DOF) to perform complicated tasks such as assembly. In order to physically touch the object and execute tasks in micro/nano world, the manipulator needs to be equipped with a proper end effector such as a gripper or cutting tool. Since the minimum size of the object to be handled is determined by the size of the tool tip, a precise tool suited for the manipulation must be developped. In order to overcome an accident of tool deformation and even continue the task, tool changeability is also necessary.

2.2.2 *Difference of Physical Laws*

In the micro world, the dominant principles of physics are different from that of the macro world or human world. The mass of the target object is so small that the forces like adhesive force due to the water layer adsorbed onto the surfaces, electrostatic force, van der Waals force, or even radiation force of light exceed the gravity in nano manipulation world[13]. Since humans operate based on the model of macro world physics, reliable task cannot be easily executed by teleoperation.

To realize automatic task execution by a robot, we have to construct a model of the task in the micro/nano world in terms of microphysics. Based on the model of micro/nano task, we can realize an algorithm for automatic operation to execute micro/nano manipulation.

Indeed, the laws of the dynamics in micro/nano manipulation are so complicated that we may not succeed in constructing the model which completely describes the task. In such case, we cannot predict what is going to happen in the next moment. Therefore, visual recognition of the circumstances and the status of the task execution is required.

2.2.3 *Uncertainty*

There is always a nonnegligible mismatch between the ideal status and the actual status in micro/nano manipulation world. Errors which are not dominant factors to prevent reliable manipulation in macro world often cause serious problems in the micro world. Some examples include errors caused by 1) positioning inaccuracy of the object, 2) inaccuracy of the manipulator itself, and 3) backlash of the actuators. While the thermal expansion by temperature change, the vibration and the deformation by external forces due to finite rigidity are troublesome also in our macro world, the displacement due to these disturbances can be fatal in micro/nano manipulation.

These would be solved by the improvement of mechanical design of the manipulator, and also by visual compensation of the position and the orientation.

Another important problem is contamination. We often find surprisingly huge dust particles in micro/nano manipulation world. Prevention and recognition of them and adaptive operation is requisite.

2.3 Micro world and Macro World

2.3.1 *Informational Mismatch*

Unfortunately, we can neither see the object nor feel the force applied to the needle directly by our unaided receptors. Therefore, an interface between the micro and macro world is necessary in micro/nano manipulation. It is only through such interface devices that we can recognize the situation of the task and execute the task. Even to be able to write a program of automantic task execution, we need the interface devices which connect the micro world and our world.

2.3.2 *Positional Mismatch*

In addition to the information interface between micro and macro world stated above, we also need positional interface function between micro and macro world. Even though we try hard to place the target object as close as possible to the center of the field of view by human hand, there still remains a need for mechanical adjusting of the position by fairly large distance in terms of micro/nano world. It is because humans are not accurate enough in terms of micro/nano world.

Therefore the manipulator must have wide translation range simultaneously with high accuracy in positioning to compensate the inaccurate positioning capability of the human.

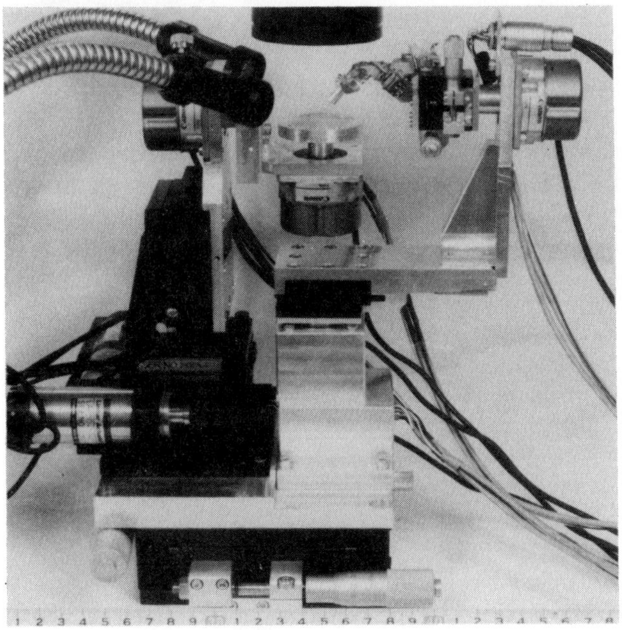

Fig. 2 Concentrated Motion Manipulator

3 Examples of Micro/Nano Manipulation System

3.1 Manipulation System

3.1.1 *Concentrated Motion Manipulator*

Figure 2 shows the micro handling system[14,15] which consists of a 'Concentrated Motion Manipulator', a worktable, and optical microscopes. In micro object handling, it is most important that operations are always executed within the field of view of a microscope. To ensure the task execution under microscope monitoring, the tool tip must always stay within the visual field. The 'Concentrated Motion Manipulator' is the manipulator where the Degrees of Freedom (DOF) are so designed that all rotational centers of the DOFs coincide with each other at the tool tip.

The handling system succeeded in executing several micro manipulation such as assembly of a micro torii(Shinto shrine gate) and a micro air turbine (dia. 500μm) as assembly of small mechanical parts, and folding a origami (paper folding) airplane as handling of flexible object.

3.1.2 *Concentrated Visual Fields Manipulation System*

Figure 3 shows a 'Concentrated Visual Fields Manipulation System[14]' which is suitable for micro object handling. The 'Concentrated Visual Fields Manipulation System' features the following: (1) multiple microscopes, (2) concentrated fields of view, (3) rotational DOF for tilting manipulators and a worktable.

To obtain operation information from multiple directions, multiple microscopes are needed. To obtain an accurate three dimenional position in the working space, one microscope should be located perpendicularly to the others. To observe in the same space at the same time, all fields of view are concentrated. Thus all optical axis of the microcope interect at one point in space in this Concentrated Visual Fields Manipulation

System.

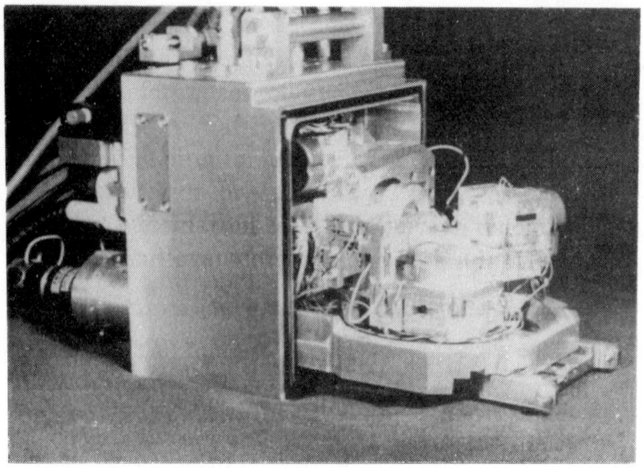

Fig. 3 Concentrated Visual Fields Manipulation System

The constructed micro handling system conists of two microscopes (a scanning electron microscope and an optical microscope) and two manipulators (main and sub) and a worktable. The main manipulator is equipped with two rotational DOF, and three highly precise translational DOF. The worktable is comprised of three translational DOF with both coarse and highly precise movement.

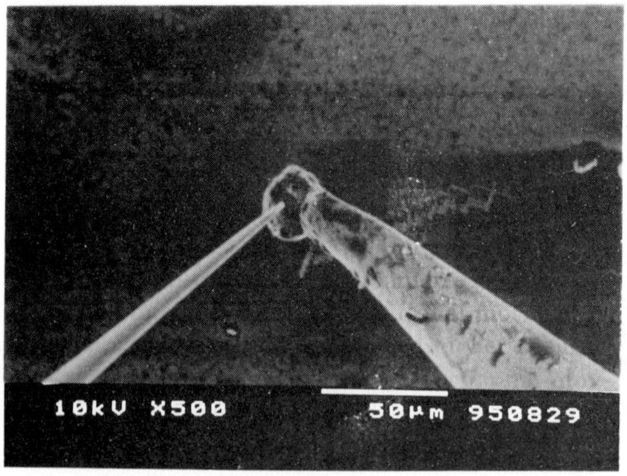

Fig. 4 Preliminary experiment on skill of incling tool

Figure 4 show an experiment in which a micro object is placed on the worktable by a primary and a secondary manipulator. In this experiment, the tungten needle is attached to the primary manipulator and a drawn glass pipe is attached to the secondary manipulator. The tool tip diameters of the primary and secondary manipulators were 20 μm less than 2 μm respectively. The contact area between the tungten needle and the object is wider than the one between the glass pipe and the object. Reliable manipulation skill of "place" is realized by utilizing the differefent van der Waals forcees between two tools of manipulators in the following procedures: (1) The primary manipulator picked the object; (2) the primary manipulator placed the object on the worktable; (3) the secondary manipulator fixed the object on the worktable as shown in Figure 4; (4) the primary manipulator moved away from the object; (5) the secondary manipulator was successfully moved away from the object.

3.2 Visual Feedback Control

To realize accurate micro manipulation and to cope with the uncertainty of the micro/nano world and inaccuracy of the manipulator, two types of Hand-Eye system , in micro and nanoworld were constructed.

3.2.1 *Hand-eye system in micro world*

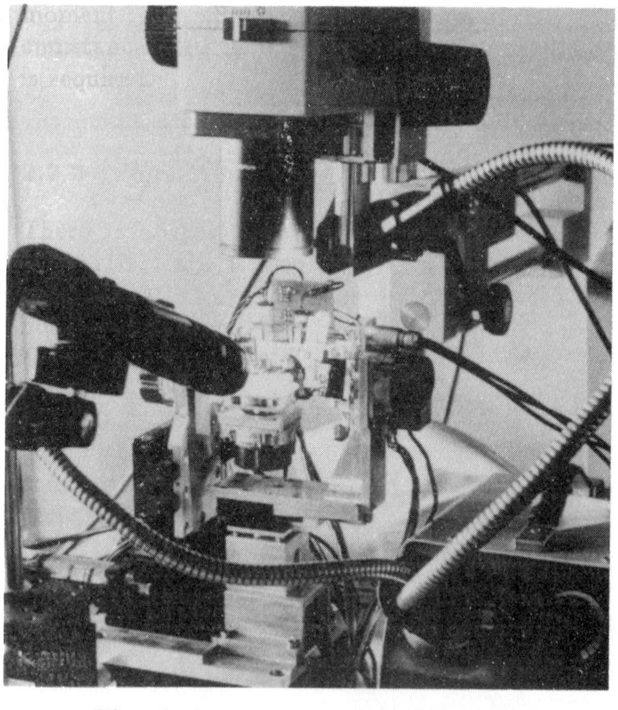

Fig. 5 Hand-eye system of micro drop application

A micro hand-eye system of micro drop application is realized for repairing micro pattern utilizing visual control as shown in Figure 5. The diameter of the micro drop ranges from 10μm to 100μm. The material of the micro drop is metal paste. A constructed prototype system is equipped with needle shaped tool, real time visual processor and a packege of task execution environment called "task package." The experiment of micro drop application were demonstarted.

3.2.2 *Hand-eye system in nano world*

The first nano manipulation system with visual feedback control is realized. The system, named a Nano-Hand-Eye System[16,17,18], has the following three main functions: manipulation (hand of 10nm positional accuracy) , visual monitoring (eye of scanning electron microscope), and integration of the hand and the eye. Micro engraving of thin line of 3μm width and 15μm length by visual feedback control is preformed with a tungsten needle on the solder substrate. The width and the direction of the line are visually controlled.

3.3 Teleoperation System

3.3.1 *Teleoperation system featuring Manual Task Execution Posture*

Figure 6 shows a micro-teleoperation system[19,20] that is inherently suitable for micro-object handling since it mitigates several drawbacks associated with human interface devices. The system uses an upward-facing CRT display monitor mounted beneath a table while looking down at the monitor with a touch sensor screen situated between the operator's arms. To execute these tasks, microscope-observed images of the target objects are utilized with the touch sensor screen and a specially designed, pencil-shaped master manipulator having a force-feedback function. The system incorporates the following characteristic features: (1) intuitive teleoperation is possible because coincidence of the operating and monitoring point is provided, (2) the pencil-shaped master manipulator enables sensitive bilateral micro-teleoperation because of its light weight and small inertia for movement, and (3) with the side of the right hand fixed in a pen-holer grip, the operator uses the master manipulator to generate command with a wide dynamic range of force feedback, i.e., the operator can simultaneously feel force feedback during operation from both pressure/slip sensory receptors of the fingertips and sensory receptors which detects stretching of the finger mustles. By utilizing the system, a micro tiny Japanese doll (Kokeshi) is sculptured.

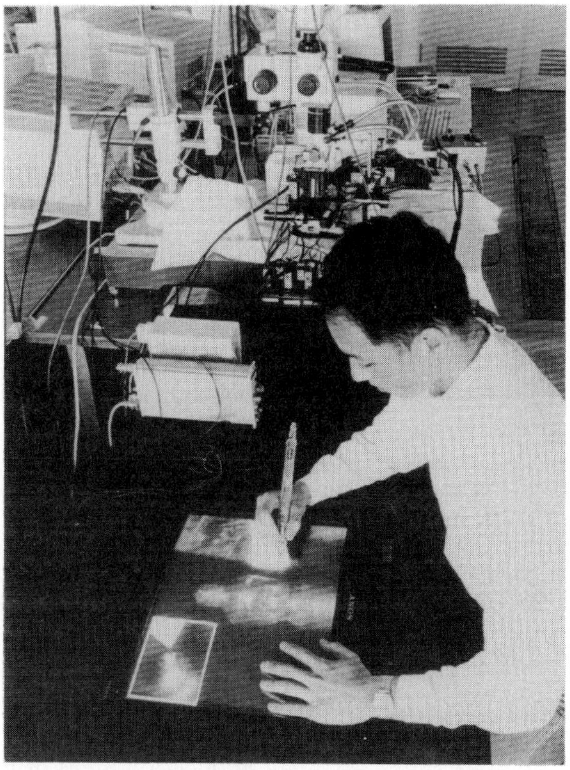

Fig. 6 Micro-teleoperation system

3.3.2 *Handling of fine particles in SEM environment*

Fabrication of three-dimensional (3D) micro structures such as optical quantum effect devices is realized by a "Pick-and-Place Forming Method[21]": a fabrication method of 3D micro structures by heaping up fine particles typically ranging from 10nm to 1μm in diameter with micro manipulator under scanning electron microscope monitoring. Particles are picked up by van der Waals force, and pick-and-place control is performed by the optimization of material combination and the trajectory of tool movement. This fabrication method has the following advantageous features: 1) excellent forming resolution as small as 10nm, 2) wide range of material selection, 3) freedom of the 3D structure, and 4) quick fabrication. Pyramidal 3D structure made of 35 polymer particles is successfully fabricated.

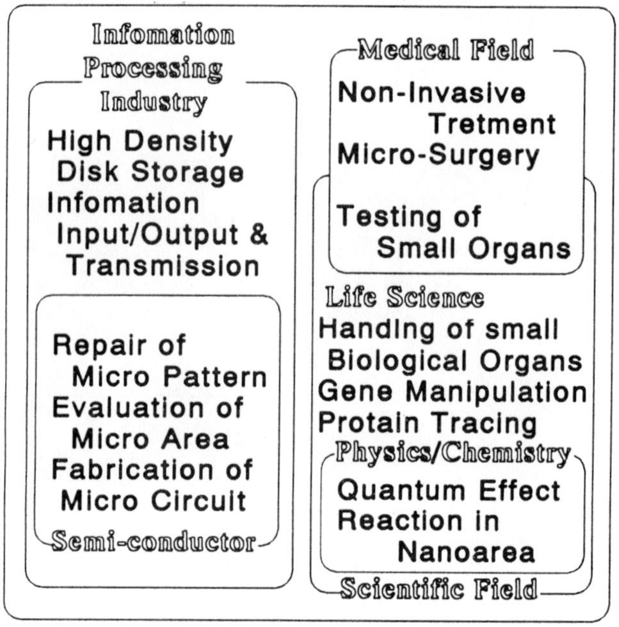

Fig. 7 Application areas

4 Application Field and Research Scope

4.1 Application Field

Figure 7 shows the main application areas of micro/nano manipulation. As for the industrial application, information processing and semiconductor related areas are the main domain where micro/nano manipiulation technologies are applied because we can expect drastical cost down of information processing aparatus and thus information processing is the leading thrust of technological inovation in the coming years. In these areas, a high density disk storage or information input/output instruments requires micro manipulation technology of assembly of microhead and components.

As for medical or life science application, micro manipulation for non-invasive operation or handling of cells and gene are required. Micro/nano manipulation offers an essential tool to cure in medical fields and to investigate biology because our body consists of cells diameter of which are aroud micro meters.

4.2 Research Scope

Figure 8 shows the list of manipulation related verbs from scientific dictionary. It is arranged according to their levels[22]. The list clearly gives the overview of the manipulation world. The top level corresponds to the genre of the robot such as an assembly robot or a construction robot. Each verb in the second level corresponds to individual robot or machine, i.e., a welding robot or a cutting machine. The lowest level is related to servoing aspect of the robot such as to moving or pushing. Inbetween these levels, we have several levels which deal with flexible object, powder or liquid. The second lowest level, say level 5, is the level of manipulation science. For example, insertion is always an important research topics in robotics. Current industrial robot technology is based upon these verbs in our macro world. Robotics so far has been related to these verbs in macro world. We should make the robot to execute these verbs in micro world and nano world. We also have to make new verbs in micro and nano world. These efforts will drive us to generate new industry in the future.

5 Summary

Micromachines, micro systems or MEMS (Micro Electrical Mechanical Systems) have been paid keen attention to realize advanced machines which could not be realized so far. With micro component or endeffector, a robot comes to be able to execute micro/nano manipulation tasks. The following are examples of micro/nano manipulation tasks: assembly of micro machines and quantum devices, repair of micro patterns, micro surgery, physical measurement of small biological organs such as ears and cells and physical handling of cells. An intelligent robot to handle such very small objects are keenly demanded. Micro manipulation and nano manipulation can be done under an optical microscope or a scanning electron microscope respectively.

In this paper,the the fundamental difficulties and necessary functions for micro/nano manipulation to cope with the difficulties are presented. As concrete example of the art of the micro maipulation technology, the paper presentes a 'Concentrated ViewiField Manipulation System' which is suitable for micro object handling under either optical microscope or scanning electron microscope. Such micro and nano task execusion of assembly of micro-air turbine, micro paper folding of Origami, construction of optical quantum effect device by heaping up sub-micro particles are presened. The paper also describes micro tele-operational system featuring 'Manual Task Execution Posture' together with task execution example of sculpturing micro tiny Japanese doll (Kokeshi).

The paper also covers the promising application areas of 1) information related instruments such as disk drive, 2) semiconductor related tasks as repair of micro patterns, 3) medical field such as key hole surgery and 4) bio/life science and physical science fields. Lastly the research scope of the micro/nano manipulation is presented by showing the list of verbs from dictionary.

In the trend toward miniaturization in many fields, above stated micro/nano manipulation technology will surely play an important role in manufacturing smaller functional devices or in revealing the nature of much smaller world. A lot of research activities should be devoted to mechanism, control, intelligence for task execution and friendly human-robot interaction in micro and nano meter manipulation world.

Verb 1 :Task

Analyze, Assemble
Build, Construct,
Disassemble,
Machining, Measure
Operate, Produce,
Repair,Test

Verb 4 :Handling of Liquid & Powder

Clean, Distribute, Draw,
Fill, Gather, Laddle,
Lubricate, Mix, Paint,
Plaster, Pour, Pump,
Scoop, Spray, Sift, Stuff,
Wipe, Write

Verb 2 :Machining

Cut, Cut-out, Dig, Drill,
Drive-nail, File, Grind,
Plane, Polish, Powder,
Punch, Saw, Scratch,
Sharpen, Shave,
Squeeze, Unnail, Weld,
Whet,Whittle,

Verb 5 :Change of Status

Attach, Arrange, Combine,
Deposit, Extract, Fly,
Hang, Insert, Interconnect,
Lean, Lock, Pack, Pile,
Place, Pull, Put-on, Rotate,
Screw, Set, Separate,
Transfer, Throw, Unscrew,
Unlock, Unpack

Verb 3 :Handling of Flexible Object

Bend, Fold, Knead,
Saw, Spread, Squeeze,
Tie, Tear-off, Wash,
Wind, Wire, Wrap

Verb 6 :Simple Motion

Fit, Grasp, Impact,
Incline, Lift, Move,Pick,
Pull, Push, Release,
Support, Shake, Slide,
Strike, Swing, Turn,
Twist, Vibrate

Fig. 8 Manipulation related verbs

Aknowledgement

We would like to thank Professor Hatamura and nanorobot group of the University of Tokyo for their valuable contributions and discussions. This reseearch is supported by a Grant-in-Aid for Developmental Scientific Research of the Ministry of Education, Sience and Culture.

References

[1]T.Sato, K.Koyano, M.Nakao, and Y.Hatamura, "Novel Manipulator for Micro Object Handling as Interface between Micro and Human Worlds," in Proceedings of the 1993 IEEE/RSJ International Conference on Intelligent Robots and Systems, pp. 1674-1681, Yokohama, Japan, July 26-30, 1993.

[2]I.W.Hunter, S.Lafontaine, P.M.F.Nielsen, P.J.Hunter, and J.M.Hollerbach, "Manipulation and Dynamic Mechanical Testing of Microscopic Objects Using a Tele-Micro-Robot System", IEEE Control Systems Magazine, pp. 3-9, Feb. 1990.

[3]R.A.Russel, "Development of a Robotic Manipulator for Micro-Assembly Operations", in Proceedings of the 1993 IEEE/RSJ International Conference on Intelligent Robots and Systems, pp. 471-474, Yokohama, Japan, July 26-30, 1993.

[4]T.Sato, J.Ichikawa, M.Mitsuishi, H.Miyazaki, and Y.Hatamura, "Micro-Teleoperation with Manual Task Execution Posture", IEEE Control Systems Magazine, pp. 22-29, Feb. 1995.

[5]M.Mitsuishi, K.Kobayashi, T.Nagao, Y.Hatamura, T.Sato, and B.Kramer, "Development of Tele-Operated Micro-Handling/Machining System Based in Information Transformation," in Proceedings of the 1993 IEEE/RSJ International Conference on Intelligent Robots and Systems, pp. 1473-1478, Yokohama, Japan, July 26-30, 1993.

[6]Y.Yokokohji, N.Hosotani, and Y.Yoshikawa, "Analysis of Maneuverability and Stability of Micro-Teleoperation Systems," in Proceedings of 1994 IEEE International Conference on Robotics and Automation, pp. 237-243, San Diego, U.S.A., May 8-13, 1994.

[7]M.Washizu, "Manipulation of Biological Objects in Micromachined Structures", in Proceedings of IEEE Micro Electro Mechanical Systems Workshop 1992, pp. 196-201, Travemunde, Germany, Feb. 4-7, 1992.

[8]H.Misawa, M.Koshioka, K.Sasaki, N.Kitamura, and H.Masuhara, "Laser Trapping, Spectroscopy, and Ablation of a Single Latex Particle in Water", Chemistry Letters, pp. 1479-1482, 1990.

[9]D.M.Eigler and E.K.Schweitzer, "Positioning Single Atoms with a Scanning Electron Microscope", Nature, 344, pp. 524-526, April, 1990.

[10]Y.Hatamura and H.Morishita, "Direct Coupling System between Nanometer World and Human World", in Proceedings of IEEE Micro Electro Mechanical Systems Workshop 1990, pp. 203-208, Napa Valley, U.S.A., Feb. 1990.

[11]S.Johansson, "Hybrid Techniques in Microrobotics", in Proceedings of 1st IARP Workshop on Micro Robotics and Systems, pp. 72-83, Karlsruhe, Germany, June 15-16, 1993.

[12]Y.Hatamura, M.Nakao, and T.Sato, "Construction of Nano Manufacturing World", MICRO SYSTEM Technologies '94, vde-verlag, pp. 37-51 (1994).

[13]R.S.Fearing, "Grasping of Micro-Parts in Air", in Tutorial Textbook of 1994 IEEE International Conference on Robotics and Automation, pp. 1-7, San Diego, U.S.A., May 8-13, 1994.

[14]K.Koyano, T.Sato, and H.Miyazaki, "Construction of Concentrated Visual Field Manipulation System for Ultra Micro Object Handling", in Proceedings of the 72nd JSME Spring Annual Meeting, VI, pp. 281-284, Tokyo, Japan, March 29-31, 1995.

[15]T.Sato, K.Koyano, M.Nakao and Y.Hatamura,"Micro Handling Robot for Micro Machine Assembly – A Vision/Operation/Motion Cofocused Manipulation System",1st IARP Workshop on Micro Robotics and Systems, pp. 138-148(1993)

[16]H.Miyazaki, T.Kameya, T.Sato, Y.Hatamura,"Construction of Ultra Micro Manipulation System based on Visual Control – Realization of Nano-Hand-Eye System–,"1994 IEEE Symposium on Emerging Technologies & Factory Automation, pp. 74-77(1994 November)

[17]T. Sato, T. Kameya, H. Miyazaki, and Y. Hatamura, "Hand-eye System in Nano Manipulation World", 1995 IEEE International Conference on Robotics and Automation, Vol.1, pp.59-66 (May. 1995).

[18]T. Sato, T. Kameya, H. Miyazaki, Y. Hatamura, and H. Morishita, "Autonomous Task Execution under SEM Observation by Nano Hand-Eye System", The 72nd JSME Spring Annual Meeting (VI), pp. 293-296 (Mar. 1995).

[19]T.Sato, J.Ichikawa M.Mitsuishi, Y.Hatamura, "Micro Teleoperation Concentrating Visual and Force Information at Operator's Hand", 3rd Int. Symp. on Experimental Robotics (ISER) Kyoto, Japan, pp. 118-124(28-30-Oct-1993)

[20]T.Sato, J.Ichikawa, M.Mitsuishi and Y.Hatamura, "A New Micro-Teleoperation System Employing a Hand-Held Force-Feedback Pencil", IEEE Robotics and Automation '94, pp. 1728-1733(May-1994)

[21]H. Miyazaki, "Fabrication of Three-Dimensional Micro Structures by Heaping up Fine Particles", The 72nd JSME Spring Annual Meeting (VI), pp. 273-276 (Mar. 1995).

[22]K.Takase, "ETL Report on Advanced Robot Project", pp.13 (1985).

Hybrid Robots Consisting of Mechanical Parts and Living Organisms for Microrobot Application

Isao Shimoyama, Hirofumi Miura, Yoshihiko Kuwana, Yushi Sayama

Department of Mechano-Informatics, The University of Tokyo

7-3-1 Hongo, Bunkyo-ku, Tokyo 113, Japan

e-mail: {isao, miura, kuwana, sayama}@leopard.t.u-tokyo.ac.jp

http://www.leopard.t.u-tokyo.ac.jp

1 Introduction

Structure of a 1 mm long microrobot that is our final goal may be different from that of a conventional robot due to scale effects. For example, rotational joints do not work well in microrobots because friction becomes dominant compared to gravity. Since insects have obtained suitable structures and functions for their scale through natural selection, they are good models for microrobots.

We've built some mechanical structures that mimic insects. However, mechanical components that current micromachining can fabricate are not so excellent as living organisms. For example, muscles generate power more efficiently, and antennae detect pheromone more sensitively. If parts of living organisms can be used for hybrid systems instead of mechanical components, the possibility of microrobots will extend.

The hybrid systems contribute to natural sciences including neuroethology. Most biologists are interested in analysis. However, emerging behavior cannot be explained clearly by molecular level analysis. Engineering synthesis may help to understand nature.

This paper shows fundamental attempts to make hybrid robots in which artificial sensors are replaced with natural antennae. First, it is shown that signals from an antenna are easily obtained with electrodes connected to both ends of an antenna. In our experiment, antennae of a silkworm moth (Bombyx mori) are used as sensors that can detect molecules of pheromone.

Concerning intelligence of microrobots, the number of neural cells in humans is 10^{10}, while the number of neural cells in insects is 10^5. Since the size of neural cells is limited by molecules, there are no tiny living organisms with a substantial amount of neural cells. The behavior of an insect is based on reflex. Also, a similar situation will be seen in a microrobot because the area available for intelligence is not large.

Currently, it is impossible to use a real nervous system for microrobots because a nervous system cannot be taken out of the body without damage. In addition, a nervous system cannot be directly connected with artificial sensors and actuators, either. Recurrent artificial neural networks are applied in this paper. The connection of the neural network is obtained by genetic algorithm (GA) and heuristics based on analytical results. As a result, a mobile robot with antennae follows a stream of pheromone.

2 A mobile robot with antennae of a silkworm moth

A male silkworm moth follows a pheromone path to locate a female. Entomologists and neuroethologists characterize this behavior as follows: when a moth is inside the pheromone stream, in which pheromone molecules (Bombykol) arrive at the antennae frequently, the moth goes straight upwind in surges. If the moth is out of the stream without pheromone stimulation, it exhibits side-to-side casting behavior to search out pheromone plumes. Until reaching a female, the moth shows tonic casting behavior as well as phasic surge behavior depending on the existence of pheromone. Details of the behavior, however, are still under investigation[6][7].

Silkworm moth antennae generate signals only if pheromone from a female silkworm moth arrives at the antennae. Figure 1 shows a silkworm moth and its antenna. The antenna is 5 mm in length and is covered with hair.

We've fabricated a pheromone sensor shown in Fig. 1 [8]. It employs silver (Ag) electrodes inserted into both ends of the antenna directly. In order to evaluate the pheromone gas sensors, dose response and life time were measured. Dose response curves were obtained by plotting the relationship between the concentration vs. amplitude of EAG (electroantennogram) at flow velocities of 0.375 m/s, 0.75 m/s and 1.125 m/s. Pheromone stimulates the sensor for 0.5 sec at 1/60Hz.

Sensitivity of the pheromone sensor was evaluated by determining the threshold using the dose response curves. The pheromone concentration is written as 10^{-10}g/10μl (i.e. artificially synthesized pheromone 10^{-10} g in weight is dissolved in 10 μl n-hexane). The resolution of this living sensor is 10^{-10}g/10μl, or 10^{-4} ppm.

The life time of this living sensor is defined as the time when the response by periodic stimulation becomes compatible to the background noise. The life time was measured at about 80 minutes experimentally.

Two pheromone sensors were mounted on the front of a mobile robot, as shown in Fig. 2. Pheromone is vaporized by hand using a syringe ahead of the robot. Only the female pheromone stimulates the pheromone sensors. Air without pheromone does not affect the sensors, therefore, this robot can trace a pheromone path just like a male moth.

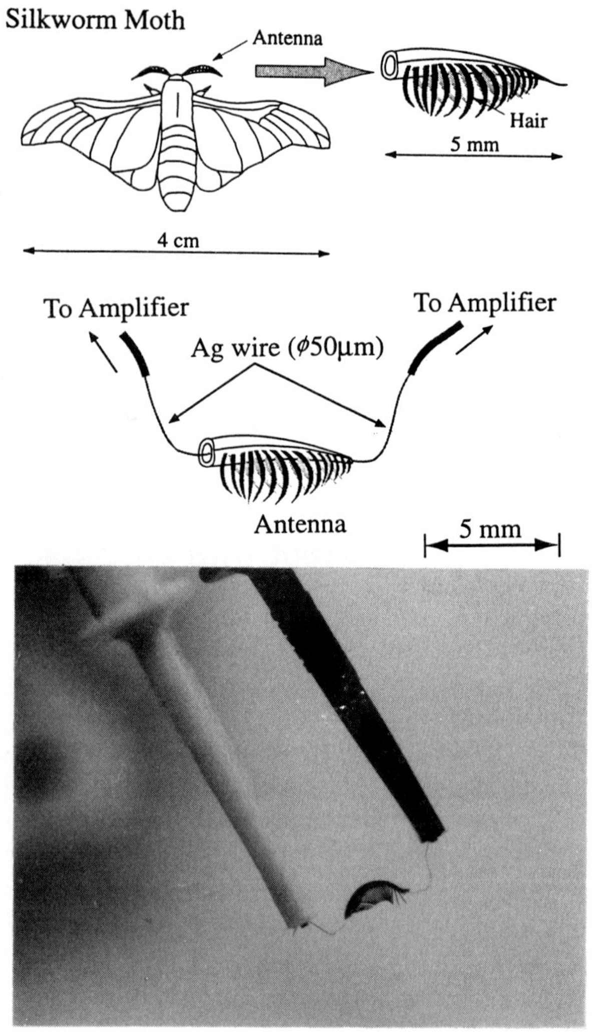

Fig. 1. Antennae of a silkworm moth can be used as sensitive pheromone gas sensors.

3 The oriented walk by recurrent neural networks

To demonstrate a moth-like behavior with pheromone gas sensors, both simulations and experiments were carried out with a mobile robot. The autonomous pheromone tracing mobile robot consists of the pheromone sensors, instrument amplifiers, an AD converter, a micro-processor, two DC motors, and batteries. During the experiment, pheromone gas flows out of a pipe at speed of 1 m/s from in front of the robot. The pheromone stream is calibrated as the gray areas in the experimental results by using a visualization technique with

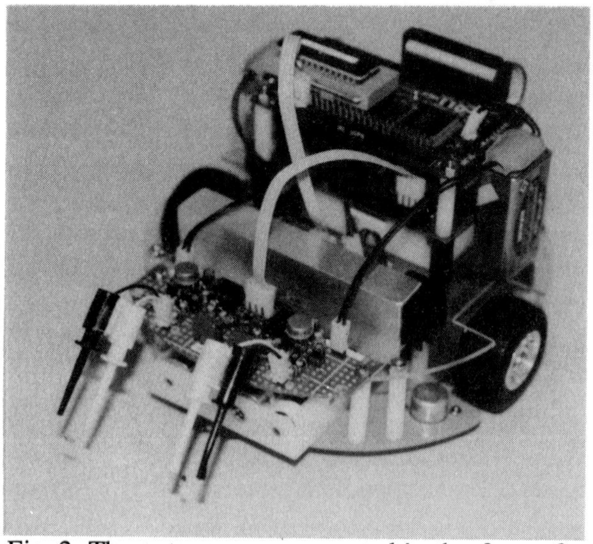

Fig. 2. The antennae are mounted in the front of a mobile robot. The robot moves to the pheromone source by responding to the pheromone distribution.

smoke, as shown in Fig.6. During the simulations, we assumed that pheromone expands inside a trapezoid, as shown in Fig. 4. Recurrent neural network based control was examined to achieve moth-like behavior.

Kanzaki[1]-[4] analyzed moth behavior in a simplified environment. Surging, casting, and turning motions appear in this order after an inactive moth is stimulated by pheromone. The direction of rotation is determined by the side of the initially stimulated antenna. For example, when the right antenna is stimulated, the moth moves forward with surges, zigzags right-left-right-left, and finally rotates clockwise. Since the direction of rotation is determined by the side of the antennae, Kanzaki proposed that there is a memory mechanism like a flip-flop in the neural system.

Based on his results, we assumed that the moth behavior emerges by interaction with the environment where pheromone is distributed complexly in spite of the simple neural system. To confirm our assumption, we conducted two experiments. Both programs are based on recurrent neural networks (RNN) with feedback loops that contribute memory to the system.

The necessary information for the oriented walk may be the existence of pheromone. A real moth also uses visual information, though. Our recurrent neural network includes two input and two output neurons. The input neurons receive pheromone. The output neurons drive two motors that determine the orientation of the mobile robot. All neurons are connected each other in both directions.

In the first experiment, the weights among neurons are determined by genetic algorithm (GA), as shown in Fig. 3. Pheromone tracing with the recurrent network is simulated in the field shown in Fig. 4. Pheromone is distributed uniformly inside the trapezoid, and since both antennae detect pheromone at an equal probability, the robot cannot reach the pheromone source only by comparing the difference of pheromone concentration between both antennae.

The weights between the neurons are encoded into a GA gene. The fitness of the pheromone tracing is defined as the nearest distance between the robot and the pheromone source. Crossover and mutation are applied to optimize the connections of the neurons. An average of five trials is taken as fitness of the pheromone tracing. In each trial, the

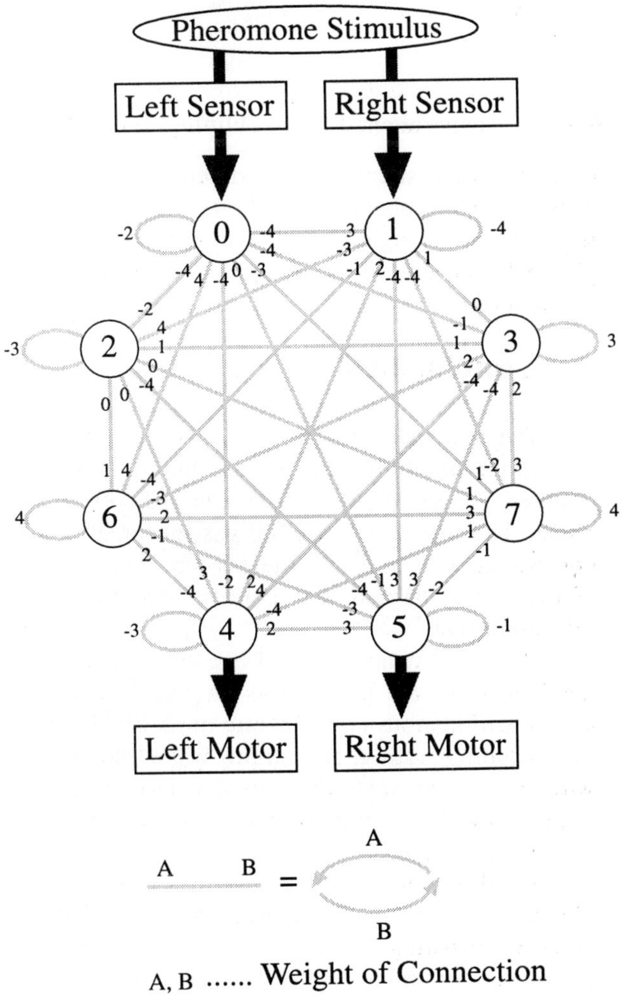

Fig.3. GA-optimized RNN.

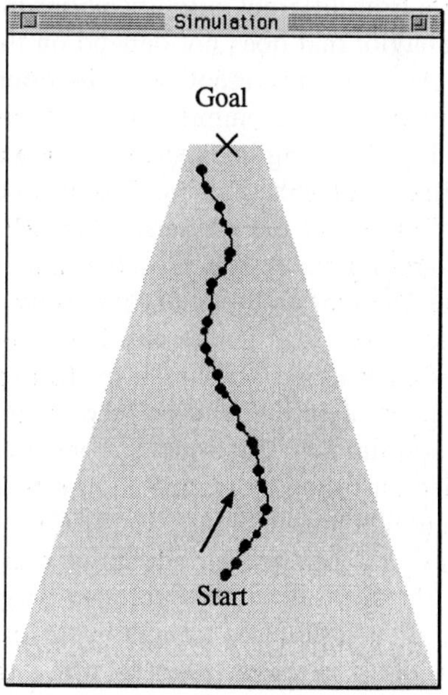

Fig.4. A simulation result. (GA-optimized RNN)

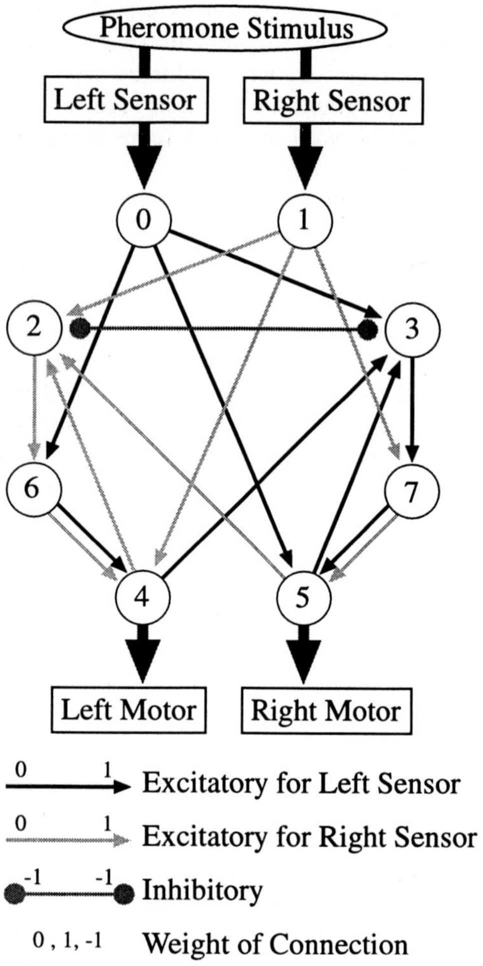

Fig.5. Flipflop-type RNN.

robot starts from different initial positions in order to obtain behavior that does not depend on the initial position. As a result, a robot with GA connections moves toward the pheromone source efficiently as shown in Fig. 4. During an experiment, however, the robot deflects out of the stream. This may occur due to the difference of the environments. When connections are determined by GA, the model should agree with the real environment precisely.

Another set of connections has been determined by examining the results by Kanzaki. One pheromone stimulation excites the sequence of casting, and turning. This sequence is achieved by the connections shown in Fig. 5. Figure 6 shows the experimental results. Contrary to the GA case, the robot traces the pheromone stream in experiment, although the simulation does not provide good performance. In addition, the turning behavior appears to detect the pheromone plumes when the robot

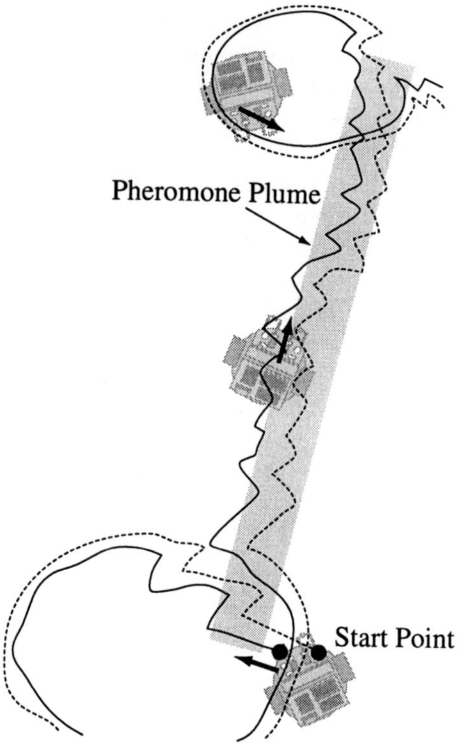

Fig.6. An experimental result. (Flipflop-type RNN)

goes out of the stream in experiment. The overall behavior of the robot looks quite similar to that of a real moth. Only eight neurons can generate a complex behavior due to the complex real pheromone distribution.

4 Conclusions

The antennae of a moth can be used as gas sensors for mobile robots for a duration of about 80 minutes. The sensitivity is 10^{-4} ppm, 10,000 times higher than that of an artificial gas sensor. This robot may be able to detect the gas pipeleaks due to earthquake damage if pheromone is mixed in with the natural gas.

Recurrent neural networks contribute memory mechanism to the pheromone tracing. The small numbers of neurons can generate moth-like behavior, including casting, and turning while interacting with the environment. The turning behavior in particular looks like a suitable tactic for small intelligence when losing pheromone.

Finally, electric stimulation has driven muscles as we want in our experiment[5]. The hybrid system consisting of antennae, muscles, and an electronic controller may be feasible empirically.

References

1. Kanzaki R., Ikeda A. and Shibuya T., (1994) Morphological and physiological properties of pheromone-triggered flip-flopping descending interneurons of the male silkworm moth, Bombyx mori. J. Comp. Physiol. A 175:1-14.
2. Kanzaki R., (1994) Olfactory processing pathways of the insect brain. Lateral accessory lobe system in the protocerebrum produces olfactory flip-flopping signals in Bombyx mori. Olfaction and Taste 11: 831-834.
3. Arbas E. A., Willis M. A. and Kanzaki R., (1993) Organization of goal-oriented locomotion: pheromone-modulated flight behavior of moths. In "Biological Neural Networks in Invertebrate Neuroethology and Robotics" (Eds. Ranall D. Beer, Roy E. Ritzmann, Thomas McKenna) pp.159-198. Academic Press Inc.
4. Kanzaki R. and Ikeda A., (1994) Morphology and physiology of pheromone-triggered flip-flop descending interneurons of the male silkworm moth, Bombyx mori. Olfaction and Taste 11: 851.
5. Yoshihiko Kuwana, Naomitsu Watanabe, Isao Shimoyama, Hirofumi Miura, (1994) Behavior Control of Insects by Artificial Electrical Stimulation, Distributed Autonomous Robotics Systems, Springer-Verlag : 291-302.
6. Tristram Wyatt, (1994) Moth flights of Fancy, Nature, Vol. 369, No. 6476, pp. 98 - 99.
7. Agenor Mafra-Neto and Ring Cardé, (1994) Fine-scale Structure of Pheromone Plumes Modulates Upwind Orientation of Flying Moths, Nature, Vol. 369, No. 6476, pp. 142- 144.
8. Yoshihiko Kuwana, Isao Shimoyama, Hirofumi Miura, (1995) Steering Control of a Mobile Robot Using Insect Antennae, IROS'95, pp.530-535.

PART 16

INDUSTRIAL SYSTEMS ISSUES

SESSION SUMMARY

K.H. Robrock
European Commission, Rue de la Loi 200, B1049 Brussels, Belgium

1. Localisation of Autonomous Guided Vehicles (*Hugh Durrant-Whyte*)

The presentation had reviewed the state of the art of localisation of autonomous guided vehicles (AGVs). The presentation had in particular addressed the problems encountered when applying a combination of sensory information coming from different sensors in conjunction with Kalman filtering techniques applied to robot localisation and map building of the environment.

The subsequent discussion addressed the following issues:

- Model Convergence:

  The proposed model of converging on an exact model of the environment does not depend on continuous observability, i.e. a system where landmarks are disappearing and reappearing. It can be proofed that the system will increase monotonically in convergence of the map regardless of the path used and regardless of how the beacon is viewed. The importance of the result obtained is that one can exploit it to actually develop tractable algorithms which guarantee convergence even if the result obtained is not an exact one. In its qualitative part the system resembles closely the model of a stretching rubber sheet.

  These systems are useful to and begin to be used in geodesy.

- GPS: Experiences with GPS have shown that this localisation tool alone is not sufficiently reliable over the envisaged long period of time. Whereas the exact source of the problems was not entirely clear, this was an experience made by various groups using it.

- Kalman Filtering: To apply Kalman filtering one needs to build an internal model of the world. Irrespective of how good the model, at some point in time the real world may not conform to the model built. Then the process of Kalman filtering becomes "brittle" i.e. it may fail catastrophically. In this case one needs to use and interpret as much as possible sensory information available. In particular, one may make use of a sequence of sensory information over time, e.g. first identify scene beacons, then apply multiple hypothesis filtering algorithms to track in time different features previously identified.

2. Robots Transportation, and Society (*Ch. Thorpe* and *D. Pomerleau*)

This presentation highlighted the technical issues behind automated highway vehicle operations and the societal and institutional context related to it.

The main point on the discussion following the presentation was the balance and trade-off between travelling of persons and acceptance of telecommunications as a replacement in interpersonal communications. This was considered as a broader issue, where certain travel could be reduced in the light of advanced and available communication means. On the other hand, for many occasions the presence of people was indispensable or undesirable and therefore a general reduction or growth of one sector over the other was considered difficult to forecast.

In the same context, smart highway cars were not the only competitor to telecommunications. In fact, transportation of humans was a question of a larger network and an infrastructure of several inter-operable subsystems, such as trains, highway cars, urban car rental systems, etc...

3. Standard Design for Intelligent Robot and its Application (*T. Oomichi*)

The presentation dealt with the requirements for intelligent robots in practical (industrial) applications, and a standard design system for intelligent robots was proposed.

As for the concrete systems under development at Mitsubishi Heavy Industries Ltd, it was expected that these could be introduced into industrial use as of the following year, although that would depend on a number of conditions to be fulfilled.

4. Manufacture of Industrial Robots (*H. Woern*)

This presentation dealt with the technological requirements for robots in industrial applications.

The first question subsequent to the presentation concerned the use of robots to assemble robots. The main point here is that although this is possible from a technical point of view, the numbers of robots typically produced are not large enough to reach economies of scale.

Off line programming and calibration of industrial robots was another point of discussion. It was considered that from an industrial point of view problems concerning integration and robot languages used and geometrical accuracy's requirements were not satisfactorily solved yet and therefore not applied on an industrial scale.

Although the presentation had been focused on "classical robots" in industry (as understood from the request), the emphasis on the industrial application of robots was now more on a holistic approach to entire manufacturing cells and shopfloor systems where the robot is an integrative element of the whole system.

Localization of Autonomous Guided Vehicles

Hugh Durrant-Whyte, David Rye, Eduardo Nebot
University of Sydney, Sydney NSW 2006 (Australia)
hugh,rye,nebot@tiny.me.su.oz.au

1 Introduction

This paper reviews and describes the state-of-the-art in localization of autonomous guided vehicles (AGVs). The localization problem, the ability to accurately sense and estimate the location of a platform, lies at the heart of almost all AGV applications. Solving the localization problem is an essential precursor to more complex AGV tasks such as planing and task execution.

Localization is essentially a problem in estimation: observations of the local environment of the AGV are made from which the relative or absolute location of the platform is computed. The most useful tool in solving such estimation problems is the Kalman Filter. The Kalman filter is simply a linear recursive estimator which employs an explicit model of the dynamics of a process and an explicit model of how the observations made are related to this process to compute estimates for a state which are"optimal" in a minimum mean squared error sense. The explicit use of a process and observation model make the Kalman filter algorithm and its derivatives particularly suited to solving the localization problem. In this paper, familiarity with the basic Kalman filter algorithm is assumed. Our review of the localization problem then follows a case-based study of different position estimation algorithms, starting from simple beacon-based positioning algorithms and proceeding to techniques for simultaneous map-building and localization. In developing these cases, we highlight a number of important issues; in the appropriate use of different sensors, the development of useful vehicle models and in assuring reliability in localization systems.

2 The Basic Problem: Beacon-Based Localization

The simplest way of obtaining information from which a vehicle location may determined is through the observation of a number of fixed and known points of reference (beacons) in its working environment. Given a number of beacon observations, the determination of position is, in principle, relatively straight forward. A wide range of beacon-based systems have been used in AGV localization problems. In particular, the use of lasers and reflective beacons is now common place in factory automation systems. Beacons based on micro-wave tags, ultra-sonic transducers and radar reflectors have also been used. The use of GPS is also increasingly common in outdoor navigation problems.

This section serves to set the base-line for the rest of this paper by describing the operation and implementation of beacon-based localization systems. The basic tools used in beacon-based systems are exploited in subsequent sections describing natural-beacon navigation and map-building.

2.1 The Kalman Filter Algorithm

In AGV systems, as well as in maritime and aerospace areas, the Kalman filter (KF) algorithm is synonymous with the localization problem. Although the KF has many limitations, its proven ability to deal with the many-fold problems involved in complex localization problems in a simple and elegant manner, make it an indispensable starting point for any discussion on the subject. In this paper we briefly list some features of the Kalman filter algorithm which are essential to a description of the localization problem.

The Kalman filter algorithm is a recursive linear estimator which exploits a model of the process under consideration and a model of the observations made of this process to provide an estimate of the state of interest. In localization problems, it is normal to start with a discrete time process model describing the motion of the platform as

$$\mathbf{x}(k) = \mathbf{f}(\mathbf{x}(k-1), \mathbf{u}(k)) + \mathbf{w}(k), \quad (1)$$

where $\mathbf{x}(k)$ is the state (location of the vehicle) at time k, $\mathbf{u}(k)$ is the average input (drive signals) to the vehicle at time k, $\mathbf{f}$ is a (kinematic) function describing the transition of the vehicle between time $k-1$ and k, and $\mathbf{w}(k)$ is a random vector describing the uncertainty in motion of the vehicle in time. Observations are made of the location of the vehicle at time k (usually by taking measurements of the location of beacons placed around the environment at known locations) according to an observation Equation

$$\mathbf{z}(k) = \mathbf{h}(\mathbf{x}(k)) + \mathbf{v}(k), \quad (2)$$

where $\mathbf{z}(k)$ is the observation made at time k, $\mathbf{h}$ is a function (typically also of beacon location) relating observations to current vehicle location, and $\mathbf{v}(k)$ is a random vector describing errors in the observation process.

The Kalman filter recursively computes an estimate of vehicle state through a linear combination of a prediction based on the process model

$$\hat{\mathbf{x}}(k \mid k-1) = \mathbf{f}(\hat{\mathbf{x}}(k-1 \mid k-1), \mathbf{u}(k)) \quad (3)$$

with a correction based on the difference between true and predicted observation as:

$$\hat{\mathbf{x}}(k \mid k) = \hat{\mathbf{x}}(k \mid k-1) + \mathbf{W}(k)\left[\mathbf{z}(k) - \mathbf{h}(\hat{\mathbf{x}}(k \mid k-1)\right], \quad (4)$$

where $\mathbf{W}(k)$ is gain determined by the relative confidence in vehicle prediction and observation.

The basic Kalman filter algorithm provides a number of features:

1. Recursive, incremental, incorporation of information into the vehicle location estimate

2. An on-going estimate of the confidence in location estimate through the state covariance matrix

3. An ability to describe a variety of different sensor observations in a single coherent form.

4. An ability to estimate states which are not observed directly.

The algorithm also has a number of important limitations which need to be considered when designing a localization system:

1. Accurate models of both vehicle and observations are required.

2. The true noise terms must be reasonably well behaved and bounded by the first and second moments of the assumed noise model.

3. Linearization assumptions are necessary for the linear predictor-corrector form of the update equation.

4. The analytic form of the models employed makes it difficult to capture many physical properties of the environment and sensors.

Practically, the Kalman filter turns out to work remarkably well, despite the fact that many of the assumptions required often break down, and has seen wide-spread application as a localization algorithm in many application domains over the last thirty years.

In this paper we will not develop the mathematics of the Kalman filter further as this can readily be found in textbooks and papers on the subject. Rather we will focus on describing various working systems, refering to the features and limitations of the basic algorithm (and references describing them) as they arise.

2.2 Lasers and Radars

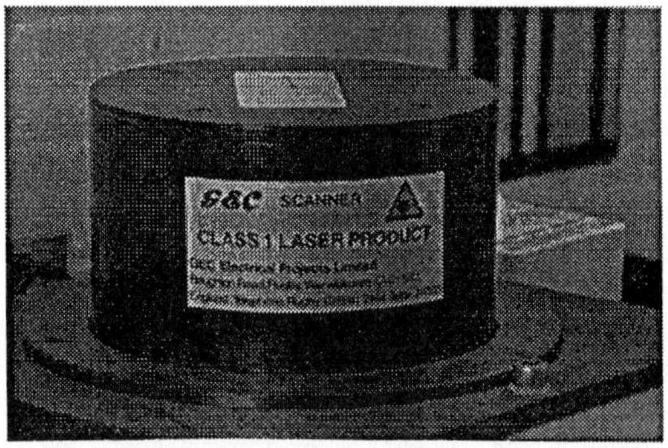

Figure 1: The GEC laser scanner as employed by the Caterpillar Self-Guided Vehicle system. The scanner rotates at a rate of approximately 2 Hz. The laser is used to detect the presence of a reflective target at a known location. Barcoded targets were used in the original system to distinguish between beacons. Simple reflectors are now more common as issues of map management and data association have been solved. The bearing-to-target measurement is produced from the scanner encoder on detection of a large reflection by the laser. Accuracies of the order of a few μradians are typical.

A wide range of beacon-based navigation systems based on scanning laser technology are now in industrial use. In its simplest form a laser is scanned through the horizontal plane and is used to detect the existence reflective beacons placed at known locations around the environment. Typically, only a bearing-to-beacon measurement from the scanner encoder is made, the laser simply being used to detect the existence of a beacon. The first laser scanners used in vehicle navigation were developed by GEC for Caterpillar in 1984 (Figure 1 shows a Caterpillar laser scanner). Since then a number of other manufacturers, notably NDC, have produced similar navigation systems. The integrated localization system is used to provide continuous 2D position and orientation estimates of vehicle location to be used in path control and tasking.

The main advantage of laser localization systems is in price and simplicity. The use of lasers is now wide-spread, providing a low-cost method for estimating vehicle locations. A number of basic extensions to the principle have been made over the past few years. In particular, methods for estimating and measuring both range and bearing have been developed, and sensors capable of scanning in both azimuth and elevation have been developed. The main disdvantage with laser-based systems is lack of range and attenuation, particularly in outdoor environments. Although lasers are increasingly used outdoors (notably in total-station surveying equipment), operation is typically limited to less than 50m (often less than 15m), and to fair weather conditions. Lasers are also adversely effected by direct sunlight and to air-borne particulate contamination.

Millimeter-wave radar (MMWR) sensors are becoming increasingly common particularly as cruise-control sensors in automotive research. The provide a good solution to the short-range and attenuation problems of lasers particularly in outdoor environments. MMWR typically operate at either 94GHz (military band) or at 77Ghz (automotive band), with wave-lengths of the order of 3mm. MMWR technology lies between laser and conventional radar; the sensors are relatively small and attenuation is much lower than that for lasers, but accuracies are much higher than conventional frequency radar. Typical operating ranges (in all weather) are around 200m, with 2Km being an upper limit. Beam widths are typically about 1° for a 100mm antenna, and range resolution is about 10cm. In localization problems, MMWR is used in much the same way as lasers; a beam is scanned in the horizontal plane with bearing measurements taken from the scanner encoder. Range measurements are often based on frequency-modulation (bandwidths of 600MHz are possible at 77GHz), but some time-of-flight devices are also used. MMWR units are becoming the sensor of choice in hostile outdoor autonomous vehicle localization, in applications such as stevadoring (see Figure 2), mining and agriculture.

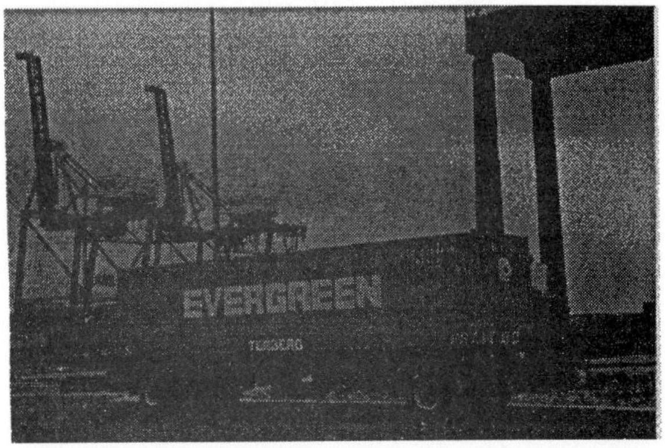

Figure 2: This outdoor cargo handling robot vehicle uses 77GHz millimeter wave radar sensors for localization. The big advantage of radar sensors is that they provide information over very long ranges in poor operating environments. The radars on this vehicle provide the range and bearing to a number of radar beacons placed about the environment at known locations. Detection distances are up to 200m, bearing resolution is approximatley 1°, range accuracy is 10cm and the scan rate is 6Hz. The radars are housed below the main deck at both front and rear.

2.3 GPS: As good as it looks ?

GPS has made an enormous impact on the development of outdoor navigation systems. GPS is essentially a beacon-based navigation system in which the satellites serve as uniquely identified beacons. In principle, reception of timing signals from four or more Navstar satellites allows the location of a vehicle to be obtained in absolute coordinates anywhere in the world at any time. Clearly this is quite useful ! Basic CA band GPS provides a positional accuracy of approximately 100m. A key problem however is that the errors in GPS signals (which are dominated by artificially generated SA errors) are not random; rather the position solution drifts over time. Indeed, it may be demonstrated that the

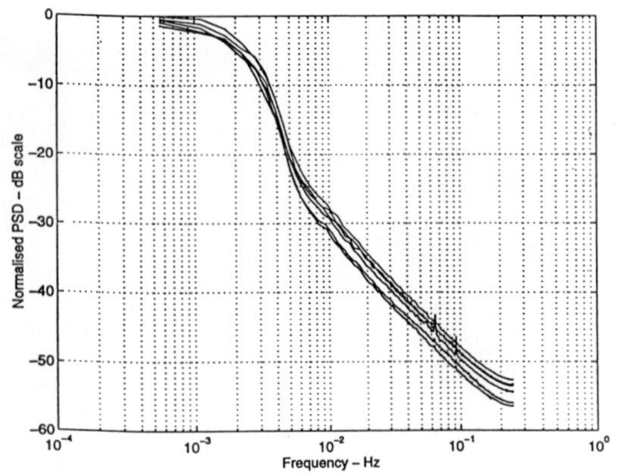

Figure 3: The Power Spectral Density of GPS pseudo-range signals from 16 of 26 satellities. The data was obtained over a six-month period from a stationary receiver. The data is the directly measured CA band pseudo-range signals. The three most important points to note are that a) the error in measurements is predictable; b) the error is the same for all satellites; c) the error is accurately modeled by a shaping filter with two oscillatory poles and one stable zero.

signals exhibit second-order Gauss-Markov error (Figure 3). Further, it may be demonstrated that all satellites have essentially the same error characteristic. The correlated form of timing signal errors means that simpple averaging of data (in a conventional Kalman filter) will not provide particulalrly good results. However, once the statistical characteristics of these erros are determined, this can be exploited in the design of the navigation system.

A number of basic variants to GPS also exist. Notably Differential GPS, in which the signals are corrected by reference to signals received from a stationary base station, and PDGPS, in which the P band carrier is also tracked (not decoded) to obtain accurate time-of-flight data. PDGPS can, in principle obtain positional accuracies of better than 2cm.

However, for autonomous vehicles, GPS has a number of basic problems. First, it requires line-of-sight. This is often difficult to guarantee all the time. Second, it is simply not sufficiently reliable in a stand-alone mode. In four years of using GPS, we have never managed to obtain a position solution sequence for a period of longer than four hours. This does not, however preclude the use of GPS as part of a suite of sensors, which may together have the ability to overcome the problem of reliability. But without this GPS is not a satisfactory solution to outdoor autonomous navigation problems.

2.4 The Research Problems in Beacon-Based Methods

Despite the relative maturity of the technology, there are still a number of important research problems remaining in beacon-based navigation methods. In particular, there is a problem associated with the occurrence of false alarms; data that appears to be a beacon, but which is not. An inability to deal with this problem in a robust manner can cause catastrophic failure. A second issue which arises, in part, because of the maturity of the technology is the long-term reliability of localization and navigation systems; how can such a system be guaranteed to operate without failure for all time ? This particular issue is becoming increasingly important as beacon-based technology moves to larger, more sophisticated platforms, operating in hazardous environments. There are essentially no known methods of guaranteeing the reliablity of an *autonomous* guidance system.

3 Natural Beacons and *a priori* Maps

The next logical extension to beacon-based localization techniques is to use an *a priori* map of the environment and natural features in place of artificial beacons. This is often termed the natural or geometric beacon method [9]. The overall technique is very similar in philosophy to the artificial beacon method. A vehicle process model in the form of Equation 1, is used to predict vehicle locations from encoder measurements. Observations are made of the environment using on-board sensors in the form

$$\mathbf{z}(k) = \mathbf{h}(\mathbf{x}(k), \mathbf{p}_i) + \mathbf{v}(k), \tag{5}$$

where the $\mathbf{p}_i$ now consist of the parameters describing the geometric features under observation. The observation equation is thus used to predict measurements made by the vehicle sensors, from a given vehicle location $\mathbf{x}(k)$, on the basis of a previously stored map described by the $\mathbf{p}_i$ parameters. Once an observation prediction is found which matches the measurement made by the actual sensors, the vehicle location may be updated. Figure 4 shows a sequence of snapshots from a natural-beacon navigation system using sonars.

One of the most problematic elements of natural-beacon navigation is to be able to relate the actual measurements made by a sensor to some piece of underlying geometry in the environment. The sensor model described by Equation 5 is typically not a simple geometric function, rather it incorporates some understanding of the physics of the transduction process. For example, in the case of sonar, some knowledge of beam-width and specularity effects is essential in determining how measurements are related to the true range and bearing to objects. This is an area of intense research activity in both sonar and other sensor areas [6, 1].

There has been a great deal of work in natural-beacon based navigation systems over the past ten years [4, 10]. Gradually, ideas in modeling have improved both the genericity and reliability of the techniques. Examples of systems which are tending to the levels of reliability found in artificial beacon systems are now becoming common. Figure 5 shows one such system developed at Oxford University over the past six years. The aim of the system was to develop demonstrably reliable natural-beacon localization in a variety of real application environments such as hospitals, offices, and factories.

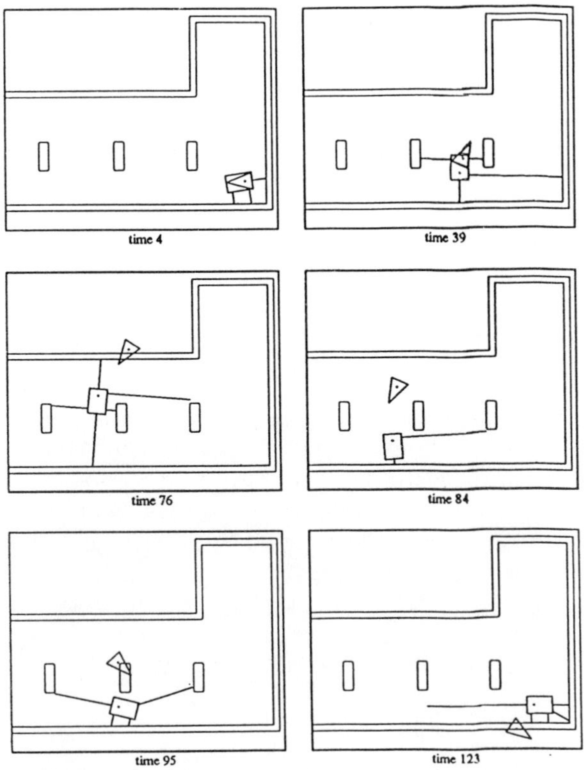

Figure 4: A series of snapshots from a localization run. The predicted (encoders) only vehicle location is shown together with the sensor updated location obtained from a set of stationary sonars. At each time-step all six sonars are fired. The figures show only those range measurements which are validated against the predicted map of the environment. Only these measurements are used to update the vehicle location.

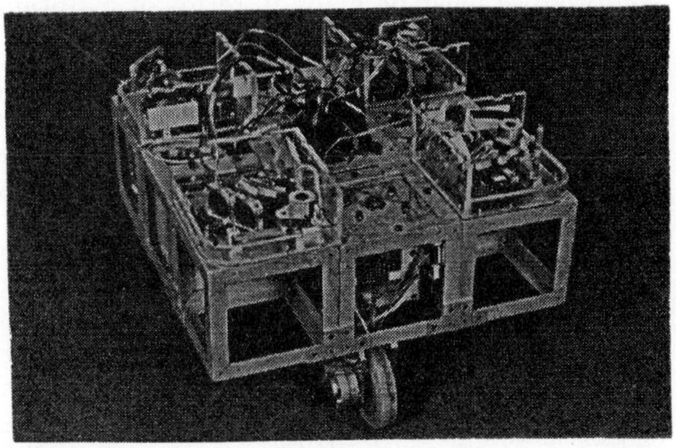

Figure 5: The OxNav robot; a vehicle design to experiment with duration localization based on natural beacons only. This is one of six vehicles built, each focused on a different application environment. Note the modular construction of drive and sensor units to facilitate reconfiguration for different applications.

Data association is a critical issue in most beacon-based localization systems. Data association is simply the problem of assigning measurements made by on-board sensors to beacons modeled as part of the navigation map. In artificial-beacon localization systems a number of simple techniques can be used to ensure correct association (either by uniquely marking beacons, or by using a nearest-neighbour algorithm). In natural-beacon localization systems, the problem is significantly harder as the possibility of incorrect association with a neighbouring feature is significant. In such cases, beacons can clearly not be marked, nor is it always the case that the nearest beacon or feature to the measurement is the correct association. The problem is further compounded by the large levels of false alarms, measurements made of unmodeled features, obtained in such systems. In natural-beacon localization systems, significantly more sophisticated data association schemes must be developed. One such technique which has seen significant application in mobile robot localization problems is the Multiple Hypothesis Tracking (MHT) algorithm [13]. The MHT algorithm allows tentative assignments of measurements to targets to be made, so generating many possible localization hypotheses. These hypotheses are tracked over time to refine and select a correct hypothesis [8, 3, 11]. Figure 6 outlines how the algorithm works with a single measurement multiple-feature assignment. In principle, the MHT algorithm is optimal (provided that no definite decisions are made !) In practise however, the number of hypotheses grows exponentially and so some compromise must be made in terms of maintaining only those most likely hypotheses. Techniques for making assignments, pruning and managing the hypothesis growth have increasingly been developed to the point where the MHT technique is now a viable method of ensuring appropriate associations are made in a robust manner.

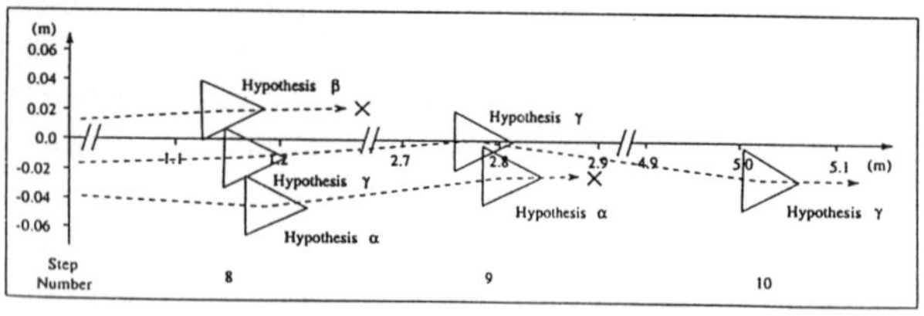

Figure 6: The operation of the Multiple-Hypothesis Tracking filter. Many hypotheses of vehicle location may be maintained simultaneously. The correct hypothesis is determined by tracking over time.

The OxNav vehicle (Figure 5) overcomes many of the data association problems in general natural-beacon localization by employing a novel 'focus of attention' sonar. The sonar, shown in Figure 7, actively locks on to and tracks individual features in the environment as the vehicle moves. This immediately reduces the data association problem to one of initial assignment. The process of disambiguation of features over time becomes significantly easier as the problem of temporal association is essentially eliminated. The technique is borrowed from methods in track-while-scan radars and has much in common with focus of attention ideas now current in computer vision. The ability to lock-on and track also makes the problem of data association in the simultaneous localization and map building problem significantly easier as the temporal tracking of features, often needed for identification purposes, is accomplished automatically.

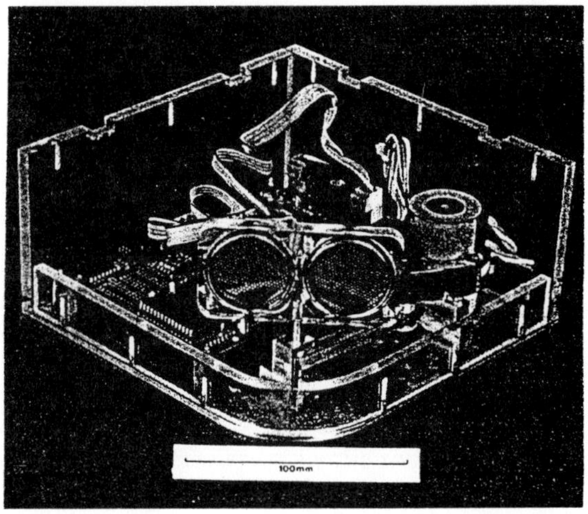

Figure 7: Active scanning sonars (RCD trackers). In this sensor, one transducer transmits and both receive an echo. By measuring the differential time-of-flight between the receivers, an accurate bearing to a feature (sonar reflector) can be obtained. The scanning unit then steers the sensor unit so that the differential time-of-flight is zero. This effectively means that the transducers are always aimed exactly at the feature of interest. The range update rate from the sensor is as high as 30Hz. As the vehicle moves, the sensor remains locked on the target feature, thus providing relative motion information from which vehicle location may be computed. This scheme avoids many complex data association problems by ensuring that the same target remains in view for long periods.

4 Simultaneous Localization and Map Building

The ideal localization system would allow the robot to start at a fixed location with no map of the surrounding environment, and from here to incrementally both build a map and use this map to locate its position relative to the start point. This is known as the simultaneous localization and map building (SLAM) problem. The SLAM problem is complex both as a mathematical exercise and in practical realization. It has received considerable attention over the years from a number of researchers [14, 5]. The essential problems are reasonably well known, however solutions have remained elusive, and a *significant* implementation has still not been attempted. We describe here the essential elements of the SLAM problem and some recent work in obtaining solutions to this problem.

4.1 The Essential Problem

There are two essential elements to the SLAM problem. The first, and most critical, is as follows: At the start of operation, both the location of the vehicle $\mathbf{x}(k)$ and the location of the features in the environment $\mathbf{p}_i(k)$ are unknown. Measurements are made of the *relative* location of the vehicle with respect to the features in the environment. The state to be estimated now consists of the combined state vector including both vehicle location and the location of all beacons: $[\mathbf{x}^T(k), \mathbf{p}_1^T, \cdots, \mathbf{p}_n^T]$. It is easy to show that the covariance matrix associated with this state vector has non-zero cross correlations between vehicle and beacon estimates, and between every pair of beacon estimates. This means that the filter used in a SLAM problem must use the complete vehicle and beacon state vector. As the filter complexity increases with the cube of the dimension of the state vector, the general SLAM estimator is computationally intractable. If it is assumed that the beacon estimates are uncorrelated with each other and the vehicle location estimate, then the map that is built will tend to diverge and become mutually inconsistent. This problem has been known for some time by various researchers [14, 10]. The second basic element of the SLAM problem is the correct identification and association of observed features in the environment. This problem is made particularly difficult by the fact that a feature identity can almost never be determined from a single observation; the feature must be tracked over space and time to provide a correct and unambiguous identification. Not knowing both the identity and association of an observation makes the data association problem very complex. In the limited case where the vehicle location may be determined independently from some other source (using either a second on-board absolute sensor or some tracking system), then this data association problem can again be solved using the MHT algorithm. Figure 8 shows an example of this case and clearly demonstrates the quality of map which can be obtained.

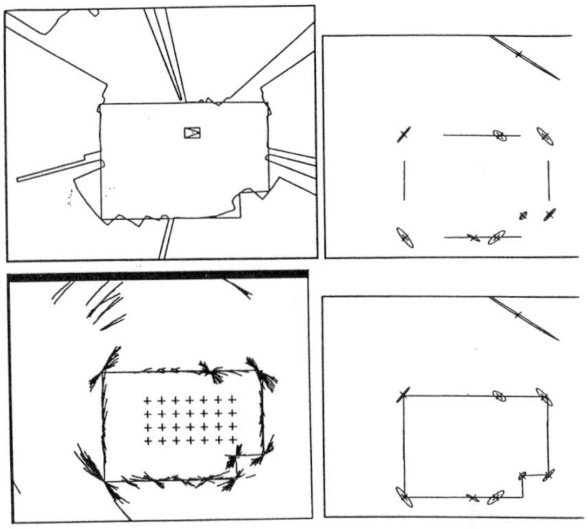

Figure 8: The process of map building using a scanning sonar. The first step in the map-building process is the identification of sensor measurements with some underlying geometric feature in the environment. This can only be done by integrating measurements of the feature over time from multiple viewpoints. Once an identification has been made, the feature may be localized. The second step in the process is to use localized feature information to provide a map from which the vehicle location may be determined.

4.2 Form of the Solution

There are a number of ways of approaching a solution to the general SLAM problem. One method adopted in the OxNav project is to actively manage the sensors on board to decouple the vehicle localization problem from the map building problem. The OxNav vehicle uses four scanning sonars mounted on the corners of the vehicle. Rear-mounted sensors are used to provide localization information from previously mapped beacons. Front-facing sonars are then used to locate and identify new features during vehicle approach. As features are mapped, the front sensors pass information to the rear sensors so that newly mapped features can be used for vehicle localization. Practically, the effect of this policy is to reduce the degree of correlation between the vehicle location estimate and the beacon location estimates. It does not eliminate the problem altogether, but by careful selection of motion trajectories (so as to triangulate the mapped area) it allows a reduced filter to maintain convergence.

The general SLAM problem has until recently eluded systematic analysis. However, some of our recent work has exposed more of the essential structure behind the problem of beacon correlation. The key new result is the following; it may be demonstrated that, regardless of the evolution of $\mathbf{x}(k)$,

$$\lim_{k\to\infty} \det\left(\mathrm{E}\begin{bmatrix} \tilde{\mathbf{p}}_i\tilde{\mathbf{p}}_i^T & \tilde{\mathbf{p}}_i\tilde{\mathbf{p}}_j^T \\ \tilde{\mathbf{p}}_j\tilde{\mathbf{p}}_i^T & \tilde{\mathbf{p}}_j\tilde{\mathbf{p}}_j^T \end{bmatrix}\right) = \mathbf{0},$$

and

$$\mathrm{E}\left[\tilde{\mathbf{p}}_i\tilde{\mathbf{p}}_i^T\right], \quad \mathrm{E}\left[\tilde{\mathbf{p}}_j\tilde{\mathbf{p}}_j^T\right], < \infty.$$

These equations mean that the determinant of all covariance matrices constructed from two (and by implication, more than two) beacons will converge to singular matrices, and that covariance matrix of a single beacon will be bounded, but may not converge to a fixed value. Thus, although it is not possible to say anything concrete about the absolute location of a beacon, the relative location of all pairs of beacons will become *completely* correlated. The implication of this is that it is possible to build an asymptotically perfect map of the environment, but the map itself cannot be absolutely registered with the true environment.

There are two important implications of this result. First, a rather elegant structure for the map building problem emerges as shown in Figure 9. The vehicle and map are defined with respect to a base coordinate system. The landmarks are defined with respect to the map base. As the vehicle travels through the map three quantities affect the observations made: the location of the vehicle the location of the map and the location of the landmarks within the map. The map begins as a deformable structure (imagine something like a rubber sheet), as more observations are made the map becomes more 'rigid' until one can think of it as a solid plate. The location of this plate with respect to the base coordinate system is not known with precision. The location of the vehicle with respect to the base coordinate system when observing elements of the map can clearly never be known with more precision than is the location of the map. However, the vehicle may know where it is on the 'plate' as well as if it had a perfect map. It is possible to build a formal model of this structure using simple Bayesian arguments. The second important consequence of the result is that it shows that, in the limit, the filter need not store all information as cross-correlations and landmark variances may be generated through knowledge of the fact that the landmark location estimates are fully correlated. This in turn leads to a number of methods for simplifying the map building problem. Such algorithms are currently being developed and validated.

A number of related problems may now also be precisely defined and analysed using this structure. In particular, the effect of adding external measurements of vehicle location (from a GPS or inertial unit for example), and the effect of adding external terrain registration information may be considered. In general, the effect of external sources of information is to reduce local state variance. However, the cross-correlation between vehicle locations does not change. A little thought shows that indeed, the cross correlations in transient map-building can never decrease.

We believe this result to be of profound significance. Not only does it, for the first time, elucidate mathematically the fundamental structure of this well known problem, but it also provides a computational model for its solution. Further, the result has wide application in other types of mapping problem, particularly in geophysical mapping and terrain modeling.

Although the structure of the SLAM problem is now better understood, it still remains to exploit this fully in the development of computationally tractable algorithms. Of note in this area are techniques being developed for obtaining always conservative bounds on cross-covariances using reduced filters. It also remains to demonstrate a localization system capable of implementing these ideas over significant temporal and spatial conditions. It is reasonably clear that this can only be done through the additional use of other sensors providing independent vehicle location or

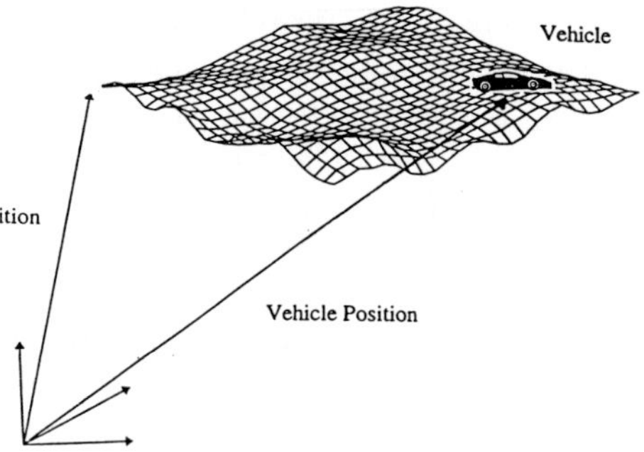

Figure 9: The esssential structure of the simultaneous map-building and localization problem. As the vehicle moves through the environment the quatity which can be determined with precision is the relation between features or natural beacons. This means that, in principle, a perfect map of the environment may be constructed: except that the absolute location of the map will not be known with precision. The three elements to be estimated therefore are the vehicle location, the relative location of features within the map and the map location itself.

map data information.

5 Combinations of Sensors

Although many localization systems use only a single source of sensor information, it is clearly sensible in many instances to use multiple sources of data, particularly where increased reliability and range of operation is required. The use of multiple sources of localization data is common in areas such as aerospace and maritime navigation for these reasons. At the most trivial level, the encoder information typically obtained from most AGV drive systems is usually used in conjunction with past sensor data to provide improved vehicle prediction information. However a number of other sensor combinations also make sense, particularly in more challenging outdoor applications. These include the use of inertial sensors, independent ground-speed sensors, multiple positioning sensors, and even traction force sensing. Inertial sensing in particular is commonly used in vehicle localization systems beacuse of its ability to provide high frequency high bandwidth data without reference to an external environment. The use of multiple sensors in localization necessitates the careful design of an overall filter structure to exploit the best features of each sensor in a system. This and the implications on system performance and reliability are discussed in this section.

5.1 Inertial Aiding

Inertial sensing units are widely used in the aerospace industry to provide localization information. A typical inertial navigation system (INS) comprises three orthogonal gyroscopes and three (sometimes nine) accelerometers. Integration of gyroscopic information provides attitude data, double integration of acceleration information provides three dimensional position. The most important feature of an INS is that if measurements are taken in broad-band noise, then integration to yield attitude and position will result in estimates that drift in time without bound. Consequently an INS system is only ever used in conjunction with some absolute position sensing system which can correct INS drift when required.

Aerospace INS systems are both accurate and expensive. The two main challenges in robotics applications are to obtain reasonable performance at reasonable price, and to work out how best to use an INS platform in conjunction with conventional robotic sensors. The increasing use of inertial sensors (gyros and accelerometers) in automotive applications has made the first of these problems much more tractable over the past five years. INS systems can now be constructed from low-cost solid-state components and by careful modeling of the error sources in both gyros and accelerometers three-dimensional position and attitude information can be obtained to a useful degree of accuracy [2].

The problem of how best to use an INS system in concert with other position and velocity sensors has been studied extensively in marine and aerospace areas. It makes sense to build on these studies in developing appropriate aided INS filter structures in robotics applications. As a general principle, a Kalman filter acts like a low-pass filter, smoothing input observations contaminated with broad-band noise to provide an estimate of a signal that is assumed band-limited. However, INS information is typically most valuable at high frequencies (and least valuable at low frequencies), so it makes little sense to pass the output of an INS system through a Kalman filter as the high value information will be attenuated. To overcome this a complimentary filter structure is often used in INS aided navigation systems. As Figure 10 shows, a complimentary filter works by passing only the error between INS indicated position and direct position sensing through a Kalman filter. Estimates of INS indicated error are therefore obtained and used to correct the signals coming from the INS in either a feed-forward or feed-back manner (the feed-forward structure is shown in the figure). Thus the inputs from the position sensor are filtered by the low-pass action of the Kalman filter in the normal way, while the subtraction of the error estimates from the feed-through INS signals result in the INS input being high-passed through to the estimate. Thus low-frequency information is obtained from the positional sensor, while high-frequency information is obtained from the INS system [12].

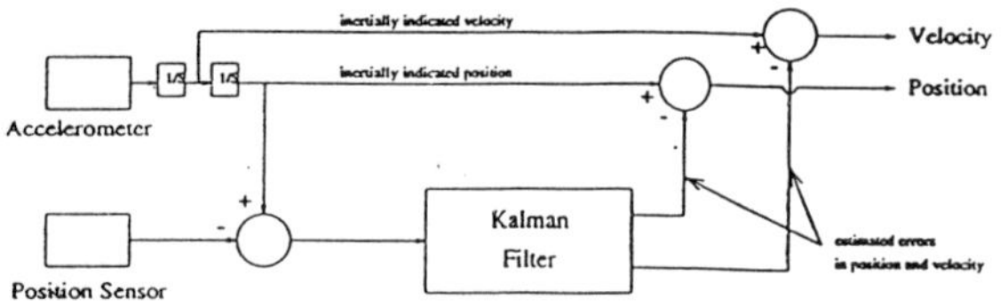

- Reliability
- Increased Range and Scope of Operations
- The Example of INS

Figure 10: The structure of the feedforward complimentary filter used in inertialy aided navigation systems. The main positional sensor is differenced from the inertialy indicated position and processed by the Kalman filter acting as a low-pass filter. The inertialy indicated position is fed through and corrected by the low pass filter, and so appears as a high-pass signal at the output. The complimentary filter thus combines low frequency information from a position sensor with high freaquency information from the inertial system.

5.2 Sensor Suite Design

The use of INS systems, and more generally further positional and velocity sensors, argues that a systematic approach to design of sensor suites and filter structures is essential. Further, the nature of the complimentary filter suggests that a good way of approaching the problem would be to analyse the system in the *frequency domain.* This particularly makes sense when complex statistical sensor models are employed as these are most readily described in terms of Power Spectral Densities (PSDs) (Figure 3 for example), which are easily understood and manipulated in the frequency domain. Figure 11 outlines how this works in practise; a low-frequency signal from a GPS sensor is combined with an INS system in complimentary form. The gap between the GPS correlation roll-off and the cut-in from the INS needs to be filled by a velocity or delta-ranging sensor. Once a suite which covers all reasonable vehicle motion characteristics (frequencies) has been developed, then a systematic analysis of error budgets over the domain of interest can be undertaken to generate either a specification for sensor accuracies and band-widths or to predict overall system performance from a given set of sensors.

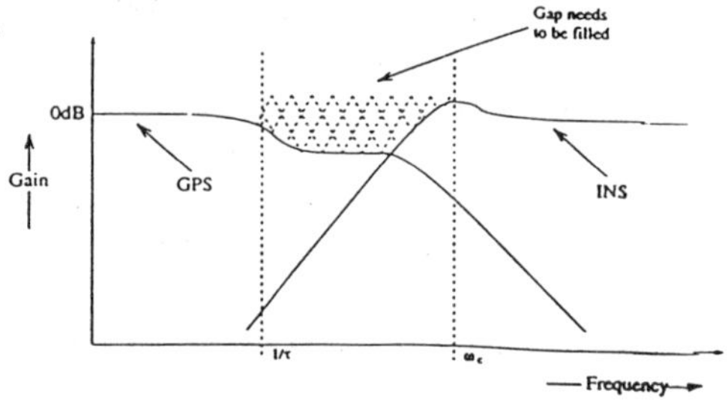

- Sensor Suite Design
- Methods in The frequency Domain
- Building Quantitative Reliability Models

Figure 11: The figure shows a filtered GPS signal that rolls of in the frequency domain at two points, the first due to the correlation times of the SA signal, the second due to the roll-off action of the filter itself. An INS is placed in a complimentary structure with the GPS unit to provide high-frequency information. This leaves a gap in response characteristics which needs to be filled by some other sensor if uniform performance over all vehicle motion frequencies is to be obtained. A sensor providing velocity or delta-ranging information would provide such information.

5.3 Reliability

The reliability and overall robustness of a localization system is essential in any real application. It is problem for which few (if any) systematic procedures exist and which very little work has so far been accomplished. Yet the reliability of *autonomous* systems is a fundamental problem which will need addressing as technologies are brought to market. The use of multiple sensors is clearly likely to help in this regard although non of the techniques described above offers any real answers as to how issues such as fault detection and recovery, varying environmental effects on sensors, physical properties of the device and algorithm failures may be overcome in a robust manner. Here we simply highlight the issue of reliability as one of the key areas of future research in Autonomous Vehicle systems.

6 Conclusions and Current Research

There are a number of key areas in which current research is now focused. The basic techniques for localization are now well known and widely used in the robotics community. Some key issues however remain to be fully resolved. Most notably, a significant demonstration of a solution to the SLAM problem is a key priority. This is particularly true in outdoor environments. There is also a substantial interest in the use of Qualitative navigation techniques and the combination of these with more traditional filter-based approaches. Studies of more complex filter structures is also becoming more important as new and more challenging applications are found. There is also a general feeling that to make much further progress in the area of localization, more must be done to develop and demonstrate real applications. This is particularly motivated by the issue of reliability. We briefly look at some of these issues in this section.

6.1 Further Progress in the SLAM problem

The generalized SLAM problem is still far from a definitive solution. Recent progress in determining the central structure of the problem may help considerably in determining the type of algorithms that need to be developed. Fundamentally, the full correlation result suggests two immediate courses of action. First to use the final steady-state solution as an upper bound on the calculation of map parameters, and second is to use additional sensors to decouple the two main parts of the problem (vehicle location and map estimates) so that an iteratively stable and convergent solution can be obtained. The prospect of a fully integrated bootstrap terrain-aided navigation system is enough to make the solution and demonstration of this problem a key priority.

6.2 Qualitative Localization Methods

Although the Kalman filter has proved both versatile and effective in localization problems, there are many facets of the general navigation problem which are not amenable to so simple an algorithm. In particular, the filters requirement for explicit models of the world make it fragile in complex operating domains. Qualitative localization schemes are a possible solution to some of these problems [7]. They offer the prospect of describing a world in purely qualitative, sensor-centric, terms so eliminating the fragility problems associated with conventional localization filters. Figure 12 outlines the type of process involved in qualitative navigation. One key element of qualitative techniques is the simple lack of constraint obtained from basic reasoning about relationships between sensor observables and an environment. There is also a basic dilemma in dealing qualitatively with observations which are intrinsically numeric. Practically this means that qualitative techniques alone are unlikely to provide all the answers. The combination of numeric filtering methods with qualitative techniques awaits further research.

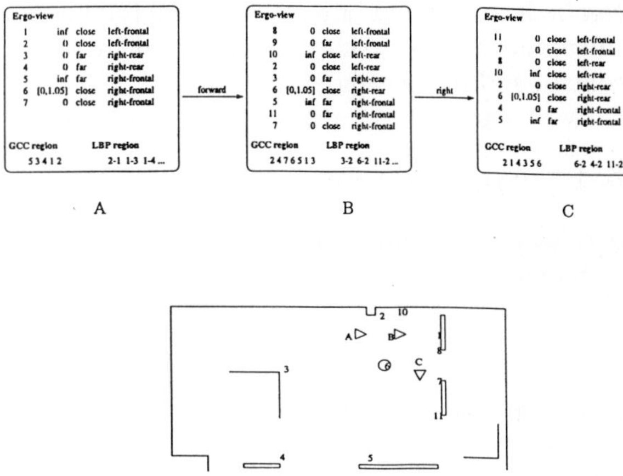

Figure 12: Navigation using qualitatitve landmarks. In a many applications it would make sense to simply use environment structure rather than detailed geometric descriptions to navigate.

6.3 Better Models

The reliance of the Kalman filter on explicit models of vehicle and sensors leads inexorably to the question of how good such models should be. Better sensor models can make a substantial difference to the quality of estimates obtained. In principle, better vehicle models can also make a significant difference to the quality of estimates. Quantifying the effects of different models on system performance is a significant area of research in the control community. Of secondary interest in mobile vehicle problems is the fact that better vehicle models can provide a much better understanding of the way in which the AGV operates. In particular, by using the observability property of the Kalman filter (in which estimates of unobserved states can be made), detailed models of platform motion can be used to estimate both the forces acting on the drive system (wheels or tracks) and also the geometry and stiffness properties of the terrain over which the vehicle is traveling. This in turn can lead to both a significantly improved understanding of vehicle motion characteristics (particularly important in outdoor and high-speed applications), and also to provide estimates from which better traction and steer control algorithms may be developed (Figure 13 shows one such vehicle model). The central research question here is how such models can be incorporated into conventional filter structures and how they may be exploited in vehicle control and navigation.

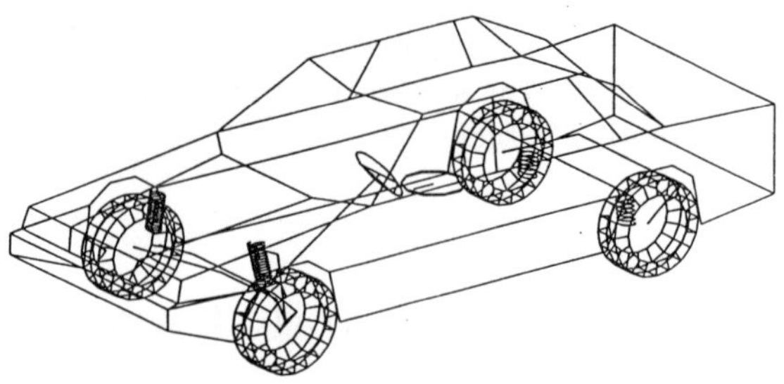

Figure 13: A full dynamic model of a conventional road vehicle (Rover 820si) developed using the ADAMS dynamic modeling package. The model includes both inertial properties of the vehicle body, steering geometry, suspension and handling system. The model also includes a full finite-element model of the tyre system. The model is used to develop vehicle process models capable of estimating forces on tyres and both geometric and material properties of the terrain over which the vehicle is driving.

6.4 Applications

Serious and significant application of localization technology is, in my view, the most essential step that research in this area must take over the next five years. We have generally come to the end of the "gee-whizz no hands" period and now need to develop and support real applications. This does not of course mean that all problems are solved; far from it; rather that key ideas and new theories need to be developed and motivated by real problems.

In terms of applications, technology based on artificial beacons is still state-of-the-art. However, the use of this technology in wider application areas is still of significant research interest, principally because it eliminates a basic problem (position sensing) from many other critically important research questions such as vehicle modeling, control, safety and reliability. Natural beacon techniques are now ripe for application, particularly when used in conjunction with beacon-based techniques. The real applications here are likely to be incremental, building on existing manufacturing AGV systems to provide additional competence. The SLAM problem is still very much a research issue but limited forms of it are likely to be developed, again when used in conjunction with sparse artificial beacon information.

Manufacturing AGV systems are a non-starter when it comes to new types of techniques. This is simply because existing systems work ! The traditional focus of the mobile robot community on indoor vehicles is also likely to be a non-starter simply on grounds of cost (although some serious real implementation work is being conducted around the world so this can not be ruled out entirely). Strangely, it is likely to be outdoor or traditionally harsh environment areas that are likely to be the most receptive and simplest to apply this type of technology. There are a number of reasons for this, first the cost of existing vehicles in these applications is high and so the additional cost of sensing and computers is not prohibitive. Second, the environments, although harsh, are often amenable to environment engineering to reduce the more difficult aspects of the localization problem. Further it is often possible to eliminate people from the environment (so eliminating a major problem). Finally, there is often a push for automation technology for reasons of both safety and productivity. Thus, the technology may be more readily developed and the user may also be more ready to accept the costs and risk involved. Figure 14 one of our current target applications.

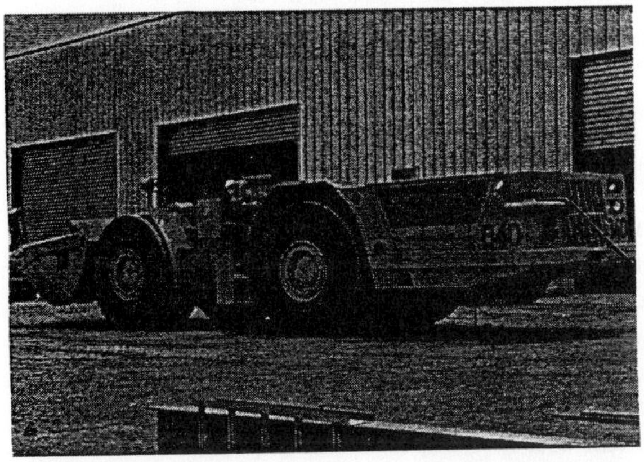

Figure 14: A Load Haul Dump (LHD) truck used for underground mining. The vehicle traverses underground passages, hauling ore from an open stope to the ore-pass. An LHD must be at least partially automated for reasons of safety. A navigation system is an essential element in their efficient use.

References

[1] A sonar sensor for accurate 3D target localization and classification. In *Proc. IEEE Int. Conf. Robotics and Automation*, 1995.

[2] B. Barshan and H.F. Durrant-Whyte. Inertial navigation systems for mobile robots. *IEEE Trans. Robotics and Automation*, 11(4), 1995.

[3] I.J. Cox and J.J. Leonard. Modeling a dynamic environment using a Bayeisan multiple hypothesis approach. *Artificial Intelligence*, 1994.

[4] J. Crowley. World modeling and position estimation for a mobile robot using ultra-sonic ranging. In *Proc. IEEE Int. Conf. Robotics and Automation*, pages 674–681, 1989.

[5] P. Grandjean and A.R. de Saint Vincent. 3-D modeling of indoor scenes by fusion of noisy range and stereo data. In *Proc. IEEE Int. Conf. Robotics and Automation*, pages 681–689, 1989.

[6] R. Kuc and V.B. Viard. A physically based navigation strategy for sonar-guided vehicles. *Int. J. Robotics Research*, 10(2):75–87, 1991.

[7] B. Kuipers and Y. Byun. A robot exploration and mapping strategy based on a semantic strategy of spatial representations. *Journal of Robotics and Autonomous Systems*, 8:47–63, 1991.

[8] J.J. Leonard, H.F. Durant-Whyte, and I.J. Cox. Dynamic map-building for an autonomous mobile robot. *Int. J. Robotics Research*, 11(4), 1992.

[9] J.J. Leonard and H.F. Durrant-Whyte. Mobile robot localization by tracking geometric beacons. *IEEE Trans. Robotics and Automation*, 7(3):376–381, 1991.

[10] J.J. Leonard and H.F. Durrant-Whyte. *Directed Sonar Navigation*. Kluwer Academic Press, 1992.

[11] D. Maksarov and H.F. Durrant-Whyte. A multiple hypothesis approach to map building and maintenance. *IEE Trans. Control Theory and Applications*, 142.

[12] P.S. Maybeck. *Stochastic Models, Estimaton and Control, Vol. I.* Academic Press, 1979.

[13] D.B. Reid. An algorithm for tracking multiple targets. *IEEE Trans. Automatic Control*, 24(6), 1979.

[14] R.C. Smith and P. Cheesman. On the representation of spatial uncertainty. *Int. J. Robotics Research*, 5(4):56–68, 1987.

Hardware Design Methods and Proposals for Open System Architecture of Intelligent Robots

Takeo Oomichi, Tomoyoshi Ibe, Taku Sasaki
Takasago Research and Development Center, Mitsubishi Heavy Industries, ltd.
2-1-1 SHINHAMA, ARAI-CHO, TAKASAGO, HYOGO PRF., 676, JAPAN
TEL +81-(0)794-45-6741 FAX+81-(0)794-45-6081 e-mail:oomichi@trdc.tksg.mhi.co.jp

1. Introduction

About 25 years have passed since industrial robots made their initial debut on factory production lines. However, even though performance has steadily improved, robots have been unable to forge into new markets. It is recognized that the lack of robotic intelligence is a key barrier to many new applications. A great deal of research surrounding robotic intelligence has been conducted to date. Even so, large numbers of practical, commercial, intelligent robots do not yet exist. We are addressing the need for additional robotic intelligence and have developed a number of robots and conducted many field trials.

For example, many of the robots that we have developed, such as those for inspection of electric power generating facilities, have to execute irregular tasks which require some intelligence. Research and development activities to date include a multi-fingered dual-arm manipulator which uses intelligent tele-operation to combine the intelligence of the robot and operator [1,2], a hybrid walking and rolling robot which effectively coordinates its "eyes", hands, and legs [3], an inspection robot for nuclear reactor pressure vessels [4], and various other robots for applications that are difficult or impossible for humans to execute.

Perspective gained through these research activities has indicated that, although an intelligent robot has a very complex system configuration, the final performance is almost solely dependent on each of the architectural elements. Furthermore, an enormous amount of money, time, and labor are required to develop a single intelligent robot. Even so, most of the behaviors which we term "intelligent" are in fact extremely simple for a human to achieve. Intelligent robots which are commercially available have been produced for specific applications and end users; typically the architecture is not "open" which limits the ability to realize the robot's full potential. We have concluded that intelligent robots will never leave the realm of academic interest as long as this situation persists.

In order to rapidly develop intelligent robots for practical use, it is important that specialists in each robotics field can clearly define functional interfaces and then separately conduct their research. For example, applying intelligence to task planning will become easier once there is a ready supply of proven hardware, sensors, and other robotic parts. That is, interfaces will be well understood and the research can focus on intelligence. A remaining challenge however is that the architecture of intelligent robots cannot readily be defined making it difficult to determine whether a design of intelligent robot can be put into practice or not. In an attempt to break out of this situation and accelerate intelligent robot development, the authors compiled and analyzed numerous past examples and grouped certain system elements as "Core Elements" (Figure 1), defined a hierarchy (Figure 2), conceived the idea of "openness" at the hardware interfaces at each level, and produced standardized designs.

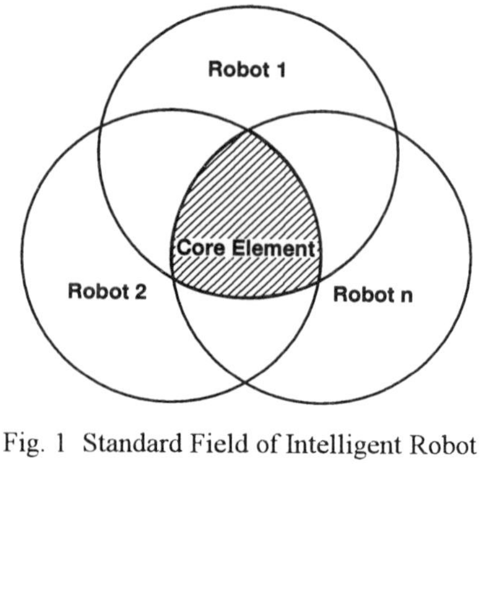

Fig. 1 Standard Field of Intelligent Robot

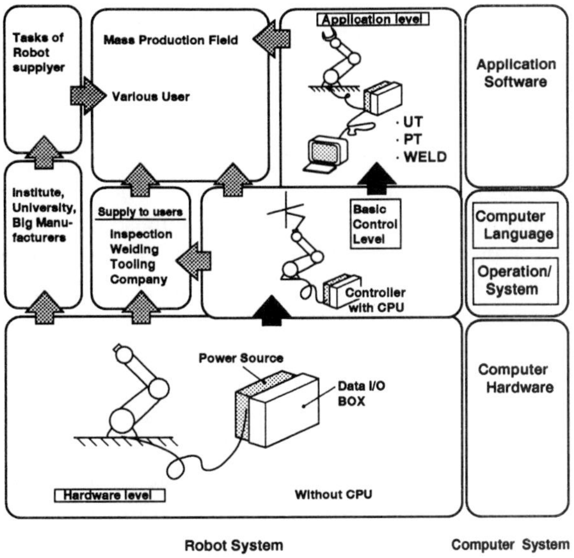

Fig. 2 Concept of Hierarchical System for Intelligent Robot Design

The standardized designs are termed SDIR-1 for "Standard Design for Intelligent Robot Version 1". To demonstrate the validity of this method and put it to practical use, modular components were developed using each standardized design and the components applied to several robotic applications.

2. System Concept and Open System Architecture

2.1 Overall Structure and Hardware

Conventionally robots have been designed by determining the target task and then making use of existing components to create a design. This method results in a lack of flexibility and versatility in the completed robot, thereby greatly diminishing the potential of the robot. Moreover, the completed robot is usually "closed" to the end user and even the smallest modification can require significant effort and perseverance.

This situation led to the concept of designing an "open" system which could be readily understood by a large number of end users. Although it is impossible to make a system fully open in every aspect, we propose to achieve openness by dividing the system into a hierarchy of levels and making interfaces between "open". The outline of this approach is shown in Figure 2. Additional hierarchies and openness occur within each level, as described later.

(1) Hardware Level

This level is principally composed of the robot hardware, that is, the robot mechanism, servo drivers, and signal I/O circuits. Using the analogy of a personal computer, the Hardware Level would be equivalent to the computer hardware (processor, memory, disk drives).

(2) Basic Control Level

This level comprises the essential control algorithms for intelligent robot control, the computer which executes the control algorithms, and communication interfaces between the control hardware and the computer. The control algorithms are divided into motion control which are, for example, path-generation algorithms, hybrid control of position and force, and compliance control. Using the personal computer analogy, the Basic Control Level would be equivalent to the operating system and high level computer language such as C.

(3) Application Level

This level, which consists of software algorithms, may be the most "intelligent" component of the robotic system. Much like application software that fully utilizes the hardware and operating system within a personal computer, the Application Level fully utilizes the elements within the robot's hardware level and basic control level. To avoid the problem of the robot's hardware and software only being usable for the initial target task, connection interfaces for future hardware, computers and control algorithms must be allowed. Under this concept, connection interfaces are open at all levels of the hierarchy making the high-level architecture completely unrestricted. As shown in Figure 2, each level can be developed and supplied by the robot supplier (as in the conventional system) or end users can further develop specific areas and in turn supply these components to other "higher level" end users. This method is extremely flexible and allows the most suitable people for development to work effectively and efficiently Furthermore, as each developed system is open, this method is expected to accelerate the development and application of intelligent robots.

2.2 Intelligent Robot Hardware Requirements

Intelligent robot applications are very broad thus making generalized requirement definitions difficult. Following is a list of proposed hardware requirements for intelligent robots:

(1) architecture to enable intelligent activities (e.g., many degrees of freedom and high positioning accuracy);
(2) capability for information input and output to support intelligent behavior (e.g., I/O for multiple sensor data);
(3) means for expansion or modification as required (e.g., provision of dual arms, and adaptability to other task-specific hardware);
(4) mobility capability to allow intelligent activity (e.g., flexibility of movement and high traversability).

Since the final intention of a robot is mechanical movement, the hardware architecture is extremely important. A correct match of software and hardware must be considered at the design stage to achieve a practical robot.

While it cannot be said that all of the robots designed by the authors achieve a high level of intelligence, many of the robots satisfy the above hardware requirements.

Analysis of the robots that we have developed indicates that the following elements are essential in the hardware architecture:

(1) joints (e.g., actuators, motors, reduction gears, position detectors);
(2) signal and power cables for connecting moving and stationary components;
(3) sensors with various types of I/O;
(4) communications systems and computer interfaces (e.g., high-speed serial links).

This is shown graphically in Figure 1. That is, there exist common architectural elements required by all robots in this broad class (termed "Core Elements"), and special architectural elements required by robots for specific application tasks. While special robot hardware designs may exist that share no Core Elements with other applications, they are considered rare and are excluded from this discussion.

3. Hardware-Level Standard Design Examples and Module Development

Using the hierarchical concept outlined above, we conducted trials by designing and developing component modules to illustrate its practicality and to understand additional benefits of the concept such as cost and weight savings.

The most important point in realizing this concept is the convenience gained by standardization. Most end users will have an interest in how to utilize the proposed method for their specific needs rather than developing openness and standardization for others. Therefore, some strategy is required to continuously maintain openness and standardization with the steady advance of technology.

3.1 Hierarchy on the Hardware Level

Figure 3 shows an example of hierarchical open hardware system for a manipulator. The hierarchy contains the mechanism level composed of drive mechanisms and actuators, the servo level composed of velocity inputs, the I/O level made up of the I/O interfaces to the servo drivers and sensors, and the communications level which transmits information between the robot and computer. Openness is designed in the connections between each of the levels including the connection between the communications interface and the basic control level.

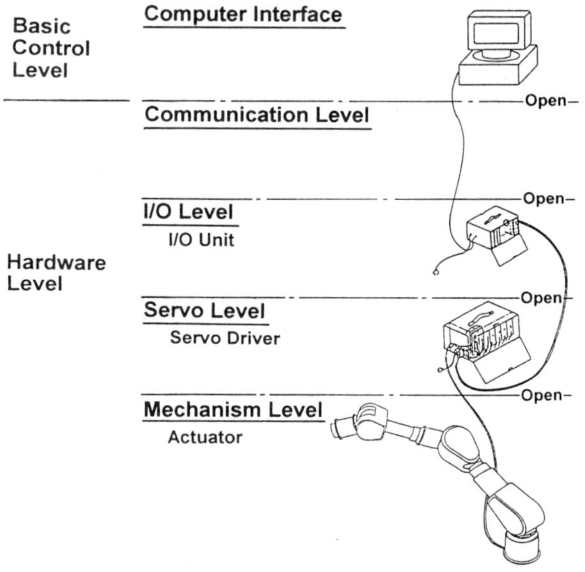

Fig. 3 Hierarchical Open Hardware System

3.2 Modular Design Elements

Figure 4 shows the system design based on the concept overviewed in Section 3.1. A manipulator is shown for ease of understanding, but application of this concept is not restricted to manipulators. Modular architectural elements were developed concurrently.

That is, we developed:

(1) rotating and pivoting joint modules incorporating high-torque AC servo motor and reduction gears which have a hollow central cavity appropriate for wire routing (Fig. 5);
(2) minimum weight and volume AC servo motor driver (Fig. 6);
(3) integrated I/O board (Fig. 7);

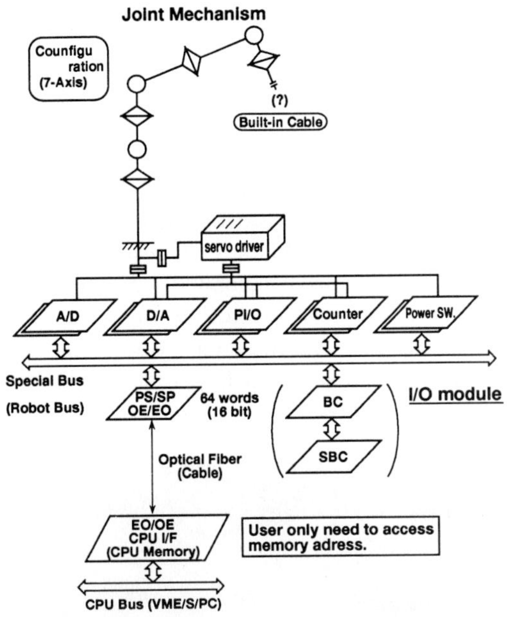

Fig. 4 Manipulator System Design Resulting from Use of SDIR-1 Concept

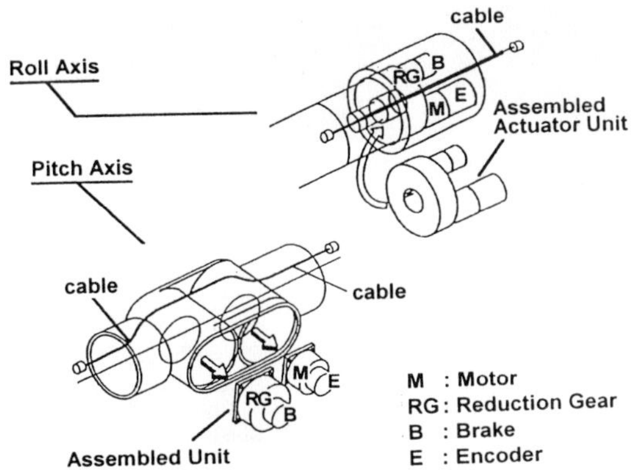

Fig. 5 Module Joint Mechanism

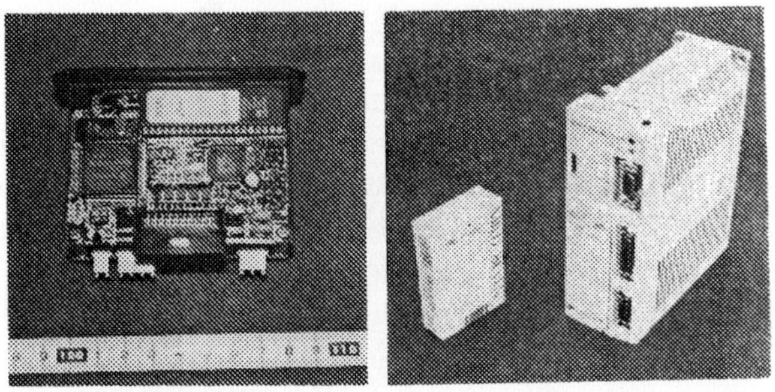

Compact AC Servo Driver
Volume : 200CC
Output : 200W
Weight : 1.9N

Market Type AC Servo Driver
Volume : 1150CC
Output : 200W
Weight : 11.8N

Fig. 6 Commercial Type AC Servo Driver

(4) multi-channel high-speed parallel-to-serial conversion IC and board; and
(5) computer interface which generalizes across many computer operating systems.

Combining these modules allows intelligent robot hardware to be easily produced.

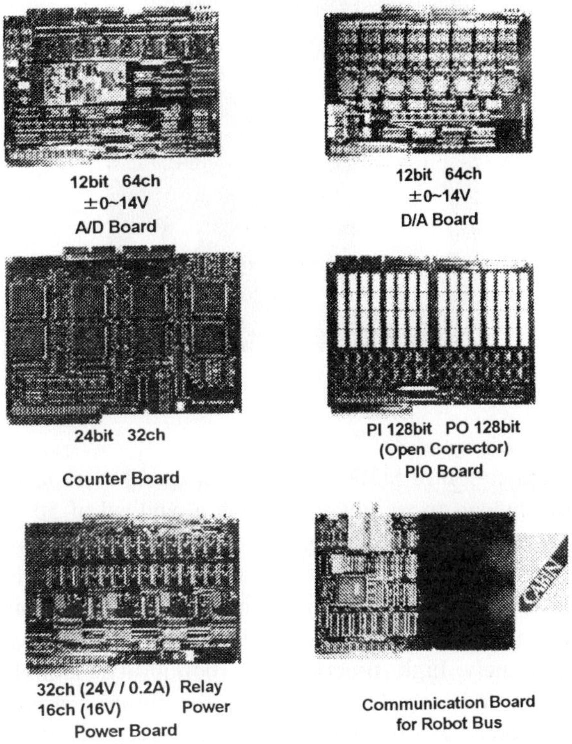

Fig. 7 I/O Boards

4. Application Examples

To allow robot developers make full use of their specific technical abilities, SDIR-1 quickly and accurately sets the required interface and peripheral conditions to support achievement of the principal development objective. In addition, the application of SDIR-1 is also expected to bring additional advantages related to functionality, development time, and cost. We are aggressively promoting this application of this concept. We now illustrate the effect of using the SDIR-1 concept on several robots which we have recently developed.

4.1 Lightweight Portable 7-axis Manipulator

The portable general-purpose intelligent arm is a commercially available 7-axis manipulator based on the open system (see Figure 8). Not only is it based on the open system, but this 7-axis manipulator also applies SDIR-1 to offer a higher level of system integration. This manipulator has a load capacity of 50 N.

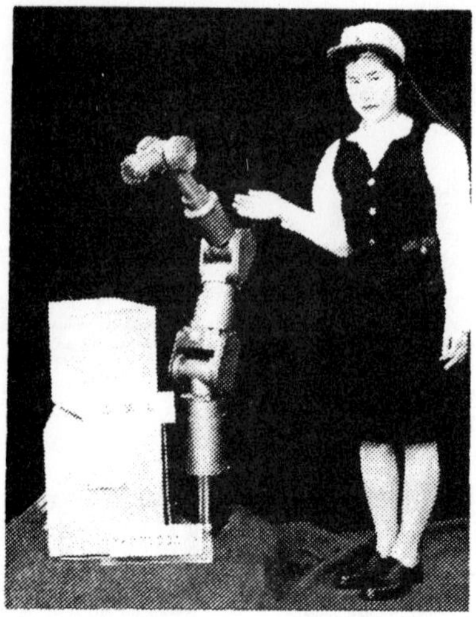

Fig. 8 7-Axis Manipulator

(1) Lightweight
Application of SDIR-1 achieved a world-class level of payload to weight ratio, with a payload of 50 N for a manipulator weight of 200 N. Furthermore, the lightweight control box (200 N) dramatically improves the portability of the manipulator. This illustrates how SDIR-1 can be applied to achieve extremely high functionality (portability, mountability on vehicles, etc.). In the future, carbon reinforced plastic (CFRP) could be used to further reduce the manipulator weight to 150 N though at several times the present cost.

(2) Redundant Axes
The robot has seven degrees of freedom, with a combination of rotary and pivoting joints. A single redundant joint allows more intelligent grasping and manipulation tasks than are possible with a standard six degrees of freedom manipulator.

(3) Data I/O
The control box includes 32 channels of A/D conversion and only requires 7 channels for manipulator operation. The extra channels are available for additional sensors and tooling resulting in a large reduction in overall system costs.

(4) Performance
The use of SDIR-1 for development of the lightweight manipulator resulted in the following advantages compared to conventional development:

(1) design time approx. 1/3
(2) overall development time approx. 1/3
(3) assembly/disassembly time approx. 1/10

A manipulator with world-class performance was produced in a short time.

4.2 Oil-Tank Floor Plate Blasting Robot
This is an example of SDIR-1 applied to a mobile robot (see Figures 9 and 10). SDIR-1 was applied to the wheel drives, steering mechanism, communications and I/O modules. Consequently, only the design of the drive component frame and the main frame remained, resulting in a significant decrease in the development time and costs compared with oil tank blasting robots built before the advent of SDIR-1.

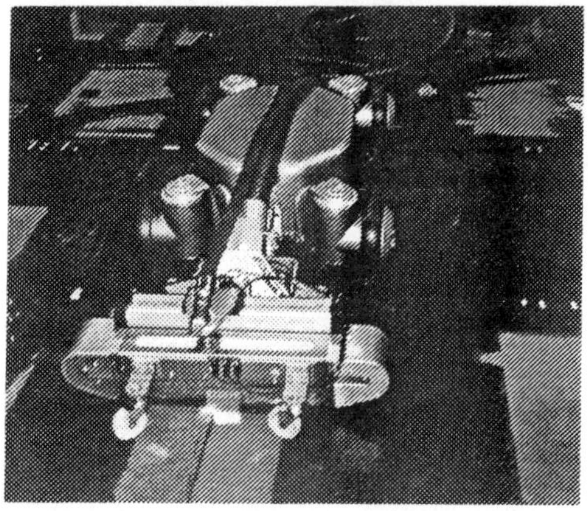

Fig. 9 Oil Tank Floor Plate Blasting Robot

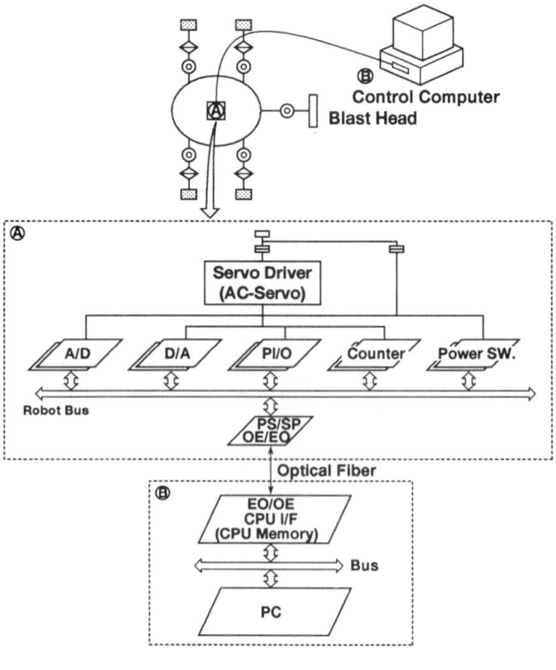

Fig. 10 Oil Tank Blasting Robot Modular System

4.3 Outdoor Patrol Robot

The outdoor patrol robot is a high capability four-wheeled robot based on the oil-tank blasting robot design (see Figure 11). It also has a faster and larger version of the 7-axis manipulator appended to the body. The open system allowed it to be easily produced by combining components developed under SDIR-1. The same multi-channel interface board previously described was also used.

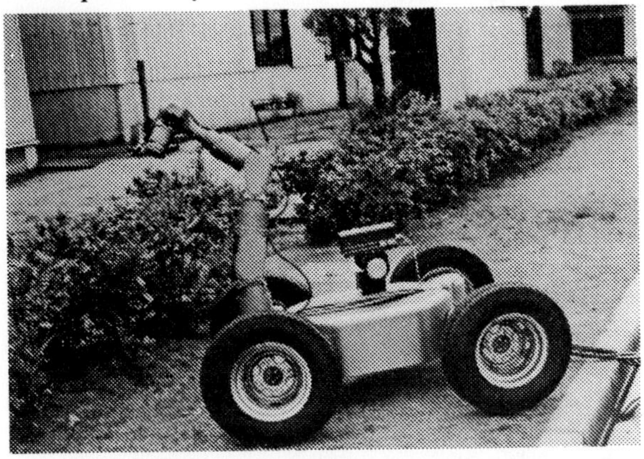

Fig. 11 Outdoor Patrol Robot

5. Future Research

Use of the SDIR-1 concept to date has shown that much intelligent robot hardware can be designed in significantly shorter periods and at significantly lower costs than with conventional methods. We are considering new standardized hardware designs to achieve even better results in the future.

5.1 Advanced SDIR-1 (SDIR-2)

The following problems exist with SDIR-1:

(a) large numbers of internal wires increase assembly costs and reduce reliability;

(b) separation between robot controller and the main body restricts system integration.

The following design concepts have been proposed to solve these problems:

(1) Unrestricted robot LAN and intelligent connectors

Reduction in internal robot wiring may be achieved by using a robot LAN. This LAN must not detract from the following robot features:

(a) real-time communications

(b) reliability

(c) expandability and flexibility

The following measures have been proposed to meet these aims:

(a) increasing data sampling and communications speeds to approximately 10 times that of conventional robots;

(b) introduction of a redundant communication system;

(c) elimination of the communications control system;

(d) producing equipment required for (a), (b), and (c) on chips with connectors;

Figure 12 shows an actual example which includes the following features:

(a) Three independent communications lines of approximately 20 Mbps capacity.

(b) Communication system based on collision detection.

(c) Ultra-compact intelligent connectors (see Figure 13). The intelligent connector transmits physical data such as sensor signals, servo commands, etc.

(d) Data transmission with sampled time.

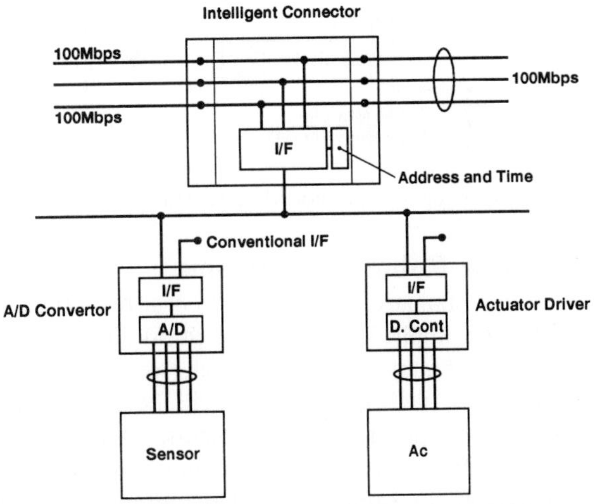

Fig. 12 Functional Concept of Intelligent Connector

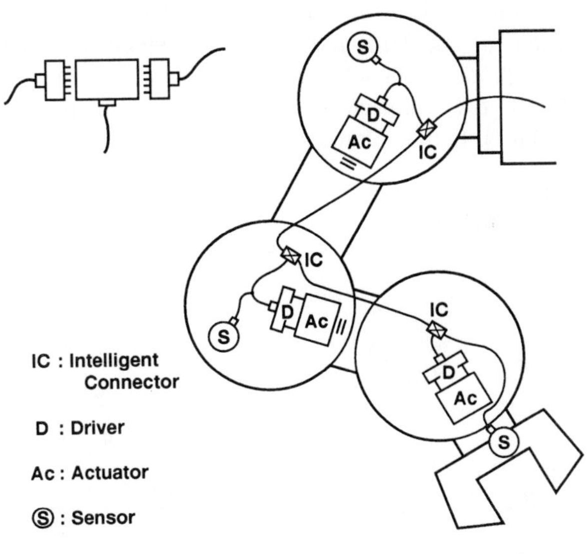

Fig. 13 SDIR-2 (shown with manipulator example)

(2) Built-in Servo Driver
Conventionally, all servo motor drivers have been located external to the robot's links and connected to the joint motors and sensors by numerous power and signal cables. This configuration greatly complicates configuration design and creates a large wiring bundle which must pass through many of the robot's links. A large reduction in servo motor driver size and brake power module size which will allow drivers to be located adjacent to the actuators. It is suggested that dramatic improvements in system integration could be achieved using hardware based on this design concept (see Figure 14).

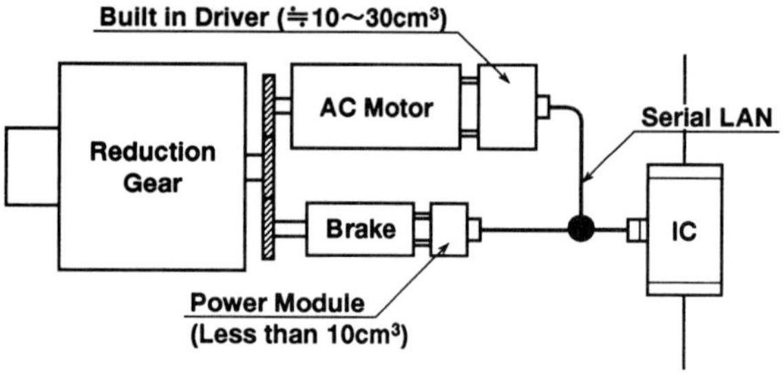

Fig. 14 Concept of Actuator Module with Built in Driver

5.2 New Robot Mechanisms

Although standardization leads to cost reductions it cannot be applied to robots with highly specific functional requirements. However, we believe that progress in intelligent control will help to reduce the need for specialized mechanisms. One example is the holonic mechanism (see Figure 15), whereby modules of the same type and function are simply linked to provide a high degree-of-freedom, very capable mechanism. These mechanisms are based upon a standard joint module which can be massproduced to achieve a radical decrease in cost. Intelligent connectors and built-in drivers are essential to the holonic mechanism.

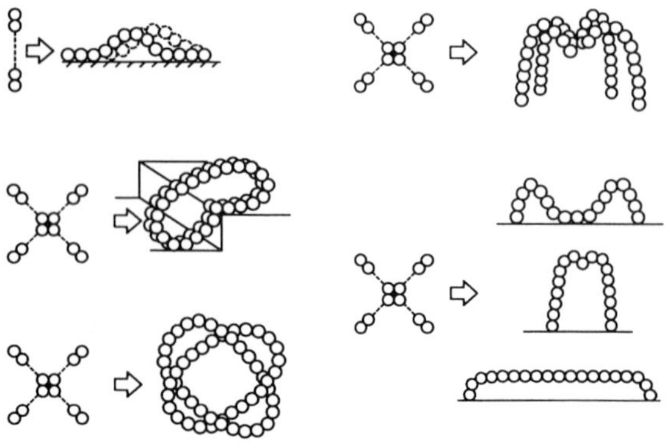

Fig. 15 Holonic Mechanism and Form Changing Control

This work was performed under the management of the Micromachine Center as the Industrial Science and Technology Frontier Program, "Research and Development of Micromachine Technology", of MITI supported by New Energy and Industrial Technology Development Organization.

5.3 Cost Predictions Using Standardized Design

Figure 16 shows costs predictions which may be realized with the introduction of standard design procedures and standardized components.

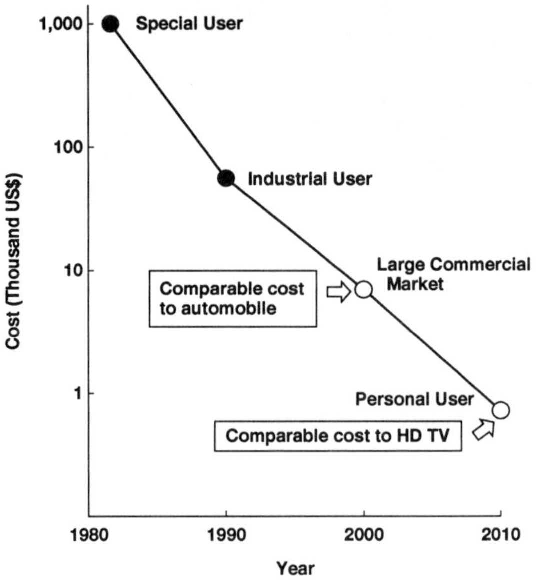

Fig. 16 Cost Decrease of 7 DOF Intelligent Manipulator

For example, an intelligent-robot-compatible manipulator with seven degrees of freedom cost approximately one million dollars ten years ago. The introduction of standard design SDIR-1 brought the cost down to about 1/10 of that figure, and the introduction of the proposed SDIR-2 concepts are predicted to reduce the cost further by a factor of 1/3 to 1/10. SDIR-2 would bring the cost of an intelligent robot to about the cost of an automobile, making intelligent robot hardware much easier to obtain. This is expected to lead to a rapid increase in the development of intelligent software. The increased demand for robots will require increased robot production, which can be expected to further reduce robot costs. We believe that this cycle will provide a strong impetus to improving the practicality of intelligent robots, which has remained stagnant for so long.

6. Conclusions

We have made the following proposals for promoting the practicality of intelligent robots and investigated their validity:

(1) Proposed robot design based on an open-system hardware architecture (SDIR-1).

(2) Used robot system examples developed with SDIR-1 to investigate the usefulness of the hardware modules, and in particular, considered the effect on development time and cost.

(3) Described the advantages of an advanced design approach (SDIR-2), which addresses the shortcomings of SDIR-1. We also showed the results that may be expected through the use of SDIR-2.

(4) Overall, we showed that open hardware architectures will lead to lower costs and generally contribute to the improved practicality of intelligent robots.

7. References

[1] K. Oonishi, T. Oomichi. Multi-fingered Manipulator Operation System for Demonstration Equipment, Japan Robot Institute Journal 1991; Vol. 9: No. 5: 88-97

[2] T. Oomichi et al. Intelligent Control of Four Fingered Manipulator System, '91 International Symposium on Advanced Robot Technology 1991; 589-596

[3] T. Oomichi et al. Industrial Robots Using Coordinated Vision, Hands, and Legs Proceedings of 30th Measurement and Automation Control Institute Symposium (SICE) 1991; 815-816

[4] T. Oomichi et al. Underwater Maintenance Robot, Proceedings of 30th Measurement and Automation Control Institute Symposium (SICE) 1991; 817-818

[5] T. Oomichi et al. Development of OS-independent Interface Elements, Proceedings of 8th Japan Robot Institute Symposium 1990; 133-134

Robots, Transportation, and Society

Chuck Thorpe and Dean Pomerleau Carnegie Mellon University

1 Introduction

It has long been a goal of robotics researchers to build smart vehicles: automobiles and buses and trucks that could avoid accidents, decrease congestion, and decrease driver effort. General Motors built self-driving cars in the 1950's; Ohio State did land mark work on wire-guided cars in the 1960's; Moravec's thesis work in the 1970's was motivated by this topic; and numerous groups in the 80's and 90's in Japan, the US, and Europe have built some sort of self-driving car.

The final goal is still elusive, for several reasons. Very good solutions have been found to parts of the problem, such as following good roads in benign weather conditions; but there are still significant challenges to making those approaches work in rain, snow, and dark of stormy night. The automotive industry is also much more price sensitive than we are used to dealing with in other areas. Even advanced safety features need to be priced lower than $1000 to be marketable. Such innovations as air bags, anti-lock brakes, traction control, even the 5 m.p.h. bumper, all took many years to introduce and often needed legislation before companies were willing to put them on vehicles. Other parts of the problem still elude technical solutions, such as coordinating vehicle motion in heavy traffic. But even more significant than the technical obstacles may be the non-technical issues. Some groups have serious questions about how intelligent vehicles would affect the environment. Concerns include increased travel and its effect on pollution and fuel use, changes in land use patterns, privacy, and liability.

CMU is involved in two major efforts to build intelligent vehicles, both funded by the US Department of Transportation. The first is Run-Off-Road Collision Countermeasures. This program does not intend to produce automated vehicles, but rather will build automated assistants that watch a person drive and provide warnings if the vehicle is in danger of departing its lane. The second program in which we are a participant is the Automated Highway System. The goal of this project is to create the specifications for vehicles and roadways that will provide hands-off, feet-off, computer controlled driving with safety and throughput much higher than today's. Our DOT work is interesting technically, and it is personally rewarding to apply our research to problems involving personal safety and broad social applicability. Dealing with the non-technical issues is often much more frustrating, but is also fascinating and educational.

Section 2 of this paper outlines the Collision Countermeasures and Automated Highway Systems pro grams. Section 3 discusses some of our work in lane sensing, obstacle detection, human factors, and vehicle positioning. Section 4 discusses the societal and institutional context for this research. Section 5 gives some conclusions and some thoughts for how we should organize our research and education for the future.

2 The Promise of Smart Cars

Current automobiles, buses, and trucks, and the high ways they run on, are remarkable systems. But in several ways they are nearing an evolutionary dead end. Congestion continues to worsen: in the US, vehicle miles traveled grow at about 4% per year, while new road construction lags far behind, and the Interstate highway system is nearly complete. Increased congestion adds to pollution, wastes valuable time, and increases driver frustration. Safety has made dramatic improvements, but is leveling off. Safer cars and roads have decreased US fatalities from a high of nearly 60,000 deaths per year to under 40,000[1][2]. But only about 2.5% of the remaining accidents are due to vehicle malfunctions, which suggests that improving vehicle safety will have diminishing returns. Add to this the increasing graying of the US population, and it becomes clear that improved safety will have to come from new ideas. Fortunately, the technology of smart vehicles is maturing, and the government agencies are preparing to exploit those results. In the US DOT, there are two main relevant agencies, each with a slightly different approach to the problems. The National Highway Safety Administration (NHTSA) is responsible for vehicle safety, and concentrates on vehicle-centered solutions. The Federal Highway Administration (FHWA) is responsible for highway construction, and is more interested in solutions that involve the infrastructure. Problems of

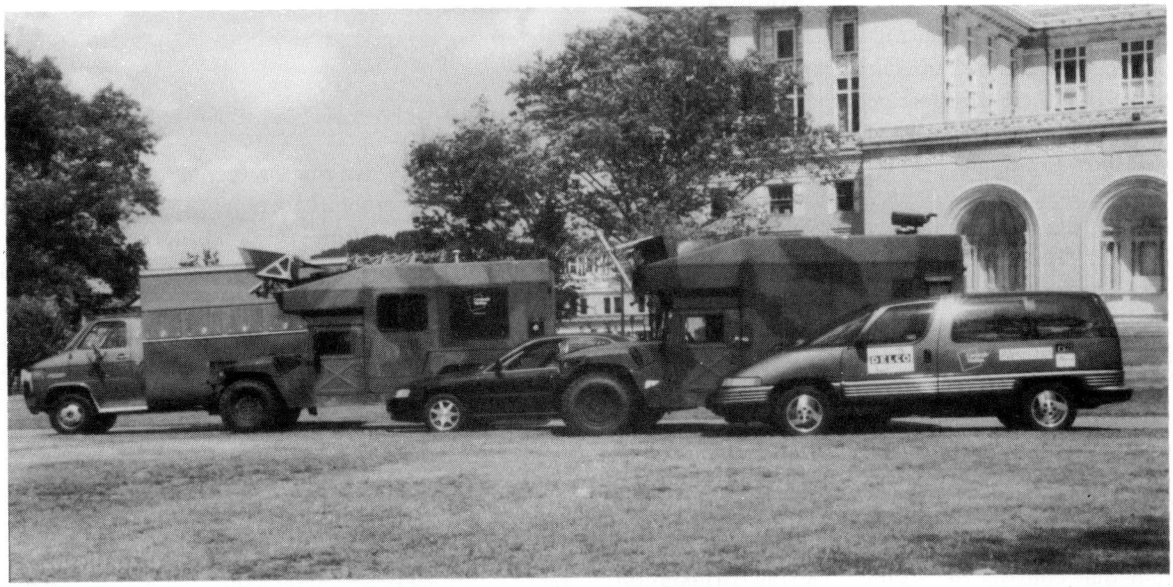

Figure 1: The CMU Navlab family in chronological order.

congestion and environ mental effects are also more likely to be part of FHWA's programs.

Our own work with intelligent vehicles at CMU has roots completely separate from DOT. The Navlab and its predecessors have all been funded by the military. The Navlab itself is funded by ARPA, as part of the Autonomous Land Vehicle and Unmanned Ground Vehicle programs. Related work in Image Understanding and high-performance computing has also been funded by ARPA, while the first contracts in the Robotics Institute for mobile robots were funded by the Office of Naval Research. More recently, robotics efforts have been funded by NASA, NSF, and private industry.

In the last two years, we have begun applying our DOD-sponsored technology to DOT programs. Our NHTSA research is part of their collision counter measures program. This program is aimed at using smart vehicle technology not to control vehicles directly, but to watch drivers' performance and to issue warnings if they are in danger. Our particular contribution is in run-off-road collision countermeasures. In this type of crash the vehicle departs the road typically because of driver inattention, intoxication, drowsiness, or excessive speed. Over 15,000 fatalities per year are caused by these single vehicle accidents[1]. The first part of this program was to examine the accident statistics to determine why these accidents happen, and to develop crash event time lines so that opportunities for warnings and for active vehicle control corrections could be modeled in detail. Simultaneously, we have been assessing the state of the art in technology for detecting potentially hazardous situations. We have also begun human factors experiments in a driving simulator, to determine what kind of warnings will produce the desired safe response from drivers. The next two stages of this program will select the most promising technology and warning methods, and will refine the experiments. This program is ambitious, because this kind of accident occurs on a wide variety of roads. Ideally, the countermeasures should work on all 4 million miles of US roads, ranging from unmarked rural roads to urban freeways.

Our preliminary results indicate that several technologies hold promise for preventing crashes. The first uses a video camera and image processing techniques to detect the vehicle's position on the road, and warn the driver when the vehicle begins to drift out of its lane [3]. In experiments on a driving simulator, it was found that such a system has the potential to prevent up to 75% of roadway departure crashes caused by driver inattention [4]. The second technology we are investigating utilizes a satellite positioning system and a digital map to warn drivers when they are approaching a sharp curve at too high speed. This system should be capable of preventing many of the 42.4% of fatal single vehicle crashes that occur on curves [5].

Our second DOT program is equally ambitious, but for very different reasons. The concept of the Auto mated Highway System (AHS) involves fully auto mated driving. After pulling onto a special lane on the highway, the driver could sit back, relax and let her specially equipped vehicle do the driving for her. The

Automated Highway System vision includes: Safety: Significantly safer than today; collision free in the absence of malfunctions Throughput: Significant increase in vehicles per hour per lane Inclement weather operation: The AHS must be able to operate at or exceed the performance levels of manual systems in the range of weather conditions that are typical in the continental U.S. Enhanced Mobility: Substantially better trip time predictability and use by impaired, aged, or less experienced drivers Improved User Comfort and Convenience: Less strain on users, higher quality trip Reduced Fuel Consumption and Emissions: Improved fuel economy and reduced emissions per vehicle mile traveled Characteristics of the AHS include affordability, user desirability, consistency with surrounding non-AHS roadways, evolvability, and ability to accommodate all normal types of vehicles.

The team that is building the AHS is a consortium that consists of Bechtel, Parsons-Brinkerhoff, Cal trans, GM, Hughes, Delco, MM, PATH, CMU, DOT. The consortium is run as an open research group. We are actively seeking outside involvement. We intend to cooperate as much as possible with our European colleagues, to learn from their Prometheus work, and with our Japanese friends, to learn from their many projects including PVS.

3 Technical Aspects and CMU Contributions

3.1 Lane Tracking

There are several daunting challenges to building intelligent vehicles. The most obvious difficulty is in lane tracking. One of the most appealing solutions is to use vision, with a forward-looking camera, to mimic human road following. There are at least ten groups doing related work, including:[3]

- General Motors
- Ford Motor Company (Jaguar Division)
- Daimler-Benz and their colleagues at the Military University of Munich
- The National Institute of Standards and Technology (NIST)
- Helsinki University
- Rockwell International
- University of Maryland
- Carnegie Mellon University
- IMRA America
- Toyota
- Tokyo University

Unfortunately, for our comparative evaluations, it has been very difficult to obtain examples of most of these systems. Some rely on proprietary software which cannot be released, others depend on special-purpose hardware that would be difficult to replicate, and many of the rest are no longer active projects. For now, we are evaluating the performance of our own vision system, ALVINN[6], RALPH[7], and AURORA[8], and are waiting for another DOT-sponsored program to evaluate the Rockwell system. We are actively seeking ways to involve other interested groups in the ongoing comparisons. Other possibilities for lane tracking involve a radar looking sideways (used by Freightliner), "magnetic nails" and magnetometers (developed at Berkeley), tracking laser reflectors (for example IMRA America), and tracking radar retroreflectors (as developed at Oxford and Ohio State).

3.2 Obstacle Detection

A second major challenge is in obstacle detection. Automotive radars are now on the market. In the US, every Greyhound intercity bus was equipped with a VORAD radar. Delco has also recently introduced radars both for personal cars and for school buses. Other companies, including Leica and Mazda, have shown prototype laser- based ranging devices. Both the ladar and radar sensors are useful in detecting slow-moving vehicles and providing warning, or even in Intelligent Cruise Control in which the sensors can be used to automatically slow the trailing vehicle to match the speed of the leading vehicle. Video sensors (shape from motion, stereo vision) also show promise, at least on days with good illumination and no precipitation.

But the problems are by no means solved. On dry pavement, a vehicle in good mechanical condition takes approximately 60 meters to stop from 100 kph. Sensors designed for adaptive cruise control may not have that range. Worse, when the pavement is wet or the tires are in bad condition, the stopping distance may be much greater. And given standard highway curvatures (radii of 900 ft are common in the US), a single-beam radar may look in the wrong lane, causing both false alarms from roadside obstacles and missing detections from vehicles around a curve. Moreover, automotive radar and ladar are usually tuned to detect other vehicles. In some rural parts of the US, over 20% of all crashes involve hitting animals[9]. Other roadway obstacles include tree branches, rock slides, and loose objects falling off preceding vehicles. Any sensor that depends on objects being nearly the size of a car, or having a strong radar reflection, may miss other obstacles that can cause dangerous accidents.

We are building a radar sensor which will be able to operate under all weather conditions at ranges up to 200 meters. It consists of a linear receiver array, using wavefront reconstruction techniques to obtain range and bearing of objects within the sensors field of view. This will give us the ability to discriminate many targets at different ranges, and up to three separate targets at any one range. In addition to building the sensor and low-level processing, we are integrating the radar into the rest of the vehicle systems. If we are running a lane-tracking vision system, we will combine the lane map with the radar obstacle map to place vehicles in their proper lanes. If we cannot see lanes, but have a detailed road map and accurate vehicle position, we can still associate radar returns with lanes. Finally, even if we have no map and no vision system, we can use the vehicle's current steering direction as a cue to where the road should be headed, and use that to filter out obstacle away from the vehicle's intended trajectory.

3.3 Other Technologies

Measuring vehicle position is important for many applications. For warning of an upcoming curve, the accuracy needs to be within several meters. For detecting a vehicle drifting out of its lane, or for automatic control to keep a vehicle in its lane, the accuracy needs to be within a meter. The most promising technology for position measurement may be GPS. We are closely tracking developments in differential corrections, including a company that broad casts differential corrections by satellite so there is no need for a local ground station. More ambitious developments include carrier phase GPS, which is beginning to produce accuracies of a few cm on one of our vehicles, and pseudolites, which provide GPS coverage in the urban canyons of big cities, where most satellites are occluded.

Once the system knows the position of the vehicle, lane boundaries, and other vehicles, there is still a significant challenge to producing smooth control. Many AHS scenarios call for vehicle to operate with only a few meters separation. Disturbances from wind gusts, hills, rough-running engines, and even the aerodynamics of the vehicle in front, all call for careful control system design. Our colleagues in the AHS consortium are concentrating on detailed models of engine dynamics and air flow[10]. At CMU we are working on adaptive and learning control systems.

3.4 Promise of AHS

If all these technological obstacles are overcome, the potential performance gains appear to be impressive For example, it could be technically feasible to have vehicles traveling 145 kph (90 mph) on a dedicated lane. These vehicles could be spaced with a headway (distance from front of one car to the front of the next car) of 8 m, calculated as 5 m average car length plus 3 m separation. That would give a total lane capacity of 18,000 cars per hour, roughly 9 times the capacity of the best manually operated lanes.

But there are remaining technical problems. First of all, it is unlikely that most AHS roadways will sup port 90 mph driving. Existing roads have much lower design speeds, and in most scenarios the AHS will be deployed along the existing rights of way. This may limit top speeds to closer to today's highway speeds, at least in urban US settings.

Second, the theoretical simulations break down once entrances and exits are introduced. If vehicles are packed together tightly, any new vehicle entering the lane will cause some vehicle to slow to let it in, which in turn causes a shock wave of slower driving to propagate backward from the entrance. Leaving gaps for slower vehicles, plus using more realistic top speeds, limits the overall throughput for a feasible system to closer to 4000 to 6000 vehicles per lane per hour.

But besides the challenges of sustained driving, there are other problems involved with checking in and out of the system. Check in involves deciding if the vehicle is properly equipped and functioning within specifications. Does the vehicle have enough fuel to reach its destination? What is the condition of its brakes, actuators, and engine? If there is an external payload, could it fall off?

Check out involves testing the driver's readiness to resume manual driving. Has the driver fallen asleep? More subtly, after automated driving with extremely short spacing, has the driver become habituated to tailgating that would be dangerous for a person to try? All these questions are under study in driving simulators.[11]

4 Institutional and Societal Issues

Beyond the technical issues, there are complex societal issues that will affect how or if an automated highway is built. The first issue is concern for increased travel, and attendant increases in fuel consumption and tailpipe emissions. At first, the auto mated highway system seems to be environmentally-friendly. Fewer traffic jams mean fewer vehicles idling and fewer high-throttle starts, which in turn should give reduced emissions and better fuel efficiency. The concern of some environmental groups is that a more efficient highway system will encourage people to travel more, and to live further from work, and will therefore undo any benefit. The analyses are fairly complex. For instance, the major effect of the AHS may not be to encourage more travel, but rather to encourage people to travel when they want to, rather than to schedule their travel to avoid rush hour. Moreover, vehicles traveling with only a few meters spacing have significantly lower wind resistance, which in turn gives much better fuel economy. A second major environmental concern involves land use and the effect of an AHS on surrounding roads. Building an AHS may affect urban development pat terns, leading to increased urban sprawl and changes in settlement patterns. Furthermore, the increased capacity of an AHS may overload cross streets. Unless care is taken in integrating the AHS with existing roadways, more vehicles may be able to exit a particular location than are able to flow into the surrounding arterials. This effect could be compounded, by traffic being diverted from parallel roads to the AHS, and therefore further overloading the feeder arterials. Conversely, the increased flow on the AHS should decrease flow on the parallel routes, making the surface streets and non-AHS freeways easier to use.

Other concerns include privacy (does the system know where you have been?); liability (if the system malfunctions, who gets sued?); maintainability (state departments of transportation are not used to dealing with advanced electronics); and social equity (are AHS's more likely to be built in socially disadvantaged communities?). Again, the analysis is often complex. For instance, as more and more people move onto an AHS, the surrounding roadways will become less and less congested, so there is an increasing benefit to the non-AHS drivers, while the AHS itself will become more crowded, and may offer a decreasing benefit.

Some of these issues are not specific to AHS, but are encountered any time a new road is built or lanes are added. Other issues have non-technical solutions, such as the use of tolls to reduce travel demand or subsidies to encourage more use of mass transit. It is also possible that the AHS may offer specialized solutions. For instance, the accurate lane-keeping that will be possible with automated driving will make it possible for inductive power pickups for roadway powered electric vehicles, which would have an enormous impact on environmental issues. The AHS will also make some transit operations much more competitive, such as high-speed dedicated bus lanes on narrow rights of way, and may change the balance between single occupancy vehicles and mass transit.

Several approaches are being developed to deal with these issues. First, the precursor studies leading up the AHS have examined a number of comparable systems to see how their introductions were accepted into society and influenced the way we live and work. Comparable systems studied ranged from the US Supersonic Transport, as an example of problems with a high-tech government funded program, to the elevator, as an example of how perceptions of safety influenced people's willingness to use a new system. Second, the National Automated Highway Systems Consortium includes seats on its management committee for "stakeholders" from interested groups, both users of the system and others who are interested in the design and evaluation process. The NAHSC is inviting representatives from the environ mental groups, the American Automobile Association, state highway organizations, trucking operators, and so forth to participate.

Third, a sizable component of the Consortium's effort will be spent on analysis and simulations. This will include both micro-scale simulations of individual vehicles driving on the AHS, and larger-scale analyses of emissions, traffic flow, and so forth. Fourth, midway through the AHS development effort, the Consortium will do a major demonstration of proof of technical feasibility. The 1997 demo will show all the major components of one possible AHS design. It is scheduled early enough in the program that test results and feedback from interested stake holders can be incorporated into the ongoing design efforts.

5 Conclusions

The AHS program is an interesting example of the robotics programs in which researchers are being invited to participate. As our research begins to have near-term practical applications, the size of the pro gram changes, the mix of people and institutions working together gets more complicated, and new issues need to be addressed. For the AHS, there are still interesting and important technical challenges. But it is not enough to discuss the interesting theory, or to show relevant analysis. In order to convince the public that the AHS is feasible and desirable, we have to build and demonstrate real integrated systems.

At the same time, we need to be much more attuned to non-technical public concerns than at least some of us have been in the past. Issues of cost, user inter face, and high reliability are traditional engineering concerns that are not usually dealt with in University research. But these are at least technical issues, for which our training and background have given us some tools. The thornier issues of public policy, environmental debates, social equity, and liability are not issues usually addressed by roboticists. We need to find partners who can help us understand and address these questions. We need to understand how to deal with those stakeholders not as adversaries, but as representatives of other issues and constituents that deserve to be heard. And we perhaps need to rethink our educational process, so the PhD's we are producing are better equipped to understand and deal with this sort of issue. Success, in AHS and in many other robotics programs, certainly depends on technical competence. But changing the paradigms of society, in such an integral area as transportation, demands additional competencies and sensitivities.

6 Bibliography

1. National Highway Traffic Safety Administration. "Fatal Accident Reporting System 1991", NHTSA Technical Report Number DOT HS 807 954

2. Wassim, G.N. Mironer, M.S. and Fraser, L.C. "Analysis of Target Crashes and ITS Countermeasure Actions", Proc. of 1995 ITS America Annual Meeting. Washington, DC, March 15-17, pp. 931-940.

3. Pomerleau, D., Kumar, B., Thorpe, C., Everson, J., Kopala, E., and Lazofson, L. (1995 August). Run-off-road collision avoidance using IVHS countermeasures: Task 3 report - Volume 1. (Contract No. DTNH22-93-C-07023). Pittsburgh, PA: Carnegie-Mellon University Robotics Institute.

4. Tijerina, L., Jackson, J. L., Pomerleau, D. A., Romano, R. A., Peterson, A. and Thorpe, C. (1995, August). Run-off-road collision avoidance using IVHS countermeasures: Task 3 - Volume II. (Contract No. DTNH22-93-C-07023). Pittsburgh, PA: Carnegie-Mellon University Robotics Institute.

5. Pape, D., Narendran, V., Hadden, J., Everson, J., Koenig, M., Pomerleau, D. and Thorpe, C. (1995, August). Run-off-road collision avoidance using IVHS countermeasures: Task 4 - Volume I. (Contract No. DTNH22-93-C-07023). Pittsburgh, PA: Carnegie-Mellon University Robotics Institute.

6. Pomerleau, D.A. (1993) *Neural network perception for mobile robot guidance*, Kluwer Academic Publishing.

7. Pomerleau, D. (1995) RALPH: Rapidly Adapting Lateral Position Handler. *Proc. of 1995 Symposium on Intelligent Vehicles*, Ypsilanti, Michigan, Sept. 25-26, 1995.

8. Chen, M., Pomerleau, D., Jochem, T. (1995) AURORA: A Vision-Based Roadway Departure Warning Systems. To Appear in *Proc. of IEEE Int. Conf. on Intelligent Robots and Systems.* Pittsburgh, PA, August, 1995.

9. Hughes, W. and Saremi, R. (1995) Investigation of Crashes with Animals. Federal Highway Administration Highway Safety Information System Report FHWA-RD-94-156.

10. Choi, S., and Hedrick, J. (1995) Comparison of Engine Models and Distance Sensors for Vehicle Longitudinal Control: Analysis and Tests Results in PATH. Proc. of 1995 ITS America Annual Meeting. Washington, DC, March 15-17, pp. 61-76.

11. Levitan, L., Plocher, T.A., and DeMers, R.E. (1995) Driver Task Analyses and Driver-System Interface Designs for the Automated Highway System. Proc. of 1995 ITS America Annual Meeting. Washington, DC, March 15-17, pp. 159-170.

Participants

Dr. Rachid Alami
LAAS-CNRS
7, avenue du Colonel Roche
31077 Toulouse Cedex
France
alami@laas.fr

Prof. Suguru Arimoto
Faculty of Engineering
University of Tokyo
Bunkyo-ku
Tokyo
113 Japan
arimoto@arimotolab.t.u-tokyo.ac.jp

Prof. Minoru Asada
Mech. Eng. for Computer-Controlled Machinery
Osaka University
Suita
Osaka 565
Japan
asada@robotics.ccm.eng.osaka-u.ac.jp

Prof. Ruzena Bajcsy
GRASP Laboratory
Depart. of Computer and Info. Science
University of Pennsylvania
Philadelphia, PA 19104-3897
USA
bajcsy@central.cis.upenn.edu

Prof. Massimo Bergamasco
Scuola Superiore S. Anna
PERCRO
via Carducci, 40
56100 Pisa
Italy
bergamasco@sssup1.sssup.it

Prof. Antonio Bicchi
Centro "E. Piaggio"
Universita di Pisa
Via Diotisalvi, 2
56126 Pisa
Italy
bicchi@anaxagoras.piaggio.ccii.unipi.it

Dr. Bob Bolles
AI Center, M/S EK290
SRI International
333 Ravenswood Avenue
Menlo Park, CA 94025
USA
bolles@ai.sri.com

Prof. J. Michael Brady
Department of Engineering Science
Oxford University
19 Parks Road
Oxford OX1 3PJ
UK
jmb@robots.oxford.ac.uk

Dr. Joel Burdick
Dept. of Mechanical Engineering
CalTech
Mail Code 104-44
Pasadena, CA 91125
USA
jwb@robby.caltech.edu

Prof. Alicia Casals
Universitat Politecnica de Catalunya
Dpt. d'Enginyeria de Sistemes,
Automatica i Informatica Industrial
C/.Pau Gargallo, 5
08028 Barcelona
Spain
casals@esaii.upc.es

Dr. Raja Chatila
LAAS-CNRS
7, avenue du Colonel-Roche
31077 Toulouse Cedex
France
raja@laas.fr

Dr. Peter Corke
CSIRO
Division of Manufacturing Technology
P.O.Box 883
Kenmore 4069
Australia
P.Corke@janus.cat.csiro.au

Prof. Paolo Dario
Scuola Superiore S. Anna
ARTS Lab.
Via Carducci 40
56100 Pisa
Italy
dario@sssup1.sssup.it

Prof. Joris De Schutter
Katholieke Universiteit Leuven
Faculty of Engineering
Dept of Mechanical Engineering
Division of Production Eng.
Machine Design and Automation
Celestijnenlaan 300 B
B-3001 Leuven
Belgium
deschutter@mech.kuleuven.ac.be

Prof. Dr.-Ing. E.D. Dickmanns
Universität der Bundeswehr
München
Institut für Systemdynamik und
Flugmechanik
Werner-Heisenberg-Weg 39
D-8014 Neubiberg
Germany
ld2brbeh@kommsrv.rz.unibw-muenchen.de

Prof. Dr.-Ing. R. Dillmann
Universität Karlsruhe
Fakultät für Informatik
Institut für Prozessrechentechnik und Robotik
Postfach 69 80
Kaiserstr. 12/Geb. 40.28
D-76128 Karlsruhe
Germany
dillmann@ira.uka.de

Prof. Hugh Durrant-White
The University of Sydney
Department of Mechanical &
Mechatronic Engineering
Sydney NSW 2006
Australia
hugh@tiny.me.su.oz.au

Dr. Jan-Olof Eklundh
Dept Numerical Analysis and Computing Science
RIT (KTH)
S-100 44 Stockholm
Sweden
joe@bion.kth.se

Prof. Michael Erdmann
Doherty Hall 4301G
Robotics Institute
CMU
5000 Forbes Avenue
Pittsburgh, PA 15213
USA
me@cs.cmu.edu

Dr. Bernard Espiau
INRIA Rhône Alpes
Bât. HITELLA, bureau 27
46, avenue Félix Viallet
38031 Grenoble Cedex 1
France
bernard.espiau@imag.fr

Dr. Olivier Faugeras
INRIA
Centre de Sophia Antipolis
Avenue Emile Hugues
06565 Valbonne Cedex
France
Olivier.Faugeras@sophia.inria.fr

Prof. R.S. Fearing
Dept. of EE&CS
University of California
Berkeley - CA 94720
USA
ronf@eecs.berkeley.edu

Dr. Wendelin Feiten
Siemens AG, ZFE
Otto-Hahn-Ring 6
D-81730 München
Germany
wendelin.feiten@zfe.siemens.de

Dr. Paolo Fiorini
Jet Propulsion Laboratory
California Institute of Technology
4800 Oak Grove Drive
Pasadena, CA 91109-8099
USA
fiorini@telerobotics.jpl.nasa.gov

Dr. Georges Giralt
LAAS-CNRS
7, avenue du Colonel-Roche
31077 Toulouse Cedex
France
giralt@laas.fr

Dr. Hideki Hashimoto
Hashimoto Lab.
Institute of Industrial Science
University of Tokyo
7-22-1 Roppongi, Minato-ku
Tokyo 106
Japan
hideki@ics.iis.u-tokyo.ac

Prof. Vincent Hayward
McGill University
Center of Intelligent Machines
3480 University Street
Montreal
Quebec H3A 2A7
Canada
hayward@mcrcim.mcgill.edu

Prof. Shigeo Hirose
Dept of Mechano-Aerospace Engineering
Tokyo Institute of Technology
2-12-1 O-Okayama
Meguro-ku
Tokyo 152
Japan
hirose@mes.titech.ac.jp

Prof. Dr.-Ing. Gerhard Hirzinger
DLR Oberpfaffenhofen
FF-DR
D-82234 Wessling
Germany
df28@master.df.op.dlr.de

Prof. Neville Hogan
MIT
Mechanical Engineering Department
545 Technology Square
Cambridge, MA 02139
USA
neville@mit.edu

Prof. John Hollerbach
University of Utah
3190 Merrill Engineering Building
Salt Lake City UT 84112
USA
jmh@cs.utah.edu

Dr. Katsushi Ikeuchi
Computer Science Dept.
Carnegie Mellon University
5000 Forbes Avenue
Pittsburgh, PA 15213-3890
USA
ki@ius.cs.cmu.edu

Prof. Masayuki Inaba
Dept. of Mechano-Informatics
University of Tokyo
7-3-1 Hongo, Bunkyo-ku
Tokyo, 113
Japan
inaba@jsk.t.u-tokyo.ac.jp

Prof. Hirochika Inoue
Department of Mechano-Informatics
University of Tokyo
7-3-1 Hongo, Bunkyo-ku
Tokyo, 113
Japan
inoue@jsk.t.u-tokyo.ac.jp

Prof. Jehuda Ish-Shalom
Columbia University
6 Oak Place
Croton-on-Hudson
NY 10520-9771
USA
jehuda@photon.poly.edu

Prof. Ray A. Jarvis
Dept. of Electrical Engineering
Monash University
Wellington Road
Clayton, Victoria 3168
Australia
ray.jarvis@eng.monash.edu.au

Prof. Takeo Kanade
Director
The Robotics Institute
Carnegie Mellon University
5000 Forbes Avenue
Pittsburgh, PA 15213-3890
USA
kanade@cs.cmu.edu

Prof. Oussama Khatib
Computer Science Dept.
B2, Cedar Hall
Robotics Laboratory
Stanford University
Stanford, CA 94305
USA
khatib@cs.stanford.edu

Prof. Daniel E. Koditschek
EECS Dept. Room 148
Advanced Technology Laboratory
College of Engineering
University of Michigan
1101 Beal Avenue
Ann Arbor, MI 48109-2110
USA
kod@eecs.umich.edu

Dipl.-Ing. Ralf H. Koeppe
Institute of Robotics and System Dynamics
Post Box 11 16
D-82230 Wessling
Germany
Ralf.Koeppe@dlr.de

Dr. Vijay Kumar
GRASP Laboratory
University of Pennsylvania
Philadelphia, PA 19104-3897
USA
kumar@central.cis.upenn.edu

Dr. Yasuo Kuniyoshi
Autonomous Systems Section
Intelligent Systems Division
Electrotechnical Laboratory
1-1-4 Umezono
Tsukuba
Ibaraki 305
Japan
kuniyosh@etl.go.jp

Prof. Jean-Claude Latombe
Dept of Computer Science
Stanford University
Stanford, CA 94305-2140
USA
latombe@coyote.stanford.edu

Dr. Christian Laugier
LIFIA - INRIA Rhône-Alpes
46, avenue Félix Viallet
38031 Grenoble Cedex
France
Christian.Laugier@imag.fr

Prof. Sukhan Lee
JPL
California Institute of Technology
4800 Oak Grove Drive
Pasadena - CA 91109-8099
USA
shlee@pollux.usc.edu

Prof. Matthew T. Mason
Robotics Institute
CMU
5000 Forbes Avenue
Pittsburgh, PA 15213
USA
Matt_Mason@cs.cmu.edu

Dr. Larry Matthies
Jet Propulsion Laboratory
California Institute of Technology
4800, Oak Grove Drive
Pasadena, California 91109
USA
lhm@robotics.jpl.nasa.gov

Prof. Hirofumi Miura
The University of Tokyo
Dept. of Mechao-Informatics
Hongo 7-3-1, Bunkyo-ku
Tokyo 113
Japan
miura@leopard.t.u-tokyo.ac.jp

Prof. Yoshihiko Nakamura
University of Tokyo
Dept of Mechano-Informatics
Faculty of Engineering
Hongo 7-3-1, Bunkyo-ku
Tokyo, 113
Japan
nakamura@mech.t.u-tokyo.ac.jp

Dr. Shree Nayar
Columbia University
Dept of Computer Science
New York, NY 10027
USA
nayar@cs.columbia.edu

Mr. Takeo Oomichi
Mitsubishi Heavy Industry
Takasago R&D Center
2-1-1 Shinhama, Arai-cho
Takasago-si
Hyogo
Japan
oomichi@trdc.tksg.mhi.co.jp

Prof. Richard (Lou) Paul
CIS Dept.
University of Pennsylvania
Room 301C
3401 Walnut Street
Philadelphia, PA 19104-6228
USA
lou@grasp.cis.upenn.edu

Prof. Marc H. Raibert
EECS Dept, Room 765
MIT AI Lab.
545 Technology Square
Cambridge, MA 02139
USA
raibert@ai.mit.edu

Dr. Karl-Heinz Robrock
Commission of the European
Communities
Office BU 3100 2058
200, rue de la Loi
B-1049 Brussels
Belgium

Prof. Bernard Roth
Dept of Mechanical Engineering
Stanford University
Stanford, CA 94305
USA
roth@Flamingo.stanford.edu

Prof. Kenneth Salisbury
MIT
AI Laboratory
545 Technology Square
Cambridge, MA 02139
USA
jks@ai.mit.edu

Dr. Tomomasa Sato
RCAST
The University of Tokyo
4-6-1, Komaba, Meguro-ku
Tokyo
Japan 153
tomo@lssl.rcast.u-tokyo.ac.jp

Prof. Isao Shimoyama
Dept. of Mechano-Informatics
Faculty of Engineering
The University of Tokyo
7-3-1 Hongo, Bunkyo-ku
Tokyo 113
Japan
isao@leopard.t.u-tokyo.ac.jp

Prof. Yoshiaki Shirai
Dept. of Mechanical Engineering
for Computer-Controlled Machinery
Faculty of Engineering
Osaka University
2-1, Yamadaoka, Suitasi
Japan, 565
shirai@ccm.osaka-u.ac.jp

Prof. Bruno Siciliano
Universita Degli Studi di Napoli Federico II
Dipartimento di Informatica e Sistemistica
Via Claudio 21
80125 Napoli
Italy
siciliano@na.infn.it

Dr. Charles Thorpe
Robotics Institute
CMU
5000 Forbes Avenue
Pittsburgh, PA 15213-3890
USA
thorpe@cs.cmu.edu

Prof. Carlo Tomasi
Robotics Laboratory
Computer Science Department
Stanford University
Stanford, CA 94305
USA
tomasi@cs.stanford.edu

Dr. Masaru Uchiyama
Dept of Aeronautics and Space Engineering
Faculty of Engineering
Tohoku University
Aramaki-Aza-Aoba
Aoba-ku
Sendai 980-77
Japan
uchiyama@space.mech.tohoku.ac.jp

Dr. Heinz Woern
KUKA (Fa. Keller und Knappich)
Blücherstr. 144
D-81165 Augsburg
Germany
Fax : (49) 821 797 1087

Dr. Kazuya Yoshida
Department of Aeronautics &
Space Engineering
Tohoku University
Aoba, Sendai 980-77
Japan
yoshida@space.mech.tohoku.ac.jp

Prof. Tsuneo Yoshikawa
Dept of Mechanical Engineering
Kyoto University
Kyoto
Japan
yoshi@mech.kyoto-u.ac.jp